KU-683-594

THE DINOSAURIA

Edited by

DAVID B. WEISHAMPEL

PETER DODSON

HALSZKA OSMÓLSKA

UNIVERSITY OF CALIFORNIA PRESS

Berkeley / Los Angeles / Oxford

University of California Press
Berkeley and Los Angeles, California

University of California Press, Ltd.
Oxford, England

Copyright © 1990 by The Regents of the University of California

First Paperback Printing 1992

Library of Congress Cataloging-in-Publication Data

The Dinosauria / edited by David B. Weishampel, Peter Dodson, Halszka Osmólska.
    p. cm.
    Includes bibliographical references
    ISBN 0-520-06727-4
    1. Dinosaurs. I. Weishampel, David B., 1952–   . II. Dodson, Peter.
III. Osmólska, Halszka.
    QE862.D5D513   1990
    567.9′1—dc20   89-20516
                                            CIP

Printed in the United States of America

The paper used in this publication meets the minimum
requirements of American National Standard for Information Sciences—Permanence of Paper for Printed Library
Materials, ANSI Z39.48–1984. ∞

To **RICHARD OWEN,**
who started it all,

and to **EDWIN H. COLBERT,**
who lit so many candles.

Contents

Abbreviations for Figures

Cranial

acc.op.: accessory opening
an: angular
aofe: antorbital fenestra
aofo: antorbital fossa
ar: articular
asc: anterior semicircular canal
ash: mesial shelf on the crown margin
b: secondary ridge(s)
boc: basioccipital
bpt: basipterygoid
bptp: basipterygoid process
bs: sinuslike opening in the basisphenoid and basioccipital
bsp: basiphenoid
bst: basisphenoid tubera
c: tertiary ridge(s)
ca: cartilage
CBL: cerebral hemispheres (general region of)
ch: choana
ci: crista interfenestralis
CLM: cerebellum (region of)
cor: coronoid
cpr: crista prootica
ct: crista tuberalis
d: dentary
dpl: ductus perilymphaticus
de: ductus endolymphaticus
ect: ectopterygoid
en: external naris
ept: epipterygoid
e: ethmoid
exoc: exoccipital
fj: jugular foramen
fl: floccular lobe
flr: foramen lacerum rostralis

flc: foramen lacerum caudalis
fm: foramen magnum
fme: fissura metotica
fo: foramen ovale
fp: footplate
fr: frontal
fs: fossa subarcuata
hsc: horizontal semcircular canal
in: internal nares
iptv: interpterygoid vacuity
itf: infratemporal fenestra
j: jugal
l: lacrimal
lc: lacrimal canal
ldp: lateral depression
lf: lacrimal foramen
lg: lagena
lsb: laterosphenoid buttress
lsp: laterosphenoid
mo: medulla oblongata
mx: maxilla
mxpr: toothlike process of maxilla
na: nasal
no: nodes
o: orbit
oc: occipital
ol: olfactory lobe
op: opisthotic
OPT: optic lobes (region of)
osp: orbitosphenoid
ot: olfactory tract
pal: palatine
palr: palatine ramus
pap: palpebral
par: parietal
pd: predentary
peg: palatine peg

pit: pituitary fossa
pitf: pituitary foramen
pmx: premaxilla
pmf: premaxillary foramen
pmlf: premaxillary-lacrimal foramen
po: postorbital
poc: paroccipital process
pof: postfrontal
pra: prearticular
prf: prefrontal
pro: prootic
ps: parasphenoid
psc: posterior semicircular canal
pt: pterygoid
ptf: pterygoid flange
ptr: pterygoid ramus
q: quadrate
qc: quadrate condyle
qj: quadratojugal
qr: quadrate ramus
r: rostral
res: resorption facet
rep. cr. replacement crown
rp: Rathke's pouch
rq: right quadrate (appears to be reconstructed
   on the left side)
S: cranial sinus
S': maxillary sinus
s: sacculus
sa: surangular
scl: sclerotic ring
set: sella turcica
soc: supraoccipital
spl: splenial
sq: squamosal
st: stapes
stf: supratemporal fenestra
t: tabular
t. gr: groove for replacement tooth crown
V: palatal vault
v: vomer
vc: vidian canal (for internal carotid a.)

Soft Anatomy
Muscle terms designate muscle or attachment site

maem: M. adductor mandibulae externus medialis
maep: M. adductor mandibulae externus
   profundus
maes: M. adductor mandibulae externus
   superficialis
mamc: M. adductor mandibulae caudalis
mdm: M. depressor mandibulae
mlp: M. levator pterygoidei
mpd: M. pterygoideus dorsalis
mpp: M. protractor pterygoidei

mps: M. pseudotemporalis
mpv: M. pterygoideus ventralis
c.n. I–XII: exits for cranial nerves
vii: palatine branch of c.n. VII
ab: A. (arteria) basilaris
ar. pal: A. palatinus
ic: internal carotid a.
lhv: lateral head vein
mpa: median palatine artery
oa: orbital artery
pa: palatine artery
rc: ramus cranialis of internal carotid a.
sl: sinus longitudinalis
ts: transverse sinus
vcc: V. (vena) cerebralis caudalis
vcd: V. capitis dorsalis
vcm: V. cerebralis medius
vcms: V. cerebralis medius secunda
v. par: V. parietalis

Axial Skeleton

at: atlas
ax: axis
c3: third cervical vertebra
ca1: first caudal vertebra
cav: caudal vertebra(e)
cedi: centrodiapophyseal lamina
ch: body of the chevron
chr: chevron rod
cv: cervical vertebra(e)
d1: first dorsal vertebra
di: diapophysis
dv: dorsal vertebra(e)
hypa: hypantrum
hypo: hyposphene
ifhy: infrahyposphenal lamina
ifpo: infrapostzygapophyseal lamina
ifpr: infraprezygapophyseal lamina
lasp: laterospinal lamina
lifpo: lateroinfrapostzygapophyseal lamina
nas: neural arch sutural surface
pa: parapophysis
padi: paradiapophyseal lamina
pat: proatlas
pl: pleurocoel
po: postzygapophysis or postzygapophyseal facet
podi: postzygodiapophyseal lamina
por: postzygapophyseal rod
posp: postspinal lamina
pr: prezygapophysis or prezygapophyseal facet
prcep: precentroparapophyseal lamina
prdi: prezygodiapophyseal lamina
prr: prezygapophyseal rod
prsp: prespinal lamina
rs1, rs2 etc.: first, second sacral rib
s1, s2, etc.: first, second sacral vertebra

sp: neural spine
sudi: supradiapophyseal lamina
supo: suprapostzygapophyseal lamina
supr: supraprezygapophyseal lamina
sy: sacral yoke
tps: sutural surface for ventral surface of the transverse process of the neural arch
tp: transverse process
zy: zygantrum

Appendicular Skeleton

a: astragalus
aap: ascending process of astragalus
ac: acetabulum
af: facet for the ascending process of the astragalus
an: antitrochanter
brs: brevis shelf
c: calcaneum
ca: cartilage
cb: carpal block
ccl: contact for clavicle
cn: cnemial crest
co: coracoid
ctgr: crista tibiofibularis groove
dc: distal carpals
dp: deltopectoral crest
dt: distal tarsals
fe: femur
ffi: fibular facet (on tibia)
fgt: greater trochanter of femur
fh: femoral head
fi: fibula
fic: fibular crest
fis: fibular sulcus
flt: lesser trochanter of femur
flts: trochanteric shelf of femur
f4t: fourth trochanter of femur
gl: glenoid cavity
h: humerus
i: intermedium
il: ilium
ilp: iliac peduncle
is: ischium
isp: ischiac peduncle
k: knob on metatarsal II (m. tibialis anticus?)
lc: lateral condyle
lepc: lateral epicondyle

mc: metacarpal
mec: medial condyle
mepc: medial epicondyle
mt: metatarsal
obf: obturator foramen
obp: obturator process
obr: obturator ridge
olp: olecranon process
pf: pubic fenestra
ph: phalanx/phalanges
prp: prepubic process
pu: pubis
pup: pubic peduncle
ra: radius
rae: radiale
sc: scapula
sgb: supraglenoid buttress
ste: sternal
t: tibia
tdp: tibial descending process
tmt: tarsometatarsal
ul: ulna
ule: ulnare
z: zygosphene
I, II,....V: first, second,...fifth digit of manus/pes

Muscles

mcf: M. caudifemoralis
mcfb: M. caudifemoralis brevis
mcfl: M. caudifemoralis longus

General Terms

Such terms should be preceded by the term for the bone of which it is a part.

e.g. **adp: dorsal process of astragalus**

ilcp: cranial process of ilium

..cp: cranial process
..dp: dorsal process
..hs: horizontal sheet
..ls: lateral sheet
..ms: medial sheet
..vp: ventral process
..cdp: caudal descending process
..cp: caudal process

Foreword

This book was conceived early in 1984, precipitated by our interest in producing a single comprehensive text on dinosaurs. The editors first met in Tübingen, Federal Republic of Germany, in September 1984, where we drew up chapter topics and made a list of prospective authors for each chapter. Shortly thereafter, we solicited authorship and came up with what we unabashedly consider among the cream of dinosaur specialists throughout the world. Each of the editors agreed to write a sample chapter, which we then submitted to prospective publishers along with a proposed outline and list of authors. Our contract was signed with the University of California Press in December 1985, and soon thereafter we authorized our contributors to begin their manuscripts in earnest. These manuscripts arrived during winter and spring 1987. They were then submitted to outside review.

The second meeting of the editors took place in August 1987, at the Tyrrell Museum of Palaeontology, Drumheller, Alberta, Canada, where we assessed the first chapter drafts and their reviews. Accordingly, we returned the contributions to authors for revision. These second drafts were returned over the course of winter 1988.

The final meeting of the editors was in April 1988 at the Zakład Paleobiologii in Warsaw, Poland, where we met as participants in the Interacademy Exchange Program of the U.S. National Academy of Science and the Polish Akademia Nauk. During our enjoyable meeting, the last drafts of the manuscripts were evaluated and integrated. The scope of several chapters was modified, and a new chapter was solicited from Ralph Molnar. Once again, we returned manuscripts to the authors for fine tuning and preparation for final submission to the University of California Press in June 1988.

In addition to the authors themselves and their respective institutions, the editors would like to thank the following for their various contributions in the form of information, criticism, advice, and technical assistance: J. B. Weishampel, D. Dodson, L. Szabelski, L. M. Witmer, D. R. Bolen, P. J. Currie and B. Kowalchuck, E. M. Conner, G. Mann, R. E. Chapman, P. Steigerwald, G. Olshevsky, M. Godinot, H.-D. Sues, J. O. Farlow, N. J. Mateer, S. Phipps, Z. Kielan-Jaworowska, W. A. Clemens, Jr., and J. A. Lillegraven.

We are especially grateful to the U.S. National Academy of Science, in particular to M. S. Iovine, D. A. Berrien, G. E. Schweitzer, and A. Cleveland of the Office of Soviet and East European Affairs, for making possible the Weishampel and Dodson visit to Warsaw through their Interacademy Exchange Program and to the Polish Academy of Science for their hospitality during their visit to the Zakład Paleobiologii. We also thank G. Waxmonsky, Scientific Attaché to the U.S. Embassy in Warsaw, for conveying the Polish manuscripts to the United States. In addition, we thank the director of the Tyrrell Museum of Palaeontology, E. M. Koster, for making possible Osmólska's visit to Drumheller in 1988.

We want to extend special thanks to the following for acting as chapter reviewers: M. Balsai, M. J. Benton, J. F. Bonaparte, E. Buffetaut, E. H. Colbert, P. J. Currie, Dong Zhiming, J. O. Farlow, A. R. Fiorillo, C. A. Forster, W. B. Gallagher, P. M. Galton, J. K. Hammond, W. Langston, Jr., T. M. Lehman, J. S. McIntosh, B. K. McNab, R. E. Molnar, E. Nicholls, D. B. Norman, J. H. Ostrom, K. Padian, M. A. Raath, D. M. Raup, D. A. Russell, P. C. Sereno, J. R. Spotila, H.-D. Sues, and L. M. Witmer.

The editors and authors acknowledge the help of the following people in preparing their chapters: W. Ackersten, J. D. Archibald, H. Armstrong, D. Baird, M. J. Benton, D. S. Berman, C. C. Black, J. R. Bolt, J. F. Bonaparte, D. Brinkman, E. Buffetaut, K. Carpenter, A. J. Charig, S. Chatterjee, D. J. Chure, E. H. Colbert, G. del Corro, A. W. Crompton, P. J. Currie, M. R. Dawson, T. A. Demere, R. Denton, Dong Zhiming, J. G. Eaton, A. G. Edmund, P. Ellenberger, R. Farrar, A. Fiorillo, R. C. Fox, E. S. Gaffney, Z. B. de Gasparini, D. D. Gillette, C. Gow, J. A. Hopson, N. Hotton III, L. L. Jacobs, H. Jaeger, S. L. Jain, J. A. Jensen, J. W. Kitching, J. Lamb, W. Langston, Jr., the late A. F. de Lapparent, P. L. Larson, R. Lavocat, G. D. Leahy, T. M. Lehman, G. Leonardi, J. A. Lillegraven, C. Llompart, J. H. Madsen, Jr., N. J. Mateer, C. McGowan, J. S. McIntosh, M. C. McKenna, W. E. Miller, A. C. Milner, R. E. Molnar, M. E. Nelson, E. Nicholls, D. B. Norman, G. Olshevsky, J. H. Ostrom, K. Padian, J. M. Parrish, R. Pascual, H. P. Powell, R. Purdy, M. A. Raath, D. A. Russell, A. Sahni, D. Sloan, R. Sloan, K. Stadtman, M. Tanimoto, P. Taquet, T. Tokaryk, A. C. Walker, K. W. Westphal, V. Wheeler, B. Whitman, R. Wild, M. E. Williams, C. Yoon, and G. Zbyszewski.

Last, we are grateful to the publishing staff of the University of California Press, particularly to Elizabeth Knoll, Bettyann Kevles (formerly with the University of California Press), Shirley Warren, Diana Feinberg, and Matthew Jaffe, for their enduring help and encouragement on *The Dinosauria*.

David B. Weishampel
Peter Dodson
Halszka Osmólska

Warsaw, April 23, 1988

Preface

When *The Dinosauria* went to press in 1988, we were full of optimism that our already active field of research would continue to burgeon. As we briefly note here, the 150th anniversary of Sir Richard Owen's first conceptualization of Dinosauria (Anderson 1991, see also Modern Geology, volume 16 [1991]) has indeed seen an ever-increasing front of new investigations of these animals.

Where we once lacked general references to dinosaur evolutionary biology, the transition from the decade of the 1980s to the '90s has seen a virtual avalanche of scholarly volumes detailing these animals. Following on the heels of the publication of *The Dinosauria, Dinosaur Systematics: Approaches and Perspectives* (Carpenter and Currie 1990) provides a host of information on the taxonomy of the major dinosaur taxa. Together with *A Bibliography of the Dinosauria (Exclusive of Aves) 1677–1986* (Chure and McIntosh 1989), *The Age of Dinosaurs* (Padian and Chure 1989) and *A Revision of the Parainfraclass Archosauria Cope, 1869, Excluding the Advanced Crocodylia* (Olshevsky 1991), these texts could easily be considered sister volumes to *The Dinosauria*. Other texts of a more functional or paleobiological nature include *Dynamics of Dinosaurs & Other Extinct Giants* (Alexander 1989) and *Dragons, Spitfires, & Sea Dragons* (McGowan 1991). In the field of dinosaur ichnology, two new texts (*Dinosaur Tracks* [Thulborn 1990], *Tracking Dinosaurs* [Lockley 1991]) complement well *Dinosaur Tracks and Traces*, the Gillette and Lockley (1989) volume that only just barely got into the earlier edition of *The Dinosauria*. Finally, three new books discuss issues that straddle aspects of taxonomy, phylogeny, footprints, biomechanics, and faunistics. These include *Paleobiology of the Dinosaurs* (Farlow 1989), *Dinosaurs from China* (Dong and Milner 1988), *Predatory Dinosaurs of the World* (Paul 1988), and *Dawn of the Age of Dinosaurs in the American Southwest* (Lucas and Hunt 1989).

The past four years have provided over three dozen new taxa and/or taxonomic alterations within Dinosauria. Most important among them are *Emausaurus ernsti*, a new taxon from the Early Jurassic of northeastern Germany (Haubold 1991) that provides important information on basal thyreophorans, and *Seismosaurus halli*, from the Morrison Formation of New Mexico, which assuredly "pushes the envelope" of body size among the already gigantic sauropods (Gillette 1991). Other significant taxonomic studies include Evans and Milner (1989) on juvenile prosauropods, Gow (1990) on heterodontosaurids, and Milner and Evans (1991) on maniraptoran theropods from Portugal.

Equally important aspects of the primary literature involve a variety of phylogenetic, diversity, extinction, and paleobiological studies. Among these are Sereno's (1991) phylogenetic research on the ornithischian *Lesothosaurus diagnosticus*. This work further substantiates the basal position of this species within the ornithischian clade, and provides complementary information to that included in the Weishampel and Witmer chapter in this volume. Other treatments and discussions of the phylogeny of Dinosauria include Dodson (1990*a*) on ceratopsids, Forster (1990) on *Tenontosaurus tilletti* and its relationship within Euornithopoda, Chatterjee (1991), Witmer (1991), Ostrom (1991), Martin (1991), Tarsitano (1991), and Sereno and Rao (1992) on basal birds and their theropod relationships, and Weishampel and Heinrich (1992) on euornithopod systematics and the completeness of their fossil record.

Several new and some old dinosaurs have altered our ideas about their reproduction and ontogeny, paleoecology, and evolutionary biology. Computer-based morphometric studies (e.g., Chapman 1990, Chapman and Brett-Surman 1990, Weishampel and Chapman 1990) are providing important information on changes in skeletal form during ontogeny, phylogeny, and taphonomic alteration. Research continues apace on life-history patterns in hypsilophodontids and other dinosaurs (Winkler and Murry 1989, Coombs 1989, Lambert 1991, Heinrich et al. in press), works that inform us about the patterns of shifting reproductive strategies and evolutionary dynamics (Weishampel and Horner in press). Related studies of fossil eggs attributed to dinosaurs include the recent work of Hirsch (1989), Kurzanov and Mikhailov (1989), Hirsch and Quinn (1990), Mikhailov (1991), and Sabath (1991). These papers discuss, among other topics, the classification and distribution of amniotic eggs (including those of dinosaurs) and the ecophysiology of eggs and nests. Among studies of feeding repertoires and strategies, it has been shown that theropods made little use of bone in their diets (Fiorillo 1991*a*). Four recently-published faunal studies include Brinkman et al. (1990) on the Judith River Formation of Alberta, Fiorillo (1991*b*) on the same unit in Montana, Jerzykiewicz and Russell (1991) on the geology, paleoecology, biostratigraphy, and faunas of Mongolia, and Buffetaut and Le Loeuff (1991) on the Late Cretaceous dinosaur faunas of Europe and their stratigraphic context. New discoveries from the Upper Triassic and Upper Cretaceous of Antarctica (Anonymous 1991, Hooker et al. 1991) have added important new information on the high-latitudinal distribution of dinosaurs beyond what had been previously known from the north slope of Alaska and the south coast of Australia. And studies of new and old dinosaur material from Europe are beginning to provide novel views on insular evolutionary dynamics across Europe during the Late Cretaceous (e.g., Weishampel et al. 1991, Le Loeuff 1991).

The application of skeletochronology to growth series of dinosaurs (Chinsamy 1990, 1991, in press) has promise to provide information on the relationship of bone histology to ontogeny and perhaps to the origin of archosaurian endothermy. Paladino et al. (1990) and Spotila et al. (1991) have analyzed dinosaur metabolism using a model of gigantothermy based on the metabolic rates of leatherback turtles, concluding that this strategy combines the thermal flexibility of endothermy in living mammals and birds with the economy of reptilian-grade ectothermy.

The diversity of all or part of Dinosauria across the Mesozoic has most recently been analyzed by Weishampel and Norman (1989), Dodson (1990*b*), and Dodson and Dawson (1991). The first study attempts to understand the pattern of herbivore diversity (Ornithischia, Sauropodomorpha, contemporary non-dinosaurian herbivores) as a consequence of events within the contemporary plant realm. In contrast, the latter two studies emphasize the patterns and gaps in our knowledge of the stratigraphic distribution of dinosaurs, the historical and geographic trends in the study of dinosaurs, and suggest an empirical basis for estimating dinosaur diversity. The problem of the Cretaceous-Tertiary boundary is itself a microcosm of the overall temporal distribution of dinosaurs. This problem has been most vigorously attacked through the work of Archibald and Bryant (1990) and Sheehan et al. (1991). The former emphasize the continuity of non-dinosaurian faunal components across the boundary and conclude that terrestrial ecosystems were not reorganized by a devastating bolide impact thought to hit the Earth at this time. The latter census disarticulated dinosaur material from the Hell Creek Formation as it approaches the Cretaceous-Tertiary boundary in Montana and North Dakota. Detecting no trend of decrease in diversity at the familial level over the final two million years of the Cretaceous, Sheehan et al. conclude that their findings are compatible with a catastrophic explanation for the disappearance of dinosaurs.

In addition to these rapidly expanding aspects of research in dinosaur biology, the political world of 1988 has also changed considerably. The first to occur was the reunification of the German Democratic Republic with the Federal Republic of Germany (under the latter name) in 1990. As part of further geopolitical changes in eastern Europe as well as in Asia, events begun in August 1991 that culminated on New Year's Day 1992 rendered the Union of Soviet Socialist Republics nonexistent. In its place, the former Soviet Republics came into their own. At the same time, several of the Yugoslavian Republics claimed their own independence. In view of these changes, but cognizant of the unsettled nature of boundaries and political subdivisions, we have altered information on the geographic context of dinosaur localities, both new and old, in the Dinosaurian Distribution chapter and in the tabular entries in individual taxonomic chapters. In addition, entries for new dinosaur genera and species and/or altered taxonomic referrals in the tables within this text are followed by a ●. Many of these new entries will assuredly figure in future revisions of the taxa under consideration in this book, but are here reported only in the context of their original publication or preliminary evaluation by chapter authors. There has been no attempt to provide textual discussions of new genera and species. Bibliographic citations for these new tabular entries are provided in the preface bibliography. In addition to the tabular entries, new taxa are also added to appropriate faunal lists in the Dinosaurian Distribution chapter if there was a prior entry for its geographic and/or stratigraphic unit. Otherwise, information on provenance, geochronology, and material is provided only within the tables (viz. *Emausaurus ernsti, Tarascosaurus salluvicus*).

Mesozoic geochronology has changed somewhat over the course of the past four years, if changes between the first and second editions of *A Geologic Time Scale* (Harland et al. 1990) are any measure. Although more than 10% of stage boundaries are unchanged between 1984 and 1990 and an additional 18% fall within ± 1 million years of previous estimates, the remaining boundary estimates are increased (sometimes considerably) over previous values. We suggest that our readers adjust for these changes in conversions from the standard European stages to absolute geochronology.

Finally, we would like to again thank our many colleagues and friends for supplying us with information on new dinosaur studies, new taxa (especially via George Olshevsky, Tracy Ford, and Mike Brett-Surman), new phylogenies, and new controversies that come out of this ever-growing research field.

BIBLIOGRAPHY

Alexander, R. M. 1989. Dynamics of Dinosaurs & Other Extinct Giants. Columbia Univ. Press, New York. 167 pp.

Anderson, A. 1991. BA meets in Plymouth on dinosaurs' birthday. Science 253: 1089–1090.

Anonymous. 1991. Dinosaur remains discovered in Antarctica. Episodes 14: 150.

Antunes, M. T., and Sigogneau-Russell, D. 1991. Nouvelles données sur les Dinosaures de Crétacé supérieur du Portugal. C. R. Acad. Sci. Paris 313: 113–119.

Archibald, J. D., and Bryant, L. J. 1990. Differential Cretaceous/Tertiary extinctions of nonmarine vertebrates; evidence from northeastern Montana. Geol. Soc. Am. Sp. Paper 247: 549–562.

Bakker, R. T., Galton, P. M., Siegwarth, J., and Filla, J. 1990. A new latest Jurassic vertebrate fauna, from the highest levels of the Morrison Formation at Como Bluff, Wyoming. Part IV. The dinosaurs: a new *Othnielia*-like hypsilophodontoid. Hunteria 2: 8–13.

Bonaparte, J. F. 1991. Los vertebrados fosiles de la Formacion Rio Colorado, de la ciudad de Neuquen y Cercanias, Cretacico superior, Argentina. Rev. Mus. Argent. Cienc. Nat. "Bernardino Rivadavia" Paleontol. 4: 15–123.

Buffetaut, E., and Le Loeuff, J. 1991a. Une nouvelle espèce de *Rhabdodon* (Dinosauria, Ornithischia) du Crétacé supérieur de l'Hérault (Sud de la France). C. R. Acad. Sci. Paris 312: 943-948.

Buffetaut, E., and Le Loeuff, J. 1991b. Late Cretaceous dinosaur faunas of Europe: some correlation problems. Cret. Res. 12: 159-176.

Carpenter, K., and Currie, P. J. (eds.). 1990. Dinosaur Systematics: Approaches and Perspectives. Cambridge Univ. Press, New York. 318 pp.

Carroll, R. L., and Galton, P. M. 1977. 'Modern' lizard from the Upper Triassic of China. Nature 266: 252-255.

Chapman, R. E. 1990. Shape analysis in the study of dinosaur morphology. In: Carpenter, K., and Currie, P. M. (eds.). Dinosaur Systematics: Approaches and Perspectives. Cambridge Univ. Press., New York, pp. 21–42.

Chapman, R. E., and Brett-Surman, M. K. 1990. Morphometric observations on hadrosaurid ornithopods. In: Carpenter, K., and Currie, P. J. (eds.). Dinosaur Systematics. Approaches and Perspectives. Cambridge Univ. Press, New York, pp. 163-178.

Chatterjee, S. Cranial anatomy and relationships of a new Triassic bird from Texas. Phil. Trans. Roy. Soc. Lond. B 332: 277–342.

Chinsamy, A. 1990. The physiological implications of bone histology of *Syntarsus rhodesiensis* (Saurischia: Theropoda). Palaeontol. afr. 27: 77–82.

Chinsamy, A. 1991. The bone histology and possible growth strategy of the prosauropod *Massospondylus carinatus.* In: Kielan-Jaworoska, Z., Heintz, N., and Nakrem, H. A. (eds.). Fifth Symp. Mesozoic Terr. Ecosyst. Biota. Contrib. Paleontol. Mus. Univ. Oslo, p. 13.

Chinsamy, A. in press. Bone histology and growth trajectory of the prosauropod dinosaur *Massospondylus carinatus* (Owen). Modern Geol.

Chure, D. J., and McIntosh, J. S. 1989. A Bibliography of the Dinosauria (Exclusive of Aves) 1677–1986. Mus. West. Colorado Paleontol. Ser. 1: 1–226.

Coombs, W. P. 1989. Modern analogs for dinosaur nesting and parental care. Geol. Soc. Am. Sp. Paper 238: 21–53.

Currie, P. J., Rigby, J. K., Jr., and Sloan, R. E. 1990. Theropod teeth from the Judith River Formation of southern Alberta, Canada. In: Carpenter, K., and Currie, P. J. (eds.). Dinosaur Systematics: Approaches and Perspectives. Cambridge Univ. Press. New York, pp. 107–125.

Dodson, P. 1990a. On the status of the ceratopsids *Monoclonius* and *Centrosaurus.* In: Carpenter, K., and Currie,

P. J. (eds.). Dinosaur Systematics: Perspectives and Approaches. Cambridge Univ. Press, New York. pp. 231–243.

Dodson, P. 1990b. Counting dinosaurs: how many kinds were there? Proc. Natl. Acad. Sci. 87: 7608-7612.

Dodson, P., and Dawson, S. D. 1991. Making the fossil record of dinosaurs. Modern Geol. 16: 3–15.

Dong, Z. 1989. [On a small ornithopod (*Gongbusaurus wucaiwanensis* sp. nov.) from Kelamaili, Junggar Basin, Xinjiang, China.] Vert. PalAs. 27: 140–147. (In Chinese with English summary)

Dong Z. 1990. [On remains of the sauropods from Kelamaili region, Junggar Basin, Xinjiang, China.] Vert. PalAs. 28: 43–58. (In Chinese with English summary)

Dong Z., and Milner, A. C. 1988. Dinosaurs from China. British Mus. Nat. Hist., London. 114 pp.

Evans, S. E., and Milner, A. R. 1989. *Fulengia,* a supposed early lizard reinterpreted as a prosauropod dinosaur. Palaeontology 32: 223–230.

Farlow, J. O. (ed.). 1989. Paleobiology of the Dinosaurs. Geol. Soc. Am. Sp. Paper 238: 100 pp.

Fiorillo, A. R. 1991a. Prey bone utilization by predatory dinosaurs. Palaeogeogr. Palaeoclimatol. Palaeoecol. 88: 157–166.

Fiorillo, A. R. 1991b. Taphonomy and depositional setting of Careless Creek Quarry (Judith River Formation), Wheatland County, Montana, U.S.A. Palaeogeogr. Palaeoclimatol. Palaeoecol. 81: 281–311.

Forster, C. A. 1990. The postcranial skeleton of the ornithopod dinosaur *Tenontosaurus tilletti.* J. Vert. Paleontol. 10: 273–294.

Gillette, D. D. 1991. *Seismosaurus halli,* gen et sp. nov., a new sauropod dinosaur from the Morrison Formation (Upper Jurassic/Lower Cretaceous) of New Mexico, USA. J. Vert. Paleontol. 11: 417–433.

Gillette, D. D., and Lockley, M. G. 1989. Dinosaur Tracks and Traces. Cambridge Univ. Press, New York. 454 pp.

Gow, C. E. 1990. A tooth-bearing maxilla referable to *Lycorhinus angustidens* Haughton, 1924 (Dinosauria, Ornithischia). Ann. S. Afr. Mus. 90: 367–380.

Harland, W. B., Armstrong, R. L., Cox, A. V., Craig, L. E., Smith, A. G., and Smith, D. G. 1990. A Geologic Time Scale 1989. Cambridge Univ. Press, Cambridge. 263 pp.

Haubold, H. 1991. Ein neuer Dinosaurier (Ornithischia, Thyreophora) aus dem unteren Jura des nördlichen Mitteleuropa. Rev. Paléobiol. 9: 149–177.

Heinrich, R. E., Ruff, C. B., and Weishampel, D. B. in press. Femoral ontogeny and locomotor biomechanics of *Dryosaurus lettowvorbecki* (Dinosauria, Iguanodontia). Zool. J. Linn. Soc.

Hirsch, K. F. 1989. Interpretations of Cretaceous and pre-Cretaceous eggs and shell fragments. In: Gillette, D. D., and Lockley, M. G. (eds.). Dinosaur Tracks and Traces. Cambridge Univ. Press, New York, pp. 89–97.

Hirsch, K. L., and Quinn, B. 1990. Eggs and eggshell fragments from the Upper Cretaceous Two Medicine Formation of Montana. J. Vert. Paleontol. 10: 491–511.

Hooker, J. J., Milner, A. C., and Sequeira, S. E. K. 1991. An ornithopod dinosaur from the Late Cretaceous of west Antarctica. Antarct. Sci. 3: 331–332.

Hunt, A. P. 1989. A new ?ornithischian dinosaur from the Bull Canyon Formation (Upper Triassic) of east-central New Mexico. In: Lucas, S. G., and Hunt, A. P. (eds.). Dawn of the Age of Dinosaurs in the American Southwest. New Mexico Museum of Natural History, Albuquerque, pp. 355–358.

Hunt, A. P., and Lucas, S. G. 1991. *Rioarribasaurus*, a new name for a Late Triassic dinosaur from New Mexico (USA). Paläont. Z. 65: 191–198.

Jensen, J. A. 1988. A fourth new sauropod dinosaur from the Upper Jurassic of the Colorado Plateau and sauropod bipedalism. Gr. Basin Nat. 48: 1121–1145.

Jerzykiewicz, T., and Russell, D. A. 1991. Late Mesozoic stratigraphy and vertebrates of the Gobi Basin. Cret. Res. 12: 345–397.

Kurzanov, S. M. 1990. Novyy rod prototseratopsid iz pozdnego mela Mongolii. [A new protoceratopsid genus from the Late Cretaceous Mongolia.] Paleontol. Zh. 1990: 91–97. (In Russian).

Kurzanov, S. M., and Mikhailov, K. E. 1989. Dinosaur eggshells from the Lower Cretaceous of Mongolia. In: Gillette, D. D., and Lockley, M. G. (eds.). Dinosaur Tracks and Traces. Cambridge Univ. Press, New York, pp. 109–113.

Kurzanov, S. M., and Osmólska, H. 1991. *Tochisaurus nemegtensis* gen. et sp. n., a new troodontid (Dinosauria, Theropoda) from Mongolia. Acta Palaeontol. Polonica 36: 69–76.

Lambert 1991. Altriciality and its implications for dinosaur thermoenergetic physiology. N. Jb. Geol. Paläont. Abh. 182: 73–84.

Le Loeuff, J. 1991. The Campano-Maastrichtian vertebrate faunas of southern Europe and their relationships with other faunas in the world; palaeobiogeographical implications. Cret. Res. 12: 93–114.

Le Loeuff, J., and Buffetaut, E. 1991. *Tarascosaurus salluvicus* nov. gen., nov. sp., dinosaure théropode du Crétacé supérieur de Sud de la France. Geobios 25: 585–594.

Li K. 1988. Researching on *Omeisaurus luoquanensis* Li, sp. nov. In: He X.-L., Li K., and Cai K.-J. The Middle Jurassic Dinosaur Fauna from Dashanpu, Zigong, Sichuan. Sichuan Publ. House Sci. Technol., Chengdu, People's Republic of China, pp. 94–105, 133. (In Chinese with English summary)

Lockley, M. 1991. Tracking Dinosaurs. Cambridge Univ. Press, New York. 238 pp.

Lucas, S. G., and Hunt, A. P. (eds.). 1989. Dawn of the Age of Dinosaurs in the American Southwest. New Mexico Museum of Natural History, Albuquerque. 414 pp.

Martin, L. D. 1991. Mesozoic birds and the origin of birds. In: Schultze, H.-P., and Trueb, L. (eds.). Origins of the Higher Groups of Tetrapods. Comstock Publ. Assoc., Ithaca, New York, pp. 485–540.

Mathur, U. B., and Srivastava, S. 1987. Dinosaur teeth from Lameta Group (Upper Cretaceous) of Kheda District, Gujarat. J. Geol. Soc. India 29: 554–566.

McGowan, C. 1991. Dinosaurs, Spitfires, & Sea Dragons. Harvard Univ. Press, Cambridge. 365 pp.

Mikhailov, K. E. 1991. Classification of fossil eggshells of amniotic vertebrates. Acta Palaeontol. Polonica 36: 193–238.

Milner, A. R., and Evans, S. E. 1991. The Upper Jurassic diapsid *Lisboasaurus estesi*—a maniraptoran theropod. Palaeontology 34: 503–513.

Nessov, L. 1985. [Rare bony fishes, land lizards and mammals from shoals and coastal plain zone of the Kysylkum Cretaceous.] Ezhegod. Vsesoyuzn. Obschchest. 28: 199–219. (In Russian)

Nessov, L. A., Kaznyshkina, L. F., and Cherepanov, G. O. 1989. Dinozavri-Tseratopsii i Krokodili Mezozoica Sredney Azii. In: Bogdanova, T. N., and Khozatsky, L. I. (eds.). [Theoretical and Applied Aspects of Modern Paleontology]. 'Nauka' Publ., Leningrad, pp. 144–154. (In Russian)

Olshevsky, G. 1991. A Revision of the Parainfraclass Archosauria Cope, 1869, Excluding the Advanced Crocodylia. Publ. Req. Res., San Diego. 196 pp.

Ostrom, J. H. 1991. The question of the origin of birds. In: Schultze, H.-P., and Trueb, L. (eds.). Origins of the Higher Groups of Tetrapods. Comstock Publ. Assoc., Ithaca, New York, pp. 467–484.

Padian, K., and Chure, D. J. (eds.). 1989. The Age of Dinosaurs. Paleontol. Soc. Short Courses Paleontol. 2: 210 pp.

Paladino, F. V., O'Connor, M. P., and Spotila, J. R. 1990. Metabolism of leatherback turtles, gigantothermy, and thermoregulation of dinosaurs. Nature 344: 858–860.

Paul, G. S. 1988. Predatory Dinosaurs of the World. Simon and Schuster, New York. 464 pp.

Peng, G.-Z. 1990. [A new species of small ornithopod type from Zigong, Sichuan.] J. Zigong Dinosaur Mus. News 1990: 19–27. (In Chinese)

Powell, J. E. 1987. The Late Cretaceous fauna of Los Alamitos, Patagonia, Argentina. Part VI—The titanosaurids. Rev. Mus. Argent. Cienc. Nat. "Bernardino Rivadavia" Paleontol. 3: 147–153.

Sabath, K. 1991. Upper Cretaceous amniotic eggs from the Gobi Desert. Acta Palaeontol. Polonica 36: 151–192.

Seiffert, J. 1973. Contribuicao para o conhecimento de Fauna do Kimeridgiano do Mina de Lignito Guimarota (Leiria, Portugal). II parte. Upper Jurassic lizards from central Portugal. Mem. Serv. Geol. Portugal 22: 7–88.

Sereno, P. C., 1991. *Lesothosaurus*, "fabrosaurids", and the early evolution of Ornithischia. J. Vert. Paleontol. 11: 145–167.

Sereno, P. C., and Rao C.-G. 1992. Early evolution of avian flight and perching: new evidence from the Lower Cretaceous of China. Science 255: 845–848.

Sheehan, P. M., Fastovsky, D. E., Hoffman, R. G., Berghaus, C. B., and Gabriel, D. L. 1991. Sudden extinction of the dinosaurs—latest Cretaceous, upper Great Plains, USA. Science 254: 835–839.

Spotila, J. R., O'Connor, M. P., Dodson, P., and Paladino, F. V. 1991. Hot and cold running dinosaurs: body size, metabolism and migration. Modern Geol. 16: 203–227.

Tarsitano, S. 1991. *Archaeopteryx*: quo vadis? In: Schultze, H.-P., and Trueb, L. (eds.). Origins of the Higher Group of Tetrapods. Comstock Publ. Assoc., Ithaca, New York, pp. 541–576.

Thulborn, T. 1990. Dinosaur Tracks. Chapman and Hall, London. 410 pp.

Weishampel, D. B., and Chapman, R. E. 1990. Morphometric analysis of *Plateosaurus* from Trossingen (Baden-Württemberg, Federal Republic of Germany). In: Carpenter, K., and Currie, P. J. (eds.). Dinosaur Systematics: Perspectives and Approaches. Cambridge Univ. Press, New York, pp. 43–51.

Weishampel. D. B., Grigorescu, D., and Norman, D. B. 1991. The dinosaurs of Transylvania: island biogeography in the Late Cretaceous. Natl. Geogr. Res. 7: 68–87.

Weishampel, D. B., and Heinrich, R. E. 1992. Systematics of Hypsilophodontidae and basal Iguanodontia (Dinosauria: Ornithopoda). Hist. Biol. 6: 159–184.

Weishampel, D. B., and Horner, J. R. in press. Life history syndromes, heterochrony, and the evolution of Dinosauria. In: Carpenter, K., Horner, J. R., & Hirsch, K. (eds.). Dinosaur Eggs and Babies. Cambridge Univ. Press, New York.

Weishampel, D. B., and Norman, D. B. 1989. Vertebrate herbivory in the Mesozoic; jaws, plants, and evolutionary metrics. Geol. Soc. Am. Sp. Paper 238: 87–100.

Wild, R. 1991. *Janenschia* n. g. *robusta* (E. Fraas 1908) pro *Tornieria robusta* (E. Fraas 1908) (Reptilia, Saurischia, Sauropodomorpha). Stuttgarter Beitr. Naturk. Ser. B 173: 1–4.

Winkler, D. A., and Murry, P. A. 1989. Paleoecology and hypsilophodontid behavior at the Proctor Lake dinosaur locality (Early Cretaceous), Texas. Geol. Soc. Am. Sp. Paper 238: 55–61.

Witmer, L. M. 1991. Perspectives on avian origins. In: Schultze, H.-P., and Trueb, L. (eds.). Origins of the Higher Groups of Tetrapods. Comstock Publ. Assoc., Ithaca, New York, pp. 427–466.

Yadagiri, P. 1988. A new sauropod *Kotasaurus yamanpalliensis* from Lower Jurassic Kota Formation of India. Rec. Geol. Surv. India 11: 102–127.

Zhang Y. 1988. The Middle Jurassic dinosaur fauna from Dashanpu, Zigong, Sichuan. Vol. III. Sauropod dinosaurs (I). Sichuan Publ. House Sci. Tech., Chengdu, China, pp. 1–89. (In Chinese with English summary)

THE DINOSAURIA

Introduction

DAVID B. WEISHAMPEL

PETER DODSON and

HALSZKA OSMÓLSKA

As in so many other scientific disciplines, the study of dinosaurs is in a state of unprecedented activity. Newly discovered dinosaurs are being described at the rate of one every seven weeks; half of all currently recognized genera have been described since 1970. These dinosaurs are being described both from terrains with a long history of paleontological exploration and from those where intensive paleontological activities are a more recent development. During the past two decades, research in Mongolia by Mongolian, Polish, and Soviet scientists has revealed astonishing diversity of Late Cretaceous dinosaurs, particularly among small carnivorous dinosaurs. Chinese scientists have been uncovering a very rich fossil record that spans the entire age of dinosaurs. There have been very important new discoveries in Argentina. At last, the remains of a dinosaur have been found in Antarctica (Gasparini et al. 1987). During the past two decades, our knowledge of many groups of dinosaurs that were poorly known has burgeoned (e.g., pachycephalosaurians, oviraptorids, dromaeosaurids) and some new and quite unexpected groups have been discovered (e.g., segnosaurs). Public interest in dinosaurs is greater than ever and is nourished by popular books (e.g., Norman 1985) and

works of art (e.g., Czerkas and Olson 1987) of exceptional quality.

Lacking in this veritable revolution in the study of dinosaurs has been any sort of authoritative professional overview of the subject. By default, comprehensive reviews of dinosaurs have come from workers outside the field, notably Romer (1956, 1966) and Steel (1969, 1970). The intention of *The Dinosauria* is to provide the first comprehensive, critical, and authoritative review of the taxonomy, phylogeny, biogeography, paleobiology, and stratigraphic distribution of all dinosaurs. It is true that the increase in our knowledge of dinosaurs has necessitated a reassessment of every aspect of our understanding of them. This book therefore treats a breadth of topics from dinosaur origins and interrelationships to dinosaur energetics, behavior, and extinction.

To accomplish these ambitious goals, it was necessary to solicit exceptional authors. Our contributors collectively have a depth and breadth of hands-on experience with various groups of dinosaurs that cannot be matched by any single person. The fossil record of dinosaurs is too extensive for any one person to command. We have striven for international authorship, if for no other reason than that many of the specialists on dinosaurs reside in distant corners of the world. We are especially pleased to have both Eastern and Western contributors to *The Dinosauria*. We were pleasantly surprised that consensus across the continents and oceans has proved to be so readily forthcoming.

A BRIEF HISTORY OF DINOSAUR STUDIES

Paleontology is a quintessentially historical subject. If the existence of dinosaurs in strata is an objective fact, then our current understanding of dinosaurs is contingent on the history of their discovery and exploration. How would our views have differed if complete skeletons had been found earlier, or if the study of dinosaurs had begun in China, Mongolia, or South America? There are several excellent reviews of the history of paleontology in general (e.g., Desmond 1982; Rudwick 1985) and of dinosaurs in particular (e.g., Colbert 1961, 1983; Wilford 1986). One of the earliest documented dinosaur discoveries was of a distal femur of a megalosaur from Oxfordshire, England, which was described by Robert Plot in 1677 as "the Thigh-bone of a Man or at least of some other Animal" (Delair and Sarjeant 1975). The first modern discovery of a dinosaur seems to be that of a tooth-bearing jaw discovered by Buckland near Stonesfield in Oxfordshire, England, in 1818 and described by Buckland as *Megalosaurus* in 1824. The description by Mantell of *Iguanodon*, based on several teeth from the Tilgate Forest, followed in 1825. Subsequent discoveries in England and Europe (*Hylaeosaurus, Thecodontosaurus, Plateosaurus, Poekilopleuron, Cetiosaurus*) were similarly fragmentary, but by 1841, Owen (1842b) coined the term *dinosaur* ("terrible lizard") in recognition of a tribe of extinct reptiles that represented the apotheosis of Mesozoic terrestrial evolution and the nearest approach of reptiles to the mammalian condition, not only in limbs and posture but, by inference, in the condition of lungs and circulation. Vertebrate fossils from the lower Mesozoic Karroo Beds of South Africa were shipped to London in quantity beginning in the 1850s. Dinosaurs were a minor component compared to mammallike reptiles, but among the former were the prosauropods *Massospondylus carinatus* described by Owen (1854) and *Euskelosaurus browni* described by Huxley (1866).

The first partial skeleton of a dinosaur which supported an interpretation of bipedal posture (and which incidentally permitted the first skeletal reconstruction and mount) was that of *Hadrosaurus foulkii*, found in New Jersey in 1858 and described by Joseph Leidy of Philadelphia. The tiny and nearly complete skeleton of *Compsognathus longipes* from the Upper Jurassic Solnhofen Limestone of Bavaria was described by Wagner in 1859. Dinosaurs continued to come to light in Europe, but, apart from *Compsognathus*, were primarily solitary, incomplete specimens. Dinosaurs were found in abundance for the first time in 1877 in the Upper Jurassic Morrison Formation at three localities in Colorado and Wyoming, providing the initial evidence for faunas dominated by huge sauropods and stegosaurs. In 1878, the mass accumulation of *Iguanodon* in a coal mine at Bernissart, Belgium (Casier 1978; Norman 1986) came to light. Between 1889 and 1892, J. B. Hatcher collected thirty-two partial or complete skulls of the horned dinosaur *Triceratops* from the latest Cretaceous Lance Formation of Wyoming. Marsh (1896) summarized the known record of dinosaurs. By this time, fewer than 20 percent of the dinosaurs we know today had come to light, but the fundamentals of a classification system had been created and a fair grasp of diversity achieved. Marsh presented complete skeletal reconstructions of twelve dinosaurs, including four European forms, and boasted (p. 238), "The remarkable discoveries in North America, however, have changed the whole subject, and in the place of fragmentary specimens many entire skeletons of dinosaurian reptiles have been brought to light, and thus definite information has replaced uncertainty."

Late Cretaceous dinosaurs were discovered in Alberta, Canada, as early as 1884. The earliest Canadian monograph was that by Lambe (1902). The celebrated Canadian dinosaur rush began in 1910, as parties from the American Museum of Natural History led by Barnum Brown were soon challenged by parties from the Geological Survey of Canada led by C. H. Sternberg and later by C. M. Sternberg. Other institutions, notably the Royal Ontario Museum and the University of Alberta, also collected specimens in the Alberta fossil beds along the Red Deer River valley. These beds, among the richest in the world (Russell 1967; Dodson 1971; Béland and Russell 1978), have continued to produce fossils to the present (e.g., Currie 1987b).

For four years beginning in 1909, Upper Jurassic beds at Tendaguru, Tanzania, produced huge quantities of sauropods and other dinosaurs for expeditions from the Humboldt Museum in Berlin under the direction of W. Janensch. These beds were later worked by expeditions from the British Museum (1924-1929). The mass burial of *Plateosaurus* at Trossingen, Germany, was worked intermittently from 1911 to 1932, resulting in the recovery of about 100 skeletons or partial skeletons (Weishampel 1984*b*). In 1922, the first of three expeditions from the American Museum of Natural History reached Mongolia and discovered rich Upper Cretaceous faunas that included nests and eggs of *Protoceratops andrewsi*. Subsequent work by paleontologists from Mongolia (R. Barsbold, A. Perle), the Soviet Union (A. K. Rozhdestvensky, S. M. Kurzanov), and Poland (T. Maryańska, H. Osmólska) has continued the momentum. Dinosaur paleontology in China dates from 1929, with the description by the Swede, C. Wiman, of several dinosaurs, including the sauropod *"Helopus"* (= *Euhelopus* Romer, 1956). Another Swede, B. Bohlin (1953) described a fauna from Gansu. The Soviet, A. N. Riabinin (1930*a*, 1930*b*) described dinosaur faunas in the region of the Amur River. The great Chinese paleontologist, C. C. Young, beginning in 1931, had a long and distinguished career as a student of Chinese dinosaurs. Since 1970, Chinese paleontologists have described many new fossils throughout the Mesozoic (review by Zhen et al. 1985).

In South America, the monograph by Huene (1929*a*) on the Cretaceous dinosaurs of Argentina is noteworthy. South America, especially Brazil and Argentina, has been the site of much work in the past two decades. A valuable summary was done by Bonaparte (1978*a*), but much new work has appeared since then. A number of prosauropods and primitive ornithischians were described from South Africa during the interval between 1911 and 1924, especially by Broom, Haughton, and Hoepen. Many of the former have now been synonymized with *Massospondylus* by Cooper (1981*a*). Dinosaur paleontology was revitalized in southern Africa with the discoveries of the early ornithischians, *Heterodontosaurus tucki* (Crompton and Charig 1962) and *Fabrosaurus australis* (Ginsburg 1964) and of the small theropod, *Syntarsus rhodesiensis* (Raath 1969).

RECENT ISSUES IN THE STUDY OF DINOSAURS

Expeditions in the 1960s, notably the Yale University expeditions to the Lower Cretaceous Cloverly Formation of Wyoming and Montana led by J. H. Ostrom and the Polish-Mongolian expeditions to Upper Cretaceous beds of the Gobi Desert led by Z. Kielan-Jaworowska and R. Barsbold, laid the groundwork for the revolution that has occurred in the study of dinosaurs since 1969. The two Yale monographs (Ostrom 1969*b*,

1970*a*) introduced an entirely new dinosaur fauna, among which was *Deinonychus antirrhopus*, an important small theropod, from a hitherto poorly known stratigraphic interval (Albian-Aptian). Results of the Polish-Mongolian expeditions are contained in a lengthy series of monographs in *Palaeontologia Polonica*. Results of Soviet-Mongolian work of the 1970s and 1980s are published in the Joint Soviet-Mongolian Paleontological Expedition, Transactions. Unexpected diversity in many groups of dinosaurs has been found, especially pachycephalosaurs (Maryańska and Osmólska 1974; Perle et al. 1982) and segnosaurs (Perle 1979; Barsbold and Perle 1980, 1983) but also including ornithomimids (Osmólska et al. 1972), small theropods (Barsbold 1974; Kurzanov 1981; Osmólska 1987), ankylosaurs (Maryańska 1977), protoceratopsids (Maryańska and Osmólska 1975), and hadrosaurids (Maryańska and Osmólska 1981*a*). Finally, instead of fossils from around the world ending up in a few centers of study (e.g., London, Berlin, New York) the past two decades have seen the emergence of vigorous national schools of dinosaur study around the world (e.g., Mongolian People's Republic, People's Republic of China, Argentina, South Africa, Australia, Canada) which have immeasurably increased our understanding of dinosaur distribution and diversity.

As our understanding of alpha and beta diversity of dinosaurs has increased, conceptual issues have come to the fore during the past two decades. Dinosaur paleobiology has emerged as a vigorous discipline. Perhaps its roots might be traced to the description of *Deinonychus antirrhopus* (Ostrom 1969*b*), with attendant revelations about dinosaur behavior. *D. antirrhopus* was an active, large-brained predator whose discovery played a pivotal role in the development of the concept of endothermic dinosaurs (e.g., Ostrom 1970*b*; Bakker 1971*c*; Thomas and Olson 1980). Forever destroyed is the concept of dinosaurs as sluggish behemoths that were evolutionary failures (Bakker 1968, 1971*c*). Other workers demonstrated the potential for sexual display (Farlow and Dodson 1975; Dodson 1975, 1976; Hopson 1975*b*), agonistic intraspecific behavior (Molnar 1977; Sues 1978*b*), and vocal communication (Weishampel 1981*a*). The recognition of juvenile dinosaurs has emerged as a major theme of study (Dodson 1975, 1976; Maryańska and Osmólska 1975; Bonaparte and Vince 1979; Horner and Makela 1979; Coombs 1980*b*, 1982; Carpenter 1982*b*; Horner 1984*a*, 1987; Fiorillo 1987; Horner and Weishampel 1988; Dodson and Currie 1988). The study of egg physiology (Seymour 1979), the inference of sophisticated maternal behavior (e.g., Horner and Makela 1979; Horner 1984*a*), and the inference of colonial nesting and nest-site fidelity (Horner 1982) have provided profound insight into the social behavior and life histories of certain dinosaurs. Other workers have integrated dinosaurs into an evolutionary ecological context, particularly with regard to major biotic shifts in the Mesozoic (e.g., the radiation of angiosperms; Bakker 1978; Weishampel 1984*a, c;* Norman and Weishampel 1985; Wing and Tiffney 1987; Coe et al. 1987; Farlow 1987*b*).

The accelerating pace of dinosaur studies during the past two decades and the nascent impact of cladistics have called into question some of the most fundamental systematic tenets of our science. The major themes in dinosaur systematics were set in the late nineteenth century with the recognition of a diphyletic Dinosauria consisting of the Saurischia and Ornithischia (Seeley 1887*b;* Huene 1914*c,* 1914*d*), each seemingly independently derived from thecodonts (Charig et al. 1965). The composition and relationships of the Archosauria have now been profoundly redefined (Benton 1985*b*). In recent years, dinosaur monophyly was first proposed by Bakker and Galton (1974), who allied ornithischians with a prosauropod grade of saurischians. The year 1984 marked the first dramatic impact of cladistics on dinosaur systematics (Benton 1984*a;* Sereno 1984; Norman 1984*a,* 1984*b;* Paul 1984*a*). Successive studies (Benton 1984*a;* Gauthier and Padian 1985; Gauthier 1986; Brinkman and Sues 1987) have corroborated monophyly in the Dinosauria and have outlined various internal monophyletic taxa.

While it once seemed an easy matter to classify animals as ornithischian or saurischian, recent reconsiderations of dinosaur relationships have forced the classical bipartite division to be abandoned (table 1). Indeed, the very definition of the Dinosauria has become subject to revision (e.g., Benton 1986, this vol.; Gauthier 1986; Brinkman and Sues 1987). A list of subtle technical characters separates nondinosaurs from primitive dinosaurs. It becomes clear that, as for all major taxa, the term *dinosaur* has a significant arbitrary component. There is a monophyletic assemblage that includes all animals we choose to call dinosaurs, but where we choose to draw the boundary between dinosaurs and nondinosaurs is a subjective matter, not a given fact of nature. It would not be incorrect to choose characters that would permit pterosaurs to be included within the Dinosauria (pterosaurs are regarded by Gauthier and Padian [1985] and Gauthier [1986] as a close outgroup of dinosaurs). Nor would it be incorrect to choose characters that would eliminate

TABLE 1 Dinosauria

```
Dinosauria
    Staurikosaurus
    Herrerasauridae
    Saurischia
        Theropoda
            Ceratosauria
            Tetanurae
                Carnosauria
                Coelurosauria
                    Ornithomimosauria
                    Maniraptora
                        Elmisauridae
                        Oviraptorosauria
                        Troodontidae
                        Dromaeosauridae
        Sauropodomorpha
            Prosauropoda
            Sauropoda
        Segnosauria
    Ornithischia
        Lesothosaurus
        Genasauria
            Thyreophora
                Scutellosaurus
                Scelidosaurus
                Stegosauria
                Ankylosauria
                    Nodosauridae
                    Ankylosauridae
            Cerapoda
                Ornithopoda
                    Heterodontosauridae
                    Hypsilophodontidae
                    Iguanodontia
                        Tenontosaurus
                        Dryosauridae
                        Camptosaurus
                        Iguanodontidae
                        Probactrosaurus
                        Hadrosauridae
                            Telmatosaurus
                            Gilmoreosaurus
                            Tanius
                            Claosaurus
                            Secernosaurus
                            Tsintaosaurus
                            Hadrosaurinae
                            Lambeosaurinae
                Marginocephalia
                    Stenopelix
                    Pachycephalosauria
                        Homalocephalidae
                        Pachycephalosauridae
                    Ceratopsia
                        Psittacosauridae
                        Neoceratopsia
                            Protoceratopsidae
                            Ceratopsidae
                                Centrosaurinae
                                Chasmosaurinae
```

Staurikosaurus, Ischisaurus, and *Herrerasaurus* from the Dinosauria, leaving intact the long-held (precladistic) view that all dinosaurs belong either to the Ornithischia or to the Saurischia. It is, however, a matter of consensus among dinosaur paleontologists that *Lagosuchus* and pterosaurs are nondinosaurs and that *Staurikosaurus, Herrerasaurus,* and *Ischisaurus* are dinosaurs that predate the split between the Saurischia and the Ornithischia. Another area that has been affected by cladistic thought in the Dinosauria concerns the relationship of birds to dinosaurs. It has become clear from the work of Ostrom (1975*a*, 1975*b*, 1976*b*, 1985) that birds are descended from small theropods. In cladistic practice, this necessitates that birds be recognized within the Dinosauria, specifically, within the Maniraptora (Gauthier 1986). While we readily accept that birds are present-day dinosaurian descendants, we have not provided discussion of avian taxonomy, phylogeny, and biology. Instead, we urge readers seeking information on birds to look elsewhere.

Although the major arrays of dinosaurs have been subjected to cladistic analysis (Maryańska and Osmólska 1985; Cooper 1985; Gauthier 1986; Sereno 1986), much of the fundamental analysis at lower taxonomic levels by appropriate specialists has not yet been done. Accordingly, we have challenged each author to apply cladistic analysis to the group in question. In this, we feel we have been at least partially successful.

Cladistic thought has emphasized several differing interpretations of dinosaurian relationships as well. In one example, Currie (this volume) makes the case that the small and poorly known theropod, *Chirostenotes pergracilis,* is an elmisaurid, while Ostrom (this volume) claims it as a dromaeosaurid. Additional and better material will assuredly aid in this animal's proper assignment. Three more fundamental issues include the monophyly of prosauropods, the relationships of the Abelisauridae, and the monophyly of the Deinonychosauria. Benton (this vol.) follows Gauthier (1986) in his discussion of sauropodomorph relationships, noting in particular the paraphyletic nature of prosauropods relative to the monophyletic Sauropoda. The latter is considered to be composed of a number of small basal taxa and two large crown clades. In contrast, Galton (this vol.) and McIntosh (this vol.) argue that the Prosauropoda is monophyletic, having only a basal relationship with the Sauropoda, and that the Sauropoda itself is composed of as many as five distinct clades. The differences in views are based principally on the generality of both cranial and postcranial characters, often notoriously missing in pivotal taxa. The abelisaurids represent a new and unusual group. Thought to be closely allied with carnosaurs, they nonetheless show a number of features that are apparently derived with respect to some members of the Ceratosauria (Molnar, this vol.). More work on basal theropod taxa may disentagle the relationships of this newcomer to dinosaur diversity. Last, Osmólska (1981, this vol.) and to an extent Gauthier (1986, Appendix) have questioned the unique relationship between the Troodontidae and Dromaeosauridae, the two taxa thought to comprise the Deinonychosauria. Colbert and Russell (1969), Barsbold (1976*b*), Gauthier (1986), and Os-

trom (this vol.) argue that the two families form a monophyletic unit that may be considered the sister taxon to birds. It is principally the construction of the enlarged ungual of the second pedal digit that is used to establish monophyly of the Deinonychosauria. In contrast, this same feature is seen as a convergence among troodontids and dromaeosaurids when viewed against other aspects of cranial and postcranial anatomy (specifically, the otherwise primitive nature of the pes).

FORMAT AND AIMS OF *THE DINOSAURIA*

The first section of this volume relates issues of general concern about dinosaurs as a whole. What constitutes the Dinosauria? What are the interrelationships of dinosaurian groups? What is the stratigraphic and geographic distribution of dinosaurs? What can be said about dinosaur behavior and physiology? And what do we know about their extinction?

Section II of *The Dinosauria* is devoted to individual dinosaur taxa, beginning with primitive groups, notably *Staurikosaurus* and the Herrerasauridae, and ending with the Neoceratopsia. Each of the chapters gives a comprehensive account of each group under consideration with a uniform format. The authors consider all dinosaur species-level taxa and critically review the status of all proposed species. For higher taxa, diagnostic features, anatomy, phylogenetic relationships, and paleoecology are emphasized. Each taxonomic chapter includes a table summarizing species-level taxonomy, geographic and stratigraphic occurrence, age, and the nature of the material on which the species is based. Also included is a listing of *nomina dubia*. The chief criterion for *nomina dubia* is whether or not further material can conceivably be referred to the taxon. *Nomina nuda* were ignored completely, because they have no scientific status until specific material is associated with a name. Each chapter thus is not only a review but an original comprehensive synthesis by an appropriate specialist. Such treatments highlight both the strengths and the weaknesses of the dinosaurian fossil record.

Ages of dinosaur occurrences are related to European marine stages and corresponding ages (fig. 1) as far as has been possible, although few of the dinosaurs are in marine beds and correlation with the European sequence may be uncertain on other continents. Often, age assignments are little more than reasonable esti-

Cretaceous	Late	Maastrichtian	65
		Campanian	73
		Santonian	83
		Coniacian	87.5
		Turonian	88.5
		Cenomanian	91
	Early	Albian	97.5
		Aptian	113
		Barremian	119
		Hauterivian	125
		Valanginian	131
		Berriasian	138
Jurassic	Late	Tithonian	144
		Kimmeridgian	150
		Oxfordian	156
	Middle	Callovian	163
		Bathonian	169
		Bajocian	175
		Aalenian	181
	Early	Toarcian	188
		Pliensbachian	194
		Sinemurian	200
		Hettangian	206
Triassic	Late	Rhaetian	213
		Norian	219
		Carnian	225
			231

Fig. 1. Geologic time scale used throughout the text (from Harland et al. 1982).

mates about the contemporaneity with the classic European stages.

A special terminology has been developed for Upper Cretaceous strata of the Western Interior of North America, where terrestrial sediments interbed with marine rocks containing ammonites. The terms *Aquilan, Judithian, Edmontonian,* and *Lancian* were proposed by Russell (1964) for successively younger biostratigraphic units based on contained terrestrial faunal assemblages, both vertebrate and invertebrate. Explicitly extending the North American Cenozoic land mammal age concept, Lillegraven and McKenna (1986) formally defined the Aquilan, Judithian, and Lancian ages, based on mammalian faunal assemblages (the Edmontonian age was regarded as based on too few taxa to be defined at present). The four ages, which correspond to early to middle Campanian, late Campanian, early to middle Maastrichtian, and late Maastrichtian, respectively (Eaton 1987), have considerable utility for dinosaur workers, at least in the northern half of the Western Interior (e.g., Russell 1984c; Lehman 1987). The well-known dinosaur faunas of the latter

three ages are distinct at generic levels from one another. Extrapolations of the Late Cretaceous land mammal ages outside of the northern part of the Western Interior have not yet been successfully applied elsewhere in the world.

Authors were challenged to apply cladistic methods of analysis to their groups, including the explicit identification of outgroups, and most were successful in this regard. We have avoided the use of quotation marks around paraphyletic taxa by making explicit statements about their status and then referring to them in the vernacular (e.g., iguanodonts, prosauropods). Similarly, metataxa as defined by Gauthier (1986) are avoided.

We have taken a slightly unconventional stance in reporting taxonomic authorship in the text and tables in *The Dinosauria*. In addition to providing the usual taxon-author combination prescribed by the International Code of Zoological Nomenclature, we have also included the date citation reflected in the Bibliography. That is, *Epanterias amplexus* Cope 1878b refers to a specific bibliographic entry in this book. We feel that this addition increases the information content of each taxonomic citation and consequently reduces ambiguity.

In a work such as *The Dinosauria*, it is important to strive for an internally consistent anatomical nomenclature that may set the standard for future work. In human anatomy, such a standard has been set by the Nomina Anatomica (1983). This document has had a great influence on anatomical nomenclature in general and on mammalian nomenclature in particular. This is despite the fact that humans are a peculiarly modified bipedal species and that human anatomists have compounded problems of comparison among species by adopting as standard posture the figure with forearms supinated. An attempt to standardize a rational anatomical nomenclature for all tetrapods is the Nomina Anatomica Veterinaria (1983). Although the nomenclature of the NAV applies specifically to domesticated animals, the general principles articulated by the NAV directly address the problems caused by indiscriminately imposing human terms on other vertebrates.

We adopt the following aspects of the NAV. The terms *ventral* (toward the belly), *dorsal* (toward the back), *cranial* (toward the head), and *caudal* (toward the tail) replace the human terms *anterior, posterior, superior,* and *inferior,* which are ambiguous when applied to tetrapods. The term *cranial* has no meaning on the head itself, hence the appropriate term is *rostral* (toward the tip or rostrum of the head). The terms *medial* (toward the midline) and *lateral* (away from the midline) are familiar and common to both the NA and the NAV. Since the tail can be considered an appendage, the terms *proximal* (toward the mass of the body) and *distal* (away from the mass of the body) are used to designate regions of the tail (e.g., proximal, middle, and distal caudal vertebrae). In addition, *proximal* and *distal* are used to designate segments of a limb or of elements within a limb (e.g., proximal shaft of femur; distal end of radius). For the manus and pes, *palmar* and *plantar*, respectively, are used to designate the surface in contact with the ground and *dorsal* for the opposite surface. An alternative set of terms can be used to designate surfaces on the limb or manus/pes: *extensor* or *flexor* surfaces. The dental nomenclature specified by Edmund (1969) is used with regard to the orientation of teeth within the jaws. The edge of a tooth toward the symphysis or premaxillary midline is the *mesial* edge, while the edge away from the symphysis along the tooth row is *distal*. The surface of a tooth toward the lip or cheek (i.e., away from the oral cavity) is *labial* or *buccal*, respectively, while the surface toward the tongue is *lingual*.

We see the immediate pool of vertebrate paleontologists, students and faculty alike, who are involved with dinosaurian or archosaurian research, as our principal readership. In addition, we hope that this book will provide an indispensible data base for invertebrate paleontologists and paleobotanists, geologists, biostratigraphers, paleoecologists, and organismal biologists. Lecturers in paleontology, interested amateurs, and evolutionary biologists should also find material of great interest. Last, we hope that those whose very important mission it is to inform the general public about dinosaurs will also find this book valuable.

It is our wish that this volume represent a benchmark in the study of dinosaurs. By achieving such a remarkable degree of international cooperation, it may mark the beginning of a new age in the science of dinosaurs. We believe there is much here that will be of lasting value. But the speed with which this book becomes obsolete will be a measure of the growth and the intellectual vigor of the field of dinosaur studies. Perhaps we will have played some small part in this exciting enterprise.

SECTION I

DINOSAUR RELATIONSHIPS, BIOLOGY, AND DISTRIBUTION

1

Origin and Interrelationships of Dinosaurs

MICHAEL J. BENTON

The Dinosauria form a well-defined group within the Archosauria and, with birds included within Dinosauria (as they must be on cladistic grounds), the group is clearly one of great significance among terrestrial vertebrates. Here, I shall review current evidence on the relationships of the Dinosauria as a whole within the Archosauria and then attempt to extract a pattern of relationships of the major taxa within the Dinosauria. The advent of cladistic analyses of archosaurs in the past ten years has revolutionized our views of their relationships, and attention will be focused on such studies.

CLADISTIC ANALYSES OF DINOSAURIA

Until recently, dinosaurian systematists were not always clear in distinguishing derived from nonderived characters, and this led to a great deal of confusion in trying to establish phylogenetic schemes that could be compared directly with one another. The work in the 1980s has established a large number of conclusions that were either suspected by only a few experts or were directly denied. Below are a few of the major conclusions and an indication of how attitudes have changed in the past few years by reference to some standard publications that represent generally held views. This is not meant as a direct criticism of the quoted authors, since they represent the views of most experts of their day.

1. The Archosauria is monophyletic. This view has been held generally for a long time, although the Archosauria have been regarded as hard to define anatomically (e.g., Romer 1956, 1966).

2. The Archosauria splits into two main evolutionary lines, one leading ultimately to crocodilians and the other to birds. This split was hinted at by Bonaparte (1975a), Krebs (1976), Cruickshank

(1979), and Chatterjee (1982), but the basal archosaurs were still generally left in a broad grouping, termed the Pseudosuchia, which included unstated ancestors of crocodilians, pterosaurs, dinosaurs, and birds.

3. The dinosaurs are closely related to ornithosuchids, *Lagosuchus,* and Pterosauria. These relationships along the bird-dinosaur line of archosaurs were not suspected by earlier authors (e.g., Romer 1956, 1966; Bonaparte 1975a, 1982b; Krebs 1976; Cruickshank 1979; Chatterjee 1982) until the work of Bakker and Galton (1974) and Gauthier (1984, 1986; Gauthier and Padian 1985).

4. The Dinosauria is monophyletic. Formerly, the origins of dinosaurs were usually seen as polyphyletic with as few as three or as many as six ancestors (Romer 1966; Reig 1970; Charig 1976; Krebs 1976; Cruickshank 1979; Thulborn 1980; Chatterjee 1982). Bakker and Galton (1974) and Bonaparte (1976) argued for dinosaurian monophyly before such views became generally accepted.

5. The Dinosauria falls into three main monophyletic groups: Theropoda, Sauropodomorpha, and Ornithischia. This view has been generally held for some time (Charig 1976, p. 87), although Cruickshank (1979) split up the Sauropodomorpha, and he and Chatterjee (1982) split up the Theropoda.

6. The Theropoda includes the ancestors of birds, which were small theropods similar to dromaeosaurids or troodontids. This view, dating from the nineteenth century, was generally denied (e.g., Romer 1956, 1966; Walker 1972; Tarsitano and Hecht 1980; Gardiner 1982). John Ostrom's work in the 1970s (e.g., 1976b), in which he detailed the anatomical similarities between theropods and birds, formed the basis for the present cladistic analyses, but such views were initially slow to gain acceptance and are still controversial (see papers in Hecht et al. 1985).

7. Within Sauropodomorpha, the Prosauropoda is a paraphyletic grade group that divides into outgroups of the Sauropoda. This view was hinted at by Colbert (1964a), Charig et al. (1965), and others but denied by Cruickshank (1979), although the precise relationships of sauropodomorph subgroups have not yet been worked out.

8. Within Ornithischia, there are two major groups, consisting of Ornithopoda (plus Ceratopsia plus Pachycephalosauria), and Thyreophora (principally Stegosauria plus Ankylosauria). This major split has been denied by some authors, (e.g., Thulborn 1971b), suspected by others, and ignored by most (e.g., Romer 1966) in the absence of strong evidence either way.

These major conclusions, and many others, are detailed below and in subsequent chapters of this volume.

THE ARCHOSAURIA

Among living vertebrates, birds and crocodilians are linked as sister groups within the Archosauria by most authors. Although seemingly very different kinds of animals, these two groups share numerous derived characters of the skull, postcranial skeleton, and soft parts (reviewed by Gauthier 1986; Benton and Clark 1988) that are absent in other living vertebrates. It should be noted, however, that Gardiner (1982) and Løvtrup (1985) have denied the existence of the Archosauria by pairing Aves with Mammalia and then those two with Crocodylia. Their postulated synapomorphies have been rejected by several authors (Benton 1985b; Kemp 1988; Gauthier, Kluge, and Rowe 1988a, b), and they have not gained general acceptance.

Molecular data on tetrapod phylogeny are equivocal regarding the relationships of birds and crocodilians. Some analyses do pair these two groups (e.g., Goodman et al. 1982; Stapel et al. 1984), but many tend to link birds and mammals more closely (e.g., Maeda and Fitch 1981; Bishop and Friday 1988). However, other protein sequence analyses give every other imaginable pairing of tetrapod groups, and their significance is debatable (Benton 1985b; Bishop and Friday 1988). The hard-part autapomorphies of Archosauria include:

1. possession of an antorbital fenestra

2. reduced postfrontal

3. fused or absent postparietals

4. laterosphenoid ossification in the braincase

5. laterally compressed serrated teeth

6. loss of trunk intercentra

7. ectepicondylar foramen absent on humerus

8. fourth trochanter on the femur

Archosaurs as Diapsids

The archosaurs, although formerly regarded as an independent reptilian subclass by Romer (1966), are now placed by nearly all biologists and paleontologists in the Diapsida (which also includes *Sphenodon*, lizards, snakes, and various extinct forms). The archosaurs form part of the Archosauromorpha, a branch of the diapsid reptiles, which includes *Trilophosaurus*, the Rhynchosauria, and the Prolacertiformes as successive outgroups to the Archosauria (Gow 1975; Brinkman 1981; Benton 1983*b*, 1984*a*, 1985*b*; Evans 1984, 1988; Gauthier 1984).

A few other recent theories of archosaur origins have suggested rather different, nondiapsid origins for the Archosauria, but these theories were based on noncladistic analyses of relationships. Romer (1966) suggested that they arose directly from captorhinomorphs, while Reig (1970) derived them from varanopsid pelycosaurs. Other authors (e.g., Hughes 1963; Cruickshank 1972; Gow 1975; Carroll 1976) suggested that the ancestor of the archosaurs was a Permian diapsid like *Youngina*, but that reptile is now recognized as belonging to the lepidosauromorph branch of diapsids (Currie 1982; Benton 1983*b*, 1984*a*, 1985*b*; Evans 1984; Gauthier 1984).

Limits of the Archosauria

There is a semantic question regarding the composition of Archosauria. That is, at which node should the name be applied? The "traditional" Archosauria (Benton 1984*c*, 1985*b*; Paul 1984*a*; Benton and Clark 1988) consists of *Proterosuchus* and more recently evolved relatives (fig. 1.1). The "crown-group" Archosauria (Gauthier 1986) consists of all the descendants of the closest common ancestor of the living forms (i.e., the Crocodylia and Aves). This latter arrangement corresponds to the Ornithosuchia and Crocodylotarsi but excludes a number of basal forms (fig. 1.1). There is no firm way of deciding which view is advisable. The traditionalists argue the need for stability in taxonomy and the convenience of having all taxa, including plesiomorphous ones, within a well-established taxon. The "crown-groupers" can support their case by pointing out that all archosaurs in their interpretation, even the fossil ones, can be said to have all the hard- and soft-part synapomorphies of the living forms. I use the traditional interpretation of the bounds of Archosauria rather than the crown-group Archosauria.

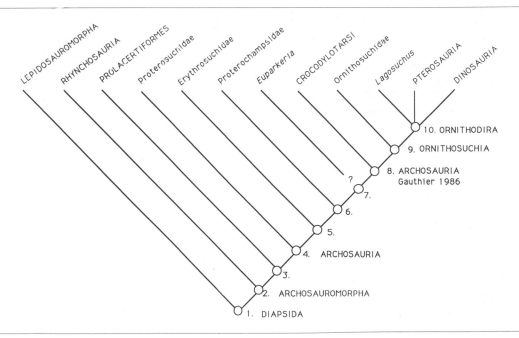

Fig. 1.1. Cladogram illustrating postulated relationships among the major archosauromorph and archosaur groups. Based on cladistic analyses by Benton (1984*a*, 1985*b*) and Benton and Clark (1988), with information from Gauthier (1986) and Evans (1988).

1. DIAPSIDA (Benton 1985*b*; Evans 1988): Supratem-

Figure 1.1, continued

poral fenestra; suborbital fenestra; cervical vertebrae longer than mid-dorsals.

2. ARCHOSAUROMORPHA (Benton 1985b; Evans 1988): Premaxilla extends up behind the naris; nares elongate and close to the midline; quadratojugal mainly behind the infratemporal fenestra rather than below it; tall quadrate; pineal foramen reduced or absent; tabulars absent; paroccipital process touches suspensorium; slender stapes without a foramen; vertebrae not notochordal; transverse processes on dorsal vertebrae project as distinctive narrow elongate processes; cleithrum absent; no entepicondylar foramen in the humerus; no foramen in carpus between ulnare and intermedium; lateral tuber on calcaneum; complex concave-convex articulation between the astragalus and calcaneum; fifth distal tarsal lost; pedal centrale displaced laterally; metatarsal V hooked in one plane only; elongate metatarsal IV.

3. UNNAMED GROUP: Long snout and narrow skull; nasals longer than frontals; posttemporal fenestrae small or absent; recurved teeth; extensive participation of the parasphenoid/basisphenoid in the side wall of the braincase; long, thin, tapering cervical ribs with two or three heads and a cranial dorsal process.

4. ARCHOSAURIA: Possession of antorbital fenestra; postfrontal reduced; postparietals fused or absent; caudal border of infratemporal fenestra bowed; marginal teeth laterally compressed; presence of an ossified laterosphenoid; no ectepicondylar groove or foramen on humerus; possession of a fourth trochanter on femur.

5. UNNAMED GROUP: Loss of the supratemporal; possession of a lateral mandibular fenestra; coronoid reduced or absent; presacral intercentra absent behind the axis; ossified portion of the scapula very tall and narrow (at least twice as tall as width of base); coracoid small, and glenoid faces largely backward; deltopectoral crest extends at least one-fourth of the way down the shaft of the humerus; distal end of the humerus is narrower than the proximal end; pelvis markedly three-rayed with a long, downturned pubis and ischium; iliac blade has a small cranial process; pubis has a strongly downturned cranial tuber in lateral view; ischium has a large caudoventral process (the ischium is longer than the iliac blade); tarsus contains only four elements; metatarsals II, III, and IV subequal in length, with III the longest; loss of cranioproximal "hook" on metatarsal V; fewer than four phalanges in pedal digit V.

6. UNNAMED GROUP: Parietal foramen absent; otic notch well developed; possession of thecodont dentition; ribs all one- or two-headed; hindlimbs under the body (semierect or erect gait); possession of "crocodiloid" tarsus (foramen lost, and rotation between astragalus and calcaneum possible); possession of dermal armor with two osteoderms per vertebra.

7. UNNAMED GROUP: Antorbital fenestra large and lying in a depression; nasals run forward between the nares; diapophysis placed fairly high on the neural arch of cervical vertebrae; parapophysis transfers to the neural arch in cranial dorsal vertebrae; diapophysis and parapophysis fuse in the caudal dorsal vertebrae and the ribs become single-headed.

8. UNNAMED GROUP (= Archosauria of Gauthier 1986): Parietals send caudal processes onto the occiput; discrete postparietal and exoccipitals absent beyond juvenile stages of development; pterygoids meet medially in the palate; palatal teeth absent.

9. ORNITHOSUCHIA (Gauthier 1986): Septomaxilla absent (parallelism in Suchia); squamosal reduced and descending ramus gracile (also in *Euparkeria*); manual digit I short and equipped with a diverging claw; no puboischiadic plate, and much reduced contact between pubis and ischium (parallelism in Suchia); pubis long, narrow, and subvertically oriented (parallelism in Suchia); pubis longer than the ischium (parallelism in Suchia); possession of a lesser trochanter; fourth trochanter a sharp flange; shaft of femur bowed dorsally; prominent cnemial crest on tibia (also in *Gracilisuchus*); ventral flange of astragalus is absent (also in *Euparkeria*); digit V of the foot is reduced (shorter than I) (parallelism in Suchia).

10. ORNITHODIRA (Gauthier 1986): Presacral vertebral column is divided into three regions (cervical, cervical-thoracic, lumbar); centra steeply inclined in at least cervicals 3–6; zygapophyses of the middle and distal caudals inclined caudoventrally; loss of the interclavicle (possibly also in *Postosuchus*); acetabulum perforated to some extent (parallelism in *Postosuchus* and Crocodylomorpha); supraacetabular crest on ilium (parallelism in *Saurosuchus*, *Postosuchus*, Crocodylomorpha); pubis more than three times the width of the acetabulum (parallelism in *Saurosuchus*, *Postosuchus*, Crocodylomorpha); fourth trochanter a winglike process; fourth trochanter runs down one-third to one-half the length of the femur shaft (parallelism in *Erythrosuchus* and *Chanaresuchus*); distal end of femur forms two subterminal condyles; knee articulates at 90°; stance digitigrade (parallelism in *Gracilisuchus*, *Postosuchus*, Crocodylomorpha); mesotarsal ankle joint with astragalus and calcaneum fused to the tibia; calcaneum with no tuber at all; ascending process of astragalus fits between the tibia and fibula; metatarsals II–IV closely bunched as a unit; metatarsals II–IV elongate and the foot functionally tridactyl.

THE RELATIONSHIPS OF THE TRIASSIC ARCHOSAURIA

"Thecodontians"

The archosaurs radiated extensively during the Triassic period, and several distinctive lineages arose. It has been widely accepted that all of the later archosaur groups arose from within the Thecodontia, a paraphyletic group that excludes three or four descendant clades: crocodilians, pterosaurs, dinosaurs, and birds. The informal term *thecodontian* will be used here to refer to all Late Permian and Triassic archosaurs that do not fall into these last three named groups.

There is no currently accepted classification of the thecodontians. Most authors have favored a basic tripartite division: the Proterosuchia (for the Proterosuchidae and Erythrosuchidae and, at times, the Rauisuchidae and Proterochampsidae), the Parasuchia (for the Phytosauridae), and the Pseudosuchia (for everything else, including some early crocodylomorphs) (e.g., Romer 1956; Reig 1970; Krebs 1976). Others have separated the actosaurs as the Aetosauria (Romer 1966, 1972*b*; Sill 1974; Bonaparte 1975*a*; Charig 1976; Thulborn 1980), some of the early crocodylomorphs as the Sphenosuchia (Bonaparte 1982*b*), ornithosuchids as the Ornithosuchia (Chatterjee 1982), and rauisuchids and poposaurids as the Rauisuchia (Chatterjee 1982). These divisions are abandoned here, except for the Rauisuchia and a revised Pseudosuchia.

Recent cladistic analyses of the Triassic archosaurs (e.g., Benton 1983*b*, 1984*a*, 1984*c*, 1985*b*; Gauthier 1984, 1986; Paul 1984*a*; Gauthier and Padian 1985; Benton and Clark 1988; Benton and Norman, in prep.) have produced similar cladograms that broadly resemble that shown in fig. 1.1. The controversial points are noted briefly below.

The Basal Archosaurs

The oldest known archosaur is *Archosaurus* from the latest Permian of the U.S.S.R. It is known from only fragmentary remains, but these show at least one diagnostic archosaurian character (the presence of an antorbital fenestra). *Archosaurus* is generally classed as a proterosuchid, a group best represented by *Proterosuchus* from the Early Triassic of South Africa, a 1.5-m long-snouted aquatic form (fig. 1.2a). *Proterosuchus* possesses all eight archosaur autapomorphies noted above.

Next on the cladogram is *Erythrosuchus* (fig.

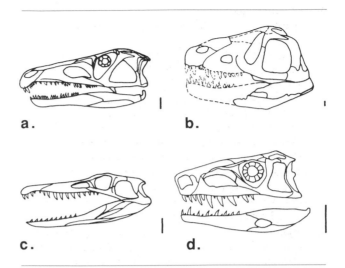

Fig. 1.2. Skulls of primitive archosaurs: a. *Proterosuchus;* b. *Erythrosuchus;* c. *Chanaresuchus;* d. *Euparkeria.* The skulls are drawn to uniform length; scale = 20 mm. Based on several sources.

1.2b), another thecodontian from the Early Triassic of South Africa. *Erythrosuchus* and its relatives were up to 5 m long, and they were clearly the top carnivores of their day. Their synapomorphies compared to *Proterosuchus* include loss of the supratemporal, presence of an external mandibular fenestra, absence of most presacral intercentra, triradiate pelvis, four tarsal elements, and metatarsals II–IV subequal in length, among other synapomorphies with later archosaurs (Benton and Clark 1988).

The Proterochampsidae of the Middle Triassic of Argentina, typified by *Chanaresuchus* (fig. 1.2c), a superficially crocodilelike fish-eater, share numerous derived characters of skull and postcranial skeleton with later forms. These include the first appearance of true thecodont dentition, the loss of the parietal foramen, and the crocodiloid tarsus, which allowed rotation between the astragalus and calcaneum.

Euparkeria and Ankle Structure

Euparkeria, a small animal from the Early Triassic of South Africa, is close to a major split in archosaur evolution that gave rise to crocodiles, on the one hand, and birds and dinosaurs, on the other. The detailed relationships of *Euparkeria* are uncertain, however. Some authors (e.g., Ewer 1965; Cruickshank 1979) place it among the basal archosaurs in association with *Erythrosuchus,* while others (Thulborn 1980; Brinkman 1981; Chatterjee 1982) regard it as more derived. Currently,

Euparkeria is regarded either as the sister taxon of all later archosaurs (= crown-group Archosauria) or as a member of the line leading to dinosaurs (fig. 1.2d).

This dichotomy of opinions hinges around the ankle structure (fig. 1.3), which has been used by several authors (e.g., Bonaparte, 1975a, 1982b; Cruickshank 1979; Thulborn 1980; Chatterjee 1982) as a general guide to archosaur relationships. Early archosaurs have a primitive mesotarsal (PM) ankle in which bending is along a simple hinge between the astragalus-calcaneum unit and the rest of the foot. Later forms have a crocodiloid ankle in which part of the line of bending runs between the astragalus and the calcaneum. However, there are two types of crocodiloid ankle, the crocodile-normal (CN) ankle found in crocodilians, phytosaurs, aetosaurs, and rauisuchians

and the crocodile-reversed (CR) ankle of ornithosuchids, an outgroup to the Dinosauria. In the CN ankle, a peg on the astragalus fits into a socket on the calcaneum, while in the CR ankle, the reverse is the case. Dinosaurs, pterosaurs, and birds have a fourth, modified ankle called advanced mesotarsal (AM) in which the astragalus and calcaneum are firmly attached to the tibia and fibula and the line of bending is between the astragalus-calcaneum and the rest of the foot. The AM ankle differs fundamentally from the PM in that the astragalus is a broad element, the calcaneum is much reduced, and both elements are virtually fused to each other and to the tibia.

Some authors (Gauthier 1986; Parrish 1986) regard the ankle of *Euparkeria* as CR and thus close to ornithosuchids and dinosaurs, while others (Cruickshank and Benton 1985) see it as merely a generalized crocodiloid type, or modified primitive mesotarsal (MPM) type, with no special CR features. Other characters of the skull and postcranial skeleton are roughly equally balanced between rival placements of *Euparkeria*. In addition to its supposed CR ankle, *Euparkeria* shares a reduced gracile squamosal and the absence of a ventral flange on the astragalus with ornithosuchians (Gauthier 1986). In addition, it lacks occipital processes of the parietal, discrete postparietal and exoccipitals, and palatal teeth and possesses medial contact of pterygoids, characters of Archosauria (Benton and Clark 1988).

The Crocodylotarsi

The crocodilian line of archosaurs, characterized by the possession of the CN ankle and other synapomorphies (viz. Gauthier and Padian 1985; Gauthier 1986; Benton and Clark 1988), includes phytosaurs as the basal group and aetosaurs, rauisuchians, and crocodylomorphs (crocodilians and crocodilianlike forms) as successively higher taxa within the lineage (fig. 1.1).

Phytosaurs are a well-defined group of long-snouted, 2- to 4-m-long animals from the Late Triassic of Europe, North America, and parts of Asia. Although superficially crocodilelike (fig. 1.4a), their aquatic and fish-eating adaptations evolved convergently.

The remaining Crocodylotarsi form a group termed the Suchia (Krebs 1976), which appears to divide into the Pseudosuchia (aetosaurs and rauisuchians) and the Crocodylomorpha. Aetosaurs were 1- to 3-m-long herbivores of the Late Triassic. They had characteristic blunt snouts, peglike teeth (fig. 1.4b), and heavily armored bodies. Rauisuchians, from the Middle and Late Triassic, include large quadrupedal

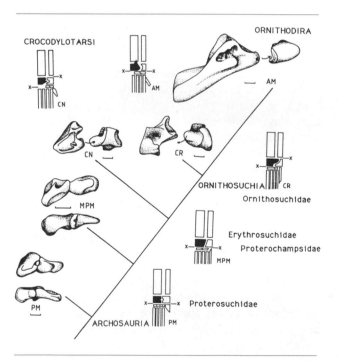

Fig. 1.3. Ankle structures of the archosaurs illustrated on a cladogram depicting broad relationships. For each ankle pattern, a cranial view of the astragalus (left) and calcaneum (right) complex is shown. An additional proximal view is shown for the PM and MPM types (upper drawing). For each ankle type, a diagram of the lower leg, ankle, and foot is shown (astragalus shaded black) with the main hinge line (x − x). The ankles are PM (primitive mesotarsal) of *Proterosuchus*, MPM (modified primitive mesotarsal) of *Chanaresuchus*, CN (crocodile-normal) of *Neoaetosauroides*, CR (crocodile-reversed) of *Riojasuchus*, and AM (advanced mesotarsal) of a prosauropod dinosaur. Based, in part, on Cruickshank and Benton (1985).

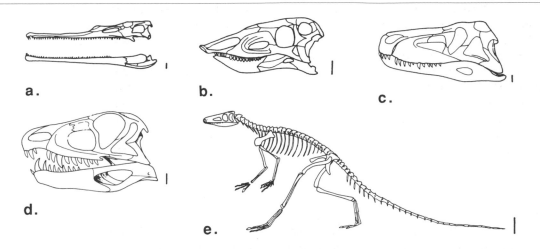

Fig. 1.4. Skulls of Triassic archosaurs: a. *Parasuchus;* b. *Stagonolepis;* c. *Saurosuchus;* and d. *Ornithosuchus.* Skele-ton of e. *Lagosuchus.* Scale = 20 mm. Based on several sources.

carnivorous forms (fig. 1.4c), up to 5 m long, which had a specialized erect gait where the pelvis was tipped almost horizontally, and the femur fitted into the acetabulum like a vertical pillar. Poposaurids were bipedal carnivores whose relationships are still unclear. They might be rauisuchians, but they also seem to share some characters with crocodylomorphs (Gauthier 1986; Benton and Clark 1988). The crocodylomorphs arose in the Late Triassic initially as small bipedal terrestrial forms. Only later, in the Jurassic, did they adopt more "crocodilian" habits of life in fresh and salt water.

The Ornithosuchia

The dinosaur-bird line, the Ornithosuchia (Gauthier 1986), consists of those archosaurs with CR and AM ankles (figs. 1.1, 1.3). The group is characterized by numerous synapomorphies that relate largely to the acquisition of erect gait (long narrow pubis and ischium with reduced contacts, lesser trochanter and sharp fourth trochanter on femur, prominent cnemial crest on tibia, digit V of foot reduced) and includes the Ornithosuchidae, *Lagosuchus,* and the Pterosauria as relatively more intimate outgroups of the Dinosauria (including Aves).

The Ornithosuchidae, from the Late Triassic of Scotland (*Ornithosuchus*) and Argentina (*Riojasuchus, Venaticosuchus*) were 1- to 3-m-long facultatively bipedal carnivores. The skull (fig. 1.4d) shows a characteristic bulbous snout and a gap in the tooth row between the premaxilla and maxilla. The CR ankle is another synapomorphy of the family. Superficially, the

ornithosuchids are very dinosaurlike, and they demonstrate the ornithosuchian synapomorphies of the hindlimb just noted as well as some features of the skull (septomaxilla absent; squamosal reduced) and hand (manual digit I short and equipped with a diverging claw).

The remaining ornithosuchians, *Lagosuchus,* the Pterosauria, and the Dinosauria have been termed the Ornithodira (Gauthier 1986), and they possess a large number of synapomorphies of the vertebral column (distinctive cervical, cervical-thoracic, and lumbar regions), shoulder girdle (loss of interclavicle), pelvis (partially to fully perforated acetabulum, supraacetabular crest on the ilium, elongate pubis), hindlimb (femur is shorter than tibia; fourth trochanter is a winglike process very low on the femur, proximal head of the femur is inturned, distal end of the femur is split into two condyles, and knee articulates as a straight hinge), and foot (digitigrade stance, AM ankle joint with ascending tibial process on astragalus, metatarsals II−IV are elongate and closely bunched as a unit, and the foot is functionally tridactyl).

Lagosuchus, a slender, long-limbed animal from the Middle Triassic of Argentina (fig. 1.4e) shares all of these synapomorphies with the Dinosauria as well as with the Pterosauria. It is known from half a dozen skeletons that show postcranial features well, but the skull is incompletely known. *Lagosuchus* may be primitive to the pterosaur-dinosaur split since it appears to lack several of their synapomorphies (loss of postfrontal, caudal zygapophyses nearly vertical, no more than four phalanges in manual digit IV, proximal head of femur fully offset; Benton and Norman in prep.). However, *Lagosuchus* shares some apparent synapomorphies with the Dinosauria which are absent in the Pterosauria, such as the caudoventrally facing glenoid facet on the scapulocoracoid, the reduced subcircular coracoid, the shortened forelimbs, and the brevis shelf on the caudal portion of the ilium (Gauthier 1986). Most of these relate to the flight specializations of pterosaurs, but it is still hard to sort out the relationships of the three ornithodiran taxa. The earlier view of Wild (1978*b*) and Benton (1982, 1984*a*, 1985*b*), that Pterosauria was the sister group of Archosauria, was proposed because they lacked seven archosauromorph synapomorphies as well as archosaurian and ornithosuchian characters such as the external mandibular fenestra, the nasals extending between the external nares, the pterygoids meeting medially, the loss of palatal teeth (*Eudimorphodon* has teeth on the pterygoid: Wild 1978*b*), the open acetabulum, the elongate pubis and ischium, the fourth trochanter on the femur, and the reduced pedal digit V (not in early pterosaurs). The view of Gauthier and Padian (1985), however, is accepted here since it is more parsimonious.

THE RELATIONSHIPS OF THE MAJOR DINOSAURIAN GROUPS

Monophyly of the Dinosauria

Until recently, nearly all authors assumed that dinosaurs were a polyphyletic group with at least two, and more probably three or four, separate origins from different thecodontian groups (Romer 1966, 1968, 1972*b*; Reig 1970; Charig 1976; Thulborn 1980; Bonaparte 1982*b*; Chatterjee 1982). Exceptions were Bakker and Galton (1974) and Bonaparte (1976), who speculated that the Dinosauria is a true clade. Recent cladistic analyses (Benton 1984*b*, 1984*c*, 1986; Gauthier 1984, 1986; Paul 1984*a*, 1984*b*; Benton and Cruickshank 1985; Cooper 1985; Gauthier and Padian 1985; Benton and Clark 1988), however, all agree that the Dinosauria is monophyletic on the basis of numerous autapomorphies:

1. elongate vomers that reach caudally at least to the level of the antorbital fenestra (Gauthier 1986).

2. three or more sacral vertebrae (paralleled in the crocodylotarsan *Postosuchus* and the Ornithosuchidae; this character is uncertain in basal dino-

saurs, and may apply to a higher node in the cladogram; *Lagosuchus* and *Lagerpeton* have only two sacrals: Arcucci 1986).

3. scapulocoracoidal glenoid facing fully backward

4. low deltopectoral crest that runs one-third or one-half of the way down the shaft of the humerus.

5. three or fewer phalanges in the fourth digit of the hand (Gauthier 1986)

6. largely to fully open acetabulum

7. fully offset proximal head of femur with a distinct neck and ball

8. greatly reduced fibula

9. well-developed ascending process of astragalus

It is worth noting that many of the characters that seem fully dinosaurian, but which are omitted here (e.g., the supraacetabular crest, elongate pubis, enlarged fourth trochanter, AM tarsus, digitigrade stance, etc.), were already present in the lineage since the origin of all ornithodirans.

The Basal Dinosaurs

There are a number of Late Triassic dinosaurs that do not fit into any of the major clades within the Dinosauria, and they are assumed here to be primitive outgroups to the main dinosaurian clade. These include *Herrerasaurus*, *Staurikosaurus*, and possibly a number of other poorly represented taxa (e.g., *Aliwalia*, *Ischisaurus*, *Spondylosoma*). In the past, attempts have been made (Colbert 1970; Cooper 1981*a*; Galton 1977*a*, 1985*a*) to place *Herrerasaurus* and *Staurikosaurus* in the Saurischia (prosauropod, theropod, or something in between: see Sues, this vol.).

It seems likely that *Herrerasaurus* and *Staurikosaurus* are successively closer outgroups of the main dinosaurian assemblage (fig. 1.5; Brinkman and Sues 1987). Galton (1985*a*) erected the Infraorder Herrerasauria for *Herrerasaurus* and *Staurikosaurus* and *Aliwalia*, while Gauthier (1986) included *Herrerasaurus*, *Staurikosaurus*, and *Ischisaurus* in the Herrerasauridae, but the authors just cited could find no potential synapomorphies of such a grouping. *Herrerasaurus* has all the dinosaurian characters noted above, although characters 1, and 3 through 5 cannot be determined from published accounts. It lacks an ornithodiran character, however, since its femur is apparently longer than its tibia (Galton 1977*a*). *Staurikosaurus* also shares the dinosaurian autapomorphies (1 and 5 cannot be deter-

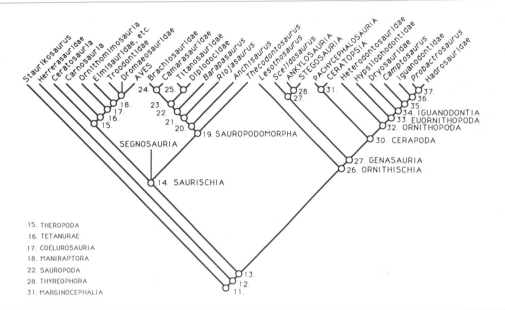

Fig. 1.5. Cladogram depicting phylogenetic relationships of the major groups of the Dinosauria, with particular focus on the early forms. Based on Gauthier (1986), Sereno (1986), and other sources (see text). Character information for nodes 15–18, 20–25, 27–37 is omitted since it is given in the text. Additional character information for certain nodes is given in other chapters of this volume: 28, 29 (Thyreophora; Weishampel, this vol.); 31 (Marginocephalia; Dodson, this vol.) 32–37 (Ornithopoda; Weishampel, this vol.).

11. DINOSAURIA: Elongate vomers that reach caudally at least to the level of the antorbital fenestra; three or more sacral vertebrae (paralleled in the crocodylotarsan *Postosuchus* and the Ornithosuchidae; this character is uncertain in basal dinosaurs); glenoid facing fully backward; low deltopectoral crest that runs one-third or one-half the way down the shaft of the humerus; three or fewer phalanges in the fourth digit of the hand; largely to fully open acetabulum; fully offset proximal head of femur with a distinct neck and ball; greatly reduced fibula; well-developed ascending process of astragalus.

12. UNNAMED GROUP: Elongate S-shaped neck; reduced contact between the pubis and the ischium.

13. UNNAMED GROUP: Presence of three or more sacral vertebrae (may move lower in cladogram); lesser trochanter on the femur is a spike or a crest; transversely expanded distal end of the tibia.

14. SAURISCHIA: Temporal musculature extending on to frontal; lateral overlap of quadratojugal on to the caudal process of the jugal; elongate caudal cervicals giving a relatively long neck; axial postzygapophyses set lateral to the prezygapophyses; epipophyses present on the cranial cervical postzygapophyses; presence of accessory intervertebral articulations (hyposphene-hypantrum) in dorsal vertebrae; manus more than 45%

of the length of the humerus + radius; distinctly asymmetrical manus, with digit II the longest; proximal ends of metacarpals IV and V lie on the palmar surfaces of digits III and IV in the hand, respectively; heavy pollex with a very broad metacarpal.

15. THEROPODA

16. TETANURAE

17. COELUROSAURIA (*sensu* Gauthier 1986)

18. MANIRAPTORA

19. SAUROPODOMORPHA: Relatively small skull (approx. 5% of body length); dentary curved down at the front; lanceolate teeth with coarsely serrated crowns; at least ten cervicals (each about twice as long as high), forming a very long neck; one to three extra sacral vertebrae, all modified from caudals; enormous pollex with an enlarged claw; absence of claws on manual digits IV and V; iliac blade with a reduced caudal process and a short cranial process (the brachyiliac condition); fused, deep, apron-like pubes that are twisted proximally; very large obturator foramen in pubis; femur longer than tibia; ascending process of astragalus keys into tibia, which has a matching descending process.

20. UNNAMED GROUP

21. UNNAMED GROUP

22. SAUROPODA

23. UNNAMED GROUP

24. CAMARASAURS

25. TITANOSAURS

26. ORNITHISCHIA: Rostral tip of the premaxilla toothless and roughened; horizontal or broadly arched palatal process of premaxilla; maxilla excluded from the margin of the external naris by a large lateral process of the premaxilla which meets the nasal; reduced antorbital fenestra; ventral margin of the antorbital fenestra parallels the maxillary tooth row; palpebral in the orbit; pre-

Figure 1.5, continued

frontal with a long caudal ramus that overlaps the frontal; subrectangular quadratojugal lying behind the infratemporal fenestra; elongate and massive quadrate; predentary bone at the front of the mandible; dorsal border of the coronoid eminence formed by the dentary; mandibular condyle set below the tooth rows (paralleled in most sauropodomorphs); buccal emargination of the jaws, suggesting the possession of cheeks in life; cheek teeth with low triangular crowns with a well-developed cingulum; crowns of cheek teeth with low and bulbous base; enlarged denticles on the margins; adjacent crowns of both maxillary and dentary teeth overlap (paralleled in part in some sauropodomorphs); recurvature absent in maxillary and dentary teeth; maximum tooth size near the middle of the maxillary and dentary tooth rows; at least five sacral vertebrae; gastralia absent (paralleled in Sauropoda); ossified tendons at least above the sacral region; opisthopubic pelvis, with small cranial process; ilium with lateral swelling of the ischial tuberosity; iliac blade with a long and thin cranial process and a deep caudal pro-

cess; pubic symphysis restricted to its distal end; ischial symphysis restricted to its distal end; pubis with an obturator notch rather than a foramen; obturator foramen formed between the pubis and ischium; distal puboischial symphysis; pendent fourth trochanter on the femur; fringe-like lesser trochanter on the femur; fifth digit of the foot reduced to a small metatarsal with no phalanges.

27. CERAPODA: Spout-shaped mandibular symphysis; entire margin of the antorbital fossa is sharply defined or extends as a lateral wall enclosing the fossa.
28. THYREOPHORA
29. EURYPODA
30. CERAPODA
31. MARGINOCEPHALIA
32. ORNITHOPODA
33. EUORNITHOPODA
34. IGUANODONTIA
35. ANKYLOPOLLEXIA
36. STYRACOSTERNA
37. UNNAMED GROUP

---

mined), although it is not clear whether it had three sacral vertebrae (Colbert 1970) or two (Galton 1977a). In comparison with *Herrerasaurus*, *Staurikosaurus* has possibly acquired two further synapomorphies of later dinosaurs—an elongate S-shaped neck and a reduced contact between the pubis and the ischium. Brinkman and Sues (1987) and Sues (this vol.) place *Herrerasaurus* above *Staurikosaurus* in the cladogram since it has indications of a "twisted tibia" seen in later dinosaurs, but the twist is only about 60° (Galton 1977a: 238), compared to a dinosaurian 90°.

The "True" Dinosaurs

All other dinosaurs appear to form a clade that may be characterized by at least three synapomorphies: the presence of three or more sacral vertebrae (? two in *Staurikosaurus*, see above), unknown in *Herrerasaurus* (? three; Benedetto 1973); a spike or crestlike lesser trochanter on the femur (only a bump in the forms so described), and the transversely expanded distal end of the tibia. The sacral vertebral character may shift down the cladogram, as noted above, depending on new studies of the early dinosaurs.

The "true" dinosaurs are divided into three monophyletic groups: Ornithischia, Theropoda, and Sauropodomorpha. There are three possible arrangements of these three taxa. The generally accepted view seems

to be that the Theropoda and Sauropodomorpha pair off as sister groups (together these make a monophyletic Saurischia), and this is defended here. Alternative views are discussed later.

The Saurischia

The evidence for monophyly of the Saurischia includes the following autapomorphies (Gauthier 1986):

1. temporal musculature extending on to the frontal

2. lateral overlap of the quadratojugal on to the caudal process of the jugal

3. elongate caudal cervicals giving a relatively long neck

4. axial postzygapophyses set lateral to the prezygapophyses

5. epipophyses present on the cranial cervical postzygapophyses

6. presence of accessory intervertebral articulations (hyposphene-hypantrum) in dorsal vertebrae

7. manus more than 45 percent of the length of the humerus and radius together

8. distinctly asymmetrical manus, with digit II the longest

9. proximal ends of metacarpals IV and V lying on the palmar surfaces of digits III and IV in the hand, respectively

10. heavy pollex with a very broad metacarpal

Within the Saurischia (*sensu* Gauthier 1986), most authors would accept the major division into the Theropoda and Sauropodomorpha.

The Theropoda

Several cladograms of Theropoda are now available (Padian 1982; Paul 1984*a*; Thulborn 1984*a*; Gauthier 1986) which hinge around the key recognition by Ostrom (1976*b*) that birds are derived from theropods close to dromaeosaurids. The Theropoda is characterized (Osmólska, this vol.; Gauthier 1986) on the basis of a number of synapomorphies of the skull and postcranial skeleton (Gauthier 1986):

1. reduced overlap of dentary onto postdentary bases and reduced mandibular symphysis

2. lacrimal exposed on the skull roof

3. extra fenestra in the maxilla

4. vomers fused rostrally

5. expanded ectopterygoid with a ventral fossa

6. first intercentrum with large occipital fossa and small odontoid notch

7. second intercentrum with broad crescentic fossa for reception of first intercentrum

8. presacral vertebrae with pleurocoels (openings to hollow centrum)

9. at least five sacral vertebrae

10. transition point in tail, with marked changes in the form of the processes

11. enlarged distal carpal I overlapping the bases of metacarpals I and II

12. digit I of hand absent or reduced to a vestige

13. digit IV of hand absent or reduced

14. elongate penultimate phalanges in hand

15. digit III of hand with short first and second phalanges

16. unguals of hand enlarged, compressed, sharply pointed, strongly recurved, and with enlarged flexor tubercles

17. long preacetabular process on the ilium

18. pronounced brevis fossa on caudal part of the ilium

19. femur convex cranially

20. fibula closely appressed to tibia and attached to a tibial crest

21. narrow, elongate metatarsus

22. digit IV of foot reduced

23. digit V of foot represented by a very reduced metatarsal

24. metatarsal I reduced, and does not contact tarsus, being attached halfway, or further, down the side of metatarsal II

25. thin-walled, hollow, long bones

The theropod clade is very well characterized, and few authors have ever doubted its validity. The inclusion of birds has, however, been controversial, but current evidence now strongly supports this view (e.g., Ostrom 1976*b*; papers in Hecht et al. 1985; Gauthier 1986). The problems arise when attempts are made to subdivide the Theropoda. Some subgroups seem to be clearly set off as monophyletic taxa, among them, ornithomimosaurs, deinonychosaurs (but see below), and carnosaurs. Others are much less clearly supported by synapomorphies. A number of the standard theropod subgroups, such as Romer's (1966) Coelurosauria, Procompsognathidae, and Coeluridac, for example, are clearly paraphyletic or polyphyletic collections of superficially similar taxa (Paul 1984*a*; Thulborn 1984*a*; Gauthier 1986). Norman (this vol.) discusses many of the problematic taxa of theropods that are hard to fit into a cladistic scheme.

Gauthier (1986) presented a cladogram of the Theropoda in which he recognized five main clades of theropods: Ceratosauria, Carnosauria, Ornithomimidae, Deinonychosauria, and Aves (his Avialae), with a number of unplaced taxa (fig. 1.5).

The Ceratosauria, including a range of small to medium-sized Late Triassic to Late Jurassic taxa, such as *Ceratosaurus, Coelophysis, Dilophosaurus,* and *Syntarsus,* are characterized by loose attachment of the premaxilla to the maxilla, a thyroid fenestra in the pubis, a narrow boned pubis, a trochanteric shelf on the femur, and other features (Rowe and Gauthier, this vol.). They appear to form the sister-group to all other theropods, termed the Tetanurae (Gauthier 1986).

The Tetanurae, including carnosaurs, ornithomimosaurs, deinonychosaurs, and birds (fig. 1.5), is defined by a number of skull and skeletal characters, including a greatly enlarged supplementary maxillary fenestra, termination of the maxillary tooth row in front of the orbit, a straplike scapula, a coracoid that tapers behind, a hand that is more than two-thirds the length of the arm (humerus and radius), absence of manual digit IV in adults, an obturator process on the ischium, a winglike lesser trochanter on the femur, a tall and broad ascending process of the astragalus, and numerous others.

The Carnosauria, the often very large carnivorous dinosaurs of the Jurassic and Cretaceous, are distinguished from other dinosaurs by their deep orbits, narrow frontals and parietals, greatly reduced hand, and other features (Molnar et al., this vol.).

The sister group to the Carnosauria is the Coelurosauria (*sensu* Gauthier 1986), including all other theropods and birds, which shows a number of autapomorphies: a subsidiary fenestra between the pterygoid and palatine, cervical ribs fused to the centra, sternal plates fused together, elongate forelimb, fourth trochanter reduced or absent, and a much enlarged ascending process of the astragalus, among others.

The basal coelurosaurs are the Ornithomimosauria, characterized by toothless jaws, a beaklike snout, all digits of the manus about equal in length, and slender hindlimbs with a very elongate metatarsus with metatarsal III strongly pinched between metatarsals II and IV (Gauthier 1986; Barsbold and Osmólska, this vol.). Ornithomimosaurians are the sister group of all remaining theropods and birds, the latter termed the Maniraptora by Gauthier (1986). The Maniraptora are characterized by reduction or loss of the prefrontal, prominent axial epipophyses, specialized ventral processes (hypapophyses) on cervicothoracic vertebrae, proximal position of transition point in tail, subrectangular coracoid, elongate forelimb, ulna bowed posteriorly, semilunate carpals, very thin metacarpal III that bows laterally, very low pubic process of the ilium, reversed pubis, shortened ischium, lesser trochanter nearly confluent with proximal head of the femur, absence of a fourth trochanter, and digit IV of the foot longer than digit II as well as other synapomorphies (Ostrom 1976b; Gauthier 1986). The Maniraptora includes the Dromaeosauridae, Troodontidae, and Aves (*Archaeopteryx* and all other birds [= Avialae of Gauthier 1986, who used a crown-group interpretation of Aves]), and probably also a number of other poorly known taxa such as *Coelurus, Ornitholestes, Microvenator, Saurornitholestes, Hulsanpes*, the Caenagnathidae, the Elmisauridae, and *Compsognathus*, as basal outgroups (Gauthier 1986), but their exact order is uncertain be-

cause of incompleteness of the specimens (see also Norman, this vol.; Currie, this vol.). The Oviraptoridae is united with the Caenagnathidae to form the Oviraptorosauria, which itself is treated here as a basal maniraptoran taxon (Barsbold et al., this vol.).

The relationships of the Dromaeosauridae, Troodontidae, and Aves are much debated at present, and a resolution may be difficult because of incomplete material. One view is that the Dromaeosauridae and Troodontidae form a taxon Deinonychosauria, which is the sister group of Aves (Gauthier 1986), and other authors (Osmólska, this vol.) have also suggested that the Oviraptorosauria and Ornithomimosauria might be related as well. The dromaeosaurids and troodontids share a number of modifications to the foot, in particular an enlarged second digit with a very large sicklelike claw presumably used to slash prey animals. Of the seventeen maniraptoran characters listed by Gauthier (1986), dromaeosaurids share five uniquely with Aves, but the status of three of these, the vertebral features, is uncertain in troodontids. Troodontids share one of the others (absence of fourth trochanter), but, significantly, lack the fifth (reversed pubis: Barsbold 1983b; Gauthier 1986: 47), which suggests that they may be a more distant outgroup to Aves than the Dromaeosauridae. Paul (1984a) and Currie (1985, 1987a) have suggested, however, that troodontids might be the closest outgroup to Aves on the basis of a number of postulated synapomorphies (Currie 1987a)—periotic pneumatic cavities, pneumatic cavities associated with the internal carotid, a more medial position for the quadrate condyle than that seen in the larger theropods, a fenestra pseudorotunda, loss of interdental plates, and the presence of a constriction between the crown and root of the teeth. Dromaeosaurids apparently lack the last two features, but the status of the first three cannot be assessed until good dromaeosaurid braincases become available. The taxa Dromaeosauridae, Troodontidae, and Aves are left as an unresolved trichotomy (fig. 1.5) for the present.

Paul (1984a) and Thulborn (1984a) offered rather different cladistic analyses of Theropoda. Paul (1984a) made *Archaeopteryx* the outgroup of Deinonychosauria and birds, since it lacked ten shared characters of the latter two groups. Gauthier (1986) discounts these as inclusive at a lower level in the cladogram or as convergences. Thulborn (1984a) placed dromaeosaurids and *Archaeopteryx* below tyrannosaurids, troodontids, ornithomimids, and birds in his cladogram since they lack a tarsometatarsus with an intercotylar prominence and a hypotarsus (a spur on the tarsometatarsus), metatarsal III that is pinched proximally, and a straplike coracoid. The distribution of these characters is uncertain in all the relevant taxa, and they appear to be greatly outweighed by the synapomorphies of *Archaeopteryx* and other birds given by Gauthier (1986).

The Sauropodomorpha

The moderate to large herbivores of the Late Triassic and Early Jurassic (the paraphyletic prosauropods) and the large to gigantic herbivores of the Early Jurassic to Late Cretaceous (the sauropods) form a well-defined second major saurischian clade called the Sauropodomorpha. The Sauropodomorpha is characterized (Dodson, this vol.) by a number of synapomorphies (Benton and Norman in prep.):

1. relatively small skull (about 5% of body length)

2. ventrally deflected front of the dentary (Paul 1984*a*)

3. lanceolate teeth with coarsely serrated crowns (Gauthier 1986)

4. at least ten cervicals (each about twice as long as high), forming a very long neck

5. one to three extra sacral vertebrae, modified from dorsals and caudals

6. enormous pollex with an enlarged claw

7. absence of claws on manual digits IV and V

8. iliac blade with a reduced postacetabular process and a short preacetabular process (the brachyiliac condition: Colbert 1964)

9. fused, deep, apronlike pubes that are twisted proximally

10. very large obturator foramen in the pubis

11. elongate femur (longer than tibia)

12. ascending process of astragalus that keys into tibia, the latter having a matching descending process (Charig et al. 1965)

Various prosauropods form successive outgroups to the Sauropoda (fig. 1.5), roughly in a sequence from *Efraasia* and *Thecodontosaurus* at the base, through *Anchisaurus*, *Plateosaurus*, *Massospondylus*, *Melanorosaurus*, *Riojasaurus*, *Vulcanodon*, and *Barapasaurus* (Gauthier 1986; Galton, this vol.). *Anchisaurus* is advanced over *Efraasia* and *Thecodontosaurus* in the possession of an

even more robust manual digit I, wide-based neural spines on proximal caudals, an arched dorsal margin of the ilium, and a completely open acetabulum (Gauthier 1986). *Riojasaurus* and higher sauropodomorphs share a compressed internarial process of the premaxilla and large nares, a mandibular condyle placed below the level of the tooth row, robust forelimbs, a broad pes, and numerous other synapomorphies (Gauthier 1986).

The placement of the Sauropoda as the most derived relatives of a paraphyletic Prosauropoda is questioned by Dodson (this vol.) and by McIntosh (this vol.). They suggest that the sauropodomorphs may have branched into two major lineages in the Late Triassic, the Prosauropoda and the Sauropoda. This solution could imply that the sauropods were primitively quadrupedal (e.g., Charig et al. 1965). It would also solve some problems with the scheme presented in figure 3.5. For example, the prosauropods share serrated leaf-shaped teeth and other dental, cranial, and skeletal characters seemingly like those of ornithischians (see below) but quite unlike those of sauropods. Of the twelve synapomorphies of the Sauropodomorpha listed above, three (nos. 2, 4, 8) do not apply to the earliest sauropods.

The Sauropoda (McIntosh, this vol.) are defined by about forty synapomorphies (Gauthier 1986) if the line is drawn below *Vulcanodon* and *Barapasaurus*. These features include the shortening of the caudal portion of the skull, deeply excavated nasals, reduced postorbital, lower temporal fenestra partly beneath the orbit, absence of an epipterygoid, absence of an external mandibular fenestra, twelve or more cervicals, cavernous vertebrae, five or six sacrals, stout metacarpals, a manus with reduced digits, a massive pubis, massive and vertical limbs, solid long bones, and a stout and broad pes.

Sauropods fall into two major lineages, according to Gauthier (1986): the camarasaurs (Camarasauridae, Brachiosauridae) and the titanosaurs (Titanosauridae, Diplodocidae). The camarasaurs are distinguished by a strongly arched internarial bar of the premaxilla, a snout that is sharply demarcated from the rest of the skull, an ischium that extends well posteriorly and twists to become more horizontal distally, and a relatively deep puboischial contact (Gauthier 1986). The Camarasauridae has as possible autapomorphies the slender ascending process of the maxilla and a jugal that is excluded from the lower rim of the skull (McIntosh, this vol.). The Brachiosauridae shares elevated nasals, a relatively elongate forelimb (humerus : femur ratio = 0.90−1.05), and an ilium with a broad anterior lobe (McIntosh, this vol.).

The titanosaurs, if they include *Euhelopus* as a basal outgroup, have as synapomorphies a quadrate that slopes up and back from the mandibular condyle, neural spines that are slightly to deeply bifurcate (also in *Camarasaurus*), and the incorporation of three or more trunk vertebrae into the cervical series (Gauthier 1986). If *Euhelopus* is excluded (McIntosh, this vol.), the titanosaur synapomorphies also include the long broad snout, premaxilla and maxilla extending dorsally to the level of the orbit, internasal processes of premaxilla and nasal reduced or absent, and external nares confluent high on the skull, reduced lacrimal, reduced dorsal process of the quadratojugal and the rostral process contacts the maxilla beneath the orbit, very long basipterygoid processes, elongate pencillike teeth at the very front of the jaws, very tall sacral neural spines, very long tail with long series of cylindrical distal caudals, relatively short forelimbs, and metatarsal IV longer than metatarsal III (Gauthier 1986). The Titanosauridae have as synapomorphies the possession of body armor (also in other sauropods?) and the biconvex first caudal centrum with the others procoelous (Gauthier 1986). The Diplodocidae show modified haemal arches with fore and aft processes, deeply cleft V-shaped neural spines in the shoulder region, and distally expanded ischia (Gauthier 1986).

These synapomorphies are tentative and may be heavily modified when a full cladistic analysis of sauropodomorphs is carried out. McIntosh (this vol.), for example, does not support Gauthier's (1986) analysis of sauropod relationships, preferring a division into two different lineages, camarasaurids plus diplodocids, and titanosaurids plus brachiosaurids. The first lineage has as potential synapomorphies the presence of bifid neural spines, forked chevrons, short metacarpals, absence of the calcaneum, and complete loss of the internal trochanter on the femur, while the titanosaurids and brachiosaurids have low sacral spines, a relatively short radius and ulna, and ischia that meet one another edge to edge distally. Within Sauropoda, then, many taxa are hard to place, and resolution of the cladogram must await fuller information on certain Asian and Gondwanan taxa and a successful splitting up of the basal paraphyletic Cetiosauridae.

The Segnosauria

Segnosaurus and its allies were initially classified (Perle 1979) as aberrant, probably herbivorous, saurischians (Barsbold and Maryańska, this vol.). Paul (1984a, 1984b) placed the Segnosauria midway between prosauropods and ornithischians on the basis of a cladistic analysis, with the assumption of a non-monophyletic Saurischia. *Segnosaurus* shares with the Ornithischia a toothless beak, a diastema, cheeks, an opisthopubic pelvis, and a tibia that partly articulates with the pes behind the astragalus. In view of the stronger evidence for a monophyletic Saurischia (see above), *Segnosaurus* might occupy a rather different position on a reanalyzed cladogram of dinosaurs. Indeed, Gauthier (1986) suggests relationship with the sauropodomorphs, and thus interprets the "ornithischianlike" features as convergences, while Dodson (this vol.) and Barsbold and Maryańska (this vol.) regard the Segnosauria as part of an unresolved trichotomy with the Theropoda and Sauropodomorpha.

The Ornithischia

The monophyly of the Ornithischia has been accepted for over a century (Seeley 1888a), and recent cladistic analyses (Maryańska and Osmólska 1984a, 1985; Norman 1984a, 1984b; Sereno 1984, 1986; Cooper 1985; Gauthier 1986) have identified a large number of autapomorphies:

1. rostral tip of the premaxilla toothless and roughened (but not in *Technosaurus*)

2. horizontal or broadly arched palatal process of premaxilla

3. maxilla excluded from the margin of the external naris by a large lateral process of the premaxilla which meets the nasal

4. reduced antorbital fenestra

5. ventral margin of the antorbital fenestra that parallels the maxillary tooth row

6. palpebral in the orbit

7. prefrontal with a long caudal ramus that overlaps the frontal

8. subrectangular quadratojugal lying behind the infratemporal fenestra

9. elongate, massive quadrate

10. predentary bone at the front of the mandible

11. dorsal border of the coronoid eminence formed by the dentary

12. mandibular condyle set below the tooth row (paralleled in most sauropodomorphs)

13. buccal emargination of both upper and lower jaws, suggesting the possession of cheeks

14. cheek teeth with low triangular crowns with a well-developed cingulum beneath

15. crowns of cheek teeth with low and bulbous base; enlarged denticles on the margins

16. adjacent crowns of both maxillary and dentary teeth overlapping (paralleled in part in some sauropodomorphs)

17. recurvature absent in maxillary and dentary teeth

18. maximum tooth size near the middle of the maxillary and dentary tooth rows

19. at least five sacral vertebrae

20. gastralia absent

21. ossified tendons at least above the sacral region

22. opisthopubic pelvis; pubis with small prepubic process

23. ilium with lateral swelling of the ischial tuberosity

24. iliac blade with a long and thin preacetabular process and a deep caudal process

25. pubis with an obturator notch, rather than a foramen; obturator foramen formed between the pubis and ischium

26. distal pubic and ischial symphyses

27. pubic symphysis restricted to its distal end

28. ischial symphysis restricted to its distal end

29. pendant fourth trochanter on the femur

30. fringelike lesser trochanter on the femur

31. digit V of the foot reduced to a small metatarsal with no phalanges

The basal ornithischian *Pisanosaurus,* which is based on incomplete material (Bonaparte 1976; Weishampel and Witmer, this vol.), shows the dinosaurian characters 2, 8, and 9, with 6 doubtfully present (see above). The two additional features seen in *Staurikosaurus* may also be present, and it shares the possession of tibial torsion of "about 90°" (Bonaparte 1976) with all later dinosaurs. *Pisanosaurus* has generally been regarded as the first ornithischian, whether as a basal ornithopod, a fabrosaur, a hypsilophodontid, or a heterodontosaurid. The arguments have centered largely on the nature of the teeth, which display ornithischian attributes 11 through 14 and 17, 18, and possibly 19 (Weishampel and Witmer, this vol.). Other ornithischian autapomorphies cannot be assessed in *Pisanosaurus:* the diagnostic predentary element and opisthopubic pelvis are only assumed by Bonaparte (1976). *Pisanosaurus* is presently regarded as a basal ornithischian.

CLADES WITHIN THE ORNITHISCHIA

Certain groups within the Ornithischia are apparently well supported by synapomorphies, and these will be noted briefly before a review of current views of their overall relationships.

The earliest ornithischians, the Fabrosauridae, are generally regarded as phylogenetically the most primitive (Weishampel and Witmer, this vol.), but they are hard to define cladistically as a family, since only *Lesothosaurus* is reasonably complete. The characters for the family noted by Galton (1978) and Cooper (1985) are primitive.

The Stegosauria (Galton, this vol.) also appear to be a monophyletic clade, all members sharing a large oval fossa in the pterygoquadrate wing, tall neural arches in middle and caudal dorsal vertebrae, a broad cup-shaped laterally facing acetabular surface on the pubis, loss of pedal digit I, prominent upright midline plates on the neck and back grading backward into spines, a lateral spine over each shoulder, and lack of ossified epaxial tendons (Sereno 1986).

The Ankylosauria (Coombs and Maryańska, this vol.) are similarly well defined as a monophyletic group by as many as twenty-six synapomorphies (Sereno 1986), including the rectangular occiput, closed antorbital and supratemporal fenestrae, contact of the quadratojugal and postorbital, fusion and dermal sculpturing of the dorsal skull roof, at least three dorsal vertebrae incorporated into the sacrum, fused scapula and coracoid, closed acetabulum, and dorsal and lateral armor of bone plates.

The Heterodontosauridae (*Abrictosaurus, Heterodontosaurus, Lycorhinus*) are distinguished from other ornithischians by the presence of three premaxillary teeth with no distinction between the root and the crown, reduced mesial two dentary teeth with no tubercles, wedge-shaped predentary, proximal head of humerus offset medially, relatively long manus, metacarpals with blocklike proximal ends, slender fibula, and extensor pits at the distal ends of proximal phalanges of pes digits II-IV (Sereno 1986; Weishampel and Witmer, this vol.).

The Hypsilophodontidae (*Hypsilophodon, Othnielia, Yandusaurus, Zephyrosaurus*) are also regarded as a monophyletic taxon by most authors (Sues and Norman, this vol.) on the basis of their shared narrow interorbital position of the frontal, premaxillary diverticulum, and steeply sloping ventral braincase region (Sereno 1986). If *Thescelosaurus* is also included in this clade, they all share a short scapula, partial ossification of sternal segments of the ribs, a rod-shaped prepubic process, and ossified hypaxial tendons in the tail (Sereno 1986).

Among other ornithopods, the Dryosauridae and Iguanodontidae are problematic. These are regarded as monophyletic groups by Norman (1984*a*) and Milner and Norman (1984) but less confidently by Norman (1984*b*). In this latter account, and Norman and Weishampel (this vol.), the two families are placed in an "iguanodontoid" clade, characterized by a large, median vertical ridge on the crowns of maxillary teeth, contact between the premaxilla and lacrimal, a distinct notch in the jugal wing of the quadrate, and a rodlike decurved ischium with a distal foot. The genera *Camptosaurus, Iguanodon*, and *Ouranosaurus* are grouped on the basis of multiple ridging of their dentary tooth crowns, and their shared robust manus, with fused carpals, metacarpal I fused to the radiale, and development of a spurlike manual digit I. The last two genera are further paired (Iguanodontidae) on the basis of the spiked pollex, the platelike first phalanx of manual digit I, the short postpubic ramus, and the reduced pedal digit I. Most authors now accept that the families Dryosauridae and Iguanodontidae, as commonly understood, are paraphyletic, forming a sequence of outgroups to Hadrosauridae, in a sequence from *Tenontosaurus*, through *Dryosaurus, Camptosaurus, Iguanodon*, and *Ouranosaurus*, as closest outgroup (Sereno 1984, 1986), or *Probactrosaurus* as closest outgroup (Sues and Norman, this vol.; Norman and Weishampel, this vol.).

The Hadrosauridae have been regarded as monophyletic by most authors (e.g., Norman 1984*a*, 1984*b*; Sereno 1984, 1986; Cooper 1985) on the basis of their well-developed dental magazines, lozenge-shaped teeth, displacement of the antorbital fenestra, loss or fusion of the palpebral, loss of the paraquadrate foramen, loss of manual digit I, loss of metatarsals I and V, and a large antitrochanter on the ilium (Milner and Norman 1984; Weishampel and Horner, this vol.). In addition, a large number of synapomorphies are shared with the successive dryosaur and iguanodont outgroups (Sereno 1986; Weishampel and Norman, this vol.). Horner (1985, in press), however, has questioned this view, arguing for close relationships of the lambeosaurine hadrosaurs with *Ouranosaurus*. His fourteen postulated synapomorphies (chap. 2) are fewer than those supporting Sereno's (1986) scheme, and some are probably plesiomorphous (e.g., high neural spines and massive ischium in lambeosaurines and *Ouranosaurus*, versus low neural spines and slender ischium in hadrosaurines and *Iguanodon*).

The Pachycephalosauria (Maryańska, this vol.) are defined by the thickened frontoparietal skull roof, broad expansion of the squamosal on the skull roof, tubercular ornamentation of the skull, elongate sacral ribs, very long ribs on proximal caudal vertebrae, relatively short forelimb, slender scapula, reduced deltopectoral crest, pubis nearly excluded from the acetabulum, and sixteen more postulated synapomorphies (Sereno 1986).

Finally, the Ceratopsia (Dodson, this vol.) all share a triangular skull in dorsal view, a median rostral bone, a tall snout with external nares highly placed, broad parietals, and other cranial features. The characteristic horns and frills are, of course, absent in psittacosaurs and are thus synapomorphous higher in the cladogram, defining the Neoceratopsia (Dodson and Currie, this vol.). Within the Neoceratopsia, two families, the Protoceratopsidae and Ceratopsidae, have generally been recognized. However, the diagnostic characters of Protoceratopsidae are primitive, or common to all neoceratopsians, and Sereno (1986) has argued that the Protoceratopsidae is a paraphyletic group.

RELATIONSHIPS WITHIN THE ORNITHISCHIA.

Although the monophyly of the Ornithischia is not questioned, relationships of the taxa within that clade are controversial. So far, five cladograms of the Ornithischia have been published, by Maryańska and Osmólska (1984*a*, 1985), by Cooper (1985), by Norman (1984*a*, 1984*b*; Milner and Norman 1984; Norman and Weishampel 1985), by Sereno (1984, 1986), and by Weishampel and Witmer (this vol.). Additional comments along these lines have also been made by Gauthier (1986, 44). These cladograms agree generally in some features—the grouping of Ankylosauria, Stegosauria, and possibly *Scelidosaurus* and *Scutellosaurus* as Thyreophora (see Coombs et al., this vol.) and the grouping of most ornithopods in the sequence of the Hypsilophodontidae, to the Dryosauridae, to the Iguanodontidae, and finally to the Hadrosauridae (see above). However, the placement of the Fabrosauridae, Heterodontosauridae, Pachycephalosauridae, and Ceratopsia is disputed. The four main proposals for ornithischian phylogeny by Maryańska and Osmólska (1985), Cooper (1985), Norman (1984*a*), and Sereno (1986) are compared in figure 1.6.

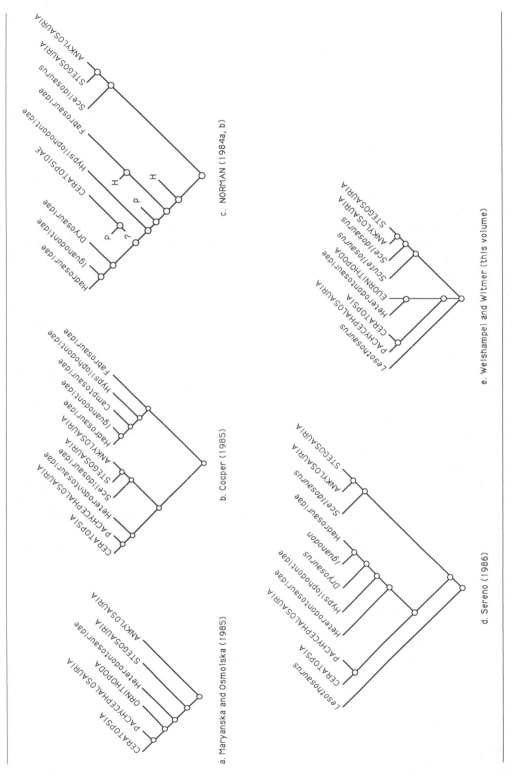

Fig. 1.6. Five cladograms depicting alternative hypotheses of the phylogenetic relationships of the Ornithischia: a. after Maryańska and Osmólska (1985); b. after Cooper (1985); c. after Norman (1984*a*, 1984*b*); d. after Sereno (1986); e. after Weishampel and Witmer (this vol.). In c, H = Heterodontosauridae; P = Pachycephalosauria.

The ankylosaurs, stegosaurs, *Scelidosaurus,* and possibly *Scutellosaurus,* all appear to fall into a clade termed the Thyreophora (Norman 1984*a;* Sereno 1984, 1986; Coombs et al., this vol.). If *Scutellosaurus* is omitted, this clade has as synapomorphies the sinuous curve of the dentary tooth row, the incorporation of a supraorbital bone into the skull roof between the frontal, postfrontal, and postorbital, the relative robustness of the medial portion of the quadrate condyle, a much shortened basisphenoid, and a tall midpalatal keel formed from the pterygoid and vomer. With the addition of *Scutellosaurus* (Thyreophora *sensu* Sereno 1986), the synapomorphies are transversely broad jugal-orbital bar, and parasagittal row of low-keeled scutes on the dorsal body surface with additional rows of lateral low-keeled scutes. Note that Maryańska and Osmólska (1984*a,* 1985) did not support Thyreophora but made Ankylosauria the sister group of the Stegosauria and all other ornithischians that share a perforate acetabulum. They noted that acceptance of this postulated synapomorphy was contingent on the acceptance of a nonmonophyletic Dinosauria. Since dinosaurian monophyly is now generally accepted, polarity of the acetabulum reverses and the imperforate acetabulum of the Ankylosauria becomes the derived state. This shift has the effect of falsifying the basal dichotomies in figure 1.6*a.*

The composition of the Ornithopoda is disputed. Santa Luca (1980) recognized the problem of including basal taxa such as fabrosaurs and heterodontosaurs. He suggested that there were two main ornithopod clades, one consisting of hypsilophodonts and fabrosaurs and one of iguanodonts and hadrosaurs. This view has not been accepted by later authors. Norman accepts the Fabrosauridae as the basal taxon of the Ornithopoda + Ceratopsia, while Sereno (1984, 1986) and Gauthier (1986) see this as a paraphyletic group. They regard *Lesothosaurus* as the only adequately known relatively complete fabrosaur and place it as the sister group of all later ornithischians, the Genasauria. The latter was ascribed a number of synapomorphies by Sereno (1986), which Weishampel and Witmer (this vol.) accept. In contrast, Maryańska and Osmólska and Cooper include *Lesothosaurus* in their Ornithopoda. The Heterodontosauridae have been interpreted as the sister group to the Cerapoda (Norman 1984*a,* 1984*b;* Maryańska and Osmólska 1985), to the Marginocephalia (Cooper 1985), or to the Ornithopoda (Sereno 1986; Weishampel and Witmer, this vol.).

All of the above authors, except for Cooper, group the Ornithopoda and Ceratopsia together, although their interpretations of Ornithopoda differ, and some also include Pachycephalosauria. This clade has been termed Cerapoda by Sereno (1986), who notes several synapomorphies: premaxillary/maxillary diastema, asymmetrical enamel on the cheek teeth, no more than five premaxillary teeth, a fingerlike lesser trochanter, a fully open acetabulum with no supraacetabular crest, and other features. Cooper (1985) places Ceratopsia in a line with heterodontosaurs and Thyreophora, separate from his Ornithopoda.

The placement of the Pachycephalosauria and Ceratopsia is still rather uncertain. Maryańska and Osmólska (1984*a,* 1985), Sereno (1984, 1986), Cooper (1985), and Gauthier (1986) all pair the Ceratopsia with the Pachycephalosauria, in a clade that Sereno (1986) names the Marginocephalia. The basis for this relationship includes the narrow parietal shelf that extends over the occipital elements, involvement of the squamosal in this shelf, a reduced premaxillary contribution to the palate, and a relatively short pubis (Sereno 1986). Norman (1984*a*) notes an alternative position for the Pachycephalosauria in association with basal ornithopods (fig. 1.6c), but this is based only on the primitive nature of their respective dentitions. He includes the Ceratopsia within his Ornithopoda, possibly grouped with the Hadrosauridae, Iguanodontidae, and Dryosauridae (fig. 1.6c), since they all share a reduced antorbital fenestra, maxillary teeth with sharp median ridges, distal dentary teeth lying medial to the coronoid process, and a strengthened predentary-dentary joint. Later, Norman (1984*b*) excluded the Ceratopsia and Pachycephalosauria from the Ornithopoda because they lack an obturator process on the ischium.

Alternative Arrangements within Dinosauria

The monophyly of Saurischia has been defined above, but a number of authors have presented a radical alternative in which the Sauropodomorpha is paired with the Ornithischia and the Theropoda are separated as the sister group of this new taxon. The third possible hypothesis of relationships, a sister group relationship between theropods and ornithischians, has not been seriously proposed and is very hard to support (Gauthier 1986). The pairing of sauropodomorphs and ornithischians was hinted at by Bakker and Galton (1974) and Bonaparte (1976), while Paul (1984*a,* 1984*b*) and Cooper (1985) have presented this view cladistically. Cooper (1981*a,* 1985), Paul (1984*a,* 1984*b*), and Sereno (1984) noted several apparent synapomorphies shared by the Sauropodomorpha (or Prosauropoda alone) and the Ornithischia:

1. tooth crown a transversely flattened blade with denticulate margins; a leaf-shaped spatulate tooth

2. differentiation of the tooth crown shape from the front backward, the mesial teeth being peglike and the distal ones more leaf-shaped

3. teeth set at an angle to the longitudinal axis of the jaws, and the distal edge of each tooth overlapping the anterior edge of the tooth behind

4. depressed jaw joint

5. reduced parietal

6. low occiput

7. elongate vomer

8. elongate preacetabular process on ilium

The dental similarities between sauropodomorphs and ornithischians (characters 1–3), in particular those in primitive forms, are striking. There are differences in the cheek teeth in both groups: the crowns are elongate in sauropodomorphs, but low in ornithischians, and the serrations are fewer and larger in ornithischians than in sauropodomorphs (Charig 1976), but this does not demonstrate that both forms were independently derived (Gauthier 1986). Characters 4–7 are less clear-cut (Gauthier 1986). The depressed jaw joint (character 4) is typical of many herbivores, and it is not seen clearly in basal sauropodomorphs like *Anchisaurus* and *Thecodontosaurus*. The parietals are reduced (character 5) in basal theropods. The occiput is not markedly lower (character 6) in sauropodomorphs or ornithischians than it is in theropods. The vomer is just as long (character 7) in theropods as it is in sauropodomorphs or ornithischians. Last, the elongate iliac prong (character 8) is seen only in *Anchisaurus* and *Ammosaurus* among sauropodomorphs and not in the basal forms *Efraasia* and *Thecodontosaurus*. The three dental characters (1–3) stand as the only potential synapomorphies in favor of a clade consisting only of the Sauropodomorpha and the Ornithischia, and they are far outweighed by the autapomorphies of the Saurischia.

THE FUTURE

The cladistic revolution in classification of the Dinosauria has only just begun, with publications dating essentially only from 1984. Already, however, a number of areas of stability have seemingly become clear: the monophyly of Archosauria; the split of Archosauria into a "bird line" and a "crocodilian line"; the close association of Ornithosuchidae, *Lagosuchus,* Pterosauria, and Dinosauria; the monophyly of Dinosauria, Sauropodomorpha, Theropoda, and Ornithischia; the positioning of Aves within Theropoda; and some general features of the relationships of major groups within Ornithischia (Ornithopoda, Thyreophora).

A great deal of work is still required, however, to address a broad range of key problems at the base of cladogram—the position of *Euparkeria,* the exact relationships of *Lagosuchus,* Pterosauria, and Dinosauria, and the arrangement of basal dinosaurian taxa. All of these problems are presently hard to solve because of poorly preserved material (many key taxa lack crucial evidence of distinguishing characters) and because of the great subsequent modifications of pterosaurs, and many dinosaurian groups, which may conceal potential synapomorphies.

Within Dinosauria, the controversy over the origin of birds has finally resolved itself in favor of Theropoda, but the relationships of theropod groups to birds, and to each other, are still unresolved. Key questions include the definitions of Dromaeosauridae and Troodontidae, and their relationships to Aves, the relative position of maniraptoran taxa that vie for positions on the cladogram close to Aves, whether below or within that taxon. Basal theropods, such as Ceratosauria and Carnosauria may be easier to resolve because the material is often of better quality. The arguments over the relationship of birds to terminal theropods will continue for some time since the material is often poor, and the characters (largely braincase and vertebral features) require exquisitely preserved specimens for determination.

The Sauropodomorpha still require a phylogenetic analysis. Some general outlines toward a cladogram are noted by Gauthier (1986: 44–45) and by Dodson, Galton, and McIntosh (this vol.). The idea of a paraphyletic Prosauropoda forming outgroups to a monophyletic Sauropoda seems clear, but the exact sequencing of prosauropod taxa and the relationships within Sauropoda have not yet been analyzed in detail. Crucial in the latter will be the breakup of the basal Cetiosauridae into several lineages.

The Ornithischia have so far, among dinosaurs, been favored with the greatest number of cladistic analyses, involving as many as ten authors to date. These studies have shown general agreement over the division of the Ornithischia broadly into an ornithopod clade and a thyreophoran clade. Basal taxa, such as fabrosaurs and scelidosaurs, have historically been hard to place, partly because of limited fossil material and partly because of incomplete descriptions of what exists. The relationships of ceratopsians and pachy-cephalosaurs, whether within the ornithopod clade or as an outgroup to it, are still to be decided. Other problems concern the genetic-level phylogeny within Ornithopoda (the relationships of dryosaurs, iguano-donts, and hadrosaurids) and within the armored or-nithischian clades.

Dinosaur systematists have made significant ad-vances in recent years as a result of the application of cladistic methods. For the first time, paraphyletic basal taxa have been widely abandoned, problems in phy-logenetic placement have been explicitly stated, new characters have been analyzed, and some novel, but convincing, patterns of relationship have emerged. The revolution we are currently witnessing marks the be-ginning of an exciting new era of clarity and precision in attempts to unravel the patterns of dinosaurian evo-lution. Whether the conclusions shown here are right or wrong, it is now possible to specify problems and conflicts, to sort out significant character transforma-tions, and to erect a firm baseline for robust and inno-vative evolutionary discussion, which is, after all, why we are in the business of paleobiology.

2

Dinosaur Paleobiology

P. DODSON

The study of the paleobiology of dinosaurs has constituted a major component of dinosaur research over the past two decades. The desire to understand dinosaurs as living creatures is compelling, despite the inherent problems of studying creatures that have not lived and breathed for at least 65 million years. Nonetheless, dinosaurs lived in a world with the same physical laws and physiological and ecological constraints as do living animals. Advances in the study of dinosaur paleobiology thus go hand in hand with advances in our general understanding of the world we live in.

It is noteworthy that all studies of dinosaur paleobiology make reference to living animals. This tradition dates from the early days of paleontology, for Richard Owen (1842*b*), as he was coining the term *dinosaur*, inferred that dinosaurs possessed circulatory and respiratory systems comparable to those of living mammals. One striking feature of studies of dinosaur paleobiology is the breadth of the animal models used to infer aspects of dinosaur biology: mammals, birds, and living reptiles.

Does it follow that, depending on the model used, dinosaurs must necessarily be regarded as Mesozoic elephants, ostriches, crocodiles, or turtles? Emphatically not! To maintain this position is to suggest that those animals living today embody all possible conditions of physiology, ecology, behavior, and so forth. Dinosaurs clearly represented unique combinations of all these, plus characters that did not survive the close of the Mesozoic.

The challenge of paleobiological research is to learn as much as possible about the dinosaurs as living animals while avoiding the twin hazards of forcing them into the mold of living animals and giving unbridled rein to fanciful speculation. The subject is vast. We have chosen to highlight only three among the many possible topics. These are dinosaur behavior (Walter P. Coombs, Jr.), dinosaur thermal physiology (James O. Farlow), and dinosaur extinction (Peter Dodson and Leonid P. Tatarinov).

Part I   BEHAVIOR PATTERNS OF DINOSAURS

WALTER P. COOMBS, JR.

What is behavior? Behavior encompasses a variety of phenomena including running ability, food-gathering strategies, circadian rhythms, circannual rhythms, social structure, reproductive behavior, and parental care of young. Topics within the general area of behavior are difficult to discuss independently; for example, communication may be considered in connection with social structure, agonistic behavior, and parental care, and running ability and foraging strategy overlap with physiology. Three topics of current interest are reviewed here: (1) gregarious behavior, (2) agonistic behavior, and (3) nesting and parental care.

There are three major approaches to analysis of dinosaur behavior:

1. The first principles or uniformitarian approach uses stable natural laws to establish functional interpretations of biological adaptations and interrelations; for example, utilization of lever mechanics and mass/strength scaling to interpret muscle-bone systems or application of the Laws of Thermodynamics to analyze energy flow in predator-prey systems.

2. Pattern matching is the most common analytic method in all the sciences. As applied to dinosaurs, it encompasses comparison with (1) the closest living relatives (i.e., birds or crocodilians); (2) the closest living morphologic equivalent (birds, crocodilians, and mammals are all used); (3) the closest living ecological equivalent (large mammals and ratites are commonly used); and (4) the closest behavioral equivalent (crocodilians, birds, and mammals are all used). Mathematical rigor can be brought to pattern matching by computing formulas from measurements of living animals, then comparing dinosaurs; for ex-

ample, calculating an encephalization quotient as an indicator of "intelligence" or using relative hip height and stride length to estimate running speed of trackways (Alexander 1976; Hopson 1980a; Farlow 1981; Kool 1981; Russell 1980; Thulborn 1982, 1984b; Russell and Seguin 1982).

3. Speculation (= scenario building, story telling, creative logic, etc.) is the easiest, most colorful, most untrustworthy, most maligned, but sometimes most fruitful, approach. It has considerable heuristic value, but because of the editorial policies of technical journals, many familiar speculations about dinosaur behavior have been most fully developed in semipopular publications or only exist as unpublished common knowledge. Speculative interpretations of dinosaur behavior have become more commonplace in recent years.

Although these three analytic methods are logically distinct, in practice they are used simultaneously and are frequently inextricably intertwined.

Gregarious Behavior

Formation of monospecific aggregates is a ubiquitous phenomenon commonly associated with (1) a breeding season or location, (2) communal feeding grounds (especially for herbivores), and (3) nomadic or migratory behavior. Aggregates may be seasonal or perennial, mono- or bisexual, coincidental or socially cohesive. Among modern ungulates, gregarious behavior and concomitant sociality is governed by the selective constraints imposed by avoiding predators and by availability of food, which in turn is influenced by foraging strategy and habitat parameters.

Ungulate gregariousness is promoted by (1) large body size, (2) unselective feeding habits, (3) open habitats, (4) strong climatic seasonality, and (5) erratic spatial or temporal availability of forage (Jarman 1974; Wilson 1975). Small forest-dwelling herbivores with selective browsing diets rarely form large social groupings. Large open-ground herbivores that graze unselectively on low-quality forage commonly form herds of complex structure and may cooperatively defend their young by assuming a living wall or rump-bump ring formation (e.g., *Taurotragus, Ovibos,* and *Loxodonta africana;* Jarman 1974; Wilson 1975; MacDonald 1984; Vaughan 1986). These species may have altruistic behavior. Dinosaurs are predominantly large herbivores that are ecologically equivalent to modern ungulates. Many dinosaurs have adaptations applicable to social behavior (see discussion of agonistic behavior, below), and they are diversified with a long evolutionary history. It is inevitable that some dinosaurs established massed monospecific aggregates on a seasonal or perennial basis. Therefore, the question of dinosaur gregarious behavior centers on which species were social, whether sociality was seasonal or perennial, and also on the nature of social interactions.

The interpretation of social interactions among members of a group is greatly altered if the group is called a school, flock, pack, pride, or herd (Coombs 1975). Some terms (notoriously, "herd") cover a diversity of social structures (Wilson 1975; MacDonald 1984). Modern ungulate herds are commonly associations of female-offspring units from which adult males are excluded except during breeding season. Males are generally solitary, although they may form temporary, unstable bachelor herds. Gregariousness of modern species may vary with climatic fluctuations or local biome structure, and closely related species may differ in degree of gregariousness (based on information from Bourliere 1964; Tener 1965; Bere 1966, 1970; Wilson 1975; Nowak and Paradiso 1983; MacDonald 1984; Vaughan 1986).

Evidence for dinosaur gregarious behavior comes from thanatocoenoses, ichnocoenoses, and analoglogical inferences (Ostrom 1986).

Dinosaur Thanatocoenoses

Near-monospecific dinosaur thanatocoenoses have traditionally been interpreted as evidence of herding behavior. The *Iguanodon* assemblage at Bernissart has been interpreted as a catastrophic kill of a herd in a chasmlike opening, possibly driven by a menacing predator (e.g., Moody 1977; McLoughlin 1979; Stout

and Service 1981), but reexamination of the stratigraphic relationships of individual skeletons has shown that the assemblage is accretional, not catastrophic (Norman 1985). The near-monogeneric assemblage of *Plateosaurus* at Trossingen is composed of adult and subadult animals and has been interpreted as a catastrophic kill of a herd in midmigration (Huene 1928). However, the stratigraphic distribution of the skeletons reveals a complex history, with erratic addition of both solitary individuals and small groups (Weishampel 1984b; Norman 1985). Thus, the "herding" interpretation for *Iguanodon* and *Plateosaurus* must be modified. The dispersion pattern may have been a mixture of small assemblages and solitary individuals (isolated bulls?), or alternatively, herding may have been seasonal, not permanent.

Recently discovered and still incompletely described mixed assemblages of hadrosaurs and near-monospecific assemblages of ceratopsians in western North America form bone beds that contain thousands of bones (data and following discussion based on Currie and Dodson 1984; Currie 1986, 1987b, pers. comm.). The fauna is commonly mixed, but in a small fraction (less than 10%) of these assemblages, a single ceratopsian (*Centrosaurus, Styracosaurus,* or other) constitutes 95 percent of the fauna. These near-monospecific thanatocoenoses may be catastrophic rather than accretional samples, although the nature of the stratigraphy, bone preservation, and burial appear to be similar in the mixed and monospecific beds. The massed aggregate interpretation is supported by the distribution of age categories, especially the high incidence of animals apparently in their prime reproductive years, based on the theory that accretional assemblages will contain disproportionate numbers of very young and very old individuals (Shipman 1981). The animals may have been killed during a massed migration. Among modern ungulates, both seasonal migrators (e.g., *Connochaetes*) and species that have erratic mass dispersals (e.g., *Antidorcas*) suffer heavy mortality at river crossings, an ideal setting for formation of a monospecific thanatocoenosis (Bere 1970). A perplexing feature of monospecific bone beds is that virtually all the material is disarticulated and broken, and complex scenarios have been proposed to explain such assemblages. Following an initial catastrophe that killed the herd, there was prolonged exposure of the carcasses to decay, scavenger dismemberment, and general bioturbation. Later the bones were swept up and buried by normal fluvial processes.

The best-known monospecific predator thanatocoenoses are the Ghost Ranch *Coelophysis* quarry and the Cleveland-Lloyd *Allosaurus* quarry (Colbert 1947, 1948*a*, 1961, 1974; Madsen 1976*a;* Ostrom 1986). The *Coelophysis* quarry has numerous skeletons of all age categories, with many prime age individuals. The skeletons are complete, articulated, and piled one atop the other as if some catastrophe overtook a group of animals (Colbert 1947, 1948*a*, 1961, 1974). The *Allosaurus* quarry is similar to ceratopsian bone beds in that the skeletons are extensively disarticulated, skulls are represented almost exclusively by isolated elements, and a wide size range of individuals is represented (Madsen 1976*a*).

There are several possible explanations for these predator assemblages. Some (many?) predatory dinosaurs may have normally formed hunting aggregates (packs or prides). Cooperative hunting exists when and if it results in an increased per diem food intake by participating individuals. Lions hunt Thompson's gazelles in pairs but hunt zebra and wildebeest in packs of three to five because the average daily catch per lion is greater than for lions that hunt the respective prey species in either smaller or larger packs (Schaller 1972; Caraco and Wolf 1975). Thus, socialization is partially governed by food supply. Small predators generally take prey over which they have an advantage in both body mass and speed. Large predators take prey more nearly their own body mass or larger that may have greater speed and endurance than the predators and may aggressively defend themselves, especially if armed with horns (Bere 1970: 69). If the most readily available prey species is too large, too swift, or too well defended to be easily taken by a solitary hunter, cooperative hunting may develop, and the most socialized modern carnivores, especially those that cooperate in hunting and sharing food, tend to be relatively large and take prey as large or larger than themselves (Wilson 1975). Pack hunting increases the maximum size of prey and permits exploitation of a wider range of prey sizes (Earle 1987). Based on the relative sizes of predator and prey calculated for modern mammals, theropods must have hunted in packs in order to tackle sauropods and some large ornithischian herbivores such as *Triceratops*. However, solitary predators or mated pairs are the most common patterns among modern Carnivora. Cooperative predation has been suggested for several carnivorous dinosaurs (Ostrom 1969*b*, 1986; Farlow 1976*a*, 1987*b*), although many were probably solitary hunters. *Tyrannosaurus* teeth are transversely broader and more rounded in section than teeth of *Albertosaurus* or *Allosaurus,* and the serrations of *Tyrannosaurus* teeth form a faint ridge on a rounded surface rather than a sharp cutting blade on a narrow edge (Molnar and

Farlow, this vol.). Thus, while *Albertosaurus* teeth look suitable for slicing meat off a large carcass, *Tyrannosaurus* teeth are more like the stab and swallow teeth of a modern killer whale (*Orca*). Tyrannosaurids therefore may have killed relatively small prey that were swallowed whole and were not major predators on adult hadrosaurids or ceratopsians.

Another possibility is that massed assemblages of predatory dinosaurs are a consequence of habitat preferences (Farlow 1987*b*), or they may have formed during breeding season in the manner of sea lions and other mammalian marine predators. In such modern predator assemblages, feeding is commonly curtailed or postponed and the animals survive for extended periods on reserves built up prior to the breeding season (e.g., Bourliere 1964; Van Gelder 1969; Garrick et al. 1978; MacDonald 1984). If dinosaurs were predominantly endothermic, a massed assemblage of *Allosaurus*-sized predators might place an unsupportable pressure on the local herbivore population, and temporary fasting for the sake of success at breeding would provide a solution. Although no modern predators appear to utilize leks, male mammalian herbivores (e.g., *Kobus kob*) and many birds that display on leks forsake or curtail feeding (Welty 1982). Predators might also aggregate during a low activity, low food supply season, akin to denning by snakes.

Finally, these monospecific carnivore thanatocoenoses may record a unique set of circumstances, a "catastrophe" that drove together animals that would normally be dispersed. In this case, the quarries are recording an environmental perturbation rather than normal behavior. The cannibalism evident in the *Coelophysis* assemblage might indicate that immature and adult animals did not normally associate but were driven together by whatever circumstances led to the mass killing and burial.

Dinosaur Ichnocoenoses

Footprints are difficult to identify and are commonly assigned to a general category of dinosaurs (e.g., "sauropod prints") rather than being ascribed to a particular genus or species based on skeletal remains. Doubt remains on correct assignment of some tridactyl-bipedal prints to carnivores rather than herbivores (detailed review by Farlow 1987*b*; see also, Thulborn and Wade 1979). The apparent excess representation of carnivorous dinosaur trackways (Leonardi 1984) may be in part an artifact reflecting this problem, although in some cases, habits and habitat selection by carnivorous species may be responsible (Farlow

1987*a*). Two trackway patterns are pertinent to the question of dinosaur gregariousness: (1) multiple trackways with random orientations (= unstructured) and (2) multiple trackways with a coordinate orientation (Ostrom 1986).

UNSTRUCTURED ICHNOCOENOSES

Most ichnocoenoses have no coordinate directional orientation. Analysis with a compass-rose diagram may reveal a preferred orientation, but if there were true herding organization it should be apparent on inspection. The lack of coordinate orientation in trackways cannot be taken as evidence against herding, however. Flocks of birds, herds of ungulates, and packs of carnivores that have strong internal cohesion will mill about randomly on feeding grounds, near watering holes, or when at leisure. Of special importance is the relative frequency of ichnospecies among trackways that lack coordinate orientations. Ichnocoenoses typically include one to four taxa each represented by numerous trackways and one or more other taxa represented by one or a few trackways. Near-monospecific ichnocoenoses are common rather than exceptional. Thus, multiple individuals of one dinosaur taxon commonly occupied small areas simultaneously. Gregarious behavior of some form is the inescapable inference.

STRUCTURED ICHNOCOENOSES

Instances of multiple trackways, apparently of single taxa, all heading in approximately the same direction have been taken as evidence of gregariousness and "flocking" since their earliest discovery (Hitchcock 1848, 1858). The famous sauropod ichnocoenosis at Davenport Ranch, Texas, was from the outset interpreted as a "herd" (Bird 1944, 1985; Farlow, in Bird 1985; Farlow 1987*a*; Lockley 1986*a*, 1987) and later as a "structured herd" with young toward the center and large adults acting as guards around the periphery (Bakker 1968; critique by Ostrom 1986). Reanalysis of the Davenport Ranch site indicates the sauropods crossed the area in small groups without the massed herd structure hypothesized earlier, and the largest footprints, which are slightly over double the linear dimensions of the smallest prints (lengths, respectively: 78 cm and 35 cm), are commonly near the center of an ichnocoenosis, possibly indicating that large sauropods were "leading" small herds (Lockley 1986*a*, 1987). There is consistency in velocities computed for trackways within what has been regarded as a set (Lockley 1987), although there is no way to know the time scale within which the overstepping trackways were made. The largest prints at Davenport Ranch are of modest size compared to sauropod prints from nearby localities

(Farlow pers. comm.), perhaps indicating some sauropod herds were sibling groups (*sensu* Coombs 1982). Other cases of multiple sauropod trackways all traveling in the same direction are known (Ishigaki 1985, pers. comm.; Lockley et al. 1986; Lockley 1986*a*, 1987; Farlow 1987*a*). Sauropods thus normally formed monospecific aggregates, and these aggregates, sometimes small (3 or 4 individuals) but possibly much larger (20 or more?), typically traveled in concert.

In early Jurassic rocks near Mt. Tom, Massachusetts, there are numerous near-parallel, tridactyl-bipedal trackways called *Eubrontes* which have long been interpreted as a flock or herd (Hitchcock 1848, 1858; Ostrom 1972, 1986). *Eubrontes* prints are generally regarded as pertaining to a theropod, although they have also been called herbivore tracks (cf. Lull 1904, 545, to Lull 1915*a*, 184; see also, Coombs 1980*a*). The trackways at Mt. Tom are arrayed in side-by-side fashion with little overstepping, an unexpected pattern for a herd. Ripple marks indicate a downslope direction approximately perpendicular to the preferred orientation of the trackways. Therefore, topographic control of the direction the animals were traveling is possible, yet 86 percent of the trackways are going in the same direction (Ostrom 1972, 1986). The "herding" interpretation of the Mt. Tom site thus remains viable. Similar side-by-side arrays of tridactyl-bipedal trackways are known (e.g., Leonardi 1984).

The Lark and Seymour quarries of Australia are mixed ichnocoenoses that contain 3,300 footprints of small tridactyl bipeds, both carnivores ("coelurosaurs") and herbivores ("ornithopods"), preserved in marginal lacustrine sediments (Thulborn and Wade 1979, 1984). Footprints of a much larger tridactyl biped are interpreted as a large predator ("carnosaur") that drove the smaller animals to flight. This ichnocoenosis has been called a "stampede."

Other cases of multiple near-parallel trackways have been reported and in some cases have been interpreted as indicating herding behavior (e.g., Shikama 1942; Colbert 1961, 1983; Ostrom 1972; Charig 1979; Lockley et al. 1983, 1986; Lockley 1987; Farlow 1987*a*), so the phenomenon, if not common, is nevertheless widespread temporally and geographically. There is one reported case of coordinate orientation in trackways interpreted as being made by juvenile dinosaurs (Currie and Sarjeant 1979). This ichnocoenosis presumably indicates a cohort or creche (= sibling group *sensu* Coombs 1982). Some cases of parallel trackways appear to be topographically controlled, that is, the animals were walking parallel to a shoreline (e.g., Lockley 1986*a*, 1987). A bimodal orientation distribution with several trackways at 180 degrees to each other, as would be predicted with shoreline control of direction, has been reported for several ichnocoenoses (Lockley 1986a, 1986b, 1987; Farlow 1987a). Unimodal directionality could also be a consequence of migratory behavior.

Analog-Logical Inferences

Migration is a cyclic, commonly circannual movement of organisms away from and back to a particular location, distinct from both nomadism, which is a noncyclic wandering that does not lead back, except coincidentally, to the starting point, and emigration, which is massed dispersal commonly induced by environmental/population stress (e.g., *Antidorcas* and *Lemmus*). True migratory behavior is rare among modern reptiles (reported for marine Chelonia but not for crocodilians), is common among birds, and is more common among marine mammals than ungulates. Migrators may travel individually, in female-offspring units, in family groups, in small flocks/herds, or in superflocks/herds. Cyclic fluctuations of forage in regions with strong climactic seasonality (summer/winter or rainy/dry) and return to traditional nesting grounds or rookeries are the primary inducements to migratory behavior (Alcock 1984; Drickamer and Vessey 1986). Based on the provenance of exotic gastroliths, a migration of up to 1,000 km has been hypothesized for some sauropods (Janensch 1929c; Russell et al. 1980). During the Cretaceous, dinosaurs were living at high latitudes in the northern hemisphere (Davies 1987).

Migratory behavior has been suggested for some large herbivores found at extremely high latitudes where, even allowing for the probable absence of a northern ice cap through most of the Mesozoic, reduced solar insolation must have lowered temperatures and curtailed primary productivity for several months (Colbert 1968; based on footprints identified as "*Iguanodon*"). A flora dominated by conifers could offer a residual browse of low quality for part of the winter. Migratory behavior has been suggested for *Plateosaurus* as a means to locate water during the dry season in an arid environment (Huene 1928). Annual sojourns to remote, "upland" nurseries from lowland feeding grounds have been hypothesized for several dinosaurs (see below). If ichnocoenoses with coordinate orientation are produced by individuals or a series of small groups rather than by massed herds, then migratory behavior may explain the common directionality. Modeling dinosaurs as inertial homeotherms leads to the

theory that near-continuous migratory behavior would generate heat by muscular activity at the same time the animals followed north-south displacements of physiologically optimum isotherms (Hotton 1980). Migrations of as much as 3,200 km are possible at walking speeds below velocities computed from trackways. Depending on relative newborn size (RNS), hatchlings either migrated with adults or hibernated over winter the first year. Hibernation and aestivation have been proposed and refuted for some dinosaurs (Cys 1967; Hotton 1980; Thulborn 1978; Hopson 1980b).

Summary

Data pertinent to gregarious behavior in dinosaurs give complexly mixed information. Some generalities may be made: (1) dinosaurs commonly formed monospecific assemblages; (2) massed assembly at breeding season was probably a ubiquitous phenomenon among dinosaurs; (3) some dinosaurs traveled in "herds," in the broadest sense of that word, or perhaps in "flocks" (see Wilson 1975; Ostrom 1986); (4) some dinosaurs were herding by the early Jurassic, leaving open the possibility for development of complex social systems over the next 140 million years (my), although these need not be more complex than what evolved in birds over 150 my; (5) speculations on dinosaur defense strategies, such as rump-bump ring formation, are reasonable if not rigorously demonstrable (dissenting opinion on ring defense by Paul 1987b); (6) some dinosaurs may have been migratory; and (7) available evidence for hibernation and aestivation is equivocal.

AGONISTIC BEHAVIOR

Agonistic behavior (= intraspecific aggression) encompasses threat, combat, appeasement, submission, territoriality, and dominance (= establishment of hierarchies), all of which are commonly linked to sexuality, especially obtaining and maintaining control of females by males. Also included are defense of young from conspecifics and parental discipline of young.

Communication

Contentious interactions between conspecifics commonly involve display or ritualized combat to avoid potentially injurious battles, although deadly fights do occur. Communication of such information as size, state of health, aggressiveness, dominance, proprietary territoriality, appeasement, and submission are critical to ritualization of agonistic interactions. Modern crocodilians and birds, both reasonable analogs for dinosaurian behavior, share a large number of communication modalities. Many birds, mammals, and lepidosaurian reptiles, but only rarely crocodilians, develop physical structures (feathers, horns, antlers, inflatable sacks, tusks, color patterns) that enhance visual agonistic displays and sometimes actual combat. Threat behavior and structural modifications used to defend territory and/or establish dominance are also commonly used in modified, mollified form to attract females, a form of positive-feedback sexual selection. Among the many dinosaurian features reasonably interpreted as visual or vocal display/threat adaptations or actual weapons are the following (see review by Molnar 1977):

1. Bright color patterns have been inferred for Dinosauria from color vision capacity of crocodilians and birds (Walls 1942; Russell 1977), extensive use of color by birds, well-developed birdlike eyes of dinosaurs evidenced by sclerotic rings (Russell 1980), and the scalation pattern of some dinosaur fossils (Osborn 1912a). Reference to modern analogs suggests small dinosaurs are better candidates than large species for bright color patterns (Hallett 1987; Paul 1987b).

2. Dinosaur tails could be used for display especially if unusually long (Diplodocidae), laterally compressed and deep (some Hadrosauridae), or decorated with armor plates (Stegosauridae, Ankylosauridae). Modern mammals, birds, crocodilians, lacertilians, and ophidians all use their tails for communication and/or display. Dinosaur tails were apparently carried off the ground by most species, as evidenced by scarcity of tail drags associated with trackways (e.g., Lockley 1987; Paul 1987b), and some dinosaurs have obvious adaptations to enhance the tail as a weapon (Stegosauridae, Ankylosauridae), indicating that many dinosaurs may have swung their tails in both intra- and interspecific combat (Colbert 1968; Bakker 1968, 1986; Coombs 1975, 1979; Carpenter 1982a).

3. In addition to their tails, sauropods probably used their elongate neck for display and/or combat. Some species (e.g., Camarasaurus, Brachiosaurus) may have had an expandable "proboscis" for display (Coombs 1975; Bakker 1986; disputed by Paul 1987b). Sauropods probably had giraffe-style neck sparring and reared up in a tripodal stance to utilize the large claws on the manus for combat (Bakker 1968, 1971a, 1971b, 1986; Wilson 1975; Coombs 1975; Paul 1987b).

4. Stegosaur plates, flattened body, and elongate hindlimbs are all suitable for flank display, with tail spikes for combat (Davitashvili 1961; Farlow et al. 1976; Buffrénil et al. 1986).

5. Hadrosaurid crests and complex narial passages could serve for both vocal and visual display, and their laterally compressed tails could be used for visual display and combat (Abel 1924; Nopcsa 1929b; Wiman 1931; Davitashvili 1961; Ostrom 1961a, 1964; Dodson 1975; Hopson 1975b; Molnar 1977; Farlow 1983; Weishampel 1981a).

6. Iguanodon thumb spikes could be used in close-quarters combat but would have necessitated some curious posturing to be effective for threat display (Owen 1872; Lydekker 1890c; Colbert 1968; Norman 1980, 1985; Paul 1987b; Hallett 1987, fig. 9).

7. Heterodontosaurus tusks, so similar to those of living Tragulidae and Moschidae, could be similarly used for display and combat (Molnar 1977; Norman 1985).

8. Ankylosaur spikes, spines, "horns" at the posterolateral corners of the skull, and tail clubs could all be used for display. Nodosaurids may have fought by hooking shoulder spikes (see restoration of Edmontonia by Carpenter, p. 21 in Czerkas and Olson 1987). Ankylosaurids could utilize their tail clubs for combat, and their broad triangular skulls, which are flat toward the rear, could be used for head-to-head shoving tests of dominance (Carpenter 1982a, pers. comm.).

9. Ceratopsian frills and horns have obvious display and combat value. Development of accessory epoccipital ornamentation of the frill enhanced full frontal displays in several genera (e.g., Anchiceratops, Arrhinoceratops, Pentaceratops, and Styracosaurus). Genera with long brow horns (e.g., Triceratops and Torosaurus) may have locked these together for wrestling. Premortem puncture wounds on frills and damaged horns attest to occasionally violent tests of dominance. Pachyrhinosaurus could have charged and battered head-on with less danger of lethal injury (Lull 1933; Col-

bert 1948a, 1961; Kurzanov 1972; Farlow and Dodson 1975; Molnar 1977; Paul 1987b, fig. 22).

10. Pachycephalosaur enlarged, thickened crania may have been used for display and combat (Galton 1970c, 1971c; Molnar 1977; Sues 1978b).

11. Dromaeosaurid saber-claws could have been used like the spurs of fighting cocks. Their complexly reinforced tails, in addition to having display value, may have served as a "third leg" during intraspecific combat (Ostrom 1969b, 1986).

12. *Acrocanthosaurus, Hypacrosaurus, Ouranosaurus* and *Spinosaurus* had saillike expansions supported by elongate neural spines for display. Enhancing coloration was probably present. A dorsal frill may have been present in many hadrosaurids (Paul 1987b).

13. Many carnivorous dinosaurs probably utilized an open mouth display of teeth (not ornithomimids), and several species had cranial crests (e.g., *Ceratosaurus* and *Ornitholestes*). Cranial crests of *Dilophosaurus* may have had enhancing coloration for display (Paul 1987b).

The agonistic functions hypothesized for the structures listed above in no way obviates their use for other behavioral or physiological purposes. Structural modifications of modern tetrapods typically have multiple functions, and arguments that dinosaurs utilized certain structures for one purpose to the exclusion of others are born in ignorance of behavior patterns in extant tetrapods.

Territoriality

Defense of a geographic plot against encroachment by certain conspecifics and sometimes other species is a ubiquitous phenomenon among vertebrates and animals generally, to the extent that it must have been common among dinosaurs. Most evidence for dinosaur territoriality is indirect, inferential, and contingent on assumptions about other behavior patterns.

Openground, colonially nesting birds such as gulls, murres, and guillemots establish and defend small territories within rookeries from conspecifics. Pugnacious species commonly establish a defense ring whose diameter is established by the maximum reach of the female when sitting on the nest. Nests of two ornithopod species have been found stacked at several horizons, suggesting repeated utilization of a traditional

rookery (term originally applied to dinosaurs nesting grounds by Lull 1904). On a single bedding plane, nests of both a medium-sized hadrosaurid and a hypsilophodontid are spaced at distances that correspond to total body length for the respective species (Horner 1982, 1984a, 1987), a pattern reminiscent of colonial nesting in some marine birds. When nesting in colonies, the nile crocodile (*Crocodylus niloticus*) establishes relatively uniform spacing of nests apparently without persistently straddling the nest, although such behavior is known (Cott 1961; Coombs in prep.).

If some dinosaurs established or traveled in massed herds (e.g., Sauropoda, Hadrosauridae, and Ceratopsidae), then personal-space mobile territories were probably maintained by males at least during the breeding season, although not through the rest of the annual cycle. Using crocodilians and birds as analogs leads to the prediction that some male dinosaurs established display-mating territories that in many cases also included nesting sites subsequently defended by either or both members of a mated pair. Lekking is also possible for some dinosaurs, although it is not especially common among modern tetrapods.

Unlike display-mating-nesting territories, which are established and defended seasonally, feeding territories are generally permanent and are defended all year throughout the life of an individual. Large herbivores commonly have traditional home ranges and migratory paths, but feeding grounds are usually only incidentally included in a male's breeding territory, a type of resource defense polygyny. However, many carnivores, whether gregarious or solitary, defend hunting territories (Wilson 1975). If some theropods (e.g., *Coelophysis, Deinonychus, Allosaurus*) hunted in packs, then they may have cooperatively defended large hunting territories in the manner of wild dogs and some large cats. Solitary dinosaurian hunters (e.g., tyrannosaurids?) would be expected to defend territories insofar as spacing of such predators is contingent on territorial behavior.

PARENTAL CARE

In current ethological theory, parental care of offspring is evaluated in terms of parental investment versus increased parental fitness (Trivers 1972, 1974). Rigorous analysis of the relentless circumstances that lead to greater parental investment (= extended parenting) in living organisms requires longitudinal studies that are necessarily impossible for dinosaurs. At the outset, therefore, it must be accepted that investigation of selective forces or "advantages" of parental care in dinosaurs will be inferential, hypothetical, or simply ignored.

Endothermy and Parental Care

The special nutritional and physiological demands of endothermic embryos makes extended parenting almost universal among warm-blooded organisms. The young require food for both growth and heat production when their physiological systems are incompletely formed, when they lack adult insulation, and when they have an unfavorable body mass/surface area relationship. Insulating juvenile down, shed with maturity, has been suggested for hatchling dinosaurs although without an evidential basis (Paul 1987b). Insulating feathers are not critical to full endothermy in hatchling birds (Pettingill 1970). The embryo, fetus, and newborn/hatchling of endothermic organisms are commonly semiectothermic, relying on their parents to supply heat either directly (body heat) or behaviorally (solar or fermentation heat for incubation; Pettingill 1970). The time taken by hatchling birds to reach physiological maturity (= independent thermoregulatory capacity) strongly influences the duration of brooding by parents (Clark 1964). Female placental mammals are locked into a system of prolonged embryo and newborn dependence from which the female cannot escape. Most female and male birds also provide extended care to embryos and hatchlings, although there are exceptions. Among both birds and mammals, males have several options and may either assist in raising offspring or may seek additional mates depending on the species and the ecological/selectional milieu (Van Tyne and Berger 1959). Although some modern endotherms have escaped all parental care (some Megapodiidae: Frith 1962; Coombs in prep.), it is nevertheless reasonable to suppose that if dinosaurs were endothermic, extended parental behavior was practiced by many species.

Dinosaurs should show a diversity of parenting behavior because (1) modern amniotes have a diversity of parenting behaviors; (2) dinosaurs were a diversified group; and (3) dinosaurs had a very long evolutionary history (Horner 1987). The question of parental care should therefore be approached on a species by species basis. Generalizing from a unique case to all members of a familial level group should be done with caution.

Parental investment in young can be divided into two components: (1) care prior to hatching (i.e., nest construction and guarding), and (2) care after young emerge (classic parenting).

Eggs and Nests

Taxonomic assignment of fossil eggs is certain only when they are directly associated with embryos and hatchlings (e.g., *Maiasaura* and *Orodromeus*: Horner and Makela 1979; Horner 1982, 1984a; Horner and Weishampel 1988) but is highly probable when there are large numbers of associated skeletons, tiny juveniles to adults, of a single dinosaurian species (e.g., *Protoceratops andrewsi*: Brown and Schlaikjer 1940a; Dodson 1976). When neither of these conditions is met, opinions on identification of eggs vary widely. Eggs from the Maastrichtian of France and Spain traditionally assigned to the sauropod *Hypselosaurus priscus* have also been ascribed to the carnosaur *Megalosaurus* (Voss-Foucart 1968) and have been interpreted as representing as few as one (Erben 1972, 1975; Erben et al. 1979) to as many as nine or more species (Dughi and Sirugue 1957, 1958a, 1958b, 1966, 1976). Eggs originally identified as dinosaurian (e.g., Young 1965) were later called avian (Dughi and Sirugue 1976), then yet again ascribed to dinosaurs (Erben et al. 1979). Within a single modern crocodilian species, egg size varies over as much as a 2:1 ratio in both mass and linear dimensions (Deitz and Hines 1980; Ferguson 1985), so size may not be a reliable taxonomic character for dinosaur eggs. Microscopic analysis of eggshells from the Maastrichtian of France indicates at least four species represented by the eggs (Williams et al. 1984). Among dinosaurs, the hadrosaurid *Maiasaura*, the ceratopsian *Protoceratops*, and the hypsilophodontid *Orodromeus* are known to be oviparous. Viviparity or ovoviviparity has been suggested for some dinosaurs (e.g., Desmond 1975; Charig 1979; Bakker 1986), although most authorities regard dinosaurs as predominantly or entirely egg-laying (e.g., Hotton 1980). It may be impossible for viviparity to evolve in species with truly cleidoic eggs (Packard et al. 1977).

Many authorities have commented on the scarcity of juvenile dinosaurs and eggs, based on unstated assumptions about age-specific distributions of dinosaur populations (most recent review by Carpenter 1982b). Among the possible explanations are the following:

1. The scarcity reflects the true age-specific distribution in long-lived species with few juveniles (extreme K-strategy; Richmond 1965).

2. Dinosaur rookeries and nurseries were remote from adult feeding grounds or were in "uplands" where fossilization is unlikely (e.g., Sternberg 1955; see below).

3. Juvenile bones are subject to scavenger destruction, rapid weathering, and disintegration.

4. Juveniles were more commonly swallowed whole by predators, hence their skeletons were less commonly available for fossilization.

5. Acid conditions in soil destroyed the eggs and many juvenile bones (Carpenter 1982*b*).

6. Tiny juvenile bones and eggs have been overlooked by collectors (collecting bias).

7. Small bones have not been recognized as juveniles of larger species (taxonomic bias: Dodson 1975).

8. Truly cleidoic eggs (= heavily calcified) are a late development in dinosaur evolution. Earlier, noncleidoic eggs (= "leathery") are unlikely to fossilize (Erben et al. 1979). This explanation seems increasingly implausible as more eggs are found.

9. Many female parents ate eggshells after the young hatched as in some modern crocodilians (Deitz and Hines 1980).

Modern Analogs for Nesting Behavior

The best modern analogs for comparison with dinosaur nesting/parenting behavior include certain birds, especially ratites and members of the family Megapodiidae, and crocodilians (Coombs in prep.). Megapode and crocodilian species share a variety of nesting/parenting behaviors. In both groups, some species excavate holes in sandy areas and bury their eggs while other species construct elevated nests of sand and vegetation or of vegetation only. Solar or fermentation heat is commonly used for incubation, and the parent attends or guards the nest. Incubation temperatures for crocodilians and megapodes differ by only a few centigrade degrees (ibid.). The parent may alter a nest to enhance development, especially in the Megapodiidae. In crocodilians but not megapodes, the young vocalize just prior to hatching and the parent commonly assists the young, especially by digging them out of the nest and transporting them to a nearby body of water. Nest failure rates of crocodilians are high if the parent is not in attendance. Among birds and crocodilians, the female, male, or both may care for young before and/or after hatching. Behavior of young after hatching is highly variable. Hatching ratites, megapodes, and crocodilians are precocial and capable of self-feeding. Nudifuguous hatchlings of megapode birds emerge at irregular intervals and are immediately independent of their parents, being fully feathered and capable of flight (Frith 1956, 1957, 1962; Welty 1982). Crocodilians typically form creches or pods under guidance of the female parent but will also form such creches spontaneously or, depending on the species, may disperse if the female is absent (Cott 1971; Garrick and Lang 1977; Garrick and Garrick 1978; Garrick et al. 1978; Deitz 1979, Deitz and Heinz 1980; Magnusen 1980; Webb et al. 1983; Whitaker and Whitaker 1984; Ferguson 1985; Lang et al. 1986; Vliet pers. comm.; Coombs in prep.).

Dinosaur Nests

The most important dinosaur nests known at present are those of *Protoceratops, Maiasaura, Orodromeus,* and nests from the Maastrichtian of France. Setting aside for the moment the nests from the French Maastrichtian, these nests share several features: (1) multiple or more or less closely spaced nests, that is, an apparent rookery; (2) a series of bedding planes that contain nests, that is, prolonged utilization of the rookery; (3) skeletal remains of a wide range of age categories—embryo or hatchling to large adults—within or near the rookery; (4) nests roughly circular and excavated into sandy material; (5) eggs positioned with some care, commonly in circles, uniformly spaced, and in some cases apparently "planted," especially *Orodromeus* eggs that have the more sharply pointed end buried downward (Horner and Makela 1979; Horner 1982, 1984*a*, 1987).

Some eggs from the Maastrichtian of France have been described as being laid singly or in paired rows on the surface of the ground (Dughi and Sirugue 1966), and other dinosaur eggs have been described as arranged in rows rather than in circular nests (Horner 1986). Nests containing two to eight eggs have also been described from this area, and one nesting ground has several different kinds of nests (Lapparent 1959; Plaziat 1961; Freytet 1965; Erben 1972; Erben et al. 1979; Kerourio 1981*a;* Breton et al. 1986). Unstratified filling is common in excavated nests, suggesting the eggs were partially or completely buried by the female. To provide sufficient gas exchange properties, the enclosing sediment and/or vegetation must have been well aerated. Shell thickness and porosity indicates that all dinosaur eggs were at least partially buried (Seymour 1979; Williams et al. 1984).

Well-sorted, well-drained fluvial sands would be especially suited for buried dinosaur nests, but such sediments have limited distribution. Consequently, female dinosaurs probably sought out traditional nesting grounds isolated within or remote from the adult home range (remote nesting grounds originally hypothesized by Lull 1904, 476; see also, Matthew 1915a, and Sternberg 1949, 1955; see Seymour 1979 and Carpenter 1982b, on selection of substrate properties for nesting). *Protoceratops, Maiasaura,* and *Orodromeus* have multiple nests on sequentially stacked layers, as would be predicted from the traditional nesting ground hypothesis. *Protoceratops* nests are associated with large numbers of skeletons, tiny juveniles to large adults, which might indicate that the adult home range was relatively small and encompassed the nesting ground (Brown and Schlaikjer 1940a; Dodson 1976). *Maiasaura* rookeries may have been somewhat remote from the home range occupied by adults during most of the year, based on the observation that eggshells and juveniles are the most common fossils in the Two Medicine Formation, but adults are more common in the Judith River Formation (Horner and Makela 1979). However, adult *Maiasaura* have not been found in more distant localities that have yielded numerous hadrosaurs (e.g., Dinosaur Provincial Park, Alberta, Canada).

Orientation of eggs with the long axis vertical, found both among dinosaurs and modern megapode birds (Horner 1987; Frith 1956, 1957, 1962), may reduce breakage when the eggs are stepped on insofar as cleidoic eggs are generally stronger down the long axis than across the short, near circular axis (Romanoff and Romanoff 1949). In megapodes, however, the vertical orientation is a coincidental consequence of egg laying (Fleay 1937; Frith 1962; Coombs in prep.). Nest construction in *Maiasaura* and *Orodromeus* is similar to that of some Megapodiidae (Horner 1984a, 1987; Coombs in prep.). *Protoceratops, Maiasaura, Orodromeus,* and some French Maastrichtian eggs are arranged in a spiral or in concentric circles, a pattern that has been erroneously compared to some megapodes (e.g., Horner 1987). Megapodes do not arrange their eggs in any regular pattern (Frith 1962; Coombs in prep.).

Posthatching Parental Care

Classic parenting is very difficult to demonstrate conclusively for dinosaurs. The occasional association of adult and juvenile skeletons does not necessarily in-

dicate parental care, insofar as such an association would be expected even if the eggs were abandoned after laying in the manner of most chelonians. However, the large number of juvenile and adult *Protoceratops* associated with nests and eggs makes interaction between hatchlings and adults inevitable and some form of parental attention to young probable. The complete growth series of *Protoceratops* skeletons found near nests indicates that much of the life cycle, including maturation from hatchling to adult, took place in the near vicinity of the rookery. Juvenile *Maiasaura* far larger than hatchlings are found in or near nests that contain broken eggs known to pertain to the same species (Horner 1982, 1984a). It has been suggested that the female parent brought food to the nest or led young on foraging expeditions out from and back to the nest (Horner 1982, 1984a; interpretation revised by Horner 1987).

Nest site fidelity by juveniles is strictly associated with parental guidance among modern amniotes. The thanatocoenosis itself, complete clutches of eggs or large numbers of same-aged juveniles dead in a single nest, suggests death following loss of an attendant parent. Unattended crocodilian nests have a higher failure rate than those guarded by the female parent, and parental assistance at hatching may be absolutely necessary for some crocodilians (Deitz and Hines 1980; Coombs in prep.). Another suggestive thanatocoenosis is "Egg Island," a nest locality with multiple nests stacked stratigraphically as well as on single bedding planes, in this case belonging to the hypsilophodontid *Orodromeus,* with the nest sites in terrestrial sediments surrounded by lake sediments (Horner 1984a, 1987). The physical dimensions of the island nesting ground make near-nest self-foraging by the young unlikely and import of food by a parent a virtual necessity.

The maximum hatchling size in oviparous species is limited by the rate of gas exchange in ever-larger eggs (Seymour 1979). An absolute upper size limit theoretically exists, and in dinosaurs with very large adults, the size differential between hatchlings and adults (= RNS; Coombs 1982) creates logistic problems for parental care. If sauropods were oviparous, hatchlings of larger species would have total body lengths, nose to tail tip, scarcely greater than the diameter of an adult's hind footprint. There are three alternative solutions: (1) young were independent from the moment of hatching as in megapode birds (Frith 1956, 1962; Coombs in prep.); (2) at least initially, hatchlings were relatively immobile, that is, they stayed in one place (the nest?) where an attendant parent was careful not to step; or (3) sauropods and perhaps other large dinosaurs were viviparous and the newborn were relatively large (see viviparity, above).

Ichnocoenoses and Parenting

As with bones, footprints of juvenile dinosaurs are quite rare (Leonardi 1981a), but this apparent scarcity may be in part an artifact of taxonomic bias. Despite lingering uncertainties about interpretation of ichnites, some useful or at least suggestive data has been gathered. The sauropod "herd" at Davenport Ranch has individuals of several size categories traveling in the same direction at approximately the same speed (Farlow 1987a; Lockley 1986a, 1986b, 1987). Within one set of trackways, the smallest prints are about half the size of the largest and have been interpreted as pertaining to subadults, although these are far larger than the expected size of hatchlings on the presumption that sauropods were oviparous.

The largest sauropod footprints in the general area come from a different locality along the Paluxy River (Farlow 1987a, pers. comm.). Interpretation of this trackway site is at best ambiguous. The modest range in footprint sizes is not convincing evidence of parental care, but assuming the footprints were made by a group rather than a series of individuals, the ichnocoenose does suggest a mixed-age "herd" (see also, Lockley et al. 1983). Tiny early Jurassic trackways called *Selenichnus breviusculus* (stride length = 100– 120 mm) have been interpreted as pertaining to juvenile dinosaurs, possibly the same genus that made the much larger *Grallator* footprints (Coombs 1982). All known trackways of *Selenichnus* are isolated individuals; neither multiple tiny individuals nor single slabs containing both *S. breviusculus* and larger prints of similar form ("parent") are known. Although inconclusive, this evidence suggests early independence for the *Selenichnus* track maker (i.e., it was highly precocial). If the *Selenichnus* track maker were a small theropod (Lull 1915a, 207), then this footprint case might support a theory that there was little or no posthatching association of juveniles and adults in some carnivorous dinosaurs (dissenting opinion by Paul 1987b). The apparent cannibalism of *Coelophysis* mentioned above would also support such an interpretation, although male lions practice infanticide on young that are not their own despite their general tolerance of young (i.e., lions are generally "good parents": Schaller 1972). Trackways of small, tridactyl, bipedal dinosaurs from the Cretaceous, thought to be prints of juveniles, do appear to be traveling in unison, which suggests formation of sibling groups (Currie and Sarjeant 1979), but footprints of adults were not found sufficiently close to support a theory of posthatching parental care.

Summary

Combining available data on dinosaur nests with data for megapodes and crocodilians results in the following generalities about dinosaur nesting/parenting behavior. Dinosaur nesting and parenting was diversified, both among different species and through the long evolutionary history of dinosaurs, and there could be intraspecific variation in some behavior patterns. Dinosaurs commonly excavated nests in soils with distinctive properties at traditional nesting grounds that might be used simultaneously by many individiuals. Alternatively, nests may have been constructed of plant litter. Clutches might be laid at one time or over an extended period by one female, or several females might share a communal nest. Nests primarily provided temperature stability (= little daily fluctuation) and elevated humidity. Solar or fermentation heat may have been important to incubation. Nests may have been guarded and the nest temperature may have been monitored by an attendant parent. Some dinosaurs may have used body heat for incubation or at least insulated eggs with their body at night. Attendant parents may have assisted young at hatching. Some but not all dinosaur hatchlings were precocial, capable of walking, running, and self-feeding, and some were probably independent as soon as they hatched, especially among the predatory species. In other species, juveniles formed creches and were at least attended or guarded by a parent after hatching. Parental transport of food to the nest during the immediate posthatching period is possible for some species. Formation of mixed-age aggregates (flocks or herds) composed of juveniles, subadults, and adults was probably common among herbivorous species. Either complete independence of precocial hatchlings or viviparity is more likely among the physically largest dinosaurs, especially the sauropods.

Part II   DINOSAUR ENERGETICS AND THERMAL BIOLOGY

JAMES O. FARLOW

Barring the invention of a time machine (fig. 2.1), no one will ever measure the body temperature or respiration rate of a living dinosaur. This has not kept paleontologists from speculating about the physiology of these spectacular animals, however. From the time that dinosaurs were first recognized as a discrete group of reptiles, interpretations of their life processes have been a prominent aspect of paleoecological and even systematic treatments of these creatures (Desmond 1979; Bakker and Galton 1974).

Until the 1960s, most workers reconstructed the physiology of dinosaurs as basically similar to that of typical reptiles (e.g., Colbert et al. 1946, 1947), although reservations were occasionally expressed: "They were scaly or armored, this we know; whether they were cold-blooded or not is a debated question. Such activity as they must have shown seems to hint at a possibility of warm blood, . . . but in the tropical climate which their known habitat implies there was little more need of a heat-maintaining mechanism than there is in modern cursorial lizards" (Lull 1925: 502).

In 1965, L. S. Russell cautiously argued for a degree of homeothermy, or even endothermy, in dinosaurs. A few years later, the case for dinosaurian endothermy was more forcefully stated by Ostrom (1970b) and Ricqlès (1974, 1980 [and earlier papers cited therein, but also see Ricqlès 1983]) and especially by Bakker (1980, 1986 [and earlier papers cited therein]) in a series of technical and popular works. Bakker's vigorous efforts on behalf of this theory prompted numerous responses, pro and con, culminating in a symposium that summarized the various arguments (Thomas and Olson 1980).

A high, constant body temperature and/or the rapid metabolic rates that are sometimes needed to make this warm-bodied homeothermy possible have a number of known or postulated benefits: increasing the rate of assimilation of nutrients; permitting intense and extended bouts of aerobic activity, with more rapid repayment of oxygen debts generated during intervals of anaerobic exertion; enhancing muscular activity and transmission and integration of nerve impulses; extending the range of thermal environments in which an organism can operate; allowing a faster growth rate and perhaps a higher reproductive potential (Bennett and Ruben 1979; Bennett and John-Alder 1984; Smith 1979; McNab 1980, 1987; Block and Carey 1985; McCosker 1987). These benefits are frequently outweighed by their metabolic or other costs, and so the majority of animal species, including most reptiles, are bradymetabolic ectotherms, particularly in situations where energy conservation is of paramount importance (Pough 1983).

Paleontologists interested in reconstructing the energetic and thermal strategies of dinosaurs have considered a number of skeletal features and aspects of the occurrence of dinosaur remains that have been thought to be correlated in some way with the body temperatures and/or metabolic rates of the great reptiles. These include such things as dinosaur limb proportions and postures, levels of activity inferred from trackways, feeding adaptations, bone histology, size and other features of the brain reconstructed from endocranial casts, putative specialized heat-exchange devices, inferences about the design of the vascular system, the relative abundance of carnivores and herbivores in dinosaur faunas, the latitudinal occurrence of dinosaur fossils, the rate of evolution of dinosaur taxa, and the presumed evolutionary relationship between theropod dinosaurs and birds (see the above-cited references).

Unfortunately, the strongest impression gained from reading the literature of the dinosaur physiology controversy is that some of the participants have behaved more like politicians or attorneys than scientists,

An instant later, both Professor Waxman and his time machine are obliterated, leaving the coldblooded/warmblooded dinosaur debate still unresolved.

Fig. 2.1.

passionately coming to dogmatic conclusions via arguments based on questionable assumptions and/or data subject to other interpretations. Many of the arguments have been published only in popular or at best semitechnical works, accompanied by rather disdainful comments about the stodgy "orthodoxy" of those holding contrary views; what began as a fresh way of considering paleontological problems has degenerated into an exercise in name-calling. All of this has made the whole field of dinosaur studies suspect in the minds of many scientists. I will nonetheless attempt an objective review of the topic and offer my own speculations about dinosaur energetics and thermal biology.

GROWTH, BODY SIZE, ACTIVITY LEVELS, AND HOMEOSTASIS

Although the energy required to synthesize a given amount of biomass is the same for ectothermic and endothermic animals, during the ontogenetic interval of most rapid growth, bradymetabolic ectotherms divert a greater proportion of their assimilated energy into growth than do tachymetabolic endotherms (Wieser 1985). Once adulthood is reached, the lower maintenance costs of ectotherms permit these animals to channel a larger fraction of their assimilated energy into reproduction than is possible for endotherms. As a result, the production efficiency (production/assimilation ratio) is usually much higher in populations of bradymetabolic animals than in tachymetabolic populations (Humphreys 1979; Pough 1980, 1983; Lawton 1981; Lavigne 1982). Wieser (1985), however, argued that ectotherms are *forced* by their low metabolic power inputs into having high production efficiencies to ensure rapid enough reproduction and offspring growth rates for animals to complete their life cycles in an ecologically feasible length of time. Tachymetabolic endotherms, with their higher food intake rates, can put a large amount of energy per unit time, but a low proportion of energy in terms of their entire energy budget, into supporting rapid reproduction and juvenile growth rates. If Wieser's interpretation of cause and effect is correct, the high metabolic rates of endotherms and the high production efficiencies of ectotherms can be seen as contrasting strategies for solving the same problem.

Reptilian growth rates are highly variable, allowing these animals to survive during periods of low resource availability but also to take advantage of increases in their food supply (Andrews 1982). Thus, the growth rates of well-fed captive reptiles can greatly exceed those of free-living conspecifics (Coulson et al. 1973; Blake and Loveridge 1975; Andrews 1982; Zug et al. 1986). Even though reptiles, following the pattern seen in other ectotherms, channel substantial proportions of their assimilated or metabolizable energy into production (Pough 1980, 1983) and as much as 10 percent of their metabolizable energy into growth alone during their periods of most rapid growth (Congdon et al. 1982; Nagy 1983; Anderson and Karasov 1988), the growth rates of reptiles in the wild are an order of magnitude less than those of mammals and birds (Case 1978a, 1978b). It seems, then, that given sufficient food supplies, the endothermic strategy for supporting rapid growth is more effective than the tactic employed by ectotherms; this contributes to one of the benefits of tachymetabolic endothermy, an increase in the intrinsic rate of natural increase—a boon especially exploited by the eutherians (McNab 1980, 1984, 1986, 1987; Henneman 1983, 1984—but see Hayssen 1984; Padley 1985). Even among reptiles, those individuals or species with higher metabolic rates than other reptiles grow more rapidly (Nagy 1983, fig. 2.3; Rhodin 1985).

Case (1978*b*) suggested that if dinosaurs grew at rates similar to those of living reptiles, large forms would have required several decades to reach sexual maturity, necessitating a very high juvenile survivorship. Dunham et al. (in press), however, argued that a juvenile survivorship high enough to make such low growth rates possible would be highly unlikely and that dinosaurian growth rates must have been faster than those extrapolated from living reptiles.

The widespread (but not universal) occurrence of fibrolamellar primary bone in dinosaurs suggests that the growth rates of these animals were indeed faster than those of most living reptiles (Ricqlès 1980; Reid 1987*a*); even the occasional occurrences of zonal bone in dinosaurs can be interpreted in terms of fairly rapid growth rates (Reid 1987*a*). This, in turn, suggests that the metabolic rates of dinosaurs may well have been greater than those of living reptiles. Although warm temperatures appear to enhance reptilian growth rates (Andrews 1982), inertial homeothermy alone may not be sufficient to cause a shift from the deposition of zonal to fibrolamellar bone, based on observations of alligators grown under uniformly warm conditions (M. W. J. Ferguson, in Reid 1987*a*); however, Ricqlès (1983) cited studies (e.g., Buffrénil 1980) in which rapidly growing, young, farmed crocodilians did show fibrolamellar bone.

If inertial homeothermy coupled with the higher production efficiencies characteristic of ectotherms were generally sufficient to permit fast enough growth rates for the formation of fibrolamellar bone in *noncaptive* reptiles, then this should be the major primary cortical bone tissue seen in most large reptiles known to live (or have lived) in warm climates, and such is not the case. This suggests some physiological difference between most dinosaurs and most living reptiles, but the occurrence of dinosaurian fibrolamellar bone does not necessarily mean that these reptiles had tachymetabolic rates comparable to those of modern birds and mammals; "dinosaurian compact bone was . . . not invariably mammal-like, and sometimes showed a mixture of reptilian and mammal-like patterns" (Reid 1987*a*: 141). Furthermore, fibrolamellar bone also occurs in primitive therapsids that give no other indications of elevated metabolic rates (Reid 1987*a*).

More controversial is the physiological interpretation of a secondary hard tissue, Haversian bone. The frequent occurrence of dense Haversian bone in dinosaurs is consistent with their having had rapid metabolic rates (Ricqlès 1980), but it is possible that Haversian bone could be developed to a significant extent in large, long-lived animals whose thermal strategy was based on inertial homeothermy in warm climates (Reid 1987*a*).

Mathematical models of the biophysical energy budgets of large reptiles (Spotila 1980; Haack 1986; Tracy et al. 1986; Turner and Tracy 1986; Dunham et al., 1989), together with field studies of the thermoregulation of such animals (McNab and Auffenberg 1976; Smith 1979; McNab 1983), indicate that large size alone is indeed a successful strategy for maintaining fairly constant, high body temperatures in mild environments. Spotila (1980) even suggested that large bradymetabolic dinosaurs with thick subcutaneous fat layers (as in leatherback turtles; Spotila and Standora 1985) could have survived in cold climates (cf. McNab 1983: 17). From the standpoint of thermoregulation alone, there would seem no need to invoke tachymetabolism to account for the success of dinosaurs during the Mesozoic, and rapid metabolic rates might even result in deleterious heat loads in very large animals living under the thermal regimes thought to have prevailed during the age of reptiles (Spotila 1980).

But large eutherian mammals—and not big ectotherms—have dominated most tropical environments since the extinction of the dinosaurs (Bakker 1980), and some extinct terrestrial eutherians were as large or larger than most dinosaurs. The huge rhinoceros *Indricotherium* (*Baluchitherium*; Granger and Gregory 1936) may have weighed about 20,000 kg (based on measurements of limb bone midshaft circumferences [data from M. McKenna pers. comm.] and the regression equation of Anderson et al. [1985]; see also Economos 1981). Rearranging Economos's (1981) regression of shoulder height against mass in living mammals, one can estimate that the extinct elephant *Mammuthus trogontherii*, with a shoulder height of about 4.5 m (Kurtén 1968), weighed some 11,000 kg. If large mammals can be endothermic without lethal heat stress, why not dinosaurs?

Spotila (1980) concluded that the heat production rates of large mammals and reptiles should differ by less than the thermal outputs of small mammals differ from those of small reptiles. Interestingly enough, the exponent of mass to which the metabolic rates of reptiles scale seems to be higher than those of mammals and birds (see equations 1 and 2 below, and Farlow 1980; Andrews and Pough 1985; Hayssen and Lacy 1985; Chappell and Ellis 1987; McNab 1987; see also some, but not all, of the equations relating field metabolic rate to mass in endotherms as opposed to iguanids in Nagy 1987), suggesting the possibility of just such a convergence at large sizes (McNab [1983], however, thought it unlikely that these differences in exponents have biological significance).

Spotila (1980: 249) reported that the metabolic rate of an elephant is only "somewhat" greater than that expected for an equally large reptile. While one might still consider this a difference between "tachymetabolism" and "bradymetabolism," Spotila's models suggest that the distinction between endothermy and ectothermy, so useful in characterizing the thermoregulatory strategies of little animals, is less meaningful when applied to dinosaur-sized creatures. If Spotila is correct, *Indricotherium* and other very large terrestrial mammals, even more than modern elephants, may not have been tachymetabolic endotherms in the usual sense. Unfortunately, *Indricotherium* is just as extinct as any dinosaur, making a conclusive test of this interpretation impossible.

Modern birds and mammals have significantly larger brains than modern fishes (except some elasmobranchs; Northcutt, 1977), amphibians, and reptiles (Jerison 1973; Martin 1981) of similar size. The reasons for this size difference are uncertain. Hopson (1977, 1980*a*) suggested that the higher food requirements of tachymetabolic endotherms force them to spend a great deal of their time in foraging and that this, in turn, requires large brains, to integrate the great input of sensory information and to control appropriate motor responses, associated with such high-level activity. Martin (1981) proposed that brain size in neonate vertebrates is constrained by maternal metabolic rate and the species' mode of reproduction (viviparity or oviparity); the rapid metabolic rates of tachymetabolic endotherms permit development of large brains in their offspring, whatever the ecoevolutionary circumstances may be that make such big brains advantageous. Armstrong (1983) concluded that brain size in mammals is controlled by the total metabolic input (approximated as body size × basal metabolic rate) supporting this energetically demanding organ; Hofman (1983) suggested that the oxygen consumption of the brain was a linear product of an animal's weight and the ratio of brain size to body weight raised to the 0.732 power.

Most dinosaurs (as well as pterosaurs and cynodonts) have reconstructed brain/body size ratios similar to those expected for comparably sized reptiles, but at least some small and large theropods have larger brains than expected, that of *Troodon* (*Stenonychosaurus*) falling within the range of brain/body size ratios seen in modern birds (Hopson 1977, 1980*a*; Martin 1981). Hopson (1980*a*) concluded that dinosaurs, apart from small theropods, were not as tachymetabolic as living birds and mammals but probably more tachymetabolic

than living reptiles, and Martin (1981) endorsed this conclusion for theropods. This argument was challenged by Bakker (1980), who found no relationship between the encephalization quotient and the ratio of observed metabolic rate to that predicted from the Kleiber equation in birds and mammals. Indeed, Martin (1981: 59) noted that "for placental mammals . . . individual values for brain size may vary by a factor of 5 on either side of the value for any given body weight . . . and this variation is not matched by equivalent variation in basal metabolic rate, so different grades of relative brain size among mammals must be influenced by other factors" (see also Hofman 1983: 497). Even so, the large overall difference in the brain and body size relationships between modern tachymetabolic and most bradymetabolic vertebrates, "noise" in these relationships notwithstanding, suggests that brain size is in some way related to an animal's metabolic rate. But without definite knowledge of the causal relationship, we cannot make conclusive interpretations of the physiological meaning of brain size in extinct vertebrates (the situation is analogous to earlier controversies over the relationship between sprawling vs. upright posture and metabolic rate). Hopson's and Martin's interpretations of brain size seem reasonable, but the small brain sizes of most dinosaurs may not necessarily mean lower metabolic rates than in theropods. The large brain sizes of theropods are consistent with a degree of tachymetabolism; because the brain is an energetically demanding organ, a large brain may require a rapid metabolic rate. However, the large brains seen in many elasmobranchs (not all of which are endothermic) indicate the need for caution in coming to the conclusion that even small theropods had rapid metabolic rates.

Bakker (1986) proposed in a popular account that the elevated metabolic rates of tachymetabolic endotherms require them to forage at higher average speeds than bradymetabolic animals. A compilation of estimated speeds based on trackways of living and fossil vertebrates (Bakker 1987) suggested that Paleozoic amphibians and reptiles moved at slower speeds than Tertiary and modern mammals, but that therapsids, thecodonts, and dinosaurs had speeds comparable to those of mammals. Bakker interpreted this to mean that therapsids and archosaurs were driven by the need to satisfy food requirements similar to those of mammals.

Bakker's argument is interesting, but not yet compelling. His basic premise—that endotherms walk at faster speeds than ectotherms—is plausible, but competing hypotheses come to mind. First of all, Bakker implicitly assumes that all, or at least most, bouts of walking reflect foraging behavior, which may or may not be true. Second, ectotherms may forage at the same speeds as endotherms of the same size but less often. If ectotherms indeed generally walk more slowly than endotherms, this might be as much a function of their limb mechanics as their metabolic rates (although these variables may themselves be correlated). Even if Bakker's assumptions about the relationship between walking speed and metabolic rates are valid, I am not convinced that the trackway data are adequate to warrant his conclusion that trackmaker speeds indicate dinosaurian tachymetabolism. Most dinosaurs and many thecodonts and therapsids were larger and/or longer-legged animals than their undoubtedly bradymetabolic predecessors, and the same is probably true of most of the mammals vis-à-vis the Paleozoic reptiles and amphibians in Bakker's trackway sample. All else being equal, a big animal will walk more rapidly than a smaller creature. Bakker's case would be more compelling if he had demonstrated that mammals and dinosaurs and other putatively tachymetabolic extinct reptiles had faster speeds than their larger size and more sophisticated locomotor adaptations alone would imply.

Activity levels and metabolic rate may be correlated in other ways, however. In modern reptiles, the capacity for sustained power input at high activity levels is rather limited (Bennett 1982, 1983; Coulson 1984), and several authors have championed the theory that a primary factor in the evolution of tachymetabolic endothermy in birds (and perhaps dinosaurs) and mammals was selection for greater aerobic endurance (Bennett and Ruben 1979, 1986; Taigen 1983; Carrier 1987—but see Pough and Andrews 1984).

Most modern reptiles are fairly small vertebrates that are well adapted to energy conservation (Pough 1980, 1983), and it is possible that interpretations of vagility based on these small animals underestimate the potential for activity of very large ectotherms with high body temperatures (cf. Spotila 1980; Bennett 1982; Wright 1986). Auffenberg (1981) reported that Komodo dragons (oras) generally walk and forage at speeds of about 4.8 km/h. Estimates of the walking speeds of medium-sized to large bipedal dinosaurs based on fossil trackways generally fall in the range of 5 to 10 km/h (Bakker 1987; Farlow 1987a); given the larger size and longer legs of most bipedal dinosaurs as compared with oras, these estimated walking speeds would not seem beyond the capacity of hypothetical ectothermic dinosaurs.

The maximal aerobic speeds of modern tachymetabolic endotherms are considerably greater than those of living reptiles (see above references), and during sprints at top speed or other intense activities, modern reptiles are forced to rely on anaerobic power

sources. However, this does not necessarily preclude impressive bursts—or sometimes even rather extended bouts—of activity. Garland (1984) reported that a 230 g *Ctenosaura similis* sprinted at 34.6 km/h. Large saltwater crocodiles vigorously resisting capture can struggle violently for a half-hour or more, developing high levels of blood lactate; most crocodiles recover at least partially from acid-base disturbance after two hours of rest (Bennett et al. 1985). Webb and Gans (1982) reported that *Crocodylus johnstoni* gallop at speeds of as much as 17 km/h. Komodo dragons will trot at speeds of 8 to 10 km/h and when frightened run at 14 to 18.5 km/h; one ora maintained a speed of about 14 km/h over a distance of somewhat over half a kilometer (Auffenberg 1981).

Although the advantages of tachymetabolism for vigorous, sustained activity are clear, there seem to be other ways (even if less effective) of accommodating a respectable activity level. Some varanids and other lizards have a relatively high factorial aerobic scope, even though their standard metabolic rates are no higher than those of other lizards (Taigen 1983; Bickler and Anderson 1986). The activity levels of varanids are made possible by a sophisticated heart structure (Burggren 1987), a high hematocrit, and a capacity for efficient oxygen transfer from lungs to blood to body tissues (Bennett 1973; cf. blue marlin [Dobson et al. 1986]).

Medium-sized theropods at least seem to have been able to run at speeds of 35 to 40 km/h (Farlow 1981), and there is no way of knowing whether these were their top speeds. Again, given the trackmakers' size and long hindlimbs, these speed estimates derived from trackways do not seem out of line with expectations based on living reptiles. There is no way of knowing how long or far hypothetical bradymetabolic theropods could have run at these estimated speeds, but the observations on large living reptiles suggest the possibility that ectothermic theropods could have moved at faster than maximal aerobic speeds for some time before having to stop and rest. Although the adaptations for vagility seen in theropods and some other dinosaurs are consistent with the hypothesis that these reptiles had high capacities made possible by tachymetabolism, the known trackway evidence does not demand such an explanation.

DINNERS AND DINERS

Bakker (1980, and earlier papers cited therein) has forcefully championed the use of reconstructed predator/prey biomass ratios in determining whether theropods and other kinds of extinct meat-eaters were ectotherms or endotherms. The details of his argument have varied, with predator/prey ratios being interpreted in terms of the energy flux to predator populations (Bakker 1972) or the efficiency with which those populations channeled ingested or assimilated energy into production (Bakker 1980). Bakker compiled a large set of reconstructed predator/prey ratios for fossil vertebrate faunas as well as comparative data for the predator/prey ratio in modern animal communities; he concluded that predator/prey biomass ratios declined from 20 to 60 percent in Early Permian tetrapod faunas to about 10 to 20 percent with the advent of advanced mammallike reptiles in the Late Permian and to even lower values with the evolution of dinosaurs and their eventual replacement by mammals. Bakker interpreted the decline of predator/prey ratios in terms of increasing levels of tachymetabolism of vertebrate carnivores, with flesh-eating dinosaurs, like mammals, having had rapid metabolic rates.

Bakker's conclusions have been challenged by other workers (Charig 1976; Farlow 1976b, 1980, 1983; Tracy 1976; Hotton 1980; Béland and Russell 1979, 1980; Reid 1987b), and his arguments require a number of assumptions that are seldom made explicit. He has never published the data base of specimen counts and animal mass estimates on which his predator/prey ratios for fossil faunas are based, but he has put some of those data in an informally distributed, unpublished manuscript. I will assume that his counts are accurate and his mass estimates reasonable.

Farlow (1976b) published an equation relating food consumption in reptiles and amphibians to body mass:

$$\text{food consumption} = 0.84 * \text{mass}^{**}0.84$$

$$\text{(equation 1)}$$

(Food consumption in Watts [after Peters 1983 with the addition of a few new data points], mass in kg; $r^{**}2$ [after logarithmic transformation of variables] = 0.945, p < 0.0001; s.e. of exponent = 0.03; 95% c.i. of exponent = 0.79−0.90; 95% c.i. of constant = 0.69−1.04; n = 56; the data on which this and the following equations are based are available on request.)

Similarly, for mammals and birds:

$$\text{food consumption} = 10.96 * \text{mass}^{**}0.70$$

$$\text{(equation 2)}$$

(Food consumption and mass in the same units as above, from Farlow 1976b after Peters 1983 with the addition of new data; $r^{**}2$ [after logarithmic transformation of variables] = 0.951, p < 0.0001; s.e. of exponent = 0.01; 95% c.i. of exponent = 0.68−0.72; 95% c.i. of constant = 10.2−11.8; n = 271.)

For mammals, the relationship between annual population biomass turnover (the production/biomass ratio) and individual adult mass is:

$$P/B = 0.94 * \text{mass}^{**}(-0.28) \qquad \text{(equation 3)}$$

(P/B in $\text{year}^{**}(-1)$, mass in kg; $r^{**}2$ [after logarithmic transformation of variables] = 0.793, p < 0.0001; s.e. of exponent = 0.02; 95% c.i. of exponent = −0.31 to −0.24; 95% c.i. of constant = 0.80−1.10; n = 55; data modified from Farlow 1976b; see also Banse and Mosher 1980; Peters 1983.)

Bakker (1972) argued that a similar relationship should apply to terrestrial ectotherms (cf. Peters 1983), but for very slowly growing forms this may not be true (Coe et al. 1979).

From the standpoint of a predator population, the productivity of prey species has two components: the biomass that is actually eaten by the carnivores and "waste." Waste includes the prey biomass lost to the predators for one reason or another (animals killed by other kinds of predators; creatures killed by disease or environmental catastrophes and buried or consumed by scavengers or decomposers before the predators of interest could locate their carcasses) as well as uneatable portions of prey carcasses. The higher the proportion of prey productivity that is lost to the predators in question as waste, the larger the prey biomass required to support the predator population. Thus, the smallest possible prey biomass needed to maintain a predator population, given a particular predator food requirement and prey biomass turnover rate, would be present in a situation where the predator population consumed all of the production of its prey populations with no carcass waste. To calculate the maximum possible predator/prey biomass ratio, then, we make the assumption (undoubtedly unrealistic in all or nearly all cases) that the

predator consumption rate = prey productivity rate.

Because the

prey turnover rate = prey productivity/prey biomass,

then the

prey turnover rate = predator consumption rate/ prey biomass

if the predator population consumes 100 percent of the prey population's production.
Consequently,

$$\text{prey biomass} = \frac{\text{predator consumption rate}}{\text{prey turnover rate}}$$

Because we are interested in the predator biomass relative to prey biomass, not the absolute value of either of these variables, we can calculate the prey biomass that would support a predator biomass equal to the average individual mass of predators in that population. We must assume that the per unit mass food consumption rate of the predator population is the same as that of a single predator of average mass. The maximum possible predator/prey biomass ratio for a predator population with a given average animal mass can then be calculated as:

$$\text{maximum predator/prey ratio} = \frac{\text{predator mass}}{(\text{predator consumption rate/prey turnover rate})}$$

Using equations 1–3, we can calculate the maximum possible predator/prey biomass ratios for modern and fossil faunas and compare them with the observed values (fig. 2.2); Vézina (1985) independently formulated a similar algorithm for estimating maximum predator/prey biomass ratios for various kinds of predator/prey complexes. For an ectothermic predator consuming prey of about the same size, the maximum possible predator/prey ratio seems to be about 20 percent. If the predator is very much larger (as in Bakker's interpretation of the Geluk fauna) and/or has a lower population turnover rate than its putative prey, the ratio can be substantially higher. Rattlesnakes and gopher snakes feeding on juvenile ground squirrels and cottontail rabbits in a southwestern Idaho community had a predator/prey biomass ratio of 8 to 11 percent; however, if prey biomass is prorated in terms of the amount actually consumed by the snakes, the ratio may be as high as 20 to 50 percent (the actual ratios would be somewhat lower, as not all prey species were censused); the high ratio presumably reflects rapid population turnover of the small mammals (ratios calculated from data in Diller and Johnson 1988). In the extreme case of poikilothermic predators (arachnids, fishes, amphibians, reptiles) feeding on small-bodied prey (insects, macrobenthic invertebrates, zooplankton) with very high population biomass turnover rates relative to those of their predators, the predator/prey biomass ratio can be very high, even approaching 100 percent (Farlow 1980; Vézina 1985); for frogs and lizards eating insects, the ratio ranges from 7 to 60 percent (Vézina 1985).

For a large endothermic predator preying on comparably sized animals, the maximum ratio seems to be about 1 to 2 percent, but again, the ratio can be higher if the prey species are considerably smaller animals than their predators; a population of ocelots in a Peruvian rain forest had a biomass of about 8 percent that of their small-mammal prey (Emmons 1987). Similarly, in a llanos community in Venezuela (Masaguaral), the ratio of crude biomass of small carnivores (felids, grisons, crab-eating foxes, tayras) to that of potential small-mammal prey (mass ≤ that of the predators) was about 4 to 6 percent (Eisenberg et al. 1979).

For modern large-vertebrate communities, the observed predator/prey ratios are in most cases substantially less than the maximum possible values, particularly in the case of the Komodo dragon and its prey (fig. 2.2). This result is expected, given the unrealistic assumption, made in calculating maximum ratios, that predators consume all of their prey species' productivity. Vézina (1985) similarly noted that observed biomasses of ectothermic and endothermic insectivores, and endothermic vertebrate-eaters, are usually less, relative to prey biomasses, than theoretical maximum values calculated in his model. The situation is quite different for fossil faunas that include the two groups of predators about whose bradymetabolism or tachymetabolism we can be reasonably sure, the pelycosaurs and mammals. In both groups, the observed predator/prey ratios are generally well above their theoretical maxima (fig. 2.2).

Bakker (1980) claimed that the predator/prey ratios of modern large-mammal communities are substantially lower than those of Tertiary mammalian faunas due to the geologically recent extinction of large-bodied carnivores thought to have been capable of killing elephant-sized prey and to human disturbance of "natural" communities. In addition, Bakker (1986) asserted that large-mammal predator/prey ratios for open habitats (e.g., the savannas of East Africa) should be less than those seen in forested areas with more cover (like the environments reconstructed for many Tertiary ecosystems); in open situations, predators should find it more difficult to stalk and ambush prey. Furthermore, Bakker proposed that in savanna systems, periodic droughts kill so many prey animals that a substantial proportion of the average herbivore biomass is unavailable to carnivores, whose population sizes have to be adjusted to the lowest seasonal levels of prey densities.

Even though Bakker (1980, 1986) presented no data in support of his assertions, they probably have some merit. African lions do have difficulties capturing gazelles in plains environments (Borner et al. 1987). Important prey species of African savanna carnivores are migratory, while the predators are not, and presumably must adjust their populations to the availability of nonnomadic prey (McNaughton and Georgiadis 1986). Many African game parks are intensively managed or otherwise disturbed, and large carnivores in these and other ecosystems have suffered at human hands throughout history (Sunquist 1981).

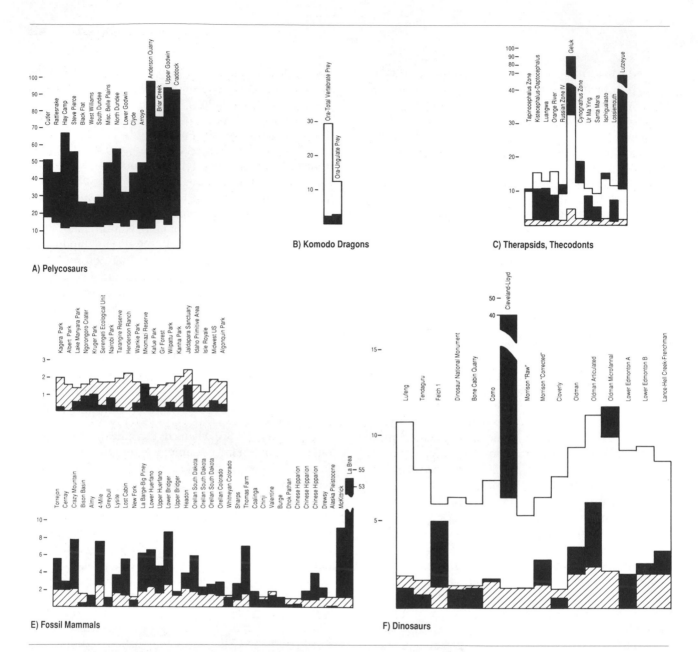

Fig. 2.2. "Observed" and theoretical maximum predator/prey biomass ratios for modern and fossil vertebrate faunas. Calculation of maximum ratios is based on the assumptions and equations given in the text, with the additional assumption that the energy content of animal tissue approximates $7 \times 10^{**}6$ joules/kg (Peters 1983, 33). Black = observed ratio, white = maximum ratio assuming bradymetabolic ectothermy on the part of the predators, cross-hatched = maximum ratio assuming mammalian and avian tachymetabolic endothermy on the part of the predators. For each fauna, the lower (lowest) of the two (or three) predator/prey ratios appears in front of the higher ratio(s); if the observed ratio is the same as a theoretical maximum ratio, only the observed ratio is shown. Modern mammal fauna predator and prey average weights and observed predator/prey ratios from Farlow (1976b). Komodo dragon data from Auffenberg (1981), as modified by Farlow (1983). Data for fossil faunas from Bakker (ms.), with the following additions: Alaska Pleistocene—data from Guthrie (1968); Morrison "Raw" data from Coe et al. (1987); Morrison "Corrected" data set

attempts to account for the taphonomic bias against smaller animals, using data from Behrensmeyer et al. (1979) to determine the relationship between the ratio of the number of living animals to the number of skeletons (corrected for observer bias in locating small carcasses in the field) and body size, a calculation slightly different from a similar approach in Coe et al. (1987); Oldman Articulated data from Béland and Russell (1978), with additional data from Dodson (1983); Oldman Microfaunal data from Dodson (1987). In the data from Bakker (ms.), an attempt to account for cannibalism and predation by one carnivore species on another was made by including the biomass of predator species, along with that of prey species, as prey biomass (i.e., the predator/prey ratio was calculated as predator biomass/[prey biomass + predator biomass]). This results in a lower ratio than the simple ratio of carnivores to their prey, particularly in assemblages with a relatively large number of predators. This modification was not made in calculating predator/prey ratios with data from other sources.

Even so, Bakker's hypotheses are unlikely to account entirely for the discrepancy in predator/prey biomass ratios between modern and ancient mammalian faunas. Some of the modern mammalian faunas for which predator/prey ratios are presented in figure 2.2 represent forest—not savanna—situations. For a Peruvian tropical rain forest community uninhabited by humans, Emmons (1987) reported a big cat (jaguar and puma)/large mammal prey biomass ratio of 0.7 percent, slightly greater than the actual predator/prey ratio. Eisenberg et al. (1979) tabulated the biomass of nonvolant mammals of a Venezuelan semideciduous to deciduous forest (Guatopo National Park); for jaguars and their potential large-mammal prey (mass > 1 kg, excluding tapirs; cf. Emmons 1987), the predator/prey crude biomass ratio was about 0.2 percent, and the ratio of small carnivore (felids, tayra) crude biomass to that of potential small-mammal prey was about 1.6 percent. At Masaguaral (llanos), the ratio of puma crude biomass to that of potential large-mammal prey was about 0.1 percent (Eisenberg et al. 1979). In Royal Chitawan National Park in Nepal (moist deciduous and riverine forest and grassland), the biomass ratio of tigers to ungulates (excluding rhinoceros, gaur, and domestic livestock) was 0.2 to 0.3 percent (Sunquist 1981). Some of the "homeothermic carnivores" for which Vézina (1985) reported biomasses relative to those of their prey were Temperate Zone raptors and mustelids. The consistently low (≤ 1–2%) predator/prey biomass ratio reported for tachymetabolic predators and their prey (when carnivores and their victims are of nearly comparable size), regardless of habitat (East 1984), degree of human disturbance, or the absolute size of the animals involved, suggests that the predator/prey ratios seen in modern large-mammal faunas represent trophodynamic equilibria rather than environmental or historical artifacts (cf. Emmons 1987: 279).

To generate higher predator/prey ratios for ancient in comparison to modern mammalian communities, the feeding rates of the carnivores in the former would need to have been well below those of modern endotherms (which seems unlikely) or the turnover rates of the prey species of ancient faunas higher than those in modern communities, or both. Bakker (1980) proposed that a greater predator pressure in the Tertiary than at present resulted in higher turnover rates of prey populations. If a greater diversity of mammalian carnivores did in fact result in a relatively heavier predation pressure on Tertiary mammals than experienced by present-day large herbivore populations (as opposed to an aggregate predation pressure of similar magnitude relative to prey population sizes being exerted by a greater number of carnivore species, with each preda-

tor species having relatively smaller populations than seen in modern large-animal communities), this could be true, to a point. However, the biomass turnover rate of mammals is affected more by body size and its influence on life history features than by predation or other agents of mortality, although the latter do have some influence (Blueweiss et al. 1978; Western 1979, 1983; Laws 1981; Lavigne 1982; Peters 1983; Stearns 1983; Calder 1984; Schmitz and Lavigne 1984; Harvey and Zammuto 1985; Allainé et al. 1987; Dickie et al. 1987; Thompson 1987). It is doubtful that biomass turnover rates of Tertiary mammals were enough higher than those of modern mammals to account for the higher predator/prey ratios reported for premodern than contemporary mammalian faunas.

The very high predator/prey ratios reported for sphenacodonts and their large prey (e.g., *Edaphosaurus, Eryops,* and *Diadectes*) are even more out of line with theoretical expectations of maximum possible ratios, and the discrepancy would be even greater if the biomass turnover rates of the ectothermic prey of pelycosaurs were lower than those of comparably sized mammals (contrary to the assumption made in my calculations), as seen in modern tortoises (Coe et al. 1979). This suggests that a significant component of pelycosaur diets is not represented in Bakker's tabulations or that carnivorous pelycosaurs are overrepresented in Permian faunas, or both. Although Bakker (1980, 1986) discounted fishes and small amphibians as *Dimetrodon* prey, it is likely that some of these animals were important constituents of the pelycosaur's diet, particularly in the later part of the Early Permian (Olson 1966, 1983). A single 5-kg *Diplocaulus* would have been large enough to satisfy the daily food requirements of a 100-kg *Dimetrodon,* if equation 1 satisfactorily predicts the predator's consumption rate, and bones of the small amphibian occur in coprolites reasonably attributed to *Dimetrodon* (Olson 1954, 1983). Even if the standing biomass of fishes and small amphibians and reptiles of Early Permian communities was less than that of large tetrapods, the presumed higher biomass turnover rates of the smaller vertebrates may have made them equally or even more attractive to *Dimetrodon* as a food source. Because *Dimetrodon* was a large animal whose remains are readily recognized and well preserved in a variety of sedimentary situations (Olson 1954), it is probably overrepresented relative to at least its smaller prey species in fossil collections (Olson 1983, pers. comm.; N. Hotton pers. comm.). Whether *Dimetrodon* is also overrepresented relative to potential large prey is arguable; if such overrepresentation does exist, it is probably due more to differences in behavior of the sphenacodont and its prey than to size-related taphonomic biases. Curiously, in one assemblage, the abundance of *Dimetro-*

don relative to other large tetrapods is closer to expectations for a bioenergetically "workable" system (Sander 1987), but whether this death assemblage more accurately reflects the living community or is itself indicative of some taphonomic artifact is hard to say.

The interpretation of Early Permian tetrapod communities presented here is consistent with my earlier suggestion (Farlow 1976*b*) that carnivores are frequently overrepresented in fossil faunas, an idea supported by Béland and Russell (1979, 1980) but not Bakker (1980). Overrepresentation of carnivores would also seem to be a plausible explanation for the excess of Tertiary carnivores as compared with their theoretical maximum abundance relative to prey, in striking contrast to the situation seen in modern large-mammal communities, but this is only a hypothesis. In a taphonomic study of the accumulation of large vertebrate carcasses in Amboseli National Park, Kenya (Behrensmeyer et al. 1979), the predator/prey ratio estimated from the bone assemblage seemed to be roughly comparable to that of the living fauna (A. K. Behrensmeyer, pers. comm.), contrary to my hypothesis. The apparent invariance of fossil mammal predator/prey ratios in different modes of faunal preservation (Bakker 1980) suggests that any oversampling of carnivores is not due to sedimentological processes. Conceivably, differences in behavior between carnivores and their prey could lead to an excess of predators in fossil faunas (Béland and Russell 1980: 99), as suggested above for sphenacodonts, but there are few data on this subject.

All of these considerations complicate interpretation of the predator/prey ratios of reptiles whose degree of tachymetabolism is in question: therapsids, thecodonts, and dinosaurs. In most fossil faunas involving these animals, the observed predator/prey ratio is well above the theoretical maximum for tachymetabolic animals, and so one is tempted to conclude, based on the relative magnitudes of observed and maximum predator/prey ratios seen in modern animals, that dinosaurs, thecodonts, and therapsids are unlikely to have been endotherms. However, given the greater values of observed than theoretical maximum ratios in faunas where the main predators were mammals or pelycosaurs, and the possibility that this is evidence for overrepresentation of carnivores in fossil faunas, one could argue that carnivorous archosaurs and therapsids are similarly overrepresented; particularly in the dinosaur faunas, the difference between the observed and maximum values of the predator/prey ratio is similar to that seen in the fossil mammalian faunas, and in some instances, the observed dinosaurian ratio is lower than the maximum value for an endothermic predator/prey complex. Furthermore, there are reasonable grounds for thinking that the advanced carnivorous therapsids, at least, were truly tachymetabolic endotherms, their seemingly high predator/prey ratios notwithstanding (Bennett and Ruben 1986).

Another complication is that there can be degrees of tachymetabolism, with different categories of endotherms having different metabolic rates and food requirements at a given body size; such differences are masked by the summary equation (2) given above (in particular, a separate allometric equation for the feeding rate of mammals has a higher exponent of body mass and predicts greater food consumption rates for large animals, than the equation for birds; see Nagy 1987 for a similar result). Carnosaurs may well have been tachymetabolic but at some intermediate level between the metabolic rates of living reptiles and the most tachymetabolic modern endotherms. Reconstructed predator/prey ratios may lack the resolution to distinguish among these possibilities.

It must be emphasized again that the calculation of the theoretical maximum predator/prey ratio is based on the assumptions (among others) that all of the important prey and predator species in the fauna have been identified, that the predators consumed all of the production of potential prey species, and that the latter were their only source of prey—that is, that the predators were food-limited and the prey populations limited by predation. One or more of these assumptions may not be true in many, or even most, cases (Farlow 1976*b*, 1980, 1983; cf. Baird and Horner 1979: 23; McNaughton and Georgiadis 1986:54; Emmons 1987:279).

Predators may kill carnivores of their own or other species, a behavior seen in crocodilians and other reptiles (Kellogg 1929; Cott 1961; Guggisberg 1972; Messel et al. 1981; Polis and Myers 1985; Delany and Abercrombie 1986; Woodward et al. 1987). Intraspecific agonistic or cannibalistic encounters in Komodo dragons were thought by Auffenberg (1981) to be the chief agent of mortality in that species and may account in part for its low observed, as opposed to theoretical maximum, predator/prey ratio (fig. 2.2).

The small theropod *Coelophysis* has been implicated in cannibalism (Colbert 1983:170–171), and this behavior may have occurred in larger carnivorous dinosaurs as well. Given their likely mobility, carnosaurs may have ranged over wide areas and perhaps frequently engaged in oralike aggressive encounters (either as solitary animals or as groups). Preserved predator/prey ratios might then reflect the land-tenure patterns of carnosaurs more than their food requirements. If these reptiles were highly territorial and their territories fairly large—but not so large as to make the chances of meeting interlopers remote—an overdispersed pattern of individuals or groups might have made the predator/prey ratio in any particular area less than the carnivores' metabolic needs alone would dictate.

Thus, the attempt to determine rates or patterns of energy flow through carnosaur populations by means of predator/prey ratios turns out to be more complicated than initially supposed. The conclusions of Bakker (1980, 1986) and Béland and Russell (1979, 1980) notwithstanding, predator/prey ratios do not provide unambiguous evidence of either tachymetabolism or bradymetabolism in carnivorous dinosaurs.

DISCUSSION AND SPECULATION

Given the occurrence of elevated metabolic rates (or at least the potential for tachymetabolism) in such disparate organisms as sea turtles, pythons, lamnid sharks, tunas, and even certain insects and plants (McNab 1983; Standora and Spotila 1982; Carey et al. 1985; Emery 1985, 1986; Spotila and Standora 1985; Lutcavage and Lutz 1986), as well as birds and mammals, it would not be surprising if at least some dinosaurs had been capable of rapid metabolic rates during some phase of their lives.

Of the various lines of evidence, bone histology provides the most compelling case for dinosaurian tachymetabolism, but even this evidence must be interpreted with caution (Ricqlès 1983; Reid 1987a). The widespread occurrence of fibrolamellar bone in dinosaurs implies rapid growth, and the question then becomes whether bradymetabolic ectotherms can grow rapidly enough to form this type of primary bone; the infrequent occurrence of fibrolamellar bone in most living and fossil reptiles, and the differences in growth rate between living mammals and birds, on the one hand, and reptiles, on the other (Case 1978a), suggest that most ectothermic reptiles cannot grow fast enough under natural conditions to deposit fibrolamellar bone. This, in turn, permits speculation that the metabolic rates of many (most?) dinosaurs were greater than those of most other reptiles.

Then we must ask whether the metabolic rates of dinosaurs were as rapid as those of living birds and mammals (and if so, which birds and mammals) or whether dinosaurian metabolic rates had elevations intermediate between those of modern reptiles and tachymetabolic endotherms (Hopson 1980a; Ricqlès 1983; Reid 1987a). A related question is whether dinosaurian tachymetabolism was uniform among the various taxa or whether the elevation of metabolic rates differed among the many kinds of dinosaurs. At present, these questions cannot be answered with certainty.

At small body sizes, birds and mammals must either elevate their metabolic rates above those expected on the basis of the Kleiber relation or undergo daily torpor; the boundary in the relationship of basal metabolic rate versus body mass between those small endotherms that do and do not enter obligatory torpor defines a "minimal boundary curve for endothermy" that can be extrapolated to larger body masses (McNab 1983). A condition of "minimal" or "marginal" endothermy exists for those larger fishes and reptiles whose metabolic rates are greater than extrapolated from the minimal boundary curve. Very large animals can achieve endothermy quite cheaply at metabolic rates considerably less than those of birds and mammals (cf. Smith 1979). Such marginal endothermy would be difficult to distinguish from inertial homeothermy (McNab 1983). If thermoregulation were the only consideration, then, it is unlikely that large dinosaurs would have been very tachymetabolic (Spotila 1980).

If any dinosaurs were tachymetabolic, I suspect that the primary benefit of an elevated metabolic rate was either the capacity for a rapid growth rate (Ricqlès 1980; Reid 1987a) or an increase in the potential for aerobic activity (Bennett and Ruben 1979; Carrier 1987), or both. Apart from the obvious benefit of reducing the interval of vulnerability of juveniles to predation and reducing the need for an unrealistically high juvenile survivorship, a faster growth rate could have shortened the length of time before the first breeding (Case 1978b), perhaps contributing to a higher reproductive potential—even if dinosaurs, like marsupials, were unable to capitalize on these potential advantages of rapid metabolism to the extent supposedly seen in modern eutherians (McNab 1986b, 1987).

As stressed by Pough (1980, 1983) and McNab (1983), a rapid metabolic rate is not always advantageous, and in some ecoevolutionary contexts an endotherm might find it better to have a lower metabolic rate than the maximal level that would otherwise be biochemically possible. In mammals, for example, the basal metabolic rate of animals of a given size varies with diet, activity level, and the climatic or microclimatic regime in which the animal lives (McNab 1986b, 1987).

It is likely that many large dinosaurs (particularly sauropods), living in a warm environment, would have found mammalian- or avian-level tachymetabolism thermally stressful, unless, as already suggested, the differences in levels of metabolism of endotherms and typical reptiles are in fact reduced at very large sizes—

in which case, the distinction between endothermy and ectothermy would be meaningless. Furthermore, even if the dangers of overheating could be alleviated, sauropods and perhaps other large dinosaurs might have found it difficult to find and process food quickly enough to sustain very rapid metabolic rates (Weaver 1983). Farlow (1987b) suggested that metabolic rates lower than those expected for mammals of comparable size might even have been advantageous to herbivorous dinosaurs feeding on poor-quality fodder.

If any dinosaurs were endotherms, that endothermy was probably based on metabolic rates that varied with the body size, diet, and thermal regime of particular species (Regal and Gans 1980). Bone histology suggests that dinosaurian metabolic capacities were probably greater than those of most or all modern reptiles, but I think it likely that only the smaller dinosaurs, and/or perhaps those living in cooler climates, had metabolic rate elevations approaching those of modern birds and mammals.

If most dinosaurs were in fact "intermediate-level" endotherms in the sense suggested here, their metabolic rates may even have varied substantially over ontogeny or on a seasonal basis. Some small birds and mammals shut down their thermogenic machinery in response to nocturnal or seasonal cold as an energy-saving strategy (McNab 1983), perhaps the ecological equivalent of the reduced metabolic rates of small reptiles at low temperatures (Pough 1980, 1983). Although white sharks seem to have the cardiac and hematological capacity to raise their body temperatures by increasing their metabolic rates (Carey et al. 1985; Emery 1985, 1986; McCosker 1987), those rates are nonetheless sometimes rather low (Carey et al. 1982). Conceivably, white sharks only increase their metabolism as needed for activity (as in green and leatherback turtles? [Spotila and Standora 1985]) or digestive purposes; if so, this might explain why the growth rate of these fishes is not especially fast (Cailliet et al. 1985).

Perhaps many dinosaurs had fairly rapid metabolic rates as juveniles (but perhaps not as tachymetabolic as those of modern birds and mammals), the better to facilitate fast growth and earlier maturation. In some (particularly smaller) forms, relatively high metabolic rates may have been retained in adulthood; this may have permitted the evolution of large brains in at least some theropods (Hopson 1977, 1980a; Martin 1981). In most large-bodied dinosaurs, the growth and metabolic rates presumably declined as the adult size and/or sexual maturity was approached (but not to the point where new primary bone was deposited in zonal instead of fibrolamellar fashion, except in special situations; cf. Buffrénil et al. 1986), with the routine level of adult standard metabolism reflecting a balance among the perhaps conflicting demands of thermoregulation, activity, and reproduction. In some dinosaurs, the best solution may have been minimal endothermy, perhaps even approaching inertial homeothermy, and in others, a higher level of tachymetabolic endothermy.

The optimal metabolic level may have shifted seasonally, with the degree of tachymetabolism increasing during the winter season and declining (particularly in large adults) during the summer months. A more rapid metabolic rate during the cooler season might have contributed to the ability of some dinosaurs to endure high-latitude winters, given sufficient food supplies (Brouwers et al. 1987; Parrish et al. 1987). A lower summer metabolic rate may have been large dinosaurs' equivalent of the torpor of small tachymetabolic endotherms, but employed to reduce heat stress rather than to conserve energy.

Ontogenetically and seasonally variable metabolic rates may even have allowed dinosaurs to exploit the contrasting benefits of both ectothermy, as practiced by modern amphibians and reptiles, and endothermy, as employed by modern birds and mammals. During periods of food shortage, dinosaurian metabolic rates could have become more bradymetabolic; growing juveniles, faced with unexpected famine, may even have been able to avoid starvation by tamping their metabolic fires until conditions improved, giving these animals even more growth rate flexibility than seen in modern reptiles (Andrews 1982). "Heterometabolically" endothermic dinosaurs may not have shown the extreme specialization for either the high production efficiencies and low metabolic rates of modern ectotherms or the low production efficiencies and rapid metabolic rates typical of birds and mammals and may have been able, in a manner not seen in any living vertebrates, to channel energy alternatively into elevated metabolism or production as ecologically appropriate.

The scenario outlined here sounds so splendid as to make one wonder why it is not employed by all endotherms, particularly the large mammals. It may be that some large mammals do have more seasonably variable metabolic rates than commonly believed; the physiological ecology of so few species has been studied in sufficient detail that generalization is risky (B. K. McNab pers. comm.). If, however, modern endotherms are indeed less seasonably heterometabolic than suggested here for dinosaurs, this may mean—to modify a suggestion made by McNab (1983, 1987)—that the phylogenetic passage of birds and mammals through animals of small body size somehow "fixed" their basal metabolic rates in such a way as to make heterometabolism to the degree proposed here for dinosaurs impossible, apart from the torpor seen in many small-bodied birds and mammals. Furthermore, eutherians may have become so specialized in their linkage of reproductive potential to tachymetabolism as to make reversion to the strategy of dinosaurs (and therapsids?) difficult. Even allowing for the depauperate nature of the modern mammalian megafauna (Bakker 1980), the majority of mammalian genera and species have always been small-bodied animals (Hotton 1980), while dinosaurs as a group opted for larger sizes; this in itself suggests some difference in the overall energetic strategies of the two groups.

All of this is admittedly speculative, and there is scant evidence to support this hypothesis of dinosaurian heterometabolism, except that the scenario seems consistent with what is known about the size-related energetic and thermal strategies of living vertebrates and with the somewhat variable development of dinosaurian primary bone (Ricqlès 1980, 1983; Reid 1987a). The chief difference between my interpretation of dinosaurian endothermy and that of most other workers is my emphasis on temporally variable levels of metabolic rate and on the possibility of interspecific differences in the degree of tachymetabolism among even closely related taxa of dinosaurs.

I predict that if and when unambiguous histological or other evidence of continuously rapid metabolic rates in dinosaurs becomes available, it will be best developed (particularly in small-bodied forms) in dinosaurs from high-latitude or other thermally seasonal situations (unless, of course, such environments were occupied only during the warm season) and poorly developed in large-bodied taxa from low-latitude faunas. My hypothesis of dinosaurian heterometabolism might also account for the occurrence of cyclical growth patterns in the primary bones of some dinosaurs (Ricqlès 1983; Reid 1987a), but there is at present no way of proving that these cycles were annual or—if they were annual—of determining the season of most active growth.

If my hypothesis becomes a viable interpretation of dinosaurian physiology, it will explain how dinosaurs achieved their evolutionary success with a metabolic strategy that was neither that of typical ectothermic reptiles nor that of endothermic birds and mammals, but could have been a fully adaptive evolutionary intermediate between the two contrasting energetic modes. Dinosaurs would be seen neither as typical tachymetabolic endotherms in the modern sense nor as "good reptiles." Perhaps we will have to call them *damn* good reptiles.

Part III   DINOSAUR EXTINCTION

PETER DODSON

LEONID P. TATARINOV

The disappearance of the dinosaurs is perhaps the most widely appreciated of all extinctions (fig. 2.3); indeed, in the public perception, the term *dinosaur* is paradigmatic for the concept "extinct." The problem of dinosaur extinction has a lengthy history of idiosyncratic explanations, the majority of which have been sterile, neither convincing in themselves nor stimulating further research. In the past, questions of both the ascen-

dance of dinosaurs in the Triassic and their extinction at the end of the Cretaceous have been explained by scenarios of competitive replacement of older groups by younger innovative ones (e.g., Davitashvili 1969, 1978; Gabunia 1969, 1984). Such explanations are rejected currently by most authors (e.g., Benton 1983a; Elliott 1986). Preferred hypotheses link extinctions with abiotic factors (e.g., Stanley 1987) or even global

"The picture's pretty bleak, gentlemen. ... The world's climates are changing, the mammals are taking over, and we all have a brain about the size of a walnut."

Fig. 2.3.

catastrophes of cosmic origin (Raup 1986). It is only since about 1980 that systematic integrated programs of research have focused on the extinction at the Cretaceous/Tertiary (K/T) boundary. The stimuli for renewed interest have been twofold: the discovery of an iridium anomaly at the K/T boundary, interpreted as evidence for an impact by an asteroid or comet (Alvarez et al. 1980), and the detection of a supposed periodicity in mass extinctions during Phanerozoic time (Raup and Sepkoski 1984) with its implication of an astronomical agent.

The end of the Cretaceous was a time of crisis in life history, and the character of the extinctions determined the course of the biosphere to the present time. Many major features of extinction are noteworthy. New groups of organisms may appear millions of years before the extinction of old ones, and old groups disappear much more rapidly than new ones radiate, leading to empty ecological space for a period of some millions of years. Ecosystem models (e.g., Zherychin 1978,

1987; Ponomarenko and Popov 1980) describe tendencies for increasing specialization of dominant types in stable ecosystems and subsequent destabilization of ecosystems when new types appear, aided by physical disturbance (climate, tectonics, etc.). Such models have been applied to the replacement of the Mesozoic cycad-broadleafed conifer flora by the modern angiosperm flora with its inferred effect on dinosaurs (Krassilov 1981, 1985, 1987; Wing and Tiffney 1987). The ecosystem model for angiosperm ascendance would seem to lack general explanatory power for the terminal Cretaceous extinctions, however, and attempts to extend its influence into the sea by means of chemical products of angiosperm decomposition are unconvincing.

Although it is generally granted that the problem of dinosaur extinction is coextensive with the study of the K/T boundary, there are two challenges to the concept that dinosaurs became extinct on the Cretaceous side of the boundary. On the one hand is the claim that dinosaur fossils are found in rocks of early Paleocene age in Montana and elsewhere, including New Mexico, South America, Europe, India, and China (Sloan et al. 1986; Rigby et al. 1987; Van Valen 1988). On the other hand is the claim that dinosaurs are not extinct at all. This claim is a popular one in fiction (e.g., Jules Verne, Edgar Rice Burroughs, Conan Doyle), but it is also maintained by strict adherents of phylogenetic systematics: if birds are descendants of small meat-eating dinosaurs, then birds must be considered a subgroup of the Dinosauria (Bakker and Galton 1974; Gauthier 1986; Cracraft 1986; Sullivan 1987). The issue of the survival of dinosaurs into the Paleocene is an interesting one, and such claims are not inherently improbable (indeed, discovery of Paleocene dinosaurs at low latitudes may be *predicted* if global climatic cooling is postulated as a cause of dinosaur extinction). However, convincing documentation of Paleocene dinosaurs lies in the future. Eaton et al. (1989) document dinosaur teeth and other Cretaceous fossils as sedimentary particles in Paleocene and Eocene deposits, and Buffetaut et al. (1980) report a dinosaur tooth in Miocene sediments. The cladistic claim that dinosaurs did not become extinct but survive as birds, while logically consistent, has little impact: the animals *universally understood* to be dinosaurs did become extinct at or near the K/T boundary.

It is probably fair to say that a consensus has not yet emerged on some of the most fundamental aspects of the problem of dinosaur extinction. Some of the fundamental questions are:

1. Was extinction of the dinosaurs sudden and catastrophic (measured in hours, days, weeks, or months), or was it slow and gradual (100,000 years, a quarter of a million years, or even a million years or longer), culminating a long period of decline in diversity and biomass?

2. Can dinosaur extinction be considered by itself, or must it be considered in conjunction with extinctions of all contemporaneous organisms, plant and animal, vertebrate and invertebrate, on land and sea?

3. Can the terminal Cretaceous extinction be considered by itself, or must it be considered as part of a nomothetic pattern of mass extinctions in earth history?

4. Are terrestrial agents sufficient to account for extinctions, or should astronomical causes be sought?

It is perhaps useful in conceptualizing extinction theories to distinguish terrestrial from astronomical or cosmic processes and biological from physical processes (table 2.1). However, ultimate geological or cosmic causes inevitably affect more volatile surficial physical processes (e.g., climate, weather) that determine physical aspects of the biosphere. Thus, multiple

pathways of causation may link the cosmic realm to the earthbound and the physical world to the biological. It is a truism that no one isolated cause can explain the complex phenomenon of mass extinction.

FACTUAL BASIS FOR DISCUSSION OF DINOSAUR EXTINCTIONS

Dinosaurs existed from the Carnian to the Maastrichtian, but the known distribution of dinosaurs through the Mesozoic is very uneven. Despite a duration of approximately 150 million years, the fossil record of dinosaurs is strongly biased toward the recent part of their range: about 50 percent of all genera currently known come from the Upper Cretaceous, and 40 percent come from the Campanian or Maastrichtian stage. Dinosaurs appear to have been at the peak of their diversity within 10 million years of their disappearance and were little diminished well into Maastrichtian time (Russell 1984b).

The Maastrichtian stage lasted 8 million years (Harland et al. 1982; Lillegraven and McKenna 1986). Simply recording that dinosaurs or any other group became extinct in the Maastrichtian (e.g., Russell 1982) provides evidence for neither tempo nor mode of extinction. Maastrichtian dinosaurs are known from Asia, Europe, and North and South America. But wherever else dinosaurs may have existed at this time, incontrovertible fossils of latest Maastrichtian age (= Lancian; see introduction for discussion of North American land mammal ages) which potentially document the crucial K/T boundary events are currently known only from Wyoming, Montana, Alberta, and Saskatchewan (Clemens 1986; Archibald 1987b), although other finds may yet be corroborated. There appears to be a strong bias in the sedimentary record against later Maastrichtian marine sediments. Kauffman (1984) estimates that 90 percent of exposed sections contain a hiatus of 2 to 5 million years duration. A similar bias exists on land. A eustatic drop in sea level of 100 to 150 m (Schopf 1982; Kauffman 1984; Hut et al. 1987; Haq et al. 1987) is a major physical consideration for the Maastrichtian. At the very least, this event caused a major loss of stratigraphic resolution of extinction events; likely, it was a contributor to extinctions both in the sea and on land.

Few dinosaur genera spanned more than one

TABLE 2.1 Classification of Extinction Processes

	TERRESTRIAL	COSMIC
BIOLOGICAL	competition with mammals egg predation by mammals origin of angiosperms poisoning by plants hyperthermic sterility alteration of sex ratios thinning of egg shells disease/epidemics faunal disequilibrium	viruses from space "little green men"
PHYSICAL	climatic cooling cold and darkness volcanism continental drift sea level changes mountain building paleomagnetic reversal	supernova comet/asteroid impact crossing the galactic plane poison gases from comet orbital eccentricity

stage. Bakker (1986) estimates that the average duration of a dinosaur species was two to three million years and of a genus, 5 to 6 million years. Indeed, given the typical rate of turnover of dinosaur genera, extinctions of considerable magnitude occurred throughout the history of dinosaurs. Only at the end of the Maastrichtian was there a failure of *origination* of new taxa (Bakker 1977). Although disappearance of the dinosaurs is a spectacular aspect of the Cretaceous extinction, it actually represented the loss of rather few species and is only a small aspect of the entire extinction (Schopf 1982; Padian and Clemens 1984; Archibald 1987*b*). The K/T extinction saw the disappearance of about 15 percent of all families of animals (Sepkoski 1982; Benton 1985*a*).

Maastrichtian extinctions were highly selective, as is the case with all mass extinctions (Jablonski 1986). All land vertebrates greater than 25 kg in body weight disappeared. This obviously included dinosaurs, but even small and young dinosaurs below this size limit disappeared. Marsupial mammals in North America almost disappeared and have never recovered their Cretaceous diversity (Padian and Clemens 1984), although they thrived in South America during the Cenozoic. Nondinosaurian reptiles (pterosaurs, ichthyosaurs, plesiosaurs, mosasaurs) disappeared. Ichthyosaurs, however, had disappeared by the early Late Cretaceous (Baird 1984), and mosasaurs, plesiosaurs, and pterosaurs appear to have dwindled greatly by the beginning of the Maastrichtian (Sullivan 1987). Mosasaurs reached the K/T boundary in New Jersey at least (Baird 1986*a*). Marine invertebrates were hard hit, but planktonic and sessile benthonic organisms suffered more than mobile benthonic ones (McKinney 1987). Tropical families were affected disproportionately as compared to extratropical ones, and endemics were more extinction-prone than were geographically widespread organisms (Jablonski 1986). Selective extinction of planktonic animals and sessile filter-feeders has been postulated, accompanied by survival of benthic marine scavengers, deposit feeders, and scavengers (Sheehan and Hansen 1986). Planktonic foraminiferal extinction is regarded as catastrophic, with a decrease from thirty-six species to one species across the boundary (Padian et al. 1984; Lipps 1986). More generally, there was a two-thirds reduction in diversity of all marine microplankton which affected larger benthonic foraminifera and calcareous nannoplankton as well. Rudistid and inoceramid bivalves, hermatypic corals, ammonites, and belemnites (Kauffman 1984) disappeared, but declines in diversity throughout the Maastrichtian can be traced (Ward 1983; Ward and Signor 1983; Ward et al. 1986; Hut et

al. 1987). At Zumaya, Spain (Ward et al. 1986), the last dwarfed ammonites occur 12.5 meters below the K/T boundary, while at Stevns Klint, Denmark, seven species of ammonites remain to the end (Alvarez and Kauffman et al. 1984).

In contrast, many organisms were little affected by the K/T extinctions. Placental mammals diversified, and multituberculates suffered little (Padian and Clemens 1984). Turtles appear to have peaked in diversity in the Maastrichtian, but suffered only modest losses (16%) across the boundary (Hutchison and Archibald 1986). In fact, continuity of nondinosaurian lower vertebrates, including champsosaurs and eusuchian crocodiles as well as turtles, across the boundary is the norm, with extinction rates at the generic level of only 26 to 31 percent (Archibald 1987*b*). Marine fishes and lizards suffered the greatest extinctions. Birds diversified into the Paleocene (Feduccia 1980), although archaic forms such as enantiornithines and hesperornithiforms disappeared. Bryozoans, brachiopods, bivalves, gastropods, echinoderms, crustaceans, and benthic foraminifera were relatively little affected.

Plant macrofossils show continuity across the K/T boundary (Hickey 1981, 1984; Tschudy and Tschudy 1986), but there is now extensive evidence for ecological disruption and reorganization of communities. The *Aquilapollenites* palynofloral province disappeared in North America but persisted in Japan and the Orient (Saito et al. 1986). In megathermal and mesothermal regions of western North America, broadleafed evergreens were dominant in the Late Cretaceous; above 66°N, the deciduous habit prevailed (Wolfe and Upchurch 1986). Immediately above the boundary, an abundance of fern spores has been documented in New Mexico (Pillmore et al. 1984), Montana (Tschudy et al. 1984), Saskatchewan (Nichols et al. 1986), and Japan (Saito et al. 1986). This may record an episode of recolonization of a disturbed land surface by opportunistic ferns. Such recolonization by ferns has been observed on the flanks of the Mexican volcano El Chichon following the 1982 eruption (Spicer et al. 1985). As a floral equilibrium was reestablished in the early Paleocene, a general pattern of extinction of evergreens and survival of deciduous plants is noted (Wolfe 1987; Wolfe and Upchurch 1986, 1987). A change from subhumid conditions in the Maastrichtian to humid conditions in the Paleocene is inferred, as expressed by the prevalence of drip points on leaves. The polar deciduous forest was disrupted but suffered little extinction, and no change to deciduous leaf habit is recorded in the Tertiary of South America.

INTERPRETATION OF MAASTRICHTIAN EXTINCTIONS

The discovery of an iridium anomaly at the K/T boundary (Alvarez et al. 1980) and its confirmation at several scores of sites, both marine and terrestrial, around the world (Alvarez and Alvarez et al. 1984) strongly invite interpretation of extinction caused by asteroid or comet impact. The recognition of an apparent 26-million-year periodicity of mass extinctions in Phanerozoic time (Raup and Sepkoski 1984, 1986, 1988; Sepkoski and Raup 1986) has earned further support for astronomical catastrophes. By inference, the ultimate agent of extinction must be celestial, as it is difficult to conceive of an earthbound agent operating with such a lengthy periodicity. The scenario of months of darkness, stormy weather, freezing temperatures, cessation of photosynthesis, dying off of large animals, hibernation of small ones, and so forth (giving rise to the concept of nuclear winter; Turco et al. 1983; Ehrlich et al. 1983), is a compelling one. But does such devastation explain more than is observed?

The reality of the 26-million-year periodicity of mass extinctions has been vigorously challenged on a variety of grounds, both statistical and biostratigraphic (e.g., Hoffman 1985; Lutz 1986; Quinn 1987; Patterson and Smith 1987; Stigler and Wagner 1987, 1988). Critical arguments have been countered with further data and further analyses (Raup and Sepkoski 1986, 1988). The statistical significance of the periodic extinctions may be sustained (e.g., Gilinsky 1986; Kitchell and Estabrook 1986), and in any case, the concept of periodic extinctions has considerable heuristic value (Raup 1986). Nonetheless, caution is always in order. The proposed periodic mass extinctions events are very uneven in their intensity. The Late Permian Guadalupian extinction 253 million years ago has long been recognized as the greatest Phanerozoic mass extinction (e.g., Newell 1967), and the Maastrichtian extinction event is perhaps the fifth greatest (Sepkoski 1982). In contrast, the mid-Miocene event is a very weak extinction (Sepkoski and Raup 1986), and the hypothesized Aptian and Bajocian events are reflected not at all in family-level data and but weakly in generic data (Raup and Sepkoski 1986). The iridium event has not been associated convincingly with other mass extinctions.

Before too much is attributed to a hypothesized impact or similar catastrophe, it should be reiterated that the Maastrichtian stage lasted approximately 8 million years. Organisms that dwindled over a significant portion of the Maastrichtian ought not to be adduced in support of a catastrophe. As yet, there has been too little resolution of the pattern of diversity of each of the affected groups within the Maastrichtian. The extinctions of ammonites, inoceramids, and rudistids were not catastrophic (Ward 1983; Stanley 1984a, 1984b; Ward et al. 1986); the extinction of planktonic foraminifera appears to have been (Lipps 1986). Kauffman (1984) stressed the synergistic roles of rapid, large-scale changes of sea level, water temperature, oxygen content, and circulation pattern. These factors caused extinctions throughout the Cretaceous and were particularly severe in the late Maastrichtian. The Maastrichtian stage was a "noisy" interval in earth history. It is now clear that much extinction occurred during 1 to 3 million years of deterioration prior to the plankton catastrophe that marks the K/T boundary. Indeed, the term *stepwise mass extinction* has now been introduced in recognition that detailed microstratigraphy of boundary sections requires modification of a single, all-encompassing, impact-related catastrophe (Hut et al. 1987).

DINOSAUR EXTINCTIONS

Sampling Problems

Do dinosaurs fit the pattern of a decline over 1 to 5 million years, or was their disappearance sudden and catastrophic? The problem of resolution is as critical here as it is for any group of fossils. In terms of numbers of species, dinosaurs were in full flower in the Maastrichtian around the world. Thus, their disappearance, whether sudden or gradual, was not predicted on the basis of their previous evolutionary history; it was an enigma (Russell 1979). It is currently only in the northern portion of the Western Interior of North America that well-characterized dinosaur faunas (i.e., those in which taxa are represented by taxonomically diagnostic skeletal material) approach the K/T boundary, which is in turn overlain by lower Paleocene sediments (Clemens 1986; Archibald 1987b).

Russell (1984b) emphasizes problems of sample size in determining diversity trends. The best-sampled dinosaur fauna, that of the Campanian Judith River Formation of Alberta, Canada (Dodson 1983), is also the most diverse dinosaur fauna known from any time or place. Russell argues that the youngest dinosaur fauna, the late Maastrichtian Lancian fauna, is less well sampled and therefore is likely to underrepresent diversity. He infers that the diversity of Lancian dinosaurs was probably about as high as that of Judithian dinosaurs.

Comparisons of diversity between formations may be confounded by differences in taphonomy. The Campanian Judith River Formation and the early Maastrichtian Horseshoe Canyon Formation preserve a mosaic of fluvial and floodplain sediments deposited on the lower coastal plain close enough to the epicontinental seaway for tidal influence to be detected (Russell 1977; Russell and Chamney 1967; Dodson 1971, 1983, 1987; Koster 1984; Koster and Currie 1986). Such sediments preserve complete skeletons of dinosaurs with regularity. But withdrawal of the epicontinental sea conditioned a different environmental setting for the Hell Creek Formation of Montana. Floodplain sediments with incipient soils supported seasonally dry, lowland, subtropical forests; channels are subordinate but provide abundant vertebrate microfossils (Fastovsky 1987; Fastovsky and Dott 1986; Fastovsky and Sweeney 1987; Retallack and Leahy 1986; Retallack et al. 1987; Sloan et al. 1986; Rigby et al. 1987). The dinosaur record of the Hell Creek Formation contrasts markedly from that of the Judith River or the Horseshoe Canyon Formation. The rarity of articulated skeletons (not a single complete skeleton of *Triceratops* has ever been found) is indicative of the taphonomic difference that makes faunal comparison, and thus the analysis of diversity trends, so difficult. Even if numbers of genera of Judithian and Lancian dinosaurs (beta diversity) are similar, it seems very unlikely that alpha diversity (i.e., within community) was similar. Diversity in the Judith River Formation is indicated by the relative abundance of five different taxa (*Stegoceras, Corythosaurus, Lambeosaurus, Centrosaurus*, and *Chasmosaurus* are more or less equally likely to be found, and *Albertosaurus* and *Gryposaurus* are about half as likely to be found as the preceding). It is impossible to predict the identity of the next recovered specimen. The Lance-Hell Creek faunas, however, are dominated by *Triceratops*, with *Edmontosaurus* a distant second. *Triceratops* is one of the most abundant dinosaurs in museum collections, with perhaps 50 skulls. Brown (1917) reported seeing more than 500 partial skulls of *Triceratops* in Montana during seven years of fieldwork (although these might include *Torosaurus* and possibly other taxa as well). Indices of faunal diversity (e.g., Ricklefs 1979) express the relative contribution of common versus rare species to community diversity. Bakker (1986) reported a diversity index of 3.8 for the Judith River fauna and an index of 1.4 for the *Triceratops*-dominated Lancian fauna (although he did not specify his census data base). When between-formation comparisons of diversity are made using skulls and/or articulated specimens, it appears that dinosaur diversity declines between the Judithian and the Lancian (Bakker 1986), but when combined sources are used (beta diversity), including taxa named only on the basis of isolated teeth, dinosaur diversity appears to be as high in the Hell Creek, at least in its early part, as in previous stratigraphic intervals (e.g., Russell 1984*a*, 1984*b*; Sloan et al. 1986).

Faunal Trends within the Lancian

There is little information on dinosaur faunal trends within the Lancian. Such data are of critical importance in establishing the tempo and mode of dinosaur extinction. Carpenter and Breithaupt (1986) report that nodosaurids became extinct in the lower part of the Hell Creek Formation, presumably in the early part of late Maastrichtian time (although Russell [1987] suggests that apparent absence of nodosaurids from latest Cretaceous sediments may reflect their preference for wetter lowland habitats). Sloan et al. (1986) report a decline in the density of dinosaur skeletons or large bones in last 27 m of the Hell Creek Formation, with the lowest density in the top 9 m of section. The purported decline in diversity and numbers of dinosaurs took place over a period of 300,000 to 700,000 years. Numbers of genera of dinosaurs within the Hell Creek Formation decline from 19 at the beginning to 12 genera in the upper 16 m of the formation to seven genera at the very top. It can be argued that an apparent decline in preserved specimens in Garfield and McCone counties of Montana need not imply that dinosaurs became extinct everywhere. Most sediments do not have dinosaur fossils or any other fossils in them. It is neither an a priori expectation nor an empirical observation that the last sediments laid down in the age of dinosaurs had dinosaur fossils in them. For the previous 160 million years, dinosaurs had apparently dwindled and become "extinct" in formation after formation, always to reappear again.

If the dinosaurs were truly dwindling in numbers and diversity as a result of climatic stress (Axelrod and Bailey 1968), it would be predicted that their ranges should contract progressively toward the equator, and Paleocene dinosaurs would not be shocking. Our knowledge of Paleocene sediments in tropical areas is certainly too incomplete to rule out such a possibility, however unlikely it may seem. Paleocene dinosaurs have been claimed from Montana (Rigby et al. 1987), New Mexico, Brazil, Bolivia, Peru, India, China, and Europe (Sloan et al. 1986; Van Valen 1988), but such claims have always been met with skepticism (e.g., Retallack and Leahy 1986; Bryant et al. 1986; Argast et al. 1987; Eaton et al. 1989). In all cases, materials are disarticulated and fragmentary. In Montana, the specific claim is made that the last dinosaur remains occur in sediments deposited 40,000 years after the end of the Cretaceous (Rigby et al. 1987). Whether or not any individuals straggled across the K/T boundary, there seems no doubt that in an ecological sense, the dinosaur chronofauna terminated at the end of the Cretaceous.

THE ASTEROID AND THE DINOSAUR

Physical evidence for an iridium anomaly is so widespread at the K/T boundary that it cannot be ignored. The "iridium event" has to be accounted for. It is widely accepted that the best source of enrichment for iridium is extraterrestrial, that is to say, a comet or an asteroid (Alvarez et al. 1980, Alvarez and Alvarez et al. 1984). In addition to iridium, shocked quartz grains have been reported from the boundary clay in Montana, New Mexico, and Europe (Bohor et al. 1984). There is some evidence that such fractured grains would result only from impact, not volcanism (Owen and Anders 1988). Abundant soot aggregates have been reported at the K/T boundary in Denmark and in New Zealand (Wolbach et al. 1985, 1988) which may have been the result of great fires on land following the impact. This evidence is difficult to evaluate, however, because its significance depends on the rate of sedimentation in the boundary clay and on the distribution of soot particles throughout the sedimentary column. (Carpenter [1987] noted the importance of charcoal [fusinite] from fire in Late Cretaceous terrestrial deposits of the Western Interior.) Consistent with the idea of widespread fire is the discovery of an anomalous and short-lived abundance of fern spores (the "fern-spore spike") immediately above the K/T boundary clay in New Mexico, Colorado, Montana, Saskatchewan, and Japan (Orth et al. 1981; Pillmore et al. 1984; Tschudy et al. 1984; Nichols et al. 1986; Saito et al. 1986). This may document a disruption of terrestrial plant communities with initial recolonization by ferns, such as has been recorded in Mexico following recent volcanism (Spicer et al. 1985). However, ferns may also be a normal component of coal-swamp succession.

An alternative view to extraterrestrial impact which has gained some support holds that major terrestrial volcanism tapping deep magma sources could be a source of iridium enrichment (Officer and Drake 1983, 1985; Officer et al. 1987; Hallam 1987). Much interest has centered on the vast flood basalts compris-

ing the Deccan Traps of India, which erupted at or near the K/T boundary (Cox 1988; Duncan and Pyle 1988; Courtillot et al. 1988). The eruptions, comprising perhaps several hundred separate flows over half a million years (Cox 1988), would entail release of tremendous quantities of gases, including CO_2 (McLean 1985) and sulfuric acid (Rampino and Volk 1988). Sequellae include climatic warming due to the greenhouse effect (McLean 1978), acid rain, and acidification and death of the ocean (the Strangelove ocean of Hsü and McKenzie 1985). The episodic nature of the eruptions lends itself better to the interpretation of stepwise extinction rather than to a single punctuated catastrophe, although the iridium signature is more compatible with a single event. Rampino and Strothers (1988) detect a 32-million-year periodicity of flood basalt volcanism over a 250-million-year period and postulate that the same comets that deposit iridium on earth may trigger the flood basalts.

Whether a catastrophe was generated by extraterrestrial impact or by massive volcanism, the immediate effect must have been on weather. Either scenario would produce dust clouds in the atmosphere, with attendant darkness and cold stormy weather (Budyko 1984). When the skies cleared and the weather ameliorated a few months of years later, the dinosaurs were gone. It seems tenable to attribute the final disappearance of dinosaurs to an asteroid strike. However, given the massive level of survival of so many organisms (i.e., roughly 85% of marine families and 86% of nonmarine vertebrates), including such thermophilic organisms as palms, reef-forming corals, crocodiles, cheloniid, and toxochelyid sea turtles [Sullivan 1987]), it is appropriate to question how dinosaurs had become so vulnerable. At present, there is at least a strong suspicion that dinosaurs had been dwindling for several hundred thousand years at least. Climatic stress resulting from cooling temperatures and deteriorating equability, exacerbated by the draining of the epeiric sea, is a candidate (Axelrod and Bailey 1968; Stanley 1987), although the relationship is complex. The Cretaceous was generally warm and frost-free (Budyko 1980), except for high latitudes in the Maastrichtian (Spicer and Parrish 1986; Wolfe and Upchurch 1987). Climatic cooling was initiated in the early Late Cretaceous and continued into the early Maastrichtian, but warming in the late Maastrichtian reversed the trend (Wolfe and Upchurch 1987). Cooling during the K/T boundary crisis is inevitably invoked (e.g., Krassilov 1985). One specific suggestion of the adverse effect of unstable temperatures on dinosaurs follows from the observation that in many living reptiles, especially turtles and crocodiles, sex is determined by incubation temperatures in the nest (Ferguson and Joanen 1982; Bull 1983). If this were also true for dinosaurs, they would have been particularly vulnerable to fluctuations in environmental temperatures, particularly if nesting grounds were in upland, less thermally buffered environments (Horner 1984a, 1984b). Thus, the last populations of dinosaurs may have been unisexual (Paladino et al., 1989).

As far as we know, dinosaur extinction was a worldwide phenomenon. Direct evidence for this event, however, comes principally from two counties in Montana. Much fundamental work remains to be done even within the Hell Creek Formation in Montana. It

may be expected that the question of gradual versus sudden disappearance of dinosaurs will be resolved by careful stratigraphic survey work throughout the whole formation. Interpretation of a worldwide phenomenon on the basis of data from a single geographic area is obviously very vulnerable. It is to be hoped that new data from localities on other continents will be forthcoming. Whatever catastrophes may have taken place in the geological past, their influence on biosphere would have been weaker than the one that would occur today by the agency of thermonuclear war, the danger of which creates a threat to the existence of life on Earth today.

3

Dinosaurian Distribution

DAVID B. WEISHAMPEL

INTRODUCTION

The extent to which dinosaur material is known throughout the world is often not well appreciated. Dinosaurs have been found on every continent of the world, from the northern slopes of Alaska, in the United States, to the Antarctic Peninsula. They are also known for virtually the entirety of the Mesozoic. This chapter is intended to summarize all dinosaur localities that are now known, on a worldwide basis. Localities are annotated for (1) geographic location, (2) lithostratigraphic unit (bed, member, formation, group, series, svita), (3) species composition, and (4) stratigraphic horizon (to stage where possible). Geographic location is based on first-order political subdivisions (states, departments, provinces, counties, etc.) of the countries in which dinosaur material has been found. The only exception to this "rule" is Antarctica, for which there are no official political subdivisions. Although political subdivisions of worldwide geography is arbitrary with respect to Mesozoic occurrences, it is equanimous with present-day political geography. Age assignments are based as far as can be ascertained on the standard European stages (see introduction) but are often limited to "best guesses" about the temporal

relationship between terrestrial and marine strata. Faunal composition is given to species-level taxa where possible, based on information presented in this volume. Finally, localities yielding skeletal remains as well as those that have produced footprints and/or trackways have been included (although ichnological designations are not used). The same applies to those localities that have yielded eggs and/or eggshell fragments (again, no ootaxa are indicated; nests, isolated eggs, and eggshell fragments are listed only as "dinosaur eggshell") and to those that have yielded coprolites. For reasons of length, no attempt has been made to characterize the distribution of faunas within lithologic units, nor are the sedimentology, depositional environments, or taphonomic aspects of localities indicated, although each of these sources of information is important in its own right to an understanding of dinosaurian distribution.

Symbols used in the maps relate to the eight series of the Mesozoic for which there are dinosaur remains: ★: Late Triassic; □: Early Jurassic; ○: Middle Jurassic; ⬠: Late Jurassic; ●: Early Cretaceous; ☆: Late Cretaceous; and ⬠: uncertain age. The first map gives a number of more poleward localities, while successive maps provide locations elsewhere in the world.

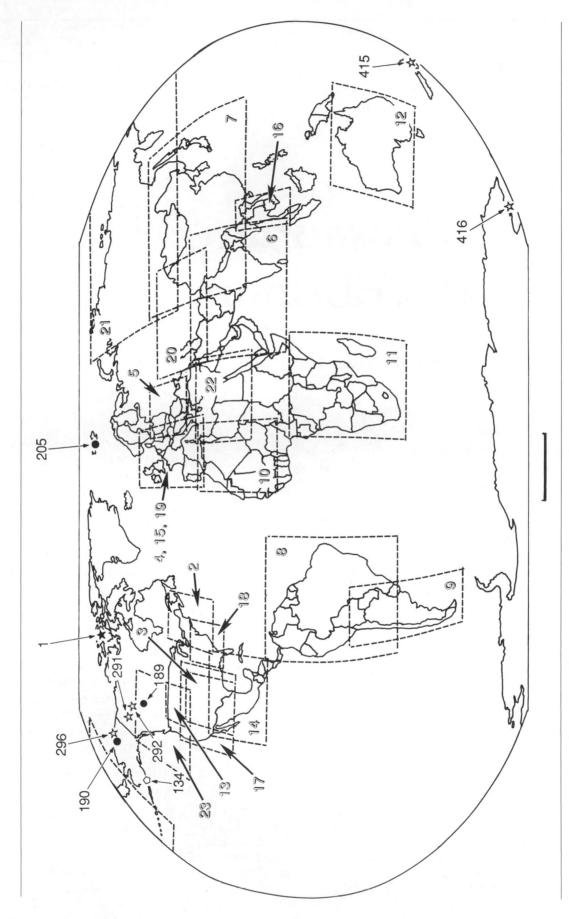

Fig. 3.1. Global index map. Shadowed numbers within rectangles indicate figures for regional coverage. Bold numbers indicate more poleward localities not included in subsequent figures. Scale = 3000 km.

LATE TRIASSIC

Late Triassic dinosaurs are known from between 75 and 100 locations in North America, Europe, Asia, South America, Africa, and Australasia.

North America

Information on Late Triassic dinosaurs in North America comes to us primarily through the work of E. H. Colbert on the early dinosaur faunas of the southwestern United States and also from the recent research by L. L. Jacobs, P. A. Murry, R. A. Long, J. M. Parrish, K. Carpenter, S. Chatterjee, K. Padian, and T. Rowe again on the Upper Triassic of the southwestern United States, and by P. E. Olsen and D. Baird in the eastern United States and Maritime Canada. Figures 3.1, 3.2, and 3.3 illustrate these Late Triassic locations.

1. Northwest Territories, Canada

   Heiberg Formation (Adams 1875; Lydekker 1889b; Galton and Cluver 1976)
   > ?Theropoda indet. (= *Arctosaurus osborni*)
   > Age: Carnian-Rhaetian (Douglas et al. 1963)

2. Nova Scotia, Canada

   a. Lower Wolfville Formation (Carroll et al. 1972; Baird pers. comm.)
   > Prosauropod indet.
   > Ornithischia indet.
   > Age: Carnian (Klein 1962)

   b. Upper Wolfville Formation (Baird pers. comm.)
   > Ornithischian footprints
   > Age: Norian (Baird pers. comm.)

   c. Blomidon Formation (Olsen et al. 1982)
   > Theropod footprints
   > Age: Norian (Olsen et al. 1982)

3. Utah, United States

   Chinle Formation (Lockley 1986a)
   > Theropod footprints
   > Age: late Carnian-early Norian (Olsen et al. 1982)

4. Arizona, United States

   Chinle Formation (Jacobs 1980; Jacobs and Murry 1980; Padian 1986; Rowe pers. comm.)
   > Theropoda
   >> Ceratosauria
   >>> *Coelophysis* sp.
   >>> cf. *Syntarsus* sp.
   >>> Undescribed ceratosaur (Rowe pers. comm.)
   >> Prosauropoda
   >>> ? Anchisaurid indet.
   >> Ornithischia indet.
   > Age: late Carnian-early Norian (Olsen et al. 1982)

5. New Mexico, United States

   a. Chinle Formation (Colbert 1974, 1989; Welles 1984)
   > Theropoda
   >> Theropoda indet. (= *Coelophysis longicollis* —exact horizon and locality unknown; *Tanystrophaeus willistoni*)
   >> Ceratosauria
   >>> *Coelophysis bauri*
   >> Ornithischia indet. (= *Revueltosaurus callenderi*)
   > Age: Carnian (Olsen et al. 1982)

   b. Dockum Group (Baird 1964; Carpenter and Parrish 1985; Parrish and Carpenter 1986)
   > Theropoda indet. (= procompsognathid indet.)
   > Ornithischian footprints
   > Age: Carnian-?Rhaetian (Olsen 1980a)

6. Colorado, United States

   Popo Agie Formation (Lockley 1986a)
   > Theropod footprints
   > Age: Carnian (Pipiringos and O'Sullivan 1978)

7. Texas, United States

   Dockum Group (Jacobs and Murry 1980; Chatterjee 1984, 1986; Murry 1986)
   > Theropoda indet. (= *Coelophysis* cf. *bauri*)
   > Theropod footprints
   > Ornithischia *sedis mutabilis*
   >> *Technosaurus smalli*
   > Age: Carnian-?Rhaetian (Olsen 1980a)

8. North Carolina, United States

   Chatham Group (Galton pers. comm.)
   > Ornithischia indet.
   > Age: Carnian-?Norian (Olsen 1980a)

9. Virginia, United States

   Cow Branch Formation (Olsen et al. 1978; Baird pers. comm.)
   > Theropod footprints
   > Ornithischian footprints
   > Age: Carnian (Olsen et al. 1978)

10. Pennsylvania, United States

    a. Lockatong Formation (Bock 1952; Baird pers. comm.)
    > Theropod footprints
    > Ornithischian footprints
    > Age: late Carnian (Olsen 1980a)

    b. New Oxford Formation (Huene 1921; Galton 1983a)
    > ?Prosauropoda indet. (= *Palaeosaurus fraserianus*)
    > Ornithischia indet. (= *Thecodontosaurus gibbidens*)
    > Age: Carnian (Reeside et al. 1957)

    c. Gettysburg Shale (Baird pers. comm.)
    > Ornithischian footprints
    > Age: Carnian (Reeside et al. 1957)

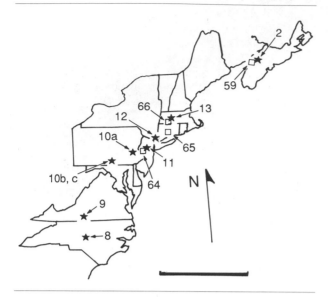

Fig. 3.2. Late Triassic- and Early Jurassic-age locations in eastern Canada and the United States. Scale = 600 km.

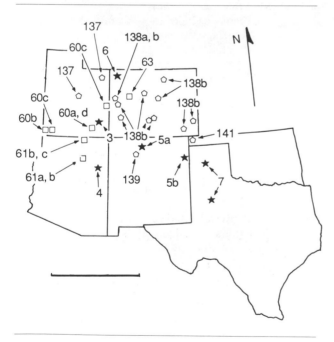

Fig. 3.3. Late Triassic- and Early Jurassic-age locations in the southern United States. Scale = 600 km.

11. New Jersey, United States

   a. Lockatong Formation (Bock 1952)
     Theropod footprints
     Age: late Carnian (Olsen 1980*a*)

   b. Passaic Formation (Baird 1957, pers. comm.)
     Theropod footprints
     Ornithischian footprints
     Age: late Carnian-Rhaetian (Olsen 1980*a*)

12. New York, United States

   Upper Stockton beds (Olsen 1980*a*)
     Theropod footprints
     Age: late Carnian (Olsen 1980*a*)

13. Massachusetts, United States

   Longmeadow Sandstone (Ostrom 1972)
     Theropod footprints
     Age: Rhaetian (Ostrom 1972)

Europe

The history of Late Triassic dinosaurs is very well known from Europe, in particular, in the classic Upper Triassic sequence of the Neckar Valley of Baden-Württemberg, Federal Republic of Germany. These German localities were originally worked by E. Fraas but most successfully by F. von Huene and later by R. Seemann. Additional work on European Upper Triassic sites include that by O. Jaekel at Halberstadt in the Federal Republic of Germany, by B. Peyer at Hallau and H. Rieber and M. Saunders at Frick, both in Switzerland, by C. Duffin and co-workers in Belgium, by J. Pidancet, S. Chopard, P. Ellenberger, and N. Theobald and co-workers in France, by A. D. Walker and M. J. Benton at Elgin in Scotland, and by K. A. Kermack, P. M. Robinson, D. I. Whiteside, N. C. Fraser, and G. M. Walkden on the fissure fillings at Bristol in England. Locations for Upper Triassic dinosaurs are given in figures 3.4 and 3.5.

14. Grampian, Scotland

   Lossiemouth Sandstone Formation (Benton 1984*a*; Benton and Walker 1985)
     Theropoda indet. (= *Saltopus elginensis*)
     Age: Carnian-Norian (Warrington et al. 1980)

15. Mid-Glamorgan, Wales

   Rhaetic Beds (Newton 1899; Galton 1985*e*)
     Theropoda
       ?Carnosauria
         ?*Megalosaurus cambrensis* (= *Zanclodon cambrensis*)
     Age: Rhaetian (Warrington et al. 1980)

16. South Glamorgan, Wales

    a. Unnamed unit of Mercia Mudstone Group (Sollas 1879: Tucker and Burchette 1977; Delair and Sarjeant 1985)
        Theropod footprints
        Age: Norian (Delair and Sarjeant 1985)

    b. Unnamed unit of Fissure Fills (Kermack 1984)
        Theropoda indet.
        Prosauropoda
            Thecodontosauridae
                Thecodontosaurus antiquus
        Age: ?Norian (Fraser et al. 1985)

17. Somerset, England

    Westbury Formation (Seeley 1898; Galton 1985*f*)
        Prosauropoda
            Thecodontosauridae
                Thecodontosaurus sp.
            Melanorosauridae
                Camelotia borealis (= *Avalonia sanfordi*)
        Age: Rhaetian (Warrington et al. 1980)

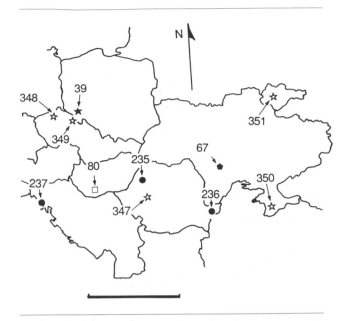

Fig. 3.5. Late Triassic- to Late Cretaceous-age locations in Czechoslovakia, Poland, Hungary, Yugoslavia, Romania, Ukraine, and Kurskaya Oblast, Russia. Scale = 600 km.

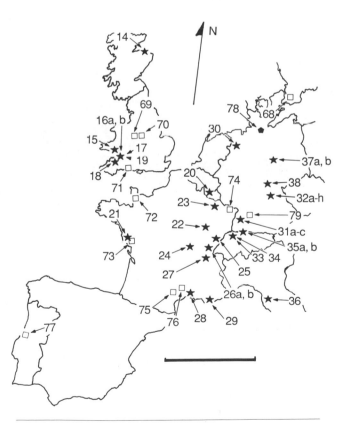

Fig. 3.4. Late Triassic- to Middle Jurassic-age locations in Scotland, England, France, Belgium, Portugal, Federal Republic of Germany, Switzerland, and Italy. Scale = 600 km.

18. Avon, England

    Magnesian Conglomerate (Riley and Stutchbury 1836, 1840; Galton 1984*b*)
        Prosauropoda
            Thecodontosauridae
                Thecodontosaurus antiquus
                Thecodontosaurus sp.
        Age: ?Norian-Rhaetian (Warrington et al. 1980)

19. Gloucestershire, England

    Unnamed unit of Fissure Fills (Fraser and Walkden 1983)
        Theropoda indet.
        Age: late Norian (Fraser and Walkden 1983)

20. Province du Luxembourg, Belgium

    ?HVL, S1, S2, UDK 1 (Duffin et al. 1983)
        Prosauropod indet.
        Age: Rhaetian (Duffin et al. 1983)

21. Département de la Vendée, France

    Niveaux 1–7 (Bessonat et al. 1965; Lapparent and Montenat 1967)
        Theropod footprints
        Age: Rhaetian (Lapparent and Montenat 1967)

22. Département de la Haute-Marne, France

"Grès de l'Infralias" (Sauvage 1907; Huene 1907–08; Galton 1985*e*)
   ?Theropoda indet. (= *Plateosaurus elizae*)
   Age: Rhaetian (Huene 1907–08)

23. Département de la Meurthe-et-Moselle, France

Grès à Avicula contorta (Laugier 1971; Buffetaut 1985; Buffetaut and Wouters 1986)
   Theropoda indet. (= procompsognathid indet.)
   Prosauropoda
      Plateosauridae
         Plateosaurus sp.
   Age: Norian (Buffetaut 1985; Buffetaut and Wouters 1986; Rhaetian, Sigogneau-Russell 1983)

24. Département de la Saône-et-Loire, France

Unnamed unit (Lapparent 1967)
   Prosauropoda
      Anchisaurid indet.
   Age: Rhaetian (Lapparent 1967)

25. Département du Doubs, France

Unnamed unit (Lapparent 1967)
   Prosauropoda
      Plateosauridae
         Plateosaurus engelhardti (= *P. poligniensis*)
   Age: ?Norian (= Keuper; Lapparent 1967)

26. Département du Jura, France

a. Marnes irisées supérieures (Pidancet and Chopard 1862; Theobald et al. 1967)
   Prosauropoda
      Plateosauridae
         Plateosaurus engelhardti
   Age: late Norian (i.e., equivalent to the Knollenmergel; Brenner 1973)

b. Unnamed unit (Henry 1876; Galton 1985*e*)
   Theropoda indet. (= *Megalosaurus obtusus*)
   Age: Rhaetian (Lapparent 1967)

27. Département de l'Ain, France

Unnamed unit (Lapparent and Lavocat 1955)
   Prosauropoda
      Thecodontosauridae
         ?*Thecodontosaurus* sp.
   Age: late Triassic (= Keuper; Lapparent and Lavocat 1955)

28. Département du Gard, France

Unnamed unit (Ellenberger 1965)
   Theropod footprints
   Age: late Triassic (= Keuper; Ellenberger 1965)

29. Département du Var, France

Unnamed unit (Ellenberger 1965)
   Theropod footprints
   Age: Rhaetian (Ellenberger 1965)

30. Niedersachsen, Federal Republic of Germany

Knollenmergel (Huene 1907–08; Weishampel and Westphal 1986)
   Prosauropoda
      Plateosauridae
         Plateosaurus cf. *engelhardti*
   Age: late Norian (Brenner 1973)

31. Baden-Württemberg, Federal Republic of Germany

a. Unterer Stubensandstein (Berckhemer 1938; Galton 1984*b*)
   Prosauropoda
      Plateosauridae
         Sellosaurus gracilis
   Age: middle Norian (Brenner 1973)

b. Mittlerer Stubensandstein (Huene 1907–08, 1932; Galton 1973*b*, 1985*a*, 1985*b*; Galton and Bakker 1985)
   Theropoda *incertae sedis*
      "*Halticosaurus*" *orbitoangulatus*
      Theropoda indet. (= *Halticosaurus longotarsus, Dolichosuchus cristatus, Procompsognathus triassicus, Tanystrophaeus posthumus*)
   Prosauropoda
      Plateosauridae
         Sellosaurus gracilis (including *Efraasia diagnostica*)
         Plateosaurid indet.
   Age: middle Norian (Brenner 1973)

c. Knollenmergel (Huene 1932; Seemann 1933; Weishampel and Westphal 1986; Galton 1985*e*)
   Prosauropoda
      Plateosauridae
         Plateosaurus engelhardti (including *P. erlenbergiensis, P. quenstedti, P. fraasianus, P. plieningeri, P. robustus, Pachysaurus giganteus, P. ajax, P. wetzelianus*)
   Age: late Norian (Brenner 1973)

d. Rhätsandstein (Huene 1907–08; Galton 1985*e*)
   Theropoda indet. (= *Megalosaurus cloacinus, Plateosaurus ornatus*)
   Age: Rhaetian–?early Hettangian (Clemens 1980)

32. Bayern, Federal Republic of Germany

a. Benkersandstein (Rehnelt 1952)
   Theropod footprints
   Age: early Carnian (Laemmlen 1956)

b. Unterer Gipskeuper (Rehnelt 1950)
   Theropod footprints
   Age: early Carnian (Warrington et al. 1980)

c. Ansbachersandstein (Heller 1952; Olsen and
Baird 1986)
  Theropod footprints
  Ornithischian footprints
  Age: late Carnian (Laemmlen 1956)

d. Blasensandstein (Weiss 1934)
  Theropod footprints
  Age: early Norian (Laemmlen 1956)

e. Plattensandstein (Kuhn 1938)
  Theropod footprints
  Age: early Norian (Laemmlen 1956)

f. Oberer Semionotensandstein (Beurlen 1950;
Heller 1952)
  Theropod footprints
  Age: middle Norian (Laemmlen 1956)

g. Feuerletten (Dehm 1935; Ulrichs 1966)
  Prosauropoda
    Plateosauridae
      Plateosaurus engelhardti
  Age: late Norian (Ulrichs 1966)

h. Unnamed unit (Kuhn 1958*a, b*)
  Theropod footprints
  Age: Rhaetian–?Sinemurian (= Rätolias,
  Kuhn 1958*a*)

33. Kanton Baselland, Switzerland

Knollenmergel (Huene 1932)
  Prosauropoda
    Plateosauridae
      Plateosaurus engelhardti (=
      Gresslyosaurus ingens)
  Age: late Norian (Brenner 1973)

34. Kanton Aargau, Switzerland

Obere Bunte Mergel (Rieber 1985*a*, 1985*b*)
  Prosauropoda
    Plateosauridae
      Plateosaurus cf. *engelhardti*
  Age: late Norian (i.e., equivalent to the
  Knollenmergel; Brenner 1973)

35. Kanton Schaffhausen, Switzerland

a. Zanclodonmergel (= Knollenmergel; Peyer 1944)
  Prosauropoda
    Plateosauridae
      Plateosaurus sp.
  Age: late Norian (Brenner 1973)

b. Unnamed unit (Peyer 1944; Tatarinov 1985)
  Theropoda
    ?Megalosaurid indet.
  Prosauropoda
    Plateosauridae
      Plateosaurus sp.
  ?Ornithopoda
    ?Heterodontosaurid indet. (=
    ?*Abrictosaurus* sp.)
  Age: ?late Norian–Hettangian (Clemens 1980)

36. Regione Toscana, Italy

Unnamed unit (Huene 1941; Tongiorgi 1980)
  ?Theropod footprints
  Age: late Carnian (Tongiorgi 1980)

37. Sachsen-Anhalt, Federal Republic of Germany

a. Knollenmergel (Jaekel 1914; Kuhn 1939)
  Theropoda indet. (= *Halticosaurus longotarsus*)
  Prosauropoda
    Plateosauridae
      Plateosaurus engelhardti (including cf.
      Palaeosaurus ?*diagnosticus*)
  Age: late Norian (Brenner 1973)

b. Unnamed unit (Jaekel 1914)
  Theropoda indet. (= *Pterospondylus trielbae*)
  Age: Rhaetian (Huene 1932)

38. Thüringen, Federal Republic of Germany

Knollenmergel (Rühle von Lilienstern et al. 1952;
Welles 1984)
  Theropoda
    Ceratosauria
      Liliensternus liliensterni (=
      Halticosaurus liliensterni)
  Prosauropoda
    Plateosauridae
      Plateosaurus engelhardti (including *P.
      longiceps, P. torgeri*)
  Age: late Norian (Brenner 1973)

39. Gorny Slask, Poland

Lissauer Breccia (Huene 1932; Osmólska pers.
comm.)
  Theropoda indet. (= *Velocipes guerichi*)
  Age: Norian (Wild pers. comm.)

Asia

Upper Triassic sites that have yielded dinosaur remains are restricted to Andhra Pradesh, India, and Yunnan, People's Republic of China. The Indian fauna, typically yielding early theropod and prosauropod remains, was early studied by R. Lydekker and F. von Huene, and later by P. Robinson, S. Chatterjee, T. S. Kutty, S. L. Jain, and T. Roy Chowdhury. C.-C. Young originally studied the Yunnan material, to be followed by Sun A.-L. and co-workers from the Institute of Vertebrate Paleontology and Paleoanthropology. Figures 3.6 and 3.7 provide information on the locations for Upper Triassic dinosaurs of Europe.

40. Madhya Pradesh State, India

Tiki Formation (Chatterjee and Majumdar 1987)
  Theropoda indet.
  Age: Carnian (Chatterjee 1978*a*)

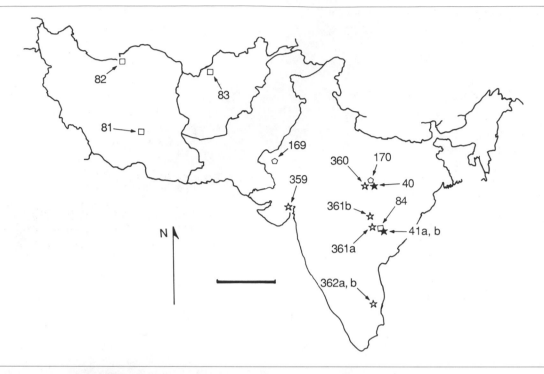

Fig. 3.6. Late Triassic- to Late Cretaceous-age locations in Iran, Afganistan, and India. Scale = 600 km.

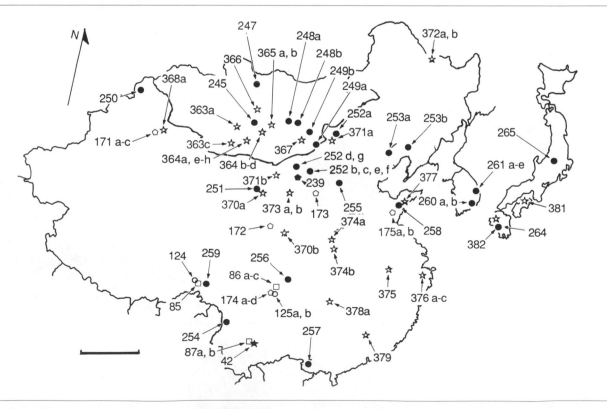

Fig. 3.7. Late Triassic- to Late Cretaceous-age locations in the Mongolian People's Republic, People's Republic of China, and Japan. Scale = 600 km.

41. Andhra Pradesh State, India

    a. Maleri Formation (Jain 1983; Chatterjee 1987)
        Theropoda *incertae sedis*
           Walkeria maleriensis
        Prosauropoda
           Prosauropoda indet. (= *Massospondylus hislopi*)
           Anchisaurid indet.
        Age: Carnian–middle Norian (Kutty 1969)

    b. Dharmaram Formation (Kutty 1969; Kutty and Roy-Chowdhury 1970)
        Prosauropod indet. (?2 species)
        Age: late Norian–Rhaetian (Kutty 1969)

42. Yunnan, People's Republic of China

    Dull Purplish Beds of the Lower Lufeng Series (Fengjiahe Formation; Young 1941*a,* 1941*b,* 1948*a,* 1948*b,* 1951; Sun et al. 1985)
        Theropoda *incertae sedis*
           Lukousaurus yini
        Theropoda indet. (= *Sinosaurus triassicus*)

        Prosauropoda
           Plateosauridae
               Lufengosaurus huenei (including *Gyposaurus sinensis, L. magnus*)
           Yunnanosauridae
               Yunnanosaurus huangi (including *Y. robustus*)
        Age: ?Rhaetian–Pliensbachian (Zhen et al. 1985; Dong pers. comm.)

South America

The earliest known dinosaur material is from South America. Collection and work on these early forms have been carried out by a number of paleontologists, among them, F. von Huene, O. A. Reig, G. Leonardi, J. F. Bonaparte, and R. M. Casamiquela. Late Triassic dinosaur localities of South America are given in figures 3.8 and 3.9.

Fig. 3.8. Late Triassic- to Late Cretaceous-age locations in Colombia, Peru, Bolivia, and Brazil. Scale = 600 km.

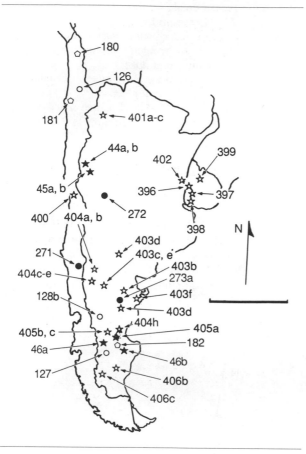

Fig. 3.9. Late Triassic- to Late Cretaceous-age locations in Uruguay, Chile, and Argentina. Scale = 600 km.

43. Estado do Rio Grande do Sul, Brazil

Santa Maria Formation (Huene 1942; Colbert 1970; Bonaparte 1973; Galton 1977a)
> Dinosauria
>> *Staurikosaurus pricei*
> Dinosauria *incertae sedis*
>> *Spondylosoma absconditum*
> Age: ?late Ladinian or ?early Carnian (Galton 1977a)

44. Provincia de La Rioja, Argentina

a. Ischigualasto Formation (Casamiquela 1967b; Bonaparte 1973, 1976)
> Ornithischia *sedis mutabilis*
>> *Pisanosaurus mertii*
> Age: Carnian (Bonaparte 1973; Stipanicic 1975)

b. Los Colorados Formation (Bonaparte 1973, 1978b; Galton 1985f)
> Theropoda indet.
> Prosauropoda
>> Melanorosauridae
>>> *Riojasaurus incertus* (including *Strenusaurus procerus*)
>> Plateosauridae
>>> *Coloradisaurus brevis*
> Age: Norian (Bonaparte 1973; Stipanicic 1975)

45. Provincia de San Juan, Argentina

a. Ischigualasto Formation (Reig 1963; Bonaparte 1973; Novas 1987; Brinkman and Sues 1987)
> Dinosauria
>> cf. *Staurikosaurus* sp.
> Herrerasauridae
>> *Herrerasaurus ischigualastensis*
>> *Ischisaurus cattoi*
> ?Theropoda *incertae sedis*
>> *Frenguellisaurus ischigualastensis*
> Age: Carnian (Bonaparte 1973; Stipanicic 1975)

b. Quebrada del Barro Formation (Bossi and Bonaparte 1978)
> Prosauropoda
>> Melanorosauridae
>>> *Riojasaurus incertus*
> Age: Norian (equivalent to Los Colorados Formation; Bossi and Bonaparte 1978)

46. Provincia de Santa Cruz, Argentina

a. "Complejo Porfirico" (Leonardi 1989)
> Theropod footprints
> Age: Late Triassic (Leonardi 1989)

b. El Tranquilo Formation (Casamiquela 1980; Bonaparte 1973; Bonaparte and Vince 1979)
> Prosauropoda
>> Plateosauridae
>>> ?*Plateosaurus* sp.
>>> *Mussaurus patagonicus*
> Dinosaur eggshell
> Age: ?Norian (Bonaparte pers. comm.)

Africa

The majority of knowledge concerning the Late Triassic and Early Jurassic comes from the famous Stormberg Series, Karroo Basin, of South Africa and Lesotho, primarily through the early efforts of S. H. Haughton and E. C. N. van Hoepen and the recent work of F. and P. Ellenberger, K. A. Kermack, J. W. Kitching, C. Gow, J. Attridge, and A. W. Crompton. Additional localities from the Upper Triassic of Africa

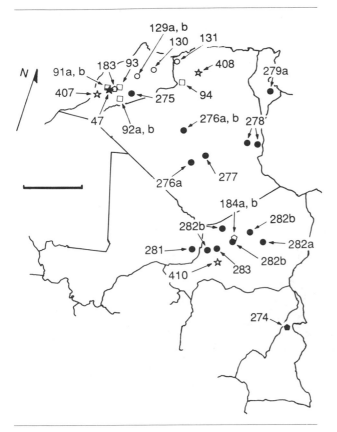

Fig. 3.10. Late Triassic- to Late Cretaceous-age locations in Morocco, Algeria, Tunisia, Niger, Mali, and Cameroun. Scale = 600 km.

are known from Morocco through work by J.-M. Dutuit and from Zimbabwe via studies by M. R. Cooper. These locations are given in figures 3.10 and 3.11.

47. Province de Marrakech, Morocco

    Argana Formation (Dutuit 1972, 1974; Biron and Dutuit 1981)
        Theropoda indet.
        Theropod footprints
        Prosauropoda
            Thecodontosauridae
                Azendohsaurus laaroussii
        Ornithischia indet. (= *Azendohsaurus laaroussii* partim; Dutuit pers. comm., in Galton 1984*b*)
        Ornithischian footprints
        Age: middle Carnian (Cousminer and Manspeizer 1976)

48. Matabeleland South, Zimbabwe

    Mpandi Formation (Cooper 1980)
        Prosauropoda
            Plateosauridae
                Euskelosaurus cf. *browni*
        Age: Norian (Cooper 1980)

49. Transvaal, South Africa

    Springbok Flats Member of Bushveld Sandstone (Haughton 1924)
        Prosauropoda
            Prosauropoda indet. (= *Gryponyx transvaalensis, Gigantoscelus molengraaffi*)
            Plateosauridae
                Euskelosaurus browni
        Age: ?late Carnian or ?early Norian (i.e., ?= Lower Elliot Formation, Kitching pers. comm.)

50. Orange Free State, South Africa

    Lower Elliot Formation (Galton and Cluver 1976; Kitching and Raath 1984; Olsen and Galton 1984)
        Prosauropoda
            Prosauropoda indet. (= *Eucnemesaurus fortis*)
            Plateosauridae
                cf. *Euskelosaurus browni*
        Age: late Carnian or early Norian (Olsen and Galton 1984)

51. Cape Province, South Africa

    Lower Elliot Formation (Haughton 1924; Kitching and Raath 1984; Olsen and Galton 1984; Galton and Heerden 1985; Galton 1985*f*)
        Dinosauria *incertae sedis*
            Aliwalia rex
        Prosauropoda
            Prosauropoda indet. (= *Orinosaurus capensis*)
            ?Prosauropod footprints
            Massospondylidae
                Massospondylus sp.
            Melanorosauridae
                Melanorosaurus readi
                Melanorosaurus sp.
            Plateosauridae
                Euskelosaurus browni (including *Plateosaurus stormbergensis, P. cullingworthi, E. africanus*)
                Euskelosaurus sp.
        ?Ornithischia indet.
        Age: late Carnian or early Norian (Olsen and Galton 1984)

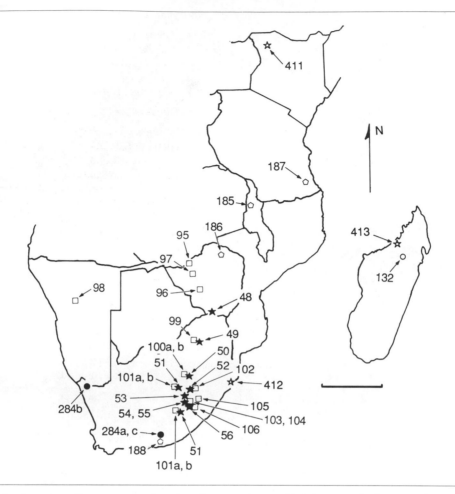

Fig. 3.11. Late Triassic- to Late Cretaceous-age locations in Kenya, Tanzania, Zimbabwe, Namibia, Lesotho, South Africa, and Madagascar. Scale = 600 km.

52. Leribe District, Lesotho

   Lower Elliot Formation (Huene 1932; Kitching and Raath 1984; Galton and Heerden 1985)
   Prosauropoda
   Melanorosauridae
   Melanorosaurus sp.
   Plateosauridae
   Euskelosaurus browni (= *Basutodon ferox*)
   Blikanasauridae
   Blikanasaurus cromptoni
   Age: late Carnian or early Norian (Olsen and Galton 1984)

53. Maseru District, Lesotho

   Lower Elliot Formation (Ellenberger 1970; Kitching and Raath 1984; Olsen and Galton 1984)
   Theropod footprints
   Prosauropoda
   Prosauropod footprints
   Melanorosauridae
   cf. *Melanorosaurus* sp.
   Age: late Carnian or early Norian (Olsen and Galton 1984)

54. Mafeteng District, Lesotho

   Lower Elliot Formation (Ellenberger 1970; Kitching and Raath 1984)
   Prosauropoda
   Plateosauridae
   Euskelosaurus sp.
   Age: late Carnian or early Norian (Olsen and Galton 1984)

55. Mohales Hoek District, Lesotho

   Lower Elliot Formation (Ellenberger 1970, 1972; Kitching and Raath 1984; Olsen and Galton 1984)
   Theropod footprints
   Prosauropoda
   Prosauropod footprints
   Plateosauridae
   Euskelosaurus sp.
   Melanorosauridae
   cf. *Melanorosaurus* sp.
   Age: late Carnian or early Norian (Olsen and Galton 1984)

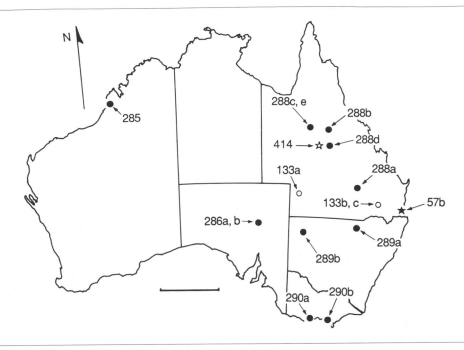

Fig. 3.12. Late Triassic- to Late Cretaceous-age locations in Australia. Scale = 600 km.

Australasia

Only two locations in Queensland, Australia, have yielded supposed Upper Triassic dinosaur material; both seem to be isolated occurrences (fig. 3.12).

57. Queensland, Australia

   a. Unnamed unit (Seeley 1891; actual locality and horizon unknown; not figured)
       Prosauropoda indet. (= *Agrosaurus macgillivrayi*)
       Age: ?Late Triassic (Molnar 1982)
   b. Blackstone Formation (Molnar 1982)
       Theropod footprints
       Age: Late Triassic (Molnar 1982)

EARLY JURASSIC

A slightly poorer record of dinosaurs is available from deposits of Early Jurassic age than from those of the Late Triassic. Dinosaurs from this time are known from North America, Europe, Asia, South America, and Africa.

North America

Like the Late Triassic, North American Early Jurassic dinosaurs are known primarily from the maritimes of Canada (through work by D. Baird, P. E. Olsen, N. Shubin, and H.-D. Sues), from the southwest of the United States (through studies by E. H. Colbert, S. P. Welles, F. A. Jenkins, A. W. Crompton, T. Rowe, and K. Padian), and again from the eastern seaboard of the United States (primarily by P. E. Olsen). (An important fauna from Tamaulipas State, Mexico [Clark and Hopson 1985] is presently under study by J. Hopson, J. Clark, and D. Fastovsky [Locality 58]. It is either Early or Middle Jurassic in age (Hopson pers. comm.) and appears to have yielded heterodontosaurid material among other terrestrial vertebrates.) North American localities for the Lower Jurassic are given in figures 3.1, 3.2, 3.3, and 3.13.

59. Nova Scotia, Canada

   McCoy Brook Formation (= Scots Bay Formation; Olsen et al. 1982; Sues et al. 1987; Olsen et al. 1987; Baird pers. comm.)
       Theropoda indet.
       Theropod footprints
       Prosauropoda
           Anchisauridae
               cf. *Anchisaurus* sp.
           Plateosauridae
               cf. *Ammosaurus* sp.
       Ornithischia indet. (?*Scutellosaurus* sp.)
       Ornithischian footprints
       Age: Hettangian (Olsen et al. 1982)

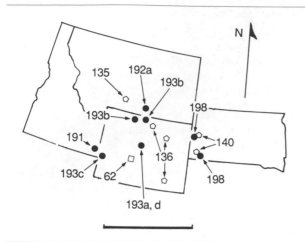

Fig. 3.13. Early Jurassic- to Early Cretaceous-age locations in the northwest-central United States. Scale = 600 km.

60. Utah, United States

 a. Wingate Sandstone (Brady 1960)
 ?Theropod footprints
 Age: Hettangian-Sinemurian (Olsen et al. 1982)

 b. Moenave Formation (Miller et al. 1989)
 Theropod footprints
 Prosauropod footprints
 Age: Hettangian (Miller et al. 1989)

 c. Kayenta Formation (Stokes and Bruhn 1960; Lockley 1986a)
 ?Theropod footprints
 Age: Pliensbachian (Olsen et al. 1982)

 d. ?Navajo Sandstone (Stokes 1978)
 Theropod footprints
 Age: Toarcian (Olsen et al. 1982)

61. Arizona, United States

 a. Moenave Formation (Morales 1986, pers. comm.)
 Theropod footprints
 Ornithischian footprints
 Age: Hettangian (Miller et al. 1989)

 b. Kayenta Formation (Colbert 1981; Welles 1984; Attridge et al. 1985; Morales 1986; Rowe 1989)
 Theropoda
 Theropod footprints
 Ceratosauria
 Dilophosaurus wetherilli
 Syntarsus kayentakatae
 Prosauropoda
 Massospondylidae
 Massospondylus sp.
 Ornithischia indet.
 Ornithischian footprints

 Ornithopoda
 Undescribed heterodontosaurid (Attridge et al. 1985)
 Thyreophora
 Scutellosaurus lawleri
 Age: Hettangian (Imlay 1952; ?Sinemurian, Rowe and Gauthier this vol.)

 c. Navajo Sandstone (Camp 1936; Baird 1980; Galton 1971a)
 Theropoda
 Theropod footprints
 Ceratosauria
 Segisaurus halli
 Prosauropoda
 Prosauropod footprints
 Plateosauridae
 Ammosaurus cf. *major*
 Plateosaurid indet.
 Age: Toarcian (Olsen et al. 1982; ?Sinemurian-Pliensbachian, Rowe and Gauthier, this vol.)

62. Wyoming, United States

 Nugget Sandstone (Stokes 1978)
 Theropod footprints
 Age: Toarcian (Olsen et al. 1982)

63. Colorado, United States

 Navajo Sandstone (Faul and Roberts 1951)
 Theropod footprints
 Age: Toarcian (Olsen et al. 1982)

64. New Jersey, United States

 a. Feltville Formation (Olsen 1980b, 1980c)
 Theropod footprints
 Ornithischian footprints
 Age: early Hettangian (Olsen 1980a)

 b. Towaco Formation (Olsen 1980b, 1980c)
 Theropod footprints
 Ornithischian footprints
 Age: Hettangian (Olsen 1980a)

 c. Boonton Formation (Olsen 1980b, 1980c)
 Theropod footprints
 Ornithischian footprints
 Age: Hettangian-Sinemurian (Olsen 1980a)

65. Connecticut, United States

 a. Shuttle Meadow Formation (Olsen 1980b, 1980c)
 Theropod footprints
 Ornithischian footprints
 Age: Hettangian (Olsen 1980a)

 b. East Berlin Formation (Olsen 1980b, 1980c)
 Theropod footprints
 Ornithischian footprints
 Age: Sinemurian (Olsen 1980a)

c. Upper Portland Formation (Olsen 1980*b*, 1980*c*)
    Theropoda indet. (= *Podokesaurus* sp.)
    Theropod footprints
    Prosauropoda
        Prosauropod indet.
        Anchisauridae
            Anchisaurus polyzelus
        Plateosauridae
            Ammosaurus major
    Ornithischian footprints
    Age: Pliensbachian or Toarcian (Olsen 1980*a*)

66. Massachusetts, United States

a. Portland Formation (Olsen 1980*a*)
    Theropod footprints
    Prosauropoda
        Anchisauridae
            Anchisaurus polyzelus
    Ornithischian footprints
    Age: Pliensbachian-Toarcian (Olsen 1980*a*)

b. Unnamed unit (= ?Portland Formation; Galton 1980*c*)
    Theropoda indet. (*Podokesaurus holyokensis*)
    Age: ?Pliensbachian-Toarcian (Olsen 1980*a*)

Europe

Much of what we know about the Early Jurassic of Europe comes from the early work by R. Owen and A. Smith Woodward in England, E. Bölau and C. Pleijel in Sweden, A. F. de Lapparent, C. Montenat, and P. Ellenberger in France, R. Wild in the Federal Republic of Germany, and most recently L. Kordos in Hungary. (The indeterminate theropod *Macrodontophion* [Zborzewski 1834] is now known to be from the Jurassic of Nikolayevskaya Oblast, Ukraine [Locality 67; Kurzanov pers. comm.], but a better age assignment is not possible at this time. In addition, "*Laelaps*" *gallicus* [a possible carnosaur; Cope 1867] comes from the Normandy region of France. Its age is at present unknown [possibly Jurassic] and consequently its location has not been plotted.) Lower Jurassic dinosaur localities of Europe are given in figures 3.4 and 3.5.

68. Malmohus Lan, Sweden

Unnamed unit (Bölau 1952, 1954; Pleijel 1975)
    ?Theropoda indet.
    Theropod footprints
    Age: Early Jurassic (Bölau 1954)

69. Warwickshire, England

Lower Lias (Woodward 1908*b*; Huene 1932)
    Theropoda indet. (= *Sarcosaurus andrewsi*)
    Age: Sinemurian (Cope et al. 1980*b*)

70. Leicestershire, England

Lower Lias (Andrews 1921; Huene 1932)
    Theropoda
        Ceratosauria
            Sarcosaurus woodi
    Age: Sinemurian (Cope et al. 1980*b*)

71. Dorset, England

Lower Lias (Owen 1861*a*, 1863; Lydekker 1888*b*; Rixon 1968)
    Theropoda
        ?Carnosauria
            Magnosaurus lydekkeri
    Thyreophora
        Scelidosaurus harrisonii
        Undescribed thyreophoran
    Age: Sinemurian (Cope et al. 1980*b*)

72. Département de la Manche, France

Couches d'Airel (Larsonneur and Lapparent 1967)
    Theropoda indet. (= *Halticosaurus* sp.)
    Age: Hettangian-Sinemurian (Medus 1983)

73. Département de la Vendée, France

Niveaux 8–13 (Bessonat et al. 1965; Lapparent and Montenat 1967)
    Theropod footprints
    Age: Hettangian (Lapparent and Montenat 1967)

74. Département de la Moselle, France

Unnamed unit (Huene 1926*a*)
    Theropoda indet. (= *Megalosaurus terquemi*)
    Age: Hettangian (Lapparent 1967)

75. Département de l'Aveyron, France

Unnamed unit (Ellenberger and Fuchs 1965)
    Theropod footprints
    Age: ?Pliensbachian (= Lotharingian; Ellenberger and Fuchs 1965)

76. Département de la Lozère, France

Unnamed unit (Thaler 1962; Ellenberger 1965)
    Theropod footprints
    Age: Hettangian (Ellenberger 1965)

77. Província do Beira Litoral, Portugal

Unnamed unit (Lapparent and Zbyszewski 1957)
    ?Thyreophora indet. (= *Lusitanosaurus liasicus*)
    Age: ?Sinemurian (Lapparent and Zbyszewski 1957)

78. Schleswig-Holstein, Federal Republic of Germany

Unnamed unit (Huene 1966)
    Theropoda indet.
    Age: Toarcian (= Lias epsilon; Geyer and Gwinner 1962)

79. Baden-Württemberg, Federal Republic of Germany

> Posidonienschiefer (Wild 1978a)
>> Sauropoda
>>> Vulcanodontidae
>>>> *Ohmdenosaurus liasicus*
>>> Age: middle Toarcian (Geyer and Gwinner 1962)

80. Megye Baranya, Hungary

> Mecsek Coal Formation (Kordos 1983)
>> ?Ornithischian footprints
>> Age: Hettangian (Kordos 1983)

Asia

Fewer than ten locations are known to have yielded Lower Jurassic dinosaur material in Asia. These include sites in Iran (worked principally by A. F. de Lapparent), India (primarily through studies by S. L. Jain and coworkers from the Indian Statistical Institute), and the People's Republic of China (again by C.-C. Young, Dong Z.-M., and co-workers at the Institute of Vertebrate Paleontology and Paleoanthropology). The last has produced the most extensive information on Early Jurassic dinosaurs from this continent. Asian dinosaur localities from the Lower Jurassic are given in figures 3.6 and 3.7.

81. Ostan Kerman, Iran

> Shemshak Formation (Lapparent and Davoudzadeh 1972)
>> Theropod footprints
>> Age: ?Hettangian-Pliensbachian (= Lias; Lapparent and Davoudzadeh 1972)

82. Ostan Mazandaran, Iran

> Unnamed unit (Lapparent and Sadat 1975)
>> Theropod footprints
>> Age: ?Hettangian-Pliensbachian (= Lias; Lapparent and Sadat 1975)

83. Faryab Velayat, Afghanistan

> Unnamed unit (Lapparent and Stocklin 1972)
>> ?Theropod footprints
>> ?Sauropod footprints
>> ?Ornithopod footprints
>> Age: ?Hettangian-Pliensbachian (= ?Lias; Lapparent and Stocklin 1972)

84. Andhra Pradesh State, India

> Kota Formation (Jain 1980)
>> Theropoda indet.
>> Sauropoda
>>> Vulcanodontidae
>>>> *Barapasaurus tagorei*
>>>> *Kotasaurus yamanpalliensis*
>> Age: Early Jurassic (Jain 1980)

85. Xizang Zizhiqu, People's Republic of China

> Daye Group (Wang et al. 1985)
>> Theropoda
>>> Undescribed ?megalosaurid
>> Prosauropoda
>>> Undescribed anchisaurid
>> ?Thyreophora
>>> Undescribed ?scelidosaur
>> Age: Early Jurassic (Wang et al. 1985)

86. Sichuan, People's Republic of China

> a. Zhenzhunchong Formation (Chen et al. 1982; Dong et al. 1983; Dong 1984a)
>> Prosauropoda
>>> Plateosauridae
>>>> cf. *Lufengosaurus huenei* (= cf. *L. magnus*)
>>> Anchisaurid indet.
>> Age: Toarcian-Bajocian (Wang and Sun 1983; Early Jurassic, Dong et al. 1983)

> b. Da'znzhai Formation (Dong et al. 1983)
>> Sauropoda
>>> ?Vulcanodontidae
>>>> *Zizhongosaurus chuanchengensis*
>> Age: Early Jurassic (Dong et al. 1983)

> c. Ziliujing Group (Wang et al. 1985)
>> Sauropoda
>>> Undescribed brachiosaurid
>> Age: Early Jurassic (Wang et al. 1985)

87. Yunnan, People's Republic of China

> a. Dark Red Beds of the Lower Lufeng Series (= Fengjiahe Formation; Young 1941a, 1941b, 1948a, 1951, 1982b; Simmons 1965; Galton 1985f; Chen et al. 1982; Sun et al. 1985)
>> Theropoda *incertae sedis*
>>> *Lukousaurus yini*
>> Theropoda indet. (= *Sinosaurus triassicus*)
>> Prosauropda
>>> Plateosauridae
>>>> *Lufengosaurus huenei* (including *Gyposaurus sinenesis*, *L. magnus*, *Tawasaurus minor*, *Fulengia youngi*)
>>> Yunnanosauridae
>>>> *Yunnanosaurus huangi* (including *Y. robustus*)
>> Ornithopoda
>>> Heterodontosaurid indet. (= *Dianchungosaurus lufengensis*)
>> ?Thyreophora
>>> *Tatisaurus oehleri*
>> Age: Early Jurassic (Olsen et al. 1982; Hettangian-Pliensbachian, Zhen et al. 1985)

> b. Fengjiahe Formation (Zhen et al. 1986)
>> Theropod footprints
>> Age: Early Jurassic (Zhen et al. 1986)

South America

Three locations in South America have produced dinosaur ichnofaunas, all from the Botucatú Formation of Brazil (worked by G. Leonardi; fig. 3.8).

88. Estado do São Paulo, Brazil

   Botucatú Formation (Leonardi 1977, 1980a, 1981b, 1989; Leonardi and Godoy 1980)
       Theropod footprints
       ?Ornithopod footprints
       Age: Hettangian-Toarcian (= Liassic; Leonardi 1989)

89. Estado do Paraná, Brazil

   Botucatú Formation (Leonardi 1980a)
       Theropod footprints
       Age: Hettangian-Toarcian (= Liassic; Leonardi 1989)

90. Estado do Rio Grande do Sul, Brazil

   Botucatú Formation (Leonardi 1989)
       Theropod footprints
       Age: Hettangian-Toarcian (= Liassic; Leonardi 1989)

Africa

Like the Upper Triassic of Africa, the Lower Jurassic is best known from the Karroo Basin of southern Africa, primarily through the efforts of S. H. Haughton, E. C. N. van Hoepen, K. A. Kermack and co-workers, F. and P. Ellenberger, J. Attridge, C. Gow, J. W. Kitching, A. W. Crompton, and A. J. Charig. Additional information on the Lower Jurassic of Africa comes from work by J. Jenny and co-workers in Morocco and M. A. Raath, M. R. Cooper, and G. Bond in Zimbabwe. Figures 3.10 and 3.11 provide information on these localities.

91. Province d'Ouarzazate, Morocco

   a. Imi-n-Ifri Formation (Monbaron et al. 1985)
       Ornithopod footprints
       Age: Sinemurian (Monbaron et al. 1985)

   b. Aganane Formation (Jenny and Jossen 1982)
       Theropod footprints
       ?Sauropod footprints
       ?Stegosaur footprints
       Age: Pliensbachian (Jenny and Jossen 1982)

92. Province de Marrakech, Morocco

   a. Aganane Formation (Jenny and Jossen 1982)
       Theropod footprints
       ?Sauropod footprints
       ?Stegosaur footprints
       Age: Pliensbachian (Jenny and Jossen 1982)

   b. Unnamed unit (Jenny et al. 1980; Taquet 1985)
       Theropoda indet.
       Sauropoda indet.
       Age: Toarcian (Jenny et al. 1980)

93. Province de Beni Mellal, Morocco

   Aganane Formation (Jenny and Jossen 1982)
       ?Sauropod footprints
       ?Stegosaur footprints
       Age: Pliensbachian (Jenny and Jossen 1982)

94. Wilaya Saida, Algeria

   Unnamed unit (Bassoulet 1971)
       Theropod footprints
       Age: Early Jurassic (Taquet 1977)

95. Mashonaland North, Zimbabwe

   Vulcanodon Beds (Raath 1972; Cooper 1984)
       Sauropoda
         Vulcanodontidae
           Vulcanodon karibaensis
       Age: ?Hettangian (Raath 1972)

96. Matabeleland North, Zimbabwe

   Forest Sandstone (Raath 1969)
       Theropoda
         Ceratosauria
           Syntarsus rhodesiensis
       Prosauropoda
         Massospondylidac
           Massospondylus carinatus
         Plateosaurid indet.
       Age: Hettangian-Sinemurian (Olsen and Galton 1984)

97. Midlands, Zimbabwe

   Gokwe Formation (Bond and Bromley 1970)
       Theropoda indet.
       Sauropoda indet.
       Age: Early Jurassic (Molnar 1980b)

98. Otjiqarongo, Namibia

   Etjo Sandstone (Huene 1925; Gürich 1926)
       Theropod footprints
       Age: ?Sinemurian (i.e., = Clarens Formation, Kitching pers. comm.)

99. Transvaal, South Africa

   Zoutpansberg Member of Bushveld Sandstone (Haughton 1924)
       Prosauropoda
         Massospondylidae
           Massospondylus carinatus
       Age: ?Sinemurian (i.e., ?= Lower Clarens Formation, Kitching pers. comm.)

100. Orange Free State, South Africa

    a. Upper Elliot Formation (Haughton 1924;
       Kitching 1979, 1981)
          Prosauropoda
              Prosauropoda indet. (= *Leptospondylus*
                  capensis, Pachyspondylus orpenii)
              Massospondylidae
                  Massospondylus carinatus (including
                    M. browni, Gryponyx taylori, G.
                    africanus, Thecodontosaurus
                    skirtopodus, Aetonyx palustris)
          Ornithopoda
              Heterodontosauridae
                  Lycorhinus augustidens (= *Lanasaurus*
                    scalpridens)
          Dinosaur eggshell
          Age: Hettangian-?Sinemurian (Olsen and
             Galton 1984)

    b. Clarens Formation (Haughton 1924)
          Prosauropoda
              Prosauropoda indet. (= *Thecodontosaurus*
                  minor)
           Massospondylidae
              Massospondylus carinatus (including
                Aristosaurus erectus,
                Thecodontosaurus dubius, Gyposaurus
                capensis)
          Age: Sinemurian (Olsen and Galton 1984)

101. Cape Province, South Africa

    a. Upper Elliot Formation (Haughton 1924; Charig
       et al. 1965; Thulborn 1974; Hopson 1975*a*; Gow
       1975; Santa Luca et al. 1976; Raath 1980;
       Kitching and Raath 1984)
          Theropoda
              Ceratosauria
                Syntarsus rhodesiensis
          Prosauropoda
              Massospondylidae
                Massospondylus carinatus (including
                M. browni, M. harriesi,
                Thecodontosaurus browni, T.
                skirtopodus, T. minor)
              Massospondylus sp.
          ?Ornithischia indet.
          Ornithopoda
              Heterodontosauridae
                Heterodontosaurus tucki
                Lycorhinus angustidens
                Abrictosaurus consors
          Age: Hettangian-?Sinemurian (Olsen and
             Galton 1984)

    b. Clarens Formation (Haughton 1924; Crompton
       and Charig 1962; Kitching and Raath 1984)
          Prosauropoda
              Prosauropoda indet. (= *Hortalotarsus*
                skirtopodus)
           Massospondylidae
              Massospondylus carinatus

         Ornithopoda
             Heterodontosauridae
                Heterodontosaurus tucki
              Heterodontosaurid indet. (=
                Geranosaurus atavus)
         Age: Sinemurian (Olsen and Galton 1984)

102. Leribe District, Lesotho

    Upper Elliot Formation (Haughton 1924; Ellenberger
    1970; Kitching and Raath 1984; Olsen and Galton
    1984)
          Theropod footprints
          Prosauropoda
              Prosauropod footprints
              Massospondylidae
                Massospondylus carinatus (= *Gryponyx*
                sp., *Aetonyx* sp.)
          Ornithischian footprints
          Age: Hettangian-?Sinemurian (Olsen and
            Galton 1984)

103. Mafeteng District, Lesotho

    Upper Elliot Formation (Ginsburg 1964; Galton
    1978; Kitching and Raath 1984; Olsen and Galton
    1984; Santa Luca 1984)
          Theropod footprints
          Prosauropoda
              Prosauropod footprints
              Massospondylidae
                cf. *Massospondylus* sp.
          Ornithischia
              Lesothosaurus diagnosticus
              Undescribed fabrosaur (Santa Luca 1984)
          Ornithischia indet. (= *Fabrosaurus australis*)
          Ornithischian footprints
          Age: Hettangian-?Sinemurian (Olsen and
            Galton 1984)

104. Mohales Hoek District, Lesotho

    Upper Elliot Formation (Ellenberger 1970; Kitching
    and Raath 1984; Olsen and Galton 1984)
          Theropod footprints
          Prosauropoda
              Prosauropod footprints
              Massospondylidae
                cf. *Massospondylus* sp. (= cf. *Gryponyx* sp.)
          ?Ornithischia indet.
          Ornithischian footprints
          Age: Hettangian-?Sinemurian (Olsen and
            Galton 1984)

105. Qachas Nek District, Lesotho

    Upper Elliot Formation (Thulborn 1970*a*, 1974;
    Hopson 1975*a*; Kitching and Raath 1984)
          Prosauropoda
              Massospondylidae
                cf. *Massospondylus* sp.

Ornithopoda
  Heterodontosauridae
    Abrictosaurus consors
Age: Hettangian-?Sinemurian (Olsen and
  Galton 1984)

106. Quthing District, Lesotho

Upper Elliot Formation (Ellenberger 1970; Kitching
and Raath 1984; Olsen and Galton 1984)
  Theropod footprints
  Prosauropoda
    Prosauropod footprints
    Massospondylidae
      Massospondylus carinatus (including
        *Thecodontosaurus browni, Gryponyx
        taylori*)
  Ornithopoda
    Heterodontosauridae
      Heterodontosaurus sp.
Age: Hettangian-?Sinemurian (Olsen and
  Galton 1984)

MIDDLE JURASSIC

The fossil record of dinosaurs during the Middle Jurassic is generally poor. Scattered localities are now
known in all continents with the exception of Antarctica, but by far the best record comes from England
and France.

North America

The record of Middle Jurassic dinosaurs in North
America is restricted to an ichnofauna from near Ar-
teaga, Mexico, described by I. Ferrusquia-Villafranca
and co-workers (fig. 3.14).

107. Estado de Michoacán, Mexico

Unnamed unit (Ferrusquia-Villafranca et al. 1980)
  Theropod footprints
  Ornithopod footprints
  Age: Middle Jurassic (Ferrusquia-Villafranca
    et al. 1980)

Europe

Middle Jurassic dinosaurs are best known from
Europe (fig. 3.15). English localities are known principally through the work by R. Owen, W. Buckland,
A. S. Woodward, H. G. Seeley, A. N. Leeds, J. B. Delair,
and W. A. S. Sarjeant, French localities by way of the
work of R. Hoffstetter, and J. Pivetaut, and Portuguese
localities through work by W. G. Kühne and R. A.
Thulborn. (Theropod and ornithopod footprints have

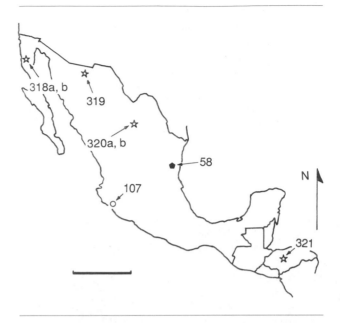

Fig. 3.14. Middle Jurassic- to Late Cretaceous-age locations in Mexico and Honduras. Scale = 600 km.

been described from the Provincia de Oviedo, Spain by
Garcia-Ramos and Valenzuela [1977a], but the locality
can only be dated to Middle-Late Jurassic and hence
appears here [Locality 108].)

109. Highland, Scotland

Lealt Shale Formation (Andrews and Hudson 1984;
Delair and Sarjeant 1985)
  Ornithopod footprints
  Age: Bathonian (Harris and Hudson 1980)

110. North Yorkshire, England

a. Lower Deltaic Series (Sarjeant 1970)
  Theropod footprints
  Age: Upper Aalenian (Sarjeant 1970)
b. Inferior Oolite (Brodrick 1909)
  Theropod footprints
  Age: Aalenian-Bajocian (Anderton et al. 1979)
c. Upper Estuarine Series (Sarjeant 1970)
  Theropod footprints
  Age: early-middle Bathonian (Cope et al.
    1980a)

111. West Yorkshire, England

Inferior Oolite (Owen 1840–45)
  Sauropoda
    Cetiosauridae
      Cetiosaurus medius (= *C. hypoolithicus*)
    Diplodocidae
      Cetiosauriscus longus
  Age: Bajocian (Cope et al. 1980a)

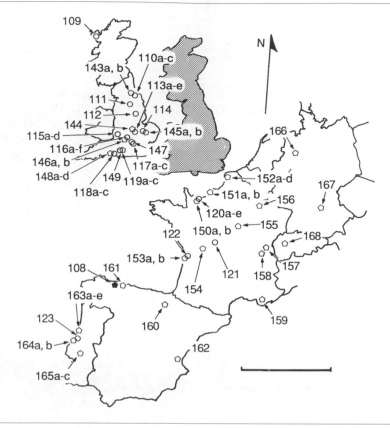

Fig. 3.15. Middle and Late Jurassic-age locations in Scotland, England (removed slightly to the west), France, Spain, Portugal, Federal Republic of Germany, and Switzerland. Scale = 600 km.

112. Nottinghamshire, England

Great Oolite (Owen 1842*a, b*)
Sauropoda
Cetiosauridae
Cetiosaurus oxoniensis
Age: Bathonian (Anderton et al. 1979)

113. Northamptonshire, England

a. Northampton Sands Formation (Reid 1984*c*)
Sauropoda
?Brachiosaurid indet.
Age: early Aalenian (Cope et al. 1980*a*)

b. Inferior Oolite (Benton pers. comm.)
Theropoda
?Carnosauria
?*Megalosaurus bucklandii*
Age: Aalenian-Bajocian (Cope et al. 1980*a*)

c. Forest Marble (Phillips 1871)
Sauropoda
Diplodocidae
Cetiosauriscus glymptonensis
Age: late Bathonian (Cope et al. 1980*a*)

d. Great Oolite (Benton pers. comm.)
Sauropoda
Cetiosauridae
Cetiosaurus oxoniensis
Age: Bathonian (Anderton et al. 1979)

e. Lower Oxford Clay (Hulke 1887; Nopcsa 1911*a*)
Stegosauria
Stegosauridae
Lexovisaurus durobrivensis (including *Omosaurus leedsi* partim)
Age: middle Callovian (Galton 1985*c*)

114. Cambridgeshire, England

Oxford Clay (Hulke 1887; Lydekker 1889*b*, 1889*c*; Nopcsa 1911*a*; Charig 1980; Galton 1980*c*, 1983*b*)
Sauropoda
Diplodocidae
Cetiosauriscus stewarti
Ornithopoda
Hypsilophodontid indet.
?Iguanodontid indet. (= *Camptosaurus leedsi*)

Stegosauria
    Stegosauridae
        Lexovisaurus durobrivensis (=
            Stegosaurus priscus)
Ankylosauria
    Nodosauridae
        Sarcolestes leedsi
Age: middle-late Callovian (Cope et al. 1980*a*)

115. Gloucestershire, England

  a. Inferior Oolite (Benton pers. comm.)
    Theropoda
        ?Carnosauria
            Megalosaurus bucklandii
    Age: Aalenian-Bajocian (Cope et al. 1980*a*)

  b. Chipping Norton Formation (Galton et al. 1980)
    Theropoda
        ?Carnosauria
            Megalosaurus bucklandii
    Sauropoda
        Cetiosauridae
            Cetiosaurus oxoniensis
    Stegosauria
        Stegosaurid indet. (= *Omosaurus vetustus*)
    Age: Bathonian (Anderton et al. 1979)

  d. Forest Marble (Benton pers. comm.)
    Theropoda
        ?Carnosauria
            Megalosaurus bucklandii
    Sauropoda
        Cetiosauridae
            Cetiosaurus oxoniensis
    Age: late Bathonian (Cope et al. 1980*a*)

116. Oxfordshire, England

  a. Sharp's Hill Formation (Galton and Powell 1983)
    Stegosauria
        Stegosaurid indet. (= *Omosaurus vetustus*)
    Age: early Bathonian (Cope et al. 1980*a*)

  b. Chipping Norton Formation (Owen 1842*a*;
    Phillips 1871; Hulke 1887; Galton and Powell
    1983; Benton pers. comm.)
    Theropoda
        Theropoda indet. (= *Scrotum humanum*)
        ?Carnosauria
            Eustreptospondylus oxoniensis
            Megalosaurus bucklandii
    Sauropoda
        Cetiosauridae
            Cetiosaurus oxoniensis (including *C.
               giganteus*)
    Stegosauria
        Stegosaurid indet. (= *Omosaurus vetustus*)
    Age: early Bathonian (Cope et al. 1980*a*)

  c. Stonesfield Slate (Huxley 1869*b;* Huene 1926*a*,
    1932; Galton 1975, 1980*c*)
    Theropoda
        ?Carnosauria
            Iliosuchus incognitus
            Megalosaurus bucklandii
    Ornithopoda
        Hypsilophodontid indet.
    Age: middle Bathonian (Cope et al. 1980*a*)

  d. Forest Marble (Phillips 1871)
    Sauropoda
        Cetiosauridae
            Cetiosaurus oxoniensis
    Stegosauria
         Stegosaurid indet. (= *Omosaurus vetustus*)
    Age: late Bathonian (Cope et al. 1980*a*)

  e. Cornbrash Formation (Galton and Powell 1983;
    Benton pers. comm.)
    Theropoda
        ?Carnosauria
            Megalosaurus bucklandii
    Stegosauria
        Stegosaurid indet. (= *Omosaurus vetustus*)
    Age: late Bathonian (Cope et al. 1980*a*)

  f. Oxford Clay (Benton pers. comm.)
    Theropoda
        ?Carnosauria
            Eustreptospondylus oxoniensis
    Stegosauria
        Stegosaurid indet. (= *Omosaurus vetustus*)
    Age: middle-late Callovian (Cope et al. 1980*a*)

117. Buckinghamshire, England

  a. Great Oolite (Phillips 1871)
    Sauropoda
        Cetiosauridae
            Cetiosaurus oxoniensis
    Age: late Bathonian (Cope et al. 1980*a*)

  b. Forest Marble (Delair and Sarjeant 1985)
    Theropod footprints
    Age: late Bathonian (Cope et al. 1980*a*)

  c. Middle Oxford Clay (Phillips 1871; Nopcsa 1906;
    Huene 1926*a*; Walker 1964)
    Theropoda
        ?Carnosauria
            Eustreptospondylus oxoniensis
    Age: late Callovian (Walker 1964)

118. Dorset, England

  a. Inferior Oolite (Huene 1923; Waldman 1974)
    Theropoda
        ?Carnosauria
            Magnosaurus nethercombensis
            Megalosaurus hesperis
    Age: Aalenian-Bajocian (Cope et al. 1980*a*)

b. Forest Marble (Benton pers. comm.)
  Theropoda indet. (= *Megalosaurus* sp.)
  Age: late Bathonian (Cope et al. 1980*a*)

c. Lower Oxford Clay (Lydekker 1888*b;* Benton
  pers. comm.)
  Sauropoda
    Cetiosauridae
      Cetiosaurus sp.
  Stegosauria
    Stegosauridae
      Lexovisaurus durobrivensis
  Age: middle Callovian (Galton 1985*c*)

119. Wiltshire, England

a. Inferior Oolite (Delair 1973)
  Sauropoda
    Cetiosauridae
      Cetiosaurus sp.
  Age: Aalenian-Bajocian (Cope et al. 1980*a*)

b. Great Oolite (Benton pers. comm.)
  Sauropoda
    Cetiosauridae
      Cetiosaurus oxoniensis
  Age: late Bathonian (Cope et al. 1980*a*)

c. Forest Marble (Owen 1841, 1875; Delair 1973;
  Benton pers. comm.)
  Theropoda
    Theropoda indet. (= *Megalosaurus* sp.)
    ?Carnosauria
      ?Megalosaurus bucklandii
  Sauropoda
    Sauropoda indet. (= *Cardiodon rugulosus*)
    Brachiosauridae
      Bothriospondylus robustus
  Age: late Bathonian (Cope et al. 1980*a*)

120. Département du Calvados, France

a. Calcaire de Caen (Eudes-Deslongchamps 1838)
  ?Theropoda *incertae sedis*
    Poekilopleuron bucklandii
  Age: early Bathonian (Lapparent 1967)

b. Unnamed unit (Lapparent 1967)
  Theropoda indet. (= *Megalosaurus* sp.)
  Age: Bathonian (Lapparent 1967)

c. Marnes de Dives (Taquet and Welles 1977)
  Theropoda
    ?Carnosauria
      Piveteausaurus divesensis
  Age: late Callovian (Taquet and Welles 1977)

d. Marnes d'Argences (Hoffstetter 1957)
  Stegosauria
    Stegosauridae
      Lexovisaurus durobrivensis
  Age: middle Callovian (Lapparent 1967)

e. Unnamed unit (Gaudry 1890; Lapparent 1967;
  Taquet pers. comm.)
  Theropoda indet. (listed as ornithopod indet.
    in Lapparent 1967)
  Age: ?Middle Jurassic (Taquet pers. comm.)

121. Département de l'Indre, France

Unnamed unit (Lapparent 1967)
  Theropoda
    ?Carnosauria
      Megalosaurus bucklandii
  Sauropoda
    Brachiosauridae
      Pelorosaurus sp.
  Age: Bathonian (Lapparent 1967)

122. Département de la Charente-Maritime, France

Unnamed unit (Lapparent 1967)
  Theropoda indet. (= *Megalosaurus* sp.)
  Age: Bathonian (Lapparent 1967)

123. Província do Beira Litoral, Portugal

Unnamed unit (Thulborn 1973*a*)
  Theropoda indet.
  Ornithischia indet. (= *Alocodon kuehnei*)
  Age: late Callovian (Thulborn 1973*a*)

Asia

Three locations are known for the Middle Jurassic of the People's Republic of China (fig. 3.7). The faunas from Sichuan have yielded the most diverse, complete, and abundant dinosaur material that is known during this time span. It has the further advantage of being exploited currently by Dong Zhiming and co-workers of the Institute of Vertebrate Paleontology and Paleoanthropology and by He Xinlu and co-workers of the Chengdu College of Geology. Largely unreported, the Xizang Zizhiqu fauna is yielding new and very important Middle Jurassic dinosaur material.

124. Xizang Zizhiqu, People's Republic of China

Dapuka Group (Wang et al. 1985)
  Theropoda
    Undescribed ?megalosaurid
    Undescribed sauropod
  Age: Middle Jurassic (Wang et al. 1985)

125. Sichuan, People's Republic of China

a. Xiashaximiao Formation (He 1979; Dong et al.
  1983; Zhang et al. 1984; He et al. 1984; He and
  Cai 1984; Dong and Tang 1985)

Theropoda
    Theropoda indet. (= *Chuandongocoelurus primitivus*)
    ?Carnosauria
        Gasosaurus constructus
        Kaijiangosaurus lini
    Theropoda *incertae sedis*
        Xuanhanosaurus qilixiaensis
Sauropoda
    Sauropoda indet.
        (= *Protognathosaurus oxyodon*)
    Cetiosauridae
        Shunosaurus lii
        Datousaurus bashanensis
        Omeisaurus jungusiensis
        Omeisaurus tianfuensis
        Omeisaurus luoquanensis
Ornithischia indet. (= *Xiaosaurus dashanpensis*)
Ornithopoda
    Hypsilophodontidae
        Yandusaurus hongheensis (including *Y. multidens*)
Stegosauria
    Huayangosauridae
        Huayangosaurus taibaii
Age: Bathonian-Callovian (Dong and Tang 1984; Bajocian, Chen et al. 1982)
b. Kuangyuan Series (Young 1944)
    Theropoda indet.
    Sauropoda indet. (= *Sanpasaurus yaoi* partim)
    Ornithopoda indet. (= *Sanpasaurus yaoi* partim)
    ?Ornithopoda indet. (= gen. indet. *imperfectus*)
    Stegosauria
        Stegosaurid indet.
    Age: Bathonian-Callovian (Dong pers. comm.)

South America

Work by A. Cabrera, R. M. Casamiquela, and most recently, by J. F. Bonaparte has introduced a few Middle Jurassic sites in South America. These are restricted to northern Chile and southern Argentina (fig. 3.9).

126. Provincia de Antofagasta, Chile

    Unnamed unit (Biese 1961; Casamiquela 1967a)
        Theropoda indet.
        Age: early Callovian (Biese 1961)

127. Provincia de Santa Cruz, Argentina

    Middle "Complejo Porfirico" (Casamiquela 1964b)
        Theropod footprints
        Age: Callovian-Oxfordian (Leonardi 1989)

128. Provincia de Chubut, Argentina

    a. Cerro Carnerero Formation (Cabrera 1947; Casamiquela 1963; not figured)

Sauropoda
    Cetiosauridae
        Amygdalodon patagonicus
Age: Bajocian (Bonaparte 1979a)
b. Cañadon Asfalto Formation (Bonaparte 1979b, 1986c)
    Theropoda
        Allosauridae
            Piatnitzkysaurus floresi
    Sauropoda
        Cetiosauridae
            Patagosaurus fariasi
        Brachiosauridae
            Volkheimeria chubutensis
    Age: Callovian (Bonaparte 1986c)

Africa

Five Middle Jurassic dinosaur locations are known from Africa (figs. 3.10, 3.11): two in Morocco and Algeria from the early work of A. F. de Lapparent and the continuing studies of M. Monbaron and P. Taquet, and one in Madagascar from the fieldwork of L. Ginsburg.

129. Province de Beni Mellal, Morocco

    a. Guettioua Sandstone (Monbaron and Taquet 1981)
        Sauropoda
            Cetiosauridae
                Cetiosaurus mogrebiensis
        Age: late Bathonian (Monbaron and Taquet 1981)

    b. Unnamed unit (Boucart et al. 1942; Monbaron 1978; Debenath et al. 1979; Taquet 1985)
        Sauropoda
            ?Cetiosaurid indet.
        Age: Bathonian (Taquet 1985)

130. Province de Fes, Morocco

    Unnamed unit (Lapparent and Lavocat 1955)
        ?Theropoda indet. (= *Megalosaurus mersensis*; may be teleosaurid crocodilian; Chabli 1985)
        Sauropoda
            Cetiosauridae
                Cetiosaurus mogrebiensis
        Age: late Bathonian (Monbaron and Taquet 1981)

131. Wilaya Tlemcen, Algeria

    Unnamed unit (Lapparent and Lucas 1957)
        Sauropoda indet.
        Age: middle Callovian (Lapparent and Lucas 1957)

132. Faritany Majunga, Madagascar

    Isalo Formation (Taquet 1977; Bonaparte 1986b)
        Sauropoda
            Brachiosauridae
                Bothriospondylus madagascariensis
                Lapparentosaurus madagascariensis
        Age: Bathonian (Taquet 1977)

Australasia

Australasian dinosaur remains from the Middle Jurassic are known from only three locations in Queensland, Australia (fig. 3.12).

133. Queensland, Australia

    a. ?Injune Creek Beds (Longman 1926, 1927)
        Sauropoda
            Cetiosauridae
                Rhoetosaurus brownei
        Age: ?Bajocian (Molnar pers. comm.)
    b. Walloon Group (Molnar 1982)
        Theropod footprints
        ?Sauropod footprints
        ?Stegosaur footprints
        Age: Middle Jurassic (Molnar 1982)
    c. Unnamed unit (Molnar 1982)
        Theropod footprints
        Age: Middle Jurassic (Molnar 1982)

LATE JURASSIC

More than any other interval of dinosaur history, the Late Jurassic represents the public's conception of dinosaur life: long-necked sauropods, predatory theropods like *Allosaurus*, and low-browsing stegosaurs and camptosaurs. These animals and their relatives are known more extensively during the Late Jurassic than previously, particularly in North America but also in Europe, Asia, South America, and Africa. There is no present record of Late Jurassic dinosaurs in Australasia or Antarctica.

North America

With the exception of undescribed Alaskan footprints, all but one of the Upper Jurassic localities are known from the Western Interior of the United States and are restricted to the Morrison Formation (figs. 3.1, 3.3, and 3.13). Our knowledge of the Morrison fauna is from the often arduous work of North America's pioneers in vertebrate paleontology. Names such as O. C. Marsh, S. W. Williston, and J. B. Hatcher; E. D. Cope; H. F. Osborn and B. Brown; E. Douglass and C. W.

Gilmore; E. S. Riggs; and J. W. Stovall are intimately associated with the exploitation of the Upper Jurassic fossil fields of the western United States (see Ostrom and McIntosh 1966 and Colbert 1968 for the history of dinosaur discoveries in the Morrison Formation). Work continues on Late Jurassic dinosaur faunas by J. A. Jensen, W. L. Stokes, J. H. Madsen, Jr., G. Callison, M. G. Lockley, W. E. Miller, and D. D. Gillette.

134. Alaska, United States

    Naknek Formation (Baird pers. comm.)
        Theropod footprints
        Ornithopod footprints
        Age: middle Kimmeridgian-late Tithonian
            (Imlay 1952, Baird pers. comm.)

135. Montana, United States

    Morrison Formation (McIntosh 1981; Horner pers. comm.)
        Sauropoda
            Sauropoda indet.
            Diplodocidae
                Diplodocus sp.
            Camarasauridae
                Camarasaurus grandis
                Camarasaurus sp.
        Ornithischia indet.
        Stegosauria
            Stegosaurid indet.
        Age: Kimmeridgian-Tithonian (Dodson, Behrensmeyer, and Bakker 1980)

136. Wyoming, United States

    Morrison Formation (Marsh 1878a, 1878b; Mook 1917a; Brown 1935; Ostrom and McIntosh 1966; Galton 1982c)
        Theropoda
            Allosauridae
                Allosaurus fragilis
            Maniraptora *sedis mutabilis*
                Coelurus fragilis
                Ornitholestes hermanni
        Sauropoda
            Cetiosauridae
                cf. *Haplocanthosaurus priscus*
            Cetiosauridae *incertae sedis*
                "*Apatosaurus*" *minimus*
            Camarasauridae
                Camarasaurus grandis
                Camarasaurus supremus
                Camarasaurus lentus
                Camarasaurus sp.

Diplodocidae
Diplodocus carnegii
Diplodocus hayi
Diplodocus sp.
Apatosaurus excelsus
Apatosaurus sp.
Ornithopoda
Hypsilophodontidae
Othnielia rex
Drinker nisti
Hypsilophodontid indet. (= *Nanosaurus agilis, Laosaurus celer*)
Dryosauridae
Dryosaurus altus
Camptosauridae
Camptosaurus dispar
Camptosaurus amplus
?*Camptosaurus* cf. *depressus*
Stegosauria
Stegosauridae
Stegosaurus armatus (= *S. ungulatus*)
Stegosaurus stenops
Stegosaurus longispinus
Stegosaurus sp.
Age: Kimmeridgian-Tithonian (Dodson, Behrensmeyer, and Bakker 1980)

137. Utah, United States

Morrison Formation (Gilmore 1914*a*, 1936; Holland 1915*b*; Ellinger 1950; Stokes 1944; Galton 1982*a*, 1983*e*; Galton and Jensen 1973*b*; Sheperd et al. 1973; Madsen 1974, 1976*a*, 1976*b*; Berman and McIntosh 1986, Hirsch et al. 1989)
Theropoda
Theropoda *incertae sedis*
Marshosaurus bicentesimus
Theropoda indet.
Ceratosauria
Ceratosaurus nasicornis
Allosauridae
Allosaurus fragilis
?Carnosauria
Stokesosaurus clevelandi
Maniraptora *sedis mutabilis*
Coelurus fragilis
Ornitholestes hermanni
Sauropoda
Sauropoda indet.
Cetiosauridae
?*Haplocanthosaurus* sp.
Diplodocidae
Apatosaurus louisae
Barosaurus lentus
Diplodocus longus
Diplodocus carnegii
Diplodocus sp.
Dystrophaeus viaemalae
?Diplodocid indet.

Brachiosauridae
Brachiosaurus cf. *altithorax*
Camarasauridae
Camarasaurus lentus (including *C. annae*)
Camarasaurus sp.
Ornithopoda
Hypsilophodontidae
Othnielia rex
Dryosauridae
Dryosaurus altus
Camptosauridae
Camptosaurus dispar
Camptosaurus sp.
Stegosauria
Stegosauridae
Stegosaurus armatus (= *S. ungulatus*)
Stegosaurus stenops
?*Stegosaurus longispinus*
Stegosaurus sp.
Stegosaurid indet. (= *Stegosaurus affinis*)
Dinosaur eggshell
Age: Kimmeridgian-Tithonian (Dodson, Behrensmeyer, and Bakker 1980)

138. Colorado, United States

a. Entrada Sandstone (Eriksen 1979*a*, 1979*b*)
Theropod footprints
Age: Oxfordian (Hintzc 1973)

b. Morrison Formation (Marsh 1884*a*, 1884*d*; Riggs 1901*a*, 1901*b*, 1903*a*, 1903*b*, 1904; Hatcher 1903*c*; Galton and Jensen 1979*b*; Callison and Quimby 1984; Jensen 1985*a*, 1985*b*; Lockley 1986*a*; Lockley et al. 1986; McIntosh and Williams 1988; Hirsch pers. comm.; Armstrong pers. comm.)
Theropoda
Theropoda indet. (= *Labrosaurus sulcatus*)
Theropod footprints
Ceratosauria
Ceratosaurus nasicornis
Allosauridae
Allosaurus fragilis
Undescribed allosaurid (Jensen 1985*b*)
?Carnosauria
Torvosaurus tanneri
Ornithomimosauria *incertae sedis*
Elaphrosaurus sp.
Ornithomimosaur indet.

Sauropoda
Sauropod footprints
Sauropoda indet. (= *Atlantosaurus montanus*)
Cetiosauridae
Haplocanthosaurus priscus
Haplocanthosaurus delfsi
Cetiosauridae *incertae sedis*
"*Morosaurus*" *agilis*
Diplodocidae
Diplodocus longus
Diplodocus lacustris
Diplodocus sp.
Apatosaurus excelsus (= *Brontosaurus amplus*)
Apatosaurus ajax
Apatosaurus sp.
Amphicoelias altus
Supersaurus vivianae
Undescribed diplodocid (Jensen 1985*b*)
Brachiosauridae
Brachiosaurus altithorax
Ultrasauros macintoshi
Dystylosaurus edwini
Camarasauridae
Camarasaurus supremus
Camarasaurus grandis
Camarasaurus lewisi
Ornithischia indet.
Ornithopoda
Hypsilophodontidae
Othnielia rex
Hypsilophodontid indet. (= *Nanosaurus agilis*)
Dryosauridae
Dryosaurus altus
Camptosauridae
Camptosaurus dispar
Iguanodontid undescribed (Armstrong pers. comm.)
?Ornithopoda indet. (= *Tichosteus lucasanus, T. aequifacies*)
Ornithopod footprints
Stegosauria
Stegosauridae
Stegosaurus armatus
Stegosaurus stenops
Stegosaurus sp.
Stegosaurid indet. (= *Hypsirhophus discurus*)
Dinosaur eggshell
Age: Kimmeridgian-Tithonian (Dodson, Behrensmeyer, and Bakker 1980)

139. New Mexico, United States

Morrison Formation (McIntosh 1981; Rigby 1982; Gillette 1987, pers. comm.)
Theropoda
Allosauridae
Allosaurus fragilis
Sauropoda
Camarasauridae
Camarasaurus cf. *supremus*
Diplodocidae
Seismosaurus halli
Age: Kimmeridgian-Tithonian (Dodson, Behrensmeyer, and Bakker 1980)

140. South Dakota, United States

Morrison Formation (Marsh 1899; Lull 1919)
Theropoda
Theropod footprints
Allosauridae
Allosaurus fragilis
Sauropoda
Diplodocidae
Barosaurus lentus
Age: Kimmeridgian-Tithonian (Dodson, Behrensmeyer, and Bakker 1980)

141. Oklahoma, United States

Morrison Formation (Stovall 1938; Langston 1974; Lockley 1986*a*)
Theropoda
Ceratosauria
Ceratosaurus sp.
Allosauridae
Allosaurus fragilis
Sauropoda
Sauropod footprints
Diplodocidae
Apatosaurus excelsus
Ornithopoda
Ornithopod footprints
Camptosauridae
Camptosaurus dispar
Stegosauria
Stegosauridae
Stegosaurus sp.
Age: Kimmeridgian-Tithonian (Dodson, Behrensmeyer, and Bakker 1980)

142. Texas, United States

?Morrison Formation (McIntosh 1981; not figured)
Sauropoda
Camarasauridae
Camarasaurus sp.
Age: Kimmeridgian-Tithonian (Dodson, Behrensmeyer, and Bakker 1980)

Europe

Scattered localities pertaining to the Late Jurassic are known throughout Europe (fig. 3.15): from England, established by J. W. Hulke, R. Owen, A. Leeds, R. Lydekker, H. G. Seeley, and most recently, by J. B. Delair; from France, by E. Sauvage, F. Nopcsa, and A. Bidar and co-workers; from Portugal, by A. F. de Lapparent and G. Zbyszewski and by W. G. Kühne and R. A. Thulborn; from the Federal Republic of Germany, by H. Friese and M. Kaever and A. F. de Lapparent; and from Romania, through the recent important work by T. Jurcsak and E. Popa.

143. North Yorkshire, England

   a. Corallian Oolite Formation (Seeley 1875; Galton 1983*b*)
      Theropoda
         ?Carnosauria
            Megalosaurus bucklandii
      Stegosauria
         Stegosaurid indet. (= *Omosaurus phillipsi*)
      Age: early-middle Oxfordian (Cope et al. 1980*a*)

   b. ?Lower Calcareous Grit (Seeley 1875; Galton 1980*e*)
      Ankylosauria indet. (= *Priodontognathus phillipsii*)
      Age: ?early Oxfordian (Cope et al. 1980*a*)

144. Northamptonshire, England

   Kimmeridge Clay (Hulke 1887)
      Sauropoda
         Brachiosauridae *incertae sedis*
            "*Ornithopsis*" *leedsi*
      Stegosauria
         Stegosauridae
            Lexovisaurus durobrivensis
      Age: Kimmeridgian (Cope et al. 1980*a*)

145. Cambridgeshire, England

   a. Ampthill Clay (Seeley 1869; Galton 1980*c*)
      Ankylosauria indet. (= *Cryptosaurus eumerus*)
      Age: late Oxfordian (Cope et al. 1980*a*)

   b. Kimmeridge Clay (Reynolds 1939; Galton 1985*c*)
      Sauropoda
         Brachiosauridae *incertae sedis*
            "*Cetiosaurus*" *humerocristatus*
            Ischyrosaurus manseli
         Sauropoda indet. (= *Gigantosaurus megalonyx*)
      Stegosauria
         Stegosauridae
            Dacentrurus armatus
      Age: Kimmeridgian (Cope et al. 1980*a*)

146. Oxfordshire, England

   a. Corallian Oolite Formation (Benton pers. comm.)
      Sauropoda
         Cetiosauridae
            Cetiosaurus sp.
      Age: early-middle Oxfordian (Cope et al. 1980*a*)

   b. Kimmeridge Clay (Galton and Powell 1980)
      Ornithopoda
         Camptosauridae
            Camptosaurus prestwichii
      Age: Kimmeridgian (Cope et al. 1980*a*)

147. Buckinghamshire, England

   Hartwell Clay (Lydekker 1893*b*; Woodward 1895)
      Theropoda indet. (= *Megalosaurus* sp.)
      Sauropoda
         Brachiosauridae
            ?*Pelorosaurus* sp. (?= *Oplosaurus* sp.)
      Age: late Kimmeridgian (Cope et al. 1980*a*)

148. Dorset, England

   a. Corallian Oolite Formation (Huene 1923; Walker 1964; Benton pers. comm.)
      Theropoda *incertae sedis*
         Metriacanthosaurus parkeri
      Stegosauria
         Stegosauridae
            Dacentrurus armatus
      Age: early-middle Oxfordian (Cope et al. 1980*a*)

   b. Kimmeridge Clay (Hulke 1874; Hoffstetter 1957; Galton 1975; Benton pers. comm.)
      Sauropoda
         Brachiosauridae *incertae sedis*
            "*Cetiosaurus*" *humerocristatus*
            Ischyrosaurus manseli
      Ornithopoda
         Hypsilophodontid indet.
      Stegosauria
         Stegosauridae
            Dacentrurus armatus
      Age: Kimmeridgian (Cope et al. 1980*a*)

   c. Portland Stone (Benton pers. comm.)
      Sauropoda
         Brachiosauridae
            ?*Pelorosaurus* sp. (= *Ornithopsis* sp.)
      Age: late Tithonian (= middle Portlandian; Cope et al. 1980*a*)

   d. Middle Purbeck Beds (Delair 1980; Delair and Sarjeant 1985)
      Theropoda
         Theropoda indet. (= *Nuthetes destructor* partim)
         Theropod footprints

Thyreophora *incertae sedis*
    Echinodon becklesii
Ornithischian footprints
Ankylosauria
    Nodosaurid indet.
Age: late Tithonian (= late Portlandian; Casey
    1973)

149. Wiltshire, England

Kimmeridge Clay (Owen 1877*a*,1877*b*; Galton
1975, 1980*c*, 1980*d*, 1985*c*)
    Theropoda indet. (= ?*Megalosaurus insignis*)
    Sauropoda
        Brachiosauridae
            Bothriospondylus suffosus
    Ornithopoda
        Hypsilophodontid indet.
    Stegosauria
        Stegosauridae
            Dacentrurus armatus (= *Omosaurus*
                armatus, including *O. hastiger*)
        Stegosaurid indet.
    Ankylosauria
        Nodosaurid indet.
    Age: early Kimmeridgian (Cope et al. 1980*a*)

150. Département du Calvados, France

a. Unnamed unit (Douville 1885; Lapparent 1967)
    Theropoda indet. (= *Streptospondylus cuvieri*)
    Age: Oxfordian (Lapparent 1967)

b. Les Sables de Glos supérieurs (Buffetaut et al.
    1985)
    Theropoda indet.
    Sauropoda
        ?Camarasaurid indet.
    Age: late Oxfordian (Buffetaut et al. 1985)

151. Département de la Seine-Maritime, France

a. Argiles d'Octeville (Nopcsa 1911*b*; Lapparent
    1967; Galton and Boine 1980)
    Stegosauria
        Stegosauridae
            Dacentrurus armatus (= *D. lennieri*)
    Age: early Kimmeridgian (Nopcsa 1911*b*;
        Lapparent 1967)

b. Unnamed unit (Eudes-Deslongchamps 1870)
    Theropoda indet. (= *Megalosaurus insignis*)
    Age: middle-late Kimmeridgian (Buffetaut
        pers. comm.)

152. Département du Pas-de-Calais, France

a. Unnamed unit (Sauvage 1897)
    Theropoda
        Theropoda indet. (= *Teinurosaurus sauvagei*)
        Ornithomimosauria *incertae sedis*
            ?*Elaphrosaurus* sp.
    Sauropoda indet. (= *Iguanodon praecursor*,
    Neosodon [no specific name], *Morinosaurus*
    typus)
    Age: Kimmeridgian (Galton 1982*c*)

b. Unnamed unit (Sauvage 1888)
    Theropoda indet. (= ?*Megalosaurus insignis*)
    Age: late Kimmeridgian-late Tithonian
        (Lapparent 1967)

c. Unnamed unit (Galton and Powell 1980)
    Ornithopoda
        Iguanodontid indet.
    Age: Kimmeridgian-Tithonian (Sauvage 1888;
        Galton and Powell 1980)

d. Unnamed unit (Sauvage 1876*b*; Moussaye 1884;
    Lapparent 1967)
    Sauropoda
        Brachiosauridae *incertae sedis*
            "*Cetiosaurus*" *humerocristatus*
    Age: late Tithonian (Buffetaut pers. comm.)

153. Département de la Charente-Maritime, France

a. Unnamed unit (Lapparent 1967)
    Theropoda indet. (= *Megalosaurus* sp.)
    Age: late Tithonian (= Portlandian; Lapparent
        1967)

b. Unnamed unit (Lapparent and Oulmi 1964)
    Theropod footprints
    Age: late Tithonian (Lapparent and Oulmi
        1964)

154. Département de la Vienne, France

Unnamed unit (Lapparent 1967)
    Theropoda indet. (= *Megalosaurus* sp.)
    Age: Oxfordian (Lapparent 1967)

155. Département de la Nievre, France

Unnamed unit (Lapparent 1967)
    Theropoda indet. (= *Megalosaurus* sp.)
    Age: Kimmeridgian (Lapparent 1967)

156. Département de la Meuse, France

Unnamed unit (Galton and Powell 1980)
    Ornithopoda
        Iguanodontid indet.
    Age: middle Oxfordian (= Rauracian; Huene
        and Maubeuge 1954)

157. Département du Jura, France

Unnamed unit (Lapparent 1943; Enay 1980)
    Theropoda indet. (= ?*Megalosaurus insignis*)
    Sauropoda
        Brachiosauridae
            Bothriospondylus madagascariensis
    Age: late Oxfordian-Kimmeridgian (Lapparent
        1943)

158. Département de l'Ain, France

Calcaires Lithographiques de Cerin (Bernier 1984;
Bernier et al. 1984)
    Theropod footprints
    Age: late Kimmeridgian (Bernier 1984;
        Bernier et al. 1984)

159. Département du Var, France

    Lithographic Limestone (Bidar et al. 1972)
       Theropoda
             Maniraptora *sedis mutabilis*
                Compsognathus longipes (= *C. corallestris*)
       Age: Kimmeridgian (Bidar et al. 1972)

160. Provincia de La Rioja, Spain

    "Terenes Marl" (Mensink and Mertmann 1984)
       Theropod footprints
       Age: Kimmeridgian (Suarez Vega 1974)

161. Provincia de Oviedo, Spain

    Unnamed unit (Garcia-Ramos and Valenzuela 1977*b*)
       Theropod footprints
       Ornithopod footprints
       Age: Late Jurassic (Garcia-Ramos and Valenzuela 1977*b*)

162. Provincia de Valencia, Spain

    Arroyo Cerezo (Lapparent et al. 1965)
       Theropod footprints
       Age: ?Kimmeridgian (Lapparent et al. 1965)

163. Província do Beira Litoral, Portugal

    a. Unnamed unit (Lapparent and Zbyszewski 1957)
       Theropoda indet. (= ?*Megalosaurus insignis*)
       Age: Oxfordian (= late Lusitanian; Lapparent and Zbyszewski 1957)

    b. Unnamed unit (Lapparent et al. 1951)
       Theropod footprints
       Age: Oxfordian (= late Lusitanian; Lapparent et al. 1951)

    c. Unnamed unit (Lapparent and Zbyszewski 1957)
       Theropoda indet. (= *Megalosaurus pombali*)
       Sauropoda
             Camarasauridae
                Camarasaurus alenquerensis (= *Apatosaurus alenquerensis*)
             Brachiosaurid indet.
       Stegosauria
             Stegosauridae
                Dacentrurus armatus (including *Astrodon pusillus*)
       Age: Oxfordian-Kimmeridgian (Lapparent and Zbyszewski 1957)

    d. Unnamed unit (Thulborn 1973*a*)
       Sauropoda
             Sauropoda indet.
             Brachiosauridae
                ?*Bothriospondylus* sp.
       Ornithopoda
             Hypsilophodontid indet. (= *Phyllodon henkeli*)
       Age: early Kimmeridgian (Thulborn 1973*a*)

    e. Unnamed unit (Lapparent and Zbyszewski 1957)
       Stegosauria
             Stegosauridae
                Dacentrurus armatus
       Age: Kimmeridgian (Lapparent and Zbyszewski 1957)

164. Província do Ribatejo, Portugal

    a. Unnamed unit (Lapparent and Zbyszewski 1957)
       Sauropoda indet. (= "*Cetiosaurus*" *humerocristatus*)
       Age: Oxfordian-Kimmeridgian (Lapparent and Zbyszewski 1957)

    b. Unnamed unit (Lapparent and Zbyszewski 1957)
       Theropoda indet. (= ?*Megalosaurus insignis*)
       Age: early Kimmeridgian (Lapparent and Zbyszewski 1957)

165. Província do Estremadura, Portugal

    a. Unnamed unit (Lapparent and Zbyszewski 1957)
       Theropoda indet. (= ?*Megalosaurus pombali*)
       Sauropoda
             Camarasauridae
                Camarasaurus ?alenquerensis (= *Apatosaurus alenquerensis*)
             Brachiosaurid indet.
       Stegosauria
             Stegosauridae
                Dacentrurus armatus
       Age: Oxfordian-Kimmeridgian (Lapparent and Zbyszewski 1957)

    b. Unnamed unit (Thulborn 1973*a*; Galton 1980*d*)
       Theropoda
             Maniraptora
                Lisboasaurus estesi
       Ornithischia indet. (= *Trimucrodon cuneatus*)
       Ornithopoda
             Hypsilophodontidae
                Hypsilophodon sp.
             Iguanodontid indet.
       Age: Kimmeridgian (Thulborn 1973*a*)

    c. Unnamed unit (Lapparent and Zbyszewski 1957, Galton 1980*a*)
       Theropoda indet. (= ?*Megalosaurus insignis*)
       Sauropoda
             Brachiosauridae
                Brachiosaurus atalaiensis
       Stegosauria
             Stegosauridae
                Dacentrurus armatus
       Ankylosauria
             Nodosauridae
                Dracopelta zbyszewskii
       Age: Kimmeridgian (Galton 1980*a*)

166. Nordrhein-Westphalen, Federal Republic of Germany

    Malm (Friese 1972, 1979; Kaever and Lapparent 1974)
       Theropod footprints

Sauropod footprints
Age: early Kimmeridgian (Kaever and
Lapparent 1974)

167. Bayern, Federal Republic of Germany

Ober Solnhofen Plattenkalk (Wagner 1861; Ostrom
1978*a*)
Theropoda
Maniraptora *sedis mutabilis*
Compsognathus longipes
Age: ?Kimmeridgian (Ostrom 1978*a*)

168. Kanton Bern, Switzerland

Unter-Virgula-Schichten (Greppin 1870; Huene
1922)
Theropoda indet. (= *Megalosaurus meriani*)
Sauropoda
Diplodocidae
Cetiosauriscus greppini (= *Ornithopsis
greppini*)
Age: ?Tithonian (= Upper Malm; Huene
1922)

Asia

Asian records of Late Jurassic dinosaurs have
been scattered and often fragmentary (figs. 3.6, 3.7,
3.16). Most of the material comes from the People's
Republic of China, originally studied by C.-C. Young.
Dong Z.-M. and associates at the Institute of Vertebrate
Paleontology and Paleoanthropology are currently
studying further Late Jurassic dinosaurs from Sichuan,
much of which provide us with a new perspective on
the fauna of this time interval. Dinosaur faunas from
Thailand have also been described by E. Buffetaut and
R. Ingavat.

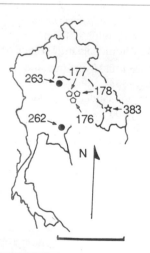

Fig. 3.16. Late Jurassic- to Late Cretaceous-age locations
in Thailand and Laos. Scale = 600 km.

169. Rajasthan State, India

?Jabalpur Beds (Mathur et al. 1985; Sahni pers.
comm.)
Dinosauria indet.
Age: Late Jurassic (Chatterjee and Hotton
1986)

170. Madhya Pradesh State, India

Jabalpur Formation (Chatterjee pers. comm.)
Sauropoda indet.
Age: Late Jurassic (Chatterjee pers. comm.)

171. Xinjiang Uygur Ziziqu, People's Republic of China

a. Shishugou Formation (Young 1937)
Sauropoda
Camarasauridae
Tienshanosaurus chitaiensis
Age: Oxfordian (Dong pers. comm.)
b. Keilozo Formation (Dong 1977; pers. comm.; =
Huoyanshan Formation)
Theropoda
Allosauridae
cf. *Szechuanosaurus campi*
Sauropoda indet. (= *Chiayusaurus lacustris*)
Age: Oxfordian-Kimmeridgian (Dong pers.
comm.)
c. Unnamed unit (Dong 1973*b*)
Theropoda indet.
Age: ?Late Jurassic (Dong 1973*b*)

172. Gansu, People's Republic of China

Hantong Formation (Young 1958*b*; Dong pers.
comm.)
Sauropoda
Diplodocidae
Mamenchisaurus constructus
Age: Oxfordian (Dong pers. comm.)

173. Shanxi, People's Republic of China

Unnamed unit (Kuhn 1958*b*)
Ornithopod footprints
Age: Late Jurassic (Kuhn 1958*b*)

174. Sichuan, People's Republic of China

a. Kyangyan Series (Young 1942*b*)
Theropoda
Theropoda indet. (= *Sinocoelurus fragilis*)
Allosauridae
?*Szechuanosaurus campi*
Age: Tithonian (Dong pers. comm.)
b. Shangshaximiao Formation (Camp 1935; Young
1939; Dong et al. 1983; He et al. 1984)
Theropoda
Theropoda indet.
Allosauridae
Szechuanosaurus campi (including *S.
yandonensis*)
?Carnosauria
Yangchuanosaurus shangyouensis
Yangchuanosaurus magnus

Sauropoda
  Cetiosauridae
    Omeisaurus changshouensis
    Omeisaurus fuxiensis
  Diplodocidae
    Mamenchisaurus constructus
    Mamenchisaurus hochuanensis
Ornithischia indet. (= *Gongbusaurus shiyii*)
Stegosauria
  Stegosauridae
    Tuojiangosaurus multispinus
    Chialingosaurus kuani
    Chungkingosaurus jiangbeiensis
    Chungkingosaurus sp.
    Stegosaurid indet.
  Age: Late Jurassic (Dong et al. 1983;
    Oxfordian, Dong pers. comm., Bathonian-
    Callovian, Chen et al. 1982)
  c. Penglaizhen Formation (Zhen et al. 1983)
    Ornithischian footprints
    Age: Late Jurassic (Zhen et al. 1983)
  d. Lower Chiating Series (Young 1960)
    Ornithischian footprints
    Age: Late Jurassic (Young 1960)

175. Shandong, People's Republic of China
  a. Meng-Yin Formation (Wiman 1929; Mateer and
    McIntosh 1985)
      Sauropoda
        Camarasauridae
          Euhelopus zdanskyi
      Stegosauria
        Stegosaurid indet.
    Age: Kimmeridgian (Dong pers. comm.; early
      Tithonian, Chen et al. 1982)
  b. Ch'ing-Shan Formation (Wiman 1929)
      Sauropoda indet.
    Age: Kimmeridgian (Dong pers. comm.)

176. Changwat Khon Kaen, Thailand
  Sao Khua Formation (Ingavat et al. 1978; Ingavat
    and Taquet 1978; Buffetaut 1982; Buffetaut and
    Ingavat 1983, 1984, 1986a)
      Theropoda
        Theropoda indet. (= compsognathid
          indet.)
        Carnosauria *incertae sedis*
          ?Spinosaurid indet. (= *Siamosaurus
            suteethorni*)
      Sauropoda indet.
    Age: Late Jurassic (Buffetaut and Ingavat
      1985)

177. Changwat Udon Thani, Thailand
  Sao Khua Formation (Buffetaut 1982; Buffetaut and
    Ingavat 1986b)
      Sauropoda indet.
    Age: Late Jurassic (Buffetaut and Ingavat
      1985)

178. Changwat Kalasin, Thailand
  Sao Khua Formation (Buffetaut 1982)
      Sauropoda indet.
    Age: Late Jurassic (Buffetaut and Ingavat
      1985)

South America

Two of the four Late Jurassic South American faunas (figs. 3.8, 3.9) are represented by dinosaur footprints or trackways from Brazil, Chile, and Argentina. These have been described primarily by G. Leonardi, R. M. Casamiquela, and O. Galli and R. Dingman. The third, from Colombia, is represented by possible cetiosaurid material described by W. Langston, Jr., and J. W. Durham. The fourth comes from a report by G. C. Diaz and Z. B. de Gasparini of an undescribed dinosaur fauna from northern Chile.

179. Departamento de Magdalena, Colombia
  "Giron" formation (Langston and Durham 1955)
      Sauropoda
        ?Cetiosaurid indet.
    Age: ?Late Jurassic (Bonaparte 1979a)

180. Provincia de Tarapaca, Chile
  Chacarilla Formation (Galli and Dingman 1962)
      Theropod footprints
      Ornithopod footprints
      ?Stegosaur footprints
    Age: ?Oxfordian (Diaz and Gasparini 1976)

181. Provincia de Antofagasta, Chile
  Unnamed unit (Diaz and Gasparini 1976)
      Undescribed Dinosauria
    Age: Late Jurassic (Diaz and Gasparini 1976)

182. Provincia de Santa Cruz, Argentina
  La Matilde Formation (Casamiquela 1964b)
      Theropod footprints
    Age: Oxfordian (Stipanicic and Bonetti 1970a,
      1970b)

Africa

The Tendaguru fauna of Tanzania is perhaps the best known of all dinosaur faunas from Africa. This site was originally worked by E. Fraas from the königlichen

Naturalienkabinett Stuttgart (now Staatliches Museum für Naturkunde, Stuttgart), but most extensively between 1909 and 1912 by W. Janensch, E. Hennig, and H. Reck under the auspices of the Humboldt-Universitat Museum für Naturkunde (for further information on the German expedition to Tendaguru, see Hennig 1912). From 1924 to 1929, Tendaguru was worked by an expedition from the British Museum (Natural History), headed successively by W. E. Cutler, F. W. H. Migeod, and J. Parkinson (Parkinson 1930). Although less spectacular, work by D.-M. Dutuit and A. Ouazzou in Morocco, L. Ginsburg and co-workers in Niger, and S. M. Haughton in Malawi sites in Africa. Additional Late Jurassic to Early Cretaceous African material is known from South Africa, originally studied by R. Broom and later by P. M. Galton and W. A. Coombs, Jr., and T. H. V. Rich, R. E. Molnar, and P. V. Rich. Upper Jurassic dinosaur locations in Africa are given in figures 3.10 and 3.11.

183. Province de Marrakech, Morocco

> Unnamed unit (Dutuit and Ouazzou 1980)
>> Sauropod footprints
>> Age: Late Jurassic or Early Cretaceous (Dutuit and Ouazzou 1980)

184. Département d'Agadez, Niger

> a. Grès d'Assaouas (Ginsburg et al. 1966)
>> Theropod footprints
>> Sauropod footprints
>> Age: Late Jurassic-Hauterivian (Ginsburg et al. 1966; Taquet 1976)

> b. Argiles de l'Irhazer (Ginsburg et al. 1966)
>> Theropod footprints
>> Ornithopod footprints
>> Age: Late Jurassic-Hauterivian (Ginsburg et al. 1966; Taquet 1976)

185. Northern Province, Malawi

> Chiweta Beds (Haughton 1928)
>> Sauropoda
>>> Titanosauridae
>>>> *Tornieria dixeyi*
>> Sauropoda indet.
>> Age: Late Jurassic (Haughton 1928)

186. Mashonaland North, Zimbabwe

> Kadsi Formation (Armstrong et al. 1967; Raath 1967)
>> Sauropoda
>>> ?Brachiosaurid indet.
>> Age: Late Jurassic (Molnar 1980b)

187. Mkoa wa Mtwara, Tanzania

> Tendaguru Beds (Janensch 1914, 1929b, 1935–36, 1955; Hennig 1924; Swinton 1950)
>> Theropoda
>>> Theropoda indet. (= *Ceratosaurus roechlingi, Megalosaurus ingens, Labrosaurus stechowi*)
>>> Ornithomimosauria *incertae sedis*
>>>> *Elaphrosaurus bambergi*
>>> Allosauridae
>>>> ?*Allosaurus tendagurensis*
>> Sauropoda
>>> Diplodocidae
>>>> *Dicraeosaurus hansemanni*
>>>> *Dicraeosaurus sattleri*
>>>> *Barosaurus africanus*
>>>> *Barosaurus gracilis*
>>> Brachiosauridae
>>>> *Brachiosaurus brancai* (including *B. fraasi*)
>>> Titanosauridae
>>>> *Janenschia robusta*
>> Ornithopoda
>>> Dryosauridae
>>>> *Dryosaurus lettowvorbecki*
>> Stegosauria
>>> Stegosauridae
>>>> *Kentrosaurus aethiopicus*
>> Dinosaur eggshell
>> Age: Kimmeridgian (Aitken 1961)

188. Cape Province, South Africa

> Upper Kirkwood Formation (Broom 1904; Galton and Coombs 1981; Rich et al. 1983)
>> Theropoda indet.
>> Sauropoda
>>> Sauropoda indet. (= *Algoasaurus bauri*)
>>> Camarasaurid indet.
>>> Brachiosauridae
>>>> ?*Pleurocoelus* sp. (including ?*Astrodon* sp.)
>> ?Ornithischian indet.
>>> Ornithopoda
>>>> ?Iguanodontid indet.
>> Stegosauria
>>> Stegosauridae
>>>> *Paranthodon africanus*
>> Age: middle Tithonian-early Valanginian (Rich et al. 1983)

EARLY CRETACEOUS

Dinosaur fossils of Early Cretaceous age are known from scattered localities in North America, Europe, Asia, South America, Africa, and Australasia.

North America

Important Lower Cretaceous localities are found primarily in the western interior of the United States, thanks to the early work by B. Brown and P. Kaisen of the American Museum of Natural History and later work by J. H. Ostrom of Yale University and J. A. Dorr, Jr., of the University of Michigan. Additional North American records include an extensive ichnofauna from British Columbia, Canada, originally described by C. M. Sternberg of the National Museum of Canada (later worked by A. G. Edmund of the Royal Ontario Museum and most recently by P. J. Currie of the Tyrrell Museum of Palaeontology), a new ichnofauna from Arkansas described by D. D. Gillette and D. A. Thomas (New Mexico Museum of Natural History), new material described by D. B. Weishampel (The Johns Hopkins University) and P. R. Bjork (South Dakota School of Mines) of *Iguanodon* from the Lakota formation of South Dakota, intriguing but poorly preserved material from the Arundel Formation of Maryland (studied by O. C. Marsh, R. S. Lull, and C. W. Gilmore of Yale University and the U.S. National Museum, respectively), and important Comanchean footprint and skeletal material from Oklahoma and central Texas, described in detail by W. Langston, Jr., of the Texas Memorial Mu-

seum. These Lower Cretaceous locations are plotted in figures 3.1, 3.13, 3.17, and 3.18.

189. British Columbia, Canada

   Gething Formation (Sternberg 1932b; Currie and Sarjeant 1979)
   Theropod footprints
   Ornithopod footprints
   ?Ceratopsian footprints
   Age: Barremian-Aptian (Stott 1975)

190. Alaska, United States

   Chandler Formation (Parrish et al. 1987)
   Ornithopod footprints
   Age: Albian-Cenomanian (Parrish et al. 1987)

191. Idaho, United States

   Wayan Formation (Dorr 1985)
   Ornithopoda
      Ornithopoda indet.
      Iguanodontia
         Tenontosaurus sp.
   Ankylosauria
      ?Nodosaurid indet.
   Dinosaur eggshell
   Age: Albian (Dorr 1985)

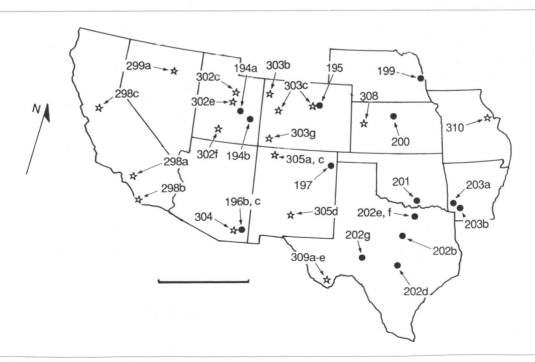

Fig. 3.17. Late Jurassic- to Late Cretaceous-age locations in the southwestern and south-central United States.

Scale = 600 km.

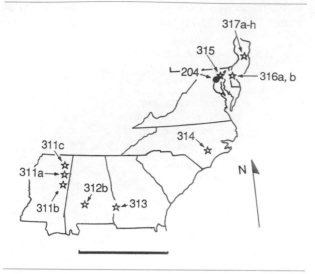

Fig. 3.18. Late Jurassic- to Late Cretaceous-age locations in the eastern and southeastern United States. Scale = 600 km.

192. Montana, United States

a. Cloverly Formation (Ostrom 1970*a*)
    Theropoda
        Theropoda indet.
        ?Maniraptora *sedis mutabilis*
            Microvenator celer
        Ornithomimid indet. (= *?Ornithomimus velox*)
        Dromaeosauridae
            Deinonychus antirrhopus
    Sauropoda
        Brachiosaurid indet. (= *?Pleurocoelus* sp.)
    Ornithopoda
        Hypsilophodontidae
            Zephyrosaurus schaffi
        Iguanodontia
            Tenontosaurus tilletti
    Ankylosauria
        Nodosauridae
            Sauropelta edwardsi
    Age: Aptian-Albian (Ostrom 1970*a*; late Aptian, Mateer et al. in press)

b. Kootenai Formation (Horner pers. comm.; not figured)
    Dinosauria indet.
    Age: Aptian (Cobban and Reeside 1952)

193. Wyoming, United States

a. Mowry or Thermopolis Shale (Marsh 1889*a*; Lull 1921; Coombs 1978*a*)
    Ankylosauria
        Nodosauridae
            Nodosaurus textilis
    Age: late Albian (Eicher 1960)

b. Cloverly Formation (Ostrom 1970*a*)
    Theropoda
        Theropoda indet.
        ?Maniraptora *sedis mutabilis*
            Microvenator celer
        Ornithomimid indet. (= *?Ornithomimus velox*)
        Dromaeosauridae
            Deinonychus antirrhopus
    Sauropoda
        Sauropoda indet.
        Brachiosauridae
            Pleurocoelus sp.
    Ornithopoda
        Iguanodontia
            Tenontosaurus tilletti
    Ankylosauria
        Nodosauridae
            Sauropelta edwardsi
    Age: Aptian-Albian (Ostrom 1970*a*; late Aptian, Mateer et al. in press)

c. Thomas Fork Formation (Dorr 1985)
    Dinosauria indet.
    Dinosaur eggshell
    Age: Albian (Dorr 1985)

d. ?Thermopolis Shale (Williston 1905)
    Ankylosauria
        Nodosauridae
            Nodosaurus textilis
    Age: Albian (Eicher 1960)

194. Utah, United States

a. Kelvin Formation (Jensen 1970)
    Dinosaur eggshell
    Age: Berriasian-Albian (Hintze 1973)

b. Cedar Mountain Formation (Bodily 1969; Jensen 1970; Galton and Jensen 1979*b*; Nelson and Crooks 1987; Madsen pers. comm.; Stadtman pers. comm.; Nelson pers. comm.; Britt pers. comm.)
    Theropoda
        Theropoda indet.
        Troodontid indet.
        Dromaeosauridae
            cf. *Deinonychus* sp.
            Dromaeosaurid indet.
    Sauropoda indet.
    Ornithischia indet.
    Ornithopoda
        Iguanodontia
            Tenontosaurus tilletti
        ?Iguanodontidae
            Undescribed ?iguanodontid (?2 species; Britt pers. comm.)
            ?Iguanodontid indet. (= *Iguanodon ottingeri*)
        ?Hadrosaurid indet.

Ankylosauria
Nodosauridae
Sauropelta edwardsi (= *Hoplitosaurus*
sp.)
Ankylosaurid indet.
Dinosaur eggshell
Age: Albian (Tschudy et al. 1984)

c. ?Dakota Formation (Marsh 1899; Carpenter pers.
comm.; not figured)
Sauropoda
Diplodocidae
?*Barosaurus lentus*
Ornithopoda
Hypsilophodontid indet.
Iguanodontid indet.
Age: late Aptian-early Cenomanian (Tschudy
et al. 1984)

195. Colorado, United States

Dakota Formation (Johnson 1931; Mehl 1931;
Lockley 1985)
Theropod footprints
Ornithopod footprints
Age: late Aptian-early Cenomanian (Lucas et
al. 1986)

196. Arizona, United States

a. Dakota Formation (unpublished; not figured)
Ornithopod footprints
Age: late Aptian-early Cenomanian (Tschudy
et al. 1984)

b. Unnamed unit (Miller 1964; Galton and Jensen
1979*b*)
Ornithopoda
Iguanodontia
Tenontosaurus sp.
Age: Aptian-Albian (Galton and Jensen
1979*b*)

c. Unnamed unit (Miller 1964)
?Stegosauria
?Stegosaurid indet.
Age: Early Cretaceous (Miller 1964)

197. New Mexico, United States

Dakota Formation (Thomas and Gillette 1985)
Theropod footprints
Ornithopod footprints
Age: late Aptian-early Cenomanian (Tschudy
et al. 1984)

198. South Dakota, United States

Lakota Formation (Lucas 1901; Anderson 1939;
Galton and Jensen 1979*b*; Bjork 1985*a*; Weishampel
and Bjork 1989)
Theropod footprints

Ornithopoda
Hypsilophodontid indet. (= *Hypsilophodon
wielandi*)
Iguanodontidae
Iguanodon lakotaensis
Camptosaurus depressus
Ankylosauria
Nodosauridae
Hoplitosaurus marshi
Age: Barremian (Sohn 1979)

199. Nebraska, United States

Dakota Sandstone (Barbour 1931; Galton and
Jensen 1979*b*)
Ornithopoda
?Hadrosaurid indet.
Age: late Aptian-early Cenomanian (Tschudy
et al. 1984)

200. Kansas, United States

Dakota Sandstone (Eaton 1960; McAllister 1989)
?Ornithopod footprints
Ankylosauria
Nodosauridae
Silvisaurus condrayi
Age: late Aptian-early Cenomanian (Tschudy
et al. 1984)

201. Oklahoma, United States

Antlers Formation (Larkin 1910; Langston 1974;
pers. comm.)
Theropoda
Allosauridae
Acrocanthosaurus atokensis
Sauropoda indct.
Sauropod footprints
Ornithopoda
Iguanodontia
Tenontosaurus cf. *tilletti*
Age: late Aptian-middle Albian (Langston
1974)

202. Texas, United States

a. Bissett Formation (Langston pers. comm.; not
figured)
Ornithopoda
Iguanodontidae
cf. *Iguanodon* sp.
Age: ?Berriasian-early Albian (Langston pers.
comm.)

b. Twin Mountains Formation (Jacobs pers. comm.)
Theropoda
Undescribed dromaeosaurid
Ornithopoda
Undescribed iguanodontid
Age: Aptian (Jacobs pers. comm.)

c. Bluff Dale Sandstone (Langston pers. comm.; not figured)
Sauropoda
Brachiosauridae
Pleurocoelus sp.
Age: late Aptian (Langston 1974; pers. comm.)

d. Glen Rose Formation (Langston 1974; Farlow 1987*a*)
Theropod footprints
Sauropoda
Sauropod footprints
Brachiosauridae
Pleurocoelus cf. *nanus*
Ornithopod footprints
Age: late Aptian-early Albian (Langston 1974)

e. Antlers Formation (Langston 1974)
Theropoda
Allosauridae
Acrocanthosaurus atokensis
Ornithopoda
Iguanodontia
Tenontosaurus cf. *tilletti*
Age: late Aptian-middle Albian (Langston 1974)

f. Paluxy Formation (Langston 1974)
Theropoda
Theropoda indet.
?Dromaeosaurid indet.
Ornithomimidae
?Ornithomimus sp.
Sauropoda
Brachiosauridae
Pleurocoelus cf. *nanus*
?Pleurocoelus sp.
Ornithischia indet.
Ornithopoda
Iguanodontia
Tenontosaurus cf. *tilletti*
Ankylosauria
Nodosauridae
?Nodosaurus sp.
Age: early-middle Albian (Langston 1974)

g. Edwards Formation (Hawthorne 1987)
Dinosauria footprints
Age: middle Albian (Stephenson et al. 1942)

h. Paw Paw Formation (unpublished; not figured)
Ankylosauria
Nodosaurid indet.
Age: middle Albian (Stephenson et al. 1942)

203. Arkansas, United States

a. DeQueen Formation (Gillette et al. 1985)
Sauropod footprints
Age: Albian (Gillette et al. 1985)

b. Trinity Formation (Quinn 1973)
Theropoda
Ornithomimid indet.
Age: Early Cretaceous (Quinn 1973)

204. Maryland, United States

Arundel Formation (Lull 1911*a*, 1911*b;* Gilmore 1921; Ostrom 1970*a;* Galton and Jensen 1979*b*)
Theropoda
Theropoda indet. (= *Allosaurus medius, Coelurus gracilis, Creosaurus potens*)
Ornithomimid indet. (= *Coelosaurus affinis*)
Sauropoda
Brachiosauridae
Pleurocoelus altus (including *Astrodon johnstoni*)
Pleurocoelus nanus
Ornithopoda
Iguanodontia
Tenontosaurus sp.
Ankylosauria indet. (= *Priconodon crassus*)
Age: Hauterivian-Barremian (Ostrom 1970*a*)

Europe

The Wealden and Wealden equivalents have produced the greatest number of dinosaur faunas of the European Early Cretaceous (figs. 3.1, 3.5, 3.19, and 3.20). Important localities include those of England (through the work of G. A. Mantell, R. Owen, W. Buckland, H. G. Seeley, J. W. Hulke, and recently, W. T. Blows and A. J. Charig), Belgium (L. Dollo), France (H.-E. Sauvage, A. F. de Lapparent and V. Stchepinsky), Spain (J. V. Santafe and co-workers, J. L. Sanz and A. Buscalioni, and R. Estes and B. Sanchiz), and Federal Republic of Germany (E. Koken, K. Oekentorp, H. Hölder, and D. B. Norman).

205. Svalbard, Norway

Festningen Sandstone (= Helvetiafjellet Formation; Lapparent 1962; Edwards et al. 1978)
Theropod footprints
Ornithopod footprints
Age: Barremian (Lapparent 1962)

206. Bedfordshire, England

Wealden (Seeley 1874; Benton pers. comm.; Norman pers. comm.)
Theropoda
Theropoda *incertae sedis*
Altispinax dunkeri
Theropoda indet. (= *?Megalosaurus bucklandii*, reworked from Potton Sands)

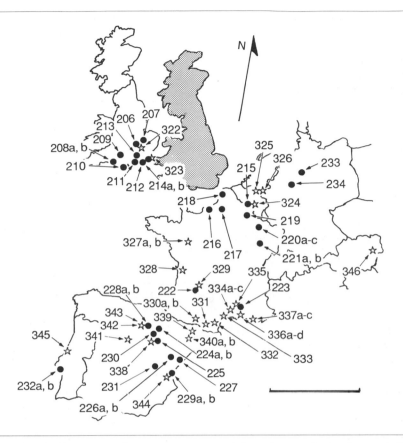

Fig. 3.19. Early and Late Cretaceous-age locations in England (removed slightly to the west), the Netherlands, Belgium, France, Spain, Portugal, Federal Republic of Germany, and Austria. Scale = 600 km.

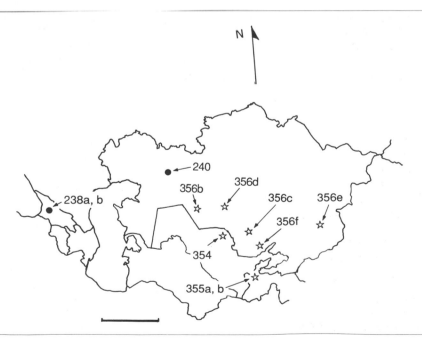

Fig. 3.20. Early and Late Cretaceous-age locations in Georgia, Kazakhstan, Uzbekistan, and Tadzikistan, central Asia. Scale = 600 km.

Ornithopoda
  Iguanodontidae
    Iguanodon cf. *atherfieldensis*
Stegosauria
  Stegosauridae
    Craterosaurus pottonensis (reworked
      from Potton Sands)
Ankylosauria
  Nodosauridae
    ?*Hylaeosaurus* sp.
  Age: Valanginian-Barremian (Rawson et al.
    1978)

207. **Cambridgeshire, England**

Cambridge Greensand (Seeley 1879; Lydekker
1888a)
  Sauropoda
    Titanosauridae
      Macrurosaurus semnus
    ?Iguanodontid indet. (= *Anoplosaurus*
      curtonotus, A. major, Acantholis
      macrocercus partim, *A. stereocercus*
      partim, *Eucercosaurus tanyspondylus*)
    Hadrosaurid indet. (= *Trachodon*
      cantabrigiensis)
  Ankylosauria
    ?Ankylosauria indet. (= *Acantholis*
      eucercus)
    Nodosauridae
      Acantholis horridus (including *A.*
      macrocercus partim, *A. platypus*
      partim, *A. stereocercus* partim)
  Age: late Albian (reworked into Cenomanian;
    Rawson et al. 1978)

208. **Dorset, England**

a. Upper Purbeck Beds (Owen 1874; Damon 1884)
  Theropoda
    Theropoda *incertae sedis*
      Altispinax dunkeri
  Ornithopoda
    Iguanodontidae
      Iguanodon hoggi
  Ankylosauria
    Nodosauridae
      ?*Hylaeosaurus* sp.
  Age: Valanginian-Barremian (Rawson et al.
    1978)

b. "Wealden Beds" (= Lower Greensand; Delair
1966)
  Theropoda indet. (= *Megalosaurus* sp.)
  Ornithopoda
    Iguanodontidae
      Iguanodon sp.
  Age: Barremian-early Aptian (Rawson et al.
    1978)

209. **Wiltshire, England**

Upper Purbeck Beds (Huddleston 1876; Delair 1973)
  Theropoda indet.
  Age: Berriasian (Rawson et al. 1978)

210. **Isle of Wight, England**

Wealden (Hulke 1874, 1881; Lydekker 1889d;
Galton 1971c; Charig 1980; Delair and Sarjeant
1985; Blows 1987; Norman pers. comm.)
  Theropoda
    Theropoda *incertae sedis*
      Altispinax dunkeri
    Theropoda indet. (= *Calamospondylus*
      oweni, C. foxi, Thecospondylus daviesi,
      Megalosaurus sp.)
    Theropod footprints
  Sauropoda
    Diplodocid indet.
    Brachiosauridae
      Pelorosaurus conybearei (including
      Ornithopsis hulkei, Bothriospondylus
      magnus, Ornithopsis eucamerotus)
      Pleurocoelus valdensis
    Camarasauridae
      Chondrosteosaurus gigas
    Titanosauridae
      ?*Macrurosaurus semnus* (=
      Titanosaurus lydekkeri)
    Titanosauridae *incertae sedis*
      "*Titanosaurus*" *valdensis*
  Ornithopoda
    Ornithopod footprints
    Hypsilophodontidae
      Hypsilophodon foxii
      Hypsilophodontid indet. (=
      Camptosaurus valdensis)
    Dryosauridae
      Valdosaurus canaliculatus
    Iguanodontidae
      Iguanodon bernissartensis (including *I.*
      seeleyi)
      Iguanodon atherfieldensis (including
      Vectisaurus valdensis,
      Sphenospondylus gracilis)
  Pachycephalosauria
    Pachycephalosauridae
      Yaverlandia bitholus
  Ankylosauria
    Nodosauridae
      Hylaeosaurus armatus (including
      Polacanthoides ponderosus,
      Polacanthus foxii)
  Age: Valanginian-Barremian (Rawson et al.
    1978)

211. West Sussex, England

Wealden (Harrison and Walker 1973; Galton 1981a;
Norman pers. comm.)
    Theropoda
        Theropoda *incertae sedis*
            Altispinax dunkeri
        Theropoda indet. (= *Wyleyia valdensis,*
            Megalosaurus oweni)
    Sauropoda
        Cetiosauridae
            Cetiosaurus oxoniensis
        Brachiosauridae
            Pelorosaurus conybearei
        Camarasauridae
            Chondrosteosaurus gigas
    Ornithopoda
        Dryosauridae
            Valdosaurus canaliculatus
        Iguanodontidae
            Iguanodon anglicus (including
                Cetiosaurus brevis, Streptospondylus
                major)
            Iguanodon atherfieldensis
            ?Iguanodon bernissartensis
    Stegosauria
        Stegosauridae
            Craterosaurus pottonensis
    Ankylosauria
        Nodosauridae
            Hylaeosaurus armatus
    Age: Valanginian-Barremian (Rawson et al.
      1978)

212. East Sussex, England

Wealden (Lydekker 1888a, 1889b; Delair and
Sarjeant 1985; Norman pers. com.)
    Theropoda
        Theropoda *incertae sedis*
            Altispinax dunkeri
            Becklespinax altispinax
        Theropoda indet.
    Sauropoda
        Brachiosauridae
            Pelorosaurus conybearei
            Pleurocoelus valdensis
    Ornithopoda
        Ornithopod footprints
        Hypsilophodontidae
            Hypsilophodon sp. (?= *H. foxii*)
        Iguanodontidae
            Iguanodon atherfieldensis
            Iguanodon dawsoni
            Iguanodon fittoni (including *I.*
              hollingtoniensis)
    Ankylosauria
        Nodosauridae
            Hylaeosaurus armatus
    Age: Valanginian-Barremian (Rawson et al.
      1978)

213. Surrey, England

Wealden (Delair and Sarjeant 1985; Charig and
Milner 1986; Benton pers. comm.; Norman pers.
comm.)
    Theropoda
        Theropoda *incertae sedis*
            Baryonyx walkeri
    Ornithopoda
        Ornithopod footprints
        Iguanodontidae
            Iguanodon atherfieldensis
    Age: Valanginian-Barremian (Rawson et al.
      1978)

214. Kent, England

  a. Wealden (Seeley 1882; Norman pers. comm.)
        Theropoda *incertas sedis*
            Valdoraptor oweni
        Theropoda indet. (= *Thecospondylus horneri*)
        Sauropoda
            Brachiosauridae
                Pelorosaurus conybearei
        Ornithopoda
            Iguanodontidae
                Iguanodon atherfieldensis
        Age: Valanginian-Barremian (Rawson et al.
          1978)
  b. Lower Greensand (Owen 1884)
        Sauropoda
            Brachiosauridae
                Pelorosaurus mackesoni (= *Dinodocus*
                  mackesoni)
        Ornithopoda
            Iguanodontidae
                Iguanodon atherfieldensis
        Age: Aptian (Rawson et al. 1978)

215. Province de la Hainaut, Belgium

Wealden (Dollo 1882, 1883a, 1883b, 1884, 1909,
1923; Dames 1884; Norman 1986)
    Theropoda
        Theropoda *incertae sedis*
            Altispinax dunkeri
    Ornithopoda
        Iguanodontidae
            Iguanodon bernissartensis
            Iguanodon atherfieldensis
    Age: Barremian/Aptian (Taquet 1976; Norman
      1980, pers. comm.)

216. Département de la Seine-Maritime, France

Unnamed unit (Buffetaut 1984)
    Sauropoda indet.
    Age: Albian (Buffetaut 1984)

217. Département de l'Oise, France

Unnamed unit (Lapparent 1946, 1967)
    Sauropoda indet.
    Age: Albian (Lapparent 1967)

218. Département du Pas-de-Calais, France

    a. Unnamed unit (Lapparent 1967)
        Theropoda indet. (= *Megalosaurus insignis*)
        Age: Valanginian (Lapparent 1967)

    b. Unnamed unit (Sauvage 1876*a*)
        Theropoda *incertae sedis*
            Erectopus superbus
        Age: Albian (= Gault; Huene 1932)

219. Département des Ardennes, France

    Unnamed unit (Lapparent 1967)
        Ankylosauria
            Nodosauridae
                Hylaeosaurus armatus
        Age: Albian (Lapparent 1967)

220. Département de la Meuse, France

    a. Unnamed unit (Lapparent and Stchepinsky 1968)
        Ornithopoda
            Iguanodontid indet.
        Age: Barremian (Lapparent and Stchepinsky 1968)

    b. Unnamed unit (Lapparent 1967)
        Theropoda indet. (= *Megalosaurus* sp.)
        Age: Albian (Lapparent 1967)

    c. Unnamed unit (Sauvage 1882; Huene 1926*a*, 1932)
        Theropoda
            Theropoda *incertae sedis*
                Erectopus superbus (= *Megalosaurus superbus* partim)
                Erectopus sauvagei (= *Megalosaurus superbus* partim)
        Age: Albian (= Gault; Huene 1932; Lapparent 1967)

221. Département de la Haute-Marne, France

    a. Unnamed unit (Lapparent and Stchepinsky 1968)
        Ornithopoda
            Iguanodontid indet.
        Age: Hauterivian (Lapparent and Stchepinsky 1968)

    b. Unnamed unit (Lapparent and Stchepinsky 1968)
        Ornithopoda
            Iguanodontid indet.
        Age: Aptian (Lapparent and Stchepinsky 1968)

222. Département du Dordogne, France

    Craie Tufau de Perigueux (Gervais 1859)
        Sauropoda *incertae sedis*
            ?*Aepysaurus* sp.
        Age: Early Cretaceous (unconfirmed)

223. Département du Vaucluse, France

    Grès vert (Gervais 1852)
        Sauropoda *incertae sedis*
            Aepysaurus elephantinus
        Age: Aptian-?Albian (Lapparent 1967; Buffetaut pers. comm.)

224. Provincia de La Rioja, Spain

    a. Enciso Formation (Aguirrezabala et al. 1985; Casanovas-Cladellas, Perez-Lorente et al. 1985; Sanz et al. 1985)
        Theropod footprints
        Ornithopod footprints
        Age: ?late Valanginian (Salomon 1982)

    b. Unnamed unit (Viera and Aguirrezabala 1982)
        Theropod footprints
        Ornithopod footprints
        Age: Barremian (Viera and Aguirrezabala 1982)

225. Provincia de Soria, Spain

    Unnamed unit (Aguirrezabala and Viera 1983)
        Theropod footprints
        Ornithopod footprints
        Age: Barremian (Aguirrezabala and Viera 1983)

226. Provincia de Teruel, Spain

    a. Las Zabacheras Beds (Sanz 1982; Estes and Sanchiz 1982; Sanz et al. 1984; Sanz et al. 1987; Buscalioni and Sanz 1987)
        Theropoda indet.
        Sauropoda
            Camarasauridae
                Aragosaurus ischiaticus
            Brachiosauridae
              cf. *Pleurocoelus* sp. (= cf. *Astrodon* sp.)
              Brachiosaurid indet.
        Ornithischia indet. (= ?*Echinodon* sp.)
        Ornithopoda
            Hypsilophodontidae
                Hypsilophodon foxii
            Dryosauridae
              cf. *Valdosaurus* sp.
            Iguanodontidae
                Iguanodon cf. *atherfieldensis*
                Iguanodon bernissartensis
        Age: early Barremian (Sanz et al. 1987)

    b. Unnamed unit (Casanovas-Cladellas et al. 1984)
        Theropod footprints
        Age: Early Cretaceous (Casanovas-Cladellas et al. 1984)

227. Provincia de Castellón, Spain

Capas Rojas of Morella (Santafe et al. 1979, 1982;
Sanz et al. 1983)
    Theropoda indet. (2 species)
    Sauropoda
        Brachiosaurid indet.
    Ornithopoda
        Hypsilophodontidae
            Hypsilophodon sp.
        Iguanodontidae
            Iguanodon bernissartensis
    Ankylosauria
        ?Nodosaurid indet.
    Age: Aptian (= early Bedoulian; Santafe et al.
    1982)

228. Provincia de Burgos, Spain

a. Unnamed unit (Sanz 1983)
    Ankylosauria
        Nodosaurid indet.
    Age: Barremian-Albian (Sanz 1983)
b. Unnamed unit (unpublished)
    Ornithopoda
        Iguanodontid indet.
    Age: ?Early Cretaceous (unconfirmed)

229. Provincia de Valencia, Spain

a. Unnamed unit (Bataller 1960)
    Theropoda indet (= *Megalosaurus* cf. *dunkeri*)
    Sauropoda
        Cetiosauridae
            Cetiosaurus sp.
    Age: Valanginian-Barremian (i.e., = Wealden;
    Bataller 1960)
b. Unnamed unit (Sanz 1985)
    Sauropoda
        Brachiosaurid indet.
    Age: Hauterivian-Aptian (Sanz 1985)

230. Provincia de Logroño, Spain

Unnamed unit (Casanovas-Cladellas and Santafe-
Llopis 1971, 1974; Viera and Angel Torres 1979)
    Theropod footprints
    Ornithopod footprints
    Age: Early Cretaceous (Casanovas-Cladellas
    and Santafe-Llopis 1971)

231. Provincia de Cuenca, Spain

Unnamed unit (Sanz 1985)
    Ornithopoda
        Iguanodontidae
            Iguanodon bernissartensis
    Age: Valanginian-Hauterivian (Sanz 1985)

232. Província do Estremadura, Portugal

a. Unnamed unit (Antunes 1976; Maderia and Dias
1983)
    Theropod footprints
    Sauropod footprints
    Ornithopod footprints
    Age: Hauterivian (Antunes 1976)
b. Unnamed unit (Lapparent and Zbyszewski 1957)
    Ornithopoda
        Iguanodontidae
            Iguanodon cf. *atherfieldensis*
    Age: Aptian-Albian (Lapparent and
    Zbyszewski 1957)

233. Niedersachsen, Federal Republic of Germany

Obernkirchen Sandstein (= Wealden; Meyer 1857;
Koken 1887; Ballerstedt 1922; Nopcsa 1923*b*; Kuhn
1958*b*; Sues and Galton 1982)
    Theropoda
        Theropoda *incertae sedis*
            Altispinax dunkeri
        Theropod footprints
    ?Ornithischian footprints
    Ornithopoda
        Ornithopod footprints
        Iguanodontidae
            Iguanodon sp.
    Marginocephalia
        Stenopelix valdensis
    Age: Berriasian (Schmidt 1969)

234. Nordrhein-Westphalen, Federal Republic of Germany

Wealden (Huckreide 1982; Hölder and Norman
1986; Norman 1987*b*)
    Theropoda indet. (= *Megalosaurus* sp.)
    Ornithopoda
        ?Hypsilophodontid indet. (=
            Hypsilophodon sp.)
        Iguanodontidae
            Iguanodon atherfieldensis
            Iguanodon bernissartensis
    Age: Aptian (Huckreide 1982)

235. Judethean Bihor, Romania

Bauxite of Cornet (Jurcsák and Popa 1979, 1983;
Jurcsák 1982; Jurcsák and Kessler 1985)
    Theropoda indet. (= *Aristosuchus* sp.)
    Ornithopoda
        Hypsilophodontidae
            Hypsilophodon sp.
            Hypsilophodontid indet. (= cf.
               Thescelosaurus sp.)
        Dryosauridae
            Valdosaurus canaliculatus
            ?Dryosaurid indet. (= *Dryosaurus* sp.)
        Iguanodontidae
            Iguanodon sp. (= *I. mantelli*)

Age: Berriasian-Hauterivian (Patrulius et al. 1983)

236. Judethean Tulcea, Romania

    Unnamed unit (Simionescu 1913)
        Theropoda indet. (= *Megalosaurus* cf. *superbus*)
        Age: Valanginian-Barremian (Simionescu 1913)

237. Istra, Croatia

    Unnamed unit (Bachofen-Echt 1925)
        Theropod footprints
        Age: Berriasian-Hauterivian (= Neocomian; Bachofen-Echt 1925)

238. Georgia

    a. Unnamed unit (Gabuniya 1951; Kurzanov pers. comm.)
        Theropod footprints
        Ornithopod footprints
        Age: Berriasian-Hauterivian (= Neocomian; Rozhdestvensky 1973*b*)

    b. Unnamed unit (Gabuniya 1951; Mateer pers. comm.)
        Undescribed Ornithopoda
        Age: early Hauterivian (Kotetishvilii 1986)

Asia

Asiatic records of Early Cretaceous dinosaurs compare well with those of the Late Cretaceous on this continent as well as elsewhere (figs. 3.7, 3.16, 3.20, 3.21). Studies of Early Cretaceous dinosaurs in Asia include those by A. K. Rozhdestvensky, C.-C. Young, Dong Z.-M., H. F. Osborn, W. Granger, B. Brown, and E. Schlaikjer, and most recently, H. M. Kim. (A series of ornithopod trackways is reported from Xinjiang Uygur Zizhiqu, People's Republic of China [Gao et al. 1981], but they are dated only as Cretaceous in age [Locality 239].)

240. Aktyubinskaya Oblast, Kazakhstan
    Neocomian Sands (Riabinin 1931*b*)
        Theropoda indet. (= *Embasaurus minax*)
        Age: Berriasian-Hauterivian (= Neocomian; Riabinin 1931*b*)

241. Yakutskaya Avtonomnaya Oblast, Russia

    Sangarskaya Svita (Rozhdestvensky 1973*b*)
        Ankylosauria
            Ankylosaurid indet.

Age: Berriasian-Hauterivian (= Neocomian; Rozhdestvensky 1973*b*)

242. Gorno-Altayaskaya Avtonomnaya Oblast, Russia

    Shestakovskaya Svita (= Ilekskaya Svita; Rozhdestvensky 1955*b*, 1960)
        Ceratopsia
            Psittacosauridae
                Psittacosaurus mongoliensis
        Age: Berriasian-Hauterivian (= Neocomian; Rozhdestvensky 1955*b*, 1960, 1973*b*)

243. Buriat-mongolskaya Avtonomnaya Oblast, Russia

    a. Unnamed unit (Rozhdestvensky 1968*b*, 1970*b*; Kurzanov pers. comm.)
        Theropoda indet. (= ?*Therizinosaurus* sp.)
        Sauropoda indet.
        Age: Early Cretaceous (Kurzanov pers. comm.; Osmólska pers. comm.)

    b. Turginskaya Svita (Riabinin 1914)
        Theropoda
            Allosauridae
                ?*Chilantaisaurus sibiricus*
        Age: Berriasian-Hauterivian (= Neocomian; Osmólska pers. comm.)

244. Khovd, Mongolian People's Republic

    Tsagantsabskaya Svita (Osmólska pers. comm.; not figured)
        Sauropoda indet.
        Ceratopsia
            Psittacosaurid indet.
        Age: Valanginian (Osmólska pers. comm.)

245. Bayankhongor, Mongolian People's Republic

    Unnamed unit (Kalandadze and Kurzanov 1974)
        Ceratopsia
            Psittacosauridae
                Psittacosaurus mongoliensis
        Ankylosauria
            Ankylosaurid indet.
        Age: Aptian-Albian (Shuvalov 1974)

246. Omnogov, Mongolian People's Republic

    Sainshandinskaya Svita (Osmólska pers. comm.; Kurzanov pers. comm.; not figured)
        Dinosaur eggshell
        Age: Albian-Cenomanian (Kurzanov pers. comm.; Osmólska pers. comm.)

247. Ovorkhangai, Mongolian People's Republic

    Khukhtekskaya Svita (= Ondai Sair and Oshih formations; Osborn 1923*a*, 1924*c*; Rozhdestvensky 1952*a*; Martinsson and Shuvalov 1973; Tumanova 1985)

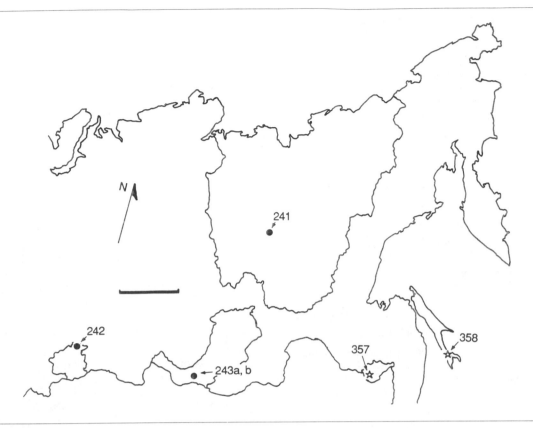

Fig. 3.21. Early and Late Cretaceous-age locations in Ya-kutskaya Avtonomnaya Oblast, Buriat-mongolskaya Auto-nomnaya Oblast, Gorno-Altajaskaya Avtonomnaya Oblast, Yevrayaskaya Avtonomnaya Oblast, and Sachalinskaya Oblast, Russia, north-central and eastern Asia. Scale = 600 km.

Theropoda indet. (= *Prodeinodon mongoliensis*)
Sauropoda indet. (= *Asiatosaurus mongoliensis*)
Ornithopoda
    Iguanodontidae
        "Iguanodon" orientalis
        Iguanodontid indet.
Ankylosauria
    Ankylosauridae
        Shamosaurus scutatus
Ceratopsia
    Psittacosauridae
        Psittacosaurus mongoliensis (including
          Protiguanodon mongoliensis,
          Psittacosaurus protiguanodonensis)
    Age: Aptian-Albian (Shuvalov 1974)

248. Dundgov, Mongolian People's Republic

  a. Khulsyngolskaya Svita (Shuvalov 1974;
     Osmólska pers. comm.)

Theropoda
    Dromaeosaurid indet.
Ceratopsia
    Psittacosauridae
        Psittacosaurus mongoliensis
    Age: Aptian-Albian (Osmólska pers. comm.)

b. Shinekhudukskaya Svita (= Khukhtekskaya
   Svita; Barsbold and Perle 1984)
   Theropoda
      Harpymimidae
         Harpymimus okladnikovi
   Ornithopoda
      Iguanodontidae
         "Iguanodon" orientalis
   Ceratopsia
      Psittacosauridae
         Psittacosaurus mongoliensis
   Age: Aptian-Albian (Barsbold and Perle 1984;
      Hauterivian-Barremian, Shuvalov 1982)

249. Dornogov, Mongolian People's Republic

a. Baiying Bologai Formation (Gilmore 1933*b;*
Coombs 1987)
Ornithopoda
?Hadrosaurid indet.
?Ankylosauria
?Ankylosaurid indet.
Age: ?early Berriasian (Morris 1936)

b. Khukhtekskaya Svita (= Upper Dzunbayinskaya
Svita; Barsbold and Perle 1983; Osmólska pers.
comm.)
Theropoda
Theropoda indet.
Troodontid indet.
Ankylosauria
Ankylosauridae
Shamosaurus scutatus
Ceratopsia
Psittacosauridae
Psittacosaurus mongoliensis
Age: Aptian-Albian (Osmólska pers. comm.)

250. Xinjiang Uygur Zizhiqu, People's Republic of China

Lianmugin Formation (Dong 1973*a,* pers. comm.;
Sereno and Chao 1988; Shen and Mateer in press)
Theropoda
Theropoda indet. (= *Phaedrolosaurus
ilikensis, Tugulusaurus faciles*)
?Carnosauria
Kelmayisaurus petrolicus
Sauropoda indet. (= *Asiatosaurus mongoliensis*)
Ornithopoda
Hadrosaurid indet.
Stegosauria
Stegosauridae
Wuerhosaurus homheni
Ceratopsia
Psittacosauridae
Psittacosaurus xinjiangensis
Age: ?Valanginian-Albian (Shen and Mateer in
press)

251. Gansu, People's Republic of China

Unnamed unit (Bohlin 1953)
Theropoda indet.
Sauropoda
Camarasauridae
?*Chiayusaurus* sp.
Sauropoda *incertae sedis*
?*Mongolosaurus* sp.
Ankylosauria indet. (= *Stegosaurides excavatus*)
Age: ?Early Cretaceous (?equivalent to the
Xinminpu Group; Shen 1981)

252. Nei Mongol Zizhiqu, People's Republic of China

a. ?Bayanhua Formation (Mateer pers. comm.; =
On Gong Formation, Gilmore 1933*b;* "Tairum
Nor" formation, Reeds 1932; Gilmore 1933*b;*
"Shirigu" formation, Reeds 1932)
Dinosauria indet.
Theropoda
Ornithomimid indet.
Sauropoda *incertae sedis*
Mongolosaurus haplodon
Age: Berriasian-Albian (Mateer pers. comm.)

b. Dashuigou Formation (Chow and
Rozhdestvensky 1960)
Ornithopoda
Probactrosaurus alashanicus
Probactrosaurus gobiensis
Age: Aptian-Albian (Rozhdestvensky 1966)

c. Lisangou Formation (Young 1931; Cheng 1983)
Ceratopsia
Psittacosauridae
Psittacosaurus guyangensis
Psittacosaurus osborni
Age: ?Aptian-Albian (Tan 1983)

d. Xinpongnaobao Formation (Young 1931)
Ceratopsia
Psittacosauridae
Psittacosaurus osborni
Age: ?Aptian-Albian (Dong pers. comm.)

e. Unnamed unit (Hu 1964)
Theropoda
Allosauridae
Chilantaisaurus tashuikouensis
Age: Aptian-?Albian (Dong pers. comm.,
contra Dong 1980)

f. Unnamed unit (Hu 1964)
Theropoda
Allosauridae
Chilantaisaurus maortuensis
Age: Albian (Dong pers. comm.)

g. Unnamed unit (Bohlin 1953)
Theropoda indet. (= *Prodeinodon* sp.)
Ankylosauria indet. (= *Sauroplites scutiger*)
Ceratopsia
Psittacosauridae
?*Psittacosaurus mongoliensis*
Age: ?Early Cretaceous (?equivalent to the
Xinminpu Group; Shen 1981)

253. Liaoning, People's Republic of China

a. Jiufotang Formation (Sereno et al. 1988)
Ceratopsia
Psittacosauridae
Psittacosaurus meileyingensis
Age: Early Cretaceous (Sereno et al. 1988)

b. Heichengtzu Series (Yabe et al. 1940; Young 1960)
Theropod footprints
Age: ?Early Cretaceous (Yabe et al. 1940; Young 1960)

254. Yunnan, People's Republic of China

Upper Jingxing Formation (Yeh 1975)
Theropoda indet. (= *Tyrannosaurus lanpingensis*)
Age: Berriasian-Hauterivian (Chen 1983)

255. Shanxi, People's Republic of China

Unnamed unit (Young 1958b)
Theropoda
Allosaurid indet.
Sauropoda indet.
Age: Early Cretaceous (Mateer pers. comm.)

256. Sichuan, People's Republic of China

Jiaguan Formation (Zhen et al. 1987)
Theropod footprints
Ornithopod footprints
Age: Aptian-Albian (Hao and Guan 1984)

257. Guangxi Zhuangzu Zizhiqu, People's Republic of China

Napai Formation (Hou et al. 1975; Dong 1980)
Theropoda indet. (= *Prodeinodon kwangshiensis*)
Sauropoda indet. (= *Asiatosaurus kwangshiensis*)
Ornithopoda indet.
Age: ?Aptian-Albian (?equivalent to Oshih Formation; Early Cretaceous, Dong pers. comm.)

258. Shandong, People's Republic of China

Qingshan Formation (Young 1958a; Chao 1962)
Ceratopsia
Psittacosauridae
Psittacosaurus sinensis
Psittacosaurus youngi
Age: Aptian-Albian (Mateer et al. MS)

259. Xizang Zizhiqu, People's Republic of China

Unnamed unit (Hao et al. 1986)
Undescribed theropods (?2 species)
Undescribed sauropods
Age: Early Cretaceous (Hao et al. 1986)

260. Kyongsang Pukdo, South Korea

a. Dogyedong Formation (Kim 1983, 1986)
Theropoda indet. (= ?*Deinonychus* sp.)
Sauropoda indet. (= *Ultrasaurus tabriensis*)
Ornithopoda
?Hypsilophodontid indet.
Age: Aptian-Albian (Choi 1985)

b. Chilgog Formation (Kim 1983)
Sauropoda indet.
Age: Berriasian-Hauterivian (Choi 1986)

261. Kyongsang Namdo, South Korea

a. Iljig Formation (Kim 1983)
?Sauropoda indet.
Age: Berriasian-Hauterivian (Choi 1986)

b. Hasandong Formation (Kim 1983)
Sauropoda indet.
Age: Aptian-Albian (Choi 1986)

c. Nagdong Formation (Kim 1983)
Sauropoda indet.
Age: Aptian-Albian (Choi 1986)

d. Jindong Formation (Kim 1983)
Theropod footprints
Sauropod footprints
Ornithopod footprints
Age: ?Albian (Choi 1985, 1986)

e. Haman Formation (Kim 1983)
Ornithopod footprits
Age: Aptian-Albian (Choi 1986)

262. Changwat Nakhon Ratchasima, Thailand

Khok Kruat Formation (Buffetaut 1982; Buffetaut and Ingavat 1986a)
Theropoda indet.
Age: ?Aptian-Albian (Buffetaut pers. comm.)

263. Changwat Loei, Thailand

Phu Phan Formation (Buffetaut, Ingavat et al. 1985)
Theropod footprints
Age: Early Cretaceous (Sattayarak 1983)

264. Kumamoto Prefecture, Japan

Mihunuye Formation (Mateer pers. comm.)
Theropoda indet.
Age: ?Albian (Mateer pers. comm.)

265. Gumma Prefecture, Japan

Sebayashi Formation (Matsukawa and Obata 1985; Mateer pers. comm.)
Theropoda indet.
?Theropod footprints
Ornithopod footprints
Age: Aptian-Cenomanian (Matsukawa and Obata 1985; Mateer pers. comm.)

South America

Twelve South American dinosaur locations are presently known from Lower Cretaceous strata (figs. 3.8, 3.9). Four of these are known to include a variety of skeletal remains, including new and very interesting saurischian material from southern Argentina described by J. F. Bonaparte of Museo Argentino de Ciencias Naturales. The remaining Early Cretaceous dinosaur occurrences consist of a wide range of ichno-faunas currently under study by G. Leonardi of Brazil.

266. Estado do Goiás, Brazil

    Corda Formation (Leonardi 1980*b*)
      Sauropod footprints
      Ornithopod footprints
      Age: pre-Aptian (Leonardi 1989)

267. Estado do Bahia, Brazil

    Bahia Series (Mawson and Woodward 1907)
      Theropoda indet.
      Ornithopoda
         Iguanodontid indet.
      Age: Aptian (Taquet 1977)

268. Estado do Ceará, Brazil

    a. Antenor Navarro Formation (Leonardi 1989)
      Theropod footprints
      Age: pre-Aptian (Leonardi 1989)

    b. Santana Formation (Leonardi 1981*b*)
      ?Ornithischia indet.
      Age: Aptian (Leonardi 1981*b*)

269. Estado do Paraíba, Brazil

    a. Antenor Navarro Formation (Leonardi 1981*b*)
      Theropod footprints
      Sauropod footprints
      Ornithopod footprints
      Age: pre-Aptian (Leonardi 1989)

    b. Sousa Formation (Leonardi 1979*b*)
      Theropod footprints
      Sauropod footprints
      Ornithopod footprints
      ?Ankylosaur footprints
      Age: pre-Aptian (Leonardi 1989)

    c. Piranhas Formation (Leonardi 1981*b*)
      Theropod footprints
      Ornithopod footprints
      Age: pre-Aptian (Leonardi 1989)

270. Estado do Paraná, Brazil

    Caiua Formation (Leonardi 1989)
      Theropod footprints
      Age: Early Cretaceous (Leonardi 1989)

271. Provincia de Malleco, Chile

    Banos del Flaco Formation (Casamiquela and Fasola 1968)
      Ornithopod footprints
      Age: Hauterivian (Diaz and Gasparini 1976)

272. Provincia de San Luis, Argentina

    La Cantera Formation (Lull 1942; not Paganzo Beds III, Bonaparte pers. comm.)
      Theropod footprints
      Age: Berriasian-Hauterivian (= Neocomian, Leonardi 1989)

273. Provincia de Chubut, Argentina

    a. Gorro Frigio Formation (Corro 1966, 1975; Bonaparte and Gasparini 1978; Bonaparte 1985)
      Theropoda
         Theropoda indet. (= *Megalosaurus inexpectatus*)
         Abelisauridae
            Carnotaurus sastrei
      Sauropoda
         Brachiosauridae
            Chubutisaurus insignis
      Age: Albian (Bonaparte 1985)

    b. La Amarga Formation (Bonaparte pers. comm.; not figured)
      Sauropoda
         Diplodocid indet.
      Age: Hauterivian (Bonaparte pers. comm.)

Africa

The African record of the Lower Cretaceous has yielded a variety of dinosaur fossils, primarily through the work of C. Deperet and J. Savornin, R. Lavocat, A. F. de Lapparent, P. Taquet, and M. Monbaron in the "Continental intercalaire" and Elrhaz Formation of Morocco, Algeria, Tunisia, Libya, Mali, and Niger, as well as the Kalahari Deposits of South Africa. (Brunet et al. [1986] noted a Cretaceous dinosaur locality in North Cameroun Oriental Province, Cameroun, but do not indicate the unit from which it comes, whether it is from Lower or Upper Cretaceous, or the material that

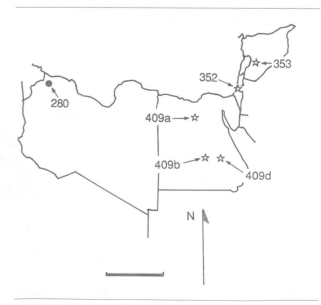

Fig. 3.22. Early and Late Cretaceous-age locations in Libya, Egypt, Israel, and Syria. Scale = 600 km.

comes from it [Locality 274].) Figures 3.10, 3.11, and 3.22 indicate the positions of these localities.

275. Province de Kasr-es-Souk, Morocco

Tegana Formation (Lavocat 1954; Monbaron 1978)
    Theropoda
        Carnosauria *incertae sedis*
           Carcharodontosaurus saharicus
        Ornithomimosauria *incertae sedis*
           Elaphrosaurus sp.
    Sauropoda
        Brachiosauridae
           Rebbachisaurus garasbae
    Age: Albian (Lapparent 1960*a*)

276. Wilaya Adrar, Algeria

  a. "Continental intercalaire" (Deperet and Savornin 1925; Lapparent 1960*a*; Bassoulet and Iliou 1967)
    Theropoda
        Theropoda indet. (= *Elaphrosaurus iguidiensis*)
        Carnosauria *incertae sedis*
           Carcharodontosaurus saharicus
    Sauropoda
        Sauropoda indet.
        Brachiosauridae
           Rebbachisaurus tamesnensis
    Age: Albian (Lapparent 1960*a*)
  b. Unnamed unit (Broin et al. 1971; Taquet 1977)
    Undescribed theropods (2 species)
    Undescribed sauropod
    Age: Albian (Broin et al. 1971)

277. Wilaya Tamenghest, Algeria

"Continental intercalaire" (Lapparent 1960*a*)
    Sauropoda
        Brachiosauridae
           Rebbachisaurus tamesnensis
    Age: Albian (Lapparent 1960*a*)

278. Wilaya Wargla, Algeria

"Continental intercalaire" (Lapparent 1960*a*)
    Theropoda
        Theropoda indet. (= *Elaphrosaurus iguidiensis*)
        Carnosauria *incertae sedis*
           Carcharodontosaurus saharicus
    Sauropoda
        Brachiosauridae
           Rebbachisaurus tamesnensis
           Brachiosaurus nougaredi
    Age: Albian (Lapparent 1960*a*)

279. Gouvernerat de Medenine, Tunisia

  a. "Continental intercalaire" (Lapparent 1960*a*; Schluter and Schwarzhans 1978)
    Theropoda
        Theropoda indet. (= *Elaphrosaurus gautieri*)
        Carnosauria *incertae sedis*
           Carcharodontosaurus saharicus
        cf. *Carcharodontosaurus* sp.
    Sauropoda
        Sauropoda indet.
        Diplodocidae
           Rebbachisaurus tamesnensis
    Ornithopoda
        Iguanodontid indet.
    Age: early Aptian (Taquet 1976)
  b. Chenini Formation (Bouaziz et al. 1988; not figured)
    Theropoda
        Carnosauria *incertae sedis*
           Carcharodontosaurus saharicus
        ?Spinosauridae
           Spinosaurus sp.
    Sauropoda indet.
    Age: early Albian (Bouaziz et al. 1988)

280. Muhafazat Gharyan, Libya

"Continental intercalaire" (Lapparent 1960*a*)
    Theropoda indet. (= *Elaphrosaurus iguidiensis*)
    Age: Albian (Lapparent 1960*a*)

281. Région de Gao, Mali

"Continental intercalaire" (Lapparent 1960*a*)
    Sauropoda indet.
    Ankylosauria
        Nodosaurid indet.
    Age: Albian (Lapparent 1960*a*; Petters 1981)

282. Département d'Agadez, Niger

  a. Elrhaz or Echkar Formation (Lapparent 1960*a*; Taquet 1976; Galton and Taquet 1982)
    Theropoda
        Theropoda indet. (= *Elaphrosaurus iguidiensis*)
        ?Spinosaurid indet.
    Sauropoda
        Diplodocid indet.
    Ornithopoda
        Hypsilophodontidae
           Valdosaurus nigeriensis
        Iguanodontidae
           Ouranosaurus nigeriensis
           Undescribed iguanodontid (?2 species)
    Age: late Aptian (Taquet 1976)

b. "Continental intercalaire" (Lapparent 1960a)
Theropoda
Theropoda indet. (= *Inosaurus tedreftensis,
Elaphrosaurus iguidiensis, E. gautieri*)
Carnosauria *incertae sedis*
Bahariasaurus ingens
Carcharodontosaurus saharicus
Sauropoda
Sauropoda indet.
Brachiosauridae
Pleurocoelus sp. (= *Astrodon* sp.)
Diplodocidae
Rebbachisaurus tamesnensis
Titanosauridae
Aegyptosaurus baharijensis
Ankylosauria
Nodosaurid indet.
Age: Albian (Lapparent 1960a; Petters 1981)

283. Département de la Tahoua, Niger

Farak Formation (Greigert et al. 1954; Lapparent
1960a; Taquet 1976)
Theropoda
Carnosauria *incertae sedis*
Bahariasaurus ingens
Sauropoda
Diplodocidae
Rebbachisaurus tamesnensis
Titanosauridae
Aegyptosaurus baharijensis
Age: ?late Albian-early Cenomanian (Taquet
1976; Petters 1981)

284. Cape Province, South Africa

a. Sundays River Formation (Mateer 1987)
Theropoda indet.
Age: ?Berriasian-Barremian (McLachlan and
McMillan 1976)

b. Kalahari Deposits (Haughton 1915; Cooper
1985)
Ornithopoda
Dryosaurid indet. (= *Kangnasaurus
coetzeei*)
Age: ?Early Cretaceous (Cooper 1985)

c. Enon Formation (Mateer 1987)
Theropoda indet.
Age: ?Early Cretaceous (possibly Late Jurassic;
McLachlan and McMillan 1976)

Australasia

Since the late 1970s, Australasia claims nearly all
major dinosaur groups known to exist during this time
interval elsewhere in the world. This rise in information on Australasian dinosaurs (fig. 3.12) is primarily the result of the efforts of A. Bartholomai,
R. E. Molnar, R. A. Thulborn, and T. H. Rich.

285. Western Australia, Australia

Broome Sandstone (Colbert and Merriless 1967)
Theropod footprints
Age: Berriasian-Hauterivian (= Neocomian;
Colbert and Merriless 1967)

286. South Australia, Australia

a. Maree Formation (Molnar and Pledge 1980)
Theropoda *incertae sedis*
Kakuru kujani
Age: Aptian (Carr et al. 1979)

b. ?Cadn-Owie Formation (Molnar 1980b)
Ornithopoda indet.
Age: ?Aptian (Molnar 1980b)

287. Northern Territory, Australia

Unnamed unit (Molnar pers. comm.; not figured)
Ornithopoda
?Iguanodontid indet.
Age: Early Cretaceous (Molnar pers. comm.)

288. Queensland, Australia

a. Bungil Formation (Molnar 1980a)
Ankylosauria
Nodosauridae
Minmi paravertebra
Age: Aptian (Exon and Vine 1970)

b. Unnamed unit (possibly Toolebuc or Allaru
Formation, Molnar 1982, pers. comm.)
Sauropoda indet.
Age: Albian (Molnar 1980b)

c. Toolebuc or Allaru Formation (Molnar 1980b,
1984a, pers. comm.)
Ornithischian indet.
Ornithopoda
Ornithopoda indet.
Iguanodontia *incertae sedis*
Muttaburrasaurus sp.
?Ankylosauria indet.
Age: Albian (Molnar 1980b)

d. Mackunda Formation (Bartholomai and Molnar
1981)
Ornithopoda
Iguanodontia *incertae sedis*
Muttaburrasaurus langdoni
Age: Albian (Vine and Day 1965)

e. Allaru Formation (Longman 1933; Molnar
1980b)
Sauropoda *incertae sedis*
Austrosaurus mckillopi
Ornithopoda indet.
Age: Albian (Molnar 1982)

289. New South Wales, Australia

a. Griman Creek Formation (Huene 1932; Molnar
1980b, pers. comm.; Molnar and Galton 1986)
Theropoda *incertae sedis*
Rapator ornitholestoides

Theropoda indet. (= *Walgettosuchus woodwardi*)
Sauropoda
    Brachiosaurid indet.
Ornithopoda
    Hypsilophodontidae
        Fulgurotherium australe (partim)
        Hypsilophodontid indet.
    Iguanodontia *incertae sedis*
        Muttaburrasaurus sp.
    Age: Albian (Molnar 1980*b*)

  b. Coreena Formation (Molnar 1980*b*)
    Theropoda indet.
    Age: Albian (Molnar 1980*b*)

290. Victoria, Australia

  a. Otway Group (Rich and Rich 1989)
    Dinosauria indet.
    Theropoda indet.
    Ornithopoda
        Hypsilophodontidae
            Leaellynasaura amicagraphica
            Atlascopcosaurus loadsi
            Fulgurotherium australe
            Hypsilophodontid indet.
    Dinosaur footprints
    Age: late Aptian-early Albian (Rich and Rich 1989)

  b. Strzelecki Group (Woodward 1906; Molnar et al. 1981, 1985; Rich and Rich 1989)
    Theropoda
        Allosauridae
            ?*Allosaurus* sp.
    Ornithopoda
        Hypsilophodontidae
            Fulgurotherium australe
            Hypsilophodontid indet.
    Age: Valanginian-Aptian (Rich and Rich 1989)

LATE CRETACEOUS

Dinosaurs are best known from the Late Cretaceous, particularly from the Campanian and Maastrichtian and principally from Canada and the United States in North America and the Mongolian People's Republic and the People's Republic of China in eastern Asia. Additional Late Cretaceous material is known from Europe, South America, Africa, Australasia, and Antarctica.

North America

The greatest diversity and abundance of dinosaurs are known from the Late Cretaceous of North America (figs. 3.1, 3.2, 3.3, 3.14, 3.17, 3.18, 3.23). Histories of the great Late Cretaceous dinosaur discoveries

of the late 1800s and early 1900s exist in the writings of Sternberg (1909, 1917), Russell (1967), and Colbert (1968). The likes of J. Leidy, E. D. Cope, O. C. Marsh, J. B. Hatcher, G. M. Dawson, L. M. Lambe, W. A. Parks, B. Brown, C. H. Sternberg and his sons (George, Charles M., and Levi), W. E. Cutler, and C. W. Gilmore were among the first to collect and study Late Cretaceous dinosaur material from Canada and the United States. In recent years, L. S. Russell, W. Langston, Jr., D. A. Russell, R. C. Fox, J. E. Storer, T. Tokaryk, and P. J. Currie have continued collecting operations in Canada, particularly southeastern Alberta, while in the Western Interior and Alaska of the United States, W. A. Clemens and co-workers from the University of California at Berkeley, M. C. McKenna and co-workers from the American Museum of Natural History, D. Gabriel and co-workers from the Milwaukee Public Museum, L. Van Valen, R. E. Sloan, J. K. Rigby, Jr., J. R. Horner, K. Carpenter, P. J. Bjork, D. B. Weishampel, A. R. Fiorillo, and P. Dodson carry on the tradition first established by Cope and Marsh.

Upper Cretaceous exposures of the San Juan Basin of New Mexico, the Big Bend area of Texas, the states of the Gulf Coast, and along the eastern seaboard of the United States have also yielded important dinosaur material. These areas were initially studied by J. Leidy, B. Brown, C. H. Sternberg, and C. W. Gilmore and more recently by W. Langston, Jr., W. A. Clemens, J. G. Armstrong–Ziegler, E. H. Colbert, S. G. Lucas, T. M. Lehman, N. J. Mateer, D. Parris, and W. B. Gallagher, among others.

Several small but important Late Cretaceous dinosaur faunas from Mexico and Honduras have been documented. They have been studied by W. J. Morris, R. E. Molnar, G. E. Murray and others, and K. L. Seymour and A. Leitch.

291. Yukon Territory, Canada

    Bonnet Plume Formation (Russell 1984*c*)
        Ornithopoda
            Hadrosaurid indet.
        Age: Maastrichtian (Rouse and Srivastava 1972)

292. Northwest Territories, Canada

  a. Summit Creek Formation (Russell 1984*c*)
    Ceratopsia
        ?Ceratopsid indet.
    Age: late Maastrichtian (Yorath and Cook 1981)

  b. Kanguk Formation (D. A. Russell pers. comm.)
    Ornithopoda
        Hadrosaurid indet.
    Age: Maastrichtian (D. A. Russell pers. comm.)

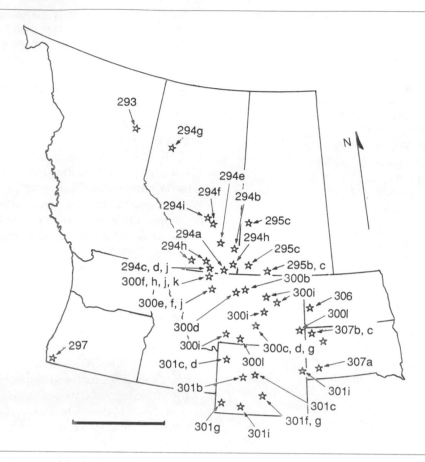

Fig. 3.23. Late Cretaceous-age locations in southwestern Canada and the northwestern United States.

293. British Columbia, Canada

Dunvegan Formation (Storer 1975)
Theropod footprints
Age: Cenomanian (Storer 1975)

294. Alberta, Canada

a. Milk River Formation (Russell 1935, 1964)
Theropoda
Tyrannosaurid indet.
?Dromaeosaurid indet.
Ornithopoda
Hadrosaurid indet.
Hadrosaurine indet. (= cf.
Kritosaurus sp.)
Ceratopsia
Ceratopsid indet. (= cf. *Brachyceratops* sp.)
Ankylosauria
Nodosaurid indet. (= cf. *Palaeoscincus*)
Age: early Campanian (Fox 1978)

b. Judith River Formation (Brown 1933; Russell
1964; Dodson 1971, 1983, 1984; Currie 1985,
1986, 1987b, 1987c)

Theropoda
Theropoda *incertae sedis*
Richardoestesi gilmorei
?Carnosauria
Aublysodon sp.
Tyrannosauridae
Albertosaurus libratus
Daspletosaurus torosus
Undescribed tyrannosaurid (Currie
pers. comm.)
Tyrannosaurid indet.
Ornithomimidae
Struthiomimus altus
Dromicieomimus samueli
Ornithomimus edmontonensis
Ornithomimid indet.
Elmisauridae
Chirostenotes pergracilis (=
Macrophalangia canadensis)
?*Chirostenotes* sp.
Elmisaurus elegans (= *Ornithomimus
elegans*)
Caenagnathidae
Caenagnathus collinsi
Caenagnathus sternbergi

Troodontidae
    Troodon formosus (= *Stenonychosaurus inequalis, Pectinodon bakkeri*)
    ?*Troodon* sp.
    Troodontid indet. (= ?*Paronychodon lacustris*)
Dromaeosauridae
    Dromaeosaurus albertensis
    Dromaeosaurus sp.
    Saurornitholestes langstoni
    ?Dromaeosaurid indet.
Segnosauria
    cf. *Erlikosaurus* sp.
Ornithopoda
    Ornithopod footprints
    Hadrosauridae
        Brachylophosaurus canadensis
        Gryposaurus notabilis
        "*Kritosaurus*" *incurvimanus*
        Prosaurolophus maximus
        ?*Maiasaura* sp.
        Corythosaurus casuarius
        Lambeosaurus lambei
        Lambeosaurus magnicristatus
        Parasaurolophus walkeri
        Undescribed lambeosaurine (Currie pers. comm.)
        Hadrosaurid indet. (= *Cionodon stenopsis, Pteropelyx grallipes, Trachodon altidens, Trachodon marginatus, Trachodon mirabilis, Trachodon selwyni*)
Pachycephalosauria
    Pachycephalosauridae
        Stegoceras validum
        Gravitholus albertae
        Pachycephalosaurus sp.
        Ornatotholus browni
        Undescribed pachycephalosaurid (Currie pers. comm.)
Ceratopsia
    Ceratopsidae
        Anchiceratops ornatus
        Chasmosaurus belli (including *C. brevirostris*)
        Chasmosaurus russelli
        Chasmosaurus canadensis (including *C. kaiseni*)
        Chasmosaurus sp.
        Monoclonius crassus (including *M. lowei*)
        Monoclonius sp.
        Centrosaurus apertus (including *C. flexus, C. cutleri, C. longirostris, Monoclonius dawsoni*)
        Centrosaurus sp.
        Styracosaurus albertensis (including *S. parksi, Monoclonius nasicornus*)
        Ceratopsid indet. (= ?*Ceratops* sp.)

Ankylosauria
    Nodosauridae
        Edmontonia rugosidens
        Edmontonia longiceps
        Panoplosaurus mirus
        ?*Panoplosaurus* sp.
        Nodosaurid indet. (= *Palaeoscincus costatus*)
    Ankylosauridae
        Euoplocephalus tutus
    Age: late Campanian (Dodson 1971, Fox 1978; middle-late Campanian, Currie, pers. comm.)
c. Belly River Formation (Currie pers. comm.; including Allison Formation, Russell 1949; Horner and Weishampel in prep.)
    Theropoda indet.
    Ornithopoda
        Hypsilophodontid indet. (= *Laosaurus minimus*)
        Hadrosaurid indet.
    Ceratopsia
        Protoceratopsidae
            Montanoceratops sp.
    Dinosauria eggshell
    Age: late Campanian (Russell 1949)
d. Two Medicine Formation (Currie pers. comm.)
    Theropoda
        Tyrannosauridae
            Albertosaurus sp.
        Troodontidae
            Troodon sp.
        Dromaeosauridae
            Dromaeosaurus sp.
            Saurornitholestes sp.
    Ornithopoda
        Hypsilophodontid indet.
    Ankylosauria
        Nodosauridae
            Edmontonia sp.
    Age: late Campanian (Currie pers. comm.)
e. Bearpaw Shale (Brinkman pers. comm.)
    Ornithopoda
        Hadrosaurid indet.
    Age: late Campanian (Cobban and Reeside 1952)
f. Horseshoe Canyon Formation (Sternberg 1926*a*; Russell and Chamney 1967; D. A. Russell 1969, 1970*b*, 1972; Wall and Galton 1979)
    Theropoda
        Theropod footprints
        Tyrannosauridae
            Albertosaurus sarcophagus
            Daspletosaurus sp.
        Ornithomimidae
            Ornithomimus edmontonensis
            Struthiomimus altus
            Dromiceiomimus brevitertius
            Ornithomimid indet.

Elmisauridae
Chirostenotes pergracilis
Troodontidae
Troodon formosus
Dromaeosauridae
cf. *Dromaeosaurus* sp.
Saurornitholestes sp.
?Dromaeosaurid indet.
Ornithopoda
Hypsilophodontidae
Parksosaurus warreni
Hadrosauridae
Edmontosaurus regalis
Saurolophus osborni
Hypacrosaurus altispinus
Pachycephalosauria
Pachycephalosauridae
Stegoceras edmontonense
Ceratopsia
Ceratopsidae
Anchiceratops ornatus (including *A. longirostris*)
Arrhinoceratops brachyops
Pachyrhinosaurus canadensis
Ceratopsid indet.
Ankylosauria
Nodosauridae
Edmontonia longiceps
Ankylosauridae
Euoplocephalus tutus
Age: early Maastrichtian (Fox 1978)
g. Wapiti Formation (Tanke 1988, pers. comm; Currie pers. comm.)
Theropoda
Tyrannosauridae
Albertosaurus sp.
Dromaeosauridae
Saurornitholestes sp.
Ornithopoda
Hadrosaurid indet.
Ceratopsia
Ceratopsidae
Pachyrhinosaurus sp.
Age: Campanian-Maastrichtian (Currie pers. comm.)
h. St. Mary River Formation (Langston 1960*b*, 1976)
Theropoda
Theropod indet.
Theropod footprints
Tyrannosauridae
?*Albertosaurus* sp.
Tyrannosaurid indet.
Coeluridae
Coelurid indet.

Ornithomimidae
Ornithomimid indet.
?Troodontidae
?*Troodon* sp.
Ornithopoda
Ornithopod footprints
Hadrosauridae
Edmontosaurus cf. *regalis*
Hadrosaurid indet.
Ceratopsia
Ceratopsidae
Pachyrhinosaurus canadensis
Anchiceratops sp. (= ?*A. ornatus*)
Ankylosauria
Nodosauridae
Edmontonia cf. *longiceps*
Dinosauria eggshell
Age: early Maastrichtian (Langston 1976)
i. Scollard Formation (Sternberg 1949; L. S. Russell 1964, 1987; Russell and Chamney 1967)
Theropoda
Tyrannosauridae
cf. *Tyrannosaurus rex*
Tyrannosaurid indet.
Ornithomimidae
Ornithomimid indet.
Troodontidae
cf. *Troodon* sp.
Ornithopoda
Hypsilophodontidae
Thescelosaurus neglectus (including *T. edmontonensis*)
?Iguanodontid indet.
Hadrosauridae
Edmontosaurus regalis
Edmontosaurus annectens
Hadrosaurid indet.
Pachycephalosauria
Pachycephalosauridae
?*Pachycephalosaurus* sp.
Ceratopsia
Protoceratopsidae
Leptoceratops gracilis
?*Leptoceratops* sp.
Ceratopsidae
Triceratops horridus (? = *T. albertensis*)
cf. *Torosaurus* sp.
Ceratopsid indet.
Ankylosauria
Ankylosauridae
Ankylosaurus magniventris
Age: late Maastrichtian (Russell 1964; Fox 1977)

j. Willow Creek Formation (Abler 1984; Horner
pers. comm.)
- Theropoda
  - Tyrannosauridae
    - *Tyrannosaurus rex*
    - Tyrannosaurid indet.
- Ornithopoda
  - Hadrosaurid indet.
- Ceratopsia
  - Protoceratopsidae
    - ?*Leptoceratops* sp.
- Age: late Maastrichtian (Schmidt 1978)

295. Saskatchewan, Canada

a. Judith River Formation (Tokaryk pers. comm.;
not figured)
- Theropoda
  - Tyrannosauridae
    - ?*Albertosaurus* sp.
  - Ornithomimid indet.
  - Elmisauridae
    - cf. *Chirostenotes* sp. (= cf. *Macrophalangia* sp.)
  - Dromaeosauridae
    - *Saurornitholestes* sp.
    - *Dromaeosaurus* sp. (?2 species)
- Ornithopoda
  - Hadrosauridae
    - Hadrosaurine indet.
    - Lambeosaurine indet.
- Ceratopsia
  - Undescribed ceratopsid
- Age: late Campanian (Dodson 1971; Fox 1978)

b. Whitemud Formation (Broughton et al. 1978;
Broughton 1981)
- Dinosaur coprolites
- Age: late Maastrichtian (Irish and Havard 1968)

c. Frenchman Formation (Russell 1964; Tokaryk
1985, 1986*a*, 1986*b*, pers. comm.)
- Theropoda
  - Tyrannosauridae
    - *Tyrannosaurus rex*
    - ?*Tyrannosaurus* sp.
  - Ornithomimid indet. (= *Ornithomimus* sp.)
  - Elmisauridae
    - cf. *Chirostenotes* sp. (= cf. *Macrophalangia* sp.)
  - Caenagnathidae
    - ?*Caenagnathus* sp.
  - Dromaeosauridae
    - *Dromaeosaurus* sp.
    - *Saurornitholestes* sp.
  - Troodontidae
    - *Troodon* sp. (including *Stenonychosaurus* sp.)
- Ornithopoda
  - Hypsilophodontidae
    - *Thescelosaurus neglectus*
  - Hadrosauridae
    - *Edmontosaurus saskatchewanensis*
    - *Edmontosaurus annectens*
    - Hadrosaurid indet.
- Ceratopsia
  - Ceratopsidae
    - *Triceratops horridus* (= *T.* cf. *prorsus*)
    - *Triceratops* sp.
    - *Torosaurus latus*
    - Ceratopsid indet.
- Ankylosauria
  - Ankylosauridae
    - cf. *Ankylosaurus* sp.
- Age: late Maastrichtian (Russell 1964)

296. Alaska, United States

Prince Creek Formation (Brouwers et al. 1987;
Parrish et al. 1987; Upper Colville Group, Davies
1987)
- Theropoda
  - Theropoda indet.
  - Undescribed tyrannosaurid
- Ornithopoda
  - Undescribed hadrosaurid
  - Hadrosaurid indet.
- Ceratopsia
  - Ceratopsidae
    - *Pachyrhinosaurus* sp.
    - Ceratopsid indet.
- Age: middle Maastrichtian (Frederiksen 1986; early middle Maastrichtian, Brouwers et al. 1987)

297. Oregon, United States

Cape Sebastian Sandstone (Leahy pers. comm.)
- ?Ornithopoda
  - ?Hadrosaurid indet.
- Age: late Campanian (Bourgeois and Dott 1985)

298. California, United States

a. Ladd or Williams Formation (Morris 1973*b*)
- Ornithopoda
  - Hadrosaurid indet.
- Age: ?Campanian (Popenoe et al. 1960)

b. Point Loma Formation (Morris 1982; Demere
pers. comm.)
- Ornithopoda
  - Hadrosaurid indet.
- Ankylosauria
  - Nodosaurid indet.
- Age: late Campanian-early Maastrichtian (Nilsen and Abbott 1981)

c. Moreno Formation (Morris 1973*b*, 1982)
 Ornithopoda
  Hadrosauridae
   Hadrosaurid indet.
   Hadrosaurine indet. (= ?*Saurolophus* sp.)
  Age: Maastrichtian (Anderson 1958)

299. Nevada, United States

a. Newark Canyon Formation (Smith and Ketner 1976)
  Theropoda indet.
  Ankylosauria
   Ankylosaurid indet.
  Age: ?Late Cretaceous (Smith and Ketner 1976)

b. Unnamed unit (Carpenter pers. comm.; not figured)
  Theropoda indet.
  Ornithopoda
   ?Hadrosaurid indet.
  Ceratopsia
   Protoceratopsidae
    ?*Leptoceratops* sp.
  Ankylosauria indet.
  Age: Late Cretaceous (Carpenter pers. comm.)

300. Montana, United States

a. Telegraph Creek Formation (unpublished; not figured)
  Ornithopoda
   Hadrosaurid indet.
  Age: late Santonian (Cobban 1955)

b. Eagle Sandstone (Marsh 1890*b*)
  Theropoda
   Tyrannosauridae
    "*Ornithomimus*" *grandis*
  Age: early Campanian (Sahni 1972)

c. Claggett Shale (Dodson pers. comm.; Fiorillo in prep.)
  Ornithopoda
   Hadrosauridae
    ?Hadrosaurine indet.
  Age: early Campanian (Sahni 1972)

d. Judith River Formation (Sahni 1972; Dodson 1984, 1986; Horner 1988, pers. comm.)
  Theropoda
   Theropoda indet. (= *Laelaps explanatus*, *Laelaps laevifrons*, *Aublysodon lateralis*)
   ?Carnosauria
    Aublysodon mirandus
   Tyrannosauridae
    Albertosaurus libratus
    Albertosaurus sarcophagus
   Ornithomimidae
    Ornithomimus sp.
    Ornithomimid indet. (= *Ornithomimus tenuis*)

Troodontidae
 Troodon formosus
 Troodontid indet. (= *Paronychodon lacustris*, *Zapsalis abradens*)
Dromaeosauridae
 Dromaeosaurus albertensis
Ornithischia indet. (Galton pers. comm.)
Ornithopoda
 Hypsilophodontidae
  Thescelosaurus neglectus
 Hadrosauridae
  ?*Prosaurolophus maximus*
  Brachylophosaurus goodwini
  ?*Corythosaurus casuarius*
  Hadrosaurid indet. (= *Dysganus encaustus*, *Diclonius pentagonus*, *Diclonius calamarius*, *Diclonius perangulatus*, *Hadrosaurus breviceps*, *Hadrosaurus paucidens*)
Pachycephalosauria
 Pachycephalosauridae
  Stegoceras validum
  Stegoceras sp.
  Pachycephalosaurus wyomingensis
Ceratopsia
 Ceratopsidae
  Monoclonius crassus
  Avaceratops lammersi
  Undescribed ceratopsid (Horner pers. comm.)
  Ceratopsid indet. (= *Monoclonius recurvicornis*, *Monoclonius sphenocerus*, *Monoclonius fissus*, *Ceratops montanus*, *Dysganus peiganus*, *Dysganus bicarinatus*, *Dysganus encaustus* partim, *Dysganus haydenianus*)
Ankylosauria
 Ankylosauria indet. (= *Palaeoscincus costatus*)
 Nodosauridae
  Edmontonia longiceps
  Edmontonia rugosidens
Dinosaur egg shell
Age: late Campanian (Sahni 1972)

e. Lower Two Medicine Formation (Gilmore 1930; Horner pers. comm.)
  Theropoda
   Theropoda indet.
   Dromaeosaurid indet.
  Ornithopoda
   Hadrosauridae
    Undescribed hadrosaurine (Horner pers. comm.)
    Hadrosaurid indet.
  Ceratopsia
   Undescribed protoceratopsid (Horner pers. comm.)

Undescribed ceratopsid (Horner pers.
comm.)
Dinosaur egg shell
Age: ?late Santonian-early Campanian
(Horner pers. comm.)

f. Upper Two Medicine Formation (Gilmore 1930;
Horner and Makela 1979; Horner 1983, pers.
comm.; Horner and Weishampel 1988)
  Theropoda
      ?Carnosauria
          Aublysodon mirandus
      Troodontidae
          Troodon sp.
      Dromaeosaurid indet.
  Ornithopoda
      Hypsilophodontidae
          Orodromeus makelai
          Undescribed hypsilophodontid
      Hadrosauridae
          Maiasaura peeblesorum
          Undescribed hadrosaurine (Horner
              pers. comm.)
          Undescribed lambeosaurine (Horner
              pers. comm.)
          Hadrosaurid indet.
  Ceratopsia
      Protoceratopsid indet. (= ?*Leptoceratops*
          sp.)
      Ceratopsidae
          Brachyceratops montanensis
          Styracosaurus ovatus
          ?*Centrosaurus* sp.
          Chasmosaurus sp.
  Ankylosauria
      Nodosauridae
          Edmontonia rugosidens
      Ankylosauridae
          Euoplocephalus tutus
  Dinosaur eggshell
  Age: middle-late Campanian (Horner pers.
      comm.)

g. Lower Bearpaw Shale (Horner 1979)
  Ornithopoda
      Hadrosauridae
          Undescribed hadrosaurine (=
              Hadrosaurus notabilis)
          Lambeosaurine indet. (=
              ?*Lambeosaurus magnicristatus*)
  Ankylosauria
      Nodosauridae
          Edmontonia sp.
  Age: late Campanian (Rice and Shurr 1983)

h. Horsethief Sandstone (Horner pers. comm.)
  Ornithopoda
      Hadrosaurid indet.
  Age: early Maastrichtian (Viele and Harris
      1965)

i. Livingston Formation (McMannis 1965; Skipp
and McGrew 1972; Horner pers. comm.)
  Theropoda
      Tyrannosauridae
          ?*Tyrannosaurus rex*
          Tyrannosaurid indet.
  Ornithopoda
      Hadrosaurid indet.
  Ceratopsia
      Protoceratopsidae
          ?*Leptoceratops* sp.
  Age: ?late Campanian-late Maastrichtian
      (Cobban and Reeside 1952)

j. St. Mary River Formation (Brown and Schlaikjer
1942; Weishampel and Horner 1987;
Weishampel in prep.)
  Theropoda
      Tyrannosaurid indet.
      Undescribed ornithomimid
      Ornithomimid indet.
      ?Dromaeosaurid indet.
  Ornithopoda
      Hadrosauridae
          Hadrosaurine indet.
          Undescribed lambeosaurine
          Lambeosaurine indet.
  Ceratopsia
      Protoceratopsidae
          Montanoceratops cerorhynchus
      Ceratopsid indet.
  Dinosaur eggshell
  Age: late Maastrichtian (Cobban and Reeside
      1952)

k. Willow Creek Formation (Russell 1968; may be
St. Mary River Formation)
  Ornithopoda
      Hypsilophodontidae
          ?*Thescelosaurus* sp.
  Age: late Maastrichtian (Honkala 1955)

l. Hell Creek Formation (Molnar 1978, 1980d;
Estes et al. 1969; Lupton et al. 1980; Brett-
Surman and Paul 1985; Sloan et al. 1986;
Carpenter and Breithaupt 1986; Currie 1987a;
Griffin et al. 1987; Horner pers. comm.)
  Theropoda
      Theropoda indet.
      Undescribed theropod (Sloan et al. 1986)
      Theropoda *incertae sedis*
          Avisaurus archibaldi
      ?Carnosauria
          Aublysodon mirandus
          Aublysodon molnari nanotyrannus
      Tyrannosauridae
          Tyrannosaurus rex
          Nanotyrannus lancensis
          Albertosaurus megagracilis
          Tyrannosaurid indet.

Ornithomimidae
　　Ornithomimus sp.
　　Ornithomimid indet.
Elmisauridae
　　?*Chirostenotes* sp.
Troodontidae
　　Troodon formosus (=
　　　　Stenonynchosaurus inequalis,
　　　　?*Saurornithoides* sp.)
　　Troodontid indet. (= *Paronychodon*
　　　　lacustris)
Dromaeosauridae
　　Dromaeosaurus sp.
　　?*Velociraptor* sp.
　　Dromaeosaurid indet. (?2 species)
Ornithischia indet. (Galton pers. comm.)
Ornithopoda
　　Hypsilophodontidae
　　　　?*Thescelosaurus garbanii*
　　　　?*Thescelosaurus* sp.
　　Hadrosauridae
　　　　Edmontosaurus regalis
　　　　Edmontosaurus annectens
　　　　Anatotitan copei
　　　　?*Parasaurolophus walkeri*
Pachycephalosauria
　　Pachycephalosauridae
　　　　Pachycephalosaurus wyomingensis
　　　　Stegoceras edmontonense
　　　　Stygimoloch spinifer (including
　　　　　　Stenotholus kohleri)
　　　　Stegoceras sp.
Ankylosauria
　　Ankylosauria indet.
　　Nodosauridae
　　　　?*Edmontonia* sp.
　　Ankylosauridae
　　　　Ankylosaurus magniventris
　　　　Ankylosaurid indet.
Ceratopsia
　　Ceratopsidae
　　　　Triceratops horridus (= *T. prorsus, T.
　　　　　　serratus*)
　　　　Triceratops sp.
　　　　Torosaurus latus
　　　　Ceratopsid indet. (= *Triceratops*
　　　　　　maximus, Ugrosaurus olsoni)
Age: late Maastrichtian (Cobban and Reeside
　　1952; Baadsgaard and Lerbekmo 1980)

301. Wyoming, United States

　a. Pierre Shale (Carpenter pers. comm.; not figured)
　　Ornithopoda
　　　　Undescribed hadrosaurid
　　Age: Campanian (Cobban and Reeside 1952)

　b. Meeteetse Formation (Keefer and Troyer 1956;
　　Keefer 1965)
　　　　Dinosauria indet.
　　Age: late Campanian (Lillegraven and
　　　　McKenna 1986)

　c. "Mesaverde" Formation (Breithaupt 1985)
　　　　Theropoda
　　　　　　Tyrannosauridae
　　　　　　　　Albertosaurus sp.
　　　　　　Troodontid indet. (= *Paronychodon*
　　　　　　　　lacustris)
　　　　　　Dromaeosauridae
　　　　　　　　Dromaeosaurus sp.
　　　　Ornithopoda
　　　　　　Hypsilophodontidae
　　　　　　　　cf. *Thescelosaurus* sp.
　　　　　　Hadrosauridae
　　　　　　　　Edmontosaurus sp.
　　　　Ankylosauria
　　　　　　Nodosauridae
　　　　　　　　Edmontonia sp.
　　　　Ceratopsia
　　　　　　Ceratopsid indet.
　　Age: late Campanian (Breithaupt 1985)

　d. Harebell Formation (McKenna 1980)
　　　　Ceratopsia
　　　　　　Protoceratopsid indet. (= ?*Leptoceratops*
　　　　　　　　sp.)
　　　　Ankylosauria
　　　　　　?Nodosaurid indet.
　　Age: Maastrichtian (McKenna 1980)

　e. Pinyon Conglomerate (McKenna and Love 1970)
　　　　Ceratopsia
　　　　　　Protoceratopsidae
　　　　　　　　Leptoceratops sp.
　　Age: Maastrichtian (McKenna and Love 1970)

　f. Medicine Bow Formation (Lull 1933)
　　　　Ceratopsia
　　　　　　Ceratopsidae
　　　　　　　　Triceratops sp.
　　Age: Maastrichtian (Cobban and Reeside
　　　　1952)

　g. Evanston Formation (Tracy and Oriel 1959;
　　Rubey et al. 1961; Lucas 1981; Breithaupt 1985)
　　　　Sauropoda
　　　　　　Titanosauridae
　　　　　　　　?*Alamosaurus* sp.
　　　　Ceratopsia
　　　　　　Ceratopsidae
　　　　　　　　Triceratops horridus (= *T. flabellatus*)
　　Age: late Maastrichtian (Cobban and Reeside
　　　　1952)

h. Ferris Formation (Bowen 1918; Lull 1933)
    Ceratopsia
        Ceratopsidae
            Triceratops sp.
    Age: late Maastrichtian (Cobban and Reeside 1952)

i. Lance Formation (including questionable Lance strata; Marsh 1889*a*; 1889*b*, 1889*c*; Hatcher 1893; Hatcher et al. 1907; Bowen 1915; Gilmore 1915*a*; Estes 1964; Ostrom 1978*b*; Carpenter 1982*b*; Breithaupt 1982, 1985; Carpenter and Breithaupt 1986; Currie 1987*a*; Carpenter pers. comm.)
    Theropoda
        Theropoda indet. (= *Laelaps cristatus*)
        ?Carnosauria
            Aublysodon mirandus
        Tyrannosauridae
            ?*Albertosaurus sarcophagus*
            Albertosaurus sp.
         cf. *Tyrannosaurus rex*
            Tyrannosaurus sp.
            ?*Dryptosaurus* sp.
            Tyrannosaurid indet.
        Ornithomimidae
            Ornithomimus sp.
            Ornithomimid indet. (= *Ornithomimus sedens*)
        Elmisauridae
            ?*Chirostenotes* sp.
        Troodontidae
            Troodon formosus (= *Pectinodon bakkeri, Stenonychosaurus inequalis,* ?*Saurornithoides* sp.)
            Troodontid indet. (= *Paronychodon lacustris*)
        Dromaeosauridae
            ?*Velociraptor* sp.
            Dromaeosaurus sp.
            Dromaeosaurid indet.
    Sauropoda
        Diplodocid indet. (may not come from Lance Formation; Russell 1984*c*)
    Ornithopoda
        Hypsilophodontidae
            Thescelosaurus neglectus
            ?*Thescelosaurus* sp.
        Hadrosauridae
            Edmontosaurus regalis
            Edmontosaurus annectens
            ?*Lambeosaurus* sp.
            Hadrosaurid indet.

Pachycephalosauria
    Pachycephalosauridae
        Pachycephalosaurus wyomingensis
    cf. *Pachycephalosaurus* sp.
        Stygimoloch spinifer
Ceratopsia
    Protoceratopsidae
        Leptoceratops gracilis
    Ceratopsidae
        Triceratops horridus (including *T. calicornis, T. elatus, T. hatcheri, T. obtusus, T. eurycephalus, T. flabellatus, T. serratus, T. prorsus*)
        Triceratops sp.
        Torosaurus latus (including *T. gladius*)
        Ceratopsid indet. (= *Agathaumas sylvestris, Bison alticornis, Triceratops sulcatus, Triceratops ingens*)
Ankylosauria
    Ankylosauria indet. (= *Palaeoscincus latus*)
    Nodosauridae
        Edmontonia sp.
        Nodosaurid indet.
    Ankylosauridae
        Ankylosaurus cf. *magniventris*
        Ankylosaurid indet.
Age: late Maastrichtian (Cobban and Reeside 1952)

302. Utah, United States

a. Wahweap (Horner pers. comm.; Carpenter pers. comm.; not figured)
    Dinosauria indet.
    Theropoda indet.
    Ornithopoda
        Hadrosaurid indet.
    Dinosaur eggshell
    Age: middle-late Campanian (Molenaar 1983)

b. Mesaverde Formation (Peterson 1924; Brown 1938; Lockley 1986*a*)
    Theropod footprints
    Ornithopoda
        Ornithopod footprints
        ?Iguanodontid indet.
    Ceratopsian footprints
    Age: late Campanian (Molenaar 1983)

c. Castlegate Sandstone (Lull and Wright 1942)
    Ornithopoda
        Hadrosaurid indet.
    Age: late Campanian (Fouch et al. 1983)

d. Straight Cliffs Formation (Carpenter pers. comm.; not figured)
    Ornithopoda
        Hadrosaurid indet.
    Age: early Santonian-late middle Campanian (Molenaar 1983)

e. North Horn Formation (Gilmore 1946*b;* Jensen 1966; Lawson 1976)
    Theropoda
        Tyrannosaurid indet.
    Sauropoda
        Titanosauridae
            Alamosaurus sanjuanensis
    Ornithopoda
        Hadrosaurid indet.
    Ceratopsia
        Ceratopsidae
            Torosaurus latus (= *T. utahensis*)
            Ceratopsid indet.
    Dinosaur eggshell
    Age: Maastrichtian (Hintze 1973)

f. Kaiparowits Formation (Gregory 1950, 1951; DeCourten 1978; Weishampel and Jensen 1979; DeCourten and Russell 1985)
    Theropoda
        Ornithomimidae
            Ornithomimus velox
    Ornithopoda
        Hadrosauridae
            Parasaurolophus sp.
            Hadrosaurine indet.
            Hadrosaurid indet.
    Ankylosauria
        Nodosaurid indet.
    Ceratopsia
        Ceratopsidae
            cf. *Triceratops* sp.
    Age: Maastrichtian (Lohrengel 1969)

303. Colorado, United States

a. Iles Formation (Mateer pers. comm.; not figured)
    Dinosauria indet.
    Age: late Campanian (Mateer pers. comm.)

b. Williams Fork Formation (Archibald 1987*a*)
    Theropoda indet.
    Ornithischia indet.
    Age: Maastrichtian (Kiteley 1983)

c. Mesaverde Formation (Peterson 1924; Brown 1938; Russell and Béland 1976; Lockley et al. 1983)
    Theropod footprints
    Ornithopod footprints
    Age: Campanian (Russell and Béland 1976)

d. Laramie Formation (Carpenter 1979; Carpenter and Breithaupt 1986; Lockley 1986*a;* Carpenter pers. comm.)
    Theropoda
        Theropoda indet. (= *Ornithomimus minutus, Tripriodon caperatus*)
        Theropod footprints
        Tyrannosauridae
            cf. *Tyrannosaurus rex*
        Troodontid indet. (= *Paronychodon* sp.)
        ?Dromaeosaurid indet.

    Ornithopoda
        Ornithopod footprints
        Hypsilophodontidae
            Thescelosaurus neglectus
            Thescelosaurus sp.
        Hadrosauridae
            Edmontosaurus regalis
            Edmontosaurus annectens
    Ankylosauria
        Nodosauridae
            ?*Edmontonia* sp.
    Ceratopsia
        Ceratopsian footprints
        Ceratopsidae
            Triceratops horridus (including *T. alticornis*)
            Triceratops sp.
            Torosaurus latus
    Age: late Maastrichtian (Carpenter 1979)

e. Arapahoe Formation (Carpenter pers. comm.; not figured)
    Ceratopsian indet.
    Age: late Maastrichtian (Cobban and Reeside 1952; Carpenter pers. comm.)

f. Denver Formation (Carpenter pers. comm.; not figured)
    Theropoda
        Theropoda indet.
        ?Carnosauria
            Aublysodon mirandus
        Ornithomimidae
            Ornithomimus velox
            Ornithomimid indet.
    Ornithopoda
        Hadrosaurid indet. (= *Cionodon arctatus*)
    Ankylosauria
        Nodosaurid indet.
    Ceratopsia
        Ceratopsidae
            Triceratops sp.
            Ceratopsid indet (= *Polyonax mortuarius, Triceratops galeus*)
    Age: late Maastrichtian (Cobban and Reeside; Carpenter pers. comm.)

g. Cimarron Ridge Formation (Lee 1912; Reeside 1924; Lehman 1987)
    Dinosauria indet.
    Age: late Maastrichtian (Lehman 1987)

304. Arizona, United States

Unnamed unit (Miller 1964)
    Theropoda indet.
    Ornithopoda
        Hadrosaurid indet.
    Age: Late Cretaceous (Miller 1964)

305. New Mexico, United States

a. Fruitland Formation (Gilmore 1916, 1919*a,* 1935; Osborn 1923*b;* Ostrom 1961*b;* Armstrong-

Ziegler 1978, 1980; Rowe et al. 1981; Lucas 1981; Currie 1987*a;* Lucas et al. 1987; Hall pers. comm.)
  Theropoda
    Theropoda indet.
    Tyrannosauridae
      ?*Albertosaurus libratus* (= *Deinodon horridus*)
      ?*Albertosaurus* sp.
      Undescribed tyrannosaurid (Hall pers. comm.)
    Ornithomimidae
      cf. *Ornithomimus* sp.
      Ornithomimid indet.
    Troodontid indet. (= *Paronychodon lacustris*)
    ?Dromaeosaurid indet.
  Ornithopoda
    Ornithopod footprints
    Hadrosauridae
      Parasaurolophus cyrtocristatus
      ?*Corythosaurus* sp.
      Hadrosaurine indet. (= ?*Kritosaurus navajovius*)
  Pachycephalosauria
    Pachycephalosaurid indet.
  Ankylosauria
    Ankylosaurid indet.
  Ceratopsia
    Ceratopsidae
      Pentaceratops sternbergii (including *P. fenestratus*)
      ?*Chasmosaurus* sp.
      Ceratopsid indet (= ?*Monoclonius* sp.)
  Age: Campanian (Fassett and Hinds 1971)
b. Menefee Formation (unpublished; not figured)
  Ornithopoda
    Hadrosaurid indet.
  Age: Campanian (Cobban and Reeside 1952)
c. Kirtland Shale (Gilmore 1916, 1919*a*, 1935; Osborn 1923*b;* Wiman 1930, 1931, Mateer 1981; Lucas 1981; Currie 1987*a;* Lucas et al. 1987)
  Theropoda
    ?Carnosauria
      Aublysodon mirandus
    Tyrannosauridae
      Albertosaurus libratus
      Tyranosaurid indet.
    Ornithomimidae
      cf. *Struthiomimus* sp.
    Troodontidae
      Troodon sp. (= ?*Saurornithoides* sp.)
    ?Dromaeosaurid indet.
  Sauropoda
    Titanosauridae
      Alamosaurus sanjuanensis

  Ornithopoda
    Hypsilophodontidae
      ?*Parksosaurus* sp.
    Hadrosauridae
      Parasaurolophus tubicen
      Hadrosaurine indet. (= *Kritosaurus navajovius*)
  Ceratopsia
    Ceratopsidae
      Pentaceratops sternbergii (including *P. fenestratus*)
      Pentaceratops sp.
      cf. *Torosaurus latus* (= *T. utahensis*)
      cf. *Chasmosaurus* sp.
      Ceratopsid indet.
  Ankylosauria
    Nodosauridae
      ?*Edmontonia* sp.
      ?*Euoplocephalus* sp.
    Ankylosauridae
      Ankylosaurid indet.
  Age: Maastrichtian (Russell 1975; Mateer 1981)
d. McRae Formation (Lozinsky et al. 1984; Lehman pers. comm.; = Red Beds of New Mexico; Lull 1933)
  Theropoda
    Theropoda indet.
    Tyrannosaurid
      Tyrannosaurus rex
  Sauropoda
    Titanosauridae
      Alamosaurus sp.
  Ankylosauria indet.
  Ceratopsia
    Ceratopsidae
      Triceratops sp.
      ?*Torosaurus* sp.
  Age: late Maastrichtian (= Lancian; Lozinsky et al. 1984)

306. North Dakota, United States

  Hell Creek Formation (Lull and Wright 1942; Lull 1933; Fastovsky pers. comm.; Williams pers. comm.; Farrar pers. comm.)
  Theropoda
    Theropoda indet.
    Tyrannosaurid indet.
    Ornithomimidae
      cf. *Struthiomimus* sp.
      ?*Ornithomimus* sp.
  Ornithopoda
    Hypsilophodontidae
      cf. *Thescelosaurus* sp.
    Hadrosauridae
      Edmontosaurus regalis
      Hadrosaurid indet.

Pachycephalosauria
    Undescribed pachycephalosaurid
      (Fastovsky pers. comm.)
Ceratopsia
    Ceratopsidae
      Triceratops sp.
Ankylosauria indet.
Age: late Maastrichtian (Cobban and Reeside
    1952)

307. South Dakota, United States

  a. Pierre Shale (Wieland 1903; Gregory 1948)
    Ornithopoda
      Hadrosaurid indet. (= *Claosaurus affinis*)
    Age: Campanian (Horner 1979)

  b. Lance Formation (Lull and Wright 1942; Bakker
    1988)
    Ornithopoda
      Hadrosauridae
        Anatotitan copei
        ?*Edmontosaurus regalis*
        Hadrosaurid indet. (= *Thespesius*
          occidentalis)
    Ankylosauria
      Nodosauridae
        Denversaurus schlessmani
    Age: late Maastrichtian (Cobban and Reeside
      1952)

  c. Hell Creek Formation (Colbert and Bump 1947;
    Morris 1976; Bjork 1985*b*; Carpenter and
    Breithaupt 1986; Horner pers. comm.)
    Theropoda
      Tyrannosauridae
        Tyrannosaurus rex
    Ornithopoda
      Hypsilophodontidae
        ?*Thescelosaurus* sp.
      Hadrosauridae
        Edmontosaurus regalis
        Edmontosaurus annectens
        Anatotitan copei
    Pachycephalosauria
      Pachycelphalosauridae
        Pachycephalosaurus wyomingensis
    Ankylosauria
      Nodosauridae
        Edmontonia sp.
    Ceratopsia
      Ceratopsidae
        Triceratops horridus
        Torosaurus latus
    Age: late Maastrichtian (Cobban and Reeside
      1952)

308. Kansas, United States

Niobrara Chalk (Marsh 1890*a*; Mehl 1936)
    Ornithopoda
      Hadrosauridae
        Claosaurus agilis
    Ankylosauria
      Ankylosauria indet. (= *Hierosaurus*
        sternbergi)
      Nodosauridae
        Nodosaurus textilis (= *Hierosaurus*
          coleii)
    Age: early Campanian (Cobban and Reeside
      1952)

309. Texas, United States

  a. Pen Formation (Lehman 1985*b*)
    Ornithopoda
      Hadrosauridae
        ?*Gryposaurus* sp. (= *Kritosaurus* sp.)
    Age: Cenomanian-early Campanian (Lehman
      1985*a*)

  b. San Carlos Formation (Lehman 1985*b*)
    Theropoda
      Tyrannosaurid indet.
    Ornithopoda
      Hadrosauridae
        ?*Gryposaurus* sp. (= *Kritosaurus* sp.)
    Pachycephalosauria
      Pachycephalosauridae
        Stegoceras sp.
    Ankylosauria
      Nodosaurid indet.
    Ceratopsia
      Ceratopsid indet.
    Age: early-late Campanian (Lehman 1985*a*)

  c. Aguja Formation (Lull and Wright 1942; Davies
    pers. comm.; Lehman 1985*b*, 1989, pers. comm.)
    Theropoda
      Tyrannosaurid indet.
      Ornithomimid indet.
    Ornithopoda
      Ornithopoda indet.
      Hadrosauridae
        ?*Gryposaurus* sp. (= *Kritosaurus* sp.)
        Hadrosaurine indet. (= *Kritosaurus*
          cf. *navajovius*)
        Lambeosaurine indet.
    Pachycephalosauria
      Pachycephalosauridae
        Stegoceras sp.
    Ceratopsia
      Ceratopsidae
        Chasmosaurus mariscalensis
      Ceratopsid indet.

Ankylosauria
  Nodosauridae
    Edmontonia cf. *rugosidens*
  Nodosaurid indet.
Age: early-late Campanian (Lehman 1985*a*)
d. Javelina Formation (Lawson 1976; Lehman 1985*b*; Davies pers. comm.)
  Theropoda
    Tyrannosauridae
      Tyrannosaurus rex
    Tyrannosaurid indet.
    Ornithomimid indet.
  Sauropoda
    Titanosauridae
      Alamosaurus sanjuanensis
  Ornithopoda
    Hadrosauridae
      cf. *Edmontosaurus* sp. (Davies pers. comm.)
  Ankylosauria
    Nodosaurid indet.
  Ceratopsia
    Ceratopsidae
      cf. *Torosaurus latus* (= cf. *T. utahensis*)
    Ceratopsid indet.
Age: late Maastrichtian (Lehman 1985*a*)
c. El Picacho Formation (Lehman 1985*b*)
  Theropoda
    Tyrannosaurid indet.
    Ornithomimid indet.
  Sauropoda
    Titanosauridae
      Alamosaurus sanjuanensis
  Ornithopoda
    Hadrosauridae
      ?*Gryposaurus* sp.(= *Kritosaurus* sp.)
  Ankylosauria
    Nodosaurid indet.
  Ceratopsia
    Ceratopsid indet.
Age: late Maastrichtian (Lehman 1985*a*)

310. Missouri, United States

Ripley Formation (Gilmore and Stewart 1945; Horner 1979)
  Ornithopoda
    Hadrosaurid indet. (= *Neosaurus missouriensis*)
Age: ?early Maastrichtian (Horner 1979)

311. Mississippi, United States

a. Basal Eutaw or Upper McShan Formation (Carpenter 1982*c*)
  Dinosauria indet.
  Theropoda
    Ornithomimid indet.
  Ornithopoda
    Hadrosaurid indet.
  Age: late Coniacian-?early Santonian (Carpenter 1982*c*)
b. Eutaw Formation (Kaye and Russell 1973)
  Ornithopoda
    Hadrosaurid indet.
  Age: late Santonian (Kaye and Russell 1973)
c. Selma Group (Carpenter 1982*c*)
  Ornithopoda
    Hadrosaurid indet.
  Age: Campanian (Carpenter 1982*c*)

312. Alabama, United States

a. Upper Eutaw Formation (Lamb pers. comm.; not figured)
  Theropoda indet.
  Ornithopoda
    Hadrosauridae
      Hadrosaurine indet.
  Ankylosauria
    ?Ankylosaurid indet.
  Age: late Santonian or early Campanian (Lamb pers. comm.)
b. Mooreville Chalk (Langston 1960*a*; Carpenter pers. comm.; Lamb pers. comm.)
  Theropoda
    Theropoda indet.
    Tyrannosauridae
      Albertosaurus sp.
    Ornithomimid indet. (= *Coelosaurus* cf. *antiquus*)
  Ornithopoda
    Hadrosauridae
      Lophorhothon atopus
    Hadrosaurid indet.
  Ankylosauria
    Nodosaurid indet.
  Age: early Campanian (Langston 1960*a*)
c. Demopolis Chalk (Lamb pers. comm.; not figured)
  Theropoda
    ?Tyrannosaurid indet.
  Age: late Campanian (Lamb pers. comm.)
d. Ripley Formation (Lamb pers. comm.; not figured)
  Ankylosauria
    ?Nodosaurid indet.
  Age: early Maastrichtian (Lamb pers. comm.)

313. Georgia, United States

Blufftown Formation (Reinhardt et al. 1981)
  Ornithopoda
    Hadrosaurid indet.
  Age: early Campanian (Reinhardt et al. 1981)

314. North Carolina, United States

Black Creek Formation (Baird and Horner 1979)
    Theropoda
        Theropoda indet. (= ?*Dryptosaurus* sp.)
        Ornithomimid indet. (= *Coelosaurus* cf.
            antiquus)
    Ornithopoda
        Hadrosauridae
            ?*Lophorhothon atopus*
            Hadrosaurid indet. (= *Hypsibema
                crassicauda*)
    Age: Campanian (Brett and Wheeler 1961)

315. Maryland, United States

Severn Formation (Baird 1986*b*)
    Theropoda
        Ornithomimid indet. (= *Coelosaurus
            antiquus*)
    Ornithopoda
        Hadrosaurid indet.
    Age: Maastrichtian (Baird 1986*b*)

316. Delaware, United States

a. Merchantville Formation (Baird pers. comm.)
    Theropoda
        Ornithomimid indet.
    Ornithopoda
        Hadrosaurid indet.
    Age: Campanian (Baird pers. comm.)

b. Marshalltown Formation (Baird pers. comm.)
    Theropoda
        Ornithomimid indet.
    Ornithopoda
        Hadrosaurid indet.
    Age: Campanian (Olsson 1980)

317. New Jersey, United States

a. Raritan Formation (Horner 1979)
    ?Theropoda indet. (Baird pers. comm. in
        Gallagher 1984)
    Ornithopod footprints
    Age: Cenomanian (Olsson 1980)

b. Potomac Formation (Baird pers. comm.)
    Theropoda
        Tyrannosauridae
            cf. *Albertosaurus* sp.
    Age: Cenomanian (Jordan 1983)

c. Merchantville Formation (Baird pers. comm.)
    Ornithopoda
        Hadrosaurid indet.
    Age: Campanian (Olsson 1980)

d. Woodbury Formation (Gallagher 1984; Baird
    pers. comm.)
    Ornithopoda
        Hadrosauridae
            Hadrosaurus foulkii
            Hadrosaurid indet.
    Age: Campanian (Olsson 1980)

e. Marshalltown Formation (Baird pers. comm.)
    Dinosauria indet.
    Ornithopoda
        Hadrosaurid indet.
    Age: Campanian (Olsson 1980)

f. Navesink Formation (Baird pers. comm.)
    Theropoda
        Theropoda indet.
        Ornithomimid (= *Coelosaurus antiquus*)
    Ornithopoda
        Hadrosaurid indet. (= *Hadrosaurus minor,
            Hadrosaurus cavatus*)
    Ankylosauria indet. (Baird pers. comm. in
        Gallagher 1984)
    Age: Maastrichtian (Olsson 1980)

g. Mount Laurel/Wenonah Formation (Baird and
    Horner 1977; Baird pers. comm.)
    Theropoda
        Ornithomimid indet. (= *Coelosaurus
            antiquus*)
    Ankylosauria indet.
    Age: Maastrichtian (Olsson 1980)

h. New Egypt Formation (Baird pers. comm.)
    Theropoda
        Theropoda *incertae sedis*
            Dryptosaurus aquilunguis
        Ornithomimid (= *Coelosaurus antiquus*)
    Ornithopoda
        Hadrosauridae
            "*Hadrosaurus*" *minor*
            Hadrosaurid indet.
    Ankylosauria
        Nodosaurid indet.
    Age: Maastrichtian (Olsson 1980)

318. Estado de Baja California Norte, Mexico

a. "El Gallo" formation (Morris 1971, 1973*a*, 1981;
    Leitch pers. comm.)
    Theropoda
        Tyrannosauridae
            cf. *Albertosaurus* sp.
            ?Tyrannosaurid indet.
        Ornithomimid indet.
        Troodontidae
            Troodon formosus
        Dromaeosauridae
            Saurornitholestes sp.
            Dromaeosaurid indet.
    Ornithopoda
        Hadrosauridae
            ?*Lambeosaurus laticaudus*
            ?*Lambeosaurus* sp.
            Hadrosaurine indet.
    Ankylosauria
        Ankylosauria indet.
        Nodosauridae
            cf. *Euoplocephalus* sp.

Ceratopsia
Ceratopsid indet.
Age: Campanian (Morris 1981)
b. "La Bocana Roja" formation (Molnar 1974)
Theropoda
?Carnosauria
Labocania anomala
Age: ?Campanian (Molnar 1974)

319. Estado de Sonora, Mexico

Snake Ridge Formation (Lull and Wright 1942)
Ornithopoda
Hadrosaurid indet.
Age: ?Maastrichtian (unconfirmed)

320. Estado de Coahuila, Mexico

a. Cerro del Pueblo Formation (Leitch and Seymour pers. comm.; Murray et al. 1960)
Theropoda
Tyrannosaurid indet.
Ornithomimid indet.
Dromaeosaurid indet.
Ornithopoda
Hadrosauridae
Hadrosaurine indet.
Ankylosauria
Ankylosaurid indet.
Ceratopsia
Ceratopsid indet. (= *Monoclonius* sp.)
Age: Campanian (Murray et al. 1960)
b. Soledad Beds (Janensch 1926; Lull 1933)
Ceratopsia indet.
Age: late Maastrichtian (= Laramie Formation, Janensch 1926)

321. Departemento do Comayagua, Honduras

Lower Valle de Angeles Redbeds (Horne et al. 1974; Finch 1981)
Ornithopoda
Hadrosaurid indet.
Age: Cenomanian (Horne pers. comm.; Finch 1981)

Europe

European Late Cretaceous dinosaurs are as yet rather poorly known. However, areas in southern France, eastern Austria, and the Transylvania region of Romania have produced remarkable dinosaur faunas thanks to the early work of F. Nopcsa, L. Dollo, A. F. de Lapparent, E. Bunzel, and H. G. Seeley, and more recently of E. Buffetaut, P. Taquet, A. W. F. Meijer, P. Kerourio, M. L. Casanovas-Cladellas, J. V. Santafe-Llopis, J. L. Sanz, A. Buscalioni, W. Brinkmann, D. Grigorescu, D. B. Weishampel, and D. B. Norman. These Late Cretaceous faunas of Europe are indicated in figures 3.5 and 3.19.

322. Hertfordshire, England

Lower Chalk (Newton 1892)
Ornithopoda
Hadrosaurid indet. (= *Iguanodon hilli*)
Age: Cenomanian (Rawson et al. 1978)

323. Kent, England

Upper Greensand (= Chalk Marl; Huxley 1867)
Ankylosauria
Nodosauridae
Acanthopholis horridus
Age: Cenomanian (Rawson et al. 1978)

324. Province de Namur, Belgium

Glauconie argileuse (Dollo 1883c, 1903)
Theropoda indet. (= *Megalosaurus lonzeensis*)
Ornithopoda
Iguanodontia *incertae sedis*
Craspedodon lonzeensis
Age: Santonian (Taquet 1976)

325. Province de Limburg, Belgium

Maastricht Beds (Buffetaut et al. 1985)
Ornithopoda
Hadrosaurid indet. (? = *Orthomerus dolloi*)
Age: Maastrichtian (Felder 1975a, 1975b)

326. Provincie Limburg, The Netherlands

Maastricht Beds (Seeley 1883; Buffetaut et al. 1985)
Theropoda indet. (= *Megalosaurus bredai*)
Ornithopoda
Hadrosaurid indet. (= *Orthomerus dolloi*)
Age: Maastrichtian (Felder 1975a, 1975b)

327. Département du Maine-et-Loire, France

a. Unnamed unit (Buffetaut et al. 1980)
Ornithopoda
Iguanodontid indet.
Age: Cenomanian (Buffetaut et al. 1980; reworked into Miocene)
b. Unnamed unit (Lapparent and Lavocat 1955)
Dinosauria indet.
Age: Cenomanian (Lapparent and Lavocat 1955)

328. Département de la Charente-Maritime, France

Unnamed unit (Lapparent 1967)
Theropoda indet. (= *Megalosaurus* sp.)
Age: Cenomanian (Lapparent 1967)

329. Département de la Dordogne, France

Craie tufau (Buffetaut pers. comm.)
Sauropoda
?Titanosaurid indet.
Age: Late Cretaceous (Buffetaut pers. comm.)

330. Département de la Haute-Garonne, France

a. Unnamed unit (Lapparent 1947)
Ankylosauria indet. (= *Rhodanosaurus lugdunensis*)
Age: ?Maastrichtian (Lapparent 1947)

b. Calcaires "nankin" (Paris and Taquet 1973)
Ornithopoda
Hadrosaurid indet.
Age: middle Maastrichtian (Paris and Taquet 1973)

331. Département d'Ariège, France

Grès de Labarre (Villatte et al. 1985)
Sauropoda
Titanosauridae
Hypselosaurus priscus
Ornithopoda
?Iguanodontia *incertae sedis*
Rhabdodon priscus
Age: Maastrichtian (Villatte et al. 1985)

332. Departement de l'Aude, France

Marnes Rouges inferieures (Lapparent and Lavocat 1955; Lapparent 1967; Breton et al. 1986; Beetschen 1985)
Ornithopoda
?Iguanodontia *incertae sedis*
Rhabdodon priscus
Dinosaur eggshell
Age: Maastrichtian (= Danian; Lapparent and Lavocat 1955)

333. Département de l'Herault, France

Grès de Saint-Chinian (Lapparent 1947; Buffetaut et al. 1986)
Theropoda
Theropoda indet. (= *Megalosaurus pannoniensis*)
?Dromaeosaurid indet.
Sauropoda
Titanosauridae
Hypselosaurus priscus
Ornithopoda
?Iguanodontia *incertae sedis*
Rhabdodon priscus
Hadrosauridae
Telmatosaurus transsylvanicus
Ankylosauria indet. (= *Rhodanosaurus lugdunensis*)
Age: Maastrichtian (Lapparent 1947)

334. Département du Gard, France

a. Unnamed unit (Lapparent and Lavocat 1955)
Theropoda indet. (= *Megalosaurus* sp.)
Age: Cenomanian (Lapparent and Lavocat 1955)

b. Unnamed unit (Buffetaut et al. 1986)
Theropoda
?Dromaeosaurid indet.
Age: Maastrichtian (Buffetaut et al. 1986)

c. Unnamed unit (Buffetaut et al. 1986)
Theropoda
?Dromaeosaurid indet.
Age: Late Cretaceous (Buffetaut et al. 1986)

335. Département du Vaucluse, France

Unnamed unit (Lapparent 1967)
Sauropoda indet.
Age: Cenomanian (Lapparent 1967)

336. Département des Bouches-du-Rhône, France

a. Couches de Rognac (Lapparent 1947; Buffetaut et al. 1986)
Theropoda
?Dromaeosaurid indet.
Sauropoda
Titanosauridae
Hypselosaurus priscus
Ornithopoda
?Iguanodontia *incertae sedis*
Rhabdodon priscus
Ankylosauria indet. (= *Rhodanosaurus lugdunensis*)
Age: Maastrichtian (Lapparent 1947)

b. Unnamed unit (Kerourio 1982)
Dinosaur eggshell
Age: late Campanian (Kerourio 1982)

c. Unnamed unit (Kerourio 1981b)·
Dinosaur eggshell
Age: early Maastrichtian (Kerourio 1981b)

d. Unnamed unit (Kerourio 1981b, 1982; Williams et al. 1984)
Dinosaur eggshell
Age: late Maastrichtian (Williams et al. 1984)

337. Département du Var, France

a. Grès à Reptiles (Lapparent 1947; Buffetaut pers. comm.)
Theropoda
Theropoda indet. (= *Megalosaurus pannoniensis*)
?Abelisaurid indet.
Sauropoda
Titanosauridae
Titanosaurus indicus
Hypselosaurus priscus
Ornithopoda
?Iguanodontia *incertae sedis*
Rhabdodon priscus
Hadrosauridae
Telmatosaurus transsylvanicus
Age: Maastrichtian (= Rognacien, Lapparent 1947)

b. Unnamed unit (Kerourio 1982; Williams et al. 1984)
Dinosaur eggshell
Age: early Maastrichtian (Williams et al. 1984)

c. Unnamed unit (Williams et al. 1984)
Dinosaur eggshell
Age: late Maastrichtian (Williams et al. 1984)

338. Provincia de Soria, Spain

Unnamed unit (Bataller 1960)
   Sauropoda
      Titanosauridae
         Hypselosaurus sp.
   Ornithopoda
      ?Iguanodontia *incertae sedis*
         Rhabdodon priscus
   Age: Maastrichtian (Bataller 1960)

339. Provincia de Lérida, Spain

Lower "Garumnian" Beds (Llompart and Kraus 1982)
   Ornithopoda
      ?Iguanodontia *incertae sedis*
         Rhabdodon priscus
   Age: Maastrichtian (Llompart and Kraus 1982)

340. Provincia de Lléida, Spain

a. Unnamed unit (Llompart et al. 1984)
   ?Sauropod footprints
   Ornithopod footprints
   Age: Maastrichtian (Llompart et al. 1984)

b. Unnamed unit (Lapparent and Aguire 1956; Brinkmann 1984; Casanovas-Cladellas, Santafe-Llopis et al. 1985*a*, 1985*b*; Casanovas et al. 1987)
   Sauropoda
      Titanosauridae
         Titanosaurus cf. *indicus*
         Hypselosaurus sp.
   Ornithopoda
      ?Iguanodontia *incertae sedis*
         Rhabdodon priscus
      Hadrosauridae
         ?*Telmatosaurus transsylvanicus*
         Hadrosaurid indet. (= *Orthomerus* sp.)
   Age: Maastrichtian (Battaller 1960)

341. Provincia de Segovia, Spain

Unnamed unit (Sanz 1985; Sanz and Buscalioni 1987)
   Theropoda indet.
   Sauropoda indet.
   Ornithischia indet.
   Age: Campanian-Maastrichtian (Sanz and Buscalioni 1987)

342. Contrée de Treviño, Spain

Unnamed unit (Sanz 1985)
   Sauropoda
      Titanosaurid indet.
   Age: Maastrichtian (Sanz 1985)

343. Provincia de Alava, Spain

Unit S1U3 (Astibia et al. 1987)
   Ornithopoda indet.
   Age: Maastrichtian (Astibia et al. 1987)

344. Provincia de Valencia, Spain

Unnamed unit (Sanz 1985)
   Ankylosauria
      Nodosaurid indet.
   Age: Santonian-Maastrichtian (Sanz 1985)

345. Província do Beira Litoral, Portugal

Unnamed unit (Lapparent and Zbyszewski 1957; Antunes and Pais 1978)
   Theropoda indet. (= ?*Megalosaurus pannoniensis*)
   ?Ceratopsia
      ?Ceratopsid indet.
   Age: late Campanian-?Maastrichtian (Lapparent and Zbyszewski 1957; Antunes and Pais 1978)

346. Niederöstereich, Austria

Gosau Formation (Bunzel 1871; Seeley 1881)
   Theropoda indet. (= *Megalosaurus pannoniensis*)
   ?Ornithischia *incertae sedis*
      Oligosaurus adelus
   Ornithopoda
      ?Iguanodontia *incertae sedis*
         Rhabdodon priscus
   Ankylosauria
      Ankylosauria indet. (= *Struthiosaurus austriacus*, *Danubiosaurus anceps* partim, *Leipsanosaurus noricus*, *Hoplosaurus ischyrus*, *Rhadinosaurus alcinus*)
      Nodosauridae
         Struthiosaurus transilvanicus (= *Danubiosaurus anceps* partim)
   Age: Campanian (Jeletzky 1960; Thenius 1974)

347. Judethean Hunedoara, Romania

Sinpetru Beds (Nopcsa 1915, 1923*a*; Huene 1929*b*, 1932; Harrison and Walker 1975; Grigorescu 1983*a*, 1983*b*, 1984; Grigorescu et al. 1985; Weishampel, Norman and Grigorescu in prep.)
   Theropoda
      Theropoda indet. (= *Megalosaurus hungaricus*)
      Undescribed theropod (Grigorescu pers. comm.)
      Troodontid indet. (= *Bradycneme draculae*, *Heptasteornis andrewsi*, *Elopteryx nopcsai*)
      ?Dromaeosaurid indet.
   Sauropoda
      Titanosauridae
         Magyarosaurus dacus
         Magyarosaurus transsylvanicus
         Magyarosaurus hungaricus
         Titanosaurid indet.

Ornithopoda
?Iguanodontia *incertae sedis*
Rhabdodon priscus
Rhabdodon sp.
Hadrosauridae
Telmatosaurus transsylvanicus
Ankylosauria
Nodosauridae
Struthiosaurus transilvanicus
?Dinosaur coprolites
Age: late Maastrichtian (Grigorescu 1983*a*, 1987)

348. Severocesky Kraj, Czechoslovakia

Korycaner Formation (Fritsch 1905)
?Dinosauria indet. (= *Iguanodon exogirarus*)
Age: Cenomanian (Fritsch 1905)

349. Vychodocesky Kraj, Czechoslovakia

Preiesener Formation (Fritsch 1905)
?Dinosauria indet. (= *Albisaurus scutifer*)
Age: ?Coniacian-?Santonian (= ?early Senonian; Fritsch 1905)

350. Krymskaya Oblast, Ukraine

Unnamed unit (Riabinin 1945)
Ornithopoda
Hadrosaurid indet. (= *Orthomerus weberi*)
Age: Maastrichtian (Jeletzky 1962)

351. Kurskaya Oblast, Russia

Unnamed unit (Kiprijanow 1883)
Theropoda indet. (= *Poekilopleuron schmidti*)
Age: Cenomanian (Rozhdestvensky 1973*b*)

Asia

The Asian record of Late Cretaceous dinosaurs (figs. 3.6, 3.7, 3.16, 3.20, 3.22) includes the famous discoveries in the Mongolian Gobi Desert by American, Russian, Polish, and Mongolian research teams, beginning in the early 1920s and continuing to the present day. Dinosaur material has also been recovered from early Late Cretaceous horizons of Kazakhstan by teams from the former U.S.S.R. under the supervision of A. K. Rozhdestvensky, as well as from central India amassed by C. A. Matley and F. von Huene and later by K. N. Prasad, S. L. Jain, A. Sahni, P. Yadagiri, and from scattered localities in the People's Republic of China, principally Shandong, studied by C. Wiman and later by C.-C. Young and Hu C.-C.

352. Mechooz Yerushalayim, Israel

Unnamed unit (Avnimelech 1962*a*, 1962*b*)
Theropod footprints

Age: early Cenomanian (Avnimelech 1962*a*, 1962*b*)

353. Muhafazat Dimashq, Syria

Unnamed unit (Hooijer 1968)
Theropoda indet.
Age: ?Cenomanian (Hooijer 1968)

354. Navoiskaya Oblast, Uzbekistan

Beleutinskaya Svita (Kurzanov 1976*a*)
Theropoda
Theropoda *incertae sedis*
Itemirus medullaris
Ornithomimid indet.
Ornithopoda
Hadrosaurid indet. (= *Cionodon kysylkumensis*)
Age: Turonian (Rozhdestvensky 1964*b*)

355. Leninabadskaya Oblast, Tadzikistan

a. Unnamed unit (Zacharov 1964; Rozhdestvensky 1973*b*)
Theropod footprints
Age: Cenomanian (Rozhdestvensky 1973*b*)
b. Yalovachskaya Svita (Rozhdestvensky 1977)
Theropoda
Tyrannosaurid indet.
Ornithomimid indet.
Ornithopoda
Hadrosaurid indet.
Age: Turonian-early Santonian (Rozhdestvensky 1977)

356. Sydarinskaya Oblast, Kazakhstan

a. Yalovachskaya Svita (Rozhdestvensky and Khozatsky 1967; not figured)
Theropoda indet.
Sauropoda indet.
Ornithopoda
Hadrosauridae
Lambeosaurine indet. (?= *Jaxartosaurus* sp.)
Age: Turonian-early Santonian (Rozhdestvensky 1977)
c. Darbazinskaya Svita (Riabinin 1931*b*, 1937, 1938; Rozhdestvensky 1968*a*; Kurzanov pers. comm.)
Sauropoda
Titanosauridae *incertae sedis*
"*Antarctosaurus*" *jaxartensis*
Ornithopoda
Hadrosauridae
Jaxartosaurus aralensis (including "*Procheneosaurus*" *convincens*)
Hadrosaurid indet. (= *Cionodon*

kysylkumensis, Bactrosaurus
prynadai)
Age: Turonian-Santonian (Rozhdestvensky
1968*a;* Kurzanov pers. comm.)

f. Unnamed unit (Nurumov 1964; Kurzanov pers.
comm.)
Dinosaur eggshell
Age: ?Late Cretaceous (Rozhdestvensky
1973*b*)

356. Kzyl-Ordaskaya Oblast, Kazakhstan

b. Beleutinskaya Svita (Rozhdestvensky 1968*a;*
Kurzanov pers. comm.)
Theropoda
Tyrannosaurid indet.
?Therizinosaurid indet.
Sauropoda indet.
Ornithopoda
Hadrosauridae
Aralosaurus tuberiferus
Hadrosaurid indet.
Ankylosauria
Ankylosaurid indet.
Age: Turonian-Santonian (Rozhdestvensky
1968*a*)

d. Bostobinskaya Svita (Shilin and Suslov 1982;
Kurzanov pers. comm.)
?Ornithopoda
?Hadrosaurid indet. (= *Arstanosaurus*
akkurganensis)
Age: Santonian-Campanian (Shilin and Suslov
1982)

e. Manrakskaya Svita (Bazhanov 1961; Kurzanov
pers. comm.)
Dinosaur eggshell
Age: Late Cretaceous (Rozhdestvensky 1973*b*)

357. Yevreyskaya Avtonomnaya Oblast, Russia

Tsagayanskaya Svita (Riabinin 1930*b;* Kurzanov
pers. comm.)
Theropoda
?Tyrannosaurid indet.
Ornithopoda
Hadrosauridae
Hadrosaurine indet. (= *Trachodon*
amurensis)
Age: ?Coniacian-?Maastrichtian (=
?Senonian; Rozhdestvensky 1973*b*)

358. Sakhalinskaya Oblast, Russia

Mh7 or Mh6 of the Miho Group (Nagao 1936;
Matsumoto and Obata 1979)
Ornithopoda
Hadrosauridae
Nipponosaurus sachalinensis
Age: late Santonian or early Campanian
(Matsumoto and Obata 1979)

359. Gujarat State, India

Lameta Formation (Infratrappean Limestone;
Dwivedi et al. 1982; Mohabey 1983, 1986;
Srivastava 1983; Srivastava et al. 1986; Mathur and
Srivastava 1987)
Theropoda
Theropoda indet. (= *Megalosaurus* sp.)
?Abelisauridae
Majungasaurus crenatissimus
Sauropoda
?Sauropod footprints
Titanosauridae indet. (= *Titanosaurus*
rahioliensis)
Dinosauria eggshell
Age: middle-late Maastrichtian (Sahni pers.
comm.)

360. Madhya Pradesh State, India

Lameta Formation (Das Gupta 1931; Huene and
Matley 1933; Chatterjee 1978*b;* Berman and Jain
1982; Jain and Sahni 1983)
Theropoda
Theropoda indet. (= *Orthogoniosaurus*
matleyi, Jubbulpuria tenuis, Laevisuchus
indicus, Lametasaurus indicus partim,
Coeluroides largus, Dryptosauroides
grandis, Ornithomimoides mobilis, O.
barasimlensis)
?Abelisauridae
Indosaurus matleyi
Indosuchus raptorius
?Carnosauria
Compsosuchus solus
Ornithomimidae
Ornithomimid indet.
Sauropoda
Titanosauridae
Titanosaurus indicus
Titanosaurus blanfordi
?*Titanosaurus* cf. *madagascariensis*
?*Titanosaurus* sp.
?*Antarctosaurus* sp.
Titanosaurid indet. (= *Lametasaurus*
indicus partim)
Titanosauridae *incertae sedis*
"Antarctosaurus" septentrionalis
Ankylosauria indet. (= *Brachypodosaurus*
gravis; may be stegosaurid, Galton 1981*a*)
?Ankylosauria indet. (= *Lametasaurus*
indicus partim)

Dinosaur eggshell
Age: middle-late Maastrichtian (Sahni pers. comm.)

361. Maharashtra State, India

a. Lameta Formation (Lydekker 1879; Prasad and Verma 1967; Berman and Jain 1982; Sahni, Ranna, and Prasad 1984; Jain and Sahni 1985)
   Sauropoda
      Titanosauridae
         Titanosaurus indicus
         Titanosaurus blanfordi
         Titanosaurus cf. *madagascariensis*
         Antarctosaurus sp.
   Dinosaur eggshell
   ?Dinosaur coprolites
   Age: middle-late Maastrichtian (Sahni pers. comm.)

b. Takli Formation (Lydekker 1890*b*; Huene and Matley 1933; Sahni, Rana, Kumar, and Loyal 1984)
   Theropoda indet. (= *Massospondylus rawesi*)
   Dinosaur eggshell
   Age: late Maastrichtian (Sahni, Rana, Kumar, and Loyal 1984)

362. Tamil Nadu State, India

a. Trichinopoly Group (Anonymous 1978; Yadagiri and Ayyasami 1979)
   Theropoda indet.
   Stegosauria
      Stegosauridae
         Dravidosaurus blanfordi
   Age: Coniacian (Sastry et al. 1969)

b. Aviyalur Group (Huene and Matley 1933; Rao 1956; Prasad 1968)
   Theropoda indet.
   Sauropoda
      Titanosauridae
         Titanosaurus indicus
   Stegosauria
      Stegosaurid indet.
   Age: Maastrichtian (Sastry et al. 1969)

363. Bayankhongor, Mongolian People's Republic

a. Nemegtskaya Svita (Barsbold 1983*b*; = Beds of Bugeen Tsav, Gradzinski et al. 1977)
   Theropoda
      Tyrannosauridae
         Tarbosaurus bataar
      Ornithomimidae
         Anserimimus planinychus
      Oviraptoridae
         Ingenia yanshini
      Troodontidae
         Saurornithoides junior
      Dromaeosauridae
         Adasaurus mongoliensis

Ornithopoda
   Hadrosauridae
      Saurolophus angustirostris
Ankylosauria
   Ankylosauridae
      Tarchia gigantea
Age: ?late Campanian or early Maastrichtian (Gradzinski et al. 1977; Maastrichtian, Fox 1978; Lillegraven and McKenna 1986)

b. Beds of Nogon Tsav (Kurzanov 1976*b*; not figured)
   Theropoda
      Tyrannosauridae
         Alioramus remotus
   Age: ?Maastrichtian (Kurzanov 1976*b*)

c. White Beds of Khermeen Tsav (Gradzinski et al. 1977) (should be listed as 364.j.)
   Theropoda
      Theropoda indet.
      Theropoda *incertae sedis*
         Therizinosaurus cheloniformis
      Tyrannosauridae
         Tarbosaurus bataar
   Ornithopoda
      Hadrosauridae
         Saurolophus angustirostris
   Ankylosauria
      Ankylosauridae
         Tarchia gigantea
   Age: ?late Campanian or early Maastrichtian (Gradzinski et al. 1977)

364. Omnogov, Mongolian People's Republic

a. Baynshirenskaya Svita (Perle 1977, 1979; Barsbold and Perle 1980, 1983; Kurzanov 1976*b*; Kurzanov and Tumanova 1978; Tumanova 1987)
   Theropoda
      Tyrannosauridae
         Alectrosaurus olseni
      Garudimimidae
         Garudimimus brevipes
      Segnosauria
         Segnosaurus galbinensis
         Erlikosaurus andrewsi
   Ornithopoda
      Hadrosaurid indet.
   Ankylosauria
      Ankylosauridae
         Amtosaurus magnus
         Talarurus plicatospineus
   Age: Cenomanian-Turonian (Barsbold 1983*a*)

b. Djadochta Formation (Gradzinski et al. 1977)
   Theropoda
      Theropoda indet.
      Tyrannosauridae
         ?*Tarbosaurus* sp.
      Oviraptoridae

    Oviraptor philoceratops
   Troodontidae
    Saurornithoides mongoliensis
   Dromaeosauridae
    Velociraptor mongoliensis
  Sauropoda indet.
  Ornithopoda
   Hadrosaurid indet.
  Ceratopsia
   Protoceratopsidae
    Protoceratops andrewsi
  Ankylosauria
   Ankylosauridae
    Pinacosaurus grangeri
  Dinosaur eggshell
  Age: ?late Santonian or early Campanian
   (Gradzinski et al. 1977; late Campanian or
   early Maastrichtian, Fox 1978; Lillegraven
   and McKenna 1986)

c. Beds of Toogreeg (Gradzinski et al. 1977)
  Theropoda
   Dromaeosauridae
    Velociraptor mongoliensis
  Ceratopsia
   Protoceratopsidae
    Protoceratops andrewsi
  Age: ?late Santonian or early Campanian
   (Gradzinski et al. 1977)

d. Beds of Alag Teg (Gradzinski et al. 1977)
  Theropoda indet.
  Sauropoda indet.
  Ceratopsia
   Protoceratopsidae
    ?Protoceratops andrewsi
  Ankylosauria
   Ankylosauridae
    Pinacosaurus grangeri
  Age: ?late Santonian or early Campanian
   (Gradzinski et al. 1977)

e. Djadochtinskaya Svita (Kurzanov 1987)
  Theropoda
   Avimimidae
    Avimimus portentosus
   Ornithomimidae
    Gallimimus sp.
  Sauropoda
   Diplodocidae
    Quaesitosaurus orientalis
  Age: Coniacian–early Campanian (Barsbold
   1983*a*)

f. Barun Goyot Formation (Gradzinski et al. 1977;
  Osmólska 1982)
  Theropoda
   Theropoda indet.
   Dromaeosauridae
    Hulsanpes perlei
    Velociraptor sp.

  Sauropoda indet.
  Pachycephalosauria
   Pachycephalosauridae
    Tylocephale gilmorei
  Ceratopsia
   Protoceratopsidae
    Breviceratops kozlowskii
  Ankylosauria
   Ankylosauridae
    Saichania chulsanensis
    Tarchia kielanae
  Age: ?middle Campanian (Gradzinski et al.
   1977; Maastrichtian, Fox 1978; Lillegraven
   and McKenna 1986)

g. Red Beds of Khermeen Tsav (Gradzinski et al.
  1977)
  Theropoda
   Theropoda indet.
   Oviraptoridae
    Ingenia yanshini
    Conchoraptor gracilis
   Dromaeosauridae
    Velociraptor sp.
  Sauropoda indet.
  Ceratopsia
   Protoceratopsidae
    Bagaceratops rozhdestvenskyi
  Ankylosauria
   Ankylosauridae
    Saichania chulsanensis
  Dinosaur eggshell
  Age: ?middle Campanian (Gradzinski et al.
   1977)

h. Nemegt Formation (Gradzinski et al. 1977)
  Theropoda
   Theropoda indet.
   Theropoda *incertae sedis*
    Therizinosaurus cheloniformis
   Tyrannosauridae
    Tarbosaurus bataar
    "Gorgosaurus" novojilovi
    Tyrannosaurid indet.
   Coelurosauria *incertae sedis*
    Deinocheirus mirificus
   Ornithomimidae
    Gallimimus bullatus
   Elmisauridae
    Elmisaurus rarus
   Oviraptoridae
    Oviraptor mongoliensis
    Oviraptorid indet.
   Troodontidae
    Saurornithoides junior
    Borogovia gracilicrus
    Tochisaurus nemegtensis
  Sauropoda
   Camarasauridae
    Opisthocoelicaudia skarzynskii

Diplodocidae
Nemegtosaurus mongoliensis
Ornithopoda
Hadrosauridae
Hadrosaurid indet.
Saurolophus angustirostris
Barsboldia sicinskii
Pachycephalosauria
Pachycephalosauridae
Prenocephale prenes
Homalocephalidae
Homalocephale calathocercos
Ankylosauria
Ankylosauridae
Tarchia gigantea
Age: ?late Campanian or early Maastrichtian (Gradzinski et al. 1977; Maastrichtian, Fox 1978; Lillegraven and McKenna 1986)
i. Sheeregeen Gashoon Formation (= Baynshirenskaya Svita; Maryańska and Osmólska 1975; not figured)
Theropoda
Theropoda indet.
Ornithomimid indet.
Ornithopoda
Hadrosaurid indet.
Ankylosauria
Nodosauridae
Maleevus disparoserratus
Ceratopsia
Protoceratopsidae
Microceratops gobiensis
Age: Cenomanian–Turonian (Barsbold 1983*a*)

365. Ovorkhangai, Mongolian People's Republic

a. Unnamed unit (Perle et al. 1982)
Pachycephalosauria
Homalocephalidae
Goyocephale lattimorei
Age: late Santonian or early Campanian (Osmólska pers. comm.)
b. "Djadochtskaya" Svita (Kurzanov 1987)
Theropoda
Avimimidae
Avimimus portentosus
Age: ?Coniacian–early Campanian (Barsbold 1983*a*)

366. Dundgov, Mongolian People's Republic

Dohoin Usu Formation (Gilmore 1933*b*)
Theropoda indet.
Ornithopoda
Hadrosaurid indet.
Age: ?Cenomanian (i.e., equivalent to Iren Dabasu Formation; Morris 1936)

367. Dornogov, Mongolian People's Republic

Baynshirenskaya Svita (Maleev 1952*a*, 1952*b*; Barsbold 1972)
Theropoda indet.
Segnosauria
Enigmosaurus mongoliensis
Segnosaurus galbinensis
Ornithopoda
Hadrosaurid indet.
Ankylosauria
Ankylosauridae
Talarurus plicatospineus
Age: Cenomanian-Turonian (Shuvalov 1976)

368. Xinjiang Uygur Ziziqu, People's Republic of China

a. Subashi Formation (Dong 1977; Chen et al. 1982)
Theropoda
Theropoda *incertae sedis*
Shanshanosaurus huoyanshanensis
Tyrannosauridae
?*Tarbosaurus bataar* (= *Tyrannosaurus turpanensis*)
Sauropoda indet. (= *Nemegtosaurus pachi*)
Ankylosauria indet.
Dinosaur eggshell
Age: ?Campanian-Maastrichtian (Shen and Mateer in press)
b. Wulungo Formation (Wu 1973, 1984; not figured)
Ornithopoda
Hadrosaurid indet. (= *Jaxartosaurus fuyunensis*)
Age: Late Cretaceous (Wu 1973)

369. Xizang Zizhiqu, People's Republic of China

Quwu Formation (Zhen et al. 1989; not figured)
Dinosauria footprints
Age: Late Cretaceous (Mateer pers. comm.)

370. Gansu, People's Republic of China

a. Minhe Formation (Bohlin 1953; Maryańska 1977; Mateer pers. comm.)
Theropoda
Theropoda indet.
Ceratopsia
Protoceratopsidae
Microceratops gobiensis (= *M. sulcidens*)
Protoceratops andrewsi
?Pachycephalosauria indet. (= *Heishansaurus pachycephalus*)
?Ankylosauria indet. (= *Peishansaurus philemys*)
Age: Campanian-Maastrichtian (Mateer pers. comm.)

b. Unnamed unit (Xie 1980)
Theropoda indet.
Sauropoda
Titanosaurid indet.
Ornithopoda
Hadrosaurid indet.
Age: Campanian-Maastrichtian (i.e.,
contemporary with the Subashi Formation,
Dong 1977; Chen et al. 1982)

371. Nei Mongol Zizhiqu, People's Republic of China

a. Iren Dabasu Formation (= Iren Nor Formation,
Erlien Dabsu Formation; Gilmore 1933a, 1933b;
Rozhdestvensky 1966; Weishampel and Horner
1986)
Theropoda
Tyrannosauridae
Alectrosaurus olseni
Ornithomimidae
Archaeornithomimus asiaticus
?Troodontidae
?Saurornithoides sp.
Ornithopoda
Hadrosauridae
Gilmoreosaurus mongoliensis
Bactrosaurus johnsoni
Age: early Late Cretaceous (Maryańska and
Osmólska 1981a; Cenomanian,
Rozhdestvensky 1977; Maastrichtian, Chen
1983)

b. Minhe Formation (Bohlin 1953; Maryańska
1977; Mateer pers. comm.)
Theropoda
Theropoda indet.
Dromaeosauridae
cf. *Velociraptor mongoliensis*
Sauropoda indet.
Ceratopsia
Protoceratopsidae
Microceratops gobiensis
Protoceratops andrewsi
Pachycephalosauria
Pachycephalosauridae
"*Troodon*" *bexelli*
Ankylosauria
Nodosaurid indet.
Age: Campanian-Maastrichtian (Mateer pers.
comm.)

372. Heilongjiang, People's Republic of China

a. Tsagayanskaya Svita (Riabinin 1930a)
Ornithopoda
Hadrosauridae
Tanius sinensis
Hadrosaurine indet. (= *Trachodon*
amurensis)
Hadrosaurid indet.
Age: ?Coniacian-?Maastrichtian (=
?Senonian; Rozhdestvensky 1973b)

b. Unnamed unit (Riabinin 1930a)
Theropoda
Tyrannosauridae
?Tarbosaurus bataar (= *Albertosaurus*
periculosus)
Ornithopoda
Hadrosaurid indet. (= *Saurolophus*
kryschtofovici)
Age: Late Cretaceous (unconfirmed)

373. Ningxia Huizu Zizhiqu, People's Republic of China

a. Unnamed unit (Young 1935b; Maryańska 1977)
Ankylosauria
Ankylosauridae
Pinacosaurus grangeri (= *Pinacosaurus*
ninghsiensis)
Age: ?late Santonian or ?early Campanian
(i.e., ?contemporary with Djadochta
Formation; Maryańska 1977)

b. Unnamed unit (Zhao and Ding 1976)
Dinosauria eggshell
Age: ?early Late Cretaceous (Zhao and Ding
1976)

374. Henan, People's Republic of China

a. Qiuba Formation (Dong 1979)
Theropoda
Tyrannosauridae
Tyrannosaurus luanchuanensis
Age: Campanian (Dong pers. comm.)

b. Nanchao Formation (Zhao 1979; Tong and Wang
1980)
?Ornithopod footprint
Dinosaur Eggshell
Age: Late Cretaceous (Zhen et al. 1985)

375. Anhui, People's Republic of China

Xiaoyan Formation (Hou 1977)
Pachycephalosauria
Homalocephalidae
Wannanosaurus yansiensis
Age: Campanian (Dong pers. comm.)

376. Zhejiang, People's Republic of China

a. Tangshang Formation (Dong 1979; Mateer pers.
comm.)
Segnosauria
"*Chilantaisaurus*" *zheziangensis*
Dinosaur eggshell
Age: Santonian-Campanian (Dong 1979)

b. Quxian Formation (Mateer 1989; not figured)
Dinosaur eggshell
Age: Late Cretaceous (Mateer 1989)

c. Wangi Formation (Zhen et al. 1989; not figured)
Dinosauria footprints
Age: Late Cretaceous (Mateer pers. comm.)

377. Shandong, People's Republic of China

Wangshi Series (Wiman 1929; Chow 1951; Young
1958a; Dong 1978)
Theropoda
Theropoda indet. (= cf. *Szechuanosaurus
campi*)
Tyrannosauridae
Chingkankousaurus fragilis
Ornithopoda
Hadrosauridae
Tanius sinensis (including *T.
chingkankouensis*)
Shantungosaurus giganteus
Tsintaosaurus spinorhinus (including
Tanius laiyangensis)
Hadrosaurid indet.
Lambeosaurine indet.
Pachycephalosauria *incertae sedis*
Micropachycephalosaurus hongtuyanensis
Dinosaur eggshell
Age: Campanian-middle Maastrichtian
(Mateer pers. comm.)

378. Hunan, People's Republic of China

a. Red Beds (Zeng and Zhang 1979)
Dinosaur eggshell
Age: Late Cretaceous (Zeng and Zhang 1979)
b. Jinjiang Formation (Zheng 1982; not figured)
Dinosaur footprints
Age: middle Late Cretaceous (Mateer pers.
comm.)

379. Guandong, People's Republic of China

Nanxiong Formation (Zhao 1975; Dong 1979)
Theropoda
Theropoda indet.
Tyrannosauridae
Tarbosaurus bataar
Tyrannosaurid indet.
Segnosauria
Nanshiungosaurus brevispinus
Ornithopoda
Hadrosauridae
Hadrosaurine indet. (=
Microhadrosaurus nanshiungensis)
Dinosaur eggshell
Age: late Campanian-Maastrichtian (Dong
1979, pers. comm.; Mateer pers. comm.)

380. Shanxi, People's Republic of China

Zhumabao Formation (Mateer pers. comm.; not
figured)
Ceratopsia
Protoceratopsidae
Microceratops sp.
Age: early Late Cretaceous (Mateer pers.
comm.)

381. Tokushima Prefecture, Japan

Akazaki Formation (Mateer pers. comm.)
Ornithopoda
Hadrosaurid indet.
Age: ?Maastrichtian (Mateer pers. comm.)

382. Kyushu Prefecture, Japan

Mitsuse Formation (Matsuo 1967; Obata et al. 1972)
Ornithopoda
Hadrosaurid indet.
Age: early Santonian or late Coniacian
(Hinokuma 1963; Matsuo 1967)

383. Savannakhet Khoueng, Laos

Unnamed unit (Hoffet 1942, 1943a, 1943b)
Sauropoda
Titanosauridae
Titanosaurus falloti
Ornithopoda
Hadrosaurid indet. (= *Mandschurosaurus
laosensis*)
Age: Coniacian-Maastrichtian (= Senonian;
Hoffet 1943a, 1943b)

South America

The South American record of Late Cretaceous
dinosaurs (figs. 3.8 and 3.9) consists of a number of
early discoveries by F. Ameghino, R. Lydekker, and F.
Huene in Argentina as well as recent important research in Bolivia by F. de Broin and co-workers, in
Peru by B. Sigé and co-workers, in Brazil by L. I. Price,
and in Argentina by J. F. Bonaparte, J. Powell, G. del
Corro, and R. M. Casamiquela.

384. Departamento de Tolima, Colombia

"Ortega" formation (Langston 1965)
Theropoda indet.
Age: Maastrichtian (Langston 1965)

385. Departamento de Potosi, Bolivia

El Molino Formation (= Toro Toro Formation;
Branisa 1968; Muizon et al. 1983; Marshall et al.
1985; Leonardi 1989)
Theropod footprints
Sauropod footprints
?Ornithopod footprints
?Ceratopsian footprints
Age: Maastrichtian (Muizon et al. 1983;
Leonardi 1989)

386. Departamento de Cochabamba, Bolivia

Santa Lucia Formation (Leonardi 1989)
Theropod footprints
Ornithopod footprints
Age: Maastrichtian (Leonardi 1989)

387. Departamento de Puno, Peru

Vilquechico Formation (Sigé 1968)
Ornithopoda indet.
Dinosaur eggshell
Age: Late Cretaceous (Grambast et al. 1967)

388. Departamento de Amazonas, Peru

Rouges Basales (Mourier et al. 1986)
Theropoda indet.
Sauropoda
Titanosaurid indet.
Age: late Santonian-Campanian (Mourier et al. 1986)

389. Estado do Matto Grosso, Brazil

Bauru Formation (Price 1960)
Theropoda indet.
Age: Campanian-Maastrichtian (Petri and Campanha 1981)

390. Estado do Goiás, Brazil

Bauru Formation (Bonaparte 1978a; not figured)
Theropoda indet.
Sauropoda
Titanosauridae
Antarctosaurus brasiliensis
cf. *Titanosaurus* sp.
Age: Campanian-Maastrichtian (Petri and Campanha 1981)

391. Estado do Minas Gerais, Brazil

Bauru Formation (Price 1951; Bonaparte 1978a, pers. comm.; Powell 1986)
Theropoda indet. (2 species)
Sauropoda
Titanosauridae
Antarctosaurus brasiliensis
cf. *Titanosaurus* sp.
Undescribed titanosaurid (Powell 1986)
Titanosaurid indet.
Dinosaur eggshell
Age: Campanian-Maastrichtian (Petri and Campanha 1981)

392. Estado do São Paulo, Brazil

Bauru Formation (Arid and Vizotto 1972; Powell 1986)
Theropoda indet.
Sauropoda
Titanosauridae
Antarctosaurus brasiliensis
cf. *Titanosaurus* sp.
Titanosaurid indet.
Age: Campanian-Maastrichtian (Petri and Campanha 1981)

393. Estado do Paraná, Brazil

Bauru Formation (Bonaparte 1978a; not figured)
Theropoda indet.
Sauropoda
Titanosauridae
cf. *Titanosaurus* sp.
Age: Campanian-Maastrichtian (Petri and Campanha 1981)

394. Estado do Maranhão, Brazil

a. Unnamed unit (Price 1947)
Sauropoda indet.
Age: ?Late Cretaceous (Price 1947)
b. Itapecuru Formation (Price 1947)
Theropoda indet.
Sauropoda indet.
Age: Cenomanian-Santonian (Petri and Mendez 1981)

395. Estado do Amazonas, Brazil

Sucunduri Formation (Price 1960)
Theropoda indet.
Age: Late Cretaceous (Price 1960)

396. Departamento Paysandú, Uruguay

Guichon Formation (Huene 1934b, Bonaparte 1978a)
Theropoda
?Ornithomimid indet.
?Ornithopoda
?Iguanodontid indet.
Age: Coniacian-Maastrichtian (= Senonian; Bonaparte 1978a)

397. Departamento Palmitas, Uruguay

Asencio Formation (Huene 1929a; Bossi 1966)
Sauropoda
Titanosauridae
Saltasaurus australis
Laplatasaurus araukanicus
Antarctosaurus wichmannianus
Argyrosaurus superbus
Age: Late Cretaceous (Bossi 1966)

398. Departamento Colonia, Uruguay

Asencio Formation (Mones 1980)
Dinosaur eggshell
Age: Late Cretaceous (Mones 1980)

399. Departamento Tacuarembó, Uruguay

Asencio Formation (Mones 1980)
Dinosaur eggshell
Age: Late Cretaceous (Mones 1980)

400. Provincia de Coquimbo, Chile

   Vinita Formation (Casamiquela et al. 1969)
   Sauropoda
       Titanosauridae
           Antarctosaurus cf. *wichmannianus*
       Titanosaurid indet.
   Age: Campanian-Maastrichtian (Gasparini
       1979)

401. Provincia de Salta, Argentina

   a. Los Blanquitos Formation (Powell 1978, 1979,
      1986)
      Theropoda *incertae sedis*
         Unquillosaurus ceibalii
      Theropoda indet.
      Sauropoda
         Sauropoda indet.
         Titanosauridae
           Laplatasaurus cf. *araukanicus*
           ?*Titanosaurus* sp.
         Titanosaurid indet.
      Age: ?Campanian (Bonaparte et al. 1977)

   b. Yacoraite Formation (Bonaparte and Bossi 1967;
      Alonso 1980; Leonardi 1989)
      Theropoda indet.
      Theropod footprints
      Ornithopoda
         Ornithopod footprints
      Hadrosaurid indet.
      Age: Maastrichtian (Bonaparte et al. 1977)

   c. Lecho Formation (Bonaparte and Powell 1980;
      Brett-Surman and Paul 1985; Powell 1986)
      Theropoda
         Theropoda *incertae sedis*
           Avisaurus archibaldi
           Noasaurus leali
         ?Abelisaurid indet.
      Sauropoda
         Titanosauridae
           Saltasaurus loricatus
      Age: ?late Campanian-Maastrichtian
         (Bonaparte et al. 1977; Powell 1986)

402. Provincia de Entre Ríos, Argentina

   Puerto Yerua Formation (= Tacuarembo Formation;
   Huene 1929a; Powell 1986)
      Sauropoda
         Titanosaurid indet. (= *Argyrosaurus*
           superbus)
      Age: Late Cretaceous (Powell 1986)

403. Provincia de Río Negro, Argentina

   a. Angostura Colorada Formation (Powell 1986; not
      figured)
      Sauropoda
         Titanosaurid indet.
      Age: Campanian (Powell 1986)

   b. Los Alamitos Formation (Bonaparte et al. 1984;
      Powell 1986)
      Theropoda indet.
      Sauropoda
         Titanosauridae
           Aeolosaurus rioneginus
      Ornithopoda
         Hadrosauridae
           "*Kritosaurus*" *australis*
      Age: late Campanian-early Maastrichtian
         (Bonaparte et al. 1984)

   c. Coli Toro Formation (Casamiquela 1964a; Powell
      1986)
      Sauropoda
         Titanosaurid indet.
      Ornithopoda
         Hadrosaurid indet.
      Age: Campanian-Maastrichtian (Powell 1986)

   d. Allen or Rio Colorado Formation
      (undifferentiated; Lydekker 1893a; Huene
      1929a; Bonaparte and Gasparini 1978;
      Bonaparte and Novas 1985; Bonaparte pers.
      comm.)
      Theropoda
         Theropoda indet.
         Abelisauridae
           Abelisaurus comahuenis
           ?Abelisaurid indet.
      Sauropoda
         Titanosauridae
           Antarctosaurus wichmannianus
           Undescribed titanosaurid (=
             Laplatasaurus araukanicus; Powell
             1986)
           Saltasaurus australis (including
             Microcoelus patagonicus, Titanosaurus
             nanus, ?*Loricosaurus scutatus*)
           Saltasaurus robustus
           ?*Titanosaurus* sp.
         cf. *Macrurosaurus* sp.
      Age: early Maastrichtian (Bonaparte and
         Novas 1985)

   e. ?Bajo Barreal Formation (= "San Jorge"
      formation; Brett-Surman 1979)
      Ornithopoda
         Hadrosauridae
           Secernosaurus koerneri
      Age: ?Maastrichtian (Brett-Surman 1979)

404. Provincia de Neuquén, Argentina

   a. Unnamed unit (Leonardi 1989)
      Theropod footprints
      Age: ?Campanian-Maastrichtian (= late
         Senonian; Leonardi 1989)

   b. Rio Limay Formation (Leonardi 1989)
      Theropod footprints
      Ornithopod footprints
      Age: ?pre-Maastrichtian (Leonardi 1989)

c. Rio Colorado Formation (Bonaparte and Gasparini 1978; Powell 1986)
Theropoda indet. (2 species)
Theropoda *incertae sedis*
Alvarezsaurus calvoi
Velocisaurus unicus
Sauropoda
Titanosauridae
cf. *Argyrosaurus* sp.
Titanosaurus sp.
Undescribed titanosaurid (Powell 1986)
Titanosaurid indet.
Age: ?Campanian (Malumian et al. 1983)

d. Rio Neuquén Formation (Huene 1929*a;* Bonaparte and Gasparini 1978)
Theropoda indet.
Sauropoda
Titanosauridae
Antarctosaurus wichmannianus
Antarctosaurus giganteus
Titanosaurus sp.
Age: ?Campanian-Maastrichtian (Digregorio 1978)

e. Unnamed unit (Tapia 1918; Huene 1929*a*)
?Ceratopsia indet. (= *Notoceratops bonarellii*)
Age: Late Cretaceous (Bonaparte pers. comm.)

405. Provincia de Chubut, Argentina

a. Laguna Palacios Formation (Bonaparte 1979*a*)
Theropoda indet.
Sauropoda
Titanosauridae
Argyrosaurus superbus
Age: Coniacian-Maastrichtian (= Senonian; Bonaparte 1979*a*)

b. Bajo Barreal Formation (Bonaparte and Gasparini 1978; Martinez et al. 1986; Powell 1986)
Theropoda
Theropoda indet.
Abelisauridae
Xenotarsosaurus bonapartei
Sauropoda
Titanosauridae
Antarctosaurus wichmannianus
Argyrosaurus superbus
Laplatasaurus sp.
Undescribed titanosaurid (Powell 1986)
Titanosaurid indet.
Sauropoda *incertae sedis*
Campylodoniscus ameghinoi
Age: Campanian-Maastrichtian (Bonaparte pers. comm.)

c. Castillo, Bajo Barreal, and Laguna Palacios Formations (undifferentiated; Huene 1929*a;* Bonaparte pers. comm.)

Theropoda
?Abelisaurid indet.
Sauropoda
Titanosauridae
Laplatasaurus araukanicus
cf. *Laplatasaurus* sp.
Antarctosaurus cf. *wichmannianus*
Argyrosaurus superbus
Argyrosaurus sp.
Sauropoda *incertae sedis*
Campylodoniscus ameghinoi
Age: Campanian-Maastrichtian (Bonaparte pers. comm.)

d. Cerro Castillo Formation (Corro 1974; Bonaparte pers. comm.)
Theropoda indet. (= *Megalosaurus chubutensis, M. inexpectatus*)
Age: Late Cretaceous (Bonaparte pers. comm.)

e. Unnamed unit (Woodward 1901; Bonaparte pers. comm., not figured)
Theropoda *incertae sedis*
Genyodectes serus
Age: Late Cretaceous (Bonaparte pers. comm.)

406. Provincia de Santa Cruz, Argentina

a. Chorrillo Formation (Powell 1986)
Theropoda indet.
Sauropoda
Titanosaurid indet.
Age: Maastrichtian (Powell 1986)

b. Cardiel Formation (= "Guaranitic" formation; Ameghino 1898, 1899; Huene 1929*a;* Molnar 1980*b;* Powell 1986; Bonaparte pers. comm.)
Sauropoda
Titanosaurid indet.
Sauropoda indet. (= *Clasmodosaurus spatula*)
?Ornithopoda indet. (= *Loncosaurus argentinus*)
Age: Late Cretaceous (Bonaparte pers. comm.)

c. Unnamed unit (Huene 1929*a*)
Sauropoda
Titanosauridae
Argyrosaurus superbus
Age: Late Cretaceous (Bonaparte pers. comm.)

d. Unnamed unit (Dibenedetto pers. comm. in Bonaparte and Gasparini 1978; not figured)
Sauropoda
Titanosauridae
Argyrosaurus superbus
Age: Late Cretaceous (Bonaparte and Gasparini 1978)

Africa

Among the best African Late Cretaceous dinosaurs come from the Western Desert of Egypt; these were collected and studied by E. Stromer. Unfortunately, this material was destroyed at the close of World War II. Additional African sites of significance include Morocco (worked by A. F. de Lapparent and co-workers), Niger (worked by F. de Broin and co-workers), and Madagascar (worked by M. C. Deperet, A. Thevenin, R. Lavocat, I. Obata and co-workers, and P. Taquet and co-workers. Late Cretaceous dinosaur sites from Africa are provided in figures 3.10 and 3.11.

407. Province d'Agadir, Morocco

    Unnamed unit (Ambroggi and Lapparent 1954)
        Theropod footprints
        Age: Maastrichtian (Ambroggi and Lapparent 1954)

408. Wilaya Laghouat, Algeria

    Unnamed unit (Bellair and Lapparent 1949)
        Theropod footprints
        Age: Cenomanian (Bellair and Lapparent 1949)

409. Marsa Matruh, Egypt

    a. Baharija Formation (Stromer 1934*a*)
        Theropoda
            Theropoda indet. (= cf. *Erectopus sauvagei*)
            Carnosauria *incertae sedis*
                Bahariasaurus ingens
                Carcharodontosaurus saharicus
            ?Carnosauria
                Spinosaurus aegyptiacus
                Spinosaurus sp.
            ?Ornithomimosauria
                Elaphrosaurus sp.
        Sauropoda
            Sauropoda indet.
            Diplodocidae
            cf. *Dicraeosaurus* sp.
            Titanosauridae
               Aegyptosaurus baharijensis
        Age: ?early Cenomanian (Said 1962)
    b. Variegated Shale of the Nubian Formation (Awad and Ghobrial 1966; Hotton pers. comm. in El-Khashab 1977)
        Ornithischia indet.
        Age: pre-Maastrichtian Late Cretaceous (El-Khashab 1977)
    c. Nubian Sandstone (Stromer and Weiler 1930; not figured)
        Theropoda indet.
        Age: ?Campanian (Said 1962)
    d. Upper Cretaceous Phosphates (Stromer and Weiler 1930; Molnar 1980*b*)
        Theropoda *incertae sedis*
            Majungasaurus crenatissimus
        Age: Campanian-Maastrichtian (Said 1962)

410. Département d'Agadez, Niger

    Unnamed unit (Broin et al. 1974)
        Theropoda
            Carnosauria *incertae sedis*
               Bahariasaurus ingens
        Sauropoda
            Titanosaurid indet.
        Age: ?Coniacian-?Turonian (= early Senonian; Broin et al. 1974)

411. Rift Valley Province, Kenya

    Turkana Grits (Arambourg and Wolff 1969; Harris and Russell MS)
        Theropoda
            Theropoda indet.
            Theropoda *incertae sedis*
               cf. *Spinosaurus* sp.
            Undescribed theropod
        Sauropoda
            Sauropoda indet.
            Titanosauridae
               cf. *Titanosaurus* sp.
        ?Ornithopoda indet.
        Age: ?Late Cretaceous ("middle" Cretaceous; Harris and Russell MS)

412. Natal, South Africa

    Unnamed unit (Kennedy et al. 1987)
        Sauropoda
            ?Titanosaurid indet.
        Age: early Santonian (Kennedy et al. 1987)

413. Faritany Majunga, Madagascar

    Grès de Maevarano (= Berivotro Formation; Deperet 1896; Thevenin 1907; Lavocat 1955*a;* Russell et al. 1976; Sues and Taquet 1979; Sues 1980*b*)
        Theropoda *incertae sedis*
            Majungasaurus crenatissimus
        Sauropoda
            Titanosauridae
               Titanosaurus madagascariensis
        Pachycephalosauria
            ?Pachycephalosauria indet. (= *Stegosaurus madagascariensis*)
        Pachycephalosauridae
            Majungatholus atopus
        Age: Campanian (Obata and Kanie 1977)

Australasia

A single Late Cretaceous locality is known from Queensland, Australia (in large part an important ichnofauna being worked by R. A. Thulborn and Mary Wade; fig. 3.12). Another site comes from North Island, New Zealand (described by R. E. Molnar and R. J. Scarlett; fig. 3.1).

414. Queensland, Australia

Winton Formation (Coombs and Molnar 1981; Thulborn and Wade 1984)
Theropod footprints
Sauropoda
Sauropod footprints
Sauropoda *incertae sedis*
Austrosaurus sp.
Ornithopod footprints
Age: Cenomanian (Thulborn and Wade 1984)

415. North Island, New Zealand

Mata Series (Molnar 1980*c*; Scarlett and Molnar 1984; Wiffen and Molnar MS)
Theropoda indet.
Ornithopoda
Hypsilophodontid indet.
Age: Campanian-Maastrichtian (= Haumurian; Wellman 1959)

Antarctica

Until 1986, no dinosaur remains had ever been found on the continent of Antarctica. That situation changed with new discoveries made by workers from the Instituto Antartico Argentino. A preliminary description of this material has been published by Gasparini et al. (1987); this site is indicated in figure 3.1.

416. Antarctica

Santa Marta Formation (Gasparini et al. 1987)
Ankylosauria
Ankylosaurid indet.
Age: Campanian (Olivero et al. 1986)

SECTION II

DINOSAUR TAXONOMY

4

Staurikosaurus and Herrerasauridae

H.-D. SUES

INTRODUCTION

The stratigraphically oldest unquestionable dinosaurian remains known to date have been recovered from a series of continental sediments of Late Triassic age in South America (Huene 1942; Reig 1963; Colbert 1970; Galton 1977a; Bonaparte 1978a; Brinkman and Sues 1987). Previously, Huene (1907–08, 1932) described a large number of allegedly dinosaurian teeth and postcranial bones from the Middle and lower Upper Triassic of central Europe and the British Isles, but most, if not all, of these specimens have subsequently been reassigned to other reptilian groups (Colbert 1970; Wild 1973; Chatterjee 1985; Galton 1985a). The Santa Maria Formation of southern Brazil, the oldest known dinosaur-bearing deposit in the world, may be latest Ladinian or earliest Carnian in age, whereas the Ischigualasto Formation of northwestern Argentina, which has yielded skeletal remains referable to several taxa of early dinosaurs, is apparently coeval with or slightly younger than the Santa Maria beds ("Ischigualastian," probably Carnian; see Bonaparte 1982a and Barbarena et al. 1985).

Of the Carnian dinosaurs from South America, only *Staurikosaurus* Colbert, 1970, from the upper part of the Santa Maria Formation (*Scaphonyx* assemblage zone *sensu* Barbarena et al. 1985) is comparatively well known (Colbert 1970; Galton 1977a). Postcranial remains of cf. *Staurikosaurus* sp. from the Ischigualasto Formation have recently provided additional anatomical information (Brinkman and Sues 1987). *Staurikosaurus*, nominal and only known genus of the Staurikosauridae Galton, 1977a, is more primitive than any other described dinosaurian taxon in the more or less circular outline of the distal articular end of the tibia and the presence of only two sacral vertebrae.

Spondylosoma Huene, 1942, from the *Dinodontosaurus* assemblage zone (*sensu* Barbarena et al. 1985) of the Santa Maria Formation may also belong to the Dinosauria. Galton (pers. comm.) is currently revising the material assigned to this taxon, which includes teeth, several vertebrae, and fragments of girdle and limb bones. *Spondylosoma* differs from *Staurikosaurus* in the structure of the vertebrae (Colbert 1970) and the form of the pubis (Galton pers. comm.), but its dinosaurian affinities have yet to be firmly established.

Herrerasaurus Reig, 1963, from the Ischigualasto Formation is more derived than *Staurikosaurus* in the moderate transverse expansion of the distal end of the tibia and in the presence of three sacral vertebrae (Reig 1963; Benedetto 1973). The skeletal material referable to this genus is currently being studied by Novas (pers.

143

comm.). A second dinosaurian taxon from the Ischigualasto Formation, *Ischisaurus* Reig, 1963, is very similar to *Herrerasaurus* in a number of skeletal features and may prove referable to this group (see below).

ANATOMY OF *STAURIKOSAURUS*

Colbert (1970) and Galton (1977*a*) have described the skeletal structure of *Staurikosaurus pricei* in detail. Their accounts, checked against the original material, and additional data from a specimen of cf. *Staurikosaurus* sp. from the Ischigualasto Formation (Brinkman and Sues 1987) form the basis for the following osteological characterization. The postcranial material of cf. *Staurikosaurus* sp. was found together with a nearly complete skull, which apparently belongs to the same individual and is currently being examined by Sues and Brinkman.

Staurikosaurus is a medium-sized (c. 2m long) bipedal form with long but relatively slender limb bones. The mandible of the holotype of *S. pricei* is almost as long as the femur, indicating a proportionately large head. The dentary is fairly deep but thin and holds 13 or 14 teeth. The retroarticular process is well developed.

The vertebral column consists of 9 or 10 cervical, 15 dorsal, 2 sacral, and at least 40 caudal vertebrae. The cervical centra are proportionately short and gently amphicoelous. Marked angulation of the articular faces only occurs on the centrum of the presumed third cervical. The third to fifth cervicals are the longest, much as in other ornithodirans except Saurischia (Gauthier 1986). The cranial cervical vertebrae lack epipophyses. Centra of dorsal vertebrae are slightly amphicoelous and constricted at midlength. V-shaped buttresses support well-developed diapophyses. The neural spines of the dorsal vertebrae are short and massive. The first sacral vertebra is very massive. The caudal vertebrae have slightly amphicoelous centra and poorly developed facets for chevrons. Transverse processes are developed up to about the twentieth caudal.

The scapular blade is very slender and not expanded proximally. The coracoid is large and platelike. The humerus bears a prominent deltopectoral crest; the humeral articular ends are distinctly expanded. The ilium (fig. 4.1A) is distinguished from that of other known dinosaurs except *Herrerasaurus* by the extensive development of the medial wall of the acetabulum. The iliac blade is short and deep, and its acetabular portion is roofed by a prominent supraacetabular rim. The preacetabular portion of the blade is deep, whereas the

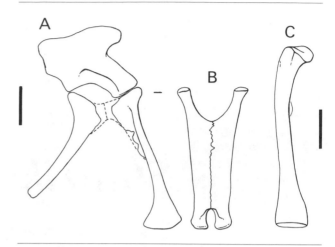

Fig. 4.1. *Staurikosaurus pricei*, postcranial elements (after Colbert 1970, Galton 1977*a*). A, reconstruction of pelvic girdle in right lateral view. B, pubic "apron" in cranial view. C, right femur in cranial view. Scale = 5 cm.

postacetabular portion is apparently short. The pubis (fig. 4.1A–B) is long, approximately two-thirds the length of the femur. The pubes contact each other along an extensive median symphysis forming a broad "apron." The distal end of the pubis is expanded craniocaudally. The ischium is broad proximally but becomes constricted distally to form a narrow blade for much of its length. The limb-bones are hollow but have fairly thick walls. The robust femur has an S-shaped shaft. The proximal articular head is set off craniomedially at a distinct angle (fig. 4.1C). The lesser trochanter is developed as a slight ridge, whereas the fourth trochanter is prominent and placed relatively close to the proximal end of the femur. The distal femoral condyles occupy subterminal positions. Thus, the knee joint was distinctly flexed, rendering the columnar attitude of the hind-limb restored by Colbert (1970) incorrect. The tibia and fibula are slightly longer than the femur. The tibia is robust and straight and bears a distinct cnemial crest. Its proximal articular end is expanded craniocaudally. The distal end is subcircular in outline and notched for the reception of the central ascending process of the astragalus.

ANATOMY OF HERRERASAURIDAE

Herrerasaurus differs from *Staurikosaurus* in its considerably larger size (length of femur 47.3 cm vs. 22.9 cm in the respective holotypes) and in a number of features of the postcranial skeleton (Reig 1963; Benedetto 1973). *Ischisaurus* Reig, 1963, appears to be very similar to *Herrerasaurus* in several features. It is smaller (length of femur 33.5 cm vs. 47.3 cm in the respective holotypes) and more lightly built than the latter. Fragmentary cranial material is known but has not been described in detail for both *Herrerasaurus* and *Ischisaurus*. According to Reig (1963), the maxilla and slender dentary of *Ischisaurus* each hold 15 to 16 laterally compressed teeth and the premaxilla holds 4 teeth.

The cervical and dorsal vertebrae have relatively short centra and craniocaudally short neural spines. *Herrerasaurus* has three sacral vertebrae.

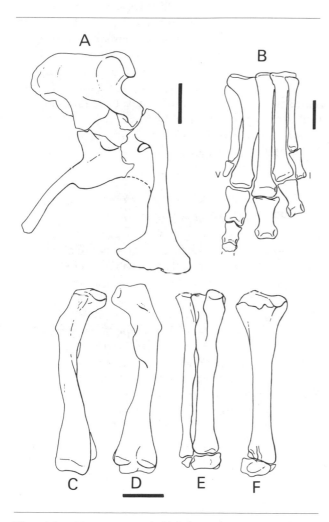

Fig. 4.2. *Herrerasaurus ischigualastensis,* postcranial elements (after Reig 1963). A, pelvic girdle in right lateral view. B, right pes in dorsal view. C-D, right femur in C, cranial, and D, caudal views. E, right fibula, tibia, astragalus, and calcaneum in extensor view. F, right tibia and astragalus in medial view. Scale = 10 cm (A, C-F), and 5 cm (B).

The humerus is slightly less than half as long as the femur in *Ischisaurus* and slightly more than half in *Herrerasaurus* (Reig 1963). In the former, the humerus bears a prominent deltopectoral crest and has expanded articular ends. The ulna of *Ischisaurus* has a distinct olecranon process.

The ilium of *Herrerasaurus* has a moderately elongate postacetabular process. The ischium and pubis together form a deep subacetabular region. The pubis of *Herrerasaurus* is very long, equaling 87 percent of the length of the femur in the holotype. The pubis of a juvenile specimen possibly referable to this genus is slightly retroverted (Paul pers. comm.). The femora of *Herrerasaurus* and *Ischisaurus* are gently S-shaped and have rather proximally placed fourth trochanters. The femur in the former genus is longer than the relatively stout tibia and bears an aliform fourth trochanter (fig. 4.2C–D). The lesser trochanter forms a distinct ridge. The distal end of the tibia of *Herrerasaurus* (fig. 4.2E) is moderately expanded mediolaterally, and the tibia shows axial torsion of about 60° (Benedetto 1973). The calcaneum is small, whereas the astragalus is broad. The pes of *Herrerasaurus* (fig. 4.2B) retains proportionately long first and fifth metatarsals. Metatarsal III equals almost half the length of the femur.

DIVERSITY OF HERRERASAURIDAE

Herrerasaurus and *Ischisaurus* are both monotypic genera. Both have been described only in a very preliminary fashion by Reig (1963). Consequently, assessments of their affinities to other Dinosauria and even to each other have to await a detailed reexamination currently being undertaken by Novas (pers. comm.).

Galton (1985*a*) has described a new genus and species, *Aliwalia rex*, on the basis of the proximal and distal articular ends of a large left femur from the lower Elliot Formation (upper Carnian or lower Norian) of southern Africa and considered it a possible herrerasaurid. Huene (1906) first commented on this specimen and incorrectly referred it to the prosauropod *Euskelosaurus*. The lesser trochanter of the femur in *Aliwalia* forms a prominent crest, and the aliform fourth trochanter is placed close to the proximal end. Galton also tentatively referred an incomplete left maxilla of a large carnivorous archosaur, apparently from the same locality as the femoral fragments, to *Aliwalia*, although he did not, in fact, establish the dinosaurian nature of this jaw fragment. Aside from the position of the fourth trochanter, *A. rex* shares no derived characters with

Herrerasaurus in the few comparable features known to date and should be considered Dinosauria *incertae sedis* at present. Galton (1985a) also noted resemblances between the femoral fragments of *Aliwalia* and a previously undescribed proximal portion of a left femur from the middle Stubensandstein (Norian) of southern Germany, but the latter specimen is again too fragmentary to permit any systematic assessment.

PHYLOGENETIC POSITION OF *STAURIKOSAURUS* AND HERRERASAURIDAE

Although our knowledge of the early dinosaurian radiation is still very imperfect, a consensus has been reached by a number of recent authors (Benton 1985b; Gauthier 1986; Brinkman and Sues 1987) that Saurischia and Ornithischia do indeed form a monophyletic grouping, Dinosauria. Ornithosuchidae and *Lagosuchus* are placed as successively closer sister taxa to this clade (Gauthier 1986).

Within this context, *Staurikosaurus*, *Herrerasaurus*, and *Ischisaurus* are the stratigraphically oldest and most primitive taxa of the Dinosauria known to date. A precise phylogenetic assessment of these genera is very difficult as they are incompletely known and/or inadequately described. The acetabula of both *Staurikosaurus* and *Herrerasaurus* are semiperforate and differ from those of other Dinosauria in the extensive development of the medial wall. It is unclear whether the deep subacetabular contact between pubis and ischium is derived for *Herrerasaurus* because the proximal portions of the pubes and ischia are incomplete in the known *Staurikosaurus* specimens. *Staurikosaurus* is more primitive than all other dinosaurs in having but two sacral vertebrae. The lesser trochanter of the femur in both forms is ridgelike, reminiscent of the condition in early sauropodomorphs. The pes of *Herrerasaurus* has a primitive appearance by retaining proportionately long first and fifth metatarsals.

Brinkman and Sues (1987) have placed *Staurikosaurus* and Herrerasauridae as successive outgroups to a clade comprising Saurischia *sensu* Gauthier (1986) (i.e., Theropoda and Sauropodomorpha) and Ornithischia. This arrangement differs from that by Benedetto (1973) who drew attention to a number of similarities between *Herrerasaurus* and *Staurikosaurus* and included both genera in the same family within the Sau-

rischia. Emphasizing the differences between the two genera, Galton (1977a) regarded *Staurikosaurus* as a very primitive theropod but preferred to classify it, along with *Herrerasaurus*, as Saurischia *incertae sedis.* More recently, Galton (1985a) referred both genera to the Theropoda as a newly created infraorder Herrerasauria but did not enumerate shared derived features in support of this reassessment of their relationships. Most of the similarities between *Staurikosaurus*, *Herrerasaurus*, and various primitive representatives of the two major saurischian clades, respectively, are symplesiomorphies and thus of little if any phylogenetic significance. *Staurikosaurus* is more primitive than the Saurischia as defined by Gauthier (1986) in the relative elongation of the third to fifth cervical vertebrae, the absence of accessory intervertebral articulations, and the absence of epipophyses on the cranial cervical vertebrae. The moderate transverse expansion of the distal articular end and consequent axial torsion of the tibia and the presence of three sacral vertebrae align *Herrerasaurus* more closely with more advanced dinosaurs. A forthcoming detailed study by Novas will undoubtedly permit more refined phylogenetic assessments of this genus and the apparently related *Ischisaurus*.

PALEOECOLOGY AND BIOGEOGRAPHY

As a consequence of the fragmentary nature of the material referred to *Staurikosaurus* and Herrerasauridae, very little information exists on the paleobiology of these dinosaurs. *Staurikosaurus* has been interpreted by Colbert (1970) as an agile, bipedal predator on the basis of the relative proportions of the hindlimb and the structure of the lower jaws. *Herrerasaurus* was a larger, presumably more ponderous carnivorous dinosaur. Both dinosaurs coexisted with large rauisuchian archosaurs. The latter were more abundant than the former and probably were the top carnivores in their respective communities (Bonaparte 1982a). These tetrapod assemblages were dominated by rhynchosaurs and herbivorous synapsids, which must have formed the main prey of the rauisuchians and early dinosaurs.

Staurikosaurus and Herrerasauridae are presently only definitely known from the Upper Triassic of South America. See table 4.1 for their geographic and stratigraphic distribution.

TABLE 4.1 *Staurikosaurus* and Herrerasauridae

	Occurrence	Age	Material
Dinosauria Owen, 1842			
Staurikosaurus Colbert, 1970			
S. pricei Colbert, 1970	Santa Maria Formation, Rio Grande do Sul, Brazil	Carnian	Incomplete skeleton
Herrerasauridae			
Herrerasaurus Reig, 1963			
H. ischigualastensis Reig, 1963	Ischigualasto Formation, San Juan, Argentina	Carnian	3 partial skeletons, isolated postcrania
Ischisaurus Reig, 1963			
I. cattoi Reig, 1963	Ischigualasto Formation, San Juan, Argentina	Carnian	2 partial skeletons
?Dinosauria incertae sedis			
Aliwalia Galton, 1985			
A. rex Galton, 1985	Lower Elliot Formation, Cape Province, South Africa	Late Carnian-early Norian	Femoral fragments
Spondylosoma Huene, 1942			
S. absconditum Huene, 1942	Santa Maria Formation, Rio Grande do Sul, Brazil	Carnian	Postcranial elements, teeth

THEROPODA

HALSZKA OSMÓLSKA

The Theropoda, the sister group to the Sauropodomorpha (see Dodson, this vol.), is the most differentiated among saurischian dinosaurs. Although monophyly of the Theropoda has rarely been doubted, the interrelationships among theropodan groups have often been disputed.

The Theropoda was originally erected by Marsh (1881c) to include all then-known Triassic dinosaurs and the Jurassic and Cretaceous carnivorous dinosaurs. This taxon embraced the forms that are now assigned to prosauropods and those we now consider proper theropods. Later on, Huene (1914c) proposed an alternative classification of the Saurischia and distinguished the Coelurosauria, which he considered as the basal unit that gave rise to all other saurischians. Subsequently, Huene (1920) erected Carnosauria for the large carnivorous saurischians, those forms more derived than the coelurosaurs. Since then, the two latter groupings were conventionally used by authors (e.g., Romer 1956) to include the "small theropods" and the "large theropods," respectively. The inadequacy of such a division of the Theropoda, in view of their observable differentiation, was indicated by Colbert and Russell (1969) and by Ostrom (1969b, 1978a), among others. Application of cladistic methods (e.g., Gauthier 1986) has brought about a significant rearrangement of phylogenetic inferences that may ameliorate the confusion of the past.

The Gauthier hypothesis on the phylogenetic relationships within the Theropoda is followed here, with some minor changes. Of these changes, the Oviraptorosauria is included and the Deinonychosauria is separated into the Troodontidae and Dromaeosauridae. Gauthier (1986, 18) included "birds and all saurischians that are closer to birds that they are to sauropodomorphs" within the Theropoda.

As in most dinosaur systematics, the Theropoda is considered in this volume as a sister taxon to the Sauropodomorpha. The former shares with the latter group several derived characters (Gauthier 1986): the reduced maxillary process of the premaxilla, invasion of the temporal musculature onto the caudodorsal surface of the frontal, the neck forming at least 33 percent the length of the presacral vertebral column, placement of the postzygapophyses more laterally than the prezygapophyses, accessory hyposphene-hypantrum articulations between the trunk vertebrae, the manus 40 percent or more of the combined length of the humerus and radius, and the pollex more robust than other manual digits, the former bearing a larger ungual.

The Theropoda is a monophyletic group characterized by members having a skull with the lacrimal broadly exposed on the roof, an accessory fenestra within the antorbital fossa, fusion of the vomers rostrally, an expanded and ventrally concave ectopterygoid. The presacral vertebrae are pleurocoelous, and the sacrum includes at least five vertebrae. The manus has a reduced or absent metacarpal and digit V and/ or IV. The penultimate manual phalanges are elongate, the unguals are enlarged, curved, and transversely compressed. The ilium has a somewhat elongate preacetabular process and pronounced brevis fossa. The femur is slightly bowed but almost nonsigmoidal, and the fibula is appressed to the tibia along the fibular crest. The pes has a compact, narrow, and elongate metatarsus with metatarsal V reduced to a splint of bone and the first metatarsal separated from the tarsus.

The Theropoda includes the Ceratosauria Marsh, 1884b (see Rowe and Gauthier, this vol.), and Tetanurae Gauthier, 1986.

The apomorphies of the Tetanurae include lack of an enlarged dentary fang, the large, circular maxillary fenestra, a tooth row that is entirely antorbital, the straplike scapula, contact of at least the proximal half of metacarpal I with metacarpal II, absence of the fourth manual digit, an obturator process on the ischium, footlike expansion of the distal end of the pubis, a tall ascending process on the astragalus, and a proximally compressed metatarsal III.

The monophyletic Tetanurae includes the Carnosauria Huene, 1920 (see Molnar, Kurzanov, and Dong, this vol.), and the Coelurosauria Huene, 1914c.

As redefined by Gauthier (1986), the Coelurosauria is no longer a basal theropod group but more derived than the Carnosauria. This concept of the Coelurosauria differs significantly from that prevailing during the last five decades.

Some important derived characters of the Coelurosauria are an accessory fenestra between the pterygoid and palatine, a deeply excavated ventral fossa on the ectopterygoid, proximal cervicals that are broader than deep cranially, fusion of the cervical ribs to centra in adults, elongate forelimbs that exceed half the length of the hindlimbs, and an elongate, gracile manus with long metacarpals II and III, and second and third digits. Within the manus, the combined lengths of phalanges 1 and 2 of the third digit are less than or equal to that of the penultimate phalanx; the fourth digit is weakly developed or absent.

The Coelurosauria (*sensu* Gauthier 1986) includes the Ornithomimosauria (see Barsbold and Osmólska, this vol.) and Maniraptora.

The Maniraptora forms a distinctive and well-differentiated group of the Coelurosauria. They are characterized by a reduced or absent prefrontal, a reduced carpus (the distal carpal is the largest, semilunate and closely covering metacarpal I). Metacarpal III and the third digit are very thin, while the ischium is at most threequarters of the pubis length with a distally placed, flangelike process.

The Maniraptora as conceived by Gauthier (1986) includes the Deinonychosauria Colbert et Russell, 1969, and birds (including *Archaeopteryx* as well as several taxa that display maniraptoran characters but that are mostly too imperfectly preserved to determine their mutual relationships). Of these taxa, *Coelurus, Microvenator, Ornitholestes, Compsognathus,* and *Hulsanpes* are separately treated in this volume by Norman, Elmisauridae by Currie, and *Saurornitholestes* is considered among the Dromaeosauridae (see Ostrom, this vol.). The Caenagnathidae is considered in this volume (see Barsbold, Maryańska, and Osmólska) as a family of the Oviraptorosauria. The latter is a most specialized group not only among the maniraptorans but also among theropods in general, and, at present, its relationship to other maniraptorans cannot be resolved.

The Deinonychosauria is often recognized as a monophyletic taxon (Gauthier 1986; Ostrom, this vol.). According to the original definition of the group, deinonychosaurians include the Dromaeosauridae and Troodontidae (see Ostrom, and Osmólska and Barsbold, respectively, this vol.). Monophyly of the Deinonychosauria, however, was questioned by Osmólska (1981) and lately also by Gauthier (1986). These two families share only one character that could be considered derived, the modified second pedal digit. However, this specialized digit is associated with the very derived metatarsus in troodontids, while the metatarsus in dromaeosaurids displays the primitive condition. It is thus considered probable here that specialization of the second pedal digit might appear convergently. It should be mentioned that Barsbold (1983a) also separated the Troodontidae (= Saurornithoididae Barsbold, 1974) from the Dromaeosauridae, based on the absence of the retroverted pubis in troodontids. The last findings (Currie 1985; Barsbold et al. 1987) suggested that the Troodontidae are best located among maniraptorans. Their relationship within this taxon is still unclear, however.

5

Ceratosauria

TIMOTHY ROWE

JACQUES GAUTHIER

INTRODUCTION

Ceratosauria is a newly diagnosed monophyletic taxon consisting of *Ceratosaurus nasicornis* and its closest relatives among Theropoda (Gauthier 1984, 1986; Gauthier and Padian 1985; Rowe 1989). Ceratosauria is the plesiomorphic sister taxon of all other theropods (fig. 5.1). It includes many of the earliest well-known theropods and by far the most abundant theropods in Late Triassic and Early Jurassic vertebrate faunas.

The oldest record of ceratosaurs is from the Carnian (Late Triassic) of North America. At their first appearance in the fossil record, ceratosaurs are already highly specialized. The oldest taxon is among the most derived ceratosaurs, possessing twenty apomorphies derived within the group and suggesting that a considerable portion of ceratosaur history does not appear in the fossil record. It follows that the origin of Theropoda and the divergence of Ceratosauria and Tetanurae (fig. 5.1) also occurred prior to the Carnian and that theropods may be expected in Middle Triassic sediments. Ceratosaurs are known in North America, Europe, and southern Africa (table 5.1). The youngest record of Ceratosauria is from the Late Jurassic.

Ceratosauria is distinguished from other theropods by ten derived characters of the vertebral column, pelvis and hindlimb, while additional synapomorphies in the skull, vertebral column, and hindlimb diagnose taxa within it. Ceratosaurs represented by adequate samples are also distinguished by a marked dimorphism that affects much of the skeleton of mature individuals. A robust form is characterized by a relatively long skull and neck, thick limbs, and powerfully developed muscle attachments such as the humeral epicondyles, olecranon process of the ulna, and lesser (anterior) trochanter of the femur. In contrast, a gracile form preserves throughout ontogeny a number of juvenile attributes, including a relatively shorter skull and neck, and slender limbs that lack hypertrophied muscle attachments. Gracile forms are not simply juveniles, however, because their characteristic structure may be observed in fully adult individuals (see below).

Ceratosaurs are also distinctive in their convergent acquisition with birds of extensive fusions in the sacrum and pelvis, tarsus, and metatarsus, between bones that were presumably separate throughout life in Theropoda ancestrally. Those features would be considered evidence of close relationship between ceratosaurs and birds if it were not for the overwhelming evidence that birds lie deeply internested within coelurosaurian theropods (Gauthier 1984, 1986) and that birds lack all other ceratosaur apomorphies. Nonetheless, if the convergent resemblances are taken together with the many homologous features that birds and ceratosaurs inherited from the ancestral theropod (Gauthier 1984, 1986), the resemblance is so striking that one might

TABLE 5.1 Ceratosauria

	Occurrence	Age	Material
Theropoda Marsh, 1881b			
Ceratosauria Marsh, 1884a			
Ceratosaurus Marsh, 1884a			
C. nasicornis Marsh, 1884a	Morrison Formation, Garden Park Quarry No. 1, near Canyon City, Colorado, Cleveland-Lloyd Quarry Utah, United States	Kimmeridgian-Tithonian	5 individuals, including nearly complete adult skeleton
Sarcosaurus Andrews, 1921			
S. woodi Andrews, 1921	Barrow-on-Soar, Leicestershire, England	Early Jurassic (?Sinemurian)	Partial adult pelvis, femur, isolated vertebrae
Segisaurus Camp, 1936			
S. halli Camp, 1936	Navajo Sandstone, Segi Canyon, Arizona, United States	Early Jurassic (?Sinemurian-Pliensbachian)	Partial subadult postcranial skeleton
Dilophosaurus Welles, 1970			
D. wetherilli (Welles, 1954) (= *Megalosaurus wetherilli* Welles, 1954)	Kayenta Formation, Gold Springs, Moenkopi Wash, Landmark Wash, Arizona, United States	Early Jurassic (?Sinemurian-Pliensbachian)	2 associated subadult skeletons, partial skeleton, 4 other fragmentary individuals
Liliensternus Welles, 1984			
L. liliensterni (Huene, 1934a) (= *Halticosaurus liliensternus* Huene, 1934a)	Knollenmergel, Sachsen-Anhalt, Federal Republic of Germany	Late Norian	2 partial subadult skeletons
Coelophysis Cope, 1889c (= *Rioarribasaurus* Hunt et Lucas, 1991 •)			
C. bauri (Cope, 1889c) (= *Coelurus bauri* Cope, 1887, *Rioarribasaurus colberti* Hunt et Lucas, 1991)	Chinle Formation, Ghost Ranch, New Mexico, Petrified Forest National Park, Arizona, United States	late Carnian-early Norian	Several hundred individuals, juvenile to adult, including nearly complete articulated skeletons
Syntarsus Raath, 1969			
S. rhodesiensis Raath, 1969	Forest Sandstone, Matabeleland North Zimbabwe; Upper Elliot Formation Cape Province, South Africa	Early Jurassic (?Hettangian-Sinemurian)	At least 30 individuals, juvenile to adult, partially articulated skeletons
S. kayentakatae Rowe, 1989	Kayenta Formation, Rock Head, Willow Springs, Arizona, United States	Early Jurassic (?Sinemurian-Pliensbachian)	16 individuals, juvenile to adult, including fully articulated adult

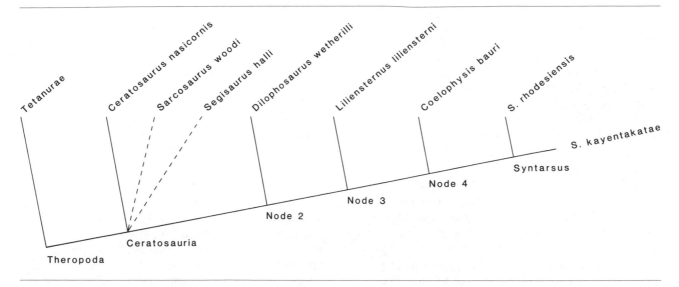

Fig. 5.1. Phylogeny of Ceratosauria (see text for character data).

conceivably mistake a fragmentary small ceratosaur for a bird. This is very likely the explanation for recent reports in the news media of a Triassic "bird" from Texas.

We distinguish between four ontogenetic stages in the following discussion, based largely on the degree of suture and fusion between elements of the postcranial skeleton. These are embryos, which are prehatching individuals; juveniles, which include hatchlings to nearly fully grown specimens (i.e., subadults); subadults, which are very near maximum size as indicated by some but not all of the developmental events marking the cessation of growth; and adults, in which the cessation of growth is marked by fusion between the axial intercentrum and atlantal centrum, fusion of this compound structure to the axial centrum, fused scapulocoracoid, fused sacral and pelvic elements, and neural arches fused to their centra. As used here, these terms make no reference to sexual maturity, which may or may not coincide with a particular stage in skeletal ontogeny (see Gauthier 1986).

Most taxa included within Ceratosauria have had long histories of problematic and controversial taxonomic assignment (see Ostrom 1981; Padian 1986). Many were placed in admitted taxonomic wastebaskets such as Podokesauridae and Megalosauridae, groups that were established on phenetic resemblance and stratigraphic distribution and that have been subject to continual revision and instability. In contrast to such assemblages, Ceratosauria is diagnosably monophyletic; it is not a catchall for early or primitive theropods.

The name Ceratosauria was resurrected from the work of Marsh (1884a), who first coined Ceratosauridae as a redundant means of emphasizing the dis-

tinctness of *Ceratosaurus nasicornis*. He later coined Ceratosauria for a more inclusive group comprising Ceratosauridae and Ornithomimidae (Marsh 1884b). Some authors subsequently recognized either one or both of Marsh's names, though disagreeing on which taxa should be placed within them (compare Gilmore 1920; Huene 1932, 1948, 1956; Welles 1984). Others have not recognized either term (e.g., Lydekker 1888b; Cope 1892a; Abel 1919; Romer 1933, 1956, 1966; Lapparent and Lavocat 1955; Colbert 1964a; Steel 1970; Charig 1979). We employ the name Ceratosauria for the group including *Ceratosaurus nasicornis, Dilophosaurus wetherilli, Liliensternus liliensterni, Coelophysis bauri, Syntarsus rhodesiensis, Syntarsus kayentakatae, Segisaurus halli, Sarcosaurus woodi*, and all other taxa stemming from their most recent common ancestor (fig. 5.1).

We follow the phylogenetic conclusions of Gauthier (1984, 1986) in our choice of outgroups used to determine polarity of character transformation in diagnosing Ceratosauria and assessing relationships within it. We have used Tetanurae, Sauropodomorpha, Ornithischia, and Herrerasauridae (of doubtful monophyly) as consecutively more distant outgroups for comparison with ceratosaurs. The analysis summarized below is based on methods discussed at length elsewhere (Gauthier 1984, 1986; Gauthier, Kluge, and Rowe 1988a; Gauthier, Estes, and de Queiroz 1988; Rowe 1988).

DIAGNOSIS

Ceratosauria is diagnosed by the following: (1) two pairs of pleurocoels in the cervical vertebrae; (2) transverse processes of dorsal vertebrae strongly backturned and triangular when viewed from above; (3) pubic plate perforated by a large, circular pubic fenestra lying below the obturator foramen; (4) sacral transverse processes, sacral ribs, neural arches, and spines completely fused to each other and sacral ribs fused to the ilium in adults (convergent with ornithuran birds); (5) pubis, ischium, and ilium fused together in adults (convergent with ornithuran birds); (6) trochanteric shelf present on lesser trochanter of femur in robust individuals; (7) a sulcus excavated into the base of the crista tibiofibularis; (8) astragalus and calcaneum fused to each other and to the tibia in late ontogeny (convergent with ornithuran birds); (9) ascending process of astragalus directed vertically, subparallel with tibial shaft, and largely overlapped rostrally by the fibula; and (10) distal tarsals II and III fused to their respective metatarsals by late ontogeny (convergent with ornithuran birds).

ANATOMY

Systematic analysis of the distributions of characters discussed below has been presented elsewhere (Gauthier 1984, 1986; Rowe and Gauthier MS). Principal descriptions of ceratosaurs are by Marsh (1884*a*, 1884*b*, 1884*c*, 1892*a*, 1896) and Gilmore (1920) for *Ceratosaurus nasicornis;* Camp (1936) for *Segisaurus halli;* Andrews (1921) for *Sarcosaurus woodi;* Huene (1934*a*) and Welles (1984) for *Liliensternus* (= *Halticosaurus*) *liliensterni;* Raath (1969, 1977, 1980, 1985) for *Syntarsus rhodesiensis;* Welles (1984) for *Dilophosaurus wetherilli;* Rowe (1989) for *Syntarsus kayentakatae;* and Colbert (1989) for *Coelophysis bauri.*

Skull and Mandible

Apart from *Ceratosaurus nasicornis,* ceratosaurs are distinguished from other theropods by a deep incisure, the "subnarial gap" of Welles (1984), that separates the premaxilla and maxilla, interrupting an otherwise continuous alveolar margin (fig. 5.2). An enlarged dentary tooth fits into the gap (see below). A similar gap occurs in all early archosauriforms (Gauthier 1984) but appears to have been lost in dinosaurs, in that it is absent in all saurischians except some ceratosaurs and all ornithischians save for the early ornithopod *Heterodontosaurus.* Welles (1984) interpreted the gap as an indication of an extraordinarily loose attachment of the premaxilla to the snout in *Dilophosaurus wetherilli.* However, in articulated material it can be seen that the palatal process of the premaxilla interdigitates with a rostral maxillary process, effecting a firm junction.

There is no bony premaxillary symphysis. As in tetanuran theropods, the alveolar border is continuous in *Ceratosaurus nasicornis,* presumably reflecting the ancestral condition for Theropoda.

The maxilla forms the rostral and ventral borders of a large antorbital fenestra that, in *Coelophysis* and *Syntarsus,* exceeds 25 percent of total skull length. Whether there is a maxillary fenestra rostral to the antorbital fenestra remains open to question because the maxilla in this region is exceedingly fragile and often broken. It is clear, however, that ceratosaurs lack the circular, caudally placed maxillary fenestra diagnostic of tetanuran theropods. In articulated specimens, the caudal end of the maxilla abuts the base of the lacrimal, excluding the jugal from the antorbital fenestra. The maxilla has a weakly interdigitating contact with the jugal. The structure of this region varies in restorations of *Dilophosaurus* (Welles 1984) and *Syntarsis rhodesiensis* (Raath 1977), but these interpretations are as yet unconfirmed by articulated specimens (see below). In *Liliensternus, Syntarsus,* and *Coelophysis,* the maxilla is strengthened by a longitudinal alveolar ridge that lies just above the tooth row.

The lacrimal encloses the antorbital fenestra caudally and dorsally and, as in theropods ancestrally, is expanded onto the skull roof. In *Dilophosaurus wetherilli* and *Syntarsus kayentakatae,* it is elaborated into a delicate longitudinal crest that projects above the skull roof. The nasals may also contribute extensively to these crests, but sutures in currently known material are obscure.

The nasals are very thin bones, even in *Ceratosaurus nasicornis* in which they support a median nasal "horn." Based on newly discovered material, J. Madsen (pers. comm.) has determined that the horn forms by the fusion of separate left and right nasal protuberances. The nasals abut against the prefrontals and meet the frontals in a squamous articulation in *Coelophysis* and *Syntarsus* and in a more strongly interdigitating suture in *Ceratosaurus nasicornis.* In both species of *Syntarsus,* the nasal, prefrontal, and frontal bound a diamond-shaped opening, termed the nasal fenestra (Raath 1977).

The frontals merely abut with each other on the midline in *Syntarsus* and *Coelophysis* but appear fused in *Ceratosaurus nasicornis.* The frontal has a complex interdigitation with the postorbital and is tightly sutured or fused to the parietal above the orbit in *Ceratosaurus* and behind it in *Coelophysis* and *Syntarsus.* The parietals are flared caudally, where they contribute substantially to the occipital plate and meet on the midline above the supraoccipital. The parietal abuts against the laterosphenoid, prootic, opisthotic, and supraoccipital.

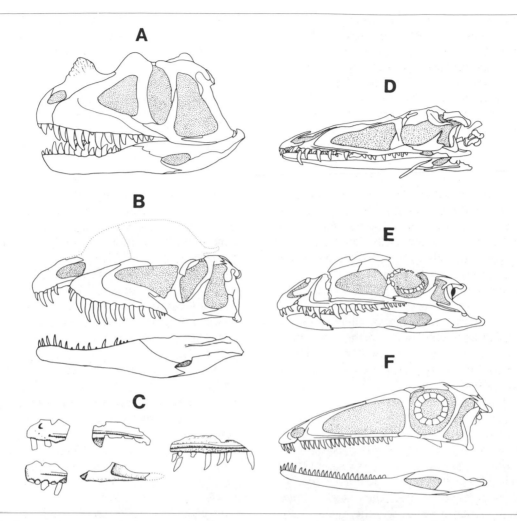

Fig. 5.2. Skulls. A) *Ceratosaurus nasicornis* (right side reversed to appear as left); B) *Dilophosaurus wetherilli;* C) *Liliensternus liliensterni,* partial left and right maxillae; D) *Coelophysis bauri;* E) *Syntarsus kayentakatae;* F) *Syntarsus rhodesiensis.* Credits for figs. 5.2–5.10: *Ceratosaurus nasicornis* (after Gil-more 1920); *Segisaurus halli* (after Camp 1936); *Sarcosaurus woodi* (after Andrews 1921); *Dilophosaurus wetherilli* (after Welles 1984a); *Liliensternus liliensterni* (after Huene 1934a); *Syntarsus rhodesiensis* (after Raath 1977, 1985); *Syntarsus kayentakatae* (after Rowe 1989).

As in other early theropods, the sutures of the cheek and suspensorium in adequately preserved ceratosaurs show virtually no interdigitation. The quadrate and quadratojugal may fuse along part of their contact, however, and are often preserved together even in young, disarticulated specimens.

The ceratosaur palate is poorly known, and virtually all useful data on it come from *Syntarsus rhodesiensis* and *Coelophysis bauri.* As in Theropoda ancestrally, the vomers, preserved only in *Coelophysis,* are fused rostrally. The vomer, maxilla, and palatine form the border of the anterior palatal vacuity. The ectopterygoid also reflects the ancestral theropod state in being expanded and bearing a fossa on its ventral surface. The palatines and pterygoids remain separated on the midline for most or all of their length. Their articulation with the basipterygoid processes is synovial, perhaps affording a degree of at least passive mobility.

The braincase is preserved in *Ceratosaurus nasicornis, Dilophosaurus wetherilli, Coelophysis bauri,* and both species of *Syntarsus.* Comparative study of this region is difficult, however, because most specimens are crushed and obscured to varying degrees by adjacent bones, and few systematic data have as yet been recovered. The braincase is formed by the laterosphenoid, prootic, opisthotic, fused basioccipital, exoccipitals and supraoccipital, and the fused basisphenoid and parasphenoid, which together form a rigid enclosure around the caudal, ventral, and lateral parts of the brain. Welles (1984) noted the possible presence of an "epiotic" in the dorsolateral braincase wall of *Dilophosaurus*, but such an element has not been observed in other ceratosaurs. Whether an ossified orbitosphenoid enclosed the neurocranium rostrally, as is the case in at least some other theropods (e.g., *Allosaurus*; Gilmore 1920), is unknown in ceratosaurs. The parietals form the roof of the neurocranium and abut without interdigitation or fusion to the bones of the neurocranium.

The paroccipital process is formed largely by the opisthotic, with a small contribution by the prootic. The opisthotic in the distal part of the process is rather spongy bone enclosed by a thin cortex. In *Syntarsus rhodesiensis*, there is a deep excavation in the prootic beneath the foramina for C. nn. V and VII which might be pneumatic; the distribution of this structure in other ceratosaurs is unknown. There are no other pneumatic depressions or openings in the bone.

Without knowledge of an ossified orbitosphenoid, little can be said of the pathways of the first four cranial nerves. The trigeminal nerve exited the neurocranium laterally through a notch in the rostrolateral edge of the prootic near the junction of the basisphenoid and prootic. The exit for c. n. VII is through a foramen perforating the prootic in *Syntarsus*, but in *Dilophosaurus*, Welles (1984) illustrated it at the junction of the prootic and basisphenoid while noting that preservation renders interpretation of this feature equivocal. The tympanic cavity is deeply recessed beneath the paroccipital process and is divided into a separate rostral and caudal chamber by a delicate oblique septum, evidently a part of the opisthotic. The rostral chamber houses only the fenestra ovalis, and the caudal chamber probably marks the location of the metotic fissure and passage of c. nn. IX–XI. C. n. XII emerged ventrolaterally from a series of small foramina in a deep excavation immediately rostral to the occipital condyle.

The occipital condyle is large and spherical and is formed largely by the basioccipital, as in dinosaurs generally. The exoccipitals fuse laterally with the opisthotic and dorsally with the supraoccipital. The bulk of the occipital plate is formed by the supraoccipital and parietals, providing attachment for powerful craniocervical muscles. As in other dinosaurs, the posttemporal fenestra is either reduced to a tiny foramen or is absent altogether. Colbert (1989) reported an open fenestra in *Coelophysis bauri*, but this has not been not confirmed in undistorted, articulated adult material.

The dentaries have no bony symphysis and only a very short overlap with the postdentary bones. An enlarged alveolus and swelling near the front of the dentary marks the presence of an enlarged dentary tooth. This is apparently a reversal to the ancestral condition for archosauriforms that occurred within ceratosaurs. The postdentary bones overlap but do not interdigitate with one another. Unlike tetanurans, ceratosaurs retain the ancestral archosauriform condition of a dorsal overlap of the surangular onto the dentary.

A single pair of hyoid rods has been recovered in *Coelophysis* and in both species of *Syntarsus*. Marsh (1884*a*, 1896) reported four hyoid bones in *Ceratosaurus*, but as Gilmore (1920, 92) described, only a single pair is actually preserved.

The premaxillary and two most mesial dentary teeth are subcircular in cross section with few or no serrations, whereas the maxillary and more distal dentary teeth are laterally flattened and strongly serrated. The largest maxillary tooth lies in or near the fourth alveolus, and the teeth gradually diminish in size on either side of it. The maxillary tooth row in all articulated specimens is plesiomorphic in extending distally to beneath the orbit, whereas it lies entirely rostral to the orbit or is absent in tetanurans. The tooth rows in *Dilophosaurus wetherilli* (Welles 1984) and *Syntarsus rhodesiensis* (Raath 1977) were reconstructed as lying antorbitally (fig. 5.2), but this configuration, which differs from other ceratosaurs, has yet to be confirmed in articulated material.

Ceratosaurus nasicornis has three premaxillary teeth, while *Dilophosaurus*, *Coelophysis*, and both species of *Syntarsus* have four. The number of maxillary and dentary teeth increases with age in the various species, and variations in numbers of teeth reported in the literature may simply reflect observation of individuals of differing ontogenetic stage. Taxa of larger adult size consistently have fewer teeth than smaller taxa. In *Ceratosaurus* there are 15 maxillary teeth, in *Dilophosaurus* Welles (1984) reported 12 (although 14 are reconstructed in his fig. 4), in *Syntarsus rhodesiensis* there are up to 20, in *Syntarsus kayentakatae* there are 18, and in *Coelophysis bauri* there are up to 26. Counts for the dentary are as follows: *Ceratosaurus*, 15; *Dilophosaurus*, 17 or 18; *Coelophysis*, 27; *Syntarsus rhodesiensis*, 25.

Postcranial Skeleton
Axial Skeleton

Twenty presacral vertebrae are preserved in *Ceratosaurus nasicornis*, but Gilmore (1920) reported that there are several breaks in the series and that more were almost certainly present in life. In *Dilophosaurus wetherilli*, there are twenty-four presacral vertebrae, whereas in *Coelophysis bauri* and *Syntarsus rhodesiensis* there are twenty-three. The remaining ceratosaurs are too fragmentary to permit a presacral count. (See fig. 5.3)

As in other theropods, there is no proatlas arch.

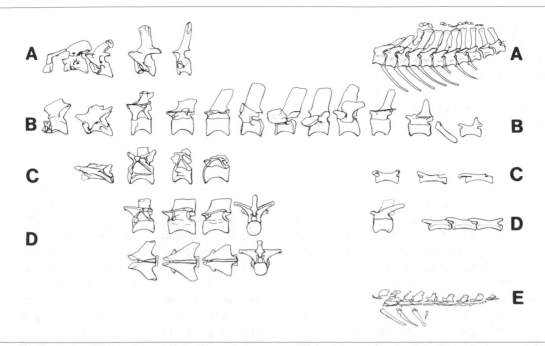

Fig. 5.3. Vertebral column. A) *Ceratosaurus nasicornis* cervicals 1, 2, 3, 6, trunk vertebra 3, partial caudal series; B) *Dilophosaurus wetherilli* cervicals 2, 6, cranial trunk vertebra, midtrunk vertebra, last two trunk vertebrae (the last of which is probably the first sacral in adults), three sacrals, caudals 3, 6, 16; C) *Liliensternus liliensterni* midcervical, trunk vertebrae 1, 3, 9, distal caudals; D) *Syntarsus rhodesiensis* three cranial trunk vertebrae in lateral, dorsal, cranial, and caudal views; one proximal and three distal caudals; E) *Segisaurus halli* partial caudal series. Illustration sources listed in fig. 5.2.

There is a deep fossa in the atlantal intercentrum that received a spherical occipital condyle. The atlantal neural arches articulate with the atlantal intercentrum but fail to meet each other above the neural canal. The atlantal centrum is fused to the axis in adults, forming a robust odontoid process. The axial intercentrum is also fused in adults to the axial centrum. The axial transverse process is greatly reduced or absent. In *Ceratosaurus nasicornis*, it consists of a small swelling, evidently a nonarticulating remnant of the diapophysis, and even this is absent in other ceratosaurs. All ceratosaurs lack a spine table on the axis that is diagnostic of tetanurans (Gauthier 1986).

The postaxial cervical vertebrae have low, elongate neural arches and unlike the condition in tetanurans, at least some retain ventral keels. Like other theropods, however, the parapophysis lies entirely on the centrum over most of the neck (see below), and the centra are opisthocoelous. Their front and rear articular surfaces are strongly offset, giving the neck a long, S-shaped curve.

Ceratosaurs have distinctive separate cranially and caudally placed pleurocoels in most postatlantal cervical vertebrae. *Ceratosaurus nasicornis* retains axial pleurocoels, but they are absent in *Dilophosaurus, Coelo-*

physis, and *Syntarsus.* In the latter two taxa, the pleurocoels are confined to vertebrae three through ten. In *Dilophosaurus,* there are also shallow depressions in the first few dorsal vertebrae below the neurocentral suture at the rostral end of the centra. Welles (1984) referred to these as pleurocoels, but their shallowness distinguishes them from cervical pleurocoels and from the trunk pleurocoels of tetanurans. *Dilophosaurus* is also distinctive in that the pleurocoels in cervicals four and six communicate with each other (preservation leaves this uncertain in the other cervicals), whereas in *Coelophysis* and *Syntarsus,* all cervical pleurocoels are closed pockets. Which of these states is primitive is currently unclear. Although *Ceratosaurus* certainly has cervical pleurocoels, it is not known whether the right and left sides communicate.

In adequately preserved ceratosaurs, ribs are present on all cervical vertebrae. The atlas and axis ribs, known only in *Coelophysis* and *Syntarsus,* entirely lack the tuberculum and articulate with very poorly developed parapophyses; these ribs are not preserved in *Ceratosaurus,* in which the parapophysis is more strongly developed. The weakness of the axial parapophysis led Raath (1977) to speculate that the axial rib was absent in *Syntarsus rhodesiensis,* but it is preserved in articulation in *S. kayentakatae*. The third cervical rib has a small capitulum and tuberculum and, in *S. kayentakatae* (unknown in other ceratosaurs), retained a potentially mobile articulation throughout life. The remaining cervical ribs are strongly bifurcate and fused to their centra in adults. In *Coelophysis* and *Syntarsus*, the ribs are unique in having greatly elongated shafts that parallel the notochordal axis. Some cervical ribs extend caudally for a distance equivalent to the length of four centra. As a result, the cervical ribs overlap to form delicate bundles of three or four that extend along most of the neck. It is not known whether this condition occurred in other ceratosaurs.

Division of the presacral column into cervical and trunk regions is imprecise because "cervicalization" of the cranial few dorsal vertebrae has increased the number of vertebrae that function in the neck, and there is no criterion or set of criteria that will consistently separate the two regions between adjacent vertebrae in all taxa. *Dilophosaurus* retains the primitive condition for saurischians in having four or five "transitional" vertebrae at the base of the neck, beginning with the ninth postcranial vertebra, in which the parapophysis straddles the neurocentral suture. This condition occurs in the twelfth vertebra of *Ceratosaurus*, but other vertebrae near the base of its neck are not preserved. *Coelophysis bauri* also appears to preserve the ancestral condition. *Syntarsus rhodesiensis* is reported to lack transitional vertebrae (Raath 1977). However, on at least the rostralmost trunk vertebra (fig. 5.3D), the parapophysis straddles the neural arch-centrum suture, as in ceratosaurs generally. The base of the neck is not preserved in other ceratosaurs.

The transverse processes of the dorsal vertebrae are unique in being strongly backturned and triangular when viewed from above (*Ceratosaurus nasicornis* is unknown in this respect). Proximally, each transverse process extends the entire length of the centrum, while distally the process slants backward, being widest along its caudal edge. Over most of the dorsal series (all in *Syntarsus* and *Coelophysis*), both articulations for the dorsal ribs are on the neural arch, with the parapophysis lying higher on the more caudal vertebrae. Dorsal ribs have a thin web of bone connecting a well-developed capitulum and tuberculum. The tuberculum diminishes in relative size toward the sacrum. The thoracic and abdominal cavities were enclosed ventrally by a network of gastralia.

There are five vertebrae in the adult sacrum, as in Theropoda ancestrally. The neural arches, transverse processes, and sacral ribs coalesce into continuous sheets of bone that extend the length of the sacrum, the suture between the sacral ribs and ilium is completely obliterated, and the centra fuse so completely that only

a swelling marks the point at which two vertebrae meet. The result is a synsacrum, another feature analogous with ornithurine birds. Growth series of *Syntarsus* (Raath 1977) and *Coelophysis bauri* (Colbert 1989) preserve young individuals in which there are only four sacrals. Welles (1984) reported only four unfused sacral vertebrae in *Dilophosaurus wetherilli*, one of many indications that the currently known associated skeletons are immature.

A deep ventral groove on the proximal caudal vertebrae distinguishes *Ceratosaurus* among Late Jurassic theropods (B. Britt pers. comm.). We have since found this groove in *Coelophysis* and *Syntarsus* but have not been able to determine its distribution elsewhere among ceratosaurs. Insofar as this groove appears to be absent outside of Ceratosauria, it may be another synapomorphy of the group. Like other theropods, the ceratosaur tail has a "transition point" at which the distal part becomes stiff and acts as a unit whose movement is largely confined to a vertical plane by elongation of the zygapophyses. In addition, the transverse processes are abruptly reduced at the transition point. In ceratosaurs, the transition point lies about midway back, contrasting with tetanuran theropods in which it lies in the proximal third of the tail.

Appendicular Skeleton

The shoulder girdle is plesiomorphic in retaining an expanded distal scapular blade and a rounded coracoid caudoventral to the glenoid. The two bones are separate in relatively young specimens (fig. 5.4A), but they later suture (e.g., *Liliensternus*, fig. 5.4B) and are completely fused in fully mature individuals (e.g., *Syntarsus kayentakatae*, fig. 5.4C). In *Dilophosaurus*, the distal scapula bears a unique, squared, distal expansion. A long left clavicle is preserved in *Segisaurus*, the only ceratosaur in which this element has been identified. The ceratosaur shoulder girdle contrasts with that of tetanurines, in which the distalmost scapula becomes narrow and straplike and the caudal end of the coracoid becomes sharply pointed or otherwise elongated caudoventral to the glenoid. Marsh (1892*a*, 1896) illustrated a distally narrowed scapula and unfused coracoid in *Ceratosaurus nasicornis*, as did Gilmore (1920). However, Gilmore (1920, 102) noted accurately that the distal portion of the scapula is not preserved and that the scapula and coracoid are fused.

In the large samples of *Syntarsus* and *Coelophysis*, cross sections of nearly all skeletal elements can be observed, and the humerus, radius, ulna, metacarpals, and phalanges are hollow and thin-walled. These samples also reveal a dimorphism in forelimb structure. A robust form has relatively long forelimbs, a prominent

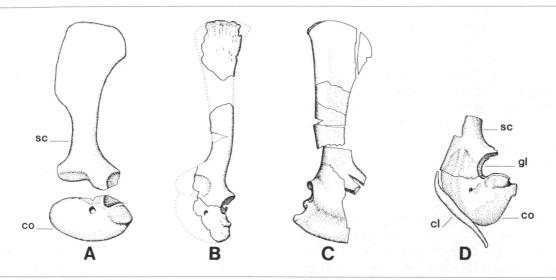

Fig. 5.4. Left scapulocoracoid. A) *Dilophosaurus wetherilli;* B) *Liliensternus liliensterni;* C) *Syntarsus kayentakatae* left scapulocoracoid in medial view; D) *Segisaurus halli* scapulocoracoid and clavicle. Illustration sources listed in fig. 5.2.

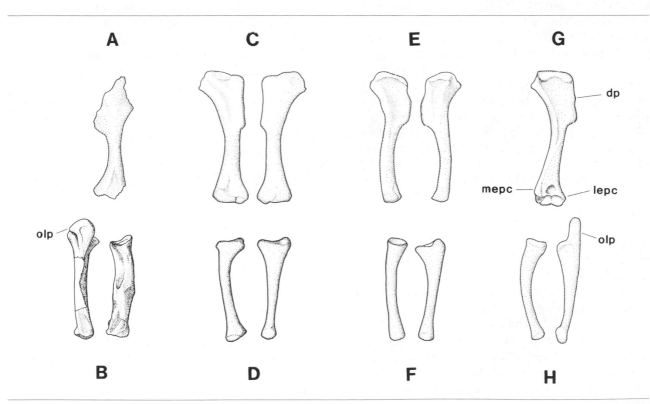

Fig. 5.5. Left humerus, radius, ulna. A) *Segisaurus halli* left humerus in oblique caudal view; B) *Ceratosaurus nasicornis* ulna and radius in caudal view; C) *Dilophosaurus wetherilli* humerus in cranial and caudal views; D) *Dilophosaurus wetherilli* radius in lateral view and ulna in medial view; E) *Lilien-sternus liliensterni* humerus in cranial and caudal views; F) *Liliensternus liliensterni* radius and ulna, both in lateral view; G) *Syntarsus rhodesiensis* humerus in cranial view; H) *Syntarsus rhodesiensis* radius and ulna, both in lateral view. Illustration sources listed in fig. 5.2.

deltopectoral crest, a relatively broad humeral head, expanded distal epicondyles, and a prominent olecranon process (fig. 5.5G—H). In the gracile morph, the forelimb is relatively shorter and has a weakly developed humeral head, deltopectoral crest, epicondyles, and olecranon (fig. 5.5E—F). The radius is slightly shorter than the ulna and is bowed in both morphs, evidently strengthening pronation of the hand (Raath 1977).

The manus (fig. 5.6) is less than two-thirds of the combined lengths of the radius and humerus. As in other theropods, ceratosaurs lack any vestige of an ossified fifth digit. The distal end of metacarpal I is offset, and the ability of digit I to oppose digits II and III appears fundamental to the organization of the hand (Galton 1971*b*). This is further reflected in an effective strengthening of digit I by the proximal quarter of metacarpal I, which is closely appressed against the base of II, and the two metacarpals are united in the wrist by the expansion of distal carpal I over their bases. However, the contact between the bases of metacarpals I and II is confined to their proximal third instead of extending half the length of metacarpal I, and metacarpal III has not been displaced to the palmar surface of II, as in tetanurans. Also unlike tetanurans, ceratosaurs retain a persistent ossified vestige of metacarpal IV and one or two of its phalanges. The entire digit IV is shorter than metacarpals II and III and was probably bound entirely, or very nearly so, within the palmar aponeurosis. Ceratosaurs share with tetanurans very short first and second phalanges of digit III and elongated penultimate phalanges of all three functional digits. The large unguals are laterally compressed, strongly recurved, and bear large flexor tubercles.

The ilium, ischium, and pubis are fused in adult ceratosaurs (fig. 5.7). This is another striking resemblance between ceratosaurs and ornithuran birds. Nevertheless, based on other systematic data (Gauthier 1984, 1986), these similarities are most simply regarded as homoplastic. The ceratosaur ilium reflects the ancestral theropod condition in having an enlarged iliac blade cranial to the acetabulum and a deep fossa at the base of the blade for the origin of M. caudofemoralis brevis caudal to the acetabulum (fig. 5.7). Ceratosaurs lack a second fossa along the base of blade, cranial to the acetabulum, which probably provided the site of origin of M. cuppedicus (Rowe 1986) in tetanurans. *Syntarsus* and *Coelophysis* are derived in that the site of origin of M. iliofemoralis on the lateral face of the ilium forms a deep fossa bounded caudally by a sharply defined rim (fig. 5.7D—E).

The ceratosaur pubis is perforated by two fenestrae. There is a relatively small obturator foramen, which opens ventrolaterally through the upper portion of the delicate, rarely preserved, puboischiadic plate. Unlike other dinosaurs, there is also a large pubic fenestra that lies ventral to the obturator. The pubis is relatively long, narrow, and in smaller taxa bowed forward (fig. 5.7). The pubis is plesiomorphic in lacking the distal expansion or pubic "foot" seen in tetanurans. A foot was restored by Marsh (1892*a*, 1896) and Gilmore (1920) in illustrations of *Ceratosaurus nasicor-*

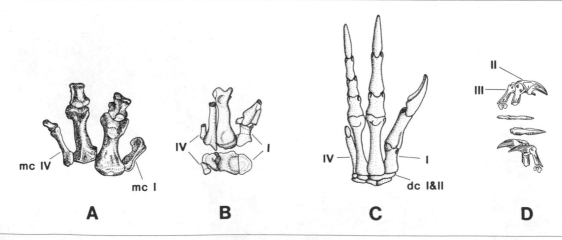

mc IV

mc I

IV   I

IV   I

dc I&II

II

III

A   **B**   **C**   **D**

Fig. 5.6. Left manus. A) *Ceratosaurus nasicornis* metacarpus and proximal phalanges in dorsal view; B) *Dilophosaurus wetherilli* metacarpus in dorsal and proximal views; C) *Syn-* *tarsus rhodesiensis* manus in dorsal view; D) *Segisaurus halli* distal phalanges of digits II and III. Illustration sources listed in fig. 5.2.

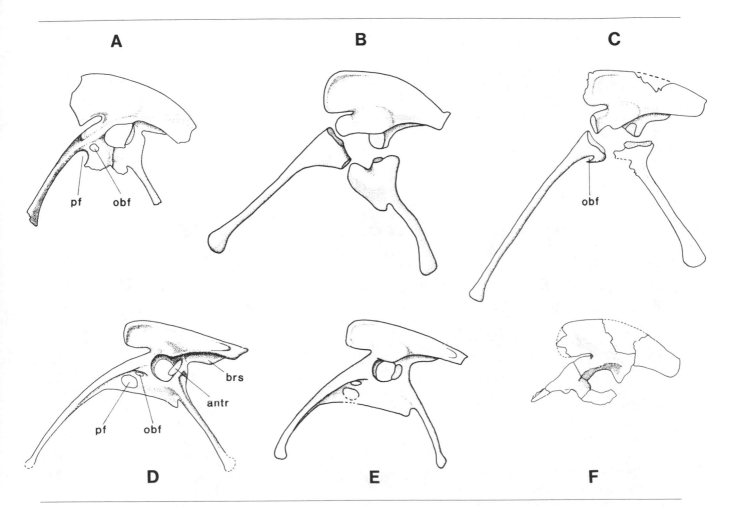

Fig. 5.7. Pelvis. A) *Ceratosaurus nasicornis*, B) *Dilophosaurus wetherilli*, C) *Liliensternus liliensterni*, D) *Coelophysis bauri*, E) *Syntarsus rhodesiensis*, F) *Sarcosaurus woodi*. Illustration sources listed in fig. 5.2.

nis. But as Gilmore described (p. 108), the restorations were all based on *Allosaurus*, and no such foot is actually preserved.

The ischium is plesiomorphic in lacking the obturator process characteristic of Tetanurae ancestrally. The ischium in the single known specimen of *Segisaurus halli* is fenestrated proximally, and the shaft is flattened and laterally expanded in a unique condition. Given the degree of crushing clearly evident in other bones, the flattening of the shaft may be artifactual.

The femur, tibia, metatarsals II–IV, and phalanges are hollow and thin-walled. In both robust and gracile morphs, the lesser (anterior) trochanter develops at the insertion of M. puboischiofemoralis internus pars dorsalis on the dorsal surface of the femur (Rowe 1986). The trochanter is relatively small and occupies the ancestral position for dinosaurs on the dorsum of the femur just distal to the head (fig. 5.8). In tetanuran theropods and ornithischians, however, it

is enlarged and displaced laterally and proximally, lying adjacent or co-ossified with the greater trochanter. The femur of robust ceratosaurs is derived in having a pronounced trochanteric shelf (Andrews 1921) that wraps from the base of the lesser trochanter laterally around the femur and continues as a distinct insertion scar onto the cranial aspect of the shaft (fig. 5.8F–I). The trochanteric shelf is absent in gracile individuals (fig. 5.8B–C).

The distal end of the femur in adequately preserved ceratosaurs bears a distinctive, deep groove at the base of the crista tibiofibularis (fig. 5.8J). This groove lies adjacent to the articular facet for the fibula, but the fibula evidently did not articulate within it. Although Welles (1984, 169) reported the groove to be absent in *Liliensternus liliensterni*, the condition of the specimen leaves this open to interpretation.

The fibula in mature individuals bears a deep sulcus on the medial surface of its proximal end which

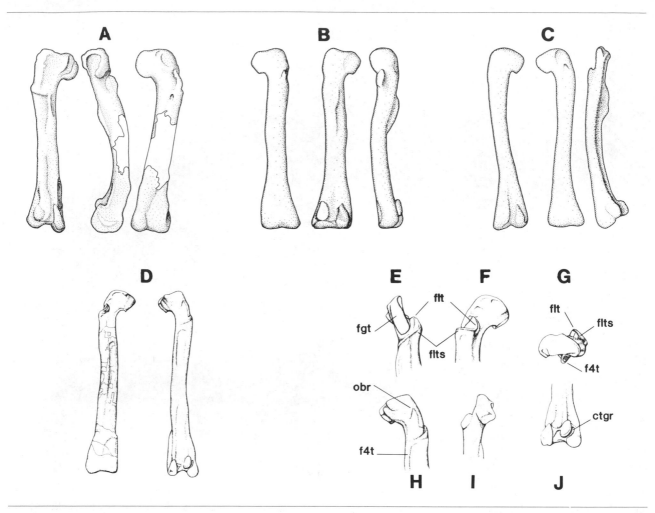

Fig. 5.8. Femur. A) *Ceratosaurus nasicornis* (left femur) in caudal, medial, and cranial views; B) *Dilophosaurus wetherilli* (left femur) in cranial, caudal, and lateral views; C) *Liliensternus liliensterni* (left femur) in caudal, cranial, and lateral views; D) *Syntarsus kayentakatae* (right femur) in cranial and caudal views; E–J) *Syntarsus kayentakatae* (right femur) proximal end in E) lateral view, F) cranial view, G) proximal view, H) caudal view, I) medial view, J) distal end in caudal view. Illustration sources listed in fig. 5.2.

probably was the site of origin of part of the pedal flexor musculature (fig. 5.9G–H). The sulcus opens caudally and extends down the fibular shaft to near the top of the crista fibularis of the tibia. The sulcus is absent in juvenile specimens of *Syntarsus kayentakatae,* and its apparent absence in *Dilophosaurus wetherilli* is probably yet another reflection of the immaturity of known material (fig. 5.9C).

In juvenile specimens, the tibia, astragalus, and calcaneum are often separate elements, but in adults, they fuse to form a tibiotarsus. The ascending process of the astragalus evidently also forms from a separate ossification center (Welles 1983). Ceratosaurs are convergent with ornithurans in these respects, but significant differences in the ankle also exist. For example,

the ceratosaur ascending process is very small and is largely covered by the flared distal end of the fibula (compare figs. 5.8I and 5.8J), misleading some authors to believe that it is entirely absent (e.g., Colbert 1964*b*), whereas in tetanurans, the ascending process forms a broad sheet that extends far up the tibia. The relative timings of fusion among tarsal elements appear to vary in different individual ceratosaurs. Colbert (pers. comm.) pointed out that fusion of the tibiotarsus in *Coelophysis* and *Syntarsus* is not well correlated with size, because it is fused in some small specimens while the astragalocalcaneum and tibia remain separate in a few much larger specimens. Nevertheless, in individuals in which the pectoral and pelvic elements are fused, the tibiotarsus is fused as well.

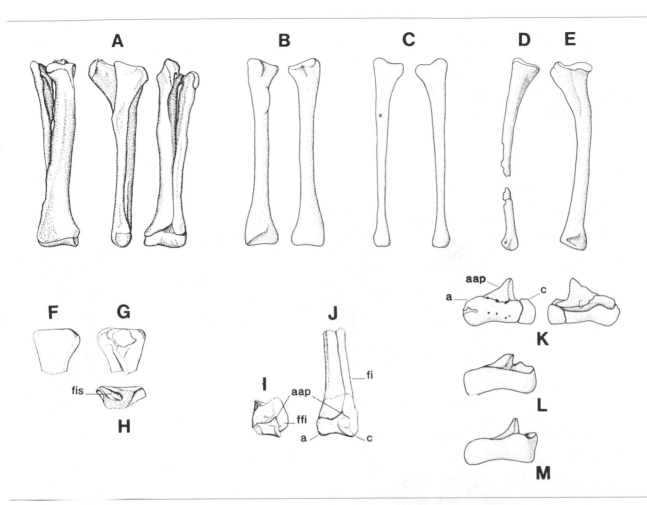

Fig. 5.9. Left tibia, fibula, tarsus. A) *Ceratosaurus nasicornis* tibia, fibula, and proximal tarsus in caudal, lateral, and cranial views; B) *Dilophosaurus wetherilli* tibia in cranial and caudal views; C) *Dilophosaurus wetherilli* fibula in lateral and medial views. D) *Liliensternus liliensterni* fibula in lateral view; E) *Liliensternus liliensterni* tibia in craniomedial view; F–H) *Syntarsus kayentakatae* proximal end of fibula in F) lateral view, G) medial view, and H) cross section; I–J) *Syntar-* *sus kayentakatae,* I) distal end of tibia with fused astragalus (calcaneum broken and fibula removed) and J) distal end of tibia, fibula, and tarsus; K) *Dilophosaurus wetherilli* astragalus and calcaneum in cranial and caudal views; L) *Liliensternus liliensterni* astragalocalcaneum in cranial view; M) *Coelophysis bauri* astragalocalcaneum in cranial view. Illustration sources listed in fig. 5.2.

Ceratosaurs are also convergent with ornithurans in having a tarsometatarsus that fused relatively early in ontogeny. Distal tarsals II and III are fused to the proximal ends of their respective metatarsals, but, unlike ornithurans, distal tarsal IV remained separate throughout ontogeny (Fig. 5.10D). There is no fifth distal tarsal. In both species of *Syntarsus,* the proximal ends of metatarsals II and II are themselves fused, and in *Ceratosaurus nasicornis,* metatarsals II, III, and IV are fused along most of their length, although the more distal fusion may be pathologic. Fusion of the distal tarsals and metatarsals also occurs in Aves, *Elmisaurus*

rarus, and elsewhere in Theropoda (Osmólska 1981; McGowan 1984; Brett-Surman and Paul 1985). In light of what is now known of theropod phylogeny, these resemblances are most simply regarded as convergent with ceratosaurs (Gauthier 1986). Metatarsals I and V are separate in all ceratosaurs. The first metatarsal has lost its contact with the ankle, as in all adult theropods, but unlike tetanurans, it retains a long, delicate dorsal process and its primitive position on the medial side of metatarsal II. The fifth metatarsal is splint-like, and though it lies close to the ankle, it fails to contact the tarsus.

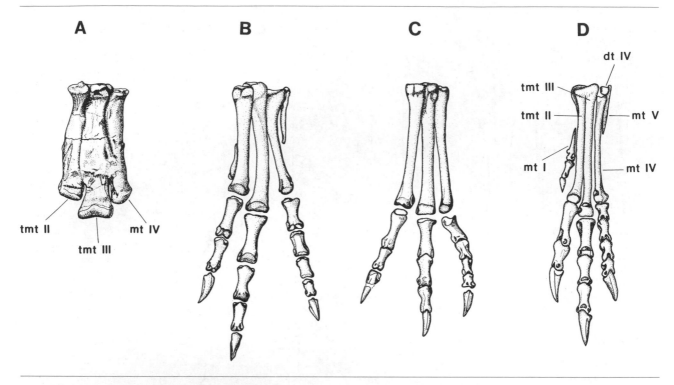

Fig. 5.10. Pes. A) *Ceratosaurus nasicornis*, B) *Dilophosaurus wetherilli*, C) *Liliensternus liliensterni*, D) *Syntarsus rhodesiensis*. Illustration sources listed in fig. 5.2.

Bone Histology

Bone histology was described by Raath (1977) for several postcranial elements of *Syntarsus rhodesiensis* based on his own observations and those of A. de Ricqlès. In a femur and tibia of undisclosed ontogenetic stage, Raath reported the bones as thin-walled, with a large medullary cavity devoid of bony trabeculae. The cortex is formed of dense, fibrolamellar tissue organized mostly in a plexiform pattern and in some areas in a less orderly reticular pattern. The cortex is well vascularized, but in these specimens, there is no evidence of cyclical deposition and only minor development of Haversian systems. However, sections through several other tibiae, ribs, and metatarsals of larger individuals showed variable degrees of lamination at the peripheries of the cortex and a "surprising" degree of osteoporosis. Fibrolamellar bone tissues are typical of large birds and most medium- to large-sized mammals, and they reflect rapid, continuous deposition of dense primary bone (Ricqlès 1980; Reid 1984a, 1984b). The presence of this tissue in *Syntarsus* and other dinosaurs is not necessarily indicative of endothermy as in extant dinosaurs (i.e., birds), but it does reflect a physiology more derived than other archosaurs because it supports rapid, continuous growth and indicates an elevated basal metabolic rate.

SYSTEMATICS AND EVOLUTION OF CERATOSAURIA

Ceratosaurus nasicornis (Marsh 1884a) is the sister taxon of other ceratosaurs (fig. 5.1). We note that its phylogenetic position is discordant with its placement in time (Late Jurassic); all other ceratosaurs are more derived yet they occur earlier in time. This conflict presumably reflects inadequacy in both the density of our temporal sample and the completeness of our morphological data. Although abundant material is now known for the Late Triassic *Coelophysis* and Early Jurassic *Syntarsus*, the remaining ceratosaurs are known from a few, incomplete specimens. Moreover, the derived nature of the earliest ceratosaurs implies an earlier unrecorded history, and virtually their entire Middle Jurassic history is also unrecorded.

Ceratosaurus nasicornis is diagnosed by its median nasal "horn," fusion between metatarsals II–IV, and epaxial osteoderms (Gilmore 1920). A nasal horn also occurs in *Proceratosaurus bradleyi,* but the latter taxon bears a large, circular maxillary fenestra and an antorbital maxillary tooth row, diagnostic characters of Tetanurae (Gauthier 1986), and it lacks the diagnostic characters of Ceratosauria. It is simplest to conclude that a nasal horn evolved convergently in these taxa. The type specimen of *C. nasicornis* is a fully grown robust individual but slightly smaller than *Dilophosaurus wetherilli.* Other material from Utah preserves the nasal horn and osteoderms and is roughly one and a half times the size of the type (Madsen and Stokes 1963; J. Madsen, pers. comm.), making *Ceratosaurus* the largest ceratosaur. The size of this material raises an unexplored possibility that there may be more than one diagnosable taxon within *Ceratosaurus.*

Additional material attributed to *Ceratosaurus* includes a fragmentary specimen from the Morrison Formation of Oklahoma (Stovall 1938). While this referral might be correct, currently known material does not include any diagnostic characters beyond those of Theropoda ancestrally. Two additional species of *Ceratosaurus* have been named, *Ceratosaurus ingens* (Janensch 1920) and *Ceratosaurus roechlingi* (Janensch 1925*b*), for material collected from the Tendaguru deposits of Tanzania. These specimens preserve no diagnostic attributes beyond those of Theropoda ancestrally, and the names are best considered *nomina dubia.* Nevertheless, a femur figured from the Tendaguru collection does appear to preserve the trochanteric shelf and crista tibiofibularis sulcus (Janensch 1925*b*, Pl. V, fig. 2a–c), although these structures are not explicitly described. Ceratosauria thus appears to be represented in the Late Jurassic of Tanzania, but this is not evidence that *Ceratosaurus* itself was present.

The remaining ceratosaurs are grouped together (fig. 5.1, Node 2) based on the presence of the subnarial gap, loss of the axial diapophysis, reduction of the axial parapophysis, and loss of axial pleurocoels. This group may also be diagnosed by expanded, triangular dorsal transverse processes (transverse processes are not preserved in *Ceratosaurus nasicornis*).

Dilophosaurus wetherilli is the plesiomorphic sister taxon of other members of this group (fig. 5.1, Node 2). *Dilophosaurus wetherilli* (Welles 1954, 1970) is diagnosed by its highly arched, paired nasolacrimal crest, cruciform distal extremities of the cervical neural spines, and the square distal expansion of its scapular blade. The type specimen (Welles 1984) is a juvenile to subadult gracile individual, and a fragmentary robust skeleton of roughly the same ontogenetic stage is also known. In addition, there is a partial skeleton of a young juvenile (Welles 1984) and fragments of at least three additional subadults or adults.

Liliensternus liliensterni, Coelophysis bauri, and *Syntarsus* appear more closely related to each other than to *Dilophosaurus* or *Ceratosaurus* (fig. 5.1, Node 3) based on the shared presence of the alveolar ridge of the maxilla. These taxa are also considerably smaller than their nearest plesiomorphic relatives. *Dilophosaurus* and *Ceratosaurus* are just over 6 meters long, while *Liliensternus, Coelophysis,* and *Syntarsus* are about half that size. The diversity of this ceratosaur subgroup may increase with the description of recently collected specimens from the Upper Triassic Chinle and Dockum formations of the southwestern United States (pers. obs.) and material being studied by D. Kermack from near the Triassic-Jurassic boundary of South Glamorgan, South Wales (A. Milner, pers. comm.; see also Kermack 1984). Material collected by P. Olsen (pers. comm.) from the Upper Triassic of Nova Scotia may also belong in this group.

Liliensternus liliensterni is known from two incomplete subadult skeletons, both gracile individuals, that evidently came from a single locality. Apart from the characters grouping them within Ceratosauria, these specimens together do not appear to share any derived characters that permit *Liliensternus* to be diagnosed as a monophyletic taxon. Until such data come to light, or until other data show *Liliensternus* to be paraphyletic, we recommend that the name be restricted to these two specimens. The skeletons were originally described by Huene (1934*a*) under the name *Halticosaurus liliensterni,* but Welles (1984, 166) later erected the name *Liliensternus* to emphasize their distinctness from *Halticosaurus longotarsus* (Huene 1908) and *H. orbitoangulatus* (Huene 1932). This decision appears justified in that the material of *Liliensternus liliensterni* does preserve data permitting its placement as a derived member within Ceratosauria, whereas no data could be found to suggest that either *H. orbitoangulatus* or *H. longotarsus* is a ceratosaur.

Coelophysis bauri Cope (1887, 1889*c*) and *Syntarsus* Raath (1969) appear more closely related to each other (fig. 5.1, Node 4) than to *Liliensternus, Dilophosaurus,* or *Ceratosaurus.* Evidence for this includes the shared presence of an antorbital fenestra that is greater than 25 percent of the length of the skull (the condition of the antorbital fenestra is indeterminable in *Liliensternus*), greatly elongated cervical ribs (the cervical ribs are not adequately known in other ceratosaurs), a distinct caudal rim on the M. iliofemoralis fossa, and the obturator ridge of the femur.

The type specimen of *Coelophysis bauri* is a deficient specimen that preserves no diagnostic attributes beyond those of Theropoda ancestrally (Padian 1986). However, in lieu of data to the contrary, we follow the convention of using this name for the large sample from Ghost Ranch, New Mexico as well as material from Petrified Forest National Park, Arizona. Several hundred individuals of *Coelophysis bauri* have now been recovered from the Ghost Ranch Quarry and they include a range of ontogenetic stages from young juveniles to adults of both robust and gracile morphs. However, even with this large sample *Coelophysis*, like *Liliensternus*, is as yet undiagnosed.

Two Connecticut Valley specimens, both probably from the Lower Jurassic Portland Formation, have also been assigned to *Coelophysis*. These are the long lost type and only specimen of *Podokesaurus holyokensis* Talbot (1911), which has been referred as a separate species (Colbert 1964*b*), and a second unnamed specimen consisting of a natural cast of a partial tibia, pubis, and rib (Colbert and Baird 1958). These specimens might be *Coelophysis*, but their resemblance is confined to characters present in Theropoda ancestrally. We agree with Olsen (1980*a*) and Padian (1986) that the name *Podokesaurus holyokensis* be restricted to the type specimen and that both of these specimens be assigned to Theropoda *incertae sedis*.

Syntarsus includes two species (fig. 5.1), *S. rhodesiensis* Raath (1969) and *S. kayentakatae* Rowe (1989). *Syntarsus* is diagnosed by the nasal fenestra and fusion of the proximal ends of metatarsals II and III. Material for both species includes a range of ontogenetic stages from young juveniles to adults (e.g., Gow and Raath 1977; Raath 1980).

Syntarsus rhodesiensis may be diagnosed by elongation of its antorbital fenestra to 40 percent of the skull length and by a ventral extension of the lacrimal that overlaps the jugal and reaches the alveolar margin in a condition unlike any other archosaur (Raath 1977). It is significant that these characters were described from disarticulated material, because the length of the antorbital fenestra is dependent on the position of the lacrimal being correctly interpreted. This interpretation also places the maxillary tooth row entirely in front of the orbit, in contrast to other ceratosaurs. Discovery of articulated cranial material will be of the utmost interest.

Syntarsus kayentakatae is diagnosed by a paired naso-lacrimal crest and fusion of the fibula to the calcaneum in adults (Rowe 1989). The crest is similar to that in *Dilophosaurus* and this, together with the co-occurrence of these taxa in the Kayenta Formation,

raises the question of whether *S. kayentakatae* might be a young *Dilophosaurus*. However, the type skeleton of *S. kayentakatae* is an adult that is about half the size of the juvenile type skeleton of *Dilophosaurus*. Moreover, *S. kayentakatae* shares a number of derived characters with *S. rhodesiensis*, *Coelophysis bauri*, and *Liliensternus liliensterni* which are absent in *Dilophosaurus*.

The phylogenetic positions of *Sarcosaurus woodi* Andrews (1921; see also Huene 1932, 49–51, pl. 2) and *Segisaurus halli* Camp (1936) are particularly uncertain, which is not surprising because they are far less complete than any other named ceratosaurs. Both are based on isolated, deficient specimens that preserve no data bearing on their positions within Ceratosauria. *Sarcosaurus* consists only of a broken pelvis, femur, and an isolated broken centrum. The ilium is roughly 20 cm long, roughly midway in size between that of *Dilophosaurus* and the *Coelophysis-Syntarsus* group. The ilium and pubis are firmly sutured, indicating a mature specimen, and the trochanteric shelf of a robust morph is present. The ilium lacks a distinct caudal rim on the M. iliofemoralis fossa, and the femur lacks the obturator ridge, indicating that *Sarcosaurus* lies outside of the *Coelophysis-Syntarsus* group (fig. 5.1, Node 4), but little more can be said of it. Because *Sarcosaurus* has no apparent diagnostic characters of its own, we suggest that the name be restricted to this specimen. Huene (1932) erected a second species, *S. andrewsi*, based on an isolated right tibia. This uninformative specimen may be a theropod, but it preserves no data to support its referral to either *Sarcosaurus* or Ceratosauria. It is best regarded as a *nomen dubium*.

Segisaurus halli Camp (1936) is based on a partial postcranial skeleton. A separate family has often been designated to accommodate this specimen because it was thought to have solid bones (e.g., Steel 1970). Welles (1984) suggested that the specimen is a secondary filling in a natural mold. Nevertheless, close inspection leaves little doubt that the preserved material is bone. Moreover, although most of the long bones are crushed, cavities in several vertebral centra and the humerus are visible, and we see no reason to doubt that the other long bones and vertebrae were hollow, as they are in theropods generally. The specimen is a robust subadult individual. *Segisaurus halli* is unique in having a large foramen in the ischium beneath the acetabulum. Flattening of the shaft of the ischium may also be diagnostic, but it is difficult to be certain that this is not due to crushing. *Segisaurus* is small, like the *Liliensternus-Coelophysis-Syntarsus* group, and is roughly contemporaneous with *Syntarsus*. However, morphological data fail to place it unequivocally with that group.

FUNCTION AND BEHAVIOR

The presence of noninterdigitating sutures between many of the bones in the skull has often been interpreted as forming a system of levers driven by jaw muscles, an enhancement to predation (e.g., Bradley 1903; Versluys 1910; Frazzetta 1962; Iordansky 1968). Bakker (1986) largely rejected a lever system but argued that mobility between the skull elements in *Ceratosaurus* was of a degree sufficient to permit swallowing of prey items larger than its own head. Welles (1984) also largely rejected a lever system and interpreted the potential mobility as an indication of weakness, arguing that the premaxillary attachment in *Dilophosaurus* (and by implication other ceratosaurs) confined it to mere scavenging.

A potential for movement between some adjacent bones may exist, but evidence for an elaborate, interconnected lever system, which was based on squamate models that have largely proved false, now appears perilous (Smith 1982). Any movement that did occur was probably more passive than active, a result of bending of the skull to accommodate stresses imposed while biting and manipulating prey items. The fibrous connections between the skull elements provide toughness while affording a degree of flexibility to the jaw apparatus. Bakker's view is close to this interpretation, but there is no evidence for the degree of mobility or size of prey items swallowed that he envisioned. Stomach contents are known in a few specimens of *Coelophysis* and *Syntarsus* and consist of relatively much smaller animals.

Ceratosaurs retain powerfully built forelimbs with strong, raptorial, grasping hands. They were cursorial, obligate bipeds. Ceratosaurs were digitigrade, and their digits terminate in sharp claws. There is little reason to doubt that they could successfully pursue fast, agile prey and use all of their limbs to aid in subduing and dismembering it before ingestion.

Ceratosaurus nasicornis, Dilophosaurus wetherilli, and *Syntarsus kayentakatae* have autapomorphic cranial ornaments that are so delicately built that they were probably used for display purposes alone. The epaxial osteoderms of *Ceratosaurus nasicornis* might also share this function. These, together with the presumed sexual dimorphism and evidence of gregarious activity (see below), suggest complex social behavior in ce-

ratosaurs. This is not surprising given the widespread occurrence of such structures and behavior in extant crocodilians and birds and the evidence for it elsewhere among extinct dinosaurs (e.g., Ostrom 1971, 1986).

PALEOECOLOGY

Ceratosaurs are consistently represented in terrestrial faunas from the Late Triassic through the Late Jurassic. Because they cover a wide size range, and in the Kayenta Formation at least two different-sized taxa occur together, ceratosaurs possibly played a variety of roles within their more general carnivorous habitude. Ceratosaurs have an essentially global distribution, which is not surprising given their cursorial skeletons. They persist in a variety of faunal associations over a time span of nearly 80 million years and are preserved in a diversity of depositional settings. One might speculate that they were capable of a broad range of climatic and dietary tolerances. Summaries of the faunal contexts in which ceratosaurs have been discovered are available in Jacobs and Murry (1980) for the Chinle Formation; Raath (1980) for the Forest Sandstone; Kitching and Raath (1984) for the Elliot Formation; Colbert (1981); Attridge et al. (1985), Jenkins et al. (1983), and Clark and Fastovsky (1986) for the Kayenta Formation; Dodson, Behrensmeyer, Bakker, and McIntosh (1980) for the Morrison Formation; and Russell et al. (1980) for the Tendaguru.

TAPHONOMY

Several ceratosaurs have been discovered in "mass burials," in which numerous individuals are preserved together, largely exclusive of other taxa. These burials include localities in Zimbabwe and South Africa which preserve numerous individuals of *Syntarsus rhodesiensis* (Gow and Raath 1977; Raath 1980; J. Kitching, pers. comm.; M. Raath, pers. comm.). Two localities in the Kayenta Formation of Arizona preserved at least three and eleven individuals of *Syntarsus kayentakatae* (Rowe 1989), and another locality in the same formation preserved three individuals of *Dilophosaurus wetherilli* (S. Welles, pers. comm.). The renowned Ghost Ranch Quarry has yielded several hundred individuals of *Coelophysis bauri* (precise estimates await final preparation of the many huge blocks removed from the quarry), and more material has yet to be excavated. *Liliensternus liliensterni* occurred at a locality that yielded at least two specimens (Huene 1934*a*).

At most of these localities, some degree of skeletal articulation is preserved. At the Ghost Ranch Quarry, a few nearly complete skeletons have been recovered, but most of the skeletons are partially dissociated. In all but one ceratosaur mass burial, a range of ontogenetic stages is represented. Relatively young individuals are much more common than adults; in some cases, adults are entirely absent.

A diversity of depositional environments has preserved ceratosaur mass burials. The Ghost Ranch Quarry is in a meter-thick mudstone of fluvial origin; a more precise taphonomic account of this remarkable accumulation has yet to be undertaken. *Dilophosaurus* and *Syntarsus kayentakatae* burials are both preserved in overbank deposits, and one mass burial of the latter is in a channel sandstone. Raath (1980) reported *Syntarsus rhodesiensis* burials in thin fluvial lenses within aeolian deposits. As Welles (1984) pointed out, the low taxonomic diversity of the mass burials and their occurrence in a variety of depositional environments suggest that group behavior contributed to the sedimentological mechanisms that were involved in these accumulations.

6

Carnosauria

RALPH E. MOLNAR

SERGEI M. KURZANOV

DONG ZHIMING

INTRODUCTION

Carnosaurian dinosaurs were the largest carnivorous animals to inhabit the land. Some of these, such as *Tyrannosaurus rex*, are known to almost every schoolchild. Unfortunately, our knowledge of many of these forms does not compare to their popularity. Indeed, few of the species are represented by complete skeletons.

It is clear that carnosaurs existed throughout the Jurassic and Cretaceous, but it is not obvious that they were the predominant carnivores of the tetrapod food chains on all continents through all this time. Good specimens thus far have been found almost entirely in only two areas, eastern Asia and western North America. Stratigraphically, good specimens so far are restricted to the Upper Jurassic and the Campanian and Maastrichtian.

Almost all of the work on carnosaurs has been taxonomic and descriptive; there has been very little work on the functional morphology or paleoecology of these animals. Our understanding of the lives of these largest of all land carnivores is still to come. Thanks to the studies of these well-preserved specimens, however, we can sketch out the lines of carnosaurian evolution.

DIAGNOSIS

The Carnosauria can be diagnosed on the following features: an enlarged foramen or foramina opening laterally at the angle of the lacrimal, a deep surangular, reduction of the axial neural spines, chevrons with craniodorsally projecting prongs, an abrupt expansion of the glenoid from the narrow parallel or subparallel scapular blade, a narrow brevis shelf on the ilium, a femoral head that is inclined upward from the greater trochanter and proximal end of the shaft, a winglike lesser trochanter, and a strongly developed extensor groove on the distal femur.

ANATOMY

Skull and Mandible

Well-preserved skulls (fig. 6.1) are known for *Allosaurus fragilis, Nanotyrannus lancensis, Albertosaurus libratus, Daspletosaurus torosus, Tarbosaurus bataar,* and *Tyrannosaurus rex* (Osborn 1912b; Madsen 1976a; Gilmore 1946c; Lambe 1917c; D. A. Russell 1970a;

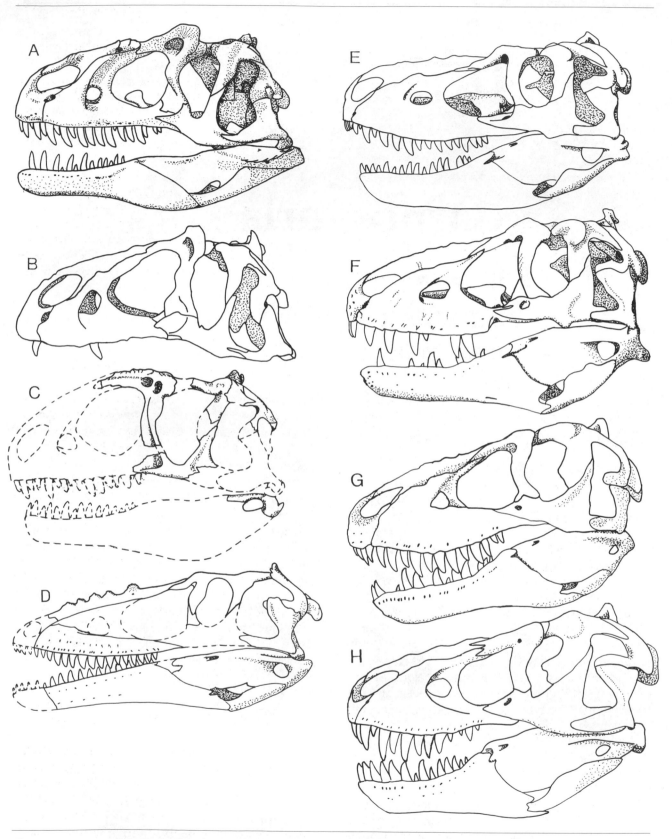

Fig. 6.1. Carnosaur skulls in lateral view. A and B, *Allosaurus fragilis* (A, composite). C, *Acrocanthosaurus atokensis*. D, *Alioramus remotus*. E, *Albertosaurus libratus*. F, *Daspletosaurus torosus*. G, *Tarbosaurus bataar*. H, *Tyrannosaurus rex*. (After Osborn 1912*b*; Stovall and Langston 1950; D. A. Russell 1970*a*; Kurzanov 1976*b*.)

Maleev 1974). Other taxa are represented by less complete skulls (*Alectrosaurus olseni* and *Alioramus remotus;* Perle 1977; Kurzanov 1976*b*). Most cranial elements are known from isolated and articulated bones. The palate has been described and illustrated for *Tyrannosaurus rex, Daspletosaurus torosus,* and *Allosaurus fragilis* (Osborn 1912*b;* D. A. Russell 1970*a;* Madsen 1976*a*). Braincases are not well known but have been described for *Acrocanthosaurus atokensis, Allosaurus fragilis, Carcharodontosaurus saharicus, Chilantaisaurus maortuensis, Piatnitzkysaurus floresi, Tarbosaurus bataar,* and *Tyrannosaurus rex* (Stovall and Langston 1950; Madsen 1976*a;* Stromer 1931; Hu 1964; Bonaparte 1986*c;* Maleev 1974; Osborn 1912*b*). More descriptive work is needed for an understanding of both palatal and braincase structure in carnosaurs.

The snout is long in *Alectrosaurus olseni* and *Alioramus remotus* and relatively short in *Tyrannosaurus rex.* The nasals and maxillae are more elongate than the parietals and frontals dorsally and the jugals and quadratojugals laterally. The snout is laterally compressed, narrower dorsally than at the palate, and the postorbital region of the skull usually only slightly broader than the base of the snout (fig. 6.2). Only in *Tyrannosaurus rex* and *Nanotyrannus lancensis* does the postorbital region extend obviously beyond the width at the base of the snout. The cranial, and some postdentary, elements have extensive internal chambers. A variety of cranial ornamentation is known in carnosaurs: almost every taxon represented by well-preserved skull material shows some form of ornamentation.

The rostrocaudally elongate nares are subterminal and bounded by the premaxilla, maxilla, and nasal. The orbits are laterally directed (and somewhat rostrally in *Nanotyrannus lancensis* and *Tyrannosaurus rex*), often broader dorsally than ventrally, and bounded by the lacrimal, postorbital, jugal, and primitively the prefrontal. In some forms, the orbits are constricted ventrally by expansion of the ventral bar of the lacrimal (*Allosaurus fragilis*) or by expansion of the postorbital (*Tarbosaurus bataar, Tyrannosaurus rex*). The position of the lacrimal canal high on the rostral margin of the orbit and the constriction of the ventral part of the orbit by the postorbital in tyrannosaurids suggest that the eyeball was located in the dorsal portion of the orbit. The large antorbital fenestra is bounded predominantly by the maxilla as well as by the lacrimal and, caudoventrally, by the jugal. The trapezoidal infratemporal fenestra, narrow dorsally and broad ventrally, opens into the adductor chamber. The infratemporal fenestra is bounded by the postorbital, squamosal, quadratojugal, and jugal and is larger than the supratemporal fe-

nestra. The supratemporal fenestra is situated dorsally, opening into the adductor chamber and bounded by the frontal, parietal, supraoccipital, squamosal, and postorbital. Much reduced posttemporal fenestrae pierce the occipital face at the junction of the opisthotic and squamosal, although these are sometimes lost (Gilmore 1920). The choanae are about halfway from the tip of the premaxillae to the occipital condyle and are surrounded by the maxilla, palatine, and vomer. The adductor chamber opens through the palate (subtemporal fenestra) behind the ectopterygoid, bounded also by the jugal, quadratojugal, quadrate, and pterygoid.

The short premaxilla is rectangular in lateral view but triangular from below. Its body and dorsal process together form the rostral half of the external naris, and its caudal process contacts the nasal and maxilla. It holds four to (rarely) five teeth. Near the toothrow are numerous labial foramina, probably for cutaneous nerves, irregularly arranged.

The maxilla is a large triangular element with a rostrally directed apex and deeply embayed from behind by the antorbital fenestra. A ramus extends above the antorbital fenestra to form almost its entire dorsal margin. The body, deep to accommodate the teeth, lies below the fenestra. The smaller maxillary (or second or rostral antorbital) fenestra is present between the antorbital fenestra and external naris. The very shallow excavation surrounding the antorbital fenestra has become deepened into an antorbital recess in *Tyrannosaurus rex.* The maxilla houses several internal chambers. The two found in *Allosaurus fragilis* open both rostrally and caudally and may have accommodated olfactory epithelium. The maxilla bears from 12 to 18 teeth (in *Tyrannosaurus rex* and *Alectrosaurus olseni,* respectively): tooth number may vary from individual to individual but is not related to age. Numerous foramina bordering the ventral margin are sometimes arranged in rows (*Alioramus remotus*). Caudally, the maxilla contacts the jugal, while dorsally, it is bordered by the nasal and lacrimal. The medial palatal process contacts the palatine.

The elongate nasals are medially fused, save at the rostral and caudal ends. They are often rugose, and *Alioramus remotus* has a double row of five vertical blades along the nasals which is continuous with the lacrimal and postorbital rugosities. The nasals are penetrated by many foramina, probably for branches of the superior nasal artery; these suggest a horny cover was present. The dorsal and ventral rostral processes form the caudal perimeter of the external naris.

The inverted L-shaped lacrimal carries the nasolacrimal canal near the top of the orbit. This element

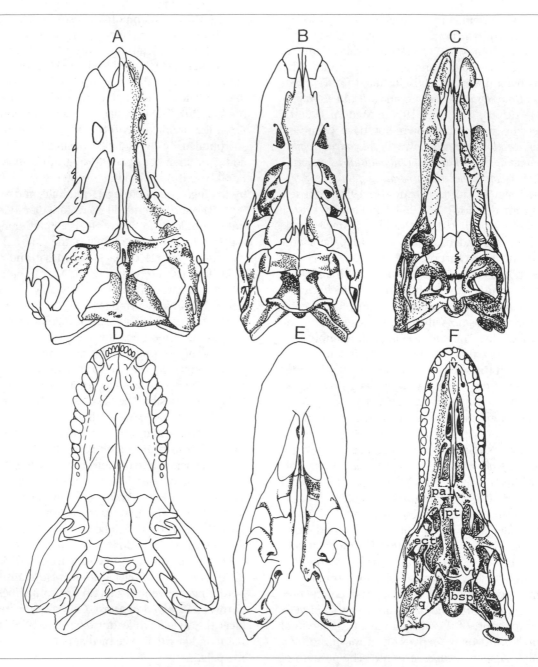

Fig. 6.2. Carnosaur skulls in dorsal (A–C) and ventral (D–F) views. A and D, *Tyrannosaurus rex*. B and E, *Daspleto-saurus torosus*, C and F, *Allosaurus fragilis*. (After Osborn 1912*b*; D. A. Russell 1970*a*; Madsen 1976*a*.)

forms the caudodorsal segment of the margin of the antorbital fenestra. The dorsal ramus contains internal chambers. In most tyrannosaurids, both ramus and chambers are enlarged compared those of the allosaurids, but the chamber of *Alioramus remotus* is small. One (tyrannosaurids) or two (*Acrocanthosaurus atokensis*) prominent apertures pierce the lacrimal just behind the caudodorsal angle of the antorbital fenestra. Rugosi-

ties or other ornamentation are often present. *Allosaurus fragilis* and *Daspletosaurus torosus* have lacrimal crests, while *Albertosaurus libratus* has a less prominent lacrimal rugosity.

The postorbital is a triradiate element, with its shorter third process extending medially to the frontal. It is fused with the postfrontal in adults. The descending process bounds the orbit caudally, to form half of the orbital rim, and reaches the jugal. The caudal process contacts the squamosal. A rugosity is located on the thickened dorsal rim above and behind the orbit in some taxa. This is prominent in tyrannosaurids but also found in *Allosaurus fragilis* (Madsen 1976a) and *Acrocanthosaurus atokensis* (Stovall and Langston 1950). In *Tarbosaurus bataar* and *Tyrannosaurus rex,* the descending process becomes widened and constricts the orbit. Medially, this process bears a socket for the laterosphenoid; in *Alioramus remotus,* due to the lengthened snout, this socket is on the caudal face of the postorbital.

The hollow squamosal is a complex wedge-shaped element with three or four processes. The ventral concavity for the quadrate may have permitted motion (see Molnar and Farlow, this vol.) and the ventral process reaches the quadratojugal below. In *Alioramus remotus,* the squamosal contacts the laterosphenoid. The squamosal often bears a smooth scar for the external mandibular adductor muscle on its lateral face just above the infratemporal fenestra.

The quadratojugal is a thin L-shaped element whose ascending process is closely applied medially to the lateral margin of the body of the quadrate. This process reaches the squamosal dorsally. In *Alioramus remotus, Allosaurus fragilis, Tarbosaurus bataar,* and *Tyrannosaurus rex,* the paraquadrate foramen lies at the junction of the quadrate with the quadratojugal. Through this foramen probably passed the anastomosing ramus of the caudal condylar artery and the trigeminal nerve branch serving the jaw joint. In tyrannosaurids, the infratemporal fenestra is invaded caudally by processes of the squamosal and quadratojugal which form a flange projecting from the rear margin of that fenestra. The triradiate jugal bounds the orbit and infratemporal fenestra ventrally and in some forms contains extensive internal chambers that open laterally via a large foramen at the caudoventral angle of the antorbital fenestra.

The small triangular prefrontal is often greatly reduced and lies between the frontal, nasal, and lacrimal (and can be overlain by the latter elements). It forms part of the orbital margin in *Alioramus remotus* and *Allosaurus fragilis.* It has not been observed in *Tarbosaurus bataar.*

The short, wide platelike frontal roofs the cavity for the forebrain and is firmly sutured medially. The frontal is invaded caudally by an extension of the supratemporal fenestra, which is shallow and limited in *Allosaurus fragilis* but deep and extensive in *Tyrannosaurus rex.* This element is thickened in *Carcharodontosaurus saharicus* and fused with the parietal in *Acrocanthosaurus atokensis.*

The parietal roofs the endocranial cavity caudally and forms two dorsal crests. The sagittal crest separates the supratemporal recesses. It is dorsally flat in *Allosaurus fragilis* where the recesses are not confluent and obtusely angled in tyrannosaurids where they are. Caudally, the parietals are twisted laterally to form the vertical plates or alae that make up the supraoccipital crest. This crest rises above the supraoccipital bone and, often with a dorsal rugosity, forms a rostrally directed V in dorsal view. Fossae for the ligamentum nuchae are situated dorsally on the caudal face of the supraoccipital crest. A pineal foramen is absent, although the parietal roof is thin over a dorsal projection of the endocranial cavity. This is thought by Hopson (1979) to have been occupied by cartilage.

In the braincase, a thin rostrally projecting parasphenoid (fig. 6.3) is usually present, although because of its delicate nature, it is often incompletely preserved or lost. The parasphenoid forms the rostral wall of the hypophysial fossa, is fused with the basisphenoid, and contacts the orbitosphenoid and laterosphenoid dorsally.

An ossified sphenethmoid is preserved in at least *Acrocanthosaurus atokensis, Albertosaurus libratus,* and *Tyrannosaurus rex.* It is a boxlike element housing the passages for the olfactory nerves. A median septum separates the two passages.

The orbitosphenoid (fig. 6.3) lies just behind the sphenethmoid and bears the optic foramen. This bone is roughly crescentic in *Allosaurus fragilis* but subquadrate in tyrannosaurids. Because of the expansion of the rostral part of the braincase, it is almost horizontal and is sutured to the frontal laterally and the laterosphenoid caudally.

The tall laterosphenoid (fig. 6.3) forms much of the lateral wall of the braincase and dorsolaterally contacts the postorbital by an apparently immobile condyloid joint. The laterosphenoid floors the endocranial cavity and forms the dorsal part of the sella turcica. The oculomotor and trochlear nerves pass through the base of the pila antotica of the laterosphenoid (*Tyrannosaurus rex:* Osborn 1912b) or laterosphenoid orbitosphenoid junction (*Albertosaurus libratus, Daspletosaurus torosus:* D. A. Russell 1970a). With a single exception (see below), the trigeminal foramen penetrates the laterosphenoid. The laterosphenoid was previously identified as the alisphenoid (e.g., Gilmore 1920) or orbitosphenoid (Osborn 1912b).

Caudal to the laterosphenoid, partly overlapping the opisthotic, is the prootic (fig. 6.3). In *Acrocanthosaurus atokensis, Allosaurus fragilis,* and *Tyrannosaurus rex,* the prootic is penetrated by the fenestra ovale and

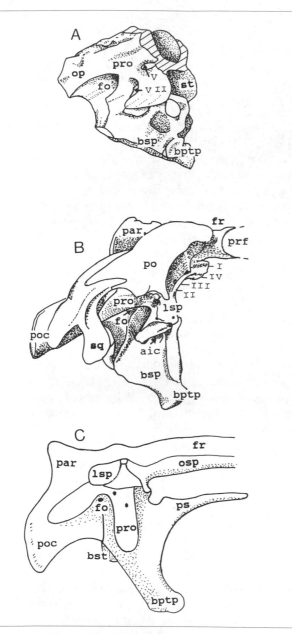

Fig. 6.3. Carnosaur braincases in lateral view. A, *Piatnit-zkysaurus floresi.* B, *Allosaurus fragilis.* C, *Tarbosaurus bataar.* (After Osborn 1912*b*; Barsbold 1983*a*; Bonaparte 1986*c*.)

jugular foramen and, with the opisthotic, roofs the sta-pedial canal (Madsen 1976*a*; Stovall and Langston 1950). In *Alioramus remotus,* this element is greatly ex-panded rostrally and houses the trigeminal foramen.

The opisthotic completes the lateral wall of the braincase and bears the foramen for the facial nerve. This element extends caudolaterally as a vertical plate to form, with the exoccipital, the paroccipital process. This process is markedly declined in *Allosaurus fragilis*

and *Chilantaisaurus tashuikouensis* but projects horizon-tally in *Tyrannosaurus rex;* the paroccipital process is hollow in that species. The glossopharyngeal, vagus, accessory, and hypoglossal nerves exit the endocranial cavity below and lateral to the foramen magnum, probably through the exoccipital component of the paroccipital process.

The occipital elements (fig. 6.4) are often fused, most extensively in the tyrannosaurids. The supraoc-cipital forms a vertical rectangle above the foramen magnum in *Allosaurus fragilis* and *Chilantaisaurus maor-tuensis* but is broadened in tyrannosaurids. The fora-men magnum is bordered by the basioccipital, opis-thotic, exoccipital, and supraoccipital in *Alioramus remotus, Allosaurus fragilis,* and *Chilantaisaurus maor-tuensis;* the supraoccipital is excluded from the foramen magnum in the other tyrannosaurids and *Acro-canthosaurus atokensis.*

The basioccipital forms from half to all of the oc-cipital condyle and most of the occipital plate below the condyle. There is some lateral contribution from the exoccipitals in *Allosaurus fragilis* and more exten-sive contribution in tyrannosaurids.

The braincase is floored by the basisphenoid. The ventrolaterally directed basipterygoid processes extend from the base of the basisphenoid to contact the ptery-goids at well-defined articular surfaces. In the basisphe-noid below the trigeminal foramen (of the laterosphe-noid) is the entrance to the internal carotid canal. The basisphenoid is excavated into the deep ventrally open-ing sphenoidal sinus (or basicranial fontanelle) rostral to other, separate internal chambers (fig. 6.4), which extend into the occipital condyle, at least in *Allosaurus fragilis* (Madsen, 1976*a*). The basisphenoid region, ros-trocaudally elongate in *Allosaurus fragilis,* becomes shortened and much expanded transversely in tyran-nosaurids.

The quadrate is a robust vertical column termi-nating ventrally in two mandibular condyles—the lat-eral usually larger than the medial—separated by a helical groove. Although apparently solid in *Allosaurus fragilis,* the quadrate is hollow in *Tyrannosaurus rex* (Molnar 1984*b*). A thin vertical platelike pterygoid process extends forward to the basipterygoid process and lies freely on the similarly thin platelike quadrate process of the pterygoid, which together form the me-dial wall of the adductor chamber.

The pterygoid is a rostrocaudally long horizontal plate, bearing the large quadrate process and a small ectopterygoid process. Rostrolaterally, the pterygoid contacts the palatine and the ectopterygoid, which overlies the pterygoid. At the base of the quadrate pro-

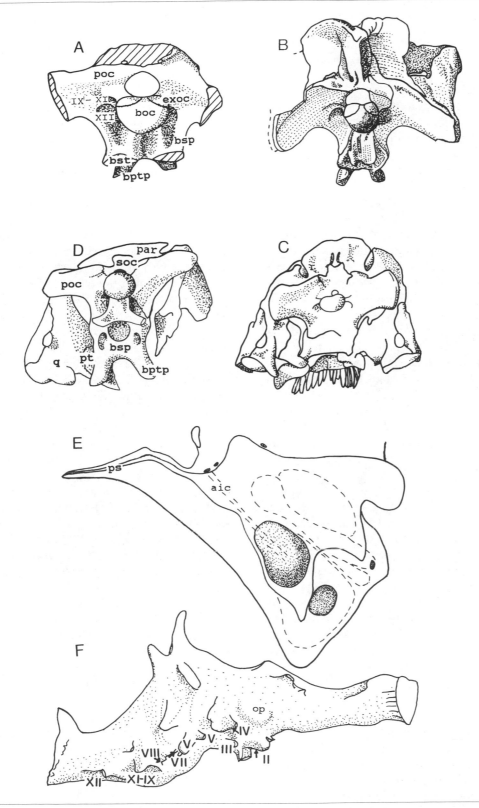

Fig. 6.4. Carnosaur skulls and braincases in caudal view. A, *Piatnitzkysaurus floresi*. B, *Allosaurus fragilis*. C, *Tyrannosaurus rex*. D, *Alioramus remotus* (in caudoventral view). E, Basisphenoid-basioccipital region of skull of *Albertosaurus* showing extent of sinuses. F, Endocranial mold of *Tyrannosaurus rex* in lateral aspect. Rostral to right. (After Osborn 1912*b;* Gilmore 1920; D. A. Russell 1970*a;* Kurzanov 1976*b;* Bonaparte 1986*c.*)

cess is the pocket for the reception of the basipterygoid process. All pterygoid contacts are overlaps and seemingly mobile. The interpterygoid vacuity separates the two pterygoids on the midline.

The ectopterygoid is a triradiate element in early taxa (*Allosaurus fragilis*), which attains a triskelion-like form (or triradiate form with hooklike extensions) in tyrannosaurids. It is inflated and deeply excavated by internal chambers. Its lateral process reaches the jugal, and its wider medial process contacts the pterygoid.

The roughly triangular platelike palatine extends medially from the palatal process of the maxilla to the midline. The palatine forms the base of the preorbital septum and the caudal margin of the choana. Rostromedially, each palatine rises to meet the other medially, thus vaulting the palate. In *Allosaurus fragilis*, the palatine is tetraradiate with dorsal processes projecting medially to contact the vomer and pterygoid. These are separated from ventral processes that contact the jugal and maxilla, by a medially open excavation. This excavation seemingly becomes enclosed to give an inflated and deeply excavated element in *Tyrannosaurus rex*.

The lanceolate vomer lays below and against the palatines or pterygoids caudally. A pit on the caudal face of the premaxillary symphysis receives the rostral tip of the head. Although slender in the other taxa, the head is much broadened in *Tyrannosaurus rex* (Osborn 1912*b*).

The long flattened epipterygoid links the quadrate process of the pterygoid to the laterosphenoid—contacting it just below the metoptic foramina. It is a thin triangular bone in *Alioramus remotus* and *Allosaurus fragilis* but quite robust in *Tyrannosaurus rex*.

The endocranial cavity is short, but tall, especially near the inner ear. The floor of the cavity is flexed upward between the jugular and trigeminal foramina. Its maximum height, at the acoustic foramen, probably results from the presence of a reduced pineal projection. Just rostral to the trigeminal foramen, the endocranial cavity is widened to accommodate the optic lobes or cerebrum. The intracoustic recess is next to, but not joined with, the inner ear. This recess penetrates the paroccipital process, and its caudal wall also forms the rostral wall of the paroccipital cavity. The intracoustic recess probably housed the flocculus.

Endocranial molds (fig. 6.4) are known for *Allosaurus fragilis, Carcharodontosaurus saharicus, Tarbosaurus bataar,* and *Tyrannosaurus rex* (Madsen 1975*a*; Stromer 1931; Maleev 1965; Osborn 1912*b*). That of *C. saharicus* represents only the dorsal portion of the endocranial cavity. The molds are basically similar in form: all show a transversely expanded rostral region, followed by a transversely constricted but dorsoventrally expanded region, tapering toward the foramen magnum (fig. 6.4; Hopson 1979). The rostral region presumably accommodated the forebrain, while the dorsoventrally expanded region housed the mid- and hindbrains. The region of the medulla is shorter in *A. fragilis* than in *T. rex*.

In *T. rex*, convexities (low on the transversely expanded portion just rostral to the dorsoventrally expanded segment) indicate either the optic lobes (Osborn 1912*b*) or caudolateral parts of the cerebrum (Hopson 1979). The molds of *C. saharicus, Tarbosaurus bataar,* and *Tyrannosaurus rex* constrict markedly rostrally (more so in the former two than the latter) only to expand again at the olfactory lobes. Indication may be seen of the pituitary fossa in *A. fragilis* and *Tyrannosaurus rex,* but the fossa itself was not molded. The cranial nerves have been identified on the molds of *Allosaurus fragilis, Tarbosaurus bataar,* and *Tyrannosaurus rex* (fig. 6.4). The mold of *Tyrannosaurus rex* differs from the others known in having three dorsal extensions from the hindbrain region. These are absent in *A. fragilis* and *Tarbosaurus bataar,* while *C. saharicus* seemingly has but two. These may represent branches of the longitudinal venous sinus like those of crocodilians (Hopson, 1979).

The relative brain size (expressed as the encephalization quotient) in carnosaurs (based on *A. fragilis* and *Tyrannosaurus rex*) is high compared to that of most other dinosaurs, but within the range of those of modern reptiles (Jerison 1973; Hopson 1980*a*). However, a more recent study (Martin 1981), which proposed a refined method for estimating brain size from body size, concluded that the brain of *Allosaurus fragilis* is larger than expected from estimates of body size.

The structure of the inner ear is preserved in *Allosaurus fragilis, Tarbosaurus bataar,* and *Itemirus medullaris*. *I. medullaris* is not unanimously accepted as a carnosaur; however, the conservative nature of the ear structure suggests that the ear is similar among all theropods. Hence *Itemirus* provides a useful supplement to the information gained from *Tarbosaurus* (the inner ear of *Allosaurus* has been figured by Madsen [1976*a*] but not yet described). The vestibule is beneath the medial part of the intracoustic recess. In the caudal portion of the rostral sinus of the vestibule is the conical depression of the lagena. Just above it is the fenestra ovale. The endolymphatic duct is displaced to the rim of the foramen magnum. The acoustic nerve probably sent two branches into the osseous labyrinth. The semicircular canals surround the intracoustic recess, housed within its side and caudal walls, and floor. Although the inner ear of carnosaurs shows no major differences from those of other theropods, it should be noted that the lagena is small and less developed than in crocodiles, the semicircular canals are very tall, and the vestibule is large.

The stapes is known only in *Allosaurus fragilis* (Madsen 1976*a*). It is a slender 1-cm-long rod that is expanded at both ends. The nearly square flat footplate is inclined to the shaft, and there was no ossified extra columella.

The laterally compressed mandible is robust rostrally but composed of less robust postdentary elements around the mandibular fossa. The postdentary region is deeper than the dentary. An external mandibular fenestra is present, bounded by the dentary, surangular, and angular.

The dentary is a stout tooth-bearing, barlike element, deepening caudally into a thin vertical plate. It bears from 12 to 18 teeth (in *Allosaurus fragilis* and *Alectrosaurus olseni*, respectively). The dentary alveoli are separated by thin partitions, which medially expand into sagittally oriented plates. Each plate is separated from the next by a notch or slot adjacent to each tooth. Along the bases of the plates is a fissure, probably for a branch of the inferior alveolar artery. These plates (and similar plates of the maxilla) are the interdental plates. A narrow bandlike supradentary medially overlies these plates along the dorsal margin of the dentary in *Allosaurus fragilis, Tarbosaurus bataar,* and *Tyrannosaurus rex.* The symphysial contact is usually loose and consists of a medial flat facet at the rostral extremity of the dentary. In the region of the 13th to 16th tooth is the entrance of the alveolar canal, through which passed the mandibular nerve. The dentary is straight in dorsal view in all forms save *Allosaurus fragilis* where it is laterally bowed.

Caudal to the dentary is the large, flat, triangular splenial that bears a small rostral mylohyoid foramen in *Allosaurus fragilis* and a large one in tyrannosaurids.

A crescentic prearticular sheathes the articular caudally and forms the caudal, ventral, and cranioventral margins of the Meckelian fossa. The craniodorsal margin of this fossa is formed by the small, flat, triangular coronoid.

The surangular is a thin, vertical plate forming the lateral wall of the Meckelian fossa. That portion just beneath the mandibular glenoid is pierced by a small foramen in *Allosaurus fragilis* and a large one in *Acrocanthosaurus atokensis* and tyrannosaurids. Cutaneous branches of the mandibular nerve and artery likely passed through this foramen as well as through the external mandibular fenestra. A flat or slightly convex facet on the dorsal margin of the surangular was apparently for attachment of the superficial mandibular adductor musculature. Just below and parallel to this facet is a bar on the lateral surface of the surangular. The surangular forms the lateral part of the mandibular glenoid.

The comma-shaped angular forms the caudoventral margin of jaw and floors the Meckelian fossa. Its rostral process lies between the splenial and dentary.

The robust, tetrahedral articular bears the mandibular glenoid and supports the short retroarticular process. In tyrannosaurids, the retroarticular process is lost, and the body of the articular is hollow and opens dorsolaterally via the foramen aerosum. The dorsally directed, transversely wide glenoid fossa is divided by an oblique ridge into a concave medial and flat lateral part. When articulated with the surangular, the glenoid form matches closely that of the articular condyles of the quadrate.

Allosaurus fragilis has a small element, not known in other species, the antarticular, at the medial end of the articular-surangular junction (Madsen 1976*a*). It probably acted as a stop to prevent caudal disarticulation of the jaw joint.

Carnosaur teeth are usually trenchant, recurved, and bladelike, with serrations on mesial and distal carinae. These serrations extend almost the entire length of the crown. Teeth are present in the premaxillae, maxillae, and dentaries only. The roots are massive and long. Maxillary and dentary (and often premaxillary) teeth usually are recurved. Dental differentiation is obvious, although not sufficiently studied to date. The distal maxillary and dentary teeth are shorter and more strongly compressed than those more mesially placed. They have recurved mesial margins but straight or only slightly recurved distal margins. Premaxillary teeth are often incisiform, more rodlike than the maxillary, and have offset carinae.

Tyrannosaurid teeth show three types of wear: slight rounding of the tip, development of flat oblique facets adjacent to the tip both labially and lingually, and abrasion of the serrations both mesially and distally. We have observed only the first two types in *Tyrannosaurus rex,* although the third type is common in other tyrannosaurids. The facets do not correspond in the upper and lower dentitions, so unless replacement accounts for the absence of corresponding facets, they were not formed in tooth-to-tooth contact. An unusual tooth form is found in *Carcharodontosaurus saharicus*. These teeth have broad, faceted crowns that are not (or are only slightly) recurved. The facets seem to be intrinsic to the tooth form and not to have resulted from wear. In addition to the serrations, these crowns have much broader marginal crenulations (Lapparent 1960*a*).

Ceratobranchials have been reported only in *Nanotyrannus lancensis* (Gilmore 1946*c*). They are elongate rods, with a curved segment near the middle of the shaft.

Postcranial Skeleton
Axial Skeleton

In carnosaurs, 9 to 10 cervicals, 13 to 14 dorsals, 5 sacrals, and 30 to 50 caudals are present. The vertebrae often exhibit pleurocoels and accessory articulations.

The atlas is reduced to the neurapophyses, crescentic intercentrum, and odontoid. The latter two are sometimes fused to the axis, as in *Piatnitzkysaurus floresi,* and sometimes free, as in *Allosaurus fragilis.* The axis (fig. 6.5) has a robust cylindrical intercentrum in *Allosaurus fragilis,* but the intercentrum is axially compressed to be twice as high as long in tyrannosaurids. The constricted cuplike axial pleurocentrum often has pleurocoels. A well-developed axial neural arch is present, but the neural spine is reduced. Prezygapophyses are present, and postzygapophyses are well developed and bear epipophyses in *Allosaurus fragilis, Szechuanosaurus campi,* and *Tyrannosaurus rex.*

The other cervical centra are opisthocoelous but are axially compressed and only weakly opisthocoelous in tyrannosaurids. The cervical centra have pleurocoels. Although parallel, the articular faces are inclined, not perpendicular, to the long axis of the centrum so that in lateral aspect the cranial face appears offset well above the caudal face (fig. 6.5). This condition is slightly developed in *Piatnitzkysaurus floresi* but prominent in later taxa; it is lost at the base of the neck. The neural spines, zygapophyses, and diapophyses are well developed in carnosaurs. Accessory lamellae are present, linking the pre- and postzygapophyses and linking both zygapophyses to the transverse process at least in cranial cervicals. These are prominent in the earlier forms (*Allosaurus fragilis, Acrocanthosaurus atokensis, Piatnitzkysaurus floresi, Szechuanosaurus campi*) but reduced in tyrannosaurids. The transverse processes are directed ventrally at about 45° cranially in the series but become laterally directed caudally. Cranial cervical neural spines may be caudally inclined or perpendicular to the axis of the centrum. Those of tyrannosaurids have a zygosphene-zygantrum articulation.

Cervical ribs are well developed and extend four central lengths in *Tyrannosaurus rex* (Osborn 1916) but seem to have been shorter in *Allosaurus fragilis.* They do not vary in length along the column and have a slightly developed tuberculum.

Regional differences in vertebral form have received little attention, but such information is necessary in interpreting incomplete specimens. These differences in cervical form have been reported only for *Allosaurus fragilis* (Gilmore 1920). The epipophyses increase in prominence to cervical 6, then decrease again with the change from wide to narrow neural spines. The diapophyses lengthen, become more robust, and assume a horizontal orientation while the neural spines become axially narrower and transversely wider.

Dorsal vertebrae exhibit prominent neural spines and transverse processes (fig. 6.5). In *Allosaurus fragilis,* the dorsals differ from cervicals by having a shorter and amphiplatyan centrum (save for dorsals 1 to 3), higher parapophyses, and longer and more strongly inclined transverse processes (Gilmore 1920; Madsen 1976*a*). In *Allosaurus fragilis* only, the cranial few dorsals are opisthocoelous, while tyrannosaurids and *Szechuanosaurus campi* have only amphicoelous or amphiplatyan dorsals. The dorsals are sometimes deeply excavated, almost pleurocoelous in tyrannosaurids, but more often cranially than caudally. In those taxa lacking deep excavations of the dorsals caudally, the centra are often markedly narrowed just below the level of the neural canal (pleurocentral excavations), as in *Allosaurus fragilis* and *Piatnitzkysaurus floresi.* The dorsal centra are constricted, most markedly in *Allosaurus fragilis,* which has expanded flangelike central articular faces. Dorsal centra are short, becoming progressively longer caudally in tyrannosaurids. The characteristic (and apparently plesiomorphic) N-shaped pattern of accessory lamellae supporting the transverse process is found in at least the cranial dorsals and often throughout the dorsal column. The diapophyses are borne on prominent transverse processes that are often, although not universally, directed dorsally (those caudal to dorsal 3). The zygapophyses are markedly elevated in some forms (e.g., *Szechuanosaurus campi*) and supplemented by hyposphene-hypantrum articulations in others. These extend caudally from dorsal 5 in *Allosaurus fragilis* and are found on all tyrannosaurid dorsals. Dorsal spines are elongate in *Acrocanthosaurus atokensis.* The neural spines often bear the ossified attachments of the interspinous ligaments (as in *Albertosaurus libratus*) and sometimes show expanded rugose summits (as in *Allosaurus fragilis*). The neural canal is expanded in the cranial dorsals of *Allosaurus fragilis,* probably in conjunction with the innervation of the forelimbs. In tyrannosaurids, which have reduced forelimbs, there is no such expansion.

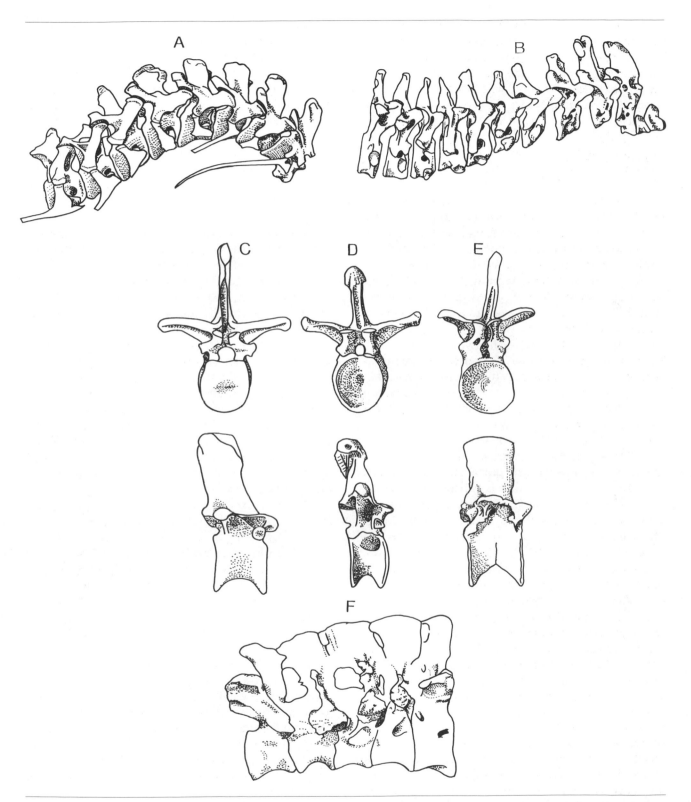

Fig. 6.5. Cervical vertebral series. A, *Allosaurus fragilis*. B, *Tyrannosaurus rex*. Dorsal vertebrae, cranial and lateral views. C, *Piatnitzkysaurus floresi*. D, *Allosaurus fragilis*. E, *Szechuanosaurus campi*. F, Sacrum of *Tyrannosaurus rex*, lateral view. (After Osborn 1906; Gilmore 1920; Dong et al. 1983; Bonaparte 1986c.)

Passing caudally in *Allosaurus fragilis*, the dorsals become larger, lose the ventral keel, develop increasingly flangelike ends of the centra (unique to the genus *Allosaurus* among carnosaurs), lose pleurocoels (at dorsal 5) but develop increasingly deeper pleurocentral excavations, horizontal transverse processes, and higher neural spines (Gilmore 1920; Madsen 1976*a*). The zygapophyses become progressively more horizontal, indicating that was the plane of the predominant motion near the sacrum. The cranial four dorsal centra are longer than high, but the more caudal centra reverse this condition.

Dorsal ribs increase in length to rib 5 or 6, then rapidly decrease. A laterally compressed neck is directed perpendicular to the shaft. The ribs are well developed and broadly curved cranially, implying a barrellike trunk, especially in tyrannosaurids. Caudally, the ribs become more strongly curved, and the trunk becomes shallower and narrower and is nearly triangular just in front of the pelvis.

The sacrum consists of five vertebrae (fig. 6.5). These include one dorsosacral and two caudosacrals together with the primitive two sacrals. The sacrals were pleurocoelous in *Allosaurus fragilis* and tyrannosaurids. The sacral ribs below and transverse processes above formed robust vertical buttresses between the sacrum and the ilia. The sacral neural spines are fused, fusion apparently commencing (phylogenetically) from the summit of the spines and progressing downward.

The number of caudals is variable. The caudals are divisible into a proximal and distal series characterized by the presence or absence of transverse processes. The transition point is at about caudal 15 in tyrannosaurids and about caudal 27 in *Allosaurus fragilis*. The transition point is also marked by some elongation of the centra and a sharp decrease in spine height. Caudal centra are amphicoelous or amphiplatyan and usually lack offset faces (as are present in the cervicals), although such were present in *Bahariasaurus ingens*, *Carcharodontosaurus saharicus*, and weakly in *Allosaurus fragilis* and *Acrocanthosaurus atokensis*. Centra bear chevron facets at both proximal and distal edges, save for the first caudal. Zygapophyses are well developed throughout the proximal and most of the distal series and are moderately to quite elongate (e.g., *Allosaurus fragilis* and tyrannosaurids). Transverse processes are broad in most forms and usually project perpendicular to the direction of the column. Proximal caudals have prominent vertical or distally inclined neural spines, with a markedly angulate proximal margin in some forms (*Allosaurus fragilis*). A small accessory spine proximal to the neural spine is retained in the distal region of the proximal series in *Allosaurus fragilis* and *Ac-*

rocanthosaurus atokensis. Tyrannosaurids exhibit neither angulate nor accessory spines. Chevrons are well developed, often elongate, and in tyrannosaurids an axially extended "skidlike" blade is developed in the distal part of the series (foreshadowed in *Allosaurus fragilis*). Zygapophyses and chevrons are retained throughout the caudal series and diminish in size toward the tail tip, while the neural spines and transverse processes are reduced in size progressively and are absent distally. In cross section, the tail is a vertical ellipse proximally, due to the long spines and chevrons, and becomes progressively more circular distally.

Sequential changes in caudal form have been reported only for *Allosaurus fragilis* (Gilmore 1920; Madsen 1976*a*). Caudal centra become both shorter and lower distally but decrease in height more rapidly than length so the centra become elongate. The angle of the prezygapophysial facets increases from about 40° to nearly 90°, so that flexibility decreases distally, while the chevrons become more curved.

Tapering rodlike gastralia have been reported extending transversely across the belly below the dorsal ribs. Madsen (1976*a*) has suggested that those attributed to *Allosaurus fragilis* actually pertained to ceratosaurs, but this has not gained general acceptance. There were extensive gastralia in *Albertosaurus libratus*, *Tarbosaurus bataar*, and *Tyrannosaurus rex*. In *Tarbosaurus bataar*, there are 12 pairs, and in *Albertosaurus libratus*, there are 19 (Lambe 1917*c*), consisting of lateral and medial segments.

Appendicular Skeleton

The scapula consists of a bandlike blade (fig. 6.6) and an acromial expansion that broadly joins the coracoid, sometimes being fused. The thin, flattened blade usually has parallel or dorsally divergent margins and is curved to fit the rib cage. The ventral expansion consists of the acromial process and a buttress supporting both glenoid and coracoid articular surfaces. This acromial expansion is well developed in tyrannosaurids. The slightly concave glenoid faces caudoventrally and slightly laterally. Both scapula and coracoid contribute equal parts to the glenoid, save in tyrannosaurids where the scapula contributes more of the glenoid.

The coracoid is an oval plate one-third to one-quarter of the scapular length. The coracoid foramen is large and usually near the center of scapulocoracoid contact. In allosaurids, there is a tubercle for M. biceps on the lateral side just caudal to the foramen; in tyrannosaurids, this is absent. The coracoid has a caudally directed angle or tip.

The forelimb is short relative to the hindlimb and

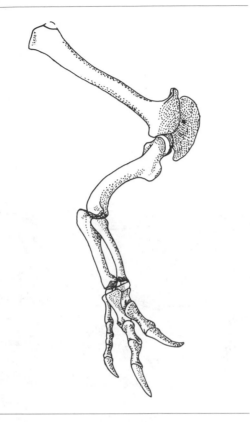

Fig. 6.6. Right forelimb and shoulder girdle of *Allosaurus fragilis.* (After Gilmore 1920.)

to the scapula. The ratio of femoral to humeral length ranges from 1.4 in *Szechuanosaurus campi* to 3.7–4.0 in *Tarbosaurus bataar* and *Tyrannosaurus rex,* averaging 2.7 among carnosaurs. The ratio of scapula-to-humerus length ranges from 1.3 in *Piatnitzkysaurus floresi* and *Szechuanosaurus campi* to 2.9 in *Tarbosaurus bataar* and among carnosaurs averages 1.9. The antebrachium is shorter than the humerus.

The robust humerus (fig. 6.7) has a prominent deltopectoral crest oriented parallel or slightly inclined to the medial margin of the humeral shaft. This crest rarely reaches and never much surpasses midshaft, nor does it extend to the proximal end of the lateral (external) tuberosity. The deltopectoral crest is markedly reduced in tyrannosaurids save for *Tyrannosaurus rex.* Proximally, the humerus is much expanded by the deltopectoral crest and the well-developed medial (internal) tuberosity. This tuberosity is placed about 10 percent of the length of the shaft distal from the head in *Szechuanosaurus campi* and *Tarbosaurus bataar,* about 15 percent in *Chilantaisaurus tashuikouensis, Daspletosaurus torosus,* and *Tyrannosaurus rex,* and intermediate in *Allosaurus fragilis.* The shaft is slightly sigmoidal and expanded distally to support the radial and ulnar condyles. These face slightly cranially, as a result of the distal curvature of the humeral shaft, and slightly medially. Thus, at rest, the elbow would have been slightly flexed and the antebrachium turned inward. The distal

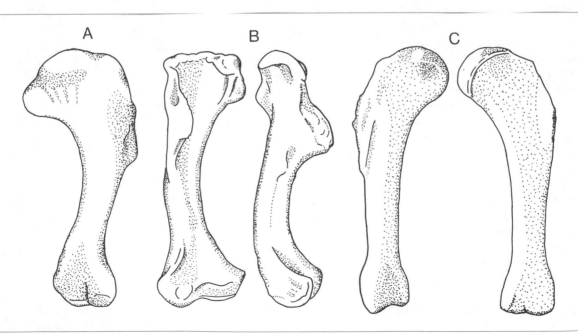

Fig. 6.7. Humerus. A, *Piatnitzkysaurus floresi.* B, *Allosaurus fragilis.* C, *Tarbosaurus bataar.* (After Maleev 1974; Galton 1982c; Bonaparte 1986c.)

humerus often bears a shallow olecranon and cuboid fossa. In *Allosaurus fragilis,* the plane of the distal end of the humerus makes a 50° angle with that of the proximal end.

The humerus of *Chilantaisaurus tashuikouensis* is longer relative to the hindlimb elements than in any other carnosaur, with a femorohumeral ratio of 2.0 (Hu 1964). It is about twice the relative size of that of *Allosaurus fragilis* but is basically an enlarged version of that humerus. It has, however, a more strongly developed deltopectoral crest, extending forward from the medial margin of the shaft.

The cylindrical radius is usually only slightly expanded proximally and distally.

The ulna tapers distally from its proximal expansion that supports both the olecranon and the humeral articular surface. The ulna is half as long as the humerus in tyrannosaurids and somewhat longer in other carnosaurs.

The carnosaur manus is primitively tridactyl (fig. 6.6) but didactyl in tyrannosaurids. Four small plate-like carpals, the proximal radiale and ulnare, and two distal elements (designated 1 and 2) are present in *Allosaurus fragilis* and tyrannosaurids, although some specimens of *Tarbosaurus bataar* have an intermedium.

Metacarpal II is usually longest and most robust and metacarpal I also robust but no more than half as long as metacarpal II (fig. 6.8). Metacarpal III is slightly reduced in allosaurids and markedly so in tyrannosaurids where it does not contact the carpus. The distal articular surfaces of metacarpals I and III diverge away from the axis of the metacarpus, as do their digits.

The phalanges are robust and have smooth ginglymoid articular surfaces (fig. 6.8). As in other theropods, the penultimate phalanges are longer than the others, and the basal phalanx of digit I is the longest. The phalanges of digit III are the smallest. The phalangeal formula is 2-3-4-0-0 for *Allosaurus fragilis* and much reduced (2-3-0-0-0) in tyrannosaurids. The ungual phalanges are more laterally compressed than those of the pes and more curved and bilaterally symmetrical. Prominent grooves are present on each ungual phalanx for the claw sheath, and a flexor tubercle is on the ventral face immediately distal to the interphalangeal articulation. This tubercle is most strongly developed on digit III and least strongly on digit I in *Allosaurus fragilis.* Only two functional digits are present in tyrannosaurids, although a splintlike metacarpal III is retained.

The digits are traditionally identified as I, II, and III, although in the absence of a generally accepted morphological series from pentadactyl ancestors to tri-

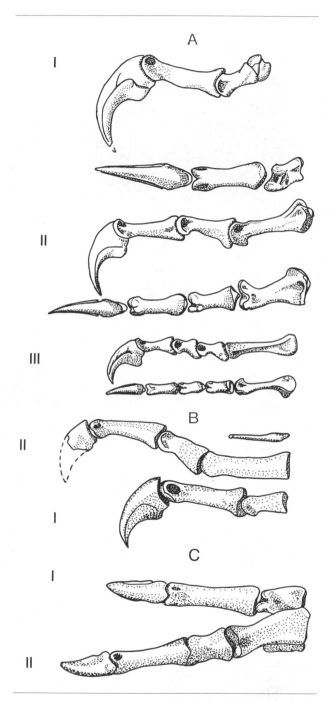

Fig. 6.8. Manus. A, *Allosaurus fragilis,* digits in medial and dorsal view. B, *Albertosaurus libratus,* digits in medial view. C, *Tarbosaurus bataar,* dorsal view. (After Lambe 1917*c*; Gilmore 1920; Maleev 1974; Madsen, 1976*a*.)

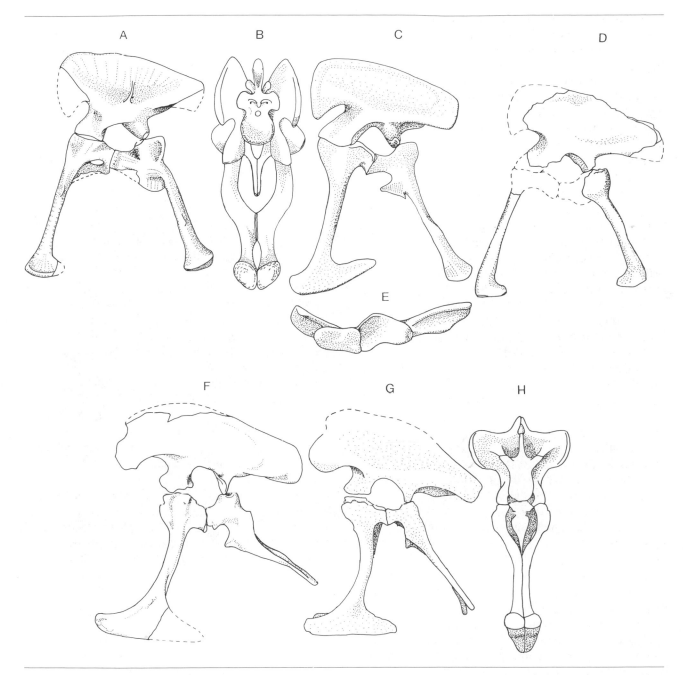

Fig. 6.9. Pelvis. A, *Piatnitzkysaurus floresi*. C, *Allosaurus fragilis*. D, *Szechuanosaurus campi*. F, *Tyrannosaurus rex*. G, *Tarbosaurus bataar*. All in lateral view. B, *Allosaurus fragilis*. H, *Tarbosaurus bataar*. Both in cranial view. E, Ilium of *A. fragilis*, ventral view. (After Osborn 1906; Gregory 1951; Maleev 1974; Dong et al. 1983; Bonaparte 1986c.)

dactyl carnosaurs, this identification must be considered tentative. Reasons for identifying the digits as II, III, and IV are given by Thulborn and Hamley (1982).

The long bladelike ilium (fig. 6.9) has a deep preacetabular and long postacetabular process, both laterally compressed. The preacetabular process approaches closely the sacral neural spines, while the usually narrower postacetabular processes gradually diverge. Projecting internally from the latter is the brevis shelf, which is deflected medially to form a ventrally open trough behind the acetabulum for attachment of M. caudifemoralis brevis. The dorsal margins of the ilia of

Allosaurus fragilis stand well apart, while those of tyrannosaurids incline medially to nearly touch each other over the sacral neural spines. The pubic peduncle is much more massive than the ischial. The dorsal margin of the acetabulum is often laterally expanded into a marked rim or lip, as in *Piatnitzkysaurus floresi* and *Allosaurus fragilis*. The cranial extremity of the preacetabular process in tyrannosaurids supports a broad ventral projection. This projection, often described as hooklike, is supported at its base by a medial shelf contacting the cranial sacral ribs. The depth, and hence overall proportions, of the supracetabular blade of the ilium varies from taxon to taxon, being deeper (relative to the acetabular diameter) in *Allosaurus fragilis* and *Szechuanosaurus campi* than in *Piatnitzkysaurus floresi*. That species, in turn, exhibits a deeper blade than tyrannosaurids. The width of the brevis shelf also varies among taxa, being relatively broader in *Tyrannosaurus rex* than in *Allosaurus fragilis*. In some forms, the ilium has a central, vertical ridge (*Piatnitzkysaurus floresi* and tyrannosaurids). The ilium of *Piatnitzkysaurus floresi* houses internal chambers (Bonaparte 1986c).

The rodlike, cranioventrally projecting pubis (fig. 6.9) is expanded distally. This expansion is drawn out both caudally and cranially into what is called a "foot" or "boot"; this is most prominent in tyrannosaurids (fig. 6.9). The shafts of both pubes fuse medially. Proximally, the pubis broadens into an acetabular portion that houses an obturator foramen in *Piatnitzkysaurus floresi* (Bonaparte 1986c). It is sometimes assumed that the ventrally open notch of this portion of the pubis (as found in *Allosaurus fragilis*) is an incompletely closed obturator foramen, although the possession of dual foramina in this portion of the pubis in more plesiomorphic theropods (ceratosaurs) suggests caution in this interpretation. Primitively, the pubes meet along the midline, but in later forms, this junction persists only distally (fig. 6.9).

The rodlike ischium (fig. 6.9) projects caudoventrally. Distally, it is often expanded, but never to the extent of the pubis, and contacts the ischum of the other side. Proximally, the ischium expands into a flattened acetabular portion often with an cranioventrally open notch between the pubic process and a platelike obturator process, as in *Allosaurus fragilis, Szechuanosaurus campi,* and tyrannosaurids. This obturator process is trapezoidal in earlier forms (e.g., *Allosaurus fragilis* and *Acrocanthosaurus atokensis*) but is triangular in tyrannosaurids. The ischium of tyrannosaurids shows no distal expansion but does have a characteristic semicircular muscle scar (apparently for M. ischiocaudalis)

along its caudal margin behind the obturator process. Like the pubes, the ischia primitively are fused medially along much of their length, but in later forms, this fusion is absent proximally.

The femur (fig. 6.10) is usually the longest element of the hindlimb, but in tyrannosaurids, it is nearly equal in length to the tibia. The femur is robust and distally curved so as to be concave caudally even in large carnosaurs such as *Tyrannosaurus rex*. In cranial view, the shaft is straight. The well-defined head projects medially where it is expanded to give a constricted neck connecting it to the proximal shaft. The head is oval or hemispherical; it is never compressed. The greater trochanter is lower than the femoral head and extends dorsally from the shaft lateral to the head. Between the greater trochanter and the head, the proximal surface of the femur is cylindrical. The lesser trochanter projects cranially from the craniolateral margin of the neck. This trochanter is lower than the greater trochanter in *Allosaurus fragilis* and material referred to *Bahariasaurus ingens* but is as high as the greater trochanter in tyrannosaurids. The neck is slightly more slender than the shaft in some forms (e.g., *Allosaurus fragilis*). The robustness of the shaft increases with the size of the animal. The often prominent fourth trochanter is just proximal to midshaft. Well-developed distal condyles are present, with the lateral condyle inset from the lateral edge of the shaft. This somewhat medial placement leaves a vertical, caudally facing shelf external to the lateral condyle. The medial condyle is confluent with or projects slightly from the medial edge of the shaft. It is only rarely slightly inset (viz. specimens referred to *Bahariasaurus ingens;* Stromer 1934a). The medial condyle is larger than the lateral. Both condyles project caudally. The flexor sulcus is always well marked, while the extensor groove is deep and prominent. M. tibialis cranialis scar on the cranial face of the shaft just above the tibial condyle ranges in development from obscure (*Piatnitzkysaurus floresi*) to well defined (*Allosaurus fragilis* and tyrannosaurids).

The tibia is robust (fig. 6.11), never fused with the fibula, and has a distinct cnemial process (also termed a tuberosity). This process is large to very large and curves laterally toward the fibula. In proximal view, a blunt but distinct process projects cranially from the lateral face of the tibia in the plane of the fibula in *Allosaurus fragilis* and tyrannosaurids (fig. 6.23). The tibial head is slightly offset medially from the shaft. A prominent lateral (or fibular) crest is found proximally on the tibia. The tibial shaft is usually straight in lateral view (slightly curved in *Chilantaisaurus tashuikouensis*)

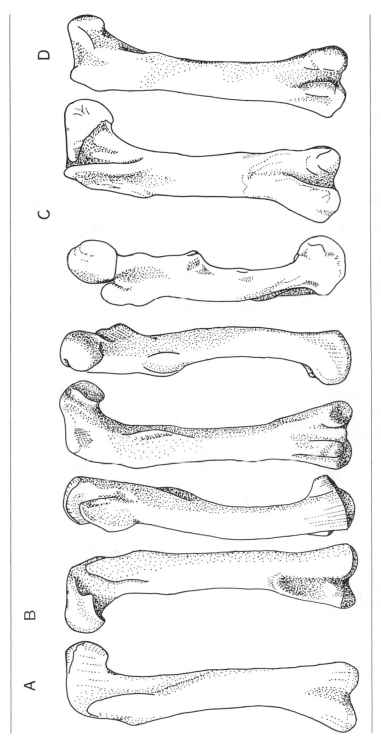

Fig. 6.10. Femora. A, *Piatnitzkysaurus floresi*, cranial view. B, *Allosaurus fragilis*, cranial, lateral, caudal, and medial views. C, *Tyrannosaurus rex*, medial and cranial views. D, *Tarbosaurus bataar*, caudal view. (After Osborn 1916; Maleev 1974; Madsen 1976*a*; Bonaparte 1986*c*.)

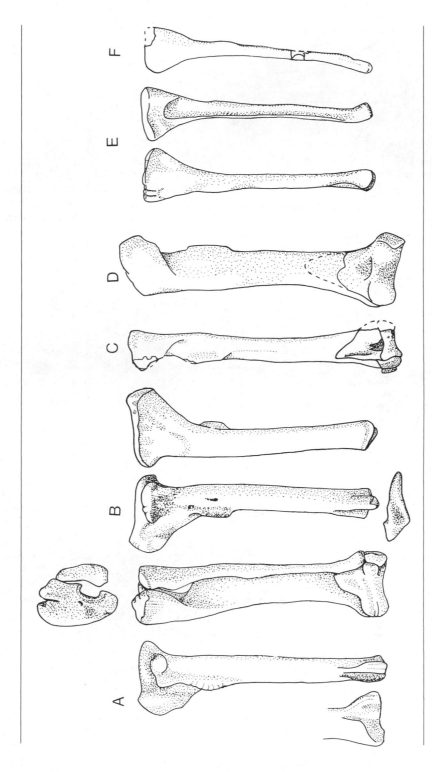

Fig. 6.11. Tibiae. A, *Piatnitzkysaurus floresi*, medial and cranial views. B, *Allosaurus fragilis*, proximal and cranial views with fibula and tarsals, medial, lateral, and distal views. C, *Alectrosaurus olseni*, cranial view with tarsals. D, *Tarbosaurus bataar*, cranial view with fibula and tarsals.

Fibulae. E, *Allosaurus fragilis*, lateral and medial views. F, ?*Nanotyrannus lancensis*, lateral view. (After Gilmore 1933*a*; Maleev 1974; Madsen 1976; Molnar 1980; Bonaparte 1986*c*.)

and often broader mediolaterally than craniocaudally. It is markedly widened distally to form the malleoli. Distally, the tibia is craniocaudally compressed.

The fibula is slender (fig. 6.11), expanded proximally into a medially concave triangular plate and often slightly expanded distally. The proximal expansion is usually restricted to the dorsal fifth of the shaft. The fibular shaft is straight (save in *Acrocanthosaurus atokensis* and *Nanotyrannus lancensis*). There is a distinct cranial tubercle in tyrannosaurids, *Acrocanthosaurus atokensis*, and material referred to *Bahariasaurus ingens* and *Carcharodontosaurus saharicus*. The distal articular face, for the calcaneum, is round.

Carnosaurs have two rows of tarsals—the astragalus and calcaneum proximally and two or three distal tarsals. The astragalus is hemicylindrical in form, with a flat ascending process (Welles and Long 1974). This process is 20 percent as long as the tibia in allosaurids but as much as 33 percent as long as the tibia in tyrannosaurids. The calcaneum is a small, disklike element. At least in *Allosaurus fragilis* and *Szechuanosaurus campi*, it bears a medial process that occupies a corresponding recess in the lateral face of the astragalus (Welles and Long 1974), making this joint immobile. Two subrhomboid platelike distal tarsals (III and IV) have been described for *Allosaurus fragilis* (Madsen 1976*a*), while *Albertosaurus libratus* retained three distal tarsals, II, III, and IV (Lambe 1917*c*).

The birdlike foot is functionally tridactyl (fig. 6.12). The metatarsus is robust, with a short, reversed digit I. Metatarsals II and IV are often subequal in length, but metatarsal II is slightly longer. Metatarsals I and V are much shorter. Metatarsal I is reduced proximally but with a well-developed distal articular surface. It contacts metatarsal II caudomedially at midshaft and bears a reduced digit. Metatarsal V is a small, distally flat bone that is closely appressed to metatarsal IV caudolaterally near its proximal end. Metatarsal III is attenuated proximally and overlapped by metatarsals II and IV in tyrannosaurids (fig. 6.13). In *Chilantaisaurus tashuikouensis*, metatarsal III is much compressed caudally at its proximal end, suggesting that this might represent an intermediate stage in the evolution of the proximally attenuated metatarsal III of tyrannosaurids. The tyrannosaurid metatarsus is longer, more than half as long as the tibia, than the allosaurid metatarsus, which is half as long as the tibia. The shaft of metatarsal III in *Allosaurus fragilis* shows a distinct medial shoulder distally, and that of tyrannosaurids exhibits both medial and lateral shoulders just below the attenuated proximal portion of the shaft. Metatarsals II and IV

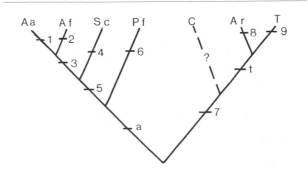

Fig. 6.13. Carnosaur relationships. The Baharija taxa are too incomplete for inclusion here. Taxa: Aa, *Acrocanthosaurus atokensis;* Af, *Allosaurus fragilis;* Sc, *Szechuanosaurus campi;* Pf, *Piatnitzkysaurus floresi;* Ar, *Alioramus remotus* (and possibly *Alectrosaurus olseni*); T, Tyrannosauridae; *Albertosaurus, Daspletosaurus, Tarbosaurus* and *Tyrannosaurus* not here differentiated. C represents the possible position of *Chilantaisaurus.* The features are discussed in the text.

have distal articular surfaces that usually are elongate and condylar, while metatarsal III has a nearly rectangular ginglymoid distal articular surface. Metatarsal II can be distinguished from metatarsal IV, in that the distal articular surface of metatarsal II is nearly square in outline and that of metatarsal IV is narrower and sometimes triangular (fig. 6.12).

The phalanges are robust, and all digits have claws. The phalangeal formula is 2-3-4-5-0. Digit III is longest. Digit I is poorly developed, not long enough to touch the ground, and is directed slightly caudally. The phalanges are wider and shorter than those of the manus. The unguals are usually stouter, less symmetrical, and less laterally compressed than those of manus. Each bears well-defined grooves for the claw sheath. The prominence of the proximal flexor tubercles varies from taxon to taxon.

EVOLUTION

Forty percent of dinosaurs are known from only the final two stages (Campanian and Maastrichtian) of the Mesozoic. Even though there are twenty-one species of carnosaurs from a period of about 150 million years (table 6.1), eleven derive from the final 10 million years. Thus, in addition to the Late Cretaceous tyrannosaurids, our knowledge rests essentially on *Allosaurus fragilis, Piatnitzkysaurus floresi,* and *Szechuanosaurus campi.*

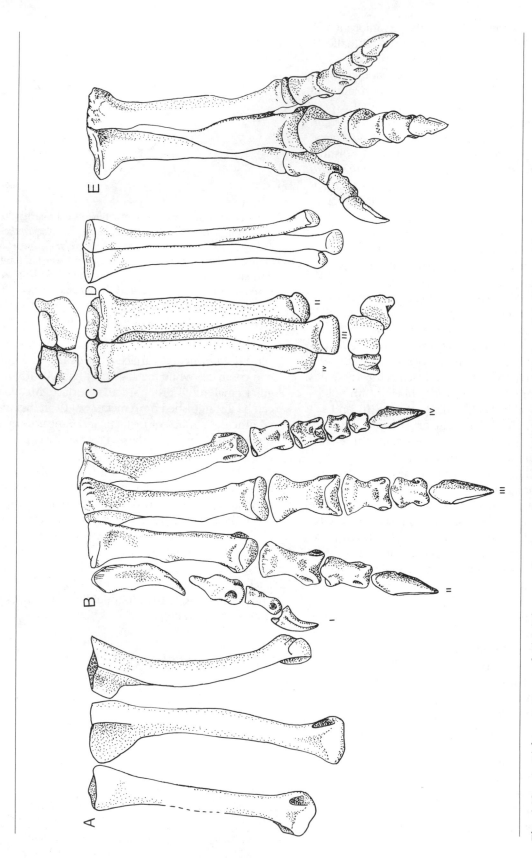

Fig. 6.12. A, Right metatarsus of *Piatnitzkysaurus floresi*. B, Right pes of *Allosaurus fragilis*. C, Left metatarsus of *Albertosaurus libratus*, proximal, cranial, and distal views. D, Left metatarsus of *Alectrosaurus olseni*, cranial view. E, Right pes of *Tarbosaurus bataar*, cranial view. (After Lambe 1917c; Maleev 1974; Madsen 1976a; Perle 1977; Bonaparte 1986c.)

TABLE 6.1 Carnosauria

	Occurrence	Age	Material

Theropoda Marsh, 1881b
  Carnosauria Huene, 1920
    Allosauridae Marsh, 1879a
      Acrocanthosaurus Stovall et Langston, 1950			

A. atokensis Stovall et Langston, 1950	Antlers Formation, Oklahoma, Texas, United States	late Aptian–early Albian	2 partial skulls with associated postcrania
Allosaurus Marsh, 1877b (= *Antrodemus* Leidy, 1870, *Creosaurus* Marsh, 1878a, *Epanterias* Cope, 1878b, *Labrosaurus* Marsh, 1879a, *Saurophagus* Ray, 1941)			
A. fragilis Marsh, 1877b (including *Poicilopleuron valens* Leidy, 1870, *Laelaps trihedrodon* Cope, 1877a, *Creosaurus atrox* Marsh, 1878a, *Epanterias amplexus* Cope 1878b, *Hypsirhophus discurus* Cope 1878a *partim*, *A. lucaris* Marsh, 1878a, *A. ferox* Marsh, 1896, *?Saurophagus* Marsh, 1896, *?Saurophagus maximus* Ray, 1941, *?Labrosaurus ferox* Marsh, 1884a)	Morrison Formation, Colorado, Montana, New Mexico, Oklahoma, South Dakota, Utah, Wyoming, United States	Kimmeridgian–Tithonian	At least 3 complete skulls, many partial skull and skull elements, many partial and complete skeletons representing at least 60 individuals
?A. tendagurensis Janensch, 1925	Tendaguru Beds, Mtwara, Tanzania	Kimmeridgian	Tibia
Chilantaisaurus Hu, 1964			
C. maortuensis Hu, 1964	Unnamed Unit, Nei Mongol Zizhiqu, People's Republic of China	Albian	Partial skull and postcranial skeleton
C. tashuikouensis Hu, 1964	Unnamed Unit, Nei Mongol Zizhiqu, People's Republic of China	Aptian–Albian	Fragmentary postcranial skeleton
?C. sibiricus (Riabinin, 1914) (= *Allosaurus sibiricus* Riabinin, 1914)	Turginskaya Svita, Buriat-mongolskaya Avtonomnaya S.S.R., Union of Soviet Socialist Republics	Berriasian–Hauterivian	Metatarsal
Piatnitzkysaurus Bonaparte, 1979b			
P. floresi Bonaparte, 1979b	Cañadon Asfalto Formation, Chubut, Argentina	Callovian	2 fragmentary skulls with associated postcrania
Szechuanosaurus Young, 1942b			
S. campi Young, 1942b (including *S. yandonensis* Dong, Chang, Li, et Zhou, 1978, *Chienkosaurus ceratosauroides* Young, 1942b *partim*)	Shangshaximiao Formation, ?Kyangyan Series, Sichuan, Keilozo Formation, Xinjiang Ugyur Zizhiqu, People's Republic of China	Oxfordian–Tithonian	Incomplete skeleton, skeletons, isolated teeth
Tyrannosauridae Osborn, 1905 *Albertosaurus* Osborn, 1905 (= *Gorgosaurus* Lambe, 1914c, *?Deinodon* Leidy, 1856)			
A. libratus (Lambe, 1914c) (= *Gorgosaurus libratus* Lambe, 1914c, including *Laelaps incrassatus* Cope 1876a, *?Deinodon horridus* Leidy, 1856, *?Laelaps falculus* Cope, 1876a, *?Laelaps hazenianus* Cope, 1876b, *?Dryptosaurus kenabekides* Hay, 1899, *?Gorgosaurus sternbergi* Matthew et Brown, 1923)	Judith River Formation, Alberta, Canada; Judith River Formation, Montana Fruitland Formation, Kirtland Shale, New Mexico, United States	late Campanian–Maastrichtian	9 partial skulls, 3 partial skeletons, 3 fragmentary skulls with associated postcrania, 6 skulls, and postcranial skeletons
A. sarcophagus Osborn, 1905 (including *A. arctunguis* Parks, 1928)	Judith River Formation, Montana, ?Lance Formation, Wyoming, United States; Horseshoe Canyon Formation, Alberta, Canada	late Campanian–early-Maastrichtian	2 fragmentary skulls with associated postcranial skeletons, fragmentary postcranial material
A. megagracilis Paul, 1988 •	Hell Creek Formation, Montana, United States	late Maastricthian	Fragmentary skull with associated postcrania

Table 6.1, continued

Alectrosaurus Gilmore, 1933a			
A. olseni Gilmore, 1933a	Iren Dabasu Formation, Nei Mongol Zizhqu, People's Republic of China; Baynshirenskaya Svita, Omnogov, Mongolian People's Republic	Late Cretaceous (?Cenomanian, ?Maastrichtian)	Skull and associated partial postcrania, fragmentary postcranial skeleton
Alioramus Kurzanov, 1976b			
A. remotus Kurzanov, 1976b	Beds of Nogon Tsav, Bayankhongor, Mongolian People's Republic	early Maastrichtian	Fragmentary skull with associated postcrania
Chingkankousaurus Young, 1958a			
C. fragilis Young, 1958a	Wangshi Series, Shandong, People's Republic of China	Campanian–early Maastrichtian	Scapula
Daspletosaurus Russell, 1970a			
D. torosus Russell, 1970a	Judith River Formation, Alberta, Canada	late Campanian	3 partial skulls, complete skull with associated partial postcrania, 3 partial skeletons
Nanotyrannus Bakker, Currie, et Williams, 1988			
N. lancensis (Gilmore, 1946c) (= *Gorgosaurus lancensis* Gilmore, 1946c)	Hell Creek Formation, Montana, United States	late Maastrichtian	Complete skull, skull with associated fragmentary postcranial remains
Tarbosaurus Maleev, 1955c			
T. bataar (Maleev, 1955a) (= *Tyrannosaurus bataar* Maleev, 1955b, including *Tarbosaurus efremovi* Maleev, 1955a, *Gorgosaurus lancinator* Maleev, 1955a, ?*Albertosaurus periculosus* Riabinin, 1930a, ?*Tyrannosaurus turpanensis* Zhai, Zheng, et Tong, 1978)	Nemegtskaya Svita, White Beds of Khermeen Tsav, Bayankhongor, Nemegt Formation, Omnogov, Mongolian People's Republic; Subashi Formation, Xinjiang Uygur Zizhiqu, ?unnamed unit, Heilongjiang, People's Republic of China	?late Campanian or early Maastrichtian	At least 5 skulls and associated postcranial skeletons representing at least 30 individuals
Tyrannosaurus Osborn, 1905 (= *Dynamosaurus* Osborn, 1905, ?*Manospondylus* Cope, 1892b)			
T. rex Osborn, 1905 (= *Dynamosaurus imperiosus* Osborn, 1905, including ?*Manospondylus gigas* Cope, 1892b)	Scollard Formation, Willow Creek Formation Alberta, Frenchman Formation, Saskatchewan, Canada, Hell Creek South Formation, Livingston Formation, Montana, Hell Creek Formation, South Dakota, Lance Formation, Wyoming, Laramie Formation, Colorado, Javelina Formation, Texas, McRae Formation, New Mexico, United States	late Maastrichtian	3 complete skulls, 5 partial skulls, skull with associated partial postcrania, 2 partial skulls with complete postcrania
T. luanchuanensis Dong, 1979	Quiba Formation, Henan, People's Republic of China	Campanian	Teeth and associated postcrania
Unnamed tyrannosaurid (= "*Ornithomimus*" *grandis* Marsh, 1890b)	Eagle Sandstone, Montana, United States	early Campanian	Pes
Unnamed tyrannosaurid (= "*Gorgosaurus*" *novojilovi* Maleev, 1955b)	Nemegt Formation, Omnogov, Mongolian People's Republic	?late Campanian or early Maastrichtian	Skull with associated postcrania
Carnosauria incertae sedis			
Bahariasaurus Stromer, 1934a			
B. ingens Stromer, 1934a	Baharija Formation, Marsa Matruh, Egypt; Farak Formation, Tahoua, "Continental intercalaire", unnamed unit, Agadez, Niger	?late Albian or ?early Cenomanian	Isolated postcranial elements
Carcharodontosaurus Stromer, 1931			
C. saharicus (Deperet et Savornin, 1927) (= *Megalosaurus saharicus* Deperet et Savornin, 1927, *Megalosaurus africanus* Huene, 1956)	Baharija Formation, Marsa Matruh,, Egypt; Tegana Formation, Ksar-es-Souk, Morocco; Chenini Formation,	late Aptian–early Cenomanian	Fragmentary skull with associated postcrania, isolated teeth, isolated postcrania

Table 6.1, continued

"Continental intercalaire", Medinine, Tunisia; "Continental intercalaire", Adrar, Tamenghest, Wargla, Algeria; "Continental intercalaire", Gharyan, Libya; "Continental "intercalaire", Agadez, Niger

Possible Carnosauria

Aublysodon Leidy, 1868b
   A. mirandus Leidy, 1868b (including *A. amplus* Marsh, 1892d, *A. cristatus* Marsh, 1892d)
Judith River Formation, Two Medicine Formation, Hell Creek Formation, Montana, Denver Formation, Colorado, Lance Formation, Wyoming, Kirtland Shale, New Mexico, United States
late Campanian-Maastrichtian
Isolated teeth, partial skull

   A. molnari Paul, 1988 •
Hell Creek Formation, Montana, United States
late Maastricthian
Fragmentary skull

Compsosuchus Huene, 1932
   C. solus Huene, 1932
Lameta Formation, Bara Simla Hill, Madhya Pradesh, India
middle-late Maastrichtian
Vertebra

Eustreptospondylus Walker, 1964
   E. oxoniensis Walker, 1964
Chipping Norton Formation, Oxford Clay, Oxfordshire, Middle Oxford Clay, Buckinghamshire, England
Callovian
Disarticulated skull and skeleton, referred limb elements

Gasosaurus Dong et Tang, 1985
   G. constructus Dong et Tang, 1985
Xiashaximiao Formation, Sichuan, People's Republic of China
Bathonian-Callovian
Postcranial skeleton

Iliosuchus Huene, 1932
   I. incognitus Huene, 1932
Stonesfield Slate, Oxfordshire, England
middle Bathonian
2 ilia

Kaijiangosaurus He, 1984
   K. lini He, 1984
Xiashaximiao Formation, Sichuan, People's Republic of China
Bathonian-Callovian
Vertebrae

Kelmayisaurus Dong, 1973
   K. petrolicus Dong, 1973
Lianmugin Formation, Xinjiang Uygur Zizhiqu, People's Republic of China
?Valanginian-Albian
Maxilla and dentary

Labocania Molnar, 1974
   L. anomala Molnar, 1974
"La Bocana Roja" formation, Baja California, Mexico
?Campanian
Fragmentary skull and very fragmentary postcrania

Unnamed possible carnosaur (= *"Laelaps" gallicus* Cope, 1867)
unnamed unit, Normandy, France
unknown age (?Jurassic)
Vertebrae, pubis, hindlimb elements

Magnosaurus Huene, 1932
   M. nethercombensis (Huene, 1926b) (= *Megalosaurus nethercombensis* Huene, 1926b)
Inferior Oolite, Dorset, England
Aalenian-Bajocian
Dentaries, vertebrae, pubis, femora, tibiae

Megalosaurus Buckland, 1824
   M. bucklandii Ritgen, 1926
Chipping Norton Formation, Stonesfield Slate, Cornbrash Formation, Oxfordshire ?Inferior Oolite, Northamptonshire, Inferior Oolite, Chipping Norton Formation, Forest Marble, Gloucestershire, Great Oolite, Forest Marble, Wiltshire, Corallian Oolite Formation, North Yorkshire, England; unnamed unit, Indre, France
Bathonian, ?Oxfordian
Dentary, referred material including teeth and postcranial remains

Table 6.1, continued

M. hesperis Waldman, 1974	Upper Inferior Oolite, Dorset, England	Aalenian-Bajocian	Skull elements
?M. cambrensis (= *Zanclodon cambrensis* Newton, 1899)	Rhaetic Beds, Mid-Glamorganshire, Wales	Rhaetian	Mold of dentary
Piveteausaurus Taquet et Welles, 1977			
P. divesensis (Walker, 1964) (= *Eustreptospondylus divesensis* Walker, 1964)	Marnes de Dives, Calvados, France	late Callovian	Braincase
Spinosaurus Stromer, 1915			
S. aegyptiacus Stromer, 1915	Baharija Formation, Marsa Matruh, Egypt	?early Cenomanian	Jaw fragments, vertebrae, hindlimb elements
Stokesosaurus Madsen, 1974			
S. clevelandi Madsen, 1974	Morrison Formation, Utah, United States	Kimmeridgian-Tithonian	Ilium
Torvosaurus Galton et Jensen, 1979a			
T. tanneri Galton et Jensen, 1979a	Morrison Formation, Colorado, United States	Kimmeridgian-Tithonian	Forelimb elements
Yangchuanosaurus Dong, Chang, Li, et Zhou, 1978			
Y. shangyouensis Dong, Chang, Li, et Zhou, 1978	Shangshaximiao Formation, Sichuan, People's Republic of China	Late Jurassic	Skeleton lacking fore limbs, pes, and distal caudal series
Y. magnus Dong, Zhou, et Zhang, 1983	Shangshaximiao Formation, Sichuan, People's Republic of China	Late Jurassic	Skull, vertebrae, pelvis, femur

Nomina dubia	Material
Aublysodon lateralis Cope, 1876a	Isolated teeth
Megalosaurus hungaricus Nopcsa, 1901	Tooth
Megalosaurus lydekkeri Huene, 1926	Tooth
Megalosaurus meriani Greppin, 1870b	Tooth
Megalosaurus pannoniensis Seeley, 1881	Isolated teeth
Megalosaurus superbus Sauvage, 1882 *partim*	Isolated teeth
Nuthetes destructor Owen, 1854 *partim*	Fragmentary dentary, postcranium
Sarcosaurus andrewsi Huene, 1932	Fragmentary pelvis, femur, vertebrae
Scrotum humanum Brookes, 1763	Fragmentary femur
Streptospondylus cuvieri Owen, 1842a	Vertebrae, limb elements
Tyrannosaurus lanpingensis Yeh, 1975	Tooth

In determining the relationships among the carnosaurs, the Ceratosauria, Herrerasauridae, and *Staurikosaurus, Lagosuchus,* Ornithosuchidae, Poposauridae, and Rauisuchidae were used in outgroup comparisons as successively more distant taxa. In all cases, assignments have been based on character states present in carnosaur taxa, regardless of potential but undemonstrated parallelisms in yet unknown portions of the skeleton. In this way, even a low number of character states in fragmentary material was used to give a tentative taxonomic assignment and evolutionary placement.

Two hypotheses of carnosaur origin have gained substantial support. Huene (1914c) proposed that carnosaurs (from which he excluded both *Ceratosaurus* and tyrannosaurids, considering these coelurosaurs) together with sauropods and prosauropods formed a natural group, called pachypodosaurs. Carnosaurs specifically were derived from prosauropods. This hypothesis has recently received support from Galton and

Jensen (1979a) based on examination of *Torvosaurus tanneri,* here considered too incompletely known to be considered a carnosaur.

The second hypothesis, that of Colbert (1964a), unites coelurosaurs (*sensu* Huene 1914c) and carnosaurs into a monophyletic group, the Theropoda (Marsh 1881b). The carnosaurs were derived along with the later coelurosaurs from the Late Triassic Podokesauridae. This family is not here accepted, many of its members now being placed in the Ceratosauria (see Rowe and Gauthier, this vol.).

A third hypothesis was proposed by Walker (1964) and recently resurrected by Chatterjee (1985). This is that carnosaurs are direct descendants of certain advanced thecodontians (ornithosuchids in Walker's scheme, poposaurids in Chatterjee's). Walker's proposal did not receive widespread support, because ornithosuchids have a crocodilian tarsus, quite unlike the theropod tarsus. Bonaparte (1984) showed that the structure and function of the entire hindlimb was quite different in theropods and rauisuchians (in which he included poposaurids) and that it was unlikely that either taxon was derived from the other. The similarities that exist among carnosaurs, ornithosuchians, poposaurids, and rauisuchians are all linked to prey handling in archosaurians that prey on large animals and hence are likely to be parallelisms.

Recently, a fourth hypothesis has been based on what is interpreted as a quadrupedal carnosaur, *Xuanhanosaurus qilixiaensis,* from the Middle Jurassic of China (Dong 1984*b*). This is the proposal that carnosaurs are derived from unspecified quadrupeds. This scheme is inconsistent with acceptance of coelurosaurs, ornithosuchids, or poposaurids as carnosaur ancestors, although perhaps consistent with prosauropod ancestry. Unfortunately *X. qilixiaensis* is based on an incomplete specimen, and thus its relationship to carnosaurs as here defined remains problematic.

Gauthier (1986) recently presented a revised classification of saurischians, integrating them with their presumed descendants, birds. This classification divided theropods (including birds) into four groups: *Procompsognathus, Liliensternus,* Ceratosauria, and Tetanurac. The Tetanurae were further subdivided into sister groups, the Carnosauria and Coelurosauria (including birds).

All of the tetanuran diagnostic characters are shown by the carnosaurs with the exception of a spine table on the axial neural spine. A spine table is a distal dilation of the spine to form a horizontal plate at its tip. This is absent in *Allosaurus fragilis* (Madsen 1976*a*) and *Piatnitzkysaurus floresi* (Bonaparte 1986*c*), although the neural spine is broad, and we have not seen an axial spine table in tyrannosaurids. If by "spine table" Gauthier refers to the broadened axial neural spine, then carnosaurs have all of the tetanuran diagnostic characters.

Gauthier gave 11 diagnostic features of carnosaurs. Of these, the dorsoventrally long orbit, reduced mandibular fenestra, craniodorsally expanded ilium, opisthocoelous cervicals and cranial dorsals, and robust postcranial skeleton are probably related to large size. Other features, such as the keyhole-shaped orbit, are not universal among carnosaurs, being absent in *Albertosaurus libratus, Alectrosaurus olseni, Alioramus remotus,* and *Daspletosaurus torosus.* Nor are narrow and very short frontals and parietals universal or even widespread among carnosaurs. The frontals of *Allosaurus fragilis* (Madsen 1976*a*), *Daspletosaurus torosus* (D. A. Russell 1970*a*), and *Tyrannosaurus rex* (Osborn 1912*b*) are as wide transversely as long, so they cannot be described as narrow. Those of *Acrocanthosaurus atokensis* (Stovall and Langston 1950) and *Albertosaurus libratus* (Russell 1970*a*) are substantially longer than wide and hence cannot be termed very short. Carnosaur parietals are both short and narrow compared with their frontals. The remaining features are found in carnosaurs, although some—lateral surangular shelf below the jaw articulation and reduced manual digits II and III—are found among other theropods as well (e.g., *Dilophosaurus wetherilli* and *Ingenia yanshini,* respectively).

The features used to diagnose the Carnosauria are noted in Diagnosis, above. In carnosaurs, the lateral face of the lacrimal bears one or two openings, enlarged over usual size of neurovascular foramen found here primitively (fig. 6.14). This opening communi-

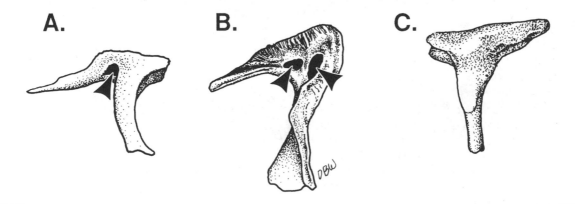

Fig. 6.14. Left lateral view of the lacrimal of the carnosaurs *Albertosaurus libratus* (A) and *Allosaurus fragilis* (B) and the dromaeosaurid *Deinonychus antirrhopus* (C). The arrows indicate prominent apertures found in carnosaurs. (After Gilmore 1920; Ostrom 1969*b*; D. A. Russell 1970*a*.)

cates with an internal chamber and is not associated with the nasolacrimal duct. This opening is greatly enlarged in *Allosaurus fragilis* but less so in the tyrannosaurids. An enlarged lacrimal aperture is also found in *Ceratosaurus nasicornis*, but in view of the absence of any such enlarged openings in primitive ceratosaurs such as *Coelophysis bauri* and *Syntarsus rhodesiensis*, this is here considered to be an independent acquisition. Other theropods have either no foramen or only a small neurovascular foramen on the lacrimal.

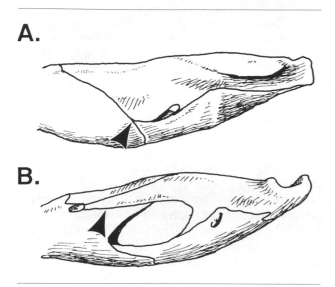

Fig. 6.15. The jaw joint, showing the deep rostral portion of the surangular (arrow) in *Allosaurus fragilis* (A) contrasted with the shallow rostral portion (arrow) in *Yangchuanosaurus shangyouensis* (B). (After Madsen 1976a; Dong et al. 1983.)

The portion of the surangular rostral to the mandibular fenestra is deep in carnosaurs. It often exceeds in depth the tooth-bearing portion of the dentary (fig. 6.15). This condition reflects the increased depth of the caudal region of the mandible. In tyrannosaurids, the depth has so increased that the ventral margin of the surangular has lost the embayment that forms the dorsal margin of the mandibular fenestra in earlier taxa (such as *Allosaurus fragilis*). In other theropods, the cranial portion of the surangular is markedly shallower than the caudal portion, as in *Yangchuanosaurus shangyouensis*.

The axial neural spine includes at most only a low vertical component projecting dorsally from the arch among carnosaurs (fig. 6.16). This differs from the prominent vertical spine in other theropods (e.g., *Gallimimus bullatus*), sometimes with a spine table (e.g., *Baryonyx walkeri*).

In addition, the chevrons of carnosaurs have a pair of small craniodorsally projecting processes (Stovall and Langston 1950), extending from the cranial margin just below the proximal articular facet (fig. 6.17). In tyrannosaurids, similar processes sometimes occur on the distal margin. Other theropods either lack such prong-like processes (e.g., *Gallimimus bullatus*) or have more obtuse "shoulders" in this position (e.g., *Deinonychus antirrhopus*).

The glenoid (or "acromial") portion of the scapula in carnosaurs is expanded to more than twice the minimal width of the blade. The margin of this portion curves away from the cranial margin of the blade forming an abrupt expansion (fig. 6.18), making an angle that ranges from about 50° to almost 90°. In *Piatnitzkysaurus floresi*, which exhibits the gentlest expan-

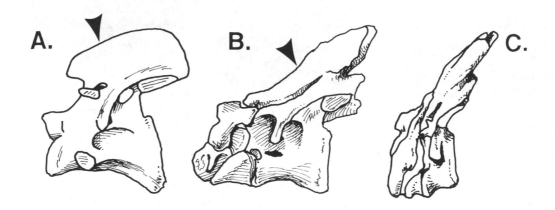

Fig. 6.16. Development of the axial neural spine (arrow) in the ceratosaur *Ceratosaurus nasicornis* (A) and two car- nosaurs *Allosaurus fragilis* (B) and *Tyrannosaurus rex* (C). (After Osborn 1906; Gilmore 1920.)

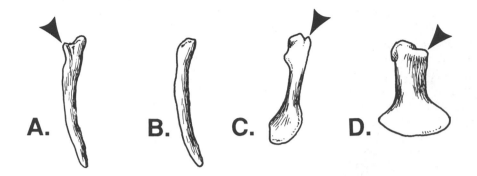

Fig. 6.17. The cranial process of the chevrons in the carnosaur *Allosaurus fragilis* (A; left lateral view), the ornithomimid *Gallimimus bullatus* (B; left lateral view), the carnosaur *Tyrannosaurus rex* (C; right lateral view), and the dromaeosaurid *Deinonychus antirrhopus* (D; right lateral view). Arrows point to the cranial process (A, C) and shelf (B, D). (After Gilmore 1920; Ostrom 1969*a*; Osmólska et al. 1972.)

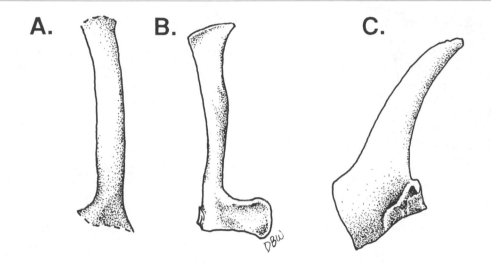

Fig. 6.18. Development of the glenoid region of the scapula in the dromaeosaurid *Deinonychus antirrhopus*, with a low glenoid region (A; right lateral view); the carnosaur *Tyrannosaurus rex*, with a very prominent glenoid region (B; right lateral view); and the ornithomimid *Ornithomimus edmontonensis*, with a low glenoid region (C; left lateral view). (After Osborn 1906; Sternberg 1933*a*; Ostrom 1969*b*.)

sion of the glenoid region, there is an angle of about 50°. Other theropods exhibit a gentler expansion of the glenoid region (e.g., *Ceratosaurus nasicornis*). Certain ornithomimids (*Ornithomimus edmontonensis*) exhibit a similar but less abrupt dilation of the glenoid region.

The brevis shelf of the ilium is, at least primitively, narrow in carnosaurs (fig. 6.19). Even when broadened, as in *Tyrannosaurus rex*, it is still proportionately narrower than among other theropods. In other theropods, this shelf is broad (e.g., *Ceratosaurus nasicornis*) and is sometimes the broadest portion of the ilium (e.g., *Syntarsus rhodesiensis*).

In carnosaurs, the neck of the femur inclines upward, so that the head rises above the greater trochanter and proximal end of the shaft (fig. 6.20). This feature is most marked in tyrannosaurids. In other theropods, the neck of the femur is either approximately horizontal (*Erectopus sauvagei*) or declined (*Dilophosaurus wetherilli*). The lesser trochanter of carnosaurs projects craniodorsally from the proximal portion of the femoral shaft (fig. 6.21). The dorsal end of the trochanter is angulate in lateral view and separated from the shaft by a V-shaped notch. The trochanter as a whole has the form of a parallelogram attached to the

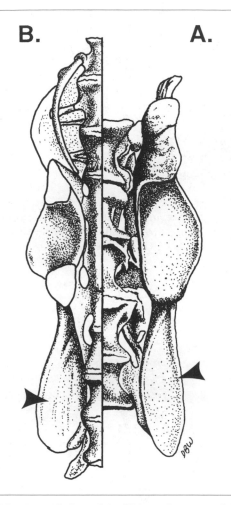

Fig. 6.19. Ventral view of the ilium and sacrum of the carnosaur *Allosaurus fragilis* (A) and the ceratosaur *Ceratosaurus nasicornis* (B). Note the difference in development of the brevis shelf (arrows). (After Gilmore 1920.)

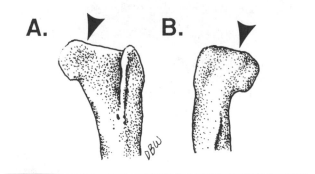

Fig. 6.20. Elevated femoral head in a carnosaur *Tarbosaurus bataar* (A; left femur) and declined head in the ceratosaur *Dilophosaurus wetherilli* (B; right femur). (After Maleev 1974; Welles 1984.)

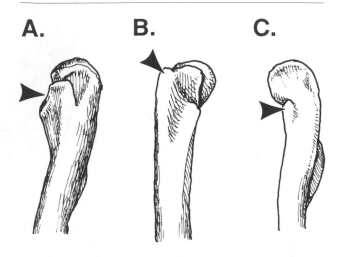

Fig. 6.21. Form of the lesser trochanter (arrow) in the carnosaur *Allosaurus fragilis* (A) compared with the dromaeosaurid *Deinonychus antirrhopus* (B) and the ceratosaur *Dilophosaurus wetherilli* (C). (After Gilmore 1920; Ostrom 1969*b*; Welles 1984.)

shaft by one edge; this has been termed a "winglike" form (Gauthier 1986). A somewhat similar form is found in *Coelurus fragilis* and *Microvenator celer*, where, however, the free margin of the trochanter is rounded and the notch U-shaped. Other theropods show a variety of different forms. Last, in distal aspect, the femur of carnosaurs has a V-shaped extensor groove comparable in depth to the flexor sulcus (fig. 6.22). Only *Gallimimus bullatus* among other theropods has a comparably deep groove. Such a groove is not known in the ancestral ornithomimid *Elaphrosaurus bambergi* and so is here considered an independent acquisition. The remainder of theropods have either a shallow extensor groove (e.g., *Erectopus sauvagei*) or none at all (e.g., *Liliensternus liliensterni*).

The first definite carnosaur is *Piatnitzkysaurus floresi* from the Middle Jurassic of Argentina. As carnosaurs are here understood, there is nothing known about *P. floresi* that would bar it from being the common ancestor of the later carnosaurs. Above *P. floresi*, carnosaurs comprise at least two families (fig. 6.13), the Allosauridae from the Jurassic and Early Cretaceous and the Tyrannosauridae from the Late Cretaceous. Successive evolutionary stages seem to have been represented by the genera *Piatnitzkysaurus*, *Allosaurus* (and *Szechuanosaurus*), *Acrocanthosaurus*, *Bahariasaurus* (and *Carcharodontosaurus*), and by the Tyrannosauridae.

The Allosauridae is characterized by the following features: a complex maxillary sinus, laterally, dor-

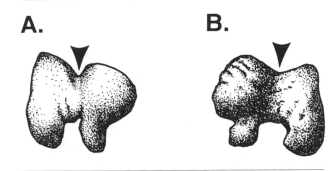

A.   **B.**

Fig. 6.22. The extensor groove of the femur in the carnosaur *Allosaurus fragilis* (A; right femur) and the noncarnosaurian theropod *Erectopus sauvagei* (B; left femur). (After Sauvage 1882; Gilmore 1920.)

sally, and medially perforate (Madsen 1976*a*); fusion of interdental plates; 14 to 16 maxillary teeth; surangular aperture small (but larger than a neurovascular foramen); medially directed notch at top of orbit bounded rostrally by lacrimal and caudally by frontal; splenial aperture small; at least some cervicals opisthocoelous; transverse processes of middle caudal series placed near middle of centrum, rather than distally on centrum; glenoid expansion of scapula makes angle of 50° to 60° with axis of scapular blade; internal tuberosity of humerus prominent; distal end of radius inclined at approximately 45° to axis of shaft; length of ilium equivalent to seven centra (comprising the caudalmost dorsals, sacrals, and the most proximal caudals); ilium extends from cranial end of caudalmost dorsal to caudal end of proximalmost caudal; cranial portion of distal pubic expansion about half as long as the caudal portion; ischium with obturator process trapezoidal in lateral view; astragalus with lateral notch receiving process of calcaneum (Welles and Long 1974); and ascending process of astragalus arising from lateral half of astragalar body (Welles and Long 1974). Madsen (1976*a*) listed nine mandibular features that characterize *Allosaurus fragilis* and may well be characteristic of all allosaurids: evenly rounded rostral margin, symphysis flat medially, fused interdental plates, rounded lingual bar, bowed ventral margin, alveoli extending to rostral extremity, uniform narrow Meckelian groove, evenly rounded lateral surface, and replacement teeth hidden.

Allosaurids may be traced back to the Callovian *Piatnitzkysaurus floresi*. *P. floresi* shares with later forms (*Allosaurus fragilis*, *Acrocanthosaurus atokensis*, and tyrannosaurids) extension of the supratemporal recesses onto the frontals, a marked notch separating the axial postzygapophysis from the neural spine, cervical and cranial dorsal pleurocoels, shortened dorsal centra, and

a rodlike pubis. These features, which persist or are more strongly developed with time in this lineage, link *Piatnitzkysaurus* to *Allosaurus* and tyrannosaurids.

Late Jurassic allosaurids include *Allosaurus fragilis* and *Szechuanosaurus campi*. *S. campi* shares certain characteristic features with *A. fragilis* (Dong et al. 1983), including shortened dorsal centra and moderately broad and high ascending process of the astragalus. The astragalus has the characteristic notch for the calcaneum (reported by Welles and Long [1974] in *A. fragilis*): the medial end of the astragalar body is inflated into a pronounced rim, unique among carnosaurs. *S. campi* shows a less advanced pelvis than *A. fragilis*, especially in the lesser development of the distal pubic expansion. The limb elements of *S. campi* are longer and more slender than those of *A. fragilis* and *P. floresi*.

Although *Allosaurus* is the best-known Jurassic carnosaur genus (Gilmore 1920; Madsen 1976*a*), its evolutionary relations to tyrannosaurids have not been explicitly given in the literature (e.g., Osborn 1912*b*). The most recent treatments (Walker 1964; Chatterjee, 1985) considered *Allosaurus* not in the line of ancestry of tyrannosaurids. *Allosaurus* is here considered to be close to the ancestry of tyrannosaurids in view of the following shared features: shortened and vertical quadrate, parietals with vertical lateral edges and extending dorsal to supraoccipitals to exclude them from the dorsal margin of the supraoccipital crest, surangular aperture, elongation of the distal chevrons, craniocaudal reduction of the ischial peduncle of the ilium, distal pubic expansion extended both cranially and caudally, angulate obturator process of the ischium, prominent attachment on the femur of M. femorotibialis, and the small process of the tibia extending parallel to the head of the fibula (fig. 6.23). The possession of five premaxillary teeth, declined paroccipital processes, and laterally bowed dentaries suggests that *A. fragilis* was not a direct ancestor of tyrannosaurids.

Chilantaisaurus (Hu 1964) is known from two species, from the Dazhuigou Formation, at Lake Chi-

Fig. 6.23. The tibia of *"Gorgosaurus" novojilovi* in proximal view, showing the cranial projection (arrow) found in *Allosaurus fragilis* and tyrannosaurids. (After Maleev 1974.)

lantai, near Alxa, Inner Mongolia, China. The humerus of *C. tashuikouensis,* is relatively twice as long relative to its femur as that of *Allosaurus fragilis.* Digit I of the manus bears an unusually large ungual phalanx. *C. tashuikouensis* shows the following carnosaurian features: elevated femoral head, winglike lesser trochanter, and marked extensor groove of femur.

The relationship of *C. tashuikouensis* with a second species, *C. maortuensis,* is unclear, as little is preserved that is common to both. Hu assigned them to the same genus because of similarities in the caudals. *C. maortuensis* has a reduced axial neural spine.

The phylogenetic position of *Chilantaisaurus* is clarified by noting its possession of declined paroccipital processes, also found in *A. fragilis,* and of a shortened axis, also found in tyrannosaurids. The short quadrate, centrally placed caudal transverse processes are found in both allosaurids and tyrannosaurids, while the caudal attenuation of the proximal part of metatarsal III seems intermediate between the nonattenuate condition of *Allosaurus fragilis* and the attenuation found in tyrannosaurids (fig. 6.24). Thus, at least in metatarsal morphology, *Chilantaisaurus* represents a form intermediate between typical allosaurids and tyrannosaurids. Rozhdestvensky's (1970*b*) contention that this genus was synonymous with *Alectrosaurus* is not here accepted in view of the absence of marked at-

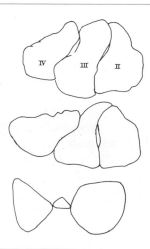

Fig. 6.24. The form of the proximal faces of the metatarsals in three carnosaurs. Above, *Allosaurus fragilis,* left metatarsus; center, *Chilantaisaurus tashuikouensis,* right metatarsus, showing the reduction in size of the caudal portion of metatarsal III; and, below, those of *Albertosaurus libratus,* right metatarsus. Those of *A. libratus* are shown sectioned just below the proximal surface. (After Lambe 1917*c;* Gilmore 1920; Hu 1964.)

tenuation of metatarsal III, as found in *Alectrosaurus olseni.*

The Albian allosaurid *Acrocanthosaurus atokensis,* in addition to the features shared with *Allosaurus* (Stovall and Langston 1950), shows the prominent jugal foramen, broadened postorbital bar, progression of pleurocoels into the caudal dorsal series, and fibular process all found in tyrannosaurids. Other taxa that share features with both allosaurids and tyrannosaurids are the Baharija carnosaurs from Egypt, *Bahariasaurus ingens* and *Carcharodontosaurus saharicus.* These forms, known from unassociated material, have been usually attributed to the Megalosauridae, although Lapparent (1960*a*) advocated their affinities with the Tyrannosauridae. The material referred to *B. ingens* and *C. saharicus* shares with tyrannosaurids the following synapomorphic features: supratemporal recesses confluent over the parietals, possession of amphicoelous rather than opisthocoelous cranial dorsal centra, neural spine central (rather than distal) in distal caudals, nearly perpendicular expansion of glenoid margin of scapula from that of scapular blade, apparently subtriangular obturator process of ischium, and cranial tubercle of fibula. Other features are shared only with allosaurids. *C. saharicus* has autapomorphic states (broad teeth that are not recurved, low preacetabular process of ilium) that suggest some independent evolution in Gondwanaland after its ancestors arrived.

These four genera, *Acrocanthosaurus, Bahariasaurus, Carcharodontosaurus,* and *Chilantaisaurus,* each have features shared with both allosaurids and tyrannosaurids. Thus, familial allocation is not immediately obvious. In the case of *A. atokensis,* the shared features may well be convergent, reflecting its large size. The broad postorbital bar and caudal dorsal pleurocoels are likely size related, and the fibular process, presumably muscle attachment, may indicate a reorganization of crural muscles related to size. If so, there is no reason to assign *A. atokensis* to the Tyrannosauridae. The Baharija taxa are difficult to treat because they are represented by unassociated elements; thus, both species will be treated together. With the exception of the dorsal central form, all of the features shared with tyrannosaurids are probably related to muscle attachments. In view of the phylogenetic lability of such features, and the incompleteness of the specimens, the Baharija taxa will not be assigned to a family here. The shortened axis of *C. maortuensis* may also be a size-related feature, convergent with tyrannosaurids. It seems unlikely, however, that proximal attentuation of metatarsal III is size related, and this may indicate closer relationship to tyrannosaurids than to allosaurids.

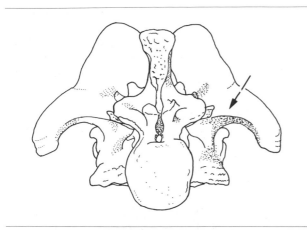

Fig. 6.25. The sacrum and ilia of *Tyrannosaurus rex* in cranial aspect, showing the medial, horizontal shelf of tyrannosaurids (arrow). (After Osborn 1916.)

The Tyrannosauridae is characterized by the following features: prominent jugal foramen present; extensions of the quadratojugal and squamosal together form a winglike process intruding into the infratemporal fenestra; rostral plate of vomer broad and rhomboid in form; prefrontal excluded from orbital margin; notch at top of orbit reduced or closed; frontals broader than long; supratemporal recesses confluent over parietals giving a blunt sagittal crest; splenial aperture large; surangular aperture large; retroarticular process lost; axial centrum short; distal caudal neural spines axially elongate; distal chevrons craniocaudally elongate; margin of glenoid expansion of scapula making approximately right angle with axis of blade; manus didactyl; horizontal medial shelf on broad ventral wing of preacetabular process of ilium (fig. 6.25); cranial portion of distal pubic expansion subequal in length to caudal portion; rodlike slender ischium with broad triangular obturator process; semicircular scar presumably for M. ischiocaudalis on caudal margin of ischium; lesser trochanter higher than greater trochanter; metatarsal III proximally attenuated but not entirely excluded from cranial face of metatarsus. In addition, Currie (1987c) gives a suite of characteristic features of the frontal: frontal overlapped rostrally by two or three nasal processes; prefrontal occupies pit on dorsolateral surface of frontal; orbital margin much reduced; frontal postorbital contact expanded, with buttress at rostral end, and frontal overlapped by postorbital caudally; sagittal crest extends onto frontals; dorsum of frontals flat between orbits; and supratemporal recess deeply invades frontals and is rostrally bounded by low sinuous ridge.

The Tyrannosauridae, although monophyletic, is not morphologically uniform. *Albertosaurus, Daspletosaurus, Nanotyrannus, Tarbosaurus,* and *Tyrannosaurus* are basically similar in form. Obviously different are two Asiatic forms, *Alectrosaurus olseni* (Gilmore 1933a; Perle 1977) and *Alioramus remotus* (Kurzanov 1976b). These appear to be slender forms, with gracile limb elements. The skulls are lower than those of the first group, and the tooth count is not reduced. *Alectrosaurus* and *Alioramus* show the following diagnostic tyrannosaurid features: squamosal-quadratojugal wing, large surangular aperture, confluent supratemporal recesses, and proximally attenuated third metatarsal. Another Asiatic species, *Chingkankousaurus fragilis* (Young 1958a), seems to be a tyrannosaurid on the basis of its very slender scapular blade.

The evolution of carnosaurs from *Piatnitzkysaurus* to tyrannosaurids involves increasing body and relative cranial size, increasing size of the preacetabular portion of the ilium, elaboration of the distal expansion of the pubis into a "foot" equally long both cranially and caudally, increasing slenderness of the ischium with concomitant increase in size of the obturator process, and proximal attentution of metatarsal III. The skull showed an increase in relative size of the adductor chamber and the attendent expansion, and ultimately confluence, of the supratemporal recesses across the parietals, increased development of cranial sinuses, shortening of the quadrate relative to the overall height of the postorbital region, reduction in tooth number, increase in tooth size relative to skull length, and development and subsequent increase in size of the surangular aperture, among other features. Two persistently primitive features are retained in tyrannosaurids: horizontal paroccipital processes and weakly opisthocoelous cervical and amphicoelous or amphiplatyan cranial dorsal vertebrae. The paroccipital processes in *Allosaurus fragilis* and *Acrocanthosaurus atokensis* are declined, and the cervical and cranial dorsals are opisthocoelous. The tyrannosaurid condition appears to be related to the shortening and stiffening of the column in conjunction with increased body size in the line leading to tyrannosaurids and thus may be a reversal of this trend.

POSSIBLE CARNOSAURS

The term "carnosaur" has too often been used simply to designate any large theropod. Although carnosaur taxa may be distinguished from one another by differences in many skeletal elements, isolated elements or

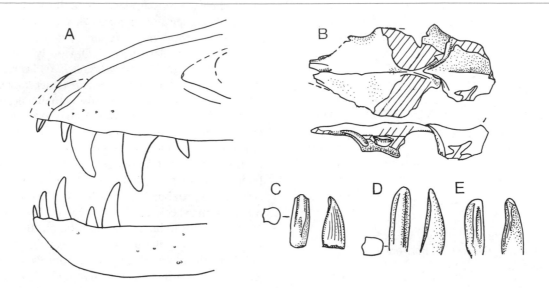

Fig. 6.26. *Aublysodon*. A, Reconstructed snout. B, Skull roof in dorsal and lateral views. C, Tooth of *A. mirandus*, lateral and distal views and section. D and E, Other teeth. (After Marsh 1892*d*; Molnar 1978.)

fragmentary specimens even when clearly distinct may be difficult or impossible to relate to well-known taxa. In consequence, a number of taxa have been referred to the Carnosauria which cannot unequivocally support this reference. The following taxa have been only tentatively associated with the Carnosauria.

AUBLYSODON Leidy 1868*b:* (fig. 6.26) Leidy (1856) proposed *Deinodon horridus* for a suite of isolated teeth of different forms, which he took to pertain to a single taxon. According to Hay (1899), Cope (1866) restricted the name *Deinodon* to those teeth not bladelike in form, although this is not clear from Cope's text. Leidy (1868*b*) then restricted the name to the blade-like teeth—just those excluded by Hay (1899)—and named the remaining three teeth *Aublysodon mirandus*. These teeth were of D-shaped cross section, and Leidy pointed out that such teeth were unknown in *Laelaps* and *Megalosaurus*. The smallest of these teeth, which lacked serrations, he thought might belong to "a different animal." Carpenter (1982*b*) chose as lectotype for *Aublysodon mirandus* this smallest tooth. An incomplete theropod skull discovered in 1966, lacking the premaxillae, had appressed against it a premaxillary tooth that matched the lectotype of *A. mirandus*. This skull, the "Jordon theropod" (Molnar 1978) from the Maastrichtian Hell Creek Formation of Garfield Co., Montana (USA), may belong to *Aublysodon* (fig. 6.26A, B).

The lectotype tooth came from the Campanian Judith River Formation, near Judith River, Montana (USA). Similar teeth have also been found in the Denver, Hell Creek, Kirtland, Lance, and Two Medicine formations. The lectotype tooth lacks serrations, and the flat distal face bears a rounded vertical ridge. Except for lacking serrations, it resembles the premaxillary teeth of tyrannosaurids.

Molnar (1978) discussed the affinities of the "Jordan theropod" skull, concluding it most likely pertained to a dromaeosaurid. However, the resemblances are plesiomorphies (such as the elongate contact between frontal and prefrontal) or otherwise dubious. The confluence of the supratemporal recesses across the parietals and other features of the frontals suggest a relationship to the tyrannosaurids, although the frontals are long. Molnar and Carpenter (in press) concluded that *Aublysodon* is a primitive genus related to tyrannosaurids, while Currie (1987*c*) believed it is a tyrannosaurid.

CHILANTAISAURUS? SIBIRICUS is based on a single incomplete metatarsal IV from the Turga Formation, at Tarbagatay Mines, Transbaikal, U.S.S.R., described by Riabinin (1914; as *Allosaurus [?] sibiricus*). In distal aspect, this metatarsal is almost identical with that of *C. tashuikouensis* in form and proportions of the distal condyle. Since both differ from other theropods in these features, and in view of the similarity in age and locality, it is here questionably referred to *Chilantaisaurus*.

COMPSOSUCHUS Huene 1932: This is one of a number of theropod genera from the Lameta Formation of Bara Simla Hill, near Jabalpur, India, established on isolated elements. The single species, *C. solus,* is based on an axis with articulated axial intercentrum; the specimen appears to be lost. This axis resembles that of *Allosaurus fragilis* in the position of the upper pleurocoel, in having an axial intercentrum cylindrical (rather than tapered) in ventral view, an axial pleurocentrum less than twice the length of the axial intercentrum, and a broad neural canal. These are apomorphies shared with *Allosaurus fragilis* and suggest that *Compsosuchus* is an allosaurid.

EUSTREPTOSPONDYLUS Walker 1964: This genus is known from a single species, *E. oxoniensis,* represented by a single well-preserved skeleton from the Middle Oxford Clay (Callovian) of Summertown, England. Prior to Walker (1964), the specimen was referred to *Streptospondylus cuvieri,* so this is the name used in the older literature (save for Owen 1842*b*).

Much of the skull is present but disarticulated. The specimen seemingly represents an immature individual, for although the scapula is fused to the coracoid (Nopcsa 1906), the postfrontal and postorbital are not fused, nor are the sacrals (Walker 1964).

The incompleteness of the other European carnosaurs makes comparison unrewarding—*E. oxoniensis* may be compared with *A. fragilis,* however. Among the similarities are a lacrimal ornamentation, an enlarged lacrimal foramen, strongly opisthocoelous cervicals, and a notch for the calcaneum in the astragalus (Welles and Long 1974). It differs from *A. fragilis* (Walker 1964) in that the frontals are longer than wide, the prefrontals are farther forward and farther apart, and the dorsal portion of the orbit is less compressed (this last perhaps a juvenile character). The reportedly heterocoelous sacrals (Huene 1923) are better described as nearly amphiplatyan. Other differences are dubious (e.g., the curved ischium: Walker 1964). Walker reported that the pubis is more similar to that of *A. fragilis* than previously believed in that it has an obturator notch (rather than a foramen) and an incipient distal "foot." Certain carnosaurian features, such as the reduced atlantal neural spine and winglike lesser trochanter, are clearly present.

On balance, *E. oxoniensis* seems related to the ancestry of the carnosaurs and is probably it-

self a carnosaur. It is advisable, however, to await further study before determining its detailed affinities.

GASOSAURUS Dong et Tang 1985: This genus is from the Lower Shaximiao Formation (Middle Jurassic) of Dashanpu quarry at Zigong, China. The single species, *G. constructus,* is represented by a single specimen (fig. 6.27), 3.5 meters long.

G. constructus retains platycoelous cervicals, a primitive feature. The femoral head extends medially more than in other taxa (this may be pathological or due to distortion). The lesser tro-

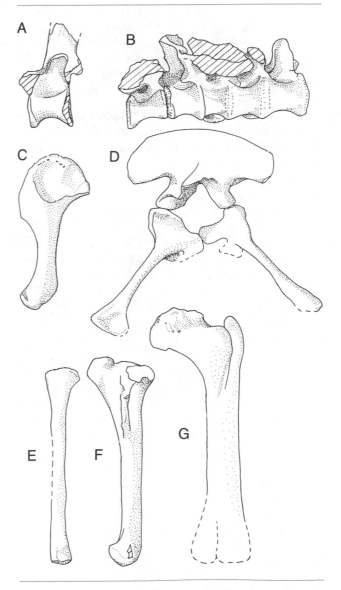

Fig. 6.27. *Gasosaurus constructus.* A, Dorsal vertebra. B, Sacrum, ventrolateral view. C, Humerus. D, Pelvis. E, Fibula. F, Tibia. G, Femur. (After Dong and Tang 1985.)

chanter rises well above the proximal end of the shaft to leave a sulcus between it and the femoral head. The ilium is short with a strongly-arched dorsal margin.

The only carnosaur feature apparent is the elevation of the femoral head. *G. constructus* may be a carnosaur, but further information is needed to confirm this conjecture.

ILIOSUCHUS Huene 1932: The single species, *I. incognitus*, is known from two isolated ilia found in the Bathonian Great Oolite near Stonesfield, England. These ilia exhibit a well-developed vertical ridge extending from the acetabular region. In all instances where ilia with such a ridge are convincingly associated with other skeletal material, that material is carnosaurian. In no case is such an ilium known from a skeleton not carnosaurian. This suggests that *I. incognitus* is a carnosaur, but in the absence of further characters, this referral must be considered tenuous.

KAIJIANGOSAURUS He 1984: *K. lini* (fig. 6.28) is from the Lower Shaximiao Formation of Sichuan, China. The type specimen consists of seven platycoelous cervical vertebrae. Additional associated elements from several individuals represent much of the postcranial skeleton. The scapular blade is narrow and may expand abruptly at the glenoid region, but unfortunately this area is broken. The femoral head appears not to be elevated, but apparently the calcaneum contacts the astragalus at a peg and socket articulation (He 1984). The noncarnosaurian characters are plesiomorphic, which together with the carnosaurian features present suggest that *K. lini* may be related to the ancestry of carnosaurs. It is possible that this genus is synonymous with *Gasosaurus*, which comes from the same formation.

KELMAYISAURUS Dong 1973a: This genus is known from a single species, *K. petrolicus*, from the ?Valanginian-Albian Lianmugin Formation, near Wuerho, China. It is represented by an incomplete left maxilla and dentary. Using the mandibular characters given by Madsen (1976a) to distinguish *Allosaurus fragilis* from *Megalosaurus bucklandii*, *K. petrolicus* agrees with *M. bucklandii* in four, with *A. fragilis* in one (fused interdental

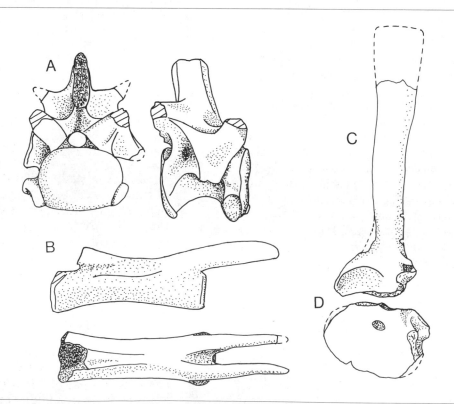

Fig. 6.28. *Kaijiangosaurus lini*. A, cervical vertebra. B, caudal vertebra. C, scapula. D, coracoid. (After He 1984.)

plates), and four cannot be determined (forms of symphysis and of lateral surface, exposure of replacement teeth, and cranial extent of alveolar row). This suggests a closer relationship to *M. bucklandii* than to *A. fragilis*.

LABOCANIA Molnar 1974: The single species, *L. anomala,* is based on a frontal, quadrate, incomplete maxilla and dentary, a chevron, proximal ischium, metatarsal II, and other fragments. The specimen was found in the Campanian "La Bocana Roja" formation, near Arroyo del Rosario, Baja California, Mexico. It has several unusual cranial features: the frontals are very thick compared with those of most other theropods, the quadrate condyle lacks the helical groove found in other theropods, the quadrate is hollow, and the dentary has an external longitudinal shelf. *L. anomala* shows several tyrannosaurid features: the hollow quadrate, a triangular obturator process of the ischium, and a characteristic semicircular muscle scar on the caudal margin of the ischium. This scar is found also in ornithomimids, which may be excluded as a possible identification in view of the maxillary and dentary teeth. The chevron has the cranial prongs characteristic of carnosaurs. However, *Labocania* is a very derived genus, in the respects given above quite unlike any tyrannosaurid. While some relationship with tyrannosaurids may exist, its nature cannot be detailed. In the absence of more complete material, its reference to the carnosaurs must remain hypothetical.

"LAELAPS" GALLICUS (Cope 1867): This name was proposed for theropod material (vertebrae, pubis, tibia, astragalus, and calcaneum) from Normandy, France, illustrated by Cuvier (1812). The tarsals show the oldest example of a separate but interlocked astragalus and calcaneum (Welles and Long 1974), characteristic of allosaurids. Strongly opisthocoelous cervicals and dorsals are also present. This species may be an allosaurid; it is not *Dryptosaurus*.

MAGNOSAURUS Huene 1932: *M. nethercombensis* (= *Megalosaurus nethercombensis* Huene 1926*a*) is based on dentaries, a dorsal and caudal, pubis, femora, and tibiae. All are incomplete, the femora being represented mainly by casts of the internal cavities. It derives from the Bajocian Inferior Oolite of Nethercomb, England. Waldman (1974) considers that the genus *Magnosaurus* is not distinguishable from *Megalosaurus;* however,

he indicates that the distal tibia is not compressed. Since the tibiae referred to *Megalosaurus* are compressed distally, this genus is here tentatively retained. The dentary agrees with that of *Megalosaurus bucklandii* in six of the nine features in which that species differs from *Allosaurus fragilis* (Madsen 1976*a*), so *M. nethercombensis* would share the taxonomic assignment of *Megalosaurus*.

Of the other species, *M. lydekkeri* (= *Megalosaurus lydekkeri* Huene 1926*b*), is based on a tooth from the Lower Lias of Lyme Regis, England, that is not now determinable. *M. woodwardi* Huene 1932 may be referrable to the ceratosaur *Sarcosaurus andrewsi*.

MEGALOSAURUS Buckland 1824: *Megalosaurus* is based on *M. bucklandii* (fig. 6.29), consisting of an incomplete right dentary from the Bathonian Stonesfield Slate, of Stonesfield, England.

Madsen (1976*a*) has shown that the type dentary of *M. bucklandii* differs substantially from that of *Allosaurus fragilis,* so that reference to the same family is dubious, a conclusion also reached by Waldman (1974). There is much referred material. Only two of the other species are based on material comparable to the type of *M. bucklandii*.

Megalosaurus hesperis (Waldman 1974) consists of parts of both premaxillae, the right maxilla (fig. 6.29), the vomer, both dentaries, and a surangular from the Bajocian Upper Inferior Oolite of Sherborne, England. The dentaries of *M. hesperis* resemble those of *M. bucklandii* in six of the nine characters listed by Madsen (1976*a*), differ in one (alveoli extending to extreme mesial margin), and two (form of symphysis and of lateral surface) cannot be determined. This species is plausibly retained in the genus *Megalosaurus*.

Megalosaurus cambrensis (= *Zanclodon cambrensis* Newton 1899) is based on a natural mold of a left dentary (fig. 6.29) from Glamorganshire, Wales. Despite its considerably greater age (Rhaetian), it shows several resemblances to *M. bucklandii*. Of the nine points listed by Madsen (1976*a*), it agrees in six and the remaining three (mesial extent of alveolar row, form of symphysis and of lateral surface) cannot be determined. Of the features in which it agrees with *M. bucklandii,* three (angular rostral margin, separate interdental plates, and exposed replacement teeth) seem to be shared, derived features. This suggests that this is correctly referred to the genus *Megalosaurus*. However, as its distinction rests largely on its much greater age, further material is necessary to establish if this is a valid species.

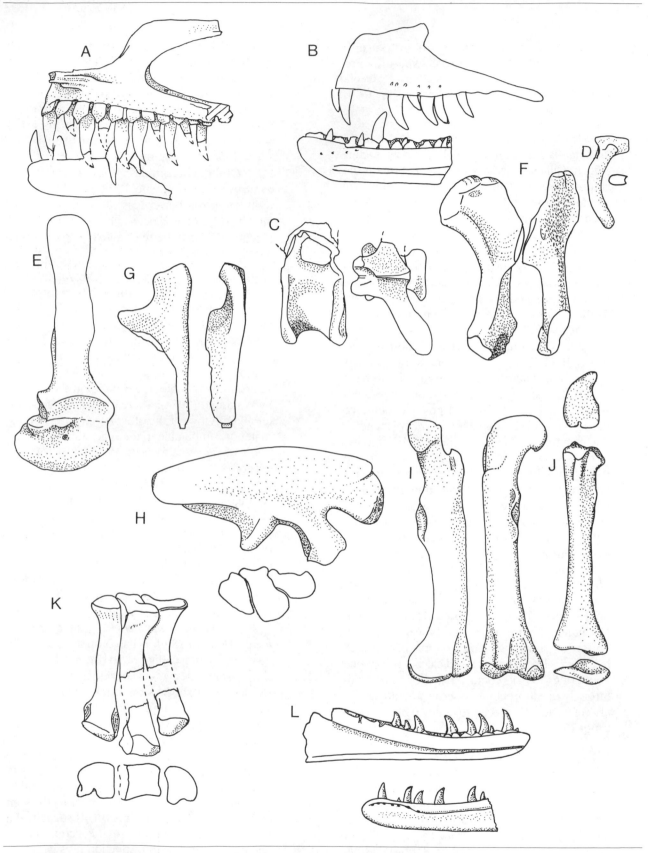

Fig. 6.29. A, Maxilla and incomplete dentary (associated) of *Megalosaurus hesperis,* medial view. B-K, *Megalosaurus bucklandii.* B, Maxilla and dentary (not associated). C, Caudal vertebra. D, Chevron. E, Scapulocoracoid. F, Humerus. G, Proximal pubis. H, Ilium. I, Femur. J, Tibia. K, Metatarsus. L, Dentary of *Megalosaurus cambrensis.* (After Huxley 1869*b;* Owen 1884; Newton 1899; Huene 1926*b;* Walker 1964; Charig 1979.)

Because no diagnostic carnosaurian characters are yet apparent in the dentary, taxonomic assignment of the genus *Megalosaurus* must depend on referred material and on the most complete referred species, *M. hesperis*. *M. hesperis* does show one carnosaurian character, a deep surangular (Waldman 1974), and no diagnostic characters of any of the other theropod groups. Its resemblance to *M. cambrensis* suggests this is a primitive species. As there is no indication that more than a single species of large theropod is found in the Stonesfield Slate, it is worthwhile to examine the referred postcranial material. The ilium shows a narrow brevis shelf, the femoral head is elevated, and the lesser trochanter is winglike. However, the scapula lacks the abrupt acromial expansion and the extensor sulcus of the femur is absent.

Megalosaurus seems related to carnosaurs, but further study is necessary to clarify its relationships.

PIVETEAUSAURUS Taquet and Welles 1977: A braincase from the marine upper Callovian of Vaches Noir, France, is the type specimen of the single species *P. divesensis* (= *Eustreptospondylus divesensis* Walker 1964). It is the size of that of an adult *Allosaurus fragilis*. Taquet and Welles (1977) list the points in which it differs from *A. fragilis*, *Ceratosaurus nasicornis*, and *Eustreptospondylus oxoniensis*. They conclude that while the differences from *C. nasicornis* are the most significant, *P. divesensis* is not confamilial with *A. fragilis* but most similar to *E. oxoniensis*, to which it had previously been referred (Walker 1964). This genus and species would presumably share the taxonomic status of *E. oxoniensis*, that is, related to the ancestry of the carnosaurs in a manner not yet clear.

SPINOSAURUS Stromer 1915: *S. aegyptiacus* was found in the Cenomanian Baharija Formation, near Baharija Oasis in central Egypt (fig. 6.30). It included an incomplete maxilla, a dentary, sple-

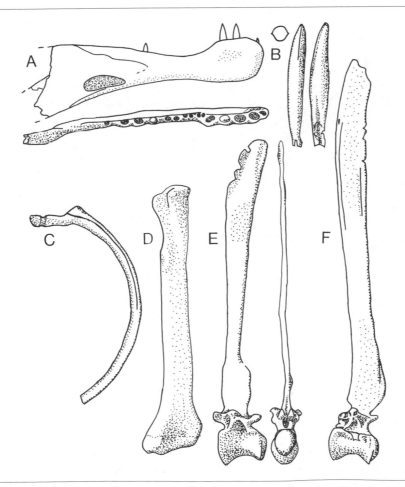

Fig. 6.30. *Spinosaurus aegyptiacus.* A, Rostral mandible. B, Isolated tooth. C, Dorsal rib. D, Tibia (from so-called *Spino-* *saurus* B). E and F, Dorsal vertebrae. (After Stromer 1915.)

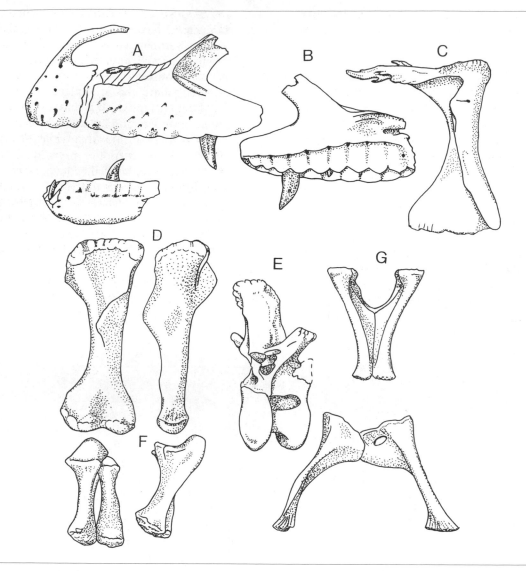

Fig. 6.31. *Torvosaurus tanneri.* A, Maxilla, premaxilla, and incomplete dentary, lateral view. B, Maxilla, medial view. C, Lacrimal. D, Humerus. E, Caudal dorsal vertebra. F, Radius and ulna. G, Pubis and ischium. (After Galton and Jensen 1979a; Jensen 1985a.)

nial, cervicals, dorsals, and caudals. Additional material ("*Spinosaurus* B") consisted of vertebrae, ribs, iliac and femoral pieces, a tibia, incomplete fibula, and phalanges. All material was destroyed during World War II.

    S. aegyptiacus has several unique characters. The teeth are conical, straight, and lack serration (Taquet 1984). Most obvious is the great elongation of the dorsal neural spines to eleven times the depth of the centrum. The only carnosaur feature is that in "*S. B*", the extensor groove of the femur is deep. Like that of tyrannosaurids, the splenial has an enlarged aperture. The referred chevrons lack cranial processes, however. It is not demon-

strably a carnosaur but is perhaps derived from *Eustreptospondylus oxoniensis,* in which the cranial dorsals are also strongly opisthocoelous.

STOKESOSAURUS Madsen 1974: Like *Iliosuchus, Stokesosaurus* is based on an isolated ilium, with a prominent ridge extending vertically from the acetabular region. The single species, *S. clevelandi,* derives from the Morrison Formation of Utah (USA). Unlike *Iliosuchus,* other material (premaxilla) has been referred to this taxon. Galton (1976b) referred this species to *Iliosuchus* but in light of further evidence, later withdrew this suggestion (Galton and Jensen 1979a). For the reasons given for *Iliosuchus, Stokesosaurus* may be a carnosaur, but further material is necessary for any certainty.

TORVOSAURUS Galton and Jensen 1979*a*: *T. tanneri* comes from the Morrison Formation of Colorado (USA). It comprises a humerus, radius, and ulna, although more material is referred (fig. 6.31), including much of the facial skeleton and vertebral column as well as the pelvis, tibia, fibula, tarsus, and metatarsus (Jensen 1985*a*).

T. tanneri has characters unexpected in a Late Jurassic theropod. The lacrimal is slender, especially the dorsal ramus, with only a small horn core. The antebrachial elements are relatively short, about half the length of the humerus, and massive in appearance. Both pubes and ischia are in contact ventrally along their entire length. The lacrimal and puboischial structures are plesiomorphic, and while the relatively massive antebrachial elements do not seem plesiomorphic for any recognized group of theropods, they are otherwise known only in Middle Jurassic genera (e.g., *Kaijiangosaurus, Poekilopleuron*).

Among the carnosaurian attributes of *T. tanneri* are the enlarged lacrimal apertures, strongly opisthocoelous cervicals, cranially projecting prongs on the chevrons, and narrow brevis shelf. Most other carnosaurian features cannot be discerned from the preserved material; however, a central vertical ridge of the iliac blade is absent as is a calcaneal notch in the astragalus.

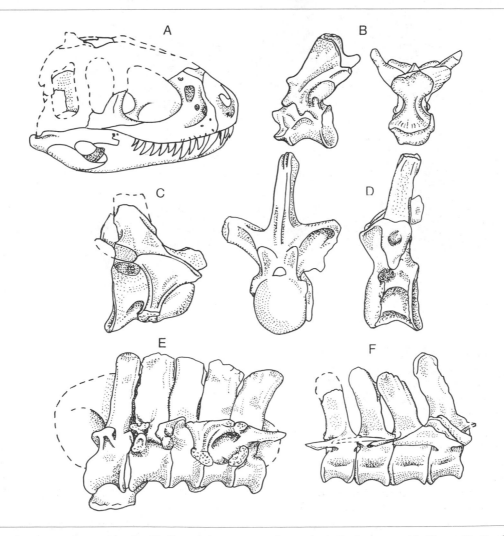

Fig.6.32. *Yangchuanosaurus magnus.* A, Skull and jaws. B, Atlas-axis vertebrae. C, Caudal cervical vertebra. D, Dorsal vertebra. E, Sacrum with ilium. F, Caudal vertebrae. (After Dong et al. 1983.)

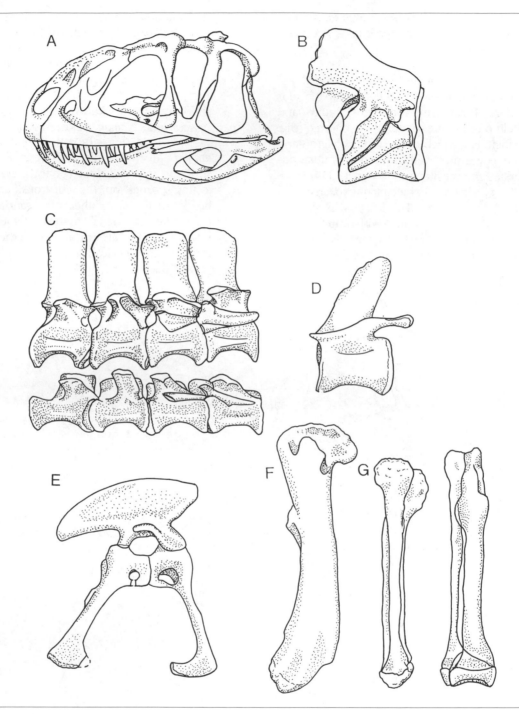

Fig. 6.33. *Yangchuanosaurus shangyouensis.* A, Skull and jaws. B, Axis, with rib and fragment of cervical 3. C, Caudal dorsal vertebrae. D, Proximal caudal vertebra. E, Pelvis. F, Femur. G, Tibia and fibula. (After Dong et al. 1983.)

These are plesiomorphic characters, and the distribution of the iliac ridge in carnosaurs is uncertain. The absence of the calcaneal notch, already known in the Oxfordian *Eustreptospondylus oxoniensis*, suggests that although *T. tanneri* seems

related to the ancestry of the carnosaurs, it was less closely related than was *E. oxoniensis*.

YANGCHUANOSAURUS Dong, Chang, Li, et Zhou 1978: *Yangchuanosaurus* is known from two species from the Upper Shaximiao Formation, near Chongqing, China. The eight-meter-long *Y. shangyouensis* (Dong et al. 1978; fig. 6.32), is known from a skeleton lacking forelimbs, hind feet, and distal tail, while *Y. magnus* (Dong et al. 1983) is represented by a snout, jaws, vertebrae, pelvis, and femur (fig. 6.33).

The following carnosaurian characters are present in at least one of the species: enlarged lacrimal aperture, reduced axial neural spine, narrow brevis shelf, and winglike lesser trochan-

ter. One carnosaurian character is clearly absent: the surangular is shallow. This appears to be a plesiomorphic feature not barring *Yangchuanosaurus* from a position in or near the ancestry of the better-known carnosaurs.

Yangchuanosaurus, *Torvosaurus*, and *Eustreptospondylus* may form a natural group related to each other and to the ancestry of the Carnosauria. Further work is necessary on these forms to determine if this suggestion is viable.

7

Carnosaur Paleobiology

R. E. MOLNAR

JAMES O. FARLOW

Few dinosaurs have captured the public imagination like the great meat-eaters. Fictional carnosaurs regularly menace jungle explorers or spar with colossal apes. As the largest terrestrial predators in earth history, carnosaurs have figured in scientific speculation about the role of large carnivores in ecological communities; interpretations of carnosaurs in such discussions range from the sluggish, thermodynamically strapped giants of Colinvaux (1978) and Halstead and Halstead (1981) to the multiton speedsters of Bakker (1986) and Paul (1987a, 1987b). Such divergent opinions, apart from inspiring wonder that the same animals were being discussed, allow latitude for speculation regarding carnosaur paleobiology. We will offer several such speculations, based on the fossil record and on comparison with living vertebrates. These interpretations seem plausible, but it must be emphasized that the plausibility of a hypothesis does not guarantee its correctness, an unfortunate fact of life often overlooked.

   This chapter will commence with a survey of the geographic and taphonomic occurrences of carnosaur finds, and then discuss functional anatomical aspects of carnosaurs, and their paleoecological and behavioral implications.

BIOGEOGRAPHY

The Jurassic fossil record indicates that carnosaurs were found at both eastern (China) and western (USA) extremities of Laurasia and in the South American and probably the African portions of Gondwanaland. Middle Jurassic carnosaurs are known only from Argentina (*Piatnitzkysaurus floresi*), but because faunas of this age are incompletely known, this apparent geographic restriction is of dubious significance. Late Jurassic carnosaurs are known from North America (*Allosaurus fragilis*) and China (*Szechuanosaurus campi*). The African species, *A. tendagurensis* (Janensch 1925), is from the Tendaguru Formation of Tanzania. It is based on a tibia, an element that seems not to exhibit distinctive allosaurid features, but in view of the existence of isolated elements from Tendaguru that do show allosaurid features, this species is here regarded as probably allosaurid.

   By the Early Cretaceous, the breakup of Pangaea would likely have progressed sufficiently for regional differentiation to be expected in the evolving lines of carnosaurs. However, Early Cretaceous carnosaurs are insufficiently known to show more than that they were still present at both extremities of Laurasia and were widespread in Gondwanaland. There is incomplete material from North America (*Acrocanthosaurus atokensis*) and Asia (*Chilantaisaurus maortuensis* and *C. tashuikouensis*). A single element from the Albian Strzelecki Group of Victoria (Australia) is reported to derive from *Allosaurus* sp. (Molnar et al. 1981, 1985). Presumably, they also inhabited at least Africa in view of the probably carnosaur material from Tanzania and the later occurrence of carnosaurs in the Egyptian Cenomanian.

Both *Bahariasaurus ingens* and *Carcharodontosaurus saharicus*, from the Baharija Formation of Egypt, show greater similarity to the Tyrannosauridae than do earlier carnosaurs. The unusual dental and iliac form of *C. saharicus* suggests that these forms had diverged from the lineage that led to tyrannosaurids and were not directly ancestral.

Late Cretaceous carnosaurs (i.e., tyrannosaurids) are very well known but for the Campanian and Maastrichtian only, from Asia (*Alectrosaurus, Alioramus, Chingkankousaurus,* and *Tarbosaurus*) and North America (*Albertosaurus, Daspletosaurus,* and *Tyrannosaurus*). The Asian tyrannosaurids include more plesiomorphic taxa (*Alectrosaurus, Alioramus*) than in North America. Fragmentary material from India previously referred to tyrannosaurids (*Indosaurus matleyi* and *Indosuchus raptorius* from the Lameta Formation of Madhya Pradesh) is possibly referable to the superficially similar abelisaurids. However, one Indian species (*Compsosuchus solus,* also from the Lameta Formation of Madhya Pradesh) seemingly represents an allosaurid. Thus, Late Cretaceous carnosaurs, with the exception of the two Egyptian and the doubtful Indian taxa, seem restricted to the Asiamerican landmass of Laurasia. There is as yet no convincing evidence of carnosaurs from the Euramerican region of Laurasia.

While this record does not yet permit confident assessment of the evolutionary biogeography of carnosaurs, it is possible that there was a Gondwanaland-Laurasian split between carnosaurs in Laurasia and other large theropods, such as the abelisaurids, in Gondwanaland in the Late Cretaceous.

TAPHONOMY

While tyrannosaurids are known from several articulated specimens, of which a substantial proportion are more than 60 percent complete, allosaurids are represented by less, usually less complete, material. *Albertosaurus libratus, Daspletosaurus torosus, Tarbosaurus bataar,* and *Tyrannosaurus rex* are each represented by skeletons at least 70 percent complete. *Szechuanosaurus campi* and *Piatnitzkysaurus floresi* are known from a few partial skeletons, and *Acrocanthosaurus atokensis, Bahariasaurus ingens, Carcharodontosaurus saharicus, Chilantaisaurus maortuensis,* and *Chilantaisaurus tashuikouensis* are known from more incomplete material. *Allosaurus fragilis* is represented by several partial skeletons with much disarticulated material from the Cleveland-Lloyd quarry, Utah (USA).

The difference in completeness between allosaurid and tyrannosaurid specimens is probably attributable simply to the difference in ages: the Late Jurassic (from which most of the allosaurid specimens derive) is about twice as old as the Late Cretaceous. This age difference probably also accounts for other differences in quality of preservation such as the proportion of taxa represented by cranial material or immature specimens. Skulls are known for almost all genera of tyrannosaurids, as opposed to only one genus of allosaurid (*Allosaurus fragilis*). Other allosaurids (*Acrocanthosaurus atokensis, Chilantaisaurus maortuensis, C. tashuikouensis,* and *Piatnitzkysaurus floresi*), however, are represented by braincases or isolated cranial elements. Only *Allosaurus fragilis* among allosaurids is represented by immature specimens, while several tyrannosaurids (*Albertosaurus libratus, Daspletosaurus torosus, Tarbosaurus bataar, Tyrannosaurus rex*) include some immature specimens, as does *Bahariasaurus ingens* (Stromer 1934*a*). Of these latter, only *Albertosaurus libratus* and *Daspletosaurus torosus* include relatively complete immature skeletons.

Several tyrannosaurid specimens (especially *Albertosaurus libratus* and *Tarbosaurus bataar;* Matthew and Brown 1923; Gradzinski et al. 1968) are preserved in the so-called opisthotonic posture. This is not a reflection of death by poisoning but of drying of the carcass for some period prior to burial (Matthew and Brown 1923).

There is no evidence for mass accumulations in which several articulated individuals are found together presumably having died more or less simultaneously. Carnosaur specimens are almost always found individually or when associated include individuals of different taxa, as at an occurrence near Jordan, Montana (USA), in the Hell Creek Formation where a specimen of *Tyrannosaurus rex* was associated with one referred to *Nanotyrannus lancensis* (Molnar 1980*d*). The only possible mass accumulation is that of *Allosaurus fragilis* at Cleveland-Lloyd Quarry in the Morrison Formation of central Utah (USA), but there is yet no evidence for the deaths of the individuals at one time.

The unique association at Cleveland-Lloyd (Madsen 1976a) is worth a brief description. Elements pertaining to at least forty-four individuals of *A. fragilis* have been found scattered and almost always disarticulated. A detailed scenario of the mode of formation of this deposit has not yet been formulated. The orientation of the elements when uncovered suggests that they were scattered and became oriented before burial. This is consistent with the occurrence of breakage and loss of delicate structures, which suggests that scavenging took place. Non-dinosaurian tetrapods are very rare, as are invertebrate and plant fossils, but such invertebrates and plants as have been found indicate deposition in a shallow, quiet, freshwater body.

The striking feature of this deposit is the relative abundance of *A. fragilis;* there are about twice as many specimens of *A. fragilis* as all other dinosaurs represented. Osmólska (1980) reported that tyrannosaurids were as common as the most common herbivore species in the Nemegt Formation of Mongolia. These are the only instances in which the usually rare carnosaur specimens outnumber those of herbivores. There is no evidence for any consistent association of carnosaurs with specific taxa of other dinosaurs.

Elsewhere than at Cleveland-Lloyd, *A. fragilis* is found preserved equably without regard to major lithofacies type (Dodson, Behrensmeyer, Bakker, and McIntosh 1980). These types include floodplain, lacustrine, and channel deposits. *Albertosaurus libratus,* however, is preferentially preserved in channel sands (Dodson 1971), and all other carnosaurs for which information is available seem preserved in channel deposits.

No convincing evidence exists that carnosaur remains were often transported to the site of preservation, as apparently were the remains of pachycephalosaurs. There is evidence, however, that some carnosaurs did not usually occupy the regions in which they are preserved. Remains of *A. libratus* are three times as abundant as those of *D. torosus* (D. A. Russell 1970a) in the Judith River Formation of Alberta (Canada). In addition, specimens of immature *A. libratus* are substantially more common than are immature *D. torosus.* This is consistent with the interpretation that *A. libratus* inhabited the region of deposition, while *D. torosus* may not have.

Evidence for predation or scavenging, with the exception of that at Cleveland-Lloyd, has never been reported. The evidence, absence of mass accumulations and common occurrence in floodplain deposits, is consistent with an interpretation of carnosaurs, especially tyrannosaurids, as uncommon lone hunters (as opposed to pack hunters) of the floodplains.

FUNCTIONAL ASPECTS OF CARNOSAUR ANATOMY

Cranial Anatomy

CRANIAL FUNCTION

Cranial kinesis in *Allosaurus fragilis* was recognized early in the century (e.g., Matthew and Brown 1922) based on the work of Versluys (1910). From work on a skull, Versluys concluded that a movable joint existed between the lacrimals and prefrontals. Movement at this joint, together with rotation of the quadrate on the squamosal (streptostyly) and flexion at an unspecified region in the skull roof, permitted slight elevation and depression of the snout on the braincase (fig. 7.1). Madsen's study (1976a) confirms the apparently mobile nature of several of the cranial joints in *A. fragilis.* In tyrannosaurids, the situation is obscure. Certain joints appear to be mobile or "loose" (not closely conformable), but these joints are not distributed in such a way as to divide the skull into mobile segments as in snakes. These seem to involve joints that resist tensional forces, where passive motion may be permitted so long as the elements do not become separated. Thus, cranial kinesis appears unlikely; however, slight motion between various elements may have occurred (Osborn 1912b; D. A. Russell 1970a).

A spheroid occipital condyle, always projecting well back of the occipital face, is found in carnosaur skulls. This implies considerable mobility of the skull on the cervical column. Strong muscle scars on the occipital face of the skull corroborate this mobility and may account for the extensive development of the su-

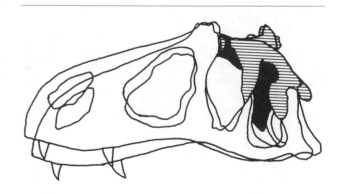

Fig. 7.1. The scheme of cranial kinesis proposed for *Allosaurus fragilis* by Versluys. The unshaded snout segment was free to move slightly up and down on the shaded braincase segment. (After Versluys 1910.)

praoccipital crest (to provide increased attachment area with the evolution of increased head size).

The craniomandibular joint is remarkably invariant in form in carnosaurs, with a helical groove in the quadrate condyle closely fitting a corresponding ridge in the articular glenoid. This ridge and groove are so oriented that the mandibles are displaced slightly laterally when the jaw is opened.

Assertions that an intramandibular joint is found in carnosaurs are widespread (Walker 1964; Madsen 1976a), but no study of its mechanism has yet appeared. Such assertions are based on observation of the "loose" articulations of the mandibular elements, especially those at the dentary-postdentary contact. The symphysial junction likewise is believed "loose" (Osborn 1906; Madsen 1976a), as the symphysial facet is flat and smooth with no sign of a rigid junction between the jaws.

Cranial musculature has not been reconstructed for carnosaurs, except for the early attempt of Adams (1919). Walker (1964) contended that in tyrannosaurids M. pterygoideus dorsalis was strongly developed but relatively shorter than in more primitive archosaurs. This would give a stronger but slower bite. Comparison of the muscle scars in both *Allosaurus fragilis* and *Tyrannosaurus rex* shows good agreement with those of modern crocodilians. It may be concluded that the pattern of the mandibular adductors was essentially similar in these two groups.

SENSORY FUNCTION

Certain features of the carnosaur endocranial molds suggest the degree of sensory development. Both olfactory and optic lobes are prominent, suggesting well-developed visual and olfactory perception. Walker (1964) pointed out that the rostrolateral orientation of the orbits in *T. rex* may have indicated "some degree of overlapping of the fields of vision," that is, frontal vision. Such orbital orientation is also found in *Nanotyrannus lancensis* but not in other carnosaurs. This orientation has been taken (e.g., Paul 1987a) to indicate "binocular vision," presumably, from the context, in the sense of stereopsis. However, there are birds with strongly overlapping visual fields which lack stereoscopic vision (Konishi and Pettigrew 1981; Pettigrew 1986), which can only be verified by behavioral or neuroanatomical studies (Pettigrew 1986). Thus, while *N. lancensis* and *T. rex* may well have had stereopsis, possession of overlapping visual fields alone does not reliably imply this.

Nonetheless, the possession of frontal vision by only two species of carnosaurs is of considerable interest. If these species did have stereopsis, what function could it have served? It confers distance perception, of great use to raptors, allowing them to pick up prey from the ground without crashing (Pettigrew 1986); this is not relevant to carnosaurs. However, stereopsis allows more than distance perception: it also allows the penetration or "breaking" of camouflage (Julesz 1971; Konishi and Pettigrew 1981; Pettigrew 1986). If the two tyrannosaurids did have stereopsis, then both of these functions would have been available to them.

Postcranial Anatomy: Stance and Locomotion

STANCE

Virtually without exception, carnosaurs have been recognized as bipedal animals. The extreme shortness of the forelimb relative to the hindlimb—in *Allosaurus fragilis,* the humerus is about 36 percent the length of the femur—and the much smaller cross-sectional area of the forelimb elements makes clear that the forelimb did not participate in locomotion. The manus bears compressed claws unlike those of the pes, and permitted motions at manual joints are not compatible with locomotor function.

This much agreed, however, further details of the stance have been contentious. The traditional pose with the vertebral column inclined to an angle of at least 45° with the ground and with the tail trailed on the ground is not universally accepted (fig. 7.2). This view was based on the work of Cope (1868) on *Dryptosaurus aquilunguis,* supported by analogy with living macropodid marsupials. However, the center of mass of the body must lie in the immediate area of the hips (Abel 1930; Alexander 1985), and many early reconstructions did not take this into account. The trunk can be balanced only if it is held so steeply as to bring the center of mass over the hind limbs, or if its weight is counterbalanced by the tail, which, to do so, must be raised from the ground. There is no consensus regarding posture, although many workers now accept that the dorsal column was probably held nearly horizontally and the weight of the trunk balanced by that of the tail.

Lambe (1917c) proposed large birds as a more applicable model for carnosaur locomotion but retained the traditional pose. Newman (1970) suggested that the presacral column was held more or less horizontally by analogy with birds and that the tail was

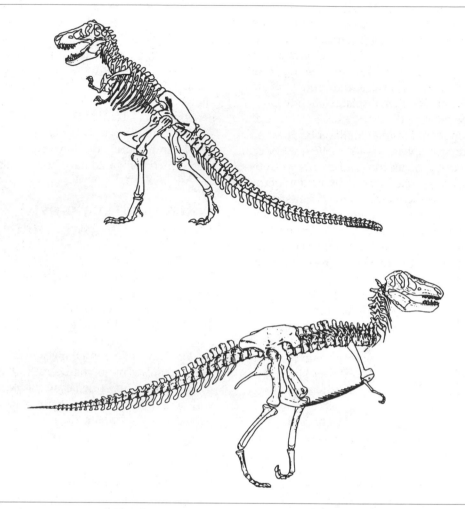

Fig. 7.2. The traditional (above) and more recently proposed (below) stance of carnosaurs, here *Tyrannosaurus rex*. The traditional stance strongly inclines the presacral vertebral column, with the tail on the ground, while the more recent stance has the presacral column more nearly horizontal, with the tail elevated. (After Osborn 1916 and Newman 1970.)

elevated from the ground to act as a counterbalance. Tarsitano (1983) argued that muscular mechanics at the hip dictated that the vertebral column be held at an angle of 20° to the horizontal.

None of these arguments is compelling. As Charig (1972) pointed out, modern birds, even large ratites, are not obviously good analogues of the much larger and tailed carnosaurs. Tarsitano (1983) assumed that M. puboischiofemoralis—which he believed to be, along with M. iliotibialis, the main femoral protractor—could not lift and protract the femur to the level of the pubis, although the muscle extends from the femur to the caudal portion of the dorsal column. Depending on the length of the muscle, it seems that pro-

traction of the femur beyond the level of the pubis is possible provided the column is not too strongly elevated. Indeed, Tarsitano himself figured the femur as being lifted and protracted slightly beyond the level of the pubis. He is likely correct, however, in asserting that the femur could not be drawn to the level of the vertebral column, but then as Paul (1987*b*) pointed out, this is unnecessary even assuming a horizontal vertebral column. Only Alexander (1985) examined the position of the center of mass with respect to the hind feet. He concluded that the posture proposed by Newman was consistent with placement of the center of mass over the feet.

More substantial arguments reconstruct the

stance from the structure and placement of the joint surfaces of the acetabulum and proximal femur. In all carnosaurs, the pubic peduncle of the ilium is massive (Lambe 1917c). In allosaurids, the supraacetabular crest is located not over the center of the acetabulum as expected if the long axis of the ilium were held horizontal but more cranially, extending onto the pubic peduncle. This may indicate that in standing, the ilium transferred the body weight to the femur at or near the pubic peduncle, which in turn would imply a more or less inclined presacral column (unless the column were arched). A necessary assumption is that the region of the acetabulum showing greatest development was that where the body-weight was transferred to the femur; however, stresses at the acetabulum incurred during locomotion would have been substantially greater than those from weight support. Hotton (1980) has argued that such stresses were caudally, not cranially, directed and so the great development of the pubic peduncle would not reflect these stresses and may indicate a stance with at least the iliosacral region inclined.

OTHER ASPECTS OF VERTEBRAL FUNCTION

The structure of the vertebral column provides no clue as to the stance habitually adopted by the animal. Madsen (1976a) recognized that the articulated cervical column of *Allosaurus fragilis* forms a sigmoid, almost swanlike, curvature. This raises the head well above the level of the cranial dorsal column. A similar curvature is found in the cervical column of *Tyrannosaurus rex* (Osborn 1916) but is absent from *Acrocanthosaurus atokensis* (Madsen 1976a). Madsen suggested that in *Allosaurus fragilis*, this curvature provided sufficient leverage for the cervical epaxial musculature to support the head without hypertrophy of the neural spines, but lacking such curvature, *Acrocanthosaurus atokensis* developed a hypertrophied epaxial musculature and with it elongate dorsal neural spines.

Tarsitano (1983) questioned the existence of this curvature in *T. rex* and illustrates the neck as projecting directly forward from the dorsal column. Examination of cervical form in both *Allosaurus fragilis* (Madsen 1976a) and *T. rex* (Osborn 1906) shows that when both centra and zygapophyses are articulated such a curvature results (fig. 7.3). Arranging the cervicals as pictured by Tarsitano results in disarticulation of the centra and zygapophyses. How these vertebrae could be arranged to form a straight line was never explained or clearly illustrated.

Allosaurids are characterized by strongly opisthocoelous cervicals, which permit great mobility of the

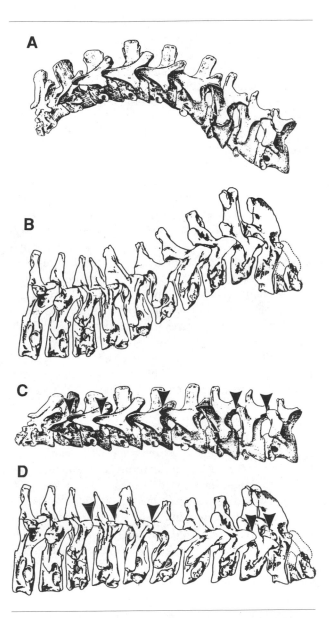

Fig. 7.3. The cervicals of *Allosaurus fragilis* (a) and *Tyrannosaurus rex* (b) showing the curvature that results from articulating both centra and zygapophyses. For *A. fragilis*, cranial is to the left; for *T. rex*, to the right. C and d. Attempt to articulate the cervicals of *A. fragilis* (c) and *T. rex* (d) in a straight line as recommended by Tarsitano (1983). Central articular surfaces are matched, but this results in poor or no articulation at some zygapophysial joints and overlap by axial neural spine of that of cervical 3. Arrows mark disarticulated zygapophysial joints. (After Osborn 1906 and Madsen 1976a.)

neck. Huene (1923) contended that the opisthocoelous cervicals and cranial dorsals functioned in disjointing the prey. He visualized carnosaurs as grasping the prey with the hind foot and, having seized the flesh in the teeth, violently dorsiflexing the presacral column to raise the head and pull off the grasped food much in the fashion of a hawk or vulture.

Tyrannosaurid presacrals would be less mobile (Osborn 1906) than those of allosaurids, as indicated by the amphicoelous centra and close approach of the neural spines. This is also suggested by ossification of the interspinous ligament at its attachment to the neural spines and by evidence for ossification in the body of that ligament (Lambe 1917c). Rigidity of the dorsal column of *T. rex* is not shown by the occurrence of a vertebral pathology (Newman 1970), a congenital block vertebra, as this condition does not produce functional disability (Paul and Juhl 1966). Presumably, the tyrannosaurid dorsal column became immobilized to permit more effective support of increased body weight following on increased size, as in ungulates.

The carnosaur tail has received little attention. Lambe (1917c) pictured that of *Albertosaurus libratus* functioning as a "third leg" on which the animal could rest while standing, much like the tails of macropodids; other workers of that period neglected tail function entirely. Recently, the tail has received attention regarding its role in locomotion. Newman (1970) maintained that contraction of M. caudifemoralis in retraction of a hindlimb would have swung the tail to that side, thus giving a waddling gait (see below). Hamley (in press) has related tail rigidity and mass to stride frequency. A heavy, flexible tail would oscillate slowly and impede rapid locomotion by "dragging" down the stride frequency. In view of this, it is interesting that the division of the carnosaur tail into proximal and distal segments implies that M. caudifemoralis longus was restricted to the proximal portion of the tail (Gatesy pers. comm.). This would lighten the tail, while the elongate prezygapophyses found in allosaurids and tyrannosaurids would, at least to some extent, increase the rigidity of the tail. Thus, tail morphology in carnosaurs may reflect modification for speed.

Carnosaur presacral vertebrae, like those of other saurischians, are characterized by marked excavations, including pleurocoels. But other elements of the skeleton are also hollow including limb bones (Osborn 1906) and, in *Piatnitzkysaurus floresi,* even the ilium (Bonaparte 1986c). Such excavations in sauropod dinosaurs have been interpreted as the result of selection for metabolic economy in view of the great size of sauropods (e.g., Colbert 1961). Carnosaurs are not larger than hadrosaurids, which lack such excavations, and extensive excavations are found even in small theropods. Bonaparte (1986c) has suggested that the selective factor in carnosaurs was decreased weight permitting increased speed. Other, complementary, interpretations have also been made, such as that of Bakker (1986) that vertebral pleurocoels housed air sacs, much like those found in modern birds. Bakker suggests that they functioned to increase the respiratory capacity.

FORELIMB FUNCTION

The function of the forelimb, especially in tyrannosaurids where it has been so much reduced, has provoked much interest. In *Allosaurus fragilis,* the proximal and distal humeral articular surfaces are situated at an angle to one another, so that when the forearm was flexed, the hands were moved away from the body wall (Gilmore 1920). Prominent muscle scars imply strongly developed musculature and hence a powerful forelimb.

The manus of *A. fragilis* is relatively large, longer than the humerus (Gilmore 1920). The form of the distal joint face of metacarpal I is inclined so that the first digit—at rest directed away from the rest of the digits— moves toward the axis of the manus as the digit is flexed, indicative of a grasping function (Gilmore 1920). The claws articulate with the penultimate phalanges such that the claws are held almost perpendicular to the axes of the phalanges, especially on digits I and II. Thus, the claws could function as hooks. These considerations led Gilmore to believe the forelimb of *A. fragilis* was an effective grasping organ.

The great reduction of the forelimb in the tyrannosaurids has raised skepticism that it functioned effectively. However, the humerus of *T. rex* (the only element of the forelimb yet known) shows well-developed muscle scars. The distal face of metacarpal I in tyrannosaurids retains its inclination as in *A. fragilis.* The claws and manual articular surfaces remain well developed. Finally, the manus is 78 percent of the length of the humerus (in *Daspletosaurus torosus,* data from Russell 1970a). These suggest that forelimbs of tyrannosaurids did retain some prehensile function, even if the specific function is not yet understood.

While Osborn (1906) suggested that the forelimb was used during mating, Newman (1970) presented a cogent argument that the tyrannosaurid forelimb functioned as a brake during rising from lying prone. The initial stages of rising require the flexed hindlimb to be straightened, which would tend to force the trunk forward along the ground. While delicate balance could also overcome this effect, Russell (1970*a*) pointed out that tyrannosaurid humeri frequently seem to suffer breakage, possibly in this context. Other interpretations of such breakage, however, are possible (see below). The anatomical considerations given above suggest that braking was not the sole function of the forelimb.

ANATOMICAL ASPECTS OF LOCOMOTION

The hindlimb was clearly the major propulsive structure. Articulation of the femoral head in the acetabulum, because of the elevated head, inclines the femoral shaft laterally (Lambe 1917*c*; Gregory 1951). However, Lambe (1917*c*) reported that the femoral distal condyles are unequally developed, bringing together the distal ends of the tibiae near the midline of the body and hence beneath the center of mass, as would be expected in such large animals (Wade 1989). Trackways of large theropods are quite narrow (see photographs in Bird 1985), a conclusion not always accepted by popular illustrators (fig. 7.4).

Hotton (1980) studied the femoral articulation with the acetabulum in theropods, and his conclusions (based in part on *Allosaurus*) hold for all carnosaurs. He found that the dorsal articular face of the femur was a narrow cylindrical surface. Movement was restricted to the parasagittal plane and excursion was limited, from vertical through an arc of about 60° cranial to vertical. From these and other considerations, Hotton concluded that large dinosaurs were primarily walkers, rather than runners as is sometimes claimed. This conclusion was independently reached by Thulborn (1982) from trackway studies.

The proximal attenuation of the third metatarsal in tyrannosaurids led Coombs (1978*c*) to suggest the potential presence of elastic (or "snap") ligaments. In this case, the suggested function involved the upward displacement of the third metatarsal when the hind foot was placed on the ground; this displacement stretched the ligaments and hence stored energy in them. This energy was released with the elevation of the foot at the end of the step, and recovered to help lift the hindlimb for the recovery phase of the step cycle.

ECOLOGICAL ASPECTS OF LOCOMOTION

The single most impressive feature of the carnosaurs is their large size. As big and rather long-legged animals, these and other large theropods give the impression of having been fairly mobile. In comparison with many modern, smaller reptiles, carnosaurs are likely to have been more like intensive foragers than cruising foragers or sit-and-wait predators (terminology of Regal 1978); Pough (1983, 165) suggested that "intensive foraging may be energetically more efficient for large lizards than for small ones," but it may be risky to extend this generalization to a comparison of theropods with lizards. The large prey animals on

Fig. 7.4. Hindlimb position in carnosaurs. The allosaurid on the left is reconstructed in accordance with anatomical and trackway data, while the *Tarbosaurus bataar* on the right is seemingly practicing a Cossack dance. This animal is re- stored with wide, ungainly transverse spacing between its hind feet, which is not in accord with evidence from large theropod trackways. (After Wade 1989.)

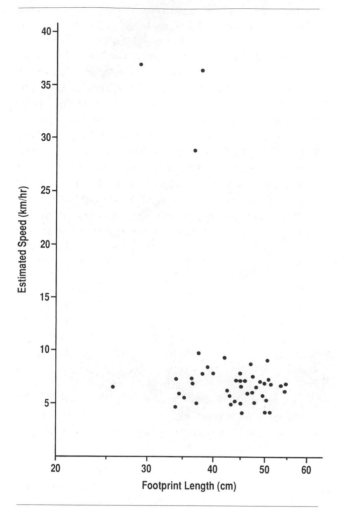

Fig. 7.5. Estimated speeds calculated from trackways attributed to theropod dinosaurs from the Lower Cretaceous (Comanchean) of Texas. Trackway data from Farlow (1987a). Speeds estimated from the equation of Alexander (1976), as modified by Thulborn (1984b). Dinosaur trackway data from other ichnofaunas are consistent with the pattern seen here, with the largest estimated speeds occurring in medium-sized dinosaurs.

which the carnosaurs presumably fed were probably widely enough dispersed to make a sit-and-wait strategy unprofitable for most meat-eating dinosaurs.

Estimates based on fossil trackways suggest that medium-sized and large theropods generally walked at speeds of 5–10 km/h (fig. 7.5; Bakker 1987). Such speeds are as fast or faster than those of modern mammals (Bakker 1987) and are consistent with the interpretation of flesh-eating dinosaurs as vagile animals, ranging widely in search of food.

More controversial are interpretations of the top speeds of the great hunters. Coombs (1978c) reviewed

the structural features that typify the best runners among living animals and used these to infer the cursorial potential of dinosaurs. Theropod adaptations related to running ability include forelimb reduction, interlocking or fused metatarsals, and a tendency to develop a symmetrical hind foot, with reduction or loss of outer digits and elongation of the middle toe. Coombs was cautious about estimating the absolute speeds attained by dinosaurs but thought it likely that the maximum speeds of carnosaurs were considerably less than those of the fastest living cursors.

Thulborn (1982) estimated the top speeds of dinosaurs by modifying a method proposed by Alexander (1976) for determining speeds from trackways. Alexander noted that living mammals shift from walking to faster gaits when the ratio of the stride length to the leg length reaches a value of 2; by substituting the quantity (2 × leg length) for stride length in Alexander's equation relating speed to leg and stride length, Thulborn calculated the speeds at which dinosaurs would shift from slower to faster gaits. Thulborn concluded that the top speeds of carnosaurs were in the range of 10 to 23 km/h, but these estimates reflect his belief that such large animals were unlikely to have accelerated much beyond the speed of the walk-trot transition.

Alexander (1985) compared the estimated bending strengths of dinosaurian limb bones with those calculated for large living mammals and the ostrich. *Tyrannosaurus rex* had a femoral bending strength somewhat greater than that of an African elephant, and Alexander cautiously suggested that this carnosaur was only a slow runner.

In contrast, Bakker (1986) claimed—but unlike Alexander presented no quantitative argument—that the hindlimbs of *T. rex* were relatively more massive than those of elephants and drew attention to the large cnemial crest on that carnosaur's tibia. Noting that living mammalian gallopers have large cnemial crests, and massive calf muscles to match, Bakker concluded that *T. rex* might have run at speeds of as much as 70 km/h. Bakker's interpretation was endorsed by Paul (1987a, 1987b), who summarized other features of the hips, limbs, and ankles which suggested cursorial potential in carnosaurs and other dinosaurs. However, the forceful limb musculature of carnosaurs does not necessarily denote great speed; conceivably, this is an adaptation associated with bearing great weight or with the stresses of subduing large prey (but see Paul 1987a for a contrary opinion).

Although on theoretical grounds one might expect larger animals to be fleeter than their smaller relatives (McMahon 1975), the biggest creatures in a particular locomotory category tend not to be the fastest (McMahon and Bonner 1983: 151). Garland (1983a) suggested the relationship between the logarithms of top running speed and of mass is curvilinear in mammals and, like Coombs (1978c), argued that maximum speeds are attained by animals of about 50–100 kg; whether the slower speeds of very large animals reflect biomechanical or energetic limits is uncertain (Peters 1983; Calder 1984). The existence of such an optimal size for cursors suggests that the biggest carnosaurs were not the fastest runners; the immense calf musculature of *Tyrannosaurus* might be an adaptation to counteract, but not entirely overcome, the limits to cursoriality imposed by large size (Coombs 1978c).

Although the trails of small bipedal dinosaurs frequently have high stride/footprint length ratios (Welles 1971; Thulborn 1984b; Thulborn and Wade 1984, Matsukawa and Obata 1985) and some trackways of large ornithopods may record running (Lockley et al. 1983), the fastest estimated speeds are for medium-sized bipedal dinosaurs (Farlow 1981; Thulborn and Wade 1984; fig. 7.5; the trail of a small theropod described by Welles may be at variance to this pattern, but its interpretation presents difficulties—Thulborn and Wade 1984; Farlow 1989). The presently available trackway evidence is consistent with the conclusion that the optimal body size of cursorial bipedal dinosaurs, like that of extant mammals, was in the middle range, although this optimal size may have been somewhat larger in dinosaurs than mammals. This interpretation must be made with caution, however, as there may be artifacts in the footprint evidence. The stride lengths of hypothetical rapidly running carnosaurs might have been too long to register on most preserved tracksite bedding surfaces. In addition, even if capable of doing so, very large and heavy animals might be loath to move at speed on the soft and often slippery substrates in which prints are formed and preserved.

Thulborn (1984b) pointed out that a walking gait for a large carnosaur would be as fast as a running gait in a smaller dinosaur. One problem with the disagreement regarding whether carnosaurs walked or ran seems to center on definitions. Some workers apparently define running in terms of speed (e.g., Bakker 1986), while others define it in terms of a suspended phase in locomotion (e.g., Thulborn 1984b). The latter is formally correct, but the former is that commonly used. Whatever their absolute speeds, carnosaurs were probably usually as fast or faster than the large herbivores on which they presumably often preyed.

Newman (1970) proposed a waddling gait for *T. rex*. He suggested that with each step the tail was flexed away from the limb being protracted, resulting in a "sinuous . . . bird-like waddling." In support, Newman adduced a trackway from the Late Jurassic of England which may not have been made by a carnosaur. However, to the extent that progression with the hind toes directed medially, in a pigeon-toed fashion, supports this idea, it is supported by data of Farlow (1987a). Newman's waddle seems not open to further independent test.

Dodson (1971: 69) suggested that carnosaurs could wade into water in pursuit of prey. Other locomotor modes have been proposed; Coombs (1980a) contended that large theropods could swim. The deep tail of *Acrocanthosaurus atokensis* was suggested specifically as an adaptation to swimming (Stovall and Langston 1950). It is reasonable to accept swimming as a part of the carnosaur locomotory repertoire; the notion of leaping is less acceptable. Originally proposed by Cope (1871) for *Dryptosaurus aquilunguis*, this mode was long rejected until the report of Bernier et al. (1984). A trackway from the Upper Jurassic of Ain, France, was interpreted as having been made by a large theropod progressing by substantial leaps. The evidence is subject to an alternate (if less spectacular) interpretation: leaping carnosaurs are not yet beyond doubt.

CARNOSAUR ONTOGENY

Anatomical changes during the growth of carnosaurs have been studied by Rozhdestvensky (1965) and Russell (1970a). The conclusions reached by Rozhdestvensky are unfortunately confused by his acceptance of "*Gorgosaurus*" *novojilovi*, a different species (Osmólska 1980), as juvenile *Tarbosaurus bataar*. Russell set out the anatomical changes with growth based largely on *Albertosaurus libratus*. He reported disproportionate growth of the supraoccipital alae of the parietals with age. He ranked the relative rates of growth of various skeletal structures, suggesting that juveniles had a relatively shorter presacral column, smaller girdles, and longer tail and hindlimb (especially distally) than adults. These trends were used to predict how a hatchling *A. libratus* would appear (fig. 7.6). The unusual features of the cranial joints in *T. rex* (mentioned above) could result from the occurrence of cranial kinesis or streptostyly in the juvenile, which was suppressed in the adult. Little work has been done on other carnosaurs; however, Steel (1970) reported, based on observations by White, that a hindlimb attributed to a juvenile *Allosaurus fragilis* had relatively larger feet than in the adult.

Although a statistical study has not yet appeared,

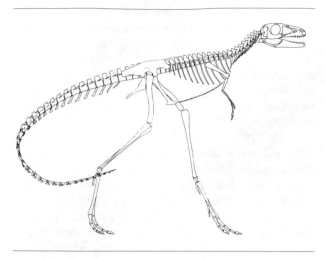

Fig. 7.6. The hypothetical hatchling of *Albertosaurus libratus* (from D. A. Russell 1970*a*).

the size distribution and range of *A. fragilis* material from the Cleveland-Lloyd Quarry led Stokes (1985) to suggest that this species grew continuously throughout life, as do modern reptiles.

Significantly, both Lawson (1976) and Molnar (1978) found that different species of tyrannosaurids show different allometric trends in homologous cranial structures. This indicates that trends found in one species cannot uncritically be assumed true for related species.

Russell (D. A. 1970: 16) reported that immature specimens of *A. libratus* are "relatively not uncommon" although very young specimens have not been found, while Madsen (1976*a*) reported that at Cleveland-Lloyd Quarry immature specimens, except for the very small, are present in some numbers. Extrapolating from such relatively small sample sizes to living population structures is risky. However, these results are consistent with any population structure that constrains the very young individuals away from the areas of deposition. This may result from parental care or a habitat removed from the areas of deposition: there is no further evidence of either (see below).

IMPLICATIONS FOR CARNOSAUR BEHAVIOR

Prey Handling

Prey handling of the largest terrestrial carnivores is a topic of interest to paleontologists and public alike. The basic question of whether carnosaurs were active predators, as often pictured, or scavengers, as some-

times suggested, is difficult to resolve from the fossil record. Lambe (1917*c*) noted the absence of heavy tooth wear in *Albertosaurus libratus* and from this suggested the carnosaur fed primarily on the soft flesh of putrefying carcasses. This interpretation, accepted by some subsequent workers (e.g., Barsbold 1983*a*), is carried to an extreme by Colinvaux (1978) and Halstead and Halstead (1981). It rests on the assumptions that putrefying carcasses were readily available and that tooth wear was usually absent: neither of these assumptions is generally warranted. The presence of soft flesh on carcasses depends on a humid climate; under dry conditions, the flesh dries and becomes hard and tough. The Morrison environment, at least, seems seasonally to have been quite dry (Dodson, Behrensmeyer, Bakker, and McIntosh 1980) and Chinese dinosaur-bearing deposits reportedly also indicate aridity (Anderson 1987). More extensive wear does occur on teeth in other carnosaur jaws, while shed crowns are often heavily worn (as Lambe acknowledged). In both placental (Walker 1981) and marsupial (W. Young pers. comm.) mammals, chipping and blunting of cusps distinguishes mammals that engage predominantly in scavenging as opposed to active predation; such work has yet to be extended to carnosaurs. However, few predators disdain feeding from a carcass already dead (Houston 1979; Auffenberg 1981), while many scavengers will sometimes attack and kill living prey (Ewer 1973).

Worn crowns suggest the teeth of carnosaurs were used in activities more strenuous than eating rotten meat. Van Valkenburgh (1988) argued that teeth (particularly canines) of mammalian carnivores are more likely to break than those of herbivores—from the mechanical stresses of killing prey. This argument may be applied to the broken and subsequently worn crowns in carnosaurs (see below) as evidence of active hunting.

Not surprisingly, there is a positive correlation between the size of a predaceous vertebrate and the average size of its prey (Gittleman 1985; Vézina 1985; Pianka 1986). Colinvaux (1978) was troubled by the absence of carnosaur-sized predators in the adaptive radiation of mammalian carnivores. This was one reason why he interpreted large carnivorous dinosaurs as sluggish creatures. However, herbivorous dinosaurs were generally larger than their modern mammalian counterparts, and this size difference probably accounts for the much larger size of their carnosaur predators than of mammalian carnivores.

In many food chains, the body size of consumers increases with each higher trophic level (Colinvaux 1978). Herbivorous zooplankters are larger than the phytoplankters they eat: herbivorous insects are often eaten by invertebrate or vertebrate predators bigger than they are, which in turn fall victim to still larger meat-eaters. Tracy (1976) proposed that something similar occurred in dinosaur communities, with carnosaurs preying on smaller theropods or other carnivores, rather than directly on herbivorous dinosaurs. Given that carnosaurs, like most predators, were likely opportunistic hunters, they probably did eat smaller flesh-eaters, but it would be surprising if they specialized on such prey to the exclusion of the presumably much greater available biomass of plant-eating dinosaurs.

Assuming that at least sometimes carnosaurs did attack and kill living prey, how was this done? Large individuals of the white shark (*Carcharodon carcharias*) feed mainly on marine mammals, especially pinnipeds (Tricas and McCosker 1984; McCosker 1985). The shark attacks from behind and beneath its victim, inflicts a savage bite with its large serrate teeth, and withdraws from the stricken animal until the prey goes into shock or bleeds to death. A similar surprise attack is often used by the ora (*Varanus komodoensis*): the lizard waits beside a game trail until a deer or boar passes, then grabs its victim with a quick lunge, often killing the animal with a bite to the abdomen (Auffenberg 1981).

Paul (1987a) proposed that carnosaurs used these tactics, making a crippling or lethal hit-and-run attack, inflicting a deep bite, then retiring to await weakening of the prey from blood loss before administering the *coup de grâce*. No anatomical considerations have yet falsified this "land shark" interpretation. Bakker (1986) was puzzled by the ability of hadrosaurids, probably less fleet-footed than contemporaneous carnosaurs, to survive in the presence of the latter. Conceivably, tyrannosaurids were better sprinters than the large ornithopods but had less endurance, so that a hadrosaurid might have been able to outdistance a carnosaur if it escaped a surprise attack, but this is pure speculation. However, recent suggestions that the land shark tactic is widespread among carnivores (e.g., Diamond 1987) have been criticized for overlooking contradictory details (Bryant and Churcher 1987); this tactic may be less widely used than has been made out.

There is likewise nothing in carnosaur anatomy that rules out tactics like those used by the big cats. These dispatch prey by holding closed the nostrils and mouth or trachea of the prey (Schaller 1972) and holding on until the prey is subdued. The gape of *T. rex* was certainly large enough to grasp the snout or neck of a contemporaneous hadrosaur. Such a tactic would expose the predator to greater risk of injury during the melee, and such injuries may be reflected in healed fractures such as those of the forelimb noted by Rothschild (1988).

Similarities of tooth form and wear suggest that carnosaur biting may have resembled that of saber-toothed cats (e.g., *Smilodon fatalis*) and oras in some ways. With jaw closure, the maxillary teeth passed outside those of the dentary; portions of the prey's flesh were trapped in notches formed by the mesial and distal keels of adjacent crowns as the opposing tooth rows slid past one another. The entire lateral toothrow may have been a crude but effective equivalent of a sabre-tooth's postulated canine shear-bite (Akersten 1985) or the action of the carnassial teeth of carnivorous mammals. The recurved shape of the tooth tips forced the mesial tooth tip against the prey item during biting, perhaps accounting for the great frequency of worn apical ends of mesial carinae. The lingual flexure of the basal segments of mesial keels may have created a screwlike action, forcing fragments of meat cut during jaw closure into the carnosaur's mouth.

When feeding, large oras advance one side of the head forward over a portion of the prey carcass, slightly close the jaws, and then pull the upper and lower jaws of that side backward (Auffenberg 1981). The breadth of the ora's muzzle causes this motion to result in each tooth's following in the path cut by the tooth behind it, enlarging that cut. Auffenberg contends this action is facilitated by a rostromedial-caudolateral inclination of the cutting edges of the ora's teeth (fig. 7.7). In the jaws we have examined, this inclination is most obvious at the basal end of the distal carina; its apical end and the mesial keel are more symmetrically placed relative to the craniocaudal axis of the crown. The heights of the lateral teeth in the ora's toothrow are such that a curve connecting the tooth tips resembles the blade of a scalpel; the tallest teeth are midway along the row, and as the jaws are retracted during feeding, progressively taller teeth cut more deeply than the teeth behind them (Auffenberg 1981).

A similar scalpel-like configuration occurs in the maxillary teeth of carnosaurs, especially so in *T. rex* (fig. 7.8). The usual lingual bending of the basal mesial keel and the labial flexure of the distal keel of carnosaur lateral teeth result in an inclination of their cutting edges similar to that of the ora. As a carnosaur's jaws closed, the distal keels on the recurved lateral teeth cut backward as well as down (Bakker 1986, 260–262). If carnosaurs subsequently retracted their jaws in oralike fashion, meat would have been forced against the distal keel, and if the teeth were completely buried in the prey's flesh, tissue adjacent to the tooth tip may have slid apically along the carina where serrations probably facilitated cutting the tissue. The labial

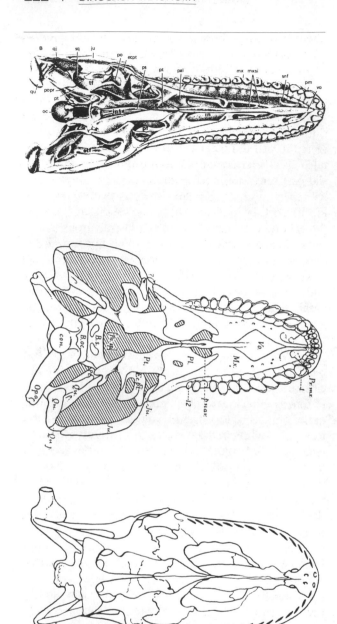

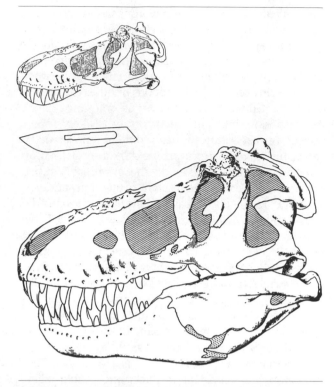

Fig. 7.8. Skull and jaws of *Tyrannosaurus rex* (from Osborn 1912). Note the relatively large size of the maxillary teeth, even in comparison with other carnosaurs. Inset: a curve connecting the tips of the teeth of the upper row forms a scalpel-like shape similar to that noted by Auffenberg (1981) for the ora.

Fig. 7.7. A comparison of muzzle shape in *Allosaurus fragilis* (from Madsen 1976*a*), *Tyrannosaurus rex* (from Osborn 1912*b*), and the ora (*Varanus komodoensis;* from Auffenberg 1981); not to scale. Note the long and narrow muzzle of *A. fragilis* as compared to the other two reptiles, the huge maxillary teeth of *T. rex,* and the inclination of the cutting edges of the ora's teeth with respect to the edges of the jaws.

inclination of the distal keel may additionally have resulted in a plowlike effect during jaw retraction, which forced pieces of flesh into the carnosaur's mouth. Pre-

sumably, these actions resulted in the flat-worn distal keels seen in many shed carnosaur crowns.

Thus far we have discussed carnosaur feeding and biting in general terms, emphasizing features that may have been common to most or all of these great meat-eaters, but there were structural variations on this theme. The muzzle of *A. fragilis* is is longer and narrower than that of *T. rex,* whose muzzle is more like that of the ora (fig. 7.7). The maxillary teeth of *T. rex* are larger relative to skull size than those of *A. fragilis,* making the Cretaceous carnosaur somewhat of a saber-toothed dinosaur. The mechanics of killing or feeding may have differed in these two carnosaurs (Paul 1987*a*).

Although the surprisingly high frequency of broken and subsequently worn teeth in mammalian carnivores suggests that tooth breakage is not as great a threat to survival as once thought (Van Valkenburgh 1988), it nonetheless seems likely that predatory mammals must exercise greater care than reptilian predators in employing their teeth against prey, given the reptilian ability to replace their teeth. Similarly, heavily worn teeth are unlikely to limit the life spans of predatory reptiles. Consequently, carnosaurs—like oras (Akersten 1985)—may have been less careful than flesh-eating mammals in deciding which portion of the potential victim to bite, perhaps biting into bone with abandon; conceivably, the wear surfaces of carnosaur premaxillary teeth were created by tooth-to-bone contact.

Paul (1987a) rightly points out that the "land shark" strategy reduces the risk of injury to the predator from resistant prey. However, carnosaur skeletons exhibit signs of more frequent injury than those of herbivorous dinosaurs. Broken ribs are found in *Tyrannosaurus rex*, fractured humeri in *Albertosaurus libratus* (Rothschild 1988; Russell 1970a), and fractured humeri and radii in *Allosaurus fragilis* (Petersen et al. 1972): all had healed. These suggest that struggles with prey did take place; it is also possible that these reflect intraspecific combat or physical competition between different species of carnosaurs.

Regardless of the strategy employed, or even whether or not carnosaurs were hunters, there is evidence of their prey. Gilmore (in Hildebrand et al. 1930) reports that one specimen of *Apatosaurus* shows several bones, particularly in the tail, with the appearance of having been "scored and bitten off." These scorings match in spacing that of the teeth of *Allosaurus*. Furthermore, the sauropod was found associated with broken carnosaur teeth, presumably from *Allosaurus*. Such evidence is certainly consistent with the notion of predation of *Allosaurus* on sauropods, or sauropod carcasses.

Sympatric predator species in modern vertebrate communities differ in preferred habitat, time, and method (e.g., solitary vs. group hunting) of foraging and the species composition or size classes of prey items (Seidensticker 1976; Bertram 1979; MacDonald 1983; Van Valkenburgh 1985; Shine 1986; Emmons 1987). Similar differences presumably occurred in Mesozoic communities, in which carnosaur species shared a potential resource base of herbivorous dinosaurs with other species of large and small theropods and predatory reptiles (Molnar 1978; Baird and Horner 1979; Farlow 1980; Dodson 1983; Coe et al. 1987). The dietary and foraging differences among these predators are difficult to reconstruct. D. A. Russell (1970a) speculated that the two Judith River tyrannosaurids of Alberta, Canada, *Albertosaurus libratus* and *Daspletosaurus torosus*, may have concentrated on hadrosaurids and ceratopsids, respectively.

Size differences among sympatric dinosaurian and other carnivores undoubtedly contributed to dietary differences, but there are confounding factors that must be considered. In modern vertebrate predators,

the size range of prey items, not just the average prey size, increases with increasing carnivore size (Vézina 1983; Gittleman 1985), resulting in some dietary overlap between large and small predators. We suspect, however, that prey size range does not increase indefinitely with increasing predator size but may reach an upper limit or even decline for enormous meat-eaters; it is difficult to imagine a carnosaur-sized flesh-eater normally being interested in mouse-sized prey. However, the occurrence of ontogenetically changing food habits in modern carnivorous reptiles (Kellogg 1929; Cott 1961; Chabreck 1972; McNease and Joanen 1977; Taylor 1979; Auffenberg 1981; Delany and Abercrombie 1986; Magnusson et al. 1987) suggests that hatchling and juvenile carnosaurs may have had considerable dietary overlap with smaller predators (Farlow 1976a, 1976b, 1980), unless parental care was elaborate enough to include provisioning of the young (Farlow 1976a; Paul 1987a).

If our interpretation of carnosaurs and other theropods as rather vagile predators is correct, these reptiles may have foraged over large areas. This in turn implies that large theropods had less specific habitat preferences—as do large modern mammals—than most living reptiles (cf. Dodson et al. 1980, on the occurrence of *Allosaurus* in all major lithofacies of the Morrison Formation) and that their diets were less constrained by habitat selection than those of modern reptiles (cf. Shine 1986; Magnusson et al. 1987).

Garland (1983b) noted that the daily movement distances of carnivorous mammals are greater than those of noncarnivorous mammals of comparable size. Similarly, carnivores have larger home ranges than noncarnivores (Peters 1983; Calder 1984). Farlow (1987a) argued that if the differences in daily movement distance and home range size seen in different trophic groups of mammals also held for dinosaurs, these differences might account for the commonly observed preponderance of theropod footprints in ichnofaunas.

Auffenberg (1980) reported that juvenile oras remain together in small groups for several months after hatching, and Auffenberg (pers. comm.) indicated that the same behavior occurs in young *Varanus grayi* and *Varanus bengalensis*. Similarly, young iguanas will associate with each other after hatching (Burghardt 1977), and the occurrence of similar gregarious behavior in young crocodilians (and protection of the young in these groups by adults) is well documented (Coombs, this vol.). It is possible that conspecific juvenile dinosaurs of at least some species also associated with each other, with or without parental guarding (Horner 1984a, 1987; Horner and Weishampel 1988; for ornithopods), and Coombs (1982) suggested that such juvenile groups might have remained together as adults (cf. Werner et al. 1987). Whether or not Coombs's hypothesis for dinosaur sociality is valid, several workers have argued on the basis of trackways and skeletal assemblages that some theropods may have been social animals, perhaps even hunting in packs (Ostrom 1969*b*, 1972; Farlow 1976*b*; Buffetaut and Ingavat 1985; Paul 1987*a*). If any carnosaurs were pack hunters, such a strategy might have permitted them to attack prey too large or dangerous for a single individual to handle (Earle 1987; Berdarz 1988), such as large sauropods or ceratopsids (the taphonomic evidence given above is consistent with solitary hunting but does not preclude pack hunting). Paleoichnological traces interpretable as records of group hunting by large theropods have been found. Trackways from the Lower Cretaceous Glen Rose Formation in Texas suggest pursuit of one or more sauropods by large theropods, possibly the contemporaneous *Acrocanthosaurus atokensis* (Langston 1983; Bird 1985). The trackways indicate that a small number (possibly four) of theropods moved in the same direction as a larger number of sauropods, after the sauropod tracks were made. However, these trackways are not as easily interpreted as may be supposed: theropod tracks also indicate movement in the antiparallel or mirror-image direction (Farlow 1987*a*). Farlow (1987*a*) concludes that while the pack-hunting scenario is consistent with the evidence, so is an interpretation of individual theropods moving independently of most of the sauropods. Evidence of similar associations (or pursuits) from the Late Cretaceous of Bolivia (Leonardi 1984) may bear on this point; equally, these may represent theropods other than carnosaurs (e.g., abelisaurids).

Other aspects of behavior

Some insight into carnosaurian behavior is made possible by the interpretation of certain anatomical structures. The existence of cranial ornamentation in all well-known carnosaur skulls suggests that such ornamentations may have been used in species-specific recognition. Each species appears to have a distinct pattern of cranial ornamentation. None appears well suited or well placed for use in combat; indeed, the lacrimal rugosities of tyrannosaurids are borne on hollow elements so delicate as to preclude such use. Among modern mammals, horns and other cranial ornamentation are restricted to herbivores that hold territories or leks (Janis 1986) and suggest that, although likely vagile (see above), carnosaurs also may have been territorial or lekking. However, modern horned mammals are not carnivorous and use their horns in combat as well as display.

Injuries may result from certain kinds of behavior and hence form a record of those behaviors. Evidence of one agonistic encounter is provided by injuries we have seen that appear to be from traumatic impacts to the jaw of one specimen of *T. rex*. These injuries may have resulted from being bitten by another individual. Other injuries seemingly inflicted by biting have been found in *A. fragilis* (Petersen et al. 1972). These suggest that carnosaurs may have been contentious animals, much like the modern Nile crocodile (Cott 1961).

Carnosaur Physiology and the Trophic Web

Carnosaurs, along with the large ceratosaurs and abelisaurids, seem to have been the top carnivores of the Jurassic and Cretaceous. Their position in the trophic web has occasioned speculation that the predator/prey ratio of the fossil assemblages in which they occur reflects their elevated metabolic rate (Bakker 1986, and work cited therein). However, fossil predator/prey ratios are not unambiguous indicators of living ratios and may reflect preservational bias or land tenure patterns (Farlow this vol.). Carnosaurs may have been tachymetabolic (elevated metabolic rates) or they may have been heterometabolic (variable metabolic rates)—both strategies have ecological advantages, in theory at least (Farlow this vol.). Or they may have been inertial homeotherms. Consistent scenarios may be proposed for any of these (and other) interpretations. The hard evidence is not yet conclusive.

8

Ornithomimosauria

RINCHEN BARSBOLD

HALSZKA OSMÓLSKA

INTRODUCTION

Members of the Ornithomimosauria were lightly built, medium to large size, cursorial theropods with rather long forelimbs provided with nonraptorial hand. They lived in the Northern Hemisphere during the Cretaceous, except for one species, known also from the Late Jurassic of Gondwanaland (table 8.1).

As indicated by their name, ornithomimosaurians resembled birds, particularly modern ground birds, in having beaklike, usually edentulous jaws and long, slender hindlimbs with an elongate, compact metatarsus. This latter feature was originally noticed by Marsh (1890b), who described the first member of this group, *Ornithomimus velox,* and compared its hindlimbs to that of the ostrich.

The name Ornithomimosauria was introduced by Barsbold (1976b) to include the Ornithomimidae but also some other related but more primitive forms: the Garudimimidae and Harpymimidae (Barsbold 1981; Barsbold and Perle 1984).

DIAGNOSIS

The Ornithomimosauria may be diagnosed by the following characters: lightly built skull with long, shallow snout, very large orbit and large antorbital fossa; broad contact between premaxilla and nasal with separation of maxilla from external naris; bulbous parasphenoid capsule open caudoventrally; jaws edentulous—but primitive forms with very reduced dentition; mandibular articulation shifted rostrally, except in primitive forms; mandible long, very shallow without coronoid prominence; neck long, constituting about 40 percent of total presacral length. Scapulocoracoid with very high acromion; scapula only slightly longer than one-third of total forelimb length. Forelimb about half the length of hindlimb; humerus slender with weak deltopectoral crest; manus tridactyl with second and third digits subequal in length and thickness; metacarpal I more than half the length of metacarpal II; manual unguals weakly curved or straight with weak flexor tubercles displaced distally from articular end; ilium hooked cranially and truncated caudally; ischium about three-quarters of pubic length with proximally placed obturator process; hindlimb long and slender with long and narrow metatarsus; pedal unguals flat ventrally, with a semicircular depression instead of flexor tubercle and with sharp outer edges produced into "spurs" close to proximal end of ungual.

ANATOMY

The skull and postcranial skeleton are known in detail only in the derived Late Cretaceous family, the Ornithomimidae, especially in the Mongolian *Gallimimus*

225

TABLE 8.1 Ornithomimosauria

Ornithomimosauria

	Occurrence	Age	Material
Theropoda Marsh, 1881b			
Ornithomimosauria Barsbold, 1976b			
Harpymimidae Barsbold et Perle, 1984			
Harpymimus Barsbold et Perle, 1984			
H. okladnikovi Barsbold et Perle, 1984	Shinekhudukskaya Svita, Dundgov, Mongolian People's Republic	Aptian-Albian	Skull with associated incomplete postcranium
Garudimimidae Barsbold, 1981			
Garudimimus Barsbold, 1981			
G. brevipes Barsbold, 1981	Baynshirenskaya Svita, Omnogov, Mongolian People's Republic	Cenomanian-Turonian	Skull with associated incomplete postcranium
Ornithomimidae Marsh, 1890b			
Ornithomimus Marsh, 1890b			
O. velox Marsh, 1890b	Denver Formation, Colorado, Kaiparowits Formation, Utah, United States	late Maastrichtian	Postcranial material
O. edmontonensis Sternberg, 1933a (including *Struthiomimus currelli* Parks, 1933)	Judith River Formation, Horseshoe Canyon Formation, Alberta, Canada	late Campanian-early Maastrichtian	Skull and associated postcranial skeleton, 2 fragmentary postcrania
Struthiomimus Osborn, 1917			
S. altus (Lambe, 1902) (= *Ornithomimus altus* Lambe, 1902)	Judith River Formation, Horseshoe Canyon Formation, Alberta, Canada	late Campanian-early Maastrichtian	Skull and complete postcrania, 2 partial skulls with associated fragmentary postcrania, 8 incomplete postcrania
Dromiceiomimus Russell, 1972			
D. brevitertius (Parks, 1926b) (= *Struthiomimus brevitertius* Parks, 1926b, including *S. ingens* Parks, 1933)	Horseshoe Canyon Formation, Alberta, Canada	early Maastrichtian	Partial skull with associated postcranium, 6 fragmentary postcrania
D. samueli (Parks, 1928b) (= *Struthiomimus samueli* Parks, 1928b)	Judith River Formation, Alberta, Canada	late Campanian	Skull with incomplete postcranium
Archaeornithomimus Russell, 1972			
A. asiaticus (Gilmore, 1933a) (= *Ornithomimus asiaticus* Gilmore, 1933a)	Iren Dabasu Formation, Nei Mongol Zizhiqu, People's Republic of China	Late Cretaceous (?Cenomanian, ?Maastrichtian)	Partial manus, metatarsus, vertebrae, limb elements
Gallimimus Osmólska, Roniewicz, et Barsbold, 1972			
G. bullatus Osmólska, Roniewicz, et Barsbold, 1972	Nemegt Formation, Omnogov, Mongolian People's Republic	?late Campanian or early Maastrichtian	2 nearly complete skeletons, complete postcranium, skull with associated fragmentary postcranium, other fragmentary postcrania
Anserimimus Barsbold, 1988			
A. planinychus Barsbold, 1988	Nemegtskaya Svita, Bayankhongor, Mongolian People's Republic	?late Campanian or early Maastrichtian	incomplete postcranium
Ornithomimosauria incertae sedis			
Elaphrosaurus Janensch, 1920			
E. bambergi Janensch, 1920	Tendaguru Beds, Mtwara, Tanzania	Kimmeridgian	Fragmentary postcranium

Nomina dubia	Material
Coelosaurus affinis Gilmore, 1920	Limb elements
Coelosaurus antiquus Leidy, 1865	Fragmentary limb elements
Elaphrosaurus gautieri Lapparent, 1960a	Isolated limb elements
Elaphrosaurus iguidiensis Lapparent, 1960a	Isolated limb elements
Megalosaurus bredai Seeley, 1883	Femur
Ornithomimus sedens Marsh, 1892b	Sacrum and fragmentary ilium
Ornithomimus tenuis Marsh, 1890b	Fragmentary metatarsal

bullatus (Osmólska et al. 1972). Other families are represented by rare, fragmentary, and still incompletely described remains. As a result, the description that follows is based largely on *G. bullatus.*

Skull and Mandible

Except for *Gallimimus bullatus* (fig. 8.1), the well-preserved and complete skull is known only in *Garudimimus brevipes.* The known skulls of *Ornithomimus edmontonensis, Struthiomimus altus, Dromiceiomimus brevitertius,* and *D. samueli* are crushed, and most of them are incomplete (fig. 8.3A–D). The skull of *Harpymimus okladnikovi* (fig. 8.6A) is still undescribed.

The skull roof is strongly narrowed across the frontals and highest between the caudal portions of the orbits, where it is convex. The frontal, at least in *Gallimimus bullatus,* seems to be separated from the orbital rim by the prefrontal. The parietals are strongly inclined caudoventrally in *Gallimimus bullatus* and *D. brevitertius,* while less so in *Garudimimus brevipes.* The nasal is very long, extending caudally to the middle of the orbits. The premaxilla and maxilla are toothless in all species except probably in *H. okladnikovi.* The ventral margins of these bones are poorly preserved in this species, and it cannot be stated whether there were teeth. The jaws were probably covered by a horny beak in all toothless ornithomimosaurians. The external naris is elongate, of moderate size, and bounded ventrally and caudally by the premaxilla and dorsally by the nasal. The jugal is usually shallow. The snout is broadly rounded rostrally in *Gallimimus bullatus* and *Garudimimus brevipes.* In other species, the shape of the beak is not known for certain, because snouts are either deformed by lateral compression or missing entirely.

In lateral aspect, the skull has a long snout, large orbit, and very short temporal region. The antorbital fossa is shallow and long and occupies more than half of the preorbital length of the skull. Within the antorbital fossa, there is one large antorbital fenestra and one or two small additional fenestrae rostral to it. The large fenestra is largest and has a subquadrate shape in *D. brevitertius.* The postorbital bar is shifted caudally (in *Garudimimus brevipes*) or caudoventrally (in all other species), and it is very close to the quadrate. The infratemporal fenestra is strongly reduced. It is small and tear-shaped in *Gallimimus bullatus,* subdivided into two small openings in *D. brevitertius* and relatively the largest in *Garudimimus brevipes.* The quadrate is very strongly inclined in *Gallimimus bullatus* (although not so strongly as is indicated on fig. 3a in Osmólska et al. 1972). As a result, in this species, the mandibular ar-

ticulation is shifted rostrally, under the caudal boundary of the orbit. The same is true for *S. altus.* In *D. brevitertius* and *Garudimimus brevipes,* the quadrate is less inclined and the jaw articulation is just behind and ventral to the orbit in the first species and well behind it in the second one. The supratemporal fenestra is reduced and placed entirely opposite the parietal portion of the skull roof. It is very short in *O. edmontonensis* and *Gallimimus bullatus* while somewhat longer in *D. brevitertius. Garudimimus brevipes* has a longer postorbital region of the skull, and its supratemporal fenestra is relatively longer.

The basioccipital caudally and the basisphenoid rostrally form stout basal tubera separated from one another by a relatively shallow sulcus. The basicranial region is pierced by several openings leading to a system of intracranial sinuses. The basisphenoid gives off two separate alae rostral to the basal tubera. These alae surround a very large opening in the base of the skull (? homologous to Rathke's pouch) which is not closed in adult ornithomimosaurians. Rostral to the opening, a strong transverse bar develops from the basisphenoid; laterally, it becomes prolonged into a stout and short basipterygoid process. The opening mentioned above leads into a large, hollow, and bulbous capsule. Most of the capsule is formed of the thin-walled parasphenoid, which rostrally forms a thin cultriform process. On each side, between the capsular and postcapsular portions of the basicranium, there is a short, vertical parabasal channel running toward the postcapsular portion of the basisphenoid. The entrance to the channel is immediately in front of the basipterygoid process. Barsbold (1983*b*) proposed that this channel developed in some theropods at the immovable basipterygoid articulation as in theriodonts and turtles (e.g., Camp and Welles 1956; Tatarinov 1966). Whether other ornithomimosaurians, beyond *Garudimimus* and *Gallimimus,* developed this peculiar basicranial structure is not directly known since lateral compression of the skulls prevents any detailed study of this region. However, indirect evidence, for example, the shape and disposition of the palatal wing of pterygoids (see below) suggest that the bulbous capsule was also present in other ornithomimosaurians. A similar capsule, although entirely closed, is also found in the Troodontidae (see Osmólska and Barsbold, this vol.).

The endocranial cavity in *D. brevitertius* was reconstructed by Russell (1972). He noted that the olfactory bulbs were short and thin and the cerebral hemispheres were deep and convex. The olfactory bulbs were relatively larger in *G. bullatus.*

There is an extensive secondary palate formed by

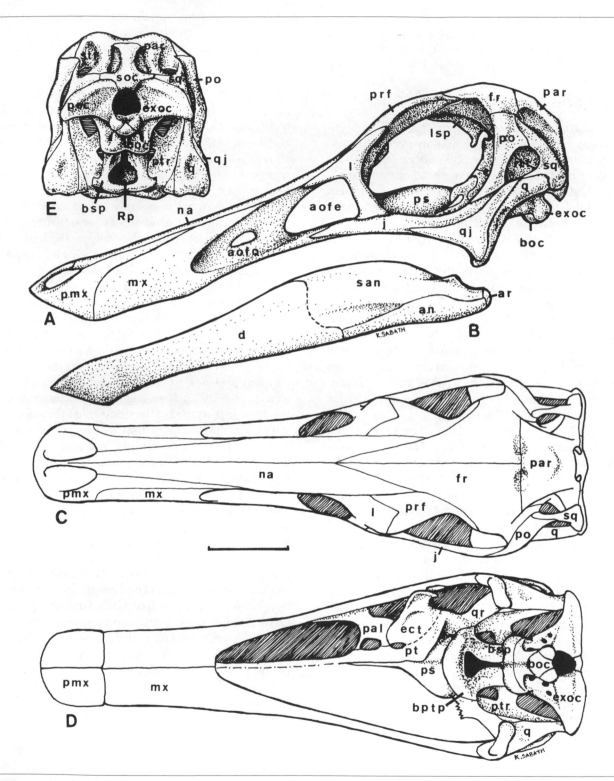

Fig. 8.1. Skull and mandible of *Gallimimus bullatus:* A. left lateral view of the skull, B. left lateral view of the mandible, C. dorsal, D. ventral, E. occipital, views of the skull. (All after Osmólska et al. 1972.) Scale = 50 mm.

the long palatal processes of the maxillae and the intervening fused vomers. The choanae are placed well caudal to the external nares. The palatal rami of pterygoids are wide and meet on the midline where they underlie the rostral portion of the parasphenoid capsule. The epipterygoid has the form of a wedgelike plate that is broader ventrally than dorsally.

The lower jaw is slender, tapers slightly forward, and is very shallow for most of its length. The surangular portion of the mandible is gently convex dorsally, but the adductor prominence is usually not pronounced. The external mandibular fenestra is small and elongate. The mandibular symphysis is relatively long and inclined caudoventrally.

The rostral half of the lower jaw is concave ventrally in *H. okladnikovi* and *D. brevitertius* but more or less straight in other species. The rostral portion of the mandible is shovellike in *Gallimimus bullatus* and *Garudimimus brevipes*. The mandibles are best known in these two species. They differ in the extent of the splenial, which reaches all the way to the symphysis in *Gallimimus* but not in *Garudimimus*. The coronoid is absent in both species, and the retroarticular process is large and has a wide, concave surface for *M. depressor mandibulae*. The medial edge of this process is enlarged to prevent excessive depression of the mandible. The mandible closes on the lingual side of the upper jaw.

The mandible is toothless in all species except *H. okladnikovi*. In the latter, there were about six small, blunt, cylindrical teeth along the very rostral part of each mandible. The lower jaw was probably covered by a horny beak, at least in the toothless forms.

Some small, recurved teeth were found separate in the same deposits that yielded the skeleton of *Elaphrosaurus bambergi*. They were considered by Janensch (1925*b*) as possibly belonging to this dinosaur. There is, however, no convincing evidence that *E. bambergi* had tooth-bearing jaws.

Neither stapes nor hyoid bones have ever been found in the ornithomimosaurians. A ring of about twenty sclerotic plates, arranged in four quadrants, is found in *D. samueli*. Judging from the very large size of the orbits, sclerotic plates probably also protected the eye in other ornithomimosaurians.

Postcranial Skeleton

Axial Skeleton

There are 10 cervicals, 13 dorsals, 5 to 6 sacrals, and about 40 caudals within the vertebral column (fig. 8.2). Most of the precaudal vertebral centra have pleurocoels in *G. bullatus*, although they are very shallow on the dorsals. Pleurocoels are lacking on the dorsals and sacrals in other species. The cervical vertebrae have platycoelous centra, and their caudal surfaces are slightly concave. The dorsal centra are platycoelous, while the caudal centra are usually amphiplatyan.

The cervicals are elongate with low neural arches and extremely low neural spines. These vertebrae increase strongly in length toward the caudal end of the series, except for two last cervicals that are somewhat shortened in relation to the preceding ones. The arches are wide, and the postzygapophyses, particularly in the second half of the series, become strongly elongate.

The centra of the dorsals lack ventral keels, except for the two most cranial ones. The neural spines of the sacrals are as high or somewhat higher than the iliac blades. These spines are long axially but are not fused with each other.

The tail is long, and its distal half differs from the proximal portion in having elongate prezygapophyses and lacking transverse processes on the caudals. The caudal centra lack pleurocoels. The transition point between the proximal and distal portions of the tail is between the 15th and 16th caudal vertebrae (in *D. brevitertius*, it is between the 12th and 13th caudals). The elongate prezygapophyses reach two-thirds the length of the preceding caudal, stiffening the distal half of tail.

All presacral vertebrae except the atlas and the last dorsal bear ribs. In the cervical region, they are fused with the vertebrae in adults. The shaft of the dorsal ribs is surmounted by a proximal expansion of bone, which is already weakly developed in the rib connected with the 10th cervical. This expansion reaches its climax on the 13th to 15th dorsal ribs.

Fifteen pairs of gastralia are known in *O. edmontonensis* and *S. altus*. Each abdominal rib consists of two overlapping segments, a dorsal and a ventral one. At the midline, each ventral segment ends in a rugose expansion, which meets its mate from the other side.

No sternum has ever been found, but the gap between the coracoids and abdominal ribs may indicate that nonossified sternals were present. Nicholls and Russell (1981) reported two elongate bony rods on each side of the rib cage in *S. altus*, at the cranial extremity of the gastralia, which they interpreted as xiphisternal structures.

Appendicular Skeleton

The scapulocoracoid (figs. 8.3G, 8.5A) differs from that of other theropods. The coracoid is usually long and shallow, except in *Gallimimus*, in which it is

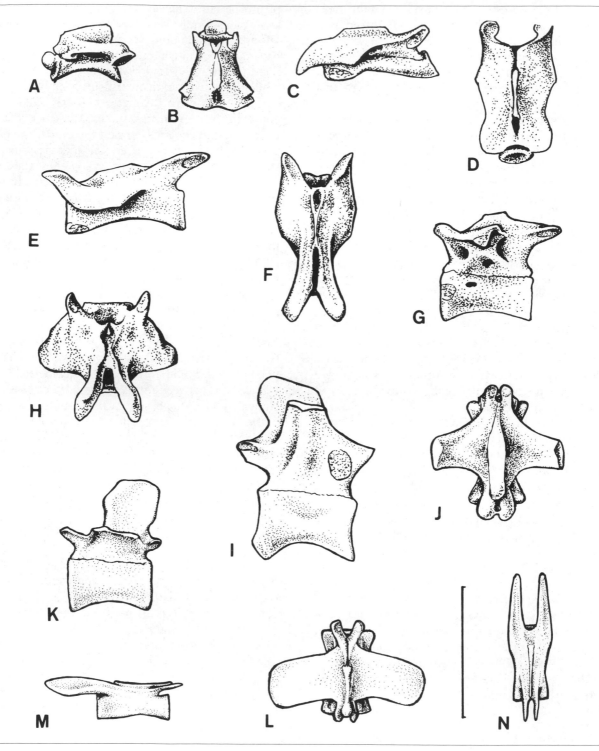

Fig. 8.2. Vertebrae of *Gallimimus bullatus:* A, axis, left lateral view, B. the same, dorsal view, C. fourth cervical, left lateral view. D. the same, dorsal view, E. eighth cervical, left lateral view, F. the same, dorsal view, G. first dorsal, left lateral view, H. the same, dorsal view, I. twelfth dorsal, right lateral view, J. the same, dorsal view, K. seventh caudal, left lateral view, L. the same, dorsal view, M. twenty-second caudal, left lateral view, N. the same, dorsal view. (All after Osmólska et al. 1972.) Scale = 50 mm.

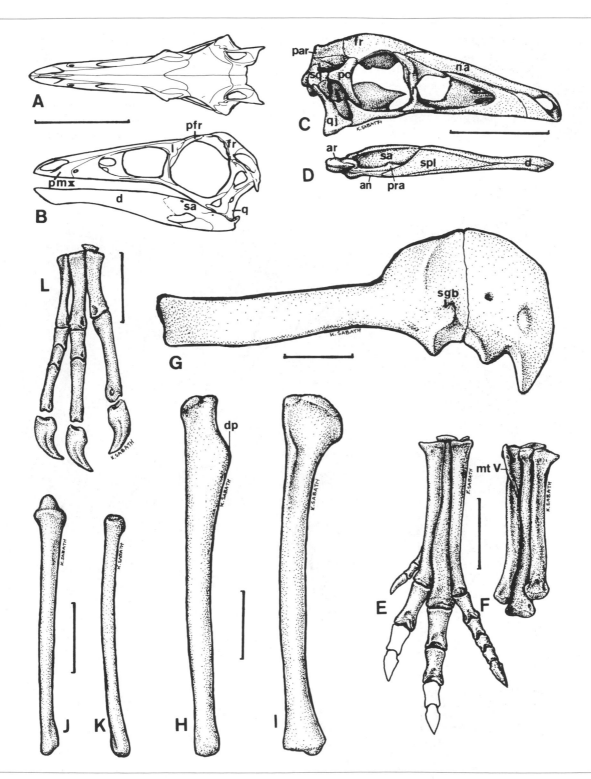

Fig. 8.3. A, B. *Dromiceiomimus brevitertius:* A. skull, dorsal view, B. skull and mandible, left lateral view. C–F, *Garudimimus brevipes:* C. skull, right lateral view, D. left mandible, medial view, E. fragmentary left pes, extensor side, F. left metatarsus, flexor side. G–L, *Gallimimus bullatus:* G. right scapulocoracoid, lateral view, H. right humerus, lateral view, I. the same, flexor side, J. right ulna, cranial view, K. right radius, cranial view, L. right manus, dorsal view. (A, B after Russell 1972; D after Barsbold 1983a; G–L after Osmólska et al. 1972.) Scale = 100 mm.

rather deep. There is an attenuated caudoventral process on the coracoid. The scapula has a high acromial process and a flange extending forward from the supraglenoid buttress (Nicholls and Russell 1985).

Contrary to Gauthier (1986), no clavicles have ever been reported in the ornithomimosaurians.

The forelimb is generally about half the length of the hindlimb (figs. 8.8, 8.9). The proportions of the

forelimb elements are highly variable (see below).

The humerus is 7 percent longer than the scapula in *O. edmontonensis* and 18 percent longer in *G. bullatus*. It equals the length of the scapula in *H. okladnikovi* and constitutes about 95 percent of the scapular length in *S. altus*. The humerus of the ornithomimosaurians (figs. 8.3H,I, 8.6B, 8.7B) is distinctive in being slender and straight. The deltopectoral crest is weak

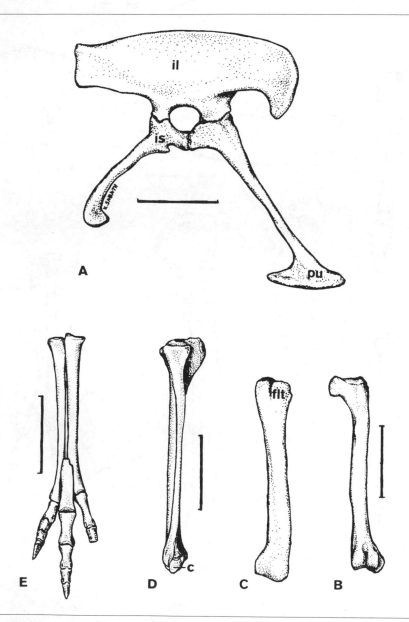

Fig. 8.4. *Gallimimus bullatus:* A. pelvis, right side, B. right femur, flexor side, C. the same, lateral view, D. right crus with proximal tarsals, lateral view, E. right pes, extensor side. (All after Osmólska et al. 1972.) Scale = 50 mm.

and proximally placed. The humeral shaft is twisted, causing the distal end to be set at an angle of 20° to 30° to the proximal end. In *H. okladnikovi*, the humerus deviates from this pattern because it is not twisted and slightly concave medially.

The radius, although always shorter than the humerus, may vary in its length, from 66 percent of humeral length in *G. bullatus* to 90 percent in *D. sam-*

ueli. The radius is as thick as the ulna in *G. bullatus* and only slightly thinner in other species (e.g., *S. altus, H. okladnikovi*). The ulna is generally slightly arched, convex toward the radius. Both these bones are tightly adherent distally in *H. okladnikovi;* in *S. altus*, they were even joined in a syndesmosis proximally and distally, preventing any relative mobility (Nicholls and Russell 1985).

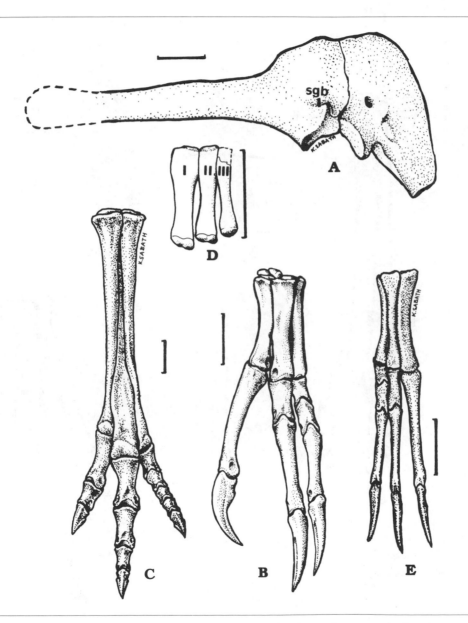

Fig. 8.5. Appendicular skeleton in the Ornithomimidae. A–C, *Struthiomimus altus:* A. right scapulocoracoid, lateral view, B. left manus, dorsal view, C. left pes, extensor side, D. left metacarpus of *Ornithomimus velox*, E. right manus of *Ornithomimus edmontonensis*, dorsal view. (A after Nicholls and Russell 1985; B, C, F after Osborn 1917; G after Sternberg 1933*a*.) Scale = 50 mm.

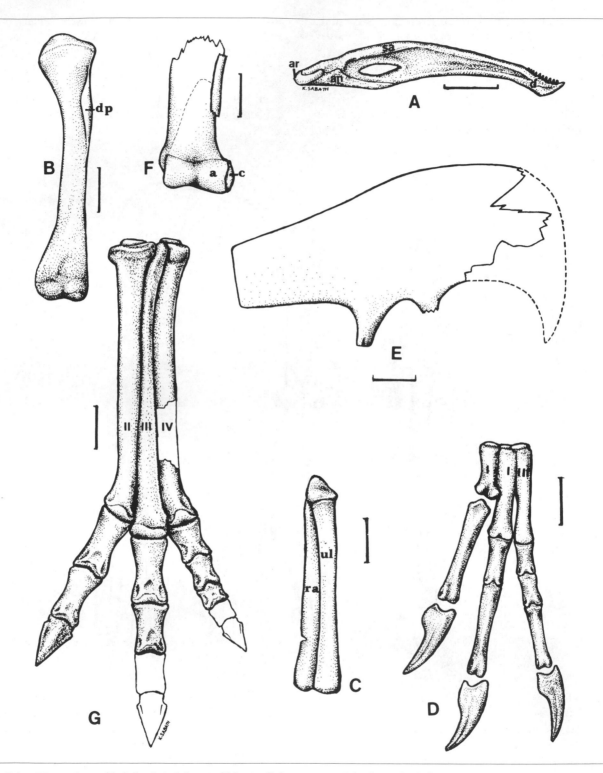

Fig. 8.6. *Harpymimus okladnikovi:* A. left mandible, medial view, B. left humerus, flexor side, C. left ulna and radius, as found, cranial view, D. left manus, dorsal view, E. fragmentary right ilium, lateral view, F. distal end of left tibiotarsus, extensor side, G. fragmentary left pes, extensor side. (A after Barsbold and Perle 1984.) Scale = 50 mm.

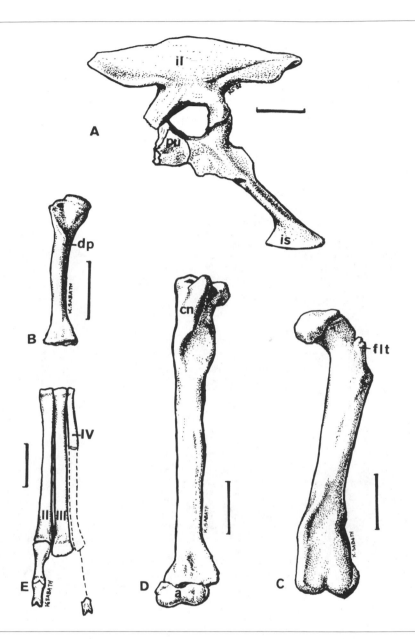

Fig. 8.7. *Elaphrosaurus bambergi:* A. fragmentary pelvis, left side, B. right humerus, flexor side, C. left femur, extensor side, D. left tibiotarsus, extensor side, E. fragmentary left pes, extensor side. (After Janensch 1925*b*.) Scale = 100 mm.

The carpals are mostly flattened bones with poorly defined articular surfaces. Five carpals are known in *S. altus.* The distal carpal represents a fusion of two or more carpals and adheres closely to metacarpals I and II. In this species, the radiale, centrale, and intermedium are concentrated above metacarpal I in a wedgelike arrangement. The pisiform is preserved adjacent to the caudal condyle of the ulna in *S. altus.* In *H. okladnikovi,* four flat distal carpals adhere to the metacarpus.

The manus is tridactyl (figs. 8.3L, 8.5B,D,E, 8.6D) and constitutes somewhat more than a third of the total length of the forelimb, except in *G. bullatus,* in which it is only a quarter that length.

All metacarpals are tightly appressed proximally, and the metacarpus is slightly arched transversely. Metacarpal I may be somewhat longer than metacarpal II, as it is in *Ornithomimus edmontonensis* and *O. velox,* or shorter, as in all other species. Its length constitutes about 95 percent of the length of metacarpal II in *S.*

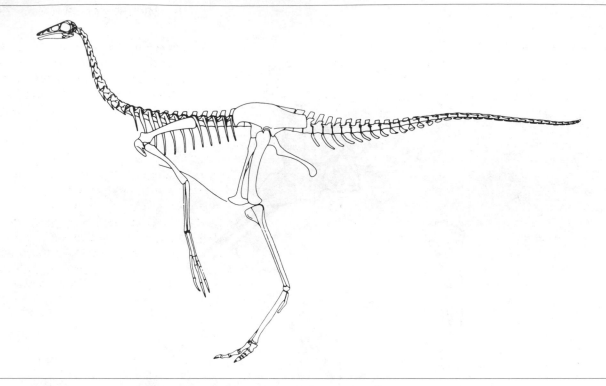

Fig. 8.8.   Restoration of the skeleton of *Dromiceiomimus brevitertius*. (After Russell 1972.)

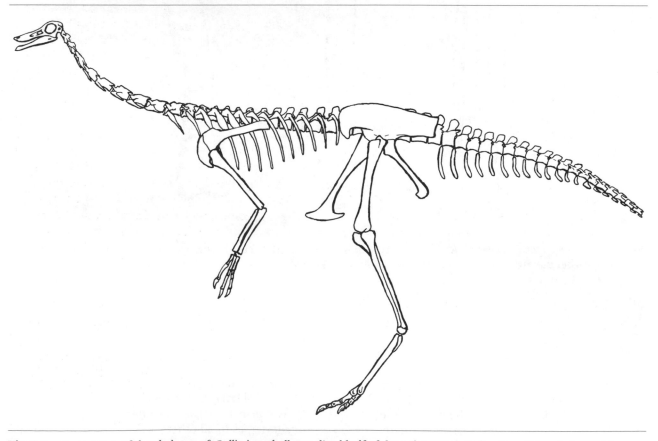

Fig. 8.9.   Restoration of the skeleton of *Gallimimus bullatus,* distal half of the tail omitted. (After Osmólska et al. 1972.)

236

altus and 85 percent in *G. bullatus* and *Archaeornithomimus asiaticus*. Metacarpal I is shortest in *H. okladnikovi*, in which it is only slightly longer than half the length of metacarpal II. Metacarpal III is usually thinner than metacarpals I and II, but in *H. okladnikovi*, this difference in the thickness is only slight. This metacarpal is about 10 percent shorter than metacarpal II in *A. asiaticus* and *G. bullatus* but somewhat longer than metacarpal II in *H. okladnikovi*. The distal articular ends of all metacarpals are rounded, nonginglymoid in all species, except *H. okladnikovi*, in which the ends of metacarpals I and II are deeply grooved.

All the interphalangeal joints are ginglymoid. Characteristic of the manus is that the third phalanx of the third digit is longer than the combined lengths of the first and second phalanges of this digit. The manual unguals are variable in forms, ranging from almost straight (in *O. edmontonensis*, *D. samueli*, and *H. okladnikovi*) to slightly recurved (in *S. altus* and *G. bullatus*). In most cases, the unguals are transversely compressed and have weak flexor tubercles that are displaced somewhat distally.

The pelvis of ornithomimosaurians (figs. 8.4A, 8.7A) is propubic and dolichoiliac. The ilium has elongate preacetabular and postacetabular processes. The preacetabular process has an undivided cranial margin, and its curved ventral portion culminates in a pointed tip. The end of the postacetabular portion is abruptly truncated. The ventral and dorsal edges of the postacetabular portion are subparallel.

The postacetabular portion of the ilium is especially long in *Gallimimus bullatus*, and its length exceeds the depth of the ilium above the acetabulum. These dimensions are about equal in all other species. The ilium is shorter than the pubis in *Garudimimus brevipes*, and the preacetabular process is relatively deep with a convex dorsal outline. The ilium is longer than the pubis in *D. brevitertius* and *Gallimimus bullatus* and about equal to that in *S. altus*: in these species, the preacetabular process is shallower and less convex dorsally than in *Garudimimus brevipes*. The preacetabular process lacks the pointed ventral tip in *E. bambergi*.

The pubis is long and slender with a moderately elongate footlike ventral symphysis. The ischium is slender and comparatively long, somewhat shorter than the pubis, and displays a small, proximal obturator process. The ventral symphysis of the ischium is expanded craniocaudally.

The hindlimb is very slender and long. The proportions of the hindlimb elements are unknown in *A. asiaticus* and *H. okladnikovi*.

The femur (figs. 8.4B,C, 8.7C) is nearly straight with a winglike lesser trochanter that does not reach the proximal end of the bone. The lesser trochanter is separated by a cleft from the weakly delimited greater trochanter. The latter is not separated from the femoral head. The fourth trochanter is developed as a faint ridge on the caudomedial side of the proximal third of the shaft. The distal end of the femur in ornithomimosaurians differs from that in other theropods in that the lateral condyle is larger than the medial condyle and the intercondylar groove is very deep and invades the extensor surface of the distal end of the femur. On the extensor face, there is a depression above the medial condyle. The craniomedial margin of the distal end of the femur is developed into a strong crest. The characteristic structure of the femur is present in *G. bullatus*, *S. altus*, and *G. brevipes* and is most probably characteristic also of other species, except *E. bambergi*. In the latter, the femur is sigmoidal, the lesser trochanter is not winglike, and the distal end is more like that of other theropods.

The tibiotarsus (figs. 8.4D, 8.7D) is usually about 10 percent longer than the femur, except in *G. brevipes*, in which the tibiotarsus is not quite as long as the femur. *D. brevitertius* has the relatively longest tibiotarsus (about 20% longer than the femur). The tibia has a fibular crest along the upper third of its shaft. The cnemial crest is strongly expanded cranially. The fibula is closely appressed to the tibia and passes onto the craniolateral edge of the tibia just below the cnemial crest of the latter. The fibula is thin for most of its length.

The astragalus forms most of the proximal tarsal articulation. Its condyles are separated distally by a broad and shallow concavity rather than a sulcus. This concavity is deeper and the lateral condyle is larger in *H. okladnikovi* than in other species. Laterally, the astragalus bears a notch for the thin, reduced calcaneum. The ascending process of the astragalus is high and shaped like an asymmetric triangle; the apex of this triangle is placed at the lateral edge of the tibia. The calcaneum has a shallow proximal notch for the fibula. On the craniomedial surface of the calcaneum, there is a small protuberance that fits into a corresponding notch on the astragalus.

The pes (figs. 8.3E,F, 8.4E, 8.5C, 8.6G, 8.7E) is tridactyl, the first and fifth digits being lost; only in *G. brevipes* is the reduced first digit still present. Whether the first digit was still present in the stratigraphically oldest species, *E. bambergi* and *H. okladnikovi*, is unknown.

The metatarsus is slender and compact and constitutes about 80 percent of the length of the femur (73% in *E. bambergi* and up to 86% in *D. brevitertius*). In *G. brevipes,* the metatarsus is only about 60 percent of the femoral length.

Metatarsal III is constricted proximally in different ways among ornithomimosaurian taxa. In all representatives of the Ornithomimidae, it is proximally restricted to the flexor side, and metacarpals II and IV contact one another proximally on the extensor surface of the metatarsus. This contact is shorter in *A. asiaticus* and in the species of *Ornithomimus* than in *S. altus* and *G. bullatus.* In ornithomimids, the metatarsus is particularly slender; its width constitutes only 11 to 13 percent of its length. There is a strongly reduced metacarpal V preserved in all adequately known species. In the earliest species, *E. bambergi, H. okladnikovi,* and *G. brevipes,* metatarsal III continues on the extensor face to the proximal end of the metatarsus and the width of the latter is 20 to 22 percent of its length. The distal articular surfaces of all metatarsals are smoothly rounded. The third digit is the longest and most robust and has nonginglymoid interphalangeal joints except between the penultimate phalanx and the ungual. The phalanges of the fourth digit are shortened. The pedal unguals are not recurved but are pointed at their tips and flat ventrally with a semicircular depression instead of the flexor tubercle. The edges of the unguals are sharp and are developed into a small "spur" on each side, close to the articular end.

SYSTEMATICS AND EVOLUTION

Recognizing the distinctiveness of his new form, *Ornithomimus velox,* Marsh (1890*b*) erected the family Ornithomimidae, also including in it two other species, *O. tenuis* and *O. grandis.* These last two, however, are *nomina nuda* since neither was described or illustrated. Moreover, according to Russell (1972, table 1) these two taxa assigned by Marsh to *Ornithomimus* do not represent ornithomimids.

Several other ornithomimid species from the North American Late Cretaceous were later described by Marsh (1892*b*), Lambe (1902), Gilmore (1920), Parks (1926*b,* 1928*b,* 1933), and Sternberg (1933*a*) as well as one species from the early Late Cretaceous of Asia (People's Republic of China) by Gilmore (1933*a*).

The African ornithomimosaurian *Elaphrosaurus bambergi* was described by Janensch (1920) from the Tendaguru beds of Tanzania (Late Jurassic), although

its ornithomimosaurian affinities were not recognized by this author. It was Nopcsa (1928*c*) who first assigned *E. bambergi* to the Ornithomimidae. Nopcsa's view was supported by Russell (1972), Russell et al. (1980), and Galton (1982*c*). These authors listed the ornithomimosaurian characters of *E. bambergi* (see below) and have considered it an ancestral species.

Russell (1972) was the first to revise the Ornithomimidae, and he erected two new genera in addition to *Ornithomimus* and *Struthiomimus: Archaeornithomimus* for the Early Cretaceous, North American *O. affinis* Gilmore, 1920 (Arundel Formation of Maryland), and the early Late Cretaceous, Asian *O. asiaticus* Gilmore, 1933*a* (Iren Dabasu Formation of People's Republic of China), and *Dromiceiomimus* for the North American *Struthiomimus brevitertius* Parks, 1926*b* (Edmonton Formation), and *S. samueli* Parks, 1928*b* (Judith River Formation), both of Alberta, Canada.

Exploration of the Cretaceous deposits of the Mongolian People's Republic by the Polish-Mongolian and Soviet-Mongolian Paleontological Expeditions between 1963 and 1981 has added three monotypic genera of ornithomimosaurian affinities: *Gallimimus* (*G. bullatus* Osmólska et al., 1972) and *Garudimimus* (*G. brevipes* Barsbold, 1981), both from the Late Cretaceous, and the Early Cretaceous genus *Harpymimus* (*H. okladnikovi* Barsbold et Perle, 1984).

While reviewing the mutual relationships among "small theropods," Barsbold (1976*b*) erected a new theropod infraorder, the Ornithomimosauria, to include at that time only the Ornithomimidae but successively two other new, monotypic families, the Garudimimidae (Barsbold 1981) and Harpymimidae (Barsbold and Perle 1984).

As the skull is inadequately known in *H. okladnikovi* and entirely unknown in *E. bambergi, A. affinis,* and *A. asiaticus,* it is impossible to state whether the distinctive characters of the Garudimimidae and Ornithomimidae (e.g., the parasphenoid capsule and parabasal channels) were already present in these early species. It is also impossible to state whether *E. bambergi* and the species of *Archaeornithomimus* had teeth. The teeth are entirely absent in representatives of both garudimimids and ornithomimids. *H. okladnikovi* still retains a few reduced teeth at the front of the dentary. These teeth, however, differ significantly from typical theropod teeth, because they are rather blunt and cylindrical. Janensch (1925*b*) considered some isolated, small, recurved, and crenelated theropod teeth as possibly belonging to *E. bambergi,* although the evidence is very weak.

Ornithomimidae Marsh, 1890*b*

The shared derived characters of the Ornithomimidae are: (1) metacarpals and digits of the manus are subequal in length; (2) the distal ends of metacarpals II and III ungrooved; (3) the combined length of the first and second phalanges of the third manual digit not greater than the length of the third phalanx of that digit; (4) metatarsus at least two-thirds the length of tibiotarsus and width 11 to 13 percent of its length; (5) metatarsal III strongly constricted along its proximal half while distally wide and slightly overlapping the adjoining metatarsals on the extensor face of the metatarsus; (6) metatarsals II and IV contacting each other proximally on the extensor face of the metatarsus; (7) pes lacking the first digit; (8) pedal digit III much longer than II and IV; (9) the first phalanx of pedal digit II is longer and more slender than that of the third digit.

The Ornithomimidae is the largest and best-known ornithomimosaurian family. It includes five genera from the Early and Late Cretaceous of North America and Central Asia (see table 8.1). These genera, as presently known, are best distinguished by the structure of the manus (Nicholls and Russell 1981; see also Russell 1972). In addition, Nicholls and Russell considered the following ornithomimid features as valid: the relative length of the presacral vertebral column to the hindlimb (combined lengths of the femur, tibia, and metatarsal III) and the relative length of the scapula to the humerus.

Ornithomimus (type species *O. velox*). The more complete of the two species assigned to *Ornithomimus* is *O. edmontonensis*. It has a length of the presacral vertebral column that is somewhat less than that of the hindlimb. The scapula is shorter than the humerus. The antebrachium is slender. Manual length is about 65 percent of the length of the femur and equals the length of the humerus. Metacarpal I is the longest, and along its entire length it adheres to metacarpal II. Metacarpals II and III are subsequal in length. All manual digits are equal in length. Their unguals are almost straight, relatively short, slightly shorter than a quarter of the length of the manus. Each ungual is shorter than its penultimate phalanx. The flexor tubercles are weakly developed. On the extensor side, the contact between metatarsals II and IV is short.

Struthiomimus is the monotypic genus represented by *S. altus*. This species is characterized by a presacral vertebral column that is longer than the hind-

limb. The scapula is longer than the humerus. The antebrachium is robust, and the radius was attached to the ulna by syndesmosis. The manus is large and powerful. The length of the manus equals the length of the humerus and is more than 70 percent the length of the femur. Of the metacarpus, metacarpal I is the shortest and, for more than two-thirds of its length, is closely applied to metacarpal II. Distally, metacarpal I diverges from the rest of the metacarpus. Metacarpals II and III are of equal length. The first digit is shorter than the remaining subequal digits. The unguals are very long (about 27% of the length of the manus), and, in comparison with other ornithomimid species, they are more robust and curved and have well-developed flexor tubercles. The unguals of the second and third digits are longer than their respective penultimate phalanges. The proximal contact of metatarsals II and IV on the extensor face of the metatarsus is extensive.

Dromiceiomimus includes two species, *D. brevitertius* and *D. samueli*, both based on incomplete specimens. The only skeletal elements in common between these two species are the caudal portion of the skull, the rostral portion of the mandible, the presacral vertebral column, and the humerus. The generic diagnostic characters were given by Russell (1972) and include, among others: the length of the presacral vertebral column is less than the length of the hindlimb (the hindlimb is lacking in *D. samueli*), the transition point in the tail between caudals 12 and 13 (the tail is unknown in *D. samueli*), the humerus is shorter than the scapula (the latter is lacking in *D. brevitertius*). According to Russell (1972), the resemblances of the cranial characters are decisive for assignment of these two species to the same genus. *D. brevitertius* has the relatively longest tibiotarsus among the ornithomimids. Its length exceeds that of the femur by 18 to 25 percent, depending on the size of the individual. Metatarsal III is excluded from the ankle joint. *D. samueli* has the relatively longest antebrachium among ornithomimids, and the length of the radius is about 92 percent of the humerus. The unguals of the manus in *D. samueli* resemble those of *O. edmontonensis* in being short and very weakly curved. As far as the preserved specimens indicate, the two *Dromiceiomimus* species differ in the length of the humerus, which is longer relative to the length of a cranial dorsal centrum in *D. brevitertius*.

Gallimimus is a monotypic genus with *G. bullatus* as the type species. *G. bullatus* has a presacral vertebral column that is equal to hindlimb length. The humerus is longer than the scapula. The manus is the shortest among ornithomimids and is 35 percent of the combined lengths of the radius plus humerus. Metacarpal I is the shortest and somewhat shorter than that in *Struthiomimus*. It adheres to metacarpal II along about its proximal half but diverges distally. Metacarpal II is the longest, and the second digit is somewhat longer than the two others. The manual unguals are short (19% of the length of the manus), somewhat curved, and bear strong flexor tubercles. On the extensor side of the metatarsus, the contact between proximal parts of metatarsals II and IV is relatively extensive.

Archaeornithomimus is a monotypic genus with one species, *A. asiaticus* (*A. affinis* is considered here a *nomen dubium*). *A. asiaticus* is represented by fragments of skeletons of several specimens (Gilmore 1933), but the skull is unknown. Within the manus, metacarpal II is the longest and metacarpal I the shortest. Metacarpal III is only very slightly shorter than metacarpal II but thinner than the latter. The proximal half of metacarpal I is tightly appressed to metacarpal II. The manual unguals are rather straight (Russell 1972; Nicholls and Russell 1981). Metatarsals II and IV contact each other proximally on the extensor face of the metatarsus but only for a very short distance, similar to that in the species of *Ornithomimus*. However, the metatarsus in *A. asiaticus* is slightly stouter than that in other ornithomimids. Its width is 18 percent of its length (in contrast to only 11 to 13 percent in other ornithomimids), and metatarsal III does not overlap the adjoining metatarsals on the extensor face of the metatarsus as it does in other representatives of the family.

A. asiaticus is the stratigraphically oldest species within the Ornithomimidae. The structure of its metatarsus is less derived than in other members of the family. It also displays a peculiar structure of the metacarpus in which metacarpal III is distinctly reduced in length and thickness, which is unique among ornithomimids. For this reason, the assignment of *Archaeornithomimus* to the Ornithomimidae may well be disputable.

Elaphrosaurus Janensch, 1920, assigned by Russell (1972) and Galton (1982c) to the Ornithomimidae, is here placed separately (see below).

Garudimimidae Barsbold, 1981

The family Garudimimidae includes but one species, *Garudimimus brevipes*. Its derived characters are: (1) the pleurocoels are lacking on the dorsal and sacral vertebrae; (2) the ilium is shorter than the pubis; (3) metatarsal III is strongly narrowed along the proximal part but visible throughout the extensor face of the metatarsus, entirely separating metatarsals II and IV; (4) the proximal phalanx of the second pedal digit is as long as that of the third digit; (5) the third pedal digit is only slightly longer than the fourth.

G. brevipes has a skull that does not differ in its basic structure from that in ornithomimid species, but the infratemporal fenestra is less reduced and the quadrate is more vertical. This species is also very primitive in many other respects among ornithomimosaurians. The ilium is shorter than the pubis. The tibiotarsus is only about 3 percent longer than the femur. The pes is tetradactyl, retaining a short first digit. The metatarsus is compact but relatively short, only slightly more than half the length of the tibiotarsus, while it is two-thirds that length in the ornithomimids. The metatarsus is also rather stout, and its width constitutes about 22 percent of its length. The pedal phalanges are short and robust. The manus is unknown.

Harpymimidae Barsbold et Perle, 1984

The family Harpymimidae includes but one species, *Harpymimus okladnikovi*. Its derived characters are (1) about six small, blunt, cylindrical teeth at the front of the dentary; (2) humerus short, about the length of the skull, and almost as long as the scapula; (3) metacarpal I only slightly more than half of the length of metacarpal II and metacarpal III is the longest.

The skull is poorly preserved in *H. okladnikovi* but has a typical ornithomimosaurian shape with a long snout, large orbits, and a shortened postorbital portion. The manus constitutes 38 percent of the forelimb. This is more than in other ornithomimosaurian species, but the large relative size of the manus only reflects the shortness of the humerus. The carpus and metacarpus are primitive in *H. okladnikovi* compared to other ornithomimosaurians. There are four unfused distal carpals. Metacarpal I is much shorter than metacarpal II and tightly appressed to the latter, and the distal articular surfaces of these two metacarpals are ginglymoid. However, in comparison with the saurischian pattern (Gauthier 1986), metacarpal I is already derived because it constitutes somewhat more than half the length of metacarpal II. The manual unguals are almost straight and have weak flexor tubercles that are displaced distally, as in other ornithomimosaurians. Their length constitutes 22 percent of the length of the manus. The ungual of digit II is about as long as its penultimate phalanx, and the unguals of the two other digits are shorter. The pes was probably tridactyl since no clear evidence of digit I has been found. The metatarsus is only moderately slender, and its width constitutes 20 percent its length. Metatarsal III uniformly narrows proximally and separates metatarsals II and IV throughout the extensor face of the metatarsus. Distally, metatarsal III does not overlap the adjoining metatarsals. The proximal phalanges of the second and third pedal digits are stout, short, and subequal in length. The pedal digits are relatively short and massive. The pedal unguals are typical for ornithomimosaurians.

Ornithomimosauria
incertae familiae

Elaphrosaurus Janensch, 1920

Elaphrosaurus includes but one Jurassic species, *E. bambergi* from the Tendaguru beds of East Africa, although additional material referred to as *E.* sp. has been described by Galton (1982c) from the Morrison Formation of North America. The systematic position of this genus within the Ornithomimosauria is still unclear. Russell (1972) and Galton (1982c) listed several characters shared between *Elaphrosaurus* and ornithomimids. Among them are the form of the presacral and proximal caudal vertebrae, the abbreviated phalanges of the fourth toe, and the straight, slender humerus with a weak deltopectoral crest. To these characters can be added proportions of the hindlimb (tibia approximately 15% longer than the femur, metatarsus approximately 73% femur length). However, the first three features are derived for ornithomimosaurians, thus primitive for the Ornithomimidae. It follows that only the hindlimb proportions are derived in this species. Russell and Galton also mentioned several characters of *E. bambergi* which are more primitive than in Ornithomimidae: shortness of ilium and hindlimbs relative to the elements of the vertebral column, the relatively short humerus, metatarsal III not shrouded proximodorsally by metatarsals II and IV (the latter two do not contact each other on the extensor face of the metatarsus), a low ascending process on the astragalus, a more distal position of the deltopectoral crest on the humerus, and the inferred teeth in this species (see above).

The manus in *Elaphrosaurus* is almost unknown. Only metacarpal I and possibly metacarpal IV are preserved (the latter as a thin and short element) in *E. bambergi.* Since metacarpal I is 20 percent of the length of the humerus, it was not particularly short. For instance, in *G. bullatus,* which has an exceptionally short manus among the ornithomimids, the length of metacarpal I is 18 percent of the humerus. In *S. altus,* the ratio is 30 percent, while in the dromaeosaurid, *Dei-*

nonychus antirrhopus, the length of metacarpal I is only 15 percent that of the humerus. Although based on limited material, these comparisons suggest that metacarpal I in *E. bambergi* may be already derived in its elongation, beyond the condition typical for most other theropod groups.

The width of the metatarsus in *E. bambergi* constitutes 22 percent of its length. It is much the same as in garudimimids, somewhat more than in the harpymimids (20%), and much more than in the ornithomimids (11–13%). Metatarsal III narrows very slightly upward, somewhat less than it does in the garudimimids and harpymimids and much less than in the ornithomimids. It is also visible throughout the entire extensor face of the metatarsus. In addition, the ilium is less derived in *E. bambergi* as compared with those in other ornithomimosaurians.

It follows that until more data on the anatomy of the species of *Elaphrosaurus* are known, this genus cannot be assigned to any established ornithomimosaurian family. Most probably, it deserves separate familial status.

The unquestionable ornithomimosaurian remains described from the Cloverly Formation by Ostrom (1970a) as *Ornithomimus* sp. do not provide any conclusive information as to their generic and familial assignment. They evidence, however, that this group of dinosaurs was already represented in North America during the Early Cretaceous.

The Ornithomimosauria (= Ornithomimidae in Gauthier 1986) plus the Maniraptora constitute the Coelurosauria Gauthier, 1986 n. comb. Gauthier's list of the shared derived characters for these two sister groups is here accepted, although the presence of a furcula and sternum should be eliminated as unique characters since they are known only in the oviraptorid *Ingenia* and not in the Ornithomimosauria.

Analysis of the distribution of the character states (mainly those concerning the proportion of the elements of the fore- and hindlimbs) among the ornithomimosaurian taxa allows a tentative reconstruction of the relationships within this group (fig. 8.10). As is expected, the Late Jurassic African *E. bambergi* is the most primitive species and displays many ancestral theropod characters (e.g., the preacetabular process of the ilium weakly extended cranioventrally, the femur is sigmoidal). The Early Cretaceous and early Late Cretaceous Asian species, *H. okladnikovi, G. brevipes,* and *A. asiaticus,* are more derived in displaying an extended preacetabular process (*H. okladnikovi, G. brevipes*) and more pinched metatarsals III. Some ancestral characters are still preserved in these species (e.g., several

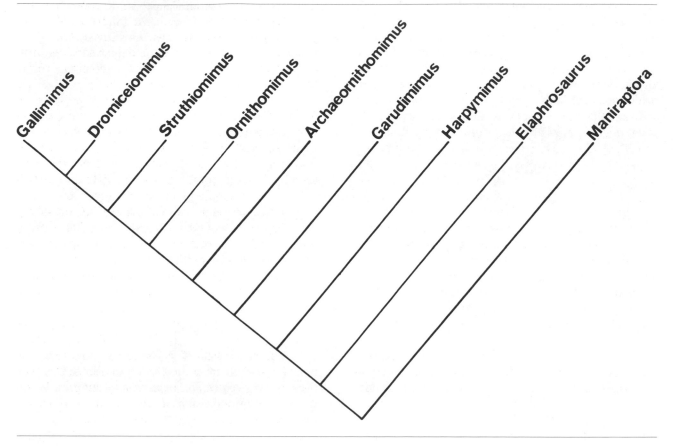

Fig. 8.10. Relationships within the Ornithomimosauria.

mandibular teeth are still present, and metacarpal I is relatively short in *H. okladnikovi;* the quadrate is little inclined and first pedal digit still occurs in *G. brevipes;* the metatarsus is relatively wide in all three species). These early species are incomplete, and their preserved skeletons are not comparable; thus, it is impossible to decide whether they are more closely related to each other than either of them is to the most derived Ornithomimidae. Among ornithomimids, *D. brevitertius,* with its long hindlimb, and *G. bullatus,* with a secondarily shortened manus, seem to be the most derived species.

The earliest Late Jurassic occurrences of the ornithomimosaurians in Africa and North America indicate that they were already widely distributed by that time. The Cretaceous record of this group in Africa is very poor and incomplete and is limited to the Early Cretaceous of North Africa (Lapparent 1960*a*). In North America, ornithomimosaurians were present during the Early Cretaceous (Gilmore 1920; Ostrom 1970*a*).

The earliest Asian ornithomimosaurian is the above mentioned *H. okladnikovi,* reported from Mongolia. The Ornithomimidae are so far restricted to the Late Cretaceous of Asia and North America. A majority of the North American species occurs in the upper Campanian and lower Maastrichtian strata, and only one valid species, the very incomplete *O. velox,* was described from the upper Maastrichtian. Nevertheless, judging by the numerous ornithomimid fragments reported from the Maastrichtian of North America (e.g., Russell 1972; Baird 1986*b*) and Europe (Lapparent 1960*a*), ornithomimosaurians lived on these continents until the end of the Mesozoic.

It is rather surprising that ornithomimosaurians have never been found south of the Gobi region, although data on other dinosaur groups in the Chinese Jurassic and Cretaceous strata are relatively abundant.

Taphonomy and Paleoecology

Ornithomimosaurians have a poor fossil record in large part because of the fragility of their hollow bones and their relatively small size. Remains of ornithomimosaurians are found both in lacustrine (*H. okladnikovi, G. brevipes*) and fluviatile (*O. edmontonensis, S. altus, D. samueli, D. brevitertius,* and *G. bullatus*) deposits, and it is likely that their natural environment may have been the peripheries of lakes and rivers. In Asia, the vertebrate fauna associated with the Early Cretaceous harpymimid shows both an aquatic component (champsosaurs) and some terrestrial components (iguanodonts, ankylosaurs, and other theropods; Kalandadze and Kurzanov 1974). The early Late Cretaceous garudimimid was associated with crocodiles and turtles but also with ankylosaurs, segnosaurs, carnosaurs, and hadrosaurids (Tsybin and Kurzanov 1979). It should be emphasized that the lacustrine environment was preferred by these Early and early Late Cretaceous primitive Asian ornithomimosaurians, while the derived ornithomimids of the late Late Cretaceous have chosen a life on the river edges, both in Asia and in North America. Ornithomimids were part of the terrestrial community that also contained large to very large dinosaurs, such as hadrosaurids, ankylosaurs, carnosaurs, and sauropods.

There were, however, a number of differences in the environmental context of ornithomimids between these two continents. In North America, ornithomimids are found mostly in deposits that correspond to lower coastal plain environments, with rich vegetation that grew under warm temperate, or possibly subtropical, climatic conditions (Dodson 1983), but they are rare (and fragmentary) in upper coastal plain habitats. Ceratopsians also occur with ornithomimids in North America. In Asia, however, horned dinosaurs are lacking from biotopes inhabited by ornithomimids. Late Cretaceous ornithomimids of Asia must have certainly lived in somewhat different environmental conditions than those of North America. Their remains have been found thus far only in Mongolia and there only in the continental interior where the climate was drier, with alternating dry and wet seasons (Gradzinski 1970; Osmólska 1980).

Ornithomimosaurian remains are rare and fragmentary in the deposits of the marginal marine environments. The only articulated specimen found in such a setting is that of *E. bambergi* from the Late Jurassic Tendaguru beds (Russell et al., 1980), but it is a very rare element of dinosaur fauna that was dominated by giant sauropods.

Ornithomimosaurians can be described as rather large (3–5 m in length and up to 2 m high at the hips) yet lightly built animals. Their long necks allowed the heads to be carried high above the shoulders. The large orbits speak for the presence of large eyes and a keen sense of vision. The lateral position of the eyes on the head and a very small overlap (if any) of the visual fields of each eye indicate that they could not see stereoscopically. But a wide range of dorsoventral and lateral movement of the cranial section of the neck ensured that ornithomimids had a broad field of observation. This was certainly a crucial condition for their safety.

Based on the presence of a large endocranial cavity, it has been suggested (Russell 1972) that ornithomimids had not only large brains but also levels of intelligence comparable to extant ratites.

Little is known of cranial kinesis in the ornithomimosaurians. Most probably, the quadrate was streptostylic and there was some mobility of the maxillary segment on the occipital segment along metakinetic and prokinetic axes. The basipterygoid joint seems to be reduced, at least in *Garudimimus brevipes* (Barsbold 1981) and *Gallimimus bullatus* (Barsbold 1983*b*), and the parasphenoid bulb, together with the closely underlying pterygoids, were most probably incorporated in the occipital segment. However, some sliding was certainly possible between the ectopterygoid and jugal, palatine, and maxilla as well as between the prefrontal and frontal, lacrimal and nasal, and, within the postorbital bar, between the postorbital and jugal. The position of the ventral zone of the palate flexion is unknown, but it must have occurred somewhere rostral to the pterygoids.

The mass of jaw adductor musculature was small, judging by the space available to the muscles in the postorbital region. Adductors (as well as the M. depressor mandibulae) were obliquely directed and inserted close to the mandibular articulation. Thus, the gape of the mouth was large but force of the muscles was weak, although the adduction of the mandible was rather quick. As a result, the slender jaws were weak and not adapted for crushing. They instead served only for procuring and transporting food, which itself may have consisted not only of soft, small items (insects, larvae, small vertebrates, eggs, fruits) as was formerly suggested (Russell 1972; Osmólska et al. 1972) but also of some relatively large objects (large gape, kinetic skull) swallowed whole. All ornithomimosaurians probably had a similar diet and feeding style, even though the harpymimids had a weak dentition along the front of jaws.

Nicholls and Russell (1985) suggested that at

least *Struthiomimus* but possibly also other ornithomimids were herbivorous. They argued that the shape of the beak (mistakingly considered by them as flat) indicates that it is likely that ornithomimids were plant-eaters and that they may have used their forelimbs for grabbing the fronds of ferns or cycads and pulling them to the mouth. In fact, the lower and upper jaws of ornithomimosaurians did not form a flattened beak comparable to that in hadrosaurids, for example. Instead, the beak, although broad, was relatively deep rostrally, at least in these species in which it was well preserved.

Both the structure of jaw apparatus and the character of skull kinesis seem to indicate that ornithomimosaurians were predominantly carnivorous, much like other theropods.

Functioning of the long ornithomimosaurian forelimbs with their large, clawed, three-fingered hands is difficult to explain. In an elegant analysis of the shoulder girdle and forelimb functions in *S. altus*, Nicholls and Russell (1985) drew attention to a number of scapulocoracoid and forelimb features present in all ornithomimids which suggest a wide range of motion of the shoulder girdle with respect to the body wall. Such mobility had the effect of increasing the forward reach of the arms. They demonstrated that the forelimbs were highly mobile in this species, yet they had a limited capability of the rotational movement. The major movements of the forelimbs were protraction and retraction at the shoulder joint and flexion and extension of the antebrachium and manus. Probably, the humerus could not have been elevated much above the horizontal.

It appears that the forelimbs in *G. bullatus* and in other ornithomimosaurians were limited in the same way. Thus, in contrast to what was formerly proposed (Russell 1972; Osmólska et al. 1972), the forelimbs in ornithomimosaurians could not have been used for raking or digging, as it would require a high degree of pronation of the limb to bring the palm parallel to the ground surface (Nicholls and Russell 1985).

As Nicholls and Russell suggested, the manus in *S. altus* formed a hook during full flexion of the digits, the second and the third digit acting in concert, the first digit somewhat opposing and converging toward the others (such a structure of the manus was a crucial argument for their suggestion that *Struthiomimus* may have been herbivorous). What *S. altus* was hooking can only be a guess.

The forelimb functions may have differed in details among ornithomimosaurians, particularly since the forelimbs in *G. bullatus* are sufficiently well known for comparing their functions with those of *S. altus*. In *G. bullatus*, the bones of the antebrachium do not seem to be as strongly united proximally and distally as they are in *S. altus*. A certain degree of rotation of the radius on ulna appears possible in this species. The much smaller manus of *G. bullatus* was unable to form the hook and hence could not have been used for clasping as in *S. altus*. Nevertheless, taking into account the ability of the limbs to reach far forward seen in both *G. bullatus* and *S. altus*, one can suggest that they may have served, for instance, to shake down the food items from the branches (fruits, small animals) or bring them from the bottom of the water basin to the surface.

The elongate crus (relative to the femur) as well as the long metatarsus are characteristic of ornithomimosaurians as a whole and indicate that these bipedal dinosaurs were capable of rather long strides. As a result, ornithomimosaurians and especially the ornithomimids were among the fleetest of all cursorial dinosaurs. Nevertheless, the estimates of running speeds of some ornithomimids (Thulborn 1982) indicate that their maximum was not very high, probably less than 60 km/h and even as low as 35 km/h.

Judging from the differences in the pelvic proportions and foot structure between the primitive and derived divisions of the Ornithomimosauria, it is clear that the harpymimids and garudimimids were somewhat distinct from the ornithomimids in terms of locomotion. As noted by Barsbold (1983*b*), the two former groups are characterized by relatively shorter metatarsi and would have been only moderately fast runners. Barsbold suggested that they were essentially wading animals, moving about on the wet, spongy substrates. Such a habit is consistent with the fact that remains of *H. okladnikovi* have been found in the swamp-lacustrine deposits of the Khukhtekskaya Svita (Kalandadze and Kurzanov 1974; Novodvorskaya 1974). In contrast, ornithomimids appear to be adapted to faster progression on hard substrates. This evidence comes, among other things, from their very long and compact metatarsi lacking the first toe. The strength and efficient mechanics of the ornithomimid hindlimb were most probably very important in accelerating these seemingly defenseless animals out of a danger. Their natural enemies might be dromaeosaurids and carnosaurs, the remains of which are commonly found in the same deposits as those of ornithomimids.

9

Elmisauridae

PHILIP J. CURRIE

INTRODUCTION

Elmisaurids are small, lightly built theropods (estimated weight 35–65 kg) of the Late Cretaceous characterized by large feet and gracile hands. The skull is at present unknown, but raptorial claws and elongate limbs suggest that they were quick and agile predators.

Chirostenotes pergracilis was described by Gilmore (1924a) on the basis of two nearly complete hands of a single individual collected from the Judith River Formation of Alberta (figs. 9.1, 9.2). A slender pair of dentaries was tentatively assigned to the species in the same paper, but there is no justification for this assignment. An isolated, articulated foot (fig. 9.2) became the type specimen of *Macrophalangia canadensis* (Sternberg 1932a). The same location produced a metatarsus that Parks (1933) described as *Ornithomimus elegans*. Various authors speculated on the possible synonymy of these three species, but until the recent discovery of a partial skeleton of *Chirostenotes pergracilis* (Currie and Russell 1988), this could not be confirmed.

Elmisaurus rarus (figs. 9.1, 9.2) was described on the basis of partial skeletons from the Nemegt Formation of the Mongolian People's Republic (Osmólska 1981). In most respects, it is very similar to *Chirostenotes pergracilis,* although the tarsals and metatarsals are coossified in the Asian form. These two species were different enough from other theropods that Osmólska (1981) established the Elmisauridae (table 9.1).

A coossified tarsometatarsus discovered recently in Alberta is close enough in morphology to suggest that *Elmisaurus* was also found in Alberta and that the type specimen *Ornithomimus elegans* is an immature specimen of *Elmisaurus* (Currie in prep.).

The skull and axial skeleton, with the exception of the sacrum, are at present unknown for the Elmisauridae. Currie and Russell (1988) speculate that *Chirostenotes* may turn out to be a caenagnathid when more complete specimens are found, whereas Ostrom (this vol.) feels that *Chirostenotes* may be synonymous with *Dromaeosaurus*.

DIAGNOSIS

The Elmisauridae is diagnosed on the basis of metatarsal III being pinched proximally between the second and fourth metatarsals but with only the proximal tip excluded from the extensor surface of the metatarsus; the second to fourth distal tarsals and metatarsals are tightly associated and tend to coossify into a tarsometatarsus; the back of the metatarsus is deeply emarginated.

Elmisaurids are also characterized by a unique suite of characters that collectively help define the family. The third digit of the manus is longer than the first digit. There is a proximodorsal "lip" on the manual unguals. The postacetabular blade of the ilium is shorter than the preacetabular blade.

TABLE 9.1 Elmisauridae

	Occurrence	Age	Material
Theropoda Marsh, 1881b			
Elmisauridae Osmólska, 1981			
Chirostenotes Gilmore, 1924a			
C. pergracilis Gilmore, 1924a (including *Macrophalangia canadensis* Sternberg, 1932a)	Judith River Formation, Horseshoe Canyon Formation, Alberta, Canada	middle Campanian–early Maastrichtian	Partial skeleton, pes, 2 manus, sacrum, isolated elements
Elmisaurus Osmólska, 1981			
E. rarus Osmólska, 1981	Nemegt Formation, Omnogov, Mongolian People's Republic	?late Campanian or early Maastrichtian	3 pedes, manus
E. elegans (Parks, 1933) (= *Ornithomimus elegans* Parks, 1933)	Judith River Formation, Alberta, Canada	middle Campanian	3 pedes

ANATOMY

Six coossified vertebrae form the sacrum in *Chirostenotes,* which is a higher number than in theropods other than troodontids, oviraptorosaurians, some ornithomimids, and birds. Sacral vertebrae are pierced by pleurocoels.

The coracoid of *Chirostenotes* is the only known element of the pectoral girdle of elmisaurids. It had a more extensive contact with the scapula than in the dromaeosaurid *Deinonychus* but is otherwise very similar in size and shape.

Metacarpal I in elmisaurids is a straight, slender bone, in contrast with the shorter, stouter version seen in oviraptorids, *Deinonychus, Ornitholestes, Dilophosaurus,* and tyrannosaurids or the longer, stouter metacarpal I of ornithomimids. *Microvenator* has a similar metacarpal I. Metacarpal II is the largest metacarpal element, whereas metacarpal III was probably very slender.

The third digit is 30 percent longer than digit I but 30 percent shorter than digit II. The longest phalanx in the hand is the second one of digit II, which is about 5 percent longer than II-1. In contrast, the first phalanx of digit I in ornithomimids and oviraptorids is longest, and the second finger is only slightly longer than the third.

The first digit is not opposable to the other fingers in elmisaurids. Digit III is very slender, and it is conceivable that the outer fingers (II, III) were syndactylous. This would only flex properly if there was some alignment between joint mcII/II-1 and mcIII/III-1 and between II-1/II-2 and III-2/III-3.

The manual unguals are strongly laterally compressed and have large, well-developed flexor tubercles. The dorsal edge of the articulation forms a prominent "lip" for the attachment of the extensor tendons.

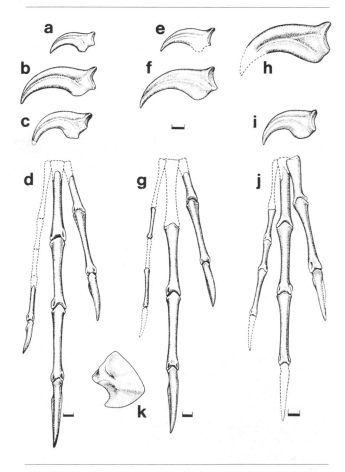

Fig. 9.1. Comparison of elmisaurid manus. a, b, c, d, manual unguals III, II, and I and reconstruction of the manus of *Chirostenotes pergracilis;* e, f, manual unguals of *Chirostenotes pergracilis* (after Currie and Russell 1988); g, reconstruction of *Chirostenotes pergracilis* manus (after Currie and Russell 1988); h, manual ungual II; i, j, manual ungual I and reconstruction of the manus of *Elmisaurus rarus* (after Osmólska 1981); k, coracoid of *Chirostenotes pergracilis* (after Currie and Russell 1988). Scale = 1 cm.

The only other theropod genera with similar unguals are *Microvenator* and *Oviraptor*.

Like most theropods, the elmisaurid ilium is dolichoiliac, having an elongate preacetabular blade that is squared off cranially. The postacetabular blade is less than 70 percent as long as the preacetabular portion of the ilium. Similar proportions are only seen in the opisthopubic dromaeosaurids, although the elmisaurid pelves are propubic.

The ischium of *Chirostenotes* is shorter than in any other known theropod and is only about 45 percent as long as the femur. There is a strong obturator process that is well separated from the pubic contact.

The hindlimb of *Chirostenotes* is long and slender. The ratio of tibia to femur is 1.2, which compares favorably with the more cursorial theropods. The femur is similar to those of other theropods. However, the base of the lesser trochanter is more distally located than in other known forms. Some of the specializations in the pelvis and femur suggest that the tail was relatively short in comparison with other theropods (Currie and Russell 1988).

In *Chirostenotes*, the cnemial crest of the tibia did not extend far along the shaft of the bone. A ridge extending distally along a third the length of the shaft clearly indicates the region of contact with the fibula, a bone that is otherwise unknown in elmisaurids.

The astragalus of *Chirostenotes* completely covered the distal end of the tibia. Its ascending process extended at least a sixth the distance up the flexor surface of the tibia. There is no pronounced horizontal groove separating the distal condyles from the ascending process. As reconstructed, the calcaneum is a small but discrete element. The distal tarsals tend to fuse to the metatarsus. Distal tarsal IV has a distinctive proximolateral projection in *Elmisaurus* specimens from Mongolia and Alberta.

The metatarsus is elongate and similar to *Ornithomimus edmontonensis* in relative length. Fusion or close association of the proximal ends of the second, third, and fourth metatarsals with the distal tarsals is the most diagnostic characteristic of elmisaurids. The lines of fusion are still visible in the type specimen of *Elmisaurus rarus* but are obliterated in other specimens from Mongolia and Alberta.

As in most theropods, the first metatarsal is divided into proximal and distal segments as the intervening shaft failed to ossify. Only the distal segment is known, and it was closely appressed to the shaft of metatarsal II. The shaft of the second metatarsal is only slightly more slender than that of the fourth. The proximal end of metatarsal II contacts metatarsal IV for a short distance on its extensor surface. Metatarsal III is

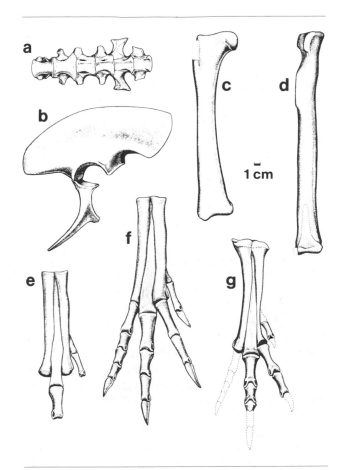

Fig. 9.2. Pelvic and hindlimb elements of elmisaurids. a, ventral view of sacrum of *Chirostenotes pergracilis* (after Currie and Russell 1988); b, ilium and ischium of *Chirostenotes pergracilis* in right lateral view, (after Currie and Russell 1988); c, d, reconstruction of right femur and tibia of *Chirostenotes pergracilis*, cranial views (after Currie and Russell 1988); e, metatarsals and phalanges of *Chirostenotes pergracilis* (after Currie and Russell 1988); f, reconstruction of pes of *Chirostenotes pergracilis* (type of "*Macrophalangia*"); g, reconstruction of pes of *Elmisaurus rarus* (after Osmólska 1981). Scale = 1 cm.

proximally pinched between the adjacent metatarsals, as in troodontids, ornithomimids, and tyrannosaurids. In *Chirostenotes*, the proximal end is diamond shaped in section, whereas it is triangular in *Elmisaurus*. The distal end of metatarsal IV is well separated from the distal end of the second metatarsal on the caudal surface, in contrast with *Troodon*, in which they almost touch (Wilson and Currie 1985). Metatarsal V is a small, boomerang-shaped bone in *Chirostenotes*.

At midlength, the flexor surface of the metatarsus is strongly arched in section, the shafts of the second and fourth metatarsals extending farther caudally than that of the third.

The phalanges of the pedal digits are more slender than those of ornithomimids, and the unguals are laterally compressed and recurved. Digits II and IV are subequal in length, and digits II, III, and IV are almost as long as their corresponding metatarsals. Phalanges II-1 and III-1 are the longest ones in the pes, III-1 being slightly shorter in *Chirostenotes*. Pedal unguals are sharply pointed but are not as trenchant as those of dromaeosaurids.

In summary, three species of elmisaurids can be easily distinguished from each other. Metatarsals II, III, and IV of *Chirostenotes pergracilis* are tightly bound proximally and may fuse in large individuals. The proximal end of metatarsal III is diamond shaped in section, and phalanx III-1 is marginally shorter than phalanx II-1. The linear dimensions of mature specimens of *Elmisaurus rarus* are about one-third smaller than *Chirostenotes pergracilis*. The distal tarsals have coossified with the proximal ends of metatarsals II, III, and IV to form a tarsometatarsus. Distal tarsal IV has a proximolateral projection, and the proximal end of metatarsal III is triangular in section. The proximal phalanges of digits II and III are subequal. *Elmisaurus elegans* is the same size as *Elmisaurus rarus* but is more gracile. A ridge along the caudolateral margin of metatarsal IV is not as strongly developed, and metatarsal II has a relatively more slender shaft in the species from Alberta.

PHYLOGENY AND EVOLUTION

Elmisaurids have been established as a separate family of theropods on the basis of incomplete but distinctive skeletons. Currie and Russell (1988) speculate on the possibility of synonymizing *Chirostenotes* and *Caenagnathus*. There are many similarities in the postcranial skeletons of the elmisaurids and *Oviraptor*, including the relative proportions of manual digits I and III, the presence of a proximodorsal "lip" on the manual unguals, the shape of the ilium, and the six sacral vertebrae with pleurocoels. These similarities suggest that the Elmisauridae is at least a sister taxon to the Oviraptorosauria. *Microvenator* (see Norman, this vol.) may be an early representative of this lineage.

PALEOECOLOGY AND BIOGEOGRAPHY

At present, elmisaurids are known only from the Late Campanian and/or Early Maastrichtian of the Northern Hemisphere. Nine articulated, partial skeletons have been collected (two of which were recovered from channel sandstones), as well as identifiable isolated bones. The sediments of the Judith River, Horseshoe Canyon, and Nemegt formations represent predominantly fluvial deposition, which favors the preservation of only large animals as articulated specimens. Small animals, such as elmisaurids, tended to be destroyed or disarticulated by the action of high energy fluvial environments or by scavengers. No cranial material can be associated with identifiable elmisaurid postcrania.

Elmisaurids are found in regions that are thought to have been warm temperate to subtropical and seasonally dry (Gradzinski 1970; Jarzen 1982).

The length of the hindlimb and its proportions suggest that *Chirostenotes* was a quick, agile animal. The manual claws were laterally compressed and recurved, indicating that elmisaurids were probably active predators. Their small size and gracility suggests that prey species were small. This is difficult to confirm without cranial material, however. The outer digit of the hand is so slender that it may have been useful for prying insects from the bark of trees if it was free or for grooming if it was syndactylous.

10

Oviraptorosauria

RINCHEN BARSBOLD

TERESA MARYAŃSKA

HALSZKA OSMÓLSKA

INTRODUCTION

The Oviraptorosauria is a group of rare Late Cretaceous theropods known so far from only seventeen specimens from the Northern Hemisphere (table 10.1). These animals were toothless, small, and cursorial, not exceeding 2 m in length, with raptorial hands and slender hindlimbs. They include two families: the Mongolian Oviraptoridae and the North American Caenagnathidae. Of the two families, the Oviraptoridae is better known, represented by several more or less complete skeletons including skulls and postcrania, some of them in an excellent state of preservation (Osborn 1924*a*; Osmólska 1976; Barsbold 1976*a*, 1976*b*, 1977*a*, 1977*b*, 1981, 1983*a*, 1986). In contrast, the Caenagnathidae is known from but two isolated mandibles (R. M. Sternberg 1940; Cracraft 1971).

Although the known postcranial skeletons of the oviraptorids have a typical theropod anatomy, the skull, mandible, and especially the mandibular joint cannot be compared with those of any dinosaur known so far.

DIAGNOSIS

The Oviraptorosauria may be diagnosed on the following characters: extensively fenestrated mandible with a medial process on the articular; dentary with two long caudal processes, the ventral process bordering the mandibular fenestra; very shallow splenial; prearticular that extends rostrally along more than half the length of the mandible; toothless jaws with rostral portions probably covered by a horny bill.

ANATOMY

The description of the cranial and postcranial anatomy that follows is largely based on the skulls and postcranial skeletons of *Oviraptor philoceratops* and *Ingenia yanshini*.

In the Caenagnathidae, the skull has never been found; thus, it is not known which of the very peculiar skull and postcranial features displayed by the Oviraptoridae as described below were characteristic of all Oviraptorosauria.

TABLE 10.1   Oviraptorosauria

	Occurrence	Age	Material
Theropoda Marsh, 1881b			
Oviraptorosauria Barsbold, 1976b			
Oviraptoridae Barsbold, 1976a			
Oviraptorinae Barsbold, 1981			
Oviraptor Osborn, 1924a			
O. philoceratops Osborn, 1924a	Djadochta Formation, Omnogov, Mongolian People's Republic	?late Santonian or early Campanian	3 specimens, including skulls, complete postcranium, partial postcranium
O. mongoliensis Barsbold, 1986	Nemegt Formation, Omnogov, Mongolian People's Republic	?late Campanian or early Maastrichtian	Skull and fragmentary postcranium
Conchoraptor Barsbold, 1986			
C. gracilis Barsbold, 1986 (including *Ingenia yanshini* Barsbold, 1981 *partim*)	Red Beds of Khermeen Tsav, Omnogov, Mongolian People's Republic	?middle Campanian	Skull and associated fragmentary postcranium, 3 fragmentary postcrania
Ingeniinae Barsbold, 1981			
Ingenia Barsbold, 1981			
I. yanshini Barsbold, 1981	Red Beds of Khermeen Tsav, Omnogov, Beds of Bugeen Tsav, Bayankhongor, Mongolian People's Republic	?middle Campanian- early Maastrichtian	6 specimens, including skulls and complete postcrania
Caenagnathidae R. M. Sternberg, 1940			
Caenagnathus R. M. Sternberg, 1940			
C. collinsi R. M. Sternberg, 1940	Judith River Formation, Alberta, Canada	late Campanian	Mandible
C. sternbergi Cracraft, 1971	Judith River Formation, Alberta, Canada	late Campanian	Mandible

Skull and Mandible

The skull of oviraptorids (figs. 10.1, 10.2) is akinetic, very short and deep, with a large orbit and large, subquadrangular infratemporal fenestra. The supraorbital fenestra is elongate, twice as long as wide. The preorbital portion of the skull is deep and strongly shortened, with a highly placed external naris and one or two antorbital fenestrae within a short, triangular antorbital fossa. The jaws form a toothless beak. The suborbital, postorbital, and infratemporal bars are thin. The bones of the skull, especially those of the skull roof, the nasal and basicranial regions, and the lateral wall of the braincase are strongly pneumatized and fused.

The premaxilla forms about half of the short snout. In *O. philoceratops*, the premaxilla displays an additional ascending process that is medial and rostral to the processes that border the elongated naris dorsally and caudally. This ascending process with its opposite fellow form the rostral margin of a large, highly pneumatized crest. In the rostral portion of the crest, there is a pair of narrow, longitudinal openings, the role of which is not known. The crest in *O. philoceratops*

is positioned above the snout, rostral to the orbit, and its margin reaches rostrally beyond the premaxilla. In *O. mongoliensis*, the crest has its highest point between the lacrimals. It is also developed caudally onto the parietals; rostrally, it slopes steeply toward the beak. In *Conchoraptor gracilis* and probably also in *Ingenia yanshini*, there is no crest and the nasal is broader and flat rostrally. Each nasal, in addition to being pneumatized, bears a large air chamber connected to the external naris, in both *O. philoceratops* and *C. gracilis*. The frontals are fused and are medially as long as the parietals. Laterally, the frontal forms the medial portion of the dorsal orbital rim. The parietals are very long and narrow between the supratemporal fenestrae. Along the line of fusion of the parietals, a low sagittal ridge is pronounced. The occiput has a crescentic shape due to the long, ventrally directed paroccipital processes. The endocranial cavity seems to be very large relative to the skull length, but the olfactory tracts are short.

The lateral wing of maxilla is very short and shallow and bears several pneumatic foramina. The prefrontal cannot be distinguished from the adjoining bones. The lacrimal does not seem to be much extended dorsally. The jugal is triradiate; its rostral pro-

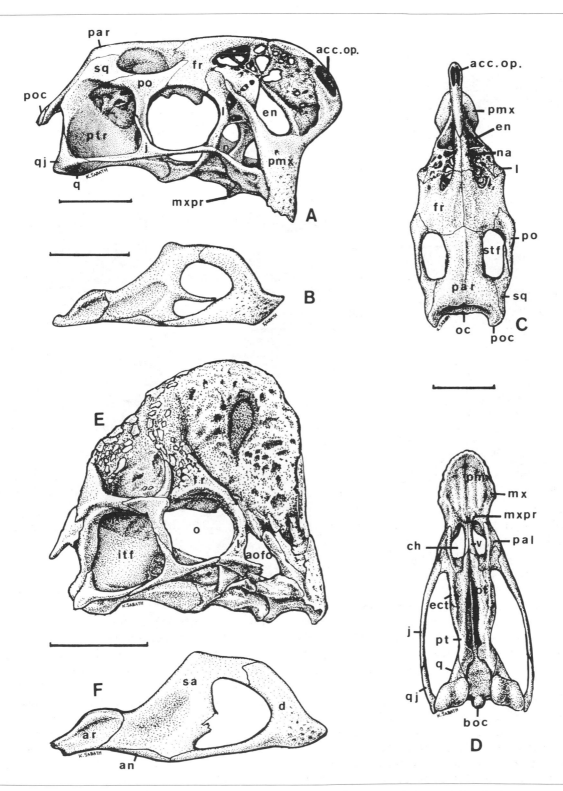

Fig. 10.1. Skulls and mandibles of the Oviraptoridae. A–D, *Oviraptor philoceratops:* A. right lateral view of the skull, B. right lateral view of the mandible, C. dorsal, D. ven-tral, views of the skull. E, F, *Oviraptor mongoliensis:* E. right lateral view of the skull, F. right lateral view of the mandible. (E, F after Barsbold 1985.) Scale = 50 mm.

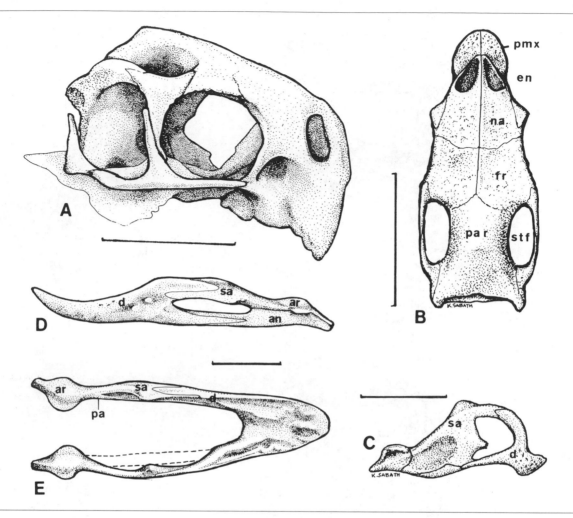

Fig. 10.2. Skull and mandibles of the Oviraptorosauria. A, B, *Conchoraptor gracilis:* A. right lateral, B. dorsal, views of the skull, C. *Ingenia yanshini,* right lateral view of the mandible, D, E, *Caenagnathus collinsi:* D. left lateral view and E. dorsal views of the mandible (A after Barsbold 1986; C after Barsbold 1977*b;* D, E after R. M. Sternberg 1940.) Scale = 50 mm.

cess is very thin and directed dorsorostrally; the dorsal process is high, overlapping posteriorly the postorbital over more than two-thirds the length of the postorbital bar; the caudal process is also thin and forms about a half of the lower temporal arcade. The quadratojugal is loosely attached to the quadrate and has a robust, dorsoventrally flattened rostral process and a very thin, craniocaudally flattened dorsal process. There is a large quadrate foramen between the quadratojugal and quadrate. The quadrate is robust and concave caudally and has a laterally expanded mandibular condyle that is divided by an oblique, shallow groove into narrow lateral and wide medial parts. The pterygoid ramus of the quadrate tightly adheres to the quadrate ramus of the pterygoid. The squamosal is extended ventrolaterally,

broadly overlapping the quadrate craniolaterally. The quadratosquamosal joint was immovable.

The palate is very peculiar in the oviraptorids, unlike that in any other dinosaur. The hard palate is formed by the premaxillae and maxillae; each maxilla bears a pair of longitudinal, rounded ridges. Caudally, the maxilla develops a ventral toothlike projection, which closely articulates with its fellow on the midline. These "teeth" are caudally buttressed by fused, laterally flattened vomers. The palatine is small and borders the small choana caudally. The quadrate ramus of the pterygoid adheres to the quadrate ventrolaterally, just above the condylar surface of the latter. More dorsally, the contact between the two bones is extensive and very tight. The palatal ramus of the pterygoid forms a robust bar and has its lateral portion reflected ventrad. The bar is continued cranially by the robust ectopterygoid, which joins the maxilla laterally, medial to the rostral end of the jugal. Along the ventral surface of the pterygoid-ectopterygoid bar, there is a longitudinal trough formed due to the ventral reflection of the lateral portion of the bar. This bar is reminiscent of that in the dicynodonts (Crompton and Hotton 1967, fig. 1) and has never been observed in other dinosaurs. Mediocaudally, the pterygoid is fused to the basipterygoid process.

Although the mandibles differ in proportions between both families of the Oviraptorosauria, the articular always displays a strong, rounded longitudinal ridge separating two stout, subcircular, and horizontal processes, the medial one usually larger. In both families, the mandibles are toothless, and there is a large external mandibular fenestra. The coronoid process is placed far forward and it is curved mediad.

There are, however, a number of differences between the two families. The mandible in the Caenagnathidae is shallow, long, and curves upward rostrally; the external mandibular fenestra is narrow and elongate; the symphysis is very long, and the coronoid process is shallow and located near the middle of the mandibular fenestra.

The mandible is similar in all species of the Oviraptoridae. It is very deep, and the external mandibular fenestra is correspondingly enlarged. This fenestra is positioned in the rostral half of the jaw and is incompletely subdivided into two smaller fenestrae by a thin, rostrally directed process of the surangular. The coronoid process is well developed and triangular and it is rough over its medial surface. With the mandible adducted, the coronoid process passed medial to the jugal bar and was located somewhat caudally and just below the rostroventral angle of the orbit, against the maxillary process of the ectopterygoid medially (Barsbold 1977a). Dorsorostrally, the dentary is deeply excavated, which resulted in a large gap between the lower and upper jaws. The gap was probably filled by a thickened portion of the horny bill during life. The symphysis is very strong and dentaries are thickened at the symphysis.

The hyoid bones, stapes, and sclerotic orbital plates are unknown in the oviraptorosaurians.

Postcranial Skeleton

Axial Skeleton

The vertebral column is basically similar to that of most other theropods, except for the tail. There are 10 cervicals, 13 dorsals, 6 or 7 sacrals, and about 40 caudals. All centra, except those in the distal half of the tail, bear pleurocoels. The dorsal centra lack a ventral keel, and there are the hyposphenes and hypantra developed between the dorsal vertebrae. The centra and prezygapophyses of the caudals are shorter than in other theropods. Ribs are present on all caudals except a few distal ones. The caudal neural spines and the chevrons are deep. The transition point is not marked in the caudal series.

The dorsal ribs are typically theropodan. There are 12 to 14 pairs of gastralia.

The sternals are known in *O. philoceratops* and *I. yanshini*. They are platelike and are sometimes fused with each other. Each sternal plate bears a groove for the coracoid on its craniolateral border.

Appendicular Skeleton

The shoulder girdle consists of a slender scapula, coracoid, and fused clavicles. The cranial edge of the nial edge of the acromial process on the scapula is thickened, forming a narrow, subtriangular articular surface for the clavicle. The clavicles are robust, fused medially without any trace of the joint surfaces, resulting in a broadly open, more or less U-shaped structure. In the midline, there is pronounced a short, stout spine on the caudal edge of this structure, which is nearly at a right angle to the external surface of the fused clavicles. The distal ends of the clavicles are flattened for contact with the scapulae. With the clavicles articulated with the scapulae, the spine is directed ventrocaudally.

The forelimb (humerus + radius) is half the length of the hindlimb (femur + tibiotarsus) in *O. philoceratops* and less than half in *I. yanshini*. The humerus is weakly twisted and has a well-developed deltopectoral crest, the apex of which is well above the midlength of the shaft in both species of *Oviraptor* and in *C. gracilis* and somewhat lower, close to the midlength, in *I. yanshini*. The upper border of the deltopectoral crest is arched between the apex and the proximal end of the humerus in *O. mongoliensis* but straight in other species. There is a well-developed medial epicondyle on the humerus of all oviraptorids. The radius is somewhat more than 70 percent the length of the humerus in *I. yanshini* and about 90 percent in the others. The ulna has a flattened distal end, which is distinctly expanded laterally. The carpus includes but two bones. The distal carpal is semilunate and immovably attached to metacarpals I and II. Along its convex proximal surface runs a broad, transversely directed groove. The proximal carpal is much smaller, roundish, and im-

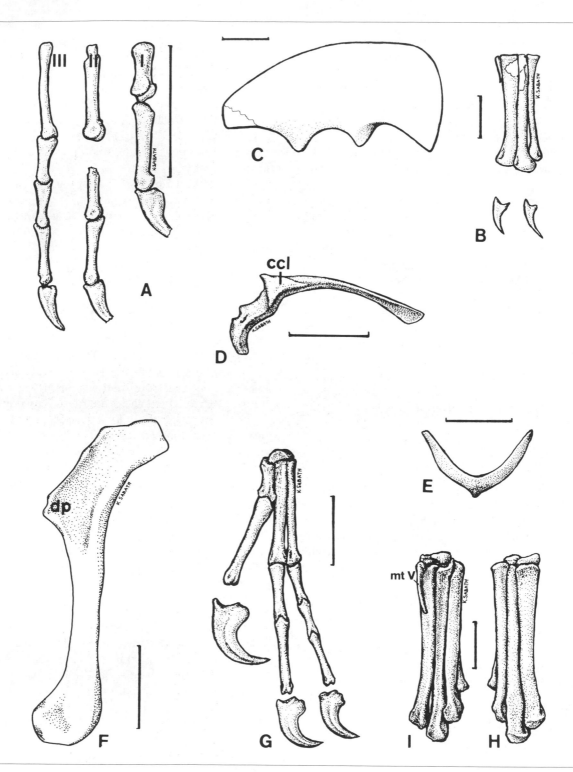

Fig. 10.3. Appendicular skeleton of the Oviraptoridae. A, B. *Conchoraptor gracilis:* A. metacarpals and digits of the right manus, lateral view, B. right metatarsus, extensor view and lateral views of pedal unguals. C. *Oviraptor mongoliensis,* right ilium, lateral view, D–I. *Oviraptor philoceratops:* D. right scapulocoracoid, cranial view, E. fused clavicles, ventral view, F. left humerus, lateral view, G. left manus, dorsal view, H. left tarsometatarsus, extensor view, I. the same, flexor view. (A after Barsbold 1986; D, E after Barsbold 1983a.) Scale = 50 mm.

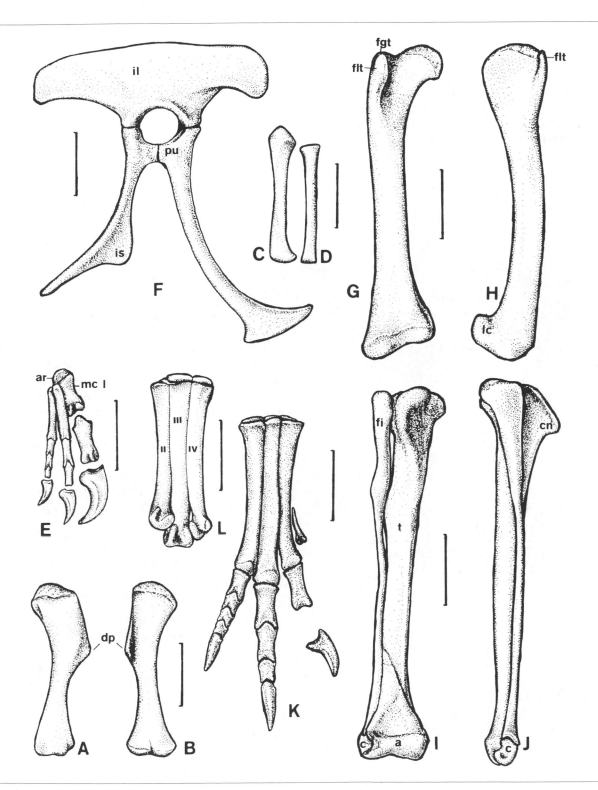

Fig. 10.4. Appendicular skeleton of *Ingenia yanshini*. Right humerus in, A. caudal view, B. cranial view, C. right ulna, cranial view, D. right radius, lateral view, E. right manus, dorsal view, F. right pelvis, lateral view, G. right femur, cranial view, H. the same, lateral view, I. right crus with proximal tarsals, extensor view, and J. the same, lateral view, K. right pes, extensor view, L. right metatarsus, flexor view. Scale = 50 mm.

movably connected with the radius. It could travel along the groove on the distal carpal.

The tridactyl manus has subequal metacarpals II and III that are about double the length of metacarpal I in *O. philoceratops* and *C. gracilis,* while only about a third longer than that in *I. yanshini.* The maximum length of the metacarpus is about half that of the humerus in both species of *Oviraptor* and *C. gracilis* and only about a third in *I. yanshini.* As a whole, the manus forms 40 percent of the total forelimb length in the oviraptorines and 30 percent in *I. yanshini.*

There are striking differences in the structure of the manual digits between *O. philoceratops* and *I. yanshini.* In the former, the digits are comparatively long and slender, and the first is shorter than the others; the penultimate phalanges are the longest; the unguals are strongly curved and have strong flexor tubercles and dorsoposterior "lips." In *I. yanshini,* the first digit is the longest and very strong; digits II and III are much shorter and thinner due to the strongly reduced phalanges, especially the penultimate ones, which are shorter than the preceding phalanx. The unguals of digits II and III are short and weakly curved with small flexor tubercles. There are no "lips" on the unguals in *I. yanshini.* The manus in *C. gracilis* is intermediate, with digital proportions as in *Oviraptor,* and small, straight unguals without "lips," more like those in *I. yanshini.*

The pelvis is propubic and narrow as in most theropods. A naturally articulated pelvis of *I. yanshini* shows that, at least in this oviraptorid species, the direction of the pubis is close to vertical rather than cranioventral as in many other theropods. The oviraptorid ilium is variable in shape, very deep with a convex dorsal margin in *O. mongoliensis,* less deep and less dorsally arched in *O. philoceratops,* and shallow with a flat dorsal margin in *I. yanshini* and *C. gracilis.* The preacetabular process of the ilium is deeper and sharply pointed cranioventrally in *Oviraptor* while shallow and obtuse in *I. yanshini* and *C. gracilis.* The postacetabular process is truncated in *I. yanshini* but not in *O. philoceratops* and *O. mongoliensis.* The brevis shelf is narrow in all these species. There is a "lip" along the cranial half of the acetabulum in oviraptorids. The pubic peduncle of the ilium is very peculiar in *O. philoceratops.* It is directed cranioventrally, rather than ventrally as in all other oviraptorids. In all oviraptorids, the pubis is concave cranially and the pubic foot is little expanded longitudinally; its caudal section is especially short. The ischium is slightly more than two-thirds the length of the pubis and has a large obturator flange that is placed approximately at midlength of the shaft.

The femur has a slightly bowed shaft. The greater trochanter is massive and well separated from the femoral head. The lesser trochanter is thick, finger-shaped, and separated from the greater trochanter by a narrow cleft and from the shaft below by a deep furrow. It extends vertically to about the level of the greater trochanter. The fourth trochanter is very weak and placed at midlength of the shaft. The distal condyles are separated by a very shallow concavity. There is a tuberlike protuberance caudally above the lateral condyle. The craniomedial edge of the distal end of the femur is sharp and crestlike. The tibia and fibula are each 20 to 30 percent longer than the femur, being proportionally the longest in *C. gracilis.* The fibula is reduced in thickness along its distal half and is closely appressed to the tibia. The astragalus displays an ascending process that covers less than a quarter the length of the tibia. The calcaneum is reduced and thin, with a craniomedial protuberance that fits into a corresponding notch on the astragalus.

The metatarsus is only moderately elongate, being somewhat less than half the length of the tibia. It is stouter in *I. yanshini* than in *O. philoceratops* and *C. gracilis.* Metatarsals I and V are strongly reduced. The differences in the lengths of the remaining metatarsals are very slight, the second being the shortest and the third the longest. Metatarsal III is only slightly thinner proximally than distally. The pes is tetradactyl with a very short first digit, and the third digit is distinctly longer than the others. The pedal unguals are compressed laterally and weakly curved.

SYSTEMATICS AND EVOLUTION

The first described representative of Oviraptorosauria, *Oviraptor philoceratops,* was assigned to the Ornithomimidae because of its edentulous jaws (Osborn 1924*a*). *Oviraptor* was also considered an ornithomimid by Romer (1956, 1966) and Steel (1970). However, that assignment was questioned by Russell (1972) who considered *O. philoceratops* not an ornithomimid. Barsbold (1976*a*) established the family Oviraptoridae, for which he later erected the infraorder Oviraptorosauria (Barsbold 1976*b*). In the meantime, Osmólska (1976) noticed a structural resemblance between the oviraptorid mandibles and those described as *Caenagnathus collinsi* and *C. sternbergi* and considered by R. M. Sternberg (1940) and Cracraft (1971) as representing birds (Caenagnathidae R. M. Sternberg, 1940). Osmólska included *Oviraptor* in the Caenagnathidae but considered this family to be theropodan. The avian status of *Caenagnathus* had been questioned before. Wetmore (1960) listed several characters supporting a relationship of *Caenagnathus* with reptiles, especially ornithomimids. Romer (1966) and Steel (1970) considered the Caenagnathidae as possible coelurosaurians. Cracraft (1971) noticed that some of reptilian characters listed by Wetmore for *Caenagnathus* (e.g., shape of the articular, forward position of the coronoid process) occur also in different bird groups. However, the presence of these characters in the undoubted Late Cretaceous theropod dinosaur, *Oviraptor*, approximately coeval with *Caenagnathus*, makes assignment of the Caenagnathidae within Oviraptorosauria more probable.

Oviraptorosauria includes two families (Barsbold 1981): the Oviraptoridae, with two subfamilies, the Oviraptorinae and Ingeniinae, and the Caenagnathidae. Following Osmólska (1976), Gauthier (1986) considered Caenagnathidae as including both *Caenagnathus* and *Oviraptor*, a view that is not accepted in this section.

As already mentioned, the oviraptorosaurian skull and postcranial skeleton are known only in the Oviraptoridae. For that reason, the two families included in the Oviraptorosauria can be only briefly diagnosed at the moment based mainly on the mandibles, which are the only common element of the Oviraptoridae and Caenagnathidae.

The derived characters of the Oviraptoridae are a short and deep mandible; a dentary with short, concave rostral portion thickened at the symphysis; two widely separated caudal dentary processes that ventrally and dorsally border a vertically enlarged external mandibular fenestra; subdivision of the external mandibular fenestra by a spinelike rostral process of the surangular; and a deep skull with a shortened snout and a palate with a pair of median, toothlike processes on the maxillae.

The differences between the Oviraptorinae and Ingeniinae, as originally conceived by Barsbold (1983), mainly concern the manus, although other differences may also be mentioned (see above.) The crest, which is an obvious feature of some oviraptorids, does not seem to have any diagnostic value at the subfamilial level. It is present in some oviraptorines (*O. philoceratops* and *O. mongoliensis*) while absent in others (*C. gracilis*); it seems to be lacking in ingeniines. As a whole, the skeleton is more robust in ingeniines, the bones being more massive than those of oviraptorines in the skeletons of comparable size.

I. yanshini was considered an ornithomimid by Gauthier (1986). However, the structure of the man-

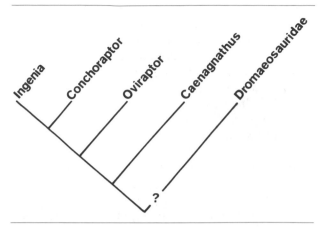

Fig. 10.5. Relationships within the Oviraptorosauria.

dible, palate, shoulder girdle, carpus, metacarpus, and pes are typical of the Oviraptoridae and quite unlike those in the Ornithomimidae.

The Caenagnathidae is monotypic and known only from the mandible. Its derived characters are the long and narrow external mandibular fenestra; the dorsal curvature of the rostral portion of the mandible; and on the dorsal surface of the symphysis, two shallow grooves along the lateral edges of the dentary. The shape of the mandible indicates that the caenagnathid skull was not shortened.

Judging from the unreduced length of the mandible, the late Campanian Caenagnathidae from Canada may be considered more primitive than the Mongolian Oviraptoridae, which range from the ?late Santonian to ?early Maastrichtian. As the oldest known oviraptorosaurian, *O. philoceratops*, already displays the highly specialized mandible and skull, one can presume that the lineage separated prior to the late Santonian.

Within the Oviraptoridae, the stratigraphically older *O. philoceratops* has a more specialized, crested skull than the middle Campanian *C. gracilis* and *I. yanshini*. However, in its postcranium, it is less advanced than *I. yanshini* in which the sacrum has seven vertebrae and the manus has distinctly shortened digits (fig. 10.5).

The Oviraptoridae display in their postcranial skeletons many characters found in theropods assigned by Gauthier (1986) to the Maniraptora.

The same structure of carpus is found in the dromaeosaurids (Ostrom, this vol.) and troodontids (Osmólska and Barsbold, this vol.). A similar, gracile metacarpal III and the third digit as seen in species of *Oviraptor* and *Conchoraptor* is found in dromaeosaurids, troodontids, *Elmisaurus*, and *Ornitholestes*. The structure of metatarsus is generally similar in dromaeo-saurids. The same form and proportions of the ischium are shared also by dromaeosaurids and troodontids, although in the former family, the pelvis is opisthopubic, while the oviraptorids have a more typical propubic configuration.

The characters of the manus in *Oviraptor* and *Conchoraptor* and of the metatarsus in all oviraptorids are primitive and might be inherited from a common coelurosaurian ancestor. Presence of the peculiar semi-lunar carpal element in the similarly reduced carpus of troodontids, dromaeosaurids, and oviraptorids supports Gauthier's (1986) suggestion that at least these three groups are closely related (Maniraptora Gauthier, 1986) since it seems probable that such a complex character is not due to convergence.

PALEOECOLOGY AND TAPHONOMY

Oviraptorosaurians constitute a rare element of the Late Cretaceous vertebrate assemblages. Most of them were found as single specimens that usually included the skull and a more or less complete postcranium. The majority of their remains come from the Mongolian red beds (*O. philoceratops, C. gracilis, I. yanshini*), which were deposited under arid or semiarid climates. Only one Mongolian species, *O. mongoliensis,* and both North American caenagnathids were found in the fluviatile deposits formed under more humid climates.

Barsbold (1977*a*, 1983*a*) has discussed in detail the probable mode of life of the oviraptorids. He drew attention to the fact that the akinetic skull, monimo-stylic quadrate, and powerful structure of the jaws indicate that their food was very hard and had to be crushed. As a result, the early suggestion that oviraptorids might specialize on dinosaur eggs (Osborn 1924*a*) was rejected. According to Barsbold (1983*a*), oviraptorids fed on pelecypods, which are very common in the freshwater deposits in the vicinity of sites where the skeletons were found. The very large infratemporal fenestrae to allow swelling of the strong adductors, the massive pterygoid-ectopterygoid structure, and the thick horny bill that probably covered the ridged palatal wings of the premaxillae and maxillae (including the median toothlike projections) speak in favor of Barsbold's hypothesis. Little can be said about feeding habits of caenagnathids. The longer and shallower mandible (and the skull) with weaker coronoid process indicates that their food had to be less resistant than that of oviraptorids and the mechanics of the jaws was different.

Accepting that oviraptorids were molluscivorous, it follows that they fed in the water. According to Barsbold, the muscular tail might have been used for aquatic propulsion. However, the typically cursorial hindlimb speaks for efficient locomotion on the land. One can thus consider that oviraptorids lived close to the shores of the inland lake basins, which themselves provided their resource base.

It should be emphasized that ingeniines may have had a somewhat different mode of life from oviraptorines, as indicated by the difference in the manus structure (although not in the jaw apparatus), which is more powerful in the ingeniines. However, the contemporaneous representatives of the two subfamilies (*I. yanshini, C. gracilis*) evidently lived on the same territory, as they are found in the same deposits and sites. In contrast, two crested *Oviraptor* species (*O. philoceratops, O. mongoliensis*) lived in different environments. The former, which is stratigraphically older, lived under more arid climatic conditions than the latter.

The function of the nasal crest in the oviraptorines is unknown. It might have served as a display structure or played a similar role to the casque of the cassowaries, which is used by these birds as an aid in passing through bushes (*fide* Sanft in Grzimek 1968).

The evident convergence in the structure of jaws and palate between oviraptorosaurians and dicynodonts has been noticed by several authors (Wetmore 1960; Cracraft 1971; Osmólska 1976). Consequently, it seems that in both groups, the specialized masticatory apparatus functioned similarly, although oviraptorosaurians are considered carnivores and dicynodonts are believed to have been herbivores.

11

Troodontidae

HALSZKA OSMÓLSKA
RINCHEN BARSBOLD

INTRODUCTION

The Troodontidae Gilmore, 1924*d*, is one of the rarest of theropod groups, known from scarcely twenty very fragmentary specimens plus isolated teeth. Troodontids come from the Cretaceous deposits of the Northern Hemisphere and their earliest record is from the Aptian or Albian of Mongolia, the latest from the late Maastrichtian of the United States (table 11.1).

Over the last decade, these dinosaurs were widely known under the name Saurornithoididae, until Currie (1987*a*) proved that the oldest available familial name for these dinosaurs is Troodontidae, introduced by Gilmore (1924*d*), however, not for theropods but for the dome-headed dinosaurs. Gilmore, on the basis of Leidy's inadequate drawing of a tooth of *Troodon formosus*, considered it identical with teeth of these ornithischians. Currie also demonstrated that *Stenonychosaurus inequalis*, one of the best-known "saurornithoidid" species, is a junior synonym of *Troodon formosus*.

Because of fragmentary preservation, the anatomy of the postcranial skeleton is poorly known in the Troodontidae, and some portions are virtually unknown (e.g., most of the vertebral column, the scapulocoracoid, and most of the forelimb).

In spite of a poor fossil record, the Troodontidae may be characterized as theropods of small adult size, with very long and slender hindlimbs and the largest relative brain size of all dinosaurs.

Although the Troodontidae shares some common characters with most other groups of "small theropods," at the moment it is impossible to state which of the latter was more closely related to troodontids. They are here tentatively considered as deinonychosaurians.

DIAGNOSIS

The Troodontidae may be diagnosed by the following features: long skull with narrow snout and large endocranial cavity; external naris bounded by the maxilla caudoventrally; basioccipital and periotic sinus systems; bulbous parasphenoid capsule; lateral depression on the lateral wall of the braincase, containing enlarged middle ear cavity; enlarged, hollow basipterygoid process; mandible with rostrally tapering dentary exposing laterally a row of foramina placed within a groove beneath the alveolar margin; as many as 25 premaxillary and maxillary teeth and 35 smaller dentary teeth; all teeth small, recurved, closely spaced; distal and often mesial edges of teeth with large, hooked

TABLE 11.1 Troodontidae

	Occurrence	Age	Material
Theropoda Marsh, 1881b			
Troodontidae Gilmore, 1924d (= Saurornithoididae Barsbold, 1974)			
Troodon Leidy, 1856 (= *Polyodontosaurus* Gilmore, 1932b, *Stenonychosaurus* Sternberg, 1932a, *Pectinodon* Carpenter, 1982b)			
T. formosus Leidy, 1856 (including *Polyodontosaurus grandis* Gilmore, 1932b, *Stenonychosaurus inequalis* Sternberg, 1932a, *Pectinodon bakkeri* Carpenter, 1982b)	Judith River Formation, Horseshoe Canyon Formation, Alberta, Canada; Judith River Formation, Hell Creek Formation, Montana, Lance Formation, Wyoming, United States	late Campanian–early Maastrichtian	At least 20 specimens, including fragmentary skulls, fragmentary postcrania, teeth
Saurornithoides Osborn, 1924a			
S. mongoliensis Osborn, 1924a	Djadochta Formation, Omnogov, Mongolian People's Republic	?late Santonian or early Campanian	Skull with fragmentary postcranial skeleton
S. junior Barsbold, 1974	Nemegt Formation, Omnogov, Mongolian People's Republic	?late Campanian or early Maastrichtian	Skull with fragmentary postcranial skeleton
Borogovia Osmólska, 1987			
B. gracilicrus Osmólska, 1987	Nemegt Formation, Omnogov, Nemegtskaya Svita, Bayankhongor, Mongolian People's Republic	?late Campanian or early Maastrichtian	Partial hind limbs
Tochisaurus Kurzanov et Osmólska, 1991 •			
T. nemegtensis Kurzanov et Osmólska, 1991 •	Nemegt Formation, Omnogov, Mongolian People's Republic	?late Campanian or early Maastrichtian	Metatarsus

Nomina dubia	Material
Bradycneme draculae Harrison et Walker, 1975	Distal tibiotarsus
Elopteryx nopcsai Andrews, 1913	Fragmentary femur
Heptasteornis andrewsi Harrison et Walker, 1975	Distal tibiotarsi
Paronychodon lacustris Cope, 1876a	Teeth
Pectinodon asiamericanus Nessov, 1985 •	Teeth
Tripriodon caperatus Marsh, 1889d	Teeth
Zapsalis abradens Cope, 1876b	Teeth

denticles that point toward the tip of the crown; distal and mesial denticles subequal in size; thin calcaneum fused with astragalus; long to very long metatarsus, with longitudinal, proximal trough along extensor face of metatarsal III; metatarsal III slightly longer than metatarsal IV and reduced to a splinter of bone for about half of its proximal length; proximal end of metatarsal III hidden on extensor side by tightly adhering ends of metacarpals II and IV; distal end of metatarsal III with tonguelike articular surface that extends proximocaudally; metatarsal IV the most robust, in caudal aspect occupying more than half the width of the metatarsus.

ANATOMY

The description that follows is based largely on two Mongolian species, *Saurornithoides mongoliensis* and *S. junior*, each represented by an incomplete postcranial skeleton with skull, and on the North American species, *Troodon formosus* represented by fragments of skulls and postcranial skeletons of about 20 specimens, plus some isolated teeth.

Skull and Mandible

The skull is long, narrow, and shallow (fig. 11.1). In lateral view, it forms a long, narrow triangle from the caudal boundary of the orbit to the tip of the snout. In dorsal aspect, it has also the form of a long, narrow triangle, tapering from the caudal edge of frontals forward. The orbits are very large and face rostrolaterally; the eyes may have been capable of stereoscopic vision as the snout in front of the orbits is narrow and shallow. The long, shallow snout is nevertheless well vaulted transversely. The external nares are small and situated subterminally. The naris is partly bounded by the maxilla caudoventrally, because the subnarial process of the premaxilla is not developed. Most of the lateral side of the maxilla is occupied by a shallow antorbital fossa with two large antorbital fenestrae, the rostral one smaller. The rostral margin of this fenestra is broadly rounded in *T. formosus* while acute in the species of *Saurornithoides*. The shaft of the lacrimal forms a comparatively strong preorbital bar descending forward to a loose contact with the jugal. The jugal is comparatively deep with a strong, caudally declined postorbital process, which in *S. junior* contacts the jugal process of the postorbital along a loose suture.

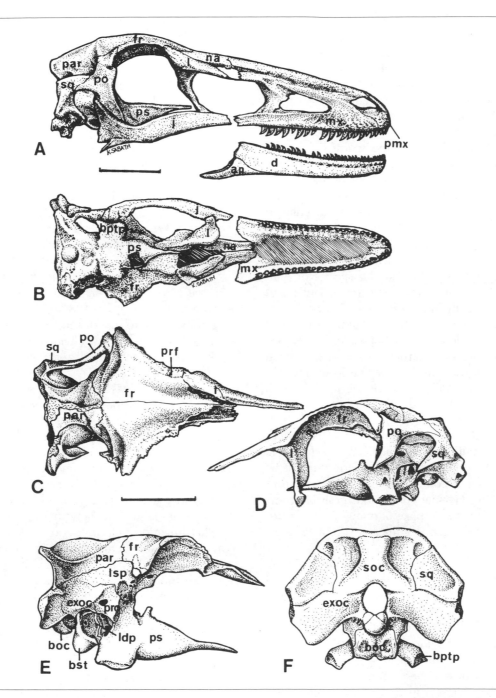

Fig. 11.1. Skulls in the Troodontidae. A, B, *Saurornithoides junior:* A. right lateral view of the skull and rostral half of the mandible, B. ventral view of the skull, C-F, *Troodon formosus:* C. fragmentary skull, dorsal view, D. the same, left lateral view, E. the same, right lateral view, and F. the same, occipital view. (C-F after Currie 1985; E, F slightly schematized.) Scale = 50 mm.

The postorbital bar is relatively wide in this species and in *T. formosus.*

The skull roof is distinctly vaulted transversely between supratemporal fenestrae; it is narrower across the parietals than across the frontals. The roof is only slightly arched and widest where the frontals contact the parietals, strongly tapering toward the V-shaped contact with the nasals. The nasal is very long. The prefrontal is narrow, fused with the frontal, excluding it from the dorsal orbital rim for about two-thirds of its length. A high sagittal crest is found on the parietals which meets a strong nuchal crest marking the caudal edge of the skull roof.

The postorbital region of the skull is relatively longer in *T. formosus* than in the species of *Saurornithoides*. The occipital region is characteristic in being sharply delimited from the skull roof by the nuchal crest. The supraoccipital has the form of an inverted Y in *T. formosus,* while it is broader and rectangular in *S. mongoliensis* and *S. junior.* The paroccipital process is deep and relatively short transversely, terminating bluntly. The basioccipital is much more strongly developed in *T. formosus* than in the species of *Saurornithoides,* the basal tubera being much larger and shifted caudoventrally in the former.

A peculiar character of the cranium, diagnostic of troodontids, is a deep lateral depression developed caudoventrally on the external surface of the lateral wall of the braincase. This depression is formed within the exoccipital-opisthotic, basioccipital, basisphenoid, and prootic. It is subdivided by a ridge in *T. formosus* and by two ridges in the species of *Saurornithoides.* In *T. formosus,* the rostral (basisphenoid-prootic) portion of the depression contained the middle-ear cavity, while the caudal (basioccipital-opisthotic) portion might be equivalent to the paracondylar pocket in *Dilophosaurus* (Currie 1985). The caudal portion was probably completely covered by a thin plate of bone in *T. formosus,* which has not yet been documented in any *Saurornithoides* species.

The medial wall of the middle-ear cavity is pierced by several foramina: two caudodorsal ones correspond to the fenestra ovalis and (?) fenestra pseudorotunda; others pass into the internal carotid artery canal, the parasphenoid capsule, and into the basipterygoid process. In the caudal portion of the lateral depression, there are also some openings leading into the sinuses in the basioccipital. Most of these foramina have been noticed by Barsbold (1974) in *S. junior.* His interpretation, however, differs from that of Currie (1985).

Another striking feature of the troodontid basicranium is the bulbous parasphenoid capsule, a hollow structure formed of the basisphenoid and parasphenoid and prolonged rostrally into the cultriform process. Caudal to the capsule, there is a distinctive rectangular platform from which stout, hollow, basipterygoid processes diverge lateroventrally in *T. formosus* and laterally in *S. junior.* Each process ends distally in an elongate, concave articular surface facing laterally and in

a smaller surface oriented caudolaterally. The spaces within the processes communicate with the parasphenoid capsule rostrally and with the lateral depression caudally.

The palate is poorly known in the troodontids, only fragments having been found in the species of *Saurornithoides.* The separation of the nasal and oral cavities was presumably very short. The palatal wings of the pterygoids were comparatively wide and closely approached each other in the midline underlying the rostral portion of the bulbous capsule in *S. junior.*

The quadrate, complete only in one specimen (indeterminate troodontid recently described by Barsbold et. al. 1987), has a shaft that is somewhat concave caudally; its dorsal condyle is well rounded. There is a quadrate foramen.

The endocranial cavity is large in the troodontids (Russell 1969) and probably corresponded entirely to the brain cavity (Hopson 1980a). The size of the brain was comparable to that of some birds. Brain weight is estimated at 37 g for an individual weighing 37 kg to 42 kg (Russell 1969; Russell and Seguin 1982; Hopson 1980a).

The mandible is known from several dentaries in *T. formosus* and in the types of *S. mongoliensis* and *S. junior.* The postdentary section of the jaw is virtually unknown, being very poorly preserved in *S. mongoliensis* and in the indeterminate troodontid mentioned above. The dentary is long and shallow. In lateral aspect, it has a narrow, triangular shape and has a distinct groove that contains a row of foramina beneath the alveolar margin. Medially, there is a well-developed Meckelian groove reaching close to the symphysis. There are no interdental plates. Dentaries were probably not very tightly bound to each other. The splenial of *S. junior* has a thick ventral margin. The caudal end of the splenial bears dorsally a concave articular surface for the angular.

Troodontid teeth can be easily distinguished from all other Cretaceous theropods by the relatively large, apically hooked denticles on the distal edge, a constriction between the root and crown (Currie 1987a), and the compressed, somewhat recurved conical crown. On the mesial edge of the premaxillary teeth, hooked denticles may also be present. The grooves between denticles are expanded into distinct pits between the bases of successive denticles.

Postcranial Skeleton

There are no complete, or even nearly complete, troodontid postcranial skeletons. The hindlimb elements, especially of the pes, have been most frequently

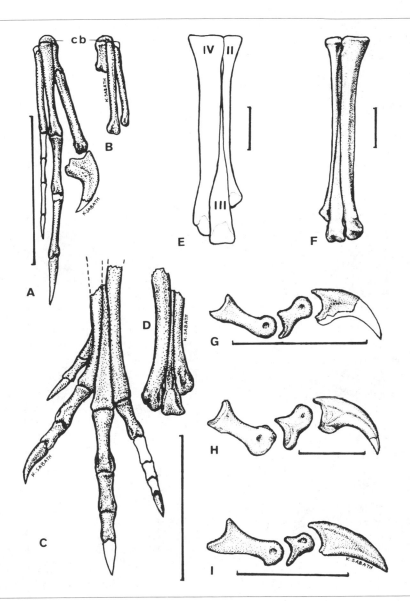

Fig. 11.2. Appendicular skeleton in the Troodontidae. A-D, indeterminate troodontid from the Early Cretaceous of Mongolia: A. fragmentary right manus, dorsal view, B. right metacarpus, palmar view, C. fragmentary left pes, extensor view, D. left metatarsus, flexor view, E. *Troodon formosus*, right metatarsus, extensor view, F. *Tochisaurus nemegtensis*, left metatarsus, extensor view, G-I, evolutionary changes of the second pedal digit in the troodontids: G. Early Cretaceous indeterminate troodontid, H. *Troodon formosus*, I. *Borogovia gracilicrus* (A-D,G, after Barsbold et al. 1987, E after Wilson and Currie 1985, F after Kurzanov 1987, H after Russell 1972, I after Osmólska 1987.) Scale = 50 mm.

found (figs. 11.2, 11.3). Fortunately, they are usually very informative taxonomically. Some of the anatomic data that follow are only known in one species.

Axial Skeleton

A very poorly preserved series of five cervicals in the Early Cretaceous indeterminate troodontid (Bars-bold et al. 1987) shows only that the vertebrae are more similar to those in the ornithomimosaurians, with presumably low neural spines, than to those in the dromaeosaurids. It is assumed (Russell 1969) that the neck was long. ˉorsals (at least the caudal ones) lack pleurocoels. The sacrum consists of six firmly fused vertebrae. Caudals of the distal half of the tail have weakly elongate prezygapophyses covering about

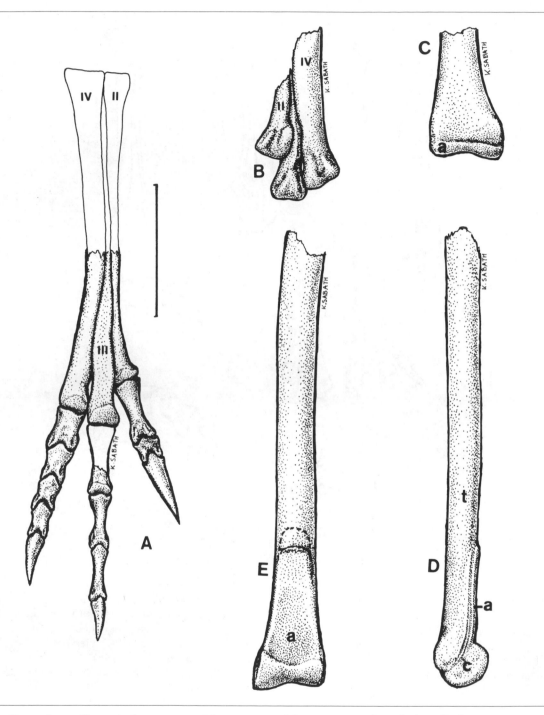

Fig. 11.3. *Borogovia gracilicrus:* A. fragmentary right pes, extensor view, B. distal end of right metatarsus, flexor view, C. distal end of right tibiotarsus, flexor view, D. distal half of right metatarsus, lateral view, E. the same, extensor view. (All after Osmólska 1987.) Scale = 50 mm.

neural spines that are divided by a longitudinal sulcus in *S. junior* or lack the spine, having only the sulcus along the dorsal surface in *T. formosus*. The transition point occurs between the 14th and 15th caudals in *S. junior*. Proximal chevrons are long ventrad; distal chevrons are very shallow and slightly elongated cra-

niocaudally. The cranial end of each distal chevron is bifurcated and embraces the caudal end of the preceding chevron. The tail thus had a certain stiffness distally, without being surrounded by long, ossified tendons characteristic of the dromaeosaurids.

Appendicular Skeleton

The scapulocoracoid is unknown. Of the fore-limb, only the ulna, a semilunate carpal, and a few manus bones were found in *T. formosus* and the semilunate carpal, the complete metacarpus, plus several phalanges and unguals in the indeterminate troodontid (Barsbold et al. 1987). The manus has a structure typical of many small theropods: metacarpal I is about one-third the length of metacarpal II, phalanx I-1 is long, and the second digit was probably the longest or subequal with the third. The latter digit is not preserved, but judging by the shaft of metacarpal III, which is only very slightly thinner than metacarpal II, the third digit was not very strongly reduced in thickness. The naturally articulated metacarpus in the indeterminant troodontid seems to indicate that it was transversely arched. Unguals of the manus are compressed and recurved with a strong flexor tubercles. From what is preserved of the manus, one can conclude that it was comparatively strong and large, equal to 50 to 60 percent of the length of the pes.

The pelvis is known from fragments in *S. mongoliensis* and the indeterminate troodontid, and it is propubic (Barsbold 1977*b*, 1983*a*, Gauthier 1986). The distal end of the pubis is broadly rounded. The ischium is about one-third the pubis length and has a large obturator process.

The femur, tibia, and fibula are incompletely preserved. The distal portions of the tibiotarsus are available in *T. formosus*, *S. junior*, and *Borogovia gracilicrus*. The tibiotarsus is characteristic in having a very reduced, thin calcaneum fused with the astragalus. The ascending process of the astragalus is high, and the astragalar condyles are well rounded, separated by deep sulcus and very prominent on the extensor side. The fibula is very strongly reduced, especially distally. The nearly complete tibiotarsus with fibula so far reported was described in *B. gracilicrus* (Osmólska 1987). In this species, the tibia is unusually long and slender as compared with other theropods, its shaft diameter being slightly more than 4 percent the estimated length of the tibiotarsus. In view of the fact that *B. gracilicrus* may be somewhat differently specialized than other troodontids, it is possible that this extreme slenderness of the tibia-fibula was not typical of the entire group. Proportions of the hindlimb bones have been estimated by Russell (1969) in *S. mongoliensis*. The tibiotarsus in this species, although quite long, seems to be shorter and less slender than that mentioned above.

The structure of the metatarsus within troodontids ranges from long and slender in *T. formosus* (diameter at the middle equals 16% the length) to very long and slender (diameter at the middle equals 10%

the length) in *Tochisaurus nemegtensis*. This later species is characteristic in being concave medially along the extensor side, especially deep proximally. The proximally concave metatarsus has also been noted by Wilson and Currie (1985) in *T. formosus;* thus, this feature may be typical of all representatives of the family. In the troodontids, metatarsal IV is the thickest, occupying caudally about two-thirds the diameter of the metatarsus. Metatarsal III is reduced to a thin splinter of bone for more than its proximal half. Metatarsals III and IV are subequal in length. The caudodistal articular end of metatarsal III is prolonged proximally as tongue-like surface. Metatarsal II is much thinner than metatarsal IV. None of the distal articular surfaces is ginglymoid, although that of metatarsal III is somewhat concave.

The second pedal digit is specialized, in having the distal articular surface on the first phalanx extending far proximally on the extensor surface and an elongated ventral heel on the proximal end of the second phalanx. This digit is usually provided with an enlarged, curved claw. The pes in *B. gracilicrus* deviates somewhat in its structure from the typical troodontids: the third digit is the weakest in the pes, the fourth digit is the strongest, and the ungual on the second digit is straight, instead of curved. The third and fourth digits are about equally strong in *T. formosus* and *S. mongoliensis.*

SYSTEMATICS AND EVOLUTION

Remains of troodontids were among the first described dinosaurs, including a single tooth described by Leidy (1856) as *Troodon formosus*, a then-presumed lacertilian. Later, Gilmore (1924*d*) considered *Troodon* to be a dome-headed dinosaur, and he erected the Troodontidae. The theropod nature of *Troodon* was firmly established by Sternberg (1945), who also speculated on the *Troodon-Stenonychosaurus* relationship (Sternberg 1951*b*). However, a redefinition of the Troodontidae as a theropod family was first suggested by L. S. Russell (1948). Successively, D. A. Russell (1969), recognizing the relationship between *Troodon formosus*, *Stenonychosaurus inequalis*, and *Saurornithoides mongoliensis*, assigned the two latter species to the Troodontidae *sensu* Russell 1948.

Regarding *T. formosus* teeth as entirely different from the teeth of *Saurornithoides mongoliensis* and *Stenonychosaurus inequalis*, Barsbold (1974) erected a new family, Saurornithoididae, for *Saurornithoides* and *Stenonychosaurus*. According to him, the monotypic

Troodontidae (*sensu* Russell 1948) were a group of uncertain affinity. Barsbold's opinion has been generally accepted.

Recently, Currie (1987*a*) undertook a thorough study of the *T. formosus*-like teeth, both isolated and associated with dentaries, from the Judith River and Horseshoe Canyon formations of Alberta. Comparing this material with the *T. formosus* holotype from the Judith River Formation of Montana, Currie demonstrated that all share a set of features, indicating a close relationship. On this basis, he synonymized *Stenonychosaurus inequalis*, as well as *Pectinodon bakkeri* described by Carpenter (1982*b*), with *T. formosus* and replaced the family name Saurornithoididae Barsbold by Troodontidae Gilmore, 1924*d*, *sensu* Russell, 1948. Importantly, Currie was able to prove that teeth in these theropods vary significantly depending on their position in the jaw and that some differences previously considered of taxonomic value are unimportant.

These few common features that are typical for the North American *T. formosus*, *Stenonychosaurus inequalis*, and *P. bakkeri* occur also in the Mongolian indeterminate troodontid (Barsbold et al. 1987), *Saurornithoides junior*, and probably also in *S. mongoliensis*. Thus, it may appear in the future that the observed tooth characters have suprageneric taxonomic value, most probably a familial one.

The systematic position of the Troodontidae within Theropoda is still unclear. The structure of the manus (Barsbold et al. 1987) with short metacarpal I, very long proximal phalanx in the first digit, and metacarpal III reduced in thickness is known in all "small theropods" excluding the derived ornithomimosaurians—the Ornithomimidae. This structure should thus be considered primitive in troodontids. Additionally, the presence of the semilunate carpal, very much like that in dromaeosaurids, oviraptorids, and *Archaeopteryx*, indicates that troodontids may share with these theropods a more recent ancestor than with the remaining theropods and that they should be included within the Maniraptora Gauthier, 1986.

Currie (1985) considered the presence of an enlarged parasphenoid capsule in troodontids and ornithomimosaurians as a common derived character of these theropods. At the same time (Currie 1985, 1987*a*), he suggested that pneumatic cavities in the basicranial region, somewhat medially placed cotylus for the quadrate, fenestra pseudorotunda, loss of interdental plates and constriction between the dental crown and root may be the common derived characters of troodontids and birds.

Although the parasphenoid capsule seems to be a unique structure unlikely to appear independently, it would be difficult to find more characters in common between troodontids and ornithomimosaurians, except those that are rather the structural consequence of the development of the capsule (e.g., the broadened basipterygoid processes and the rectangular platform between them). Thus, it seems more parsimonious to interpret the development of the capsule as a convergence.

Of the characters quoted by Currie (1985, 1987*a*) as common only to troodontids and birds, pneumatization of the otic region is found also in the oviraptorids (Osmólska 1976; see also Barsbold et al., this vol.), and the quadrate cotylus is also somewhat more medial in a dromaeosaurid, *Velociraptor*, than in other theropods (unpubl. obs.); homology of a foramen caudal to the fenestra ovalis in *T. formosus* with the fenestra pseudorotunda cannot also be proved beyond doubt (Currie 1985:1656). The loss of interdental plates and constriction between the crown and root are rather weak evidence in favor of the common ancestry of troodontids and birds advocated by Currie.

The most widely accepted opinion is that the Troodontidae is closely related to the Dromaeosauridae (in 1969, Ostrom considered the taxa now assigned to troodontids as belonging to dromaeosaurids), and they were often classified with that family within the Deinonychosauria (Barsbold 1976*b*; Osmólska 1982). This assignment is based only on the comparable specialization of the second pedal digit and may be additionally supported by some skull characters, for example, the medial position of the quadrate cotylus (at least in *Velociraptor*) and the presence of strongly pronounced nuchal and sagittal crests. However, when many other troodontid characters are considered (e.g., the bulbous parasphenoid, the lateral depression on the braincase wall, the pneumatized basicranium and otic region, the very derived, slender metatarsus), this similarity between troodontids and dromaeosaurids may indicate a less close relationship than has been suggested. In some respects (e.g., propubic pelvis, lack of ossified caudal tendons), troodontids are less derived than dromaeosaurids, and these facts, together with the differences mentioned above, cast further doubt on the monophyly of Deinonychosauria (Gauthier 1986).

There are some resemblances in the manus and metatarsus of troodontids and *Elmisaurus rarus* (Elmisauridae; see Currie, this vol.). Metacarpal I is more slender and relatively longer than in other maniraptorans, and metatarsal IV is the strongest element of the pes and close in its length to the length of metatarsal III, in both groups. However, the pes in *E. rarus* lacks the specialized second digit. Other skeletal elements are either unknown or poorly known in the elmisaurids;

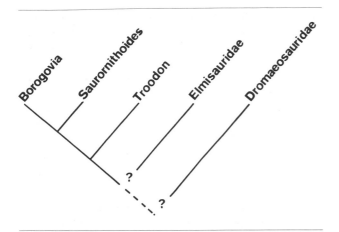

Fig. 11.4. Relationships within the Troodontidae.

thus, a closer relationship between these two families than with other maniraptorans is not sure.

As follows from the preceding discussion, the Troodontidae should be considered at the moment as maniraptorans of unresolved close relationships to any other maniraptoran family.

The three valid genera presently assigned to the Troodontidae may briefly be diagnosed as follows.

Troodon (with the only species *T. formosus*) shows very strong basal tubera, which are shifted caudo-ventrally; the basipterygoid processes are directed ventrolaterally; the temporal region is about as long as the orbit, the supraoccipital has the form of an inverted Y. Within the pedal digit II, the second phalanx is somewhat longer than half the length of the first phalanx.

Saurornithoides is characterized by the weakly developed basal tubera, laterally directed basipterygoid processes, and the rectangular supraoccipital: the temporal region is somewhat shorter than the orbit. In the pes, the third digit is as strong as the fourth: within digit II, the second phalanx is half as long as the first phalanx. As presently known, both species of this genus, *Saurornithoides mongoliensis* and *S. junior*, show only differences in the size and number of teeth. *S. mongoliensis* is smaller and has 18 maxillary and 28 dentary teeth, while *S. junior* has 20 and 35 teeth, respectively. The postcranial skeleton is very fragmentary in *S. junior*, and proportions within the pedal digits are unknown in this species.

Borogovia differs from all the troodontids in that pedal digit III is the thinnest, the second phalanx of digit II is shorter than half the length of the first phalanx, and the ungual of the second pedal digit is straight. The only species of this genus, *B. gracilicrus*, comes from a locality 100 km from where *S. junior* was found, and it is about the same age as the latter. Thus, it is possible that the two taxa may be found to be conspecific in the future.

The remaining and newly described troodontid, *Tochisaurus* is characterized by a strongly reduced second metatarsal.

The Troodontidae appear in the fossil record in the Early Cretaceous of Asia, and the oldest representative, the indeterminant troodontid described by Barsbold et al. (1987), shows characters of the dentary, teeth, and pes typical of the Late Cretaceous forms. The relatively long second phalanx in the pedal digit II indicates its primitiveness compared to the later species. The evolutionary tendency, at least in the Asian troodontid line, leads toward a shortening of that phalanx, as is documented by its extreme shortness in the youngest species. *B. gracilicrus*, from Mongolia (figs. 11.2G–I, 11.4). As far as may be judged from the preserved basioccipital portion of the braincase, the indeterminant troodontid (Barsbold et al. 1987) had not developed the lateral depression, or at least it still lacked the basioccipital portion of that depression.

PALEOECOLOGY, BEHAVIOR, AND PALEOGEOGRAPHY

Troodontids were small, lightly built theropods weighing about 50 kg. Their long legs, with slender, strongly elongated crus and metatarsus, are evidence that they were capable of very long strides and thus of more rapid progression than all other contemporaneous theropods, except, perhaps, the avimimids (Kurzanov 1987). Their hands were probably capable of rather precise movements and were comparatively strong.

The encephalization quotient of the troodontids is among the largest within known dinosaurs (Hopson 1980a). They probably had more complex perceptual abilities and more precise motor-sensory control mechanisms. It implies comparatively high activity level and may indirectly indicate endothermy. The short secondary palate, however, would be less supportive of this explanation. The large eyes and stereoscopic vision also suggest agility (Walls 1942; Russell and Seguin 1982). The enlarged middle ear cavity and the presence of a periotic sinus system indicate that troodontids had a well-developed sense of hearing and probably an advanced detection capability of low-frequency sounds; the parasphenoid capsule connected with the lateral depression and periotic sinuses might also be associated with hearing (Currie 1985). In contrast, the sense of olfaction may have been inferior to those of sight and hearing, taking into account the comparatively narrow, although long olfactory tracts and small external nares.

Delicate, long, and slender jaws, closely spaced, sharp, recurved teeth indicate carnivorous feeding habits in these dinosaurs. The relatively loose mandibular symphysis and the movable joint between the angular and splenial indicate that there was a potential for intramandibular kineticism. The prey of troodontids was relatively small and may have consisted of lizards, mammals, and hatchlings and young of other dinosaurs; it may also have included insects. It was suggested by Russell and Seguin (1982) that *T. formosus* was a crepuscular or nocturnal carnivore, preying on small mammals, which themselves were active by night. If, however, the troodontid diet included insects and small nonmammalian vertebrates, diurnal activity would seem more reasonable.

By comparison with dromaeosaurids, which have a similarly specialized second pedal digit serving as an offensive device to disembowel the prey, a similar use of this digit has generally been suggested for troodontids. This, however, is extremely unlikely: the digit was associated with the totally different, very long and slender metatarsus and the long, delicate crus in troodontids; the nonginglymoid articular joint between metatarsal II and the proximal phalanx of the second digit did not limit excursion of the claw to only one plane or the range of the second digit movements, as was the case in dromaeosaurids. In conclusion, the force that could be exerted on the second digit must have been very much weaker in troodontids than in dromaeosaurids (Osmólska 1982). Thus, the use of the second digit was most probably quite different in the two groups. This is further supported by the case of *Borogovia gracilicrus* (Osmólska 1987), which, with its straight claw, is the extreme example in the troodontid line of specialization of the second pedal digit and of the pes. By no means could the pes serve for offense or killing activity.

Remains of troodontids are extremely rare, in large part due to the fragile nature of their thin-walled bones. In Mongolia, troodontids are found in red sands and sandstones of the Djadochta Formation, deposited under arid conditions (dunes and intermittent lake sediments), as well as in the light yellowish-gray sands and sandstones of the Nemegt Formation and Bugeen Tsav beds, deposited under wetter, more seasonal, warm-temperate climatic conditions (fluviatile or flood-plain sediments). Interpretation of Lower Cretaceous deposits that have yielded the stratigraphically oldest but indeterminant troodontid (Barsbold et al. 1987) has not yet been presented. In the Djadochta Formation, in which only one specimen of *Saurornithoides mongoliensis* was found (represented by a skull and fragmentary postcranium), this material came from the assemblage dominated by a herbivore dinosaur, *Protoceratops andrewsi*. Other dinosaurian carnivores within the assemblage include *Velociraptor mongoliensis*, equally as rare as *S. mongoliensis*, although additionally represented by numerous isolated teeth. In contrast, isolated teeth of *S. mongoliensis* were extremely rare in the deposits of the Djadochta Formation. In addition to the dinosaur skeletal remains, various eggs and nests were found in the Djadochta Formation, as well as numerous small lizards and mammals: multituberculates and insectivores (Osmólska 1980). The Nemegt Formation and Bugeen Tsav beds assemblages were totally different from the Djadochta assemblage. Two troodontid species were found, each represented by single, very fragmentary skeletons: *S. junior* and *B. gracilicrus*. *Saurolophus angustirostris* was the dominant herbivorous dinosaur there, but giant carnivores (*Tarbosaurus bataar*) as well as ornithomimids (*Gallimimus bullatus*) were also numerous. Neither mammals nor lizards were found in these deposits, probably because of preservational factors.

In North America, all known troodontid remains belong to *T. formosus,* most of them found in the Judith River and Horseshoe Canyon formations of Alberta. It was also a rare element of the fauna, dominated by large herbivore hadrosaurs and to a lesser extent by ceratopsids (Russell 1977; Dodson 1983). The environmental conditions were different from those in Mongolia. The climate was more humid, because the region was closer to the seashore and covered by dense vegetation.

The only European record of troodontids comes from the Maastrichtian deposits of Romania, where indeterminate tarsometatarsi (*Bradycneme draculae, Heptasteornis andrewsi*) have been found (Harrison and Walker 1975). The state of preservation of these bones indicates that they were reworked (Grigorescu 1982), and nothing can be deduced about their anatomy or ecology.

12

Dromaeosauridae

JOHN H. OSTROM

INTRODUCTION

Among the most distinctive of the Theropoda are those taxa assigned to the Dromaeosauridae (= Dromaeosaurinae Matthew *et* Brown 1922), such as *Deinonychus antirrhopus, Velociraptor mongoliensis, Adasaurus mongoliensis,* and, of course, *Dromaeosaurus albertensis,* together with several others (Table 12.1). Included in the Dromaeosauridae are a variety of small to medium-sized, agile theropods (estimated live weight ranging from 30 to 80 kg), all of which were lightly built with obligatory bipedal posture and gait. All dromaeosaurids are distinctive in the construction of the foot with an exceptionally raptorial talon, highly mobile hand-wrist complex, and unique caudal vertebral adaptations for balance control.

One of the best-known examples of the family is *Deinonychus antirrhopus* (Ostrom 1969*a,* 1969*b,* 1974*a,* 1976*a*), represented by partial skeletons of at least six individuals from several sites in Montana (U.S.A.). Also included here are the less completely described remains assigned to *Velociraptor mongoliensis* (Osborn 1924*a;* Barsbold 1983*a*), *Dromaeosaurus albertensis* (Matthew and Brown 1922; Colbert and Russell 1969), and *Adasaurus mongoliensis* (Barsbold 1983*a*). The family appears to have ranged from Early Cretaceous (Aptian/Albian) to Late Cretaceous (middle to late Campanian and possibly Maastrichtian) in North America and Asia.

Sometimes dromaeosaurs are allied with members of the Troodontidae, formerly the Saurornithoididae of Barsbold's 1974 systematic scheme (see Currie 1987*a*), although not everyone is convinced. Together, these two constitute to many the core of the Deinonychosauria (Colbert et Russell, 1969), which are distinctive in foot structure and hand and wrist construction. In fact, the Deinonychosauria was proposed as a formal taxonomic category only because of these distinctive anatomical features (a few of which are confined to the Dromaeosauridae) that do not conform with previous diagnoses of the Theropoda subcategories "Carnosauria" and "Coelurosauria." Nor are these features found in any other theropod, although the manus configuration is similar in the Oviraptoridae. Colbert and Russell (1969) proposed the category Deinonychosauria to resolve this systematic dilemma. Barsbold (1977*b*) followed that by elevating the several other divergent theropod families to infraordinal rank, the scheme followed in this chapter. Last, it should be pointed out, however, that some authors have not supported this alliance of dromaeosaurids and troodontids (see Osmólska, this vol.).

Recently, Gauthier's (1986) cladistic analysis proposed a new alignment of the traditional theropod categories. His Maniraptora was proposed for those theropods (including birds) and all "coelurosaurs" that he considers closer to birds than to Ornithomimidae. Included here are a number of taxa (*Coelurus, Ornitho-*

TABLE 12.1   Dromaeosauridae

	Occurrence	Age	Material
Theropoda Marsh, 1881b			
Dromaeosauridae Matthew et Brown, 1922			
Adasaurus Barsbold, 1983			
A. mongoliensis Barsbold, 1983	Nemegtskaya Svita, Bayankhongor, Mongolian People's Republic	?late Campanian or early Maastrichtian	Partial skull and fragmentary postcrania
Deinonychus Ostrom, 1969a			
D. antirrhopus Ostrom, 1969a	Cloverly Formation, Wyoming, Montana, United States	Aptian-Albian	More than 8 articulated and disarticulated skeletons and skulls
Dromaeosaurus Matthew et Brown, 1922			
D. albertensis Matthew et Brown, 1922	Judith River Formation, Alberta, Canada; Judith River Formation, Montana, United States	late Campanian	Incomplete skull and postcranial fragments
Hulsanpes Osmólska, 1982			
H. perlei Osmólska, 1982	Barun Goyot Formation, Omnogov, Mongolian People's Republic	middle Campanian	Partial foot
Saurornitholestes Sues, 1978a			
S. langstoni Sues, 1978a	Judith River Formation, Alberta, Canada	late Campanian	?3 skull and postcranial fragments
Velociraptor Osborn, 1924a			
V. mongoliensis Osborn, 1924a	Djadochta Formation, Beds of Toogreeg, Omnogov, Mongolian People's Republic; Minhe Formation Nei Mongol Zizhiqu, People's Republic of China	?late Santonian or early Campanian	More than 6 partial to complete skulls and skeletons
?Dromaeosauridae			
Chirostenotes Gilmore, 1924a			
C. pergracilis Gilmore, 1924a	Judith River Formation, Horseshoe Canyon Formation, Alberta, Canada	late Campanian	2 partially complete manus

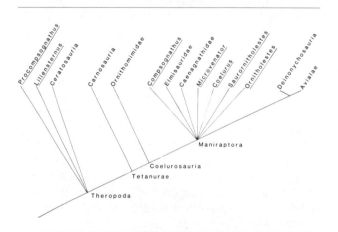

Fig. 12.1. Cladogram of proposed phylogenetic relationships among theropods and birds (from Gauthier 1986).

lestes, Microvenator, Compsognathus, and others), plus the Deinonychosauria and Avialae. The last comprises what Gauthier describes as "winged theropods" (more commonly known as birds, both living and extinct). This arrangement, as summarized in his cladogram (fig. 12.1), expresses but one current hypothesis on the phyletic affinities of deinonychosaurians (and therefore dromaeosaurids) to birds.

Most recently, Paul (1984a, 1988) has made the curious suggestion that dromaeosaurs might be the flightless Cretaceous descendants of Jurassic archaeopterygians. More pertinent here, though, is Paul's further suggestion that *Deinonychus* is a junior synonym of *Velociraptor* and that *Saurornitholestes* may be also. I do not agree that *Velociraptor* and *Deinonychus* are congeneric for anatomical and temporal reasons.

DIAGNOSIS

Diagnostic characters of the Dromaeosauridae include the following derived features: (a) cranium relatively large with long, slender pterygoids, a deeply-pocketed ectopterygoid of triradiate form, a deep maxilla defining part or all of three antorbital fenestrae; (b) maxillary and dentary teeth laterally compressed bearing both mesial and distal serrations with distinct size differences between mesial and distal serration denticles; (c) forelimb not reduced with elongated tridactyl manus bearing large raptorial claws; (d) carpals consist solely of proximal elements: a semilunate radiale with an asymmetrical proximal ginglymus, plus an oval wedge-shaped ulnare; (e) coracoid, thin semicircular plate of disproportionate size; (f) pes with three primary digits, pedal digit II highly modified for extreme hyperextension and bearing a greatly enlarged trenchant ungual; (g) pes functionally didactyl for all locomotory purposes; (h) ilium deep and short with pubic peduncle angled back and ventrally; (i) pubis long and opisthopubic at variable angles. Pubis bears a large, laterally compressed distal expansion ("boot"); (j) ischium short, thin, and tapered distally, bearing a prominent triangular obturator process near midlength; (k) distal 80 percent of the caudal vertebral series features extremely long extensions of the prezygapophyses (up to 10 vertebral segments long in some taxa); and (l) all but the most proximal chevrons develop comparable paired (on each side) long proximal bony rods analogous to the overlying vertebral rods.

Special note should be made of the following: the foot is sharply modified from the normal tridactyl theropod condition to a didactyl weight-bearing design involving only digits III and IV. The second pedal digit is specialized as an offensive device terminating in a very large and strongly recurved claw (for support of an even larger horny talon). The manus bears three long clawed fingers designed for grasping but not for support or walking. The carpals provided some degree of "pronation-supination" mobility linked with wrist flexion-extension. Finally, the modified caudal vertebral series is unique to this family in that all except the most proximal vertebrae are enclosed in bundles of long ossified rods (= bony tendons) that appear to be bony extensions of the prezygapophyses—and probably represent ossified tendinous muscle insertion attachments—that would seem to have made the tail almost completely inflexible throughout its length. The long pubis, with a large and laterally compressed distal expansion, with its opisthopubic orientation, as preserved in *Adasaurus mongoliensis, Deinonychus antirrhopus,* and *Velociraptor mongoliensis,* is yet another uniquely dromaeosaurid condition. The caudal and pubic conditions are unknown in other taxa here referred to the Dromaeosauridae.

ANATOMY

As reported elsewhere (Ostrom 1969*b*), the cranium and jaws are lightly and loosely constructed in *Dei-*

nonychus antirrhopus but may have been more firmly united in other dromaeosaurids (fig. 12.2). In *Deinonychus,* the premaxilla nearly encircles the nares, much less so in *Velociraptor,* and bears not fewer than four teeth. The maxilla is deep, defines two-thirds of the antorbital fenestra rostrally, and contains one or two auxiliary antorbital fenestrae. It is moderately robust and bears 15 deeply socketed and blade-like teeth. All other sub- and postorbital and palatal bones are thin to delicate. Except for the only known specimen of *Dromaeosaurus albertensis,* the orbit of all dromaeosaurids is large, elliptical and slightly inclined backward from the vertical. The dentary is thin transversely and narrow vertically with nearly parallel upper and lower margins, bearing up to 16 tooth positions. There is doubtful evidence of "interdental plates" in the dentary (and maxilla), despite Currie's (1987*a*) suggestion that they exist. Teeth (except mesially) are laterally compressed and strongly recurved with prominent serration denticles along the distal carina almost twice as large as the denticles of the mesial serrations on the same tooth.

Eight to nine cervical vertebrae are all strongly angled (centrum faces relative to centrum axes) forming a distinct S-shape curve to the neck; 13 to 14 dorsals, the caudal dorsals with modest pleurocoels; and 3 or 4 sacral vertebrae (fig. 12.3). All cervical (except the atlas and axis) and all but the last dorsals are platycoelous. Thirty-six to forty vertebrae in the caudal series, which is highly modified by extremely long (up to 8 or more segmental lengths) extensions of the prezygapophyseal processes above and equally long forward-projecting chevron elongations beneath (fig. 12.4). Together, these structures must have rendered the tail almost completely inflexible throughout most of its length. Only the first three or four proximal caudals were free of these ossified tendon rods. Both prezygapophyseal and chevron rods bifurcate at their bases into paired elements on each side. Dorsal ribs are of normal theropod form and are supplemented with a series of straight or sigmoid-shaped gastralia of unknown number and configuration.

The scapula is of normal theropod design with a narrow, stout blade, but the coracoid is unusually large, forming a thin concavoconvex sheet of bone of nearly triangular outline. The latter features a pronounced biceps tubercle and a relatively large coracoid foramen. The humerus is also of medium length but stout, slightly sigmoidal in lateral aspect and bearing a very large and robust deltopectoral crest. The ulna is of moderate length and slightly bowed convex externally,

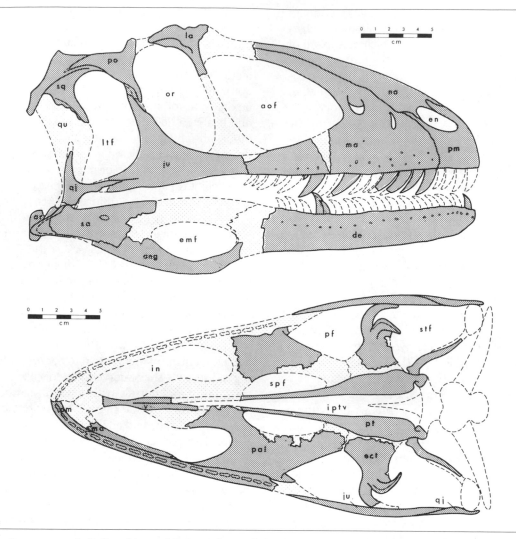

Fig. 12.2. Reconstructed skull and jaws of *Deinonychus antirrhopus*. (From Ostrom 1969*b*.)

while the radius is more slender and nearly straight. The manus is long and slender with three functional digits, of which digit II is the longest, digit III a little shorter and slightly divergent, and digit I much shorter and more robust than the others and slightly twisted from the palmar plane (fig. 12. 5). The carpus consists of two proximal elements, a distinctive semilunate radiale and a smaller suboval ulnare. The radiale articulates compactly with metacarpals I and II, but not metacarpal III, and features a specialized and distinctive asymmetrical proximal ginglymus for highly mobile articulation with the radius. This feature provided limited degrees of "pronation/supination" as the wrist was flexed or extended..

In the pelvis, the ilium is relatively short with a truncated cranial process. The ilium is canted relative to the sacral axis and bears a very robust pubic peduncle that is inclined backward. The pubis is very long and markedly opisthopubic (although exact attitude is unknown), bearing a large and stout distal expansion. The ischium is much shorter than the pubis and thin, with a pronounced obturator process and a short distal tapering.

The femur, known in only a few taxa (*Deinonychus* and *Velociraptor*), is relatively robust and shorter than the tibia, with a stout medially offset head. The shaft is hollow (but with stout walls), slightly curved craniocaudally, and bears a prominent "posterior trochanter" distal and caudal to the greater and lesser trochanters. This feature may be a unique muscle at-

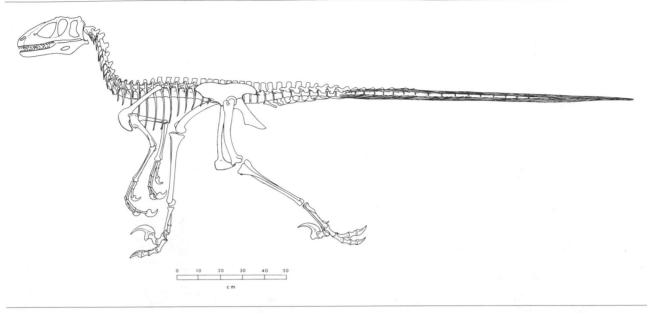

Fig. 12.3. Reconstruction of the skeleton of *Deinonychus antirrhopus* (after Ostrom 1969b). The horizontal attitude of the dorsal vertebrae is modeled after the normal posture of the vertebral column in large ratites (e.g., *Struthio*), the enlarged interspinous ligament scars on the dorsal vertebrae, and the natural curvature of the cervical series.

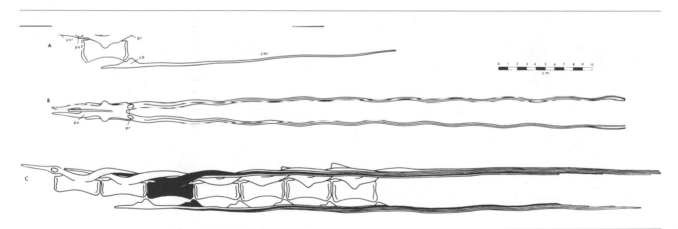

Fig. 12.4. Caudal vertebra and chevron of *Deinonychus antirrhopus* in lateral (A) and dorsal (B) views. C illustrates the relationship of a single vertebra with its bony rods to preceding (right) and succeeding (left) units.

tachment site related to the special offensive hindlimb excursion of dromaeosaurids. The tibia is stout and long with a very prominent cnemial crest, while the fibula is slender, except proximally where it is greatly expanded craniocaudally. The standard theropod mesotarsal ankle consists proximally of a small irregularly oval calcaneum and a very large roller-bearing astragulus (capable of as much as 130° to 150° of flexion/ extension) with a broad and high ascending process. Distally, two irregularly shaped and flat tarsals cap the three main metatarsals. Total hindlimb is relatively long.

The pes is of moderate length and functionally didactyl for walking (fig. 12.6). The metatarsus is also moderately long. Metatarsals II and III are strongly ginglymoid distally, but metatarsal IV is strongly convex. Metatarsal I is reduced to the distal half only, and meta-

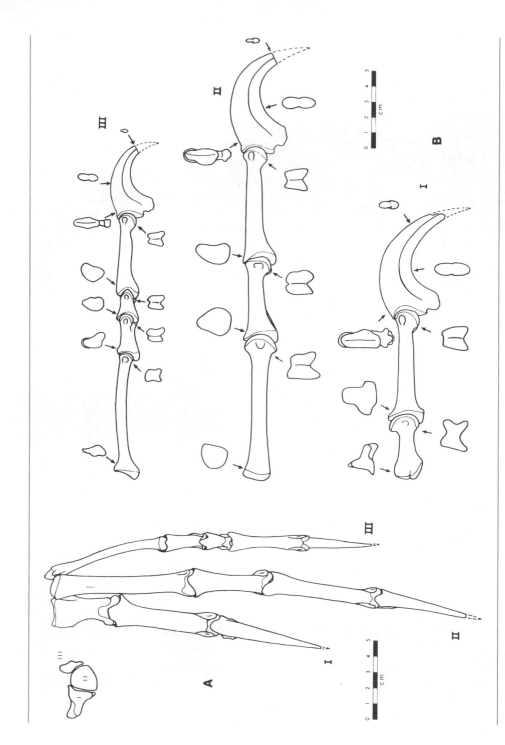

Fig. 12.5. The left manus of *Deinonychus antirrhopus* in dorsal (A) and medial (B) aspects, with proximal and distal profiles. Proximal ends of the metacarpals are outlined at upper left in A. (From Ostrom 1969b.)

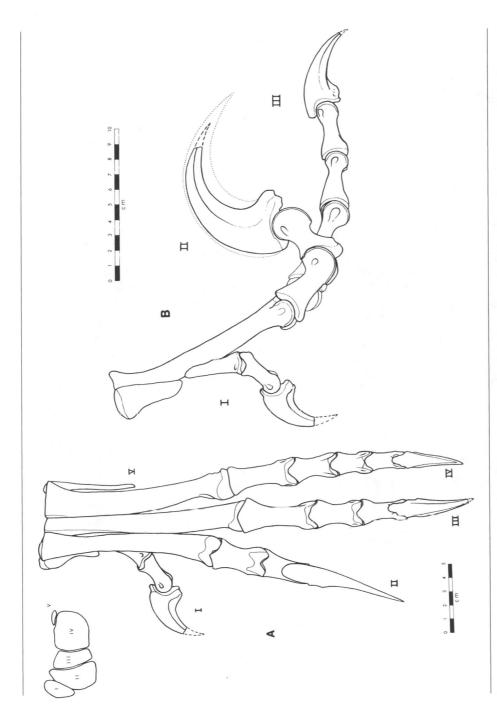

Fig. 12.6. The left pes of *Deinonychus antirrhopus* in dorsal (A) and medial (B) aspects. The dotted line of B indicates a conservative estimate of the size and shape of the horny talon on the second toe. Note the extreme contrast in form and size of the second and third toe unguals. (From Ostrom 1969b.)

tarsal V consists only of a proximal splint with no evidence of any articulating phalanges. Digit I is short, medial in position, and slightly reverted. Digits III and IV are of normal design and phalangeal formulae, but the second toe is highly derived, even though the phalangeal formula is the primitive three. All second digit articulations are deeply ginglymoid. The proximal phalanx has a prominent dorsally expanded ginglymus (for extreme extension), while the second phalanx bears a unique ventrally located proximal expansion ("heel"), presumably for attachment of a large flexor muscle mass. Distally, the second phalanx has an enlarged and ventrally expanded ginglymus for extreme flexion of the ungual. The ungual is very large, very narrow and trenchant, and strongly curved. Proximally, the ungual has a very narrow articular facet and a relatively enormous flexor tubercle. The entire second toe is designed for extreme hyperextension and powerful flexion.

EVOLUTION/SYSTEMATICS

The rarity of comparable material from different taxa makes it difficult to infer evolutionary relationships. The earliest known dromaeosaurid is *Deinonychus antirrhopus* from the Aptian/Albian Cloverly Formation of Wyoming and Montana, United States. The origin of the family is unknown, but fragmentary evidence suggests that it started in the *Coelurus-Ornitholestes* complex (see Norman, this vol.; Ostrom 1980*a*). *Coelurus* has the distinctive asymmetrical radiale characteristic of the Deinonychosauria. In *Ornitholestes,* the pes may have borne a modified second digit with an enlarged claw and thus approached the didactyl condition of deinonychosaurs. Also, the manus of *Ornitholestes* most closely resembles those of *Deinonychus* and *Velociraptor* (and also *Chirostenotes*). Neither *Coelurus* or *Ornitholestes* remains preserve any evidence of dromaeosauridlike modification of the caudal vertebrae as in *Deinonychus* and *Velociraptor*, but, nevertheless, these Late Jurassic forms are the best evidence available suggestive of the beginnings of the deinonychosaurian trend.

No evidence exists indicating when the split between the dromaeosaurids and troodontids occurred, but by Santonian or early Campanian time, the two lineages are distinct in Asia and by middle to late Campanian time, in North America. In the latter, *Dromaeosaurus* and perhaps *Saurornitholestes* represent Campanian descendants of *Deinonychus*. In Asia, *Velociraptor*

and perhaps *Adasaurus* and *Hulsanpes* represent the eastern derivatives of the dromaeosaurid lineage.

The evidence is very incomplete, but it does suggest a North American origin of the family in Late Jurassic time, followed by dispersal to Asia and subsequent diversification in both regions. Dromaeosaurids apparently became extinct by early Maastrichtian time.

Ostrom (1973, 1975*a*, 1975*b*, 1976*a*, and 1985) noted the anatomical similarities between dromaeosaurids and *Archaeopteryx*. The evidence strongly indicates a close phyletic relationship between dromaeosaurids and primitive birds, and one could argue that an (as yet unknown) Early or Mid-Jurassic dromaeosaurid was ancestral to archaeopterygians and later birds. This theme has been summarized cladistically by Gauthier (1986), Gauthier and Padian (1985), and Padian (1982). The hypothesis is not without critics (Martin 1983*a*, 1983*b*; Martin et al. 1980; Tarsitano and Hecht 1980; Hecht and Tarsitano 1982), yet it should be noted that the Dromaeosauridae may have had more than ordinary phyletic significance.

At the present time, the following taxa are assigned to the Dromaeosauridae.

Adasaurus mongoliensis   Barsbold, 1983*a*

Based on an incomplete specimen consisting of a fragmentary skull and a few associated postcranial elements including an opisthopubic pelvis, a slender metacarpal II, and a "reduced" ungual claw II (reduced as compared with the pedal claw condition in *Deinonychus;* fig. 12.7). No other diagnostic dromaeosaurid char-

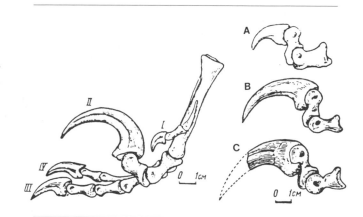

Fig. 12.7.   Pes of *Deinonychus antirrhopus* (left) and the second pedal digit of *Adasaurus mongoliensis* (A), *Velociraptor mongoliensis* (B), and an indeterminate deinonychosaur from the Mongolian People's Republic (C) (From Barsbold 1983*a*.)

acters have been reported as yet, so the familial assignment is not certain. The material is of Late Cretaceous age from the Nemegt Basin, Mongolian People's Republic.

Deinonychus antirrhopus   Ostrom, 1969

As of this writing, *Deinonychus antirrhopus* appears to be the best-known and best-reported example of the family, represented by at least six or seven partial skeletons (Ostrom 1969*a*, 1969*b*, 1974*a*, 1976*a*). All specimens preserve one or more of the diagnostic family characters, and all parts of the skeleton are known except for the braincase and skull roof. No sternum or clavicles have been found, although these elements are

present in *Velociraptor mongoliensis* (Barsbold 1983*a*). All material was recovered from Lower Cretaceous strata (Aptian/Albian), the Cloverly Formation of Montana and Wyoming (U.S.A.). (Paul [1984*a* and 1988] suggested that *Deinonychus* is synonymous with *Velociraptor*, a notion that seems most unlikely on anatomical and temporal stratigraphic grounds.)

Dromaeosaurus albertensis
Matthew et Brown, 1922

This taxon is known from a single specimen consisting of a partial skull and jaws (fig. 12.8), hyoids, associated with several foot elements, all from the Judith River Formation of Late Cretaceous age, Alberta, Canada.

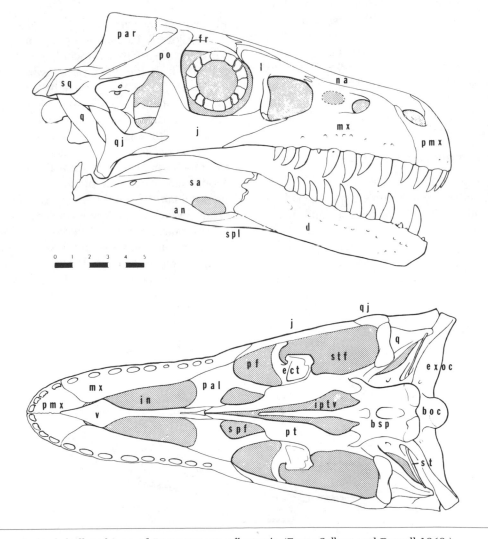

Fig. 12.8.   Reconstructed skull and jaws of *Dromaeosaurus albertensis*. (From Colbert and Russell 1969.)

Diagnostic features of pedal digit II, although incomplete, establish its affinity with *Deinonychus*. (See Colbert and Russell 1969.)

Hulsanpes perlei   Osmólska, 1982

Hulsanpes is based on a single specimen consisting of an incomplete foot from the Barun Goyot Formation (Campanian), Upper Cretaceous of the Mongolian People's Republic. It appears to display the diagnostically modified form of the second toe. Other dromaeosaurid distinctions have not been reported as yet.

Saurornitholestes langstoni   Sues, 1978*a*

S. langstoni is known from three fragmentary specimens plus the holotype. The type consists of associated cranial elements and fragmentary postcranial material. The other specimens consist of cranial elements only. All are from the Upper Cretaceous Judith River Formation (Campanian) of Alberta, Canada. The postcranial fragments are very similar to *Deinonychus*, but absence of diagnostic dromaeosaurid characters leaves the familial assignment probable but uncertified. The two associated teeth are indistinguishable from those of *Deinonychus*, though. (Paul [1984*a* and 1988] concludes that *Saurornitholestes* is a junior synonym of *Velociraptor*.)

Velociraptor mongoliensis   Osborn, 1924*a*

V. mongoliensis (fig. 12.9) is well known from at least half a dozen fragmentary to complete specimens from Upper Cretaceous strata, the Djadochta and Barun Goyot formations of Mongolia. This is the best-known example of the family after *Deinonychus antirrhopus* but not as yet completely described. Most specimens preserve one or more of the foot, hand, wrist, or caudal features that are diagnostic of the family. One or more of the specimens include a sternum (but no clavicles),

which has not yet been found or recognized in other dromaeosaurid material.

Chirostenotes pergracilis   Gilmore, 1924*a*

This taxon is based on a single specimen consisting of an incomplete but articulated pair of hands from the Upper Cretaceous Judith River Formation of Alberta, Canada. The digital proportions and hand configuration are comparable to those of *Deinonychus antirrhopus* and *Velociraptor mongoliensis*, hence the suggested family assignment. It should be pointed out here that it is quite possible that material pertaining to this species represents the unknown hands of *Dromaeosaurus albertensis* (however, see Currie, this vol.).

PALEOECOLOGY

Remains of *Deinonychus antirrhopus* occur exclusively in the Cloverly Formation of Wyoming and Montana. Virtually all specimens (including isolated teeth) were collected from the upper part, a variegated sandy bentonitic claystone that probably represents overbank floodplain deposits reflecting a lowland riparian environment. Coarse channel sands occasionally are associated with these claystones (Ostrom 1970*a*). *Velociraptor mongoliensis* remains derive primarily from the much younger Djadochta Formation of southern Mongolian People's Republic. Djadochta lithology is predominantly a poorly cemented, fine-grained arkosic sandstone with very large-scale cross-stratification and occasional caliche zone. The unit is usually interpreted as eolian dune deposits and associated playa lake sediments (Gradzinski et al. 1977). This points to at least semiarid and perhaps high plateau environs. The only certifiable remains of *Dromaeosaurus albertensis* are from the Judith River Formation of Alberta. The matrix is a

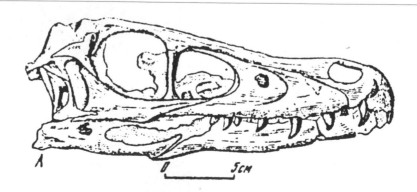

Fig. 12.9.   Skull and jaws of *Velociraptor mongoliensis* (from Barsbold 1983*a*).

drab fine-grained claystone and also may represent an overbank floodplain setting. The precise locality and stratigraphic horizon are unknown. Consequently, taphonomic details and paleontologic conditions are not available, but the deposition environment of the Judith River Formation was predominantly fluvial, indicating a lowland riverbank habitat for *Dromaeosaurus* also.

Anatomical evidence among the dromaeosaurids is unequivocal: all were active bipedal predators with moderately elongate and robust limbs, highly sectorial teeth, long raptorial hands, and specialized foot structures. Taphonomic evidence associated with the remains of *Velociraptor mongoliensis* and *Deinonychus antirrhopus* further documents the above inference of predatory habits. Dental and skeletal evidence pertaining to *Deinonychus* has been recovered from sites in exclusive association with skeletal remains of *Tenontosaurus tilletti*, a medium- to large-sized herbivorous ornithopod. A complete skeleton of *Velociraptor mongoliensis* found in the Djadochta Formation actually "embracing" its prey—a skeleton of the herbivore *Protoceratops andrewsi*—provides irrefutable evidence about the preferred prey of *Velociraptor*. At one site of *Deinonychus antirrhopus* in southern Montana, remains of not less than four individuals were recovered associated with fragmentary remains of a single large individual of *Tenontosaurus tilletti*, apparently killed during their attack on the much larger prey animal. This occurrence suggests that *Deinonychus* hunted in packs of half a dozen or more individuals—much like modern wolves and Cape hunting dogs—enabling it to prey on much larger animals. Only in the case of *Compsognathus longipes* is there comparable evidence of such specific prey preference in any other theropod (Ostrom 1978*a*).

The elongated raptorial construction of the hands establishes dromaeosaurids as obligate bipeds. A quadrupedal stance was impossible. Walking and running most likely was by alternating strides, there being no evidence of symmetrical saltatorial gait for any theropod. All known theropod footprint evidence substantiates this conclusion even though no certifiable dromaeosaurid trackways have been reported to date. The grasping power of the hands is evident in their morphology, but the derived nature of the carpals indicates greater than normal theropod grasping capabilities. The asymmetrical ginglymus of the proximal surface of the radiale caused automatic supination of the hand as the wrist was adducted or pronation as the wrist and hand was extended and abducted. The manual dexterity resulting from this distinctive feature must have increased the grasping efficiency of dromaeosaur forelimbs.

The derived state of the second digit of the pes (all joints of which feature deeply grooved/keeled ginglymoid articulations) is evident. These joints confined all interphalangeal movements in that toe to a single precise plane but permitted an extreme arc of rotation of up to 150° or more by combined extension/flexion of the two distal joints. This arc of rotation could have been further increased by movements against metatarsal II, plus flexion and extension at the ankle, which was dominated by the large astragalar roller bearing-like mesotarsal articulation with the tibia-fibula. In short, the entire foot, but especially digit II, was capable of a very large arc of excursion. Add to this the usual excursion capabilities at the knee and hip. Retraction or extension of the scimitarlike ungual (bearing a much larger horny talon) of the second toe minimized abrasion or other damage to the claw from contact with the ground during normal walking or running. Extreme flexion of the second toe coupled with full hindleg retractile stroke, produced a very long curving slash-action of the claw against the prey animal during the attack—a most effective predatory adaptation.

This unusual adaptation is even more remarkable because this all-important killing device is located on each foot of an obligate biped, which thereby required an extraordinary degree of agility, balance, and leaping ability. It is conceivable that *Deinonychus, Velociraptor, Dromaeosaurus,* and other dromaeosaurids attacked with both feet simultaneously. That possibility is given strong credence by the design of the caudal vertebral series described earlier. The extreme hypertrophy of the prezygapophyseal and chevron processes must have had several important consequences: most of the tail length (= approximately two-thirds of total animal length) was greatly stiffened; the mass of the tail was increased over the normal theropod condition; the contractile forces of the proximal caudal muscles attaching to these bony tendons were applied simultaneously to the entire tail, rather than separately to individual segments. The four bundles of ossified tendons (one bundle at each quadrant) would allow the massive tail to move as a unit in any direction necessary and thus function as a dynamic inertial stabilizer with a high moment of inertia to counteract the destabilizing effects of the violent body movements of the predator during an attack.

The inferred picture gained from this suite of unique anatomical features is that dromaeosaurids were highly active, agile, fleet-footed, and bizarre "killing machines" that must have been effective against large and small prey alike.

13

Problematic Theropoda: "Coelurosaurs"

D. B. NORMAN

INTRODUCTION

Small theropods have generally been referred to as "coelurosaurs" since the systematic review of Huene (1914c). As with all relatively small, delicate vertebrates, postmortem physical processes are very likely to lead to the rapid maceration and fragmentation of their remains. A direct consequence is that the Mesozoic fossil record is literally strewn with the imperfectly preserved remains of small theropods which have few or no diagnostic characteristics.

This situation has created some very serious problems for comparative anatomists and taxonomists. New taxa have often been created on the basis of such poor, and in many cases inadequately preserved, material. The reasons for this are quite varied. First, simply because the material is "dinosaurian" it is likely to generate a disproportionate amount of interest; second, the material may represent a new stratigraphic and/or geographic occurrence of the group; and third, historical precedent actively encourages the continuation of erecting new taxa, complete with Linnaean binomials, on

the basis of taxonomically inadequate material. Clearly none of these reasons are adequate. In ideal circumstances, taxa should only be erected and given binomial status on the basis of diagnostic characters (synapomorphies). If material is discovered which is of considerable scientific interest by virtue of its stratigraphic or geographic position, or because it represents new evidence of hitherto unappreciated faunal associations, then it would be perfectly legitimate to define the new material in general (family, ordinal status) terms rather than more specifically, thus drawing attention to important material and ideally provoking greater research.

Nearly all of the taxa described below fall into the categories that have just been outlined. For completeness of coverage in a work of this kind, it is considered essential to review these taxa and assess their status and systematic position. This is particularly so in cases where the records indeed indicate that the material may be indeterminate, even at the familial level, and yet records an important stratigraphic or geographic occurrence.

One final problem that is manifested in many groups of dinosaur is the assessment of ontogenetic status of the material under consideration. There is the clear possibility that some small theropods ("coelurosaurs"), especially when known only from fragmentary remains, may well be juvenile forms of contemporary large theropods, immature versions of merely slightly larger contemporary theropods, or valid adults of taxa of naturally small size.

Much of the description and discussion to follow will involve some consideration of the points raised above in an attempt to justify some of the suggestions that have been made in a review of this type. However, it must be emphasized that this procedure is to a large degree purely arbitrary. I have adopted three categories for the assignment of such theropod remains: first, Theropoda *incertae sedis* (material that appears to justify its specific status but cannot be assigned to any higher-level taxonomic category with confidence); second, Provisionally Theropoda *incertae sedis* (material that is of considerable interest but lacks sufficient evidence, for a variety of reasons, for confident specific designation); and third, *nomina dubia* (indeterminate material).

THE TAXA

The taxa included in this chapter (table 13.1) were originally assigned to a series of familial or higher-level taxonomic categories either by the authors of the original descriptions or by subsequent reviewers. These have been disregarded, and the taxa are arranged in such a way that well-preserved and definable taxa are presented in the first section of the taxonomic review and those that are relegated to the position of being provisionally *incertae sedis*, or clear *nomina dubia*, follow. Concluding remarks provide a discussion of the problems with any higher-level taxonomic assignments of the definable taxa in the light of recent systematic reviews of the Theropoda.

1. Theropoda *incertae sedis*

Avimimus portentosus   Kurzanov, 1981

MATERIAL:

Parts of a skull (rostral portion of the jaws, much of the braincase and part of the lower jaw). There are also significant parts of the postcranial skeleton, in-

TABLE 13.1   Problematic Theropoda: "Coelurosaurs"

	Occurrence	Age	Material
Theropoda incertae sedis			
Avimimus Kurzanov, 1981			
A. portentosus Kurzanov 1981	"Barungoyotskaya" Svita, Omnogov, "Djadochtinskaya" Svita, Ovorkhangai, People's Republic	?late Santonian or ?early Campanian	3 partial skeletons
Coelurus Marsh, 1879c			
C. fragilis Marsh, 1879c (including *C. agilis* Marsh, 1884a)	Morrison Formation, Wyoming, Utah, United States	Kimmeridgian-Tithonian	Postcranial skeleton
Compsognathus Wagner, 1861			
C. longipes Wagner, 1861 (including *C. corallestris* Bidar, Demay, et Thomel, 1972)	Ober Solnhofen Plattenkalk, Bavaria, Federal Republic of Germany; Lithographic Limestone, Canjuer, Var, France	Kimmeridgian	2 skulls and associated postcranial skeletons
Deinocheirus Osmólska et Roniewicz, 1970			
D. mirificus Osmólska et Roniewicz, 1970	Nemegt Formation, Omnogov, Mongolian People's Republic	?late Campanian or early Maastrichtian	Forelimb elements
Ornitholestes Osborn, 1903			
O. hermanni Osborn, 1903	Morrison Formation, Wyoming, Utah, United States	Kimmeridgian-Tithonian	Skull and associated postcranial skeleton
Therizinosaurus Maleev, 1954			
T. cheloniformis Maleev, 1954	Nemegt Formation, Omnogov, White Beds of Khermeen Tsav, Bayankhongor, Mongolian People's Republic	?late Campanian or early Maastrichtian	Forelimb elements
Provisionally Theropoda incertae sedis			
Alvarezsaurus Bonaparte, 1991 •			
A. calvoi Bonaparte, 1991 •	Rio Colorado Formation, Neuquen, Argentina	Coniacian	Partial postcranial skeleton
Avisaurus Brett-Surman et Paul, 1985			
A. archibaldi Brett-Surman et Paul, 1985	Hell Creek Formation, Montana, United States; Lecho Formation, Salta, Argentina	late Maastrichtian	Metatarsals
Unnamed theropod (= *"Halticosaurus" orbitoangulatus* Huene, 1932)	Mittlerer Stubensandstein, Baden-Württemberg, Federal Republic of Germany	middle Norian	Skull

Table 13.1, continued

Kakuru Molnar et Pledge, 1980			
K. kujani Molnar et Pledge, 1980	Maree Formation, Andamooka South Australia, Australia	Aptian	Hindlimb
Lisboasaurus Seiffert, 1973 •			
L. estesi Seiffert, 1973 •	Unnamed unit, Estremadura, Portugal	Oxfordian	Maxilla
Lukousaurus Young, 1948b			
L. yini Young, 1948b	Lower Lufeng Series, Yunnan People's Republic of China	Rhaetian-Pliensbachian	?3 skulls
Microvenator Ostrom, 1970a			
M. celer Ostrom, 1970a	Cloverly Formation, Montana, Wyoming, United States	Aptian-Albian	Partial skeleton
Noasaurus Bonaparte et Powell, 1980			
N. leali Bonaparte et Powell, 1980	Lecho Formation, El Brete, Salta, Argentina	?late Campanian-Maastrichtian	Isolated skull elements, vertebrae, pedal elements
Podokesaurus Talbot, 1911			
P. holyokensis Talbot, 1911	?Portland Formation, Massachusetts, United States	?Pliensbachian-Toarcian	Incomplete, fragmentary skeleton
Richardoestesia Currie, Rigby, et Sloan, 1990 •			
R. gilmorei Currie, Rigby, et Sloan, 1990 •	Judith River Formation, Alberta, Canada	late Campanian	2 dentaries with replacement teeth
Walkeria Chatterjee, 1986			
W. maleriensis Chatterjee, 1986	Maleri Formation, Godavari Valley, Andhra Pradesh, India	Carnian-middle Norian	Partial skull, vertebrae, femora, partial pes

Nomina dubia	Material
Arctosaurus osborni Adams, 1875 (doubtfully theropod)	Vertebra
Avipes dillstedtianus Huene, 1932	Metatarsals
Calamospondylus foxi Lydekker, 1889d	Vertebrae
Calamospondylus oweni Fox, 1866 (type of *Aristosuchus* Seeley, 1887a)	Vertebrae, pubes, phalanx
Chuandongocoelurus primitivus He, 1984	Vertebrae, ?scapula, pelvis, femur, tibia, fibula, partial pes
Coelurus gracilis Marsh, 1888a	Manual ungual, teeth
Coelurus longicollis Cope, 1887 (type of *Longosaurus* Welles, 1984)	Cervical vertebrae
Dolichosuchus cristatus Huene, 1932	Tibia
Euronychodon portucalensis Antunes et Sigogneau-Russell, 1991 •	Teeth
Halticosaurus longotarsus Huene, 1907	Mandibular fragment, vertebrae, humerus, ilium, femur, metatarsal
Inosaurus tedreftensis Lapparent, 1960a	Vertebrae
Jubbulpuria tenuis Huene et Matley, 1933	Vertebrae
Laevisuchus indicus Huene et Matley, 1933	Vertebrae
Megalosaurus bredai Seeley, 1883 (type of *Betasuchus* Huene, 1932)	Partial femur
Ornithomimus minutus Marsh, 1892b	Partial metatarsus
Phaedrolosaurus ilikensis Dong, 1973b	Tibia, phalanges
Procompsognathus triassicus Fraas, 1914	Partial skeleton
Pterospondylus trielbae Jaekel, 1914	Vertebra
Saltopus elginensis Huene, 1910a	Partial skeleton, including caudal part of skull
Sinocoelurus fragilis Young, 1942b	Isolated teeth
Sinosaurus triassicus Young, 1948b	Maxilla with teeth
Tanystrophaeus posthumus Huene, 1908 (type of *Tanystrosuchus* Kuhn, 1963)	Caudal vertebra
Tanystrophaeus willistoni Cope, 1887	Ilium
Teinurosaurus sauvagei (Huene, 1932) (type of *Caudocoelus* Huene, 1932)	Vertebra
Thecospondylus daviesi Seeley, 1888b (type of *Thecocoelurus* Huene, 1923)	Cervical vertebra
Thecospondylus horneri Seeley, 1882	Internal mold of sacrum
Tugulusaurus faciles Dong, 1973a	Hindlimb, rib, vertebral centrum
Velocipes guerichi Huene, 1932	?Tibia
Wyleyia valdensis Harrison et Walker, 1973	Partial humerus

cluding much of the limbs, the vertebral column, and the pelvis (figs. 13.1–13.9).

Recovered from Nemegt Basin, Mongolia (Campanian-Maastrichtian).

DIAGNOSIS:

Edentulous premaxillae, with bony denticulations (hadrosaurlike); narial opening extends to above the orbit (sauropodlike); fusion of the braincase (as in birds and some theropods), ?hypapophysis on cervical vertebrae (maniraptoran and birdlike); ilium markedly inclined medially (birdlike); retention of metatarsal V (primitive dinosaur trait); crest on shaft of ulna (unique ?birdlike); fusion of the bases of metacarpals I–III (birdlike).

DESCRIPTION:

The recent detailed description of *Avimimus portentosus* by Kurzanov (1987) forms the basis for this brief redescription of the taxon.

Cranial. The type material consists of a poorly preserved skull cap, comprising a major part of the frontoparietal plate and adjacent fragments of the neurocranium. Additional material includes a well-preserved posterior portion of the skull (fig. 13.1) which is rather birdlike in its arrangement and proportions. The sutures seem to be obliterated as a result of fusion. The occipital plate is approximately triangular in caudal aspect (fig. 13.1B), and, rostral to this, the skull roof rises sharply to form a domed frontoparietal plate; the parietal portion of the latter is waisted laterally where it

forms the inner borders of the oblique, and posteriorly directed, supratemporal fenestrae; the frontal portion of the frontoparietal plate is high and arched but curves down abruptly in the area where it would have met the rostral bones (and forms the posterior margins of the external nares—fig. 13.1, n) and has a sharply defined rim laterally where it forms the margin of the orbit. The orbit is very large and has displaced the supra- and infratemporal fenestrae; it is not certain, from the published figures, whether the infratemporal fenestra has been lost by incorporation into the orbital cavity or whether a thin postorbital bar (fig. 13.1) separates the one from the other. The suspensorium is fused, with the pterygoids meeting in the midline. There is a trochlear condyle to the quadrate and a thin straplike jugal. In addition, there is some evidence of pneumatization of the braincase (sinuslike openings in the base of the basisphenoid—fig. 13.1, bs).

Other cranial material referred to this taxon includes isolated fused premaxillae that have a broad edentulous occlusal surface, marked by irregular denticulations, and border a large nasal aperture (fig. 13.2). A tiny fragment from the tip of the lower jaw is also recorded. Finally, a small portion of the distal end of the mandible is preserved.

Postcranial: Axial. Representative parts of the axial skeleton are preserved. Cervical vertebrae (fig. 13.3A,B) are typically theropod in shape, with low, cylindrical centra, short, square neural spines, large zygapophyses and a broad, decurved, lateral diapophyseal shelf that

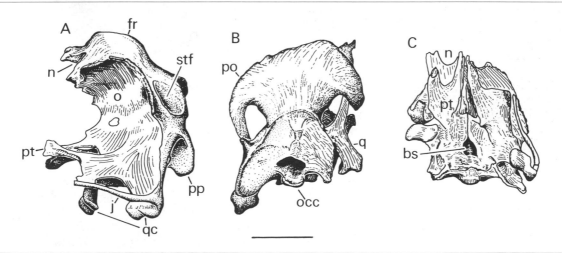

Fig. 13.1. *Avimimus portentosus* Kurzanov, 1981. Partial skull in lateral (A), dorsal (B), and ventral (C) aspects. Scale = 2 cm. (After Kurzanov 1987.)

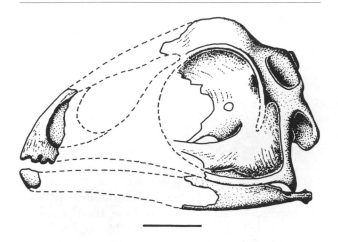

Fig. 13.2. *Avimimus portentosus* Kurzanov, 1981. Reconstruction of the skull and lower jaw in lateral aspect. Scale = 2 cm. (After Kurzanov 1987.)

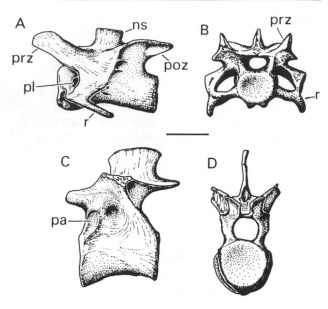

Fig. 13.3. *Avimimus portentosus* Kurzanov, 1981. Cervical vertebra in lateral (A) and cranial (B) aspects. Dorsal vertebra in lateral (C) and cranial (D) aspects. Scale = 1 cm. (After Kurzanov 1987.)

is, frequently if not invariably, fused to the tuberculum of the rib. A pleurocoel foramen is found on the lateral surface of the centrum, just behind the cranial articular surface, and in some vertebrae there may be two foramina, one above the other. The relationship of these latter vertebrae, with others (belonging to the type) that are of similar size, lack pleurocoels, have rib-head facets positioned as expected for a cervical centrum, and possess very prominent hypapophyses (fig. 13.4), is uncertain. Dorsal vertebrae (fig. 13.3C,D) have shorter and taller centra, loftier neural arches with short, square neural spines, upwardly curving diapophyseal shelves and parapophyses placed on the side of the neural arch. Sacral vertebrae appear to have numbered at least six, but no details of their form are currently available to the reviewer. Caudal vertebrae are poorly preserved.

Postcranial: Appendicular. The forelimb and girdle are unfortunately not well represented. No girdle elements are known. The humerus has the proportions of that of a typical bipedal dinosaur: quite slender, poorly defined head, modestly developed deltopectoral crest, and quite well-defined distal articular surfaces for the radius and ulna (fig. 13.5). The proximal end of the ulna is preserved and shows a roughened articular portion of the olecranon, a curved shaft, and a roughened ridge running along one edge (fig. 13.6D, cr). The only known portion of the manus is interpreted to be a fragment of the fused base of metacarpals I–III (fig. 13.6A, B, C).

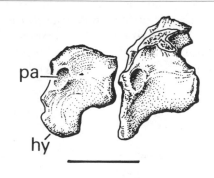

Fig. 13.4. *Avimimus portentosus* Kurzanov, 1981. Two cervical vertebrae referred to the type material of *Avimimus*, showing the well-developed hypapophyses. Scale = 2 cm. (After Kurzanov 1987.)

The hindlimb and pelvic girdle are well preserved (figs. 13.5, 13.7–13.9). The ilium has a large, broad, medially inclined blade, which, in dorsal view, has the profile of a typical theropod ilium (large rounded rostral portion and a slightly longer and more tapering caudal portion) and has a very broad brevis shelf (fig. 13.7, brs). There is a large, oblique boss immediately caudal to the margin of the large, open acetabulum.

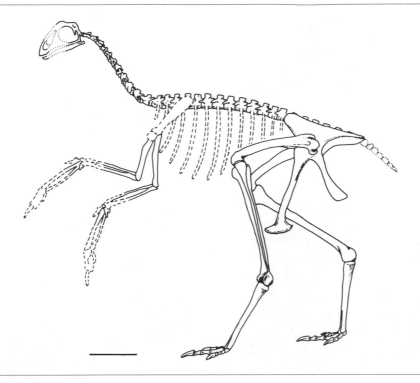

Fig. 13.5. *Avimimus portentosus* Kurzanov, 1981. Reconstruction of the skeleton. Scale = 10 cm. (After Kurzanov 1987.)

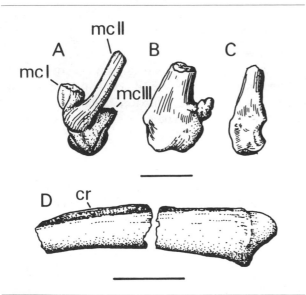

Fig. 13.6. *Avimimus portentosus* Kurzanov, 1981. Portion of the carpometacarpus (A, B, C) and the fragmentary ulna (D) in ?caudal view. Scale = 1 cm. (A-C); = 2 cm (D) (After Kurzanov 1987.)

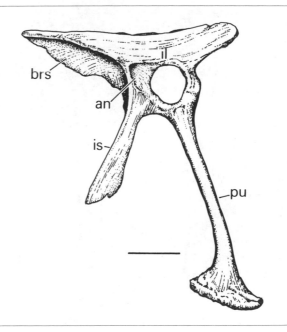

Fig. 13.7. *Avimimus portentosus* Kurzanov, 1981. Restored pelvis in lateral view. Scale = 4 cm. (After Kurzanov 1987.)

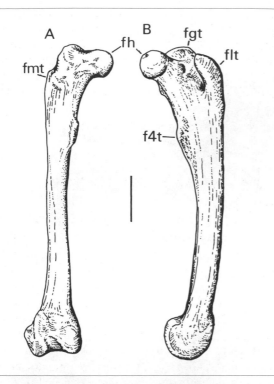

Fig. 13.8. *Avimimus portentosus* Kurzanov, 1981. Femur (left) in caudal aspect (A) and in medial aspect (B). Scale = 3 cm. (After Kurzanov 1987.)

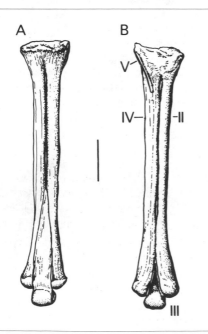

Fig. 13.9. *Avimimus portentosus* Kurzanov, 1981. Metatarsus (left) in cranial (A) and caudal (B) aspects. Scale = 2 cm. (After Kurzanov 1987.)

The pubes have slender, curved shafts and meet in a symphysis along their lower third; the distal ends have the form of a rostrocaudally expanded "foot." The ischium is poorly preserved and indicates little of its natural form. The femur (fig. 13.8) has a typically offset head, and a saddle-shaped shoulder region separates the head from the greater trochanter. Separated from the greater trochanter by a large cleft is the lesser (anterior) trochanter, which is thick, fingerlike, and curved medially toward its tip. Beneath the greater trochanter and nearer the caudal border of the femur is an additional (minor) trochanter similar to that seen in dromaeosaurids. Beneath the head of the femur on the caudomedial side of the shaft there is a crested fourth trochanter. The tibia, fibula, and proximal tarsals appear to be fused distally, while proximally, there is a prominent cnemial crest, and distal to this, on the lateral surface of the shaft there is a pronounced fibular crest (typical of many theropods). The proximal (unfused) portion of the fibula is flattened proximally and tapers rapidly distally; the fibula has a tubercle where it is attached to the fibular crest of the tibia. The pes is well preserved. The metatarsus (fig. 13.9) is long and slender, with metatarsals II and IV fused proximally and metatarsal III confined to the distal area of the metatarsus, where it is wedged between the adjacent metatarsal shafts. Unusually, there also appears to be evidence of a splintlike metatarsal V. The phalangeal formula is 0, 2, 3, 4, 0 and the unguals are narrow and pointed but only slightly decurved.

ASSESSMENT:

Avimimus is clearly an unusual theropod from the evidence of the published description possessing, as it does, a very distinctive mixture of features, some of which are seen either in other groups of dinosaurs or in birds. Listing the salient points:

i. The premaxilla is very reminiscent of the form seen in ornithopod dinosaurs (lambeosaurine hadrosaurids). The narial fossa is remarkable in its development. Surrounding the external naris, it forms a very large embayment on the lateral side of the premaxilla to the notch above the orbit.

ii. The braincase resembles that of birds in some respects (particularly in the expansion of the fronto-parietal plate, the enlarged orbits, and the reduction of the temporal region); however, many of these features also resemble the condition in sauropod braincases. The skull and lower jaw fragment of this species seem to bear some similarity to oviraptorids.

iii. The vertebral column appears to be typical of that of most small theropods, but several cervical vertebrae are recorded with the type material that would appear to be of a radically different form. Most notably, these specimens lack pleurocoels and possess large hypapophyses (typical of modern birds and incorporated in Gauthier's [1986] suprageneric taxon Maniraptora). The proportions of the tail are unknown, although Kurzanov (1987) has illustrated this form with a hypothetically short tail.

iv. The pelvis is unusual in the form of the iliac blades, which are inclined medially in a very avian fashion. The remainder of the pelvis is typically theropod as far as can be determined.

v. The hindlimb is typical of theropods in most respects. It does exhibit fusion of the distal end of the fibula with the tibia and proximal tarsals to form a typically avian tibiotarsus, but this type of formation has been observed in some theropods. The retention of a vestige of metatarsal V is very unusual in forms from the Late Cretaceous.

vi. The forelimb is perhaps the most puzzling part of the skeleton and is (typically) the least well preserved part of this animal. The humerus is typical of a bipedal dinosaur in all respects (many theropods have a far better developed deltopectoral crest). However, the interpretation of the ulna by Kurzanov (1981, 1987) as showing evidence of feather attachment is most interesting, especially when it is combined with the metacarpal fragment associated with this material. The latter is very similar in appearance to the proximal portions of the manus of a typical adult bird.

The unique combination of characters in this taxon is remarkable. Evidence can be brought forward to suggest theropod, sauropod, ornithopod, and avian affinities with some of the individual remains in this apparently associated material.

Using Gauthier (1986) as a guide it is clear that *Avimimus* displays some basal theropodan characters: spurlike metatarsal V, enlarged preacetabular portion of the ilium and large brevis fossa, and bowed femoral shaft and a fibular attachment crest on the lateral sur-

face of the tibia; it also exhibits a small number of tetanuran and ornithomimid characters and one maniraptoran feature (development of hypapophyses on cervical vertebrae). However, it seems clear that our knowledge of this form is insufficient to establish precise relations with currently recognized higher taxa.

Coelurus fragilis   Marsh 1879*c*

SYNONYMY:

Coelurus agilis   Marsh 1884*a*

MATERIAL:

Ostrom (1980*a*) has begun to revise the osteology of this genus for the first time since the detailed revision of carnivorous dinosaurs by Gilmore (1920). Instead of a single type specimen, which was in fact a composite of two partial caudal vertebrae, Ostrom has proposed that the major part of a single skeleton was recovered over a period of about a year (September 1879–September 1880) from Quarry 13 at Como Bluff, Wyoming. Type and referred material of *C. fragilis* comprises one cervical (fig. 13.10) and several dorsal vertebrae and a series of caudals. Type material of *C. agilis* comprises several cervical and dorsal vertebrae as well as one proximal caudal vertebra, numerous fore- and hindlimb bones, both pubes, parts of the ilium, and portions of the manus and pes.

The material comes from the Morrison Formation (Kimmeridgian-Tithonian) of Como Bluff, Wyoming.

DIAGNOSIS:

Small coelurosaurian theropod (*sensu* Gauthier—footed pubis and fused cervical ribs), cervical vertebrae are opisthocoelous, have shallowly convex cranial articular surfaces, and the lateral diapophyseal shelf is very broad and decurved.

DESCRIPTION:

As Gilmore (1920) supposed, the two species of *Coelurus* are identical. They were originally distinguished by size (the pubis of *C. agilis* [fig. 13.11] being considerably larger than that which might be expected to belong to *C. fragilis*), but Ostrom (1980*a*) has stated that this is incorrect. In addition, the reassignment of material allows several portions of the two type specimens to be compared directly.

Ostrom (1970*a*) indicated there are similarities between the material referred to *Calamospondylus oweni* (see below) from the Wealden (Barremian) of the Isle of Wight (southern Britain) and that of *Coelurus fragilis:*

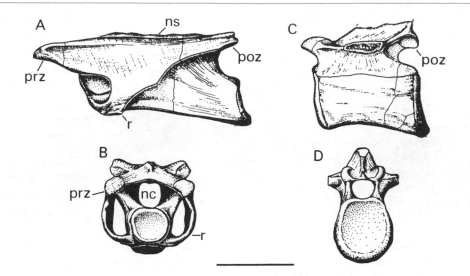

Fig. 13.10. *Coelurus fragilis* Marsh, 1879c. Cervical vertebra in lateral (A) and cranial (B) aspects and a proximal caudal vertebra in lateral (C) and cranial (D) aspects. Scale = 2 cm. (After Marsh 1881d.)

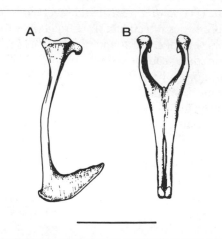

Fig. 13.11. *Coelurus fragilis* Marsh, 1879c. Pubes in lateral (A) and cranial (B) aspects. Scale = 10 cm. (After Marsh 1884a.)

notably, in the shape of the cervical vertebrae and the structure of the pubis.

Coelurus is a small, highly cursorial form, with well-developed pleurocoels in the cervical and dorsal vertebral centra and with cervicals that possess diapophyseal shelves that are broad and decurved. In addition, Ostrom (1976a) restored the forelimb of *Coelurus* with a semilunate carpal. On this basis alone, Gauthier (1986) assigned *Coelurus* to the Maniraptora;

however, since the cervicals lack hypapophyses (a key maniraptoran character), this taxon is regarded as a provisionally valid species (pending the work of Ostrom), but its relations within the Theropoda are uncertain.

Compsognathus longipes   Wagner, 1861

SYNONYMY:

Compsognathus corallestris   Bidar, Demay, et Thomel, 1972

MATERIAL:

Skeleton lacking only the distal half of the tail. Collected from Jachenhausen in the Reidenberg-Kelheim area of Bavaria, southern Germany (Ostrom 1978a). The stratigraphic horizon of this locality is dated at Kimmeridgian. Additional material referred to this species includes a less well preserved but somewhat larger skeleton from the Kimmeridgian limestones of Canjuers near Nice in southern France.

DIAGNOSIS:

Absence of mandibular fenestra, skull slender yet 30 percent of total body length. Restriction of pleurocoels to the cervical series, spool-shaped dorsal centra and the "fan-shaped" neural spines, absence of transverse processes on the caudal vertebrae, extreme shortness of the forelimb (37 percent of hindlimb length), digital formula of the manus 2-2-0-0-0, rostral expansion of the pubic foot, prominent obturator process of ischium and distal portion of ischial shaft fused to neighbor, first digit of the pes reduced (entire digit not attaining length of metatarsal II).

DESCRIPTION:

The osteology of the two known specimens of *Compsognathus* has been revised by Ostrom (1978*a*) and will be summarized in the following account (see figs. 13.12–13.15); it is paradoxically one of very few well-preserved small theropod dinosaurs and yet is not clearly attributable to one or another of the higher-level categories of theropods (see below).

The reconstruction of the holotype skeleton (fig. 13.12) of the dinosaur gives an estimated length of 700 mm; the original specimen suffers from a little post-mortem crushing and distortion, in particular, to the areas around the skull, forelimbs, and abdomen.

The skull is long and low (fig. 13.13), with a sharply tapering snout, and is very lightly constructed, there being very large openings that break up the dermal shield into a narrow median skull roof formed of the paired premaxillae, nasals, frontals, and parietals—none of which show any sign of fusion. Lateral to, and beneath, the skull roof there are a series of openings, separated by narrow spars of bone that connect with the tooth-bearing, lower margin of the skull. The most

prominent of the skull openings is the orbit. As reconstructed, this is large and subcircular and enclosed by the lacrimal rostrally, the postorbital dorsally and caudally, and the jugal ventrally. Immediately rostral to the orbit, there is a very large approximately triangular antorbital fossa enclosed by the lacrimal and maxilla; this has a substantial, subcircular antorbital fenestra, with a smaller triangular subsidiary fenestra in the rostral corner of the fossa. Near the tip of the snout there is a modest-sized elliptical narial opening. Caudal to the orbit there are two temporal fenestrae: the supratemporal fenestra is probably a parasagittally oriented ellipse, while the infratemporal fenestra is a vertical ellipse. Indeed, the temporal region of the skull is somewhat rostrocaudally compressed, which suggests that the temporal musculature was not particularly strongly developed.

The tooth-bearing bones of the upper jaw include the premaxilla, a relatively small approximately triangular bone, with a deep caudal embayment for the rostral margin of the naris. The extreme rostral tip of the premaxilla is slightly deflected downward and bears three teeth along its ventral margin, behind which there is a clear diastema before the teeth of the maxilla. The teeth of the premaxilla are long, slender, and taper to sharp points, which are somewhat recurved; the first tooth seems to have been slightly procumbent, and the

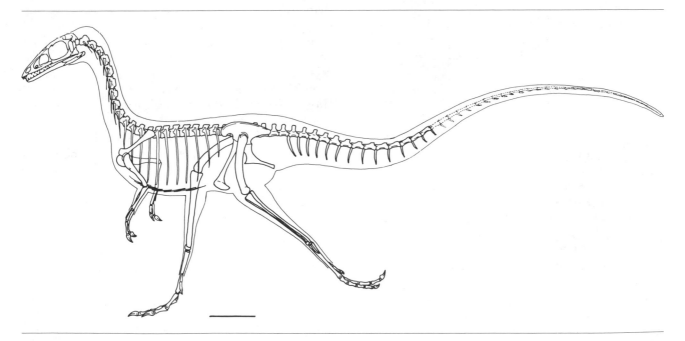

Fig. 13.12. *Compsognathus longipes* Wagner, 1861. Full skeletal restoration. Scale = 5 cm. (After Ostrom 1978*a*.)

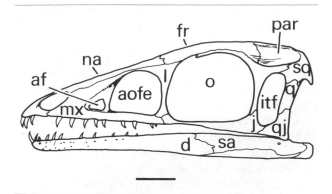

Fig. 13.13. *Compsognathus longipes* Wagner, 1861. Skull restoration in lateral aspect. Scale = 1 cm. (After Ostrom 1978a.)

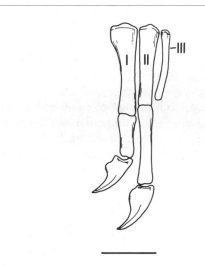

Fig. 13.14. *Compsognathus longipes* Wagner, 1861. Manus (left) in dorsal aspect. Scale = 1 cm. (After Ostrom 1978a.)

second tooth is the longest of the three. The maxilla has a large exposure on the lateral side of the snout where it forms a thin sheet of bone. Ventrally, the maxilla has a surprisingly shallow tooth-bearing margin; this apparently lacks alveoli rostrally and was described by Ostrom as extending for "approximately two-thirds of the maxilla length" (Ostrom 1978a: 82), even though teeth were illustrated extending along almost the entire length of the maxilla. It was reported that there may have been 15 or 16 maxillary teeth. Maxillary teeth are described as recurved, more laterally compressed than the premaxillary teeth, and possessing a distal carina with serrations. Distally, teeth become progressively smaller and less recurved, and mesial teeth seem to converge on the form of the premaxillary teeth. The

lower jaw is very slender, with nearly parallel upper and lower margins. There is no evidence of either a coronoid process or an external mandibular fenestra (extremely unusual for theropods). The dentary teeth mirror those described in the premaxilla and maxilla. Mesially, the teeth are long and slender with slightly recurved tips and the first is slightly procumbent, as is that of the premaxilla. More distal teeth are laterally compressed and more strongly recurved but lacked serrated distal carinae of those of the upper jaw. There are eighteen tooth positions in the dentary.

The skull appears to be very long and slender and is unusually large for the size of the animal (the ratio of skull length to presacral vertebral column [0.30] is greater than all other "coelurosaurs" and actually greater than that of the carnosaur *Allosaurus* [0.28]); the orbit is very large; the skull may have been highly kinetic (although this may well reflect the immaturity of the specimen); the mesial teeth are procumbent, and there is a distinct diastema in the upper jaw; the temporal region is constricted, and there is no evidence of a coronoid "eminence" or "process," (all of the latter points suggest a weak "bite"); and possibly associated with this, there is no evidence of either an intramandibular hinge or a mandibular fenestra. The postcranial skeleton has a vertebral column of 10 cervicals, 13 dorsals, 4 sacrals, and 30+ caudals and a well-preserved appendicular skeleton.

The cervical vertebrae are mostly well preserved, although the atlas is fragmentary. All of the remaining vertebrae are strongly opisthocoelous, are "waisted" between the articular surfaces, and have rostrally placed pleurocoels on the sides of the centrum. The neural arches of this part of the vertebral column are poorly preserved. Dorsal vertebrae are distinguished from cervicals by the absence of pleurocoels and by the change from opisthocoely to an essentially platycoelous condition; they also have rather long, slender, spool-shaped centra. Again, few details of the neural arch can be discerned. One peculiarity is the shortness but relatively great axial extent of the neural spines of dorsal vertebrae. The cranial and caudal edges of the neural spines show evidence of strong ligamentous attachment, as is typical of many theropods. The sacrum is unknown as it is obscured by the pelvis. Caudal vertebrae are known from the first sixteen of the series; proximal ones resemble those of the dorsal series in that they are long, slender, and somewhat spool-shaped. There are no pleurocoels or transverse processes, and the centra seem to be amphiplatyan. Small curved chevrons are present beneath the tail.

Ribs are present throughout the presacral col-

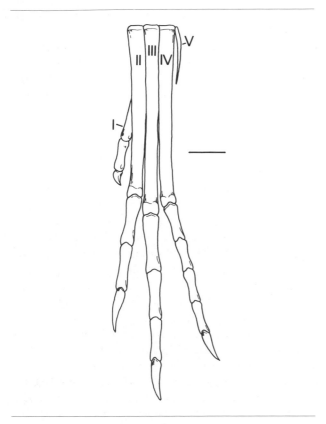

Fig. 13.15. *Compsognathus longipes* Wagner, 1861. Pes (left) in dorsal view. Scale = 1 cm. (After Ostrom 1978*a*.)

a symphysis, and the distal end is expanded to form the typical footed structure of most tetanuran theropods (although the expansion of the end of the pubis is a little atypical in that it is not noticeably expanded rostrally). The ischium is well preserved and rather short compared with most theropods; its proximal portion is narrow and expands ventrally to produce a thin but prominent obturator process, after which the ischium tapers backward as a narrow tapering cylinder. The ventral edge of the ischium caudal to the obturator process is fused to its neighbor.

The hindlimb is long and the distal elements show elongation. The femur is not well preserved; impressions indicate that it was relatively stout and curved. The tibia is considerably longer than the femur, its shaft is straight, and, as with the femur, the shaft is hollow. The fibula is long, very slender, and unfused to the tibia. The ankle is mesotarsal, but the details are now obscure (Ostrom 1978*a*). The pes (fig. 13.15) is typical of that of most theropods, showing cursorial adaptations, with long metatarsals (II–IV), a relatively shorter metatarsal I, a rudimentary metatarsal V, and slender pedal phalanges. Deep collateral ligament fossae are also to be seen on metatarsals and phalanges. The phalangeal formula, 2-3-4-5-0, is the same as that found in most theropods. Digit I is very short, with the entire digit not reaching the level of the lower end of metatarsal II, and the terminal claw is very small.

ASSESSMENT:

The small size of this dinosaur suggests that it is an immature specimen. Indicators of immaturity include the disproportionately large head and orbit and possibly the absence of detailed anatomical features in some limb bones. Immaturity may also explain why, in an otherwise little-disturbed skeleton, the skull is largely disarticulated. Corroboration may come from the Nice specimen (*C. corallestris*), which is some 50 percent larger than the original specimen, even though it appears to be identical in all other respects.

Also of note is the report of the stomach contents of the type specimen (Ostrom 1978*a*): the major part of a skeleton of a lepidosaur (*Bavarisaurus* cf. *macrodactylus*) within the abdominal cavity of the dinosaur; this may well be one of the few examples of the dietary preference of carnivorous dinosaurs.

The anatomy of *Compsognathus* is unusual in that it has a combination of features that consistently fails to fit into the systematic categories in current use. This focuses attention on the problematic status of this species in relation to such working hypotheses of relationship. Characters of *Compsognathus* appear sporadically

umn. The cervical ribs are not fused to the vertebrae and are double-headed, tapering to an extremely fine end. Fine bony gastralia are preserved.

The forelimb is stout and unsually short for small theropodans, being only 37 percent of the length of the hindlimb. The most peculiar feature of the forelimb is the manus (fig. 13.14). Two stout digits are reconstructed as well as the vestigial metacarpal of a third. By comparison with the manus structure in other theropods, Ostrom (1978*a*) concluded that the manual digits are I, II, and III. Digits I and II possessed an intermediate phalanx and an ungual each, while digit III consists of a single, slender, and bowed metacarpal (fig. 13.14). The ungual phalanges of the hand are unusual in that they are not strongly curved, as they are in a typically raptorial species, so that they were unlikely to have been used solely as piercing or cutting devices.

The pelvis and hindlimb are moderately well preserved. The pelvis is that of a conventional (nondeinonychosaurian) theropod, with a long, low, iliac blade, slightly convex dorsally, with the rostral processes longer and deeper than the caudal blade. The distal half of the pubis is fused to its neighbor, forming throughout the levels of the most recent cladogram for Theropoda (Gauthier 1986). Gauthier (1986, fig. 6) clearly placed *Compsognathus* in the Maniraptora, despite the fact that his review of theropod systematics consistently failed to find maniraptoran characters. Notwithstanding comments to the contrary (Gauthier 1986: 36), only two maniraptoran characters are found in *Compsognathus*: a bowed third metacarpal (already a vestigial bone in this species) and metatarsal IV subequal in length to metatarsal III; I do not regard either of these as convincing.

This species is clearly well defined, but its systematic position with regard to theropods generally cannot be established with precision. Using Gauthier's review, *Compsognathus* can be classified as a tetanuran theropod on the basis of a large caudally placed antorbital fenestra; maxillary teeth found only rostral to the orbit; absence of digit IV in manus; obturator process on the ischium; expanded pubic foot; and short first metatarsal.

Deinocheirus mirificus
Osmólska et Roniewicz, 1970

MATERIAL:
Pectoral girdle and forelimbs of a single specimen from the Nemegt Formation (Campanian-Maastrichtian) of Mongolia.

DIAGNOSIS:
Forelimbs of gigantic size, scapula 25 percent longer than the humerus, deltopectoral crest robust and proximally placed, manus equal to half the length of humerus and radius, manus unguals robust and strongly curved with flexor tubera close to the proximal articular surface.

DESCRIPTION:
Deinocheirus mirificus was undoubtedly a gigantic creature, with forelimbs that were 2.4 m long. The scapula is unusually long, while the coracoid is very deep dorsoventrally and has a poorly developed caudal process. The humerus is, as in most theropods, slender, somewhat twisted about its long axis, with a robust deltopectoral crest. The radius is slightly more than 50 percent of the length of the humerus, and yet the manus forms 33 percent of the total length of the forelimb. The digits are of equal length, although there is some disparity in length between the individual bones of the manus.

ASSESSMENT:
The gigantic size of the forelimb and girdle elements of *Deinocheirus* prompted Osmólska and Ronie-

wicz (1970) to assign this genus to a new family Deinocheiridae, within the Carnosauria. Somewhat later, Rozhdestvensky (1970b) noted that carnosaurs generally tend to exhibit forelimb reduction, rather than extreme elongation, and compared *Deinocheirus* to ornithomimids; this latter view has been supported by Ostrom (1972) and others (Galton 1982c). Barsbold (1976a) proposed a new infraorder, Deinocheirosauria, to include both the Deinocheiridae and Therizinosauridae. However, it would appear (see below) that such a suggestion is untenable.

In their general proportions, the limbs of ornithomimids and *Deinocheirus* are very similar. However, Nicholls and Russell (1985) noted that the scapulocoracoid of *Deinocheirus* has a more typical theropod structure than that of most ornithomimids: there is no flange on the supraglenoid buttress, the coracoid is very deep dorsoventrally, and the posterior coracoid process is rather poorly developed. In addition, the scapula is disproportionately long in *Deinocheirus* even when compared to ornithomimids. Finally, the manus unguals are raptorial in *Deinocheirus*, which is not the case in ornithomimids.

The curious combination of extraordinarily large size of the known remains of this dinosaur with the complete lack of further diagnostic material assignable to this taxon leaves its systematic position within the Theropoda in doubt. Affinities to ornithomimids are suggested, but until more material of this taxon is recovered, it must remain as a valid species of uncertain affinities (Theropoda *incertae sedis*).

Ornitholestes hermanni   Osborn, 1903

MATERIAL:
Partial skeleton, including a well-preserved skull (fig. 13.16) and an articulated partial manus of unproved association. Recovered from the Morrison Formation (Kimmeridgian-Tithonian) of Wyoming.

DIAGNOSIS (modified from Paul 1988):
Small size (1–2 m), oral margin of premaxilla short, external nares large, horn or crest over external nares formed by premaxillae and nasals, differentiation of dentition (mesial teeth conical with reduced marginal serrations compared to the distal crowns that are laterally compressed, recurved, and serrated blades), dentary tooth row shorter than maxillary row.

DESCRIPTION:
There is no detailed description of this taxon published to date. The skull is well preserved (fig. 13.16) and is typical of most theropods in having a short, box-

like caudal area with large supratemporal fenestrae and infratemporal fenestrae that are partially shielded by the quadratojugal and squamosal; in this, they tend to resemble carnosaurs. The orbit is unusually large and round, and in this respect, they more closely resemble the smaller theropodans, whereas the larger forms tend to have narrow rostrocaudally compressed (keyhole-like) orbital cavities. The antorbital fenestra is large and approximately triangular and bears a subsidiary fenestra on its rostral medial wall.

The teeth are restricted to the preorbital area of the mouth and were described by Osborn (1903, 1916) and Paul (1988) as differentiated into mesial ones, which tend to be more conical in shape and have less pronounced marginal serrations, and the distal ones, which are more typical theropod, being laterally

compressed, recurved, and markedly serrated. Osborn (1916) reconstructed the skull of this form with four particularly prominent premaxillary teeth, while Paul (1988) illustrated this species with three far smaller teeth using the same skull material. The teeth in the lower jaw are restricted to the rostral half of the dentary and would appear to work against only the rostral half of the upper dentition.

The lower jaw is robust and decurved distally. As restored by Paul, the mandible has a large fenestra, although this was not indicated on Osborn's revised figure (1916).

The postcranial skeleton (fig. 13.17) is partly preserved and was illustrated reasonably accurately by Osborn (1916). There are three cervical vertebrae (Ostrom 1980a). They are shorter and less complex than those of *Coelurus* (Osborn 1916: 254). There appear to be ten dorsal vertebrae, which are claimed to be distinct from those of the latter genus. Portions of the caudal series are also preserved. These suggest that the tail was long and whiplike and not reinforced by extensions of either the zygapophyses or chevrons.

The forelimb is represented by humeri and portions of the forearm and manus, with an additional unassociated manus also having been referred to this species (fig. 13.18). The pelvis is represented by incomplete ilia, pubes, and ischia; the pubis does not indicate whether there was a pubic foot, but the ischium does have a proximally positioned obturator process. The femur is longer than the tibia (indicating somewhat reduced cursorial capabilities). Only fragments of the foot are preserved.

ASSESSMENT:
Described in some detail by Osborn (1903, 1916), this form has been referred persistently to the genus

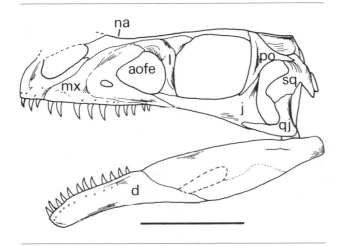

Fig. 13.16. *Ornitholestes hermanni* Osborn, 1903. Skull in lateral view, with indications of possible crest above the nares. Scale = 5 cm. (Modified from Paul 1988.)

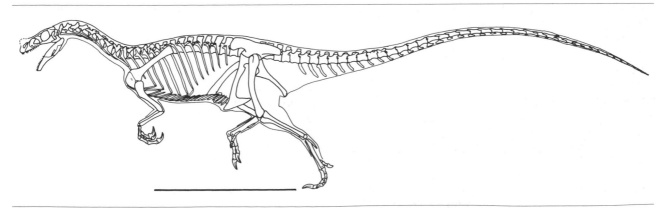

Fig. 13.17. *Ornitholestes hermanni* Osborn, 1903. Full skeleton restoration in "pursuit mode." Scale = 50 cm. (After Paul 1988.)

Fig. 13.18. *Ornitholestes hermanni* Osborn, 1903. Manus (left) in ventral (palmar) aspect. Stippling indicates restored digits. Scale = 5 cm. (After Osborn 1916.)

Coelurus following the comprehensive review by Gilmore (1920). However, the brief reappraisal of *Ornitholestes* and *Coelurus* by Ostrom (1980a) confirmed the earlier view of Osborn that this is a distinct genus. Very recently, Paul (1988) takes this distinction a stage further by proposing that *Ornitholestes* should be separated from other theropods, along with *Proceratosaurus*, into the Ornitholestinae within the Allosauridae. This view is in advance of the detailed revision of the species being undertaken by Ostrom and will be substantiated or otherwise as a result of that work.

The more conventional view of the systematic position of *Ornitholestes* is found in Gauthier (1986), who aligns this taxon with other forms in the Maniraptora (along with forms that in the recent past have tended to be referred to as "coelurosaurs" in the broadest Hueneian sense). The position of *Ornitholestes* within the Maniraptora is at best tentative since of the seventeen characters that define the node, only two clear ones appear; the forelimb comprises some 75 percent of presacral vertebral column length and the postdorsal margin of the ilium curves ventrally in lateral view. There is also a slender third metacarpal, though it is uncertain if this was bowed or not, but the remaining characters are either absent or cannot be determined.

On the basis of Gauthier's analysis, *Ornitholestes* would best be placed as a tetanuran theropod using the following characters: (i) no fanglike teeth in dentary; (ii) maxillary fenestra enlarged and caudally placed; (iii) maxillary teeth found rostral to orbit; (iv) manus two-thirds of the length of the humerus and radius

(provided that the referred manus is correctly attributed); (v) obturator process on the ischium; (vi) wing-like cranial trochanter (if Osborn's interpretation is correct). More precise relationships are not possible.

Therizinosaurus cheloniformis　Maleev, 1954

MATERIAL:

Partial forelimbs and pectoral girdles, from the Campanian of the Nemegt Basin, Mongolia.

DIAGNOSIS:

Gigantic theropods with long forelimbs, humerus not twisted, distal carpals blocklike, manus tridactyl and metacarpals of uneven length, unguals long and weakly curved and yet very narrow.

DESCRIPTION:

The proximal portion of the scapula is very wide and the distal portion short and narrow compared with most theropods; there is a very prominent acromion process, and close to the cranial edge of the proximal end there is a foramen not seen in any other theropod. The coracoid has a well-developed caudal process and a large coracoid foramen.

The humerus is massive and longer than the scapula and is curved but does not exhibit the axial twisting seen in *Deinocheirus,* and the deltopectoral occupies approximately 70 percent of the total length of the humeral shaft, even though the most prominent portion is clearly proximally positioned. The carpals are represented by two sutured bones, which are in turn sutured to the proximal surfaces of metacarpals I and II. The manus has the proportions of many nonornithomimosaurid theropods in that metacarpal II is the longest and most robust, with metacarpal I approximately half this length and metacarpal III intermediate in length and thinner than either of the others. Well-developed ginglymoid joints are developed on the distal ends of the metacarpals and on known phalanges. In the only known articulated digit (II), the first and second phalanges are shortened and support an enormously elongate ungual phalanx, which is approximately equal in length to the radius. The latter phalanx is weakly curved and strongly laterally compressed; it appears that the other two digits were similarly constructed.

ASSESSMENT:

The first remains of *Therizinosaurus* to be described were thought to belong to chelonians (Maleev 1954; Sukhanov 1964), and it was not until Rozhdestvensky (1970b) that the remains were referred to as those of theropod dinosaurs. Barsbold (1976*a*) described the first articulated remains that supported Rozhdestvensky's views.

Several gigantic and enigmatic theropods are currently known but have proved impossible to classify satisfactorily: *Chilantaisaurus, Deinocheirus,* and some undescribed forms from Kazakhstan, Zabaykalie, and Niger (Rozhdestvensky 1970*b*; Ricqlès 1967); none of these displays the enormous, slender, and weakly curved form of unguals seen in *Therizinosaurus cheloniformis.*

The unique form of the forelimb, girdle, and unguals justify the taxon as an entity; however, its relations to other taxa are entirely unknown. The appearance of gigantic long-armed theropods in the latest Cretaceous is a great mystery primarily because their remains are so distinctive and yet we know so little about the form and biology of such creatures. Theropoda *incertae sedis.*

2. Taxa of Questionable Status: Provisionally Theropoda
incertae sedis.

Avisaurus archibaldi   Brett-Surman et Paul, 1985

MATERIAL:

Left metatarsals (coossified), recovered from the Hell Creek Formation (Maastrichtian) of Montana. Two additional specimens have been referred to either as *Avisaurus* sp.—a small right metatarsal from the Lecho Formation of Argentina (previously referred to as an enantiornithid bird by Walker [1981]) and another specimen of a right metatarsus, again from the Hell Creek of Montana, which has been referred to as avisaurid.

DIAGNOSIS:

Knob on metatarsal II, extreme thinness of the shaft of metatarsal IV.

ASSESSMENT:

The type material consists of an isolated metatarsus (fig. 13.19), which is claimed to have a "uniquely divergent morphology" (Brett-Surman and Paul 1985: 137). The characters that are claimed to make this form distinctive are as follows: metatarsal III visible along its entire length (clearly found in other theropods, though rare in Late Cretaceous forms) and the broadest element proximally (not confirmed by illustrations where the situation is obscured by fusion and metatarsal is reported to be the most robust [see below]); metatarsals

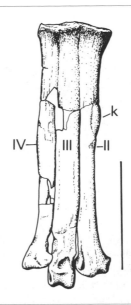

Fig. 13.19. *Avisaurus archibaldi* Brett-Surman et Paul 1985. Metatarsal of the pes (right) as preserved. Scale = 3 cm. (After Brett-Surman and Paul 1985.)

II and IV equal in length (this is not a unique character; almost all theropods have equal or subequal metatarsals II and IV); large concave articular facets proximally (also found in dinosaurs generally) and the "trochlea" of metatarsal IV vestigial, spoon-shaped, divergent, and asymmetrical (I would dispute the use of trochlea because it cannot be demonstrated that this area incorporates the distal tarsal; the area may well be irregular simply because it provides a basin for the attachment of a distal tarsal), knob on metatarsal II, metatarsal IV very flattened, metatarsal III triangular in section (condition found in wide range of theropods); metatarsal II the most robust (also seen in enantiornithine birds), and tendency to form a tarsometatarsus by fusion proximally (fusion may be observed, but there is no means of establishing a tendency on the basis of the material currently known; in fact, on the basis of the form of the proximal surface of metatarsal IV, I would argue that this specimen shows no evidence of a tarsometatarsus).

Therefore, of these characteristics, the metatarsals are unusual by virtue of the presence of a knob on metatarsal II and the extreme thinness of the shaft of metatarsal IV.

While probably theropod, it would not seem prudent at this stage to establish new higher-level taxonomic ranks for this material on the basis of what is currently known.

"Halticosaurus" orbitoangulatus   Huene, 1932

MATERIAL:

The partial and crushed skull of a theropod. The remains were recovered from the Pfaffenhofen area of Württemberg in the Stubensandstein (middle Norian).

ASSESSMENT:

Huene (1932) attempted a rather speculative reconstruction of this skull, which needs further preparation of the material so that it can be evaluated more accurately. It is possible that the material is referable to the contemporaneous *Liliensternus liliensterni* (see Rowe and Gauthier, this vol.), but this cannot be established currently.

Despite the poor state of this material and the inadequate diagnosis, it is best to retain the binomial in recognition of the potential importance of this fossil and to refer to this taxon as Theropoda *incertae sedis*. Prematurely discarding taxa, particularly those that may prove of importance to our understanding of Late Triassic faunas and early dinosaur evolution, is not fruitful at this stage.

Kakuru kujani   Molnar et Pledge, 1980

MATERIAL:

Casts (only) of the major part of the tibia, a fragment of the (?)fibula, and a phalanx. All of this material comes from the Andamooka area of South Australia, Maree Formation, dated as Aptian and opalized originally. The original material was sold by the owner.

DIAGNOSIS:

Small but very long, slender tibia, with very high, narrow, astragalus facet.

ASSESSMENT:

The tibia is long, slender, and hollow and has preserved on its distal end the outline of the astragalus attachment facet. The latter is of distinctive form as described by Molnar and Pledge (1980) in that the ascending process is high and narrow (a form that is apparently unique among theropods).

The material is of insufficient quality to make fruitful comparisons with other species. Molnar (pers. comm.) suggests that the tibia is extremely slender and equaled in this regard only by *Avimimus*. The astragalar facet may well have contours that result from postmortem distortion; however, Molnar (pers. comm.) reports an undescribed tibia and astragalus from Africa which is very similar.

It is probably best to refer to this material as Theropoda *incertae sedis*. But it is a useful record of theropods from the Aptian of South Australia.

Lukousaurus yini   Young, 1948*b*

MATERIAL:

The rostral portion of a small skull, including the lower jaw. Other material tentatively associated with this specimen includes a single tooth and a portion of a humerus.

The material was collected from the Lower Lufeng deposits (Upper Triassic—Lower Jurassic) of Yunnan, China. R. Molnar (pers. comm.) reports two further specimens from the same deposits, both of which are rostral portions of the skull.

ASSESSMENT:

The skull was described but poorly illustrated by Young (1948*b*) and is 70 mm long, extending back as far as the middle of the orbit (fig. 13.20). The snout is rather deep for its size and quite robust compared with many small theropods. The antorbital fenestra is large and triangular, and the naris is small and confined to the dorsal border of the snout. Five premaxillary teeth were described and ten maxillary tooth positions were noted (although there may well have been more in the complete skull). The teeth were described as typically theropod, being laterally compressed, recurved, and serrated.

This material is theropodlike and is of considerable interest since it is from an animal living either at the end of the Triassic or earliest Jurassic in an important geographic province.

While currently lacking a diagnosis, this material is of sufficient importance to merit being retained as a provisionally valid taxon.

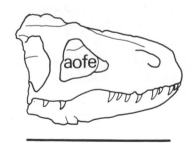

Fig. 13.20. *Lukousaurus yini* Young, 1948*b*. Skull as preserved, taken from an illustration by Young (1948*b*). Scale = 5 cm.

Microvenator celer   Ostrom, 1970*a*

MATERIAL:

A fragmentary skeleton, including parts of the skull (very fragmented), parts of 17 presacral vertebrae, 1 sacral, 9 caudals, and several fragments of the appendicular skeleton (coracoid, humerus, ulna, pubes, phalanges, femur, tibia, fibula).

From Wheatland County, Montana, Cloverly Formation (Aptian-Albian).

DIAGNOSIS:

Two sequentially arranged pleurocoels on cervical vertebrae, neural spines absent; forelimbs slender and very long; hindlimb with prominent cranial trochanter and a shallow pit instead of a crested fourth trochanter.

ASSESSMENT:

The cervical vertebrae possess two pleurocoels (one behind the other), are opisthocoelous, and have a flattened rostral surface; the neural arches lack a neural spine, are sculpted, and have long, low zygapophyses but lack the overhanging diapophyseal shelf seen in forms such as *Coelurus* and *Calamospondylus*. The dorsals retain a single pleurocoel and have low neural arches and spines. The forelimbs seem remarkably long and slender, as is the hindlimb with its cursorial proportions. The principal peculiarity is the femur, which has a prominent cranial (lesser) trochanter and a shallow pit instead of a crested fourth trochanter.

The small size of this specimen prompted some discussion from Ostrom (1970*a*) over the issue of its immaturity. Ostrom concluded that it was a mature individual, despite its small size and the fact that it had unfused neurocentral sutures (a feature that many would regard as an indicator of immaturity). Ostrom compared *Microvenator* with a variety of theropods, and concluded that it was most probably related to "coelurids," that is, *Compsognathus*, *Ornitholestes*, and *Coelurus*.

More recently, Gauthier (1986) reassessed *Microvenator* and placed it as a maniraptoran on the basis of four characters: elongate forelimb, bowed ulna, reduction of the cranial part of the pubic foot, and absence of a crested fourth trochanter. Although not a great deal is known of this specimen, the characters of the forelimb and hindlimb are suggestive not only of Maniraptora but also of the Ornithomimidae. Members of the latter family are characterized by their slender proportions, long arms, the unusually elevated cranial trochanter, and the absence or virtual absence of a crested fourth

trochanter, though none of the femoral characters are used by Gauthier in his definition of the latter family.

It has been suggested (Molnar pers. comm.) that *M. celer* has affinities with oviraptorids, but at present, I have insufficient evidence to substantiate this proposal. The taxon is tentatively regarded as Theropoda *incertae sedis*, pending further discoveries.

Noasaurus leali   Bonaparte et Powell, 1980

MATERIAL:

Includes a partial left maxilla, right quadrate, right squamosal, cervical neural arch, dorsal centrum, cervical rib, right metatarsal II, phalanx of pes, ungual phalanx (fig. 13.21).

From the El Brete area of Salta Province, Argentina, Lecho Formation (?Maastrichtian).

DIAGNOSIS:

Large antorbital fenestra, cervical neural arches with prominent cranially and caudally directed spines, and unguals with flexor pits rather than tubercles.

ASSESSMENT:

The partial remains of a small theropod, characterized by a large antorbital fossa, with a single fenestra. The ventral margin of the fossa is marked by a sharp ventral edge, which, as in some material referred to "*Marshosaurus*" Madsen 1976*b*, compares quite closely with that of dromaeosaurids. A most unusual feature of the neck is that there are sharp laterally positioned

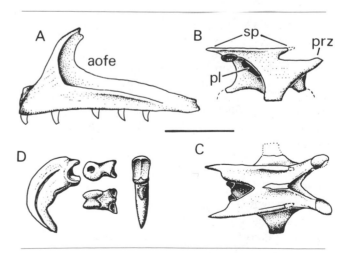

Fig. 13.21. *Noasaurus leali* Bonaparte et Powell, 1980. A, maxilla in lateral view. B, cervical neural arch in lateral view and C, in dorsal view. D, penultimate phalanx and ungual of the ?pes. Scale = 3 cm. (After Bonaparte and Powell 1980.)

spines developed from the epipophyses projecting caudally and from the lateral edges of the caudal zygapophyses projecting cranially (fig. 13.21B, C). The preserved metatarsal (?II) is very long (reminiscent of those of *Hulsanpes*—see Ostrom, this vol.), and the isolated ungual phalanx (fig. 13.21D) is very unusual in that there is a ventral pit instead of the usual flexor tubercle just anterior to the articular surface.

The material is too fragmentary to be of much value in the elucidation of either its biology or its relationships even though it is a useful geographic record of small theropods from South America in the Late Cretaceous. So far as our current understanding of Late Cretaceous theropods goes, the group that may be closest to this in general form is the troodontids (see Barsbold and Osmólska, this vol.), but the low density of teeth in the maxilla and the very unusual form of the ungual phalanx make this assignment contentious.

The creation of a new family of theropods (Noasauridae Bonaparte et Powell, 1980) on the basis of this imperfect specimen seems inappropriate at present.

Podokesaurus holyokensis   Talbot, 1911.

MATERIAL:

The type material comprises a series of caudal dorsals and distal caudals, a partial humerus, ribs, pubis, ischium, femora, left tibia and fibula, metatarsals, pedal phalanges, and possible skull fragments. The original specimens were destroyed by fire, though casts are preserved in the collections of the Peabody Museum (Yale University).

The type material was recovered from a glacially transported boulder in Holyoke, Massachusetts. Lull (1953) believed that this boulder originated within the Portland Formation (?Pliensbachian-Toarcian).

ASSESSMENT:

The original material (as well as the casts that are now the only ones available for study) was very poorly preserved. Colbert (1964*b*) in a review of early theropods proposed that the greater craniocaudal width of the neural spines of this taxon and the form of the ischium were unique to *P. holyokensis*, when compared to *Coelophysis bauri*. If these characters fall outside the normal range of variation for the latter species (Colbert in prep.) then *P. holyokensis* would appear to be a valid species.

The relationships of *P. holyokensis* have been discussed by Colbert and Baird (1958) and Colbert (1964*b*). Colbert and Baird noted similarities in the form of the hindlimb of *P. holyokensis* and *Coelophysis bauri*, while Colbert (1964*b*) used some basic anatomi-

cal, and preliminary morphometric, observations to conclude that the genera *Coelophysis* and *Podokesaurus* were synonymous, such that *P. holyokensis* became *Coelophysis holyokensis*. Unfortunately, most of the anatomical features merely provide confirmation of the dinosaurian status of these genera rather than a specific relationship between them.

The status and relationships of *P. holyokensis* are difficult to decide at present. As argued by Olsen (1980*a*), it is probably best to retain this taxon as a species of small, early theropod whose relationships have yet to be clarified. It should be regarded as Theropoda, provisionally *incertae sedis*.

Walkeria maleriensis   Chatterjee, 1987

MATERIAL:

Partial skull, the remains of 28 vertebrae, partial femora, and an astragalus.

Collected from the Upper Triassic Maleri Formation, Pranhita-Godavari Valley, Andhra Pradesh, southern India.

DESCRIPTION:

(Chatterjee, 1987: fig. 2a, b, c): The skull is 4 cm long and consists of a partial left maxilla with the distal ends of both dentaries attached. A fragmentary premaxilla lies adjacent to the distal tip of the maxilla.

The rostroventral margin of the antorbital fenestra lacks a well-marked ventral margin. The dentary rami are slender, and the teeth are slender and conical mesially and larger and laterally compressed distally; there appear to be no marginal serrations, as so commonly seen on theropod teeth.

The vertebrae are poorly preserved, similar in form to other theropods. The dorsals are small, cylindrical, and amphiplatyan, and the bases of the neural arches are excavated, but lack pleurocoels. Two caudals are described and identified because of their chevron facets.

The femoral head is strongly inturned, the lesser trochanter is positioned low (one quarter of the way down the shaft), and the fourth trochanter is large and crested, lying on the mediocaudal side of the shaft. The astragalus is low and hemicylindrical, with a low ascending process.

ASSESSMENT:

It appears to be clear that *Walkeria maleriensis* as currently defined is a small, carnivorous theropod dinosaur from the Late Triassic, with no particularly strong affinities with other known theropods below those characters that are used to define the Theropoda.

It is distinguished from other contemporary forms by the absence of a ridge below the antorbital fenestra, a long dentary symphysis, unserrated teeth, conical and procumbent mesial dentary teeth (see also *Compsognathus*), open astragalocalcaneal joint; and excavation of the bases of the neural arches of dorsal vertebrae. On this admittedly tenuous basis, it is considered best to refer this taxon to the holding category of imperfectly understood theropods.

3. *Nomina Dubia.*

Avipes dillstedtianus   Huene, 1932

MATERIAL:

The type and only material referable to this species comprises the proximal portions of three unfused metatarsals (fig. 13.22A). The specimens are very poorly preserved and embedded in small slab and are approximately 80 mm long. The locality of this material is given by Huene as Thüringen, and the horizon is reported as "Lettenkohle" at the base of the Lower Keuper. This probably corresponds to the Lettenkohlensandstein of the Lower Keuper (upper Ladinian) from Geiger and Hopping (1968).

ASSESSMENT:

This material is of a small and lightly built reptile. The fact that the metatarsals are bunched together suggests a digitigrade stance, which may in turn be interpreted as evidence of the dinosaurian status of this

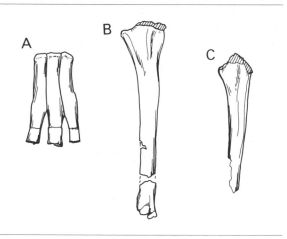

Fig. 13.22. A, *Avipes dillstedtianus* Huene, 1932—type material (metatarsal fragments). B, *Dolichosuchus cristatus* Huene, 1932—type material (tibia). C, *Velocipes guerichi* Huene, 1932—type material (mesopodial fragment?). (After Huene 1932.)

genus. This is, however, an extremely subjective line of reasoning. A variety of nondinosaurian archosaurs (saltoposuchid crocodilians, lagosuchids, and some pterosaurs) living in the Middle and Late Triassic also had a digitigrade stance.

The material attributed to this species is without doubt generically and specifically indeterminate and probably represents the remains of a digitigrade archosaur. The status of *Avipes* as an early dinosaur on the basis of the material is extremely dubious, and, furthermore, its presence in the Ladinian makes this seem even more unlikely, since there are few, if any, confirmed records of pre-late Carnian dinosaurs.

Calamospondylus oweni   Fox, 1866

MATERIAL:

The type material consists of 9 vertebrae (2 dorsals, 5 sacrals, and 2 caudals), the fused distal portions of both pubes, and an ungual phalanx (fig. 13.23). All of this material was recovered from the Wealden shales (upper Barremian) of the southern coast of the Isle of Wight, United Kingdom. It should be noted that Seeley (1887a) questioned the reference of the two sacral vertebrae to this taxon.

ASSESSMENT:

Calamospondylus oweni Fox, 1866, is clearly a senior objective synonym of *Poikilopleuron pusillus* Owen, 1876b, and *Aristosuchus pusillus* (Owen 1876b). Despite the poor original description of *Calamospondylus oweni* by Fox, this is clearly the type species of this material and relegates *Aristosuchus pusillus* and *Poikilopleuron pusillus*—names that have occurred in the recent literature on a number of occasions—into synonymy.

Additional genera and species have also become entangled with this group of taxonomic names. *Thecospondylus daviesi, Coelurus daviesi,* and *Thecocoelurus daviesi,* based, as they all are, on a partial cervical vertebra, are clearly Theropoda *incertae sedis* and should be considered *nomina dubia. Calamospondylus foxi* (renamed *Calamosaurus foxi*), based as it is on two cervical vertebrae (fig. 13.25) and a referred tibia (fig. 13.26), is not directly comparable to *Calamospondylus oweni* and should also be considered a *nomen dubium.*

Thecospondylus horneri (fig. 13.24) is a large specimen (far too large to be considered seriously as belonging to a small theropod) that is purported to be the natural cast of the internal surface of the sacrum. It is not comparable to any of the material described to date, and until sacral casts become used as systematic characters, it will remain of little value as a taxonomic entity.

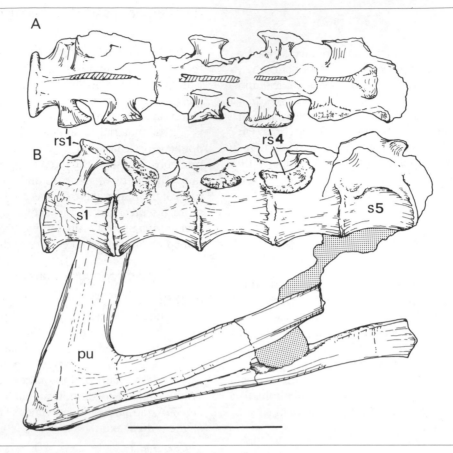

Fig. 13.23. *Calamospondylus oweni* Fox, 1866. Sacrum and pubis in dorsal (A) and lateral (B) aspects. Scale = 10 cm. (After Seeley 1887a.)

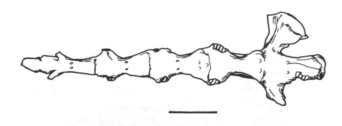

Fig. 13.24. *Thecospondylus horneri* Seeley, 1882. Sacral cavity preserved as a natural cast. Scale = 10 cm. (After Seeley 1882.)

The fossil evidence recovered to date from the Wealden shales of the Isle of Wight indicates that there is at least one small theropod. If diagnostic characters emerge for this material in the future, then it can be referred to, with some confidence, as *Calamospondylus oweni.*

Chuandongocoelurus primitivus He, 1984

MATERIAL:

Remains include several cervical (fig. 13.27) and dorsal vertebrae, fragmentary ?scapula, partial articu-

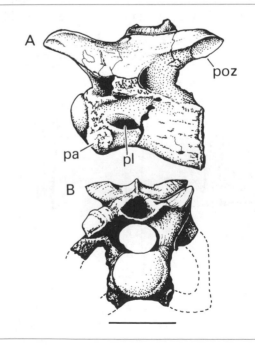

Fig. 13.25. *Calamospondylus oweni* Fox, 1866. Cervical vertebra in lateral (A) and cranial (B) aspects. Scale = 2 cm. (After Seeley 1891.)

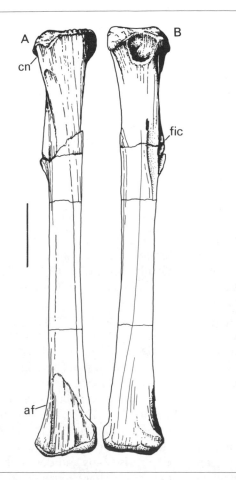

Fig. 13.26. *Calamospondylus oweni* Fox, 1866. Tibia (right) in (A) cranial and (B) caudal aspects. Scale = 2 cm. (After Seeley 1891.)

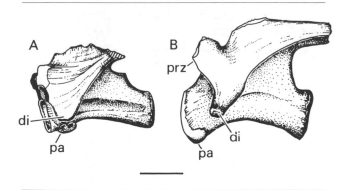

Fig. 13.27. *Chuandongocoelurus primitivus* He, 1984. Two cervical vertebrae in lateral aspect. Scale = 2 cm. (After He 1984.)

lated pelvis, femur, tibia and fibula, astragalus and calcaneum, and pes (partial).

From the Middle Jurassic of Sichuan.

ASSESSMENT:

Some of the vertebrae appear to have unfused neurocentral sutures, which is suggestive of immaturity of the individual. The hindlimb elements indicate theropod affinities for this taxon, with the femur having a simple dorsal curvature to the shaft, a prominent cranial trochanter, and a crested fourth trochanter. The ascending process of the astragalus is not very prominent and is more reminiscent of that of an ornithopod. The scapula appears to be unusually broad in lateral aspect, instead of the straplike form of theropods generally. The pes is represented by the metatarsus, which is unfused and shows no lateral compression of the proximal end of metatarsal III and is also long and slender. The metatarsals are symmetrical, with metatarsals II and IV subequal. The articulated phalanges of digit IV seem to be preserved in association with the metatarsus.

This material is insufficient to provide diagnostic characters that justify a binomial. I consider it best to refer to this taxon as a *nomen dubium* until better material is described.

Coelurus gracilis   Gilmore, 1920

MATERIAL:

Isolated ungual phalanx of theropod type.

From the Arundel Formation, Maryland. Hauterivian-Barremian.

ASSESSMENT:

Generically and specifically indeterminate.

Dolichosuchus cristatus   Huene, 1932

MATERIAL:

Left tibia (imperfect) with a well-developed fibular crest on the lateral surface of the shaft near the proximal end (fig. 13.22B). The material was recovered from Kaltenthal (near Stuttgart) in the Stubensandstein (middle Norian).

ASSESSMENT:

The material is too fragmentary to be of any use for descriptive, or, more important, comparative purposes, and should be regarded as a *nomen dubium*.

Halticosaurus longotarsus   Huene, 1907–08

MATERIAL:

Three cervical, 2 dorsal, 2 (fused) sacral, and part of a proximal caudal vertebrae; the proximal end of a humerus, a fragmentary ilium (restored in the form of a prosauropod!), the proximal end of a femur, a long and well-preserved metatarsal, and a very poorly preserved lower jaw.

This material was recovered from Pfaffenhofen in Württemberg from the Stubensandstein (middle Norian).

ASSESSMENT:

The material is very poorly preserved and could be from either a small theropod or a small prosauropod (especially if the ilium has been correctly restored). In the form of the femur, the presence of only two fused sacrals, and the absence of pleurocoels on the cervicals (perhaps the structure of the ilium as well), this species shows affinities with prosauropods. The one unusual feature is the exceptional length and slenderness of the metatarsal associated with the other remains, and in this, it does resemble a theropod.

The material is inadequate for diagnostic purposes and should be regarded as a *nomen dubium.*

Inosaurus tedreftensis   Lapparent, 1960*a*

MATERIAL:

Twenty-five small vertebrae and the proximal portion of a left tibia recovered from three localities in the Sahara Desert. The locality has been reported to be Lower Cretaceous.

ASSESSMENT:

The dorsal vertebrae are reported to be "massive," and the caudals are described as very tall, with a median ventral groove between the chevron facets.

There is no proof of association of this material, and if it belongs to a theropod, then it is of a highly unusual type. For the time being, it would be better to refer to this material as a *nomen dubium.*

Jubbulpuria tenuis   Huene et Matley, 1933

MATERIAL:

Two small vertebrae, recovered from the "Carnosaur Bed" of Jubbulpore (Madhya Pradesh) of India, Lameta Formation (Maastrichtian—Sahni pers. comm.).

ASSESSMENT:

Indeterminate small theropod dorsal vertebrae.

Laevisuchus indicus   Huene et Matley, 1933

MATERIAL:

Three small cervicals recovered from the "Carnosaur Bed" of Jubbulpore (Madhya Pradesh) of India, Lameta Formation (Maastrichtian).

ASSESSMENT:

The form of the vertebrae is reminiscent of those of *Calamospondylus,* and this is undoubtedly a small theropod species; however, it cannot be readily distinguished from the other species of small theropod that have been described from the same area on the basis of equally fragmentary but noncomparable material.

Procompsognathus triassicus   Fraas, 1913

MATERIAL:

Partial skeleton, very poorly preserved, recovered from the Stubensandstein (middle Norian) of the Pfaffenhoffen (now Heilbronn) area of in Württemberg.

ASSESSMENT:

Poor preservation makes interpretation of this material very difficult (figs. 13.28, 13.29). The material was reviewed as far as possible by Ostrom (1981).

The skull (fig. 13.28) is very crushed but gives the impression of a long, narrow-snouted form, with a large antorbital fossa, a large orbit, slender jaws with recurved teeth, a large infratemporal fenestra, and a narrower elliptical supratemporal fenestra. No mandibular fenestra seems to have been present. No pleurocoels are preserved on the vertebrae, and the dorsals tend to be rather long (fig. 13.29, dv). The pelvis and hindlimb are the most informative elements. The pubis has a rather broad, rectangular form reminiscent of that of "prosauropod" saurischians and does not have any obvious distal expansion characterizing most Jurassic and Cretaceous theropods. The femur, which is longer

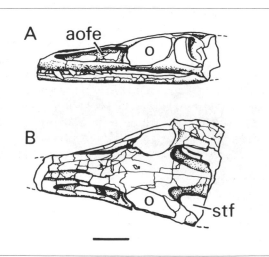

Fig. 13.28. *Procompsognathus triassicus* Fraas, 1913. Skull illustration in lateral (A) and dorsal (B) aspects. Scale = 1 cm. (After Ostrom 1981.)

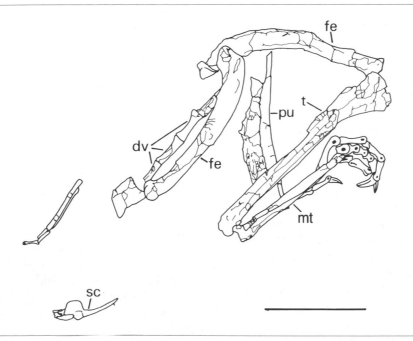

Fig. 13.29. *Procompsognathus triassicus* Fraas, 1913. Partial skeleton as seen preserved on a slab of matrix. Scale = 5 cm. (After Ostrom 1981.)

than the tibia, seems to be curved dorsally in lateral view, and the metatarsals are long and bunched together as is typical of erect-gaited archosaurs and dinosaurs generally. The foot has reduced digit I, and digit V is reduced to a short, hooked metatarsal splint.

On the basis of the structure of the pubis primarily, this form was tentatively allied to other Late Triassic/Early Jurassic forms such as *Coelophysis*, *Syntarsus*, and *Halticosaurus)*.

The structure of the pelvis in this early form is reminiscent of that of advanced archosaurs (lagosuchids) and early dinosaurs (herrerasaurids), but the preservation of this material is not sufficiently good to allow firm taxonomic assessment. Provisionally regarded as a *nomen dubium* because the supposed diagnostic characters are primitive for all theropods.

Pterospondylus trielbae   Jaekel, 1914

MATERIAL:

Single middorsal centrum, approximately twice the size of those of *Procompsognathus* from the Rhaetic of Halberstadt (FRG).

ASSESSMENT:

White (1973) referred to this material as a probable *nomen dubium*, and this was confirmed by Ostrom (1981).

Saltopus elginensis   Huene, 1910a

MATERIAL:

Remains include a partial postcranial skeleton of a small dinosaur, with a portion of the caudal part of the skull preserved. Collected from the Lossiemouth Sandstone Formation (early Norian) of Lossiemouth, Scotland.

Benton and Walker (1985) refer briefly to this rather poorly known form and amend several of Huene's original observations. The animal is reported to be a very primitive theropod, having three sacral vertebrae (rather than four) and a short ilium. The hindlimb shows cursorial proportions, and the metatarsals are bunched together to give a typically (although not exclusively) dinosaurian digitigrade stance. As reconstructed by Benton and Walker (1985), the skull is, as in *Procompsognathus*, very low and slender.

ASSESSMENT:

The type material is extremely poorly preserved and impossible to interpret unambiguously. It is not sufficiently well preserved to permit comparison with, for example, *Procompsognathus*, and it cannot be categorically denied that it may equally well be the fragmentary remains of a lagosuchid or an early pterosaur. It is considered best to regard this taxon as a *nomen dubium*.

Tanystrophaeus posthumus Huene, 1907–08

MATERIAL:

Isolated distal caudal vertebra. Collected from the Heslach area near Stuttgart, West Germany. Stuben-sandstein (middle Norian).

ASSESSMENT:

The presence of relatively long prezygapophyses is suggestive of theropod affinities for this material but is inadequate for a more precise assignment.

Tanystrophaeus willistoni Cope, 1887

MATERIAL:

The type material consists of a small, imperfectly preserved ilium. In addition, Cope (1887) referred to this taxon several dorsal vertebrae and miscellaneous bone fragments. All of the material referred to this taxon was recovered from the Chinle Formation, New Mexico, (U.S.A.)

ASSESSMENT:

The ilium of this species is clearly that of a thero-pod dinosaur; however, apart from this very general level of affinity, it is neither possible to justify the reten-tion of this species on this basis nor to refer this mate-rial with any confidence to the coeval genus *Coelophysis* (Rowe and Gauthier, this vol.), or any other approxi-mately contemporary species.

Cope (1889c) synonymized this species with *Coelophysis willistoni.* In all probability this specimen is a juvenile of *C. bauri,* as has been proposed by Colbert (1964b), but this cannot at present be proven. Proof awaits the appearance of the morphometric study of *Coelophysis* (Colbert 1989).

At present, it is considered prudent to regard *T. willistoni* Cope, 1887 as a *nomen dubium.*

Teinurosaurus sauvagei (Huene, 1932)

MATERIAL:

Isolated distal caudal vertebra of a theropod di-nosaur, collected from the area of Boulogne-sur-Mer, France (Kimmeridgian).

ASSESSMENT:

The material was originally referred to as *Igua-nodon prestwichii* by Sauvage (1897) but was subse-quently renamed *Teinurosaurus* by Nopcsa (1928c) and then renamed *Caudocoelus sauvagei* by Huene (1932) before finally being renamed *Teinurosaurus sauvagei* by Olshevsky (1978). The material is clearly theropod but must be regarded as a *nomen dubium.*

Velocipes guerichi Huene, 1932

MATERIAL:

Broken proximal end of a tibia (?fibula) (fig. 13.22C). Collected from the Keuper (late Carnian/ early Norian) of Schlesien (West Germany).

ASSESSMENT:

The material is indeterminate. *Nomen dubium.*

Wyleyia valdensis Harrison and Walker, 1973

MATERIAL:

Imperfect right humerus from the Weald Clay of Henfield, Sussex (Barremian).

ASSESSMENT:

The humerus is incomplete, lacking the distal portion of the shaft. There is a prominent deltopectoral crest and a slender shaft, which is quite thin walled. No evidence of a pneumatic foramen beneath the proximal head. Harrison and Walker (1973) compared this speci-men to remains of pterosaurs, dinosaurs (ornithopod and theropod), and birds and concluded that this was most probably a bird humerus. The comparisons to theropod dinosaurs relied largely on the published de-scription of *Deinonychus* and cannot be regarded as comprehensive. It seems most probable, in the absence of any clearly diagnostic avian characters of the prox-imal end of the humerus and in the absence of avian pneumatic openings into the humeral shaft, that this material is theropod rather than avian; it is, however, insufficiently preserved to be regarded as anything more than a *nomen dubium.*

CONCLUSION

The majority of taxa considered above (section 3) have proved to have been based on material that is clearly inadequate to allow unambiguous diagnosis in support of their taxonomic status. A smaller number of taxa (section 2) are regarded as being of considerable poten-tial interest, but currently of little practical systematic value, and have been assigned the status of Provi-sionally Theropoda *incertae sedis.* The reason for doing this is an attempt to focus attention on particular taxa as a means of stimulating future research. There still remain, however, four taxa (section 1): *Avimimus, Coelurus, Compsognathus,* and *Ornitholestes,* all of which are based on reasonably well preserved material and are justifiably regarded as valid taxa. As has been explained, none of these taxa can readily be grouped with any of the currently recognized higher taxonomic categories of theropods, even in the light of more thorough systematic reviews such as that by Gauthier (1986).

That this should be the case is not particularly surprising when the material is poorly preserved. However, when forms such as *Compsognathus longipes,* which are known from virtually complete skeletons, fall into this category of indeterminacy, it does pose severe questions of currently accepted classificatory schemes and their systematic bases. The very existence of a chapter such as this in a review of dinosaurs clearly points to the fact that we cannot be complacent about current systematic and classificatory schemes for all fossil taxa.

14

Problematic Theropoda: "Carnosaurs"

R. E. MOLNAR

Ideas of wide utility are also often widely misunderstood; this is as true in dinosaurian systematics as elsewhere. Attempts to apply Huene's (1914c) division of post-Triassic theropods into carnosaurs and coelurosaurs resulted all too often in grouping these creatures by size. The large were called carnosaurs and the small coelurosaurs, although this was not Huene's intent. Our understanding is still inadequate to properly systematize many large theropods. These range from specimens that are indeterminate to species that are clearly distinct but that cannot yet be classified. Some of these may relate to one of the known groups of large theropods: abelisaurs, carnosaurs, or ceratosaurs; others may represent new groups. Five years ago, the only group of large theropods was the Carnosauria; given this situation, it would be unwise to presume that further groups will not be discovered.

Systematic understanding of some of this material (table 14.1) is hampered by the loss or destruction of type specimens. Some of those from the Lameta Formation, near Jabalpur, India, can no longer be located, while some European material has been destroyed by

war. Understanding of such material rests entirely on the literature. In addition, much of the existent material has not been restudied since its original description, often in the last century. New work on this material in the light of more recent discoveries is highly desirable. The difficulty in placing many of these specimens taxonomically results as much from the lack of detailed description as from the incomplete or fragmentary condition of the specimens.

Abelisaurus comahuensis
Bonaparte et Novas, 1985

A skull from the Maastrichtian Allen Formation of Rio Negro, Argentina, is the basis of *Abelisaurus comahuensis*. About two-thirds of the skull is known. Its length is 85 cm, indicating that *A. comahuensis* was a large animal. The antorbital and infratemporal fenestrae are exceptionally large. Although superficially resembling tyrannosaurid skulls in some features (four premaxillary teeth, rugose nasals, postorbital intruding into orbit), diagnostic tyrannosaurid features are absent. Rather than projecting rostrally into the infratemporal fenestra, the slender descending process of the squamosal lies adjacent and parallel to the tall quadrate. The supratemporal recesses are not confluent but separated by a sagittal crest that is broad rostrally. The skull roof elements are thoroughly fused. There is no indication of an enlarged lacrimal foramen characteristic of carnosaurs, although that region is incomplete. However, the similarity of the squamosal-quadrate structure to that of *Carnotaurus sastrei* suggests that these animals are closely related (Bonaparte 1985; Bonaparte and Novas 1985). Both *A. comahuensis* and *C. sastrei* are thought to represent a theropod lineage (Abelisauridae) evolving independently of the carnosaurs, perhaps from ceratosaur ancestors.

Altispinax dunkeri   Huene, 1923

The genus *Altispinax* was originally proposed by Huene (1923, 1926*a*, 1926*b*) to accommodate *Megalosaurus dunkeri* Dames, 1884, a theropod tooth from the Wealden near Hannover, Federal Republic of Germany, as well as an articulated series of three dorsal vertebrae

from the Wealden of Battle, England, and additional isolated teeth from the Wealden of Hollington, England. In 1991, Olshevsky restricted *A. dunkeri* solely for the type tooth, as well as possibly additional teeth from the Wealden that have been referred to *A. dunkeri*, but not the dorsal vertebrae. This chapter follows Olshevsky's (1991) revision of this material.

Becklespinax altispinax   Paul, 1988

As presently recognized, *Becklespinax altispinax* consists of an articulated series of three vertebrae from the caudal region of the dorsal series (fig. 14.1). Originally figured by Owen (1884), this specimen comes from the Wealden of Battle, England. The neural spines of these vertebrae are elongate, five times as high as the centra (fig. 14.1). They bear no particular similarity to the dorsals of *Spinosaurus aegyptiacus* to which they often have been compared: in that species, the spines are craniocaudally dilated just dorsal to the arches, while in *Becklespinax*, they are constricted just above the arches. The centra exhibit no characters useful in determining their phylogenetic position, although they do represent

---

TABLE 14.1   Problematic Therpoda: "Carnosaurs"

Theropoda incertae sedis	Occurrence	Age	Material
Abelisaurus Bonaparte et Novas, 1985			
A. comahuensis Bonaparte et Novas, 1985	Allen Formation, Rio Negro, Argentina	early Maastrichtina	Nearly complete skull
Altispinax Huene, 1923			
A. dunkeri (Dames, 1884) (= *Megalosaurus dunkeri* Dames, 1884)	Wealden, Hannover, Federal Republic of Germany, Wealden, Hainaut, Belgium; Wealden, Bedfordshire, Dorset, West Sussex, East Sussex, Isle of Wight, Upper Purbeck Beds, Dorset, England	Barremian	Isolated teeth
Baryonyx Charig et Milner, 1986			
B. walkeri Charig et Milner, 1986	Wealden, Surrey, England	Barremian	Partial skull and associated postcranial skeleton
Becklespinax Olshevsky, 1991 •			
B. altispinax (Paul, 1988) (= *Acrocanthosaurus altispinax* Paul, 1988) •	Wealden, East Sussex, England	Barremian	Dorsal vertebrae
Carnotaurus Bonaparte, 1985			
C. sastrei Bonaparte, 1985	Gorro Frigio Formation Chubut, Argentina	Albian-Cenomanian	Complete skeleton and skull
Dryptosaurus Marsh, 1877a			
D. aquilunguis (Cope, 1866) (= *Laelaps aquilunguis* Cope, 1866)	New Egypt Formation, New Jersey, United States	Maastrichtian	Partial skeleton
Erectopus Huene, 1921			
E. sauvagei Huene, 1932	unnamed unit, Meuse, France	Albian	Fragmentary skeleton
E. superbus (Huene, 1932) (= *Megalosaurus superbus* Sauvage, 1882)	unnamed unit, Meuse, unnamed unit, Pas-de-Calais, France	Albian	Teeth
Frenguellisaurus Novas, 1986			
F. ischigualastensis Novas, 1986	Ischigualasto Formation, San Juan, Argentina	Carnian	Fragmentary skull, cervical and caudal vertebrae
Genyodectes Woodward, 1901			
G. serus Woodward, 1901	unnamed unit, Chubut, Argentina	Late Cretaceous	Partial skull
Indosaurus Huene et Matley, 1933			
I. matleyi Huene et Matley, 1933	Lameta Formation, Madhya Pradesh, India	middle-late Maastrichtian	Partial skull, fragmentary postcranium
Indosuchus Huene et Matley, 1933			
I. raptorius Huene et Matley, 1933	Lameta Formation, Madhya Pradesh, India	middle-late Maastrichtian	Fragmentary skull, postcranial fragments

Table 14.1, continued

Itemirus Kurzanov, 1976a			
I. medullaris Kurzanov, 1976a	Beleutinskaya Svita, Navoiskaya Oblast, Uzbekistan	late Turonian	Braincase
Majungasaurus Lavocat, 1955a			
M. crenatissimus (Deperet, 1896) (= *Megalosaurus crenatissimus* Deperet, 1896)	Grès de Maevarano, Majunga, Madagascar	Campanian	Partial mandible
Marshosaurus Madsen, 1976b			
M. bicentesimus Madsen, 1976b	Morrison Formation, Utah, United States	Kimmeridgian-Tithonian	Partial skull and postcranial skeleton
Metriacanthosaurus Walker, 1964			
M. parkeri (Walker, 1964) (= *Megalosaurus parkeri* Huene, 1923)	Corallian Oolite Formation, Dorset, England	early-middle Oxfordian	Partial postcranial skeleton
Poekilopleuron Eudes-Deslongchamps, 1838			
P. bucklandii Eudes-Deslongchamps, 1838	Calcaire de Caen, Calvados, France	early Bathonian	Forelimb elements
Proceratosaurus Huene, 1926a			
P. bradleyi (Woodward, 1910) (= *Megalosaurus bradleyi* Woodward, 1910)	Great Oolite, Gloucestershire, England	Bathonian	Partial skull and mandible
Rapator Huene, 1932			
R. ornitholestoides Huene, 1932	Griman Creek Formation, New South Wales, Australia	Albian	Metacarpal
Shanshanosaurus Dong, 1977			
S. huoyanshanensis Dong, 1977	Subashi Formation, Xinjiang Uygur Ziziqu, People's Republic of China	?Campanian-Maastrichtian	Partial skull and associated postcranial skeleton
Tarascosaurus Le Loeuff et Buffetaut, 1990 •			
T. salluvicus Le Loeuff et Buffetaut, 1990 •	Fuvélian Beds, ?Bouche-du-Rhone, France	Campanian	Femur, vertebrae
Unquillosaurus Powell, 1979			
U. ceibalii Powell, 1979	Los Blanquitos Formation, Salta, Argentina	?Campanian	Pubis
Valdoraptor Olshevsky, 1991 •			
V. oweni (Lydekker, 1889e) (= *Megalosaurus oweni* Lydekker, 1889e) •	Wealden, Kent, England	Valanginian-Barremian	Metatarsals
Velocisaurus Bonaparte, 1991 •			
V. unicus Bonaparte, 1991 •	Rio Colorado Formation, Neuquen, Argentina	Coniacian	Partial hindlimb
Xenotarsosaurus Martinez, Gimenez, Rodriguez, et Bochatey, 1987			
X. bonapartei Martinez, Gimenez, Rodriguez, et Bochatey, 1987	Bajo Barreal Formation, Chubut, Argentina	?Campanian	Vertebra, hindlimb elements
Xuanhanosaurus Dong, 1984			
X. qilixiaensis Dong, 1984	Xiashaximiao Formation, Sichuan, People's Republic of China	Bathonian-Callovian	Vertebrae, forelimb

Nomina dubia	Material
Allosaurus medius Marsh, 1888a	Tooth
Ceratosaurus roechlingi Janensch, 1925b	Quadrate, vertebrae, fragmentary fibula
Coeluroides largus Huene, 1932	Isolated vertebrae
Creosaurus potens Lull, 1911b	Vertebra
Dryptosauroides grandis Huene, 1932	Vertebrae
Embasaurus minax Riabinin, 1931a	Vertebrae
Labrosaurus stechowi Janensch, 1925b	Isolated teeth
Labrosaurus sulcatus Marsh, 1896	Tooth
Laelaps cristatus Cope, 1876b	Tooth
Laelaps explanatus Cope, 1876a	Tooth
Laelaps laevifrons Cope, 1876b	Tooth
Laelaps macropus Cope, 1868	Pedal remains
Lametasaurus indicus Matley, 1921 *partim*	Sacrum, ilia, tibia
Macrodontophion Zborzevsky, 1834 (no species name given)	Tooth
Massospondylus rawsei Lydekker, 1890b	Tooth
Megalosaurus chubutensis Corro, 1974	Tooth
Megalosaurus cloacinus Quenstedt, 1858	Tooth
Megalosaurus inexpectatus Corro, 1966	Tooth
Megalosaurus ingens Janensch, 1920	Tooth
Megalosaurus insignis Eudes-Deslongchamps et Lennier, 1870	Teeth, isolated postcrania
Megalosaurus lonzeensis Dollo, 1883c	Pedal ungual
Megalosaurus obtusus Henry, 1876	Tooth
Megalosaurus pombali Lapparent et Zbyszewski, 1957	Vertebrae
Megalosaurus terquemi Huene, 1926b	Teeth
Ornithomimoides barasimlensis Huene et Matley, 1933	Vertebrae
Ornithomimoides mobilis Huene et Matley, 1933	Vertebrae
Orthogoniosaurus matleyi Das Gupta, 1931	Tooth
Poekilopleuron schmidti Kiprijanov, 1883	Ribs, tibia fragment
Prodeinodon mongoliensis Osborn, 1924c	Fragmentary tooth
Prodeinodon kwangshiensis Hou, Yeh, et Zhao, 1975	Isolated teeth
Tomodon horrificus Leidy, 1865 (type of *Diplotomodon* Leidy, 1868a)	Tooth
Walgettosuchus woodwardi Huene, 1932	Vertebra

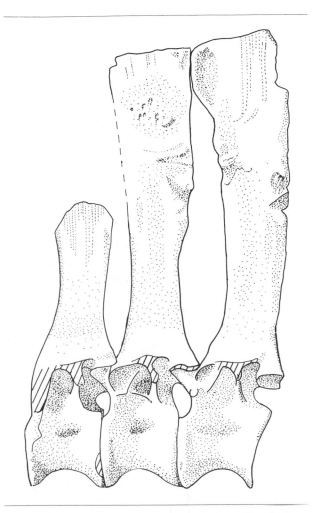

Fig. 14.1. *Becklespinax altispinax.* The three dorsals from Battle, England. (After Owen 1884.)

a valid taxon and are not referable to the better-known *Baryonyx walkeri* (Charig and Milner 1986).

Two referred isolated metatarsals described as II and IV (Lydekker 1889c) seemingly represent two fourth metatarsals, of similar diameter but different length. They presumably do not both pertain to the same species and do not demonstrably pertain to *Becklespinax altispinax.*

Baryonyx walkeri   Charig et Milner, 1986

Discovered in the Barremian Wealden Formation, in Surrey, England, was a large, reasonably complete skeleton of a previously unknown carnivorous dinosaur, *B. walkeri*, in some ways quite unlike other theropods. The teeth are only slightly recurved and rounded, not strongly compressed, in section. The snout is low

and elongate, with a spatulate tip. Apparently, the premaxillary-maxillary contact was mobile with a subrostral notch. Seven premaxillary and thirty-two dentary teeth give the highest tooth count known in theropods. The nasal supports a low longitudinal crest, and long cervicals suggest a long neck. A large manual claw inspired the media-given name "claws."

B. walkeri seems distinct from *Becklespinax* as there is no indication that *B. walkeri* had elongate dorsal neural spines. Since the specimen has yet to be completely described, its affinities are not clear, although at present there is no reason to consider it a carnosaur.

Carnotaurus sastrei   Bonaparte, 1985

Based on a skeleton lacking only the hind feet and distal tail, *Carnotaurus sastrei* is an exceptional theropod. The skull is very high and short and bears two stout horn cores that project caudaldorsally above the orbits. The postorbital region is greatly expanded transversely, as in *Nanotyrannus lancensis* and *Tyrannosaurus rex*, having the effect of directing the orbits rostrally. The medial and caudal walls of the supratemporal recess are extended upward to increase the attachment area, and presumably the volume, of the mandibular adductors and the cervical epaxial muscles. The forelimbs are remarkably abbreviate. Together with *Abelisaurus comahuensis*, *C. sastrei* is included in the Abelisauridae by Bonaparte and Novas (1985), and this assignment is supported here.

Coeluroides largus   Huene et Matley, 1933

C. largus was established for isolated incomplete vertebrae from the Lameta Formation near Jabalpur, India. The vertebrae were described as dorsals (Huene and Matley 1933) but are caudals (Welles 1984). The caudals of *C. largus* are primitive in having the transverse process linked with both pre- and postzygapophyses by laminae. They show no useful taxonomic features.

Diplotomodon horrificus   Leidy, 1868a

Diplotomodon horrificus (= *Tomodon horrificus* Leidy, 1865) is based on a single tooth, now lost, from the "Green-sand formations" (probably either the Navesink or Mount Laurel Formation) of Gloucester County, New Jersey (U.S.A.). Leidy (1865) suggested that the tooth was plesiosaurian and later (1868a) that it might be fish. Welles (1952) recognized it as theropod. Preim (1914) figured an apparently similar tooth from the Campanian phosphates of Wadi Oum Hemaiet, Egypt. Although the type tooth, broad and symmetrical in lateral view rather than recurved, is similar to some of those of *Carcharodontosaurus saharicus* (Stromer 1931), it is also similar to the tooth third from the front in the jaw fragment of *Dryptosaurus aquilunguis* (Cope 1871). It may be an isolated tooth of that species.

Dryptosauroides grandis   Huene et Matley, 1933

Another of the Lameta Formation carnosaurs from near Jabalpur, India, *D. grandis* is represented by six incomplete caudals and a referred cervical and fragmentary ribs. They were thought by Huene and Matley (1933) to resemble those of *Dryptosaurus aquilunguis*. No carnosaurian features are shown by these vertebrae.

Dryptosaurus aquilunguis   Marsh, 1877*a*

D. aquilunguis (= *Laelaps aquilunguis* Cope, 1866) was based on the first partial skeleton (fig. 14.2) of a theropod from North America—from the Maastrichtian New

Egypt Formation in Gloucester County, New Jersey (U.S.A.). The material is poorly preserved and needs restudy. The sacrals are not fused, suggesting that they derive from an immature individual (Cope 1871). Among the distinctive features are the large manual ungual I and the slender hindlimb elements (D. A. Russell 1970*a*). The taxonomic position of *D. aquilunguis* is unclear. There are features reminiscent of the tyrannosaurids: a jugal foramen (but apparently medial, not lateral; Cope 1871); rounded alveoli at the front of the maxilla (Cope 1871); and a well-defined cranial muscle scar on the distal femur (Baird and Horner 1979). However, *D. aquilunguis* is not demonstrably a carnosaur, and because of the incompleteness of the specimen, it is not possible to be certain of features such as the form of the lesser trochanter. One carnosaurian feature is absent—the extensor groove of the femur.

Other species have been doubtfully referred to

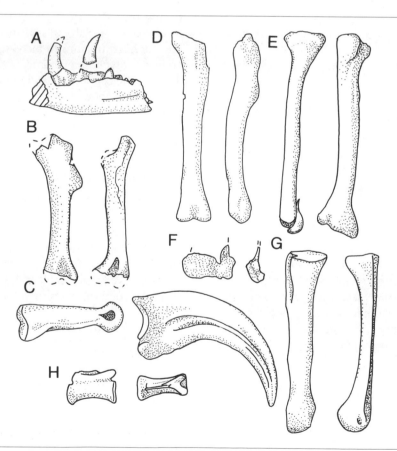

Fig. 14.2. *Dryptosaurus aquilunguis.* A, jaw fragment with teeth. B, humerus. C, manual phalanx and ungual. D, femur. E, tibia. F, astragalus. G, metatarsal IV in cranial, proximal, and distal views. H, distal caudal. Not to scale. (After Cope 1869*b*, Huene 1932.)

this genus, *D. potens* (= *Creosaurus potens* Lull, 1911*b*) and *D. medius* (= *Allosaurus medius* (Marsh, 1888*a*), from the Arundel Formation of Maryland (U.S.A.). These are based on an isolated caudal and a tooth, respectively, and may represent a single species (Gilmore 1920). Other specimens, including a very large manual ungual (Gilmore 1920), have been referred to this form. Material possibly of *Dryptosaurus* has been reported from the Campanian Black Creek Formation of Bladen and Sampson counties, North Carolina (U.S.A.) (Baird and Horner 1979).

The generic name *Dryptosaurus*, having been applied to one of the first-known North American theropod genera, temporarily gained a wider currency than warranted. Because it is still sometimes incorrectly applied, it is worth briefly outlining its history. The genus was initially named *Laelaps*, which was preoccupied. Marsh (1877*a*) suggested the replacement name. Material from the Judith River Formation of Montana (Cope 1876*a*, 1876*b*) and, later, Alberta (Lambe 1904*b*) was referred to this genus under both of these names. Later, this material was recognized to be quite different from *Dryptosaurus* (e.g., Lambe 1917*c*) so this generic name cannot be used to designate material from western North America.

Embasaurus minax   Riabinin, 1931*a*

E. minax, the sole species, is based on two dorsal centra from Mount Koi-Kara, Gurievsk (U.S.S.R.). They were found in the "Neokomian sands." One is platycoelous, the other incomplete. Platycoelous dorsal vertebrae often occur in older taxa, such as *Kaijiangosaurus lini*, and suggest that this form may have been relatively primitive. They are not known in carnosaurs so there is no reason to consider this poorly known species a carnosaur.

Erectopus sauvagei   Huene, 1923

A fragmentary skeleton of a theropod of moderate size from the Albian Gault of the Bois de la Panthière, France, was made the type of *E. sauvagei* Huene, 1932. This was associated with teeth, the type of *Megalosaurus superbus* Sauvage, 1882, which Huene deemed too large to associate with the other remains. *M. superbus* is sometimes considered a second species of *Erectopus*.

Sauvage's (1882) figures show unusual skeletal elements (fig. 14.3). The head of the femur is displaced caudally from the midline of the shaft. Sauvage (1882) wrote of, but did not figure, a presumed jaw fragment with a coronoid process. Such features are otherwise unknown in theropods so this material needs further

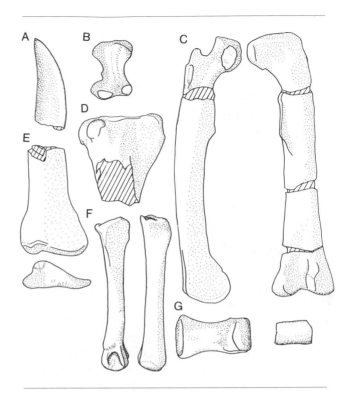

Fig. 14.3. *Erectopus*. A, a tooth of *E. superbus*, probably one of the syntypes (the remainder of the figures all refer to *E. sauvagei*). B, metacarpal I. C, femur. D, proximal tibia. E, distal tibia in caudal and distal views. F, metatarsal II. G, pedal phalanx. (After Sauvage 1876, 1882.)

study. A manus with five metacarpals was suggested for *E. sauvagei* (Huene 1926*a*); however, the manual skeleton is incomplete and this surmise is best treated with caution. The manual claws are small. Metatarsal III is half as long as the femur, indicating an elongate foot.

No carnosaurian features are evident in the femur or tibia, the only elements sufficiently complete for assessment. The femoral head seems not elevated, the lesser trochanter is not wing-like, and the extensor groove is modest. The distal tibia is almost identical with that of *Poekilopleuron bucklandii*.

Isolated teeth from Portugal (e.g., Lapparent and Zbyszewski 1957), England (Sauvage 1882), and Romania (Simionescu 1913) have also been attributed to this genus: quite possibly incorrectly. Huene (1926*b*) also attributed metatarsals and fragments from Grandpré, Ardennes, France.

Frenguellisaurus ischigualastensis   Novas, 1986

F. ischigualastensis, recently described by Novas (1986), comes from the Ischigualasto Formation (Upper Triassic) of San Juan Province, Argentina. It is presently

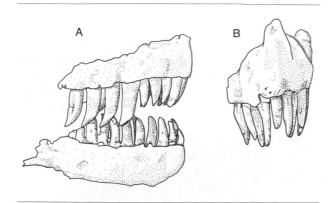

Fig. 14.4. The snout and jaws of *Genyodectes serus.* A, in lateral view, and B, snout only in rostral view. (After Woodward 1901.)

represented by an incomplete skeleton consisting of portions of the skull, the axis, a cranial cervical vertebra, and a section of distal caudal vertebrae. The maxilla bears three strongly-developed caniniform teeth, set mesially among the tooth row. The accessory antorbital fenestra is small. A slight diastema may have existed between the maxillary tooth row and the more elevated premaxillary tooth row. The infratemporal fenestra is large. Of the vertebral material, the caudal series is best known. The prezygapophyses of these caudal vertebrae are considerably elongate, extending forward as far as the middle of each preceding vertebra. Novas (1986) correctly rejected rauisuchid and ornithosuchid relationships for *F. ischigualastensis,* suggesting instead that it represents a primitive theropod. It may yet be a member of the non-saurischian, non-ornithischian dinosaur nexus, much like *Staurikosaurus* and the Herrerasauridae (Sues this volume).

Genyodectes serus   Woodward, 1901

The rostral portion of a snout and dentaries (fig. 14.4) from the "areniscas rojas" (?Upper Cretaceous) of the Grand Canyon, Argentina, is the basis of the species *G. serus.* Huene (1926*b*) suggested that a dorsal centrum from Neuquen, Argentina, might also pertain to this form.

Genyodectes is often referred to the Tyrannosauridae but shows none of the diagnostic characters of that family. It does not, as has sometimes been claimed, have premaxillary teeth of D-shaped section (Stromer 1934*b*) as do the tyrannosaurids; indeed, it shows no diagnostic carnosaur characters and must be considered Theropoda *incertae sedis.*

Indosaurus matleyi   Huene et Matley, 1933

The single species, *Indosaurus matleyi,* derives from the Lameta Formation near Jabalpur, Madhya Pradesh, India. It is based on a braincase (now lost; Chatterjee 1978) including chiefly the frontals, parietals, laterosphenoid, and basioccipital. Other cranial and postcranial elements have been assigned to this species by Walker (1964) and Chatterjee (1978). However, since at least three distinct large theropods (*Composuchus solus, I. matleyi,* and *Indosuchus raptorius*) are present from the same site, these assignments are not accepted here.

The skull roof of *I. matleyi* resembles that of *Carnotaurus sastrei* in two features. The frontals are massively thickened into what appear to be the bases of supraorbital horn cores, while the supraoccipital and sagittal crests of the parietals are markedly elevated. These features suggest that *I. matleyi* is an abelisaurid, a proposal also made on other grounds by Bonaparte and Novas (1985).

Indosuchus raptorius   Huene et Matley, 1933

I. raptorius is another Jabalpur theropod also based on cranial material, two braincases, and an isolated frontal.

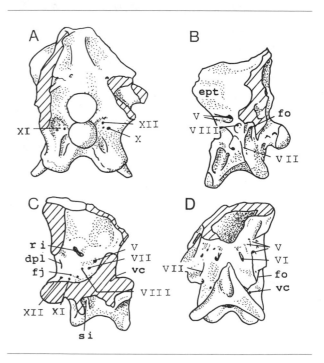

Fig. 14.5. The braincase of *Itemirus medullaris.* A, caudal view. B, lateral view. C, section. D, ventral view (rostral to left). "ri" marks the interacoustic recess and "si" the basisphenoid sinus. (After Kurzanov 1976*a*.)

This material also is lost or misplaced (Chatterjee 1978); other cranial and postcranial elements have been referred to *I. raptorius* by Walker (1964) and Chatterjee (1978). For reasons given for *Indosaurus matleyi*, these assignments are not accepted here.

The skull roof of *I. raptorius* is distinctly less massive and less elevated than that of *Indosaurus matleyi* and lacks horn cores. The supratemporal recesses approach each other closely caudally but are separated by a wedgelike block of the frontals rostrally, as in *Abelisaurus comahuensis*. This feature suggests that *I. raptorius* is an abelisaurid, also noted by Bonaparte and Novas (1985).

Itemirus medullaris   Kurzanov, 1976a

A single isolated braincase (fig. 14.5) is the type of *I. medullaris*: it is from the Late Turonian Central Kyzylkum Sands at Dzhara-Kuduk, Itemir (Uzbekistan). The specimen presumably derived from an aged individual, as almost all sutures are obiterated by fusion. Opening into the endocranial cavity from the region of the middle ear is an extensive chamber, the recessus interacusticus (fig. 14.5C). Kurzanov concluded that *I. medullaris* was the sole member of the Itemiridae, related to both the Tyrannosauridae and the Dromaeosauridae, more closely to the former than to the latter.

This braincase shows none of the specializations of ornithomimids or saurornithoidids. The basipterygoid processes are prominent and longer than the sphenooccipital tubercles but not to the extent found in *Allosaurus fragilis* and *Tyrannosaurus rex*. The basipterygoid processes bear characteristic deep lateral excavations (Fig. 14.5B). This braincase does not bear very close resemblance to those of carnosaurs, and the relationships of this taxon remain enigmatic.

Lametasaurus indicus   Matley, 1921

Under the name *L. indicus*, Matley described remains that he attributed to an armored dinosaur. They proved to pertain to three different orders: an armored form (possibly sauropod; Galton 1981a), a crocodile, and a theropod (Chakravarti 1935). The material of the latter consists of a sacrum, ilia, and a tibia.

The tibia is massive and lacks the lateral crest. The ilium appears to have a broad brevis shelf, although crushing makes this uncertain. There is no evidence suggesting that this form is a carnosaur.

Macrodontophion   Zborzevsky, 1834

This genus is founded on a slender, slightly recurved tooth from the Late Jurassic or Cretaceous of the Ukraine (U.S.S.R.). No specific name was proposed by Zborzevsky, and the tooth has not been studied since, although a figure of the specimen was reproduced by Steel (1970). From this figure, this tooth appears to taper little if at all from neck to tip, quite unlike most theropod teeth. It cannot be determined if the tooth derived from a carnosaur, or even if from a dinosaur.

Majungasaurus crenatissimus   Lavocat, 1955

This genus was proposed by Lavocat (1955) for an incomplete dentary from the Grès de Maevarano of Meravana, Madagascar. Although Lavocat did not propose a specific name for this specimen, he did assign *Megalosaurus crenatissimus* Deperet (1896) to *Majungasaurus*, and this has been accepted as the type species. Deperet's material consisted of two teeth, a caudal and two sacral centra, and a claw. Although the teeth seem to have a cross section of unusual form, none shows diagnostic features. The mandible, although incomplete, is distinctly concave dorsally, has triangular interdental plates, and a longitudinal medial shelf. Lavocat associated this mandible with Deperet's material because of the similarity of the teeth. No elements with diagnostic carnosaurian features are known for *M. crenatissimus*. However, among other theropods, only *Carnotaurus sastrei* has a similarly curved mandible. It is thus possible that *M. crenatissimus* is an abelisaurid.

Marshosaurus bicentesimus   Madsen, 1976b

Deriving from the Morrison Formation of Utah (U.S.A.), this genus is known from the single species *M. bicentesimus*. The type specimen is an isolated ilium and other elements—ilia, pubes, ischia, a premaxilla, maxillae, and dentaries—are referred. The ilium has distinctive shallow excavations in the articular face of the pubic peduncle.

The taxonomic position of *Marshosaurus bicentesimus* is unclear. The ilium appears to have a narrow brevis shelf (Galton and Jensen 1979a) as in carnosaurs, while the pubic shaft is bowed cranially as in ceratosaurs (Rowe and Gauthier, this vol.). That this pubis does pertain to this species is confirmed by the match of the convexities of its iliac articular face to the concavities of the pubic articular face of the ilium (Madsen 1976b). Furthermore, the pubis otherwise resembles that of *Coelurus agilis* (Madsen 1976b). *Marshosaurus* cannot be considered a carnosaur.

"Megalosaurus"

Many species traditionally referred to *Megalosaurus* (Meyer 1832) are here considered dubious. *Megalo-*

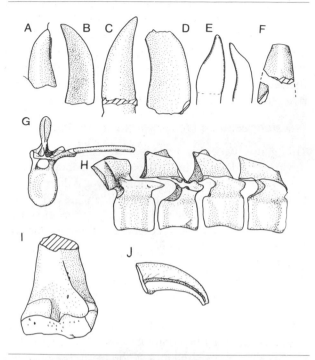

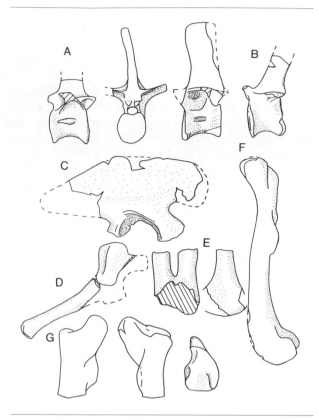

Fig. 14.6. Dubious species of *"Megalosaurus"* based on miscellaneous materials, principally teeth. A, *"M." cloacinus* from the Rhaetian of Württemberg, Federal Republic of Germany. B, *"M." dunkeri* from the Wealden of Hanover, Federal Republic of Germany. C, *"M." ingens* from the Tendaguru Beds of Tanzania. D, *"M." insignis* from the Portlandian of Boulogne, France. E, *"M." terquemi* from the Hettangian of Lorraine, France. F, *"M." obtusus* from the Rhaetian of Moissey, France. *"Megalosaurus" insignis*. G and H, dorsals, and I, distal femur of *"Megalosaurus" insignis*. J, the type ungual of *"Megalosaurus" lonzeensis*. (After Koken 1887, Sauvage 1888, Huene 1907–08, Janensch 1920, Lapparent and Zbyzewski 1957, Dollo 1883c.)

Fig. 14.7. *Metriacanthosaurus parkeri*. A, two dorsal vertebrae, one in lateral and caudal view. B, caudal vertebra. C, ilium. D, ischium. E, distal pubis(?). F, femur. G, proximal tibia in cranial, medial, and dorsal views. (After Huene 1926a, 1926b, Walker 1964.)

saurus chubutensis Corro, 1974, *Megalosaurus cloacinus* Quenstedt, 1858, *Megalosaurus inexpectatus* Corro, 1966, *Megalosaurus ingens* Janensch, 1920, and *Megalosaurus obtusus* Henry, 1876 are all based on isolated teeth (fig. 14.6), while *Megalosaurus lonzeensis* Dollo, 1903 is based on an isolated manual ungual (fig. 14.6D) and *Megalosaurus terquemi* Huene, 1926b is based on two teeth (fig. 14.6E) and an ungual. It is not clear that any of these is carnosaurian, much less referable to *Megalosaurus*.

The remaining referred species are a mixed bag. *Megalosaurus insignis* is based on teeth, an incomplete sacrum, and phalanges from the Kimmeridgian of Besancon, France (Eudes-Deslongchamps in Lennier 1870). Referred material (Lapparent and Zbyszewski 1957) includes dorsals (fig. 14.6) from the Kimmeridgian of Portugal. None of this material shows definitive

carnosaurian characters. *Megalosaurus oweni* Lydekker, 1889e, is based on a partial left metatarsus from the Wealden of England. This taxon was recently renamed *Valdosaurus oweni* by Olshevsky (1991; now separately listed in Table 14.1). *Megalosaurus pombali* Lapparent and Zbyszewski, 1957 from the Lusitanian (Oxfordian) of Porto da Barcas, Portugal, is based on associated dorsals and caudals and isolated worn centra. Pending study of the material referred to *Megalosaurus bucklandii*, these taxa cannot confidently be assigned.

Metriacanthosaurus parkeri   Walker, 1964

This monotypic genus is represented by the species, *M. parkeri* (= *Megalosaurus parkeri* Huene, 1926a), known from a single specimen from the Upper Oxford Clay of Weymouth, England. The specimen consists of dorsals, caudals, an ilium, fragmentary ischium and pubis, a femur, and a proximal tibia (fig. 14.7).

M. parkeri has been described as having elongate dorsal neural spines. Although unlike those of *Altispi-*

nax, they are not substantially longer than those of *Allosaurus fragilis.* Autapomorphies are the angled dorsal iliac margin and the distinct lateral ridge of the proximal part of the ischium (fig. 14.7D). No features allow this form to be confidently classified.

Ornithomimoides mobilis   Huene et Matley, 1933

The type species, *O. mobilis,* is based on five dorsal vertebrae from the Lameta Formation near Jabalpur, India. A second tentative species, *O barasimlensis* Huene, 1932, was based on four smaller dorsals from the same locality. The centra are hollow, with thin walls, and show distinct flanges developed at the articular ends like those of *Allosaurus fragilis.* These dorsals may pertain to a carnosaur, but this cannot be demonstrated.

Orthogoniosaurus matleyi   Das-Gupta, 1931

This genus is based on an isolated tooth. The tooth, the type of *O. matleyi,* is from the Lameta Formation from near Jabalpur, India. It is not from the same stratigraphic level as the other Jabalpur theropods but from the sauropod bed (Huene and Matley 1933). The straight, serrate distal margin meets at the tip a strongly recurved mesial margin that is not serrate. Although its general form is unusual compared to most illustrated theropod teeth, this form is common in caudal maxillary and dentary teeth, although it is unusual for such teeth to lack the mesial serration entirely. The tooth may derive from one of the other nominate Jabalpur theropods, but it cannot be shown to be carnosaurian.

Poekilopleuron bucklandii
Eudes-Deslongchamps, 1838

P. bucklandii is based on one of the first theropod skeletons found, but unfortunately much of the skeleton was destroyed prior to collection (Eudes-Deslongchamps 1838). What was collected was destroyed in World War II. *P. bucklandii* was found in the Bathonian Great Oolite near Caen, France. The species is characterized by short and massive antebrachial elements (fig. 14.8) and a relatively large manus (Huene 1923). The radius has a prominent medial process at midlength, while the second metacarpal also seemingly bears a medial process at midlength (fig. 14.8). Neither is known in other theropods. Although both manus and pes have been reported to be pentadactyl (Huene 1923), reexamination of the original figures indicates no evidence for this.

Despite the restudy of *P. bucklandii* by Huene (1926*b*), its position is unclear. There is no indication of carnosaurian characters in the chevrons. If the scapula of Huene (1926*b*, originally identified as a pubis by Deslongchamps) is correctly identified, which is not certain, then the blade is broad and unlike that of a carnosaur. There is no clear indication that *Poekilopleuron* is a carnosaur.

The name *Megalosaurus poikilopleuron* was proposed for *P. bucklandii* by Huene (1926*b*), who believed it referable to that genus. Since there is no dentary material of *P. bucklandii,* this cannot presently be determined. Other material referred to *P. bucklandii* from Boulonnais, France (Sauvage 1880), and Tilgate, England (Owen 1884), seems to be incorrectly referred.

Kiprijanov (1883) based *P. schmidti* on what may be fragments of ribs and a "humerus" (a distal tibia; Rozhdestvensky 1973*b*) from the "Sewerische Osteolithe" at Kursk, U.S.S.R. Information on this deposit is difficult to obtain, but it appears to be Cenomanian to Santonian in age. This poorly preserved tibia does not show the characteristic malleolar form of *P. bucklandii* and could belong to almost any large Late Cretaceous theropod.

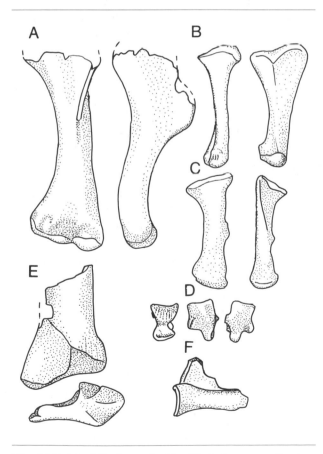

Fig. 14.8. *Poekilopleuron bucklandii.* A, humerus. B, ulna. C, radius. D, metacarpal I. E, distal tibia in cranial, caudal, and ventral views. F, astragalus. (After Eudes-Deslongchamps 1838, Hulke 1979*b*.)

Proceratosaurus bradleyi Huene, 1926*a*

A single skull with both jaws but lacking the cranial roof is the basis of *P. bradleyi* (Woodward 1910), originally described as a species of *Megalosaurus*. The specimen comes from the Bathonian Great Oolite of Gloucestershire, England. Although the dorsal portion of the skull is missing, enough remains to show that a medial horn core sat on the nasals above the nares. The skull is lightly built and the premaxillary and mesial dentary teeth are distinctly smaller than the more distal teeth of both jaws. The surangular is shallow, indicating that *Proceratosaurus* is not a carnosaur. Paul (1988) has argued that *P. bradleyi* is related to *Ornitholestes hermanni*.

Prodeinodon mongoliensis Osborn, 1924*c*

P. mongoliensis is based on a fragmentary tooth from the Lower Cretaceous Khukhtekskaya Formation at Oshi Nuru, Mongolian People's Republic. Other material has been reported from Nei Mongol Zizhigu, People's Republic of China (Bohlin 1953). Bohlin found the teeth associated with large but fragmentary limb bones that could not be collected because of bad weather. A second species, *P. kwangshiensis* (Hou et al. 1975), is based on four teeth from the Early Cretaceous at Fusui, People's Republic of China but these are not thought to be distinctive. The teeth of *Prodeinodon* are obviously theropod but have no characters by which they can be assigned at the infraordinal level.

Rapator ornitholestoides Huene, 1932

Proposed by Huene for an isolated left first metacarpal, the sole species, *R. ornitholestoides,* was named for a slight resemblance to the much smaller *Ornitholestes hermanni*. The metacarpal derives from the Albian Griman Creek Formation at Lightning Ridge, Australia. It is unusual in possessing a prominent caudomedial process, much stronger than that of *O. hermanni* (which is hardly comparable) and unlike anything seen in other theropods. There is no reason to consider *Rapator* a carnosaur.

Shanshanosaurus huoyanshanensis Dong, 1977

S. huoyanshanensis Dong, 1977, the only species, was found in the Campanian or Maastrichtian Subashi Formation, of the Turpan Basin, People's Republic of China. It is known from an incomplete skeleton, including portions of the skull (fig. 14.9). Although *S. huoyanshanensis* shows a narrow scapula, with an abrupt glenoid expansion (fig. 14.9), and a deep surangular, it lacks other features of carnosaurs. The femoral head is not elevated. Far from being opisthocoelous, the figured cervical appears gently procoelous. *S. huoyanshanensis* does have a surangular foramen and premaxillary teeth with a "tendency to [be] incisiform" (Dong 1977), but these features also appear in dromaeosaurids. The well-developed pubic "foot" is also found in abelisaurids, ornithomimids, and oviraptorids. In view of this apparent mosaic of features, Dong's assignment of this genus to a unique family is justified. However, assignment to a higher level is not now possible.

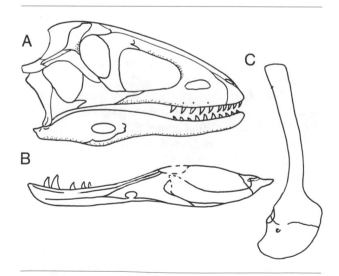

Fig. 14.9. *Shanshanosaurus huoyanshanensis.* A, reconstructed skull. B, mandible in medial view. C, scapula and coracoid. (After Dong 1977.)

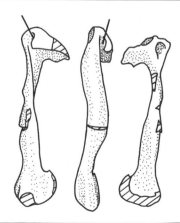

Fig. 14.10. Pubis of *Unquillosaurus ceibaḻii* in lateral, cranial, and medial views. Bar indicates the distinctive sulcus mentioned in the text. (After Powell 1979.)

Unquillosaurus ceibalii   Powell, 1979

An isolated left pubis (fig. 14.10) from the Los Blanquitos Formation of El Ceibal, Argentina, is the type specimen of *U. ceibalii*. Proximally, just below the iliac articular surface, is a marked, cranioventrally-facing sulcus, quite unlike anything known in other theropod pubes. This species seems valid but shows no similarity to any known carnosaur.

Walgettosuchus woodwardi   Huene, 1932

The unique species, *W. woodwardi*, is established on an isolated, incomplete, amphicoelous caudal. It was found in the Albian Griman Creek Formation, at Lightning Ridge, New South Wales, Australia. For reasons not now apparent, Huene believed that the caudal had possessed elongate prezygapophyses. What remains of the caudal is not distinguishable from those of allosaurids or ornithomimids and so cannot be identified as carnosaurian.

Xenotarsosaurus bonapartei
Martinez, Gimenez, Rodriguez, et Bochatey, 1987

The single species, *X. bonapartei* is based on a femur, tibiotarsus, fibula, and associated dorsal vertebra. The specimen is from the pre-Maastrichtian Upper Cretaceous Bajo Barreal Formation of Chubut, Argentina. The astragalus and calcaneum are fused, and both are fused to the tibia. The ascending process of the astragalus is relatively small and narrow. The femur has both a trochanteric shelf and a crista tibiofibularis sulcus, both also found in ceratosaurs (Rowe and Gauthier, this vol.). Martinez et al. (1987) list several resemblances of the femur and dorsal to those of *Carnotaurus sastrei*. These features may imply that *X. bonapartei* is an abelisaurid.

Xuanhanosaurus qilixiaensis   Dong, 1984*b*

Represented by two cervicals, four dorsals, and a forelimb (including the shoulder girdle), *X. qilixiaensis* was recovered from the Lower Shaximiao Formation of Sichuan, People's Republic of China. The forelimb is robust, and the hand retains four metacarpals, while the scapula has a prominent "acromial" projection quite unlike the flattened form of the glenoid region in other theropods. The cervical vertebrae are opisthocoelous. The nature of the carpal and phalangeal joints suggests that the digits had restricted mobility and that the manus was not capable of grasping. This in turn suggests, together with massive elements of the forelimb and shoulder girdle, that the forelimb was used in quadrupedal locomotion (Dong 1984*b*). The relationship of such an apparently primitive theropod to better-known taxa is uncertain.

These species are either too poorly known or too different from the better-understood theropods to be confidently classified. Four of these species, *Baryonyx walkeri*, *Itemirus medullaris*, *Shanshanosaurus huoyanshanensis*, and *Xuanhanosaurus qilixiaensis*, probably each represent distinct families. Other species, among them "*Altispinax dunkeri*," *Dryptosaurus aquilunguis*, *Erectopus sauvagei*, *Marshosaurus bicentesimus*, *Metriacanthosaurus parkeri*, *Poekilopleuron bucklandii*, *Rapator ornitholestoides*, and *Unquillosaurus ceibalii*, seem to be valid but inadequately known for classification. Others, such as *Megalosaurus lonzeensis*, *M. obtusus*, and *Walgettosuchus woodwardi*, seem too incomplete for classification and are best regarded as indeterminate.

These forms, however, highlight our lack of knowledge of the Mesozoic fauna. The discovery of such genera as *Baryonyx* and *Xuanhanosaurus* show that large theropods were more diverse than we had previously supposed.

SAUROPODOMORPHA

P. DODSON

The Sauropodomorpha consists of the Prosauropoda and the Sauropoda, collectively, the saurischian herbivores. Prosauropods, of Late Triassic and Early Jurassic age, were an ephemeral group of dinosaurs, although they were the first to attain wide geographic distribution and numerical abundance. Sauropods were very large and persisted through the entire Mesozoic. Autapomorphies of the Sauropodomorpha, including *Thecodontosaurus* and *Anchisaurus,* include robust pollex with an enlarged ungual, enlarged hallux ungual, lanceolate teeth with coarsely serrated crowns, comparatively small skull, ten or more moderately elongate cervicals, hindlimb equal to or shorter than trunk, tibia shorter than femur, and ascending process of the astragalus keyed into the descending process of the tibia (Gauthier 1986; Benton, this vol.; Galton, this vol.).

Monophyly of the Sauropoda is universally accepted. Gauthier (1986) listed more than thirty characters uniting sauropods. Sauropod origins remain clouded. It is usually assumed that sauropod ancestry lies within the Prosauropoda. As described by Galton (this vol.), melanorosaurids and blikanasaurids are very large, sauropodlike animals. What remains unclear is the level within the prosauropod assemblage at which sauropod relationships are to be found. Was there a smooth transition between the highly derived melanorosaurids or blikanasaurids and basal sauropods? If this can be demonstrated, then paraphyly of the Prosauropoda follows. If, however, it can be demonstrated that sauropods are the sister group of the Thecodontosauridae or the Anchisauridae, for example, a monophyletic Prosauropoda could be defined. The transitional status of *Vulcanodon* and *Barapasaurus* (which are regarded by Gauthier [1986] and Benton [this vol.] as outgroups to the Sauropoda but by McIntosh [this vol.]) as primitive sauropods is clearly suggestive of the paraphyly of prosauropods. However, data on critical apomorphies of the skull are totally lacking for melanorosaurids, blikanasaurids, and vulcanodontids. Certain features found among prosauropods seem autapomorphic. For instance, the deep braincase in the Plateosauridae has a steplike ventral profile, with the parasphenoid located well below the occipital condyle (Galton, this vol.). Such a feature may suggest that the sister group of sauropods lies at a more primitive level within prosauropods. It is also clear that feet of *all* prosauropods, including very primitive *Thecodontosaurus,* are more derived than the feet of any sauropod in the reduction of metatarsal V. Prosauropods also seem more derived in elongation of the cervicals than are primitive sauropods. Galton concludes that sauropods are the sister group of the prosauropods. At present, the absence of critical data on the structure of derived prosauropods and primitive sauropods seemingly precludes definitive assessment of the relationship of prosauropods and sauropods. Hence, alleged paraphyly of the Prosauropoda (e.g., Gauthier 1986; Benton, this vol.) is at best an interesting hypothesis. Resolution of this problem lies in the future.

The recently discovered Segnosauria (Barsbold and Maryańska, this vol.) presents a truly novel assortment of characters imposed on a rather primitive *Bauplan.* It is particularly enigmatic that such an experimental body design should appear in the Late Cretaceous rather than earlier in the age of dinosaurs. Initial suggestions that segnosaurs are aberrant theropods (Perle 1979; Barsbold and Perle 1980; Barsbold 1983*b*) have not been sustained. The suggestion that segnosaurs present relicts of a "prosauropod-ornithischian transition" (Paul 1984*b*) seems closer to the mark. Gauthier (1986), Benton (this vol.), and Barsbold and Maryańska (this vol.) find it likely that segnosaurs are a sister group of sauropodomorphs, although an unresolved trichotomy with both theropods and sauropodomorphs seems the most candid assessment at present (Barsbold and Maryańska, this vol.). In any event, the split between sauropodomorphs and segnosaurs seems to have been very ancient, at least before the earliest Jurassic. Segnosaurs evidently existed as undetected plesiomorphic forms until they acquired the distinctive apomorphies that characterize the known Late Cretaceous representatives. Perhaps the ancestral stock persisted in some sort of biogeographic isolation. Homoplastic characters abound, obscuring the taxonomic signal. For instance, it is presumed that loss of the mesial dentition is not indicative of ornithischian relationships and that the retroverted pubes indicate neither dromaeosaurid nor ornithischian affinities. Segnosaurs offer a powerful test of the ability of natural selection to modify body plans in accordance with the demands of particular modes of life. Thus modified, segnosaurs present a keen challenge to our methods of phylogenetic analysis.

15

Basal Sauropodomorpha— Prosauropoda

PETER M. GALTON

INTRODUCTION

Prosauropods have been found on all continents except Antarctica. These animals were small to large (maximum lengths approximately 2.5 to 10 m), bipedal, facultatively bipedal or quadrupedal, herbivorous dinosaurs with small skulls and elongate necks (fig. 15.1A,C) and are usually the most common terrestrial vertebrates in the beds in which they occur. The earliest records pertain to *Azendohsaurus* from the mid-Carnian of Morocco and isolated teeth of an indeterminate prosauropod of comparable age from Canada (Carroll et al. 1972). However, the earliest good record of the group is from the Lower Elliot Formation (late Carnian or early Norian) of South Africa (*Euskelosaurus, Melanorosaurus, Blikanasaurus*). From the middle to late Norian of Europe and South America, prosauropods are very common. It has long been assumed that prosauropods were restricted to the Triassic but the transfer of several important Upper Triassic beds that have yielded remains (e.g., upper Newark Supergroup and Glen Canyon Series of North America, Upper Elliot and Clarens formations and equivalents of southern Africa,

upper Lower Lufeng Series of the People's Republic of China) to the Lower Jurassic means that prosauropods remained important until at least the Pliensbachian or Toarcian (Olsen and Galton 1977, 1984; Olsen and Sues 1986).

The Anchisauridae and Plateosauridae were erected by Marsh (1885, 1895), but it was not until 1920 that Huene (see also Huene 1932) recognized the Prosauropoda, which he combined with the Sauropoda to form the Sauropodomorpha. The various systematic positions and constituents of the Prosauropoda have been reviewed by Colbert (1964a) and Charig et al. (1965) and summarized by Steel (1970), who also reviewed all the species of prosauropods. The Prosauropoda is considered to be a paraphyletic taxon by Gauthier (1986) and Benton (this vol.), but I consider it to be monophyletic.

DIAGNOSIS

The Prosauropoda are medium- to large-sized and bipedal to quadrupedal sauropodomorphs that can be di-

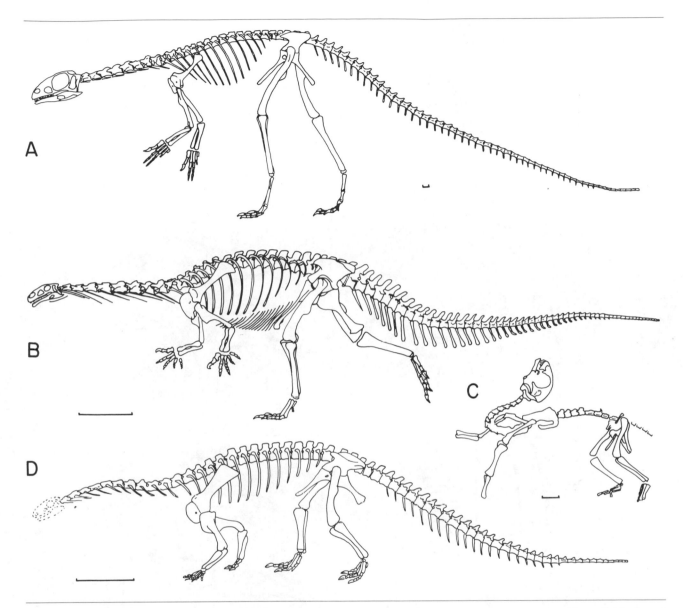

Fig. 15.1. Prosauropod skeletons: A, thecodontosaurid *Thecodontosaurus*, juvenile individual, estimated length approximately 1 m (after Kermack 1984); B, plateosaurid *Plateosaurus*, length approximately 6.5 m, maximum approximately 9 m (after Weishampel and Westphal, 1986); C, plateosaurid *Mussaurus*, extremely juvenile individual, estimated length approximately 25 cm (after Bonaparte and Vince 1979); D, melanorosaurid *Riojasaurus*, length approximately 6 m, maximum approximately 10 m (after Bonaparte 1972). Scale = 1 cm (A,C), 50 cm (B,D).

agnosed on the following basis: skull about half the length of femur, jaw articulation situated slightly below the level of the maxillary tooth row, dentition consists of small, homodont or weakly heterodont, spatulate-shaped teeth with coarse, obliquely angled marginal serrations; digit I of manus bears a twisted first phalanx and an enormous, trenchant ungual that is medially directed when hyperextended; digits II and III subequal in length and bear small, only slightly recurved ungual phalanges; digits IV and V reduced and lack ungual phalanges; typical phalangeal formula 2-3-4-(3 or 4)-3, bladelike distal parts of pubes form a broad flat apron; pes with vestigial digit V (*Herrerasaurus*, *Staurikosaurus*, and *Lagosuchus* used as outgroups; see Benedetto 1973; Bonaparte 1975b, 1978a; Brinkman and Sues 1987; Colbert 1970; Galton 1977a; Reig 1963; Sues, this vol.).

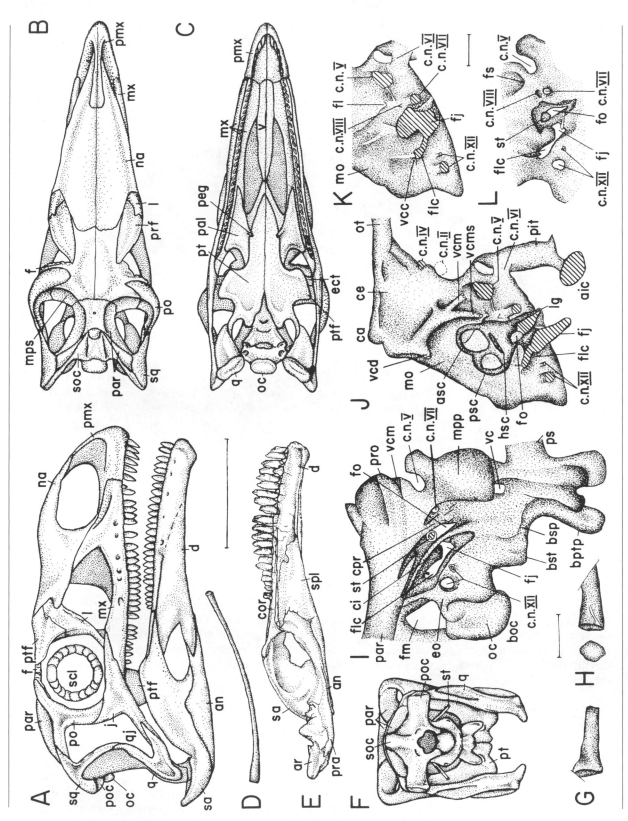

Fig. 15.2. Cranial anatomy of plateosaurid *Plateosaurus* (after Galton 1984a, 1985a). A-C, restoration of skull: A, right lateral; B, dorsal; and C, ventral *views*; D, right ceratobranchial in lateral view; E, left mandible in medial view; F, reconstruction of skull in caudal view; G,H, proximal part of stapes: G, left in caudomedial view; H, right in ventromedial view; I, braincase in right laterocaudal view to show foramina; J, K, endocranial cast in right lateral view: J, complete cast with semicircular canal restored; K, as J with inner ear and adjacent structures removed; L, left medial wall of braincase in mediodorsorostral view to show foramina. Scale = 1 cm (G,H), 4 cm (I-L), 10 cm (A-F).

ANATOMY

Skull and Mandible

Articulated and disarticulated cranial material of *Plateosaurus* (figs. 15.2, 15.4E) has been recently redescribed by Galton (1984*a*, 1985*e*). Cranial material is also illustrated for *Anchisaurus* (figs. 15.3A-C, 15.4A; Huene 1914*a*; Galton 1976*a*), *Coloradisaurus* (figs. 15.3G,H, 15.4C; Bonaparte 1978*b*), *Lufengosaurus* (fig. 15.3N-P; Young 1941*a*, 1947), *Massospondylus* (fig. 15.3J-M; Attridge et al. 1985; Cooper 1981*a*; Crompton and Attridge 1986), *Mussaurus* (fig. 15.3I; Bonaparte and Vince 1979), *Thecodontosaurus* (figs. 15.3D-F, 15.4B,D; Huene 1907–08, 1914*a*; Kermack 1984), and *Yunnanosaurus* (fig. 15.3Q,R; Young 1942*a*, 1951). The skull of *Plateosaurus* is very narrow relative to its height as in *Yunnanosaurus*. The skull is proportionally wider in other prosauropods and especially so in *Coloradisaurus* and *Massospondylus*.

The elongate dorsal and caudal processes of the premaxilla, together with the nasal, enclose the external nares in *Anchisaurus*, *Sellosaurus*, and *Plateosaurus*, whereas in *Coloradisaurus*, *Lufengosaurus*, *Massospondylus*, *Mussaurus*, and *Yunnanosaurus*, the maxilla borders this opening caudoventrally. The sutural surface between the premaxillae of *Plateosaurus* is flat except for a prominent excavation into which the rostral processes of the maxillae and vomers fit (fig. 15.2C). The dorsal process of the maxilla meets the lacrimal; it is low in *Yunnanosaurus* and high in all other prosauropods. The low lateral lamina of the maxilla extends caudally from the base of the dorsal process of the maxilla. This lamina is short and merges with the body of the maxilla in *Anchisaurus*, *Lufengosaurus*, *Sellosaurus*, *Thecodontosaurus*, and *Yunnanosaurus*. It is long in *Coloradisaurus*, *Massospondylus*, and *Plateosaurus*. The medial lamina of the maxilla backs the rostral part of the antorbital fenestra to form an antorbital fossa that is small in all prosauropods except *Coloradisaurus* and *Plateosaurus*. The maxillary tooth row terminates just rostral to the orbit in *Anchisaurus*, *Thecodontosaurus*, and *Yunnanosaurus*, underlaps the orbit to a slight extent in *Massospondylus*, and extends caudally to a level approximately below the middle of the orbit in *Coloradisaurus*, *Lufengosaurus*, *Mussaurus*, and *Plateosaurus*.

In all prosauropods except *Plateosaurus*, the length of the nasal is less than half the length of the skull roof. The nasolacrimal canal extends from the orbital margin to the rostral end of the lacrimal. The lateral lamina of the lacrimal encloses the caudolateral corner of the antorbital fenestra while the medial lamina backs the antorbital fossa. These laminae are narrow except in *Plateosaurus* (wide lateral lamina) and *Coloradisaurus* (wide medial lamina).

The prefrontal wraps around the lacrimal, overlaps the frontal, and forms the rostrolateral part of the orbital rim. The orbital part of the prefrontal is short and transversely narrow in *Coloradisaurus* and *Massospondylus*, longer but still narrow in *Anchisaurus*, *Sellosaurus*, and *Thecodontosaurus*, and elongate and transversely broad in *Plateosaurus*, *Lufengosaurus*, and *Yunnanosaurus*.

The frontal provides an important contribution to the orbital rim in *Anchisaurus*, *Coloradisaurus*, *Massospondylus*, *Sellosaurus*, and *Thecodontosaurus*. However, the frontal is mostly excluded from the orbit by the overlapping prefrontal and postorbital in *Plateosaurus* (fig. 15.2B); it is likely similar in *Lufengosaurus* and *Yunnanosaurus*. The frontal and the adjacent parts of the parietal and postorbital of *Plateosaurus* are deeply excavated to form the area of origin of the M. pseudotemporalis. This attachment area is less prominent in other prosauropods but recognizable in *Thecodontosaurus* and *Massospondylus*. The parietals are separate bones in *Anchisaurus*, *Massospondylus*, and *Thecodontosaurus* and are partly or completely fused in *Coloradisaurus*, *Lufengosaurus*, *Plateosaurus*, and *Yunnanosaurus*. A parietal foramen is present in some individuals of *Plateosaurus* and absent in others (Huene 1932; Galton 1985*a*). The tetraradiate squamosal overlaps the parietal, opisthotic, postorbital, and quadrate.

The dorsal head of the quadrate fits into a socket in the body of the squamosal. The mandibular condyle is slightly below the maxillary tooth row in *Thecodontosaurus*, *Anchisaurus*, *Massospondylus*, and *Yunnanosaurus* and is well below the level of the dentary tooth row in *Sellosaurus*, *Plateosaurus*, *Coloradisaurus*, and *Lufengosaurus*. The quadrate is rostroventrally directed in *Anchisaurus* and *Thecodontosaurus* and vertical or caudoventrally directed in the remaining genera.

The V-shaped quadratojugal overlaps the quadrate and jugal. The large pterygoid ramus of the quadrate overlaps the pterygoid medially. The rostral edge of the lateral wall of the quadrate is overlapped by the squamosal and quadratojugal so the quadrate is excluded from the border of the infratemporal fenestra. The rostral and dorsal processes make an angle of about 90° to each other in *Anchisaurus*, *Mussaurus*, and *Thecodontosaurus*; this angle is 50° to 60° in *Coloradisaurus*, *Massospondylus*, and *Yunnanosaurus* and less than 45° in *Plateosaurus* and *Lufengosaurus*. The apical part of the quadratojugal in *Plateosaurus* fits into a

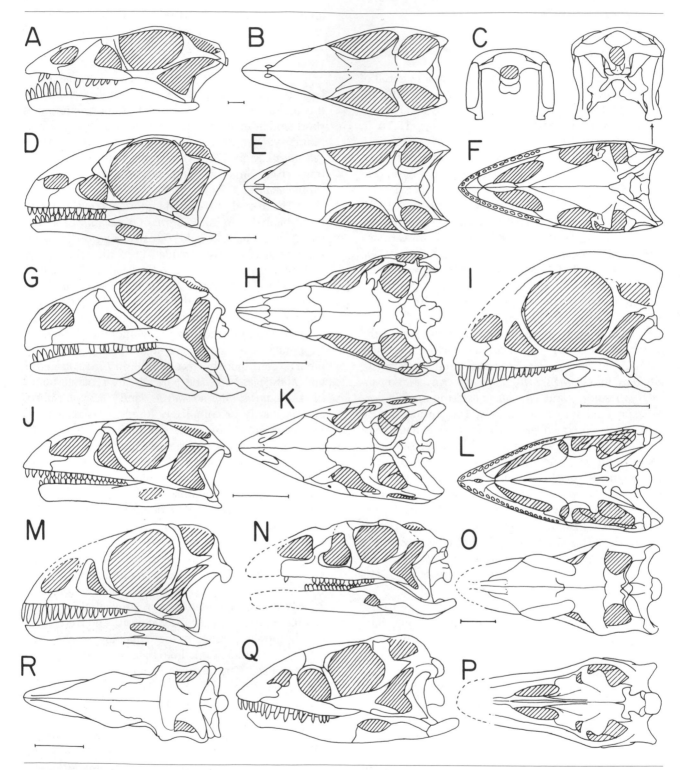

Fig. 15.3. Reconstruction of skulls of anchisaurid (A–C), thecodontosaurids (D–F), plateosaurids (G–I, N–P), melanorosaurids (J–M), and yunnanosaurids (Q,R) in left lateral (A,D,G,I,J,M,N,Q), dorsal (B,E,H,K,O), ventral (F,L,P), and caudal views (C,F). A–C, *Anchisaurus* (after Galton 1976a); D–F, *Thecodontosaurus*, juvenile (after Kermack 1984); G,H, *Coloradisaurus* (after Bonaparte 1978b); I, *Mussaurus* juvenile (after Bonaparte and Vince 1979); J–L, *Massospondylus* (after Attridge et al. 1985); M, *Massospondylus*, juvenile (after Cooper 1981a); N-P, *Lufengosaurus* (after Young 1941a, 1951); Q,R, *Yunnanosaurus* (after Young 1942). Scale = 1 cm (A–F,I,M), 5 cm.

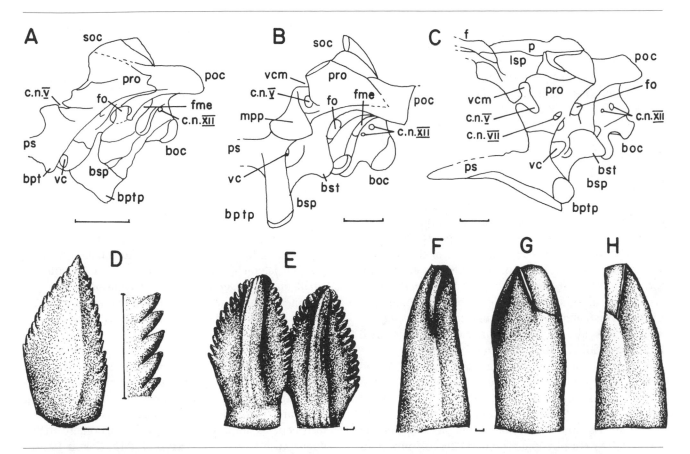

Fig. 15.4. A–C, prosauropod braincases in left lateral view: A, anchisaurid *Anchisaurus* (after Galton and Bakker 1985); B, thecodontosaurid *Thecodontosaurus* (after Galton and Bakker 1985); C, plateosaurid *Coloradisaurus* (after Bonaparte 1978*b*); D,E, right dentary teeth in buccal view: D, thecodontosaurid *Thecodontosaurus* (after Huene 1907-08); E, *Plateosaurus* (after Galton 1985*e*); F–H, yunnanosaurid *Yunnanosaurus* (after Galton 1985*d*): right maxillary tooth in F, mesial, G, labial, and H, distal views. Scale = 1 cm A–C), 1 mm (D–H).

prominent depression on the main body of the quadrate. The Y-shaped jugal overlaps the maxilla, lacrimal, and ectopterygoid and is overlapped by the postorbital and quadratojugal. The dorsal part bordering the orbit is transversely thick, whereas the ventral process is thin and slender, especially in *Anchisaurus*.

The pterygoids form the major part of the caudal palate. They meet along the midline in *Thecodontosaurus*, *Lufengosaurus*, and *Plateosaurus* whereas in *Massospondylus* they do not. The pterygoids are relatively small in *Plateosaurus* compared to those of *Lufengosaurus* and especially those of *Thecodontosaurus* and *Massospondylus*. The central part of the pterygoid is thick but constricted. From it extend the caudodorsally and laterally directed broad quadrate ramus, the caudomedially and slightly dorsally directed short process that supports a socket for the basipterygoid processes of the basisphenoid, the ventrally directed pterygoid ramus, and the rostrodorsally directed and elongate palatal ramus. The epipterygoid is known only in *Plateosaurus*; it is a slender bone that overlaps the rostrodorsal part of the quadrate ramus of the pterygoid and the ventral part of the laterosphenoid.

The bar-shaped ectopterygoid is expanded to fit laterally against the jugal and medially against the pterygoid ramus of the pterygoid. The rostrocaudally expanded part of the palatine that fits against the maxilla laterally is small in *Massospondylus*, proportionally larger in *Thecodontosaurus* and *Lufengosaurus*, and largest in *Plateosaurus*. The thin medial processes of the palatine overlap the pterygoid and converge rostrally to meet ventrally before contacting the deep vomer. The palatines occupy about half the length of the palate in *Massospondylus*, in which the vomers are presumably very short and hidden by the palatines, about a third in *Thecodontosaurus*, and about a quarter in *Plateosaurus* and *Lufengosaurus* in which the vomers are the longest. Ventromedially the palatine of *Plateosaurus* has a prominent, peglike projection that probably supported the caudal part of a soft secondary palate (also supported by the medial horizontal edge of the maxilla and the median vomers).

The supraoccipital in *Anchisaurus, Sellosaurus, Thecodontosaurus,* and *Yunnanosaurus* is steeply inclined at an angle of about 75° so that its apex is well caudal to the basipterygoid processes. In *Coloradisaurus, Lufengosaurus, Massospondylus,* and *Plateosaurus,* the supraoccipital slopes at about 45° and its apex is approximately above the basipterygoid process. The horizontal paroccipital process is presumably formed from the opisthotic. These processes are set at an angle of about 50° to 60° to the midline in all prosauropods except *Plateosaurus,* in which this angle is about 40°. The exoccipital forms the caudal part of the sidewall of the braincase. The occipital condyle is apparently formed only from the basioccipital in *Coloradisaurus,* whereas in all other prosauropods the exoccipital contributes dorsolaterally. In *Anchisaurus, Massospondylus, Sellosaurus,* and *Thecodontosaurus,* the occipital condyle is in line with the parasphenoid, but in *Anchisaurus,* the basal tubera extend farther ventrally than do the very small basipterygoid processes. In *Coloradisaurus, Lufengosaurus,* and *Plateosaurus,* the steplike ventral outline of the braincase provides a deep rostral floor to the braincase and the occipital condyle is above the level of the parasphenoid. The basipterygoid processes converge rostrally to form a V-shaped depression in all prosauropods except *Plateosaurus,* in which this area is flat because of a lamina between the processes. The parasphenoid is a slender rod, subtriangular in cross section, that extends rostrally from the basipterygoid processes and between the pterygoids. A rugose, rostrolaterally facing surface on the adjacent parts of the basisphenoid and prootic is the area of origin of the M. protractor pterygoideus in *Thecodontosaurus* and *Plateosaurus* (figs. 15.2, 15.4B). The laterosphenoid of *Sellosaurus* and *Plateosaurus* is triangular in lateral view and tapers from the prootic to the transversely expanded rostral end. The orbitosphenoid is only known for *Plateosaurus,* in which it is a rib-shaped bone situated immediately rostral to the laterosphenoid.

Information from endocranial casts (*Plateosaurus, Thecodontosaurus*) suggests that the brain (fig. 15.2J) is short and very deep with a short and deep medulla oblongata, a steeply inclined caudodorsal edge to the metencephalon, and prominent cerebral and pontine flexures. In the telencephalon, the slightly differentiated cerebral hemispheres form the widest part of the brain. These taper rostrally into elongate olfactory tracts that then widen into large olfactory bulbs. In the

diencephalon, there appear to be no dorsal diverticulum. The dorsal apex of the diencephalon probably represents a cartilaginous infilling between the ossified supraoccipital and the overlying part of the parietal. C. n. II arose from the diencephalon, the ventral part of which was within part of the pituitary fossa. The extent of the mesencephalon is uncertain because the optic lobes are not differentiated and the points of origin of c. n. III and IV cannot be determined. The metencephalon has no dorsal cerebellar expansion, but the prominent floccular lobes of the cerebellum do project caudodorsally to occupy the fossa subarcuata in the medial wall of the prootic (fig. 15.2K,L). C. n. V originates from this region, whereas c. n. VI to XII originate from the myelencephalon (fig. 15.2K). In the inner ear (fig. 15.2K), the anterior semicircular canal is the longest, the lateral canal is the shortest, the sacculus is small, and the lagena is short as in sauropods. The thin indented dorsal part of the myelencephalon becomes widest at the level of the vena cerebralis caudalis.

The dentary is more than half the length of the mandible in all prosauropods except *Thecodontosaurus.* A prominent ridge continues from the lateral surface of the overlapping surangular onto the dentary and passes diagonally across the latter, such that the more distal teeth are slightly inset. This ridge is probably the attachment site for a muscular cheek functionally analogous to that of mammals (Paul 1984*b*; Galton 1984*a,* 1985*e*) and ornithischian dinosaurs (Galton 1973*c*). Additional evidence for cheeks in prosauropods shows that the vascular foramina of the dentary and maxilla are large and few in number, rather than small and numerous as in reptiles without cheeks (Paul 1984*b*). Cheeks were probably developed to a varying degree in all prosauropods.

The prominent external mandibular fenestra is bordered by the dentary, surangular, and angular. The coronoid eminence of the surangular is low, and the jaw articulation is in line with the dentary tooth row in *Anchisaurus, Thecodontosaurus, Massospondylus, Mussaurus,* and *Yunnanosaurus.* In *Coloradisaurus, Lufengosaurus, Plateosaurus,* and *Sellosaurus,* the coronoid eminence is deeper and the jaw articulation is well offset ventrally. The medial aspect of the mandible is described only in *Plateosaurus.* The elongate intercoronoid described by Brown and Schlaikjer (1940*c*) and Galton (1984*a*) is probably the rostral part of the small coronoid that medially overlaps the dentary. The large splenial overlaps the dentary and the prearticular, more caudally overlapping the surangular and the articular, the middle part of which is constricted dorsoventrally. The funnellike caudal part of the prearticular overlaps the articular, the transversely broad central part of which articulated with the quadrate. The stout retroarticular process of the articular (and overlapping sheet of surangular) is short in *Anchisaurus* and *Massospondylus* and long in the remaining genera.

The teeth are set in sockets that are bordered lingually by small interdental plates alternating with special foramina of varying sizes. Numerous front to back wave patterns of tooth replacement (or *Zahnreihen*) with eruption in sequential tooth positions appear to pass along the jaws during the life of a reptile. Spacing between *Zahnreihen* or Z-spacing values of between 2.0 and 3.0 tooth positions result in the secondary pattern of tooth replacement that is commonly seen in reptiles—back to front waves affecting alternating tooth positions (DeMar 1972). This alternating pattern is present in *Plateosaurus*, but, because the length of the replacement wave decreases from 5 to 4 to 3 tooth positions passing caudally along the tooth row, the Z-spacing values show a corresponding increase from 2.5 to 2.66 to 3.0.

The dental formulae for the premaxilla, maxilla, and dentary are *Anchisaurus* 5, 11, ?; *Coloradisaurus* 3, 23 or 24, 22; *Lufengosaurus* 5, 19, 25; *Massospondylus* juvenile 4, 15, ?; adult from Arizona 4, 16, 20; adult from South Africa 4, 18, 17; *Mussaurus* juvenile 4, 10 to 12, ?; *Plateosaurus* 5 to 6, 24 to 30, 21 to 28; *Sellosaurus* 4 or 5, 25, 22; *Thecodontosaurus* juvenile 4, 10, 14, adult dentary 20 (see Galton 1985*b*); and *Yunnanosaurus* 4, 15, 17 or 18.

The crowns of the teeth of the premaxilla and the rostral part of the dentary of thecodontosaurids, anchisaurids, massospondylids, and plateosaurids taper apically and are slightly recurved with little mesiodistal expansion, whereas those of the maxilla and the rest of the dentary are more spatulate, transversely compressed, more expanded mesiodistally, and symmetrical in mesial and distal views. The crowns are oriented slightly obliquely relative to the long axis of the maxilla and dentary so that the distal edge of one tooth slightly overlaps the mesial edge of the tooth behind it. The base of the crown is wide. The root is circular in cross section. The buccal surface of the crown bears a central thickening, so that it is slightly more convex mesiodistally than the lingual surface, but there are no prominent vertical ridges as is seen in many ornithischians. The marginal serrations are prominent (high notches or *Spitzkerben* of Huene 1926*c*) and set at an angle of about 45° to the cutting edge (fig. 15.4D,E). The mesial edge has fewer serrations; these are found farther from the root than on the distal edge. The teeth are slender and elongate in *Azendohsaurus*, *Anchisaurus*, and juvenile individuals of *Massospondylus* and *Mussaurus*. The teeth are slender in *Coloradisaurus*, *Lufengosaurus*, *Sellosaurus*, *Thecodontosaurus*, and adult individuals of *Massospondylus* and broad in *Plateosaurus*. The teeth of the melanorsaurid *Riojasaurus* are similar (J. F. Bonaparte pers. comm.).

The teeth of the yunnanosaurid *Yunnanosaurus* are transversely asymmetrical, and the apices are directed slightly lingually in both maxillary and dentary teeth. The buccal surface is uniformly convex vertically, whereas the lingual surface is concave except near the root. The form of these teeth is similar to those referred to the Lower Cretaceous sauropods *Astrodon* and *Pleurocoelus* (Galton 1985*d*; see also McIntosh, this vol.). The teeth of *Yunnanosaurus* bear well-developed wear surfaces. The wear facet on the distal edge of the maxillary tooth is slightly curved and the smooth enamel border forms a slightly raised rim around the dentine (fig. 15.4F,G). Comparable wear in *Astrodon* was probably formed during feeding by tooth-food contact. The larger mesial wear surface is flat, obliquely inclined, and faces mesiolingually (fig. 15.4G,H). Similar surfaces occur on the maxillary teeth of the sauropod *Brachiosaurus*, thought to have resulted from wear against distobuccal surfaces of reciprocal dentary teeth (Janensch 1935-36).

The stapes is a slender bony rod, most of which is preserved in a skull of *Plateosaurus* (fig. 15.2F). The medial footplate is only slightly expanded, and there is a short diagonal ridge that passes distally onto the shaft (fig. 15.2G,H).

Only three complete sclerotic rings are known for *Plateosaurus* (fig. 15.2A; Galton 1984*a*). Each ring consists of eighteen plates that are grouped into four unequal quadrants. Otherwise, only a few plates are preserved in other prosauropod taxa (e.g., *Coloradisaurus*).

The hyoid apparatus is represented by a pair of elongate, asymmetrical, rodlike first ceratobranchials, preserved caudoventral to the mandibles in several individuals of *Plateosaurus*. These elements vary from 50 to 60 percent of the length of the mandible (fig. 15.2D).

Postcranial Skeleton

The postcranial skeleton is much more uniform than the skull among prosauropods. The main overall difference is that the bones are slender and lightly built in smaller genera (e.g., *Anchisaurus*, *Ammosaurus*, *Massospondylus*, *Sellosaurus*, and *Thecodontosaurus*), thicker and more heavily built in larger plateosaurids (e.g., *Euskelosaurus*, *Lufengosaurus*, and *Plateosaurus*), melanorosaurids, and yunnanosaurids, and very robust and ex-

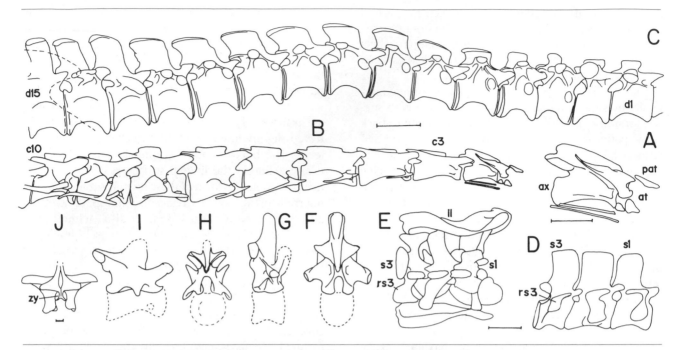

Fig. 15.5. Prosauropod vertebrae in right lateral (A–D,G,I), dorsal (E), and caudal views (F,H,J). A–E, plateosaurid *Plateosaurus* (after Huene 1907–08, 1926c, 1932): A, proatlas, atlas, and axis; B, cervical vertebrae 1 to 10 with ribs; C, dorsal vertebrae 1 to 15; D, sacrum; E, sacrum with ilia; F–I, neural arches of unnamed melanorosaurid (after Bonaparte 1986b): F,G, cranial dorsal; H,I, caudal cervical; J, massospondylid *Massospondylus*, neural arch of middorsal (after Cooper 1981a). Scale = 1 cm (J), 5 cm (A), 10 cm (B–E), scale not known for F–I.

tremely heavily built in blikanasaurids. Illustrations are given of the vertebral column (fig. 15.5), pectoral girdle and forelimb (fig. 15.6), and pelvic girdle and hindlimb (figs. 15.7, 15.8). Unless indicated to the contrary, the following references were used for particular genera: *Anchisaurus* (figs. 15.6N, 15.8B,G; Galton 1976a), *Ammosaurus* (fig. 15.8A; Galton 1976a), *Blikanasaurus* (fig. 15.8K; Galton and Heerden 1985), *Euskelosaurus* (Heerden 1979), *Lufengosaurus* (fig. 15.6P; Young 1941a, 1941b, 1947, 1951), *Massospondylus* (figs. 15.5J, 15.6Q; Cooper 1981a), *Melanorosaurus* (Heerden 1977, 1979), *Plateosaurus* (figs. 15.1B, 15.5A-E, 15.6A-K, 15.7; Huene 1907-08, 1926c, 1932), *Riojasaurus* (figs. 15.6L, 15.8C-F,L; Bonaparte 1972), *Sellosaurus* (fig. 15.8H,I; Huene 1907-08, 1932; Galton 1985b), *Thecodontosaurus* (figs. 15.1A, 15.6O; Huene 1907-08; Kermack 1984), and *Yunnanosaurus* (figs. 15.6M, 15.8J; Young 1942a, 1951).

Axial Skeleton

The complete vertebral series of *Plateosaurus* (figs. 15.1B, 15.5A-E) consists of the proatlas plus 10 cervicals, 15 dorsals, 3 sacrals, and about 50 caudals. Similar vertebral counts are found in *Sellosaurus, Lufen-gosaurus, Yunnanosaurus,* and *Riojasaurus*. In *Plateosaurus*, the proatlas consists of a pair of thin plates that overlap the zygapophysislike processes of the supraoccipital and the neural arch of the atlas (fig. 15.5A). Each neural arch of the atlas has an elongate caudodorsal process. The cervical vertebrae are low and elongate cranially but become higher and eventually slightly shorter caudally. The centra are strongly compressed transversely and markedly amphicoelous, especially cranially. The neural spine of the axis is large, while that of the third cervical vertebra is long and low. Passing caudally, the neural spines of cervical vertebra 4 to dorsal vertebra 3 become progressively shorter and taller. The neural spines of the remaining dorsal vertebrae continue to increase slightly in height and especially in length. However, the last three dorsals show a decrease in length of the neural spine. From the eighth cervical vertebra onward, lamellae are formed cranioventral and more strongly caudoventral to the diapophyses. The central part of the deep, platycoelous dorsal centra are strongly compressed transversely, and the last five dorsals have lateral pleurocentral indentations. The middle and especially the caudal dorsal vertebrae have zygosphene-zygantrum accessory articulations for the zygapophyses (fig. 15.5J). In the first six dorsal

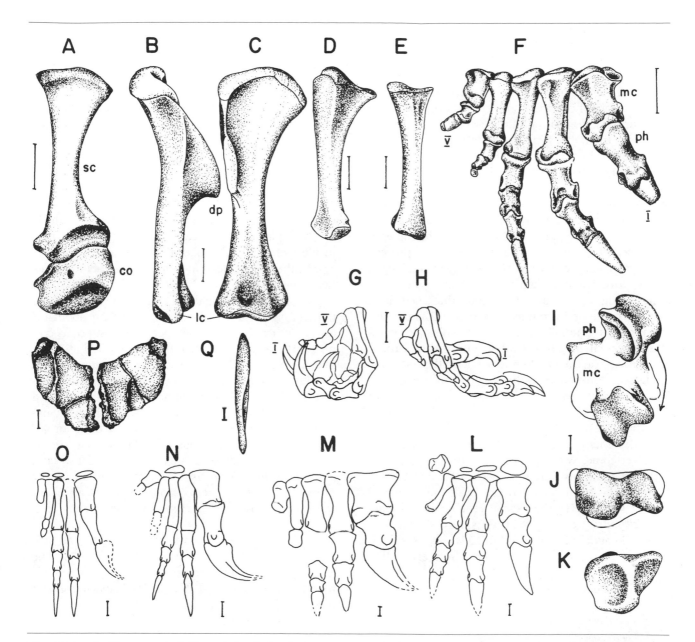

Fig. 15.6. A–K, pectoral girdle and forelimb of plateosaurid *Plateosaurus* (after Huene 1926c, 1932) all from right side: A, scapula and coracoid in lateral view; B,C, humerus in B, lateral, and C, cranial views; D, ulna in lateral view; E, radius in lateral view. F–H, manus: F, in cranial view with digit I partly flexed and digits II to V straight; G, in lateral view with digits fully flexed and H, in medial view with digits in full extension in weight supporting pose during quadrupedal walking. I–K, metacarpal I and first phalanx: I, phalanx shown in full extension and full flexion on distal end of metacarpal I; J, metacarpal I in distal view; K, phalanx in proximal view. L–O, right manus in cranial view with all digits straight: L, melanorosaurid *Riojasaurus* (after Bonaparte 1972); M, yunnanosaurid *Yunnanosaurus* (after Young 1947); N, anchisaurid *Anchisaurus* (after Galton 1976a); O, thecodontosaurid *Thecodontosaurus* (after Galton and Cluver 1976). P, plateosaurid *Lufengosaurus*, sternal plates in dorsal view (after Young 1947). Q, massospondylid *Massospondylus*, interclavicle in ventral view (after Cooper 1981a). Scale = 10 cm (A), 5 cm (B–H,P), 1 cm (I–O,Q).

vertebrae, the diapophyses are connected to the prezygapophyses cranially and to the postzygapophyses caudally by thin lamellae. Three cavities below the transverse process are defined by the two ventral lamellae. The remaining dorsals have only two cavities due to the convergence and fusion of the two cranial lamellae. In smaller prosauropods such as *Anchisaurus, Ammosaurus, Massospondylus, Sellosaurus,* and *Thecodontosaurus,* the centra of the dorsal vertebrae are proportionally lower, and the lamellae ventral to the transverse process are less prominent. Bonaparte (1986*b*) has described a caudal cervical and a cranial dorsal vertebra of an unnamed melanorosaurid which are more sauropodlike than the vertebrae of other prosauropods (see below).

Each cervical rib (figs. 15.1B, 15.5B) is thin, delicate, and over twice the length of the supporting vertebra. The atlantal rib lacks a tuberculum (fig. 15.1A). In the remaining cervical ribs, the line of the shaft is continued beyond the point of divergence of the capitular and tubercular processes as a spinous cranial process. These processes are short in *Massospondylus* and long in *Plateosaurus.* The first eight of the dorsal ribs are especially strong, and in life, the slightly thickened distal end was continued ventrally by cartilage. The remaining ribs taper gradually to a point and show a progressive decrease in length. The tuberculum forms a prominent process on the first four ribs, but on the rest, it is a small facet dorsolateral to the long capitular process.

Gastralia consisting of slender subparallel rods form a basketlike support for the ventral abdominal wall (fig. 15.1B).

The sacrum of most prosauropods consists of three vertebrae, with sacral ribs 1 and 2 much more robust than 3. However, in *Massospondylus* and *Melanorosaurus,* the last dorsal vertebra is also incorporated into the sacrum. The third sacral vertebra in *Ammosaurus, Sellosaurus,* and *Riojasaurus* is very similar to a proximal caudal vertebra, and in *Ammosaurus,* the diapophysis and the sacral rib diverge from each other distally.

The slender, oblique neural spines and transverse processes of the caudal vertebra of *Plateosaurus* progressively decrease in size distally and disappear at about caudal vertebrae 36 and 28, respectively. The axial width of the neural spines is less on proximal caudal vertebrae of *Thecodontosaurus* and *Sellosaurus* than it is in other prosauropods. Height is greater than length in the proximal centra, and more distally, the centra are lower and longer. Ventral facets on adjoining centra together support a single chevron. The first chevron is a nubbin of bone between sacral centrum 3 and the first caudal centrum. Chevrons 2 to 4 are long and slender and increase in length, but distal to this, the chevrons progressively decrease in length. In *Riojasaurus,* the first chevron is borne between caudals 3 and 4 (fig. 15.1D).

Appendicular Skeleton

The scapula is long and slender with expanded ends, especially ventrally. The suboval coracoid has a prominent notch ventral to the glenoid cavity in all prosauropods except the melanorosaurid *Riojasaurus.* In this genus, the caudal border is convex and ventrally projects as a prominent process. Huene (1926*c*) showed a slender rodlike clavicle contacting the cranial part of the scapula and coracoid in *Plateosaurus.* In *Massospondylus,* the interclavicle is long, moderately broad, and spatulate. It is found adjacent to the ventral edge of the coracoid. The sternal plates consist of a pair of thick, irregular, platelike bones caudal to the coracoids which in life were probably suboval in outline. In *Lufengosaurus,* the sternal plates are sutured together to form a heart-shaped shield (fig. 15.6P).

The proximal and distal ends of the humerus are transversely expanded, the axes of which are about 45° to each other. The large deltopectoral crest points cranially. The apex of the deltopectoral crest is at about midlength except in *Anchisaurus* and *Thecodontosaurus,* in which it is more proximally placed. The radius has an oval, saddle-shaped proximal articular surface and a flat, square obliquely facing distal articulation. The expanded ends of the ulna are at an angle of about 40° to each other. The proximal end is triangular in outline, while the distal end has a convex articular surface. In *Lufengosaurus, Yunnanosaurus,* and *Riojasaurus,* the proportions of the humerus, radius, and ulna are about the same as those of *Plateosaurus,* but the articular ends are much larger.

The structure of the manus is similar in all prosauropods (fig. 15.6F-O; Galton and Cluver 1976). The main variation is that, compared to all other prosauropods, metacarpal I is slender in *Thecodontosaurus* and metacarpals II to V are also slender in *Anchisaurus, Thecodontosaurus,* and a juvenile individual of *Sellosaurus.* The proximal carpals are rarely preserved in prosauropods and were probably represented by cartilage rather than bone. An intermedium of *Massospondylus* is pyramidal. Distal carpals I and II are preserved in most prosauropods and consist of large and small thin plates capping metacarpals I and II, respectively. Occasionally, a small distal carpal III is also preserved. Metacarpal I is short but stout, with a triangular proximal end and a very asymmetric distal ginglymus. This distal ginglymus extends well onto the extensor surface to the degree that marked hyperextension was possible. The proximal articular surfaces of the first phalanx of digit I are also unequally developed, and the rotational axis of

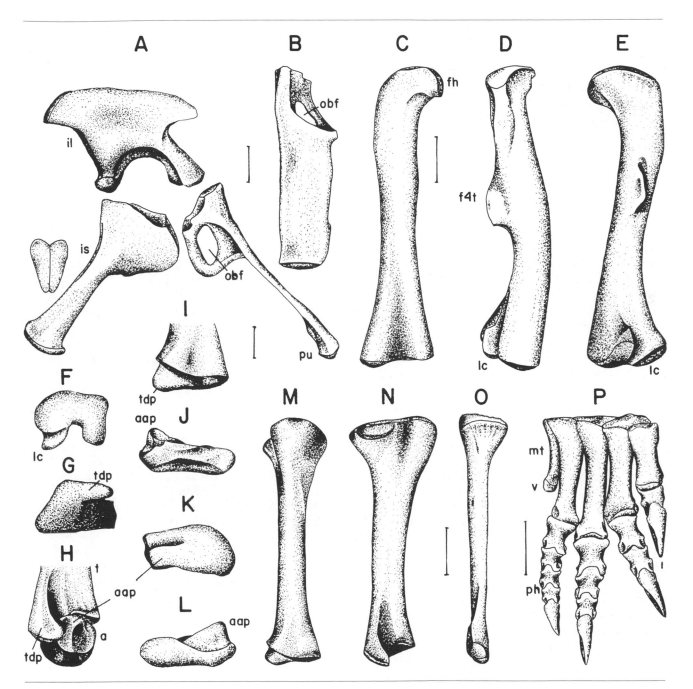

Fig. 15.7. Pelvic girdle and hindlimb of plateosaurid *Plateosaurus*, all from right side (after Huene 1926c). A, pelvic girdle in lateral view with distal end of ischia; B, pubis in dorsal view. C–E, femur in C, cranial, D, lateral, E, caudal, and F, distal views. G–I, distal end of tibia in G, distal, H, lateral (with astragalus), and I, cranial views. J–L, astragalus in J, cranial, K, proximal, and L, caudal views. M,N, tibia in M, cranial, and N, lateral views. O, fibula in lateral view. P, pes in cranial view. Scale = 10 cm (A–F, M–P), 5 cm (G–L).

the distal condyles makes an angle of about 45° with that of the proximal articulation. The very trenchant, raptorial ungual of digit I is very large, exceeding the size of the largest ungual of the pes. Metacarpal II is

slightly longer and more robust than metacarpal III. Digits II and III are subequal in length, cannot be hyperextended, and bear much smaller unguals that are only slightly recurved. Metacarpal IV is slightly shorter

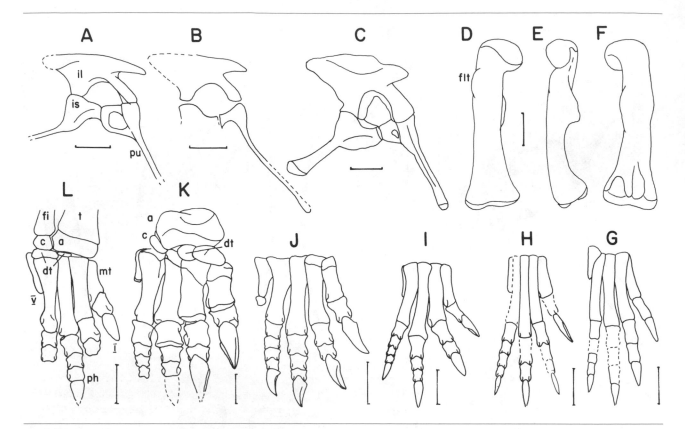

Fig. 15.8. Prosauropod pelvic girdles and hindlimbs, all from right side. A−C, pelvic girdles in lateral view: A, plateosaurid *Ammosaurus* (after Galton 1976*a*); B, anchisaurid *Anchisaurus* (after Galton 1976*a*); C, melanorosaurid *Riojasaurus* (after Bonaparte 1972). D−F, femur of *Riojasaurus* in D, cranial, E, medial, and F, caudal views (after Bonaparte 1972). G−L, pes in caudal (G) and cranial views (H−L); G, *Anchisaurus* (after Galton 1976*a*); H,I, plateosautid *Sellosaurus* (after Galton 1973*b*, 1984*b*); H, from juvenile and I, larger individuals; J, yunnanosaurid *Yunnanosaurus* (after Young 1947); K, blikanasaurid *Blikanasaurus* (after Galton and Heerden 1985); L, *Riojasaurus* (after Bonaparte 1972). Scale = 10 cm (A,B,G,H,K), 5 cm (C−F,I,J,L).

and more slender than metacarpal III and bears a vestigial series of phalanges, as does the very small metacarpal V.

The ilium is proportionally deep, except in *Ammosaurus* and *Anchisaurus*, in which the cranial process is proportionally elongate (figs. 15.7A, 15.8A-C). The long robust pubic penduncle of the ilium is large compared to the small ischial penduncle in all prosauropods except *Riojasaurus*. In the latter, the iliac portion of the acetabulum is much deeper. The brevis shelf on the ventromedial edge of the short postacetabular process is narrow, and the third sacral rib fits against this shelf while the other sacral ribs fit against the body of the ilium. The thin supraacetabular rim projects laterally and extends along the craniodorsal part of the acetabulum. The acetabulum itself is partially backed by the ilium and, to a lesser extent, by parts of the pubis and ischium. The pubis is twisted along its length

with its proximal part forming a deep, thin, and oblique subacetabular region, which bears an obturator foramen (figs 15.7A,B, 15.8A-C). This region is shallow in *Riojasaurus* and still shallower in *Anchisaurus,* in which the obturator foramen opens ventrally. The obturator foramen is small in *Massospondylus* and *Riojasaurus* and large in *Ammosaurus* and *Plateosaurus*. The main part of the pubis is transversely expanded, and, because there is an extensive median symphysis, the two pubes form a platelike apron. This apron is broad in all prosauropods except *Anchisaurus*, in which it is relatively narrow. The deep subacetabular part of the ischium is also oblique and sweeps back to a short, deep median symphysis (fig. 15.7A). The cranioventral part of the subacetabular region is low in *Ammosaurus*.

The femur of *Thecodontosaurus, Anchisaurus, Massospondylus, Yunnanosaurus,* and plateosaurids is sigmoidal in cranial and lateral views. The medially directed femoral head is not well rounded proximally and lacks a distinct neck (figs. 15.7C-E). Instead, the head merges with the shaft. The greater trochanter is represented by a thick lateral edge adjacent to the head, and the lesser trochanter consists of a low ridge craniodistal to the greater trochanter. The fourth trochanter is large with an apex that extends beyond femoral midlength; it is well removed from the medial edge of the shaft. The shaft is twisted about its long axis so the articular axis of the subequal distal condyles makes an angle of about 25° with the axis of the femoral head. The main variation on the anatomy of the femur in nonmelanorosaurid prosauropods is that the fourth trochanter is more proximally placed in smaller forms such as *Anchisaurus, Ammosaurus, Massospondylus, Sellosaurus,* and *Thecodontosaurus* and has a pendant shape in some individuals of *Massospondylus.* In melanorosaurids such as *Riojasaurus,* the femoral head is well rounded and medially directed, the lesser trochanter is prominent, the shaft is straight in caudal view, and the fourth trochanter is medial (fig. 15.8D-F; Galton 1985*f*).

The expanded ends of the fibula are set at about 40° to each other so that the distal cranial edge also faces medially. The tibia has a prominent cnemial crest, the shaft is triangular in cross section, and the distal end is transversely expanded. There is a prominent groove that deepens laterally to accommodate the ascending process of the astragalus (fig. 15.7G-N). This groove is backed by a prominent thick descending process formed by the medial malleolus of the tibia. The tibiae of smaller prosauropods such as *Anchisaurus, Sellosaurus,* and *Thecodontosaurus* are proportionally more slender, whereas those of *Riojasaurus* and *Blikanasaurus* are more robust with a much larger cnemial crest.

The calcaneum is a small, subtriangular bone that contacts the fibula proximally and the astragalus medially. The large astragalus has a craniolateral ascending process that fits into the grooved distal end of the tibia. In most prosauropods, two distal tarsals are preserved capping metatarsals III and IV. However, in *Blikanasaurus,* the two distal tarsals cap metatarsals I to III, articulating with the medial part of the astragalus, and the calcaneum contacts metatarsal IV rather than a distal tarsal (fig. 15.8K,L).

The form of the pes is uniform. Metatarsals I and V are short, but I is robust and bears a short but strong digit, whereas V is broad, thin, and only bears a vestigial phalanx. Digits II and IV are subequal. Digit III is only slightly longer. Unguals decrease in length from digits I to IV in all prosauropods except *Anchisaurus,* in which the ungual of digit I is shorter than that of digit II (fig. 15.8G). In smaller forms such as *Anchisaurus* and

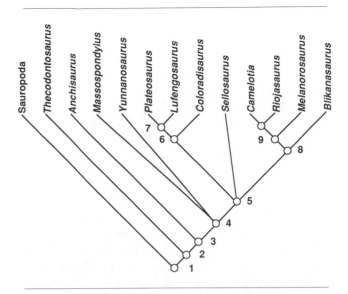

Fig. 15.9. A preliminary cladistic analysis of the Prosauropoda. See text for derived characters at Nodes 1 to 9.

Thecodontosaurus, the pes is slender, but this difference may reflect ontogenetic differences as indicated by different-sized individuals of *Sellosaurus* (fig. 15.8H,I; Huene 1907-08, 1915, 1932; Galton 1985*a*) and *Massospondylus* (Cooper 1981*a*). The pes of *Blikanasaurus* is very massive with the proximal ends of metatarsal III and especially II very expanded craniocaudally (fig. 15.8K).

SYSTEMATICS

Derived characters of the Sauropodomorpha (Node 1, fig. 15.9, primitive characters for the Prosauropoda) include the following: skull proportionally small, teeth spatulate and coarsely serrated; neck moderately elongate with 10 vertebrae; sacrum with three vertebrae with additional vertebra incorporated from tail; short metapodials; manus with very large ungual phalanx on digit I; combined lengths of femur, tibia, and metatarsal III shorter than presacral series (85% or less), tibia shorter than femur; ascending process of astragalus keys into descending process on back of tibia; pes with largest ungual on digit I and unreduced metatarsal V (*Herrerasaurus, Staurikosaurus,* and *Lagosuchus* used as outgroups; see Benedetto 1973; Bonaparte 1975*b,* 1978*a*; Brinkman and Sues 1987; Colbert 1970; Galton 1977*a*; Reig 1963; Sues, this vol.).

It is usually assumed that prosauropods of at least partly bipedal habits are ancestral to the Sauropoda (e.g., Huene 1932, 1956; Colbert 1951; Lapparent and Lavocat 1955; Romer 1956). The Melanorosauridae (fig. 15.1D) has been thought to represent an intermediate stage between the other prosauropods and sauropods (Huene 1929c, 1932, 1956; Romer 1956). Charig et al. (1965), however, argued that such a relationship is unlikely because melanorosaurids (*Melanorosaurus* actually), *Blikanasaurus,* and plateosaurids (based on small spatulate teeth; *Euskelosaurus* not referred to Plateosauridae by Heerden until 1979) occur together in the Lower Elliot Formation of South Africa. They noted that there is no evidence that the sauropod line of evolution was other than purely quadrupedal and considered that the partly bipedal prosauropods were offshoots from this main quadrupedal line that included melanorosaurids and blikanasaurids. However, the presence of prosauropods from the mid-Carnian of Africa and North America means that the quadrupedal families of sauropodomorphs could still be derived from a lineage derived from bipedal thecodontosaurids or facultatively bipedal anchisaurids (or even plateosaurids), as was presumably the case for the yunnanosaurids with their very sauropodlike teeth (fig. 15.4F-H). As Gauthier (1986) noted, additional outgroup comparisons with *Staurikosaurus* and *Lagosuchus* indicate that the ancestral saurischian was a small, erect-postured, cursorial carnivorous biped, and *Thecodontosaurus* is closer to this in its size and bipedality than any other prosauropod. The very juvenile holotype of *Mussaurus* (figs. 15.1C, 15.3I) is of interest in having a sauropodlike skull with a deep symphysial region to the dentary, a short snout, short but high premaxilla and maxilla, a high antorbital fenestra, and a large orbit. However, apart from the cervical vertebrae, which resemble those of the thecodontian *Lagosuchus,* the rest of the anatomy is typically prosauropod. Bonaparte and Vince (1979) postulated that the Sauropoda might have originated by the persistence of juvenile characters, that is, neoteny, in one of the prosauropod families, and that would explain the embryolike flexure of the skull above the braincase of sauropods (see Janensch 1935-36; Berman and McIntosh 1978). Unfortunately, the skull of *Riojasaurus* is still to be described and that of the other melanorosaurids and *Blikanasaurus* is unknown. However, as Cruickshank (1975) noted, the fifth metatarsal is unreduced in all sauropods. Hence *Riojasaurus, Blikanasaurus,* and *Thecodontosaurus* (Kermack 1984), with a reduced metatarsal V (fig. 15.8G-L), cannot be directly ancestral to sauropods. The reduced size of metatarsal V in all pro-sauropods is the derived condition, so the Sauropoda is regarded as the sister taxon to the Prosauropoda (fig. 15.9).

Using *Herrerasaurus, Staurikosaurus, Lagosuchus,* and *Euparkeria* (skull only) for outgroup comparisons (see Benedetto 1973; Bonaparte 1975b, 1978a; Brinkman and Sues 1987; Charig et al. 1976; Colbert 1970; Galton 1977a; Reig 1963), *Thecodontosaurus* is the most primitive prosauropod (figs. 15.3D-F, 15.4B,D; Huene 1907-08, 1914a; Galton and Cluver 1976; Kermack 1984), sharing with all other prosauropods (table 15.1) the derived characters listed in the diagnosis of the Prosauropoda (see above and Node 2, fig. 15.9).

Compared to *Thecodontosaurus,* the other prosauropods show various combinations of the following derived characters:

1. enlargement of external nares

2. loss of caudal contact between premaxilla and nasal

3. increase in height of dorsal process of maxilla

4. decrease in size of antorbital fenestra by development of lateral lamella on maxilla (condition in some genera uncertain)

5. decrease in size of antorbital fenestra by development of medial lamina on maxilla to form antorbital fossa

6. elongation of maxilla and dentary

7. increase in tooth count

8. loss of denticles on cheek teeth that curve lingually and have wear surfaces

9. increased transverse width of prefrontal

10. increased contribution to orbital rim of prefrontal (so decrease of contribution from frontal)

11. rostrocaudal expansion of lacrimal

12. fusion of parietals to form single bone

13. reorientation of quadrate to vertical to caudoventral direction

14. more ventrally offset jaw articulation

15. increased depth of coronoid eminence

16. less steeply inclined supraoccipital

17. deepening of rostroventral part of braincase so that occipital condyle is dorsal to level of parasphenoid

TABLE 15.1 Prosauropoda

	Occurrence	Age	Material
Prosauropoda Huene, 1920			
Thecodontosauridae Lydekker, 1890			
Azendohsaurus Dutuit, 1972			
A. laaroussii Dutuit, 1972 *partim*	Argana Formation, Atlas Mountains, Marrakech, Morocco	middle Carnian	Dentary, isolated teeth, adult
Thecodontosaurus Riley et Stutchbury, 1836			
T. antiquus Morris, 1843	Magnesian Conglomerate, Avon, England; Fissure Fillings, South Glamorgan, Wales	?Norian-Rhaetian	More than 100 disarticulated elements, skull, partial skeletons, isolated elements, juvenile to adult
Anchisauridae Marsh, 1885			
Anchisaurus Marsh, 1885 (= *Megadactylus* Hitchcock, 1865, *Amphisaurus* Marsh, 1882, *Yaleosaurus* Huene, 1932)			
A. polyzelus (Hitchcock, 1865) (= *Megadactylus polyzelus* Hitchcock, 1865, including *Anchisaurus colurus* Marsh, 1891b)	Upper (Portland) Beds, Newark Supergroup, Connecticut, Massachusetts United States	Pliensbachian or Toarcian	Nearly complete skull and skeleton, fragmentary skeleton, adult
Massospondylidae Huene, 1914b			
Massospondylus Owen, 1854 (= *Aetonyx* Broom, 1911, *Aristosaurus* Hoepen, 1920, *Dromicosaurus* Hoepen, 1920, *Gryponyx* Broom, 1911, *Gyposaurus* Broom, 1911)			
M. carinatus Owen, 1854 (including *Aetonyx palustris* Broom, 1911, *Aristosaurus erectus* Hoepen, 1920, *Dromicosaurus gracilis* Hoepen, 1920, *Gryponyx africanus* Broom, 1911, *G. taylori* Haughton, 1924, *Gyposaurus capensis* Broom, 1911, *Massospondylus browni* Seeley, 1895, *M. harriesi* Broom, 1911, *M. schwarzi* Haughton, 1924, *Thecodontosaurus browni* Huene, 1932, *T. dubius* Haughton, 1924)	Upper Elliot Formation, Clarens Formation, Orange Free State, Cape Province, Bushveld Sandstone, Transvaal, South Africa, Upper Elliot Formation, Leribe, Quthing, Lesotho, Forest Sandstone, Matabeleland North, Zimbabwe	Hettangian-Pliensbachian	At least 80 partial skeletons, isolated skull and skeletal elements, 4 skulls, juvenile to adult
Yunnanosauridae Young, 1942a			
Yunnanosaurus Young, 1942a			
Y. huangi Young, 1942a (including *Y. robustus* Young, 1951)	Upper Lower Lufeng Series, Yunnan, People's Republic of China	Hettangian-Pliensbachian	More than 20 partial to complete skeletons, 2 skulls, juvenile to adult
Plateosauridae Marsh, 1895			
Ammosaurus Marsh, 1891b			
A. major (Marsh, 1891b) (= *Anchisaurus major* Marsh, 1889a, including *Anchisaurus solus* Marsh, 1892e)	Upper (Portland) Beds, Newark Supergroup, Connecticut, Navajo Sandstone, Arizona, United States	Pliensbachian or Toarcian	4 incomplete skeletons, juvenile to adult
Mussaurus Bonaparte et Vince, 1979			
M. patagonicus Bonaparte et Vince, 1979	El Tranquilo Formation, Santa Cruz, Argentina	Norian	More than 10 fragmentary to complete skeletons, 4 skulls, juvenile to adult
Plateosaurus Meyer, 1837 (= *Dimodosaurus* Pidancet et Chopard, 1862, *Dinosaurus* Rütimeyer, 1856a, *Gresslyosaurus* Rütimeyer, 1856b, *Pachysauriscus* Kuhn, 1959, *Pachysaurops* Huene, 1959, *Pachysaurus* Huene, 1907-08)			
P. engelhardti Meyer, 1837 (including *P. erlenbergiensis* Huene, 1907-08, *P. fraasianus* Huene, 1932, *P. integer* Fraas vide Huene, 1915, *P. longiceps* Jaekel, 1914, *P. plieningeri* Huene, 1907-08, *P. reinigeri* Huene, 1907-08, *P. trossingensis* Fraas, 1914, *Dimodosaurus poligniensis* Pidancet et Chopard, 1862, *Dinosaurus gresslyi* Rütimeyer, 1856a, *Gresslyosaurus ingens* Rütimeyer, 1856b, *G. robustus* Huene, 1907-08, *G. torgeri* Jaekel, 1911, *Pachysaurus ajax* Huene, 1907-08, *P. giganteus* Huene, 1932, *P. magnus* Huene, 1907-08, *P. wetzelianus* Huene, 1932, *Zanclodon laevis* Plieninger, 1846 *partim*, *Z. plieningeri* Fraas, 1896, *Z. quenstedti* Koken, 1900)	Knollenmergel, Baden-Württemberg, Niedersachsen, Feuerletten, Bavaria, Federal Republic of Germany; Knollenmergel, Baselland, Obere Bunte Mergel, Aargau, Switzerland; Marnes irisees supérieures, Jura, unnamed unit, Doubs, France; Knollenmergel, Sachsen-Anhalt, Federal Republic of Germany	late Norian	More than 100 fragmentary to complete skeletons, isolated elements, at least 10 skulls, juvenile to adult
Sellosaurus Huene, 1907-08 (= *Efraasia* Galton, 1973b)			
S. gracilis Huene, 1907-08 (including *Palaeosaurus diagnosticus* Huene, 1932, *Sellosaurus fraasi* Huene, 1907-08, *Teratosaurus minor* Huene, 1907-08, *T. trossingensis* Huene, 1907-08, *Thecodontosaurus diagnosticus* Fraas, 1912, *T. hermannianus* Huene, 1907-08)	Unterer and Mittlerer Stubensandstein, Baden-Württemberg, Federal Republic of Germany	middle Norian	21 partial skeletons, isolated elements, 3 partial skulls, juvenile to adult

Table 15.1, continued

Coloradisaurus Lambert, 1983 (= *Coloradia* Bonaparte, 1978b)

C. brevis (Bonaparte, 1978b) (= *Coloradia brevis* Bonaparte, 1978b)	Upper Los Colorados Formation, La Rioja Argentina	Norian	Skull, adult

Euskelosaurus Huxley, 1866 (= *Plateosauravus* Huene, 1932)

E. browni Huxley, 1866 (including *E. africanus* Haughton, 1924, *Plateosaurus cullingworthi* Haughton, 1924, *P. stormbergensis* Broom, 1915,	Lower Elliot Formation, Orange Free State, Cape Province, Bushveld Sandstone, Transvaal, South Africa; Lower Elliot Formation, Leribe, Lesotho; Mpandi Formation, Matabeleland South, Zimbabwe	late Carnian or early Norian	16 fragmentary skeletons, adult

Lufengosaurus Young, 1941a

L. huenei Young, 1941a (including *L. magnus* Young, 1947, *Gyposaurus sinensis* Young, 1941b, *Fulengia youngi* Carroll et Galton, 1977 •, *Tawasaurus minor* Young, 1982b)	Upper Lower Lufeng Series, Yunnan, People's Republic of China	Hettangian-Pliensbachian	More than 30 fragmentary to complete skeletons, 2 juvenile to adult

Melanorosauridae Huene, 1929a
Camelotia Galton, 1985f

C. borealis Galton, 1985f (= *Avalonia sanfordi* Seeley, 1898 partim)	Westbury Formation, Somerset, England	Rhaetian	Vertebrae, pubis, ischium, femur, tibia, phalanges

Melanorosaurus Haughton, 1924

M. readi Haughton, 1924	Lower Elliot Formation, Cape Province, South Africa	late Carnian or early Norian	Partial skeleton, adult

Riojasaurus Bonaparte, 1969 (= *Strenusaurus* Bonaparte, 1969)

R. incertus Bonaparte, 1969 (including *Strenusaurus procerus* Bonaparte, 1969)	Upper Los Colorados Formation, La Rioja, Quebrada del Barro Formation, San Juan, Argentina	Norian	More than 20 fragmentary to complete skeletons, skull, juvenile to adult

Blikanasauridae Galton et Heerden, 1985
Blikanasaurus Galton et Heerden, 1985

B. cromptoni Galton et Heerden, 1985	Lower Elliot Formation, Leribe, Lesotho	late Carnian or early Norian	Partial hindlimb

Nomina dubia	Material
Agrosaurus macgillivrayi Seeley, 1891	Radius, manual ungual, tibia, distal caudal vertebra
Basutodon ferox Huene, 1932 (doubtfully prosauropod)	Tooth
Eucnemesaurus fortis Hoepen, 1920	Fragmentary skeleton
Gigantoscelus molengraaffi Hoepen, 1916	Distal femur
Gryponyx transvaalensis Broom, 1912	Ungual, distal metatarsal
Hortalotarsus skirtopodus Seeley, 1894b	Partial pes
Leptospondylus capensis Owen, 1854	2 caudal vertebrae
Massospondylus hislopi Lydekker, 1890b	Isolated vertebrae
Orinosaurus capensis Lydekker, 1889a	Proximal tibia
Pachyspondylus orpenii Owen, 1854	Fragmentary vertebrae
Palaeosaurus fraserianus Cope, 1878e (doubtfully prosauropod)	Tooth
Plateosaurus elizae Sauvage, 1907 (doubtfully prosauropod)	Tooth
Plateosaurus ornatus Huene, 1907-08 (doubtfully prosauropod)	Tooth
Thecodontosaurus minor Haughton, 1918	Cervical vertebra, tibia, ischium

18. increased length of neck (especially cervical vertebrae 3-8) relative to hindlimb, so neck up to twice length of femur

19. increased length of trunk relative to hindlimb, so trunk subequal to or longer than hindlimb

20. increased development of laminae ventral to transverse processes of cranial dorsal vertebrae

21. increased height of centrum relative to length in dorsal vertebrae

22. increased sacralization of sacral vertebra 3 by increased proximodistal width of neural spine and massiveness of sacral rib

23. fourth sacral vertebra incorporated from dorsal series

24. increased axial width of neural spines of proximal caudal vertebrae

25. increased length of forelimb relative to hindlimb

26. increased size and more distal position of apex of deltopectoral crest of humerus

27. increased massiveness of metacarpals, especially the first

28. increased size of ilium relative to femur

29. increased length of cranial process of ilium

30. reduction in extent of closure of acetabulum by ilium

31. ventral emargination of subacetabular part of pubis

32. increased width of apronlike part of pubis

33. ventral emargination of subacetabular part of ischium

34. femur becomes straight in craniocaudal view

35. lesser trochanter of femur becomes large and sheetlike

36. more distal position of fourth trochanter of femur

37. fourth trochanter more medially on edge of femoral shaft

38. medial expansion of caudal malleolus at distal end of tibia

39. increased massiveness of metatarsus

40. reduction in size of the ungual of pedal digit I

Thecodontosaurus is a fully bipedal animal (see above) that is regarded as the sister taxon to all the other prosauropods and the characters held in common are those listed for the Prosauropoda (see Diagnosis, above, and Node 2, fig. 15.9). Derived characters of *Anchisaurus* and the other prosauropods compared to *Thecodontosaurus* include a dentary that exceeds half the length of the mandible, a trunk that is at least subequal to the length of the hindlimb (femur + tibia + metatarsal III), and a massive first digit to the manus (Node 3, fig. 15.9; the facultatively bipedal prosauropods). Derived characters of *Massospondylus, Yunnanosaurus,* and the other prosauropods compared to *Anchisaurus* include a vertical or caudoventrally oriented quadrate, a proportional widening of the apronlike distal part of the pubis, a manus with more massive metacarpals II to V, and a pes with more massive metatarsals I to IV (Node 4, fig. 15.9, the broad-footed prosauropods of Galton and Cluver 1976). Node 4 represents an unresolved trichotomy. A derived character of *Yunnanosaurus* relative to all other prosauropods is the transversely asymmetrical teeth with very few marginal serrations and prominent wear surfaces. A derived character of the Plateosauridae, Melanorosauridae, and ?Blikanasauridae relative to *Massospondylus* and *Yunnanosaurus* is the pronounced ventral offset of the jaw articulation that is well below the line of the dentary tooth row (Node 5, fig. 15.9). Within the Plateosauridae, *Sellosaurus* is included as part of an unresolved trichotomy at Node 5 (fig. 15.9). A derived character of

Coloradisaurus, Lufengosaurus, and *Plateosaurus* relative to *Sellosaurus* is the prominent ventral step in the braincase so that the occipital condyle is well above the level of the parasphenoid (Node 6, fig. 15.9). A derived character of *Lufengosaurus* and *Plateosaurus* relative to *Coloradisaurus* is the transversely wide and elongate prefrontal (Node 7, fig. 15.9). Derived characters of *Plateosaurus* relative to *Lufengosaurus* include a large medial lamina to the maxilla and a transverse lamina between the basipterygoid processes. A derived character of the Melanorosauridae and ?Blikanasauridae relative to the Plateosauridae is a trunk that is about 140 percent of hindlimb length (Node 8, fig. 15.9, quadrupedal prosauropods). A derived character of the Melanorosauridae is a femur that is straight in craniocaudal view with a prominent lesser trochanter and a fourth trochanter that is on the medial edge of the femoral shaft (Node 9, fig. 15.9). Derived characters of *Blikanasaurus* relative to all other prosauropods include the extremely massive metatarsus (expanded both transversely and craniocaudally) and the distal tarsals contacting metatarsals I to III. The positioning of *Blikanasaurus* at Node 8 (fig. 15.9) is very tentative, given that the type specimen is only a partial hindlimb, but it seems the most reasonable place to put the genus.

Gauthier (1986) noted that among the broad-footed sauropodomorphs, *Riojasaurus, Vulcanodon* (see Cooper 1984), *Barapasaurus* (see Jain et al. 1975, 1979), and the Sauropoda (see McIntosh, this vol.) appear to be the most closely related. However, he does not list any derived characters to support this grouping and instead lists 25 derived characters for *Vulcanodon, Barapasaurus,* and the Sauropoda relative to all other sauropodomorphs, including *Riojasaurus.* Apart from the form of the femur and the increased length of the trunk relative to the hindlimb, *Riojasaurus* is extremely similar in its postcranial anatomy to *Plateosaurus* (fig. 15.1B,D; see Bonaparte 1972 and Huene 1926c) and very different from *Vulcanodon* and *Barapasaurus* (see McIntosh, this vol.). Gauthier (1986) cited the reduced size of the lesser trochanter of the femur of *Vulcanodon, Barapasaurus,* and the Sauropoda as a derived character. However, it is probably a primitive character for these taxa because the trochanter is small in all prosauropods except the Melanorosauridae, in which its large size is a derived character relative to the other prosauropods.

Bonaparte (1986b) noted that the cervical and cranial dorsal vertebrae of some prosauropods such as *Plateosaurus* and especially those of an unnamed melanorosaurid from the Late Triassic (Norian) of Argentina show a set of derived characters that approach the structure of Middle Jurassic sauropods. In the cervicals, the cited characters include (fig. 15.5B,H,I):

1. proportions and progressive slenderness toward the skull

2. median ventral keel on cranial half of centrum

3. pendant diapophyses

4. low, axially elongate neural spine that coalesces with the postzygapophyses

5. the depressed form ventral and cranial to the postzygapophyses

6. the laminae running from diapophyses toward the front and rear.

The first dorsal of the unnamed melanorosaurid shows the following additional sauropodlike characters (fig. 15.5F,G):

7. tall but axially short neural arches and spines

8. high and steeply inclined zygapophyses

9. well-developed suprapostzygapophyseal laminae forming a deep caudal axial depression between them

10. cranial side of neural spine free of laminae, long and transversely convex

11. large infrapostzygapophyseal cavities caudally and small infrazygapophyseal cavities cranially

12. diapophyses project ventrolaterally and a caudal lamina connects them to the postzygapophyses

The caudal dorsals of prosauropods (fig. 15.1D, 15.5C) do not show any approach to the corresponding vertebrae of sauropods, but Bonaparte (1986b) apparently does not have these vertebrae in his unnamed melanorosaurid.

Although the listed characters show an approach to the sauropod condition, it should be noted that characters 2, 3, 6, 7, 10, and 12 also occur in the Staurikosauridae and Herrerasauridae (Galton 1977a; Reig 1963; Benedetto 1973; Bonaparte 1978a). Hence, prosauropods show the primitive rather than the derived condition. Most of character 4 is primitive for prosauropods except that the neural spine is short in *Staurikosaurus*. The elongate form of the neural spine of prosauropods is probably related to character 1. Compared to plateosaurids and *Riojasaurus* (figs. 15.1B,D, 15.5B,C), the vertebrae of the unnamed melanorosaurid (fig. 15.5F-I) approach the more derived, sauropod condition with regard to characters 5, 8, 9,

and 11, and it definitely represents a new family of sauropodomorphs. However, Bonaparte (1986b) provided no information on the size of metatarsal V in this new family that may be on the evolutionary line leading to sauropods. However, the form of the vertebrae (fig. 15.5F-I) may be another example of the mozaic expression of sauropod-type characters in different prosauropod families, for example, the teeth of *Yunnanosaurus* (fig. 15.4F-H), the femur of melanorosaurids (fig. 15.8D-F), and the massive metatarsals I to IV of *Blikanasaurus* (fig. 15.8K).

On the basis of the preliminary cladistic analysis given above, I provisionally recognize seven families of prosauropods: the Thecodontosauridae, Anchisauridae, Massospondylidae, Yunnanosauridae, Plateosauridae, Melanorosauridae, and Blikanasauridae.

Thecodontosauridae. Recognition of this family is based on the very primitive nature of *Thecodontosaurus* as regards the form of the skull and the fully bipedal proportions of the skeleton (fig. 15.1A); all other prosauropods were either facultatively bipedal or fully quadrupedal (fig. 15.1B-D; see below). The only derived characters are the relatively long retroarticular process of the mandible (short in *Anchisaurus*, *Massospondylus*) and closure of the interpterygoid vacuity (open in *Massospondylus*).

Thecodontosaurus antiquus is a small form (length approximately 2.5 m) represented by many isolated bones and a skeleton from a juvenile individual of approximately 1 m from the Late Triassic (?Norian-Rhaetian) of Great Britain (Huene 1907-08, 1914a; Kermack 1984).

Azendohsaurus laaroussii from the Late Triassic (middle Carnian) of Morocco was described by Dutuit (1972) as an ornithischian dinosaur. One tooth crown (Dutuit 1972, fig. c; Galton 1985b) is similar to those of the fabrosaurid *Lesothosaurus diagnosticus* (see Thulborn 1970) in being markedly asymmetrical in mesiodistal views and having from five to seven very coarse marginal serrations on each cutting edge. The teeth in the dentary and a loose tooth (Dutuit 1972, figs. a, b; Galton 1985b) are from a prosauropod because the tooth crowns are symmetrical in mesiodistal views and the marginal serrations are more numerous and finer as in *Thecodontosaurus antiquus*. *A. laaroussii* is presumably a thecodontosaurid in which the central vertical ridge on the tooth crown is more prominent than in other species in the family (assuming that it is not an anchisaurid).

Anchisauridae. Derived characters of this family include the large basisphenoid tubera that project farther ventrally than the very small basipterygoid processes, the proportionally elongate cranial process of the ilium (also in *Ammosaurus major*), the ventral emargination of the proximal part of the pubis, and reduction in size of the first ungual of the pes so it is smaller than the second ungual.

Anchisaurus polyzelus is a small animal (length approximately 2.5 m; Huene 1906, 1914*a*; Lull 1953; Galton 1976*a*) from the Early Jurassic (Pliensbachian or Toarcian) of the Connecticut Valley (U.S.A.). Primitive characters include the jaw articulation of the slenderly built skull being only slightly below the level of the maxillary tooth row. The nasal is small, the lacrimal narrow, the prefrontal thin, and the distal apronlike part of the pubis is proportionally narrow.

Massospondylidae. A derived character of this family is the centrally situated and almost vertical dorsal process of the maxilla. *Massospondylus carinatus* from the Early Jurassic (Hettangian to Pliensbachian) of southern Africa is a medium-sized animal (length approximately 5 m) in which the jaw articulation is only slightly below the level of the maxillary tooth row and the coronoid eminence is low (Cooper 1981*a*). Material referred to *Ammosaurus major* from the Early Jurassic (Pliensbachian or Toarcian) of Arizona (see Galton 1976*a*) may be referable to *Massospondylus*, a complete skull of which is described from Arizona by Attridge et al. (1985).

Yunnanosauridae. Characters of the Yunnanosauridae include weakly spatulate and noticeably asymmetrical maxillary and dentary teeth, the apices of which are directed slightly medially. In addition, there are at most only a few coarse apically directed marginal denticles. It is represented only by *Yunnanosaurus huangi*, a large animal (length approximately 7 m; Young 1942*a*, 1951; Galton 1985*d*) from the Early Jurassic (Hettangian or Pliensbachian) of the People's Republic of China. The skull shows several primitive characters such as a short snout, small external nares, a low dorsal process to maxilla, a narrow lacrimal, a large antorbital opening, and a jaw articulation that is only slightly below the line of the maxillary tooth row. A derived character of the proportionally tall and narrow skull is the long and transversely wide prefrontal (also in *Lufengosaurus* and *Plateosaurus*).

Plateosauridae. This family is characterized by a ventrally offset jaw articulation that is well below the level of the dentary tooth row.

African Plateosauridae: *Euskelosaurus browni* from the Late Triassic (late Carnian or early Norian) of southern Africa has long been referred to the Melanorosauridae (including Cooper 1980), but Heerden (1979) showed that it is a large plateosaurid (length approximately 10 m). Unfortunately, the skull is unknown (the large maxilla described by Seeley 1894*a* is not prosauropod, ?herrerasaurid; Galton 1985*a*). The deltopectoral crest of the humerus of *E. browni* is sigmoidal in cranial view.

South American Plateosauridae: *Coloradisaurus brevis* from the Late Triassic (Norian) of northern Argentina is represented by a deep and wide skull with a snout that is short (Bonaparte 1978*b*). Derived characters include a wide dorsal process and a large medial lamina to the maxilla, a large medial lamina to the lacrimal, and a ventral step in the braincase so the occipital condyle is well above the parasphenoid.

Mussaurus patagonicus from the Late Triassic (Norian) of southern Argentina is represented by an extremely small skeleton (length approximately 25 to 30 cm; Bonaparte and Vince 1979). However, a series of intermediate-sized skeletons link this juvenile to larger individuals (length approximately 3 m) from the same site which were briefly described by Casamiquela (1980) as *Plateosaurus* sp. (Bonaparte, pers. comm.). The diagnosis of *Mussaurus patagonicus* must await a full description of the adult individuals, the skulls of which differ from that of *Coloradisaurus brevis* (Bonaparte, pers. comm.).

European Plateosauridae: *Sellosaurus gracilis* is known from several skeletons (length up to approximately 6.5 m) from the Late Triassic (upper Norian) of West Germany (Huene 1907-08, 1915, 1932; Galton 1973*b*, 1984*b*, 1985*a*, 1985*b*; Galton and Bakker 1985). The skull of *S. gracilis* lacks several of the derived characters of those of *Plateosaurus engelhardti*, so it is no longer referred to that genus (Galton 1985*b*). Derived characters of *S. gracilis* include a medium-sized medial lamina to the maxilla and a medium-sized lateral lamina to the lacrimal.

Plateosaurus engelhardti is the best-represented prosauropod in the world with numerous skeletons from the Late Triassic (late Norian, Rhaetian) of France, East and West Germany, and Switzerland (Huene 1907-08, 1915, 1926*c*, 1932; Galton 1984*a*, 1985*e*, 1986*a*; Weishampel and Westphal 1986). It is large (length up to approximately 9 m), and derived characters include a tall and narrow skull with an elongate snout, the maxilla has a long lateral lamina and a large, triangular-shaped medial lamina, the lacrimal has a large lateral lamina, and the prefrontal is transversely broad and ends close to the postorbital such that the frontal provides only a small portion of the orbital rim. In the braincase, the occipital condyle is above the level of the parasphenoid, and there is a vertical lamina of the basisphenoid between the basipterygoid processes.

North American Plateosauridae: *Ammosaurus major* from the Early Jurassic (Pliensbachian or Toarcian) of Connecticut, U.S.A., is an animal of approximately 4 m, the skull of which is unknown (Huene 1906, 1914*a*; Lull 1953; Galton 1976*a*). Derived characters include the proportionally elongate cranial process to the ilium and the ventral emargination of the proximal part of the ischium. A primitive character is the distal separation of the transverse process and the rib of sacral vertebra 3.

Asian Plateosauridae: *Lufengosaurus huenei* from the Early Jurassic (Hettangian or Pliensbachian) of the People's Republic of China is a medium-sized animal (length approximately 5 m), known from several articulated skeletons (Young 1941*a*, 1941*b*, 1947, 1948*a*, 1951). Derived characters of the skull include very large prefrontals and an expanded top to the dorsal process of the maxilla. Kutty (1969) reports plateosaurids from the Dharmaram Formation (Upper Triassic, Norian) of India similar to those of China, but the material is still to be described.

Melanorosauridae. Members of this family are characterized by an ischial peduncle to the ilium that is only slightly smaller than the pubic peduncle, so this part of the acetabulum is deep. In addition, the femur is straight in craniocaudal view and bears a lesser trochanter in the form of a prominent sheet, and the fourth trochanter is found on the medial edge of the shaft. The teeth of *Riojasaurus* resemble those of plateosaurids (J. F. Bonaparte pers. comm.).

The Melanorosauridae is based on *Melanorosaurus readi* Haughton, 1924, from the Upper Triassic of South Africa. Heerden (1979) regarded *M. readi* as a junior synonym of the sympatric *Euskelosaurus browni* Huxley, 1866, which he transferred from the Melanorosauridae to the Plateosauridae. However, Galton (1985*f*) pointed out that the femur of *Melanorosaurus* resembles that of *Riojasaurus* from the Upper Triassic of Argentina (fig. 15.8D-F; Bonaparte 1972), a taxon that Heerden (1979) accepted as a nonplateosaurid, because it is straight in caudal view with the fourth trochanter close to or on the medial margin of the shaft. In *Euskelosaurus* (Heerden 1979; Galton 1985*f*), the femur is sigmoidal in caudal view with the fourth trochanter well removed from the medial edge of the shaft as in other plateosaurids (e.g., *Plateosaurus*; fig. 15.7C-E). Concerning the femur of *Melanorosaurus*, Heerden (1979) noted that the straightness of the femoral shaft may be due to distortion because the proximal end, which lacks a proper caput femoris, is unlike that found in *Riojasaurus* and similar to *Euskelosaurus*. However, the femur does not appear distorted (see stereo photographs in Heerden 1977, pls. 6−8, and 1979, pls. 64, 65), and the degree of development of the caput femoris varies in different-sized femora of *Riojasaurus* (see Bonaparte 1972, fig. 68). Consequently, I agree with Heerden (1977) in retaining *Melanorosaurus* in the family Melanorosauridae along with *Riojasaurus* (Galton 1985*f*).

Riojasaurus incertus (Late Triassic, Norian, Argentina; Bonaparte 1972), *Camelotia borealis* (Late Triassic, Rhaetian, England; Seeley 1898; Huene 1907-08; Galton 1985*f*), and *Melanorosaurus readi* (Late Triassic, late Carnian or early Norian, South Africa; Haughton 1924; Heerden 1977, 1979) are large animals (length approximately 10 m, 9 m, and 7.5 m, respectively), but only the former is adequately known and the skull is still to be described (J. F. Bonaparte, pers. comm.). The lesser trochanter of the femur of *Riojasaurus incertus* is proportionally smaller than that of *Camelotia borealis* and larger than that of *Melanorosaurus readi*. The length of caudal dorsal and proximal caudal centra exceeds their height in *Riojasaurus incertus*, whereas the reverse is the case for *Camelotia borealis*. Heerden (1977) noted that the postcranial elements referred to the theropod dinosaur *Sinosaurus triassicus* (Early Jurassic, Pliensbachian, People's Republic of China) by Young (1948*b*, 1951) include a mixture of plateosaurid and melanorosaurid elements. He has not elaborated further on this, but if true, it would represent the most recent record of a melanorosaurid. Bonaparte (1986*b*) described a caudal cervical and a cranial dorsal vertebra from the Late Triassic (Norian) of Argentina and referred them to an unnamed melanorosaurid. However, these vertebrae are much more sauropodlike than are those of other melanorosaurids (see above), so these vertebrae probably represent a new family of prosauropods.

Blikanasauridae. The hindlimb in members of this family is extremely stocky. The metatarsus is proportionally very short and broad. The proximal end of metatarsal III and especially metatarsal II is expanded craniocaudally. In addition, the distal tarsals contact metatarsal I to III so the calcaneum contacts metatarsal IV rather than a distal tarsal (fig. 15.8K cf. L). Represented only by *Blikanasaurus cromptoni*, which is based on an incomplete hindlimb of a medium-sized animal (length approximately 5 m; Galton and Heerden 1985) from the Late Triassic (late Carnian or early Norian) of South Africa.

TAPHONOMY

Prosauropods often occur as complete or partial skeletons, but complete skulls are rare. The remains of these animals usually occur in marls, mudstones, and siltstones (e.g., *Euskelosaurus, Lufengosaurus, Massospondylus, Melanorosaurus, Plateosaurus, Riojasaurus, Yunnanosaurus*), in sandstones (e.g., *Anchisaurus, Ammosaurus, Massospondylus, Sellosaurus*), and in eolian sandstones (e.g., *Ammosaurus* from Arizona). Articulated material of *Massospondylus* from Zimbabwe occurs in thin, water-laid sediments within an eolian dominated basin (Raath 1980). The bones may be distorted, and surfaces are commonly sun-cracked (e.g., *Euskelosaurus, Lufengosaurus, Plateosaurus, Yunnanosaurus*). Disarticulated remains of *Euskelosaurus* and *Massospondylus* from South Africa occur in separate thin beds of mudstone (1-3 m and 2 m thick) with numerous calcareous concretions that represent conditions of subaerial exposure with the production of calcareous soil or caliche (Kitching and Raath 1984). Remains of *Thecodontosaurus* from Great Britain occur in Rhaetic fissure fills in Carboniferous limestone as mostly disarticulated bones of adults in a conglomerate (see Huene 1907-08, 1914a) and as articulated bones of juveniles in a stratified marl (see Kermack 1984).

Some prosauropods occur as isolated skeletons (e.g., *Ammosaurus, Anchisaurus, Blikanasaurus, Coloradisaurus, Euskelosaurus, Massospondylus*), whereas others occur as monospecific mass accumulations. Examples of the latter include *Lufengosaurus huenei* (Young 1951), *Mussaurus patagonicus* (Bonaparte and Vince 1979; Casamiquela 1980), *Plateosaurus engelhardti* (Huene 1907-08, 1916, 1928, 1932; Weishampel 1984b; Weishampel and Westphal 1986; Galton 1986a), *Riojasaurus incertus* (Bonaparte 1982a), *Sellosaurus gracilis* (Huene 1907-08, 1932; Galton 1985a), *Thecodontosaurus antiquus* (Huene 1907−08), and *Yunnanosaurus huangi* (Young 1951).

Detailed taphonomic information is available for *Plateosaurus engelhardti*. Included is a map of their occurrence at Halberstadt, East Germany (Jaekel 1913), and an analysis of 1930s quarry data from Trossingen, West Germany (Weishampel 1984b). On the basis of age-class and completeness distributions, Weishampel (1984b) interpreted the lower frequency and less complete remains from two marl bone beds as the result of normal background turnover in an animal that was very common in the region. However, he suggested that a concentration of more complete carcasses in the top 1 m of the lower bone bed indicates catastrophic mortality related to the overlying intermediate bed, a bone-shard dominated layer that represents a rapid mud flow.

PALEOECOLOGY

The earliest prosauropods first occur in the mid-Carnian but they are rare (Galton 1984b). However, in the Lower Elliot Formation (upper Carnian or lower Norian) of South Africa and Lesotho, *Euskelosaurus* is common (Heerden 1979; Kitching and Raath 1984), while *Melanorosaurus* and *Blikanasaurus* are very rare. Other elements of this fauna include a capitosaurid amphibian, a rauisuchian, and a traversodontid therapsid (Olsen and Galton 1984). This paleofauna is the earliest occurrence of the "Prosauropod Empire" of Benton (1983a). The numerical predominance of prosauropod specimens in the fossil record continues for the rest of the Norian and into the Early Jurassic (to the Toarcian).

Massospondylus from the Upper Elliot and Clarens formations shared the scene with primitive ornithischians, primitive theropods, protosuchid and pedeticosaurid crocodilomorphs, tritylodontid, tritheledontid, and diarthrognathid therapsids, and morganucodontid mammals (Olsen and Galton 1984). From data given by Benton (1983a), *Plateosaurus* represents at least 75 percent of the individual animals from the Knollenmergel (upper Norian) of East and West Germany, and *Lufengosaurus* (together with *Yunnanosaurus*) represents 82 percent of the individuals from the upper Lower Lufeng Series of the People's Republic of China. In both cases, prosauropods are by far the largest animals and certainly formed a majority of the original biomass. However, in the Los Colorados Formation (Norian) of Argentina, the large (up to 10 m) melanorosaurid prosauropod *Riojasaurus* constitutes about 40 percent of the collected individuals (Bonaparte 1982a), while the smaller plateosaurid *Coloradisaurus* is rare (Bonaparte 1978b). The remainder of this fauna consists of primitive theropods, stagonolepids, rauisuchians, ornithosuchids, sphenosuchid and protosuchid crocodilomorphs, and pachygenelid therapsids (Bonaparte 1982a).

Prosauropods are usually considered to represent the first radiation of herbivorous dinosaurs (Colbert 1951; Charig et al. 1965; Romer 1966; Galton 1976a; Bonaparte 1982a; Benton 1983a; Weishampel 1984c). The evidence for such a dietary habit is discussed by Galton (1984a, 1985d, 1986b), who also rejects a scavenging predatory diet with occasional cannibalism in these animals (Cooper 1981a).

Part of the reason for the debate concerning the diet of prosauropods is that their masticatory apparatus appears to be poorly adapted for dealing with resistant plant material, especially when compared with ornithischian dinosaurs and other Carnian and Norian herbivores (e.g., chelonians, procolophonids, trilophosaurids, rhynchosaurs, mammallike reptiles such as dicynodonts and tritylodontids, and the very small mammals; Galton 1986b; Crompton and Attridge 1986). Compared to the skulls of most of these herbivores, those of prosauropods are lightly constructed, with noninterdigitating, overlapping, or abutting sutures that tended to separate prior to preservation (see Galton 1985a), there is proportionally much less room for the adductor muscles (Crompton and Attridge 1986), and, apart from *Yunnanosaurus* (fig. 15.4F-H), well-developed tooth-tooth wear facets are not present on the teeth. However, the cranial characters of prosauropods clearly agree with those of herbivorous reptiles rather than carnivorous reptiles (Galton 1984a, 1985d, 1986b). The jaw articulation is set below the line of the tooth row, and the amount of ventral offset ranges from small in *Anchisaurus, Massospondylus,* and *Thecodontosaurus* to quite large in *Coloradisaurus, Plateosaurus,* and *Lufengosaurus* (figs. 15.2A, 15.3). Similarly, aetosaurs (Charig et al. 1976) and ornithischian dinosaurs (Galton 1973c) are two other groups of herbivorous reptiles with a ventral jaw articulation. It is true that transverse movement of the mandibles was not possible in prosauropods, but two functions proposed for the dorsally offset articulation of herbivorous mammals that do not involve transverse movements also apply to the ventrally offset jaw articulation of prosauropods. First, offsetting increases the angle between the lever arm of the bite force and the plane of the teeth, which is important in dealing with resistant plant material (Crompton and Hiiemae 1969). Second, it allows a more even distribution of the biting force by ensuring that the tooth rows are almost parallel at occlusion, so that contact is made along the complete length of the tooth row by a "nutcracker"-like action (Colbert 1951:89).

The coarsely serrated and spatulate crowns of the maxillary and dentary teeth of most prosauropods are transversely compressed and mesiodistally expanded, such that the maximum width of the crown is greater than that of the root (fig. 15.4D,E). This crown shape is nearly identical to that of herbivorous lizards of the families Iguanidae and Agamidae. Such an expansion and consequent reduction of the space between adjacent teeth results in a more nearly continuous cutting edge than occurs in insectivorous and carnivorous lizards (Hotton 1955; Ray 1956; Montanucci 1968; Throckmorton 1976; Auffenberg 1981). The teeth of most prosauropods also resemble those of *Iguana iguana* in the *en echelon* arrangement of the maxillary and dentary tooth crowns that bear prominent serrations set at 45° to the cutting edges (Throckmorton 1976). The edges of the flat and obliquely inclined wear surface on the maxillary tooth of *Yunnanosaurus* (fig. 15.4F-H) were self-sharpened by wear against the corresponding surface on the opposing dentary tooth much like that in the sauropod *Brachiosaurus* (Janensch 1935−36). Consequently, the series of relatively coarse, 45° inclined marginal denticles seen in other prosauropods was no longer necessary for cutting plant material, and vestiges of this system are retained on only a few teeth in *Yunnanosaurus* (Young 1942a, 1951; Galton 1985d).

The lack of wear facets on the teeth of most prosauropods indicates that the teeth did not come into contact with each other, so there was no tooth-tooth occlusion. Throckmorton (1976) showed that the teeth of *Iguana iguana* also lack wear facets and that they are used only to bite off a piece of a plant and not for oral processing. Unlike lizards, prosauropods probably had cheeks, so food was retained, and an extensive soft secondary palate in *Plateosaurus* may have allowed breathing and chewing at the same time. Oral processing probably involved a piercing and puncturing action of the teeth in most prosauropods. This also produced the tooth-food wear surfaces on the teeth of *Yunnanosaurus* in which more extensive chewing also produced tooth-tooth wear surfaces.

Once the plant food was ingested, a gastric mill provided further mechanical breakdown of the plant material. Indeed, a well-preserved gastric mill, consisting of a concentrated mass of small stones, has been found associated with the stomach contents of several specimens of *Massospondylus* (Bond 1955; Raath 1974) and one of *Sellosaurus* (Huene 1932; Galton 1973b). Modern birds, many of which are herbivorous, lack teeth but use gastroliths for the mechanical breakdown of the food (Welty 1975).

Because of their relative lack of oral processing compared to most ornithischian dinosaurs and their reliance on mechanical breakdown of food in the gizzard, prosauropods probably had low turnover times in the fermentation chamber for the digestion of poor quality plant material, although it is possible that they opted for high passage rates and concentrated on the most nutritive components of the food (Farlow 1987). If so, they probably had long large intestines and short small intestines, partitions across the colon, and a symbiotic microflora in a hindgut fermentation chamber to degrade dietary fiber as in modern herbivorus reptiles (Guard 1980; Iverson 1980, 1982; Farlow 1987*b*). The gastrointestinal situation for prosauropods is unknown, but the barrellike rib cage (see photos in Huene 1932; Weishampel and Westphal 1986) and broad, apronlike pubis indicate a capacious digestive tract.

Osmólska (1979) pointed out that the external nostrils are enlarged in most herbivorous dinosaurs including prosauropods such as *Plateosaurus*. She suggested that the nostrils housed a large lateral nasal gland that functioned as a salt gland to unload the excess potassium ions in the food of large herbivores. However, Whybrow (1981) questioned whether there was sufficient room for the nostrils to house the salt gland and its associated duct system and still perform a respiratory function.

Bakker (1978) suggested that prosauropods could assume a tripodal feeding posture in order to browse at high levels. However, as Coe et al. (1987) noted, elephants can easily stand on their hind legs but do not usually feed in this position. The long neck certainly extended the vertical feeding range so that higher vegetation could be reached, much like the situation in giraffes. Food could not have been conveyed directly to the mouth by the manus. However, the manus was probably used to hold onto branches to assist balancing while the animal reared up on its hindlimbs to reach higher levels of vegetation. In addition to foliage, prosauropods probably fed on fleshy nilssonialian fruits and possibly on bennettitalian inflorescences at lower levels (0–1 m) and, as the only Triassic browsers capable of feeding at higher levels (1–3 m), on lycopsid fructifications as well (Weishampel 1984*c*).

Among prosauropods, *Thecodontosaurus* is considered fully bipedal. The forelimbs are about half the length of the hindlimbs as in other prosauropods, but the hindlimbs are proportionally much longer (fig. 15.1A). The trunk to hindlimb ratio is 1 : 1.6 in *Thecodontosaurus* in comparison with other fully bipedal dinosaurs (1 : 1.22 to 1 : 1.9; Galton 1970*a*). The remaining prosauropods were probably only facultatively bipedal (trunk-hindlimb ratios ranging from 1 : 0.95 to 1.15). However, *Riojasaurus* (and other melanorosaurids, ?blikanasaurids) must have been fully quadrupedal (trunk-hindlimb ratio of 1 : 0.71; cf. ratios of 1 : 0.69–1 : 0.9 in undoubtedly quadrupedal dinosaurs; Galton 1970*a*). The increase in the proportional length of the trunk provided storage space for the elongate viscera necessary for a herbivorous diet.

In nonmelanorosaurid prosauropods, the hindlimb was held more upright than in "sprawlers," but the femur was probably at an angle of about 20° to the parasagittal plane with the head pointing craniomedially so the parasagittal tibia could be moved craniocaudally (Cooper 1981*a*; 30–45° according to Heerden 1979) and probably also in *Yunnanosaurus* (see Young 1942*a*, 1951). Cooper (1981*a*, 1981*b*) suggested that during retraction the femur was twisted so the head faced medially and perforated the open acetabulum. The femur was probably much closer to the vertical in melanorosaurids, in which the head is more medially set and the bone is straighter in cranial and caudal views (Bonaparte 1972; Galton 1985*f*). However, even in mammals a fully erect limb posture is only present in cursorial and graviportal mammals, and the femur is found at an angle of 20°–50° to the parasagittal plane in many other mammals (Jenkins 1971).

The locomotory capabilities of dinosaurs can be estimated by comparing plots of the hindlimb proportions (tibia : femur/metatarsal III : tibia) with those of living mammals (see Coombs 1978*c*). On this basis, small to medium-sized prosauropods such as thecodontosaurids, anchisaurids, massospondylids, *Ammosaurus*, and *Sellosaurus* were low-grade subcursorial runners, and large prosauropods such as *Lufengosaurus*, *Plateosaurus*, yunnanosaurids, and melanorosaurids were mediportal. Consequently, prosauropods were probably the slowest of the bipedal dinosaurs but better runners than most other quadrupedal dinosaurs (Coombs 1978*c*).

The constancy of the structure of the prosauropod manus correlates well with the development of an enormous, trenchant first ungual that was used for offense or defense while the animal was in a bipedal stance (Galton 1971*a*, 1971*b*). With the first digit in full extension, the ungual would clearly have been a formidable weapon (fig. 15.6H,L–O). During quadrupedal locomotion, with the digits of the manus in full extension, the weight of the forequarters was taken by digits II to IV (in particular, digit II and III), while the first ungual was held clear of the ground (fig. 15.6H). Only if the substrate was irregular or soft would the lateral surface of this phalanx have touched the ground, and even then the point of the claw would not have been damaged. Such contact is shown by the trackway of *Navahopus falcipollex* which Baird (1980) shows was made by a prosauropod walking quadrupedally. The first ungual was usually held clear of the ground as a result of two specializations of the first digit (fig. 15.6I–K). First, the ginglymus of metacarpal I extends well above the dorsum of the shaft and the articular surface extends proximally onto the dorsal surface so that the first phalanx was capable of marked hyperextension. Second, the first phalanx is twisted along its length (45° between the rotational axes). In full extension, the ungual phalanx is directed ventromedially. During flexion, the first digit of the prosauropod manus rotates laterally around its long axis so that the first ungual comes to lie more or less parallel with the unguals of digits II and III. This rotation results from the asymmetrical structure of the ginglymus of metacarpal I, the lateral condyle of which is larger than the medial condyle (fig. 15.6J). When flexed, the prosauropod manus must have acted as an efficient grasping organ, particularly during bilateral use of the hands.

The main variation in the structure of the prosauropod manus is that metacarpal I is more slender in *Thecodontosaurus* and metacarpals II to IV are more slender in *Anchisaurus* and *Thecodontosaurus*, in which the metatarsals are also more slender. Most prosauropods were quite large, and the more massive nature of the metacarpals was a graviportal adaptation that was unnecessary in the smaller forms.

16

Sauropoda

J. S. McINTOSH

INTRODUCTION

The sauropod dinosaurs (Coombs 1975; Dodson, this vol.) were gigantic quadrupedal herbivores, up to thirty meters long and 80,000 kg in weight (fig. 16.1). They appeared in the earliest Jurassic and survived until the end of the Cretaceous (table 16.1) and were the dominant terrestrial herbivores during most of that period. They were characterized by small skulls and brains, long necks and tails, and sturdy upright limbs. Some of the Late Cretaceous forms possessed dermal armor. Eggs found in close proximity to sauropod remains suggest that these animals were oviparous.

Any overall review of the sauropods is complicated by factors not applicable to most other groups of dinosaurs. There the major diagnostic characters are usually found in the skulls and perhaps the degree of specialization of the fore and hind feet. Partly because the skulls are unknown in so many important genera and partly because the need for lightening of the skeleton of these huge animals has resulted in a number of widely divergent types of highly specialized vertebrae, it is the latter elements that are often most useful for generic identification. But almost any part of the skeleton may be diagnostic in individual cases. The result has been the introduction of nearly 90 different taxa at the generic level and over 150 at the specific. The majority of these are based on small parts of the skeleton which may be diagnostic when compared to similar elements of other taxa but obviously cannot be compared to taxa based on different elements. Virtually complete skeletons are known for only five genera. Fourteen species are based solely on teeth, 25 on one or more vertebrae, and 6 on a single limb bone. Consequently, there are only about a dozen genera that may be said to be firmly established and perhaps another dozen "fairly well established." In this group of two dozen, skulls are known in less than half. In a certain sense, all the other generic taxa might be termed *nomina dubia*, but that judgment would be far too sweeping. Many of these are clearly separable from the "established genera" but not from one another. A further complication arises from the fact that until very recently virtually all the well-established genera were from an almost identical age—the latestmost Jurassic. Recent discoveries in Argentina, Morocco, Madagascar, and Texas but primarily in China provide hope that when this material is adequately described and figured, a much more coherent picture of the relationships of the various sauropod families will emerge. At this writing, only some of the Chinese and Argentine material has been described in any detail, and even most of these descriptions are preliminary with few figures.

In this review, the characters separating the different species of a given genus are not discussed, both for the sake of brevity and also because only in the more completely known Morrison and Tendaguru genera is it currently possible to attempt any sort of meaningful species analysis. Thus, the table of species (table 1) is quite uneven in that for all those better-

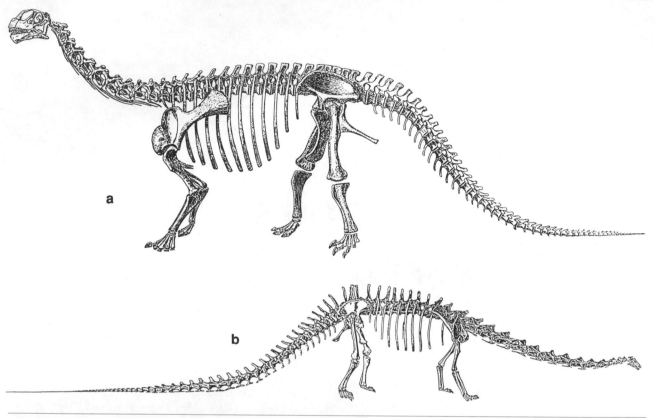

a

b

Fig. 16.1. Skeletal reconstruction. a) *Camarasaurus lentus,* (after Gilmore 1925); b) *Diplodocus carnegii* (After Hatcher 1901.)

TABLE 16.1 Sauropoda

	Occurrence	Age	Material
Sauropodomorpha Huene, 1932 **Sauropoda** Marsh, 1878b **Vulcanodontidae** Cooper, 1984 ***Vulcanodon*** Raath, 1972			
V. *karibaensis* Raath, 1972	Vulcanodon Beds, Mashonaland North, Zimbabwe	?Hettangian	Partial skeleton, scapula
Barapasaurus Jain, Kutty, Roy-Chowdhury, et Chatterjee, 1975			
B. *tagorei* Jain, Kutty, Roy- Chowdhury, et Chatterjee, 1975	Kota Formation, Andhra Pradesh, India	Early Jurassic	Scattered remains of more than 6 partial skeletons without skulls, manus, or pedes
Ohmdenosaurus Wild, 1978			
O. *liasicus* Wild, 1978	Posidonienschiefer, Baden-Württemburg, Federal Republic of Germany	middle Toarcian	Tibia, tarsus
Zizhongosaurus Dong, Zhou, et Zhang, 1983			
Z. *chuanchengensis* Dong, Zhou, et Zhang, 1983	Da'znzhai Formation, Sichuan, People's Republic of China	Early Jurassic	Dorsal vertebra, humerus, pubis
Kotasaurus Yadagiri, 1988 •			
K. *yamanpalliensis* Yadagiri, 1988 •	Kota Formation, Andhra Pradesh, India	Early Jurassic	Nearly complete skeleton without skull
Cetiosauridae Lydekker, 1888b **Cetiosaurinae** Janensch, 1929b ***Cetiosaurus*** Owen, 1841			
C. *medius* Owen, 1842b (including *C. hypoolithicus* Owen, 1842a)	Inferior Oolite, West Yorkshire, England	Bajocian	Caudal vertebrae, additional vertebrae, fragmentary limb elements
C. *oxoniensis* Phillips, 1871 (including *C. giganteus* Owen [ex Huxley 1870])	Great Oolite, Nottinghamshire, Northamptonshire, Chipping Norton Formation, Great Oolite, Forest Marble, Gloucestershire, Chipping Norton Formation, Forest Marble, Oxfordshire, Great Oolite, Buckinghamshire, Great Oolite, Wiltshire, England	Bathonian	2 partial skeletons without skulls, isolated vertebrae, limb elements

Table 16.1, continued

C. mogrebiensis Lapparent, 1955	Guettioua Sandstone, Beni Mellal, Morocco	late Bathonian	Skeleton with skull, 6 partial skeletons
Bellusaurus Dong, 1990 •			
B. sui Dong, 1990 •	Wucaiwang Formation, Xinjiang Uygur Ziziqu, People's Republic of China	Middle-Late Jurassic	17 partial skeletons, juvenile
Haplocanthosaurus Hatcher, 1903b (=*Haplocanthus* Hatcher, 1903a preoccupied)			
H. priscus (Hatcher, 1903a) (= *Haplocanthus priscus* Hatcher, 1903a, including *Haplocanthosaurus utterbacki* Hatcher, 1903a)	Morrison Formation, Colorado, Wyoming, United States	Kimmeridgian-Tithonian	2 partial skeletons lacking skulls
H. delfsi McIntosh et Williams, 1988	Morrison Formation, Colorado, United States	Kimmeridgian-Tithonian	Partial skeleton without skull
Amygdalodon Cabrera, 1947			
A. patagonicus Cabrera, 1947	Cerro Carnerero Formation, Chubut, Argentina	Bajocian	Partial skeleton
Patagosaurus Bonaparte, 1979b			
P. fariasi Bonaparte, 1979b	Cañodon Asfalto Formation, Chubut, Argentina	Callovian	As many as 12 skeletons with jaw elements
Shunosaurinae nov.			
Shunosaurus Dong, Zhou, et Zhang, 1983			
S. lii Dong, Zhou, et Zhang, 1983	Xiashaximiao Formation, Sichuan, People's Republic of China	Bathonian-Callovian	As many as 20 skeletons with 5 skulls
Datousaurus Dong et Tang, 1984			
D. bashanensis Dong et Tang, 1984	Xiashaximiao Formation, Sichuan, People's Republic of China	Bathonian-Callovian	2 partial skeletons without skulls, ?jaws
Rhoetosaurus Longman, 1925			
R. brownei Longman, 1925	?Injune Creek Beds, Queensland, Australia	?Bajocian	Partial skeleton
Omeisaurus Young, 1939 (= *Zigongosaurus* Hou, Chao, et Chu, 1976)			
O. junghsiensis Young, 1939 (= *Zigongosaurus fuxiensis* Hou, Chao, et Chu, 1976)	Xiashaximiao Formation, Sichuan, People's Republic of China	Bathonian-Callovian	Several partial skeletons
O. changshouensis Young, 1958b	Shangshaximiao Formation, Sichuan, People's Republic of China	Late Jurassic	Partial skeleton
O. fuxiensis Dong, Zhou, et Zhang, 1983	Shangshaximiao Formation, Sichuan, People's Republic of China	Late Jurassic	Basioccipital, maxillae and mandibles
O. tianfuensis He, Li, Cai, et Gao, 1984	Xiashaximiao Formation, Sichuan, People's Republic of China	Bathonian-Callovian	Skeleton with partial skull
O. luoquanensis Li, 1988 •	Xiashaximiao Formation, Sichuan People's Republic of China	Bathonian-Callovian	Parial skeleton
Cetiosauridae incertae sedis			
Unnamed cetiosaurid (= "*Morosaurus*" *agilis* Marsh, 1889a)	Morrison Formation, Colorado, United States	Kimmeridgian-Tithonian	Cranium, 3 cervicals
Unnamed cetiosaurid (= "*Apatosaurus*" *minimus* Mook, 1917a)	Morrison Formation, Wyoming, United States	Kimmeridgian-Tithonian	Sacrum, pelvis
Brachiosauridae Riggs, 1904			
Brachiosaurus Riggs, 1903a (= *Giraffotitan* Paul, 1988 •)			
B. altithorax Riggs, 1903a (= *Giraffotitan altithorax* Paul, 1988)	Morrison Formation, Colorado, Utah, United States	Kimmeridgian-Tithonian	Partial skeleton from 2 individuals
B. brancai Janensch, 1914 (including *B. fraasi* Janensch, 1914)	Tendaguru Beds, Mtwara, Tanzania	Kimmeridgian	5 partial skeletons more than 3 skulls, isolated limb elements
B. atalaiensis Lapparent et Zbyszewski, 1957	unnamed unit, Estremadura, Portugal	Kimmeridgian	Vertebrae, ischia, pubes, forelimb and hindlimb elements
B. nougaredi Lapparent, 1960a	"Continental intercalaire", Wargla, Algeria	Albian	Sacrum, forelimb elements
Bothriospondylus Owen, 1875 (?= *Marmarospondylus* Owen, 1875)			
B. suffosus Owen, 1875	Kimmeridge Clay, Wiltshire, England	early Kimmeridgian	7 dorsal and sacral centra

Table 16.1, continued

B. robustus Owen, 1875 (= *Marmarospondylus robustus* Owen, 1875)	Forest Marble, Wiltshire, England	late Bathonian	Dorsal vertebra
B. madagascariensis Lydekker, 1895	Isalo Formation, Majunga, Madagascar	Bathonian	Isolated vertebrae and limb elements of at least 10 individuals
Lapparentosaurus Bonaparte, 1986b			
L. madagascariensis Bonaparte, 1986b	Isalo Formation, Majunga, Madagascar	Bathonian	More than 4 partial skeletons lacking skulls, teeth
Volkheimeria Bonaparte, 1979b			
V. chubutensis Bonaparte, 1979b	Cañodon Asfalto Formation, Chubut, Argentina	Callovian	Partial skeleton consisting of presacral and sacral vertebrae, pelvis, hind limb
Pelorosaurus Mantell, 1850 (= *Ornithopsis* Seeley, 1870, *Dinodocus* Owen, 1884, *?Oplosaurus* Gervais, 1852)			
P. conybearei (Melville, 1849) (= *Cetiosaurus conybeari* Melville, 1849, including *Ornithopsis hulkei* Seeley, 1870, *Bothriospondylus elongatus* Owen, 1875, *B. magnus* Owen, 1875, *?Ornithopsis eucamerotus* Hulke, 1882a, *?Oplosaurus armatus* Gervais, 1852)	Wealden, Isle of Wight, West Sussex, East Sussex, Kent, England	Valanginian- Barremian	Humerus, caudal vertebrae, sacrum, pelvis, isolated dorsal vertebrae
P. mackesoni (Owen, 1884) (= *Dinodocus mackesoni* Owen, 1884)	Lower Greensand, Kent, England	Aptian	Complete and partial limb elements
Pleurocoelus Marsh, 1888a (=*?Astrodon* Johnston, 1859)			
P. nanus Marsh, 1888a	Arundel Formation, Maryland, Paluxy Formation, Glen Rose Formation, Texas, United States	Hauterivian- Barremian	Isolated remains of more than 6 individuals, including skull elements
P. altus Marsh, 1888a (?including *Astrodon johnstoni* Leidy, 1865)	Arundel Formation, Maryland, United States	Hauterivian- Barremian	Tibia, fibula
P. valdensis Lydekker, 1889e	Wealden, East Sussex, Isle of Wight, Englan	Valanginian- Barremian	Teeth, dorsal and caudal centra
Ultrasauros Jensen, 1985b •			
U. macintoshi Jensen, 1985b •	Morrison Formation, Colorado, United States	Kimmeridgian- Tithonian	Dorsal vertebra, scapulocoracoid
Dystylosaurus Jensen, 1985b			
D. edwini Jensen, 1985b	Morrison Formation, Colorado, United States	Kimmeridgian- Tithonian	Dorsal vertebra
Chubutisaurus Corro, 1974			
C. insignis Corro, 1974	Gorro Frigio Formation, Chubut, Argentina	Albian	2 partial skeletons including most limb elements and caudal vertebrae
Brachiosauridae incertae sedis			
Unnamed brachiosaurid (= "*Cetiosaurus*" *humerocristatus* Hulke, 1874)	Kimmeridge Clay, Cambridgeshire, Dorset, England; unnamed unit, Pas-de-Calais, France	Kimmeridgian	Humerus
Ischyrosaurus Hulke, 1874			
I. manseli (Hulke [ex Lydekker, 1888b]) (= *Ornithopsis manseli* Hulke (ex Lydekker, 1888b))	Kimmeridge Clay, Dorset, Cambridgeshire, England	Kimmeridgian	Humerus
Unnamed brachiosaurid (= "*Ornithopsis*" *leedsi* Hulke, 1887)	Kimmeridge Clay, Northamptonshire, England	Kimmeridgian	Pelvis
Camarasauridae Cope, 1877d			
Camarasaurinae Nopcsa, 1928c			
Camarasaurus Cope, 1877e (= *Caulodon* Cope, 1877c, *Morosaurus* Marsh 1878a, *Uintasarus* Holland, 1919)			
C. supremus Cope, 1877e (including *C. leptodirus* Cope, 1879, *Amphicoelias latus* Cope, 1877d, *Caulodon diversidens* Cope, 1877c, *Caulodon leptoganus* Cope, 1878d)	Morrison Formation, Wyoming, Colorado, New Mexico, United States	Kimmeridgian- Tithonian	At least 5 partial skeletons, including braincase and jaws
C. grandis (Marsh, 1877b) (= *Apatosaurus grandis* Marsh, 1877b, including *Morosaurus impar* Marsh, 1878a, *M. robustus* Marsh, 1878b, *Pleurocoelus montanus* Marsh, 1896)	Morrison Formation, Wyoming, Colorado, Montana, United States	Kimmeridgian- Tithonian	At least 6 partial skeletons, including 2 skulls, hundreds of postcranial elements
C. lentus (Marsh, 1889a) (= *Morosaurus lentus* Marsh, 1889a, including *C. annae* Ellinger, 1950, *Uintasaurus douglassi* Holland, 1919)	Morrison Formation, Wyoming, Utah, United States	Kimmeridgian- Tithonian	5 skeletons with skulls, hundreds of postcranial elements

Table 16.1, continued

C. *alenquerensis* (Lapparent et Zbyszewski, 1957) (= *Apatosaurus alenquerensis* Lapparent et Zbyszewski, 1957)	unnamed unit, Estremadura, Portugal	Oxfordian-Kimmeridgian	Partial skeleton lacking skull and cervical series, other material including caudal series, isolated material referred to 8 additional individuals
C. *lewisi* (Jensen, 1988) (= *Cathetosaurus lewisi* Jensen, 1988 •)	Morrison Formation, Colorado, United States	Kimmeridgian-Tithonian	Nearly complete postcranial skeleton
Aragosaurus Sanz, Buscalioni, Casanovas, et Santafe, 1987			
A. *ischiaticus* Sanz, Buscalioni, Casanovas, et Santafe, 1987	Las Zabacheras Beds, Teruel, Spain	early Barremian	Caudal vertebrae, scapula, fore limb, ischium, pubis
Euhelopus Romer, 1956 (= *Helopus* Wiman, 1929, preoccupied)			
E. *zdanskyi* (Wiman, 1929) (= *Helopus zdanskyi* Wiman, 1929)	Meng-Yin Formation, Shandong, People's Republic of China	?Kimmeridgian	Skull and partial postcranial skeleton, additional fragmentary skeleton
Tienshanosaurus Young, 1937			
T. *chitaiensis* Young, 1937	Shishugou Formation, Xinjiang, Weiwuer Ziziqu, People's Republic of China	Oxfordian	Partial postcranial skeleton
Opisthocoelicaudiinae nov.			
Opisthocoelicaudia Borsuk-Bialynicka, 1977			
O. *skarzynskii* Borsuk-Bialynicka, 1977	Nemegt Formation, Omnogov, Mongolian People's Republic	?late Campanian or early Maastrichtian	Skeleton lacking skull and cervical series
Chondrosteosaurus Owen, 1876b			
C. *gigas* Owen, 1876b (?including *Regnosaurus northamptoni* Mantell, 1848)	Wealden, Isle of Wight, England	Hauterivian-Barremian	Several cervical vertebrae
Diplodocidae Marsh, 1884d			
Diplodocinae Janensch, 1929b			
Diplodocus Marsh, 1878b			
D. *longus* Marsh, 1878b	Morrison Formation, Colorado, Utah, United States	Kimmeridgian-Tithonian	2 skulls, caudal series
D. *carnegii* Hatcher, 1901	Morrison Formation, Wyoming, Utah, United States	Kimmeridgian-Tithonian	5 skeletons without skull or manus, 2 skulls, hundreds of isolated postcranial elements
D. *hayi* Holland, 1924b	Morrison Formation, Wyoming, United States	Kimmeridgian-Tithonian	Partial skeleton with braincase
D. *lacustris* Marsh, 1884d	Morrison Formation, Colorado, United States	Kimmeridgian-Tithonian	Jaw with teeth
Barosaurus Marsh, 1890 (?= *Gigantosaurus* Fraas, 1908)			
B. *lentus* Marsh, 1890b (including *B. affinis* Marsh, 1899)	Morrison Formation, South Dakota, Utah, United States	Kimmeridgian-Tithonian	5 partial skeletons without skulls, isolated limb elements
B. *africanus* (Fraas, 1908) (= *Gigantosaurus africanus* Fraas, 1908)	Upper Tendaguru Beds, Mtwara, Tanzania	Kimmeridgian	More than 3 partial skeletons, a few skull elements, many isolated postcranial elements
B. *gracilis* Janensch, 1961	Tendaguru Beds, Mtwara, Tanzania	Kimmeridgian	Many isolated limb elements
Apatosaurus Marsh, 1877b (= *Brontosaurus* Marsh, 1879c, *Elosaurus* Peterson et Gilmore, 1902)			
A. *ajax* Marsh, 1877b (including *A. laticollis* Marsh, 1879a, *Atlantosaurus immanis* Marsh, 1878a)	Morrison Formation, Colorado, United States	Kimmeridgian-Tithonian	2 partial skeletons, braincase
A. *excelsus* (Marsh, 1879c) (= *Brontosaurus excelsus* Marsh, 1879c, including *B. amplus* Marsh, 1881b, *Elosaurus parvus* Peterson et Gilmore, 1902)	Morrison Formation, Wyoming, Utah, Oklahoma, United States	Kimmeridgian-Tithonian	6 partial skeletons without skulls, hundreds of postcranial elements
A. *louisae* Holland, 1915b	Morrison Formation, Utah, United States	Kimmeridgian-Tithonian	2 skeletons, one with skull, 3rd partial skeleton, isolated limb elements
Amphicoelias Cope, 1877d			
A. *altus* Cope, 1877d (including *A. fragillimus* Cope, 1878c)	Morrison Formation, Colorado, United States	Kimmeridgian-Tithonian	Dorsal vertebrae, pubis, femur
Supersaurus Jensen, 1985b			
S. *vivianae* Jensen, 1985b	Morrison Formation, Colorado, United States	Kimmeridgian-Tithonian	Cervical vertebra, scapulocoracoid, ischium, proximal caudal vertebra
Dystrophaeus Cope, 1877b			
D. *viaemalae* Cope, 1877b	Morrison Formation, Utah, United States	Kimmeridgian-Tithonian	Scapula, ulna, partial manus

Table 16.1, continued

Cetiosauriscus Huene, 1927			
C. *stewarti* Charig, 1980	Oxford Clay, Cambridgeshire, England	Callovian	Rear half of skeleton, dorsal series, caudal series
C. *longus* (Owen, 1842b) (= *Cetiosaurus longus* Owen, 1842b, including *Cetiosaurus epioolithicus* Owen, 1842a)	Inferior Oolite, West Yorkshire, England	Bajocian	Caudal vertebrae
C. *glymptonensis* (Phillips, 1871) (= *Cetiosaurus glymptonensis* Phillips, 1871)	Forest Marble, Northamptonshire, England	late Bathonian	Caudal vertebrae
C. *greppini* (Huene, 1922) (= *Ornithopsis greppini* Huene, 1922)	Unter-Virgula-Schichten, Bern, Switzerland	?Tithonian	At least 4 partial skeletons without skulls
Seismosaurus Gillette, 1991 •			
S. *halli* Gillette, 1991 •	Morrison Formation, New Mexico, United States	Kimmeridgian-Tithonian	Pelvis, partial vertebral column
Dicraeosaurinae Janensch, 1929b			
Dicraeosaurus Janensch, 1914			
D. *hansemanni* Janensch, 1914	Middle Tendaguru Beds Mtwara, Tanzania	Kimmeridgian	Skeleton lacking skull and fore limbs, 2 partial skeletons, isolated vertebrae and limb elements
D. *sattleri* Janensch, 1914	Upper Tendaguru Beds, Mtwara, Tanzania	Kimmeridgian	2 partial skeletons without skulls, isolated postcranial remains
Nemegtosaurus Nowinski, 1971			
N. *mongoliensis* Nowinski, 1971	Nemegt Formation, Omnogov, Mongolian People's Republic	?late Campanian or early Maastrichtian	Skull
Quaesitosaurus Kurzanov et Bannikov, 1983			
Q. *orientalis* Kurzanov et Bannikov, 1983	"Barungoyotskaya" Svita, Omnogov, Mongolian People's Republic	late Santonian or early Campanian	Partial skull
Rebbachisaurus Lavocat, 1954			
R. *garasbae* Lavocat, 1954	Tegana Formation, Ksar-es-Souk,. Morocco	Albian	Dorsal vertebrae, scapula, humerus, sacrum
R. *tamesnensis* Lapparent, 1960a	"Continental intercalaire", Adrar, Tamenghest, Wargla, Algeria; "Continental intercalaire", Medenine, Tunisia; "Continental intercalaire", Agadez, Farak Formation, Tahoua, Niger	Albian	4 teeth, abundant isolated postcranial elements
Mamenchisaurinae Young et Chao, 1972			
Mamenchisaurus Young, 1954			
M. *constructus* Young, 1954	Shangshaximiao Formation, Sichuan, People's Republic of China	Late Jurassic	Partial postcranial skeleton
M. *hochuanensis* Young et Chao, 1972	Shangshaximiao Formation, Sichuan, People's Republic of China	Late Jurassic	4 partial skeletons without skulls
Titanosauridae Lydekker, 1885			
Titanosaurus Lydekker, 1877			
T. *indicus* Lydekker, 1877	Lameta Formation, Madhya Pradesh, Maharashtra, Aviyalur Group, Tamil Nadu, India; Grès à Reptiles, Var, France; unnamed unit, Lleida, Spain	middle-late Maastrichtian	Isolated postcranial material
T. *blanfordi* Lydekker, 1879	Lameta Formation, Madhya Pradesh, Maharashtra, India	middle-late Maastrichtian	Caudal vertebrae
T. *madagascariensis* Deperet, 1896	Grès de Maevarano, Majunga, Madagascar; Lameta Formation, Maharshtra, India	Campanian	2 caudal vertebrae, scute
T. *falloti* Hoffet, 1942	unnamed unit, Khoueng, Laos	Coniacian-Maastrichtian	Caudal vertebrae, femur
Magyarosaurus Huene, 1932			
M. *dacus* (Nopcsa, 1915) (= *Titanosaurus dacus* Nopcsa, 1915)	Sinpetru Beds, Hunedoara, Romania	late Maastrichtian	Isolated postcranial material of at least 10 individuals
M. *transsylvanicus* Huene, 1932	Sinpetru Beds, Hunedoara, Romania	late Maastrichian	Isolated postcranial material
M. *hungaricus* Huene, 1932	Sinpetru Beds, Hunedoara, Romania	late Maastrichtian	Fibula

Table 16.1, continued

Laplatasaurus Huene, 1929a			
L. araukanicus Huene, 1929a	Castillo Formation, Bajo Barreal Formation, Laguna Palacios Formation, Rio Negro, Los Blanquitos, Formation, Salta, Argentina; Asencio Formation, Palmitas, Uruguay	Campanian-Maastrichtian	Caudal series, isolated limb elements
Saltasaurus Bonaparte et Powell, 1980 (= *Microcoelus* Lydekker, 1893a, *?Loricosaurus* Huene, 1929a)			
S. loricatus Bonaparte et Powell, 1980	Lecho Formation, Salta, Argentina	?late Campanian-Maastrichtian	Partial skeletons of approximately 6 individuals, including jaws and armor
S. australis (Lydekker, 1893a) (= *Titanosaurus australis* Lydekker, 1893a, including *Titanosaurus nanus* Lydekker, 1893a, *Microcoelus patagonicus* Lydekker, 1893a, *?Loricosaurus scutatus* Huene, 1929a)	Rio Colorado Formation or Allen Formation, Rio Negro, Argentina; Asencio Formation, Palmitas, Uruguay	early Maastrichtian	Vertebrae and limb elements of at least 10 individuals, skull elements
S. robustus (Huene, 1929a) (= *Titanosaurus robustus* Huene, 1929a)	Rio Colorado Formation or Allen Formation, Rio Negro, Argentina	early Maastrichtian	Vertebrae and limb elements of approximately 3 individuals
Aegyptosaurus Stromer, 1932			
A. baharijensis Stromer, 1932	Baharija Formation, Marsa Matruh, Egypt	?early Cenomanian	Partial postcranial skeleton
Alamosaurus Gilmore, 1922			
A. sanjuanensis Gilmore, 1922	Upper Kirtland Shale New Mexico, North Horn Formation, Utah, Javelina Formation, El Picacho Formation, Texas, United States	Maastrichtian	Partial postcranial skeleton, isolated postcranial remains
Macrurosaurus Seeley, 1896			
M. semnus Seeley, 1869 (including *Acantholpholis platypus* Seeley, 1869 *partim*, *?Titanosaurus lydekkeri* Huene, 1929)	Cambridge Greensand, Cambridgeshire, Wealden, Isle of Wight, England	Valanginian, late Albian	Isolated postcranial elements
Hypselosaurus Matheron, 1869			
H. priscus Matheron, 1869	Grès de Labarre, Ariège, Grès de Saint-Chinian, Couches de Rognac, Bouches-du-Rhône, Grès à Reptiles, Var, France	Maastrichtian	Isolated postcranial remains of at least 10 individuals
Argyrosaurus Lydekker, 1893a			
A. superbus Lydekker, 1893a	Castillo Formation, Bajo Barreal Formation, Laguna Palacios Formation, Rio Negro, Laguna Palacios Formation, Bajo Barreal Formation, Chubut, unnamed unit, Santa Cruz, Argentina; Asencio Formation, Uruguay	Campanian-Maastrichtian	Fore limb
Antarctosaurus Huene, 1929a			
A. wichmannianus Huene, 1929a	Castillo Formation, Bajo Barreal Formation, Laguna Palacios Formation, Allen or Rio Colorado Formation, Rio Negro, Rio Neuquen Formation, Neuquen, Bajo Barreal Formation, Chubut, Argentina; Asencio Formation, Palmitas, Uruguay; Vinitas Formation, Coquimbo, Chile	Campanian-Maastrichtian	Partial skeleton, including braincase, mandible, and most of the limbs
A. giganteus Huene, 1929a	Rio Neuquen Formation, Neuquen, Argentina	Campanian-Maastrichtian	2 femora, pubis
A. brasiliensis Arid et Vizotto, 1972	Bauru Formation, Sao Paulo, Goias, Minas Gerais, Brazil	Campanian-Maastrichtian	Fragmentary postcranial remains
Tornieria Sternfeld, 1911			
T. dixeyi (Haughton, 1928) (= *Gigantosaurus dixeyi* Haughton, 1928)	Chiweta Beds, Northern Province, Malawi	Late Jurassic	Caudal vertebrae
Janenschia Wild, 1991 •			
J. robusta (Fraas, 1908) (= *Gigantosaurus robustus* Fraas, 1908)	Upper Tendaguru Beds, Mtwara, Tanzania	Kimmeridgian	3 hind limbs, 2 fore limbs, manus, ?2 dorsal vertebrae, caudal series

Table 16.1, continued

Aeolosaurus Powell, 1987 •			
A. rionegrinus Powell, 1987 •	Los Alamitos Formation, Rio Negro, Argentina	Campanian	Caudal vertebrae
Titanosauridae incertae sedis			
Unnamed titanosaurid (= *"Antarctosaurus" septentrionalis* Huene et Matley, 1933)	Lameta Formation, Madhya Pradesh, India	middle-late Maastrichtian	Basicranium and partial postcranial skeleton
Unnamed titanosaurid (= *"Antarctosaurus" jaxartensis* Riabinin, 1939)	Dabrazinskaya Svita, Syderinskaya Oblast, Kazakhstan	Turonian-Santonian	Femur
Unnamed titanosaurid (= *"Titanosaurus" valdensis* Huene, 1929a)	Wealden, Isle of Wight, England	Valangian-Barremian	Several caudal vertebrae
Sauropoda incertae sedis			
Austrosaurus Longman, 1933			
A. mckillopi Longman, 1933	Allaru Formation, Queensland, Australia	Albian	Dorsal vertebrae, ?incomplete limb remains from several individuals
Aepysaurus Gervais, 1853			
A. elephantinus Gervais, 1853	Grès Vert, Vaucluse, France	Aptian-?Albian	Humerus
Camplodoniscus Kuhn, 1961 (= *Campylodon* Huene, 1929 preoccupied)			
C. ameghinoi (Huene, 1929a) (= *Campylodon ameghinoi* Huene, 1929a)	Bajo Barreal Formation, Chubut, Argentina	Campanian-Maastrichtian	Maxilla with teeth
Mongolosaurus Gilmore, 1933b			
M. haplodon Gilmore, 1933b	?Bayanhua Formation, Nei Mongol Zizhiqu, People's Republic of China	Berriasian-Albian	Teeth, basioccipital, 3 cervical vertebrae
Unnamed sauropod (= *"Pelorosaurus" becklesii* Mantell, 1852)	Wealden, Sussex, England	Hauterivian-Barremian	Fore limb, epidermal impressions

Nomina dubia	Material
Algoasaurus bauri Broom, 1904	Dorsal vertebra, femur, ungual
Asiatosaurus kwangshiensis Hou, Yeh, et Zhao, 1975	Teeth
Asiatosaurus mongoliensis Osborn, 1924c	2 teeth
Cardiodon rugulosus Owen, 1845	Teeth
Chiayusaurus lacustris Bohlin, 1953	Tooth
Clasmodosaurus spatula Ameghino, 1899	Teeth
Gigantosaurus megalonyx Seeley, 1869	Unassociated sacral vertebra, radius, fibula, tibia
Iguanodon praecursor Sauvage, 1876a	Teeth
Morinosaurus typus Sauvage, 1874	Teeth
Nemegtosaurus pachi Dong, 1977	Teeth
Neosodon (no species name given) Moussaye, 1885	Teeth
Protognathosaurus oxyodon Zhang, 1988 •	Dentary
Sanpasaurus yaoi Young, 1946 *partim*	Isolated postcranial elements
Titanosaurus montanus Marsh, 1877a (type of *Atlantosaurus* Marsh, 1877b)	2 1/2 sacral centra
Titanosaurus rahioliensis Mathur et Srivastava, 1987 •	Teeth
Ultrasaurus tabriensis Kim, 1983 •	Fragmentary humerus, vertebra

known forms, many species are synonymized, whereas no attempt has been made, or at this time could be made, to synonymize many of the incompletely known species. This is particularly true of the Titanosauridae, where a major reduction in the number of genera and species will eventually be necessary.

HISTORICAL REVIEW

Sauropod history begins in 1841 in England with Owen's (1841) description of *Cetiosaurus* based on scat-tered remains of caudal vertebrae and various fragments. He later (1860) referred this animal to a new suborder of crocodiles under the name Opisthocoelia. Isolated bones, including a large humerus of *Pelorosaurus* (Mantell 1850) and a humerus of *Aepysaurus* (Gervais 1852), were reported spasmodically during the ensuing years, but it was not until Phillips's description in 1871 of a large part of a skeleton of *Cetiosaurus oxoniensis* from the mid-Jurassic of England that a true understanding of these animals began to emerge. The skull was missing, and except for one dorsal and a few caudals, the vertebrae were represented only by centra.

However, most of the limb and girdle bones were present and correctly identified. Phillips did, however, misinterpret the proper orientation of the pubis and ischium, mistaking their iliac articular surfaces for those along which the two bones met.

The two decades from 1870 to 1890 saw the solutions of most of the problems concerning sauropod anatomy. In England, Seeley (1869, 1870, 1871, 1876, 1889) and Hulke (1869, 1870, 1872, 1874, 1879a, 1880, 1882, 1887), in a series of papers based largely on isolated bones, described the intricate construction of the highly specialized presacral vertebrae, and Hulke (1882) reported the correct orientation of the pubis and ischium. The year 1877 represented the beginning of a virtual explosion of discoveries in western North America of a large amount of material, some of it articulated. Collecting parties were sent to Como Bluff, Wyoming, and Morrison and Garden Park, Colorado (Ostrom and McIntosh 1966), by O. C. Marsh and E. D. Cope. Their intense rivalry for priority resulted all too often in very brief descriptions with inadequate figures of new taxa, based in many cases on specimens the greater portions of which had not yet been exhumed. The result was a plethora of inadequately based taxa, the synonymies of which are still in the process of resolution. Nevertheless, Marsh's papers were to establish firmly the identity of the group he first described as a new order of reptiles, the Sauropoda (Marsh 1878b), and later (Marsh 1879b) as a suborder of the Dinosauria. Among his major contributions were (1) the first description (1879b) of a partial sauropod skull, that of his "*Morosaurus*," and six years later of the first complete skull, that of *Diplodocus*, (2) the first published reconstruction in 1883 of the complete skeleton of a sauropod, that of *Apatosaurus* (= *Brontosaurus*), (3) the establishment (Marsh 1881e) of the sternum as paired plates, (4) the recognition of the greatly enlarged neural cavity in the sacrum (Marsh 1896). Three well-established Morrison genera were to emerge from the Marsh-Cope years: *Camarasaurus, Apatosaurus,* and *Diplodocus.*

The two decades from 1890 to 1910 were marked first by the discovery of sauropods in South America and Africa. Lydekker (1893a) described and figured a large collection from the Late Cretaceous of Argentina as belonging to a separate family, the Titanosauridae, several vertebrae of which he had years before described from India (Falconer 1868; Lydekker 1877). Other mid-Jurassic remains were reported from Madagascar (Lydekker 1895; Thevenin 1907). Beginning in 1897, parties from the American Museum of Natural History took a fine partial *Diplodocus* skeleton from Como Bluff (Osborn 1899a), a skeleton that helped correct a number of misconceptions concerning that animal. The same institution opened the nearby Bone Cabin Quarry a year later and in the next eight years took from it the second largest number of sauropod bones yielded by any American site. A skeleton of *Apatosaurus* (= *Brontosaurus*) from a neighboring quarry was to be the first *permanent* mount of a sauropod skeleton (Matthew 1905). The year 1899 saw the first of many Carnegie Museum expeditions to Wyoming, Colorado, and Utah; the superb series of specimens collected by them were to result in the classic Hatcher memoirs on *Diplodocus* (Hatcher 1901), *Haplocanthosaurus* (Hatcher 1903c), and *Apatosaurus* (= *Brontosaurus*) (Hatcher 1902). The first two of these contained more or less complete descriptions of the all-important vertebral columns, and the third contained the first description of a complete articulated sauropod manus with its single claw. The Carnegie Museum efforts culminated in 1909 with the discovery by E. Douglass of the greatest of all sauropod quarries, that at Dinosaur National Monument in Utah. This quarry was worked by the Carnegie Museum until 1923 and afterward by the Smithsonian Institution, the University of Utah, and today by the U.S. Park Service. From this site has been taken a large number of very rare sauropod skulls—nine of them more or less complete: five of *Camarasaurus*, three of *Diplodocus*, and the only known one of *Apatosaurus*. From it has also come five mounted skeletons: *Apatosaurus* in Pittsburgh, *Camarasaurus* in Pittsburgh and Washington, *Diplodocus* in Washington and Denver. Several other skeletons complete enough to mount remain, the most significant being one of the rare genus *Barosaurus*. In this era, Field Museum (Chicago) parties under Riggs exhumed the original skeleton of the long-limbed *Brachiosaurus* (Riggs 1903a, 1904) and a mountable *Apatosaurus* (Riggs 1903b) from the Grand Junction, Colorado, area.

Shortly before Douglass's discovery, E. Fraas (1908) described new sauropod taxa collected by a German expedition to East Africa. A continuation of these efforts under Janensch up until World War I resulted in vast collections and his exhaustive studies (Janensch 1914, 1922, 1929a, 1929b, 1935–36, 1939, 1947, 1950a, 1950b, 1961) of four African forms: *Brachiosaurus, Dicraeosaurus, "Barosaurus,"* and *Janenschia*. This sauropod fauna along with that from the Morrison of North America represent the best-known faunas to this date.

The 1920s and 1930s were notable chiefly for the publication of landmark memoirs: (1) on *Camarasurus* by Osborn and Mook (1921) and another on the first essentially complete skeleton of a sauropod by Gilmore (1925); (2) on *Apatosaurus* by Gilmore (1936); (3) on the Chinese *Euhelopus* (formerly *Helopus*) by Wiman (1929); and (4) on the South American titanosaurids by Huene (1929a). Other notable events included the mounting of the gigantic *Brachiosaurus* skeleton in Berlin (Janensch 1950b) and the excavation of the great Howe quarry near Shell, Wyoming, by Barnum Brown, the specimens from which are still not described.

Since World War II, most of the advances in sauropod studies have come from China and more recently from South America. C. C. Young (1935a, 1937, 1939, 1944, 1948c, 1954, 1958b: Young and Chao 1972) reported partial skeletons of new Jurassic genera from Sichuan and Xinjiang provinces, of which the most complete was the bizarre, enormously long-necked *Mamenchisaurus*. More recently, the discovery in Sichuan Province of a large number of more or less complete skeletons *with* skulls of *Shunosaurus* and several other new genera by Dong and co-workers (Dong 1973a; Dong and Tang 1984; Dong, Zhou, and Zhang 1983) has already greatly aided our understanding of sauropod phylogeny and will no doubt contribute much more when the material is fully studied. Argentina has yielded much sauropod material from all ages from Middle Jurassic to latest Cretaceous. The most significant discoveries come from the work of J. Bonaparte and co-workers (Bonaparte 1979b, 1986b, 1987; Bonaparte and Bossi 1967; Bonaparte and Gasparini 1979; Bonaparte and Powell 1980) on the cetiosaurid *Patagosaurus*, the Upper Cretaceous *Saltasaurus* with which body armor was found, and other remains currently under study. Large quantities of the best-preserved material yet found of the Late Cretaceous titanosaurids were collected in Brazil by L. I. Price, but these remains have not yet been described. Other important discoveries in recent years include (1) a partial skeleton without skull (Borsuk-Bialynicka 1977) and two isolated skulls (Nowinski 1971; Kurzanov and Bannikov 1983) from the Late Cretaceous of Mongolia, (2) as an yet undescribed *"Cetiosaurus"* skeleton with skull from Morocco (Monbaron 1983), (3) huge sauropods found by J. A. Jensen (1985b) at Dry Mesa, Colorado, and (4) a very large sauropod quarry in northern Wyoming by B. R. Erickson.

DIAGNOSIS

The Sauropoda is diagnosed on the basis of very large body size, relatively small skulls, long necks, long tails, skull with large dorsally placed nares, greatly reduced jugal usually excluded from the ventral border of the skull, large quadratojugal, relatively small endocranial capacity, highly vaulted palate with large pterygoids, presacral centra lightened by deep pleurocoels and/or cancellous bone, neural arches and spines largely reduced to a complex of thin laminae, 12 to 19 cervicals, 8 to 14 dorsals, scapula oriented more nearly horizontal than vertical, ilium with broadly expanded preacetabular process and with pubic peduncle much longer than ischial, limb bones robust and solid, no notch between head and greater trochanter in femur, carpus and tarsus reduced to one or two elements each in all but perhaps the earliest forms, metacarpals longer than metatarsals, number of phalanges greatly reduced in manus, digit I alone retaining a claw, number of phalanges reduced in digits IV and V of pes.

ANATOMY

Skull

Virtually complete skulls and mandibles are known in *Diplodocus, Camarasaurus, Brachiosaurus,* and *Shunosaurus* (the latter has received only a preliminary description) (fig. 16.2). Partial skulls and jaws are known in *Apatosaurus, Euhelopus, Dicraeosaurus, Omeisaurus, Datousaurus, Nemegtosaurus, Quaesitosaurus, Antarctosaurus,* "*Barosaurus*" *africanus,* and the as yet undescribed *"Cetiosaurus" mogrebiensis.* The other genera are known from a few skull elements or more often from nothing but possibly a few teeth. In this latter category are such important forms as *Haplocanthosaurus,* the North American species of *Barosaurus, Opisthocoelicaudia, Cetiosaurus* (limited to the English species), *Pleurocoelus, Barapasaurus, Vulcanodon, Mamenchisaurus, Alamosaurus,* and almost all the other titanosaurids. The sauropod skull is notable in that the external nares have migrated high up onto the snout and in some cases to the skull roof; the antorbital fenestra, orbit, and infratemporal fenestra have come much closer together, the orbit having been "squeezed" up between the other two; the antorbital fenestrae have been reduced in size; and perhaps most striking, the jugal is greatly reduced in size and often completely excluded from the lower rim of the skull. The greatest variation occurs in the jaws and teeth and also in the shape and placement of the external nares. Significant differences also occur in the bones of the palate, but these have been studied in detail in few forms. Diplodocids have additional antorbital fenestrae, and small fenestrae between the premaxilla and maxilla occur in at least *Diplodocus.*

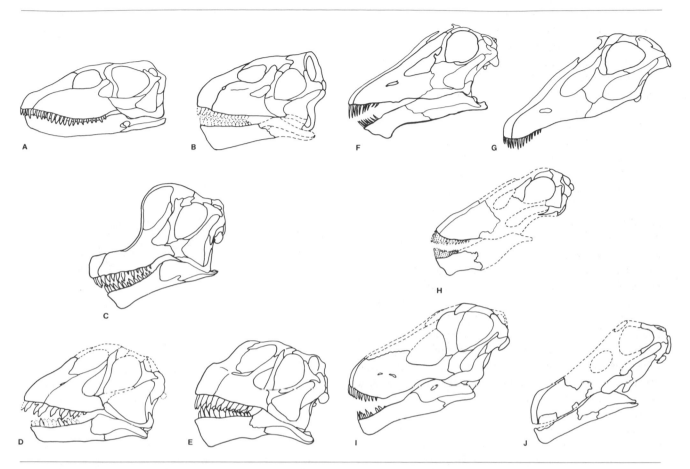

Fig. 16.2. Skull and mandible, lateral view a) *Shunosaurus lii,* b) *Omeisaurus tianfuensis,* c) *Brachiosaurus brancai,* d) *Euhelopus zdanskyi,* e) *Camarasaurus lentus,* f) *Diplodocus longus,* g) *Apatosaurus louisae,* h) *Dicraeosaurus hansemanni,* i) *Nemegtosaurus mongoliensis,* j) *Quaesitosaurus orientalis.* (After He et al. 1984 (b); Janensch 1935–36 (c, h); Mateer and McIntosh 1985 (d): Berman and McIntosh 1978 (g); Nowinski 1971 (i); Kurzanov and Bannikov 1983 (j); (a) courtesy of R. E. Sloan.)

In *Shunosaurus, Omeisaurus,* and *Euhelopus,* the external naris is oval and found on the upper part of the side of the skull, rostrodorsal to the antorbital fenestra. It is larger than the latter but smaller than the orbit. Those of *Camarasaurus* and *Brachiosaurus* are similar, but the external naris is as large as the orbit, and in *Brachiosaurus,* the external naris is located farther backward and upward, so as to lie higher than the orbit. Extreme narial shifting occurs in *Diplodocus* and *Apatosaurus,* in which the external nares, though small, lie behind the antorbital fenestra and above the orbit, with no median bar separating them from one another. *Nemegtosaurus* appears to have a similar arrangement, but the skull is damaged in this area. Furthermore, the narial region is missing in known skulls of *Dicraeosaurus, Quaesitosaurus,* and *Antarctosaurus.*

The braincase is completely ossified, and the bones of the braincase including those of the skull roof are tightly coossified. The occiput slopes downward and forward from the skull roof at an angle varying from about 15° to 40° from the vertical. The robust supraoccipital is subtriangular in caudal view with a marked crest at its apex. Ventrally, the bone rests on the prootic rostrally, on the opisthotic-exoccipital complex caudally, and in some forms at least, forms the center of the dorsal rim of the foramen magnum. Rostrally, it loosely abuts the rear wing of the parietal and, beneath that, the squamosal. Ringing the sides of the foramen magnum are the exoccipitals, so firmly coossified with the opisthotics that the suture line between them is virtually never discernible. Broad paroccipital processes of the opisthotic are directed laterally and in some genera lateroventrally. These become expanded and large at the ends where they abut the quadrate. The lower rim of the foramen magnum is formed by the large ball-shaped condyle of the basioccipital. Rostroventrally, the paired basal tubera of the basioccipital are large and bulbous in one extreme in *Camarasaurus* and smaller at the other extreme in *Dicraeosaurus*. The floor of the braincase slopes forward and strongly upward to form an angle of about 40° with the plane of the skull roof. In front of the basioccipital is the basisphenoid, which is solidly fused to the former just in front of the tubera. Opening into the braincase from below and forming a large downwardly elongated cavity in the center of the basisphenoid is the pituitary fossa. The lower part of the basisphenoid splits into the paired, diverging basipterygoid processes. These processes are short, robust, and directed downward and backward in *Camarasaurus, Brachiosaurus,* and *Quaesitosaurus* but are long and slender and directed downward and forward in *Diplodocus, Apatosaurus, Dicraeosaurus,* and *Nemegtosaurus*. They fit loosely into a socket on the dorsal edge of the pterygoid. Lying above the basisphenoid are the prootic and laterosphenoid. The parasphenoid juts forward from the top half of the rostral edge of the basisphenoid, any suture between the two bones being lost. The parasphenoid lies in a vertical plane, short and high in *Camarasaurus* and elongated and low in the diplodocids. The blocklike prootic is a short, high, and fairly broad element wedged between the laterosphenoid and opisthotic and bounded dorsally by the supraoccipital and ventrally by the basisphenoid. From its lateral face there projects a thin lamina of bone, the crista prootica, which attaches to the side of the basipteryoid process and proceeds down it for a considerable distance. In front of the prootic is the laterosphenoid, a narrower bone whose height decreases rostrally. Another thin lamina, the crista antotica, projects from the lateral face of the laterosphenoid to connect with a ridge on the medial face of the postorbital. Just in front of the laterosphenoid is the small orbitosphenoid, which inclines inward toward its mate, the two being separated by the large olfactory canal. The epipterygoid, occasionally reported in prosauropods, is absent in sauropods.

The braincase is roofed by the parietals and frontals (fig. 16.3). These elements are coossified to their mates along a tightly sinuous articulation and to one another by a similar rather straight articulation. Laterally, the parietal sends out two rami that enclose more than half of the supratemporal fenestra. The rostral ramus unites with the postorbital, the caudal with the squamosal. Caudally, the parietals unite with the supraoccipital and have a small contact surface with the squamosals. The presence of a pineal opening in the parietal has been the subject of much controversy

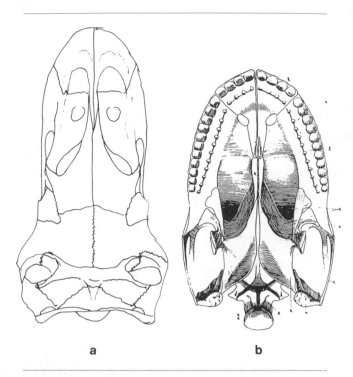

Fig. 16.3. Skull. a) dorsal view, *Camarasaurus lentus* (after Gilmore 1925), b) palatal view, *Brachiosaurus brancai* (after Janensch 1935/1936).

(Roth and Roth 1980). Some skulls of *Camarasaurus* and *Diplodocus* appear to have such an opening; others do not. It may be a juvenile character, or it may be that the very thin osseous covering of the pineal organ becomes fractured in some skulls. The frontals are flat, thin, but sturdy, subrectangular plates broader than long. The prefrontal is a relatively small blocklike element lying between the nasal and the frontal and forming a small part of the dorsal rim of the orbit. Ventrally, there is a very complex union of the prefrontal, nasal, lacrimal, and dorsal process of the maxilla. The nasal arches gently forward from the frontal and sends out two slender unequal processes. The longer one is directed forward on the midline and ends on the side of the ascending tongue of the premaxilla; the shorter process is directed ventrolaterally into the union with the prefrontal, maxilla, and lacrimal. The straight lacrimal descends from this union, where it is abutted rostrally by the dorsal process of the maxilla, above and in front by the nasal, and caudodorsally by the prefrontal. It lies between the antorbital opening and the orbit. Its base is usually somewhat expanded and set on the jugal, or in some forms it has migrated forward to the rear of the maxilla. In *Camarasaurus* and related forms, it is reduced to a complex of several thin laminae. It is pierced at midlength by a small foramen. There are no supraorbital bones, and no separate postfrontal is seen. Viewed laterally, the postorbital is triradiate with a short rostrodorsal orbital process, an even shorter caudal squamosal process, and a long, slender rostroventral jugal process. Medially, the postorbital unites with the frontal and parietal. The squamosal process is V-shaped and fits firmly into a groove on the outer side of the head of the squamosal. The jugal process separates the orbit and the infratemporal fenestra, its slender lower end joining the equally slender ascending process of the jugal in an overlapping, grooved articulation. On the medial side of the postorbital, a ridge descends down the jugal process to which is attached the crista antotica of the laterosphenoid. Viewed laterally, the squamosal is tightly curved where it is contacted by the postorbital. Medially, the squamosal successively meets the parietal, supraoccipital, and paroccipital process. The broad descending body of the squamosal is a cotylus that receives the head of the quadrate. The prequadratic process reaches the dorsal process of the quadratojugal in some forms. The quadrate has a straight shaft. The pterygoid ramus of the quadrate makes a broad squamose articulation with the quadrate ramus of the pterygoid. The lateral aspect of the quadrate shaft is lapped ventrally by the quadratojugal and dorsally by the prequadratic process of the squamosal. The mandibular condyle is expanded fore and aft but more so transversely for its articulation with the articular. In skulls of *Camarasaurus, Brachiosaurus*, and *Shunosaurus*, the quadrate is directed caudoventrally with respect to the skull roof, but in many others (e.g., *Diplodocus, Apatosaurus, Nemegtosaurus, Quaesitosaurus, Dicraeosaurus*), it is directed rostroventrally. The quadratojugal is a thin L-shaped plate, the long axis being horizontal. The angle between the rami is 90° or less in brachiosaurids and camarasaurids but much greater in diplodocids. A striking feature is its enlargement at the expense of the jugal. In most forms, the horizontal ramus extends forward beneath the infratemporal fenestra and unites with the rear of the maxilla, excluding the jugal from the lower margin of the skull. The jugal is reduced to a small lune or even crescent-shaped bone, the center abutting the rear end of the maxilla, the lower ramus following the ventral margin of the infratemporal fenestra and then narrowing to continue for a ways beneath the overlapping quadratojugal. The reduced body of the jugal articulates with the footlike base of the lacrimal, while the postorbital process of the jugal narrows as it meets the jugal process of the postorbital.

The maxilla is the largest bone in the skull. There are two major parts, a heavy horizontal tooth-bearing quadrangular body and a prominent dorsal process.

The body articulates with the quadratojugal, jugal, and sometimes the lacrimal. Near the top of the medial face of the maxilla well behind the tooth row are two roughened depressions, rostrally for the palatine, caudally for the ectopterygoid. Rostrally, the medial surface of the maxilla articulates with the vomer in a manner varying from *Camarasaurus* to *Diplodocus*. In the diplodocids, *Nemegtosaurus, Shunosaurus*, and *Euhelopus*, the dorsal process lies on the rostral half of the maxilla and terminates against the rostrodorsal margin of the lacrimal and nasal. Above it lies the external naris. In *Brachiosaurus, Camarasaurus*, and *Omeisaurus*, a recessed shelf develops above the rostral end of the main body of the maxilla beneath the external naris, and the dorsal process arises from near the center of the main body. The rostral end of the maxilla articulates along a broad, flat surface with the premaxilla, and this articulation continues along the leading edge of the dorsal process in *Diplodocus, Apatosaurus, Nemegtosaurus, Dicraeosaurus*, and to a certain degree *Euhelopus*. The premaxilla tapers upward and backward separating the external nares. A small foramen is often found on the articular margin between the maxilla and premaxilla.

The bones of the palate (fig. 16.2) exhibit considerable variation in the few genera in which they have been described. By far the largest of these are the pterygoids, triradiate nearly vertical plates. Their dorsal margins meet on the midline at an angle of about 60°, resulting in a highly vaulted palate. The quadrate ramus overlaps the pterygoid ramus of the quadrate along a broad, flat, articular surface. Near the back of the expanded upper edge of the pterygoid lies the socket for the basipterygoid process. The ectopterygoid ramus is directed ventrally to contact the ectopterygoid and palatine. The palatine ramus is the longest and is directed rostrally, meeting its fellow side by side on the midline. Rostrally, the pterygoids are wedged between the vomers. The small straplike ectopterygoid has an expanded base that firmly contacts the maxilla. As it proceeds medially, it bends so as to lie against the lateral surface of the ectopterygoid process of the pterygoid. The palatine is a somewhat larger, triangular, flat element with a robust, horizontal process that extends backward to contact the maxilla just in front of the ectopterygoid articulation. The palatine overlaps the ectopterygoid ramus of the pterygoid and runs dorsally to the caudolateral rim of the vomer. The vertical vomers are roughly triangular robust elements. Rostrally, they are wedged between the maxillae. Caudally, they diverge around the rostral tips of the pterygoids. The rim of the choana is thus irregular. A second small opening behind the choana is ringed by the maxilla, ectopterygoid, and palatine.

Mandible

The mandible is short, largely because of the shortened dentary. Viewed laterally, the dentary in front and the surangular and angular behind are about the same length. In primitive forms like *Shunosaurus*, a small circular mandibular foramen occurs on the lateral face at the junction of those elements. It has not been reported in any genera of Late Jurassic or Cretaceous time. In some later forms such as *Camarasaurus*, there is a small surangular foramen. The articular and thus the craniomandibular joint lies very low in the jaw. The adductor fossa is rimmed by the surangular laterally, the articular behind, and the thin vertically oriented prearticular, small coronoid, and intercoronoid medially. A very thin vertically oriented splenial completes the medial surface of the mandible rostrally.

The dentary occupies roughly the rostral half of the mandible. The height of the main body of the bone is greatest at the symphysis but viewed medially, decreases regularly caudally ending near the high point of the jaw, where it is overlapped by the surangular. The dentary symphysis is weak in *Camarasaurus* but very strong in *Diplodocus*. The teeth occupy much of the dentary in most skulls but are confined to the front in skulls of *Diplodocus* and *Antarctosaurus*. A row of foramina is placed beneath the teeth on the medial surface. The angular is elongate, relatively sturdy, and narrow. It tapers to a blunt termination at both ends, rostrally fitting into a slot in the dentary. Lying just above the angular is the larger but thinner surangular, which overlaps the back of the dentary to form the lateral surface of the upper part of the caudal half of the mandible. Caudally, the surangular narrows rapidly above the angular. On the medial side of the mandible, the long, slender prearticular is placed just above the medial rim of the angular. it tapers caudally, its rim deflected inwardly so that the distal ends of the surangular and prearticular cradle the massive, blocklike articular. A very thin, vertical splenial lies beneath the Meckelian canal on the medial face of the dentary, covering the lateral platelike extension of the dentary and the rostral half of the angular to form much of the lower border of the medial mandible. Behind the tooth row, the splenial narrows abruptly, where it is covered by the prearticular. An elongate thin strip of bone, the intercoronoid (complementare), lies above the splenial, covering the medial surface of the rear half of the tooth row. Between the back of the splenial and the upper end of the prearticular is the small coronoid.

Sauropod teeth clearly relate to their herbivorous habits but exhibit a considerable range of variation. The extremes are represented, on the one hand, by the robust teeth of *Camarasaurus* with their high, enlarged, lingually curved spoon-shaped crowns and, on the other hand, by the very long, slender, peglike teeth of *Diplodocus* with unexpanded cylindrical crowns. Intermediate are the teeth of the early *Shunosaurus* and the later *Pleurocoelus*, those of the former possessing the slenderness of diplodocids and the crown shape of camarasaurs. Small denticles or serrations are present on the teeth of the very early *Barapasaurus* but are absent on most other forms, a trace lingering on some of the teeth of *Brachiosaurus* and *Mongolosaurus*.

There are four teeth in the premaxilla. In early forms like *Shunosaurus*, there are as many as 21 maxillary and dentary teeth, but this number is reduced to 14 or fewer in later forms. The teeth of the diplodocids are limited to the front of the jaws. All teeth exhibit wear facets. Details of the dental batteries have been given by White (1958) for *Camarasaurus* and by Janensch (1935-36) for *Brachiosaurus*. In *Camarasaurus*, the fore and aft edges of succeeding teeth have been described as tightly appressed "so the complete battery of teeth presented a continuous cutting edge for cropping vegetation. All teeth exhibit wear facets varying from slight to a condition on which . . . half of the crown is worn away." According to White, the replacement pattern is that of the *replacement wave*, in which the order of replacement and degree of wear progresses along the jaw with the process modified in that alternate teeth constitute two separate waves that progress in opposite directions. Janensch denied two such waves of teeth in *Brachiosaurus*, but White claimed that the former's own diagrams belie this conclusion and that the *Brachiosaurus* teeth also exhibit two waves of alternate teeth progressing in opposite directions.

The stapes is a very small, straight, slender rod extending from the fenestra ovalis to the head of the quadrate.

Sclerotic rings have been reported surrounding the eyeball in at least three genera: *Diplodocus* (Holland 1924*b*), *Camarasaurus* (Gilmore 1925), and *Nemegtosaurus* (Nowinski 1971). They were likely present in all. They are best preserved in *Nemegtosaurus*, in which ten overlapping hexagonal plates complete the ring.

Hyoid elements have been reported in five genera, but in no case has the complete hyoid apparatus been found articulated and in place. Marsh (1883) mentioned two pairs of "rodlike elongate somewhat curved" bones in "*Brontosaurus*." A pair of straight rodlike bones and another shorter one were found with *Camarasaurus* (Gilmore 1925). A pair of gently

curved rods was found with *Brachiosaurus* (Janensch 1935–36). More sharply curved (~110°) rods were found with *Omeisaurus,* and a yokelike hyoid bone has been reported in *Shunosaurus.*

Axial Skeleton

The presacral formula for the Early Jurassic sauropods is not known, but the process of neck elongation among sauropods, in which additional dorsals are taken into the neck, had already begun by the Middle Jurassic. There is also the tendency in some forms to increase the number of presacrals by up to 6, invariably added to the neck. A further reduction to the number of dorsals by absorbing 1 to 2 into the sacrum could reduce their number to as low as 9. There are 4 to 6 functional sacrals, and the number of caudals varied from about 34 to more than 80.

To increase strength while minimizing increase in weight in animals the size of the sauropods, the centra are greatly lightened by the development of paired pleurocoels, which in extreme cases (e.g., the dorsals of *Camarasaurus* and *Brachiosaurus*), occupy most of the interior of the centrum. Pleurocoels are less common in the caudals, but in some forms, for example, *Diplodocus,* lightening was obtained by ventral excavation of the centrum. Alternatively, cavities in the presacral neural arches, are developed (e.g., in *Barapasaurus* and *Dicraeosaurus*). The more common method of lightening the arches in the cervicals, but more particularly in the dorsals, is to reduce them to little more than a complex of thin laminae extending from one structure to another. Differing terminologies are used for the different laminae. The most comprehensive is that of Janensch (1929*a*), who recognized no fewer than sixteen, some of which are used here.

The number of cervicals varies from 12 in the more conservative forms to as many as 19 in *Mamenchisaurus.* The intercentrum of the atlas is crescentic with well-developed facets for the articulation of a single-headed rib, which is rarely if ever fused to it. The odontoid is a blunted cone whose base is fused tightly to the axis at a very early age. The two neurapophyses fuse to the intercentrum (but not to one another) in the adult. In perhaps all genera, paired inverted triangular elements form a proatlas. The lower two sides articulate with the exoccipital and neurapophysis of the atlas, respectively. The centrum of the axis is opisthocoelous and, in most advanced forms at least, contains pleurocoels or some lateral depression. Prominent parapophyses are directed outward and in some instances backward from the cranioventral part of the lateral face of

the centrum. In several genera, a small intercentrum is placed just in front of and fused to the centrum. The neural arch is low cranially with small prezygapophyses to which the neurapophyses of the atlas connect. The arch rises to a maximum height at the rear. The spine is broad and shows no sign of bifurcation. The very short transverse processes arise at midlength of the vertebra on the lower part of the arch just above its joint with the centrum. The centra of all succeeding cervicals (fig. 16.4) are strongly opisthocoelous, increasing in length for the majority of the series before decreasing again. The cervicals usually contain pleurocoels of varying depth and complexity, those of particularly the brachiosaurids, camarasaurids, and diplodocids being subdivided in various ways. Prominent parapophyses extend from the bottom of the cranial part of the centrum, their ventral surfaces usually coplanar with ventral surface of the centrum. In several derived forms such as *Apatosaurus,* the parapophyses are directed not only outward but ventrally as well. The transverse processes increase in length backward. The articulations lack a hyposphene. Spines of the cetiosaurids, brachiosaurids, and titanosaurids are simple, but in camarasaurids and diplodocids, they are bifid, beginning sometimes as cranially as cervical 3 and persisting well into the dorsal series. Maximal bifurcation occurs near the base of the neck. This unusual development apparently arose independently in the two families and provided a channel for a heavy ropelike muscle, which connected the skull with the spines of the cervical and dorsal vertebrae and was useful in raising the skull and neck. Parts of the calcified muscle are actually preserved in one specimen of *Camarasaurus* (D. Chure pers. comm.).

Except for that of the atlas, the cervical ribs from the axis backward are double headed and firmly coalesced to the diapophysis and parapophysis at an early age. Usually, they lie a short distance below and parallel the ventral border of the centrum. In many genera, a cranial extension of the rib arises at the junction of the capitulum and tuberculum and is projected toward the cranial end of the vertebra. The main shaft of the rib is directed backward and in the caudal third of the neck of some genera, for example, *Brachiosaurus* and *Camarasaurus,* it extends caudally beneath one or two more cervicals.

The transition between the cervical and dorsal series in sauropods is more reasonably defined by a sharp transition in the type of rib borne by the vertebra. Thus, there is a transition region of two to four cranial dorsal vertebrae that bear dorsal ribs but resemble cervicals in having the parapophyses on the centrum. Additional

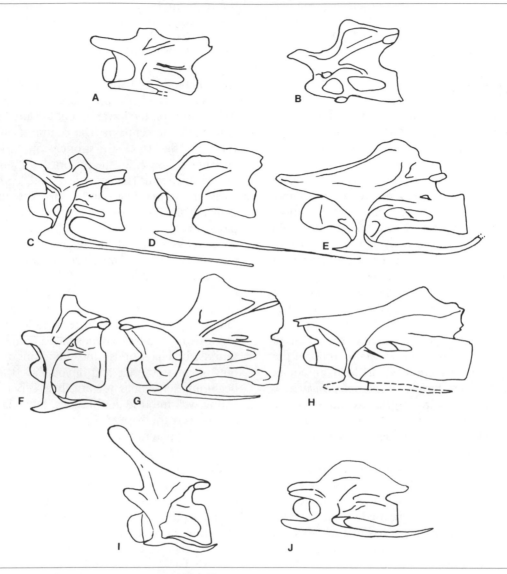

Fig. 16.4. Caudal cervical vertebrae, lateral view. a) *Cetiosaurus oxoniensis,* b) *Haplocanthosaurus priscus,* c) *Camarasaurus lentus,* d) *Euhelopus zdanskyi,* e) *Brachiosaurus brancai,* f) *Apatosaurus excelsus,* g) *Diplodocus carnegii,* h) *Barosaurus lentus* i) *Dicraeosaurus hansemanni,* j) titanosaurid from Brazil. (After Holland 1924*a* (c); Wiman 1929 (d); Janensch 1929*a,* 1950 (e, i); Gilmore 1936 (f); Hatcher 1901 (g); (j) is from a photograph courtesy of L. I. Price.)

laminae connect the prezygapophyses with one another and with the floor of the arch, and others connect the postzygapophyses to the spine. In the transition region of the cranial dorsals, the upward-facing prezygapophyses are relatively far apart and articulate with the downward-facing postzygapophyses in the normal fashion, as in the cervicals.

The number of dorsals ranges from 14 in the cetiosaurid *Haplocanthosaurus* apparently to 9 in the diplodocid *Barosaurus.* The centra are generally opisthocoelous in the cranial part of the column (fig. 16.5), a condition that persists back to the sacrum in some

forms. In others, the centra become amphiplatyan or slightly amphicoelous in the rear half of the series. The centra usually contain pleurocoels of varying depth. In most forms, the lengths of the centra remain nearly constant throughout, perhaps decreasing slightly at the end. In others, particularly the diplodocines, the cranial centra are much the longest, but after the first three vertebrae, the length decreases steadily and remains fairly uniform thereafter. The highly developed laminar structure of the neural arches has been referred to above. The transverse process is directed laterally in most forms but also has an upward component

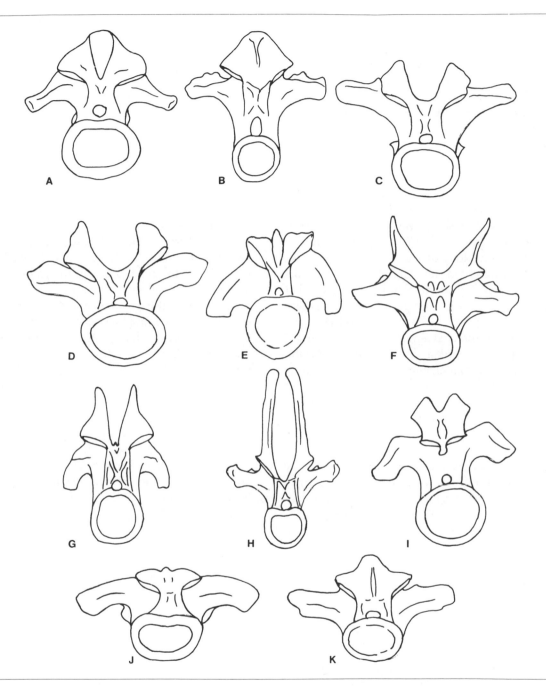

Fig. 16.5. Cranial dorsal vertebrae, cranial view. a) *Omeisaurus tianfuensis*, b) *Haplocanthosaurus priscus*, c) *Camarasaurus supremus*, d) *Opisthocoelicaudia skarzynskii*, e) *Euhelopus zdanskyi*, f) *Apatosaurus louisae*, g) *Diplodocus carnegii*, h) *Dicraeosaurus hansemanni*, i) *Mamenchisaurus hochuanensis*, j) *Janenschia robusta*, k) titanosaurid from Brazil. (After He et al. 1984 (a); Osborn and Mook 1921 (c); Borsuk-Bialynicka 1977 (d); Wiman 1929 (e); Gilmore 1936 (f); Hatcher 1921 (g); Janensch 1929*a* (h); Young and Chap 1972 (i); (k) is from a photograph courtesy of L. I. Price.)

in some cetiosaurids and *Dicraeosaurus*. The transverse process contains four laminae, of which the first is the centrodiapophysial, extending from the diapophysis caudoventrally to be joined by the infrapostzygapophysial lamina from the postzygapophysis above, where they reach the caudal end of the floor of the neural canal. The other three laminae extending from the diapophysis (prezygodiapophysial, postzygodiapophysial, supradiapophysial) are directed, respectively, forward to the prezygapophysis, backward to the postzygapophysis, and upward to the spine (Bonaparte 1986b). In the caudal dorsals, the parapophysis is located high up on the centrodiapophysial lamina where it remains for much of the series. At approximately dorsal 4, the parapophysis descends rapidly to reach the centrum; it actually is found below the pleurocoel in dorsal 1 of *Diplodocus* and *Barosaurus*. Farther back, the articulation is greatly strengthened by the development of a strong hyposphene on which the articular surface of the postzygapophyses is continued ventrally after a nearly right-angle bend. The hyposphene locks tightly into the hypantrum of the next vertebra, providing transverse stability not afforded by the conventional articulation. The neural spines vary widely from genus to genus and also from fore to rear in a given animal. Laminae are directed longitudinally to the zygapophyses and ventrolaterally to the diapophyses. The caudal dorsal spines may be short, robust, and expanded transversely in cetiosaurids, brachiosaurids, and camarasaurids or slender and elevated in diplodocids (fig. 16.6). They may be directed almost vertically, the usual condition, or inclined caudally as well as dorsally in the Titanosauridae. The cranial spines are simple or bifurcated like those of the cervicals of the corresponding forms. In brachiosaurids, the spines actually increase in height from the sacrum forward.

The first dorsal rib is the most distinctive, in being shorter and heavier and in having a larger tuberculum and a greater spread between the tuberculum and the capitulum than those following. The first several ribs increase in length rapidly, then slowly, to past the cen-

ter of the thorax, after which they decrease gradually, at the same time becoming less robust.

The sacrum of typical Late Jurassic sauropods is composed of five vertebrae, usually interpreted (Riggs 1903b) as three primary sacrals augmented on either end by a modified dorsal or dorsosacral and a modified caudal or caudosacral. In *Barapasaurus* and the Chinese cetiosaurids, there are four sacrals, sacrals I to IV. In the Late Cretaceous titanosaurids and occasionally in very old individuals of other families, a second dorsosacral is added to give a total of six sacrals. Sacral ribs are winglike vertical plates, robust below but otherwise thin. The amount of coossification between the various sacral elements varies considerably as a function of age and also of taxon. The ribs fuse to the vertebrae at a very early age, but the centra remain free. In the adult, the ribs fuse to one another, to the centra, and to the ilia to form a rigid sacrocostal yoke, the latter sharing with the ilium the acetabular surface. The centra and certain of the neural spines also fuse. Marsh (1896) has pointed out that in many Morrison forms at least the sacral centra are hollow, the cavity extending even into the fused sacral ribs. He also emphasized the greatly expanded neural canal in the sauropod sacrum. Bonaparte (1986c) has shown that this feature is already present in the Middle Jurassic *Patagosaurus*.

The long tail is one of the more obvious sauropod features, but great variation in the caudal vertebrae exists among the different genera both as to number and form. Complete tails are rare, and the exact number of caudals may well have varied by a few from indi-

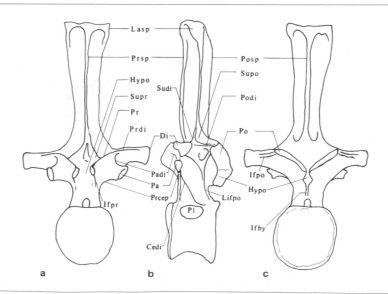

Fig. 16.6. Caudal dorsal vertebra of *Apatosaurus louisae* to identify laminae: a) cranial view, b) lateral view, c) caudal view. (After Gilmore 1936.)

vidual to individual. The early cetiosaurid *Shunosaurus* had 44 caudals and *Camarasaurus* had 53. Diplodocids, however, had many more (82 in one specimen of *Apatosaurus*). At the other extreme is the late Cretaceous *Opisthocoelicaudia*, which had only 34. The proximal centra (fig. 16.7) range from amphicoelous in cetiosaurids and brachiosaurids to slightly procoelous in *Camarasaurus*, to moderately procoelous in diplodocids, to very procoelous with deep proximal concavities and great distal ball-like convexities in titanosaurids. In contrast, *Opisthocoelicaudia* has a large proximal ball and distal concavity. Pleurocoels and ventral concavities occur in diplodocids but are absent in most other forms. The chevron facets are particularly pronounced in *Vulcanodon*, *Cetiosaurus*, *Haplocanthosaurus*, and titano-

saurids. The middle caudals are generally amphicoelous and spool shaped, the greatest variation being their relative lengths. In diplodocids, a number of elongate biconvex centra lacking arches complete the tail as a "whiplash." In general, the arches and spines are relatively simple, lacking the elaborate laminae of the dorsals and sacrals. The articulation is ordinarily simple, but a hyposphene-hypantrum articulation is found in the proximal caudals of at least *Pleurocoelus* (Langston 1974). The caudal ribs are simple in most genera, decreasing in size and finally vanishing around caudal 16. In diplodocids, however, the proximal caudal ribs are specialized, winglike thin plates similar to sacral ribs. These become gradually modified backward into those of the more conventional form.

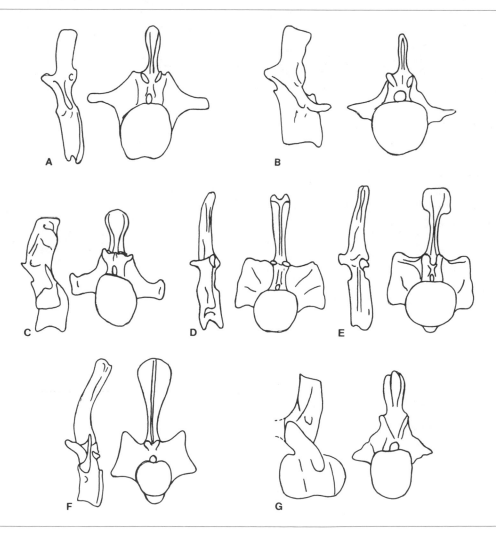

Fig. 16.7. Proximal caudal vertebra, lateral and cranial views. a) *Haplocanthosaurus priscus*, b) *Brachiosaurus brancai*, c) *Camarasaurus supremus*, d) *Diplodocus carnegii*, e) *Apatosaurus excelsus*, f) *Dicraeosaurus hansemanni*, g) *Alamosaurus sanjuanensis*. (After Hatcher 1901, 1903c (a, d); Janensch 1929a, 1950 (b, f), Osborn and Mook 1921 (c); Gilmore 1936 (e).)

Chevrons appear after the first or second caudal vertebra, at first lengthening backward but then gradually shortening until they disappear altogether near the end of the tail in short-tailed forms or much farther forward in those sauropods possessing the whiplash. The most interesting development is that of the typical *Diplodocus*-like chevrons, in which a cranial projection on midcaudal chevrons progressively expands through the more distal series. The forward part

of the chevron becomes coequal in length with the backward part and, when viewed from above, takes on a diamond shape. Distally, the right and left halves separate and form two parallel rods, each articulating with the overlying centrum. This development reaches its maximum in *Diplodocus* but occurs also in the other diplodocids. The condition was thought to be very derived (Marsh 1884*d*), but its presence in four Early and Middle Jurassic Chinese cetiosaurids indicates that the

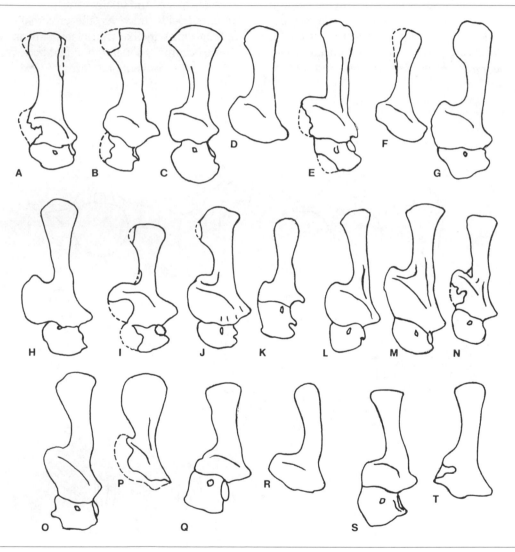

Fig. 16.8. Scapulocoracoids, lateral view. a) *Barapasaurus tagorei*, b) *Cetiosaurus oxoniensis*, c) *Haplocanthosaurus priscus*, d) *Omeisaurus junghsiensis*, e) *Tienshanosaurus chitaiensis*, f) French "*Bothriospondylus madagascariensis*," g) *Brachiosaurus brancai*, h) *Euhelopus zdanskyi*, i) *Camarasaurus? alenquerensis*, j) *Camarasaurus grandis*, k) *Opisthocoelicaudia skarzynskii*, l) *Apatosaurus louisae*, m) *Diplodocus carnegii*, n) *Barosaurus lentus*; o) *Supersaurus vivianae*, p) *Rebbachisaurus garasbae*, q) *Saltasaurus loricatus*, r) *Laplatasaurus araukanicus*, s) *Alamosaurus sanjuanensis*, t) *Antarctosaurus wichmannianus*. (After Jain et al. 1975 (a); Phillips 1870 (b modified); Hatcher 1901 (c, 1903*c*, m); Dong et al. 1983 (d); Young 1939 (e, h); Janensch 1950 (g); Lapparent and Zbyszewski 1957 (i); Marsh 1877 (j); Borsuk-Bialynicka 1977 (k); Gilmore 1922, 1936 (l, s); Jensen 1985*b* (o); Lavocat 1954 (p); Bonaparte and Powell 1980 (q); Huene 1929*a* (r, t).)

character is, if not primitive, at least a very early development. Its apparent absence in the other cetiosaurids is important in the consideration of sauropod phylogeny. In most genera, the chevrons are open above the hemal canal, but in *Diplodocus, Apatosaurus, Mamenchisaurus,* and the proximal chevrons of *Dicraeosaurus* and *Patagosaurus,* the hemal arch is closed between the two heads.

Appendicular Skeleton

The scapula and coracoid are free in the juvenile but invariably coosified in the adult. Interclavicles are absent.

The scapula is an elongated, concave medially, plate that fits against the rib cage (fig. 16.8). The proximal end is broadly expanded. The diagonal acromion separates a large fossa for the M. deltoides scapularis below from a smaller fossa. It is greatly thickened in the area of the glenoid cavity. The shaft is thin and straight, and the distal end is variably expanded. In brachiosaurids, camarasaurids, and some cetiosaurids, it is greatly splayed. The distal end is rugose for the attachment of a cartilaginous suprascapula, which is occasionally fossilized. The proper orientation of the scapula against the body was long debated. The matter was finally resolved in favor of the less vertical orientation advocated by Marsh and Hatcher, as opposed to the more vertical orientation of Osborn and Mook (1921),

by the discovery of a nearly complete articulated skeleton of a juvenile *Camarasaurus* (Gilmore 1925).

The large coracoid can be oval to quadrangular in shape. Its articular margin with the scapula may be nearly straight or take a noticeable defection near the center just above the prominent foramen. The plate is heaviest near the glenoid cavity but thins markedly forward.

Hatcher (1901) interpreted small, gently curved, rodlike elements found near the pectoral region of two *Diplodocus* skeletons as clavicles. The bone is flattened and slightly expanded at one end and is two-headed at the other. It has also been suggested that these bones might be sternal ribs, although they bear no resemblance to the supposed sternal ribs of *Apatosaurus.* Nopcsa's (1905b) suggestion that they might represent an os penis has met with no acceptance. Much larger straight, supposed clavicles have also been reported in *Shunosaurus.*

The ossified sternum consists of a pair of broad plates (fig. 16.9), whose placement has been the subject of considerable controversy. It is now generally recognized that Marsh's (1881e) interpretation in which the heavier ends lie forward (contra Hatcher [1901]) is the correct one. The plates vary considerably in shape and size and are particularly large in *Cetiosaurus, Haplocanthosaurus,* and titanosaurids. The rear edges are rugose for the attachment of cartilaginous sternal ribs. These are seldom fossilized, but several elements (Marsh

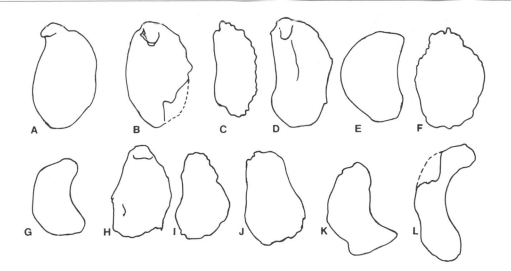

Fig. 16.9. Sternal plates, ventral view. a) *Shunosaurus lii,* b) *Cetiosaurus oxoniensis,* c) *Haplocanthosaurus delfsi,* d) *Brachiosaurus brancai,* e) *Pleurocoelus nanus,* f) *Camarasaurus lentus,* g) *Opisthocoelicaudia skarzynskii,* h) *Apatosaurus excelsus,* i) *Diplodocus carnegii,* j) *Barosaurus? africanus,* k) *Alamosaurus sanjuanensis,* l) *Saltasaurus robustus.* (After Zhang et al. 1984 (a); McIntosh and Williams 1988 (c); Janensch 1947, 1956 (d, j); Gilmore 1925, 1925 (f, k); Borsuk-Bialynicka 1977 (g); Marsh 1883 (h); Hatcher 1901 (i); Huene 1929a (1).)

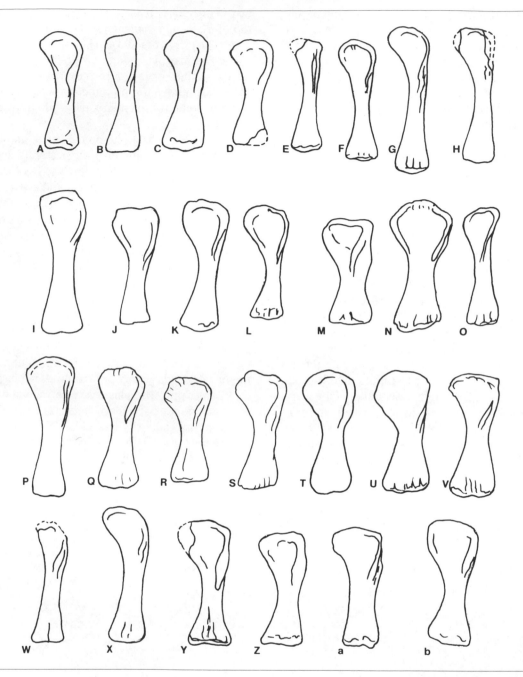

Fig. 16.10. Humeri, cranial view. a) *Cetiosaurus oxoniensis*, b) *Shunosaurus lii*, c) *Omeisaurus junghsiensis*, d) *Tienshanosaurus chitaiensis*, e) *"Cetiosaurus" humerocristatus*, f) French *"Bothriospondylus madagascariensis*,*"* g) *Brachiosaurus brancai*, h) *Pelorosaurus conybearei*, i) *Pleurocoelus nanus*, j) *Euhelopus zdanskyi*, k) *Camarasaurus? alenquerensis*, l) *Camarasaurus grandis*, m) *Opisthocoelicaudia skarzynskii*, n) *Apatosaurus louisae*, o) *Diplodocus* sp., p) *Barosaurus lentus*, q) *Barosaurus? africanus*, r) *Dicraeosaurus hansemanni*, s) *Cetiosauriscus stewarti*, t) *Mamenchisaurus* sp., u) *Janenschia robusta*, v) *"Pelorosaurus" becklesii*, w) *Aegyptosaurus baharijensis*, x) *Laplatasaurus araukanicus*, y) *Alamosaurus sanjuanensis*, z) *Saltasaurus australis*, a) *Argyrosaurus superbus*, b) *Magyarosaurus dacus*. (After Phillips 1870 (a); Dong and Tang 1984 (b); Dong et al. 1983 (c); Young 1937, 1954 (d, t); Hulke 1874 (e); Lapparent 1943 (f); Janensch 1914, 1950*b*, 1961 (g, q, r, u); Mantell 1850 (h); Wiman 1929 (j); Lapparent and Zbyszewski 1957 (k); Marsh 1877*a* (l); Borsuk-Bialynicka 1977 (m); Gilmore 1936, 1946*b* (n, y); Mook 1917*b* (o); Woodward 1905 (s); Stromer 1932 (w); Lydekker 1879, 1893*a* (x, a).)

1883) found with a skeleton of *Apatosaurus* can hardly be anything else.

The forelimb is solid and sturdy, the humerus always exceeding the radius and ulna in length. The humerus is straight and expanded at both ends, usually more so at the proximal (fig. 16.10). A prominent deltopectoral crest occupies the upper half of the lateral edge of the cranial face. The less expanded, stockier distal end exhibits small but distinct radial and ulnar condyles on its cranial side and a broad olecranon groove on the caudal one.

The ulna (fig. 16.11) is slightly longer than the radius, and the two bones range from a minimum of about three-fifths the length of the humerus in some brachiosaurids to nearly four-fifths in some diplodocids. The olecranon is rudimentary. The expanded head of the ulna has a triangular cross section, concave in front, to cradle the head of the radius, the latter lying more or less in front of the ulna rather than to the side. The radius does not cross the ulna as once thought (Osborn and Granger 1901; Hatcher 1903*d*). The two

bones are nearly straight but very slightly bowed cranially. The cross section of the radial shaft is nearly circular, often gradually broadening toward the distal end. The proximal end (fig. 16.12) is expanded, with a somewhat pointed expansion directed medially. The interosseous border runs nearly the length of the caudal face. In titanosaurids, both ends of the radius are abnormally expanded, and the distal margin is not perpendicular to the shaft but rather downwardly inclined.

Only two blocklike elements remain in the typical sauropod carpus, the larger a fusion of distal carpals I and II, the smaller a fusion of distal carpals IV and V (Osborn 1904). In *Shunosaurus*, there are three elements, the larger one covering metacarpals I and II, the smaller, metacarpals II and III. Lavocat (1955*b*) had previously reported three distal carpals in a large Middle Jurassic sauropod from Madagascar. The extreme reduction of the carpus to a single element lying above metacarpals II to IV occurs in the diplodocid *Apatosaurus*, but whether this is also true for other members of that family is not known.

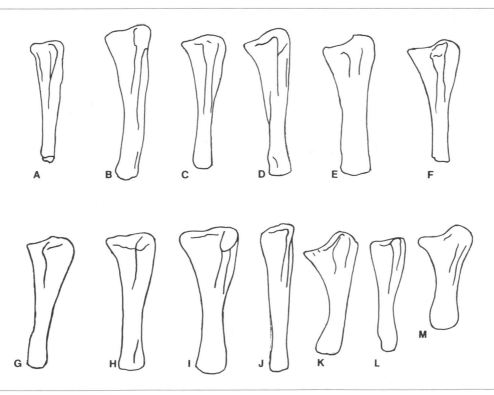

Fig. 16.11. Ulnae, craniolateral oblique view. a) *Vulcanodon karibaensis,* b) *Cetiosaurus oxoniensis,* c) French *"Bothriospondylus madagascariensis,"* d) *Brachiosaurus brancai,* e) *Chubutisaurus insignis,* f) *"Pelorosaurus" becklesii,* g) *Camarasaurus? alenquerensis,* h) *Camarasaurus grandis,* i) *Apato-* *saurus,* j) *Diplodocus* sp., k) *Janenschia robusta,* l) *Magyarosaurus dacus,* m) *Saltasaurus robustus.* (After Cooper 1984 (a); Lapparent 1943 (c); Janensch 1914, 1922, 1950*b* (d, k); Marsh 1977*a* (h); Gilmore 1936 (i); Mook 1917*b* (j).)

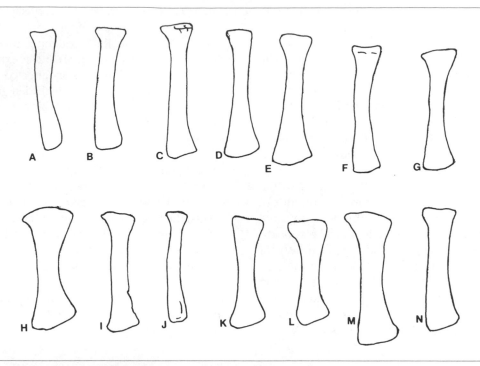

Fig. 16.12. Radius, cranial view. a) *Vulcanodon karibaensis*, b) *Cetiosaurus oxoniensis*, c) *Brachiosaurus brancai*, d) *Pleurocoelus nanus*, e) *Chubutisaurus insignis*, f) *Camarasaurus? alenquerensis*, g) *Camarasaurus grandis*, h) *Opisthocoelicaudia skarzynskii*, i) *Apatosaurus louisae*, j) *Diplodocus* sp., k) *Jan-* *enschia robusta*, l) *Saltasaurus robustus*, m) *Alamosaurus sanjuanensis*, n) "*Pelorosaurus*" *becklesii*. (After Cooper 1984 (a); Janensch 1922, 1950*b* (c, k); Borsuk-Bialynicka 1977 (h); Gilmore 1936, 1946*b* (i, m); Mook 1917*b* (j).)

All sauropods possessed a pentadactyl, digitigrade manus with five strong metacarpals but reduced number of phalanges on all digits except digit I. The pillarlike metacarpals stand vertically in a circle, metacarpal V almost reaching metacarpal I. They are of roughly equal length, but metacarpal III is usually the longest. In a few examples (e.g., *Brachiosaurus*), metacarpal II is slightly longer. The metacarpals range from long and very slender in brachiosaurids and camarasaurids to short and stocky in *Apatosaurus* to short and slender in *Diplodocus*. The metacarpus-to-humerus-length ratio varies between 0.23 to 0.25 in *Diplodocus* and *Apatosaurus* to 0.30 to 0.32 in *Camarasaurus*, *Brachiosaurus*, and *Chubutisaurus*. Only digit I retained a claw. In *Shunosaurus*, the phalangeal formula is 2-2-2-2-1. In *Brachiosaurus*, *Camarasaurus*, and *Apatosaurus*, it is 2-1-1-1-1 with the occasional inclusion of a small vestigial II-2. The inwardly directed claw on digit I is greatly reduced is *Brachiosaurus*, and Early Cretaceous presumably brachiosaurid footprints (Farlow 1987*a*) suggest an even further reduction or loss of the claw altogether.

The ilium, largest of the pelvic elements, is usually oriented with its long axis horizontal (fig. 16.13). The preacetabular process is more expanded than the postacetabular process and is inclined outward. The strong pubic peduncle, located at midlength or more forward, is much longer than the ischiadic peduncle, and its caudal face contributes significantly to the acetabular surface. The dorsal margin of the ilium is curved, and the bone here is very thin and often damaged. In *Brachiosaurus*, the preacetabular process and pubic peduncle are very expanded so that the axis of the bone is inclined upward and forward. In titanosaurids, the tip of the preacetabular process is outwardly inclined, affording a nearly horizontal orientation. Distinct vertical ridges on the medial face mark the articulations with the sacral ribs.

The pubis (fig. 16.14) is generally larger and more robust than the ischium, and the two bones are coossified along their articular margin in at least one old individual of *Apatosaurus*. Primitively, the blade forms an apron and the transverse diameter of the shaft greatly exceeds the sagittal one (e.g., *Vulcanodon*). In

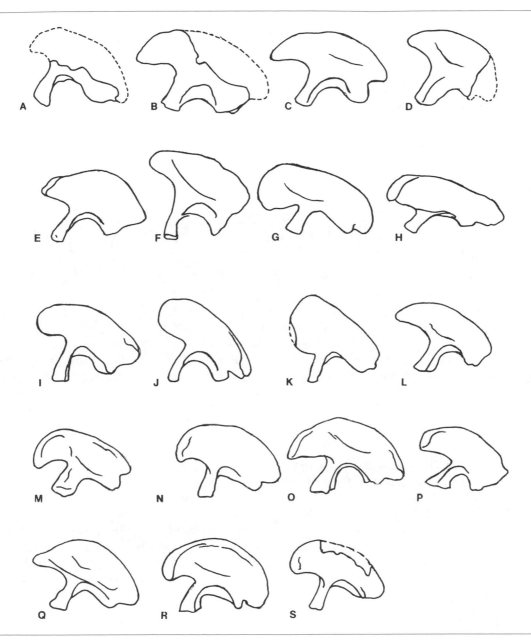

Fig. 16.13. Ilia, lateral view. a) *Vulcanodon karibaensis*, b) *Barapasaurus tagorei*, c) *Shunosaurus lii*, d) *Omeisaurus junghsiensis*, e) *Patagosaurus fariasi*, f) *Tienshanosaurus chitaiensis*, g) *Haplocanthosaurus priscus*, h) *"Apatosaurus" minimus*, i) *Volkheimeria chubutensis*, j) *Brachiosaurus brancai*, k) *Pelorosaurus conybearei*, l) *Euhelopus zdanskyi*, m) *Camarasaurus? alenquerensis*, n) *Camarasaurus grandis*, o) *Dicraeosaurus han-semanni*, p) *Mamenchisaurus hochuanensis*, q) *Apatosaurus excelsus*, r) *Diplodocus carnegii*, s) *Barosaurus lentus*. (After Cooper 1984 (a); Jain et al. 1975 (b); Dong et al. 1983 (c); Young 1937 (d, f); Bonaparte 1986b (e, i); Mook 1917a (h); Janensch 1914, 1950b (j, o); Lapparent and Zbyszewski 1957 (m); Marsh 1877a (n, q); Young and Chao 1972 (p); Hatcher 1901 (r).)

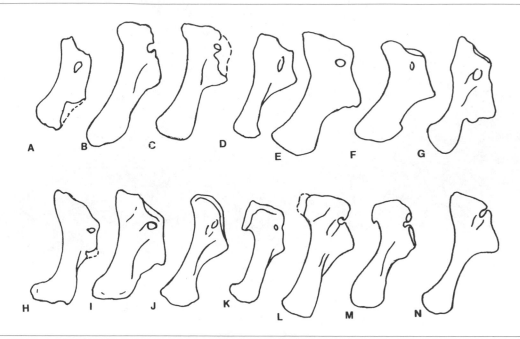

Fig. 16.14. Pubes, lateral view. a) *Cetiosaurus oxoniensis*, b) *Shunosaurus lii*, c) *Omeisaurus junghsiensis*, d) *Haplocanthosaurus priscus*, e) *"Apatosaurus" minimus*, f) *"Ornithopsis" leedsi*, g) *Brachiosaurus brancai*, h) *Camarasaurus? alenquerensis*, i) *Camarasaurus supremus*, j) *Apatosaurus excelsus*, k) *Diplodocus carnegii*, l) *Barosaurus lentus*, m) *Dicraeosaurus hans-* *emanni*, n) titanosaurid from Brazil. (After Phillips 1870 (a); Zhang et al. 1984 (b); Young 1937 (c); Mook 1917a (e); Janensch 1914, 1950b (g, m); Lapparent and Zbyszewski 1957 (h); Osborn and Mook 1921 (i); Marsh 1877a (j); (n) from a photograph courtesy of L. I. Price.)

derived forms, the shaft is less apronlike and both the shaft and distal end are much expanded. The pubic foramen appears closed in all adult forms.

The ischium (fig. 16.15) undergoes a similar transformation from the straight, elongated prosauropodlike form to a shorter bone at most equal in length to that of the pubis. The shaft is typically directed caudoventrally. The manner of articulation of the distal ends (which are sometimes fused) varies considerably. In *Vulcanodon*, *Haplocanthosaurus*, *Camarasaurus*, and *Opisthocoelicaudia*, the twisted shafts and narrow distal ends meet edge to edge. In *Brachiosaurus*, the same is true but the shafts and distal ends are broader, and the ischium is directed more nearly vertically. In diplodocids, including *Dicraeosaurus*, the distal ends are greatly expanded in both directions, and they meet more nearly side by side, the extreme development of the feature seen in *Apatosaurus*.

The three hindlimb elements are straight, sturdy bones differing from genus to genus largely in their overall robustness. The femur is always longer than the tibia and fibula, and the fibula is somewhat longer than the tibia, its lower end extending beyond that of the

tibia to fill the void left by the much reduced or absent calcaneum.

Except in some brachiosaurs, the femur (fig. 16.16) is the longest bone in the skeleton. It possesses only a hint of the sigmoid curve typical of many prosauropods, and it is expanded at both proximal and distal ends, usually a bit more so at the former. The straight shaft varies from having an almost circular cross section to an oval whose transverse diameter exceeds the craniocaudal one. Two prominent features are a ridge that rises at midlength on the medial side of the caudal face and represents the rudimentary fourth trochanter and a sharp deflection on the proximal third of the lateral margin, more prominent in brachiosaurids and titanosaurids than in other forms in which it may cease to exist. This may represent the last trace of an otherwise missing lesser trochanter (P. Galton pers. comm.). The head of the femur is almost at right angles to the shaft. It rises slightly above and is somewhat more expanded craniocaudally than the greater trochanter, which occupies the lateral half of the proximal end of the bone. There is no trace of a notch between the head and greater trochanter. The distal end

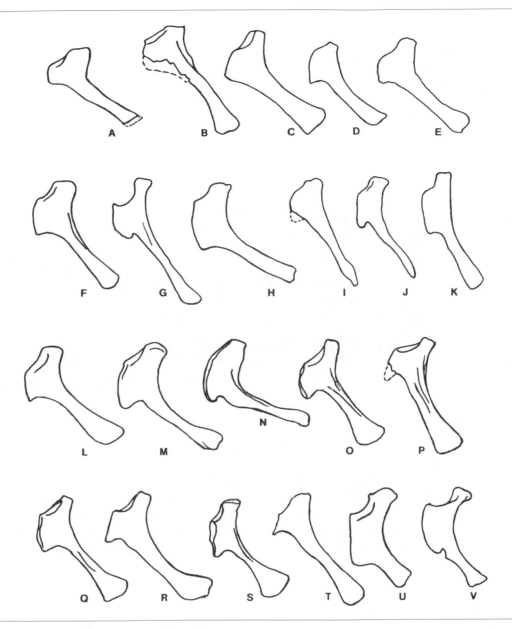

Fig. 16.15. Ischia, lateral view. a) *Vulcanodon karibaensis,* b) *Barapasaurus tagorei,* c) *Shunosaurus lii;* d) *Omeisaurus junghsiensis,* e) *Tienshanosaurus chitaiensis,* f) *Cetiosaurus oxoniensis,* g) *Haplocanthosaurus priscus,* h) *"Apatosaurus" minimus,* i) *"Ornithopsis" leedsi,* j) French *"Bothriospondylus madagascariensis,"* k) *Brachiosaurus brancai,* l) *Euhelopus zdanskyi;* (m); *Camarasaurus? alenquerensis,* n) *Camarasaurus lentus,* o) *Apatosaurus excelsus,* p) *Diplodocus carnegii,* q) *Bar-osaurus? africanus,* r) *Barosaurus lentus,* s) *Dicraeosaurus hansemanni,* t) *Mamenchisaurus hochuanensis,* u) *Alamosaurus sanjuanensis,* v) titanosaurid from Brazil. (After Cooper 1984 (a); Jain et al. 1975 (b); Zhang et al. 1984 (c); Young 1937 (d, e); Mook 1917a (h); Lapparent 1943 (j); Marsh 1877a, 1889a (n, o); Janensch 1914 (s); Gilmore 1946 (u); (v) from a photograph courtesy of L. I. Price.)

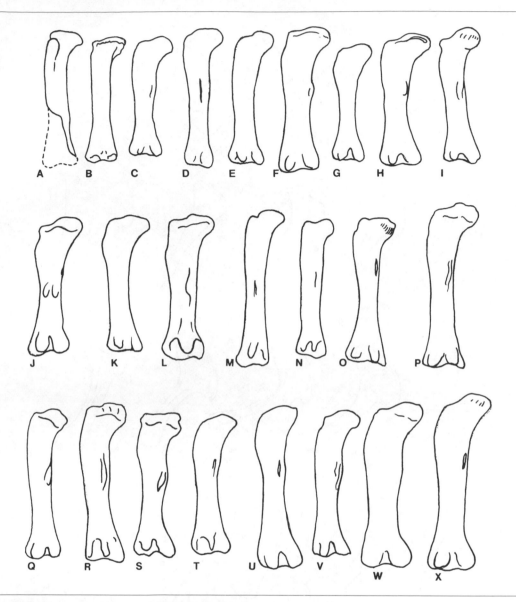

Fig. 16.16. Femora, caudal views. a) *Vulcanodon karibaensis*, b) *Barapasaurus tagorei*, c) *Cetiosaurus oxoniensis*, d) *Shunosaurus lii*, e) French *"Bothriospondylus madagascariensis,"* f) *Brachiosaurus altithorax*, g) *Pleurocoelus nanus*, h) *Camarasaurus? alenquerensis*, i) *Camarasaurus grandis*, j) *Opisthocoelicaudia skarzynskii*, k) *Euhelopus zdanskyi*, l) *Apatosaurus louisae*, m) *Amphicoelias altus*, n) *Diplodocus carnegii*, o) *Barosaurus lentus*, p) *Barosaurus? africanus*, q) *Cetiosauriscus stewarti*, r) *Dicraeosaurus hansemanni*, s) *Mamenchisaurus* sp., t) *Magyarosaurus dacus*, u) *Titanosaurus indicus*, v) *Aegyptosaurus baharijensis*, w) *Saltasaurus australis*, x) *Antarctosaurus wichmannianus*. (After Cooper 1984 (a); Jain et al. 1975 (b); Phillips 1870 (c); Zhang et al. 1984 (d); Lapparent 1943 (e); Riggs 1904 (f); Lapparent and Zbyszewski 1957 (h); Borsuk-Bialynicka 1977 (j); Gilmore 1936 (l); Woodward 1905 (q); Swinton 1947 (u); Stromer 1932 (v).)

of the femur is dominated by the two prominent condyles on the distocaudal face. The tibial condyle is the larger of the two.

The straight tibia has an expanded proximal end, on the lateral face of which is a well-developed cnemial crest that cradles the proximal end of the fibula (fig. 16.17). The shaft is slightly flattened craniocaudally. The distal end is less expanded than the proximal. A process that locks the astragalus in place is found on the caudomedial side of the distal end of the tibia.

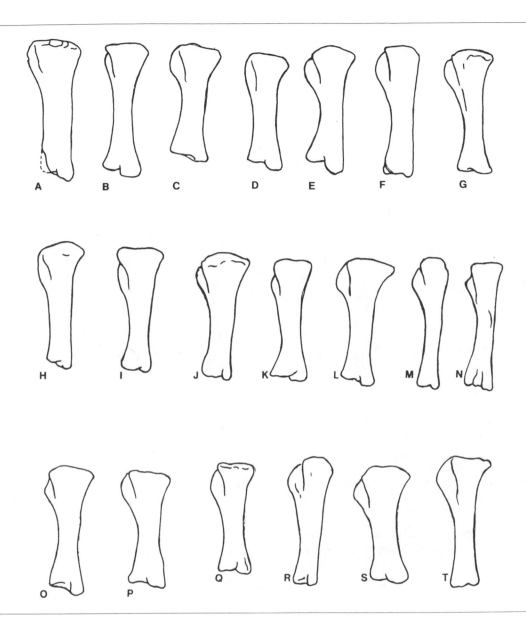

Fig. 16.17. Tibiae, caudal views. a) *Vulcanodon karibaenis*, b) *Barapasaurus tagorei*, c) *Cetiosaurus oxoniensis*, d) *Omeisaurus junghsiensis*, e) *Brachiosaurus brancai*, f) *Pleurocoelus nanus*, g) French "*Bothriospondylus madagascariensis*," h) *Chubutisaurus insignis*, i) *Euhelopus zdanskyi*, j) *Camarasaurus? alenquerensis*, k) *Camarasaurus grandis*, l) *Apatosaurus louisae*, m) *Diplodocus carnegii*, n) *Barosaurus lentus*, o) *Barosaurus? africanus*, p) *Mamenchisaurus hochuanensis*, q) *Janenschia robusta*, r) *Titanosaurus indicus*, s) *Saltasaurus robustus*, t) *Antarctosaurus wichmannianus*. (After Cooper 1984 (a); Jain et al. 1975 (b); Dong and Tang 1984 (d); Janensch 1914, 1950*b* (e, o); Lapparent 1943 (g); Lapparent and Zbyszewski 1957 (j); Gilmore 1936 (l); Young and Chao 1972 (p); Swinton 1947 (r).)

The slender straight fibula has a flattened proximal end (fig. 16.18). On the medial side of the proximal end, there is a triangular muscle scar, the hypoteneuse of which extends from the proximocaudal corner diagonally downward to the cranial border. The shaft shows another prominent muscle scar about halfway down on its external face. The medial face of the distal end is convex where it roughly abuts into a concavity on the lateral face of the astragalus.

The tarsus has been reduced to one or two proxi-

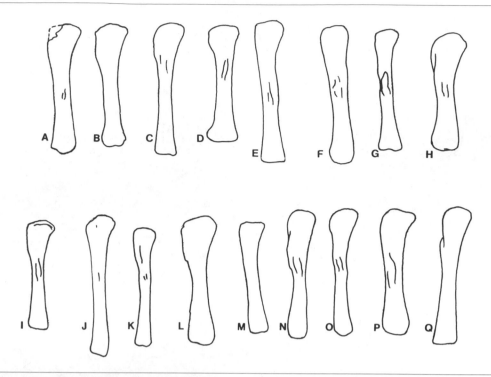

Fig. 16.18. Fibulae, lateral view. a) *Cetiosaurus oxoniensis,* b) *Omeisaurus junghsiensis,* c) French "*Bothriospondylus madagascariensis,*" d) *Brachiosaurus brancai,* e) *Pleurocoelus nanus,* f) *Camarasaurus? alenquerensis,* g) *Camarasaurus grandis,* h) *Opisthocoelicaudia skarzynskii,* i) *Apatosaurus louisae,* j) *Diplodocus carnegii,* k) *Barosaurus lentus,* l) *Mamenchisaurus hochuanensis,* m) *Tornieria robusta,* n) *Titanosaurus indicus,* o) *Magyarosaurus* sp., p) *Saltasaurus robustus,* q) *Laplatasaurus araukanicus.* (After Dong and Tang 1984 (b); Lapparent 1943 (c); Janensch 1950*b* (d); Lapparent and Zbyszewski 1950 (f); Marsh 1877*a* (g); Borsuk-Bialynicka 1977 (h); Gilmore 1936 (i); Young and Chao 1972 (l); Swinton 1947 (n).)

mal elements, a large astragalus, which articulates tightly with the distal end of tibia, and a very much reduced calcaneum, which has disappeared completely in diplodocids. The astragalus is a heavy, solid, block-like bone, broadly convex below, flat above with an excavation on the caudomedial quarter into which the downwardly directed process of the tibia fits. The lateral face is squarish but concave. The calcaneum, when present, is greatly reduced and lies just below the distal end of the fibula. It has a generally globular shape with a very rugose surface.

As with the manus, the pes is pentadactyl with a reduction in the number of phalanges of the outer digits, which is less pronounced than in the manus. Once described as plantigrade, it is now clear from footprints (Farlow 1987*a*) that semidigitigrade is a more apt description. The metatarsals are arranged in semicircular fashion, diverging outward and downward in contrast to the vertically oriented metacarpals. as revealed by footprints (Farlow 1987*a*), a heavy elephantlike pad

formed the heel. Metatarsal I is the shortest and most robust (fig. 16.19). Metatarsal II is longer and also robust. Metatarsals III and IV are long and more slender. Metatarsal V is shorter with a characteristically much larger proximal than distal end. Metatarsals II and III are the longest in some families (e.g., the Camarasauridae), III and IV are the longest in others (e.g., Diplodocidae). The phalangeal formula is usually 2-3-4-2-1, with minor variations such as the loss of IV-2 or III-3. The first three digits bore claws of decreasing size. Occasionally, an otherwise complete foot is reported lacking the claw on digit III, but this may well be due to its loss at the time of burial. Jensen (pers. comm.) has reported finding a small claw on digit IV of a *Camarasaurus* pes, but this can be no longer verified as it was supposedly lost during collection.

Dermal Armor
There is mounting evidence that all the titanosaurids possessed dermal armor. Deperet's (1896) claim

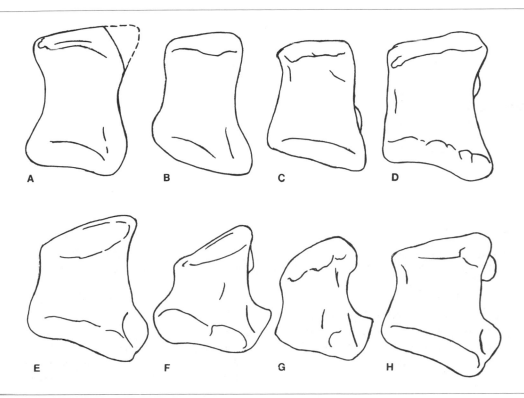

Fig. 16.19. Metatarsal I, cranial view. a) *Cetiosaurus medius*, b) *Macrurosaurus semnus*, c) *Euhelopus zdanskyi*, d) *Camarasaurus grandis*, e) *Cetiosauriscus stewarti*, f) *Apatosaurus excelsus*, g) *Diplodocus carnegii*, h) *Barosaurus lentus*. (After Marsh 1877*a* (d).)

that a scute found with titanosaur caudals in Madagascar was ignored until incontrovertible evidence of the association of several types of osteoderms with Argentinian titanosaurids was presented by Bonaparte and Powell (1980) and Powell (1980). More recently, Sanz and Buscalioni (1987) reported a third association from the Upper Cretaceous of Spain.

In *Saltasaurus* (Bonaparte and Powell 1980), the osteoderms are of two types: dermal plates 10 to 12 cm in diameter and ossicles 6 to 7 mm across. The relatively rare plates, whose position is not known with certainty, are oval to circular. Their dorsal surfaces are rugose and bear a low but sharp ridge; their ventral surfaces are described as smooth, "concave in some with a median ridge in others." The irregularly shaped ossicles, outwardly convex, are tightly packed and apparently covered the dorsal and lateral parts of the body. A third type of ossification typified by the material from Spain and Madagascar and Argentinian material referred by Huene (1929*a*) to a supposed nodosaurid *Loricosaurus scutatus* (Powell 1980) can be differentiated from the nodosaurids according to Sanz and Buscalioni by three characters: "the appearace of a

well-developed cingulum, surface texture made of nodules [without pattern], and the appearance of internal ducts."

SYSTEMATICS

Following the pioneering work of Marsh (1898*a*), most writers prior to 1929 recognized some half-dozen sauropod families. In that year, Janensch (1929*b*) proposed a new classification, which divided the group into two families based on those with broad spatulate teeth versus those with slender peg teeth. This scheme became widely adopted with variations as to the subfamilies and also the names of the families, those most commonly used being Brachiosauridae and Titanosauridae. The result of this classification was to throw together the highly specialized diplodocids with the, in some ways, more generalized Upper Cretaceous titanosaurs, two widely divergent groups, which except for the teeth, exhibit almost no resemblance in other parts of the skeleton. As discussed below, sauropod phylogeny remains very confused. Hence, it would appear prudent to reject the bifamily classification in favor of one that recognizes five or six.

Family Vulcanodontidae   Cooper, 1984

Family Characters. Teeth with coarse denticles on both edges; presacral vertebrae lacking pleurocoels; sacrum narrow with four functional sacrals; sacrocostal yoke not contributing to acetabulum; caudals deeply furrowed beneath with strong chevron facets; forelimb relatively long; pubes primitive, forming prosauropod-like apron; ischium noticeably longer than pubis; femur slender; calcaneum present; metatarsals elongated.

This family diagnosis must be regarded as tentative since it incorporates some characters possessed by only one of the two genera, *Vulcanodon* and *Barapasaurus,* assigned to the family; the condition in the other is unknown.

Vulcanodon Raath, 1972. The earliest and most primitive sauropod, from the Triassic-Jurassic boundary of Zimbabwe, is known from the pelvis, hindlimb and foot, forearm, proximal section of the tail, and fragments of the scapula, humerus, and manus. The skull and teeth are unknown.

The presacral vertebrae are totally unknown, and the preserved fragments indicate a narrow sacrum. The proximal caudal centra are amphicoelous and contain no pleurocoels but are strongly waisted. They contain a deep furrow on the ventral surface and strong chevron facets at both ends. The transverse processes are directed outward but also backward and slightly downward.

The fragmentary scapula appears to have its proximal end less expanded than in other primitive sauropods, for example, *Barapasaurus* and cetiosaurids. From an incomplete humerus, Cooper (1984) has estimated the ratio of lengths of the forearm to humerus as 0.9, similar to that given by Yadagiri et al. (1979) for *Barapasaurus.* This estimate may be too high. The latter bones are very slender but typically sauropod. The metacarpus is only a third the length of the radius, far shorter than in later forms. The pelvis is primitive. The preacetabular process of the ilium is not preserved, but the pubes more nearly resemble those of the prosauropods than of typical sauropods, with their shafts twisted to form a transverse apron similar to but even more primitive than in *Barapasaurus.* The long, straight ischium is much longer than the pubis. The shaft expands gently but continuously to the distal end, which meets its mate side by side in a manner reminiscent of diplodocids. The nearly complete, relatively slender femur is unique among sauropods in that it retains a reduced but well-marked lesser trochanter. The fourth

trochanter is also more prominent than in later forms. The tibia : femur ratio is a relatively small 0.58 (Raath 1972). The ulna : tibia ratio is a large 1.04. The astragalus is large, and there is a prominent calcaneum. The longest metatarsal is the third, 0.32 the length of the tibia and much longer than in any other sauropod.

Barapasaurus Jain, Kutty, Roy-Chowdhury et Chatterjee, 1975. All parts of the Early Jurassic Indian genus *Barapasaurus* have been recovered except for the skull and feet, making it potentially the best-known Early Jurassic sauropod, but thus far the animal has received only a very preliminary description. Thus, its inclusion in the Vulcanodontidae rather than the Cetiosauridae is tentative but appears to be justified by the narrow sacrum.

No part of the skull is known except for isolated teeth. These are spoon shaped with coarse denticles on both edges.

The vertebral formula is unknown. The cervicals and cranial dorsals are opisthocoelous; the caudal dorsals platycoelous. The presacral centra have shallow depressions on their lateral faces but no true pleurocoels. The lightening of the neural arch by thin laminae of the more advanced sauropods is little developed, but instead deep excavations occur in the side of the arch below the transverse processes (somewhat as in *Dicraeosaurus*), and there is also a very significant central hollowing of the cranial and caudal faces of the arch below the zygapophyses around the neural canal, which is itself a vertical slot in cross section. The diapophyses of the dorsals are directed laterally, and the presacral neural spines are all undivided. Four sacral vertebrae (including a dorsosacral) contribute to the sacrocostal yoke, which itself does not contribute to the acetabulum. The spines of sacrals 2 and 3 are united. The most striking feature is the extreme narrowness of the sacrum, a primitive character. The tail remains undescribed except for the observation that the caudal centra are not procoelous and do not contain pleurocoels. Nothing is known concerning the middle chevrons.

The scapula resembles somewhat that of *Cetiosaurus.* The proximal end is less expanded than in later forms, the shaft relatively broad, and the distal end moderately expanded. The forelimb has not yet been described by Jain et al. (1975, 1979), but Yadagiri et al. (1979) state that the humerus, radius, ulna, tibia, and fibula are of almost the same length and that the humerus is two-thirds the length of the femur. If these statements are accurate, they are of great significance, first, since the sauropod forearm in later forms (see, however, *Vulcanodon* above) ranges from about two-thirds to three-fifths the length of the humerus, and, second, since in the cetiosaurids the humerus is four-fifths or more the length of the femur. However, these ratios should be viewed with caution since published photographs of a mounted skeleton of *Barapasaurus* show a humerus considerably longer than the radius and ulna. The ilium is of the typical sauropod type with the expanded preacetabular process. The pubis has a slender shaft and a relatively long symphysis, resulting in a pelvic cavity smaller than in later sauropods but greater than in prosauropods. The ischium is longer than the pubis, with a slender, straight shaft and small distal expansion. The femur is straight and relatively slender and retains a central cavity but no lesser trochanter, in sharp contrast to that of *Vulcanodon*.

Ohmdenosaurus liasicus Wild, 1978a, from the Early Jurassic of West Germany may be a vulcanodontid. It is represented by a tibia and astragalus of a very primitive sauropod. The tibia is very similar to that of *Vulcanodon*. The astragalus has the same general shape, but it is more compressed vertically and possesses a concavity on the ventral surface unique among sauropods.

Zizhongosaurus chuanchengensis Dong et al., 1983, known from a dorsal arch and a fragmentary pubis and humerus, is important because it documents a primitive sauropod in the Lower Jurassic of China, but it cannot be properly evaluated based on what is now known of it. The dorsal spine is undivided and of moderate height. The diapophyses are stout and directed laterally.

Family Cetiosauridae   Lydekker, 1888b

Family Characters. Spatulate teeth smaller, more slender, and more numerous than those of the brachiosaurids and camarasaurids; cervical centra with relatively simple pleurocoels; short simple caudals with large chevron facets; moderate humerofemoral ratio.

As here constituted, the Cetiosauridae represent a family of relatively generalized sauropods, all members of which are incompletely known at this time. That they comprise a monophyletic group is far from certain. I have recognized two subfamilies, the Cetiosaurinae with conventional chevrons and the Shunosaurinae, Chinese forms with forked chevrons in the midtail region like those of *Diplodocus*. Except for a few teeth and the jaws of *Patagosaurus*, the skull is not known in the Cetiosaurinae, although a recently reported (Monbaron et al. 1981, 1983) but as yet undescribed partial skull and nearly complete skeleton from Morocco may belong to this subfamily. When the new

material from China and Morocco has been fully described, a much clearer picture of the Cetiosauridae will no doubt emerge.

SUBFAMILY CETIOSAURINAE JANENSCH, 1929b

Cetiosaurus Owen, 1841. The characterization of *Cetiosaurus* is based on the Middle Jurassic English species *C. oxoniensis* known from most parts of the skeleton except the skull. Whether the Moroccan species, *C. mogrebiensis*, belongs to this genus must await the description of the skeleton now under study (Monbaron 1983). Certain limb proportions in the two species are quite different, so the African form will be treated separately.

No part of the skull is known except perhaps for a single incomplete tooth of the spatulate variety. The heart-shaped teeth of *Cardiodon* have often been referred to *Cetiosaurus*, but this cannot be established.

The numbers of cervical and dorsal vertebrae are unknown. The cervical centra are of moderate length and contain elongated, rather shallow pleurocoels uncomplicated by dividing septa seen in many advanced forms. The neural spines are undivided. The amphicoelous or platycoelous middle and caudal dorsal centra also contain pleurocoels. The arches are high, and the spines are low. The diapophyses are directed upward as well as outward. The sacrum is not known. The relatively short caudal centra are amphicoelous, lack pleurocoels, and exhibit prominent chevron facets. The arches and spines of the proximal and middle caudals resemble those of *Camarasaurus*. Of the chevrons so far reported, none are forked as in diplodocids.

The proximal part of the scapula is less broad than in most later forms, but the distal end is widely expanded. The forelimbs are of moderate length (humerus:femur ratio = 0.80) in contrast to those of the African species discussed below. The three bones are a bit more robust than those of *Camarasaurus*, much more so than in *Shunosaurus*, *Barapasaurus*, *Diplodocus*, or *Brachiosaurus* but less so than in *Apatosaurus*, *Opisthocoelicaudia*, or *Saltasaurus*. The pubis is relatively broad, the ischium slender and only slightly expanded distally. The bones of the forearm are relatively long compared to those of the lower leg (ulna:tibia = 1.00), exceeded only by *Vulcanodon* and the skeleton from France originally referred to *Bothriospondylus madagascariensis* (Dorlodot 1934; Lapparent 1943), both of which have ratios of 1.04.

Cetiosaurus ?mogrebiensis (Lapparent 1955). The original partial skeletons of this Moroccan species were already characterized by a relatively long forelimb compared to that of the English form, but the recently reported complete skeleton raises questions concerning its systematic position. The maxilla is characterized by a broad dorsal process like those of the Chinese cetiosaurids (Monbaron 1983). There are at least thirteen dorsal vertebrae. Their centra contain large pleurocoels. The scapula is heavy with a ventral end more expanded than in *C. oxoniensis*. The forelimb is much longer than in the latter species (humerus : femur ratio is 0.97), and all three forelimb bones are considerably more slender. The bones of the forearm are relatively shorter—humerus : radius ratio is 0.62 in the Moroccan species, 0.75 in the English. The relationship of this species to *Cetiosaurus* is yet to be established.

Haplocanthosaurus Hatcher, 1903*b*. *Haplocanthosaurus* represents a holdover of the Cetiosauridae into the Late Jurassic of the United States. No part of the skull or teeth can yet be assigned with certainty to this animal, nor can any of the limb or foot bones save the femur. Otherwise, except for the latter half of the tail, the *Haplocanthosaurus* skeleton has been well described and figured. Two species are recognized: *H. priscus* and *H. delfsi*.

Gilmore (1907) has suggested that the fragment of the cranium together with cervicals 1 to 3 named *Morosaurus agilis* by Marsh (1889*a*) may belong to *Haplocanthosaurus*, and, although the suggestion has merit, future discoveries of direct association will be needed before the question is settled.

Hatcher (1903*c*) estimated the number of cervicals to be 15 (as in *Diplodocus*), but examination of the most complete neck of this form suggests that this number may be too large. The centra are relatively short with well-developed pleurocoels divided into a smaller cranial and larger caudal part by a ridge. The arches are of moderate height, rising toward the back. The spines are simple. There are 14 dorsal vertebrae. The centra are small, opisthocoelous throughout but less so caudally, with well-developed ovate pleurocoels. The arches are high with transverse processes extending upward as well as outward. The spines are short to moderate and not bifurcated. The overall height of the dorsals increases in the first third of the thorax and then remains stationary. There are the usual five sacrals containing pleurocoels. The sacral spines are relatively short, and the number that coalesce varies from three to five. The caudal centra are amphicoelous, and their ventral surfaces contain a deep longitudinal furrow. Both fore and rear chevron facets are strongly developed. The transverse processes are directed straight outward and persist for about eleven vertebrae. The arches rest on the centers of the centra, and the simple spines are directed vertically on the proximal caudals but become gradually inclined somewhat caudally on the midcaudals. The chevrons are simple and lack the forking of the Chinese cetiosaurids and diplodocids.

As in all cetiosaurids, the proximal end of the scapula is less developed than in other sauropods, and the angle of its ridge with the shaft is large. The shaft itself is longer and less broad than that of *Camarasaurus*, but the distal end is just as widely splayed, these two genera having the broadest distal expansion. The coracoid is subcircular. The ilium has a relatively short pubic peduncle. The pubis is less robust than in *Camarasaurus*, and the ischium has a straight, slender shaft slightly expanded at the distal end. The femur, the only known limb bone, is not distinctive.

Amygdalodon Cabrera, 1947. The remains assigned to this Middle Jurassic Argentinian species, *A. patagonicus*, include teeth, a cervical centrum, three dorsal centra and arches without apophyses, three caudals, ribs, most of a pubis, and the distal end of a tibia. These indicate cetiosaurid affinities but do not serve to clearly differentiate this animal from the other members of the family. The expanded tooth crowns are heart shaped, closely resembling those of the English *Cardiodon*. The entire lateral face of the elongated cervical centrum is concave, as in *Cetiosaurus*, rather than containing a sharp-lipped pleurocoel. It is relatively longer than that of the *Patagosaurus* (Bonaparte 1987). Ventrally, it has a sharp median ridge. The dorsal centra are platycoelous and contain depressions rather than true pleurocoels. They resemble those of *Cetiosaurus*, except that ventrally they are gently concave. The proximal caudal is procoelous, the median caudals amphicoelous. The pubis is relatively massive, resembling *Cetiosaurus* rather than *Barapasaurus* or *Vulcanodon*. It is a bit more robust than that of *Patagosaurus*.

Patagosaurus Bonaparte, 1979*b*. This second Middle Jurassic Argentinian cetiosaurid, *P. fariasi*, is known from pieces of the jaws, vertebrae from all parts of the column, and all the girdle and limb bones except the fibula and fore and hind feet.

The maxilla possesses a broad dorsal process, arising behind a relatively low external naris. The teeth are broad and resemble those of *Amygdalodon*. The dental formula is

$$\frac{4, 12}{?}.$$

The number of presacral vertebrae is unknown. The cervical and cervical dorsal centra are opisthocoelous, becoming amphiplatyan in the mid and cranial dorsal regions. The lateral depressions in the cervical and caudal dorsal centra are deeper than those in

Amygdalodon but do not contain the sharply lipped pleurocoels seen in *Haplocanthosaurus*. In the caudal dorsals, the depressions are less deep. All preserved neural spines remain undivided. The arches are lower and the spines higher than those of *Haplocanthosaurus*. They resemble more nearly those of *Cetiosaurus*. The transverse processes are directed upward as well as outward as in *Haplocanthosaurus* and *Cetiosaurus* and in contrast to those of *Omeisaurus*. All five sacral centra are tightly fused, but fusion of the spines occurs only in sacrals 2 and 3. The caudal centra are amphiplatyan. The distal caudals resemble most closely those of *Cetiosauriscus* with their long, backwardly directed spines. The chevrons are apparently all of the normal type. The hemal canal is enclosed dorsally.

The scapula resembles that of *Cetiosaurus* with a broad distal expansion and modest proximal expansion and with a 60 to 70 degree angle between shaft and ridge. The three bones of the forelimb are virtually identical with those of *Cetiosaurus*. The ilium is deep with smaller than normal cranial crest. The ischium is a bit more robust and more expanded distally than in *Cetiosaurus*. Tibia:femur varies from 0.68 in the juvenile to 0.58 in the adult.

SUBFAMILY SHUNOSAURINAE, NOV.

Shunosaurus Dong, Zhou, et Zhang, 1983. Nearly complete skeletons of this Middle Jurassic Chinese genus have been found and, when completely described and figured, it will become only the second sauropod (after *Camarasaurus*) to be known in its entirety.

The skull (Zhang et al. 1984) is relatively long and low with narcs facing out. The teeth are relatively small, spatulate, and with long crowns intermediate in shape between those of *Camarasaurus* and *Diplodocus*. They were more numerous than in later forms. The dental formula is

$$\frac{4, 18 \text{ or } 19}{20 \text{ or } 21}.$$

The ascending process of the maxilla is broad.

There are probably 13 cervical vertebrae. The centra are short, opisthocoelous, with simple lateral depressions and a longitudinal ventral keel. The spines of the caudal cervicals become relatively high and club shaped. According to Dong, Zhou, and Zhang (1983), the last several spines are weakly bifurcated, a feature not mentioned by Zhang, Yang, and Peng (1984) in their revised description based on more complete material. There are thirteen dorsal vertebrae and a dorsosacral, whose centra are platycoelous and solid without honeycomb structure. They possess some lateral depressions but no true pleurocoels. The neural arches are fairly high, increasing from front to rear. The transverse processes of the caudal dorsals are directed upward as well as outward. The first dorsosacral is fused to three remaining sacrals in adults, and the fairly high spines of sacrals 2 and 3 may do so as well. The caudosacral remains free, but the rib is a winglike plate. The centra from sacral 3 back through the middle caudals are short and procoelous, gradually becoming amphiplatyan. There are 43 caudals. The middle chevrons are forked.

The shaft of the scapula is long, its distal end little expanded. The forelimb is relatively short (humerus:femur ratio is 0.62). There are three carpals in the distal row, which increase in size from the medial side outward. The third metacarpal is the longest. The phalangeal formula of the manus is 2-2-2-2-?1. The pubic peduncle of the ilium is fairly long and forward of the center of the bone. The pubis and ischium resemble those of *Diplodocus*, the distal ends of the latter somewhat expanded. The femur is straight and slender. Even in this early form, there is no lesser trochanter, and the prominence on the proximal half of the lateral margin which might represent its final trace is barely perceptible. The tibia:femur ratio is 0.57, and the muscle scar on the lateral face of the fibula is weak. A small calcaneum abuts the lateral face of the astragalus just below the distal end of the fibula. Metatarsal III is the longest. The reported phalangeal formula of the pes is 2-3-3-3-2 without a well-developed claw on digit III.

Datousaurus Dong et Tang, 1984. *Datousaurus bashanensis* represents a second and much rarer sauropod from the Middle Jurassic of China. Also found isolated was the front half of a large non-*Shunosaurus* skull that Dong and Tang (1984) referred to this animal. While they are probably correct, past experience of assigning sauropod skulls to postcranial skeletons not found in direct contact suggests that the assignment be viewed with caution. Several fairly good skeletons have been found but have so far received only preliminary descriptions. Were it not for the skull, *Datousaurus* would be a prime candidate for an ancestor of the diplodocids.

The skull is short and high, with a blunt rounded muzzle reminiscent of that of *Camarasaurus*, but the nares are smaller. The jaws are massive, and the ascending process of the maxilla is more slender than in *Shunosaurus* and *Cetiosaurus ?mogrebiensis*. The teeth are large and spoon shaped, with a median ridge on the concave surface. Four or five small cusps are present on the mesial edge of the premaxillary tooth crown, and denticles are likewise present on the maxillary teeth but are lacking on the dentary teeth. The dentary is more massive than in *Shunosaurus*. The dental formula is

$$\frac{4, 10 \text{ to } 12}{10 \text{ to } 12}.$$

The braincase and palate are unknown.

There were at least twelve cervical vertebrae. They are at once distinguished from those of *Shunosaurus* by being much longer (cervical length to dorsal length 2.5 : 1) with lower spines. The pleurocoels are better developed. The neural spines are undivided throughout the presacral region. The number of dorsals is 13. The centra are platycoelous and have lateral depressions rather than true pleurocoels. The arches and spines are high, the latter widened transversely. The fragmentary sacrum is said to have four vertebrae. The caudal centra, amphiplatyan and higher than long in the proximal part of the tail, become progressively amphicoelous with a greater length-to-height ratio distally. The middle chevrons are forked.

A clavicle has been reported. The humerus is stouter than that of *Shunosaurus*, as is the shaft of the pubis. The distal end of the ischium has a moderate expansion, not as great as in the diplodocids but greater than in *Camarasaurus*. The straight femur is said to be nearly round in cross section. The fourth trochanter is located just above half length of the femur. The moderately expanded distal end of the tibia exhibits the descending astragular process of the later sauropods.

Omeisaurus Young, 1939. *Omeisaurus* is now known from most parts of the skeleton including the skull. When fully described, it will be one of the best-known sauropods. This Late Jurassic Chinese sauropod is most closely allied to, and perhaps a descendant of, *Datousaurus*. Among the several species, *O junghsiensis* and *O. tianfuensis* are most completely known.

The skull most nearly resembles that of *Shunosaurus* but is relatively shorter and higher, and the supratemporal fenestra is larger. There is a small premaxillary foramen. The dorsal process of the maxilla is still large and heavy, but above and in front of it, below the external naris, there is a shelf much like that in *Camarasaurus*, which is lacking in *Shunosaurus*. The massive dentary also more nearly resembles that of the former than that of the latter. The dental formula is reduced to

$$\frac{4, 11}{13 \text{ to } 15},$$

and the teeth are much larger than in *Shunosaurus* but still retain serrations on their mesial margins.

There are 15 to 17 cervical vertebrae. They resemble those of *Datousaurus*. The opisthocoelous centra have well-developed pleurocoels and are keeled ventrally. The spines are relatively low and are undivided in all presacrals. The cervical ribs are long and slender. There are 12 dorsals, the 13th here considered a sacral. The centra are opisthocoelous but not keeled ventrally. The pleurocoels are undeveloped in dorsal 1 but strongly developed from dorsal 4 to 12. The hyposphene-hypantrum articulation occurs first between dorsals 3 and 4 and persists to the caudal dorsals. The parapophyses migrate upward on the centrum from dorsals 1 to 3 and reach the arch on dorsal 4. The neural spines are of moderate height and decrease in transverse width from fore to rear. The typical laminae of sauropod dorsals are well developed. There are four coossified sacrals, but the caudosacral remains free. Fusion of the sacral spines is variable, ranging from 1 to 4 to merely sacral 2 and 3. Sacral centra lack pleurocoels. Sacrals 1 to 4 contribute to the typical sacrocostal yoke. The caudosacral has a large fan-shaped rib that apparently does not fuse to the yoke. The caudal centra are amphicoelous, the proximal ones only faintly so, and ridged ventrally. The proximal caudal ribs are relatively long and directed straight out, decreasing in size backward and disappearing after caudal 14. The neural spines are long and flat, directed upward in the proximal part of the tail, but they become increasingly swept backward. They are not expanded transversely. Forked chevrons begin at about caudal 16 or 17.

The proximal plate of the scapula is more developed than in many cetiosaurids; the distal end is less expanded than in *Camarasaurus*. The scapulocoracoid articulation margin is fairly straight. Dong, Zhou, and Zhang (1983) identify a rather straight bone, flattened and slightly expanded at one end, as the clavicle, but its large size (0.9 m) casts some doubt on the identification. The humerus is stouter than that of *Shunosaurus*, comparable to that of *Camarasaurus*, but with a less expanded proximal end. Comparative limb measurements approximated from Young's (1939) original material suggest a humerus : femur ratio of 0.80 to 0.85. The pubic peduncle of the ilium is placed at midlength. The shaft of the pubis is slender, the distal end somewhat expanded. The ischium resembles somewhat that of *Camarasaurus*, but the distal end is a bit more expanded. The tibia and fibula are relatively robust, the muscle scar on the latter not pronounced. Metatarsals II and III are the longest.

Rhoetosaurus Longman, 1926. Although apparently of Early Jurassic age, the large Australian form *Rhoetosaurus brownei* (Longman 1926, 1927) clearly possessed derived features separating it from the Vulcanodontidae. It is known from incomplete cervicals, dorsals, sacrals, and a series of proximal and middle caudals, several of the latter complete. From the appendicular skeleton, there are the incomplete pubes, the greater part of a femur, and the as yet undescribed tibia, fibula, and pes (Thulborn, pers. comm.).

The single incomplete opisthocoelous cervical possesses a distinct pleurocoel. The preserved dorsals all have the ball and socket articulation of the centra, which also possess well-defined pleurocoels. The arches are very high, but the spines and transverse process are incomplete. However, the arch shows the development of the thin laminae separated by cavities so typical of later sauropods. A well-developed hyposphene-hypantrum articulation is also evident. The proximal caudal centra are short and the arches and spines relatively high. The middle caudals are also short.

The incomplete pubes are notable for their very large size. The femur is quite robust, in sharp contrast to those of the vulcanodonts. The fourth trochanter is placed at midlength.

Family Brachiosauridae   Riggs, 1904

Family Characters. Skull moderately long with elevated nasals and large external nares. Elongated slender cervical ribs, neural spines in sacral region very low, tail quite short with low spines and simple chevrons; forelimb greatly elongate (humerus:femur ratio = 0.9 to 1.05); long slender metacarpals; ilium with deep preacetabular process, long pubic peduncle, longitudinal axis of ilium directed upward; ischium directed downward with twisted shaft so that unexpanded distal ends meet edge to edge; femur with a prominence on upper lateral margin.

Brachiosaurus Riggs, 1903a. *Brachiosaurus* is known from all parts of the skeleton except for the important neural arches of the caudal cervical and cranial dorsal vertebrae and much of the pes. It was one of the largest sauropods and the one in which the lengthening of the forelimb relative to the hindlimb reached its extreme. *B. altithorax* was present in the Late Jurassic of North America, *B. brancai* in eastern Africa, and *B. atalaiensis* in Europe.

The skull is relatively large, of the general camarasaur type but broader and with a more elongate snout. The jaws are robust with broad subspatulate teeth, the crowns convex laterally, concave medially, many with a median ridge. The dental formula is

$$\frac{4, 11 \text{ to } 13}{13 \text{ to } 15}$$

The maxilla was relatively longer than that of *Camarasaurus,* its ascending process rising from farther back near the center of the bone. The external nares are placed high up on the lateral face of the skull and sepa-

rated from one another by a very narrow strip of bone formed from the premaxilla and nasal. There is a pair of small foramina between the premaxilla and maxilla. Unlike most sauropods, the jugal appears to reach the lower rim of the skull, barely separating the maxilla from the quadratojugal. The quadrate is nearly perpendicular to the skull roof. The parasphenoid is more slender than in *Camarasaurus,* and the basipterygoid processes are relatively short and stout. Many elements are similar to those of *Camarasaurus,* among them the quadratojugal, squamosal, quadrate, pterygoid, and postorbital. The lacrimal and jugal are somewhat broader, the vomer smaller, and the palatine broader dorsally. The mandible resembles that of *Camarasaurus* but is relatively longer and the dentary less robust.

The number of presacral vertebrae is uncertain because of questions concerning those in the shoulder region. There are 13 cervicals and 11 or perhaps 12 dorsals. The mid and caudal cervicals are enormously elongated with deep, complex pleurocoels. They are distinguished from the equally long *Barosaurus* cervicals in that the neural arch covers the cranial four-fifths of the centrum rather than all of it, and neural spines are all undivided. The ribs of the caudal cervicals are very much elongated, continuing backward under the next cervical or two. The first few dorsals, particularly their arches and spines, are incomplete or absent in all known skeletons. All the dorsal centra are opisthocoelous with deep pleurocoels. The arches and spines next to the sacrum are very low but increase in height steadily forward to at least dorsal 4, probably reaching a maximum height at dorsal 1 or 2. The high neural spines in the shoulder region are correlated with the very long forelimbs, just as the high neural spines in the sacral region of the diplodocids are correlated with the relatively long hindlimbs. The five sacral centra are coalesced, and the sacrum is notable for the very low neural arches and spines. The tail is very short with more than 50 short caudals, the centra of which are amphicoelous and without pleurocoels. The caudal ribs are simple but relatively long at the base of the tail, being directed caudolaterally. The arches and spines are low and simple. The chevron facets are weak and the chevrons themselves simple throughout without a bridge of bone across the hemal canal.

Relative to the pelvis and hindlimb, the pectoral girdle and limb are relatively larger than in any other sauropod. The scapula is greatly expanded both proximally and particularly distally, where the end is widely splayed. The coracoid lacks the regular oval or quadrangular shape of the camarasaurids and diplodocids and had an irregular outline, the scapular border exhibiting a jog upward from rear to fore at midlength.

All the forelimb bones, but particularly the humerus, are enormously long and slender. The latter is usually a bit longer than the femur. The ulna:humerus ratio is 0.62. There are two ossified carpals. The metacarpals are greatly elongated (metacarpal III:humerus is 0.30). The phalanges are greatly reduced in size (formula is 2-1-1-1-1), and, in particular, the claw on digit I is very small. Because the forelimbs are elongated, the axis of the ilium is inclined sharply upward from rear to fore, with a greatly elongated slender pubic peduncle and a greater development of the preacetabular process. The pubes are massive and lack the ambiens process on the proximal end. The shafts of the ischia are directed more nearly downward. These shafts are of moderate breadth and do not increase or thicken toward the distal end, which meets its mate edge to edge as in *Camarasaurus*. The transverse diameter of the shaft of the femur greatly exceeds the longitudinal diameter. The upward sweep of the lateral margin exhibits a prominence at three-quarters height. The lower segment of the leg is relatively short (tibia:femur ratio = 0.58). The tibia is massive with a well-developed cnemial crest. The fibula, barely longer than the tibia, is also massive with little development of the muscle scar on its lateral face. The inwardly directed process on the distal end, which fits loosely into the lateral concavity of the astragalus, is unusually large. Several blocklike elements of varying shape have been identified by Janensch as the calcaneum. They are unlike the globular calcanea of *Camarasaurus* and *Shunosaurus* but bear some resemblance to the bone assigned by Huene (1929*a*) to the calcaneum of *Saltasaurus* (= *Titanosaurus*) *australis*. The metatarsals of the incompletely known pes appear to be relatively elongate.

Bothriospondylus Owen, 1875. The type species, *B. suffosus*, was founded on four caudal dorsal and three uncoossified sacral centra of a juvenile sauropod from the Upper Jurassic of England. The dorsal centra are gently opisthocoelous, but the cranial end is almost flat, lacking the greater convexity of *Brachiosaurus*. They contain fairly prominent pleurocoels, as do the sacrals, but one of the latter, which is greatly flattened vertically, possesses a large parapophysis on the caudal end of the centrum and a smaller one on the rear, a most unusual condition. The others suggest a juvenile brachiosaurid.

A series of vertebral centra, limb, and girdle bones of numerous adult animals from the Middle Jurassic of Madagascar have been referred to this genus as *B. madagascariensis*. In addition, Ogier (1975) referred to this species the beautifully preserved remains of three or four disarticulated juvenile skeletons collected by Ginsburg in the same stratum. These were recently referred to a new genus *Lapparentosaurus* by (Bonaparte 1986*b*) as discussed below. Of the adult material, all that is known of the skull is a single broad-crowned tooth. Presacrals are represented by several cervical centra with prominent but less complex pleurocoels than in *Brachiosaurus*, a number of dorsal centra with deep pocketlike pleurocoels of the *Brachiosaurus* or *Camarasaurus* type, and caudals similar to those of *Brachiosaurus* but perhaps a bit shorter. The limb bones resemble closely those of *Brachiosaurus*, but relative proportions are not available as these bones were not associated. Significantly, three bones are reported (Lavocat 1955*b*) in the carpus. The ischia are not as broad distally as in *Brachiosaurus*. That the Madagascar species actually belongs to *Bothriospondylus* is doubtful, particularly in light of the dubious status of the type species. However, pending a full review of the material as well as that discussed in the ensuing paragraphs and a full description of the juvenile *Lapparentosaurus* material, it is perhaps prudent to continue to use *B. madagascariensis*.

One of the best sauropod finds so far in Europe is a partial skeleton from the Upper Jurassic of France. It has also been referred (Dorlodot 1934; Lapparent 1943) to *B. madagascariensis*. Teeth, several presacrals, sacrum, limb and girdle bones (except for the radius), and partial feet were found. The limb and several of the dorsal centra are indeed very similar to those of the adult animal from Madagascar, but it would not be surprising if more complete material of the latter would show the animals to be different. The teeth are very similar to those of *Brachiosaurus*, with large subspatulate crowns, not as broad as those of *Camarasaurus*, and with a low ridge on the concave labial face. The nearly complete relatively short cervical is probably the last. Its pleurocoel is very large and deep but not subdivided. The centrum is grooved below. The vertebra appears much less broad then that of *Brachiosaurus*, but this is due, in part at least, to crushing. The middle dorsals have a less prominent cranial ball but an equally large pleurocoel. The sacrocostal yoke tapers from fore to rear as in *Brachiosaurus*, but the centra of only four vertebrae are coossified. Several of the sacrals appear to possess the unusual character of *B. suffosus* in having the parapophysis shared by two vertebrae. The leading edge of the incomplete scapula is clearly less splayed than in *Brachiosaurus*. The humerus is slender but not as extreme as in the latter. The humerofemoral length ratio is 0.90 and ulna:humerus is 0.69. The ilium, ischium, and hindlimb bones are virtually identical to those of *Brachiosaurus*. Should this animal be referred to *Brachiosaurus*? The vastly shorter humerus and sacrum mentioned above suggest that it is more prudent to leave the species in *Bothriospondylus* pending a full review.

Lapparentosaurus Bonaparte, 1986*b*. The Middle Jurassic juvenile brachiosaurid material from Madagascar has been fully described in Ogier's thesis but remains unpublished, except for preliminary statements by Bonaparte (1986*b*). It appears that *Lapparentosaurus madagascariensis* is very close to, if not the direct ancestor of, *Brachiosaurus*. Bonaparte characterized the animal by the very low neural arches and spines in the caudal dorsal and sacral regions. They are relatively flat, lacking the divergent thin laminae that characterize later forms.

Volkheimeria Bonaparte, 1979*b*. *V. chubutensis* is based on caudal dorsals, partial sacrum, pelvis, femur, and tibia from the Middle Jurassic of Argentina. The neural spines in the hip region are low and relatively flat. They are very similar to those of the closely related *Lapparentosaurus*. The ilium was high with elongated pubic peduncle. Bonaparte (1986*b*) separates *Volkheimeria* from *Lapparentosaurus* on the basis of the pubis and ischium. The former is more slender, and the shaft and distal end of the latter are less flattened.

Pelorosaurus Mantell, 1850. *Pelorosaurus conybearei* clearly documents the presence of a brachiosaurid in the Lower Cretaceous (Wealden) of England. There is no reason to believe that the caudal vertebrae found with the type humerus and included in the type by Mantell (1850), but referred to variously as *Cetiosaurus brevis* (Owen 1842*b*) and *C. conybearei* (Melville 1849) elsewhere, do not belong to the same species and not unlikely to the same individual. A number of brachiosaurid dorsal vertebrae as well as a sacrum and pelvis (Swinton 1946), usually referred to *Ornithopsis* (Hulke 1879*a* and 1880; Seeley 1870), can also safely be referred to *Pelorosaurus*. Less surely but not improbably the tooth named *Oplosaurus armatus* (Gervais 1852) also belongs here.

Pelorosaurus cannot be separated from *Brachiosaurus* with certainty from what is now known, but its younger age and smaller size lead one to suspect that future finds will prove the two animals different. Various brachiosaurid remains from the Upper Jurassic of England have been referred to *Pelorosaurus*, *Ischyrosaurus*, and *Ornithopsis*. It is not possible to assign them to either *Pelorosaurus* or *Brachiosaurus* at this time. The description below applies only to the Early Cretaceous species.

The only skull part currently referable to *Pelorosaurus* is the much-figured large spatulate tooth origi-

nally described as *Oplosaurus armatus*. It is about the same size and shape as those of *Brachiosaurus* but appears to differ in one respect. Most of the teeth of the latter have a prominent ridge down the center of the lingual face extending to the tip of the crown. The trace of a ridge on the "*Oplosaurus*" tooth dies out well up on the crown, and the face is concave most of the way to the tip.

No cervical vertebrae clearly referable to *Pelorosaurus* are known from the Wealden. Should the centra named *Chondrosteosaurus* eventually prove to belong, the separation of *Pelorosaurus* and *Brachiosaurus* would be assured, as these caudal cervicals are considerably shorter than those of the latter. There are a number of middle and caudal dorsals, which closely resemble those of *Brachiosaurus* in having opisthocoelous centra with deep pleurocoels, massive arches with similar laminae, and short massive spines. In the preserved sacrum, the dorsosacral is not coossified to the other four. Otherwise, it resembles that of *Brachiosaurus* closely. The proximal caudals discussed above resemble those of *Brachiosaurus* in their large centra and short simple arches and spines, but the centra are relatively a bit shorter, and the caudal ribs proceed directly outward rather than outward and backward. The chevrons are open above the hemal canal.

The pectoral girdle is unknown. The long, straight, slender humerus, more expanded proximally than distally, is slightly more massive than that of *Brachiosaurus*, but the two bones are strikingly similar. The bones of the forelimb and manus are unknown. The ilia associated with the sacrum mentioned above are virtually identical with those of *Brachiosaurus*, possessing the same upward and forwardly directed axis of the bone, broadly expanded cranial lobe, and long, slender, recurved pubic peduncle. The caudal border of the blades of the associated pubes are incomplete, causing the bones to appear less robust than those of *Brachiosaurus*, but they may not have been so when complete. A pair of pubes of *Pelorosaurus conybearei*, originally named *Ornithopsis eucamerotus* (Hulke 1882*a*), closely resemble those of *Brachiosaurus* and also "*Ornithopsis*" *leedsi* (Hulke 1887). The associated ischia do so as well in having a broad, well-developed head and twisted shaft with ends meeting edge to edge rather than side by side, but the shaft was shorter and broader. No bones of the hindlimb can be assigned to *Pelorosaurus*.

Pleurocoelus Marsh, 1888*a*. *Pleurocoelus nanus* probably represents a small Lower Cretaceous survivor of the brachiosaurid group. The skull is known only from a few disarticulated elements. Vertebral centra are present from all parts of the column, but only in the tail are there neural arches. Most of the limb and foot bones are known, but all are disarticulated, so relative lengths cannot be determined. The girdle bones are known only from fragments. Specimens have been recovered from Maryland and Texas in the United States and England. It is not unlikely that the tooth named *Astrodon johnstoni* (Leidy 1865) represents an adult form of *Pleurocoelus,* but this remains to be established.

The only skull elements so far collected are the supraoccipital, laterosphenoid, pterygoid, dentary, and several incomplete maxillae (Lull 1911*b*; Kingham 1962). The first three elements are of little diagnostic value. The maxilla actually resembles that of *Camarasaurus* more closely than that of *Brachiosaurus,* as the (missing) dorsal prefrontal process arises much farther forward, suggesting the short high skull of the former. The dentary, however, is lighter and less deep than in *Camarasaurus,* more nearly resembling that of *Brachiosaurus.* The number of maxillary teeth has been estimated at 9 or 10 as opposed to at least 13 in the dentary. The teeth are very characteristic, being intermediate between the broad spoon-shaped camarasaur-brachiosaur type and the slender peglike *Diplodocus* or *Apatosaurus* type. They are generally cylindrical but relatively broader than those of the latter.

The numbers of cervical and dorsal vertebrae are not known. The centra are all opisthocoelous and contain huge pleurocoels, more strongly developed than in any other reported form. The cervical centra are less elongated than in *Brachiosaurus.* It is not known if the presacral spines are bifid. The sacral centra are relatively elongated and solid but with a depression behind the articular surface for the rib. The caudal vertebrae resemble those of *Brachiosaurus* but differ in several respects. The short procoelous proximal centra have little backward convexity and do not contain pleurocoels. The caudal ribs are directed caudolaterally and dorsally as well. The neural spines are relatively low and simple. The most characteristic feature is the strong development of the hyposphene-hypantrum articulation (Langston 1974) typical of sauropod dorsals but not developed in the tail of any other known form. The middle caudals have the neural arch on the cranial half of the centrum. The hemal canal is not bridged over in the chevrons.

The backwardly directed expansion of the scapula above the glenoid fossa is less protruding than in *Brachiosaurus,* and the distal end of the bone is less splayed. The humerus is relatively long and slender but not to the degree found in *Brachiosaurus.* It was evidently shorter than the femur, but bones of an associated individual have not been found. The slender radius exhibits a distal expansion reminiscent of the titanosaurids, but the proximal end is not similarly expanded. The bone has a very sharp ridge running down the caudal (extensor) face. The metacarpals are long and slender. The femur exhibits the prominence on its upper lateral margin, typical of the brachiosaurids and titanosaurids.

Chubutisaurus Corro, 1975. This South American Lower Cretaceous form, *C. insignis,* is known from several incomplete dorsals, a series of caudals (mostly centra), fore- and hindlimb bones, metacarpals, and incomplete girdle bones. Once again, the lack of a skull and presacral vertebrae blurs its relationships to other sauropods.

The strongly opisthocoelous dorsal centrum contains a large complex pleurocoel. The caudal centra, even the most proximal ones, are short, amphicoelous, and without pleurocoels. Ventrally, they exhibit the blunted V-shape of *Apatosaurus,* but the apex of the V is not only blunted but also gently excavated to produce a shallow ventral trough. Otherwise, they resemble *Brachiosaurus.* The humerus is long and relatively slender, its proximal end squared. The humerofemoral length ratio is 0.85 while radius : humerus is 0.59, the latter a characteristic of brachiosaurids and titanosaurids. The radius and ulna are fairly robust, the distal end of the former more greatly expanded than its proximal end. The femur possesses a distinct prominence on its upper lateral margin. The metacarpals are relatively long (metacarpus : radius ratio is 0.54 compared to 0.51 in *Brachiosaurus* and *Alamosaurus,* 0.46 in *Camarasaurus,* and 0.31 in *Diplodocus*). They are stouter than in *Brachiosaurus.*

The noted characters suggest the Brachiosauridae. There are also some similarities to the titanosaurids, but the total absence of the deeply procoelous caudal centra preclude any close relationship.

Ultrasauros Jensen, 1985*b* (non Kim 1983). Recently, Jensen (1985*b*) reported three new taxa of gigantic sauropods from a single Upper Jurassic quarry in Colorado, each based on a few bones. *Ultrasauros macintoshi* based on a large cranio-middle dorsal and scapulocoracoid is a very large brachiosaurid. The dorsal resembles *Brachiosaurus* but is larger than that of either the American or African species. It is lighter than those of *B. altithorax,* but that might be due to its more cranial position in the column. The scapulocoracoid also differs from that of *Brachiosaurus* in some respects, but a proper evaluation of the genus compared to *Brachiosaurus* must await the preparation of more material.

Dystylosaurus Jensen, 1985*b. D. edwini* is based on a very large cranio-middle dorsal from the same site as the above. It is also clearly brachiosaurid. The infraprezygapophysial lamina leads down to the massive parapophysis located on the boundary of centrum and arch. Cranial to and parallel to this, a second lamina descends from the lower end of the prezygagophysis, providing the generic name. The vertebra cannot be compared to one of *Brachiosaurus* since the type of *B. altithorax* contains only the last seven dorsals, and this one is probably one or two forward from this. Compared to the seventh dorsal forward from the sacrum of *Brachiosaurus*, the *Dystylosaurus* centrum is shorter and has a smaller pleurocoel. The transverse process is similarly directed straight outward but appears more massive, in part perhaps because most of the laminae in the former have been severely damaged and are largely restored in plaster. The three major differences pertain to the spine, which in *Dystylosaurus* is broader transversely, much shorter axially, and noticeably lower than would be expected. The first of these might be explained by preservation and the second possibly by its position in the series. As to the third, if the height of the spine is really virtually complete as it appears, the vertebra can hardly belong to *Brachiosaurus*.

Family Camarasauridae   Cope, 1877*e*

Family Characters. Short sturdy basipterygoid processes, jugal excluded from lower rim of skull, midpresacral vertebrae with U-shaped divided spine. The family may be divided into two subfamilies, the Camarasaurinae and the Opisthocoelicaudiinae, the inclusion of the latter subfamily in the Camarasauridae being tentative.

SUBFAMILY CAMARASAURINAE NOPCSA, 1928*c*

Camarasaurus Cope, 1877*e*. The Upper Jurassic genus *Camarasaurus* (Cope 1877*e*, 1878*d*; Osborn and Mook 1921; Gilmore 1925) is the most common North American sauropod and the only one for which all parts of the skeleton are known. At least three species are recognized: *C. supremus*, *C. grandis,* and *C. lentus.*

The powerfully built skull is short and high, with a "bulldog" shaped muzzle. The quadrate is directed at right angles to the long axis of the skull. The basipterygoid processes of the basisphenoid are very short and stout. Both the upper and lower jaws are massive and contain large spoon-shaped teeth. The dental formula is

$$\frac{4, 10}{13}.$$

The maxilla is shorter than in *Brachiosaurus* and develops a shelf over its dorsal rostral end. This forms the floor of the external naris, which opens laterally, not as high up on the face as in *Brachiosaurus*. The ascending process of the maxilla is located more forward. The vomer is more robust.

There are 12 cervical vertebrae. Their centra are strongly opisthocoelous with very deep pleurocoels. They are of moderate length. The neural spines are bifid, beginning as far forward as cervical 4 (depending on the age of the individual), from which point the character increases backward in a prominent U-shaped cleft. The cervical ribs are very long and slender, the longest being that of cervical 9, which extends the length of three vertebrae (Holland 1924*a*). The 12 dorsal centra also contain large deep pleurocoels and are strongly opisthocoelous all the way back to the sacrum. The arches are high, but the spines, the first four of which are cleft, are short and massive. They increase moderately in height backward. The transverse processes are directed horizontally. The sacrum has the standard composition, the sacral ribs of all five vertebrae participating in the strongly fused yoke. However, the centrum of the dorsosacral is usually not coossified with that of sacral 2 even in the adult. It is not unusual for the spines of sacrals 2 to 5 or even 1 to 5 to fuse into a bony plate. The spines are more elevated than in *Brachiosaurus* but much less so than in the diplodocids. The centra, at least several of them, contain small pleurocoels. There are 53 caudal vertebrae with relatively short amphicoelous centra, which contain no pleurocoels and weak chevron facets. The proximal caudal ribs are simple, not broad plates as in the diplodocids, and are directed straight outward. They decrease in size backward and disappear at about caudal 13. The spines are simple, but in some individuals (?males) they are noticeably expanded transversely into a club. The chevrons are simple throughout, and the hemal canal is almost never bridged over with bone.

Both the proximal plate and distal end of the scapula are broadly expanded as in the Brachiosauridae. The angle between the ridge separating the two muscular fossae on the lateral face and the axis of the shaft of the bone is about 66°. The coracoid is oval, its articular margin with the scapula straight. The bones of the forelimb are more robust than those of *Brachiosaurus* and a bit more robust than those of *Diplodocus* but much more slender than those of *Apatosaurus* and *Opisthocoelicaudia*. The forelimb is relatively longer than those of diplodocids but shorter than those of *Omeisaurus, ?Euhelopus,* and brachiosaurids. The humerofemoral length ratio is about 0.76. There are two carpal bones. The metacarpals are relatively very long and slender compared to *Diplodocus* and *Apato-*

saurus (humerus : metacarpus ratio is 0.32). The pubis is relatively very robust. The slender blade of the ischium is directed backward as well as downward and twisted so as to meet its unexpanded fellow edge to edge distally. The femur shows almost no trace of a prominence on the upper lateral margin. Its head is set noticeably higher than the greater trochanter, a generic character. The tibiofemoral ratio is greater than that of brachiosaurids but less than that of diplodocids. There is a small globular calcaneum. Metatarsal III is the longest one.

In addition to the two or three valid American species, *Apatosaurus alenquerensis* Lapparent et Zbyszewski, 1957, from the Upper Jurassic of Portugal is provisionally referred to *Camarasaurus* pending the discovery of more complete material. This species is based on a large part of the skeleton lacking the skull and vertebral arches. It agrees with *Camarasaurus* in the following characters: 12 dorsals rather than 10, even the caudal ones with strong ball and socket articulation, distal end of the scapula broadly expanded, long slender humerus, distal end of ischium unexpanded. The chief hesitation with this identification lies with the last point. However, the humerofemoral ratio is not only much higher than in *Apatosaurus* (0.84 vs. 0.64) but also considerably larger than that of the American species of *Camarasaurus* (0.76). The robustness indices, that is, the ratios of least circumference of shaft to length are essentially the same in the American and European species but quite different in *Apatosaurus*. Hence the species will be referred to as ?*C. alenquerensis*.

Aragosaurus Sanz, Buscalioni, Casanovas, et Santafe, 1987. *Aragosaurus ischiaticus* from the Lower Cretaceous of Spain is known (Sanz et al. 1987; Lapparent 1960*b*) from caudal vertebrae from all parts of the tail, chevrons, scapula, forelimb, pubis, ischium, femur, and several phalanges including a claw. All the caudals are virtually identical with those of *Camarasaurus*. The distal end of the scapula is not complete, but it was apparently less splayed than those of the American or Portuguese species of *Camarasaurus*. The forelimb was relatively longer than those of *C. grandis*, *C. lentus*, and *C. lewisi* (humerus : femur is 0.82) but similar to that of ?*C. alenquerensis*. The humerus resembled that of *Camarasaurus* but was somewhat more slender and the proximal end less expanded. The ulna and radius also resembled those of that genus, but the former was again somewhat more slender (humerus : ulna is 0.73; humerus : radius is 0.71). The pubis was apparently less robust than that of *Camarasaurus*, and the shaft of the ischium was shorter. The femur was virtually identical to that of the American species, but its head was less protruding than that of the Portuguese form.

Euhelopus Romer, 1966. The known parts of *Euhelopus zdanskyi* from the Upper Jurassic of China include the partial skull, all the presacral and sacral vertebrae, the pelvis, hindlimb, and foot. In addition, a scapulocoracoid and humerus found later are supposed to belong. The skull and presacral vertebrae more nearly resemble *Camarasaurus*, while the relatively long forelimb suggests *Brachiosaurus*, but the increased number of presacral vertebrae suggests neither family. *Euhelopus* has been referred to both families represented by those genera but is here referred to the Camarasauridae.

The braincase has not been found, but most of the other skull elements are known (Wiman 1929; Mateer and McIntosh 1985). The short high skull resembles that of *Camarasaurus* but is more lightly built. The dorsal process of the maxilla is slender. The pterygoid is very large. The jaws are noticeably less massive than in *Camarasaurus*, and the spoon-shaped teeth are similar but not quite as large and have no serrations. There are four premaxillary teeth, nine in the maxilla and one more on the margin between the two bones.

There are 17 cervical vertebrae. They are opisthocoelous, of only moderate length, and contain no pleurocoels in cervicals 1 to 14, although the entire side of the centrum is concave beneath the postcentrodiapophysial lamina. Cervical 15 has a lateral depression, cervical 16 a deeper one, and cervical 17 a true pleurocoel. Ventrally, the caudal cervical centra exhibit a distinct but modest keel. The ribs in the caudal half of the neck are very long, extending beneath the centra of one or two more caudal cervicals. The dorsum of the spine of cervical 10 is flat; that of 11 shows the first trace of bifurcation; those of 11 to 15 are divided but minimally so. A short median spine appears between the two in cervicals 14 and 15 and continues on in 16, 17, and dorsal 1. There are 14 dorsals and a dorsosacral, all markedly opisthocoelous with large balls remaining on the cranial ends of the caudal dorsal centra. They all contain pleurocoels, but these are not nearly as large as in *Camarasaurus*. The first few centra continue to exhibit a slight keel. The arches and spines are of only moderate height and do not increase in height cranially as one might expect if the forelimb were unusually long. The transverse processes are directed outward and decrease in length from dorsal 2 backward. The parapophysis lies on the centrum beneath the pleurocoel in dorsal 1, beside and just in front of it in dorsals 2 and 3, just above it in 4, and finally reaches the arch in 5. All five sacrals are firmly coalesced, and it is possible that the centrum of caudal 1 is also. In addition, the distal ends of the shortened thoracic ribs of dorsals 13 and 14 unite with one another and the forward tip of the ilium. The spines of all five sacrals are united. They are relatively short. Except for the centrum of caudal 1, the tail is unknown.

A scapulocoracoid and humerus came from the same area and supposedly the same site (Young 1935a), which had yielded one of two *Euhelopus* skeletons many years before. While these bones likely belong to this species and perhaps to the skeleton, their large size ordains that this supposition be treated with some caution. The scapula has a well-developed proximal plate and long, relatively slender shaft, which expands regularly toward the distal end, but the expansion is considerably less than in *Camarasaurus* or *Brachiosaurus*. It resembles a bone figured by Dong et al. (1983) as *Omeisaurus*. The coracoid is relatively small, and its border with the scapula exhibits a marked jog upward just above the coracoid foramen. The humerus has a slender shaft, broader distal end, and much broader proximal end somewhat reminiscent of *Camarasaurus* and more nearly resembling that of *Datousaurus* but not *Brachiosaurus*. However, if the humerus does belong to the type skeleton, the ratio humerus:femur would approach that of the brachiosaurids. The lower forelimb and manus are unknown in *Euhelopus*. The pelvis more nearly resembles that of *Omeisaurus* than *Camarasaurus*. The pubis is less robust and the blade and distal end of the ischium broader than in the latter. The femur also is less robust. The tibia:femur ratio is 0.63. A bone identified as phalanx III-2 by Wiman may instead be the calcaneum. If so, it is larger than those of either *Shunosaurus* or *Camarasaurus*. Metatarsal I does not possess the distal process characteristic of diplodocids. Wiman evidently confused metatarsals III and IV, which are the longest. The metatarsofemoral ratio is 0.14.

Euhelopus represents a form with characters between the cetiosaurid *Omeisaurus* and the camarasaurid *Camarasaurus*. The skull and tooth characters and the beginnings of the neural spine bifurcation suggest that it be included in the Camarasauridae. Despite a possible high humerofemoral ratio, *Euhelopus* exhibits little similarity to brachiosaurids.

Tienshanosaurus Young, 1937. *T. chitaiensis*, a relatively small sauropod from the Jurassic of Xinjiang, China, is known from several cervical and dorsal vertebrae, a series of caudal vertebrae, chevrons, pectoral girdle, humerus, ilium, ischium, a fragmentary pubis, and hindlimb bones.

The cervicals are relatively short, and the one that has been illustrated suggests a divided neural spine, but this may be an artifact of the preservation. The cervical rib is slender and extends backward beyond the end of the centrum. The caudal dorsals are amphicoelous and contain weak pleurocoels. The caudal centra are amphicoelous and do not contain pleurocoels. The proximal spines are slender and moderately high, the caudal ribs normal. The distal caudals have a pronounced longitudinal ridge on the side of the centrum. The proximal chevrons are figured with the hemal canal bridged across dorsally.

The scapula has a broad proximal plate and a long shaft, which is little expanded distally. The angle between the axis of the shaft and the ridge above the ventral fossa is 68°. The coracoid is relatively small, and the scapulocoracoid border shows a jog above the coracoid foramen. The humerus is relatively slender and resembles that of *Camarasaurus*. The ilium is short and high with a long pubic peduncle. The ischium resembles that of *Camarasaurus* in being little expanded distally. The ratio ischium:humerus is 0.82 compared to 0.88 in *Camarasaurus*, over 0.9 in *Diplodocus*, and 1.0 or more in *Apatosaurus*. The hindlimb bones are too fragmentary to be of use.

The relationships of *Tienshanosaurus* are uncertain. The caudals immediately eliminate reference to the Brachiosauridae or Titanosauridae. If the cervical spine is indeed divided, the Cetiosauridae would also be eliminated. The long, slender, cervical ribs, simple proximal caudal ribs, short distal caudals, and shape of humerus and ischium suggest referral to the Camarasauridae. The scapula suggests a relationship with *Apatosaurus*, and the bridged-over chevrons also indicate a relationship with the Diplodocidae. The shape of the ilium, weak dorsal pleurocoels, and ridge on the lateral face of the caudals suggest neither. The provisional reference is to the Camarasauridae.

Chondrosteosaurus Owen, 1876b. *Chondrosteosaurus gigas* is based on cervical centra from the Early Cretaceous of England. The texture of the bone is coarsely cancellous in contrast to the fine spongy nature in *Cetiosaurus*. The large pleurocoel, divided in two by a vertical ridge, quite unlike the titanosaur condition, suggests the Camarasauridae or Brachiosauridae, but the latter may apparently be ruled out by the shortness of the centrum and its flatness ventrally. It is broader than high. The closest resemblance is with cervical 10 of *Camarasaurus*, and for that reason, *Chondrosteosaurus* is tentatively assigned to the Camarasauridae.

SUBFAMILY OPISTHOCOELICAUDIINAE, NOV.

Subfamily characters. Caudal vertebrae opisthocoelous, tail very short.

Opisthocoelicaudia Borsuk-Bialynicka, 1977. *O. skarzynskii* is known from an otherwise essentially complete skeleton lacking the skull and neck. Although scattered fragments suggesting that other sauropods than titanosaurids survived into the Late Cretaceous had been found previously, this specimen established the fact with certainty. However, its relationship to other sauropods is not entirely clear.

There are eleven (or possibly twelve) dorsal vertebrae. The centra are all strongly opisthocoelous, the prominent ball and socket articulation persisting back to the sacrum. They all contain prominent pleurocoels and a ventral concavity divided by a sagittal ridge. The concavity deepens toward the sacrum. The arches are relatively low, the transverse processes heavy and directed outward. The spines are short and directed caudodorsally, reminiscent of those of the titanosaurids. They are deeply divided by a broad U-shaped cleft, which in the cranial dorsals most resembles *Camarasaurus*. A notch persists to the back of the dorsal column. The parapophysis on dorsal 1 lies just below the pleurocoel. It migrates upward, reaching the base of the arch on dorsal 4. The sacrum contains six sacrals, an extra caudosacral having been added (in contrast to the titanosaurids where an extra dorsosacral has been added). The centra are all completely coossified and apparently do not contain pleurocoels. The spines are low. The tail is very short with perhaps no more than 35 vertebrae. The proximal caudal centra are opisthocoelous, the first five strongly so with a huge cranial ball. The distal ones become amphiplatyan and finally biconvex. They contain no pleurocoels. The arches and spines are very robust and low, the spines very broad transversely. The distal caudal arches lie on the front halves of the centra as in *Pleurocoelus* and the titanosaurids. The chevrons are all simple (not forked) with the hemal canal open above.

The scapulocoracoid is relatively robust; the distal end of the scapula is little expanded. The forelimb is relatively short (humerus : femur is 0.72) and massive. The humerus is very robust as in *Saltasaurus* and to a lesser degree *Apatosaurus*. The antebrachium resembles that of *Camarasaurus* but is much more robust. In *Opisthocoelicaudia*, metacarpals I and II are coequal in length, and longer than the other metacarpals; in *Camarasaurus*, metacarpals II and III are coequal and longer than all others. The metacarpus is rotated laterally. The pubis is not as massive as that of *Camarasaurus*, and the ischium is greatly reduced with a relatively short shaft more expanded distally than in the latter. The hindlimb bones are relatively robust. The fourth trochanter of the femur is situated uniquely below the middle of the shaft. The prominent deflection on the upper lateral margin is more pronounced than in *Camarasaurus*, less so than in *Brachiosaurus*. The bones of

the lower leg are relatively short (femur : tibia ratio = 0.58). The calcaneum is absent. The third metatarsal is the longest. The phalangeal formula is 2-2-2-1-0? with claws on digits I to III. It represents a greater reduction than in any other sauropod.

Family Diplodocidae   Marsh, 1884*d*

Family Characters. Skull relatively long; quadrate rostroventrally inclined; external nares opening dorsally; basipterygoid processes slender and elongate; mandible light; teeth slender, peglike; proximal caudals with procoelous centra and winglike transverse processes; middle chevrons forked; distal ends of ischia expanded, meeting one another side by side.

This family may be divided into three subfamilies: the Diplodocinae, Dicraeosaurinae, and Mamenchisaurinae. Inclusion of the latter in the family is tentative since the skull is completely unknown.

SUBFAMILY DIPLODOCINAE JANENSCH, 1929*b*

Subfamily Characters. Twenty-five presacral vertebrae; increase of number of cervicals at the expense of the dorsals; very high neural spines in sacral region; forelimbs short (humerus : femur is 0.6-0.7); metatarsals III and IV longest; no calcaneum; a small process on distal plantar edge of lateral surface of metatarsal I.

There are three well-established North American genera: *Diplodocus, Barosaurus,* and *Apatosaurus*. The English form *Cetiosauriscus* also appears to be a primitive member of the subfamily. Three other poorly represented North American genera also belong. Two of the latter are truly gigantic.

Diplodocus Marsh, 1884*d*. The genus *Diplodocus* (Osborn 1899*a*; Hatcher 1901; Mook 1917*b*; Gilmore 1932*a*) is completely known except for some details concerning the manus and the end of the tail. Three species are recognized: *D. longus, D. carnegii,* and *D. hayi*.

The snout is elongated with the teeth confined to the very front of the jaws. The nares are located on the very top of the skull and are confluent. The dorsal process of the maxilla is long and broad to its upper end. The jugal is excluded from the lower rim of the skull by the quadratojugal. The basipterygoid processes are very long. The mandible is squared in front. The peglike teeth are particularly weak. (Marsh 1884*d*; Holland 1906, 1924*b*; McIntosh and Berman 1975). Dental formula is

$$\frac{4, 9 \text{ to } 11}{10}.$$

There are 15 cervical vertebrae and 10 dorsals. All the presacral centra are opisthocoelous, the cervicals and cranial dorsals markedly so and the caudal dorsals weakly so. The cervical centra are moderately elongated with large pleurocoels divided by a complex series of laminae. Large simple pleurocoels persist throughout the dorsal series. The spines of the presacral vertebrae are bifid beginning with a small notch in cervical 3 and soon developing into a deep V-shaped cleft that reaches its extreme development at the cervicodorsal boundary. A small secondary spine develops at the base of the cleft in cervical 13 and persists through the first five dorsals. The cleft spines in the shoulder region are relatively elevated but not nearly as much as in *Dicraeosaurus* and certainly not as much so as in the single spines of *Brachiosaurus*. The cervicodorsal transition is marked by an abrupt change from the slender horizontal cervical ribs, which extend backward just to the rear of the centra, to the elongated heavy downwardly projected thoracic rib of dorsal 1. As seen in the vertebrae themselves, the transition is much more gradual. The dorsal centra decrease in length progressively from an elongated dorsal 1 to dorsal 4 (the last one with a prominent cranial ball) and henceforth remain fairly constant. The parapophysis on dorsal 1 lies on the bottom of the lateral face of the centrum well below the pleurocoel. On dorsal 2, it has progressed upward just below and in front of the pleurocoel; on dorsal 3, it lies just above the pleurocoel, reaching the border of the centrum and arch on dorsal 4 and finally the arch itself on dorsal 5. The cleft in the spine decreases progressively backward to a small notch on dorsal 8 and nothing on dorsals 9 and 10. The height of the spines increases regularly, reaching a maximum at the sacrum, where the total height of the vertebra is four times that of the centrum. The transverse processes are directed laterally. The five sacral centra are fused in adults, and their ribs enter the sacrocostal yoke. All five have deep pleurocoels, that on the caudosacral somewhat reduced. The spines of sacrals 2 and 3 are always united, that of sacral 4 sometimes entering the plate as well. A small ossicle is often found between the tips of the spines of sacrals 1 and 2. The proximal caudal centra are short, gently procoelous, and with deep pleurocoels as far back as caudal 19. The centra are excavated ventrally with two parallel grooves separated by a central ridge. Farther back, the ridge disappears and the excavation becomes a deep trough. The spines are high and slender, and the first eight or so are slightly bifid. Most characteristic are the caudal ribs attached by thin laminae to transverse processes to produce winglike structures similar to the sacral ribs. These elaborate structures, present in the most proximal caudals of other diplodocids, continue backward about to caudal 13 before gradually assuming the more normal form. They finally disappear at about caudal 20. The middle caudals are greatly elongated, and the 30 or so distal ones are reduced to biconvex rods, forming the so-called whiplash. The exact number of caudals is not known, at least numbering 70 and perhaps more than 80. The proximal chevrons are normal, but the hemal canal is always bridged over. After caudal 12, the chevrons are typically forked with the appearance of a cranial process at the front margin. Viewed from above, the chevrons assume a diamond shape with the two articular facets at the two central apices, the other two occurring where the lateral branches join cranially and caudally. This development is typical of all diplodocids and Chinese cetiosaurids but is most extreme in this genus. Farther back, the branches become detached and the chevron develops into two parallel rods.

The scapula has a large proximal plate, and the angle between the ridge on it and the shaft is more acute than in *Apatosaurus* or *Camarasaurus*. The distal end is more expanded than in other members of the family. The forelimb bones are very slender (humerus : femur ratio is 0.65; humerus : metacarpus ratio is 0.23). The structure of the carpus is not known. The metacarpals are slender but short, markedly so compared to those of *Camarasaurus*. The pubis has a relatively long, slender shaft and has a prominent hooklike process on the cranial margin for the attachment of the ambiens muscle. The ischium has a slender shaft, which expands and thickens distally, so that the meeting of the two ischia appears somewhat between the side-by-side arrangement of *Apatosaurus* and the edge-to-edge articulation of *Camarasaurus* and *Brachiosaurus*. The femur is straight and very slender, resembling a stove pipe. The tibia : femur ratio is 0.69 to 0.71, and the metatarsus : tibia ratio is 0.23. The medially directed projection of the astragalus is relatively shorter than in *Camarasaurus*. There is no calcaneum. The metatarsals are more slender than in *Apatosaurus*. Metatarsals III and IV are the longest. The phalangeal formula is 2–3–4–2–1 or 2–3–3–2–1.

Barosaurus Marsh, 1890*b*. In the North American *Barosaurus lentus* (Lull 1919), the skull, front of the neck, forearm and manus, and distal half of the tail remain unknown. The genus is very closely related to *Diplodocus*, and the limb bones are so similar as to be indistinguishable. *Barosaurus* differs from *Diplodocus* in its enormously elongated cervical vertebrae, which are relatively 33 percent longer than those of the latter.

The number of cervical vertebrae is unknown,

but with a reduction by 1 of the dorsals, it is likely that there are 16 or perhaps more. The enormously elongated cervicals are generally similar to those of *Diplodocus* if the latter were stretched. They resemble those of *Brachiosaurus* superficially but differ in several respects: the spine is bifid in the middle and caudal part of the neck, the arch virtually covers the centrum and is not set forward, the cervical ribs, although long and slender, extend only to the back end of the centrum. The bifid spines begin farther back in the column but become very deep Vs at the rear end of the neck, after which the notch disappears more rapidly than in *Diplodocus*. The small auxiliary central spine seen in that form also occurs in *Barosaurus*. There appear to be 9 dorsals, although the first one appears very much like a cervical. It is still very elongated with the parapophysis on the bottom of the centrum, but it appears that the rib (detached in known specimens) is not fused to it. That is the main reason for provisionally regarding it as a dorsal. The parapophysis remains at the bottom of the centrum on dorsal 2, moves up just in front of the pleurocoel on dorsal 3, and reaches the base of the arch on dorsal 4. Beyond the first, the dorsals closely resemble those of *Diplodocus* with the same well-defined thin laminae, opisthocoelous centra with prominent balls on the front through dorsal 5, and large pleurocoels throughout. All five sacrals enter into the formation of the yoke. The spines are not known. The caudals also resemble those of *Diplodocus* but are somewhat shorter, particularly in the midcaudal region, and the extreme specializations of that genus are less pronounced. The winglike transverse processes and caudal ribs have been modified to the more normal form by caudal 7. The proximal centra are less procoelous; the pleurocoel disappears after caudal 14, and the ventral excavation is a broad, gentle concavity. The neural spines of the proximal caudals are broad transversely, and their tops are not bifid. Few chevrons have been found. Those in the midcaudal region are of the *Diplodocus* type but are much smaller. A whiplash probably existed, but beyond caudal 30, the tail is unknown. The tail was almost certainly noticeably shorter than that of *Diplodocus*, however.

The distal end of the scapula is less expanded than in *Diplodocus*, and the angle between shaft and ridge is less acute. The humerus is similar, but the humerofemoral ratio is apparently greater (0.72). The forearm and manus are unknown. The pelvis closely resembles that of *Diplodocus*, but the ambiens process of the pubis is not as strongly developed. The hindlimb and foot bones are virtually indistinguishable from those of the latter.

Gigantosaurus africanus Fraas, 1908, from Tanzania has been referred to *Barosaurus* (Janensch 1922), but the evidence is not absolutely compelling. Nevertheless, this form is provisionally referred to as *?Barosaurus africanus*.

To the African species have been referred braincases and incomplete jaw bones not distinctive beyond the family identification. Many girdle and limb bones, some fragmentary presacrals, and some middle caudals do not help to identify the animal (or animals) beyond showing a similarity to *Barosaurus* and *Diplodocus*.

Apatosaurus Marsh, 1877*b*. The anatomy of *Apatosaurus* (Marsh 1883; Riggs 1903*b*; Gilmore 1936) is completely known except for the mandible. Overall, it is similar to that of *Diplodocus* and *Barosaurus*, except that the bones, particularly those of the neck and forelimb, are much more robust, the caudal vertebrae are shorter and less specialized, and the typical forked diplodocid chevrons less developed. Three species are recognized: *A. ajax*, *A. excelsus*, and *A. louisae*.

The skull (Berman and McIntosh 1978) is known from a single fairly well preserved specimen without lower jaw. It is remarkably similar to that of *Diplodocus* but with a slightly broader muzzle, shorter quadrate, and shorter basipterygoid processes. The teeth, insofar as they have been observed, are typically peglike.

There are 15 cervical vertebrae. The atlas bears a single headed rib. The centra are opisthocoelous with well-developed pleurocoels. In comparison to those of other diplodocids, they are short, cervicals 10 and 11 being the longest. The V-shaped cleavage in the neural spine begins at about cervical 6 and becomes quite pronounced in the rest of the cervical series. The great robustness of the cervical ribs is striking. There are 10 dorsal vertebrae, all the centra of which have well-developed pleurocoels. They are opisthocoelous, except that beyond the first, the cranial ball is greatly reduced and appears only on the upper half of the centrum. The spines increase in height to become quite high at the sacrum. The transverse processes are directed horizontally. The parapophysis lies below the pleurocoel on dorsal 1 and migrates upward on the centrum reaching the arch by dorsal 4. The five sacral centra are firmly fused in the adult and contain deep pleurocoels. As in *Diplodocus*, the sacral spines are very high, sacrals 2 and 3 fusing clear to the top. The tail contains 82 caudals (Holland 1915*a*). The first few have elaborate winglike caudal ribs, which may be perforated. The amphicoelous centra do not contain pleurocoels. The proximal ones have a rounded V-shape below in contrast to those of *Diplodocus* and *Barosaurus*. The midcaudals are elongate but not as markedly so as those of *Diplodocus* and *Barosaurus*. The neural arch decreases regularly in size and vanishes completely between caudals 35 and 40. The last 40 or so are the biconvex bony rods of the whiplash. The proximal chevrons are bridged over the hemal canal. The middle chevrons are of the forked, diplodocid type but smaller.

The scapula has a long slender blade only slightly expanded at the distal end. The angle between ridge and blade is broader than in *Diplodocus*. The coracoid is quadrangular. The bones of the forelimb are relatively short and very massive. The humerofemoral ratio is 0.64; the ulna:humerus ratio is 0.70. The expansion of the proximal end of the ulna is greater than in *Diplodocus*. The carpus (Hatcher 1902) has been reduced to a single element, which lies above metacarpals II–IV. The metacarpals are short and heavy (metacarpus:humerus is 0.25). The phalangeal formula of the manus is 2−2−1−1−1, with the small phalanx I-2 present at least in some individuals. The pelvis resembles that of *Diplodocus*, but the bones are somewhat more robust; the ambiens process on the upper border of the pubis is slightly developed but much less so than in *Diplodocus* and *Barosaurus*, and the distal ends of the ischia, which meet side by side, have been thickened dorsoventrally to a greater extent than in any other sauropod. The bones of the hindlimb are sturdy; tibia:femur is 0.60. The calcaneum is absent. The metatarsals are short and robust, with a metatarsus:tibia ratio of 0.21.

Cetiosauriscus Huene, 1927. Although the skull, cervical vertebrae, most of the dorsal vertebrae, and some of the girdle and foot bones are not known in *C. stewarti*, this form from the Upper Jurassic of England (Charig 1980) clearly belongs to the Diplodocidae, but it is less clear whether it should be assigned to the Diplodocinae or the Mamenchisaurinae. The proximal caudals have high thin spines, gently procoelous centra, and a few have the rudiments of the wing-shaped caudal ribs. In addition, there are several diplodocine characters not possessed by *Mamenchisaurus*. The humerofemoral ratio is two-thirds. There is a projecting process on the distal end of metatarsal I, and the calcaneum would appear to be absent. However, the forked chevrons resemble more closely those of *Mamenchisaurus* than those of *Diplodocus*, and the bone texture is more spongiose as in *Cetiosaurus*. *Cetiosauriscus* is thus placed in this subfamily tentatively.

Amphicoelias Cope, 1877d. *Amphicoelias altus* is known from two dorsal vertebrae, a pubis, a femur, and the arbitrarily assigned scapula, coracoid, and ulna, representing a very large diplodocid from the top of the Morrison Formation. It is closely allied to *Diplodocus* and *Barosaurus*, but the meager materials available make its identification with either very difficult. The very slender femur more closely resembles that of *Diplodocus*. The validity of *Amphicoelias* as distinct from *Diplodocus* rests on the interpretation of an erect spine in dorsal 10 of the former (Osborn and Mook 1921) compared to a caudoventrally inclined condition in the latter (Gilmore 1932). For the present, *Amphicoelias* is included in the Diplodocinae as a genus of doubtful validity.

Supersaurus Jensen, 1985b. As discussed under *Ultrasaurus* above, there are several types of gigantic sauropods from the Dry Mesa quarry, Upper Jurassic of Colorado, including the diplodocid *Supersaurus vivianae*. Only a fraction of this material has yet been prepared, but bones that can be referred to this form with reasonable certainty include the scapula-coracoid, a proximal caudal, an ischium, and probably a caudal cervical and some distal caudals.

The scapula is of the diplodocid type, but its distal end is a bit more expanded than in *Diplodocus* and considerably more so than in *Barosaurus*. It bears some resemblance to the scapula referred to *Amphicoelias*. The proximal caudal vertebra resembles more nearly *Barosaurus*. Its spine shows no sign of cleavage. The huge cervical is crushed almost flat, so much so that it is difficult to tell whether the spine is bifurcate or not. It apparently is, but there is a question as to whether Jensen's (1985b) claim that it possessed no pleurocoel, a character that would remove it from both the Brachiosauridae and the Diplodocidae, can be verified. The ischium and distal caudals bear no special characters.

Dystrophaeus Cope, 1877b. *D. viaemalae* was one of the first sauropods described from the North American Upper Jurassic (Cope 1877b; Huene 1904). Although often referred to the Cetiosauridae, it probably belongs to the Diplodocidae. Known parts include the scapula, ulna, partial radius, and metacarpals III to V. The scapula is too incomplete to be of diagnostic value. The relatively short metacarpals preclude any relationship to the Brachisauridae or Camarasauridae. The very slender ulna eliminates *Apatosaurus*. Among the established Morrison genera, this leaves *Diplodocus*, *Barosaurus*, and *Haplocanthosaurus*. The ulna and manus are unknown in the latter two genera, but as stated, the short metacarpals suggest a diplodocid rather than a cetiosaurid relationship. Whether this is *Diplodocus*, *Barosaurus*, or perhaps a third form cannot now be determined.

SUBFAMILY DICRAEOSAURINAE JANENSCH, 1929b

Subfamily Characters. Twenty-four presacrals, number of cervicals not increased at expense of dorsals;

dorsals without pleurocoels; neural spines very high, particularly in sacral region, and those of presacrals bifid to extreme degree.

Dicraeosaurus Janensch, 1914. *Dicraeosaurus* from the Upper Jurassic of Tanzania represents an aberrant diplodocid, the vertebrae of which maintain some primitive cetiosaurid characters but some very derived ones as well. The most surprising is the maintenance of only 12 cervical vertebrae, an extreme condition directly opposite to that of the other aberrant branch typified by the long-necked *Mamenchisaurus*. Two species are recognized: *D. hansemanni* and *D. sattleri*.

The skull is known from the braincase, lacrimal, and upper and lower jaws. As reconstructed by Janensch (1935-36), it is reminiscent of *Diplodocus*. The orbit has the same shape as that of the latter, but the form of most of the other openings is hypothetical. In particular, nothing is known about the external naris, although the incomplete premaxilla suggests that it may resemble that of *Diplodocus* or *Apatosaurus*. Janensch recognized a small fontanelle between the frontal and parietal on the midline and a smaller opening between the parietal and supraoccipital, which he termed the *postparietale lucke*, similar to the supposed pineal opening in *Diplodocus*. If these features, particularly the former, are standard in the adult animal, they would be important diagnostic characters. The preserved rostral parts of the premaxilla and maxilla are like *Diplodocus*, but the dentary is more massive. The teeth are slender and peglike, and the dental formula is

$$\frac{4, 12}{16}.$$

The articulated vertebral column from atlas to caudal 19 is known in one specimen, and disarticulated caudals representing the rest of the tail have been found. Janensch (1929*b*) recognized 12 cervicals and 12 dorsals, but the interpretation of 11 cervicals and 13 dorsals appears to be equally justified based on the length and orientation of the rib of presacral 13 (measured from the sacrum). The presacral centra are opisthocoelous with the typical ball and socket articulation in the cervical and cranial dorsal regions, but beyond dorsal 4, the ball becomes reduced to a gentle convexity. Notable is the weakness of the pleurocoels. In the cranial and middle cervicals, the side of the centrum is concave behind with no distinct rim of a pleurocoel. This concavity develops into a small pleurocoel beneath the diapophysis. A relatively small but more typical pleurocoel develops on the last two cervicals, only to vanish on the cranial and all later dorsals. It be-

comes replaced by a small, well-marked cavity on the lateral face of the neural arch just below the transverse process. The parapophysis is directed outward from the bottom of the centrum of the first ten cervicals in normal fashion. From cervical 11 onward, it migrates upward to reach the boundary of the centrum and neural arch by dorsal 2 and thereafter is located on the arch. The cervical centra are relatively short, the longest being cervical 8, after which they decrease in length to cervical 12, and the rest of the presacral centra are all about the same length. Beginning with cervical 3, the spines are unusually tall, gradually increasing in height throughout the presacral region, so that the height of the last dorsal is four times that of the centrum. More strikingly, the spines are deeply divided starting with cervical 3, the condition being enhanced backward to a maximum development on dorsals 3 and 4, after which the degree of division decreases rapidly, a small notch remaining in dorsal 7 and none thereafter. The caudal dorsal spines consist of four thin laminae placed at right angles to one another, producing a cross-shaped cross section with the transversely directed laminae exceeding the longitudinal directed ones. In *Dicraeosaurus*, the spine division reaches the maximum development. The slender transverse processes of the dorsal vertebrae are directed upward and outward in a manner reminiscent of cetiosaurids. The cervical ribs are short, reaching the rear of the centrum throughout most of the neck, but the last one is very long, over half the length of thoracic rib 1, and is directed downward, another feature suggesting the alternate interpretation that the thirteenth presacral forward from the sacrum is dorsal 1. The sacrum has the usual configuration, with all five contributing to the sacrocostal yoke. The spines of sacrals 2 to 4 are fused. The proximal caudal centra are procoelous, a condition that soon modifies to planoconcave and then amphicoelous. They do not contain pleurocoels or ventral excavations. The proximal centra are relatively short, the middle ones a bit longer, and the distal ones considerably longer than broad. There is whiplash of rodlike distal caudals, but the number of caudals is not known. By caudal 17, the chevrons become forked. The hemal canal is bridged over.

The distal end of the scapula was only moderately expanded, the angle between shaft and ridge fairly acute, and a small jog occurred on the scapulocoracoid border above the foramen. The forelimb was short (humerus:femur ratio is 0.62). The humerus was stouter than those of other sauropods except *Apatosaurus*, *Opisthocoelicaudia*, and *Saltasaurus* (ulna:humerus ratio is 0.66). The manus is unknown. The ilium resembles those of *Diplodocus* and *Apatosaurus*. The shaft of the pubis is rather long and slender, with a prominent hooklike ambiens process. The distal end of the ischium is broadened dorsoventrally to articulate with its mate side by side. The femur is less slender than in *Diplodocus,* and the shaft was nearly circular. The tibiofemoral ratio is 0.62. The pes is incompletely known.

Nemegtosaurus Nowinski, 1971. *Nemegtosaurus mongoliensis* is based on an isolated skull, the most complete of the known Late Cretaceous skulls. The dorsal processes of the maxillae and the rostral part of the nasals as well as the front of the braincase and a number of bones of the palate, have been damaged. Otherwise, it is nearly complete.

The skull is generally like that of the *Diplodocus* with slender peg-teeth confined to the front of the jaws. It differs from that of *Diplodocus* in the following ways: (1) it has several small intramaxillary foramina but no large second antorbital fenestra; (2) the snout is higher and much broader; (3) the infra-temporal fenestra is narrower; (4) the basipterygoid processes are perpendicular to the skull roof rather than forward and downward, and they are shorter and stouter; (5) the palatine has an expanded maxillary contact; (6) the lacrimal is more robust and much broader at its upper end; and (7) the dentary is higher at the symphysis, which is weak rather than strong. The shape and size of the external naris is completely unknown. The dental formula is

$$\frac{4,\,8}{13}.$$

Nemegtosaurus most closely resembles *Dicraeosaurus* (Nowinski 1971). However, it differs from the latter in the length and positioning of the basipterygoid processes. Serious questions remain, since none of the very distinctive dicraeosaurlike vertebrae have ever been found in the Upper Cretaceous. The possibility that the isolated skull of either this genus or of *Quaesitosaurus* belong to *Opisthocoelicaudia,* the other Late Cretaceous mongolian sauropod, is unlikely. The skulls of the former two are diplodocid; the skeleton of the latter, camarasaurid.

Quaesitosaurus Kurzanov et Bannikov, 1983. *Quaesitosaurus,* represented by a second isolated skull from the Late Cretaceous of Mongolia, resembles that of *Nemegtosaurus* in many respects. It is less complete than that of the latter. The dorsal processes of the maxillae and premaxillae are missing, as are the jugal, lacrimal, prefrontal, and part of the frontal, so no parts

of the rims of either the external nares or antorbital fenestrae are preserved, and their shapes remain hypothetical. The mandible is essentially complete.

The preserved parts of the skull resemble *Nemegtosaurus* in the following respects: a broad snout, quadrate inclined rostroventrally with respect to the axis of the mandible, relatively broad quadratojugal, very narrow infratemporal fenestra, short basipterygoid processes. The reported differences include a significantly broader snout, a shorter squamosal, which does not contact the quadratojugal, lack of parietal aperture, development of a concavity on the caudal face of the quadrate, the presence of nine rather than eight maxillary teeth, a longer tooth row in the mandible, and a prominent canal opening between the basal tubera of the basioccipital beneath the condyle, leading to the pituitary fossa. The age of the individual, state of preservation of the skull, and individual variation might account for some of these differences. The most significant characters would appear to be the quadrate concavity and especially the canal between the tubera. The presence of the suboccipital canal is, however, puzzling, for as Kurzanov and Bannikov (1983) note, it is quite unique, nothing similar occurring in any other sauropod. Other significant characters of *Quaesitosaurus* are the broad, high parasphenoid and stout basipterygoid processes, both sharply contrasting with the same elements in *Diplodocus.* The dental formula is

$$\frac{4,\,9}{13}.$$

The teeth are peglike.

Rebbachisaurus Lavocat, 1954. *Rebbachisaurus,* from the Early Cretaceous of Morocco, is distinct from any other sauropod, but its systematic position is unclear. The type specimen of *R. garasbae* consists of a scapula, a humerus, caudal dorsal vertebrae, and an unprepared sacrum. Only a very brief description has appeared (Lavocat 1954), but an exhibited dorsal in Paris allows a few further observations. Lapparent (1960*a*) has referred various remains from Niger to a second species, *R. tamesnensis,* but there is some question as to whether most of this material belongs to this genus. The discussion below is based on *R. garasbae.*

The single prepared caudal dorsal is very large and has an enormously elevated but incomplete neural spine. It most nearly resembles dorsal 12 of *Dicraeosaurus,* with the exception that it contains a deep pleurocoel in the centrum. The very thin laminae contributing to the arch, and particularly the four laminae of the spine, are very similar between the two taxa. The dorsal end of the spine is missing so it is unclear whether it expands as much transversely as that of *Dicraeosaurus,* but the slender transverse processes are similarly directed upward as well as outward. The unprepared sacrum is said to have the spines of four sacrals coalesced.

The proximal end of the scapula is incomplete, but the blade is broad, and the distal end uniquely expands into a broad racket shape. A meter-long humerus remains undescribed.

A dorsal vertebra resembling that of *R. garasbae* from the Late Cretaceous of Argentina (Nopcsa 1902) would appear to extend the range of this genus.

If Lapparent's assignment of materials to the remaining species, *Rebbachisaurus tamesnensis,* is correct, it has spoon-shaped teeth, nonprocoelous caudal vertebrae, distal whiplash caudals, long, slender, *Diplodocus*-like humeri, an ischium with a long, slender shaft slightly expanded distally, and robust metatarsals (Lapparent 1960*a*). When this material is prepared and compared to newer and better material collected by Taquet in Niger, much clarification concerning the status of *Rebbachisaurus* is certain to occur.

SUBFAMILY MAMENCHISAURINAE Young et Chao, 1972

Subfamily Characters. Number of cervical vertebrae increased to eighteen or nineteen.

Mamenchisaurus Young, 1954. *Mamenchisaurus* represents another aberrant line of diplodocids. It is known from the almost complete vertebral column and many of the limb and girdle bones, but the skull and much of the feet are lacking. Two species are recognized: *M. constructus* (Young 1954 and 1958) and *M. hochuanensis* (Young and Chao 1972).

There are 29 presacral vertebrae, of which Young and Chao (1972) took 19 to be cervicals. However, one and possibly two of the supposed cervicals are probably dorsals. The cervical vertebrae are moderately elongate, which, when coupled with their increased number, produced a very long neck. The presacrals are all opisthocoelous. The pleurocoels are apparently only rather weakly developed, especially in the cervicals and cranial dorsals. The spines of dorsals 1 to 4 (or 2 to 5) are bifurcate, more so than in *Euhelopus* but less so in the other camarasaurids and diplodocids. The dorsal spines are relatively low, and in the sacral region, the height of the vertebrae is less than three times that of the centrum. Sacral 1 (last dorsal of Young and Chao) is not fused to sacral 2. The spines of sacrals 2 to 4 are fused. The caudal centra are more strongly procoelous than those of the diplodocines, and this condition persists throughout the 35 preserved caudals. They lack

pleurocoels, and the winglike caudal ribs are only proximally present. The total number of caudals and whether the tail ends in a whiplash are not known. The chevrons begin to fork at caudal 12, and the later chevrons show that the extreme fore and aft expansion is developed more strongly than in any genus other than *Diplodocus* itself.

Nothing is known of the pectoral girdle. The forelimb is relatively long for a diplodocid (humerus : femur is 0.79). The humerus is more robust than in *Diplodocus* or *Barosaurus* but less so than in *Apatosaurus*. It approximated those of *Cetiosauriscus* and *Dicraeosaurus*. The ulna-humerus ratio is 0.72. Little is known of the manus. The ilium is short with a very robust pubic peduncle. The ischium has a straight shaft and distal end less expanded than other diplodocids. The femur is sturdy, with a near-circular cross section like that of *Dicraeosaurus.* The tibia and fibula are robust, the muscle scar on the fibula weak. The fibula : femur ratio is about 0.6, and the metatarsus : femur ratio is 0.22. If the small element is correctly identified as the calcaneum, its presence is a subfamily character.

Mamenchisaurus differs enough from the typical diplodocids to be placed in a separate subfamily but is referred to the Diplodocidae for the following reasons: divided spines of cranial dorsals, procoelous caudals, and strong development of *Diplodocus*-like chevrons.

Family Titanosauridae   Lydekker, 1885

Family Characters. Dorsals with irregularly shaped pleurocoels and spines directed strongly backward; transverse processes directed dorsally as well as laterally, very robust in shoulder region; a second dorsosacral, its rib fused to ilium; caudals strongly procoelous with a prominent ball on distal end of centrum throughout tail; caudal arches on front half of centrum; sternal plates large; preacetabular process of ilium swept outward to become almost horizontal.

Titanosaurids were by far the predominant sauropods in the Late Cretaceous, but their relationships to other families remain an enigma, in part because no good skulls are known. Titanosaurids have been found throughout the world, most prolifically in South America and India. A large number of taxa have been described, but many are probably suspect. It is expected that the systematics of these animals will improve with the publication of work by J. E. Powell (thesis, Univ. of Tucumán, 1986).

Titanosaurus Lydekker, 1877. *Titanosaurus indicus* was founded on caudal vertebrae and a slender femur from the Late Cretaceous of India. Later (Lydekker

1893a; Huene 1929a), the robust-limbed *T. australis* and *T. robustus* from Argentina were referred to the genus. The analysis below is limited to the Indian species *T. indicus* (Lydekker 1877, 1879; Swinton 1947). *T. australis* and *T. robustus* are referred to *Saltasaurus australis* and *S. robustus,* respectively. Thus limited, *Titanosaurus* is known only from caudal vertebrae and most of the limb bones of a medium-sized sauropod.

The middle caudals are strongly procoelous with large distal balls. The centra are rather higher than broad, and their flattened sides give them an almost rectangular cross section.

The limb bones are much more slender than those of *Saltasaurus* and *Alamosaurus*. The distal end of the slender radius is expanded both medially and laterally, but the distal margin is perpendicular to the axis of the bone in sharp contrast to *Saltasaurus* and *Alamosaurus*. The humerofemoral ratio is 0.74. The femur shows almost no trace of a lateral prominence. The tibiofemoral ratio of 0.65 compares favorably with that of *Tornieria* but is somewhat less than that of *Aegyptosaurus* (0.69).

Titanosaurus falloti Hoffet, 1942, from the Late Cretaceous of Laos, known from a robust femur and amphicoelous caudals, may belong to another genus and possibly another family.

Saltasaurus Bonaparte et Powell, 1980. *Saltasaurus* is a relatively small, stocky-limbed titanosaurid from the Upper Cretaceous of Argentina. It is known from a series of associated but largely disarticulated bones of a number of individuals representing all parts of the skeleton except most of the skull and many of the foot bones. Osteoderms, including both scutes and ossicles found associated by Bonaparte and Powell (1980), proved what had been suggested by Deperet (1896)—that titanosaurids had body armor.

The differences between *Saltasaurus* and *Titanosaurus australis* and *T. robustus* noted by Bonaparte and Powell (1980) are not here deemed of taxonomic importance, and hence the latter two species are tentatively referred to *Saltasaurus*.

The only skull elements so far reported are a few detached elements from the rear and roof of the skull (Huene 1929a) and perhaps a braincase (Berman and Jain 1982). They are of little diagnostic value.

The number of cervical and dorsal vertebrae are unknown. The presacral centra are opisthocoelous with a well-developed ball and socket throughout. The pleurocoels are small in the cervicals and of only moderate size in the dorsals, and their margins tend to be irregular rather than sharp lipped. The postzygapophyses of the cervicals are large and project backward well beyond the end of the centrum. The spines are low, broad, and single. They remain undivided in the dorsal region but are directed strongly caudally in the middorsal region and remain so almost to the sacrum. The sacrum incorporates a second dorsosacral, so that there are 6 strongly coossified sacral vertebrae. The first caudal is typically biconvex. The succeeding caudals are strongly procoelous, a condition that persists almost to the end of the tail. The proximal caudals possess lateral depressions and are concave ventrally. The arches and spines are sturdy and placed on the cranial half of the centrum. At the end of the tail, there is a series of rod-like caudals indicating some kind of whiplash, but its extent is probably much less than in diplodocids.

The scapula has a very stout blade and is little expanded both proximally or distally. It has a long glenoid margin, so that the large quadrangular coracoid articulates farther forward than in most sauropods. The forelimb bones are massive. The radius is noted for its expanded proximal and distal ends. In addition, the proximal end is particularly directed medially. The distal margin is not perpendicular to the axis of the bone but is directed downward medially. Little is known of the manus. The preacetabular process of the ilium projects strongly laterally. The shaft of the pubis is slender, and the distal end is expanded. The ischium is relatively short and has a well-developed head. The bones of the hindlimb are fairly stout. A prominence is present on the upper half of the lateral margin of the femur as in the brachiosaurids. Little is known of the pes.

Body armor consisting of both dermal plates and tightly packed globular osteoderms was found associated with bones of *Saltasaurus*. They are quite distinct from those of the nodosaurids to which Huene (1929a) had originally assigned them.

Laplatasaurus Huene, 1929a. From three major titanosaur quarries in Argentina, Huene (1929a) somewhat arbitrarily separated specimens representing most of the limb and girdle bones, which he took to be a new slender-limbed genus some 50 percent larger than *Saltasaurus*, naming it *Laplatasaurus araukanicus*. He found no skull material, only two or three presacral centra and twenty caudals, ten of them articulated, as well as some juvenile material to go with the limbs. Bonaparte and Gasparini (1979) selected a tibia and fibula as the type of *L. araukanicus*. While it is reasonable that some of these bones, particularly the slender limbs, belong to one different from *Saltasaurus*, it is not clear that they all do. In particular, it is important that the caudals, distinctive at the family level, are properly referred to the limb material. Pending publication of Powell's work, the tentative diagnosis of *Laplatasaurus* as envisioned by Huene (1929a) follows: much larger than *Saltasaurus*; large pleurocoels in dorsals; more slender scapula and limb bones than in the latter. In addition, the caudal centra are relatively short and the neural arches placed farther back on the centra.

Huene referred the poorly preserved titanosaurid, *Titanosaurus madagascariensis* Deperet, 1896, from Madagascar to *Laplatasaurus,* a judgment that cannot be confirmed. The importance of the specimen is that it represents the first report of body armor in a sauropod.

Alamosaurus Gilmore, 1922. The one continent on which titanosaurids were rare is North America, but, ironically, this continent has produced the most complete articulated skeleton (one quarter complete) yet found. New specimens from Texas will add materially to our knowledge of *Alamosaurus* (Langston, pers. comm.), but the only parts presently described are the pectoral girdle, forelimb and foot, sternum, ischia, and tail of quite a large animal, *A sanjuanensis* (Gilmore 1922, 1946*b*; Mateer 1976).

There is an articulated series of thirty caudal vertebrae beginning at the first. The first caudal centrum is biconvex; all others are strongly procoelous. They do not contain pleurocoels. The neural arches and spines diminish rapidly in height and are placed on the cranial half of the centrum. The prezygapophyses are long. The transverse processes disappear after caudal 8. The prominent chevron facets give the appearance of some ventral excavation, not as pronounced as in *Magyarosaurus*. The chevrons are simple. The sternal plates are very large.

The scapula is large with only a slight distal expansion. The angle between shaft and ridge on the proximal plate is large. The otherwise straight scapulocoracoid margin is broken in the middle by a prominent upward deflection. The forward rim of the coracoid projects beyond the scapula but not as markedly so as in *Saltasaurus* and *Magyarosaurus*. The forelimb is relatively long. The humerus is rather stout and greatly expanded at both ends. The ulna : humerus ratio is 0.65, and the radius : humerus ratio is 0.59, indicating a short forearm. The radius is stout and expanded at both ends, particuarly the distal. The metacarpus : humerus ratio is 0.30. The ischia are short, stout, and united edge to edge for their entire length. The shaft is broad and the distal ends quite broad but thin.

Magyarosaurus Huene, 1932. *Magyarosaurus* was created by Huene (1932) for a large collection of scattered titanosaurid bones from the top of the Cretaceous in Romania, most of them belonging to small individuals. Represented are dorsal vertebrae, many caudals, scapulae, coracoids, pubes, limb bones, and a few metapodials. As with so many of the titanosaurid collections from India, Argentina, southern France, and Spain,

these bones are totally dissociated and may represent more than one genus as shown by the slender and robust types of humeri. Much of Huene's reason for separating *Magyarosaurus* from *Titanosaurus* hinges on a comparison with *T.* (removed to *Saltasaurus*) *australis*. If missing spines of the dorsal vertebrae do indeed point forward (Huene 1932), they would provide a differentiating character, but this condition is not yet demonstrated. Most of the *Magyarosaurus* bones belong to a slender-limbed type of titanosaurid. The prezygapophyses of the proximal caudals are particularly robust. Little more can be said.

Macrurosaurus Seeley, 1869. *Macrurosaurus semnus* is known from several caudal series from the midtail region. An articulated metatarsus from the same strata in the middle Cretaceous of England, originally assigned to the nodosaurid *Acanthopholis* as *A. platypus,* likely belongs here (Seeley 1876). It is one of the oldest genera of the Titanosauridae. The proximal centra are only slightly concave ventrally, and this condition quickly gives way to a flat ventral surface. The middle caudals are more elongate than in later titanosaurids. The metatarsus is noted for the relatively short metatarsal I and long metatarsal V.

Aegyptosaurus Stromer, 1932. *Aegyptosaurus baharijensis,* based on only one specimen consisting of 3 caudal vertebrae, a partial scapula, and 9 limb bones from the Cenomanian of Egypt, is another of the oldest examples of a titanosaurid. Regrettably, this specimen was destroyed during World War II. The bones exhibit minor differences from those of other titanosaurids, but unless future discoveries of topotypes are made, it is uncertain that the validity of the genus can be established. The importance of this form lies not only with its age but also in providing comparative limb measurements of a single individual.

The best-preserved caudal is from the midcaudal series. It is typically procoelous with an arch more elevated than in *Titanosaurus, Saltasaurus,* or *Alamosaurus.* The limbs are comparable to those of *Titanosaurus* but considerably less robust, and the humerus is narrower proximally than those of *Saltasaurus, Alamosaurus,* or *Argyrosaurus.* The femur differs from that of *Titanosaurus* in the placement of the fourth trochanter, which lies just above half length rather than one-third of the way down from the head. The humerus : femur ratio is 0.78, the ulna : humerus ratio is 0.75, and the tibia : femur ratio is 0.69.

Argyrosaurus Lydekker, 1893*a*. *Argyrosaurus superbus* was founded on a gigantic forelimb and foot from the Late Cretaceous of Argentina, and largely for that reason, it has been considered a titanosaurid. The association of a femur and large titanosaurid caudal from other localities with this animal is purely conjectural. The humerus is relatively robust with a square proximal end reminiscent of *Saltasaurus* and also *Opisthocoelicaudia*. The radius and ulna are relatively more slender. The ulna:humerus ratio is 0.69, somewhat greater than that of *Alamosaurus* (0.65), comparable to that of *Opisthocoelicaudia* (0.68), but considerably less than in *Aegyptosaurus* (0.75) or *Janenschia* (0.77). The metacarpals are relatively quite long, comparable to those of camarasaurids and brachiosaurids but much longer than those of diplodocids.

Antarctosaurus Huene, 1929*a*. *Antarctosaurus wichmannianus* is known from a supposedly associated partial skeleton of a very large animal, consisting of a cranium, partial mandible, fragmentary cervical, scapula, incomplete humerus, metacarpus, incomplete ischium, femur, tibia, fibula, metatarsus, and fragments. Other specimens of a large animal (Bonaparte and Bossi 1967) have been referred to the genus.

Among the distinguishing features of the skull fragment are the very slender parasphenoid (in sharp contrast to those of *Camarasaurus* and especially *Quaesitosaurus*) and the large prefrontal, whose articular surface with the frontal occupies more than half of the rostral margin of that element and suggests that the missing nasal must be quite different from that of either *Diplodocus* or *Camarasaurus*. The mandible is square in front like that of *Diplodocus*. Largely on the basis of the similarities in the mandible and weak dentition, Huene (1929*a*) has restored the skull along the lines of *Diplodocus*, but much of this restoration is hypothetical.

A single badly preserved cervical is of no diagnostic value. Huene (1929*a*) concluded that a biconvex first caudal from the same quarry as the type might belong to this animal but more probably was that of *Laplatasaurus*.

The scapula has a long slender shaft, little expanded distally, large angle between axis of shaft and muscle ridge, and a coracoid articulation with a short but sharp central deflection. The incomplete humerus appears to be long and slender. The metacarpals are relatively slender. The ischium is very unlike that of *Alamosaurus* with a long slender shaft and not nearly as broad a distal end. The tibiofemoral ratio is 0.67. Two elements identified as astragalus and calcaneum are totally unlike those of other sauropods, a fact that casts doubt on their identification. The metatarsals are relatively stout. Metatarsal I shows no indication of the characteristic distal process of diplodocids; metatarsals III and IV are the longest. Contrasted with those of *Macrurosaurus*, metatarsal I is relatively much longer and metatarsal II much shorter.

The pair of even more gigantic femora that formed

the basis of *Antarctosaurus giganteus* (Huene 1929*a*) are relatively much more slender than those of the type species *A. wichmannianus*. Their assignment to this genus has not been definitely established. Likewise, *A. septentrionalis* (Huene and Matley 1933) from India, based on an incomplete braincase, scapulae, forelimb, and ?sternal bone, doubtfully belongs to this animal. The slender humerus does indeed resemble the fragmentary one of *A. wichmannianus*, but the shaft of the scapula is relatively much longer and of a quite different shape from that of the latter. The associated radius and ulna removed from this animal by Huene as relatively too short should be returned. The radius to humerus ratio is 0.58, compared to 0.59 in *Alamosaurus*. This is important as the radius has the typical titanosaurid expansion of the ends and confirms that *A. septentrionalis* is a true titanosaurid. The large associated caudal vertebra has a flat-sided centrum reminiscent of *Titanosaurus*. Possibly this form represents a large species of that animal.

Janenschia Wild, 1991. *Janenschia* from the Upper Tendaguru beds is usually taken to be the oldest known member of the Titanosauridae. It is known from several series of very robust limb and foot bones. Janensch (1961) also referred to the genus a pubis and an ischium from the type locality and, in addition, a long series of caudal vertebrae and two huge dorsals from other localities. The justification for the assignment of the vertebrae is that they are not similar to any of the other three established Tendaguru genera. Since it is not inconceivable that there were more than four Tendaguru sauropods, the assignment of the vertebrae should be viewed with caution, especially since the characters of the tail provide the principal evidence for calling *Janenschia* a titanosaurid.

Two very large, poorly preserved cranial dorsals are characterized by massive transverse processes and almost no neural spine at all. In this respect, they are unique. The series of 30 articulated caudals begins near the sacrum. The first five or six of these are strongly procoelous with deep cups on the cranial ends of the centra and large balls behind. The arches and spines bear some resemblance to those of *Alamosaurus*. Behind these, the caudal ball quickly disappears, and the centra maintain a gentle concavity in front and are virtually flat behind unlike those of the typical titanosaurid. The neural arches are not located on the cranial half of the centrum, and ventrally the centra are not excavated. No chevrons are known. Thus, the titanosaurid affinities of even the tail are not entirely clear.

The pectoral girdle is unknown. The limbs are among the most robust found among the Sauropoda. The proximal end of the humerus is broader than the distal. The robust radius resembles that of *Alamosaurus*. The humerofemoral ratio is 0.70 and the ulna : humerus and radius : humerus are 0.77 and 0.69, respectively (Janensch 1961). This greater disparity than usual suggests one or both may be inaccurate, a distinct possibility as both lower arm bones are poorly preserved. In an isolated manus assigned to *Janenschia*, the carpals are missing and the metacarpals are short and stout, rather more like those of *Apatosaurus* than like *Camarasaurus* or *Brachiosaurus*. Metacarpal II is the longest. The phalangeal formula is 2−2−1−1−1 (Janensch 1922) with a large claw on digit I. The pubis assigned to *Janenschia* is of the diplodocid type with relatively slender shaft and strongly developed ambiens process on the proximal end. The ischium is also diplodocid in nature but with a heavier shaft than those of that family. The most distinctive feature of the stout femur is the prominent deflection on its upper lateral margin reminiscent of that found in brachiosaurids and titanosaurids. The tibia is extremely robust and the fibular muscle scars weak. No calcaneum has been found. The metatarsals are very short and robust, but there is no trace of the diplodocid process on the lower end of metatarsal I. Metatarsals II and III are the longest. The phalangeal formula of the pes is 2−3−3−2−1 with claws on digits I to III.

While the procoelous caudals suggest *Janenschia* may have been an ancestral titanosaurid, the present evidence is not convincing. However, the Early Cretaceous species from Malawi, "*Gigantosaurus*" *dixeyi* (Haughton 1928), based on caudal, pubis, scapula, and sternal plate, is certainly a titanosaurid. That it belongs to *Janenschia* is unclear.

Hypselosaurus Matheron, 1869. *Hypselosaurus priscus* from the Late Cretaceous of southern France and Spain differs from other European titanosaurids in the robustness of its limbs. Eggs found associated have been referred to this genus (Lapparent 1947).

Sauropoda *incertae sedis*

"*Apatosaurus*" *minimus* Mook, 1917a, is based on a complete sacrum and pelvis from the Late Jurassic of Wyoming. The species cannot belong to *Apatosaurus*. The sacrum consists of 6 coossified vertebrae, a second dorsosacral having been added. The capitulum and tuberculum of its short rib are those of a typical dorsal, but the distal end is fused to the preacetabular process of the ilium. The spines of sacrals 2 to 5 are fused, slender, and of moderate height. The ilia are low, and their preacetabular processes are directed strongly laterally somewhat as in the titanosaurids. The distal end of the ischium is unexpanded.

"*Morosaurus*" *agilis* Marsh, 1889a, is based on an articulated cranium, proatlas, and cervicals 1 to 3 of a small sauropod from the Early Jurassic of Colorado. The axis has a small but distinct intercentrum fused to the rostral end of the (true) centrum (Gilmore 1907). Cervical 3 is elongate, and its spine is directed rostrodorsally. The size of the specimen and its occurrence in the same beds with several *Haplocanthosaurus* skeletons led to Gilmore's suggestion that it may belong to that animal.

"*Pelorosaurus*" *becklesii* Mantell, 1852. This species is based on one of the best specimens from the English Wealden, yet its affinities are much in doubt. It consists of the three well-preserved forelimb bones and a large patch of skin impressions of a rather small animal. The humerus is stocky. The proximal end of the ulna is broadly expanded, the distal end very little. The fairly robust radius is little expanded proximally. The ulna : humerus ratio is 0.71. The latter character and its robustness immediately excludes the animal from the genus *Pelorosaurus* and any other brachiosaurid such as *Pleurocoelus*. Huene (1932) referred it to *Camarasaurus*, but the limb is more robust than that of the latter, more nearly resembling *Apatosaurus* except for size.

Mongolosaurus Gilmore, 1933b. Based on 5+ teeth, the basioccipital, and cervicals 1 to 3, *Mongolosaurus haplodon* from the Early Cretaceous of Mongolia is difficult to classify. The basioccipital is relatively shorter than those of *Camarasaurus* and *Brachiosaurus*, and the heavy descending processes are considerably shorter than those of *Camarasaurus* and shorter dorsoventrally than those of *Diplodocus*. The teeth are unique among sauropods. They are slender and cylindrical and convex on both faces of the crown, which tapers to a slightly blunted point at the tip. Carinae develop on either edge and extend to the apex of the crown. In some of the teeth, the carina is slightly serrate.

The atlas exhibits clear facets for the articulation of cervical ribs. The odontoid is less prominent than in other species. The parapophysis of the axis occurs at midheight on the cranial end of the lateral face of the centrum, and a ridge proceeds backward from it dividing a lower and upper concavity but no true pleurocoel. Ventrally, the centrum is strongly keeled. Cervical 3 is noticeably longer than that of *Camarasaurus*. Small cavities in the side of the centrum suggest the beginnings of a pleurocoel. The spine is very low but consists of paired incipient ridges, separating it from brachiosaurids and titanosaurids. This feature and the slender teeth suggest a possible affinity with the diplodocids.

Austrosaurus Longman, 1933. *Austrosaurus mckillopi* was founded on three pairs of very incomplete dorsal vertebrae from the Early Cretaceous of Australia. The centra are all strongly opisthocoelous with large balls and cups. They contained deep pleurocoels bored straight into the centrum, not expanding inside it. The rest of the centrum is lightened by a labyrinthine complex of small intramural cavities separated by very thin plates of bone. These centra are unlike those of the camarasaurids and brachiosaurids but bear no real resemblance to other forms, either. *Austrosaurus* is sometimes referred to the generalized Cetiosauridae, but the close cancellous structure of the bones in that group is not similar either. Longman (1933) shows that there is some resemblance to the little-known *Chondrosteosaurus*.

Recently (Coombs and Molnar 1981), a number of fragmentary limb and girdle bones from the Middle Cretaceous of Australia have been provisionally referred to *Austrosaurus* because of the similarity of the texture of a dorsal centrum. The caudal centra are short and range from platycoelous to amphiplatyan to biconvex in the distal ones. The arches are set forward on the centra but not as markedly so as in the Titanosauridae. The humerus is elongate and slender, and the radius is more robust with greatly expanded distal but not proximal end. The metacarpals are long (metacarpus : radius = 0.52), as in *Brachiosaurus* (0.51), *Chubutisaurus* (0.54), and *Alamosaurus* (0.51).

Campylodoniscus Kuhn, 1961. This form is based on an incomplete maxilla from the Late Cretaceous of Argentina (Huene 1929a). It contained teeth with spatulate crowns broader than those of the diplodocids but not as broad as those of the camarasaurids and quite different from the tiny slender teeth of *Antarctosaurus* and also those assigned to *Alamosaurus* (Kues et al. 1980). In a second maxilla (Huene and Matley 1933) from the Late Cretaceous of India, the teeth are absent, but the element is high and robust, as in camarasaurids. Because of the incompleteness of the two maxillae, they cannot be compared critically and may be in no way related. However, they are important in that they demonstrate the existence of sauropods with nondiplodocid skulls in the Late Cretaceous.

Aepysaurus Gervais, 1852. Huene (1932) referred *A. elephantinus* based on a humerus from the French Aptian to the Titanosauridae because of some resemblance to the humerus of *Laplatasaurus*. This humerus is of the slender type and does bear that resemblance, but it also resembles that of *Camarasaurus* and some brachiosaurids. Its true family position must await future discoveries.

SAUROPOD RELATIONSHIPS

The origins of the sauropods are obscure. For many years, they were assumed to have arisen from bipedal prosauropod ancestors. In 1965, Charig et al. argued convincingly for a quadrupedal ancestry from animals allied to the (at the time) very imperfectly known melanorosaurid prosauropods. The description in recent years of better material of the Melanorosauridae and Blikanasauridae has led to some clarification of the relationships of these forms to the sauropods, but none have yet been found which could have been directly ancestral. In most, the reduction of digits in the fore and/or hind feet had already proceeded beyond that of the sauropods. The major points of divergence of sauropods from their presumed melanorosaurid ancestors are reduction in number of teeth and their specialization from the crenellate leaflike type to the two basic sauropod types, the broad spatulate or pencillike cylindrical one, both types without serrations or crenellations except in the earliest forms; development of the vertebrae from the *Plateosaurus* type with low flat arches and spines and no lateral cavities to those with complex cavities in both the centra and arches and reduction of the neural arches and spines to a complex of thin laminae (Bonaparte [1986b] has shown that these developments began in the cervicals and cranial dorsals of a Late Triassic Argentinian prosauropod); increase in sacral vertebrae to 4, 5, or 6 and the enlargement of the sacral neural canal; expansion of the proximal plate of the scapula to accommodate a greater mass of the M. deltoides scapularis associated with the heavier forelimb of these large quadrupeds; significant changes in the pelvis, including the development of a greatly expanded preacetabular process of the ilium, a reduction of its ischial peduncle, and the transformation of the slender pubis into a broad plate; reduction of the deltopectoral crest of the humerus; transformation of the sigmoid-curve shaped femur with prominent lesser trochanter into a straight bone with greatly reduced or, in all later forms, nonexistent lesser trochanter; reduction of the ossified carpus to 2 or 3 distal elements; reduction in number of phalanges in manus; and reduction of the ossified tarsus to a large astragalus tightly bound to the tibia and, in some cases, a much reduced calcaneum.

Although recent discoveries are beginning to clarify the problems of sauropod phylogeny, we are still very far from being able to construct a cladogram. I have chosen five dichotomies representing specializations that divide the sauropods: (1) the development of slender peg-teeth occupying the front of the jaw of diplodocids and apparently titanosaurids; (2) the development of bifid presacral spines of camarasaurids including *Opisthocoelicaudia* and of diplodocids including *Dicraeosaurus* and *Mamenchisaurus*; (3) the development of forked chevrons possessed by the diplodocids and shunosaurine cetiosaurids; (4) the abnormally long forelimbs of the brachiosaurids and abnormally short forelimbs of the diplodocids; and (5) the marked increase of five or more presacral (largely cervical) vertebrae found in the three Chinese genera *Omeisaurus, Euhelopus,* and *Mamenchisaurus.* Any attempt to reconstruct the phylogeny must recognize that one or more of these characters arose more than once.

The oldest known sauropod, *Vulcanodon,* is also the most primitive, particularly in terms of its pelvis, narrow sacrum, and relatively long forelimb. The other established Early Jurassic form, *Barapasaurus,* was probably off the line leading to later forms, because of its specialized cavities in the dorsal arches.

By the Bathonian, a split into three groups was already evident, but it is not clear whether the brachiosaurids split off first from the cetiosaurines and shunosaurines or whether the latter group split off first. The craniodorsally directed ilium, broader shafted and downwardly directed ischium, lengthening forelimb, and low-spined sacrum were already differentiating the brachiosaurids from the cetiosaurids, but the shunosaurines from China had already well developed, *Diplodocus*-like forked chevrons separating them from the English cetiosaurines (and the brachiosaurids). Since it is most unlikely that the latter character could have developed more than once, if it appeared before those characters separating the cetiosaurines and brachiosaurids, a cladistic analysis would require raising the shunosaurines to a separate family.

In the Late Jurassic, all five dichotomies were firmly established. The peg-teeth first appeared in the perhaps Kimmeridgian in the three North American diplodocids and the two or more from Tanzania including *Dicraeosaurus.* The bifid presacral spine (the first trace of which might have occurred in *Shunosaurus*) was now evident in the same group of diplodocines and also in the Chinese *Mamenchisaurus* and *Euhelopus* as well as in *Camarasaurus.* Similarities in the skull and broad teeth of *Camarasaurus* and *Brachiosaurus* suggest that their two lines divided long after the development of the forked chevrons, in which case the divided spine would have had to have developed independently in the diplodocids and camarasaurids. It would appear that *Mamenchisaurus* with well-developed forked chevrons and V-shaped bifid spines in some of its dorsals could be assigned to the diplodocid line, while *Euhelopus* with a distinctly *Camarasaurus*-like skull and dentition and minimally divided spines would go into the camarasaurid line. The argument is weakened by the total absence of skull and teeth in *Mamenchisaurus* and of the tail and chevrons of *Euhelopus.* It is further confounded by the increase of 5 or 6 presacrals in both Chinese forms, *Euhelopus* and *Mamenchisaurus,* and also in the third Chinese Late Jurassic genus *Omeisaurus* with forked chevrons and undivided spines. A final com-

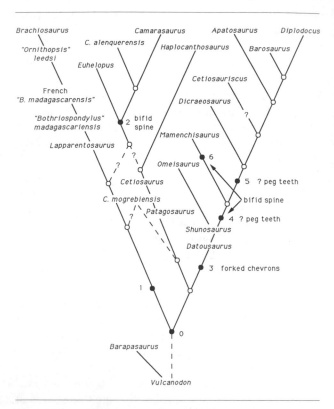

Fig. 16.20. Phylogenetic diagram of Jurassic sauropods. See text for explanation of characters.

plication arises from the fact that *Mamenchisaurus* had a fairly long forelimb, and *Euhelopus* may have had a quite long one. The pictorial representation of relationships among the sauropods is shown in figure 16.20.

A possible variation in sauropod relationships includes *Euhelopus* and *Mamenchisaurus.* If the former should turn out to have forked chevrons and the latter does not have peg-teeth, *Euhelopus* would be removed from its position in the Camarasauridae and its line connected to point 6 beneath *Mamenchisaurus.* Although this would mitigate the problem of multiple development of the increase in presacrals, it would ignore similarities in the skulls and teeth of *Euhelopus* and *Camarasaurus* and seems very unlikely. A second variation, already mentioned and much more plausible, would be for the line to *Cetiosaurus* to have its origin at point 1 on the brachiosaurid line. In any case, it seems likely that diplodocids arose from the Shunosaurinae. Whether *Shunosaurus* itself or *Datousaurus* was closer to the ancestral position must await future disclosures. A third possibility exists. *Patagosaurus* has been assumed to have had normal chevrons, but little information is available concerning the tail. If it should prove to have had forked chevrons, it would then be necessary to transfer it to the Shunosaurinae, and it would become a possible diplodocid ancestor. Indeed, Bonaparte (1986*b*) has presented evidence for this possibility based on similarities in the dorsal vertebrae of *Patagosaurus* and *Diplodocus,* namely, in "the laminae connecting the diapophyses with the different parts of the neural arch," a "cavity of the neural arch just above the neural canal," and other general proportions of these vertebrae. Finally, the position of *Janenschia* is ambiguous and will be discussed below.

Bonaparte (1986*b*) has recently discussed the evolution and classification of the Jurassic sauropods based on a careful analysis of characters of the presacral and sacral vertebrae. His conclusions differ in some respects from those presented here. The major point of disagreement relates to the position of *Haplocanthosaurus,* which he would remove from the Cetiosauridae and place, along with *Dicraeosaurus,* in a separate family—Dicraeosauridae.

As complicated and uncertain as the problems with the Jurassic phylogeny are, those involving the Cretaceous are far worse. Most easily disposed of are the Early Cretaceous brachiosaurids. *Pelorosaurus* is essentially on the same line from *Brachiosaurus. Pleurocoelus* and *Chubutisaurus* must each have diverged from that line earlier, as neither can be descended from *Brachiosaurus.* The real problem concerns the Titanosauridae. No forked chevrons have ever been found with members of the family, and most of the chevrons are present in the long tail segment of *Alamosaurus* (Gilmore 1946*b*). Since it is difficult to imagine this

character being suppressed once it had developed (or the bifid spine either), the divergence from the diplodocid line would have had to have occurred earlier than point 3. But the peglike teeth in the diplodocids did not develop until point 5 (or possibly 4). If indeed the peglike teeth in the Titanosauridae and Diplodocidae did arise independently, it would appear more reasonable that the titanosaurid line arose from cetiosaurines rather than the shunosaurines, perhaps through *Janenschia.* Although the peg-toothed *Nemegtosaurus* and *Quaesitosaurus* seem to approach the dicraeosaurines most nearly, the possibility that they might be titanosaurid does exist and would raise real problems if true. There remains the remote possibility that the titanosaurids did not have peglike teeth. *Antarctosaurus,* whose vertebrae are virtually unknown, might possibly be a diplodocid. However, this does not appear very likely as no diplodocid vertebrae have been found in the Late Cretaceous of Argentina, and only peglike sauropod teeth (just 2 or 3 of them) have been found in rocks of that age in North America (Kues et al. 1980). The problem of the titanosaurid relationships will not be solved until a complete skull is found associated with a titanosaurid skeleton, and connecting links in the Early Cretaceous will probably also be necessary. The final problem involves *Opisthocoelicaudia,* which is certainly neither titanosaurid (opisthocoelous caudals, bifid spine) nor diplodocid (simple caudals and chevrons). The most likely conclusion based on available evidence is that it arose within the camarasaurid line.

17

Sauropod Paleoecology

PETER DODSON

The sauropods were archaic dinosaurian herbivores with relatively small heads, simple teeth, and the smallest brain-to-body sizes among all dinosaurs (Hopson 1977). Other groups of dinosaurs surpassed sauropods in relative brain size, masticatory efficiency (Weishampel 1984a; Norman and Weishampel 1985), locomotor ability (Coombs 1978c), and sociality as indicated by secondary sexual display structures (Dodson 1975, 1976; Farlow and Dodson 1975). Paradoxically, sauropods also represent one of the most successful groups of dinosaurs in terms of longevity (Early Jurassic–Late Cretaceous), diversity (more valid genera and species than any other major group of dinosaurs), and distribution (known from all continents but Antarctica), not merely sheer physical bulk. By the Early Jurassic, they attained enormous body size (*Barapasaurus,* India: femur length 1.7 m; Jain et al. 1975) and persisted throughout the Mesozoic. Sauropods present fertile ground for inquiry into the biology of dinosaurs, including such topics as reproduction, growth, longevity, metabolism, and life habits generally. Anatomic, taphonomic, sedimentologic, stratigraphic, and biogeographic features all contribute to our understanding of sauropod paleoecology.

Collectively, sauropods have an excellent fossil record, but at the species level (McIntosh, this vol.), the record of sauropods is poor. Of nearly 90 proposed genera, roughly half may sustain claims of validity. Skulls are known from fewer than 12 genera, and only 5 genera are known from essentially the entire skeleton. Although taphonomic theory predicts that animals of large body size may potentially be overrepresented in the fossil record (Behrensmeyer et al. 1979; Coe et al. 1987), the probability of preservation of a *complete* specimen declines at large size, given the vagaries of both sedimentation and erosion. Body parts most likely to be lost before burial include tail, feet, and skull (Dodson 1971). The largest dinosaurs that are regularly preserved complete measure about 10 m in length (e.g., iguanodonts, hadrosaurids). Complete sauropod skeletons are exceedingly rare. However, juvenile specimens are also very uncommon. The overwhelming majority of sauropod specimens are at least 80 percent of adult size.

Owen (1859), working with very incomplete material, believed the British sauropod, *Cetiosaurus* ("whale lizard"), to have been a gigantic marine crocodile. Thus, from the beginning, sauropods were considered aquatic or marine animals (Cope 1878g; Marsh 1883; review in Coombs 1975). Although a forceful case for terrestrial sauropods was made by Riggs (1904), the persistent image of aquatic sauropods, as painted by C. Knight and Z. Burian, has been influential until the 1970s. Bakker (1971) interpreted the elephantine, columnar limbs, short, compact feet, and abbreviated narrow, slablike thorax as evidence for a terrestrial, high-browsing life-style, a view endorsed by Coombs (1975). A survey of the taphonomy of major quarries in the sauropod-dominated Upper Jurassic Morrison Formation of the western United States (Dodson, Behrensmeyer, Bakker, and McIntosh 1980; Dodson, Behrensmeyer, and Bakker 1980; Dodson et al. 1983) shows that sauropod remains are distributed across a spectrum of sedimentary environments. Although Morrison habitats included lakes and swamps, it is clear that sauropods ranged across all available habitats in the Morrison terrain. Indeed, environmental analysis of the Morrison (Peterson and Turner-Peterson 1987) show that moist habitats were subordinate to dry ones on the western American landscape at that time. Morrison sauropods probably had to cope with seasonal aridity. Comparison of the preservation of Morrison sauropods with that of large African mammals of Pleistocene age in Kenya (Behrensmeyer 1975) suggests that sauropods were as terrestrial as are Recent elephants.

The Morrison Formation is the best-known sauropod-bearing formation in terms of sedimentary environments and dinosaurian and nondinosaurian faunas. Nonetheless, sauropods are too diverse to permit generalization based on the Morrison Formation alone. In the Upper Jurassic Tendaguru Formation of Tanzania, sauropod remains are sometimes found mixed with marine invertebrates (Russell et al. 1980), and the principal bonebed concentrations were deposited not far from the sea. In the Glen Rose Limestone (Lower Cretaceous) of Texas, sauropods left abundant trackways preserved in dolomitic mudstones that represent intertidal and supratidal deposits adjacent to a lagoon (Farlow 1987a); marine invertebrates are found above and below the track-bearing strata. Lower Cretaceous Wealden sauropod remains are often found in aquatic or marine habitats (Allen 1975). Lower Cretaceous sauropods of Maryland (Lull 1911b) come from coastal deposits of the Arundel Formation. It is probably significant that the Mongolian Upper Cretaceous sauropods *Nemegtosaurus* and *Opisthocoelicaudia* come not from the Djadochta and Barun Goyot formations, which were deposited in semiarid red bed terrains, but in the pale fluvial deposits of the overlying Nemegt Formation (Osmólska 1980) which document the reestablishment of more mesic conditions. Erben et al. (1979) attribute the success of *Hypselosaurus* in the Maastrichtian of southern France to humid conditions.

All of these studies suggest that as a whole, sauropods prospered under humid conditions. Abundant sauropod trackways (e.g., Gillette et al. 1985; Lockley 1986a) confirm that sauropods readily walked on substrates that were either under water or recently emergent. Possibly the adaptation of Morrison sauropods to seasonally dry environments represents an extreme rather than a typical situation for sauropods. Nonetheless, the concepts of sauropods submerged in deep water and breathing through scarcely exposed nostrils (e.g., Hay 1908, 1910) and that of behemoths too heavy to support their weight on land may be confidently rejected (Kermack 1951).

The geographic distribution of sauropods has ecological significance. In the Morrison Formation, sauropod faunas are essentially homogeneous over distances of 1,000 km (Dodson, Behrensmeyer, Bakker, and McIntosh 1980). This distribution, coupled with evidence of seasonal aridity and absence of abundant plant remains, further suggests that Morrison sauropods may have migrated long distances in search of food (see also Calder 1984 on the positive relationship between increased body size and tendency to migrate). Another consideration favoring the concept of vagile sauropods is the impact that herds must have had on the relatively slow-growing local vegetation (Coe et al. 1987). Regular herd movements, such as those shown by modern African ungulates (Sinclair and Norton-Griffiths 1979), would provide the opportunity for depleted vegetation to recover. Even though Pangaea was still assembled in the early Mesozoic, similar dinosaurs are found on opposite sides of juxtaposed continents. For example, *Brachiosaurus* and *Barosaurus* are found in Tanzania and the western United States, separated by moderate barriers to dispersal (Galton 1977b). Long legs in dinosaurs, however, indicative of general locomotor abilities (Bakker 1971; Coombs 1978c), undoubtedly aided long-distance movements, even at very slow rates of 10 or 20 km per day (Hotton 1980; Parrish et al. 1987).

Sauropod remains are found in mass accumulations (bone beds) in the Morrison Formation of the western United States (Dodson, Behrensmeyer, Bakker, and McIntosh 1980), in the Tendaguru beds of East Africa (Russell et al. 1980), in the Kota Formation of India (Jain 1980), and in the Lower Shaximiao Formation of the Sichuan Basin in China (Xia et al. 1983). Vast sauropod footprint assemblages (e.g., Bird 1985; Lockley 1986a, 1986b; Farlow 1987a) provide complementary data that suggest that at least some sauropods were gregarious some of the time (Coombs, this vol.). However, some sauropods are never found in major bone beds. These include, for example, *Brachiosaurus* and *Haplocanthosaurus* in the United States and all of the Mongolian and European sauropods. Abundant massed remains of *Brachiosaurus* are found at Tendaguru. It is not known if there was a genuine behavioral difference between American and African species (which is not inherently improbable) or if the fossil record of *Brachiosaurus* in the Morrison is simply too poor. It is reasonable to conclude that some but not necessarily all sauropods were gregarious.

Sauropods rapidly attained large body size and dominated Jurassic terrestrial vertebrate faunas. They clearly coevolved with contemporary floras, which consisted principally of conifers, ginkgoes, cycads, ferns, and horsetails (Weaver 1983; Weishampel 1984c; Coe et al. 1987), with a high degree of success. In the Cretaceous, sauropods were replaced in North America and western Asia by derived ornithischians (Bakker 1978; Norman and Weishampel 1985; Wing and Tiffney 1987) that coevolved with the newly radiating angiosperms. However, sauropods prospered on southern continents (McIntosh, this vol.) and persisted on northern continents through the Cretaceous. In North America, sauropods appear to have become extinct at the end of the Early Cretaceous, but the appearance of *Alamosaurus* in the Maastrichtian of Texas, New Mexico, and Utah documents the reintroduction of sauropods from South America (Lehman 1987).

The long necks of sauropods invite comparison with those of giraffes, which are high browsers (Dagg and Foster 1976). *Brachiosaurus*, with its elongate forelimbs, conforms well to the giraffe model. However, such morphology is *not* typical of sauropods: most have forelimbs that are relatively short and withers that are lower than the hips. No sauropod has elongated neural spines at the base of the neck which could provide leverage for epaxial muscles for sustained elevation of the neck in a giraffelike manner. An alternative model of sauropod feeding (Martin 1987) notes the usefulness of the neck in sweeping both laterally and vertically a large volume of feeding space from a fixed locus. Sauropods had the potential to exploit a high vertical feeding range of 3.5 m (Martin 1987) to 15 m (Jensen 1985b), depending on the species and the degree of maturity of the individual. These animals competed with contemporary dinosaurian herbivores (e.g., stegosaurs and ornithopods) at lower levels (0–3 m off the ground) but were unchallenged, except interspecifically, at greater heights (Coe et al. 1987). Relative neck length varies greatly among sauropods and may offer a partial basis for niche separation. The cetiosaurids (e.g., *Cetiosaurus oxoniensis*; *Haplocanthosaurus priscus*;

Shunosaurus lii) and camarasaurids (e.g., *Camarasaurus* species) have relatively short necks (Bonaparte 1986b; Martin 1987), while the diplodocids (e.g., *Diplodocus longus*, *Barosaurus lentus*, *Mamenchisaurus constructus*) have relatively long ones (McIntosh, this vol.). In *Mamenchisaurus constructus* (Young 1954), the relatively enormous neck equaled the length of the rest of the body.

How much time sauropods typically spent with their necks elevated may be questioned. A strong nuchal ligament of elastin may have aided in maintaining the neck in a horizontal or modestly elevated position (Alexander 1985), but its inability to shorten actively would have rendered it ineffective for elevation of the neck, which can only be accomplished by muscular effort. Low neural spines of the cervical and dorsal vertebral series of the derived camarasaurids and diplodocids are conspicuously bifid (McIntosh, this vol.). It may be assumed that the broad area within the split spine was occupied by muscle plus a nuchal ligament of modest dimensions. Alexander's (1985) biomechanical analysis of the neck of *Diplodocus carnegii* suggests that the transversospinalis muscles contained within the bifurcation would have been barely able to lift the neck of this animal. Although there is no reason why these muscles could not have been supplemented by the action of the longissimus and iliocostalis muscles lateral to the spines, it appears that the margin of safety was small, at least in *D. carnegii*. Furthermore, perfusion of the brain at a pressure of 60 mm Hg in a sauropod with an elevated head would have placed tremendous and potentially very dangerous stress on the cardiovascular system. Estimates of systolic blood pressures required to perfuse the brains in *Apatosaurus, Camarasaurus, Diplodocus*, and *Brachiosaurus* standing with heads maximally elevated range from 367 to 717 mm Hg (Hohnke 1973; Seymour 1976). Such pressures, with attendant risks of cardiovascular failure, appear to be without precedent among living vertebrates (Ostrom 1980; Pedley 1987; Hargens et al. 1987 on giraffes). In giraffes, special adaptations of the vessels and fascia of the distal limb prevent severe edema. Aquatic habits would circumvent the problems associated with high blood pressure, but sauropods appear to be well adapted for a terrestrial mode of life (Bakker 1971a; Coombs 1975; Dodson, Behrensmeyer, Bakker, and McIntosh 1980). A mode of life for sauropods that required lower blood pressures, that is, habitual feeding at moderate rather than very high levels, seems inherently more likely. Instead of sustaining elevated systolic blood pressures, sauropods may have tolerated periods of transient hypoxia during bouts of high feeding. The lower their oxygen demand, the more feasible would be this strategy. In addition, specializations of the cervical and cranial vasculature are to be expected. For example, in giraffes, elasticity of the *rete mirabile* at the base of the brain prevents hypertensive failure of cerebral vasculature when the elevation of the head is suddenly changed (Dagg and Foster 1976). Additionally, sauropods may well have had a carotid sinus that served as a reservoir of arterial blood to perfuse the brain when the carotid arteries collapsed due to insufficient systolic pressure during high feeding.

It is sometimes suggested (Osborn 1899; Riggs 1904; Coombs 1975) that sauropods could extend their vertical feeding ranges by adopting a bipedal or tripodal stance. Given the shortness of the dorsal vertebral column, location of the center of gravity near the pelvis (Alexander 1985), the high sacral neural spines, and the robustness of the tail, such behavior seems well within the potential biomechanical repertoire of many sauropods. Indeed, the ability to rise and support weight on the hindlimbs was undoubtedly necessary for the procreation of the species. However, it seems unlikely that sauropods would have been inclined to do so except under conditions of stress, such as intraspecific combat (e.g., Bakker 1987), or of food shortage, when foliage accessible from normal posture was exhausted.

The teeth of primitive sauropodomorphs and of vulcanodontids are coarsely denticulate (Galton, this vol.; McIntosh, this vol.). Derived from this condition are two tooth types that are widespread among sauropods (e.g., Carey and Madsen 1972; Coombs 1975; Galton 1986b). Robust, erect, spatulate teeth, which may show heavy wear, are characteristic of *Camarasaurus* and other members of the family Camarasauridae but also of the Cetiosauridae and the Brachiosauridae (McIntosh, this vol.). Slender, procumbent, peglike teeth that are confined to the front of the jaws and typically show relatively little wear are characteristic of *Diplodocus* and other members of the Diplodocidae and Titanosauridae. Diplodocoid teeth seem suited for little but nipping prehension or rakelike stripping of succulent plant fronds and possibly straining aquatic "gruel," while camarasaur teeth indicate a modest degree of initial oral processing of coarser plants (e.g., Galton 1986b). Sauropod communities in the Morrison Formation include four or five apparently sympatric species (Dodson, Behrensmeyer, Bakker, and McIntosh 1980), among which one had spatulate teeth (*Camarasaurus*), two had slender peglike teeth (*Diplodocus, Apatosaurus*), and two lack skulls (*Haplocanthosaurus, Barosaurus*). Differences between the two food-processing

systems seem fundamental, but no ecological synthesis has been achieved on the basis of these differences.

Sauropod food requirements must have been impressive, regardless of metabolic rate. Gut capacity was enormous, and it is generally believed that sauropods employed gastric mills ("gizzard stones") to aid in the mechanical breakdown of food in the gut in lieu of effective oral processing (Brown 1907; Cannon 1907; Farlow 1987b; Stokes 1987). However logical the concept of a gastric mill may be, unequivocal evidence for this is annoyingly rare (e.g., Darby and Ojangas 1980), in keeping with the taphonomic problems discussed above. A specimen of ?*Barosaurus* from the Morrison Formation (Bird 1985) showed sixty-four polished stones between the pelvis and the ribs. Sauropods are unlikely to have possessed a rumen, as the effectiveness of the rumen for selective delay of food particles to allow microbial fermentation of cellulose decreases with increasing body size (e.g., Janis 1977; Dement and Van Soest 1985). For mammals above 1,200 kg, food is already held to the point of essentially complete digestion without a specialized foregut. Large mammals employ specializations of the hindgut (cecum) for microbial fermentation of fibrous, low-quality plant material as well as detoxification of plant toxins (allelochemicals or secondary compounds). It is probable that sauropods utilized a similar system (Farlow 1987b; Coe et al. 1987; Dunham et al. 1989). One would expect adaptations in sauropods to increase the rate of gut clearance, not to cause selective delay, because it is inefficient to obstruct the gut with materials from which no further energy can be extracted.

Estimation of food requirements for sauropods is of considerable interest. Bakker (1980) has argued that relative head size does not correlate with body size in living mammalian herbivores, and thus small relative head size does not preclude tachymetabolism in sauropods. Weaver (1983) has reviewed the caloric content of living representatives of the principal potential food sources and demonstrated that ferns and horsetails have significantly lower caloric values than do cycads or conifers; ginkgos are intermediate. McNab (1986a, 1986b) has shown that in eutherian and metatherian mammals, basal metabolic rate is controlled by diet, and fruits and leaves are associated with relatively low basal metabolic rates. Using elephants as a model, Weaver (1983) estimated that a sauropod could have gathered 200 kg of forage per day. At this rate, a 15-ton sauropod could have eaten enough to sustain an endothermic metabolism eating ginkgoes or better. A 40-ton sauropod eating only cycads or conifers could have maintained endothermy, but by this model, endothermy would have been impossible with any diet for the truly gigantic sauropods (e.g., *Ultrasauros,* Jensen 1985*b*). Whether sauropods fed selectively is doubtful. Feeding selectivity generally varies inversely with body size (e.g., Jarman and Sinclair 1979; Calder 1984; Wing and Tiffney 1987). Large herbivores cannot afford energetically the luxury of seeking out high-quality food. Instead, they are constrained to eat whatever is available and then to digest it by microbial fermentation in capacious guts (Farlow 1987*b*). Ferns were the dominant herbaceous plants of the Jurassic and Cretaceous, forming vast fern savannas and prairies (Coe et al. 1987). It seems highly probable that ferns, in spite of their nutritional shortcomings, formed a major part of the sauropod diet.

Models of heat transfer in sauropods (Spotila et al. 1973; Spotila 1980; Dunham et al. 1989) show that metabolic heat would be lost so slowly that they would have effectively been homeothermic ("inertial homeotherms") irrespective of metabolic rate. Fermentation provided a source of heat, and the term "fermentative endotherms" has been proposed (Mellett 1982; Farlow 1987*b*). It may be argued that, for a very large animal, tachymetabolism in a warm climate would cause more problems (i.e., those relating to heat loss) than it would solve (Dunham et al. 1989). Despite the enthusiastic claims of tachymetabolic endothermy for sauropods by Bakker (1971*c*, 1980), most workers (e.g., McNab 1983; Weaver 1983; Coe et al. 1987; Farlow 1987*b*; Reid 1987*a*, 1987*b*; Dunham et al. 1989) continue to regard sauropods as poor candidates for this status, having more to lose (heat exhaustion, starvation) than to gain.

Reproduction in sauropods is poorly understood. Eggs and nests are very rare for all dinosaurs. There is only a single presumed association between a sauropod and an egg, that for *Hypselosaurus,* a relatively small Maastrichtian sauropod from southern France (Dughi and Sirugue 1976; Ginsburg 1980). A very porous 1,700 cc egg from the Gobi is inferred to be sauropod (Sochava 1969) but is associated with no described genus. Flattened eggs and shell fragments from the Upper Cretaceous Infratrappean and Lameta formations, Deccan, India, have been referred to sauropods on the basis of size (15 cm diameter), spherical shape, and associated remains of the sauropods *Titanosaurus* and *Antarctosaurus* in the vicinity (Vianey-Liaud et al. 1987). The eggs of *Hypselosaurus* are the largest dinosaur eggs known, although with a volume of only about 1,900 cc (Seymour 1979), they are only about a fifth of the size of the largest known egg, that of the extinct elephant bird, *Aepyornis,* from Madagascar

(Schmidt-Nielsen 1984). *Hypselosaurus* eggs are described as having been laid in pairs along a linear transect (Dughi and Sirugue 1976), and Ginsburg (1980) facetiously suggested that they were laid by a fleeing animal. However, their porous structure is best interpreted as indicating that they were covered with mounds of vegetation (Seymour 1979). In default of further evidence, it is assumed that all sauropods (indeed all dinosaurs) were egg-layers. Sauropod eggs are not known from the great sauropod terrains of the world, but the rarity of fossil eggs generally (as may be expected from their ephemeral nature, fragility, susceptibility to predation, and ease of dissolution in acidic substrates; Carpenter 1982*b*) demonstrates that no significance can be attributed to this observation. A weak case has been made for live birth in sauropods, based principally on negative evidence, the lack of strong evidence for eggs (Bakker 1980). Discoveries of dinosaur eggs and nests are increasing, and negative evidence will become increasingly less attractive. Nothing about pelvic structure can be construed as supporting live birth in sauropods in preference to oviposition (Dunham et al. 1989).

Aepyornis eggs indicate that there is no physiological barrier to producing 10 kg, 1.5 m long hatchlings such as one must postulate for large sauropods, although there is no described evidence of sauropods this small. Age of *Hypselosaurus* at sexual maturity was calculated by Case (1978*b*) as 62 years (range 25 to 72 years) and age at adult size as 82 to 118 years. For *Brachiosaurus,* sexual maturity would have taken over a century. These estimates, however, are based on frank extrapolations of reptilian growth rates; assumptions of avian or mammalian growth rates would make an approximately ten-fold difference (Case 1978*b*). The presence of mammalianlike levels of primary and secondary vascularization of dinosaur bone in general and sauropod bone in particular (e.g., Enlow and Brown 1957; Currey 1962; Ricqlès 1980) encourages the view that sauropods grew relatively rapidly and continuously but does not necessarily confirm endothermy (Reid 1984*a*, 1984*b*, 1987*b*). Furthermore, lamellae documenting cyclical growth are reported in a sauropod humerus (Ricqlès 1983) and a pelvis (Reid 1981). Ricqlès (1983) interprets the histological evidence as consistent with inertial homeothermy coupled with incipient endothermy. However, whatever the metabolic rate of sauropods, Dunham et al. (1989) believe that sexual maturity occurred at 20 years of age or less, on the grounds that further delay of first breeding beyond that age requires highly improbable levels of juvenile survivorship. Longevity generally correlates positively with body size (e.g., Calder 1984), and it would seem that life spans on the order of 100 years should not be regarded as unreasonable, given the brackets of 70 years for elephants (Vaughan 1978) and 150 years or more for tortoises (Minton and Minton 1973).

All sauropods were large and ranged from 12 m to 30 m or more in total length. Few other dinosaurs were larger than even small sauropods (Hotton 1980). Large size with all its biological implications (e.g., McMahon and Bonner 1983; Peters 1983; Calder 1984; Schmidt-Nielsen 1984) was intrinsic to sauropod biology and was established at the outset of sauropod history. Advantages of large size include access to foods unavailable to smaller herbivores and low mass-specific metabolic rates. Lower metabolic rates permit the fermentive digestion of coarser, less nutritious food and confer relative resistance to periods of short-term food privation. The energetic cost of locomotion per unit body mass decreases with increasing size, but animals of larger body mass spend a greater proportion of their daily energy budgets on movement than do smaller ones (Garland 1983b). Although the cost of lifting a unit of mass is independent of size, metabolic rate is not. In small animals, the cost of climbing (lifting its own mass) is a small proportion of resting metabolic rate, but in a large animal, it is a much larger proportion of metabolic rate. Thus, the relative cost of climbing hills increases in large animals, which ascend slopes at lower angles (Taylor et al. 1972). Very large dinosaurs had considerable thermal stability (inertial homeothermy: Spotila 1980), but even at low metabolic rate, heat loss could have been a serious problem. The attenuated form of sauropod appendages (neck, tail, limbs) may help to maximize surface area for heat loss (Wheeler 1978; Colbert 1983).

Very large sauropods clearly approached the theoretical maximum for mobile terrestrial organisms, but with recent reports of ever-larger sauropods (Jensen 1985b; Gillette 1987), there is no reason to believe that the largest possible sauropod has yet been found. The weight of the supporting skeleton increases at a faster rate than the mass of the body (allometric coefficient 1.09; Prange et al. 1979). In a 6.6-metric-ton elephant,

the skeleton accounts for 26 percent of the body weight. Strength of the skeleton does not appear to be a limiting factor in very large animals (Schmidt-Nielsen 1984), but efficiency of enormous viscera may be. For instance, the heart of a 30-ton sauropod must have weighed at least 150 kg, and an 80-ton sauropod 400 kg (conservatively, 0.5% body mass; Schmidt-Nielsen 1984). By comparison, a 16 m fin whale (*Balaenoptera physalus*) that weighs 29 metric tons has a 150 kg heart (McAlpine 1985), and a 30 m blue whale (*Balaenoptera musculus*) that weighs 120 metric tons (Slijper 1962) may be expected to have a heart that weighs 600 kg. Whales operate effectively with sauropod-sized hearts but have the great advantage of living in a hydrostatically dense medium. Seymour (1976) predicted that sauropod hearts would have to undergo myocardial hypertrophy by a factor of eight (conservative estimate) in order for the ventricular wall to sustain stresses of 500 mm Hg in the absence of hydrostatic support. Sauropod hearts would thus range in size from 1.2 to 3.2 metric tons. What are the limits to oxygenation of the myocardium? How could the respiratory system tolerate the burden of dead air space in the elongated trachea? Giraffes reduce dead air space through tracheal stenosis (Calder 1984). What are the limits to narrowing of the trachea? Could a very large sauropod sustain a fall without breaking its bones? Is body size limited by the ability of an animal to crop and digest sufficient food in a day?

Although rare terrestrial mammals (e.g., *Baluchitherium*; Granger and Gregory 1935) attained gigantic size, sauropods as a group represent a unique experiment in life history. Wing and Tiffney (1987) argue that mammals have rarely exploited very large body size because forests are mechanically difficult habitats in which to forage and that high plant growth is more efficiently harvested by small arboreal animals. Coe et al. (1987) suggest that if body size is limited by foraging ability, the difference in body size between modern elephants and sauropods may directly reflect the maximum biomass sustained by an endotherm versus that sustained by an ectotherm. Sauropods present fertile ground for studying the limits of physiology and biomechanical engineering on land.

Segnosauria

RINCHEN BARSBOLD
TERESA MARYAŃSKA

INTRODUCTION

The Segnosauria constitutes a very rare, exclusively Cretaceous group of aberrant saurischians. They were medium- to large-sized animals with a small skull in comparison with body size, low mandibles, and short and broad tetradactyl pes with rudimentary metatarsal V. They are represented by four valid monospecific genera: *Segnosaurus* Perle, 1979; *Nanshiungosaurus* Dong, 1979; *Erlikosaurus* Barsbold et Perle, 1980; and *Enigmosaurus* Barsbold et Perle, 1983. *"Chilantaisaurus" zheziangensis* Dong, 1979, represented by very fragmentary postcranial elements, is treated here as a possible segnosaurian.

Perle (1979), Barsbold and Perle (1980), and Barsbold (1983a) considered the Segnosauria as most probably representing a separate line of predatory dinosaurs within the Theropoda. However, Perle (1981) suggested that the Segnosauria, as presently known, do not show characters that unequivocally justify their assignment to the Theropoda or any other known suborder of Dinosauria. According to Paul (1984b), segnosaurians represent surviving relics of the Triassic prosauropod-ornithischian transition. It seems more reasonable to interpret segnosaurians as an aberrant, herbivorous rather than carnivorous group of Saurischia *sedis mutabilis*, most probably more closely re-

lated to Sauropodomorpha (Gauthier 1986) than to the Theropoda.

Perle (1982) includes in the Segnosauria the Therizinosauridae and Deinocheiridae. Both are here treated as Theropoda by Norman (this vol.), although there is some possibility that the Therizinosauridae might belong to the Segnosauria.

DIAGNOSIS

Segnosaurians are characterized as follows: skull and mandible shallow and long, toothless rostrally; external naris greatly elongate; palate highly vaulted with elongate, caudally shifted vomers and palatines and rostrally reduced pterygoids; premaxillary-maxillary secondary palate well developed; basicranium and ear region strongly enlarged and pneumatized; mandible downwardly curved rostrally; six firmly coalesced sacrals with long transverse processes and sacral ribs; humerus with strongly expanded proximal and distal ends; pelvis opisthopubic, ilia broadly separated from each other; preacetabular process of ilium deep and long with pointed cranioventral extremity that flares outward at a right angle to the sagittal plane; postacetabular process very short, bearing a strong, knoblike caudolateral protuberance; astragalus with tall ascending process and reduced astragalar condyles, only partly covering the distal end of the tibia; and pedal claws comparatively large, narrow, and decurved.

ANATOMY

Skull and Mandible

The description of segnosaurian cranial structures that follows is based on the skull of *Erlikosaurus andrewsi*, which is the only known skull in the group (figs. 18.1A–F, 18.2A, B); that of the lower jaw and postcranial skeleton is based on *E. andrewsi* and *Segnosaurus galbinensis*.

The skull of *E. andrewsi* is long and low (fig. 18.1), with large, elongate external nares surrounded by a deep narial shelf. The premaxillae form a relatively wide, toothless beak that is heavily vascularized and was probably provided with a horny sheath during life. The nasal process of the premaxilla is very narrow and long. The triangular maxilla forms the caudoventral margin of the external naris and bears a long jugal ramus caudal to the last maxillary tooth. The rostralmost portion of the maxilla is edentulous. The maxillary tooth row ends at the level of the middle of the antorbital fenestra. Contrary to the suggestion of Paul (1984b), there is no lateral maxillary shelf developed in *E. andrewsi*. There are only a few large foramina along the dentigerous portion of the maxilla.

The prefrontal is small and triangular; it ventrally contacts the T-shaped lacrimal. The long and rostrally narrow nasal forms two-thirds of the dorsal border of the very elongate external naris. The jugal is long, low rostrally, and divided caudally. The caudal jugal process forms much of the ventral half of the infratemporal fenestra. The quadratojugal has a broad exposure on the lateral surface of the skull. It overlaps rostrally the caudoventral process of the jugal, forms a middle part of the caudal border of the infratemporal fenestra, and contacts the squamosal dorsally. The quadrate is vertical, with a concave caudal margin.

The rostralmost portion of the palate is formed by horizontal palatal processes of the premaxillae and maxillae. The choanae are bordered rostrally by the maxillae. Caudal to the palatal processes of the maxillae, the palate is highly vaulted. The fused vomers form a long vertical plate that is wedged rostrally between the maxillae and reaches as far caudally as the infratemporal foramen. The large and long palatine is shifted caudally, overlying the ectopterygoid ventrally.

The ectopterygoid bears a stout process ventrally, near the contact with the pterygoid. The palatal ramus of the pterygoid is very short and lies close to the parasphenoid. There is no interpterygoid vacuity. The quadrate ramus of the pterygoid is very close to the braincase. The basicranial and ear regions of the skull are swollen and strongly pneumatized. The basipterygoid process is weakly developed. The parasphenoidal rostrum is comparatively long and wedged between the vomers. The semilunate orbitosphenoid is very small, and the laterosphenoid has a strongly convex ventrolateral surface.

The supraoccipital bears a vertical median ridge. This bone extends rostrodorsally and participates in the formation of the caudomedial portion of the skull roof. The parietal invades the occiput, where it is low. The exoccipital is rather large with a long and high paroccipital process.

The lower jaw is generally shallow, and the rostral region is curved downward. Downward bending is stronger in *S. galbinensis* than in *E. andrewsi*. The dentary is very shallow for most of its length but is deeper caudally to form a slight coronoid prominence, very low in *E. andrewsi* and only slightly elevated in *S. galbinensis*. The rostralmost portion of the dentary is toothless. In *E. andrewsi*, the tooth row is bordered laterally by a shelf starting at the fifth tooth position and running to the end of the dentary, suggesting the presence of cheeks (see also Paul 1984b). This shelf is also visible in *S. galbinensis*, but here it starts first at the fourteenth tooth position and continues backward only for a half of the dentary. The large surangular overlaps the caudal aspect of the dentary. There is no coronoid and retroarticular process. The jaw articulation is set below the line of the tooth row. The external mandibular foramen is large, elongated, and bordered by the dentary, surangular and angular.

Maxillary teeth, 24 in number, are more or less uniform in shape in *E. andrewsi*. These diminish slightly in size from mesial to distal. Maxillary teeth are straight and narrow, somewhat pointed, and slightly flattened transversely. The teeth, with the exception of the five mesialmost, are closely spaced. Mesial and distal edges of each tooth are serrated. The serration is inclined to the edges.

There are 31 dentary teeth in *E. andrewsi* which are similar in shape to those of the maxilla. The first five teeth are straight, markedly larger than the others, and loosely spaced, while the distal teeth are rather closely packed and diminish in size toward the rear. They are also more mesiodistally expanded. The crown is widest halfway up the crown. The labial surface of

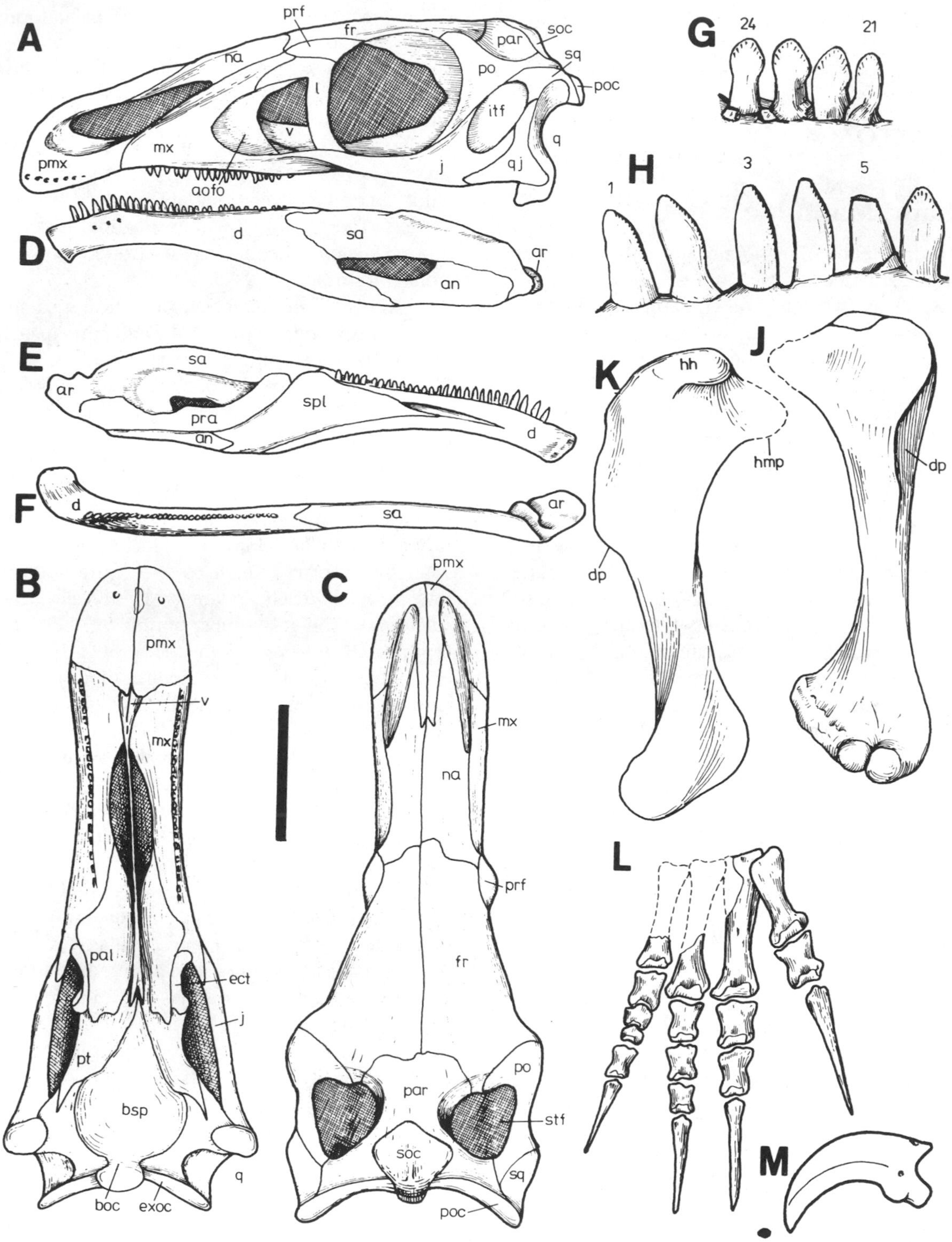

Fig. 18.1. A–M, *Erlikosaurus andrewsi:* skull in lateral (A), palatal (B), and dorsal (C) views; left mandible in labial (D), lingual (E), and dorsal (F) views; labial view; 21st–24th dentary teeth (G), left dentary teeth, first to sixth (H), lingual view; left humerus in cranial (J) and caudal (K) views; right pes (L) in dorsal view (after Barsbold and Perle 1980); pedal claw of digit I (M) in lateral view (after Perle 1981). Scale = 6 cm (A–F), 1 cm (G, H), 9 cm (J, K), 12 cm (L), 8 cm (M).

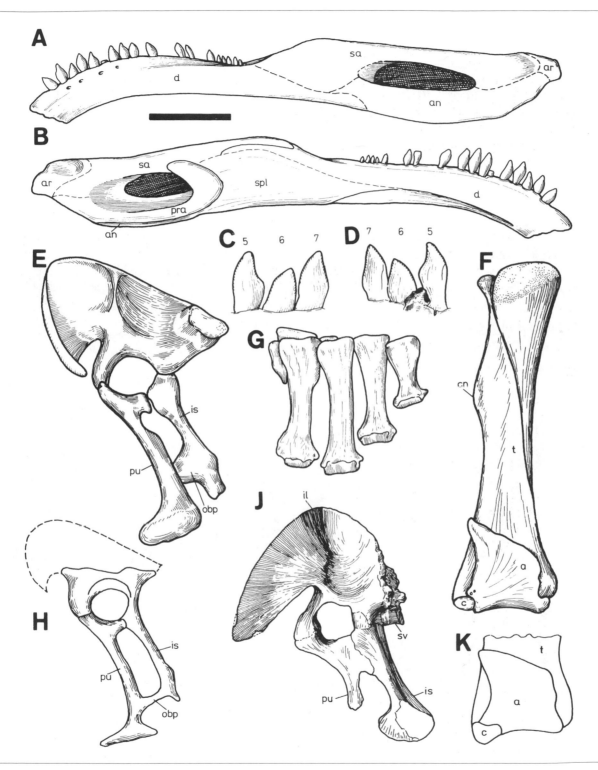

Fig. 18.2. A—G, *Segnosaurus galbinensis:* left mandible in labial (A) and lingual (B) views; 5th—7th dentary teeth in labial (C) and lingual (D) views; pelvis in left lateral view (E; after Barsbold 1983*a*); right tibia, astragalus, and calcaneum in cranial view (F; after Perle 1979); right metatarsus in dorsal view (G; after Perle 1979). H, *Enigmosaurus mongoliensis:* pel- vis in left lateral view (after Barsbold 1983*a*). J, *Nanshiungo- saurus brevispinus:* pelvis in left lateral view (after Dong 1979). K, segnosaurid indet.: distal end of left tibia with as- tragalus and calcaneum, cranial view. Scale = 6 cm (A, B), 2.5 cm (C, D), 35 cm (E), 20 cm (F), 15 cm (G), 40 cm (H), 30 cm (J), 18 cm (K).

these distal teeth is convex. Mesial and distal edges are serrated (fig. 18.1G,H).

The dentary of *S. galbinensis* bears 24 sharply pointed teeth. The mesial teeth are somewhat recurved only along their upper halves. The last few teeth are straight and markedly smaller. The transverse compression of the teeth is greater on the larger, more mesial teeth. The labial surfaces are convex, while the lingual aspects are concave. The serrations of the mandibular crowns in *S. galbinensis* are more distinct than those in *E. andrewsi*. The teeth in *S. galbinensis* are placed somewhat obliquely, so that the serrated mesial and distal edges are slightly lingual and labial, respectively. The entire dentition, with the exception of the first two teeth, is closely spaced. Wear surfaces are not observed on the teeth in *E. andrewsi,* whereas they are weakly developed on the distal edges of some dentary teeth in *S. galbinensis* (fig. 18.2C,D).

Axial Skeleton

The cervical vertebrae in *E. andrewsi, S. galbinensis,* and *Nanshiungosaurus brevispinus* are large and elongate and bear low neural spines and large pleurocoels. Cervical ribs are fused to the vertebrae. The centra of the dorsals, known only in *N. brevispinus,* are pneumatized, lacking pleurocoels. Neural spines are low, both in the dorsal and sacral vertebrae. Ossified interspinous ligaments are present between the sacral vertebrae in *S. galbinensis* and *N. brevispinus.*

Appendicular Skeleton

The pectoral girdle is preserved only in *S. galbinensis.* The scapula and coracoid are fused. The scapular blade is straight and narrow. The coracoid is large with a very small coracoid foramen and a very well developed acromial process.

The humerus, known in both *E. andrewsi* and *G. galbinensis,* is massive with strongly expanded proximal and distal ends. The humeral head is situated at about the middle of the proximal end and overhangs the caudal surface. This overhang is more pronounced in *E. andrewsi* than in *S. galbinensis.* Medial to the humeral head, an internal tubercle is well developed. The deltopectoral crest is massive and extends more than a third of the humeral length. A sharply pointed tubercle is present on the caudomedial surface of the humerus at the midlength of the bone. The radial and ulnar condyles are small, rounded, and separated by a very narrow flexor groove. Lateral and medial epicondyles are well developed; the medial epicondyle is larger (fig. 18.1J-K).

The antebrachium is known only in *S. galbinensis.* The radius is massive, and its length is approximately 60 percent of the length of the humerus. The massive ulna bears a well-developed olecranon process. The metacarpals and digits are poorly preserved. Similarly, the exact proportions of fore- to hindlimbs is unknown, because the preserved elements of the forelimbs do not belong to the same specimen as the hindlimb elements.

The pelvis, more or less completely preserved in *S. galbinensis, N. brevispinus,* and *Enigmosaurus mongolensis,* is opisthopubic, with widely separated ilia fused to the sacral diapophyses (fig. 18.2E, J, H). The long preacetabular process of the ilium with its long, pointed cranioventral end is flared at right angles to the sagittal plane. The end of the very short postacetabular process bears a strong knoblike lateral protuberance. The pubic peduncle is long; the ischial peduncle is much shorter. The footed pubis is directed caudoventrally. The shaft of the pubis is long. The ischium is shorter than the pubis and bears an obturator process.

The femur and crus are known only in *S. galbinensis* (fig. 18.2F, K). The femur is very weakly sigmoid. The femoral shaft is oval in section. The femoral head is turned in an almost right angle and separated from the proximal shaft by a distinct neck. The greater trochanter is well developed, expanded craniocaudally, and separated from the lesser trochanter by a deep cleft. The fourth trochanter, in the form of a rugose crest, is placed somewhat above the middle of the femoral shaft. The distal condyles are well separated; the lateral condyle is larger than the medial. The extensor and flexor grooves are very shallow.

The length of the tibia is more than 80 percent of the length of the femur. The tibia bears a well-developed cnemial crest, has an expanded distal end, and supports the distal end of the fibula. Closely applied to the tibia for most of its length, the fibula is very slender distally.

Both proximal and two distal tarsals (3 and 4) are preserved in *S. galbinensis.* The astragalus with reduced condyles covers only the medial part of the distal surface of the tibia and bears a tall, laterally curved and pointed ascending process. The calcaneum is suboval, wider than high, and fits into a prominent socket on the astragalus.

The pes of *E. andrewsi* and *S. galbinensis* is short and broad (figs. 18.1L, H, 18.2G). The completely preserved metatarsus of *S. galbinensis* is composed of five elements, metatarsal V being rudimentary. Metatarsals I–IV are massive with widened distal extremities. Metatarsal I is the shortest and somewhat divergent from the rest. In *S. galbinensis,* the length of the metatarsus is approximately 30 percent of the femoral length. The phalangeal formula in *E. andrewsi* is 2−3−4−5−0. In this species, the first digit is shortest, while digits II and III are the longest and subequal in length. The fourth digit is slightly shorter and the thinnest of all. The phalanges of digits I−IV are short and robust. The third phalanx of digit IV is especially short. Pedal claws are relatively large, recurved, and strongly compressed transversely in *E. andrewsi* but less compressed in *S. galbinensis*.

SYSTEMATICS AND EVOLUTION

As presently known, the Segnosauria is represented by *Erlikosaurus andrewsi*, *Enigmosaurus mongoliensis*, *Nanshiungosaurus brevispinus*, and *Segnosaurus galbinensis* (table 18.1) The Segnosauridae was established by Perle (1979) for *Segnosaurus* and was provisionally assigned to the Theropoda. Later, *Erlikosaurus* was assigned to this family (Perle, in Barsbold and Perle 1980; Perle 1981). In 1980, Barsbold and Perle erected another family within the Segnosauria, the Enigmosauridac, for a new monospecific genus, *Enigmosaurus*. Erection of this family was based solely on characters of the pubis and ischium (size and shape of the pubic foot and form of the obturator process) in *E. mongolien-*

sis, as other parts of the skeleton of this species are not preserved. For this reason, creation of the Enigmosauridae may not be justified. It seems more reasonable to maintain only one family, the Segnosauridae, for all segnosaurian genera.

All species within the Segnosauria are based on fragmentary specimens, and only few skeletal fragments are comparable. Thus, it is very difficult to determine the full range of specific and generic features that characterized those segnosaurians thus far described.

Segnosaurus, with only one species, *S. galbinensis*, from Baynshirenskaya Svita of the Mongolian People's Republic, is characterized by, among other features, mesial mandibular teeth that are markedly flattened and slightly recurved and by pedal unguals that are moderately compressed transversely.

In the only species of *Erlikosaurus*, *E. andrewsi*, from the same stratigraphic unit, the small mandibular teeth are straight and only slightly flattened, and the pedal claws are strongly compressed.

Enigmosaurus mongoliensis (from the same stratigraphic unit) is characterized by an uncompressed, narrow pubic shaft and a small and shallow obturator process on the ischium. This condition differs from *S. galbinensis*, in which the shaft of the pubis is flattened and widened, the distal foot has an ellipsoidal shape, and the obturator process is deep and subquadrate.

The remains of all three species mentioned above

TABLE 18.1 Segnosauria

	Occurrence	Age	Material
Saurischia *sedis mutabilis*			
Segnosauria Barsbold et Perle, 1980			
Segnosauridae Perle, 1979 (including Enigmosauridae Barsbold et Perle, 1983)			
Enigmosaurus Barsbold et Perle, 1983			
E. mongoliensis Barsbold et Perle, 1983	Baynshirenskaya Svita, Dornogov, Mongolian People's Republic	Cenomanian-Turonian	Incompletely preserved pelvis
Erlikosaurus Perle, 1980			
E. andrewsi Perle, 1980	Baynshirenskaya Svita, Omnogov, Mongolian People's Republic	Cenomanian-Turonian	Skull, pes, other fragmentary postcranium
Nanshiungosaurus Dong, 1979			
N. brevispinus Dong, 1979	Nanxiong Formation, Guandong, People's Republic of China	late Campanian	Vertebral column, pelvis
Segnosaurus Perle, 1979			
S. galbinensis Perle, 1979	Baynshirenskaya Svita, Omnogov, Dornogov, Mongolian People's Republic	Cenomanian-Turonian	3 specimens, including mandible, pelvis, hind limb, scapulocoracoid, incomplete fore limb, fragmentary vertebrae 2 pedal digits
Unnamed segnosaurian (= *"Chilantaisaurus" zheziangensis* Dong, 1979)	Tangshang Formation, Zhejiang, People's Republic of China	Santonian-Campanian	

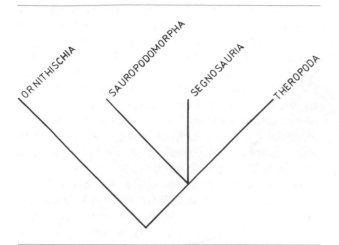

Fig. 18.3. Relationships of the Segnosauria.

come from Upper Cretaceous deposits, provisionally assigned to Cenomanian or Turonian, from the southeastern region of the Mongolian People's Republic.

Nanshiungosaurus brevispinus, originally described by Dong (1979) as a sauropod from the Maastrichtian Nanshiung Formation of northern Guandong, People's Republic of China, represents a fourth segnosaurian species. *N. brevispinus* is known to have an opisthopubic pelvis that is very similar to that of *S. galbinensis* and hence typical of segnosaurian dinosaurs. The two differ slightly in that the cranial portion of the preacetabular process of the ilium is less deep in *N. brevispinus* than in *S. galbinensis.* The structure of cervical vertebrae in *N. brevispinus* is similar to that in the Mongolian segnosaurians. However, since other known parts of the skeleton in the Mongolian segnosaurians are not duplicated in *N. brevispinus,* a more extensive comparison is not possible at this time.

There is some probability that a second Chinese species, *"Chilantaisaurus" zheziangensis,* in fact is a segnosaurian. This species was described by Dong (1979) as a theropod on the basis of very poor material (pedal digits II and III only). Because the structure of the short and massive pedal phalanges and large, decurved claws in *"C." zheziangensis* is very similar to the structure of those elements in *S. galbinensis,* the segnosaurian affiliation of this Chinese species cannot be excluded. For this reason, *"C." zheziangensis* is regarded here as a possible segnosaurian.

In 1982, Perle described a fragmentary hindlimb (including fragments of femur, tibia, astragalus, calcaneum, metatarsals, and pedal digits) from the White Beds (= Nemegtskaya Svita) of Khermeen Tsav, Mongolian People's Republic, and referred it to *Therizino-*

saurus. The sole species of *Therizinosaurus, T. cheloniformis* (see Norman, this vol.), is known only from the pectoral girdle and forelimb. Thus, the referral of the previously mentioned hindlimb to the *Therizinosaurus* is at present unjustified. However, the structure of preserved parts of the hindlimb from Khermeen Tsav, especially the short and broad metatarsus and astragalus with a laterally curved ascending process, is very similar to the structure of respective bones in Mongolian segnosaurians (fig. 18.2). For this reason, the specimen described by Perle (1982) as *Therizinosaurus* sp. is here considered as the Late Cretaceous Campanian or Maastrichtian indeterminate representative of the Segnosauria.

The interrelationships among segnosaurian species are not yet possible to resolve because of their incomplete remains. The relationship of the Segnosauria to other dinosaurs, by contrast, is controversial in part because of the unusual suite of characters found in the group.

Perle (1979), Barsbold and Perle (1980), and Barsbold (1983a) included segnosaurians in the Theropoda. This assignment was rather arbitrary, and the supposed theropod affinities of the Segnosauria never were precisely discussed. The first to compare segnosaurs with thecodonts, prosauropods, and ornithischians was Paul (1984b), who explicitly tested the theropod assignment of the Segnosauria by means of cladistic methods. He expressed the opinion that Segnosaurians show no distinctive theropod characters. Instead, he suggested that the Segnosauria are phylogenetic intermediates between prosauropods and ornithischians and constitute a late surviving relict of the Triassic prosauropod-ornithischian transition.

There are, in fact, a number of segnosaurian features that are shared uniquely with theropods (Gauthier 1986), including lack of a process of the dentary passing dorsal to the external mandibular fenestra, presence of pleurocoels in the vertebrae, the fibula closely appressed to the tibia, expanded pubic foot, and tall ascending process of the astragalus. However, other characters that the Segnosauria share with theropods (e.g., a low and long skull, slender jugal, low maxilla, broad lacrimal, absence of coronoid process) are primitive not only for all theropods but also for all dinosaurs.

Some segnosaurian characters, such as large elongate nares, deep narial shelf, deeply recessed antorbital fossa and fenestra, downwardly curved rostralmost portion of the dentary, general tooth structure, short, broad tetradactyl pes, and an incipient lateral dental shelf are comparable to those of basal sauropodomorphs. However, many derived characters of the

Segnosauria, for example, structure of pelvis, structure of palatine with caudally shifted vomers and rostrally reduced pterygoids, and edentulous premaxillae, make it impossible to include the segnosaurians among theropods or sauropodomorphs.

The suggestion of Paul (1984*b*) that the Segnosauria represent relicts of a transition between Prosauropoda and Ornithischia, having acquired an ornithischian grade cheek structure, pubis, and derived ornithischianlike food-gathering and food-processing apparatus, is here rejected. This suggestion was based on the mistaken supposition that segnosaurians had fully developed lateral maxillary and dentary shelves and a diastema between the mesialmost teeth and the main tooth row. The grade of development of the dentary shelves in segnosaurians is similar to that in prosauropods, while the incipient short cheeks covering the distalmost teeth in *E. andrewsi* and *S. galbinensis* were also present in *Plateosaurus* (Galton 1985*e*). The structure of the massive, retroverted, footed pubis and ischium with an obturator process is characteristic of the Segnosauria and more like that in some opisthopubic saurischians (e.g., dromaeosaurids) than in any ornithischian.

Gauthier (1986) suggested that segnosaurians bear a close relationship with broad-footed sauropodomorphs. This relationship seems to be more probable than others but needs a detailed analysis. For this reason, it is reasonable to consider the Segnosauria as Saurischia *sedis mutabilis*.

PALEOECOLOGY

Very rare and fragmentary skeletons of segnosaurians come only from Asia. Currie's (1987*c*) record (a frontal bone) of *Erlikosaurus* sp. from the Judith River Formation in Alberta, Canada, appears suspect since evidence from this element is not particularly compelling.

The fragmentary skeletons of *Segnosaurus galbinensis*, *Erlikosaurus andrewsi*, and *Enigmosaurus mongoliensis* were discovered in sediments included to the Baynshirenskaya Svita in the southeastern region of the Mongolian People's Republic. The remains of *S. galbinensis* and *E. andrewsi* were found in poorly cemented, gray sands with intraformational conglomerates, gravels, and gray claystones, interpreted by Tsybin and Kurzanov (1979) as deposits in small lakes. In the same sediments, skeletons of the theropods *Alectrosaurus olseni*, *Garudimimus brevipes*, the ankylosaur *Talarurus plicatospineus*, and abundant hadrosaurs and turtles were discovered. The bones of *Enigmosaurus mongoliensis* come from the red beds, which also yielded bones of sauropods, theropods, and stems of Taxodiaceae in situ (Martinsson 1975). The Chinese *Nanshiungosaurus brevispinus* comes also from red beds of north Guandong Province of the People's Republic of China.

It appears that segnosaurians were rather slow-moving animals with short and broad feet and a bulky trunk. It has been suggested by Barsbold and Perle (1980) that segnosaurians were amphibious carnivores, preying on fishes. In contrast, Paul (1984*b*) suggested these dinosaurs had a herbivorous diet. The edentulous rostral portion of the snout (possibly covered by a horny rhamphotheca), the structure of the teeth, partial development of lateral dentary shelves, and a few large nerve foramina along the flat, lateral maxillary wall above the tooth row, suggesting the presence of cheeks, speak in favor of a herbivorous diet. The other segnosaurian character that may be connected with their herbivorous habit is the massive opisthopubic pelvis. This pubic retroversion probably served to create space and as a support for a massive gut for digesting plant food, as was suggested by Paul (1984*b*).

19

Lesothosaurus, Pisanosaurus, and *Technosaurus*

DAVID B. WEISHAMPEL

LAWRENCE M. WITMER

INTRODUCTION

The animals considered in this chapter are all small (1–2 m long), facultatively bipedal, and appear to be predominantly herbivorous. The phylogenetic positions of *Pisanosaurus* and *Technosaurus*, and to a lesser extent *Lesothosaurus*, have been controversial. Recent cladistic analyses of their relationships to other ornithischians have demonstrated, however, that they are among the most primitive members of this large dinosaurian group (Sereno 1986; Gauthier 1986).

ANATOMY

Lesothosaurus diagnosticus is particularly well known in terms of skull and postcranial material (Thulborn 1970b, 1972; Santa Luca 1984; Weishampel 1984a; Crompton and Attridge 1986; Norman and Weishampel in press). Additional postcranial material thought to

pertain to a closely related taxon (Santa Luca 1984; see Systematics, below) is also discussed here. *Pisanosaurus mertii* is known from a single specimen that was described by Casamiquela (1967b) and Bonaparte (1976). The specimen is very poorly preserved and includes portions of the skull, axial skeleton, and hindlimb with a few fragments of the shoulder and pelvic girdles. Last, only the premaxilla, mandible, partial dentition, dorsal vertebra, and astragalus are known for *Technosaurus smalli* (Chatterjee 1984).

Skull and Mandible

The following description relies chiefly on the skull of *Lesothosaurus* but includes information on *Pisanosaurus* and *Technosaurus* where relevant.

The skull of *Lesothosaurus* (fig. 19.1a) is relatively high and long, certainly longer than illustrated by Thulborn (1970b, fig. 8). The external naris is relatively small, while the orbit is large. The infratemporal fenestra is oblong, broadest dorsally. There is a moderate external mandibular fenestra at the junction of the dentary, surangular, and angular.

The premaxilla of *Lesothosaurus* forms the ventral rim of the external naris. Immediately rostral to the external naris is a small foramen entering the premaxillary body. In *Technosaurus,* there is a series of five foramina set in an inclined row directly beneath the external naris. The dorsal premaxillary process in *Lesothosaurus* meets its counterpart as a distinct butt joint and contacts the nasal in a scarf joint rostral to the external naris. The short lateral process overlaps the nasal-maxilla contact for only a short distance but completely excludes the maxilla from the external naris. There are five or six teeth in the premaxilla in *Lesothosaurus.* The arcade of premaxillary teeth is interrupted by an edentulous area at the rostral end of the element. In contrast, the rostrum is not edentulous in *Technosaurus;* rather, the arcade of five teeth per premaxilla continues fully to the mesial midline. The ventral part of the medial premaxillary surface is excavated to receive the rostral process of the maxilla.

The nasal in *Lesothosaurus* forms the dorsal margin of the muzzle, the roof of the nasal cavity, and the entire dorsal margin of the external naris. Ventrally, the nasal contacts the maxilla and prefrontal. Medial to the nasal-prefrontal joint, the nasal overlaps the frontal.

The maxilla in *Lesothosaurus* forms the dorsal, rostral, and ventral margins of the antorbital fossa, such that only the caudal margin of the fossa is bounded by the lacrimal. The antorbital fenestra is located at the caudodorsal margin of the fossa near the lacrimal-maxilla junction. Rostrally, the maxilla is closely applied to the body and lateral process of premaxilla and sends a small rostral process to contact the premaxilla. Caudally, the upper surface of the maxilla is formed into an elongate and narrow facet that receives the jugal. A shallow longitudinal ridge directly beneath the antorbital fossa has the effect of slightly setting the dentition in from the side of the skull, suggesting the presence of cheeks (Paul 1984*b*, 1986; Norman pers. comm.). There is a rounded ectopterygoid shelf on the lateral surface of the maxilla in *Pisanosaurus.* There are 15 to 20 tooth positions in the maxilla in *Lesothosaurus* and at least 11 in *Pisanosaurus* (fig. 19.1b; Bonaparte [1976] estimated a full maxillary complement of 16 to 18).

The lacrimal in *Lesothosaurus* is a strutlike element between the antorbital fossa and orbit. It articulates dorsally with the prefrontal and ventrally with the jugal along short scarf joints. Along with the prefrontal, the upper portion of the lacrimal acts as a platform for the base of the palpebral.

The crescentic prefrontal contacts the nasal rostrally, the lacrimal ventrally, and the frontal medially; it also contributes the rostrodorsal margin of the orbit. The frontal forms the broad rostral portion of the skull roof and contributes the dorsal margin of the orbit. A transverse and moderately interdigitate suture is made with the parietal. The interfrontal joint is straight throughout its length. The fused, broad parietals form the caudal portion of the skull roof and slightly overhang the occiput. There is a moderate sagittal crest. Caudally, the parietal is drawn out into laterally projecting flanges that contact the squamosals.

The short medial process of the postorbital contacts the frontal and parietal along a slightly interdigitate suture. The ventral process reaches nearly to the body of the jugal to form virtually the entire postorbital bar but articulates principally with the front of the dorsal process of the jugal. The postorbital makes an extensive scarf joint with the squamosal to form the upper temporal arch.

The jugal forms the lateral wall of the adductor chamber, covering the coronoid process of the mandible. The jugal makes an open scarf articulation with the postorbital and a broad lapping articulation with the quadratojugal. The quadratojugal is large and triangular. Dorsally, the quadratojugal contacts the prequadratic process of the squamosal to exclude the quadrate from the caudal margin of the infratemporal fenestra.

The area immediately above the prequadratic process of the squamosal accommodates the attachment of M. adductor mandibulae externus superficialis. The body of the squamosal houses a lateroventrally facing cotylus for the head of the quadrate. Directly behind this cotylus, the postquadratic process contacts the lateral aspect of the paroccipital process.

Broad and somewhat flat dorsally, the quadrate head is buttressed caudally by a small protuberance on the upper part of the quadrate shaft which contacts the postquadratic process of the squamosal. The shaft of the quadrate is shallowly concave caudally, while rostrally, it is emarginated for the quadratojugal. Extensive quadrate-pterygoid contact is provided by the large, rostromedial pterygoid ramus. The mandibular condyle of the quadrate is broadened transversely to form a unicondylar to slightly bicondylar jaw joint.

Only certain aspects of the braincase and palate of *Lesothosaurus* are known. The supraoccipital is high, wide, transversely arched, and appears to contribute virtually the entire dorsal margin of the foramen magnum. It bears a rounded, vertical eminence on its caudal face. Laterally, the supraoccipital contacts the

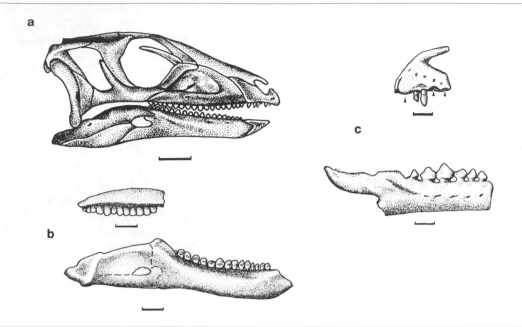

Fig. 19.1. a. Right lateral view of the skull of *Lesothosaurus diagnosticus*. b. Maxilla and dentary of *Pisanosaurus mertii* (after Bonaparte 1976). c. Premaxilla and dentary of *Tech-nosaurus smalli* (after Chatterjee 1984; arrows indicate position of empty premaxillary alveoli). Scale = 10 mm.

parietal. The exoccipitals and opisthotics are indistinguishably fused to form the paroccipital processes. The latter processes are virtually straight along their length, projecting horizontally from the lateral margins of the foramen magnum, and are weakly pendent at their termini. The basipterygoid processes are relatively short, extending rostrolaterally and slightly ventrally to terminate in blunt articular surfaces that contact the body of the pterygoid.

The vomers are long, thin, and make up the midline of the palate. On its caudolateral margin, the vomer articulates with the medial surface of the palatine. The palatine is transversely broad and contacts the maxilla. At the caudal terminus of the vomer-palatine joint, the palatine contacts the pterygoid and ectopterygoid. The pterygoid is not well known but appears to articulate extensively with the medial margins of both the ectopterygoid and palatine. The dorsal aspect of the pterygoid is excavated for articulation with the basipterygoid processes of the basisphenoid. Ventrally, there are relatively stout pterygoid flanges, while caudally, there is a robust quadrate ramus. The ectopterygoid braces the palate on the caudomedial surface of the maxilla. Medially, the ectopterygoid forms a relatively broad scarf joint with the ventrolateral surface of the quadrate ramus of the pterygoid. Rostrally, the ectopterygoid overlies the dorsal surface of the palatine.

A small spatulate predentary caps the rostral tip of each dentary. It sends short lateral processes around the lateral dentary borders and a ventral process under the full length of the mandibular symphysis. This ventral process is unilobate. Relatively large vascular foramina are found in the predentary body at the base of each lateral process.

The dorsal and ventral margins of the dentary in *Lesothosaurus*, *Pisanosaurus*, and *Technosaurus* are subparallel for most of its length, diverging caudally only with the formation of the low coronoid process (fig. 19.1). Small foramina are found near the undersurface of the lateral face of the dentary. The dentaries are closely and broadly applied to each other at the mandibular symphysis; these articulations are ovoid in shape, relatively rugose, and inclined rostrally. In *Lesothosaurus*, the dentary accommodates as many as 17 tooth positions, at least 15 positions in *Pisanosaurus*, and at least 9 positions in *Technosaurus*. There appears to be a slight rostral diastema between the first tooth and the position of the predentary. The tooth row is emarginated from the side of the mandible, especially in *Pisanosaurus*, due to a prominent lateral shelf. Ventral to this ridge are five longitudinally placed foramina in *Technosaurus*. *Lesothosaurus* and especially *Technosaurus* have a relatively large external mandibular fenestra. Casamiquela (1967*b*; followed by Bonaparte

1976) described a relatively large opening at the junction of the surangular, dentary, and angular which he ascribed to preservational destruction. However, because of its bony relationships and its distribution among other archosaurs, we cannot rule out its being the external mandibular fenestra.

The surangular contributes the caudal aspect of the coronoid process, the dorsal margin of the external mandibular foramen, the lateral wall of the glenoid, and the lateral aspect of the stout retroarticular process in all three species. Laterally, the rim of the jaw joint is buttressed by a liplike thickening of the surangular. The articular is restricted to the medial surface of the jaw joint; it also forms a portion of the medial wall of the retroarticular process. Directly in front of the articular, the prearticular is teardrop shaped and forms the medial wall of the entrance to the Meckelian canal. The long and narrow splenial covers virtually the entire medial surface of the mandibular body. It also forms the lateral, dorsal, and ventral walls of the Meckelian canal. The relatively long and low angular forms the caudoventral margin of the lower jaw, including the caudal angle of the external mandibular fenestra and a small portion of the lateral aspect of the glenoid, at least in *Technosaurus*.

The dentition of *Lesothosaurus* has been thoroughly described (viz. Thulborn 1970*b*, 1971*a*; Galton 1978; Weishampel 1984*a*; Crompton and Attridge 1986). Premaxillary teeth are relatively narrow, conical, and moderately recurved. A small cingulum marks the base of the crown. On the lingual surface of the more mesial premaxillary teeth, there is a vertical furrow and an adjacent sharp ridge toward the distal edge of the crown. More distally, the crowns become more triangular and acquire distal and then mesial and distal denticulation. The roots of the premaxillary teeth are relatively long and circular in cross section. In *Technosaurus*, the base of the straight, fusiform premaxillary crown is formed into a cingulum that separates the crown from the thick root. There is faint indication of longitudinal grooves but no denticles. Both maxillary and dentary tooth crowns are labiolingually compressed and mesiodistally expanded to assume their characteristic rhomboidal or triangular shape, usually as wide as high (fig. 19.2) Four to seven denticles mark each mesial and distal edge; these are continuous onto the buccal surface of maxillary crowns and lingual surface of dentary crowns as distinct ridges and grooves. The central eminence beneath the apex is more restricted to the midline of the crown, forming a distinct ridge. The outermost denticles (anterior and posterior accessory cusps of Chatterjee 1984) are restricted to-

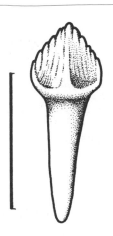

Fig. 19.2. Dentary tooth of *Lesothosaurus diagnosticus*. Scale = 5 mm.

ward the base of the crown where they are continuous with the bulbous cingulum. Wear facets can develop on either the mesial or distal halves of the crown (lingual aspect for maxillary teeth; buccal aspect for dentary teeth), depending on the age of the tooth. Each facet has a shallowly oblique orientation to one another but together form an occlusal surface that is inclined 65° to 75° to the horizontal.

Although not well known, dental morphology in *Pisanosaurus* differs significantly from that of *Lesothosaurus* and *Technosaurus*. Maxillary and dentary teeth are relatively closely packed and appear to lack a cingulum. Occlusal surfaces are continuous between adjacent teeth. For maxillary teeth, which curve lingually from their base, the occlusal plane is inclined approximately 45°. Dentary teeth vary in size (small mesially to relatively large distally). Unlike maxillary crowns, dentary crowns are relatively straight and vertical. The slightly buccally concave wear surfaces are inclined 60° to 70°.

Postcranial Skeleton

As with the skull, postcranial descriptions rely heavily on *Lesothosaurus* (fig. 19.3; see also Thulborn 1972, Santa Luca 1984), noting both *Pisanosaurus* and *Technosaurus* where relevant.

The vertebral formula of *Lesothosaurus* has not been documented, due to the relatively poor preservation of the material. The vertebrae are morphologically conservative. Centra are apparently amphicoelous to amphiplatyan throughout the column. Neurocentral

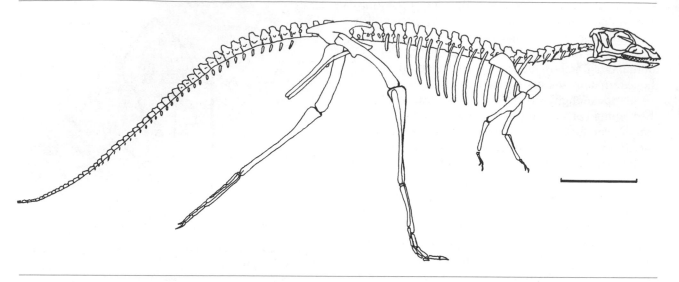

Fig. 19.3. Skeletal reconstruction of *Lesothosaurus diagnosticus* (after Thulborn 1972, Galton 1978). Scale = 10 cm.

sutures remain open. Cervical centra are excavated ventrolaterally for muscle attachment such that a median ventral keel is produced. The excavation and keel run the length of the centra in the cranial cervicals. Caudal cervicals restrict the excavation to the cranial portion of the centrum, and the keel is weaker caudally. The parapophyses migrate from a position below the neurocentral suture in the cranial cervicals to just above the suture in the caudal cervicals. The diapophyses are situated dorsal to the suture. The neural spine is represented by a low median ridge.

The three vertebrae pertaining to *Pisanosaurus* have been interpreted by Bonaparte (1976) as cervicals (originally thought to be caudals by Casamiquela [1967*b*]). They are laterally excavated, with transverse and spinous processes weakly developed or absent. A peculiarity noted by Bonaparte (1976) is that the prezygapophyses appear to project cranially, while the postzygapophyses do not even reach the caudal end of the centrum.

The dorsal centra of *Lesothosaurus* lack the well-developed lateral excavations, and the ventral keel is weak or absent. The centra are amphicoelous to amphiplatyan, spool-shaped structures. The poorly preserved series of seven dorsals in *Pisanosaurus* and the isolated dorsal of *Technosaurus* appear very similar. The transverse processes are stout in *Lesothosaurus* and long in *Technosaurus*. In both taxa, they are horizontal, especially caudally, and bear the diapophyses at their tips. The parapophyses are located dorsal to the neurocentral suture and just cranial to the base of the transverse process; they migrate caudodorsally toward the diapophyses in the caudal portion of the series. The

neural spines are strong and roughly rectangular, although Santa Luca (1984) noted that the dorsal portion of the spine diverges, producing a more trapezoidal profile. The zygapophyses are somewhat inclined in the cranial dorsals and become more horizontal caudally.

The sacrum of *Lesothosaurus* consists of five sacral vertebrae and their ribs. Sacral centra are broad with kidney-shaped amphiplatyan articular surfaces. The neural canal is excavated ventrally. A median ventral keel is variably developed. The transverse processes are broad, horizonal, and do not taper laterally. The neural spines are transitional in shape between the dorsals and caudals. The zygapophyses are close to the midline and are nearly vertical.

The number of caudal vertebrae cannot be estimated. The caudal centra are spool-shaped and become relatively lower in the distal part of the series. The articular surfaces become more circular distally. Proximal caudals bear ventral chevron facets on their distal surfaces, but chevrons appear to have been absent in the distal portion of the series. Proximal caudal neural arches exhibit tall spines and long, horizontal, transverse processes; these structures become reduced more distally in the series such that the spines become low ridges and the transverse processes are lost. The prezygapophyses are apparently longer than the postzygapophyses; their inclination decreases distally such that the most distal caudals have nearly horizontal zygapophyses.

Ribs are two-headed. The serial changes in the diapophyses and parapophyses are mirrored in the ribs: the tuberculum and capitulum approach each other in successively more caudal portions of the column. Thulborn (1972) considered the long transverse processes of the caudal vertebrae to represent fused ribs.

Ossified tendons are known in *Lesothosaurus*, where they are found in the dorsal and caudal portions of the column. Each tendon has a pointed end and an expanded end. Tendons are grouped in bundles that may have been arranged in a diamond-shaped lattice.

The scapula is an elongate straplike bone 10 to 15 percent longer than the humerus. The shaft is roughly rectangular and expands in width distally. The distal two-thirds is laterally bowed, reflecting the curvature of the rib cage. The distal surface of the scapula is pitted, indicating a cartilaginous suprascapula. The caudoventral margin of the distal end is produced into a distinct hooklike process. The proximal portion is expanded caudoventrally and exhibits a cranially directed acromion, a grooved sutural area for the coracoid, a notch leading to the coracoidal foramen, and the scapular contribution to the glenoid. The lateral surface of the proximal portion is concave for muscle attachment. The coracoid is a simple rounded rectangle and is pierced laterally by the coracoidal foramen. Its contribution to the glenoid is thick and results in a relatively deep caudoventrally directed glenoid fossa.

The humerus is a simple straight bone (fig. 19.4*b*). A tuberosity is located at the proximomedial corner of the bone, and the humeral head is more centrally placed. The shaft is transversely expanded proximally. The deltopectoral crest projects cranially from the lateral margin of the shaft; the crest is triangular and situated relatively proximally. The distal end is very simple, consisting of rounded radial and ulnar condyles. The ulnar condyle is oval with a transverse axis of rotation. The radial condyle is somewhat laterally inclined. A shallow fossa is situated between the distal condyles on both the cranial and caudal surfaces.

The radius is about 60 percent as long as the humerus. It is a straight bone with an oval cross section. Its humeral articular surface is semicircular to reniform. Its carpal articular surface is circular and slightly saddle-shaped. The ulna is slightly longer than the radius and has a triangular cross section. An olecranon is absent.

The carpus is known from only two elements; these elements are flattened plates of bone and probably represent distal carpals II and III. The manus is also incompletely known. The third metacarpal is longest, followed by the second, fourth, and first. The fourth metacarpal is much more slender than the others. The fifth metacarpal is not known. The metacarpal bases are simple ellipsoid condyles, while the heads are bicondylar trochleae. Several phalanges are preserved, but a phalangeal formula cannot be estimated. The

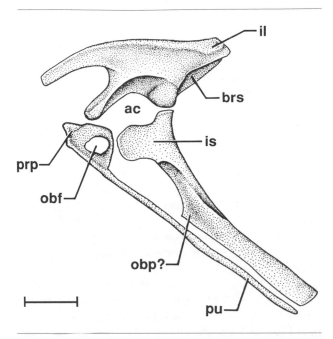

Fig. 19.4. Left lateral view of the pelvis of *Lesothosaurus diagnosticus* (modified after Thulborn 1972, Santa Luca 1984). Scale = 3 cm.

phalanges are unremarkable with the typical expanded proximal articular surfaces and distal ginglymi.

The hindlimb is very long and gracile and is much longer than the forelimb. The ilium is long and low (fig. 19.4). The preacetabular process tapers cranially and varies from being relatively straight to slightly decurved. It is much shorter than in higher ornithischians, its length being approximately equal to acetabular diameter, rather than much longer. The postacetabular process is shorter and deeper. The ilium slightly walls off the acetabulum medially; thus, the acetabulum is not completely perforate. The pubic peduncle is a strong, horizontal, rectangular process. The brevis shelf (for M. caudifemoralis brevis) is well developed. The shelf is initially horizontal where it medially undercuts the postacetabular process caudal to the ischial peduncle; it then becomes vertical where it is walled off medially. A pronounced lateral swelling of the ilium dorsal to the acetabulum, the supraacetabular flange, results in a deep acetabulum. The flange is strongest just cranial to the ischiadic peduncle. There is no antitrochanter. Medially, there are four to five facets for the sacral ribs; the precise positions of the facets vary.

In *Lesothosaurus*, the long, rodlike pubis bears a short deep prepubic process (fig. 19.4). The prepubic process consists of a laterally concave plate of bone cranial to the obturator foramen and cranioventral to the acetabulum. The pubic contribution to the acetabulum is straight and thickened. The obturator foramen is enclosed caudally by a ventral process from the acetabular margin and a dorsal process from the postpubis. Opisthopuby in *Pisanosaurus* is difficult to confirm (contra Bonaparte 1976).

The ischium is a long complex bone situated parallel to the pubis. (fig. 19.4a). The proximal end bears an iliac peduncle and a larger pubic peduncle. The peduncles are separated by the acetabular surface. The shaft of the ilium exhibits some torsion and is bowed cranially. Thulborn (1972) identified a proximally situated obturator process, but Sereno (1986) denied its presence in the same material. Sereno (1986:247), however, did note a form "of possible close affinity to *Lesothosaurus*" which does possess an obturator process. The caudal margin of the ischium is deeply grooved for muscle attachment.

The femur (fig. 19.5) is about 20 percent longer than the ilium. Proximally, the medial aspect of the femur is enlarged; there is no neck and hence no distinct head set off from the shaft. The greater trochanter is a vertical, caudolateral blade that is relatively continuous with the head. The lesser trochanter is shorter than either the head or the greater trochanter, is separated by a vertical cleft from the greater trochanter, and is directed craniodorsally from the craniolateral surface of the proximal end. There is no cranial intercondylar fossa between the distal condyles, but the caudal intercondylar fossa is deep in both *Lesothosaurus* and *Pisanosaurus*. The lateral condyle is slightly larger than the medial. The femora described by Thulborn (1972) and Santa Luca (1984) are not identical. The head is dorsal to the greater trochanter in Thulborn's specimen, while the opposite is true in Santa Luca's specimen. Although in both samples the pendent fourth trochanter projects from the caudomedial portion of the shaft and is concave medially, it is more triangular and slightly more proximally situated in Thulborn's material than in Santa Luca's. Both femoral shafts are bowed cranially, but the bowing is more pronounced in Santa Luca's specimen.

The tibia in *Lesothosaurus* is long, about 25 percent longer than the femur. Contact between the tibia and femur is horizontal and flat. The laterally compressed, robust cnemial crest intervenes between the convex medial surface and the concave lateral surface. A thickened crest runs distally from the lateral condyle. In *Pisanosaurus,* the distal end of the tibia is only slightly wider than the shaft. In contrast, the distal tibia is distinctly expanded in *Lesothosaurus*. The tibia of both taxa are twisted roughly 70° to 90°. The medial and lateral margins of the distal tibia are separated by a groove or notch on the cranial surface. The lateral margin is broader than the medial and projects more distally.

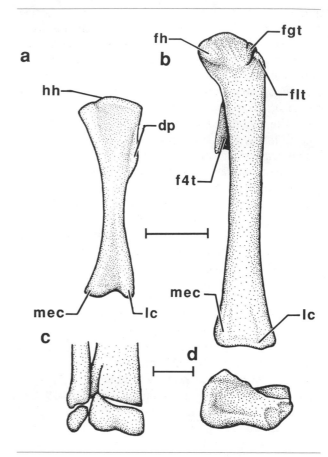

Fig. 19.5. Cranial view of left humerus of *Lesothosaurus diagnosticus* (after Thulborn 1972). Scale = 2 cm. b. Cranial view of left femur of *Lesothosaurus diagnosticus*. c. Cranial view of distal end of right tibia of *Pisanosaurus mertii* (after Casamiquela 1967*b*, Bonaparte 1976). d. Cranial view of right astragalus of *Technosaurus smalli* (after Chatterjee 1984). Scale = 1 cm.

The fibula in *Lesothosaurus* is shorter than the tibia and is expanded both proximally and distally. The proximal end of the fibula is laterally compressed with pronounced cranial and caudal processes. The shaft is slender. The semicircular distal end of the fibula lies on the cranial surface of the lateral tibial malleolus. The calcaneal articulation is concave.

In *Lesothosaurus,* the tarsus consists of astragalus, calcaneum, and at least two distal tarsals. The astragalus is attached to the medial tibial malleolus. A relatively strong ascending process of the astragalus projects proximally on the cranial face of the tibia within the intermalleolar sulcus. The ascending process is similar in *Pisanosaurus* and *Technosaurus* in that its apex is situated laterally (fig. 19.5). The calcaneum is relatively elongate proximodistally and bears a planar facet (in *Lesothosaurus*) or convexity (in *Pisanosaurus*) for the fibula. Bonaparte (1976) considered a caudal tubercle on the calcaneum of *Pisanosaurus* to be homologous to the calcaneal tuberosity of nonmesotarsal archosaurs. Based on this tubercle and the calcaneofibular articulation, he considered the ankle of *Pisanosaurus* to be "intermediate" between crurotarsal and mesotarsal. The biomechanics of such an intermediate are unclear, and the ankle of *Pisanosaurus* was certainly functionally mesotarsal. Perhaps more important, a mesotarsal ankle was already present ancestrally among dinosaurs; thus, an "intermediate" condition in *Pisanosaurus* would constitute a reversal. The distal tarsals of *Lesothosaurus* are disc-shaped; one articulates with metatarsal III and the other with metatarsal IV. A small distal tarsal is preserved in *Pisanosaurus* which bears broad and small contacts for metatarsals IV and III, respectively.

The pes is about as long as the tibia in *Lesothosaurus* and about 60 percent as long in *Pisanosaurus*. In *Lesothosaurus*, metatarsal I is unusual in that it is small, thin, and splintlike and may be fused to metatarsal II. The other three metatarsals are elongate, with metatarsal III being longest (as in *Pisanosaurus*) and II longer than IV. in *Lesothosaurus*, the bases are flat and the heads are weakly bicondylar. Metatarsals II and III are laterally compressed proximally and are closely appressed throughout most of their length. Metatarsal IV is transversely expanded proximally and is somewhat separated from the other metatarsals. The metatarsus exhibits a weak transverse arch in which the plantar surface is concave. In *Pisanosaurus*, the base of metatarsal III overlaps that of metatarsal IV. The phalangeal formula is 2−3−4−5−? (digit V is unknown.) The phalanges are generally typical. The distal phalanges are clawlike; they are slightly curved, cone-shaped with a flat plantar surface, and grooved laterally and medially.

SYSTEMATICS

As one of the best-known early ornithischians, *Lesothosaurus* once typified the fabrosaurid *Bauplan*. Furthermore, the Fabrosauridae was considered the stem taxon from which all other ornithischians descended (Thulborn 1971*b*; Galton 1972, 1978; Chatterjee 1984), and with few exceptions, they have been included as basal members of the Ornithopoda. The Fabrosauridae as conceived by Galton (1978) has to date included numerous taxa: *Fabrosaurus, Lesothosaurus, Scutellosaurus, Tawasaurus, Trimucrodon, Echinodon, Technosaurus, Nanosaurus, Alocodon, Xiaosaurus*. Additionally, Santa Luca (1984) described the postcranium of what he considered to be a new fabrosaur, and undescribed fabrosaurs have been reported by a number

of authors (Gow 1981; Callison and Quimby 1984; Sereno 1986; Crompton and Attridge 1986). However, Sereno (1984, 1986) and Gauthier (1986) denied not only the ancestral status of fabrosaurids but also their ornithopod status and even their validity as a taxon. Sereno and Gauthier are correct in stating that, as previously constituted, the Fabrosauridae is an unnatural paraphyletic assemblage. For example, *Scutellosaurus* has been assigned to the Thyreophora (Sereno 1984; Gauthier 1984; see also Coombs et al., this vol.). The same may apply to *Echinodon* (see Coombs et al., this vol.). Galton (1983*d*) assigned *Nanosaurus* to the Hypsilophodontidae. *Xiaosaurus* may be a hypsilophodontid (Dong pers. comm.) and clearly needs more study. Finally, many of the taxa can be removed from the family and classified as *nomina dubia* (*Fabrosaurus, Tawasaurus, Trimucrodon,* and *Alocodon*; see table 19.1).

The choice of taxa included in this study (table 19.1) requires a broad phylogenetic framework. Gauthier (1984, 1986; see also Benton, this vol.) presented evidence that saurischian dinosaurs are monophyletic and stand as the sister group of ornithischians, with basal dinosaurs (*Herrerasaurus, Staurikosaurus,* etc.), pterosaurs, and lagosuchids forming successively more remote outgroups. Sereno (1986) implied that the Saurischia was both monophyletic and the sister group of the Ornithischia. However, he previously suggested monophyly of prosauropods and ornithischians (Sereno 1984), and this position has had a considerable following (Bakker and Galton 1974; Bonaparte 1976; Cooper 1981*a*, 1985; Paul 1984*a*, 1984*b*; Bakker 1986). Furthermore, Paul (1984*b*) refined the concept of a Phytodinosauria (a taxon formally proposed by Bakker 1986), suggesting that segnosaurs are the sister group of ornithischians, with sauropodomorphs representing the next outgroup.

Choice of outgroup is critical in the present case in that sauropodomorphs and ornithischians share derived dental features (Sereno 1984; Gauthier 1986). We considered an informal cluster of basal sauropodomorphs (Galton 1976*a*; Cooper 1981*a*; Attridge et al. 1985; Crompton and Attridge 1986) and basal theropods (Raath 1969, 1977; Welles 1984; Padian 1986) to represent the saurischian sister group of ornithischians. Segnosaurs (Perle 1979; Barsbold and Perle 1980; Paul 1984*b*) also were considered in character polarization but not in any rigorous way. The next outgroup was a cluster of the basal dinosaurs *Herrerasaurus, Staurikosaurus,* and *Ischisaurus* (Reig 1963; Colbert 1970; Benedetto 1973; Galton 1977*a*; Brinkman and Sues 1987).

Gauthier (1986) and Sereno (1984, 1986) removed *Lesothosaurus* from the Ornithopoda and instead regarded it as the sister group of all other ornithischians (the latter named Genasauria by Sereno

TABLE 19.1 *Lesothosaurus, Pisanosaurus, Technosaurus*

	Occurrence	Age	Material
Ornithischia Seeley, 1888a			
Lesothosaurus Galton, 1978			
L. diagnosticus Galton, 1978	Upper Elliot Formation, Mafeteng District, Lesotho	Hettangian-?Sinemurian	At least 4 skulls and associated skeletal material
Undescribed fabrosaur (Santa Luca 1984)	Upper Elliot Formation, Mafeteng District, Lesotho	Hettangian-?Sinemurian	Fragmentary skull material and associated skeletons of several individuals
Pisanosaurus Casamiquela, 1967			
P. merti Casamiquela, 1967	Ischigualasto Formation, La Rioja Province, Argentina	Carnian	Fragmentary skull and skeleton
Technosaurus Chatterjee, 1984			
T. smalli Chatterjee, 1984	Dockum Formation, Texas, United States	Carnian	Isolated skull and skeletal elements

Nomina dubia	Material
Azendohsaurus laaroussii Dutuit, 1972 *partim*	Isolated teeth
Alocodon kuehnei Thulborn, 1973a	Tooth
Fabrosaurus australis Ginsburg, 1964	Fragmentary dentary with teeth
Gongbusaurus shiyii Dong, Zhou, et Zhang, 1983	2 isolated teeth
Revueltosaurus callenderi Hunt, 1989 •	Teeth
Taveirosaurus costai Antunes et Sigogneau-Russell, 1991 •	Teeth
Thecodontosaurus gibbidens Cope, 1878e	Teeth
Trimucrodon cuneatus Thulborn, 1973a	Tooth
Xiaosaurus dashanpensis Dong et Tang, 1983 (may be a hypsilophodontid)	Teeth and isolated postcranial material

[1986]). Features that are absent in *Lesothosaurus* but present in these higher ornithischians include a sharp definition to the entire margin of the antorbital fossa (Sereno 1986), reduction of the external mandibular fenestra (Sereno 1986), and elongation of the preacetabular process of the ilium. A spout-shaped mandibular symphysis (Sereno 1986) is very difficult to interpret but may be a further synapomorphy for higher ornithischians. The same applies to the reduced robustness of the pubic peduncle of the ilium relative to the ischial peduncle (Sereno 1986). We view buccal emargination of the maxilla (Sereno 1986) as a basal ornithischian trait since it is present in *Lesothosaurus* (Paul 1984b; Galton 1986b; Norman pers. comm.), *Pisanosaurus, Technosaurus,* and basal thyreophorans (*Scutellosaurus, Tatisaurus, ?Echinodon*).

It should be noted that Santa Luca (1984) and Sereno (1986) identified another unnamed genus and/or species closely related to *Lesothosaurus diagnosticus*. If true, then there is the distinct possibility that a monophyletic clade of lesothosaurs exists (Lesothosauridae Halstead et Halstead, 1981?).

Pisanosaurus is so fragmentary and so poorly preserved that its phylogenetic position is extremely problematic; it has been classified as an ornithopod (Casamiquela 1967b), a hypsilophodontid (Thulborn 1971a; Galton 1972, 1986), and a heterodontosaurid (Bonaparte 1976; Charig and Crompton 1974; Cooper

1981a, 1985; Weishampel and Weishampel 1983; Weishampel 1984a; Crompton and Attridge 1986). Unfortunately, most recent cladistic analyses (Paul 1984a, 1984b; Sereno 1984, 1986; Gauthier 1984, 1986) have omitted *Pisanosaurus* altogether.

Our analysis tentatively places *Pisanosaurus* as one of the most primitive members of the ornithischian clade. *Pisanosaurus* is certainly a dinosaur by virtue of its fully "twisted" tibia and the notch in the cranial face of the distal tibia for reception of the ascending process of the astragalus. However, relationship with higher dinosaurs is complicated by the lack of distal expansion of the tibia, approximately as much as in the basal dinosaurs *Staurikosaurus* and *Herrerasaurus*. However, in theropods, sauropodomorphs, and ornithischians, the distal tibia is greatly expanded in the transverse plane. For *Pisanosaurus* to be considered an ornithischian, its narrow distal tibia has to be considered a reversal.

Bonaparte's (1976) suggestion that five sacral vertebrae are present in *Pisanosaurus* allies it with the Ornithischia. The jaws, however, exhibit morphology that is found elsewhere only in ornithischians. Both the maxilla and the dentary display modest buccal emargination. Systematic occlusion (producing extensive wear between maxillary and dentary teeth) is unique to *Pisanosaurus* and a variety of other ornithischians. Sereno's (1986) series of ornithischian characters invite comparisons with *Pisanosaurus*. Of these, loss of recurvature of the maxillary and dentary teeth, separation of the crown and root of the teeth by a neck, maximal tooth size being in the middle of the tooth row, and the dentary forming the rostral portion of the coronoid process stand as synapomorphies of *Pisanosaurus* and Ornithischia.

A number of primitive features deny *Pisanosaurus* a relationship with higher ornithischians. The coronoid process is low as in fabrosaurs and basal thyreophorans. The medial metatarsals overlap the lateral ones; this is the condition in *Herrerasaurus* (Reig 1963), basal sauropodomorphs (Galton 1976*a*), and basal theropods (Welles 1984), but all other ornithischians show the apomorphic loss of the overlap. In *Pisanosaurus*, the apex of the ascending process of the astragalus is laterally situated near the articulation with the calcaneum. This is also the condition in basal sauropodomorphs (Galton 1976*a*; Cooper 1981*b*), basal theropods (Welles 1984; Rowe and Gauthier, this vol.), *Technosaurus* (Chatterjee 1984), and fabrosaurs (Santa Luca 1984).

Ornithischians at least at the level of the Cerapoda, however, exhibit the derived condition in which the apex of the ascending process of the astragalus has moved medially away from the calcaneal articulation (Gilmore 1915*a*; Galton 1974*a*, 1981*c*, Brown and Schlaikjer 1940*a*; Maryańska, this vol.). Thyreophorans are equivocal on this point; *Scutellosaurus* appears to exhibit a unique configuration (Colbert 1981), but it is uncertain as to whether this pertains to all thyreophorans. The situation in *Scelidosaurus* is unclear (viz. Owen 1863). Although a few features suggest relationship between *Pisanosaurus* and heterodontosaurids (i.e., the often-broad contact between mesial and distal surfaces of adjacent crowns, the presence of extensive planar tooth wear, and the degree of buccal emargination), these are considered convergences.

From this discussion, it can be concluded that *Pisanosaurus* is a member of the ornithischian clade. Its position within the clade is less clear, but it appears to be a very primitive member. Based on the primitive aspect of what postcranial material is preserved, it seems likely that *Pisanosaurus*'s position at the base of the cladogram will be corroborated by future discoveries.

Chatterjee (1984) referred *Technosaurus smalli* to the Fabrosauridae, but it appears to lack the derived characters of both fabrosaurs (as conceived here) and the larger taxon to which they belong. *Technosaurus* appears to retain important primitive characters including a complete premaxillary dental arcade, a large external mandibular fenestra, and, like *Pisanosaurus*, a laterally situated apex of the ascending process of the astragalus. The teeth resemble those of other ornithischians in being triangular and having *en echelon* emplacement but yield no taxonomically useful features. The metatarsals are not known in *Technosaurus*, and thus their amount of overlap is unknown. As a result, we consider *Technosaurus* also one of the most primitive ornithischians yet known.

PALEOECOLOGY

Because of the rarity of fossils and often imperfect nature of the material referred to *Lesothosaurus*, *Pisanosaurus*, and *Technosaurus*, very little of substance can be said about their paleoecology. The paleogeographic distribution of these animals is limited to small portions of the United States, South Africa, and Argentina. Stratigraphic ranges are restricted to the Late Triassic and Early Jurassic.

Local environments from which these animals derive represent chiefly semiarid habitats (see similar comments for heterodontosaurids, Weishampel and Witmer, this vol.). As herbivores, *Lesothosaurus*, *Pisanosaurus*, and *Technosaurus* were active foragers on ground cover and shrubby vegetation. A cornified rhamphotheca was apparently present in at least *Lesothosaurus* but perhaps not in *Technosaurus*, given its complete premaxillary dental arcade. Thulborn (1971*a*) and Weishampel (1984*a*) have suggested that *Lesothosaurus*, like all other primitive ornithischians, relied solely on simple adduction of the lower jaws to produce vertical or near vertical tooth-tooth shearing motion between bilaterally occluding maxillary and dentary teeth.

On the basis of limb and trunk proportions and the general gracile nature of their skeleton, all forms considered here were probably agile, cursorial bipeds (Thulborn 1982). Quadrupedal stance was probably used during foraging or when standing still. The femur was held in a slightly abducted position to facilitate a parasagittal gait and horizontal knee joints. Lever-arm mechanics suggest that femoral retraction was powerful. The attitude of the axial skeleton was largely horizontal. In *Lesothosaurus*, the forelimbs are quite short in comparison to the hindlimbs, and the distal portions of the forelimb are reduced. What little is known about the manus suggests that it was of little use in prehension. *Lesothosaurus* possesses a semiperforate acetabulum and lacks a femoral neck, suggesting the kinematics of hindlimb excursion were different from other ornithischians.

THYREOPHORA

DAVID B. WEISHAMPEL

Originally proposed by Nopcsa (1915) to include ceratopsians, stegosaurs, and ankylosaurs, the Thyreophora has been restricted to include stegosaurs and ankylosaurs as well as their more basal relatives, *Scelidosaurus* and *Scutellosaurus* (Norman 1984*a;* Sereno 1984, 1986; Cooper 1985; Gauthier 1986). These animals are among the most primitive ornithischians, and hence character polarity, particularly for the most basal taxa, is difficult to assess. Thyreophorans appear to be united on the basis of having a jugal with a transversely broad postorbital process and parasagittal rows of keeled scutes on the dorsal surface of the body (Sereno 1986). These characters are not present in the sister group to thyreophorans, the more derived cerapodans (ornithopods plus marginocephalians), nor are they present in *Lesothosaurus*.

Thyreophorans are now considered among the most primitive ornithischians (Norman 1984*a;* Sereno 1984, 1986; Gauthier 1986). Sereno (1986), Gauthier (1986), and Weishampel and Witmer (this vol.) regard *Lesothosaurus* as the sister taxon to all remaining ornithischians, including thyreophorans (termed the Genasauria by Sereno [1986]).

Relationships of taxa within the Thyreophora are separately discussed for basal thyreophorans(Coombs et al., this vol.), the Stegosauria (Galton, this vol.), and the Ankylosauria (Coombs and Maryańska, this vol.).

20

Basal Thyreophora

WALTER P. COOMBS, JR.
DAVID B. WEISHAMPEL
LAWRENCE M. WITMER

INTRODUCTION

The Thyreophora is a diversified group of armored, predominantly quadrupedal ornithischians, most of which fall into the Stegosauria and Ankylosauria, which are thought to be sister taxa (Eurypoda of Sereno 1986). In addition, it now appears that several taxa are more primitive members of this clade. These basal thyreophorans, *Scutellosaurus lawleri, Scelidosaurus harrisonii,* and possibly *Tatisaurus oehleri* and *Echinodon becklessii,* are the subject of this chapter (Table 20.1). The following description of *Scelidosaurus* is based primarily on Owen (1861a, 1863), Charig (1979), Norman (1985), and Sereno (1986). Described material of *Scutellosaurus* consists of parts of the skull and the majority of the postcranial skeleton (Colbert 1981). Only the dentary, fragmentary lower dentition, and a dorsal rib fragment are known for *Tatisaurus oehleri* (Simmons 1965). *Echinodon becklesii,* a possible thyreophoran, is known from fragmentary skull elements and perhaps some scutes (Owen 1861b; Galton 1978). In all cases, these taxa represent very primitive thyreophorans and hence are among the most primitive of ornithischians. *Scelidosaurus, Scutellosaurus,* and perhaps *Echinodon* bore a regular arrangement of dermal armor on their backs and were at least partially quadrupedal. All were herbivorous and ranged from one to four meters in total length.

Because the taxa considered in this chapter do not constitute a monophyletic group, diagnoses are deferred to the Systematics section.

ANATOMY

Skull and Mandible

Caudally, the skull of *Scelidosaurus* is high but tapers rostrally to produce a wedge-shaped lateral profile (fig. 20.1a). In dorsal view, the skull is widest across the jugals, from which it tapers rostrally to just in front of the orbits where the snout becomes almost parallel-sided (fig. 20.1b). Thus, the orbits face slightly forward. The skull proportions of the other taxa under consideration cannot presently be characterized.

The premaxilla, known only in a partially disarticulated skull referred to *Scelidosaurus* (Norman 1985), has several large, slightly recurved teeth. In this specimen, the premaxilla has a large lateral process that appears to have extended to the lacrimal. However, in

TABLE 20.1 Basal Thyreophora

	Occurrence	Age	Material
Thyreophora Nopcsa, 1915			
Scutellosaurus Colbert, 1981			
S. lawleri Colbert, 1981	Kayenta Formation, Arizona, United States	Hettangian	Fragmentary skull and skeleton from at least two individuals
Emausaurus Haubold, 1991 •			
E. ernsti Haubold, 1991 •	"bituminosen Schiefer, Mecklenberg-Vorpommern, Federal Republic of Germany	early Toarcian	Nearly complete skull with associated postcrania
Scelidosaurus Owen, 1860			
S. harrisonii Owen, 1861	Lower Lias, Charmouth, England	Sinemurian	Nearly complete skull with associated postcrania
Unnamed ?thyreophoran (so-called juvenile *Scelidosaurus*; Rixon 1968)	Lower Lias, Charmouth, England	Sinemurian	Articulated postcranial skeleton, juvenile
Tatisaurus Simmons, 1965			
T. oehleri Simmons, 1965	Dark Red Beds, Lower Lufeng Series, Yunnan, People's Republic of China	Early Jurassic	Isolated dentary
?Thyreophora incertae sedis			
Echinodon Owen, 1861b			
E. becklesii Owen, 1861b	Middle Purbeck Beds, Dorset, England	late Tithonian	Isolated skull elements of at least three individuals, scutes

Nomina dubia	Material
Lusitanosaurus liasicus Lapparent et Zbyszewski, 1957	Skull fragment

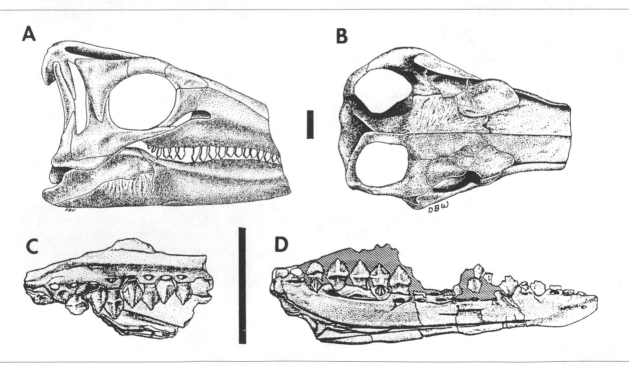

Fig. 20.1. a. Right lateral view of the skull of *Scelidosaurus harrisonii*. b. Dorsal view of the skull of *Scelidosaurus harrisonii*. c. Medial view of the right maxilla of *Scutellosaurus lawleri* (from Colbert 1981). d. Medial view of the left dentary of *Scutellosaurus lawleri* (from Colbert 1981). Scale = 20 mm.

the type of *Scelidosaurus*, the lateral process of the premaxilla must have been relatively short because there is uninterrupted contact between the nasal and maxilla rostral to the lacrimal. For this reason, pending reconstruction of the referred skull, this specimen is only tentatively accepted as pertaining to *S. harrisoni*. In *Scutellosaurus*, the premaxilla bears a field of horizontal striations laterally. There are six alveoli between the mesial midline and the base of the lateral process. The ventral border of the external naris is formed into a slight horizontal lip. A small foramen is found at the rostral margin of the external naris near the base of the dorsal premaxillary process.

The maxilla of *Scelidosaurus* and *Scutellosaurus* is triangular. In *Scutellosaurus* (fig. 20.1c), it is longer and lower, and the rostral two-thirds is shallowly excavated to form the antorbital fossa, which extends onto the body of the maxilla. The rostral edge of this fossa is interpreted as the caudal margin of the antorbital fenestra. An overhanging maxillary shelf projects lateral to the tooth row in *Scelidosaurus* and to a lesser degree in *Scutellosaurus*. In *Scutellosaurus*, the medial maxillary surface is well rounded. The maxilla makes a long butt contact with the nasal. There are at least 10 maxillary alveoli in *Scutellosaurus* and more than 19 in *Scelidosaurus*. In *Scutellosaurus*, there is a resorption pit immediately dorsal to the medial aspect of each alveolus for eruption of successive replacement teeth.

The nasal of *Scelidosaurus* is long and broad. Although poorly known, the nasal of *Scutellosaurus* appears to be relatively long and dorsally concave in transverse section. Caudally, the nasal-maxillary suture in *Scelidosaurus* is continuous with that between the prefrontal and lacrimal. The lacrimal is semirectangular in lateral view and makes up the majority of the rostral margin of the orbit. The prefrontal is slightly oval and contacts the frontal medially, the nasal rostrally, and the supraorbital caudally. The supraorbital makes up the dorsal rim of the orbit and intervenes between the prefrontal and postorbital, excluding the frontal from the orbit.

The postorbital and jugal of *Scelidosaurus* form a broad postorbital bar between the orbit and the narrow, tall infratemporal fenestra. The jugal is long and narrow at both maxillary and quadratojugal contacts. The L-shaped quadratojugal partially overlies and hides the quadrate in lateral view. The nearly vertical quadrate is slightly concave along its caudal edge. The quadrate head is lodged in a somewhat shallow squamosal cotylus. The mandibular condyle lies ventral to the level of the maxillary tooth row. The postorbital-squamosal suture is not clearly indicated.

The frontal-parietal suture appears to be interdigitate throughout its length in *Scelidosaurus*. In *Scelidosaurus*, there is a straight median suture between the frontals and between the parietals, but the frontal-parietal suture is not visible. What can be clearly identified as the frontal is approximately rectangular. In *Scutellosaurus*, the dorsal surface of the thin, triangular frontal is flat, while ventrally, it is excavated to accommodate the cerebrum and long olfactory tracts. The frontal-postorbital contact is a scarf joint. The frontal contributes to the dorsal margin of the orbit. The parietals of *Scelidosaurus* are shaped like an hourglass. The squamosal forms the slender caudal and lateral boundaries of a nearly circular supratemporal fenestra. A narrow parietal shelf separates the two supratemporal fenestrae. The occipital margin of the skull is deeply notched.

The palate of *Scelidosaurus* has been illustrated in cross section (Owen 1861a). The vomers are small, laterally compressed, and sit well dorsal to the level of the maxillary teeth but ventral to the respiratory passage.

In *Scutellosaurus*, the mandible is straight, dorsoventrally slender, and has a low coronoid process (fig. 20.1d). The dentaries of both *Scelidosaurus* and *Tatisaurus* (fig. 20.1e) converge rostrally over the course of the mesial six or so alveoli to give a slightly sinuous profile. In *Scutellosaurus*, the lateral dentary contains a few large, ovoid foramina near the tooth row, while approximately 15 small foramina are found close to the rostral end of the dentary. Although no predentary is known in any of the taxa under consideration here, prominent facets receiving the lateral and ventral predentary processes are exposed rostrally in *Tatisaurus*. In all taxa, the mandibular symphysis is broad, and the articular surfaces are very rugose and inclined rostrally. The dentary supports as many as 18 alveoli in *Scutellosaurus* and *Tatisaurus* and more than 16 in *Scelidosaurus*. In *Scutellosaurus*, a long Meckelian groove marks the ventromedial aspect of the dentary. Resorption pits marks the entrance of replacement teeth medial to the functional dentition. The broadly ovoid symphysis is upturned at the rostral end of the dentary. *Scelidosaurus* lacks an external mandibular fenestra; this condition is not yet known in the other forms. In *Scelidosaurus*, *Tatisaurus*, and *Scutellosaurus*, the dentary dentition is emarginated from the side of the mandible. The surangular is large, forms the caudal half of the coronoid process, and projects caudally as a well-developed retroarticular process. The wedge-shaped angular forms the ventral part of the caudal half of the mandible. The splenial is a narrow splinter of bone on the ventral margin of the mandible.

Premaxillary tooth morphology is not known for

either *Scutellosaurus* or *Scelidosaurus*. Maxillary and dentary tooth crowns in all forms are transversely narrow, mesiodistally expanded, and broadly triangular in shape, as wide or wider than high (only dentary dentition is known for *Tatisaurus*). The apex of the tooth caps a broadly rounded eminence that divides the lingual face of the smooth crown into subequal mesial and distal regions. The buccal aspect of the crown is flat. Mesial and distal edges each bear three to four denticles in *Scutellosaurus* or four to six denticles in *Scelidosaurus* which decrease in size toward the apex. In *Tatisaurus*, the few small marginal denticles on mesial and dental edges of the crown are well removed from the apex. The outermost denticles merge to form a slightly undulatory or rounded cingulum at the base of the crown, which in *Tatisaurus* is bulbous. In all taxa, enamel is symmetrically distributed on buccal and lingual faces of the crown. Adjacent teeth may touch each other, but there is no notching or other modification of adjoining surfaces. Roots appear to be long, straight, and aligned with the crown.

Axial Skeleton

As preserved, the vertebral count of *Scelidosaurus* is 6 cervicals, 17 dorsals, 4 sacrals, and 35 caudals (fig. 20.2a). Owen (1863) suspected that one or two cervicals were missing. The vertebral column of *Scutello-saurus* consists of 24 presacral, 5 sacral, and about 60 caudal vertebrae (fig. 20.2b). Of the presacrals, there are 6 or 7 cervicals (including the atlas and axis), 17 or 18 dorsals, and 5 sacrals. The neurocentral sutures are open in the presacral series but closed in the tail. All vertebrae are amphicoelous. Although the specimens are disarticulated, Colbert (1981) arranged the vertebrae in sequence, presumably based on size and the regional differentiation observed in other species.

In *Scutellosaurus*, the cervical centra exhibit deep lateral excavations and a medial ventral keel that becomes smaller in the caudal cervicals. The central articulations narrow ventrally. Cervical neural arches are poorly known. The length of cervical centra is as great or greater than their diameter. The transverse processes are dorsoventrally compressed and project laterally to slightly caudally in the cervicals and cranial dorsals. Parapophyses are not apparent on cervical centra. In *Scelidosaurus*, the centrum of the axis is fused to the atlas.

In both *Scutellosaurus* and *Scelidosaurus*, the dorsal centra are less constricted, more spool-shaped, lack the ventral keel, and bear longitudinal striations. Their articular surfaces are round. In *Scutellosaurus*, the neural spines are poorly known but appear to expand dorsally to assume a trapezoidal outline. The transverse processes are thinner on the cranial dorsals and become broader caudally. This broadening reflects the migration of the parapophysis from the base of the trans-

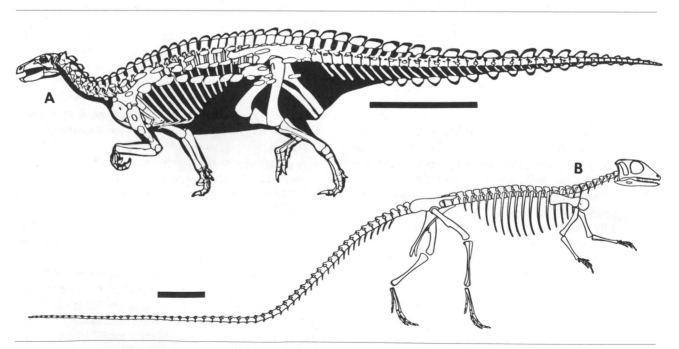

Fig. 20.2. a. Skeletal reconstruction of *Scelidosaurus harrisonii* with dermal armor (from Paul 1987*b*). b. Skeletal reconstruction of *Scutellosaurus lawleri* (from Colbert 1981). Scale = 50 cm.

verse process on the cranial dorsals to a position on the transverse process near the diapophysis on caudal dorsals. The diapophysis becomes smaller caudally as the parapophysis approaches it; in the last dorsals, only the parapophysis remains. The zygapophyses are simple and only slightly inclined. In *Scelidosaurus*, the transverse processes of cranial dorsals bear a diapophysis distally, while the parapophyses are on the flanks of the centra. The parapophyses gradually shift dorsally to the underside of the transverse processes along the dorsal series. The last three dorsals have only the diapophyses at the distant tips of the transverse processes. The last dorsal does not bear a rib. The neural spines are laterally compressed and short.

In *Scutellosaurus*, the sacral centra are broader than long with crescentic articular faces and smooth ventral surfaces. The transverse processes are broad. The postzygapophyses are small, almost vertical, and close together. In *Scelidosaurus*, the neural spines of the sacral series are in direct contact but do not fuse with one another.

In *Scutellosaurus*, the tail is about 2.5 times the presacral length as a consequence of the large number of elongate vertebrae. Individual vertebrae of the proximal two-thirds of the caudal series are longer than the presacral vertebrae. The proximal caudal centra are broad, and the distal caudals become narrower. Chevron facets are present in the proximal three-fifths of the series. The neural arches are fused to the centra throughout the tail. Neural spines are present in the proximal three-fourths of the series and are thin and distally inclined. They become progressively smaller distally. Transverse processes are relatively large in the most proximal caudals but become smaller distally and are absent in the distal two-thirds of the tail. Zygapophyses are small and close together. In *Scelidosaurus*, the caudal series is relatively short. Transverse processes are inclined 15° to 20° in the caudal dorsals. Distal caudals are long and slender.

Presacral ribs in *Scutellosaurus* reflect the serial changes noted above for the vertebrae: the head and tubercle are widely separated but not biramate in the cranial dorsals and become closer together in the caudal dorsals. As noted, the last few dorsal vertebrae have only the parapophysis, and consequently several ribs are holocephalous. In *Scelidosaurus*, the ribs of the cranial dorsals are bicipital, but those of the last few dorsals are holocephalous. The four sacral ribs are generally broad, long, appear to be fused to their transverse processes, and are progressively shorter down the sequence. Each projects straight laterally in a horizontal plane and expands distally, but adjacent sacral ribs do not contact each other.

Rodlike, laterally compressed chevrons are present intervertebrally at least to caudal 28 in *Scelidosaurus* and to caudal 30 in *Scutellosaurus*.

Colbert (1981) identified a few fragments of *Scutellosaurus* as possible ossified tendons. Elongate slender bones identified by Owen (1862) as abdominal ribs also may be ossified tendons.

Appendicular Skeleton

In *Scutellosaurus*, the scapular shaft is broad craniocaudally and is uniformly bowed to accommodate the trunk. The expanded proximal end has a reniform glenoid facet that faces ventrocaudally, a truncate acromion, and in between, a broad, triangular lateral depression. The articulation between scapula and coracoid is sinuous and congruent. The coracoid is rounded in lateral view and is pierced by a coracoidal foramen near its caudodorsal margin. The coracoidal glenoid is a rounded, caudally directed facet. In *Scelidosaurus*, the scapular blade is straight with a pronounced expansion both opposite the glenoid and distally. The coracoid is irregularly circular in lateral view.

In *Scelidosaurus*, the humerus is straight, moderately thick, and has a large deltopectoral crest extending almost to humeral midlength. In contrast, the humerus of *Scutellosaurus* is relatively slender and slightly laterally bowed, and the triangular deltopectoral crest is proximally situated. In addition, the proximal end is expanded transversely and rounded. A small head lies just lateral to a medial tuberosity. The distal end is in approximately the same plane as the proximal. The radial and ulnar condyles are similar in size. The olecranon fossa is slightly deeper than the cuboidal fossa.

The radius and ulna are incompletely known in *Scutellosaurus*, and the length of the antebrachium cannot be estimated. Both bones are straight and cylindrical. The ulna lacks a well-developed olecranon. The distal ends of the bones are about the same size and present simple rounded articulations. The radius, ulna, and manus are unknown in *Scelidosaurus*.

The carpus is unknown in *Scutellosaurus*, and the manus is represented only by a few metacarpal fragments and several phalanges. The unguals are small, pointed claws. Although incomplete, the hand apparently was moderately large.

The pelvic girdle and femur of *Scelidosaurus* are known from a partial skeleton identified as a juvenile (fig. 20.3b; Rixon 1968; Romer 1968; Charig 1972; Sereno 1986). However, the taxonomic assignment of this material is uncertain (Thulborn 1977; see below). This specimen is hereafter referred to as the unnamed thyreophoran. The postacetabular process of the ilium

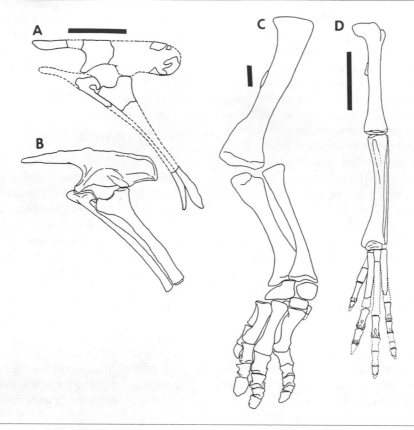

Fig. 20.3. a. Pelvic girdle of *Scutellosaurus lawleri* (after Colbert 1981). b. Pelvic girdle of undescribed thyreophoran, a specimen referred to *Scelidosaurus harrisonii* (after Charig 1972). c. Hindlimb of *Scelidosaurus harrisonii* (after Owen 1863). d. Hindlimb of *Scutellosaurus lawleri* (after Colbert 1981).

is laterally compressed, long, and near vertical in both the unnamed thyreophoran and *Scutellosaurus*, while the pronglike preacetaular process is dorsoventrally compressed and somewhat longer and shallower in the unnamed form. In *Scutellosaurus*, the dorsal margin is thickened along its length and slightly overhangs the preacetabular process (fig. 20.3a). There is no antitrochanter in either of these forms. In *Scutellosaurus*, there appears to be a small brevis shelf. In both forms, the pubic peduncle is large and somewhat inclined cranially. The ischial peduncle is smaller. Colbert (1981) described a supraacetabular flange in *Scutellosaurus*. The ilium of *Scelidosaurus* apparently diverges from the midline, insofar as the first several rib projects farther laterally than the more caudal sacral ribs. Paul (1987b) shows a very *Stegosaurus*-like ilium in a dorsal reconstruction of *Scelidosaurus*.

In the unnamed thyreophoran, *Scelidosaurus*, and *Scutellosaurus*, the rodlike ischium extends caudoventrally below the ilium and is bifurcated proximally into subequal iliac and pubic peduncles. The three forms have neither an ischial flange partially occluding the acetabulum nor an obturator process on the ischium (Charig 1972; Sereno 1986 contra Colbert 1981). The distal ends of the ischia expand into thin, bladelike structures.

The pubis is an elongate slender rod that extends down the entire length of the ischium in *Scutellosaurus* and the unnamed thyreophoran. The distal end of the postpubic process curves cranioventrally away from the ischium in *Scutellosaurus*. In the unnamed form, the prepubic process is short and blunt (Charig 1972). The obturator foramen is closed caudally by the ischium.

The femur is straight to slightly bowed in both *Scelidosaurus* and *Scutellosaurus* (fig. 20.3c, d), while that of the unnamed thyreophoran is more strongly bowed (Charig 1972). In all three taxa, the head is displaced medially and is set off from the shaft by a slight neck in *Scutellosaurus* but not in *Scelidosaurus*. The greater trochanter is at the same level as the head and is continuous with it. The lesser trochanter is prominent and well separated from the greater trochanter by a deep groove. The fourth trochanter is large, pendent, and positioned at about femoral midlength in both

Scutellosaurus and *Scelidosaurus* (Romer 1956; Charig 1972; Colbert 1981). The distal condyles are well developed and at about the same level. The lateral femoral condyle is slightly larger than the medial. There is a distinct medial supracondylar ridge. The cranial intercondylar fossa is absent, but the caudal intercondylar fossa is prominent.

The tibia of *Scutellosaurus* and *Scelidosaurus* is slightly longer than the femur. The craniocaudal expansion of the proximal end results from the considerable development of the cnemial crest. Despite this expansion, the proximal end is transversely broad. The lateral tibial condyle is much smaller than the medial condyle and bears part of the fibular articulation. The medial border of the shaft is sharp distally. The distal end of the tibia is twisted about 70° from the proximal end. The medial margin of the distal tibia is larger than the lateral and projects farther distally. The fibula is a long, slender bone that expands proximally.

Scelidosaurus has a very large, free astragalus, a large calcaneum, and two additional distal tarsals. The tarsus of *Scutellosaurus* is poorly known. The astragalus bears a low but well-developed ascending process; its apex is situated close to the medial border.

In *Scutellosaurus*, the metatarsals are moderately long. Proximally, they are closely appressed, metatarsals II-IV forming a curved row. Distally, the metatarsals apparently diverged somewhat. A deep median pit near the heads of the proximal phalanges receive a process from the second phalanges and suggest considerable extension at this joint. The unguals are clawlike. The phalangeal formula of *Scutellosaurus* is unknown. *Scelidosaurus* has four short, massive metatarsals and a vestigial metatarsal V. Phalanges are also short and dorsoventrally compressed. The unguals are blunt, dorsoventrally compressed, and lack flanking grooves. The pedal digital formula of *Scelidosaurus* is 2–3–4–5–0.

The dermal armor of *Scelidosaurus* includes numerous roughly oval plates that have an outer median longitudinal ridge or keel. Cranial to its apex, the ridge is long, upwardly convex, and obliquely slanted, whereas caudal to the apex it is shorter, straighter, and more vertically oriented. Most larger plates are excavated on the median surface. There are also numerous smaller plates, some with keels and some merely small nubbins. In situ plates indicate a pair of unique three-pointed plates just caudal to the skull, a longitudinal row of bilateral pairs of plates adjacent to the midline starting above the atlas and axis and extending to the sacral region, at least two additional longitudinal rows of plates more lateral over the thorax, and four longitudinal rows of plates down the tail (median, dorsal, median ventral, and a row on each side; fig. 20.2a).

Similarly, *Scutellosaurus* was covered in dermal armor consisting of hundreds of scutes, although the exact number and pattern of placement are unknown. Colbert (1981) divided the scutes into six classes that presumably relate to regional differentiation, but only two or three classes are markedly different. All scutes are very rugose with a high degree of pitting, and they almost always have a longitudinal keel on the dorsal surface. The scutes are excavated such that they are generally of uniform thickness. Nearly all scutes are asymmetrical, varying from low and relatively flat to elongate and sharply keeled. Others are very tall and triangular in profile. The scutes were probably located on the body cranial to the tail with the larger scutes occupying a more dorsal position. Their asymmetry suggests that none were medial structures. Most of the remaining scutes are symmetrical, long, and narrow. Colbert (1981) suggested that they belong to a median series along the tail. Last, there are transitional scutes with double keels and symmetrical bases.

SYSTEMATICS

Nopcsa's (1915, 1917, 1923b, 1928c) Thyreophora has recently been reconstituted to include stegosaurs, ankylosaurs, and a small number of basal taxa (Coombs 1972; Norman 1984a; Sereno 1984, 1986; Cooper, 1985; Gauthier 1986; Weishampel, this vol). *Scelidosaurus* has always been recognized as phylogenetically close to stegosaurs and possibly ankylosaurs. The presence of distinctive armor, a single supraorbital bone in the orbital rim, and various details of the skull identify it as the sister group to the stegosaur-ankylosaur clade (together, the Thyreophoroidea of Sereno [1986]). *Scutellosaurus* was originally placed in the Fabrosauridae (Colbert 1981), but Gauthier (1984) and Norman (1984a) considered it to be the most primitive member of the Thyreophora, an assignment accepted by later workers (Sereno 1986; Galton 1986). The most obvious thyreophoran synapomorphy of *Scutellosaurus* is the presence of armor that is very similar to that of *Scelidosaurus* (Owen 1863), basal stegosaurs, and ankylosaurs in that the scutes are dorsally keeled and ventrally excavated (Colbert 1981). The presence of armor probably represents a single derived character, because armor is absent in all outgroups.

Gauthier (1986) suggested that other thyreophoran features of *Scutellosaurus* include the elongation of the trunk relative to the hindlimb and elongation of the ilium relative to the femur. Sereno (1986) also noted the transverse breadth of the orbital bar as a further derived feature. Finally, Colbert (1981) noted that the sacrum is quite broad in *Scutellosaurus:* the sacral centra are wider than long, their transverse processes are broad, and the sacral ribs are broad and long. This broad sacrum may constitute a thyreophoran synapomorphy. A feature that is possibly unique to *S. lawleri* is that the distal ends of the pubis and ischium are pointed and diverge rather than being blunt and parallel (Colbert 1981, fig. 22c).

Tatisaurus oehleri is here assigned provisionally to the Thyreophora. Although this taxon consists of only jaws and teeth, and thus none of the features noted above (with one possible exception) can be assessed, the preserved material lacks the dental apomorphies suggesting cerapodan relationships (see Benton, this vol.; Weishampel, this vol.). *Tatisaurus* was originally referred to the Hypsilphodontidae by Simmons (1965), but one feature, the ventral deflection of the mesial end of the dentary tooth row, is shared with the *Scelidosaurus*-Stegosauria-Ankylosauria clade but not with *Scutellosaurus* or other outgroups. Sereno (1986) apparently referred to the same character when he noted the sinuous curve of the dentary tooth row. *Tatisaurus* is too incompletely preserved to determine whether it or *Scelidosaurus* is closer to the stegosaur-ankylosaur group.

Galton (1978) originally included *Echinodon* (fig. 20.1f) within his Fabrosauridae but later (Galton 1986b) considered it related to *Scutellosaurus* without defending his position. Galton (1986b) referred dermal scutes from the same locality to *Echinodon*; Owen (1861b) previously had referred these scutes to the juvenile megalosaur *Nuthetes*. Thus, *Echinodon* ultimately may prove to be a thyreophoran as well.

PALEOECOLOGY

The fossil record of *Scelidosaurus harrisonii, Scutellosaurus lawleri,* and *Tatisaurus oehleri* is patchy in both time and space but encompasses both Laurasia (United States, People's Republic of China) and Gondwanaland (South Africa) and ranges from the Late Triassic through the Early Cretaceous.

The depositional environment from which the disarticulated remains of *Scutellosaurus* have been found (Kayenta Formation of the United States) has been interpreted as relatively broad floodplain habitats drained by moderate, sediment-rich streams (Clark and Fastovsky 1986). Water appears to have been abundant, although soil formation is known at several horizons. The fauna of the Kayenta is rich in terrestrial vertebrates, including thecodonts, crocodilians, theropod di-

nosaurs, and tritylodontids. The Kayenta Formation also has a freshwater bivalve fauna, indicating well-watered conditions. Unfortunately, floral evidence is lacking from the Kayenta. The Kayenta environment was probably stable, water-rich, warm, and humid.

Scelidosaurus is known from relatively complete, articulated material from the marine Lias of England. Owen (1863) regarded *Scelidosaurus* as terrestrial or at best amphibious along the margins of streams and suggested that carcasses were washed out to sea. We have no problem with such a paleoecological interpretation.

Early thyreophorans were active foragers on shrubby vegetation within one meter of the ground. It is likely that all forms under consideration relied solely on simple adduction of the lower jaws to produce vertical or near-vertical tooth-tooth shearing motion between maxillary and dentary teeth, similar to *Lesothosaurus* (Thulborn 1971a; Weishampel 1984a).

Scutellosaurus appears to have been a bipedal herbivore (Colbert 1981), but the hindlimbs are shorter relative to the length of the preacetabular process than in species regarded as obligate bipeds (Colbert 1981). The notably elongate tail, slender hindlimb, and compact pes with slender metatarsals all suggest cursorial habits, but the lengths of the femur and metatarsal III relative to the tibia are less reflective of cursorial habits than in other bipedal ornithischians (Colbert 1981). The disproportionately long trunk and forelimbs, a moderately large manus, a wide pelvis, and, most important, extensive dermal armor, however, suggest a degree of quadrupedality. *Scutellosaurus* may have been derived from bipedal, cursorial ancestors but had adopted occasional quadrupedality, perhaps for foraging. As suggested by Colbert (1981), the species may be a harbinger of full quadrupedality in ankylosaurs and stegosaurs.

Scelidosaurus was probably predominantly quadrupedal, given the nature of their limb proportions, but the skeleton of these animals retains features of their bipedel ancestry (Thulborn 1977; Colbert 1981). *S. harrisonii* may have been capable of assuming a bipedal or tripodal stance, as has been forcefully advocated for sauropods and stegosaurs (Coombs 1975; Bakker 1986, 1987; Paul 1987b), but whether such behavior had major biological significance is debatable. Following the general pattern of all dinosaurs, the quadrupedal *Scelidosaurus* has fewer of the morphological correlates of cursorial habits than do bipedal dinosaurs (Coombs 1978c). Among quadrupeds, however, it was probably a relatively fleet animal.

21

Stegosauria

PETER M. GALTON

INTRODUCTION

The Stegosauria has been known for over 110 years, for the first 50 years consisting of all armored and quadrupedal taxa from the Jurassic and Cretaceous. However, Romer (1927) recognized the Stegosauria as a mostly Jurassic suborder of the Ornithischia on the basis of the form of the pelvic girdle and hindlimbs. Stegosaurs are medium to large quadrupedal herbivores (length up to approximately 9 m) with proportionally small heads, short and massive forelimbs, long columnar hindlimbs with a very long femur, short metacarpals and metatarsals with hooflike unguals, and an extensive system of parasagittal osteoderms consisting of vertical plates and caudodorsally inclined spines in various proportions plus a shoulder spine (fig. 21.1). The earliest records consist of fragmentary remains from the Lower Bathonian of England (Galton and Powell 1983) plus excellent articulated material including skulls of *Huayangosaurus* (figs. 21.1B, 21.3M-P) from the Bathonian-Callovian of the People's Republic of China (Zhou 1984). Stegosaurs are best represented in the Upper Jurassic with excellent articulated material from Africa (*Kentrosaurus*, fig. 21.1C; Hennig 1924), Asia (*Chungkingosaurus, Tuojiangosaurus*, figs. 21.1E, 21.3E,G.H; Dong et al. 1983), Europe (*Dacentrurus, Lexovisaurus*, fig. 21.1D,F; Galton 1985c), and North America (*Stegosaurus*, figs. 21.1A, 21.2,

21.5, 21.6D,E, 21.7, 21.8; Gilmore 1914a; Ostrom and McIntosh 1966). The most recent record is a partial skeleton from the Upper Cretaceous (Coniacian) of India (Yadagiri and Ayyasami 1979; Maastrichtian stegosaur also reported but not described).

DIAGNOSIS

The Stegosauria can be diagnosed in the following way: large oval fossa or fenestra formed by the quadrate flange of the pterygoid; coronoid eminence of mandible formed into a vertical lamina that is prolonged rostrally, tall neural arches in middle and caudal dorsals; elevated transverse processes in caudal dorsal vertebrae; broad acromial region of the scapula; two very massive compound carpals in adult; broad, short ischium with a prominence of the dorsal margin at the midlength of the shaft; pubis with broad, cup-shaped laterally facing acetabular surface; columnar femur; fourth trochanter represented by a low ridge; digit I absent in the pes, digits III and IV lack at least one or two phalanges, respectively; prominent parasagittal osteoderms with moderately sized vertical plates grading caudally into caudodorsally angled spines, including two pairs of terminal tail spines, shoulder spine ("parasacral") with expanded base; and no ossified axial tendons.

435

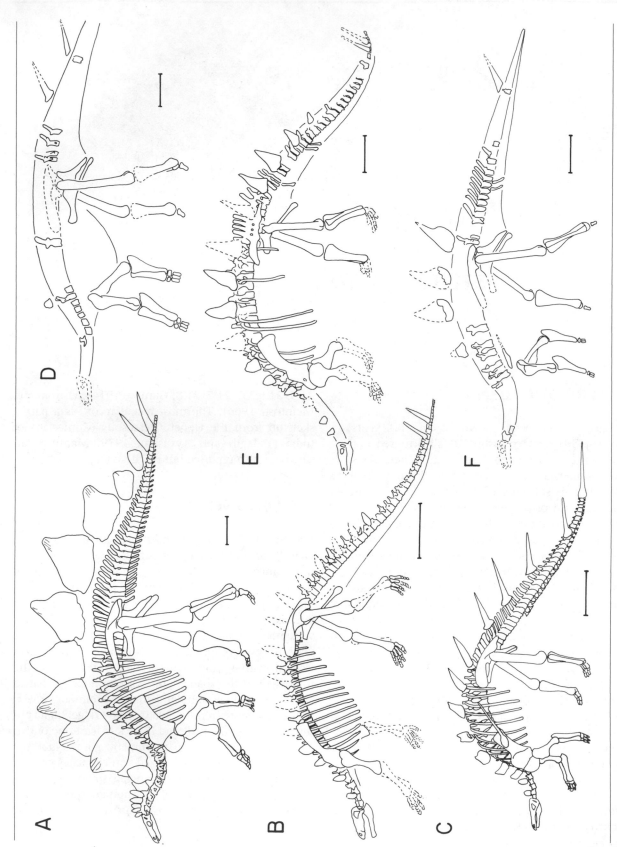

Fig. 21.1. Stegosaur skeletal reconstructions. A, *Stegosaurus* (after Czerkas 1987); B, *Huayangosaurus* (after Zhou 1984); C, *Kentrosaurus* (after Janensch 1924 and Galton 1982b); D, *Dacentrurus* (after Galton 1985c); E, *Tuojiangosaurus* (after Dong et al. 1983); F, *Lexovisaurus* (after Galton 1985c). Scale = 50 cm.

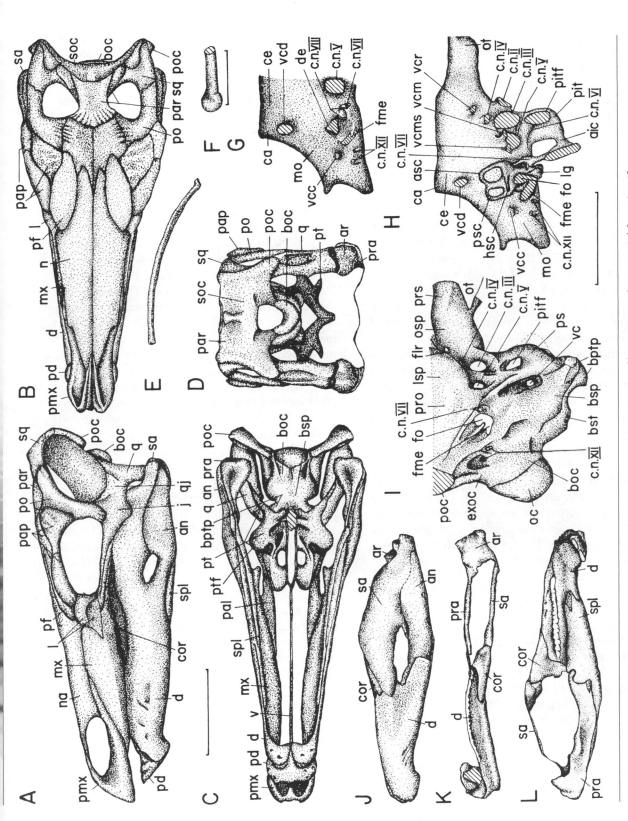

Fig. 21.2. Cranial anatomy of *Stegosaurus*. A–D, reconstruction of skull in A, left lateral; B, dorsal; C, ventral; and D, caudal views (after Galton in prep.); E, first ceratobranchial; F, proximal end of left stapes; G, H, endocranial cast (from I) in right lateral view; G, caudal part with inner ear and adjacent structures removed, and H, complete endocast; I, right ventrolateral view of braincase to show cranial foramina (for complete braincase, see McIntosh 1981, fig. 18); J–L, left mandible in J, lateral; K, dorsal; and L, medial views (after Berman and McIntosh 1985). Scale = 1 cm (F), 5 cm (G–I), 10 cm (A–E, J–L).

ANATOMY

Skull and Mandible

Cranial bones are known for only a few genera, among them *Chialingosaurus* (Dong et al. 1983), *Chungkingosaurus* (fig. 21.3E; Dong et al. 1983), *Dravidosaurus* (fig. 21.3I,J; Yadagiri and Ayyasami (1979), *Huayangosaurus* (figs. 21.3A-d, 21.4J; Dong et al. 1982; Zhou 1984; Sereno 1986), *Kentrosaurus* (figs. 21.3K, 21.4E,F; Hennig 1924, 1936; Janensch 1936; Galton 1988), *Paranthodon* (figs. 21.3F, 21.4G-I; Galton and Coombs 1981), *Stegosaurus* (figs. 21.2, 21.3L, 21.4A-D; Gilmore 1914a; Galton in prep.); and *Tuojiangosaurus* (figs. 21.3G,H, 21.4K,L; Dong et al. 1983).

In *Stegosaurus*, the caudal width is slightly greater than the height, the snout is proportionally long, low and narrow, and the mandible is deep (fig. 21.2A-D). The proportions of the preserved parts of the skulls of *Paranthodon*, *Tuojiangosaurus*, and *Dravidosaurus* are similar to those of *Stegosaurus*. In *Huayangosaurus*, however, the snout is proportionally shorter and higher and tapers more abruptly in dorsal view and the mandible is not as deep. The snout of *Chungkingosaurus* is also proportionally shorter and deeper than that of *Stegosaurus*.

Each premaxilla bears seven teeth in *Huayangosaurus*, but in *Chungkingosaurus*, *Paranthodon*, *Stegosaurus*, and *Tuojiangosaurus*, this bone is endentulous. The external nares are small in *Huayangosaurus* and *Tuojiangosaurus*, larger in *Chungkingosaurus* and *Paranthodon*, and very large in *Stegosaurus*. The caudodorsal process of the premaxilla does not reach the lacrimal, and this process is short and broad in *Huayangosaurus*, long and thin in *Chungkingosaurus*, long and very thin in *Stegosaurus*, and very long and broad in *Paranthodon*. The elongate maxilla has a prominent horizontal ridge for the area of origin of cheek musculature as in other ornithischians (Galton 1973c). This ridge marks the approximate dorsal limit of the thickened alveolar region in *Stegosaurus*, and above this ridge is a thin lateral sheet. In *Huayangosaurus*, this sheet borders a triangular antorbital fenestra (or ?fossa) that is absent in *Stegosaurus*, *Chungkingosaurus* and *Tuojiangosaurus*. In *Stegosaurus* and *Chungkingosaurus*, at least, the fenestra closure is accomplished by a lateral lamina that extends rostroventrally from the lacrimal. The nasals are the longest bone in the skull roof, especially so in *Stegosaurus*. The palpebrals of *Stegosaurus* are sculptured with an irregular series of ridges, grooves, and depressions. Three palpebrals are also present in *Huayangosaurus*

and *Tuojiangosaurus* (Dong in press). Only two palpebrals are figured for *Dravidosaurus*, but the specimen is poorly preserved. The palpebrals exclude the suboval prefrontal and the frontal from the orbital rim. The short and thick frontals are proportionally longer in *Huayangosaurus* and *Tuojiangosaurus* than they are in *Stegosaurus* and *Dravidosaurus*. The medial part of the parietal forms a flat area that is proportionally smaller in *Huayangosaurus* than it is in *Stegosaurus* and *Tuojiangosaurus*. The subtriangular squamosal roofs the caudolateral part of the supratemporal fenestra that is proportionally larger in *Stegosaurus*, *Tuojiangosaurus*, and *Dravidosaurus* than in *Huayangosaurus*. The triradiate postorbital is more slender in *Tuojiangosaurus* and *Dravidosaurus* than it is in *Huayangosaurus* and *Stegosaurus*. The rostral half of the jugal is very shallow in *Huayangosaurus*, *Stegosaurus*, and *Tuojiangosaurus*, so that the height is less than the transverse width in the orbit. However, it deepens caudally and is especially marked in *Stegosaurus*, in which the jugal forms a ventral flange. The sutural relationships of the jugal are unclear in *Stegosaurus* but were probably the same as in *Huayangosaurus*, in which the jugal overlaps a small quadratojugal that in turn overlaps the ventral part of the quadrate. The dorsal head of the quadrate of *Stegosaurus* is thin but wide and has an extensive contact on its caudomedial surface with adjacent parts of the squamosal and paroccipital process. In large individuals, this suture is firmly fused. The shaft of the quadrate is continuous with the large pterygoid flange, and there is no lateral lamina as occurs in ornithopods (Weishampel 1984a). The medial mandibular condyle is larger than the lateral condyle, so the articular surface faces lateroventrally. The quadrates of *Chialingosaurus*, *Huayangosaurus*, and *Tuojiangosaurus* resemble that of *Stegosaurus*, but that of *Kentrosaurus* has a prominent lateral notch that, along with the quadratojugal, formed a prominent paraquadrate foramen.

The palate in *Huayangosaurus* and *Stegosaurus* is divided into two arched, ventrally open passageways by a deep median keel formed by the vomers, palatines, and pterygoids. The transversely thin vomers extend for less than half of the length of the palate in *Huayangosaurus*. In *Stegosaurus*, the vomers are proportionally much longer and extend to the caudal end of the tooth row so that the pterygoids and most of the palatines are caudomedial to the teeth. The palatine and pterygoid appear to border a foramen in *Stegosaurus* but not in *Huayangosaurus*. The interpterygoid vacuity is very small because of the close proximity of the pterygoids. The complex pterygoid has rami diverging from the

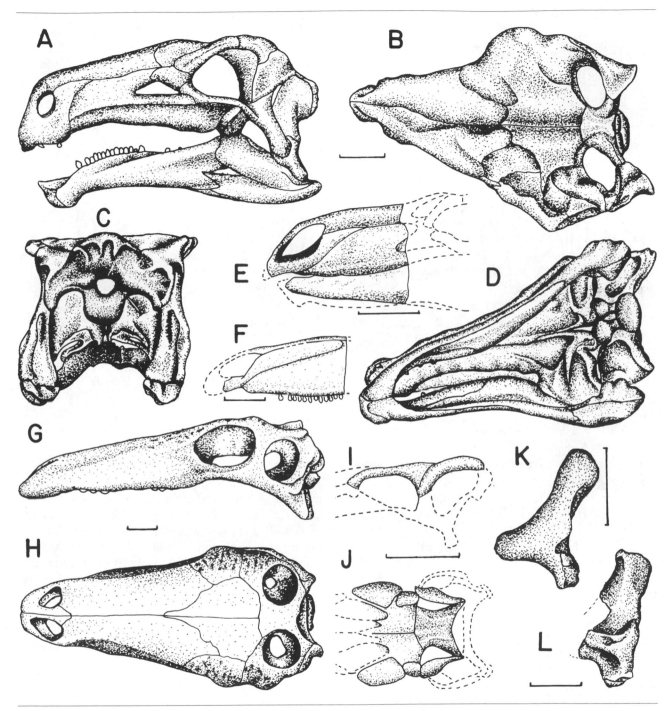

Fig. 21.3. Stegosaur skull material in left lateral (A,E−G,I), dorsal (B,J), ventral (D), and caudal views (C): A–D, *Huayangosaurus* (after Zhou 1984); E, *Chungkingosaurus* (after Dong et al. 1983); F, *Paranthodon* (after Galton 1981*a*); G,H, *Tuojiangosaurus* (after Dong et al. 1983); I,J, *Dravidosaurus* (after Yadagiri and Ayyasami 1979); K,L, left quadrates in lateral view; K, *Kentrosaurus* (after Galton 1988); L, *Stegosaurus* (after Gilmore 1914*a*). Scale = 5 cm.

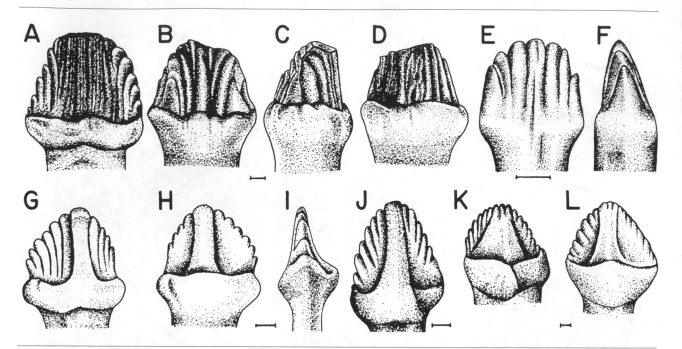

Fig. 21.4. Teeth of stegosaurian dinosaurs. A—D, *Stegosaurus* A, right maxillary tooth in lingual view; B—D, isolated cheek tooth in three views; E, F, isolated cheek tooth of *Kentrosaurus* (after Galton in press); G—I, *Paranthodon* G, H, left maxillary tooth in G, lingual, and H, buccal views (after Galton and Coombs 1981); I, isolated tooth in mesial view (after Galton and Coombs 1981); J, *Huayangosaurus* cheek tooth in buccal view (after Zhou 1984); K,L, *Tuojiangosaurus* K, right dentary tooth in buccal view; L, left maxillary tooth in buccal view. Scale = 1 mm.

body. The caudal part of the pterygoid body overlaps the rostrodorsally facing articular surface of the basipterygoid processes. The pterygoid flange projects ventrally, while the quadrate ramus is roundly bifurcate (Sereno 1986).

Several braincases and endocranial casts are known for *Stegosaurus* and *Kentrosaurus* (Galton 1988, in prep.), but the braincase is less known in *Huayangosaurus* and *Tuojiangosaurus*. The large occipital condyle is formed mostly from the basioccipital with a dorsolateral contribution from the exoccipitals (fig. 21.2D,I). The basipterygoid processes are short, directed rostroventrally with a rostrodorsally facing surface for the pterygoids. The parasphenoid is not known. The thin orbitosphenoids meet at the midline, forming the floor of the olfactory tracts (fig. 21.2H,I), are semicircular in cross section, and are slightly constricted at midlength. The laterosphenoids are short and thick, especially rostrally. The prootic is an irregular bone with an extensive ventral suture with the basisphenoid so rostrally the prootics form up to 75 percent of the width of the floor of the braincase. The exoccipitals form the caudal part of the sidewall of the braincase, and the ventral part adjacent to the foramen magnum is indented (fig.

21.2I). The paroccipital processes are deep and thick and presumably formed from the opisthotic. The supraoccipital is inclined at an angle of about 55°.

An endocranial cast of *Stegosaurus* was the first to be described for any dinosaur (Marsh 1880), but it is incomplete. A second endocast (Gilmore 1914*a*; Ostrom and McIntosh 1966), complete but badly distorted and with few anatomical details, misled Hopson (1979) in his reinterpretation of several structures of the stegosaur brain (Galton in prep.). From an undistorted braincase (fig. 21.2H), it is clear that the brain (fig. 21.2G,H) is relatively short and deep with strong cerebral and pontine flexures and a steeply inclined caudodorsal edge compared to those of ornithischian dinosaurs (see Hopson 1979). The widest part of the brain is formed by the slightly expanded cerebral hemispheres of the telencephalon. These hemispheres taper rostrally to short olfactory tracts that widen again into small olfactory bulbs. The sidewalls of the sella turcica are ossified except for a small pituitary foramen (fig. 21.2H,I). The extent of the mesencephalon is uncertain because the optic lobes are not differentiated and the points of origin of C. nn. III and IV cannot be determined. There is no cerebellar expansion in the dorsal

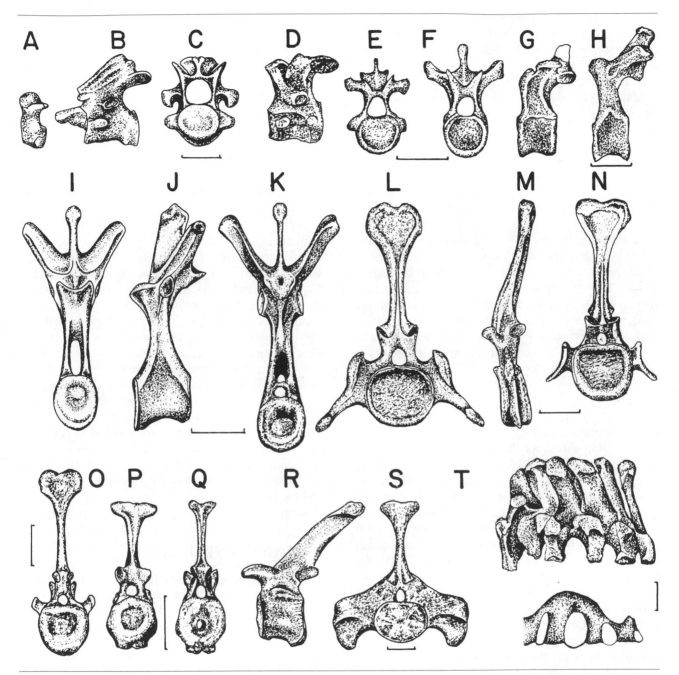

Fig. 21.5. Vertebrae of *Stegosaurus* (after Ostrom and McIntosh 1966), in left lateral (A,B,D,G,H,J,M,R,T), cranial (C,E,F,I,N−Q), and caudal views (K,L,S): A, atlas; B, axis; C,D, cranial cervical; E, middle cervical; F,G, caudal cervical; H−J, cranial dorsals; K, middorsal; L,M, caudal 3; N, caudal 5; O, proximal caudal; P, middle caudal; Q,R, caudal slightly more distal than P; S, last sacral vertebra with sacral ribs; T, neural arches of sacrum of juvenile individual with endosacral cast. Scale = 5 cm (A−D), 10 cm (remaining fig.).

region of the metencephalon. The flocullar lobes of the cerebellum are interpreted from a slight concavity, the fossa subarcuata, in the medial wall of the prootic and supraoccipital. C. n. V originates from this region of the brain stem, the caudal part of which has transversely constricted sidewalls to accommodate the inner ears. With the adjacent part of the myelencephalic walls, this area is the narrowest part of the brain. C. nn. VI to XII originate from this part of the myelencephalon, the widest part of which is slightly caudal to the vena cerebralis caudalis. The inner ear is well preserved (fig. 21.2H). The anterior semicircular canal is the longest, the lateral canal is the shortest, the lagena is moderately well developed, and the fenestra ovalis was probably filled with cartilage in life so its diameter more closely matches the size of the footplate of the stapes (fig. 21.2F).

The single predentary of *Stegosaurus* is wider than the combined width of the premaxillae, but it does not extend as far forward as the latter. The rostral surface of the predentary is rather flat. In life, the tips of the premaxillae and predentary were probably covered by a horny rhamphotheca. The rostral part of the dentary of *Huayangosaurus* (fig. 21.3A) is similar to that of other ornithischians in that the tooth row is emarginated from the side of the face. There is a prominent lateral ridge for the attachment of muscular cheeks (Galton 1973c). However, this ridge becomes a vertical lamina that increases in depth caudally, so the rear teeth are hidden from view. The ridge merges with the rostral edge of the deep coronoid eminence. Passing caudally, the surface between the tooth row and the base of the vertical lamina becomes progressively more horizontal. In *Chungkingosaurus, Kentrosaurus,* and *Stegosaurus,* that portion of the dentary lateral to the tooth row is horizontal and the teeth are hidden by a very extensive vertical lamina (fig. 21.2A,J-L). The dentary is deep but very thin except for the thick alveolar region. The coronoid forms the base of the coronoid eminence rostromedially and extends onto its lateral surface in *Stegosaurus* but not in *Huayangosaurus.* The coronoid eminence is proportionally higher in *Stegosaurus* than it is in *Huayangosaurus.* In both genera, this eminence is formed mostly by the large surangular. The surangular, angular, and dentary surround an external mandibular fenestra that is proportionally larger in juveniles (fig. 21.2A,J; Berman and McIntosh 1985). A smaller fenestra is present in *Huayangosaurus* (fig. 21.3A), and Dong et al. (1977) mention that it may also be present in *Tuojiangosaurus.* The articular is a broad, thick bone overlapped laterally by the surangular and medially by the prearticular (Galton 1982a). The retroarticular process is shorter in *Stegosaurus* than it is in *Huayangosaurus.*

There are seven premaxillary teeth in *Huayangosaurus* (Sereno 1986). These teeth have denticulate margins and a weakly developed cingulum (Dong in press). In all other stegosaurs, the premaxilla is edentulous. In *Stegosaurus,* the maxilla bears about 25 teeth, whereas the dentary count ranges from 20 to 23. In *Huayangosaurus* and *Tuojiangosaurus,* the counts are 27 for both upper and lower jaws (Dong in press), whereas the counts for *Kentrosaurus* (Hennig 1936) and *Chi-*

alingosaurus (Dong in press) appear to be lower than in *Stegosaurus.* The cheek teeth of stegosaurs are enameled on both sides. The cheek teeth of *Huayangosaurus, Tuojiangosaurus, Stegosaurus,* and *Paranthodon* have the usual *en echelon* arrangement found in all other ornithischians. Dong et al. (1983) note that the teeth of *Chialingosaurus* and *Chungkingosaurus* do not overlap. However, the three functional teeth preserved in the dentary of *Chialingosaurus* are separated from each other by empty alveoli, and the tooth rows of *Chungkingosaurus* are not illustrated. The crowns of stegosaurian cheek teeth are asymmetrical (fig. 21.4A-L). The convex mesial outline bears fewer and/or larger marginal denticles, whereas the distal outline is slightly concave with smaller denticles. The simplest crown occurs in *Kentrosaurus.* Here, there are only seven denticles and a weak cingulum. In *Huayangosaurus, Tuojiangosaurus, Stegosaurus,* and *Paranthodon,* there are more denticles and the cingulum is more prominent. The denticles continue onto the adjacent surface of the crown to a varying extent, but only in *Stegosaurus* is there a complex network of secondary ridges.

The stapes is a slender rod that is preserved in situ in *Huayangosaurus* (fig. 21.3C), and the proximal end with an expanded footplate is preserved adjacent to the fenestra ovalis in *Stegosaurus* (fig. 21.2F).

The hyoid apparatus is represented only by the first ceratobrachials, each of which is a slender curved rod, with slightly expanded ends (fig. 21.2E), that is just under half the length of the lower jaw (see Gilmore 1914a, figs. 8, 9).

To date, sclerotic rings have not been described.

Postcranial Skeleton

The postcranial skeleton of stegosaurs is better known than the skull, particularly in *Chialingosaurus* (Young 1957; Dong et al. 1983), *Chungkingosaurus* (fig. 21.6H,I; Dong et al. 1983), *Dacentrurus* (figs. 21.1D, 21.6G; Owen 1875, 1877b; Galton 1985c), *Dravidosaurus* (Yadagiri and Ayyasami 1979), *Huayangosaurus* (figs. 21.1B, 21.6A,F; Zhou 1984), *Kentrosaurus* (figs. 21.1C, 21.6B,G; Hennig 1924; Janensch 1924; Galton 1982b), *Lexovisaurus* (figs. 21.1F, 21.6C; Nopcsa 1911a; Galton 1985c; Galton et al. 1980), *Stegosaurus* (figs. 21.1A, 21.5, 21.6D,E, 21.7, 21.8; Gilmore 1914a; Ostrom and McIntosh 1966), *Tuojiangosaurus* (figs. 21.1E, 21.6L; Dong et al. 1983,) and *Wuerhosaurus* (fig. 21.6K; Dong 1973a). Illustrations are given of the vertebrae (figs. 21.1, 21.5, 21.6E-L), pectoral girdle and forelimb (figs. 21.1, 21.7), and pelvic girdle and hindlimb (figs. 21.1, 21.6, 21.8).

Axial Skeleton

The vertebral formula for *Stegosaurus* is 26 presacrals (presumed to represent 10 cervicals and 16 dorsals), 5 sacrals, and about 45 dorsals (figs. 21.1A, 21.5). In *Huayangosaurus*, it is presumed to be 8 cervicals, 17 or 18 dorsals, 4 sacrals, and 35 to 42 caudals. The neural arches of the atlas are fused to the intercentrum in *Huayangosaurus, Tuojiangosaurus, Kentrosaurus,* and *Stegosaurus*. The odontoid process is ankylosed to the centrum of the axis in *Huayangosaurus, Kentrosaurus, Lexovisaurus,* and *Stegosaurus*. Cranially, both ends of the centra are flat, then strongly excavated, especially caudally, and those are amphicoelous. In addition, the large, ventrally situated parapophyses become smaller and more dorsal in position, while the transverse processes become longer, change from a ventrolateral to a dorsolateral orientation, and originate more dorsally on the neural arch. The neural spines increase in size and become more caudal in position. The neural canal increases in size from cervical 8 to dorsal 3 to form an enlargement for the brachial nerve complex (Lull 1917).

The dorsal vertebrae of *Stegosaurus* and of most other stegosaurs are characterized by a very high neural arch, the more ventral part of which forms a solid pediclelike region above the neural canal. The marked upthrust of the transverse processes in middle and caudal dorsals is as much as 60° above the horizontal, almost equaling the narrow neural spines in height. The transverse processes unite craniomedially to form a vertical surface immediately caudal to the prezygapophyses (fig. 21.4I,J). The zygapophyses also unite ventromedially to give V-shaped prezygapophyses and triangular postzygapophyses (fig. 21.4E,G). In dorsals of *Dacentrurus*, the pedicle is low (as in *Huayangosaurus*), the transverse processes reach a maximum of 45° to the horizontal, and there is a prominent notch ventrally between the postzygapophyses (also true for *Huayangosaurus* and for cranial dorsals of *Lexovisaurus*). The centra are massive with the transverse width greater than length for the last two-thirds of the series (reverse of usual condition), and there is a prominent lateral depression. In *Kentrosaurus*, the neural canal extends into most of the pedicel region, the dorsal part of which is occupied only by a thin transverse lamina. In *Craterosaurus*, there is a very deep depression immediately caudal to the prezygapophyses (Nopcsa 1912; Galton 1981*a*).

Except for the atlas, the ribs on all the presacral vertebrae are double-headed. Cervical ribs 2 to 10 are triradiate, and, passing caudally, the shaft lengthens, as do the articular processes, especially the tubercular. Dorsal ribs 1 to 6 show a marked increase in length, ribs 7 to 11 are subequal in length, and ribs 12 to 15 show a rapid shortening and become more slender. The tubercular process is short, whereas the capitular process is long and perpendicular to the shaft to give a deep and narrow rib cage. The cross-section of the proximal half of the shaft of ribs 6 to 11 is T-shaped, with the flat surface on the outside. In *Huayangosaurus*, the cranial dorsal ribs bear a prominent hamularis process.

The sacrum of stegosaurs consists of five or six coossified centra (including one or two dorsosacrals and one caudosacral). However, in *Stegosaurus stenops*, the first sacral bears a free rib (see Gilmore 1914*a*, fig. 23), and the neural spine is not coalesced with those of the remaining sacrals. All stegosaur sacra have four pairs of robust sacral ribs that are borne by sacrals 2 to 5. In some individuals of *Chungkingosaurus, Kentrosaurus, Dacentrurus,* and *Stegosaurus*, sacral 1 may bear an additional pair of slender sacral ribs (fig. 21.6G,I; Galton 1982*a*). There is dimorphism in the number of sacral ribs in *Kentrosaurus aethiopicus* (Galton 1982*b*). Similar dimorphism is probably present in *Dacentrurus armatus* (fig. 21.6G; Lapparent and Zbyszewski 1957, pl. 34), *Chungkingosaurus jiangbeiensis* (fig. 21.5I; Dong et al. 1983, fig. 97), and *Stegosaurus stenops* (Gilmore 1914*a*; Ostrom and McIntosh 1966). In juveniles of *Stegosaurus*, the sacral ribs unite distally. However, there is a large oval sacral foramen between successive transverse processes (fig. 21.5T). Such large foramina are also present in adults of *Huayangosaurus* and *Chungkingosaurus* (fig. 21.6F,H.I). In other stegosaurs, however, the transverse processes and the adjacent parts of the sacral ribs form an almost solid dorsal plate extending from the base of the neural spines to the flat dorsal surface of the ilium with which it is continuous. Caudally, the centrum of the last sacral is markedly concave transversely. Endosacral casts show that there is an extremely large sacral enlargement, especially in the neural arches of sacrals 2 and 3 (fig. 21.5T; Hennig 1924). This enlargement is for the nerve plexus for the very large hindlimb and tail (as suggested by Lull 1910*b*, 1917, there is no "sacral brain").

The proximal caudals of stegosaurs have small neural canals, axially short and transversely circular centra, and prominent transverse processes. These processes are laterally and/or slightly ventrally directed in *Huayangosaurus, Lexovisaurus, Chialingosaurus, Dacentrurus, Kentrosaurus,* and *Wuerhosaurus* and ventrolaterally directed in *Chungkingosaurus* and *Stegosaurus* (fig. 21.5L,M). Each transverse process may bear

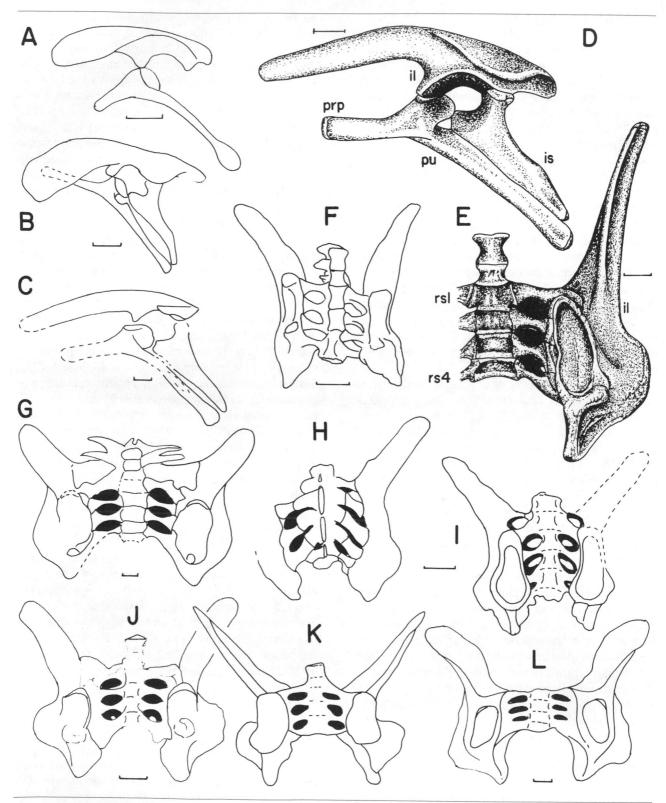

Fig. 21.6. Left pelvic girdles (A–D) and sacra with ilia (E–L) in lateral (A–D), dorsal (H), and ventral views (E–G, I–L) of huayangosaurids (A,F) and stegosaurids (B–E, G–L): A,F, *Huayangosaurus* (after Zhou 1984); B,J, *Kentrosaurus* (after Hennig 1924); C, *Lexovisaurus* (after Galton 1985c); D,E, *Stegosaurus* (after Gilmore 1914a; Ostrom and McIntosh 1966); G, *Dacentrurus* (after Galton 1985c); H,I, *Chungkingosaurus* (after Dong et al. 1983); K, *Wuerhosaurus* (after Dong 1973a); L, *Tuojiangosaurus* (after Dong et al. 1983). Scale = 10 cm.

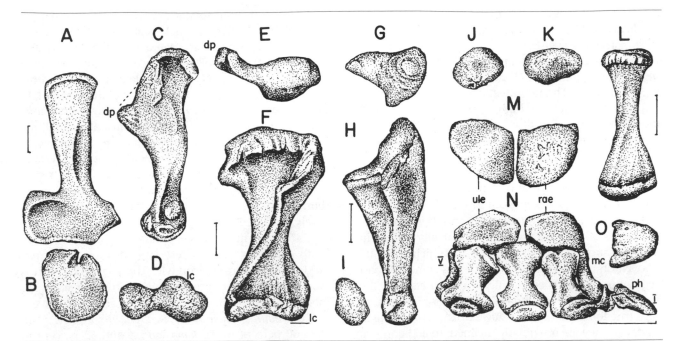

Fig. 21.7. Pectoral girdle and forelimb of *Stegosaurus*, all left, except M–O (after Gilmore 1914a; Ostrom and McIntosh 1966). A, scapula in lateral view; B, coracoid in lateral view; C–F, humerus in C, lateral, D, distal, E, proximal, and F, cranial views; G–I, ulna in G, proximal, H, lateral, and I, distal views; J–L, radius in J, proximal, K, distal, L, cranial views; M, proximal carpals in proximal view; N, articulated manus and proximal carpals in cranial view; O, manual ungual of digit I in cranial view. Scale = 10 cm.

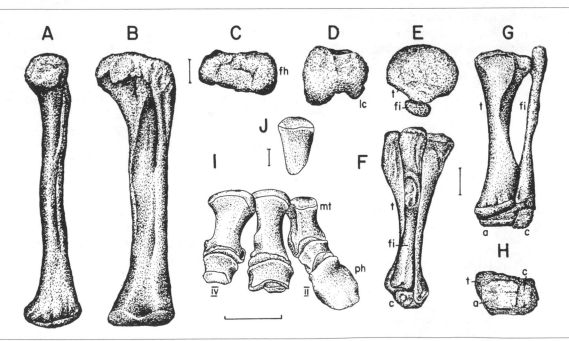

Fig. 21.8. Hindlimb anatomy of *Stegosaurus*, all left except I,J (after Gilmore 1914a, Ostrom and McIntosh 1966). A–D, femur in A, medial, B, cranial, C, proximal, and D, distal views; E–H, tibia, fibula, astragalus, and calcaneum in E, proximal, F, lateral, G, cranial, and H, distal views; I, articulated pes in cranial view, distal views; J, metatarsal V in cranial view. Scale = 1 cm (J), 10 cm (remaining fig.).

a proximodorsal projection that is small in *Dacentrurus*, *Kentrosaurus*, and *Wuerhosaurus* and large in *Stegosaurus*. The transverse processes decrease in size as they pass distally and disappear at about caudal 16 in *Stegosaurus*, caudal 11 in *Lexovisaurus*, and caudal 27 in *Kentrosaurus*. The neural spines of the most proximal caudals are low in *Huayangosaurus*, *Lexovisaurus*, and *Dacentrurus*, of medium height in *Chialingosaurus*, *Chungkingosaurus*, *Kentrosaurus*, and *Stegosaurus*, and extremely tall in *Wuerhosaurus*. The apices are simple in *Huayangosaurus*, *Chungkingosaurus*, and *Chialingosaurus*, moderately expanded transversely in *Lexovisaurus*, *Dacentrurus*, *Tuojiangosaurus*, and *Kentrosaurus*, and very expanded in *Stegosaurus* and *Wuerhosaurus*. The apices of the neural spines are dorsally convex in all stegosaurs except *Stegosaurus*, in which they are grooved on caudals 1 to 25. The neural spines of most stegosaurs are inclined distally. In *Kentrosaurus*, the neural spines distal to caudal 8 are more or less vertical (also true for caudal 30 of *Lexovisaurus*), while distal to caudal 18 they are proximally inclined. In addition, the neural arches of all caudals of *Kentrosaurus* lean proximally with respect to the long axis of the centrum, rather than normal to it as seen in other stegosaurs. The proximal chevron facet of stegosaurs is usually much smaller than the distal chevron facet and well separated from it (fig. 21.5R). In *Lexovisaurus*, however, the proximal facet is large on the proximal caudals, and both large facets on caudals 8 to 11 unite ventrally to give a V-shaped outline to the centrum in lateral view. The chevrons of most stegosaurs are slightly shorter than the corresponding neural spines and taper distally. However, those of *Kentrosaurus* are short and plough-shaped in lateral view. In *Stegosaurus*, the large hemal canal is either open or closed in the same individual.

Although there are quite a few articulated skeletons of stegosaurs, there are no traces of any ossified tendons. Given the ubiquitous occurrence of ossified tendons in all other groups of ornithischian dinosaurs, it is reasonable to assume that their absence in stegosaurs represents a secondary loss (Sereno 1984).

Appendicular Skeleton

In stegosaurs, the blade of the scapula gradually increases in width passing dorsally from its junction with the broad ventral region. In *Huayangosaurus*, the cranial edge of the blade lacks a well-defined acromial process but is set at an angle of about 150° to the small acromial region. In *Tuojiangosaurus*, *Kentrosaurus*, *Wuerhosaurus*, and *Stegosaurus*, the acromial process is prominent, the acromial region is progressively larger, and the blade-acromial angle is about 120°, 100°, 90°, and 70°, respectively. In *Huayangosaurus*, the greatest craniocaudal width of the coracoid is slightly less than the greatest width of the scapula and slightly more than the greatest dorsoventral height of the coracoid. In *Tuojiangosaurus*, *Kentrosaurus*, *Wuerhosaurus*, and *Stegosaurus*, the width of the coracoid is just above half the width of the scapula and less than the height of the coracoid (fig. 21.7A,B). Clavicles and interclavicles are unknown in stegosaurs. The only evidence for a sternum is the small triangular and asymmetrical paired bones in *Stegosaurus* which Gilmore (1914*a*) tentatively identified as sternal plates.

The humerus of stegosaurs is short but massive with expanded ends, a very large deltopectoral crest, and a large lateral epicondyle. The proximal width is about 36 percent of humeral length in *Chialingosaurus*, about 48 percent in *Huayangosaurus*, about 55 percent in *Lexovisaurus*, *Dacentrurus*, *Kentrosaurus*, and *Stegosaurus*, 60 percent in *Tuojiangosaurus*, and 62 percent in *Chungkingosaurus*. The apex of the deltopectoral crest occurs about 22 percent of humeral length (measured from the proximal humeral head) in *Huayangosaurus* and *Chungkingosaurus*, 29 percent in *Tuojiangosaurus*, and about 37 percent in *Lexovisaurus*, *Dacentrurus*, *Chialingosaurus*, *Kentrosaurus*, and *Stegosaurus*. However, in some individuals of *Stegosaurus*, this crest may reach midlength. The distal humeral width is about 32 percent of humeral length in *Tuojiangosaurus*, about 37 percent in *Lexovisaurus*, *Dacentrurus*, and *Chialingosaurus*, about 44 percent in *Kentrosaurus*, *Stegosaurus*, and *Wuerhosaurus*, and 48 percent in *Chungkingosaurus*. The radius is 66 percent of the humeral length in *Huayangosaurus*, whereas in *Kentrosaurus* and *Stegosaurus*, this ratio is about 70 percent and 73 percent, respectively. In *Huayangosaurus*, the radius is slender and the proximal end is wider than the distal end. However, the reverse is true for the more massive radii of *Dacentrurus*, *Kentrosaurus*, and *Stegosaurus*. The ulna of *Huayangosaurus* is slender, whereas that of *Lexovisaurus* and *Dacentrurus* is more expanded proximally. The ulna of *Stegosaurus* is extremely massive, especially proximally. Hence the length of the antebrachium exceeds that of the humerus, whereas the reverse is true in juveniles in which the olecranon process is less prominent.

In the depth of the carpus, the shortness of the metacarpals, and the arch of the manus, there is a striking resemblance between *Stegosaurus* and elephants (fig. 21.7N). In young *Stegosaurus*, there are four proximal carpals (see Gilmore 1914*a*, figs. 36, 37; Galton

1982a; fig. 21.7D,E), but in adults there are only two large, blocklike bones. The metacarpus consists of five short, robust metacarpals. The first phalanx of digits I and II are short and wide, with well-formed articular surfaces. Each of these digits bears a large, wide, and depressed hooflike ungual (fig. 21.7N,O; ungual II not preserved) that in life was encased by a horny hoof. The first phalanges of digits III and IV have small distal articular surfaces, each of which articulated with a reduced terminal cartilaginous ungual. Judging from the distal end of metacarpal V, at least one phalanx was present, indicating that the phalangeal formula for *Stegosaurus* is probably 2–2–2–2–?1. The few manual elements known for other stegosaurs are similar to those of *Stegosaurus*.

The preacetabular process of the dorsoventrally compressed ilium of stegosaurs is long and faces slightly dorsally, whereas the body forms a flat surface that faces mostly dorsally (fig. 21.6). It is the iliac body that fits against and is continuous with the transverse processes or the dorsal plate of the sacrum (fig. 21.6H). The preacetabular process is set at an angle of about 35° to the midline in *Huayangosaurus*. In other stegosaurs including *Stegosaurus* (see Galton 1982a), the angle is between 40° and 50°. This process is broad and tapers cranially in *Huayangosaurus*, *Chungkingosaurus*, and *Tuojiangosaurus*, broad and widens cranially in *Kentrosaurus*, and long and slender in *Lexovisaurus*, *Stegosaurus*, and *Wuerhosaurus*. The pubic peduncle is prominent in all stegosaurs. The ischial peduncle is pronounced in *Huayangosaurus* and projects ventrally. In all other stegosaurs, the ischial peduncle is small, projects very little, and lacks a supraacetabular buttress. The large antitrochanter overhangs the caudal half of the acetabulum and the adjacent part of the very short postacetabular process (fig. 21.6). This antitrochanter is small in *Huayangosaurus*, slightly larger in *Chungkingosaurus*, and very large in *Lexovisaurus*, *Dacentrurus*, *Kentrosaurus*, *Tuojiangosaurus*, *Stegosaurus*, and *Wuerhosaurus*. The open acetabulum is partly backed by thin sheets of bone from the ilium, pubis, and ischium.

The pubis of *Huayangosaurus* has a stout prepubic process (short as preserved but incomplete cranially, see Zhou 1984), a robust body, and a slender, distally expanded pubic rod (fig. 21.6A). The prepubic process is long in *Dacentrurus*, *Kentrosaurus*, *Stegosaurus*, and *Wuerhosaurus* and deeper and thinner in *Dacentrurus*, *Stegosaurus*, and *Wuerhosaurus*, and the proximal end is expanded dorsoventrally in *Dacentrurus* (fig. 21.6B-D). The distal expansion of the pubis is much less prominent in *Lexovisaurus*, *Dacentrurus*, *Kentrosaurus*, *Stego-*

saurus, and *Wuerhosaurus*. Like the pubes, the ischia contact each other only at their distal end. However, a very rugose area medially on the angled distal third of the bone (fig. 21.6D) indicates that strong ligaments spanned the symphysis. The ischium lacks an obturator process, and the distal part is unexpanded. In *Stegosaurus stenops*, the entire length of the ischium contacts the pubis ventrally (fig. 21.6D), but in *S. ungulatus*, there is a gap between these bones due to their mutual angulation; the same is true for *Dacentrurus armatus* and *Kentrosaurus aethiopicus*. The distal part of the ischium is straight and tapering in *Dacentrurus* and *Chungkingosaurus* and expanded in *Wuerhosaurus*.

In *Huayangosaurus*, the femur is only slightly longer that the humerus (115%) and the tibia (110%). The femur to humerus ratio is 147 percent in *Dacentrurus*, about 163 percent in *Chialingosaurus*, *Chungkingosaurus*, and *Kentrosaurus*, 173 percent in *Lexovisaurus*, and 180 to 230 percent in *Stegosaurus*. The femur to tibia ratio is 126 percent in *Chungkingosaurus*, about 142 percent in *Kentrosaurus*, 150 percent in *Lexovisaurus*, and 168 to 185 percent in *Stegosaurus*. The femur of stegosaurs is slender in lateral view with a very straight shaft of nearly uniform width. However, it is transversely wide. There is no constricted neck region producing a distant femoral head. In juvenile individuals, the lesser trochanter is a slender, vertical, fingerlike process, but in larger individuals it may be obliterated by proximal remodeling, particularly in *Kentrosaurus* in which the proximal end becomes extremely massive laterally. In *Chialingosaurus*, the triangular lesser trochanter has a broad base. The vestigial fourth trochanter is represented by a distinct ridge in *Lexovisaurus*, *Chialingosaurus*, *Tuojiangosaurus*, *Dacentrurus*, and *Kentrosaurus*. However, in *Chungkingosaurus* and *Stegosaurus*, the fourth trochanter is represented only by rugosities on the shaft at about midlength. The stegosaur tibia is short but massive, and, in large individuals, the distal end is fused with the fibula, astragalus, and calcaneum (fig. 21.8F-H). In small individuals of *Stegosaurus* and in large individuals of *Dacentrurus*, the distal end of the tibia is deeply excavated to receive the large proximal projection of the astragalus (see Gilmore 1914a, figs. 48, 49). The fibula is slender and only slightly expanded at both ends. The astragalus is a massive bone that caps the tibia. The calcaneum is a small bone that fits against the adjacent parts of the astragalus, tibia, and fibula. In *Stegosaurus*, the shape and relative sizes of the astragalus and calcaneum are exceedingly variable.

Distal tarsals are unknown and were presumably cartilaginous. The three robust metatarsals of the stegosaur pes are identified as II to IV by comparison with tridactyl ornithopods, in which digit loss occurs on both sides of the pes, rather than as metatarsals I to III as identified by Gilmore (1914a), who noted the resemblance of the innermost element to metatarsal I of sauropods. A vestigial metatarsal V is preserved proximally on the caudal surface of metatarsal IV in one specimen of *Stegosaurus*. Metatarsals II to IV are straight but divergent with only a small contact proximally. The metatarsals of *Lexovisaurus* and *Dacentrurus* are similar to those of *Stegosaurus*, but those of *Kentrosaurus* are more slender and even more so in *Huayangosaurus*. The proximal phalanges are short, broad, and relatively much heavier than those of the manus. The unguals are broad and depressed and in life were encased by a horny hoof. Articulated pedes of *Huayangosaurus*, *Kentrosaurus*, and *Stegosaurus* have a phalangeal formula of 0−2−3−3−0 (Dong in press), 0−2−2−2−0, and 0−2−2−2−0, respectively.

Osteoderms

The osteoderms are developed as a double series of small plates and spines in *Huayangosaurus*, *Chialingosaurus*, *Chungkingosaurus*, *Tuojiangosaurus*, and *Kentrosaurus* (fig. 21.1B,C,E). The best-known example of this configuration is *Kentrosaurus*, which is presumed to have carried 15 paired plates and spines on its back, but only the position of the terminal pair of tail spines is fixed. The neck and cranial half of the back carries six pairs of small vertical plates; the fourth pair is very thin and subrhomboidal in shape, the fifth resembles a nuchal plate of *Stegosaurus*, and the sixth is a thin oval plate with most of the outline of a spine and its base on the lateral surface. The rest of the back supports three pairs of flat spines. Last, the tail carries five pairs of large spines, three of which are stocky, steeply inclined, and possess large oval bases. The remaining caudal elements are long, horizontal spines that are supported by smaller, subcircular bases. A pair of horizontal spines, each with an extremely enlarged, medially positioned base, has long been placed on the flat dorsal surface of the ilia and the adjacent dorsal plate of the sacrum. However, these "parasacral" spines occur close to the scapula in *Huayangosaurus* (Dong in press), their site of preservation in a complete articulated stegosaur skeleton from the Upper Jurassic of Sichuan, People's Republic of China (Gao et al. 1986). Nopcsa (1911a) described a similar spine with a very small base in *Lexovisaurus* from England as a shoulder spine, and a much longer shoulder spine with a very large base is known from France (Galton et al. 1980). The paired

osteoderms of *Huayangosaurus* include small nuchal plates similar to those of *Stegosaurus*, larger subrectangular plates in the cranial dorsal region, and small spines on the rest of the back, large spines on the sacral region, and a series of small spines on the tail that has two pairs of elongate terminal spines (fig. 21.1B). In *Chungkingosaurus*, the osteoderms are large and thick, with a shape intermediate between plates and spines, and there were probably four pairs of tail spines (Dong in press). *Tuojiangosaurus* is reconstructed with 17 pairs of plates and spines, with spines predominating (fig. 21.1E). The plates on the neck are irregular to spherical, while those of the back are triangular and swordlike with a widened discoid base. The spines of the sacrum and tail are short and conical with the largest pair in the sacral region. There are two pairs of elongate terminal tail spines. Proportionally large, thin plates occur in *Lexovisaurus* (fig. 21.1D), while in *Dravidosaurus*, there are ten triangular-shaped plates with thick bases (height 50-250 mm, length 30-150 mm; Yadagiri and Ayyasami 1979; Anon. 1978). The bony plates are large, long, and rather low in *Wuerhosaurus*. However, large thin plates reach their extreme development in *Stegosaurus* (fig. 21.1A).

The osteoderms of *Stegosaurus stenops* consist of small, depressed, and subangular ossicles embedded in the throat region and scattered over the skin, a series of 17 erect and thin plates of varying sizes along the dorsum of the neck, back, and tail, and two pairs of elongated, spikelike tail spines. These dermal plates have been reconstructed in several different ways, first as a single median row by Marsh (1891c, 1896; Ostrom and McIntosh 1966) and as two rows that were either bilaterally paired throughout the entire series (Lull 1910a, 1910b) or two rows of staggered alternates (Gilmore 1914a, 1915b, 1918). Gilmore (1914a) favored the alternating arrangement because some of the plates are arranged this way in an almost complete skeleton still embedded in rock, and, in addition, no two plates have exactly the same shape or size. However, Czerkas (1987) makes a case that the 17 plates of the *Stegosaurus stenops* represent the complete series and that there is sufficient room for their bases to be arranged in a single median row (fig. 21.1A). However, plates 6 to 13 show a dorsal overlap because the maximum craniocaudal length of each plate is greater than the length of the corresponding base.

The five nuchal plates are small, thin, and vertically elongate with roughened and only slightly transversely expanded basal ends. Only the last nuchal plate has a base that is asymmetrically expanded. The depth of insertion of these plates into the thick skin is indicated by a diagonal groove along the base at which the vascular grooves of the upper surface end abruptly. The oval plates 6 through 9 on the rest of the neck and adjacent part of the back have asymmetrical and transversely wide bases, the lateral surfaces of which are equally rugose. These plates increase markedly in size caudally. The bases are often cleft along their ventral aspect and are very short axially compared to the maximum length of the plate. The wide asymmetrical bases of plates 5 to 9 would have provided the necessary support for these large plates to be angled away from the midline on the neck (allowing moderate movement in this region; Czerkas 1987). The remaining dorsal and caudal plates have transversely narrow and symmetrical bases almost as long as the maximum length of the plate. The largest plate is positioned over the base of the tail. These caudal plates are subrectangular proximally but triangular distally. Lateral surfaces are rugose and evenly covered with vascular grooves. The lack of overlap of the caudal plates enabled the tail to move laterally without hindrance (Czerkas 1987). The external surfaces of the two pairs of terminal tail spines are also covered with vascular impressions.

The dermal armor of the tail of the other species of *Stegosaurus* show some differences from that of *S. stenops*. The two pairs of tail spines of *S. longispinus* are extremely elongate and possess unexpanded bases such that the spine thickens slightly axially before tapering distally very gradually. The distal two-thirds of the tail of *S. ungulatus* carries at least four spinelike plates of different sizes, each with a very oblique flat base and sharp fore and aft edges. The spinelike plates may have been paired because there are two large plates of almost the same size (Ostrom and McIntosh 1966). An isolated pair of stocky spines with extremely large bases is probably from a very large individual of *S. ungulatus.*

SYSTEMATICS

The remains of stegosaurs are common in the Jurassic, with two well-represented genera from the Middle Jurassic (*Huayangosaurus*, People's Republic of China; *Lexovisaurus*, Europe) and six from the Upper Jurassic (*Chialingosaurus, Chungkingosaurus, Tuojiangosaurus*, People's Republic of China; *Dacentrurus*, Europe; *Kentrosaurus*, Tanzania; *Stegosaurus*, U.S.A.; table 21.1). There is a poor record of the group from the Lower Cretaceous of England (*Craterosaurus*), South Africa (*Paranthodon*), and the People's Republic of China

(*Wuerhosaurus*) plus a partial skeleton from the Upper Cretaceous of India (*Dravidosaurus*, Coniacian; Maastrichtian stegosaur mentioned but not described by Yadagiri and Ayyasami 1979). The absence of stegosaurs from Antarctica, Australia, and South America probably results from the incompleteness of the fossil record. However, Leonardi (1984) has referred earliest Cretaceous footprints from Brazil to the Stegosauria.

The most recent review of the Stegosauria is by Steel (1969), who included several genera from the Jurassic of Europe which are now referred to the basal Thyreophora (*Echinodon, Scelidosaurus*; see Coombs et al., this vol.) and to the Ankylosauria (*Sarcolestes, Priodontognathus*; see Coombs and Maryańska, this vol.). Recent cladistic analyses by Norman (1984a), Sereno (1984, 1986), Cooper (1985), and Gauthier (1986) indicate that the Stegosauria and Ankylosauria are more closely related to each other than to any other ornithischians (the Eurypoda of Sereno 1986).

Derived characters linking Stegosauria and Ankylosauria (hence also primitive characters for Stegosauria) include the following: three palpebrals form the dorsal rim of the orbit, quadrate with the shaft and pterygoid ramus in the same plane with no distinct lateral ramus and distal articular surface ventrolaterally directed; no otic notch between paroccipital process and quadrate; deep median keel formed by the vomers and pterygoids extends to caudal end of palate, vertical median portion of palatal ramus of pterygoid developed caudally; ventral part of exoccipital adjacent to foramen magnum is indented to form a recess that is overhung by the dorsal border of exoccipital plus the supraoccipital and floored by the occipital condyle; fusion of neural arches of atlas to atlantal intercentrum, proximal caudals with axially short and transversely circular centra and a small neural canal; uniform width to blade of scapula; relatively short metacarpals and hooflike unguals; ilium with greatly enlarged preacetabular process directed at least 35° lateral to the midline, a very short postacetabular process, and the dorsal edge forms a prominent antitrochanter that overlaps the acetabulum and the ischial peduncle (which lacks a supraacetabular buttress); distally, the blade of the ischium is unexpanded and does not slant ventromedially; femur with head effectively terminal and reduced lesser and fourth trochanters; tibia and fibula fuse distally and with astragalus and calcaneum in adult; spreading short metatarsals, no sigmoidal curve in shaft of metatarsal IV, loss of one phalanx on digit IV (thus four phalanges), hooflike unguals; and elevated dermal spines (mostly after Sereno 1986, *Lesothosaurus, Scutellosaurus* and *Scelidosaurus* used for outgroup

TABLE 21.1 Stegosauria

	Occurrence	Age	Material
Stegosauria Marsh, 1877c			
Huayangosauridae Dong, Tang, et Zhou, 1982			
Huayangosaurus Dong, Tang, et Zhou, 1982			
H. taibaii Dong, Tang, et Zhou, 1982	Xiashaximiao Formation, Sichuan, People's Republic of China	Bathonian-Callovian	Complete skeleton with skull complete skull, 5 fragmentary postcrania, adult
Stegosauridae Marsh, 1880			
Chialingosaurus Young, 1959			
C. kuani Young, 1959	Shangshaximiao Formation, Sichuan, People's Republic of China	Late Jurassic	Partial postcranial skeleton, adult
Chungkingosaurus Dong, Zhou, et Zhang, 1983			
C. jiangbeiensis Dong, Zhou, et Zhang, 1983	Shangshaximiao Formation, Sichuan, People's Republic of China	Late Jurassic	Incomplete skeleton with skull, 3 fragmentary postcrania, adult
Craterosaurus Seeley, 1874			
C. pottonensis Seeley, 1874	Wealden (Woburn Sands; reworked), Bedfordshire, England	Valanginian-Barremian	Incomplete dorsal vertebra, adult
Dacentrurus Lucas, 1902 (= *Omosaurus* Owen, 1875)			
D. armatus (Owen, 1875) (= *Omosaurus armatus* Owen, 1875, including *O. hastiger* Owen, 1877, *O. lennieri* Nopcsa, 1911b, *Astrodon pusillus* Lapparent et Zbyszewski, 1957)	Lower Kimmeridge Clay, Cambridgeshire, Dorset, Wiltshire, Corallian Oolite Formation, Dorset, England; Argiles d'Octeville, Seine-Maritime, France; unnamed unit, Beira Litoral, unnamed unit, Estremadura, Portugal	Oxfordian-Kimmeridgian	Nearly complete postcranial skeleton, partial postcranial skeleton, 4 sacra, femora, 16 fragmentary postcrania, juvenile to adult
Dravidosaurus Yadagiri et Ayyasami, 1979			
D. blanfordi Yadagiri et Ayyasami, 1979	Trichinopoly Group, Tirchirapalli District, Tamil Nadu, India	Coniacian	Fragmentary skeleton with partial skull, adult
Kentrosaurus Hennig, 1915 (= *Doryphorosaurus* Nopcsa, 1916, *Kentrurosaurus* Hennig, 1916b)			
K. aethiopicus Hennig, 1915	Tendaguru Beds, Mtwara, Tanzania	Kimmeridgian	2 composite mounted skeletons, 4 braincases, 7 sacra, more than 70 femora, approximately 25 isolated elements, juvenile to adult
Lexovisaurus Hoffstetter, 1957			
L. durobrivensis (Hulke, 1887) (= *Omosaurus durobrivensis* Hulke, 1887, including *O. leedsi* Seeley, 1901 *partim*, *Stegosaurus priscus* Nopcsa, 1911a)	Lower Oxford Clay, Kimmeridge Clay, Northamptonshire, Cambridgeshire, Dorset, England; Marnes d'Argences, Calvados, France	middle Callovian-Kimmeridgian	3 partial postcranial skeletons, 10 isolated elements, juvenile to adult
Paranthodon Nopcsa, 1929a			
P. africanus (Broom, 1912) (= *Palaeoscincus africanus* Broom, 1912, *Paranthodon owenii* Nopcsa, 1929, *Anthodon serrarius* Owen, 1876 *partim*)	Kirkwood Formation, Cape Province, South Africa	middle Tithonian-early Valanginian	Partial skull, teeth, adult
Stegosaurus Marsh, 1877c (= *Diracodon* Marsh, 1881b)			
S. armatus Marsh, 1877c (including *S. ungulatus* Marsh, 1879c, *S. sulcatus* Marsh, 1887, *S. duplex* Marsh, 1887, *Hypsirhophus seeleyanus* Cope, 1879)	Morrison Formation, Colorado, Wyoming, Utah, United States	Kimmeridgian-Tithonian	2 partial skeletons, 2 braincases, at least 30 fragmentary postcrania, adult
S. stenops Marsh, 1887c (including *Diracodon laticeps* Marsh, 1881b)	Morrison Formation, Colorado, Wyoming, Utah, United States	Kimmeridgian-Tithonian	Complete skeleton with skull, 4 braincases, at least 50 partial postcrania, juvenile to adult
S. longispinus Gilmore, 1914a	Morrison Formation, Wyoming, ?Utah, United States	Kimmeridgian-Tithonian	Fragmentary postcranial skeleton, adult
Tuojiangosaurus Dong, Li, Zhou, et Zhang, 1977			
T. multispinus Dong, Li, Zhou, et Zhang, 1977	Shangshaximiao Formation, Sichuan, People's Republic of China	Late Jurassic	2 partial skeletons
Wuerhosaurus Dong, 1973a			
W. homheni Dong, 1973a	Lianmugin Formation, Xinjiang Uygur Zizhiqu, People's Republic of China	?Valanginian-Albian	Partial skeleton

Nomina dubia	Material
Hypsirhophus discurus Cope, 1878a	2 dorsal vertebrae, caudal neural arch fragment
Omosaurus phillipsi Seeley, 1893	Juvenile femur
Omosaurus vetustus Huene, 1910b	Juvenile femur
Stegosaurus affinis Marsh, 1881a	Pubis

comparisons; see Thulborn 1970*b*, 1972; Colbert 1981; Owen 1861*a*, 1863; Coombs et al., this vol.).

Huayangosauridae

This family includes only *Huayangosaurus taibaii,* the sister taxon to all other stegosaurs (Sereno 1986), and it is diagnosed by the following characters: cranial dorsal ribs with a prominent hamularis process and a much enlarged distal end to the pubic rod.

Huayangosaurus taibaii is the best-represented Middle Jurassic (Bathonian-Callovian) stegosaur with several skeletons (length approximately 4.3 m) known from Sichuan, People's Republic of China (Dong et al. 1982; Zhou 1984; Sereno 1986), and it is characterized by numerous primitive characters. The skull is deep with a short snout region, seven premaxillary teeth, a small triangular antorbital fenestra, three supraorbitals, rostral part of orbit overlapping maxillary tooth row, and a small external mandibular fenestra. The vertical lamina lateral to the tooth row extends rostrally for a third of the length of the dentary. The sacrum is pierced by three large sacral foramina between adjacent transverse processes. The width of the coracoid exceeds its height and is only slightly less than the width of the ventral part of the scapula that lacks a well-defined acromial process. The ilium possesses a broad preacetabular process and, a prominent ischial peduncle and is transversely narrow. The limb bones are rather hollow, and the humerus is slender with a proximally placed deltopectoral crest. The femurohumeral ratio is 115 percent. The phalangeal formula of the pes is 0−2−3−3−0. The osteoderms consist of two rows of bony plates and spines of various shapes and two pairs of terminal tail spines plus a pair of shoulder spines (Dong in press). Plesiomorphic characters include reduction in height of the rostral part of jugal compared to that of *Tuojiangosaurus* (also low in *Stegosaurus*), loss of the paraquadrate foramen between quadrate and quadratojugal which is present in *Kentrosaurus* (also absent in *Stegosaurus*), the decreased size of the external mandibular fenestra compared to *Stegosaurus*, increased complexity of the teeth compared to those of *Kentrosaurus* with more numerous marginal denticles and a central vertical eminence, and elevation of the transverse processes of the dorsal vertebrae up to 60° compared to 45° in *Dacentrurus*.

Stegosauridae

The remaining stegosaurs are referred to the Stegosauridae that is diagnosed by the following derived characters: skull with orbit caudal to maxillary tooth row; a femorohumeral ratio of at least 145 percent (modified from Sereno 1986), width of coracoid less than its height, and only slightly more than half the width of the ventral part of scapula that has a prominent acromial process, ilium with a prominent antitrochanter, and a reduced ischial peduncle.

European Stegosauridae: *Lexovisaurus durobrivensis* is represented by a few partial skeletons (length approximately 6 m) lacking skulls from the Middle Jurassic (middle Callovian) of England and France (Nopcsa 1911*a*; Galton 1985*c*; Galton et al. 1980). Derived characters include caudal centra in the proximal third of the tail with a large proximal chevron facet that unites with the distal one to give a V-shaped centrum in caudals 7 to 11, almost solid dorsal plate to sacrum, midcaudals with vertical neural spines, an ilium with a long, thin preacetabular process, a pubis with a rugose central thickening, and osteoderms that include several very large, tall, thin plates whose height is over twice the craniocaudal length. A shoulder spine is present.

Dacentrurus armatus is represented by a few partial skeletons (length approximately 7 m) from the Upper Jurassic (Lower Kimmeridgian of England, France, and Portugal (Owen, 1875, 1877*b*; Galton 1981*b*, 1985*c*; Galton and Boine 1980; Nopcsa 1911*b*; Lapparent and Zbyszewski 1957; Zbyszewski 1946). Derived characters include dorsal vertebrae with massive centra that are wider than long and with lateral pleurocoelous depressions, almost solid dorsal plate to sacrum, deep prepubic process of pubis, and long dermal tail spines with sharp lateral and medial edges. Primitive characters include dorsal vertebrae with the pedicle of the neural arch low and transverse processes elevated at 45°, and a femur to humerus ratio of 147 percent.

Craterosaurus pottonensis is represented by the neural arch of a dorsal vertebra (Seeley 1874; Nopcsa 1912; Galton 1981*a*, 1985*c*) from the Lower Cretaceous (Aptian, reworked from ?Valanginian or Barremian) of England which is characterized by a unique, very deep excavation immediately caudal to the prezygapophyses.

Kentrosaurus aethiopicus is known from hundreds of bones, mostly disarticulated, from the Upper Jurassic (Kimmeridgian) of Tanzania (length approximately 5 m; Hennig 1915*b*, 1916*a*, 1936; Janensch 1924, 1936; Galton 1982*b*, 1988). However, much of the material mentioned by Hennig (1924) can no longer be located and is presumed destroyed. Derived characters include a large neural canal within the pedicle and bordered dorsally by a thin transverse lamina in dorsal vertebrae, almost solid dorsal plate to centrum, neural arches cranially leaning in caudal vertebrae, the transverse processes extending as far as caudal 28, and the neural spines of the distal two-thirds of the tail show a marked anticline. Additionally, the hemal arches are plough-shaped, and the preacetabular process of the ilium is wide. Primitive characters include a skull with a prominent paraquadrate foramen, simple cheek teeth whose crowns bear only seven marginal denticles, and a shoulder spine. In addition, the neck and cranial half of the back carry six pairs of erect plates (sixth is a spine-plate), there are three pairs of flat spines on the rest of back, and the tail carries five pairs of large spines including one terminal pair.

Paranthodon africanus from the Upper Jurassic or Lower Cretaceous (Middle Tithonian or Early Valanginian) of South Africa (Galton 1981*a*; Galton and Coombs 1981) has a snout region characterized by a uniquely long, broad, caudal process to the premaxilla and maxillary teeth with a very large cingulum and prominent vertical ridges on the crown.

Asian Stegosauridae: The record of stegosaurids from the Sichuan basin of the People's Republic of China is the best in the world, but much of it is still in need of detailed study.

Chialingosaurus kuani is known from one partial, medium-sized skeleton from the Late Jurassic (Oxfordian or Kimmeridgian) of Sichuan (Young 1959; Dong et al. 1983). It has a high narrow skull, a slender humerus and ulna, and a relatively small platelike spine. The only unique derived character is the lesser trochanter of the femur that is triangular with a broad base.

Chungkingosaurus jiangbeiensis is known from several medium-sized partial skeletons (length approximately 3-4 m) from the Late Jurassic (Oxfordian or Kimmeridgian) of Sichuan (Dong et al. 1983). The skull is high and narrow with a thick lower jaw. The sacrum is primitive in retaining three pairs of large foramina between the transverse processes. The humerus is primitive in retaining a very proximal deltopectoral crest but derived in having very broad ends. Another derived character is the presence of four pairs of terminal tail spines (Dong in press). The bony dermal plates are large and thick, and many are intermediate in form between plates and spines.

Tuojiangosaurus multispinus is known from two large partial skeletons (length approximately 7 m) from the Late Jurassic (Oxfordian or Kimmeridgian) of Sichuan (Dong et al. 1977, 1983). The skull has a low, elongate facial region, a reduced jugal, and three supraorbitals. There are three pairs of small, elongate bean-shaped sacral foramina between the transverse processes (Dong in press). The proximal caudals are unique in possessing craniolaterally oriented sheets on their neural spines. Osteoderms consist of 17 pairs of symmetrical bony plates and spines, some of which are spherical in the neck region. Large high spines are found in the lumbar and in the sacral region.

Wuerhosaurus homheni is represented by several fragmentary large individuals (length approximately 7 to 8 m) from the Lower Cretaceous (Valanginian or Albian) of the Dzungar Basin (Dong 1973*a*). Derived characters include the almost solid dorsal plate to sacrum, the very elongated neural spines of the proximal caudals, a dorsoventrally expanded distal end of the ischium, and osteodermal plates that are long, large, and low.

Dravidosaurus blanfordi is a small stegosaur from the Upper Cretaceous (Coniacian) of southern India (Yadagiri and Ayyasami 1979; Anon. 1978; Galton 1981*a*). It has a proportionally smaller skull than in other stegosaurs with ?two supraorbitals and a postorbital that is thin caudally. The iliac plate curves caudally and is narrow at the ischial peduncle. Osteoderms are large and triangular with thick bases, and the tail spike has a uniquely expanded middle region.

North American Stegosauridae: *Stegosaurus* is the largest stegosaur (length up to 9 m in *S. ungulatus*) and is represented by numerous remains from the Upper Jurassic (Kimmeridgian or Tithonian) of the United States, mostly from Colorado, Utah, and Wyoming (Marsh 1896; Gilmore 1914*a*, 1918; Ostrom and McIntosh 1966; Galton 1982*a*, in prep.; Czerkas 1987). The skull is characterized by large external nares, long nasals, maxillae and vomers producing an elongate snout, a ventral process to the jugal, the vertical lamina lateral to the tooth row extending almost to the rostral end of the dentary, and a wide flat parietal platform between the supratemporal fenestrae. The sacrum has an almost solid dorsal plate. The proximal caudals have an apex to the neural spine that is greatly expanded transversely and a prominent dorsal process on the transverse process. The femorohumeral ratio is 180 percent or more. The osteoderms consist mostly of a series of thin, very large plates, the length of which is about equal to the vertical height. Several of the holotypes of the different species of *Stegosaurus* are still to be described, so the synonymies given in the table are tentative. Apart from two large dorsal plates and several caudal vertebrae, the holotype of *S. armatus* is still to be prepared, but *S. ungulatus* is probably a junior synonym that is characterized by sacral centra that are broad, rounded, and without keel, with tall neural spines to the sacrals and proximal caudals that are grooved dorsally, elongate radius, ulna, and femur, the middle part of the tail bears paired caudal spine-plates and there are four pairs of terminal tail spines (figs. 21.4I-S, 21.5E, 21.6A-J, M-O; Marsh 1896; Gilmore 1914*a*; Ostrom and McIntosh 1966). In *S. stenops* (approximately 7 m in length), the sacral centra have a decided ventral keel, the neural spines of the sacrals and proximal caudals are proportionally shorter with transversely flat tops, the middle part of the tail bears thin plates, and there are two pairs of terminal tail spines (figs. 21.1A, 21.1A-D, 21.3A, 21.4A-H,T,U, 21.5D, 21.6K,L; Marsh 1896; Gilmore 1914a; Ostrom and McIntosh 1966). *S. longispinus* (length approximately 7 m) is characterized by two pairs of very elongate tail spines, the bases of which are subequal rather than the proximal pair having a much larger base.

The derived characters of the Stegosauridae are given above. Unfortunately, relationships within the family are unclear. One reason is the relative lack of cranial material that, apart from *Stegosaurus* (fig. 21.2), is very fragmented (*Tuojiangosaurus*), very incomplete (*Chungkingosaurus, Dravidosaurus, Paranthodon* [fig. 21.3E,F,I,J], *Chialingosaurus, Kentrosaurus*), or completely lacking (*Craterosaurus, Dacentrurus, Lexovisaurus, Wuerhosaurus*). The postcranial skeleton is completely known only for *Kentrosaurus* and *Stegosaurus*. The partial postcranial skeletons of *Tuojiangosaurus, Dacentrurus, Lexovisaurus* (fig. 21.1D-F), *Chungkingosaurus, Chialingosaurus, Wuerhosaurus,* and *Dravidosaurus* show a progressive decrease in completeness, *Craterosaurus* is represented by a part of a dorsal vertebra, and the postcrania of *Paranthodon* are unknown. In addition, many of the characters are proportional differences that are not clear-cut.

In comparison to *Huayangosaurus,* derived characters of the skull of *Chungkingosaurus* are the edentulous premaxillae (also in *Paranthodon, Stegosaurus, Tuojiangosaurus*) and closure of the antorbital fenestra (also in *Stegosaurus, Tuojiangosaurus*). In comparison to *Huayangosaurus, Chialingosaurus, Chungkingosaurus,* and *Paranthodon,* a derived character of the skulls of *Stegosaurus* and *Tuojiangosaurus* is an elongation of the snout region with a proportionally longer and lower maxilla. In comparison to *Huayangosaurus,* the skulls of *Tuojiangosaurus* and *Stegosaurus* show the following derived characters: a decrease in proportional size and height, the more vertical orientation of the quadrate, and an increased width of the flat parietal plate between the supratemporal fenestrae that decrease in size. However, these characters are not known for *Chialingosaurus, Chungkingosaurus,* and *Paranthodon.*

Dacentrurus, Kentrosaurus, Lexovisaurus, Tuojiangosaurus, Wuerhosaurus, and *Stegosaurus* are grouped together and apart from the other stegosaurids by Sereno (1986) on the basis of derived postcranial characters compared to *Huayangosaurus.* The characters cited include the complete or nearly complete fusion of the adjacent transverse processes of the sacral vertebra to form a solid dorsal plate, a relative lengthening of the prepubic process of the pubis, and an increase in the ratio of the lengths of the femur and humerus to at least 145 percent (modified from Sereno 1986). However,

the femur to humerus ratio is about 163 percent in *Chialingosaurus* and *Chungkingosaurus,* and, at least in *Chungkingosaurus,* the sacral foramina are large (fig. 21.6H). In addition, the prepubic process of the pubis is incomplete cranially in *Huayangosaurus* (Zhou 1984, fig. 30, pl. 10; fig. 21.2) and in *Lexovisaurus* (fig. 21.6C), and the pubis is not known for *Tuojiangosaurus, Chialingosaurus,* or *Chungkingosaurus.* Consequently, the derived postcranial characters cited to link *Dacentrurus, Kentrosaurus, Lexovisaurus, Tuojiangosaurus, Wuerhosaurus,* and *Stegosaurus* are reduced to one, the solid or nearly solid dorsal sacral plate, and this character may be size related because it strengthens the sacrum.

Primitive characters of *Dacentrurus* include middle and caudal dorsal vertebrae with the pedicel of the neural arch low and transverse processes at 45°, and a femurohumeral ratio of 147 percent. All other stegosaurids appear to be derived with respect to *Dacentrurus* in having middle and caudal dorsal vertebrae with a tall pedicle, transverse processes reaching at least 60° and a femorohumeral ratio of at least 160 percent.

Compared to *Huayangosaurus,* the different genera of stegosaurids show various combinations of the following derived characters:

1. elimination of ventral notch between prezygapophyses and between postzygapophyses of dorsal vertebrae to form a V-shaped structure in cranial or caudal views

2. more ventral orientation of the transverse processes on proximal caudal vertebrae

3. development of a proximodorsal process on transverse process of proximal caudal vertebrae

4. increase height of the neural spines of proximal caudal vertebrae

5. increased transverse thickness of the apices of neural spines of proximal caudal vertebrae

6. increased solidity of limb bones

7. increased massiveness of forelimb bones, especially both ends of the humerus, the proximal end of the ulna, and the distal end of the radius

8. enlargement of the deltopectoral crest of humerus by increased width and/or by the more distal position of the apex

9. increased length of the radius and ulna relative to that of humerus

10. increased transverse width and decreased height of ilium

11. increased angle between the preacetabular process of ilium and the midline

12. increased length of the preacetabular process of ilium

13. increased size of the antitrochanter of ilium

14. increased length of the hindlimb relative to trunk, chiefly by increased length of femur

15. loss of fourth trochanter of femur

16. increased massiveness of the metatarsals

17. reduction of the phalangeal formula of the pes from 0−2−3−3−0 to 0−2−2−2−0

18. increased size of osteoderms with an emphasis on either spines (*Tuojiangosaurus*, *Kentrosaurus*) or plates (*Lexovisaurus*, *Stegosaurus*, *Wuerhosaurus*, *Dravidosaurus*)

19. loss of the shoulder spine

20. number of terminal tail spines increased from two to four pairs.

TAPHONOMY

The European record of stegosaurs consists of single isolated carcasses that drifted downstream and disintegrated to a varying degree before being deposited in marine sediments. *Kentrosaurus* is a relatively minor element of the Tendaguru fauna of Tanzania, occurring in nearshore deposits thought to prevail in a warm climate with periodic droughts. Preservation favored medium-sized individuals (Russell et al. 1980). Quarry "St" represents an enormous concentration of partly articulated and partly sorted remains. Hennig (1924) noted that several sacra were found together in one place; in another, a row of limb bones were found; and at another, vertebrae of all types were especially numerous; and manual and pedal bones were almost completely lacking.

In the Morrison Formation, the remains of species of *Stegosaurus*, *Camptosaurus*, *Allosaurus*, *Apatosaurus*, *Camarasaurus*, and *Diplodocus* are broadly distributed across a spectrum of lithofacies that include channel sands, overbank variegated and drab mudstones, and floodplain limestone-marls (Dodson, Behrensmeyer, Bakker, and McIntosh 1980). In contrast to the other genera, *Stegosaurus* occurs more frequently in channel sands that represent a concentration of bones from animals that probably spent much of their lives in floodplain areas. *Stegosaurus* may have been somewhat separated ecologically from sauropods, perhaps inhabiting areas farther from sources of water. The remains

of *Stegosaurus* in overbank mudstones and floodplain limestone marls probably were buried where they died. *Stegosaurus* occasionally occurs as a single skeleton, but normally it is part of an accumulation of 20 to 60 skeletons of other dinosaurs distributed over a relatively small area and showing only a moderate to low degree of articulation. Morrison dinosaur carcasses (including those of *Stegosaurus*) typically decomposed in open dry areas or spent considerable time in channels prior to deposition. Quarry 13 at Como Bluff, Wyoming, United States (Ostrom and McIntosh 1966), is unique because it consists of a concentration of *Stegosaurus* and *Camptosaurus* remains with only a few *Camarasaurus* remains.

The degree of articulation of stegosaurs from Sichuan is greater than those from the Morrison Formation. Several skeletons from Sichuan include skulls (see Dong et al. 1977, 1982, 1983; Zhou 1984), so the carcasses were buried more rapidly than were those of the Morrison Formation. The Lower Shaximiao Formation in which *Huayangosaurus* occurs consists of alternating deposits of fluvial and lacustrine facies. The dinosaur fauna occurs in sandstones that were deposited in a lakeshore shallow-bank environment under low-energy conditions (Xia et al. 1984.)

PALEOECOLOGY AND BEHAVIOR

The teeth of stegosaurs are obviously those of herbivorous animals, but they are proportionally small and usually lack the prominent wear surfaces so well developed in most ornithischians. However, because they were obviously quite successful during the Middle and Late Jurassic, these animals presumably adopted a feeding strategy different from that of other ornithischians, and, as discussed by Farlow (1987*b*), there are several different possibilities. It is usually assumed that stegosaurs, because of their short forelimbs and long hindlimbs, were low browsers, feeding close to the ground. However, Bakker (1978, 1986) suggested that stegosaurs would have also reared up on the hindlimbs that, with the tail, acted as a tripod for balance while the animal was feeding on leaves high up in the trees (up to about 6 m for the largest *Stegosaurus*). However, as Coe et al. (1987) noted, elephants can easily stand on their hind legs but do not usually feed in this position. They regarded *Stegosaurus* as a very important browser near the 1 m level (see also Weishampel 1984*c*), the maximum "comfortable" height it could raise its head without rising on its hind legs. In addition to foliage, Weishampel (1984*c*) suggested that stegosaurs browsed on the fleshy parts of bennettitalian inflorescences plus the fructifications of the Nilssoniales and Caytoniales plant groups.

Forelimb and hindlimb proportions and manual and pedal morphology indicate a graviportal habit for stegosaurs (Coombs 1978c).

Marsh (1880) was the first to note that *Stegosaurus* was protected by a powerful armor that served for both defense and offense. Lull (1910a) referred to the armor plates and defensive spines, and Gilmore (1914a), while agreeing about the spines, noted that the plates were protective only to the extent of providing the animal with a more formidable appearance. Davitashvili (1961) also questioned the protective value of the plates, noting that their dorsal location along a narrow strip left the sides unprotected and that their basal insertion into the integument and their thinness meant that they were incapable of taking much weight from above. Hotton (1963) and Bakker (1986) have suggested that the horn-covered plates were movable from a recumbent position, through which they provided an effective flank defense with the largest plates over the especially vulnerable hindlimb, to a vertical position for warding off attack from above. Bakker (1986) noted that, because of the lack of ossified tendons and the presence of well-developed zygapophyses almost to the end of the caudal vertebrae, the powerful tail was more flexible than that of other ornithischians. He also noted that, as a result of the proportions of the limbs, *Stegosaurus* could pivot easily on its long hindlimbs by pushing sideways with its short but powerful forelimbs (also Lull 1910b) using the extremely large deltopectoral muscles.

The pattern of plates and spines is characteristic for each species, and it was probably important for intraspecific recognition. Davitashvili (1961) assumed that the armor was important for ritual intraspecific displays for sexual selection within a hierarchical social system, and, as Spassov (1982) noted, the armor is ideally arranged for maximum effect during lateral display during agonistic behavior. Sexual dimorphism in the sacrum is known for several species of stegosaurs and is probably reflected by the two sizes of shoulder plates in *Lexovisaurus*. However, the presence or absence of sexual dimorphism in the dorsal plates and/or spines cannot be determined for a particular species at present.

Buffrenil et al. (1984, 1986) studied the histology of a plate of *Stegosaurus* (for textural and mineralogical analysis, see Brinkman and Conway 1985). The main site of active bone deposition and plate growth was the basal third of the plate which was buried in a thick dermis rich in connective fibers. The system of anchoring Sharpey's fibers in this region are symmetrical relative to the sagittal plane of the plate, so the plate was held erect and was not moved to and from a recumbent orientation by axial or skin muscles. The thin wall of the plate consists of incompletely remodeled bone surrouding a large cancellous region containing large vascular channels. The apical two-thirds of the plate was skin covered with an extensive dermal vascular system as indicated by the external system of large dichotomizing grooves. The surfaces of both plates and spines lack any imprints that would indicate deposition of horn. Buffrenil et al. (1984, 1986) conclude that the plates could not have functioned as an armor because they do not consist of a thick, compact bone, and, because the plates could not unfold suddenly, they were probably not used as an intraspecific deterrent display structure. However, histology provides supporting evidence for the use of the plates as intraspecific agonistic (sexual display) structures and/or as themoregulatory devices. In an alternating arrangement, they would have functioned well as forced convection fins to dissipate heat (Farlow et al. 1976) or as heat absorbers (Seitel [1979] assumed that plates were movable). Buffrenil et al. (1984, 1986) emphasized the role of the richly vascularized skin supported by the scaffoldinglike plate as the heat exchange structure and note that a linear arrangement of the plates means that radiative and convective heat transfer may have been equally important. They noted that a heat-absorbing role for the plates makes sense if *Stegosaurus* was an ectotherm, whereas heat loss by radiation or forced convection would be useful if the metabolic regime of *Stegosaurus* was ectothermic (so acted as a heat dumping system) or to any degree endothermic.

The large plates of *Lexovisaurus* are very thin with an extensive series of grooves (see Galton 1985c) so they probably served a display and/or thermoregulatory function as suggested for *Stegosaurus*. However, the dorsal plates of most other stegosaurs are much thicker and do not have a well-developed series of surface grooves. Hence, they were probably used more for display. All stegosaurs have terminal caudal spines that were probably defensive, and, given their orientation, the animal probably backed toward the predator much as a modern porcupine does.

22

Ankylosauria

WALTER P. COOMBS, JR.
TERESA MARYAŃSKA

INTRODUCTION

Ankylosaurs are quadrupedal armored ornithischians primarily of Cretaceous age, although some fragmentary specimens have been found in the Jurassic. The skull is broad and dorsoventrally flattened, especially in the Ankylosauridae, and is covered by a mosaic of armor plates. Ankylosaurs retain the denticulate, leaf-shaped, noninterlocking teeth of primitive ornithischians. The ribs arch up and outward to form a broad, rounded body, especially among Late Cretaceous species. The limbs are stout, with the forelimb about two-thirds to three-quarters the length of the hindlimb, and the manus and pes are short with somewhat divergent digits. The tail accounts for about half of total body length. Body armor is variable from species to species but is generally composed of small rounded plates that form a fairly continuous shield over the dorsal surface of the body. Nodosaurids commonly have tall spikes or spines in the armor, and ankylosaurids have a club of dermal plates at the end of the tail.

Huene (1914d) used the term "Ancylosauria" in a phylogenetic diagram without diagnosis or comment, but he is not generally credited with creating the suborder. Osborn (1923a) used the term "Ankylosauria" without diagnosis or comment, and he is generally given credit for creating the suborder. Romer (1927, 1956) forcefully advocated subordinal status for ankylo-

saurs, the structure of the pelvis being critical (Romer 1968). Workers outside of North America continued to unite ankylosaurs and stegosaurs into a group variously called the Stegosauria (Hennig 1924), Thyreophora (Nopcsa 1929a; Huene 1948), or Stegosauroidea (Lapparent and Lavocat 1955). After publication of a lengthy diagnosis (Romer 1956), subordinal status for the Ankylosauria has been widely accepted.

DIAGNOSIS

The Ankylosauria can be diagnosed on the following basis: cranium low and flat, rear of skull wider than high; antorbital and supratemporal fenestrae closed; sutures between cranial bones of the skull roof obliterated in adults; maxilla with deep, dorsally arched cheek emargination; pterygoid closes the passage between the space above the palate and that below the braincase; accessory antorbital ossification(s) and postocular shelf partially or completely enclose orbital cavity; a bony median septum extends from ventral surface of skull roof into palate as fused vomers; quadratojugal contacts postorbital; dorsoventrally narrow pterygoid process of the quadrate; quadrate slants rostroventrally from underside of squamosal; mandible with coossified keeled plate along ventrolateral margin; caudal dorsal ribs tend to fuse to vertebrae both at centrum and along the transverse process and may also underlie and fuse to the expanded preacetabular segment of the ilium; at least three caudal dorsal vertebrae fuse to sacrals to form a presacral rod, and the ribs of the former contact the preacetabular process of the ilium; neural and hemal spines of distal caudals elongate along axis of tail, with contact between adjacent hemal spines; ilium rotated into horizontal plane; acetabulum closed; pubis small, with very short extension adjacent to ischium; pubis almost completely excluded from the acetabulum; no prepubic process; body covered dorsally by armor plates of three or four shapes including flat, oval to rectangular plates that bear a keel, ridge, or short spine externally; larger armor plates commonly symmetrically arranged in transverse rows.

ANATOMY

Skull and Mandible

Ankylosaur skulls are always wider than high in occipital view, and the antorbital and supratemporal fenestrae are invariably closed (figs. 22.1, 22.2, 22.3). Secondary superficial bony deposits, either coossified armor plates or bone that has accreted under the inductive influence of epidermal scutes, obscure the cranial sutural pattern of all adult ankylosaurs. Because the armor plates are the most conspicuous feature of ankylosaur skulls, these will be described before the normal bones of the skull roof. In ankylosaurids, two pairs of plates in the nasal region form the boundary of the external naris (especially in *Ankylosaurus magniventris;* Coombs 1978*a*), and another pair forms prominent caudolateral "horns" on the skull roof that may be blunt and rounded to moderately pointed (*Euoplocephalus tutus*), large and pyramidal (*Ankylosaurus magniventris*), or compressed with spike-shaped points (*Saichania chulsanensis*). In ankylosaurids, the quadratojugal lies lateral to and in approximately the same transverse plane as the quadrate. A triangular armor plate fused to the quadratojugal projects caudoventrally to hide both the entire length of the quadrate and the infratemporal fenestra in lateral view, although the fenestra is not actually closed. The medial surface of the quadratojugal/plate complex may provide extra area for the origin of the M. adductor mandibulae externus superficialis, which normally originates from the dorsal margin of the infratemporal fenestra. An unusually wide jugal together with adjacent elements similarly hide the infratemporal fenestra of the nodosaurid *Silvisaurus condrayi* (Eaton 1960). Large marginal plates on ankylosaurid skulls are generally arranged in bilateral pairs, but plates that cover the central and rostral regions of the skull roof commonly depart from perfect bilateral symmetry and may be small and numerous (e.g., *Euoplocephalus tutus* and *Ankylosaurus magniventris;* Coombs 1971, 1972, 1978*a*) or larger and fewer (e.g., *Saichania chulsanensis;* Maryańska 1977). Individual armor plates of ankylosaurid skulls tend to remain distinct in large, presumably old individuals, but the central postorbital region of the skull roof is commonly more or less smooth.

There is no nodosaurid skull in which armor plates can be detached from the skull roof and shown to be separate ossifications as opposed to accretional texturing below dermal scutes. Typically, nodosaurids have a single large plate filling the central and caudal region of the skull roof with a variable number of small plates peripherally (fig. 22.3A). Nodosaurids lack the caudolateral "horns" and the large triangular plate fused to the jugal and quadratojugal, and consequently, the quadrate, its condyle, and the infratemporal fenestra are commonly visible in lateral view (exception: *Silvisaurus condrayi;* Eaton 1960). Typically, nodosaurids have a median, roughly trapezoidal plate fused to the premaxillae at the rostral tip of the snout. Hence, the nostrils face laterally. Grooves separating the plates are deeply incised on some skulls (e.g., the type of *Panoplosaurus mirus*) but are shallow or absent on others (e.g., type of *Edmontonia longiceps*). Generally, the grooves are best developed on small skulls and absent from large skulls, suggesting an ontogenetic change, but in some cases, the prominence of the grooves may have taxonomic significance (K. Carpenter pers. comm.). Detailed differences in patterns of cranial armor have not generally been used as taxonomic characters at the generic and specific level.

Sutural boundaries of the skull elements are known from a single ankylosaurid skull that has been identified as a juvenile *Pinacosaurus grangeri* (Maryańska 1971, 1977). The subequal parietal and frontal are approximately square in outline and together are about the same size as the larger nasal. The latter extends laterally onto the side of the skull to form the dorsal margin of the naris. Caudally, it contacts the laterally elongate prefrontal along a suture that extends rostrolaterally. The premaxilla has only a short caudodorsal process so that the nasal contacts both the maxilla and the lacrimal. This *Pinacosaurus* skull has dermal plates that are suturally united to normal cranial elements in such a way as to form structural components of the skull roof and also has cranial elements that are otherwise uncommon or unknown within the Or-

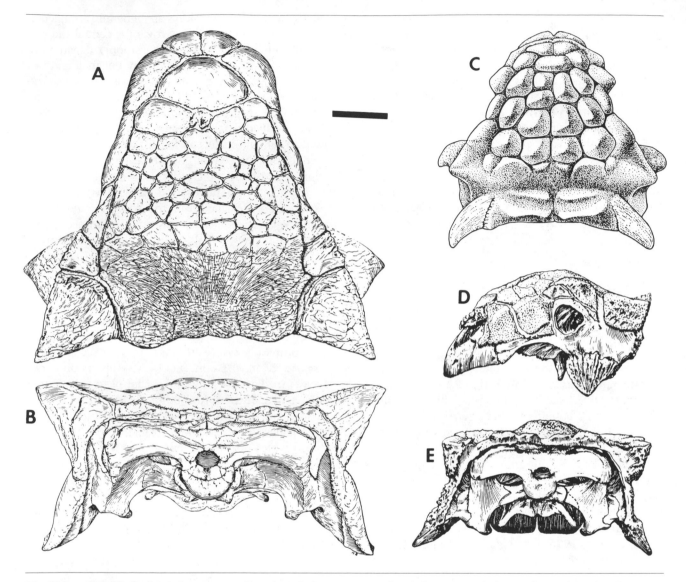

Fig. 22.1. A, B: Skull of *Ankylosaurus magniventris* (Ankylosauridae) in dorsal and occipital view. C: Skull of *Saichania chulsanensis* (Ankylosauridae) in dorsal view. D, E: Skull of *Euoplocephalus tutus* (Ankylosauridae) in lateral and occipital view. Scale = 10 cm. (C after Maryańska 1977; others after Coombs 1978*a*.)

nithischia (fig. 22.2A; Maryańska 1971, 1977). These include a "tabular" bone along the caudal margin of the skull roof (homology uncertain; Maryańska 1977) and three extra supraorbital elements dorsal to the orbit (postfrontal, presupraorbital, and postsupraorbital of Maryańska 1971, 1977; palpebrals of Coombs 1972).

Paroccipital processes of ankylosaurids project horizontally either directly laterally (*Pinacosaurus grangeri*, *Euoplocephalus tutus*, and *Ankylosaurus magniventris*, fig. 22.1) or slightly caudally (*Talarurus plicatospineus* and *Shamosaurus scutatus*). In most individuals,

the processes as well as the occipital region are hidden in dorsal view by the overhanging skull roof and the caudodorsal armor plates. Distal tips of the paroccipital processes may be either free of the dorsal tip of the quadrate (*Euoplocephalus tutus*, *Tarchia gigantea*, and *Pinacosaurus grangeri*) or the two bones may be sutured together (*Ankylosaurus magniventris* and *Saichania chulsanensis*). In nodosaurids, there is no caudal overhang of the skull roof and the paroccipital processes curve caudally and ventrally, especially at their distal tips, to form prominent projections along the occipital margin

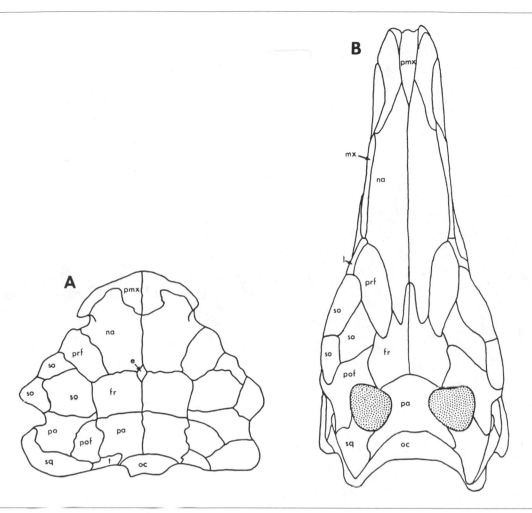

Fig. 22.2. A: Skull roof elements of *Pinacosaurus* (Ankylosauridae). B: Skull roof elements of *Stegosaurus* (Stego-sauridae). (A after Maryańska 1971, 1977; B after Gilmore 1914a.)

of the skull (fig. 22.3A). In all nodosaurids, the paroccipital process, dorsal end of the quadrate, and squamosal are fused.

Ankylosaur palatal structure is well preserved in several specimens, yet many problems remain as to the exact arrangement of bones because of extensive fusion. The solid roofing plate formed by the premaxillae within the cutting margin of the beak is in some specimens pierced by foramina of variable size and erratic distribution which have been interpreted as accommodating a vomeronasal organ (Maryańska 1977). In nodosaurids, the cutting margin of the beak curves inward as a ridge that connects to the mesial limit of the maxillary tooth row outlining an oval region commonly described as ladle- or scoop-shaped (fig. 22.3B).

Premaxillary teeth are present along this ridge in the nodosaurids *Silvisaurus condrayi* (Eaton 1960) and *Sauropelta edwardsi,* and indirect evidence suggests presence of premaxillary teeth in "*Struthiosaurus*" *transilvanicus* (Nopcsa 1915). In ankylosaurids, the cutting edge of the beak aligns with the extreme lateral margin of the lateral maxillary shelf, and there is no ridge interrupting a smooth connection between the broad premaxillary scoop and the lateral maxillary shelf (fig. 22.4). No known ankylosaurid has premaxillary teeth.

Ankylosaurs appear to have a deep buccal emargination and narrow palate because the short, broad, flat skull brings the dorsal margin of the maxillary cheek pouch almost lateral to the teeth and because the maxillary tooth row curves medially. In fact, ankylosaur

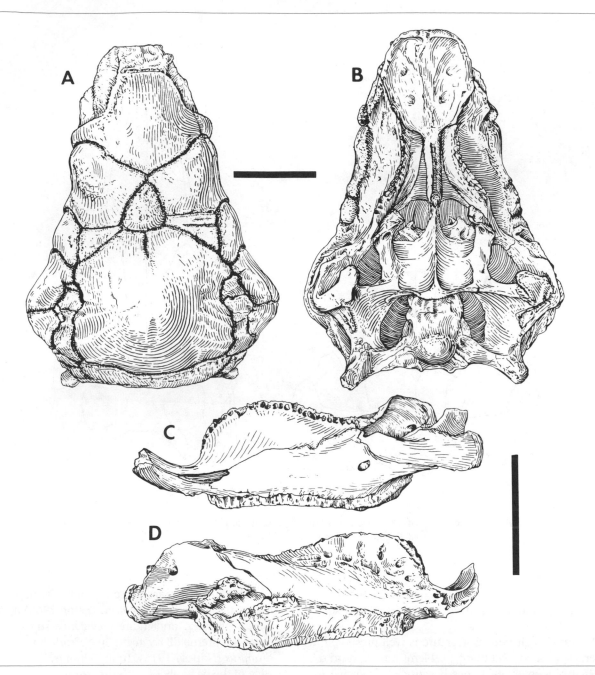

Fig. 22.3. A, B: Skull of *Edmontonia longiceps* (Nodosauridae) in dorsal and palatal view. C, D: Mandible of *Euoplocephalus tutus* (Ankylosauridae) in medial and lateral view. Scale = 10 cm. (A and B after Coombs 1978a.)

palates are not especially narrow relative to their length compared to those of other ornithischians. A conspicuous feature of the palate is the sagittal septum that divides the entire snout rostral to the braincase (figs. 22.3B, 22.4, 22.5C,D,E). This septum is formed by vomers and pterygoids caudoventrally and by premaxillae and nasals rostrodorsally, with additional contri-

butions by ethmoid ossifications (Maryańska 1977). A median vomer plate is present in many ornithischians, but its extension to the underside of the skull roof appears to be a uniquely ankylosaurian feature possibly related to the depressed shape of the skull. The caudal region of the palate is composed primarily of pterygoid and palatine bones, but sutural borders are visible on

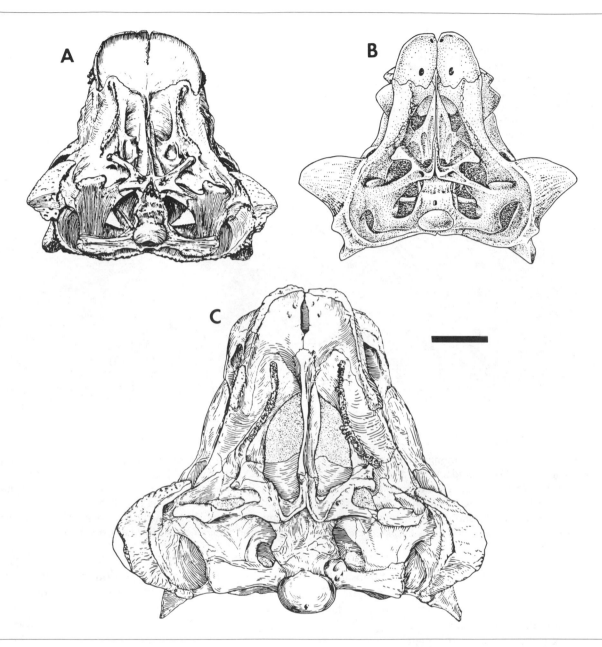

Fig. 22.4. A: Skull of *Euoplocephalus tutus* in palatal view. B: Skull of *Saichania chulsanensis* in palatal view. C: Skull of *Ankylosaurus magniventris* in palatal view (all three, Ankylo-sauridae). Scale = 10 cm. (A and C after Coombs 1978*a;* B after Maryańska 1977.)

few skulls. In ankylosaurids, the mandibular ramus of the pterygoid projects rostrolaterally, but in nodosaurids, there is a broad transverse plate at the back of the palate and the mandibular process of the pterygoid arises laterally and projects rostrally parallel to the midline. An elongate, slender, rod-shaped element found in specimens of *Saichania chulsanensis, Tarchia gigantea,*

and *Euoplocephalus tutus* has been identified as an epipterygoid (Maryańska 1977).

Ankylosaurids and nodosaurids have a bony secondary palate that appears to have developed independently in each family, as indicated by the substantially different morphology in the two families and by the fact that *Sauropelta edwardsi,* an early nodosaurid, lacks

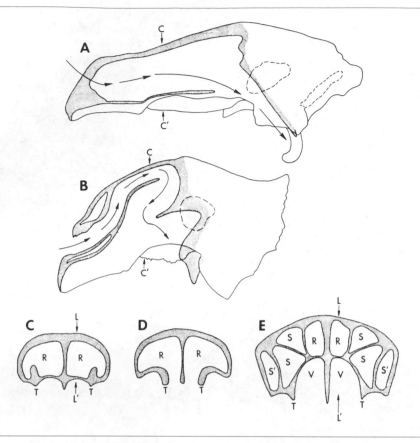

Fig. 22.5. A: Diagrammatic parasagittal section through a skull of *Edmontonia* (Nodosauridae). B: Parasagittal section through a skull of *Euoplocephalus* (Ankylosauridae). C: Cross section through the snout of *Edmontonia* (Nodosauridae). D: Cross section through the snout of *Sauropelta* (Nodosauridae). E: Cross section through a skull of *Euoplocephalus* (An-kylosauridae). Abbreviations: C–C' = approximate line of cross sections shown in C, D, E; L–L' = approximate line of parasagittal sections shown in A and B; R = respiratory path (= dorsal air chamber in Ankylosauridae); S = cranial sinus; S' = maxillary sinus; V = palatal vault. The arrows show the respiratory path. (After Coombs 1978a.)

a secondary palate (fig. 22.5D). In derived nodosaurids such as *Edmontonia longiceps* and *E. rugosidens,* the palate is a single horizontal plate that extends caudally from the premaxillae to the last two or three maxillary teeth (figs. 22.3B, 22.5A,C). This plate is formed laterally by a maxillary shelf and medially by a shelf from the vomer.

The more complex ankylosaurid secondary palate has as its major components two shelves (figs. 22.4, 22,5B,E). First, a rostrodorsal shelf projects medially to contact the sagittal septum. This shelf slants obliquely caudodorsally from the premaxillary scoop parallel to the skull roof and forms the floor of the respiratory passage (= dorsal air passage of Maryańska 1977) and parts of the cranial sinuses. Second, a caudoventral shelf begins at the caudal margin of the premaxillae but projects only slightly medially in a horizontal plane

along the lingual side of the tooth row. Just caudal to the midpoint of the tooth row, this shelf projects medially to contact the vomer and possibly the palatine. The caudoventral shelf is wider in later ankylosaurids such as *Euoplocephalus tutus* and *Saichania chulsanensis* than in the primitive ankylosaurid *Shamosaurus scutatus* (Tumanova 1983, 1986, 1987). The exact composition of the ankylosaurid palatal shelves is uncertain, but at least in *Pinacosaurus grangeri* and *Saichania chulsanensis,* the rostrodorsal shelf is composed of maxillae, vomers, and an element identified as a maxilloturbinal (Maryańska 1977). The caudoventral shelf is composed of maxillae, vomers, and palatines. The respiratory tract extends caudally above the rostrodorsal shelf to a position just rostral to the braincase where it turns ventrally to enter the cavum nasi proprium. The tract then turns rostrally into the palatal vault, then ventrally and back caudally around the wide section of the caudoventral shelf.

A unique feature of ankylosaurids is the complex of sinuses associated with the folded respiratory path (fig. 22.5B,E). Their exact number, shape, and orientation is incompletely known, but in *Ankylosaurus magniventris, Euoplocephalus tutus, Saichania chulsanensis,* and *Pinacosaurus grangeri,* some generalities can be made: (1) the true, major respiratory path (= dorsal air passage of Maryańska 1977) lies directly below the skull roof, adjacent to the sagittal septum described above; (2) at least three pairs of sinuses lie lateral to the respiratory path, and each of these arches convexly upward and extends down the entire length of the snout; (3) there may be small, flattened sinuses within the premaxillary plates that form the roof of the premaxillary scoop and also within the nasals caudodorsal to the nostrils; (4) the most lateral sinus, the maxillary sinus, lies in part above the lateral maxillary shelf; and (5) the maxillary sinus may open to the outside lateral to the external naris (*Euoplocephalus tutus* and *Ankylosaurus magniventris,* Coombs 1978*a*) or may connect with a premaxillary sinus and open ventral to the naris (*Saichania chulsanensis,* Maryańska 1977). Interconnections between the sinuses and the dorsal air passage are not well understood. The bony laminae between sinuses are in part extensions of palatal and skull roof elements, but ossifications otherwise unknown among diapsids are also present, notably, nasoturbinal and ethmoturbinal elements (Maryańska 1977).

The rostral inclination of the quadrate is a distinctly ankylosaurian feature, but there is considerable variation within the suborder. In *Tarchia gigantea,* the quadrate is almost vertical and perpendicular to the long axis of the skull. In *Euoplocephalus tutus* and *Ankylosaurus magniventris,* there is a moderate rostral inclination, while in *Panoplosaurus mirus, Edmontonia longiceps,* and *E. rugosidens,* the quadrate slants strongly rostrally. Another distinctive feature of ankylosaurs is the partial bony enclosure of the orbit. Caudal to the eye, there is a postocular shelf (Haas 1969) that separates the orbit from the mandibular adductor. This plate, composed of wings from the postorbital and jugal (Maryańska 1977), is broader and extends farther medially in ankylosaurids than in nodosaurids. The jugal component curves from an oblique to a near-horizontal orientation to form a partial floor along the ventrolateral orbital cavity. The rostral orbital wall is formed by a large lacrimal process together with accessory ossifications of uncertain homology. Within the orbit of some *Euoplocephalus tutus* skulls, there is a cup-shaped element not attached to surrounding skull bones

which has been identified as a bony eyelid (= "palpebral"; Coombs 1972). No similar element has been found in other ankylosaurs.

The tall, arched shape of the mandible is partially a consequence of the overhanging maxillary shelf and medial position of the maxillary teeth relative to the lateral margin of the snout (fig. 22.3C,D). In transverse section, the mandible is oblique such that its ventrolateral margin extends laterally beyond the overhanging maxillary shelf, while the dentary teeth lie medial to the maxillary teeth (contra Haas 1969). A large dermal plate fused to the ventrolateral margin of the mandible is uniquely ankylosaurian. The symphysis is narrow and the predentary slender with only slight development of a caudoventral process at the symphysis. Predentaries detach readily from the mandible and are commonly missing from otherwise well-preserved specimens, a phenomenon that led to the erroneous theory that ankylosaurs had no predentary ("Apraedentalia" of Huene 1948).

 Ankylosaur teeth are small relative to skull size, laterally compressed and leaf shaped with 8 to 16 or 17 marginal denticles (fig. 22.6). The crown flanks typically have vertical grooves that align with notches between adjacent marginal cusps, but in some species, there are complex grooves and ridges that have no regular relationship to marginal cusps (e.g., *Euoplocephalus tutus*). Ridges, grooves, and enamel thickness are nearly symmetrical on labial and lingual flanks of most tooth crowns, but individual teeth in a single row may be slightly asymmetrical. The basal cingulum common on nodosaurid teeth is typically higher on one crown face than the other. The small crown is aligned with a long, straight root. Neither adjacent teeth nor sequential replacement teeth interlock in any way. In all these features, ankylosaur teeth are basically primitive for ornithischians. There is considerable variation in teeth of a single individual and substantial overlap in tooth morphology among species of ankylosaurs, but it is generally possible to distinguish teeth of the two families. Nodosaurid teeth are slightly larger both absolutely and relative to skull size, and there are fewer teeth per row than in ankylosaurids (14-17 maxillary teeth in *Panoplosaurus mirus, Edmontonia longiceps,* and *E. rugosidens;* 19-24 in *Talarurus plicatospineus, Euoplocephalus tutus,* and *Saichania chulsanensis;* and 34-35 in *Ankylosaurus magniventris*). Nodosaurid teeth are also more compressed labiolingually and commonly have a conspicuous basal cingulum, while ankylosaurid teeth commonly have a swollen base without a distinct cingulum. Grooves on the crown flank are generally better developed on nodosaurid

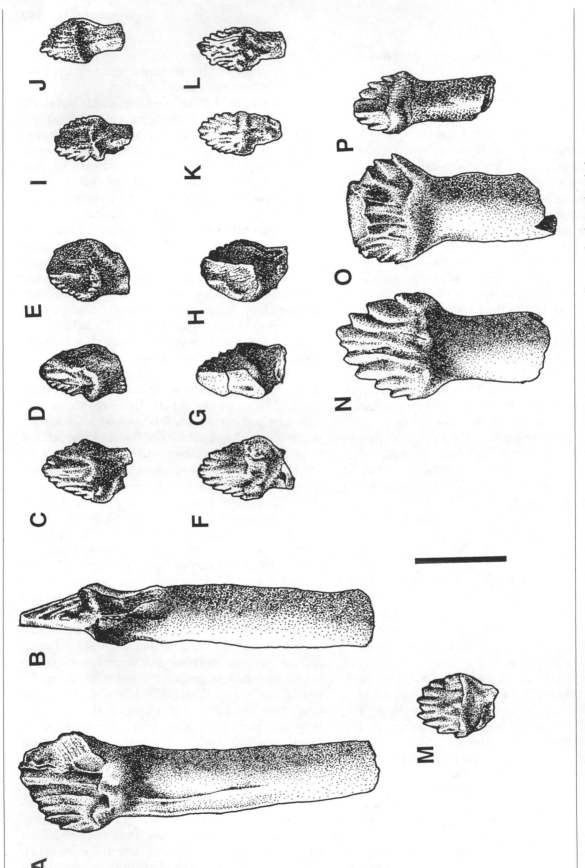

Fig. 22.6. Teeth of ankylosaurs. A, B: *Two views of a tooth of* Edmontonia rugosidens *(Nodosauridae) which show the long, straight root. In some teeth, the crown may be angled relative to the axis of the root as seen in view B and the root may be slightly curved. C, D, E: Three teeth of* Ankylosaurus magniventris *(Ankylosauridae). F, G, H: Lateral views of the same three teeth. I, J: Two maxillary teeth of* Euoplocephalus tutus *(Ankylosauridae). K, L: Opposite crown face of the same two teeth. M: Tooth of* Sauropelta edwardsi *(Nodosauridae). N, O, P: Three teeth of* Edmontonia rugosidens *(Nodosauridae) which show part of the size range that may be present in a single individual. Scale = 1 cm.*

teeth. Premaxillary teeth of the nodosaurids *Silvisaurus condrayi, Sauropelta edwardsi,* and *"Struthiosaurus" transilvanicus* are small, with smooth crown flanks and typically a single apical cusp (Nopcsa 1929; Eaton 1960; Ostrom 1970*a*). Wear surfaces on ankylosaur teeth are commonly vertical along the lingual crown flank of maxillary and the labial crown flank of dentary teeth, but there are exceptions within single tooth rows. Some teeth develop wear first along the cingulum or swollen base, but other teeth wear down the entire crown leaving the cingulum unworn. On some skulls, the wear surfaces of adjacent teeth are aligned, but on other skulls, the wear surfaces of adjacent teeth may have different angles relative to the parasagittal plane. Thus, occlusion of ankylosaur teeth appears to be imprecise and subject to individual variation, and the orientation of the wear surface may change as a single tooth is worn down.

Axial Skeleton

No ankylosaur vertebral column is sufficiently complete to provide an accurate count of the entire vertebral series, but pooling information from all specimens yields the following numbers: 7 or 8 cervicals; 9 to 13 free dorsals; a coossified sacral complex of 3 to 6 dorsals that form a presacral rod, 3 to 6 sacrals, and 1 to 3 caudals; 20 to 40 caudals including in the Ankylosauridae; at least 11 free caudals and 7 to 11 caudals fused to form a tail club "handle" (Nopcsa 1905*a*, 1928*a*; Sternberg 1921; Lull 1921; Parks 1924; Gilmore 1930; Eaton 1960; Ostrom 1970*a*; Maryańska 1977; Carpenter 1982*a*).

Contrasting sharply with that of most ornithischians, the atlas of ankylosaurs has a complete neural arch with elongate postzygapophyses extending over one-third to one-half the length of the axis neural arch (fig. 22.7). The axis bears a massive odontoid process and a heavy neural arch with a short, laterally compressed neural spine. The atlas and axis are separate in the ankylosaurids *Pinacosaurus grangeri* (Gilmore 1933*b*) and *Euoplocephalus tutus* and in the nodosaurids *"Struthiosaurus" transilvanicus* (Nopcsa 1929*a*) and *Sauropelta edwardsi* (Ostrom 1970*a*). The atlas and axis are fused in the ankylosaurid *Saichania chulsanensis* (Maryańska 1977) and in the nodosaurids *Panoplosaurus mirus* (Sternberg 1921) and *Edmontonia longiceps* (Gilmore 1930). Among more caudal cervical vertebrae, there may be centra with a cranial face that is higher than the caudal face (cervical 5? in *Edmontonia rugosidens,* Gilmore 1930; cervical 4 in *Ankylosaurus*

magniventris; all cervicals in *Talarurus plicatospineus,* Maryańska 1977), but in other taxa, the cranial face is always lower than the caudal (*Saichania chulsanensis* and *Pinacosaurus grangeri,* Maryańska 1977). Diapophyses of cervical vertebrae project straight laterally either in a horizontal plane or with a slight ventral inclination (fig. 22.7). The neural spines of cervical vertebrae differ from one taxon to another and may also change shape down the length of the cervical series.

Centra of free dorsal vertebrae have subequal length and diameter and expanded articular faces (fig. 22.7). Diapophyses are dorsoventrally compressed, bladelike, and have a strong upward inclination that in *Ankylosaurus magniventris* increases from about 30° to more than 50° caudally along the dorsal series. Ribs of caudal free dorsals commonly fuse to both the flanks of their centra and to their diapophyses, a feature in part responsible for the name "ankylosaur" (Brown 1908), but there are exceptions (e.g., *Panoplosaurus mirus:* Lambe 1919; Sternberg 1921; K. Carpenter, pers. comm.). Fusion may have an ontogenetic basis. Neural spines of free dorsals are laterally compressed, axially elongate quadrilateral blades. Along the dorsal vertebrae, the zygapophyses approach the midline and the angle between becomes more acute, until in the caudalmost free dorsals there is a single, almost U-shaped articulation formed by prezygapophyses and a matching, roughly semicylindrical surface formed by the postzygapophyes.

In the three to six dorsals that fuse to form a presacral rod (Eaton 1960; Ostrom 1970*a*; Coombs 1971, 1978*a*; = "lumbaroid vertebrae" of Nopcsa 1928*a*; = "sacrolumbars" of Gilmore 1930), the articular faces are less expanded, and the middle region of the centra is relatively wider than on free dorsals, but the approximate boundaries of individual vertebrae are still discernible (fig. 22.11). Diapophyses rapidly lose their upward inclination along the presacral rod. In the sacrum, the rib-diapophysis complex projects straight laterally with little or no upward inclination. Centra of sacral vertebrae are dorsoventrally compressed and commonly the widest vertebrae of the column. Neural arches are low and broad, and the neural canal is commonly enlarged and excavated into the dorsal surface of the centra (exception: *Polacanthus foxii;* Nopcsa 1905*a*). Neural spines in the presacral rod and sacrum are laterally compressed and form a continuous, bladelike ridge. The first one or two caudal vertebrae have massive transverse processes that project craniolaterally from the entire height of the centrum and neural arch (fig. 22.7K,L). These massive processes may contact the postacetabular region of the ilium, and the first one

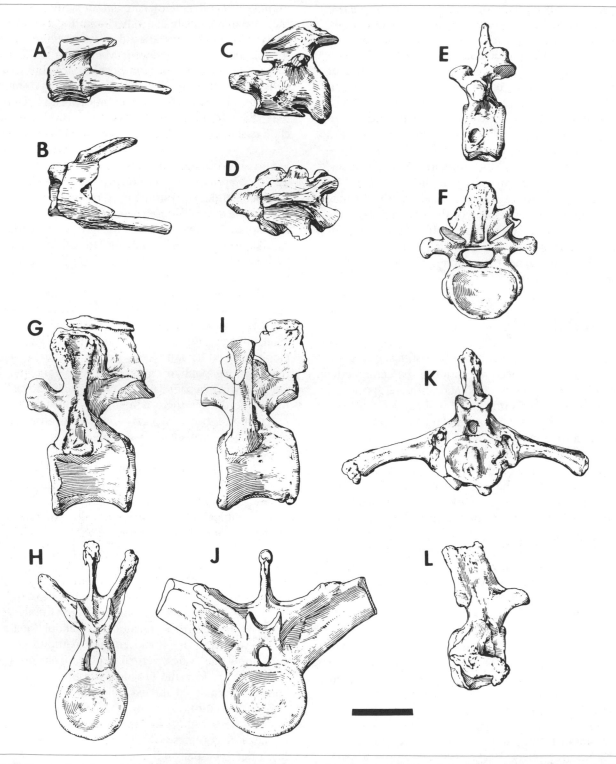

Fig. 22.7. A, B: Atlas vertebra of *Sauropelta edwardsi* (Nodosauridae) in lateral and dorsal view. C, D: Axis vertebra of *S. edwardsi* (Nodosauridae) in lateral and dorsal view. E, F: Sixth cervical vertebra of *Ankylosaurus magniventris* (Ankylosauridae) in lateral and cranial view. G, H: Sixth dorsal of *A. magniventris* (Ankylosauridae) in lateral and cranial view. I, J: Tenth dorsal vertebrae of *A. magniventris* (Ankylosauridae) in lateral and cranial view. K, L: First caudal vertebra of *Euoplocephalus tutus* (Ankylosauridae) in cranial and lateral view. Scale = 10 cm.

or two caudals may be fused into the sacral complex. Pre- and postzygapophyses are typically fused throughout the presacral rod and sacrum proper but are variably fused or free in the first one or two caudals.

Proximal caudal centra have diameters about double the length of the centrum, but smaller distal caudals are more elongate and spool shaped. Proximally, the transverse processes arise from the junction of neural arch and centrum, but more distally, the attachment tends to shift ventrally to the upper half or third of the centrum, and the processes, which are massive on the first caudal (fig. 22.7K), attenuate rapidly and are absent from the distal third of the series. In ankylosaurids, coossified vertebrae that form the

"handle" of the tail club lack neural spines, as do the ten terminal caudals of the nodosaurid *Sauropelta edwardsi* (fig. 22.8B,C). Vertebrae of the tail club "handle" have elongate prezygapophyses that overlap at least half the length of the next more proximal centrum and have postzygapophyses that fuse to form a tongue-like plate that overlaps at least half the length of the next more distal centrum (fig. 22.8A,B).

In all ankylosaurs, the height of the neural spines throughout the entire vertebral series accounts for a maximum of about half the entire height of the neural arch-spine structure, and the maximum height of the arch and spine together never exceeds twice the maximum diameter of the centrum.

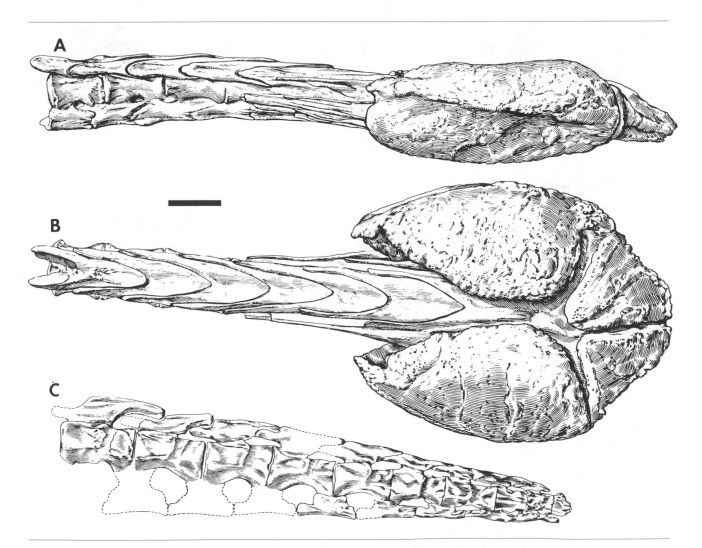

Fig. 22.8. A, B: Distal caudal vertebrae and tail club of *Ankylosaurus magniventris* (Ankylosauridae) in lateral and dorsal view. C: Distal caudal vertebrae of *Sauropelta edwardsi* (Nodosauridae) in lateral view. Scale = 10 cm. (After Coombs 1978a, 1979.)

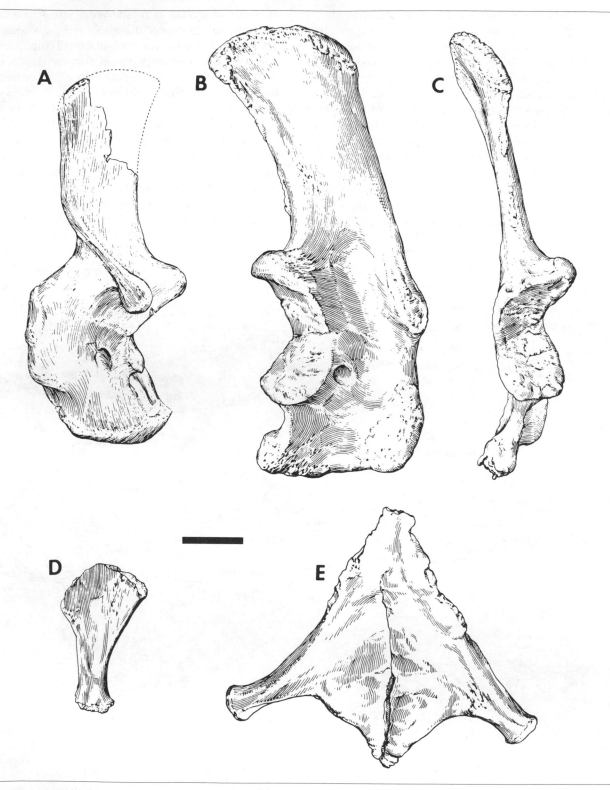

Fig. 22.9. A: Scapulocoracoid of *Sauropelta edwardsi* (Nodosauridae) in lateral view. B, C: Scapulocoracoid of *Euoplocephalus tutus* (Ankylosauridae) in lateral and caudal view. D: Sternal plate of *Sauropelta edwardsi* (Nodosauridae). E: Sternal plate of *Euoplocephalus tutus* (Ankylosauridae). Scale = 10 cm. (A, B, C after Coombs 1978*a*, 1978*b*.)

Appendicular Skeleton

Fusion of the scapula and coracoid is the usual condition in adult ankylosaurs, although the bones are separate in *Hylaeosaurus armatus,* *"Struthiosaurus" transilvanicus,* and sometimes in *Edmontonia rugosidens.* Ankylosaurids have an elongate scapula and a relatively small coracoid, the latter having a distinct corner demarcating its cranial edge from the medioventral edge (fig. 22.9). Nodosaurids have a relatively smaller scapula with a relatively larger coracoid. The glenoid tends to be more open in nodosaurids, and the cranial edge of the coracoid curves smoothly into the medioventral edge. The medioventral edge of the coracoid is commonly coarsely rugose, sometimes with a raised rim, suggesting extensive fibrous/cartilaginous connection with the sternum. The ankylosaurid scapula has a ridge-shaped scapular spine at the extreme cranial edge of the scapular blade opposite the upper half of the glenoid. A striking feature of typical nodosaurids is the caudally displaced, knoblike scapular spine and the small prespinous fossa formed cranial to it. In *Hylaeosaurus armatus,* the spine forms a transverse ridge adjacent to the scapula-coracoid junction, then curves dorsally just cranial to the glenoid, and there is no prespinous fossa. This is apparently a primitive condition in nodosaurids (Coombs 1978*b*). The poorly preserved scapula of *Hoplitosaurus marshi* may be similar to that of *Hylaeosaurus armatus* (Galton 1983*b*). Nodosaurids have a pair of flat, pyriform sternal plates. In ankylosaurids, the plates are fused along the midline to form a rhomboidal plate with caudolaterally projecting processes (fig. 22.9).

The ankylosaur humerus is short and broad, generally more massive in ankylosaurids and more elongate and slender in nodosaurids, a difference in part caused by a more distal extension of the deltopectoral crest in the former (fig. 22.10). The radial condyle of the humerus is more nearly spherical and somewhat more elevated in nodosaurids and the radius more nearly circular in cross section. In ankylosaurids, the radius is somewhat oval in cross section, and the radial condyle of the humerus is more crescentic and lies flatter against the humeral shaft. These differences suggest a slightly greater range of pronation/supination in nodosaurids. The ankylosaur ulna is short with a massive olecranon process that commonly accounts for up to one-third of the length of the ulna.

The ankylosaur manus is known from only a few reasonably complete specimens, and the phalangeal count is uncertain in all. The known metacarpal and phalangeal counts are as follows: *Saichania chulsanen-sis,* five metacarpals (Maryańska 1977); *Pinacosaurus grangeri,* five metacarpals, 2–3–3–3–2 (Maryańska 1977); *Sauropelta edwardsi,* five metacarpals, 2–3–4–3?–2? (Ostrom 1970*a*); *Panoplosaurus mirus,* three metacarpals, 2–3–3–?–0? (Lambe 1919); *Edmontonia rugosidens,* four metacarpals. Metacarpals of ankylosaurs are short and broad, contact adjacent metacarpals only proximally, and tend to diverge distally. Phalanges except the unguals are generally short, disklike elements. Unguals of adult nodosaurids are slightly longer than wide and rounded at the tip. Unguals of adult ankylosaurids are commonly wider than long, broadly curved across the tip, coarsely textured, and sometimes perforated along the margins (but note the different form of supposed juvenile unguals: Coombs 1986). In general, ankylosaurs have short, splayed digits that form a broad, graviportal type of foot (Coombs 1978*d*).

The pelvic girdle is one of the most distinctive and diagnostic postcranial features of ankylosaurs (fig. 22.11). The ilium has an abbreviated postacetabular and an hypertrophied preacetabular process that diverges cranially (Coombs 1979). Near the acetabulum, the entire iliac blade is horizontal and gently convex dorsally. Cranially, the blade twists to face ventromedially. The pubic preduncle is absent, in connection with great reduction of the pubis. The ilium forms the overhanging dorsal wall of the acetabulum. The ankylosaurid pubis is a small nubbin that does not extend ventrally adjacent to the ischium and that is fused to the anterior rim of the acetabulum. The nodosaurid pubis has a crescentic body and a short, blunt prepubic process (Gilmore 1930; Ostrom 1970*a*). The pubis extends ventrally medial to the ischium as a short, laterally compressed blade. In lateral view, the ischium of ankylosaurids has a slight curve, concave cranially, and a slight caudal slant below the acetabulum (Coombs 1979, 1986). In nodosaurids, the ischium is directed more caudally below the acetabulum, and there is a sharp ventral flexion at about midlength (Ostrom 1970*a*; Gilmore 1930). Most of the vertical wall of the acetabulum is formed by the proximal end of the ischium. In all undoubted ankylosaurs, the acetabulum is completely closed.

The femur of ankylosaurs is typically straight as seen in either cranial or lateral view (fig. 22.12). The head is directed upward but is not set off on a distinct neck. In ankylosaurids, the head is almost terminal, while in nodosaurids, it is displaced to the medial side of the shaft. The combined greater and lesser trochanters form a shoulder at the lateral margin of the proximal end of the femur and are only weakly distinguished

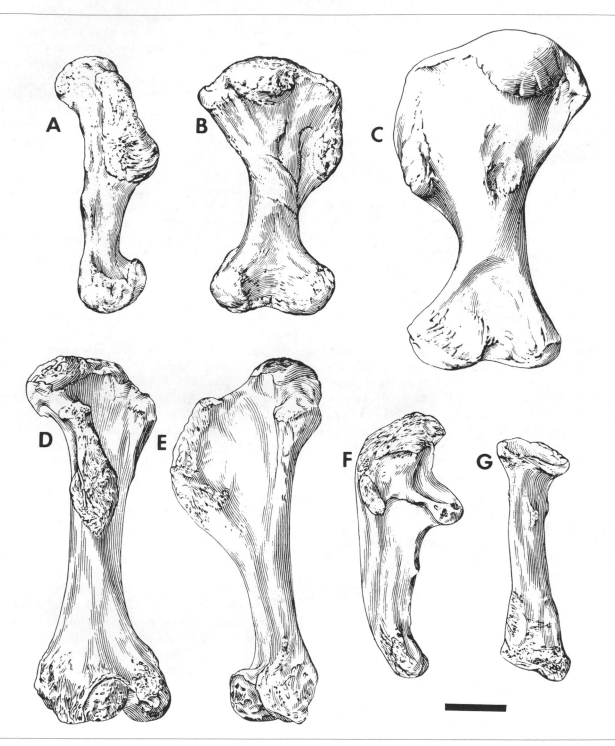

Fig. 22.10. A, B: Humerus of *Euoplocephalus tutus* (Ankylosauridae), lateral and caudal views. C: Humerus of *Ankylosaurus magniventris* (Ankylosauridae), caudal view. D, E: Humerus of *Sauropelta edwardsi* (Nodosauridae), cranial and medial views. F, G: Ulna and radius of *Sauropelta edwardsi* (Nodosauridae). Scale = 10 cm. (After Coombs 1978*a*, 1978*b*.)

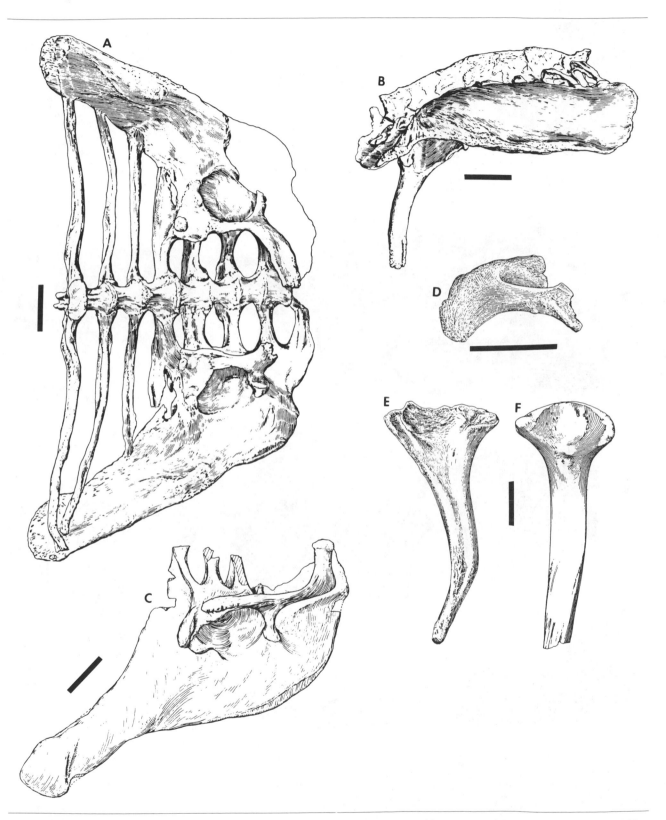

Fig. 22.11. A, B: Pelvic girdle and sacral complex of a *Euoplocephalus tutus* (Ankylosauridae), ventral and lateral views. C: Ilium and sacral ribs of *Sauropelta edwardsi* (Nodosauridae), ventral view. D: Pubis of *Edmontonia rugosidens* (Nodosauridae), medial view. E: Ischium of *Edmontonia rugosidens* (Nodosauridae), lateral view. F: Ischium of *Ankylosaurus magniventris*, lateral view. Scale = 10 cm. (D, E after Gilmore 1930; others after Coombs 1978a, 1979.)

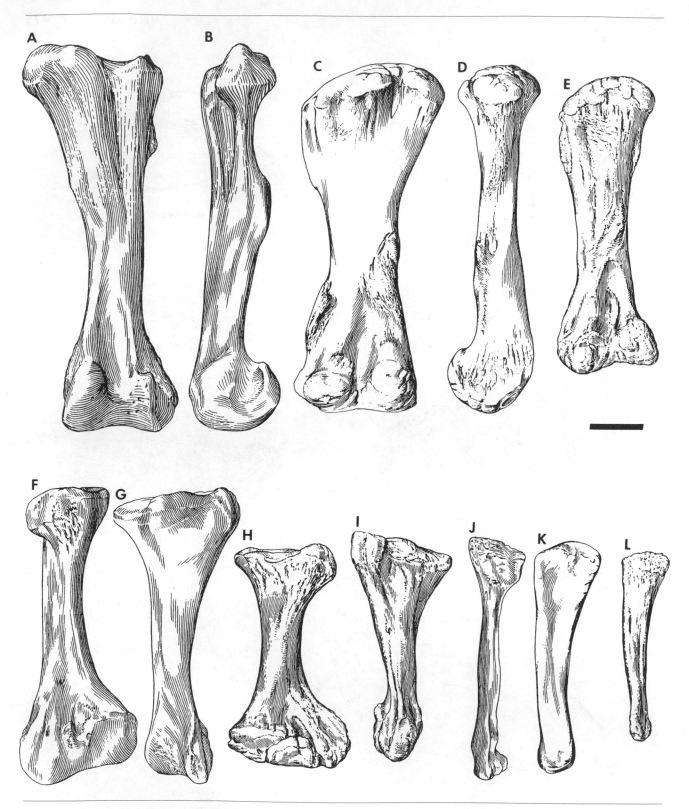

Fig. 22.12. A, B: Femur of *Sauropelta edwardsi* (Nodosauridae), cranial and medial views. C, D: Femur of *Ankylosaurus magniventris* (Ankylosauridae), caudal and medial views. E: Femur of *Euoplocephalus tutus* (Ankylosauridae), caudal view. F, G: Tibia of *Sauropelta edwardsi* (Nodosauridae), cranial and lateral views. H, I: Tibia of *Euoplocephalus tutus* (Ankylosauridae), cranial and lateral views. J: Fibula of *Sauropelta edwardsi* (Nodosauridae), cranial view. K: Fibula of *Ankylosaurus magniventris* (Ankylosauridae), cranial view. L: Fibula of *Euoplocephalus tutus* (Ankylosauridae), cranial view. Scale = 10 cm. (After Coombs 1978a, 1979.)

from the head. A separate, distinct lesser trochanter is absent from the majority of ankylosaurs (possible exceptions: *Hoplitosaurus marshi* and *"Cryptodraco" eumerus*; Galton 1980*c*, 1983*b*). In all ankylosaurs, the fourth trochanter is set closely against the femoral shaft, is never protruding and pendent, and is more proximal in nodosaurids than in ankylosaurids, being at or distal to femoral midlength in the latter.

The tibia of ankylosaurs is straight, wide craniocaudally at the proximal end, and wide transversely at the distal end (fig. 22.12). The astragalus is normally fused onto the tibia in adult ankylosaurs (Coombs 1979, 1986). The slender fibula may have a slight curvature to the shaft and may have the calcaneum coossified to its distal end (e.g., *Nodosaurus texilis* (= *Hierosaurus coleii*); Mehl 1936), although the calcaneum is indistinguishable in the majority of ankylosaurs. A separate calcaneum and astragalus have been reported in a subadult *Euoplocephalus tutus* (Coombs 1986).

The ankylosaur pes is known from several good specimens. Phalangeal formulae are as follows: *Euoplocephalus tutus*, three metatarsals, 0−3−4−5−0 (Coombs 1986); *Pinacosaurus grangeri*, four metatarsals (Maryańska 1977); *Talarurus*, four metatarsals, 2−3−4−4−0 (Maryańska 1977); *Tarchia gigantea*, four(?) metatarsals (Maryańska 1977); *Sauropelta edwardsi*, five metatarsals, 2−3−4−4/5−0 (Ostrom 1970*a*); *Nodosaurus textilis*, four or five metatarsals, 2−3−4−5−4 (Lull 1921; Mehl 1936). The last phalangeal count is suspect because no other ornithischian is known to have four phalanges in digit V of the pes. Metatarsals of ankylosaurs are more elongate and slender than the metacarpals and are less divergent distally. Pedal phalanges, including the unguals, are similar to comparable elements of the manus.

Ankylosaur armor is composed of a series of large, oval to rectangular plates arranged in pairs in roughly transverse rows with a mosaic of much smaller plates filling the spaces between larger plates (fig. 22.13). The larger plates commonly bear a longitudinal ridge or keel that tends to be higher on plates toward the flanks of the body, although in some species these ridges may be higher near the midline in some rows (e.g., *Euoplocephalus tutus;* Carpenter 1982*a*). Transverse alignment of large plates, as well as the large plates themselves, may be absent from the caudal half of the body. Nodosaurids but not ankylosaurids may have tall conical spikes or spines that have heights at least twice basal diameter. These spikes are positioned at the flanks of the body, especially over the shoulders. In some taxa, there are spines along the sides of the neck (e.g., *Sauropelta edwardsi*; Carpenter 1984). In *Edmontonia longiceps* and *E. rugosidens*, the spike-tips

project cranially. Fusion of plates over the caudal half of the body to form a solid carapace has been reported for some species (e.g., *Hylaeosaurus armatus* (= *Polacanthus foxii*); Nopcsa 1905*a*; Blows 1987). In ankylosaurids, the keeled plates are commonly excavated on the medial surface and are consequently quite thin even when there is a tall keel, but nodosaurid plates tend to be flat or only slightly cupped on their medial surface. The first and second transverse rows of keeled plates tend to unite to form a pair of yokes or half-rings around the neck. In ankylosaurids, the individual plates of these yokes are well separated, retain their oval shape, and appear to be fused onto a separate underlying transverse band of bone, but there are exceptions. In nodosaurids, the adjacent individual plates of the half-rings are larger than those of ankylosaurids and tend to abut along their common margins to become more quadrilateral than oval and to form the half-rings by fusion to adjacent plates.

SYSTEMATICS AND EVOLUTION

From the discovery of *Hylaeosaurus armatus* in 1833 up to the 1970s, there was no consensus on subdivisions within the Ankylosauria. Numerous family and subfamily level names were proposed, and genera and species were shuffled among these in a bewildering diversity of arrangements (reviews by Hennig 1924; Coombs 1971). Coombs (1978*a*) proposed that all known ankylosaurs (table 22.1) belonged in one of two families, Nodosauridae and Ankylosauridae, and pointed out the numerous differences between these groups, although formal diagnoses (Coombs 1971) were not published. This two-family system has been widely adopted (e.g., Maryańska 1969, 1971, 1977; Carpenter 1982*a*, 1984; Charig 1979, Sereno 1984, 1986; Bakker 1986). Sereno (1984, 1986) attempted a phylogenetic analysis of the two families using stegosaurs as an outgroup and arrived at 11 autapomorphies for the Nodosauridae and 32 for the Ankylosauridae. Most of these characters are among those listed by Coombs (1971, 1978*a*) but with many plesiomorphies eliminated and some new characters added. The list of diagnostic features given here for each of the families follows Sereno (1986) with some modifications.

Nodosauridae

Diagnostic features of the Nodosauridae include the following: hourglass-shaped palate; basipterygoid processes consisting of a pair of rounded, rugose stubs;

Fig. 22.13. Upper: Skull and thoracic armor of *Edmontonia rugosidens* (Nodosauridae), dorsal view. Lower: Armor of *Sauropelta edwardsi* (Nodosauridae), dorsal view. The tail vertebrae are mostly reconstructed in the lower specimen.

TABLE 22.1 Ankylosauria

	Occurrence	Age	Material
Thyreophora Nopcsa, 1915			
Eurypoda Sereno, 1986			
Ankylosauria Osborn, 1923a			
Ankylosauridae Brown, 1908			
Amtosaurus Kurzanov et Tumanova, 1978			
A. magnus Kurzanov et Tumanova, 1978 (may be hadrosaurid)	Baynshirenskaya Svita, Omnogov, Mongolian People's Republic	Cenomanian-Turonian	Fragmentary skull
Ankylosaurus Brown, 1908			
A. magniventris Brown, 1908	Hell Creek Formation, Montana, Lance Formation, Wyoming, United States; Scollard Formation, Alberta, Canada	late Maastrichtian	3 specimens, including skulls and postcranial material
Euoplocephalus Lambe, 1910 (= *Stereocephalus* Lambe, 1902, *Anodontosaurus* Sternberg, 1929b, *Dyoplosaurus* Parks, 1924, *Scolosaurus* Nopcsa, 1928c)			
E. tutus (Lambe, 1902) (= *Stereocephalus tutus* Lambe, 1902, including *Palaeoscincus asper* Lambe, 1902, *Dyoplosaurus acutosquameus* Parks, 1924, *Scolosaurus cutleri* Nopcsa, 1928a, *Anodontosaurus lambei* Sternberg, 1929b)	Judith River Formation, Two Medicine Formation, Montana, United States; Judith River Formation, Horseshoe Canyon Formation, Alberta, Canada	late Campanian-early Maastrichtian	More than 40 specimens, including 15 complete or partial skulls, isolated teeth, and 1 nearly complete skeleton with *in situ* armor
Maleevus Tumanova, 1987			
M. disparoserratus (Maleev, 1952b) (= *Syrmosaurus disparoserratus* Maleev, 1952a)	Baynshirenskaya Svita, Omnogov, Mongolian People's Republic	Cenomanian-Turonian	Fragmentary skull
Pinacosaurus Gilmore, 1933b (= *Syrmosaurus* Maleev, 1952a)			
P. grangeri Gilmore, 1933b including *P. ninghsiensis* Young, 1935b, *Syrmosaurus viminicaudus* Maleev, 1952a)	Djadochta Formation, Red Beds of Alag Teg, Omnogov, Mongolian People's Republic; unnamed unit, Ningxia Huizu Zizhiqu, People's Republic of China	?late Santonian early Campanian	More than 15 specimens, including 1 complete and 4 partial skulls, nearly complete skeleton with armor
Saichania Maryańska, 1977			
S. chulsanensis Maryańska, 1977	Barun Goyot Formation, Red Beds of Kermeen Tsav, Omnogov, Mongolian People's Republic	?middle Campanian	3 specimens, including 2 2 complete skulls, nearly complete postcranial skeleton and armor
Shamosaurus Tumanova, 1983			
S. scutatus Tumanova, 1983	Khukhtekskaya Svita, Dornogov, Ovorkhangai, Mongolian People's Republic	Aptian-Albian	3 specimens, including complete skull and jaw, partial skeleton, armor
Talarurus Maleev, 1952b			
T. plicatospineus Maleev, 1952b	Baynshirenskaya Svita, Dornogov, Omnogov, Mongolian People's Republic	Cenomanian-Turonian	More than 5 specimens, including 2 partial skulls, nearly complete skeleton and armor
Tarchia Maryańska, 1977			
T. gigantea (Maleev, 1956) (= *Dyoplosaurus giganteus* Maleev, 1956, including *T. kielanae* Maryańska, 1977)	Barungoyotskaya Svita, Nemegt Formation, White Beds of Khermeen Tsav, Omnogov, Barungoyotskaya Svita, Nemegt Formation, Nemegtskaya Svita, White Beds of Khermeen Tsav, Bayankhongor, Mongolian People's Republic	middle Campanian-early Maastrichtian	7 specimens, including complete and partial skulls with nearly complete postcranial skeleton and armor
Nodosauridae Marsh, 1890a			
Acanthopholis Huxley, 1867 (= *Syngonosaurus* Seeley, 1879, *Eucercosaurus* Seeley, 1879)			
A. horridus Huxley, 1867 (including *A. macrocercus* Seeley, 1869 *partim*, *A. platypus* Seeley, 1869 *partim*, *A. stereocercus* Seeley, 1869 *partim*)	Cambridge Greensand, Cambridgeshire, Upper Greensand, Kent, England	late Aptian-Cenomanian	Teeth, postcranial elements, armor plates, mostly fragmentary

Table 22.1, continued

Denversaurus Bakker, 1988			
D. schlessmani Bakker, 1988	Lance Formation, South Dakota, United States	late Maastrichtian	Skull, teeth, armor fragments
Dracopelta Galton, 1980a			
D. zbyszewskii Galton, 1980a	unnamed unit, Estremadura, Portugal	Kimmeridgian	Partial rib cage and armor
Edmontonia Sternberg, 1928			
E. longiceps Sternberg, 1928	Judith River Formation, Montana, United States; Judith River Formation, Horseshoe Canyon Formation, St. Mary River Formation, Alberta, Canada	late Campanian-early Maastrichtian	4 specimens, including complete skulls, much of the skeleton and armor
E. rugosidens (Gilmore, 1930) (= *Palaeoscincus rugosidens* Gilmore, 1930, type of *Chassternbergia* Bakker, 1988)	Judith River Formation, Two Medicine Formation, Montana, Aguja Formation, Texas, United States; Judith River Formation, Alberta, Canada	Campanian	4 specimens, including complete skulls, much of the skeleton and armor
Hoplitosaurus Lucas, 1902a			
H. marshi (Lucas, 1901) (= *Stegosaurus marshi* Lucas, 1901)	Lakota Formation, South Dakota, United States	Barremian	Partial postcranial skeleton with armor plates
Hylaeosaurus Mantell, 1833 (= *Polacanthus* Huxley, 1867, *Polacanthoides* Nopcsa, 1928a, *Vectensia* Delair, 1982)			
H. armatus Mantell, 1833 (= *Hylosaurus mantelli* Fitzinger, 1843, *Hylaeosaurus oweni* Mantell, 1844, *Polacanthus foxii* Hulke, 1881, *P. becklesi* Hennig, 1924, *Polacanthoides ponderosus* Nopcsa, 1929a)	Wealden (Tunbridge Wells Sand/Grinstead Clay), West Sussex, East Sussex, Wealden (Wessex Formation, Vectis Formation, ?Ferruginous Sands), Isle of Wight, England; unnamed unit, Ardennes, France	late Valanginian-Barremian	2 fragmentary postcranial skeletons with dermal isolated postcranial elements and armor plates
Minmi Molnar, 1980			
M. paravertebra Molnar, 1980	Bungil Formation, near Roma, Queensland, Australia	Aptian	Fragmentary postcranial skeleton with armor
Nodosaurus Marsh, 1889b			
N. textilis Marsh, 1889b (including *Stegopelta landerensis* Williston, 1905, *Hierosaurus coleii* Mehl, 1936)	Mowry or Thermopolis Shale, Wyoming, Niobrara Chalk, Kansas, United States	Albian, early Campanian	3 fragmentary postcranial skeletons with armor
Panoplosaurus Lambe, 1919			
P. mirus Lambe, 1919	Judith River Formation, Alberta, Canada	late Campanian	Partial skeleton with complete skull, jaws, and much of the armor, additional isolated teeth, postcranial elements and armor
Sarcolestes Lydekker, 1893c			
S. leedsi Lydekker, 1893c	Lower Oxford Clay, Cambridgeshire, England	Callovian	Partial left mandible
Sauropelta Ostrom, 1970a			
S. edwardsi Ostrom, 1970a	Cloverly Formation, Wyoming Montana, Cedar Mountain Formation, Utah, United States	late Aptian-Cenomanian	Several partial skeletons, 1 crushed skull, numerous isolated postcranial elements and armor plates
Silvisaurus Eaton, 1960			
S. condrayi Eaton, 1960	Dakota Formation, Kansas, United States	late Aptian-Cenomanian	Partial skeleton with skull
Unnamed nodosaurid (= *"Struthiosaurus" transilvanicus* Nopcsa, 1915, including *Danubiosaurus anceps* Bunzel, 1871 *partim*, *Crataeomus lepidophorus* Seeley, 1881, *C. pawlowitschii* Seeley, 1881 *partim*)	Sinpetru Beds, Hunedoara, Romania; Gosau Formation, Neue Welt, near Wiener Neustadt, Niederöstereich, Austria	Campanian-Maastrichtian	Partial skeleton with fragmentary skull and additional postcranial elements

Nomina dubia	Material
Acanthopholis eucercus Seeley, 1869 (doubtfully ankylosaurian)	2 caudal centra
Brachypodosaurus gravis Chakravarti, 1934	Humerus
Cryptosaurus eumerus Seeley, 1869 (type of *Cryptodraco* Lydekker, 1889)	Femur
Danubiosaurus anceps Bunzel, 1871 *partim* (*partim* type of *Pleuropeltus suessi* Seeley, 1881; *partim* type of *Crataeomus pawlowitschii* Seeley, 1881)	Indeterminate fragments
Hierosaurus sternbergi Wieland, 1909	6 dermal plates
Hoplosaurus ischyrus Seeley, 1881	Fragmentary humerus, scapula, coracoid, vertebrae, armor

Table 22.1, continued

Iguanodon phillipsii Seeley, 1869 (type of *Priodontognathus* Seeley, 1875)	Maxilla with several teeth
Lametasaurus indicus Matley, 1923 *partim*	Sacrum, ilia, tibia, spines, armor
Leipsanosaurus noricus Nopcsa, 1918	Tooth
Palaeoscincus costatus Leidy, 1856	Tooth
Palaeoscincus latus Marsh, 1892	Tooth
Peishansaurus philemys Bohlin, 1953 (doubtfully ankylosaurian)	Very fragmentary jaw with 1 tooth
Priconodon crassus Marsh, 1888a	Tooth
Rhadinosaurus alcinus Seeley, 1881 (doubtfully ankylosaurian)	Indeterminate fragments, ?2 humeri, ?2 vertebrae
Rhodanosaurus ludgunensis Nopcsa, 1929a	Dermal plates
Sauroplites scutiger Bohlin, 1953	?Ilium, ribs, armor
Stegosaurides excavatus Bohlin, 1953	Very fragmentary vertebrae, spine base, isolated elements
Struthiosaurus austriacus Bunzel, 1871	Fragmentary basicranium

hemispherical occipital condyle composed of basioccipital only, set off from braincase on a short neck and angled about 50° downward from line of maxillary tooth row; quadrate angled rostroventrally; skull roof with large plate between orbits, rostrocaudally narrow plate along caudal edge of skull; three scutes above the orbits and additional small scutes along the snout rostral to the orbits; scapular spine displaced toward glenoid; coracoid large and craniocaudally long relative to dorsoventral width; and ischium ventrally flexed near midlength.

At least 11 valid genera of the Nodosauridae have been reported from the Callovian (Galton 1980c, 1983b) to the Maastrichtian (Coombs 1971, 1978a; Carpenter and Breithaupt 1986), but the family is primarily Cretaceous. Nodosaurids are known from Europe, North America, and Australia. Some armor fragments from South America might be nodosaurid (described as "*Lametasaurus indicus*": Matley 1923; Huene and Matley 1933). Gasparini (pers. comm.) reports some fragments, including teeth, that appear to be nodosaurid from James Ross Island off the coast of Antarctica. Nodosaurids are not found in Asia or Africa (doubtful occurrence reported by Lapparent 1958). Galton 1980b, 1980c, 1983b, 1983c) has assigned four European Jurassic species to the Nodosauridae, although some of these are herein regarded as *nomina dubia*.

European nodosaurids are generally small animals, with a total length generally less than 3.5 m. Many species are based on very fragmentary remains, and some of these are herein regarded as *nomina dubia*. However, several taxa have been retained despite our reservations regarding the adequacy of the type specimens.

The single fragmentary mandible of late Callovian age which constitutes the type of *Sarcolestes leedsi* can be recognized as an ankylosaur by the following features: small teeth with marginal denticles; teeth extend to rostral end of dentary; short medially projecting symphyseal process; low coronoid eminence; and dermal plate welded to the lower edge of the mandible (Galton 1980c, 1983b, 1983c). Assignment to the Nodosauridae is based on the large size of the teeth relative to the mandible and the presence of basal cingula.

The Kimmeridgian *Dracopelta zbyszewskii* has typical nodosaurid dermal plates accompanied by ankylosaurian curvature of the ribs and ossified tendons (Galton 1980a, 1983b).

Hylaeosaurus armatus is a moderate to large nodosaurid (length 3 to 5 m) characterized by a scapular spine that slants obliquely across the entire scapular blade. This feature together with absence of a prespinous fossa makes *H. armatus* the most primitive known nodosaurid (fig. 22.14; Coombs 1971, 1978a, 1978b; Sereno 1986). The scapula and coracoid are separate as in "*Struthiosaurus*" *transilvanicus* (Nopcsa 1929a) and one specimen of *Edmontonia rugosidens*. Blows (1987) attempted to separate "*Polacanthus foxii*" and "*Polacanthoides ponderosus*" from *H. armatus*, but we find the characters listed unconvincing for specific, much less generic, diagnosis. Regrettably, there are no homologous elements in the type specimens of *H. armatus* and "*P. foxii*," so the question of their synonymy will never be fully resolved. "*Vectensia*" Delair (1982), based on a single dermal plate from the Isle of Wight, is another probable synonym of *H. armatus*. Fragments possibly assignable to *H. armatus* have been reported from Spain (Sanz 1983).

Assignment of *Acanthopholis horridus* to the Nodosauridae is based primarily on tooth structure and presence of tall conical spines and keeled plates that are flat or slightly convex medially. *A. horridus* appears to be a small nodosaurid, and although the species is retained here, it may be a *nomen dubium* (Coombs 1971).

The Campanian to Maastrichtian *Struthiosaurus* is in need of revision and its present status is uncertain. The type species, *S. austriacus* Bunzel (1871), is based on a small, isolated, fragmentary basicranium that has

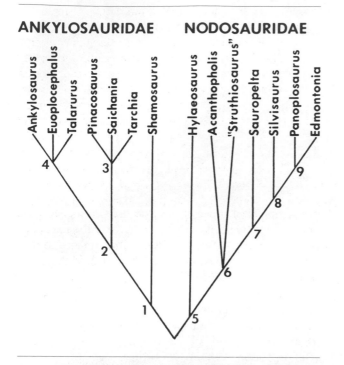

ANKYLOSAURIDAE   **NODOSAURIDAE**

Fig. 22.14. Cladogram of the Ankylosauria. Characters for the nodes are as follows: 1. Derived characters of the Ankylosauridae (see text). 2. Orbits more caudally positioned and facing more laterally; quadrate condyle below orbit; occipital condyle crescentic and not protruding on a distinct neck; squamosal horns large. 3. Rostrodorsal skull roof covered by 14 to 18 large plates; squamosal horns narrow in dorsal view; laterally flaring naris circular in rostral view; short premaxillary beak rostral to naris. 4. Rostrodorsal skull roof covered by over 30 small plates; squamosal horns broad based and pyramidal in dorsal view. 5. Derived characters of the Nodosauridae (see text). 6. Scapular spine knoblike; prespinous fossa cranial to scapular spine. 7. Size increase; scapula and coracoid fused (exception: one specimen referred to *Edmontonia rugosidens*). 8. Secondary palate present rostrally from middle of maxillary tooth row. 9. No premaxillary teeth; atlas and axis fused. The characters for node 7 are weak because of the incomplete material available for taxa below this node. *Nodosaurus* probably fits just above or just below node 8.

large ventrally projecting basisphenoid, a saurischianlike feature. *"Struthiosaurus" transilvanicus* Nopcsa (1915, 1929*a*) is based on a partial skull and skeleton of about 2 to 3 m total length. The scapulocoracoid has the displaced scapular spine and prespinous fossa diagnostic of derived nodosaurids, and the dermal armor includes typical nodosaurid spikes. The skull fragment has a closed supratemporal fenestra, narrow infratemporal fenestra, and fused paroccipital process and squamosal, all features of nodosaurids. Several other

Late Cretaceous European taxa are herein listed as synonyms of *"Struthiosaurus" transilvanicus* (table 22.1; see also Romer 1956, 1968), although the extremely fragmentary nature of the type specimens makes assignment even to the Ankylosauria tenuous in some cases. Additional fragmentary specimens from the Late Cretaceous of Europe may also be assigned to this species (e.g., Lapparent 1947, 1954).

North American nodosaurids are generally large animals with total body lengths commonly exceeding 5 m. *Sauropelta edwardsi* (Ostrom 1970*a*) is the oldest North American nodosaurid known from abundant material. It is characterized by premaxillary teeth, absence of a secondary palate, separate atlas and axis, and a distinctive pattern of armor (Carpenter 1984).

The Albian *Nodosaurus textilis* Marsh (1889*b*) has been characterized by distal limb elements that are relatively longer than proximal elements and distinctive armor (K. Carpenter pers. comm.). However, despite the familiarity of the name, the type specimen is rather poor and may not be adequate to define the taxon.

Hoplitosaurus marshi is based on badly distorted material including some undoubtedly nodosaurid elements, especially a scapulocoracoid with a displaced scapular spine (Gilmore 1914*a*; Galton 1983*b*) and large armor spikes. The presence of a lesser trochanter distinct from the greater trochanter is a feature known otherwise only in the enigmatic femur of *"Cryptodraco" eumerus* (Galton 1980*c*, 1983*b*). All other femora assigned to either the Ankylosauria or Stegosauria have the lesser trochanter fused to the greater (Sereno 1986). *H. marshi* has been regarded as a *nomen dubium* by some authors (Ostrom 1970*a*; Coombs 1971, 1978*a*) but as a valid taxon by others (Galton 1980*a*, 1980*c*, 1983*b*). *Hoplitosaurus marshi* is hesitantly retained here as a separate genus and species (table 22.1).

Silvisaurus condrayi is of medium size (length approx. 4 m), characterized by a skull about threefourths as wide as long, with a toothed premaxilla and an infratemporal fenestra that is hidden below a caudoventrally expanded jugal-quadratojugal-postorbital (Eaton 1960). The latter condition is unique within the Nodosauridae and differs from the diagnostic ankylosaurid condition in which dermal plates cover the infratemporal fenestra. Therefore, this feature is not indicative of a special relationship between *S. condrayi* and the Ankylosauridae.

Panoplosaurus mirus is a large animal (length approx. 6 m) characterized by a relatively wide skull (width exceeding four-fifths of length), external nares well separated dorsally, toothless premaxilla, and absence of tall spines in the armor (Lambe 1919; Sternberg 1921; Carpenter and Breithaupt 1986; K. Carpenter pers. comm.). The atlas and axis vertebrae are coossified.

Edmontonia longiceps is at least as large as *Panoplosaurus mirus* (length approx. 6-7 m). It is characterized by a relatively long, narrow skull with a quadrate more oblique than in *P. mirus* (fig. 22.2A). The premaxillary palate is longer than wide. As in *P. mirus,* the atlas and axis vertebrae are coossified (Sternberg 1928). The skull of *E. rugosidens* is pyriform in dorsal view, the premaxillary palate is wider than long, and the atlas and axis are coossified (Gilmore 1930). There are large, cranioventrally directed, double-pointed spines over the shoulder in *E. longiceps* and *E. rugosidens* (fig. 22.13 upper). *Edmontonia rugosidens* is known from the early Maastrichtian (Carpenter and Breithaupt 1986), making it the last of the nodosaurids.

Minmi paravertebra from the Aptian of Australia (Molnar 1980*a*) is a small nodosaurid (length approx. 2 m) and the only unequestionable ankylosaur from Gondwanaland. The outstanding feature of *M. paravertebra* is the presence of elongate bony rods, apparently modifed ossified tendons, located between the neural spines and the tips of the transverse processes. These structures have been called "paravertebrae" (Molnar 1980*a*).

Ankylosauridae

Diagnostic features of the Ankylosauridae include the following: maximum width of skull equal to or greater than length; snout arches above level of postorbital skull roof; external nares divided by premaxillary septum with ventral or lateral opening leading into a maxillary sinus; caudal process of premaxilla along margin of beak extends lateral to most mesial teeth; premaxillary palate wider than long; no ridge separating the premaxillary palate from the lateral maxillary shelf; edge of premaxillary beak continuous with lateral edge of maxillary shelf; complex secondary palate twists respiratory passage into vertical S-shaped bend; paired sinuses in the premaxilla, maxilla, and nasal; postorbital shelf composed of postorbital and jugal bones extends farther medially and ventrally than in nodosaurids; near-horizontal epipterygoid contacts pterygoid and prootic; prominent, wedge-shaped caudolaterally projecting quadratojugal dermal plate; large, wedge-shaped, caudolaterally projecting squamosal dermal plate; infratemporal fenestra, paroccipital process, quadratojugal, and all but quadrate condyle hidden in lateral view by united quadratojugal and squamosal dermal plates; sharp lateral rim and low dorsal

prominence for each lateral supraorbital element; flat lateral supraorbital margin above the orbit; skull roof covered by numerous small scutes including paired premaxillary scutes, a median nasal scute, two scutes laterally between the orbit and external naris, scutes on each supraorbital element, single scute on the squamosal and on the quadratojugal dermal plates; scute pattern poorly defined in supraorbital region; coronoid process lower and more rounded than in nodosaurids; distal caudal centra partially or complete fused and with elongate prezygapophyses that are broad dorsoventrally and elongate postzygapophyses that are united to form a single dorsoventrally flattened, tonguelike process; hemal arches of distal caudals elongate with zygapophysis-like overlapping processes and elongate bases that contact to form a fully enclosed hemal canal; fused sternal plates; deltopectoral crest and transverse axis through distal humeral condyles in same plane; postacetabular process of ilium short; ischium near-vertical below acetabulum; pubis reduced to nubbin fused to other pelvic elements; fourth trochanter on distal half of femur; terminal tail club of large pair of lateral plates and two smaller terminal plates; and ossified tendons surrounding distal caudal vertebrae.

Ankylosaurids are known only from the Cretaceous (Aptian through Maastrichtian) and only from North America and Asia. Nine monospecific genera are recognized.

The Aptian-Albian *Shamosaurus scutatus,* the earliest Asiatic ankylosaurid, is regarded as the most primitive member of the family (fig. 22.14; Tumanova 1983, 1986, 1987; Sereno 1986). It is a medium-sized animal (skull length to 360 mm, width to 370 mm) characterized by narrow premaxillary beak not covered by dermal ossifications; laterally facing orbits positioned more rostrally than in other ankylosaurids; quadrate weakly inclined rostrally; quadrate condyle positioned behind the caudal margin of the orbit; slitlike nostril facing rostrolaterally; and nostril divided by a vertical premaxillary septum. The caudal part of the skull roof is flat with weakly developed, hornlike dermal projections. The occipital condyle extends farther from the braincase than in other ankylosaurids. The rostral part of the secondary palate forms a horizontal plate that extends caudally from the premaxillary to the midpoint of the maxillary tooth row, a pattern similar to that in derived nodosaurids (e.g., *Edmontonia longiceps*). The tooth row is relatively long in *Shamosaurus scutatus,* equaling about half of total skull length. The lower jaw is relatively longer and lower than in other ankylosaurids. Nothing is known of the postcranial skeleton.

Talarurus plicatospineus is a medium-sized ankylosaurid (length approx 4-5 m) that has a relatively long narrow skull (length approx. 240 mm, width approx. 220 mm). The occipital condyle is partially visible in dorsal view, a feature unique within the Ankylosauridae. Maxillary teeth have swollen bases cut by W-shaped furrows on the labial side. The manus is pentadactyl, the pes tetradactyl. Postcranial elements are wide relative to their length, the armor plates are ribbed, and the tail club is small.

Maleevus disparoserratus and *Amtosaurus magnus* each differ from *Talarurus plicatospineus* in various details and proportions of the basicranial region (Kurzanov and Tumanova 1978; Tumanova 1987). Maryańska (1977) regarded these differences as sufficient for specific diagnosis but pointed out details in the pattern of cranial foramina that are also present in *T. plicatospineus* and no other known ankylosaurid. In deference to Tumanova, both *Maleevus disparoserratus* and *Amtosaurus magnus* are retained with reservation.

The Santonian-Campanian *Pinacosaurus grangeri* is a medium-sized animal (length approx. 5 m; skull length to 300 mm, width to 340 mm) that is represented by abundant material including an excellent skull and postcranial skeleton of a juvenile. The premaxilla is not covered by dermal ossifications, and the large oval nostril faces rostrally and is divided by a horizontal septum. At least in juveniles, the nasal area has a laterally situated third opening that leads into a premaxillary sinus. The beak is slightly wider than the distance between the caudalmost maxillary teeth. The mandibular condyle of the quadrate is positioned below the caudal margin of the orbit. The postcranial elements, especially the limb bones, are relatively slender. The manus is pentadactyl and the pes tetradactyl.

The large Mongolian ankylosaurid *Saichania chulsanensis* (length to 7 m; skull length to 450 mm, width to 480 mm) is characterized by large oval external naris situated rostrally, divided by a horizontal septum; relatively small number of scutes cover the skull roof (fig. 22.11C); premaxilla covered by well-developed dermal ossifications; width of the beak less than distance between caudalmost maxillary teeth (fig. 22.4B); exoccipital very low; quadrate fused to paroccipital process and strongly inclined rostrally; and mandibular condyle below the middle of the orbit. The postcranial skeleton is very massive. Dermal armor covers the ventral and dorsal sides of body, a feature unique within the Ankylosauria, although the armor of many taxa is incompletely known.

The youngest and largest Mongolian ankylosaurid is *Tarchia gigantea* (body length to 8 m; skull length to 400 mm, width to 450 mm). In this species, the premaxilla is only partially covered by dermal ossifications, the width of the beak is equal to the distance between the caudalmost maxillary teeth, and the premaxillary area of the palate has subequal length and width. A very high exoccipital is not fused to the quadrate, and the mandibular condyle is positioned below the caudal part of the orbit. The pes is tetradactyl and the tail club is large.

Two monospecific genera of North American ankylosaurids are recognized, following Coombs (1971, 1978a), although K. Carpenter (pers. comm.) believes that some of the species here considered junior synonyms of *Euoplocephalus tutus* may warrant retention as separate taxa (table 22.1).

Euoplocephalus tutus is a medium- to large-sized ankylosaurid (length 6-7 m; skull length to 460 mm, width to 475 mm) that is represented by more skulls than any other ankylosaur (table 22.1). Important features include the following: premaxillae not covered by dermal ossifications (fig. 22.1D); slitlike external nostril faces rostrally and divided by a vertical septum; width of beak equal to or greater than distance between caudalmost maxillary teeth (fig. 22.1D); and pes tridactyl. Numerous tail clubs associated with *Euoplocephalus tutus* material are highly variable in size and shape. *Euoplocephalus tutus* is the common ankylosaurid of Campanian and early Maastrichtian deposits of western North America.

Ankylosaurus magniventris is the last and largest of all known ankylosaurids (length approx. 8-9 m; skull length to over 750 mm, width to over 770 mm; fig. 22.1A,B). A distinctive feature is a rostral and lateral expansion of dermal armor of the nasal bones that covers the premaxillae and restricts the external nares to small circular openings placed far laterally near the premaxillary-maxillary suture at the margin of the mouth. Other features include the following: beak narrower than distance between caudalmost maxillary teeth (fig. 22.2C); premaxillary area of palate with subequal width and length; and large pyramidal squamosal dermal plates form hornlike projections at caudolateral corners of skull roof (fig. 22.1A,B). The tail club is massive (fig. 22.8A,B).

Discussion

All known ankylosaurs are highly derived, and all reasonably complete specimens can be assigned unambiguously to either the Nodosauridae or the Ankylosauridae (fig. 22.14). Therefore, there is no species that both falls within the definition of the Ankylosauria and is a sister group to the two families. Thus, the divergence of the two families and the evolution of their numerous morphologic differences are not documented by currently known fossils. Divergence of the Nodosauridae and Ankylosauridae probably took place before the Upper Jurassic (Maryańska 1977; Coombs 1978a; Galton 1980c; contra Thulborn 1977).

Both families of the Ankylosauria are very conservative in their evolution. Diagnostic characters are present in the oldest-known species of each family, except that Jurassic nodosaurids are too fragmentary to document significant morphology (Galton 1980a, 1980c, 1980e, 1983b, 1983c). *Hylaeosaurus armatus* from the earliest Cretaceous (late Valanginian) is more primitive than other nodosaurids in lacking the prespinous fossa and knoblike scapular spine of more-derived species (Coombs 1978b; Sereno 1986; Blows 1987). "*Struthiosaurus*" *transilvanicus* and *Acanthopholis horridus* may represent a lineage of small insular European nodosaurids in which premaxillary teeth and a separate atlas and axis may have been retained until very late in the Cretaceous (fig. 22.14). North American nodosaurids are generally much larger animals, among which *Sauropelta edwardsi* is primitive in having premaxillary teeth, a separate atlas and axis, and no secondary palate (Ostrom 1970a; Coombs 1978a). *Silvisaurus condrayi* also retains premaxillary teeth but has developed the characteristic bony palate of derived nodosaurids (Eaton 1960). *Panoplosaurus mirus, Edmontonia rugosidens,* and possibly *E. longiceps* lack premaxillary teeth and have a fused atlas and axis (Sternberg 1921; Gilmore 1930; Russell 1940).

Shamosaurus scutatus is the earliest ankylosaurid (Aptian-Albian), and it has several primitive characters including the following: secondary palate simpler than that of derived ankylosaurids; orbits face laterally and positioned more rostrally than in other ankylosaurids; squamosal horns weakly developed; occipital condyle separated from braincase; teeth possibly larger relative to skull than in derived ankylosaurids (Tumanova 1983, 1986, 1987; Sereno 1986). In all other respects, *S. scutatus* is a typical ankylosaurid, at least in its cranial morphology. Tumanova (1987) separated *S. scutatus* together with *Saichania chulsanensis* into a subfamily Shamosaurinae, regarded as the primitive sister group to all other Ankylosauridae, which were placed in the subfamily Ankylosaurinae (*sensu* Sereno 1986). We agree with the interpretation of *S. scutatus* but oppose subdivision of the Ankylosauridae above the generic level.

All other ankylosaurid taxa can be considered a single lineage with the following general evolutionary trends: later species such as *Ankylosaurus magniventris* are generally larger than earlier species much as *Talarurus plicatospineus* and *Pinacosaurus grangeri;* the caudoventral shelf of the secondary palate is more restricted in earlier species such as *Talarurus plicatospineus* and *Pinacosaurus grangeri* but expands in later species including *Saichania chulsanensis, Tarchia gigantea,* and *Euoplocephalus tutus;* in *Ankylosaurus magniventris,* dermal armor of the skull expands rostrally to cover the premaxillae and restrict the external nares to small circular openings that are not visible in rostral view.

There are two interpretations of ankylosaur relationships within the Ornithischia. In one system, all armored ornithischians and/or all quadrupedal ornithischians are placed in a single group, thus dividing the Ornithischia into bipedal-unarmored and quadrupedal-armored clades. The latter group has always included stegosaurs and ankylosaurs and depending on the author and date, might also include scelidosaurs, ceratopsians, and pachycephalosaurs. Names applied to the group include Scelidosauridae (Huxley 1870; Seeley 1887b), Stegosauria (Marsh 1889b; Hennig 1915a), Orthopoda (Huene 1909; Hennig 1924), Thyreophora (Nopcsa 1915, 1917, 1918, 1923c; Huene 1948), and Stegosauroidea (Lapparent and Lavocat 1955). In the second system, advocated primarily by Romer (1927, 1956, 1966), the Ankylosauria is a suborder equal in rank to the Stegosauria, Ceratopsia, and Ornithopoda. The first three suborders are envisioned as reversions to a quadrupedal stance, each independently evolved from the bipedal Ornithopoda. Recent opinion has generally favored a sister group relationship for stegosaurs and ankylosaurs (Coombs 1972; Maryańska 1977, fig. 13; Colbert 1981: 53; Norman 1984, Sereno 1984, 1986; Maryańska and Osmólska 1984a, 1985; Bakker 1986). The name generally favored for the combined Ankylosauria and Stegosauria has been Nopcsa's (1915) Thyreophora. Adopted here is Sereno's (1986) system, which places stegosaurs and ankylosaurs in a new group, the Eurypoda, which together with *Scelidosaurus harrisonii, Lusitanosaurus liasicus,* and *Scutellosaurus lawleri* constitute the Thyreophora (further discussion in Coombs et al., this vol.). Traditionally, stegosaurs and ankylosaurs have been united because of three features: both groups are quadrupedal; both groups have armor; and tooth structure is similar. Sereno (1986) lists 20 synapomorphies that unite ankylosaurs and stegosaurs, including the following: two lateral supraorbitals (= palpebrals *sensu* Coombs (1972) forming the dorsal margin of the orbit; quadrate condyle angled ventromedially; no otic notch between quadrate and paroccipital process; vertical median portion of pterygoid palatal ramus developed caudally; median palatal keel extending to caudal end of palate; symphyseal region of dentary slender relative to maximum dentary width; dorsal and ventral margins of scapular blade parallel; fourth trochanter of femur a low muscle scar or absent; hypertrophied preacetabular segment of ilium diverging from midline cranially at about 40°; postacetabular segment of ilium short; absence of distal expansion of ischium; and short, spreading metapodials. Sereno (1986) also lists as synapomorphies the fusion of the lesser femoral trochanter to the greater, but the lesser trochanter is separate in *Hoplitosaurus marshi* (Galton 1980c, 1983b). The fourth trochanter of *H. marshi* is near midlength of the femur but is too broken to determine if it is a flat scar or a protruding process.

TAPHONOMY

Most European ankylosaur specimens are isolated, fragmentary elements, with few partial skeletons, so that preservation appears to be haphazard and provides no insights into the biology of the animals. In contrast, Upper Cretaceous ankylosaurs from Asia, especially the Mongolian People's Republic, are commonly preserved as nearly complete or at least partially articulated skeletons. Except for material from Baynshirenskaya Svita, isolated bones are uncommon, and preservation of disarticulated, displaced bones in breccialike sediments indicative of long fluvial transport are almost unknown. Thus, most Mongolian ankylosaurs were buried at or near the site of death, that is, within or close to their normal life habitat. Mongolian ankylosaur-bearing strata are typically eolian deposits or stream channel sediments with rapid rates of accumulation. Data from the Djadochta Formation, Barungoyotskaya Svita, and Nemegt Formation (Gradzinski 1970; Lefeld 1971; Gradzinski and Jerzykiewicz 1974) suggest that Mongolian ankylosaurids lived under warm, semiarid to warm, and rather humid climates. Most Mongolian localities that yield ankylosaurs also produce large predators that might have fed on them. An exception is Bayn Dzak, which has yielded several specimens of *Pinacosaurus grangeri* and other medium-sized herbivorous dinosaurs but only skeletons of predators too small to attack an ankylosaur unless they hunted in packs.

North American ankylosaurs are commonly found in channel deposits either as isolated elements (especially from the Cloverly Formation) or as partial skeletons (especially from the Judith River Formation and various Maastrichtian deposits). Unlike Mongolian specimens, partial skeletons are commonly found upside down (Sternberg 1970), possibly because they bloated and floated in that position before burial. The general climate for at least Late Cretaceous ankylosaurs appears to be more humid in North American than in Asia.

PALEOECOLOGY AND BEHAVIOR

As noted above, ankylosaurs are most commonly found as individual specimens in formations that contain a variety of other dinosaurs. The Cloverly Formation of Montana and Wyoming has produced numerous specimens of *Sauropelta edwardsi*, but these are primarily scattered individual bones (Ostrom 1970a). *Euoplocephalus tutus* specimens from the Judith River Formation of Alberta are most commonly found as isolated elements or partial skeletons (Dodson 1971; Béland and Russell 1978). This mode of preservation suggests solitary habits or small group clusters, although some species may have formed larger aggregates at least intermittently.

Among dinosaurs, ankylosaurs were relatively slow-moving animals, similar to stegosaurs, more fleet than sauropods, but probably not as fast as ceratopsians or bipedal dinosaurs (Coombs 1978c). Both the heavy armor and short massive limbs make high-speed endurance running improbable (Coombs 1978b, 1978c, 1979). Analysis of condyle orientation and footprints ascribed to ankylosaurs suggests that ankylosaurs kept the forelimbs close to the midline with the elbows flexed and close against the body (Coombs 1978b; Carpenter 1984). The humerus was parasagittal when the forelimb was brought forward and the elbow extended but would angle outward as the forelimb was retracted and the elbow flexed. The hindlimbs had a near-vertical, pillarlike orientation and moved primarily in a parasagittal plane (Coombs 1979; Carpenter 1984). Some modifications of the forelimbs, including the unusually positioned scapular spine of derived nodosaurids, suggest high force delivery to the manus (Coombs 1978b), but the behavioral significance of these modifications is unclear. The ankylosaur manus has none of the typical modifications found in diggers. Minor differences in the postcranial skeletons of some genera, especially in the pectoral girdle and forelimb, suggest differences in locomotor ability. For example, *Saichania chulsanensis* appears to be a slower-moving animal than *Pinacosaurus grangeri*.

When attacked by predators, ankylosaurs primarily defended themselves passively, relying on their extensive armor and low-slung, difficult-to-overturn body conformation. Outrunning of predators seems unlikely. Ankylosaurids may have used their tail clubs for active defense by sweeping it just above the ground to strike at the fragile ankles of an attacking predator (Coombs 1979). The club may also have been used for intraspecific combat. Bakker (1986) suggested that nodosaurids used the shoulder spines actively by charging a predator, but this idea seems to us more colorful than plausible.

Ankylosaurs have long been recognized as herbivores (Leidy 1856). Ankylosaur teeth fall into type A of Hotton (1955) indicating a predominantly or exclusively herbivorous diet. The relatively tiny teeth, especially of ankylosaurids, may indicate that nonabrasive plant material was selected or that relatively little mastication was done in the mouth. The majority of plant maceration was probably accomplished by a gizzard, by fermentation, by chemical processes, or most likely by some combination of these (Coombs 1971; Bakker 1986; Farlow 1987b). In connection with lesser running ability than other ornithischians, ankylosaurs may have had a somewhat slower metabolic rate, thus reducing the per diem demand for food and obviating the need for a complex dental battery similar to that of ceratopsids and hadrosaurids. We do not accept the theory of insectivorous habits for ankylosaurs (Nopcsa 1926, 1928c; Maryańska 1977; similar opinion expressed by Carpenter 1982a). Differences in the shape of the beak, narrow in *Shamosaurus scutatus* and most nodosaurids, very broad in some ankylosaurids (e.g., *Euoplocephalus tutus*), may be correlated with differences in foraging strategies, perhaps selective versus unselective browsing (Carpenter 1982a) or perhaps differences in plant morphotypes such as upright and bushy versus low and sprawling. The horny beak is more extensive in forms with exposed premaxillae (e.g., *Euoplocephalus tutus*) than in species with armor covering the premaxillae (e.g., *Ankylosaurus magniventris*), a feature possibly correlated with some dietary difference. The well-developed hyoid apparatus and, at least in *Saichania chulsanensis*, a long entoglossal process may indicate a long flexible tongue in some species as an alternate or accessory food-gathering adaptation (Maryańska 1977).

Initial development of a secondary palate in ankylosaurs was probably a structural solution to the weakened area around the maxillary teeth caused by the near-horizontal rather than vertical orientation of the maxillary shelf. *Sauropelta edwardsi* has no secondary palate, and consequently the dental battery is not as well supported and braced internally as in other ankylosaurs (fig. 22.5D). In most ankylosaurs, the secondary palate forms a brace connecting the lingual side of the tooth row to the sagittal plate of the vomers, which in turn braces against the skull roof (fig. 22.5C,E). Ankylosaurs are occasionally found in marine deposits but not as commonly as Jurassic sauropods and not with sufficient regularity to connect the secondary palate with aquatic or amphibious habits (e.g., Mehl 1936).

There has been relatively little speculation on the function of the complex narial passages of ankylosaurids. In connection with a low head position and relatively sluggish habits, olfaction may have been important to ankylosaurs (Maryańska 1977). The complex narial passages of ankylosaurids, with development of ossified conchal and "turbinal" elements, may have supported an expanded olfactory epithelium, although there is no enlargement of the olfactory lobes in endocasts (Coombs 1978d), and modern mammals with exceptionally well developed olfaction do not necessarily have unusually large and complex narial passages (e.g., Canidae and Ursidae). Other possible functions for the sinuses include warming, moistening or filtering of inhaled air, or some combination of these functions (Maryańska 1977); housing of various glands; formation of a resonating chamber; and strengthening of the skull to compensate for its width and shallowness and the near-horizontal orientation of the maxillary shelf. Strengthening the skull would be an extension of the presumptive function of the secondary palate. These several potential functions are not mutually exclusive.

ORNITHOPODA

DAVID B. WEISHAMPEL

The Ornithopoda represents one of the largest and most long-lived ornithischian clades, known from the earliest Jurassic through the close of the Cretaceous. Ornithopods were long thought to be the stem group for virtually all other ornithischian taxa (Romer 1966, 1968; Galton 1972). However, with the segregation of some members to the Thyreophora (see Weishampel, this vol.), Pachycephalosauria (Maryańska, this vol.), Ceratopsia (Dodson, this vol; Pachycephalosauria and Ceratopsia are together considered the Marginocephalia by Sereno [1986]; see Dodson, this vol.), and more distantly related Ornithischia (e.g., *Lesothosaurus;* Weishampel and Witmer, this vol.), a monophyletic Ornithopoda has been recognized in the studies of Norman (1984*a*), Sereno (1984, 1986), Cooper (1985), and Maryańska and Osmólska (1985).

The Ornithopoda as discussed here includes some of the most diverse dinosaurian herbivore groups (Norman and Weishampel 1985). The group is known to consist of the Heterodontosauridae, Hypsilophodontidae, Dryosauridae, Iguanodontidae, Hadrosauridae, and various intercalated taxa (for details of the internal relationships of these taxa, see Weishampel and Witmer, Sues and Norman, Norman and Weishampel, and Weishampel and Horner, this volume). The monophyly of this group is well corroborated using marginocephalians, thyreophorans, and *Lesothosaurus* as successive outgroups (Sereno 1986). Hence, ornithopods are characterized by the following synapomorphies (modified from Sereno 1986): pronounced ventral offset of the premaxillary tooth row relative to the maxillary tooth row, crescentic paroccipital processes, strong depression of the mandibular condyle beneath the level of maxillary and dentary tooth rows, and caudal elongation of the lateral process of the premaxilla to contact the lacrimal and/or prefrontal, thereby covering the maxilla-nasal articulation (absent in hypsilophodontids). Sereno (1986) called this group the Euornithopoda, but it is here referred to as the Ornithopoda, reflecting the more all-encompassing status of the name and prior usage to include all taxa from heterodontosaurids to hadrosaurids (Euornithopoda may then be applied to the more restricted clade consisting of hypsilophodontians and iguanodontians *sensu* Sereno 1986).

Heterodontosaurids represent the primitive sister group of all other ornithopods (Weishampel and Witmer, this vol.). This higher monophyletic assemblage (hypsilophodontids, dryosaurids, iguanodontids, and hadrosaurids, among others) consists of those forms possessing a moderate-size antorbital fossa, loss of the external mandibular fenestra, elongation of the prepubic process beyond the preacetabular process of the ilium, and a tabular obturator process on the ischium (modified from Sereno 1986).

Remaining higher ornithopods (i.e., *Tenontosaurus,* dryosaurids, *Camptosaurus,* iguanodontids, *Probactrosaurus,* hadrosaurids; termed the Iguanodontia, Sereno 1986) can be diagnosed on the basis of loss of premaxillary teeth, eversion of the oral margin of the premaxilla, relatively small or absent antorbital fossa, enlargement of the external naris relative to the orbit, predentary with bilobate ventral process, denticulate oral margin of the predentary, reduction of manual digit III to only three phalanges, and femur with a deep flexor groove between distal condyles, among others (modified from Sereno 1986). Successive sister group relationships among iguanodontians are covered in respective chapters (Sues and Norman, Norman and Weishampel, and Weishampel and Horner, this vol.). It should be pointed out, however, that a few of Sereno's systematic assessments have been questioned (see Sues and Norman and Norman and Weishampel, this vol.). Among them are the relative positions of *Tenontosaurus, Ouranosaurus,* and *Probactrosaurus.* Seemingly, the quality and quantity of material of the first two taxa are sufficient for an unambiguous placement of these animals on a cladogram, but *Probactrosaurus* is clearly in need of further documentation and analysis, as pointed out by all authors.

Ornithopods are thought to share a close relationship with both pachycephalosaurians and ceratopsians (Cooper 1985; Maryańska and Osmólska 1985; Sereno 1986). Together, the Marginocephalia and Ornithopoda are termed the Cerapoda (Sereno 1986). This large clade of ornithischians includes those forms with an enlarged coronoid process, reduction of premaxillary teeth to five, oral dimensions of the premaxilla approximately that of the predentary, relatively large diastema between premaxillary and maxillary dentition, asymmetrical deposition of enamel on the surfaces of the cheek teeth (thicker on the lingual side of dentary teeth and on the buccal side of maxillary teeth), some differentiation of the ridging pattern on the crowns of cheek teeth, fingerlike lesser trochanter closely applied to the greater trochanter on the femur, laterally protruding ischial peduncle on the ilium, and medial position of the apex of the ascending process of the astragalus, among others (modified from Sereno 1986, with additional observations).

Because of these proximal relationships and still more distant relationships with thyreophorans and *Lesothosaurus,* the long-held view that ornithopods represent the stem group for all other ornithischians is rightfully abandoned. As such, ornithopods have a phylogenetic destiny unto themselves throughout the course of the Mesozoic.

23
Heterodontosauridae

DAVID B. WEISHAMPEL
LAWRENCE M. WITMER

INTRODUCTION

Heterodontosaurids are among the smallest ornithischians (1-2 m long) and also among the earliest (Early Jurassic: Hettangian-?Sinemurian). Virtually all material comes from South Africa, but additional material has been reported from the southwestern United States and the People's Republic of China. The skull of these animals is greatly modified in conjunction with their herbivorous habits (depressed jaw joint, robust and closely packed teeth, extensive tooth wear). Best known from the remains of *Heterodontosaurus tucki*, the Heterodontosauridae claims an additional three species. Together, these animals represent the most primitive clade among the Ornithopoda (see Weishampel, this vol.).

In general, the remains of heterodontosaurids are relatively poorly known (table 23.1). *Lycorhinus angustidens* is based on jaws and teeth, while *Abrictosaurus consors* is based on incomplete skull material and an undescribed skeleton (we regard the type material of *A. consors* as juvenile and/or female, with referred material representing an adult individual of the same species). Fortunately, *Heterodontosaurus tucki* is known from a complete skeleton and a second virtually complete skull.

DIAGNOSIS

The Heterodontosauridae can be diagnosed on the following characters: high-crowned cheek teeth, chisel-shaped crowns with denticles restricted to the apicalmost third of the crown, and presence of a caniniform tooth in both premaxilla and dentary.

ANATOMY

Skull and Mandible

Several complete skulls are known for *Heterodontosaurus tucki* (fig. 23.1a,b), while somewhat less complete material has been referred to *Abrictosaurus consors* (fig. 23.2b; Crompton and Charig 1962; Thulborn 1974; Charig and Crompton 1974; Hopson 1980b; Crompton and Attridge 1986). Partial maxillae and a fragmentary dentary, all with teeth, are known for *Lycorhinus angustidens* (fig. 23.2a, c; Gow 1975, 1990; Hopson 1975a). The following description pertains mostly to *Heterodontosaurus,* supplemented by information from *Abrictosaurus* and *Lycorhinus*.

The skull of *Heterodontosaurus* is moderately short and high, while that of *Abrictosaurus* is more truncated with a more abrupt downward slope (perhaps a juvenile condition). The orbit, infratemporal fenestra, and antorbital fossa are relatively large, while the external naris is small. A poorly preserved sclerotic ring is known in *Abrictosaurus*. The antorbital fossa in *Heterodontosaurus* is chiefly in the maxilla but also extends caudally onto the maxillary process of the jugal. There is a relatively large antorbital fossa extending rostrally across the lateral surface of the maxilla in *Lycorhinus*. Here the rostral edge of the fossa is broadly rounded and runs parallel to the premaxilla-maxilla articulation. The large antorbital fenestra of *Lycorhinus* is found caudally within the fossa. Bounded dorsally and caudally by the lacrimal, the antorbital fossa of *Heterodontosaurus* houses two foramina, the principal opening, a teardrop-shaped foramen at the back of the fossa adjacent to the maxilla-jugal contact, and a smaller rostral opening beneath the contact of the maxilla with the lateral process of the premaxilla.

The rostral region of the premaxilla is edentulous and likely supported a narrow rhamphotheca. Directly behind, the premaxilla of adult *Abrictosaurus* and *Heterodontosaurus* accommodates three teeth; in smaller *Abrictosaurus*, there are two premaxillary teeth. Caudally, the body of the premaxilla marks the beginning of a diastema that accommodates the dentary caniniform tooth. The narrow, elongate lateral process of the premaxilla reaches the junction between the prefrontal and lacrimal, thereby covering the contact between the nasal and maxilla. Limited contact takes place between the dorsal process of the premaxilla and the nasal. The nasal, in turn, roofs the nasal cavity and forms the dorsal margin of the external naris.

The maxilla contacts the lateral process of the premaxilla along a shallowly convex joint surface of the premaxilla. This process also backs the diastema between the premaxillary and maxillary dentition which received the caniniform dentary tooth. Caudally, the lateral maxillary surface has a prominent articular facet for the jugal. Eleven maxillary tooth positions are known in *Heterodontosaurus*, 12 in *Abrictosaurus*, and at least 12 in *Lycorhinus*. The tooth row is emarginated from the side of the skull by a prominent horizontal shelf. Three small foramina are found beneath this ridge, at least in *Lanasaurus*.

The lacrimal contributes the caudal and dorsal margins of the antorbital fossa where it passes medial to the maxilla. The thickened ventral margin of the lacrimal contacts the rostrodorsal aspect of the jugal in a short articulation. The base of the narrow palpebral contacts the lower region of the prefrontal, a small portion of the dorsal margin of the lacrimal, and possibly the caudal tip of the lateral process of the premaxilla.

Ventrally, the crescentic prefrontal makes a long butt joint with the lacrimal; it contacts the nasal and frontal medially. The frontal is roughly pentangular in dorsal view. Laterally, it contributes to the dorsal margin of the orbit and caudally forms a moderately interdigitate transverse suture with the parietal. The interfrontal suture is straight along its length. Laterally, an undulating suture is formed between the frontal and postorbital. The parietal is wide and makes up the caudal aspect of the skull roof. The caudal margin is incised but still covers the dorsal portion of the supraoccipital. The parietal is pulled up into a moderate sagittal crest. Short contact is made laterally with the postorbital.

The very short medial process of the postorbital meets the frontal and a small portion of the parietal

TABLE 23.1 Heterodontosauridae

	Occurrence	Age	Material
Ornithopoda Marsh, 1881b			
Heterodontosauridae Romer, 1966			
Abrictosaurus Hopson, 1975a			
A. consors (Thulborn, 1974) (= *Lycorhinus consors* Thulborn, 1974)	Upper Elliot Formation, Cape Province, South Africa; Upper Elliot Formation, Qachas Nek, Lesotho	Hettangian-?Sinemurian	2 skulls, 1 associated with fragmentary skeleton; material consisting of either male and female or adult and juvenile
Lycorhinus Haughton, 1924 (= *Lanasaurus* Gow, 1975)			
L. angustidens Haughton, 1924 (= *Lanasaurus scalpridens* Gow, 1975)	Upper Elliot Formation, Cape Province, Orange Free State, South Africa	Hettangian-?Sinemurian	Isolated dentary maxillae
Heterodontosaurus Crompton et Charig, 1962			
H. tucki Crompton et Charig, 1962	Upper Elliot Formation, Clarens Formation, Cape Province, South Africa	Hettangian-Sinemurian	2 complete skulls, 1 associated with complete skeleton, fragmentary jaw

Nomina dubia	Material
Dianchungosaurus lufengensis Young, 1982b	Fragmentary skull
Geranosaurus atavus Broom, 1911	Fragmentary jaws

a

b

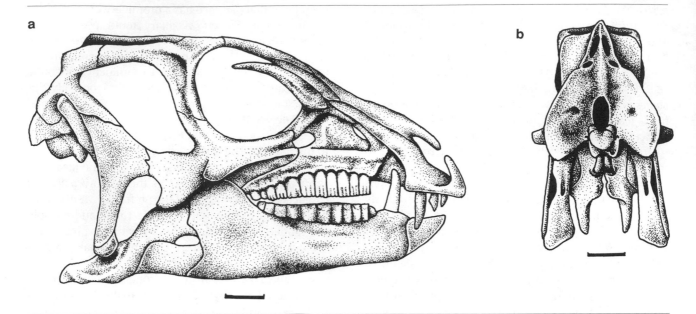

Fig. 23.1. a. Left lateral view of the skull of *Heterodontosaurus tucki*. b. Occipital view of the skull of *Heterodontosaurus* *tucki*. Scale = 10 mm.

along a coarsely interdigitate suture. The caudal process is robust and meets the squamosal along a broad squamous suture to form the upper temporal bar. The ventral postorbital process meets the jugal along an interdigitate contact to form the postorbital bar. The lateral surface of this process is marked by a prominent dorsoventrally oriented ridge reflecting the extensive attachment of M. adductor mandibulae externus superficialis.

The jugal of heterodontosaurids is an irregularly shaped bone that forms the ventral margins of the orbit and infratemporal fenestra and the lateral wall of the adductor chamber. Rostrally, the body of the jugal extensively contacts the lateral surface of the maxilla and is excavated for a caudal extension of the antorbital fossa (the latter is not known in *Abrictosaurus*). A stout jugal boss projects laterally from the face of the jugal. The caudal process of the jugal makes an interdigitate suture with the quadratojugal. Ventral to this junction, there is a prominent ventral extension of the jugal which descends across the surangular portion of the coronoid process.

The large quadratojugal forms a broad dorsal process that articulates along a short interdigitate contact with the prequadrate process of the squamosal. The quadratojugal extends nearly to the base of the

quadrate and is closely applied to virtually the entire lateral surface of the quadrate shaft.

The body of the squamosal is dominated by a caudolaterally and slightly ventrally facing cotylus that receives the quadrate head. The rostral process of the squamosal is short and robust as it contacts the postorbital. The prequadrate process is robust and meets the dorsal process of the quadratojugal to exclude the quadrate from the margin of the infratemporal fenestra. Caudally, the squamosal contributes only a short postquadratic process to the lateral aspect of the paroccipital process.

From its articulation with the squamosal, the quadrate curves rostroventrally toward its transversely expanded mandibular condyle. The lateral wall of the midshaft is excavated to form (with the overlying quadratojugal) a paraquadrate foramen. Medially, the quadrate flares into an extensive pterygoid ramus that is applied to the pterygoid. A small foramen is found on the caudal surface of the midsection of the quadrate shaft. The mandibular condyle is divided into prominent lateral and smaller medial portions.

The palate is known only in *Heterodontosaurus*. The paired crescentic vomers are known only by their caudal extremities where they are slightly separated. Each vomer has a loose ventral contact with the pal-

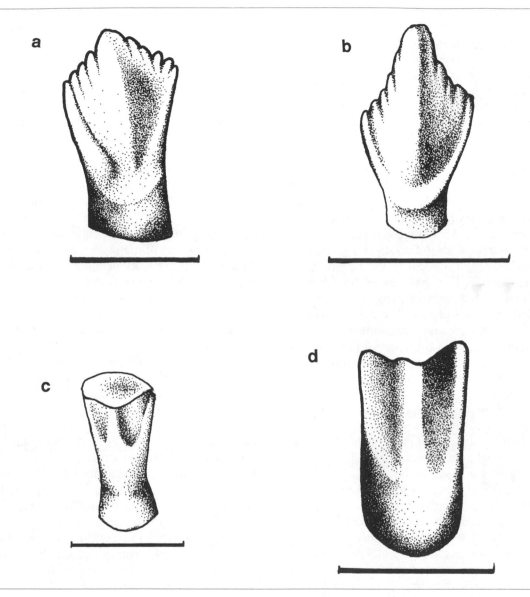

Fig. 23.2. a. Maxillary tooth of *Lycorhinus angustidens* (after Gow 1975). b. Dentary tooth of *Abrictosaurus consors* (after Thulborn 1974). c. Dentary tooth of *Lycorhinus an-* *gustidens* (after Hopson 1975*a*). d. Worn dentary tooth of *Heterodontosaurus tucki*. Scale = 5 mm.

atine and the palatine ramus of the pterygoid. The U-shaped palatine articulates extensively with the caudal process of the maxilla along a long, thin butt joint. Caudally, the palatine extends parasagittally and appears to contact the pterygoid along the rostral margin of the latter element.

The body of the pterygoid articulates with the basipterygoid process of the basisphenoid. Its quadrate ramus is formed in two parts. The more ventral, horizontal aspect is short and stout, while the more dorsal contact is high and oblique. The rostral edge of the pterygoid flange articulates with the ectopterygoid and ventrally projects nearly to the level of the mandibular condyle. The ectopterygoid is a stout buttress between the maxilla and pterygoid.

The braincase has yet to be fully described in *Het-*

erodontosaurus. Several of its features suggest that its construction is different from other ornithopods (Charig and Crompton 1974). The foramen for the middle cerebral vein is found between the parietal and the pro-otic-opisthotic complex. Caudal to this position, there is an opening marking the entrance of the posttemporal canal. This canal opens onto the occiput by way of two oval posttemporal foramina adjacent to the broad midline supraoccipital ridge or pillar. The caudal portion of the opisthotic flares laterally to form a stout, strongly pendent paroccipital process. A circular vascular foramen is found on the caudal face of each of the paroccipital processes. The parasphenoid is long and tapers across the back of the orbit. The basipterygoid process is highly declined and projects rostrally to articulate with the reciprocal, rostrolaterally facing articular surface of the pterygoid. Caudal to the basipterygoid process is the relatively large basal tuber. The foramen for c. n. V is positioned approximately at the junction of the basisphenoid and prootic. The occiput is high, particularly above the foramen magnum. The tall supraoccipital contributes the entire dorsal rim and half of the lateral margins of the foramen magnum. The spherical occipital condyle dominates the caudal aspect of the basioccipital.

The back of the scoop-shaped predentary in both *Abrictosaurus* and *Heterodontosaurus* fits against the inflated, blunt rostral surfaces of each of the dentaries, creating a spheroidal joint between them. The body of the predentary of *Abrictosaurus* has a relatively large foramen on the lateral aspect. A short ventral predentary process extends as a narrow wedge along the ventral surface of the broadly rounded mandibular symphysis. The body of the dentary diverges caudally, especially where it forms the highly elevated coronoid process. In *Abrictosaurus,* three foramina are found at the rostral end of the dentary near its contact with the predentary, but these foramina appear to be lacking in *H. tucki.* The dentition (13 dentary teeth in *Heterodontosaurus,* 14 in *Abrictosaurus*) is emarginated from the side of the dentary by a prominent lateral shelf.

The surangular forms the caudal aspect of the coronoid process, the lateral wall of the mandibular fossa, the outside of the glenoid, including a lip or buttress somewhat rostral to the articulation, and the lateral wall of the retroarticular process. The articular forms the medial portion of the jaw joint as well as the medial surface of the retroarticular process. In lateral view, the angular is high, but it is decidedly narrow in medial view, such that it contributes only a wedge to the caudoventral border of the mandible. In *Heterodontosaurus,* the relatively large external mandibular fenestra is found at the junction of the surangular, angular, and dentary. In contrast, the external mandibular fenestra appears to be lacking in *Abrictosaurus* (a juvenile condition?). The large splenial makes up most of the medial aspect of the mandible at its midsection.

In *Heterodontosaurus,* the first premaxillary tooth is small, peglike, and slightly recurved. The second tooth is larger and has a small lingual shelf but is otherwise similar to the first tooth. The third premaxillary tooth is caniniform. Small-scale serrations are found on both mesial and distal margins. All premaxillary teeth are completely enameled. Tooth wear on the inner crown surface is a consequence of occlusion with the upper border of the predentary and its rhamphotheca. The premaxillary dentition is virtually identical in *Abrictosaurus.* The chief difference is the lack of a caniniform tooth, which led Thulborn (1974) to suggest that this individual of *A. consors* was female; we suggest that this absence may well represent a juvenile condition.

The cheek teeth are relatively well known in *Abrictosaurus, Lycorhinus,* and *Heterodontosaurus* (fig. 23.2). Crowns are chisel shaped, with denticles restricted to the uppermost third of the crown. Mesial and distal margins are well separated in *Abrictosaurus,* moderately overlapping in *Lycorhinus* (based only on dentary dentition), and closely packed in *Heterodontosaurus.* Tooth replacement appears to have been continuous but at reduced rates in old age. Spacing between *Zahnreihen* varies around 3.0. In *Lycorhinus,* the teeth within the tooth row are arranged in offset triplets that reflect emplacement of teeth of roughly the same age. Enamel is continuous on all surfaces of the teeth in *Lycorhinus* and *Abrictosaurus,* but the buccal surface of maxillary teeth and the lingual surface of dentary teeth are more heavily invested. In contrast, in *Heterodontosaurus,* enamel is restricted to the buccal and lingual surfaces of maxillary and dentary crowns, respectively. The maxillary teeth of *Lycorhinus* have three to seven marginal denticles on each mesial or distal edge. The outermost distal denticle is the largest and is separated from the remainder of the crown on both buccal and lingual surfaces by a deep groove. As a result, this distal denticle merges with the basal cingulum. There is the slight suggestion of a broad central eminence on the buccal surface of the crown, but no secondary ridges appear to be present. The crowns of the maxillary dentition in *Abrictosaurus* vary from slightly higher than wide to twice as high as wide. The apex of these crowns is either symmetric or slightly displaced distally. Four to seven denticles are situated toward the apex of the crown and are supported by weak, near-vertical ridges on the buccal surface. The outermost distal ridge merges with the basal cingulum. Ornamentation is not differentiated into primary and secondary ridges. In *Heterodontosaurus*, maxillary teeth are very high crowned and more robust than in *Abrictosaurus*. A large primary ridge is symmetrically placed on the buccal surface of the crown. The outermost mesial and distal ridges are slightly less robust than the primary ridge; these outer ridges converge at the base of the crown. A single secondary ridge is developed mesially and distally midway between the primary ridge and the outer ridges. There is no cingulum, such that the crown merges with the columnar root. Extensive tooth wear is found in all heterodontosaurids. In *Lycorhinus*, the lingual surface is planed off nearly to the roots, thereby truncating virtually all marginal features with the exception of the outermost distal denticle. The resulting mesial and distal facets are of slightly different orientation but create an overall occlusal surface that is highly inclined (75°–80° below horizontal). In *Abrictosaurus*, there is a very large facet, while in *Heterodontosaurus*, there are two facets of slightly different orientation, each of which is continuous on adjacent teeth. Angle of wear varies from 30° to 40° below horizontal in *Heterodontosaurus*.

In all known heterodontosaurids, the large caniniform first dentary tooth fits into the maxillary diastema. In *Abrictosaurus*, only the sharp mesial edge of the caniniform tooth is marked with fine serrations, while in both *Lycorhinus* and *Heterodontosaurus*, both mesial and distal margins are serrated. In all species, the entire crown is thinly enameled. The only material to lack a caniniform dentary tooth is a possibly juvenile specimen of *Abrictosaurus consors*. Instead, the first dentary tooth is small, relatively broad, and recurved. Mesial and distal margins lack denticles. The second and third crowns are similar in morphology, the latter, however, being denticulate on its distal edge. A small gap separates the caniniform tooth from the remaining dentary cheek teeth. The second tooth is small, conical, and recurved, at least in *Abrictosaurus* and *Lycorhinus*. More distally, the dentary crowns are spatulate, approximately twice as high as wide. In *Abrictosaurus*, the middle of both lingual and buccal aspects of the crown is enlarged to form a broad central eminence, while in *Heterodontosaurus* and possibly *Lycorhinus*, there is clear

differentiation of a strong central primary ridge. In all heterodontosaurids, the apex of the crown tends to be symmetrical, supported by a strong primary ridge. Where they are asymmetrical with respect to the lingual surface of the tooth, the apex and primary ridge are offset somewhat mesially. The apex is surrounded by three to five denticles on each mesial and distal margin in *Abrictosaurus*, and these denticles are supported by short secondary ridges. In *Heterodontosaurus* and *Lycorhinus*, the number and configuration of secondary ridges are often poorly known because of the extreme tooth wear. Those that support the outermost mesial and distal denticles, however, merge with the base of the crown. A weakly developed cingulum is present in both *Abrictosaurus* and *Lycorhinus* but is absent in *Heterodontosaurus*. Dentary tooth wear in heterodontosaurids produces either a single facet confluent mesially and distally with adjacent teeth (in *Abrictosaurus*) or a bifaceted surface, each facet of which is continuous with a facet on neighboring crowns (in *Lycorhinus* and *Heterodontosaurus*). The angle of these wear facets increases with tooth age, ranging from 45° to 65° below horizontal.

Postcranial Skeleton

The following discussion of heterodontosaurid postcrania is based largely on *Heterodontosaurus tucki* (fig. 23.3; Santa Luca et al. 1976; Santa Luca 1980; see also Bakker and Galton 1974; Bakker 1986; Norman 1985).

There are about nine cervical vertebrae. Only the atlas is missing. The odontoid process of the axis is well developed. The sigmoidal curvature of the neck is demonstrated by the lateral profiles of the cervical centra: parallelograms cranially, rectangular in the middle of the series, and trapezoidal caudally. The centra are longer in the cranial portion of the series, but central height is relatively constant. The centra are spool-shaped with a median ventral keel. There is a lateral concavity in the centra that becomes deeper caudally, especially just caudoventral to the parapophysis. The parapophyses remain on or just below the neurocentral suture. The diapophysis is near the parapophysis cranially but migrates caudodorsally in the caudal portions of the series. The neural spine of the axis is long and nearly horizontal; at its base, it bears lateral flanges. The neural spines become weaker toward the middle of the series but are tall caudally. The prezygapophyses of the axis face laterally to receive the atlas. The other cranial

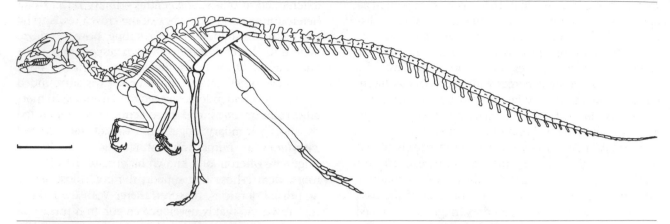

Fig. 23.3. Skeletal reconstruction of *Heterodontosaurus tucki* (modified from Santa Luca et al. 1976, Paul 1987*b*). Scale = 10 cm.

cervical zygapophyses are more horizontal, whereas the middle ones are inclined and the caudal ones more horizontal again.

The dorsal region of the vertebral column is relatively short, comprising twelve vertebrae. The cranial dorsals are transitional between the cervicals and other dorsals. The dorsal centra are spool shaped and present rectangular lateral profiles. The transverse processes are horizontal, directed caudolaterally, and lie at the level of the zygapophyses. The parapophysis moves onto the transverse process early in the series. In the middle portion of the series, the transverse processes are bifid with parapophysial and diapophysial facets on the cranial and caudal portions, respectively. The caudalmost dorsals lack bifid transverse processes, and the facets have merged. The strong neural spines are rectangular and nearly as long as the centra. The cranial dorsal zygapophyses are horizontal and become inclined to about 45° caudally.

There are six sacral vertebrae, the first four of which appear to be sacralized dorsals. The sacral centra are fused. Details are obscured by crushing and the overlying ilia. Thulborn (1974) reported that *Abrictosaurus* has four unfused sacrals; this may be a juvenile feature, consistent with the lack of caniniform teeth in the same individual.

Proximal and distal portions of the tail are preserved in *Heterodontosaurus*; an estimated 34 to 37 vertebrae are present in the caudal series. The proximal caudal centra are more concave ventrally because the hemal arch articulation is more pronounced; distally, the hemal arch articulation is reduced and the ventral border is flatter. The proximal centra also bear a prominent excavation ventral to the transverse processes; the excavation is shallower and eventually absent distally. The transverse processes are horizontal, ventral to the neurocentral suture, and directed laterodistally. The proximal transverse processes are strong, but more distal ones are weaker and the distalmost are absent. The neural spines are vertical and tall proximally but become inclined and shorter distally. Neural spines are absent distal to the tenth caudal. The zygapophyses are elevated about 45° proximally but soon become nearly vertical. Chevrons are present throughout the caudal series. Proximally, they are strong and become progressively weaker distally. The chevrons from the middle of the series are expanded distally.

The cervical ribs are all two-headed. Cranially, the capitulum and tuberculum are close together. Distally, they become more widely separated and the tubercular process becomes shorter. The capitulum and tuberculum are close together in the dorsal series. The tuberculum becomes reduced distally, and the last dorsal rib is single headed.

Ossified tendons are found in the dorsal and proximal sacral portions of the column. Ossified tendons were apparently absent from the tail.

The scapula is slightly longer than the humerus. The scapular shaft is circular in cross section proximally but expands into a flattened blade distally. The distal end is pitted, indicating presence of a cartilaginous suprascapula. The shaft is slightly bowed laterally for the rib cage. The proximal end is expanded. A well-developed acromion projects cranially. The scapular glenoid fossa is located ventrally. The triceps tubercle lies dorsocaudal to the glenoid. The coracoid is rectangular with flat proximal and distal ends. The distal portion curves strongly medially. The coracoidal glenoid fossa is directed caudoventrally. At the ventral corner of the coracoid is a tubercle for muscle attachment. Santa Luca (1980) reported a sternal plate; its morphology and attachments cannot be determined.

The humerus is strongly constructed. The shaft is very slightly twisted. A thickened boss in the middle of the proximal end represents the humeral head. There is a fossa on the cranial surface of the bone adjacent to the head. Medial to the head there is a large tuberosity. The dectopectoral crest is strong, large, and directed craniolaterally running about 40 percent of the length of the humerus. Its distal portion ends abruptly. The humerus bears well-developed distal condyles; the articular surfaces do not extend caudally, and there is no olecranon fossa. The ulnar condyle is rounded and projects distally beyond the radial condyle. A prominent entepicondyle projects medially from the shaft above the ulnar condyle. The radial condyle is less rounded and is inclined laterally somewhat. A lateral supracondylar ridge extends proximally from the radial condyle.

The radius is about 70 percent of the length of the humerus. The radial shaft is somewhat twisted. Proximal cross sections are circular and become more rectangular distally. The humeral articulation is a semicircular sulcus; a process extends proximomedially from the caudal surface and contacts the ulna. The distal articulation has planar contacts for the ulnare as well as the radiale. Proximal to the distal articulation on the caudal surface is a tubercle for muscle attachment. The ulna is about 80 percent of humeral length. The most striking feature of the ulna is the prominent olecranon that bears rugosities for the triceps insertion. The coronoid process is also well developed. The humeral articulation is somewhat trochlear. The shaft thins in the middle and expands distally. Caudally, on oblique crest runs proximomedially-distolaterally and ends on reaching the distal third. The articulation for the ulnare is transversely convex. The ulna also articulates with the pisiform.

The carpus is very well preserved and remarkable in containing nine elements. There is a proximal row (radiale, ulnare, and pisiform), a middle row ("centrale"), and a distal row (distal carpals I-V). The radiale has broad contacts with the radius and distal carpal I and smaller contacts with adjacent carpals. The ulnare

offers a concave articulation for the ulna and a planar one for the radius; it articulates broadly with the "centrale" and distal carpal IV and narrowly with distal carpal V and the pisiform. The "centrale" is problematic. The bone occupies the position of the centrale of other tetrapods—intervening between proximal and distal carpals. *Heterodontosaurus,* however, appears to lack an intermedium, a typical component of the proximal carpal row in ornithischians (e.g., *Camptosaurus,* Gilmore 1909, *Hypsilophodon,* Galton 1974*a*). The problem is whether the bone in question represents a distally displaced intermedium or a true centrale. Santa Luca (1980) listed four possible resolutions. The most likely possibility is that the intermedium has been incorporated into the ulnare (which is quite large) and the bone in question is a true centrale. Distals carpals I-III are relatively large; distal carpals IV-V are abducted.

Although Thulborn (1974) described the manus of *A. consors* as "diminutive," it is quite large in *Heterodontosaurus,* being longer than the radius or ulna and almost as long as the humerus. Metacarpal II is the longest, followed by III, I, IV, and V. The carpal articulation of metacarpals I and II is carried onto the dorsal surface, indicating that these carpometacarpal joints could be extended farther than the others. The dorsal surfaces of the bases of metacarpals I-IV bear strong lateral and medial tubercles for extensor tendons. As a result of the tight planar articulations between adjacent metacarpals and the carpals, the bases of metacarpals I-IV are almost cuboidal. Metacarpal I is unusual in that it is asymmetric, with the lateral border longer. Its metacarpophalangeal joint is also asymmetric: the medial condyle is in line with the long axis of the shaft, whereas the lateral condyle is perpendicular to it. This asymmetry would cause the digit to deviate medially on extension and laterally on flexion. Metacarpophalangeal joints II and III are symmetrical and restrict movement to simple flexion and extension. The cranial surfaces of the metacarpal heads of digits I-III bear deep pits that received processes from the proximal phalanges; this suggests that digits I-III could be hyperextended at the metacarpophalangeal joints. Metacarpal IV is reduced distal to its proximal cuboidal portion, and metacarpal V is even more reduced.

The phalangeal formula of *Heterodontosaurus* is 2−3−4−3−2. Digits IV and V are reduced. Their interphalangeal joints are very simple and lack well-defined distal condyles. Digits I-III, however, are large, complex, and exhibit elements of asymmetry. The first phalanx of digit II and the first and second of digit III are characterized by asymmetric distal articulations. The shafts are twisted such that the heads are medially rotated. Likewise, the lateral margins are longer, imparting a medial curve to the shafts. Finally, the medial condyle is larger and oriented differently from the lateral condyle. These articulations would cause lateral deviation during flexion and medial deviation during extension. The penultimate phalanges of digits I-III are symmetrical and permitted simple extension and flexion. The unguals of digits I-III are large claws; the flexor tubercles are very large for an ornithischian. The phalangeal formula of *A. consors* is reported to be 2–3–4–2–1.

The unusual manus of *Heterodontosaurus* has received considerable attention. Bakker and Galton's (1974) reconstruction of the hand showed a medially deviated pollex. The resemblance of this "twist-thumb" (Bakker 1986) to that of prosauropods was used as evidence for monophyly of prosauropods and ornithischians (Bakker and Galton 1974; Bakker 1986). Santa Luca (1980) described Bakker and Galton's figure as being "completely inaccurate" and offered a reconstruction with the pollex more in line with the other digits. Bakker (1986) responded, noting that the specimen had been damaged subsequent to his studies and repaired without the twist. Despite these discrepancies, both authors agreed that the pollex did not have straight flexion/extension but instead deviated medially during extension and somewhat laterally during flexion.

The hindlimb is long and gracile in *Heterodontosaurus*, but, owing to the considerable length of the forelimb, it is only about 37 percent longer than the forelimb. In *Abrictosaurus*, however, the forelimb is not elongate and thus the hindlimb/forelimb ratio is much higher (Thulborn 1974).

The ilium is dorsally convex (fig. 23.4). The preacetabular process is about twice as long as the postacetabular process and bears a pronounced tuberosity at its end. The brevis shelf is horizontal and shallow. The pubic peduncle is quite long and projects cranioventrally about 20°. The ischial peduncle is shorter than the pubic and rectangular in section. The supraacetabular rim is weak except caudodorsally where an antitrochanter is developed. A medial view is not available, but the first three sacral ribs and perhaps the last dorsal rib attach to the preacetabular process.

The pubis has a short prepubic process that is stout, squared-off cranially, and bears three small tubercles (fig. 23.4). At the acetabular margin of the pubis there is a small tubercle. The obturator foramen is fully within the pubis. The pubis itself is a simple, long rod.

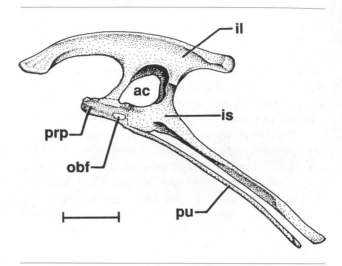

Fig. 23.4. Left lateral view of the pelvis of *Heterodontosaurus tucki* (modified from Santa Luca 1984). Scale = 3 cm.

The ischium contacts the ilium via a rather long, rectangular iliac peduncle and the pubis via a broad, spatulate pubic peduncle (fig. 23.4). There is no obturator process. A ventral concave shelf projects laterally from the shaft of the ischium; in the middle of the shaft, the shelf moves dorsally such that a laterally facing flange is developed.

The femur is relatively straight, about 16 percent longer than the ilium, and only slightly bowed cranially. The femoral head is oriented about 90° to the shaft. The greater trochanter is not well separated from the head and is on about the same level. Likewise, the lesser trochanter is not separated from the greater trochanter by a cleft but is instead a pronounced rugosity on the craniolateral surface. The fourth trochanter is pendent and somewhat styloid. There are neither cranial nor caudal intercondylar fossae between the distal condyles. The condyles are obliquely situated, with the smaller lateral condyle being distal to the larger medial condyle.

Heterodontosaurus is very unusual in having the tibia fused to both the distal fibula and the proximal tarsals, producing a rather avian tibiotarsus (or more properly, a tibiofibulotarsus). The tibia (plus astragalus) is about 30 percent longer than the femur. The medial tibial condyle is larger than the lateral (to receive the large medial femoral condyle) and is laterally compressed. The cnemial crest is well developed and bears an elongate fossa on its lateral surface. The tibial shaft is straight and is transversely expanded distally. The fibular head is laterally compressed. The shaft of the fibula attenuates distally and is fused to the tibia at its distal terminus.

The astragalus and calcaneum are fused to each other and the tibiofibula. There are three distal tarsals that are fused with each other and with the metatarsals, producing a tarsometatarsus. Distal tarsal 1 is applied to metatarsals I and II, and distal tarsals 2 and 3 contact metatarsals III and IV, respectively. The caudal surface of distal tarsal 1 is expanded into a vertical flange functionally analogous to the hypotarsal ridges of birds. The cranial surface of distal tarsal 3 is pierced by a foramen.

The pes is generally similar to other small ornithischians except that the proximal ends of the metatarsals appear to be fused to each other. Metatarsal III is the longest, followed by IV, II, I, and V. Digit V is represented by only a metatarsal splint. Digit I was too short to reach the ground during normal progression. The metatarsal bases are roughly cuboidal. The axis of metatarsophalangeal joint III is perpendicular to the long axis of the pes, whereas those of digits II and IV are deviated from this long axis. The heads of metatarsals III and IV bear deep pits on their cranial surfaces, suggesting that metatarsophalangeal joints III and IV could be extended farther than II. The phalangeal formula is 2−3−4−5−0. The interphalangeal joints are strongly trochlear and are characterized by deep pits for the collateral ligaments. The ungual phalanges are laterally compressed claws with grooves on both sides.

SYSTEMATICS

The Heterodontosauridae (fig. 23.5) includes the following three species: *Abrictosaurus consors, Lycorhinus angustidens,* and *Heterodontosaurus tucki.* Although the nominative species *H. tucki* is characterized by numerous derived features, the other members of the group are either highly fragmentary or incompletely described, making diagnosis of the group difficult. Heterodontosaurids are unique in that the denticles are restricted to the apicalmost third of the chisel-shaped crowns. Heterodontosaurid monophyly can be further characterized by the presence of a caniniform tooth in both the premaxilla and dentary. It is also probable that the heterodontosaurid clade is united by a further feature: low-angle tooth wear characterizes dentary teeth of all known taxa. This feature suggests on a biomechanical basis the presence of a spheroidal articulation between the predentary and dentary.

Within the heterodontosaurid clade, *Abrictosaurus consors* itself appears to be characterized by a few

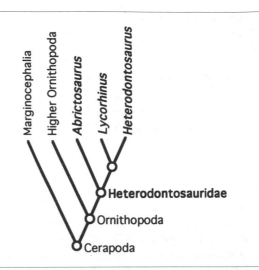

Fig. 23.5. Cladogram of the Heterodontosauridae. Characters discussed in text.

autapomorphies. The facial skeleton of *A. consors* is short and steeply sloping, which is either a uniquely derived feature or may represent a juvenile condition. Those individuals of this species that lack caniniform teeth are thought to be female (Thulborn 1974) and/or juvenile. Finally, the number of phalanges in manual digits III and IV is reduced in *A. consors* over that seen in other species.

Lycorhinus and *Heterodontosaurus* are sister taxa based on the presence of both mesial and distal serrations on the dentary caniniform tooth (Charig and Crompton 1974) and the broad contact of mesial and distal surfaces of the crown (convergently developed in *Pisanosaurus*). *Heterodontosaurus tucki* is characterized by a large number of unique features, including the presence of an accessory antorbital fenestra within the antorbital fossa and the presence of a distinctive jugal boss. Postcranially, the species is characterized by bifid transverse processes of the dorsal vertebrae; loss of ossified tendons in the tail; a well-developed entepicondyle on the humerus; fusion of the ulnare and intermedium; an elongate manus (Sereno [1986] applied this to all heterodontosaurs but Thulborn [1974] described the manus of *Abrictosaurus* as "diminutive"); cuboidal metacarpal bases (Sereno [1986] again applied this to all heterodontosaurids); lack of cleftlike separation between the lesser and greater trochanters; loss of the caudal intercondylar fossa on the femur; fusion of the distal tibia, distal fibula, astragalus, and calcaneum; a very slender fibula lacking distal expansion (Sereno 1986), which probably relates to the fusion noted above; fusion of the distal tarsals and proximal metatarsus; and finally, the presence of extensor pits on the heads of the proximal pedal phalanges (Sereno 1986).

The position of heterodontosaurids has generated a considerable diversity of opinion. Santa Luca (1980), based on the absence of the obturator process in *Heterodontosaurus,* rejected any relationship with ornithopods and considered them quite primitive ornithischians. Cooper (1985) placed heterodontosaurids as the sister group of the Marginocephalia on the basis of a prominent jugal boss and a few less convincing characters. The jugal boss in our scheme would either represent an independent acquisition or loss of the boss in higher ornithopods. Maryańska and Osmólska (1985) excluded heterodontosaurids from the Cerapoda, but the polarities of their characters reverse when basal thyreophorans are considered. We agree with Sereno (1984, 1986) that heterodontosaurids are the sister group of higher ornithopods. We accept the four synapomorphies suggested by Sereno (1986) which unite these two groups: the projection of the premaxillary tooth row below the level of the maxillary tooth row (this feature is also found in pachycephalosaurs and may be a cerapodan feature); paroccipital processes strongly pendent laterally; very tall quadrates, making the occipital profile taller than wide and depressing the jaw joint far below the tooth row; elongate lateral process of the premaxilla contacting the lacrimal and covering the contact of the nasal and maxilla (reversed in hypsilophodontids). Reduction or loss of the cingulum may also characterize this node. The last character appears to reverse in *Abrictosaurus,* but again this may be an age-related phenomenon.

PALEOECOLOGY

The semiarid depositional environment of the upper parts of the Stormberg Series (Upper Elliot and Clarens formations) of South Africa have yielded the best material pertaining to heterodontosaurids. These sediments consist of mixed flood fan and eolian dune facies (Anderson and Anderson 1970; Visser 1984; Kitching and Raath 1984). Silts and muds are well oxidized, such that little organic matter is preserved. Flood sands interspersed with eolian deposits indicate that abundant water was available only sporadically. Consequently, preservation of heterodontosaurid remains is rather poor. Isolated elements dominate, although at least one fully articulated skeleton is known. In addi-

tion, most species are not known from more than three individuals, and the majority come from single specimens. Of these, jaw elements are the most common.

Altogether, Upper Elliot and Clarens habitats seem extremely inhospitable. It is perhaps surprising then that the associated terrestrial fauna is extremely diverse and often rich, including early mammals, advanced cynodonts, the prosauropod *Massospondylus carinatus,* sphenosuchians, and terrestrial crocodilians. No fully aquatic members of the fauna are known. It is tempting to think that at least some of these remains were transported into the catchment area by floods from more benign and luxuriant environments, but no taphonomic and biogeographic work has been done on this subject.

Heterodontosaurids clearly were active herbivores, chiefly foragers on ground cover and shrubby vegetation. Browsing appears to have been concentrated within the first meter above the ground. Absence of teeth along the rostral part of the premaxilla, accompanying development of a cornified rhamphotheca, and narrowing of the oral margins of the premaxillae suggest relatively selective cropping abilities. Wear on the premaxillary teeth, produced by occlusion with the outer margins of the predentary, indicates not only piercing prehension of food but also shearing. Studies of mastication in heterodontosaurids by Weishampel (1984*a*) and Crompton and Attridge (1986) suggest that, like all other ornithischians, these animals chewed bilaterally. To combine this condition of isognathy with a relatively low angle transverse power stroke as indicated by dentary teeth, all heterodontosaurids combined vertical mandibular movement with a slight degree of medial rotation of the mandible about their long axes (Weishampel 1984*a*; Norman and Weishampel 1985; Crompton and Attridge 1986). This rotational mobility is facilitated by the restructuring of the symphysis and dentary-predentary contact into a spheroidal articulation. Otherwise, the skull was exceptionally rigid to resist deformation during chewing.

On the basis of limb and trunk proportions and the general gracile nature of their skeleton, heterodontosaurids are viewed as agile, cursorial bipeds (Thulborn 1982). Quadrupedal stance was probably used during foraging or when standing still. The femur was held in a slightly abducted position to facilitate a parasagittal gait and horizontal knee joints. Lever-arm mechanics suggest that femoral retraction was powerful. The attitude of the axial skeleton was largely horizontal. In *Heterodontosaurus,* the cranial series of dorsal vertebrae are turned downwardly, while the middle and caudal cervical vertebrae are dorsiflexed. Ossified epaxial tendons over the caudal dorsals strengthened this portion of the vertebral column, but their absence over the caudal series left the tail to move about more or less freely. The hindlimbs appear to be highly adapted for rapid progression, displaying greatly elongate distal limb segments and fusion of elements within segments. The forelimbs are also specialized, but their purported function is controversial. On the basis of the large entepicondyle, large olecranon, and ungual flexor tubercles, Santa Luca (1980) suggested that forelimb function encompassed powerful elbow extension and antebrachial and manual flexion. He further regarded quadrupedal locomotion as being important only at slow speeds. In contrast, Paul (1987b) considered *Heterodontosaurus* to be a quadrupedal galloper. Although Paul (1987b) did not cite any features, the same attributes noted by Santa Luca are equally expected in quadrupedal animals; to this list can also be added the modified wrist and hand joints that suggest hyperextension of the carpometacarpal and metacarpophalangeal joints. Neither of these ideas about forelimb function, however, take into account the morphology of the claws and the relations of the scapula, coracoid, and sternum to limit scapular rotation, both of which argue against rapid quadrupedal locomotion. We suggest that, alternatively, all of these forelimb features are consistent with a digging and tearing habitus in the manner of pangolins, aardvarks, and certain edentates (Coombs 1983). It seems possible that *Heterodontosaurus* may have used its large claws and powerful forelimbs to dig up roots and tubers or tear apart insect (e.g., termite) nests.

Aestivation and the significance of caniniform teeth comprise the sole basis for inferences about the behavioral repertoire of heterodontosaurids. Arguing that heterodontosaurids aestivated during times of environmental stress, Thulborn (1978) relied on the purported lack of tooth replacement and propalinal jaw action in these animals. In his rejoinder to Thulborn, Hopson (1980b) correctly argued that heterodontosaurids had a transverse power stroke and that tooth replacement was reduced, not lost, in these animals. Hence, there is no compelling reason to believe that heterodontosaurids engaged in aestivation.

The presence of caniniform teeth in heterodontosaurids may relate to their importance as display and combat structures in intraspecific behavior (Steel 1969; Thulborn 1974; Molnar 1977). Since caniniform teeth are thought to be present only in mature males (Thulborn 1974), then they would likely have been used not only for gender recognition but also for intraspecific combat, ritualized display, social ranking, and possibly courtship (much like that in suids and tragulid artiodactyls; Molnar 1977). Similarly, the development of the jugal boss might also be interpreted along these lines, but data are lacking from the heterodontosaurid fossil record.

24

Hypsilophodontidae, *Tenontosaurus,* Dryosauridae

HANS-DIETER SUES

DAVID B. NORMAN

INTRODUCTION

The Hypsilophodontidae is a structurally conservative assemblage of small to medium-sized (2-4 m long), bipedal ornithopod dinosaurs. These animals are known from Asia, Australia, Europe, and North America (table 24.1) and range in time from the Middle Jurassic to the Late Cretaceous. Their skeletal structure has become well known only recently as the result of a series of detailed studies by Galton (1969, 1971*d*, 1973*d*, 1974*a*, 1974*b*, 1975, 1983*d*; Galton and Jensen 1973*b*, 1979*b*).

Galton (1973*d*, 1974*a*, 1974*b*, 1980*d*) depicted hypsilophodontids as a paraphyletic "plexus" of persistently conservative forms from which several iguanodont lineages arose iteratively throughout the later Mesozoic. The taxon Hypsilophodontidae can, however, be defined as a monophyletic grouping that includes *Atlascopcosaurus, Fulgurotherium, Hypsilophodon, Leaellynasaura, Orodromeus, Othnielia, Parksosaurus,* *Thescelosaurus, Yandusaurus,* and *Zephyrosaurus* (see table 24.1).

Tenontosaurus, originally classified as an iguanodontid (Ostrom 1970*a*), has been referred to the Hypsilophodontidae by Dodson (1980). Sereno (1986) considers it the most primitive member of his clade Iguanodontia. Although *Tenontosaurus* is represented by a number of excellent specimens, a detailed anatomical survey of this taxon has not yet been published.

Galton (1977*b*, 1981*c*, 1983*d*) included *Dryosaurus* (including *Dysalotosaurus*) and *Valdosaurus* in the Hypsilophodontidae on the basis of their overall structural similarity. Milner and Norman (1984), Norman (1984*a*), and Sereno (1986) pointed out that *Dryosaurus* differs from other members of that taxon in a number of features, which it shares with the clade including *Camptosaurus, Iguanodon,* and *Ouranosaurus* (see Norman and Weishampel, this vol.). Following Milner and Norman (1984), we place *Dryosaurus* and *Valdosaurus* in a separate family, Dryosauridae, which

498

we consider the sister group of all other taxa within the Iguanodontia (Sereno 1986; Norman and Weishampel, this vol.; see table 24.1).

ANATOMY OF HYPSILOPHODONTIDAE

The following anatomical account is based principally on Galton's (1974*a*) description of *Hypsilophodon foxii* from the Lower Cretaceous (Barremian) of the Isle of Wight, England. Much of the skeletal material used in that study represents immature specimens, and this fact should be acknowledged when making comparisons with other ornithopod taxa.

Skull and Mandible

The skull of *Hypsilophodon* is relatively small and has a short snout (fig. 24.1A). The premaxilla bears five teeth. Its rostral end is edentulous and rugose; in life, it probably bore a horny beak, as did the predentary. The premaxillary teeth bit against the predentary beak. The ventral margin of the premaxilla is situated below that of the maxilla. The premaxilla does not completely separate maxilla and nasal on the side of the face. The lateral surface of the maxilla overhangs the alveolar margin to produce a pronounced recess; the tooth row of the dentary is similarly inset. An elongate rostral process of the maxilla fits into a recess on the medial surface of the premaxilla and is wedged in place by the vomer. An extensive antorbital fossa is enclosed by thin laminae of the maxilla; the antorbital fenestra is small. The outer surface of the tooth-bearing portion of the maxilla is pierced by several large foramina that communicate with the antorbital fossa. The maxilla has a short dorsal contact with the nasal and overhangs the lacrimal both laterally and medially, where it forms the inner wall of the antorbital fossa. The nasal is thin. The lacrimal fits into a slot on the nasal rostrally, and its dorsal edge bears a groove for reception of the prefrontal. The prefrontal has a thickened, rugose corner for contact with the palpebral. The palpebral is curved and rather stout. The jugal is deep and does not contact the quadrate. The large quadratojugal, which is pierced by a prominent foramen, forms the ventral margin of the lower temporal fenestra. The quadrate has a triangular proximal head, which fits into a socket in the squamosal. The pterygoid flange of the quadrate broadly overlaps a winglike process of the pterygoid.

The frontals are relatively narrow transversely, especially between the orbits (fig. 24.1B). Each frontal bears a distinct lateral peg or spike that fits into a socket on the medial surface of the triradiate postorbital.

Prootic, opisthotic, and exoccipital form the lateral walls of the braincase, and the basioccipital and basisphenoid/parasphenoid make up its floor. The braincase is roofed over behind by the supraoccipital. The laterosphenoid apparently does not fuse to the remainder of the braincase. Its rostrolateral extremity forms a synovial joint with the skull roof and is expanded into a transverse condyle. The latter fits into a well-defined depression on the ventral aspect of the frontal and postorbital, which is traversed by the suture between these two bones. The squamosal is extensively sutured to the parietal medially and to the opisthotic ventrally. The paroccipital process projects transversely. Small exoccipitals frame the foramen magnum ventrolaterally.

The palate is moderately vaulted. Rostrally, the fused vomers form a narrow median plate of bone, which curves upward and increases in depth caudally. The palatine floors the orbital socket. The triradiate pterygoid links the suspensorium to the braincase, maxilla, and rostral portion of the palate by means of thin, winglike processes. The ectopterygoid forms a sharply curved bar of bone.

The relatively robust mandible lacks an external mandibular fenestra (fig. 24.1C). The dentary has a spoutlike rostral end, which articulates with the predentary. It becomes thicker and deeper caudally. The coronoid process is moderately well developed. The jaw articulation is set below the occlusal plane of the teeth. The mandible has a sinuous outline in occlusal view, and the tooth row extends along the medial edge of the dentary.

The heterodont dentition of *Hypsilophodon* consists of five simple, slightly recurved premaxillary teeth, 10 or 11 buccolingually compressed maxillary teeth that have mesiodistally broad crowns with denticulated edges, and probably 13 or 14 dentary teeth. The first 3 or 4 dentary teeth appear to be simpler and more conical, but the crowns of the other dentary teeth are buccolingually compressed and bear denticulated edges, much like the maxillary teeth. The heavily enameled buccal faces of maxillary tooth crowns (fig. 24.1D) bear small vertical ridges. The thickly enameled lingual faces of most of the dentary teeth (fig. 24.1E) are marked by a strong vertical median ridge and several weaker secondary ridges. These teeth also typically show substantial wear, in the form of obliquely inclined, planar wear facets. A depression along the mesial edge of the root and part of the crown of a maxil-

TABLE 24.1 Hypsilophodontidae, *Tenontosaurus*, Dryosauridae

	Occurrence	Age	Material
Ornithopoda Marsh, 1881b			
Hypsilophodontidae Dollo, 1882			
Agilisaurus Peng, 1990 •			
A. louderbacki Peng, 1990 •	Xiashaximiao Formation, Sichuan, People's Republic China	Bathonian-Callovian	Nearly complete skeleton with skull
A. multidens (He et Cai, 1983) • (= *Yandusaurus multidens* He et Cai, 1983)	Xiashaximiao Formation, Sichuan, People's Republic of China	Bathonian-Callovian	Nearly complete skeleton with skull
Atlascopcosaurus Rich et Rich, 1989			
A. loadsi Rich et Rich, 1989	Otway Group, Victoria, Australia	late Aptian-early Albian	Maxilla, teeth
Drinker Bakker, Galton, Siegwarth, et Filla, 1990 •			
D. nisti Bakker, Galton, Siegwarth, et Filla, 1990 •	Morrison Formation, Wyoming, United States	Kimmeridgian-Tithonian	Partial skull and postcranial skeleton
Fulgurotherium Huene, 1932			
F. australe Huene, 1932	Griman Creek Formation, New South Wales, Australia	Albian	Femoral fragments, isolated postcranial elements, ?teeth
Gongbusaurus Dong, Zhou, et Zhang, 1983 •			
G. shiyii Dong, Zhou, et Zhang, 1983 •	Shangshaximiao Formation, Sichuan, People's Republic of China	Late Jurassic	Isolated teeth
G. wucaiwanensis Dong, 1989 •	Shishiugou Formation, Xinjiang Ugyur Ziziqu, People's Republic of China	early Late Jurassic	Dentary, sacral and caudal vertebrae, partial forelimb, complete hindlimb
Hypsilophodon Huxley, 1869			
H. foxii Huxley, 1869	Wealden Marls, Isle of Wight, England; Las Zabacheras Beds, Teruel, Spain	Barremian-Aptian	3 nearly complete skeletons, approximately 10 partial skeletons, cranial and postcranial elements
Leaellynasaura Rich et Rich, 1989			
L. amicagraphica Rich et Rich, 1989	Otway Group, Victoria, Australia	late Aptian-early Albian	Skull fragments, teeth, isolated postcrania
Orodromeus Horner et Weishampel, 1988			
O. makelai Horner et Weishampel, 1988	Two Medicine Formation, Montana, United States	late Campanian	3 partial skeletons, cranial and postcranial elements, clutch of 19 eggs containing embryos
Othnielia Galton, 1977b			
O. rex (Marsh, 1877b) (= *Nanosaurus rex* Marsh, 1877b, including *Laosaurus gracilis* Marsh, 1878b, *Laosaurus consors* Marsh, 1894)	Morrison Formation, Colorado, Utah, Wyoming, United States	late Kimmeridgian early Tithonian	2 partial skeletons, postcranial elements, teeth
Parksosaurus Sternberg, 1937			
P. warreni (Parks, 1926a) (= *Thescelosaurus warreni* Parks, 1926a)	Horseshoe Canyon Formation, Alberta, Canada	early Maastrichtian	Incomplete skeleton
Thescelosaurus Gilmore, 1913			
T. neglectus Gilmore, 1913 (including *T. edmontonensis* C. M. Sternberg, 1940a)	Lance Formation, Wyoming, Judith River Formation, Hell Creek Formation, Montana, Hell Creek Formation, South Dakota, Laramie Formation, Colorado, United States; Scollard Formation, Alberta, Frenchman Formation, Saskatchewan, Canada	?late Campanian, late Maastrichtian	Approximately 8 partial skeletons, cranial and postcranial elements, teeth
?*T. garbanii* Morris, 1976	Hell Creek Formation, Montana, United States	late Maastrichtian	Incomplete hindlimb, vertebrae
Yandusaurus He, 1979			
Y. hongheensis He, 1979	Xiashaximiao Formation, Sichuan, People's Republic of China	Bathonian-Callovian	Nearly complete skeleton with skull
Zephyrosaurus Sues, 1980a			
Z. schaffi Sues, 1980a	Cloverly Formation, Montana, United States	late Aptian	Partial skull, vertebrae
Unnamed hypsilophodontid (= "*Thescelosaurus*" sp. described by Morris, 1976)	Hell Creek Formation, South, Dakota, United States	late Maastrichtian	Skull
Unnamed taxon (tenontosaurs)			
Tenontosaurus Ostrom, 1970a			
T. tilletti Ostrom, 1970a	Cloverly Formation, Montana, Cedar Mountain Formation, Utah, ?Antlers Formation, Oklahoma, ?Antlers Formation, ?Paluxy Formation, Texas, United States	late Aptian	Approximately 27 skeletons, cranial and postcranial elements, teeth
Dryosauridae Milner et Norman, 1984			
Dryosaurus Marsh, 1894 (= *Dysalotosaurus* Virchow, 1919)			
D. altus (Marsh, 1878b) (= *Laosaurus altus* Marsh, 1878b)	Morrison Formation, Colorado, Wyoming, Utah, United States	late Kimmeridgian-early Tithonian	Nearly complete skeleton, approximately 7 partial skeletons, postcranial elements, teeth
D. lettowvorbecki (Virchow, 1919) (= *Dysalotosaurus lettow-vorbecki* Virchow, 1919)	Middle Saurian Bed, Tendaguru Beds, Mtwara, Tanzania	late Kimmeridgian	Large number of mostly disassociated cranial and postcranial elements
Valdosaurus Galton, 1977b			
V. canaliculatus (Galton, 1975) (= *Dryosaurus canaliculatus* Galton, 1975)	Wealden, Isle of Wight, West Sussex, England; Bauxite of Cornet, Bihor, Romania	Berriasian-Barremian	Femora, other postcranial elements, dentary, teeth
V. nigeriensis Galton et Taquet, 1982	El Rhaz Formation, Agadez, Niger	late Aptian	Femora

Hypsilophodontidae nomina dubia	Material
Camptosaurus valdensis Lydekker, 1889b	Fragmentary femur
Hypsilophodon wielandi Galton et Jensen, 1979b	Femur
Laosaurus celer Marsh, 1878a	Vertebrae
Laosaurus minimus Gilmore, 1924b	Partial hindlimb, vertebrae
Nanosaurus agilis Marsh, 1877d	Partial skeleton
Phyllodon henkeli Thulborn, 1973a	Teeth

Dryosauridae nomina dubia	
Kangnasaurus coetzeei Haughton, 1915	Tooth, postcranial elements

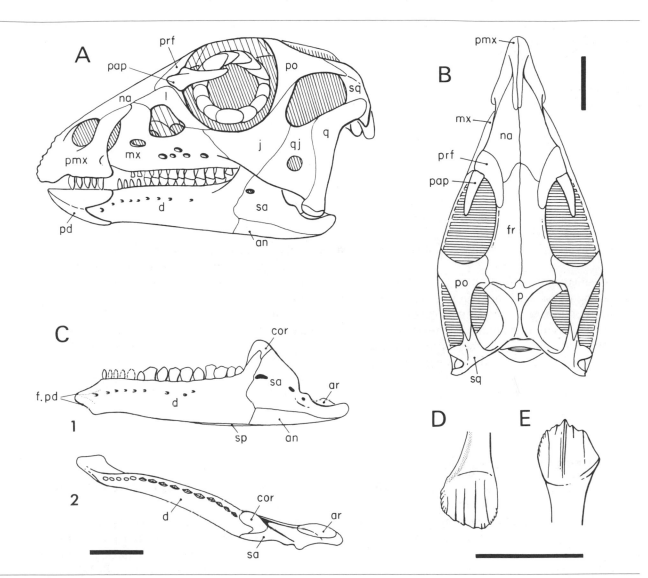

Fig. 24.1. Skull and teeth of *Hypsilophodon foxii* (after Galton 1974a). A–B, reconstruction of skull in A, left lateral, and B, dorsal views. C, left lower jaw (without predentary tooth) in C1, lateral, and C2, dorsal views. D, unworn maxillary tooth crown in labial view. E, worn dentary tooth crown in lingual view. Scale = 2 cm (A–C), = 5 mm (D–E).

lary tooth receives the distal edge of the crown of its immediate predecessor. This partially interlocking arrangement produces a continuous if uneven cutting edge.

The sclerotic ring is composed of fifteen bony plates that are distributed over quadrants of unequal size. The only hyoidal elements known to date are rodlike presumed first ceratobranchials.

Postcranial Skeleton

Axial Skeleton

Hypsilophodon has 9 cervical, 15 or 16 dorsal, 5 or 6 sacral, and about 45 to 50 caudal vertebrae. The centra of the third to seventh cervical vertebrae are opisthocoelous, whereas those of the eighth and ninth are amphiplatyan. With the exception of the first two, all cervical ribs are double-headed. All dorsal vertebrae have amphiplatyan centra, the length of which increases slightly caudad. The dorsal ribs, with the exception of the last one or two, are double-headed. The first six or seven ribs are curved, especially near their proximal ends, and have broad and flattened distal shafts. Ribs caudal to the seventh dorsal vertebra become progressively shorter and straighter. The sternal segments of the ribs are frequently ossified.

A dimorphism in the structure of the sacrum has been reported by Galton (1974a). Some specimens have 16 dorsal and 5 sacral vertebrae, but the centrum of the last dorsal is expanded caudally and forms extensive sutural contacts with the first sacral ribs. By contrast, certain other individuals have 15 dorsal and 6 sacral vertebrae; the first vertebra is a sacral because its ribs are sutured to both centrum and neural arch and contact the pubic peduncles of the ilia.

The centrum of the first caudal vertebra is opisthocoelous, but other caudal centra are amphicoelous and become progressively lower and more slender toward the distal end of the tail. The first chevron is small and borne between the first two caudal vertebrae. The second is apparently flattened craniocaudally, whereas the third tapers distally. The distal portions of the more distal chevrons are craniocaudally expanded and flattened.

Ossified tendons extend in parallel longitudinal rows along the sides of the neural spines in the dorsal vertebral column, rather than in a rhomboidal latticelike arrangement as in most iguanodont ornithopods. Relatively few tendons were present in the proximal portion of the tail, but at least the distal third of the tail is ensheathed by a large number of ossified tendons.

The presence of a light armor in *Hypsilophodon foxii* is conjectural (see Galton 1974a for further discussion).

Appendicular Skeleton

The scapula is about as long as the humerus. The broader proximal end of the humerus bears the cranially directed deltopectoral crest and is set at an angle relative to the moderately expanded distal end. The ulna bears a moderately developed olecranon process. The shaft of the radius is subtriangular in transverse section. The carpus is quite robust, with well-developed proximal carpals fixed to the distal ends of the radius and ulna (Norman, unpub. data). The manus is very small relative to the pes. Metacarpal V is apparently set off from the others to some extent; metacarpal III is the longest. The manus has a phalangeal formula of 2−3−4−3?−1+? (fig. 24.2A).

The ilium (fig. 24.2B) is low. Its preacetabular process bears a prominent ridge along its medial surface and curves outward, with the lateral surface facing somewhat dorsally. The postacetabular portion is deep. The pubic peduncle is slender; the first sacral rib fits onto the medial surface of this peduncle. The long prepubic process of the pubis has a flat, striated external surface, and its ventral edge is grooved. The expanded proximal end of the ischium is separated from the large, bladelike distal portion by a constricted shaft.

The obturator process is situated at about midlength on the shaft of the ischium. The hindlimb is longer and much more robust than the forelimb (fig. 24.3A). The femur (fig. 24.2C-D) has a large, pendent trochanter on its proximal half. The lesser trochanter is separated from the greater trochanter by a shallow cleft and is triangular in transverse section. The extensor groove on the distal articular end of the femur is weakly developed. On the flexor aspect of the distal end, the lateral and medial condyles are almost equal in size. The tibia is longer than the femur. Its proximal end is only moderately developed, and the cnemial crest is small. The fibula has a flat distal articular surface that fitted against the concave proximal surface of the calcaneum. The astragalus caps the distal articular end of the tibia and wraps around onto the extensor face of that bone to form a short ascending process. Metatarsal III is the longest, equaling more than half of the length of the femur, and most robust. Metatarsal V is a mere splint and metatarsal I is about half as long as metatarsal III. The pes has a phalangeal formula of 2−3−4−5−0 (fig. 24.2E). The hallux was not opposable, contrary to earlier assertions, and the foot is digitigrade (Galton 1971d).

SYSTEMATICS AND EVOLUTION OF HYPSILOPHODONTIDAE

We diagnose the Hypsilophodontidae Dollo, 1882, as follows: maxilla with rostral process that fits into groove on medial aspect of premaxilla; frontal with lateral peg that fits into socket on medial surface of postorbital; scapula as long as or slightly shorter than humerus; prepubic process of pubis rod-shaped, generally wider mediolaterally than deep dorsoventrally; the distal portion of the tail is ensheathed by epaxial and hypaxial ossified tendons.

Zephyrosaurus from the Lower Cretaceous of Montana, United States, is very closely related to *Hypsilophodon*. It differs from the latter in a number of cranial features including the presence of a prominent lateral boss on the jugal, the structure of the postorbital, and the presence of a small rostrolateral boss on the maxilla (Sues 1980a). Its postcranial skeleton is virtually unknown.

Orodromeus from the Upper Cretaceous of Montana is known from a remarkable growth series, including still unhatched embryos in eggs (Horner and Weishampel 1988). The presence of a lateral boss on

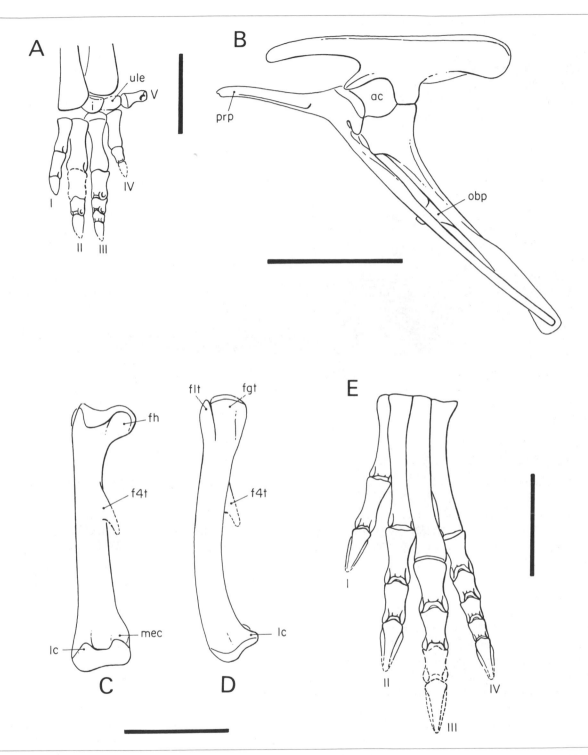

Fig. 24.2. Postcranial elements of *Hypsilophodon foxii* (after Galton 1974*a*). A, left manus in dorsal view. B, pelvic girdle in left lateral view. C–D, left femur in C, caudal, and D, lateral views. E, left pes in dorsal view. Scale = 3 cm (A), = 10 cm (B), = 4 cm (C–D), = 5 cm (E).

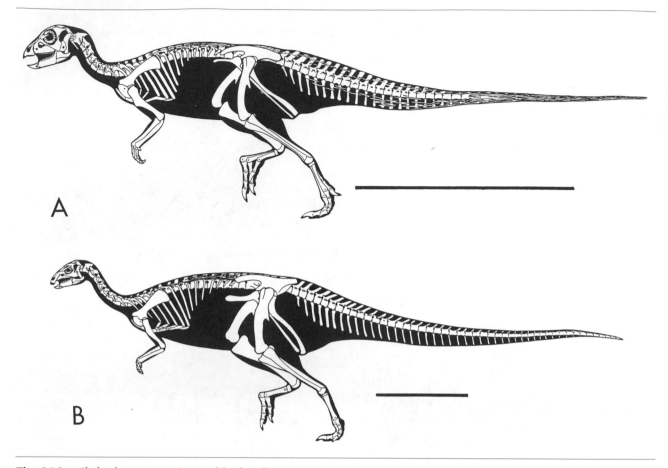

Fig. 24.3. Skeletal reconstruction and body silhouette of A, *Hypsilophodon foxii* (after Galton 1974*a*) and B, *Dryo-* *saurus altus*. Scale = 0.5 m. (Courtesy Gregory S. Paul.)

the jugal and the expanded central portion of the postorbital indicate a close relationship to *Zephyrosaurus*. The teeth of *Orodromeus* resemble those of *Fabrosaurus* very closely. The maxillary and dentary teeth are broadly triangular and often as wide as high. The crowns are subequally divided by a rounded primary ridge. The tibia of *Orodromeus* is considerably longer than the femur. "*Laosaurus*" *minimus* from the Upper Cretaceous of Alberta, Canada (Gilmore 1924*b*), is probably referable to *Orodromeus*, but this synonymy cannot be firmly established owing to the inadequate nature of the type and only known specimen.

Othnielia from the Upper Jurassic of the western United States closely resembles *Hypsilophodon* in the structure of the postcranial skeleton (Galton and Jensen 1973*b*; as *Nanosaurus*(?) *rex*) but little about its skull is presently known. It differs from *Hypsilophodon* in the greater depth of the cleft between the greater and lesser trochanters of the femur and in the more proximal position of the obturator process on the ischium.

Referred teeth have very distinctive, finely textured enamel faces (Galton 1983*d*).

Parksosaurus from the Upper Cretaceous of Alberta, Canada, is distinguished from *Hypsilophodon* by the great depth of the caudal process of the premaxilla, the presence of numerous low, rounded ridges on the thickly enameled surfaces of the cheek teeth, and the apparent expansion of the dorsal end of the jugal (Galton 1973*d*). It is more primitive than the latter in retaining an extensive sutural contact between the maxilla and nasal and a shallow lower temporal arch.

Yandusaurus from the Middle Jurassic of China also lacks prominent primary ridges on the lingual faces of the dentary teeth (He and Cai 1983). The quadratojugal is long and narrow. The lower temporal bar is moderately shallow. The obturator process is placed close to the proximal articular end of the ischium. The lesser trochanter of the femur is placed below the greater trochanter and is separated from it by an apparently shallow cleft.

Fulgurotherium from the Upper Cretaceous of Australia was originally considered a small theropod dinosaur (Huene 1932). The holotype is an abraded distal end of an opalized femur, but the recovery of additional partial femora from the same horizon established the hypsilophodontid affinities of *Fulgurotherium* (Molnar and Galton 1986). The lateral condyle of the femur is narrow. The greater and lesser trochanters are separated by a shallow cleft.

Rich and Rich (1989) have recently described a considerable number of dissociated bones of Hypsilophodontidae from the Lower Cretaceous of Australia, some of which they refer to two new genera, *Atlascopcosaurus* and *Leaellynasaura*. *Atlascopcosaurus* resembles *Zephyrosaurus* in the structure of its unworn maxillary teeth except for the more pronounced development of the primary ridge. *Leaellynasaura* has up to five ridges on both the buccal and lingual aspects of the crowns of unworn maxillary teeth. These ridges are equally well developed on both sides of the tooth crowns, unlike the condition in *Atlascopcosaurus*. Femora referred to *Leaellynasaura* by Rich and Rich (1989) are characterized by craniocaudally narrow distal ends. The fragmentary nature of the Australian hypsilophodontid material described to date does not permit a more precise assessment of the phylogenetic affinities of the two genera under discussion.

Thescelosaurus from the Upper Cretaceous of the western United States and western Canada, long regarded as a hypsilophodontid (Gilmore 1915*a*; Sternberg 1937, 1940*a*; Morris 1976), was referred to the Iguanodontidae by Galton (1974*b*) because the tibia is shorter than the femur in this genus. It differs from other hypsilophodontid ornithopods in the robustness of its skeleton (convergent on *Tenontosaurus*) but does resemble them in derived features such as the rodlike prepubic process (see also Sereno 1986). Furthermore, there are no synapomorphies supporting reference to the clade Iguanodontia. Aside from relative proportions, *Thescelosaurus* differs from other Hypsilophodontidae particularly in the greater interorbital width of the skull roof (Galton 1974*b*). Numerous ridges are developed on the thickly enameled surfaces of the maxillary and dentary teeth of *Thescelosaurus* and form two converging crescentic patterns. An incomplete skull from the Upper Cretaceous of South Dakota, identified as ?*Thescelosaurus* sp. by Morris (1976), may not be referable to this genus (P. Galton pers. comm.).

The interrelationships of the various hypsilophodontid genera remain largely unresolved owing to the lack of comparable skeletal material for several of the taxa in question, and, therefore, subfamilial groupings serve no useful purpose at present. Sternberg (1937, 1940*a*) and Sereno (1986) both noted the distinctiveness of *Thescelosaurus* from other Hypsilophodontidae.

ANATOMY OF *TENONTOSAURUS*

The following anatomical account is based on the type and only known species, *T. tilletti*. The skull and postcranial skeleton were briefly described by Ostrom (1970*a*). Subsequently, Weishampel (1984*a*) has discussed certain aspects of the dentition and jaw function and Dodson (1980) the structure of the postcranial skeleton.

Tenontosaurus has a recorded size range from 1.5 m to more than 7.5 m (Dodson 1980), and some of the features that distinguish it from the Hypsilophodontidae other than *Thescelosaurus* may simply reflect the greater body size and robustness of this genus. *Tenontosaurus* differs from the Hypsilophodontidae in the absence of premaxillary teeth, the presence of three, rather than four, phalanges in digit III of the manus, and a prepubic process that is narrower transversely than deep dorsoventrally. It shares these characters with the Iguanodontia but lacks the other diagnostic synapomorphies of that group (including *Dryosaurus*; see below) and, pending a detailed revision of this genus, is best regarded as representing an as yet unnamed higher taxon of advanced ornithopods.

Skull and Mandible

The skull (fig. 24.4) is long and moderately deep and has very large external nares and large, subrectangular orbits. The external antorbital fenestra is long and narrow. The ventral margins of the robust, edentulous premaxillae are everted to form a rounded, spatulate beak. The premaxilla does not separate the maxilla and nasal. A rostral process of the maxilla fits into a small recess on the ventral face of the premaxilla, but this recess is not floored as in *Hypsilophodon*. The maxilla is very high. The nasal is long and narrow. Both the prefrontal and lacrimal are robust. The broad and stout frontals are tightly sutured to each other. The massive parietals are fused into a single element. The palpebral is small. The nearly straight quadrate is tall dorsoventrally but narrow transversely. A large foramen pierces the quadratojugal near its sutural contact with the jugal. Little is known about the palate. The paroccipital process is hook-shaped and distinctly de-

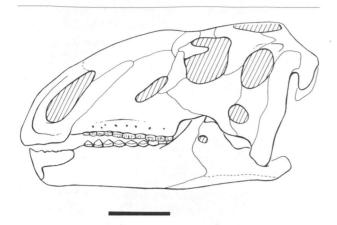

Fig. 24.4. Reconstruction of the skull of *Tenontosaurus tilletti* in left lateral view (modified from Weishampel 1984*a*). Scale = 10 cm.

flected laterally. The predentary is shallow and horseshoe-shaped and has a single ventral process; its sharp dorsal margin bears numerous toothlike projections. The dentary forms a short coronoid process, which is separated by a shallow notch from a bladelike projection of the surangular. The mandible has a long, curved retroarticular process.

The tooth crowns of both the maxillary and dentary teeth are stout and unusually broad. The enameled buccal faces of the maxillary tooth crowns bear five or six weak, subequal ridges. The enameled lingual faces of the dentary teeth bear a prominent median primary ridge; smaller subsidiary ridges are developed on either side of the primary ridge.

Axial Skeleton

The vertebral column of *Tenontosaurus* comprises 12 cervical, 16 dorsal, 5 sacral, and more than 59 caudal vertebrae (Ostrom 1970*a*). The cervical vertebrae have low neural arches and short but well-developed neural spines. The centra of the dorsal vertebrae are amphiplatyan and have subcircular articular ends. The robust neural arches are low and bear stout zygapophyses, which have articular facets inclined at about 45 degrees throughout the dorsal column. The sacral centra do not fuse to each other and have broadly rounded ventral faces. The sacral series may incorporate an additional dorsosacral and/or caudosacral vertebra. The total length of the caudal column equals two to two and a half times the combined lengths of presacral and sacral segments. The proximal caudal vertebrae have short, deep, and broad centra and nar-

row neural spines of moderate height. The tail is enclosed by ossified tendons for most of its length. The epaxial tendons extend from the fourth or fifth dorsal vertebra to the distal end of the tail. Hypaxial tendons begin at about the fifth or sixth caudal vertebra and extend to the distal extremity of the tail.

Appendicular Skeleton

The scapula is distinctly longer than the humerus. The sternal is thin, flat, and crescentic. The deltopectoral crest of the humerus is more robust and occupies a more distal position than in *Hypsilophodon*. The carpus consists of three proximal elements, of which the stout ulnare is the largest, and one or two sesamoidlike distal bones. The metacarpals have flattened proximal ends and are disposed in a spreading manner. The manus has a phalangeal formula of 2−3−3−2−2 (Dodson 1980). The terminal phalanges of digits IV and V are mere nubbins of bone.

The ilium has a concave dorsal margin and long, distinctly deflected preacetabular process. A brevis shelf is not developed. The ischium has a straight, flattened shaft and is not expanded distally. The prepubic process of the pubis is narrower transversely than deep dorsoventrally. No extensor groove is developed on the distal articular end of the femur. The tibia is slightly but consistently longer than the femur in juvenile specimens of *Tenontosaurus;* in more mature individuals, however, the tibia becomes somewhat shorter than the femur (Dodson 1980). The astragalus lacks a cranial ascending process. In addition to the astragalus and calcaneum, the tarsus comprises two distal tarsals. Metatarsal V is absent. The pes has a phalangeal formula of 2−3−4−5−0 (Dodson 1980).

ANATOMY OF DRYOSAURIDAE

The structure of the skull and postcranial skeleton of *Dryosaurus* from the Upper Jurassic of the western United States and Tanzania has been thoroughly described by Janensch (1955) and Galton (1981*c*, 1983*d*). As noted previously, *Dryosaurus* resembles hypsilophodontid ornithopods in many skeletal features but differs significantly from them in the following characters:

absence of premaxillary teeth (also in *Tenontosaurus*);

presence of prominent median vertical ridge on buccal aspect of maxillary tooth crowns;

contact between premaxilla and both prefrontal and lacrimal;

quadratojugal small, overlapped extensively by jugal and quadrate;

predentary with bilobed ventral process;

humerus relatively long and slender, with low deltopectoral crest;

ischium with proximally placed obturator process;

ischium with curved (caudodorsally convex) shaft and rostrocaudally expanded distal end.

Dryosaurus shares these characters with iguanodont ornithopods and, therefore, should not be included in the Hypsilophodontidae.

Skull and Mandible

The premaxillae do not enclose the external narial openings dorsally (fig. 24.5A-B). A long process of the premaxilla extends backward to contact both the prefrontal and lacrimal, thereby excluding any contact between the nasal and maxilla. Premaxillary teeth are absent. The rostral end of the maxilla bears a lateral process in addition to the long medial one. The frontals are relatively broad. The palpebral bridges the entire orbit

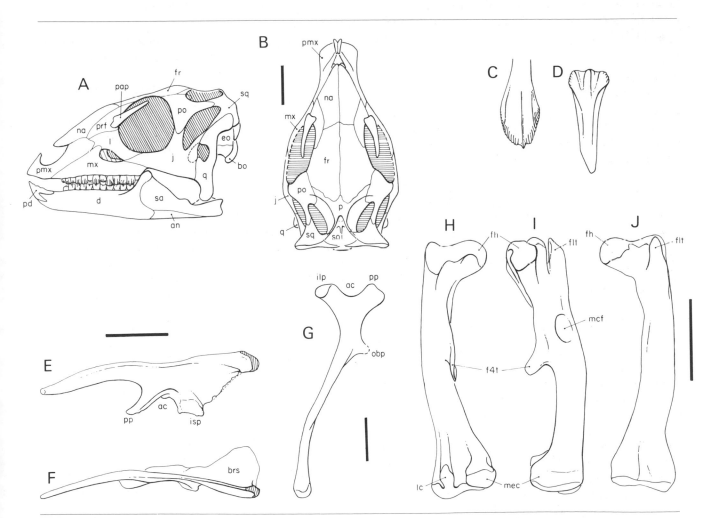

Fig. 24.5. *Dryosaurus lettowvorbecki.* A–B, reconstruction of the skull in A, left lateral, and B, dorsal views. C, maxillary tooth crown in labial view. D, dentary tooth crown in lingual view. E–F, left ilium in E, lateral, and F, dorsal views. G, right ischium in lateral view. H–J, right femur in H, caudal, I, medial, and J, cranial views. (A–F after Janensch 1955; G–J after Galton 1981c.) Scale = 3 cm (A–B), unavailable for C, D [original incorrectly given as "1/3 nat. Gr." by Janensch 1955], = 5 cm (E–G), = 10 cm (H–J).

in *D. altus*. The large jugal has a deeply V-shaped caudodorsal margin for the infratemporal fenestra. The quadratojugal is small. The rostral margin of the quadrate bears a distinct embayment, which presumably corresponds to the foramen in the quadratojugal of *Hypsilophodon*. The coronoid process of the dentary is elevated. The predentary has a bilobed ventral process, rather than a simple one as in *Hypsilophodon* and *Tenontosaurus*.

The thickly enameled buccal surfaces of the maxillary tooth crowns bear prominent primary ridges (fig. 24.5C), which are often set asymmetrically and are much more clearly defined than those on the lingual faces of the corresponding dentary tooth crowns (fig. 24.5D). Two to four secondary ridges are developed mesial to the primary ridge, and up to three are present distal to it. The secondary ridges terminate in well-developed denticles along the margins of the tooth crowns.

Postcranial Skeleton

The humerus bears a low but robust deltopectoral crest. The low ilium forms a very wide post-acetabular shelf for the origin of M. caudifemoralis brevis (fig. 24.5F). The obturator process occupies a proximal position on the ischium (fig. 24.5G). The distal shaft of the ischium in *Dryosaurus* is curved in lateral view and craniocaudally somewhat expanded. Femora of *Dryosaurus* (fig. 24.5H-J) and *Valdosaurus* have a distinct extensor groove (which is particularly well developed in the latter genus) on the distal end of the femur, a separation of the greater and lesser trochanters by a deep cleft, and a distinct pit, presumably for insertion of M. caudifemoralis longus, at the base of the fourth trochanter. The pes of *Dryosaurus* retains a vestigial metatarsal I. Its phalangeal formula is 2–3–4–5–0.

SYSTEMATICS AND EVOLUTION OF DRYOSAURIDAE

We diagnose the Dryosauridae Milner et Norman, 1984, as follows: premaxilla does not enclose external naris dorsally; ilium with wide brevis shelf; femur with deep extensor groove on distal articular end; deep pit (for insertion of M. caudifemoralis longus) developed at base of fourth trochanter of femur; metatarsal I vestigial.

Dryosaurus is known from the Upper Jurassic of the western United States (*D. altus*) and Tanzania (*D. lettowvorbecki*) (Galton 1977*b*). Galton (1981*c*) noted that *D. altus* is distinguished from *D. lettowvorbecki* only by minor skeletal differences such as the much greater length of the palpebral in the American species and the more prominent development of the olecranon process in the African species.

Valdosaurus from the Lower Cretaceous of Europe (Galton 1977*b*; Jurcsák and Popa 1979) and Niger (Galton and Taquet 1982) differs from *Dryosaurus* in the possession of a prominent extensor groove on the distal end of the femur. It is still poorly known and, consequently, very difficult to characterize.

PALEOECOLOGY AND BIOGEOGRAPHY

The dentitions of the hypsilophodontid and dryosaurid ornithopods are well suited for oral processing of plant fodder. These dinosaurian herbivores probably foraged within one to two meters above the ground and were apparently common in Late Jurassic and Cretaceous terrestrial communities.

The ecological success of the Hypsilophodontidae and other advanced ornithopods may well relate to the deployment of a transverse power stroke for effective oral processing of plant food, which is reflected by a distinctive pattern of intracranial mobility (Weishampel 1984*a*; Norman 1984*b*; Norman and Weishampel 1985; see also Sues 1980*a*). Pleurokinesis (*sensu* Norman 1984*b*) involved slight rotation of a functional unit comprising maxilla, jugal, lacrimal, palatine, pterygoid, and ectopterygoid relative to the muzzle and skull roof by means of a diagonal hinge between the premaxilla and maxilla. When the upper and lower teeth were brought bilaterally into occlusion, opposing surfaces sheared obliquely past one another, and, as a consequence, the maxillae rotated laterally and the dentaries medially. Pleurokinesis reduced torsional stresses generated during bilateral occlusion in skulls without a rigid secondary palate. At the same time, it permitted transverse excursion of the upper teeth relative to the lower within a primitively isognathous masticatory apparatus. Resistance to maxillary movement, necessary for shearing between occluding teeth, was presumably provided by the constrictor dorsalis musculature (or a ligamentous remnant thereof) connecting the palate (and hence the upper jaw) and braincase. In this way, effective shear of the food between the teeth could be achieved.

Hypsilophodon was long regarded as *the* arboreal dinosaur par excellence (e.g., Heilmann 1927), but

Galton (1971*d*, 1974*a*) has convincingly argued that it had no structural adaptations for such a mode of life. In particular, the hallux of *Hypsilophodon* was not opposable. Rather, both dryosaurids and hypsilophodontids were fast bipedal runners (fig. 24.3), judging from the significant elongation of both the tibia and metatarsal III relative to the femur and the shortness of the trunk relative to the hind leg (Galton 1974*a*).

Horner and Weishampel (1988) have presented evidence that the hatchlings of *Orodromeus* were precocial. They had well-developed limb bones and epiphyseal condyles; the latter were composed entirely of calcified cartilage. This pattern of ossification is suggestive of an activity pattern similar to that of adult individuals (which attained a length of about 2.5 m) and quite unlike the prolonged nesting behavior documented in the much larger hadrosaurid ornithopod *Maiasaura* from the same beds.

Tenontosaurus was a large, more robustly built form. Judged on the basis of the hindlimb proportions in presumed adult specimens, it was less cursorial than any of the aforementioned taxa, but this is not the case for immature individuals (Dodson 1980). Its dental specializations were essentially similar to those found in the Hypsilophodontidae, but, by virtue of its larger body size, it had a larger vertical foraging range (up to >3 m above the ground).

Skeletal remains referable to the Hypsilophodontidae are currently known from the Upper Jurassic to Upper Cretaceous of western North America, the Upper Jurassic and Lower Cretaceous of Europe, the Middle Jurassic of Asia, and the Lower Cretaceous of Australia. *Hypsilophodon* is known from the Lower Cretaceous of Europe (Galton 1974*a*) and North America (Winkler et al. 1988). *Tenontosaurus* is presently only known from the Lower Cretaceous of western North America. *Dryosaurus* and *Valdosaurus* both had intercontinental geographic ranges (Galton 1977*b*; Galton and Taquet 1982).

Both Dryosauridae and Hypsilophodontidae probably were widely distributed throughout Laurasia and Gondwana after the presumably Early Jurassic divergence between the hypsilophodontian and iguanodontian clades, and further discoveries will surely increase their respective documented geographic and stratigraphic ranges.

25

Iguanodontidae and Related Ornithopods

DAVID B. NORMAN

DAVID B. WEISHAMPEL

INTRODUCTION

Iguanodontids and closely related forms have been known since the early years of the nineteenth century, following the discovery of *Iguanodon* (Mantell 1825). *Iguanodon* was in fact one of the founding members of the Dinosauria (Owen 1842*b*). Additional iguanodonts were slowly described from this time forward (*Camptosaurus*, Marsh 1879*c*, Hulke 1880; *Probactrosaurus*, Rozhdestvensky 1966; *Ouranosaurus*, Taquet 1976; *Mutiaburrasaurus*, Bartholomai and Molnar 1981), but this increase in knowledge has produced a rather complete picture of these animals. Many were relatively large (2,500 kg in *Iguanodon bernissartensis*), mostly bipedal herbivores that are known principally from the Late Jurassic through the Early Cretaceous of both the northern and southern hemispheres.

Iguanodonts, as all the animals considered in this chapter will be referred to (i.e., species of *Iguanodon*, *Camptosaurus*, *Ouranosaurus*, *Probactrosaurus*, and *Muttaburrasaurus*), do not comprise a monophyletic group, although they have at one time or another been collectively called the Iguanodontidae (*sensu* Galton 1972). They do, however, fall between the Hypsilophodon-

tidae (Sues and Norman, this vol.) and the Hadrosauridae (Weishampel and Horner, this vol.), and more restricted subgroups among iguanodonts are monophyletic. These will be diagnosed in the Systematics section of the chapter; a diagnosis of the whole group is impossible given its paraphyletic status.

ANATOMY

The general anatomy of iguanodonts has been the subject of a number of detailed reviews (Gilmore 1909; Taquet 1976; Dodson 1980; Galton and Powell 1980; Norman 1980, 1986, 1987*a*, 1987*b*, 1987*c*; Bartholomai and Molnar 1981; Weishampel and Bjork 1989).

Skull and Mandible

The skulls of all iguandonts are very similar in many of their features (figs. 25.1-25.5). They range in size appreciably, from the smallest in *Camptosaurus prestwichii* to the largest in *Iguanodon bernissartensis*. From the tip of the snout, the dorsal border of the skull profile generally rises in a smooth curve to a point above the orbit, beyond which the skull roof forms a horizontal table immediately above the braincase. The caudal edge of the skull is overhung by the winglike, tapering paroccipital process, which encloses a narrow notch probably supporting the tympanic membrane. Rostral to the paroccipital process, the quadrate is lodged in a cup-shaped squamosal cotylus and is held in position by a short, twisted prequadrate process of the squamosal. Farther ventrally, the lateral wall of the quadrate is notched where it forms the caudal border of the paraquadrate foramen. Ventrally, the lateral wall merges with the mandibular condyle. The latter is transversely expanded and has an articular surface that is subdivided into a rounded lateral portion separated by a step from a somewhat more flattened medial portion. The pterygoid ramus of the quadrate is thin and dorsoventrally expanded, subdividing the adductor chamber along an oblique vertical plane.

The quadratojugal is generally a relatively small bone, overlapped extensively by the jugal. However, it is apparently large in *Muttaburrasaurus langdoni* (fig. 25.4), although incompletely known. In most species, the quadratojugal forms the cranial margin of the paraquadrate foramen; however, this does not appear to be the case in *Camptosaurus dispar,* in which the quadratojugal has been reconstructed lodged in the entire notch in the quadrate (Gilmore 1909). Dorsally, the quadratojugal usually forms a slender, tapering, fingerlike process that follows the rostrodorsal margin of the quadrate.

The jugal is notable for the deep excavation on its rostroventral end for reception of the thick, oblique jugal process of the maxilla. Immediately adjacent, the medial side of the jugal has a rugose, pitted surface for the extreme tip of the ectopterygoid. The dorsal margin of the jugal has two deep embayments for the orbit and infratemporal fenestra. The main body of the jugal is thin and swings from the side of the skull to clear the coronoid process of the mandible. Caudally, the jugal expands dorsoventrally and meets the quadratojugal along an extensive overlapping suture. In *Ouranosaurus,* the rostral end of the jugal is slightly deeper than in *Iguanodon* and *Camptosaurus.* The rostral end of the jugal does not appear to form the caudal margin of the antorbital fenestra as reconstructed by Taquet (1976).

The maxilla is approximately triangular in lateral view and has a long, narrow rostral extension that lies beneath the premaxilla. The maxillary part of the maxilla-premaxilla articulation is generally flat to slightly concave transversely. A stylelike dorsomedial maxillary process is lodged between the paired premaxillae. The dorsal process of the maxilla ends in a

thumblike lacrimal process. Immediately caudoventral to the lacrimal process, there is an embayment for the antorbital fenestra, somewhat larger in species of *Camptosaurus* than in either *Iguanodon* or *Ouranosaurus.* Galton and Powell (1980) reported a distinctive pit in the dorsal surface of the maxilla adjacent to the antorbital fenestra in *C. prestwichii,* but the presence of this pit has not been confirmed in *C. dispar.* A small shoulder beneath the fenestra forms the base of the prominent jugal process; the latter points obliquely ventrolaterally from the body of the maxilla. The caudal end of the maxilla is bluntly truncated where it forms a wedge between the suture for the pterygoid and ectopterygoid. The medial surface of the maxilla is essentially flat and vertical, with a slight medial overhang of the dorsal border rostrally. The most notable feature of the medial surface is the arcade of special foramina that circumscribe the alveolar parapet. The latter has a regularly scalloped margin and supports the roots of functional and successional teeth. The maxilla of *Ouranosaurus* shows slight proportional differences to that of *Camptosaurus* and *Iguanodon* in that the caudal border of the maxilla of the former is shorter and the rostral portion is deeper and appears to lack a dorsomedial process.

The ventral surface of the premaxilla is rounded and irregularly flattened laterally, and the extreme rostral margin is strongly denticulate, pitted, and somewhat curved dorsally where it is opposed by the predentary. The lateral and ventral margins of the external nares are somewhat flared, trumpet-shaped, and partially separated by a thin median septum. Flaring is broadest in *Ouranosaurus,* less so in *Camptosaurus* and *Iguanodon.* The long sloping facial region is very distinctive in *Ouranosaurus.* In particular, the premaxillae are dorsoventrally compressed and broad, such that the external nares are higher and much more prominent. Caudal to the narial opening, the lateral process of the premaxilla forms a long, slender, sloping surface on which the nasal rests and which in turn rests in a trough-shaped depression on the dorsal surface of the maxilla. The lateral process terminates caudally as an overlapping suture with the lacrimal and prefrontal (the extreme caudal tips of the lateral processes of the premaxillae in *Ouranosaurus* have been reconstructed as contacting the lacrimal alone [Taquet 1976], but this condition is conjectural since this region of the skull is not well preserved). There is evidence of a median cleft in the palatal roof of the premaxillae in *I. atherfieldensis* (Norman 1986). A prominent dorsal process of the premaxilla curves caudally over the external nares to wedge between the nasals. However, the degree of de-

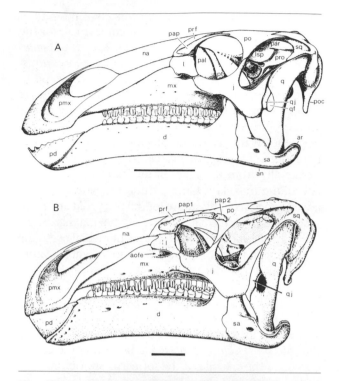

Fig. 25.1. (A) *Iguanodon atherfieldensis*. Skull in lateral view (after Norman 1986). (B) *I. bernissartensis*. Skull in lateral view. Scale = 10 cm. (After Norman 1980.)

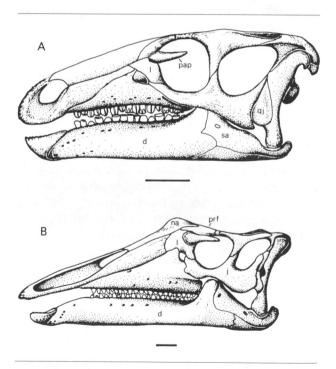

Fig. 25.2. (A) *Camptosaurus dispar*. Skull in lateral view (after Gilmore 1909). (B) *Ouranosaurus nigeriensis*. Skull in lateral view. Scale = 5 cm. (After Taquet 1976.)

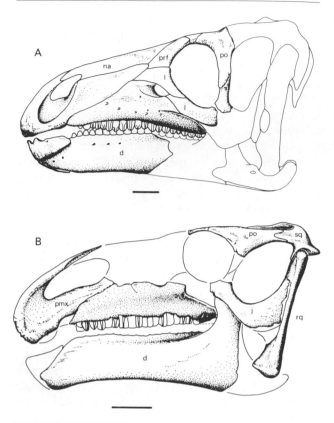

Fig. 25.3. (A) *Iguanodon lakotaensis*. Skull in lateral view. (After Weishampel and Bjork 1989.) (B) *Probactrosaurus gobiensis*. Skull in lateral view. Scale = 5 cm. (After Rozhdestvensky 1966.)

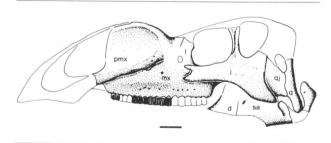

Fig. 25.4. *Muttaburrasaurus langdoni*. Skull in lateral view. Scale = 5 cm. (After Bartholomai and Molnar 1981.)

velopment of this process varies considerably among species. Short medial premaxillary processes meet immediately beneath the external nares. These processes roof the rostral portion of the oral cavity and probably provide sutural contact for the rostral portion of the palate.

The smoothly-curved nasal borders the external naris and roofs the nasal cavity. Rostrally, it tapers to a

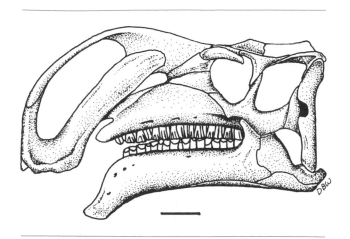

Fig. 25.5. *"Iguanodon" orientalis.* Skull in lateral view (based on photographs courtesy of S. M. Kurzanov). Scale = 5 cm.

sharp point where it is overlapped by the dorsal process of the premaxilla. In *Camptosaurus dispar,* the nasal appears to meet the short dorsal process of the premaxilla directly above the narial opening. In *Ouranosaurus nigeriensis,* the nasal is remarkable for its distinct rounded thickenings located near the frontal suture. These features, along with the extreme breadth and dorsoventral compression of the premaxillae, add considerably to the exaggerated slope of the snout. In *Muttaburrasaurus langdoni,* the nasals are equally remarkable but instead form a high arch somewhat similar to that in gryposaur hadrosaurids (Weishampel and Horner, this vol.).

The lacrimal in *I. atherfieldensis* is a small block-like bone wedged against the maxilla between the lateral wall and the lacrimal process. It is overlapped by a thin caudal extension of the premaxilla and meets the prefrontal dorsally in a loose, rostrally overlapping suture (Norman 1986). The orbital exposure of the lacrimal is narrow and has a vertically elliptical naso-lacrimal canal leading into the nasal cavity. The ventral margin of the lacrimal forms the dorsal margin of the antorbital fenestra and contacts the rostral end of the jugal. In *O. nigeriensis,* the lacrimal appears to be proportionally much larger than that of *I. atherfieldensis,* but this difference may reflect poor preservation of the prefrontal in the latter species.

The prefrontal is small and smoothly curved around the rostrodorsal margin of the orbit. At its rostral end, it is overlapped by the caudal tip of the premaxilla, above which it meets the nasal in an oblique suture rostrally and the frontal caudally. In *Ouranosaurus,* the prefrontal has been restored in a more dorsal position compared to other iguanodonts. In addition, it may have overlapped the lacrimal to a greater degree

and had a more extensive exposure on the side of the face.

The palpebral of *I. atherfieldensis* is long, tapering, and curves around the orbital margin parallel to the prefrontal, against which its base articulates exclusively. A subsidiary palpebral is positioned beyond the distal extremity of the primary palpebral, adjacent to the postorbital (Norman 1980). In *C. dispar,* the palpebral articulates with both the lacrimal and prefrontal (Gilmore 1909) and is shorter and appears to project across the orbit instead of following the orbital margin. The palpebral of *O. nigeriensis* seems smaller than that of species of *Iguanodon.* However, it may well have curved around the orbital margin in a similar manner.

In *Iguanodon,* the postorbital has a slightly everted and rugose orbital margin. Ventrally, it produces a tapering angular process that forms a loose overlapping suture with the jugal, while caudally it develops two fingerlike projections that lie against the squamosal and form the rostral part of the upper temporal arch. Medially, the postorbital is extensively sutured to the frontal along a slightly interdigitate, curved surface, and is butt-jointed to the parietal and laterosphenoid medially. The form of this bone is very similar in all other iguanodonts that are adequately known.

The frontals are fused together to form a broad sheet of bone across the roof of the skull, between the orbits and supratemporal fenestrae (fig. 25.6). The suture between the frontal and nasal is transversely broad and overlapping, interrupted laterally by notches to receive the prefrontals. There is a short orbital exposure of the frontals, shrouded (in the case of *Iguanodon*) by the palpebral. Caudally, there is a long, serrated, curved suture with the postorbital, which is continuous with a midline suture with the parietals. The ventral surface of the frontals is excavated, laterally forming the roof of the orbital cavities and medially the roof of the common passage for the olfactory nerve.

Dorsally, the fused parietal plate is generally saddle shaped, has a low sagittal crest, and caps the braincase (fig. 25.6). The size of the sagittal crest appears to vary with maturity and possibly sex. The (restored) sagittal crest of *Probactrosaurus alashanicus* is particularly prominent. The lateral wall of the parietal and the sagittal crest formed the area of attachment of much of the adductor musculature.

The squamosal forms the caudolateral corners of the skull roof. It has a forwardly directed process that lies against the postorbital. The external surface of this process has a ledge that extends to a position above the squamosal cotylus for the attachment of M. adductor mandibulae externus pars superficialis (fig. 25.7). The prequadratic and postquadratic processes

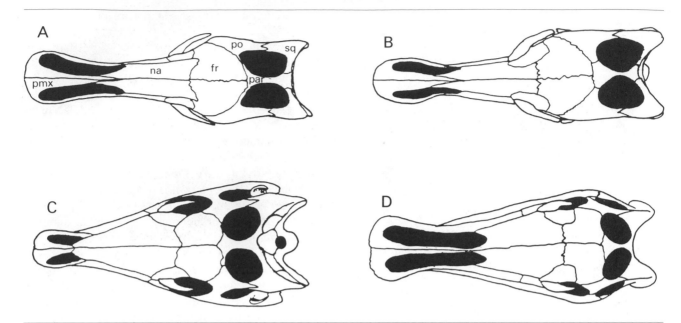

Fig. 25.6. Dorsal views of the skull of selected iguanodonts: (A) *Iguanodon atherfieldensis*, (B) *I. bernissartensis*, (C) *Camptosaurus dispar*, (D) *Ouranosaurus nigeriensis* (all after Norman 1986).

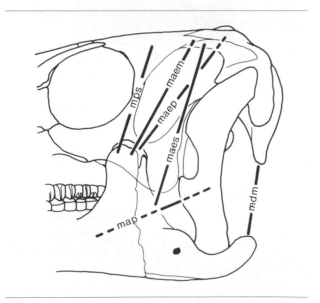

Fig. 25.7. Cranial jaw musculature restored in *Iguanodon atherfieldensis*, using an approximation to the lines of action of the various muscles.

of the squamosal restrain rotation of the quadrate head. The squamosal lies against the oblique lateral surface of the paroccipital portion of the opisthotic. The dorsomedial portion of the squamosal overlaps the shoulder of the paroccipital process as well as the supraoccipital and abuts the parietal along the dorsal margin of

the occiput in species of *Camptosaurus, Iguanodon*, and *Ouranosaurus*. In contrast, the squamosals exclude the parietals from the occiput and appear to separate the parietals for a considerable distance between the low sagittal crest in *Probactrosaurus gobiensis*.

The neurocranial region in adults (fig. 25.8) is firmly fused and consists laterally of the orbitosphenoid, laterosphenoid, prootic, and optisthotic (the latter fused with the exoccipital to form the caudal wall of the braincase as well as the paroccipital process). The crista prootica extends across the prootic and opisthotic to the base of the paroccipital process, marking the attachment site of M. levator pterygoideus (fig. 25.9). Ventrally, the braincase is comprised of the parasphenoid, basisphenoid, and basioccipital. The shallowly concave surface between the traces of the maxillary and mandibular branches of c. n. V along the lateral wall of the basisphenoid marks the site of attachment of either M. protractor pterygoideus or a ligament that extends onto the dorsal surface of the pterygoid bone. The basipterygoid process of the basisphenoid is low and oblique (Weishampel 1984a; Norman 1986), and the occipital condyle portion of the basioccipital is set on a distinct neck and tends to vary in size and shape from species to species. In some taxa, it is almost heart shaped in occipital view (e.g., *Ouranosaurus*), while in others it may be globular (e.g., *Iguanodon*).

The occiput is formed of the massive laterally projecting paroccipital processes, while between these,

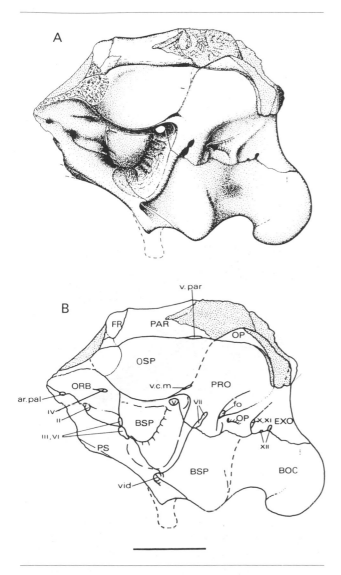

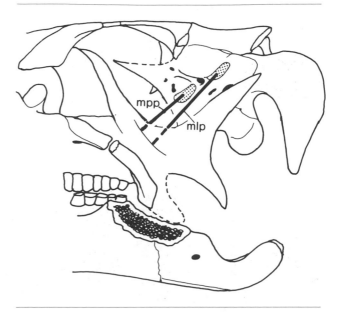

Fig. 25.9. Constrictor dorsalis musculature of *Iguanodon atherfieldensis,* based on a partially "dissected" skull, showing the pleurokinetic system.

Fig. 25.8. *Iguanodon* cf. *atherfieldensis.* Neurocranium in later view. (A) natural appearance, (B) diagrammatic interpretation. Scale = 5 cm. (After Norman 1986.)

the foramen magnum is surrounded by the exoccipitals, which together form a horizontal bar separating the foramen magnum from the more dorsally positioned supraoccipital (fig. 25.10). The supraoccipital is not well defined. It is hidden dorsally in a recess in the occipital surface beneath the parietals and between and below the shoulders of the occiput formed by the squamosals. It appears to be thin and poorly ossified in *I. atherfieldensis* but prominent in *I. bernissartensis* (Norman 1980).

In contrast to *Iguanodon, Camptosaurus dispar* has an occiput that is very reminiscent of that in hypsilophodontids in that the supraoccipital forms the dor-

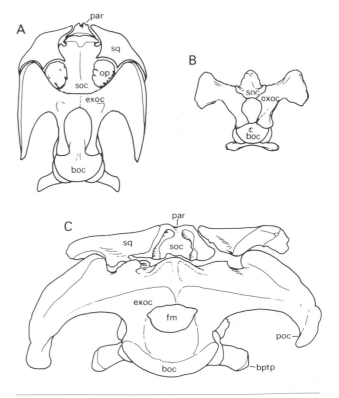

Fig. 25.10. Occipital view of the skulls of (A) *Iguanodon atherfieldensis,* (B) *Camptosaurus dispar* (after Gilmore 1909), (C) *Ouranosaurus nigeriensis* (after Taquet 1976).

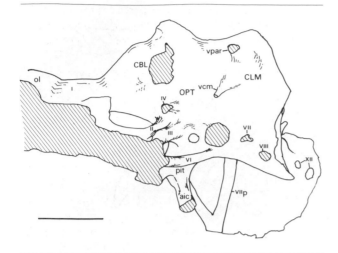

Fig. 25.11. *Iguanodon bernissartensis*. Natural endocranial cast in lateral view. Scale = 5 cm.

sal margin of the foramen magnum instead of being excluded by the exoccipital bar. *Ouranosaurus nigeriensis* has a lower and broader braincase than in any other iguanodont, being more similar to that seen in hadrosaurids. All bones within the occiput are fused, but it is reported that the supraoccipital is excluded from the foramen magnum (Taquet 1976).

Internally, the endocranial cavity of *Iguanodon* (fig. 25.11) is very similar to that seen in hadrosaurids (cf. Norman 1986; Ostrom 1961*a*). The rostral telencephalic region is broadly expanded, followed by a constriction accommodating the mesencephalon and finally the dorsoventrally expanded metencephalic and myelencephalic region that taper caudally toward the foramen magnum (fig. 25.12A, B). The area immediately beneath the mesencephalon is complicated by the presence of vessels leading to and from a large sinus on the dorsolateral wall of the endocranial cavity.

The rather bulbous telencephalon (mainly comprised of the paired cerebral hemispheres) tapers rostrally to a constricted region, the coronal sulcus, which marks the rostral termination of the cerebral expansion and separates this part of the brain from the olfactory lobes. Beneath the broadly expanded area of the cerebral hemispheres, there is a well-preserved pituitary fossa and indications of the paths of the associated vessels and nerves. All twelve cranial nerves are indicated in figure 25.12B. The osseous labyrinth is also known in *Iguanodon* (fig. 25.13; Norman 1986). The labyrinth is notable for the small size of the sacculus and the elongate lagena, a combination of features that is very similar to that in many other dinosaurs.

The palate is the least well known portion of the

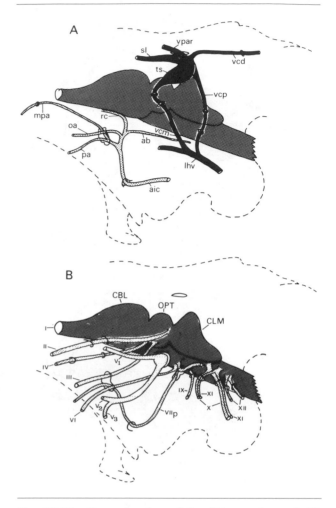

Fig. 25.12. Reconstruction of the (A) vascular and (B) neural anatomy of the endocranial region of *Iguanodon*.

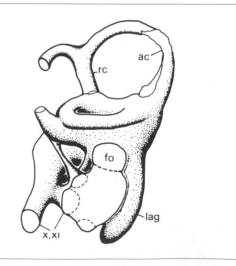

Fig. 25.13. Reconstruction of the osseous labyrinth of *Iguanodon*.

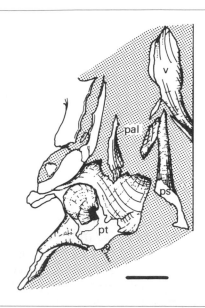

Fig. 25.14. *Iguanodon bernissartensis.* Ventral view of the palate. Scale = 5 cm.

skull in iguanodonts (fig. 25.14). In species of *Iguanodon,* the vomers appear to form deep, thin plates that are fused along their ventral edges. The rostral portion probably contacts the medial process of the premaxilla. From here, the vomers form a keel-like blade that bisects the nasal cavity, extending caudally along the midline as far as the level of the lacrimal. The dorsal portion of each vomer consists of a caudally deepening plate that diverges from its neighbor, thereby forming the medial septal walls of the nasal cavity. The caudal extremity of each vomer extends dorsally in a rising arch to meet or lie adjacent to the rostral extremity of the pterygoid and the dorsal margin of the palatine.

The palatine is rhomboidal is outline and forms the medial orbital wall. The ventral portion articulates with the caudomedial part of the body of the maxilla. From here, the palatine rises in a smooth arch toward the midline just beneath the frontals. The rostral edge of the palatine is smoothly curved and has a rounded edge. The caudal edge, by contrast, is apparently fused to the narrow, obliquely inclined rostral process of the pterygoid.

The rostral portion of the pterygoid is narrow, thin, and axially twisted. Farther caudally toward the body of the pterygoid, the ventral edge is sutured to the oblique caudal margin of the maxilla, and this area is further bound to the maxilla by the ectopterygoid. Caudal to the articulation with the basipterygoid process, the central plate of the pterygoid produces a deep, thin ramus that is forked caudally and laps against the medial side of the pterygoid ramus of the quadrate.

Ventrally beyond the caudal extremity of the maxilla, the pterygoid develops a broadly curved ventral pterygoid flange, the site of attachment of M. pterygoideus dorsalis.

The ectopterygoid is adequately known only in *Iguanodon* (Norman 1980, 1986). It appears to be a narrow strap of bone that lies in a recess on the lateral surface of the caudal end of the maxilla. The rostral end fits against the medial side of the jugal adjacent to the jugal-maxilla articulation and extends caudally across the maxilla to the pterygoid, where it seems to form part of the lateral margin of the ventral pterygoid flange.

The mandible of all iguanodonts is bowed slightly downward along its length and has a prominent coronoid process that stands lateral to the tooth row, marking the rear of the buccal cavity. The apex and caudal margin of the coronoid process is the site of attachment of virtually all of the jaw adductor musculature. The unpaired predentary has a smoothly curved buccal margin that in occlusal view is broadly horseshoe-shaped. The rostral portion of the predentary is coarsely denticulate, while more laterally and caudally, the edge comprises a flattened shelf bordered by a raised ridge. The form of the predentary varies within the group, from relatively narrow in *Camptosaurus dispar* to very broad in *Ouranosaurus nigeriensis.* The primary function of the predentary (apart from prehension of food) is to stabilize the small, horizontal dentary symphysis by wedging itself between the ends of the mandibles and clasping the dentaries with a pair of flat lateral processes. A flat and bilobate ventral process underlies the symphysis.

The dentary is long and transversely thick, more so caudally. The alveoli are positioned along the extreme medial edge. The teeth form a more or less straight array mesially. However, toward the distal end of the series the tooth row curves laterally toward the base of the coronoid process. The extreme tip of the dentary is shaped like an open trough, which is capped by the predentary. There can be a diastema between the predentary and the tooth-bearing portion of the dentary, variably developed among iguanodonts (long in *Ouranosaurus nigeriensis,* short or absent in *Iguanodon lakotaensis* and *Camptosaurus dispar*). The surangular is large and well exposed on the lateral surface of the mandible. The rostral portion of this bone is irregular and lies medial to the partially tubular caudal end of the dentary. However, dorsal to this junction, the surangular develops a complex sutural relationship with the caudal margin of the coronoid process (Norman 1980, 1986). Caudal to the area of sutural contact with the dentary, the surangular drops sharply and broadens to form the major part of the mandibular articulation.

Here the surangular consists of a broad cup-shaped depression that receives the convex lateral portion of the mandibular condyle of the quadrate. Behind the joint, there is a variably developed retroarticular process. The medial side of the retroarticular process has a shallow depression for the articular bone. Rostral and ventral to this area, there are further depressions for the prearticular, angular, and splenial. The surangular also forms the lateral margin of the adductor fossa.

The articular is small and somewhat lozenge shaped; it contributes to the medial part of the jaw joint (i.e., articulating with the medial portion of the mandibular condyle) and the retroarticular process.

The prearticular serves to at least partly sandwich the articular in position by enclosing the rostral part of the latter medially. The prearticular also forms the medial border of the glenoid. Rostrally, the prearticular is thin where it forms the medial margin of the adductor fossa and sends a spur toward the base of the coronoid process. Further rostrally, the prearticular is long and tapering beneath the arcade of special foramina medial to the dorsal port of the Meckelian canal and overlain by the splenial ventrally. It is not certain how far rostrally the prearticular extends.

The splenial lies ventral to the prearticular and with the latter covers the Meckelian canal. The rostral extent of the splenial is so far not known. Caudally, the splenial appears to lie beneath the prearticular and contacts the surangular by means of a lateral, horizontal shelf (Norman 1980).

The angular lies along the ventral edge of the mandible and is usually just visible in lateral view. Although somewhat variable among genera, the angular tends to be a long, thin bone that sits in a recess on the ventral surface of the surangular. It contacts the splenial medially in what appears to be a butt suture.

The small coronoid attaches to the medial side of the cornoid process of the dentary. Again, this bone is visible in lateral view as a small corona above the dentary portion of the process. The caudal edge of the coronoid appears (at least in species of *Iguanodon*) to be involved in a complex suture between the dentary and surangular. In addition, there is a ventrally placed bone that extends medial to the caudal alveoli and gradually tapers rostrally. Such a position is occupied by an "intercoronoid" in ceratopsids (Brown and Schlaijker 1940c) and in *Plateosaurus engelhardti* (Galton 1985e). However, in the latter examples, the "intercoronoid" may in fact be the ventral extension of the normal coronoid, rather than a separate ossification. The same may apply in iguanodonts.

The maxillary tooth count varies according to the size or age of individuals of the same species and also among taxa. Maximum counts are 23 in *Iguanodon atherfieldensis*, 29 in *I. bernissartensis*, 14 in *Camptosaurus dispar*, 22 in *Ouranosaurus nigeriensis*, and more than 23 in *Probactrosaurus gobiensis*. There is never more than one replacement tooth per tooth position, although a rudimentary second replacement tooth is reported for *P. alashanicus* (Rozhdestvensky 1966).

Maxillary crowns (fig. 25.15) are narrow and generally more lanceolate in buccal or lingual aspect

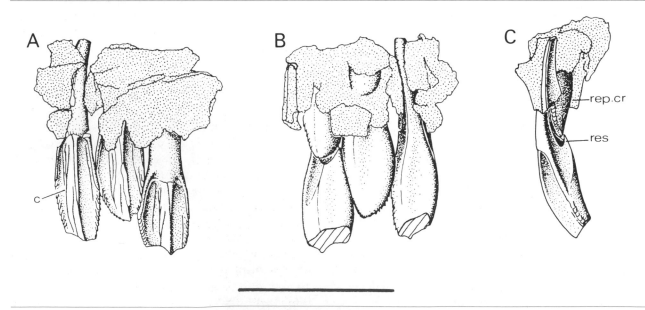

Fig. 25.15. *Iguanodon* cf. *atherfieldensis*. Maxillary teeth in a fragment of maxilla in (A) buccal, (B) lingual, and (C) distal aspects. Scale = 5 cm. (After Norman 1986.)

than those of the dentary. The buccal surface of the crown is thickly enameled and has a strongly denticulate margin. The buccal surface is dominated by a very large primary ridge arising from the base of the crown (this appears to be true of all iguanodonts here considered, including *Probactrosaurus,* despite errors in the English translation of Rozhdestvensky 1966). The remainder of the crown surface is relatively smooth apart from a few fine ridges of variable length in nearly all iguanodont species; *Rhabdodon priscus* represents one of the exceptions in having a very strong secondary ridge on either side of the primary ridge. The crown margins in iguanodont maxillary teeth are thickened and coarsely denticulate, the denticles themselves having an irregularly undulating edge. The bases of the denticulate margins are curled across the crown near the base to form a very oblique shelf that merges with the base of the primary ridge. The root is long and tapering and has shallow longitudinal grooves running along the sides to accommodate adjacent successional crowns within the alveolus.

Craspedodon lonzeensis, known only from presumed maxillary teeth (fig. 25.16), has crowns that are relatively narrow and transversely expanded. The buccal surface of the crown is marked by a very prominent primary ridge and well-formed tertiary ridges. The coarsely denticulate and highly sculptured mesial and distal margins coverge toward the base to form a complete cingulum. Finally, the crowns of *Muttaburrasaurus langdoni* lack a large primary ridge (fig. 25.4). Instead, the enameled buccal surface is covered by a series of fine vertical ridges that vary in number depending on tooth row position and size of the crown.

Dentary tooth counts are known for several species that have relatively well preserved jaws. There are 21 dentary teeth in *I. atherfieldensis,* 25 in *I. bernissar-*

tensis, 16 in *C. dispar,* 23 in *O. nigeriensis,* and 21 to 23 in *P. gobiensis.* Tooth counts are generally lower in the dentary than maxilla, since dentary crowns are broader. No more than a single replacement tooth is found per tooth position, and, as with maxillary teeth, the long tapering root has shallow vertical grooves along both mesial and distal edges for the crowns of adjacent successional teeth. The crown is transversely compressed and leaf shaped in lingual aspect (fig. 25.17). Mesial crowns tend to be more lanceolate, while those distally are somewhat broader. Mesial and distal margins are denticulate. The lingual surface is thickly enameled and traversed by a series of ridges. The largest and most clearly defined (i.e., primary) ridge is distal to the middle of the crown and extends to the distal corner of the truncated apex. Another, less well defined (i.e., secondary) ridge extends upward on the crown mesial to the primary ridge but tends to fade before reaching the apex. A variable number of tertiary ridges continue to the bases of the marginal denticles. Just below the widest part of the crown, the distal denticulate margin curls inwardly to produce an oblique cingulum.

Replacement in the lower dentition of species of *Iguanodon* yields a *Zahnreihe* pattern with Z-spacing around 2.4. Wear consists of confluent, oblique facets, often with a pronounced step at the trailing edge of the occlusal surface of each tooth (buccal for dentary teeth, lingual for maxillary teeth). Wear striae are oriented approximately transverse to the long axis of the tooth row.

Portions of the hyoid apparatus are commonly preserved between the mandibles in relatively undisturbed skulls (fig. 25.18). The paired ceratobranchials are elongate, curved rods tapering toward one end. The end of the ceratobranchial is abruptly truncated and sometimes preserves the articular surface for adjacent parts of the hyoid.

Postcranial Skeleton

Axial Skeleton

There are 10 to 11 cervical vertebrae, 17 dorsals, 6 sacrals, and more than 45 caudals in *I. atherfieldensis. I. bernissartensis* has 11 cervicals, 17 dorsals, 7 sacrals, and more than 50 caudals. *C. dispar* has 9 cervicals, 15 to 16 dorsals, 5 to 6 sacrals, and more than 40 caudals. Last, *O. nigeriensis* has 11 cervicals, possibly 17 dorsals, 6 sacrals, and more than 40 caudals.

The atlas typically comprises a crescentic intercentrum bearing a concave dorsal margin that receives the odontoid process of the axis, a smoothly concave cranial articulation for the occipital condyle and liplike facets for attachment of the single-headed first cervical rib (fig. 25.19). The neural arch is paired, and each half

Fig. 25.16. *Craspedodon lonzeensis.* Newly erupted maxillary crown in buccal view. Scale = 2 cm. (After Dollo 1883c.)

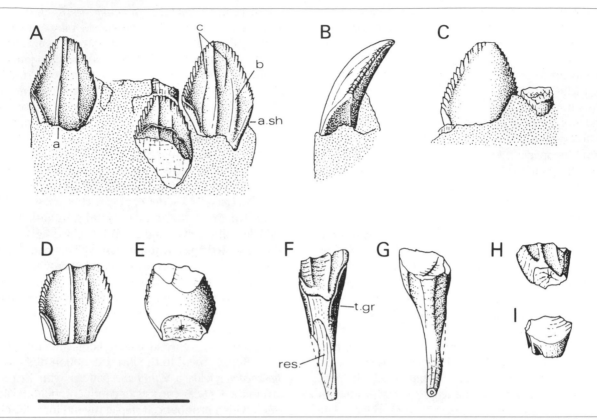

Fig. 25.17. *Iguanodon* cf. *atherfieldensis*. Dentary teeth in (A) lingual, (B) mesial, and (C) buccal views; isolated worn dentary crown in (D) lingual and (E) buccal views; isolated worn dentary tooth in (F) lingual view and (G) buccal view, showing well-developed resorption facet on the root; (H, I) worn dentary crown (root resorbed and crown heavily worn). Scale = 5 cm.

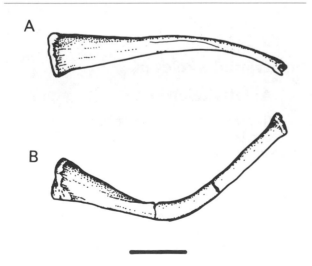

Fig. 25.18. *Ouranosaurus nigeriensis*. Hyoid in (A) dorsal and (B) lateral views. Scale = 4 cm. (After Taquet 1976.)

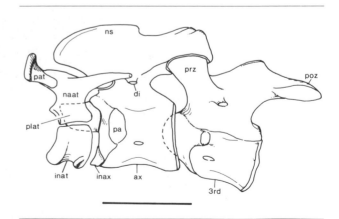

Fig. 25.19. *Iguanodon* cf. *atherfieldensis*. Reconstruction of the cranial cervical vertebral series. Scale = 5 cm. (After Norman 1986.)

has a small articular surface for the occipital condyle, above which is a narrow arch that curves over the neural canal. From the dorsolateral margin of the arch, there is a slender transverse process that is directed obliquely backward and upward; medially, the postzygapophysis is horizontal. The ventral portion of the neural arch curves forward and supports the small proatlas elements interposed between the occiput and atlas. Proatlases are rarely preserved but have been described in *I. atherfieldensis* (Norman 1986) and *O. nigeriensis* (Taquet 1976). As in all archosaurs, the atlas pleurocentrum is attached to the centrum of the axis, forming the rotational axis for the atlas-axis complex.

The axis is notable for a prominent neural spine that forms a bladelike structure above the centrum, extending forward between the separate arches of the atlas. In addition, the ventral surface of the axis pleurocentrum has a small, wedge-shaped intercentrum attached along its rostroventral margin. Both of these structures control the movement between the atlas and axis. A large parapophyseal facet is present immediately behind the rostral articular surface of the centrum, and above and somewhat caudal to this, there is a small diapophyseal facet. The caudal surface of the centrum is concave, the degree of concavity depending on the species and size of the individual. This surface receives the convex cranial surface of the third cervical centrum.

The form of the third cervical is typical for the remainder of the cervical series. The main body of the centrum has a large convex cranial articular surface

(least well developed in *C. dispar*), behind which the centrum is contracted laterally but retains a prominent ventral keel; caudally, the articular surface is strongly concave. The neural arch is divided into oblique, forwardly directed processes that form the cranial supports of the zygapophyses. Caudally, the postzygapophyses are long and curved, diverging from a low midline spine or ridge.

A number of serial changes occur down the cervical series (figs. 25.20-25.23). The neural spine becomes gradually taller, forming a hook-like process as the dorsal series is approached. The centra become shorter in axial dimensions and taller dorsoventrally. The divergence between the prezygapophyses becomes less marked. Finally, the diapophysis becomes more robust and higher on the side of the neural arch.

Ribs are found on all cervical vertebrae. The first is a simple straight rod, while the remainder are two-headed, with a robust capitulum and more slender tuberculum. The rib shaft increases in length as the dorsals are approached. *Ouranosaurus* is unusual in that it has cervical vertebrae with randomly fused ribs (Taquet 1976), probably due to the old age of the individual.

The change from cervical to dorsal vertebrae is marked by the migration of the parapophysis from the surface of the centrum onto the base of the neural arch. The first two dorsals retain many of the characters of the cervicals: the convex cranial articular surface, slight opisthocoely, and the relatively small size and hooked form of the neural spine. In more caudal members of

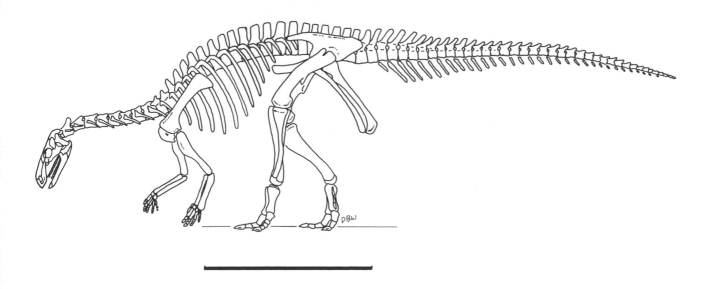

Fig. 25.20. *Camptosaurus dispar.* Skeletal restoration. Scale = 1 m.

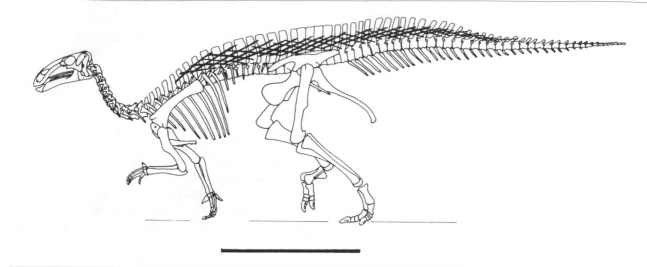

Fig. 25.21. *Iguanodon atherfieldensis.* Skeletal restoration. Scale = 1 m. (After Norman 1986.)

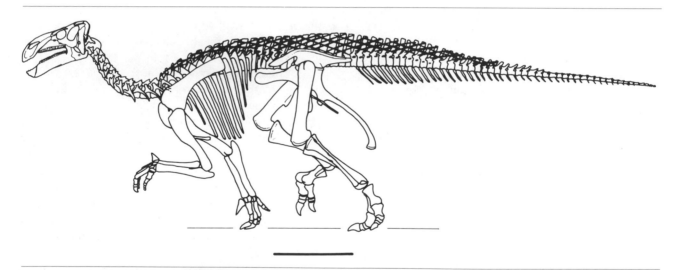

Fig. 25.22. *Iguanodon bernissartensis.* Skeletal restoration. Scale = 1 m. (After Norman 1980.)

the series, the centrum becomes taller and more laterally compressed, eventually developing a sharp ventral keel, while the articular surfaces become amphiplatyan with slight opisthocoely. The neural arch also becomes axially longer, and the parapophysis rises gradually to a position adjacent to the prezygapophysis and then migrates laterally along the cranial margin of the transverse process toward the diapophysis. The transverse processes, which are at first steeply inclined toward the neural spine and quite robust, become progressively lower until they are directed horizontally and become relatively more slender. The neural spines increase markedly in height caudally from dorsals 1 to 5, forming a row of tall, slightly inclined plates with truncated

and slightly expanded dorsal edges. The degree of development of the spines reaches its maximum in *Ouranosaurus*, where the spines are nine times the height of the centrum (fig. 25.24). The centra of the last dorsal vertebrae are axially compressed, have broad articular surfaces, tend to be opisthocoelous, and have zygapophyses that are practically horizontal. The last neural spine in the dorsal series of *Ouranosaurus* is unusual in that it has a groove down the caudal margin; this evidently fitted against the cranial margin of the spine of the first sacral vertebra. The entire dorsal series is gently convex dorsally but slightly more so along the cranial end of the column.

The dorsal vertebrae (and caudals) of *Iguanodon*

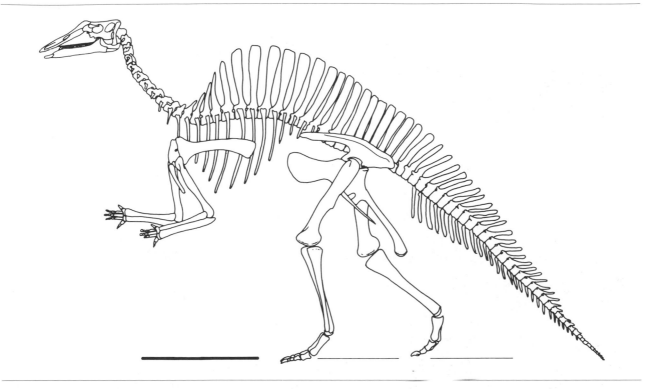

Fig. 25.23. *Ouranosaurus nigeriensis.* Skeletal restoration. Scale = 1 m. (After Taquet 1976.)

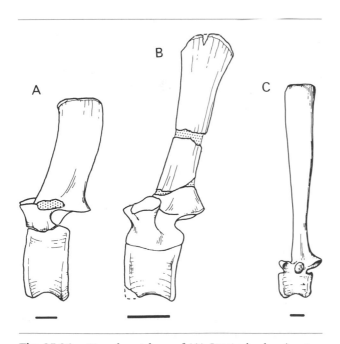

Fig. 25.24. Dorsal vertebrae of (A) *Iguanodon bernissartensis,* (B) *I. atherfieldensis* (after Norman 1986), (C) *Ouranosaurus nigeriensis.* Scale = 5 cm. (After Taquet 1976.)

fittoni are very distinctive in their very long, narrow, and steeply inclined neural spines (Norman 1987*c*). These do not in any way resemble those of *O. nigeriensis.* Another more robust contemporary species, *I. dawsoni,* has larger vertebrae that have shorter, broader neural spines and almost cuboid caudal centra. The composite reconstruction of a dorsal vertebra of *Muttaburrasaurus langdoni* resembles in its proportions that of *I. dawsoni* or a large camptosaur (Bartholomai and Molnar 1981).

The cranial dorsal ribs differ little from those of the cervical region. However, they increase in length markedly in the front of the dorsal series; the fifth and sixth ribs are the longest. Ribs 3 through 9 develop blunt distal ends that mark the articulation with the sternal ribs. Further caudally, the rib shafts become narrower and shorter as they approach the sacrum.

There are five to seven sacrals among iguanodonts (figs. 25.25-25.27). The principal load-bearing sacral ribs and vertebrae are found at the cranial end of the sacrum. The sacral ribs are massive, twisted elements forming the sacral yoke laterally, where they meet and fuse together, and have an extensive articulation with adjacent centra and neural arches. The neural arches of adjacent vertebrae are often fused together, as are the neural spines. The peg-and-socket articulation between sacral centra is known in both *Camptosaurus*

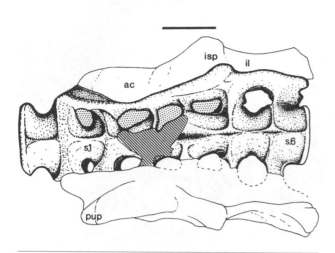

Fig. 25.25. *Iguanodon atherfieldensis.* Sacrum in ventral view. Scale = 10 cm. (After Norman 1986.)

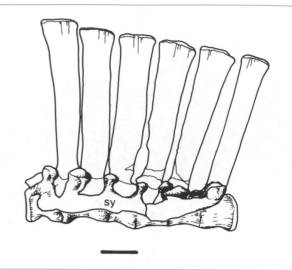

Fig. 25.26. *Ouranosaurus nigeriensis.* Sacrum in lateral view and partly restored. Scale = 3 cm. (After Taquet 1976.)

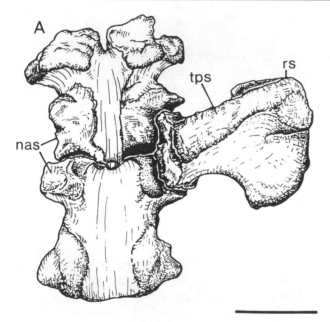

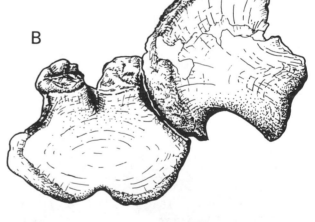

Fig. 25.27. *Iguanodon bernissartensis.* Two sacral vertebral centra and an isolated sacral rib placed in articulation. (A) dorsal and (B) cranial views. Scale = 5 cm. (After Norman 1980.)

dispar and species of *Iguanodon* and may be a general adaptation associated with strengthening the sacrocentral articulation, particularly in immature individuals.

Proximal caudals have rather tall, rectangular centra, with well-developed ribs fused to the vertebra along the line of the neurocentral contact. Neural spines are tall and obliquely inclined distally. They also tend to be narrower than the neural spines of the dorsal ver-

tebrae. The prezygapophyses are relatively small and point obliquely forward, while the postzygapophyses are small and located on the distal margin of the neural spine. Middle caudals have lost the caudal ribs and hence are hexagonal or octagonal in cross section. Centrum height is reduced. The centra of distal caudals are cylindrical and yet again reduced in height. They bear separate chevron facets on the ventral margin

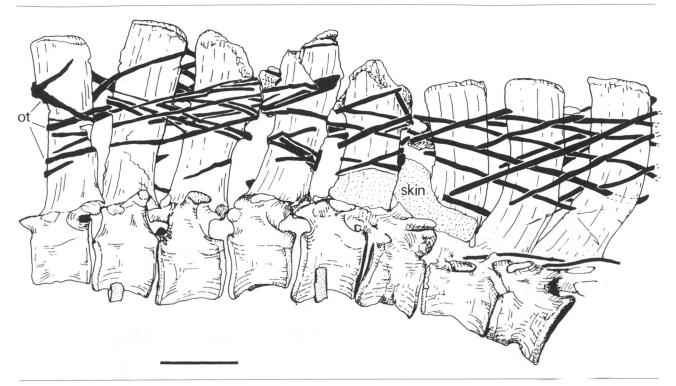

Fig. 25.28. *Iguanodon atherfieldensis.* Dorsal vertebrae in articulation, showing complexity of the ossified tendon lattice. Scale = 10 cm. (After Norman 1986.)

of the distal articular surface. The neural spines become reduced to apophyses for the support of the postzygapophyses.

Chevrons are first found between the second and third caudal vertebrae and are long, slender, and slightly curved laterally compressed rods. There is a well-developed hemal canal at the proximal end, above which is an expanded bifaceted articulation with both distal and proximal undersurfaces of adjoining vertebrae. Chevrons regularly shorten along the caudal series, mirroring the decline in height of the neural spines dorsally. Chevrons disappear toward the middle of the distal caudal series.

Ossified tendons are arrayed in a rhomboid lattice of at least three overlapping layers (fig. 25.28). Individual tendons seem to span about four to six vertebral spines. They are found along the dorsal, sacral, and proximal caudal series.

Appendicular Skeleton

The scapula of iguanodonts is typically moderately tall, longer than the humerus, and curved along its length (fig. 25.29). The distal blade and proximal end are expanded. A projection from the caudal border

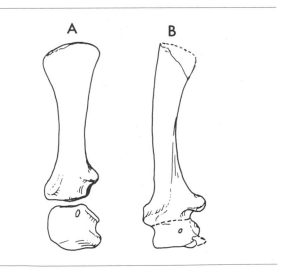

Fig. 25.29. Scapula and coracoid. (A) *Camptosaurus dispar* in lateral view (after Gilmore 1909), (B) *Ouranosaurus nigeriensis* in lateral view. (After Taquet 1976.)

of the proximal end forms the dorsal buttress of the glenoid; a cranial expansion forms the clavicular boss. The coracoid is relatively small and saucer-shaped. It is modified caudally, where there is a hook-like sternal process. A large coracoid foramen is always found close to the scapular articulation and may either be involved in the contact or remain separated from it.

The paired sternals (fig. 25.30) are in most instances hatchet shaped. The broad proximal portion of the sternal (the blade) meets its neighbor along the ventral midline, and a narrow rod-shaped process (the handle) projects obliquely caudolaterally. In *Camptosaurus dispar*, the sternal appears to differ markedly from that described above (Dodson and Madsen 1981). Here, the sternal is flat, with thickened and textured cranial and caudal ends, a smooth, thickened, and concave lateral margin, and a very thin, convex medial margin. Finally, in *I. bernissartensis*, there is a very characteristic additional intersternal ossification (Norman 1980), which is an irregular bone developed in the sternal cartilage between the coracoid and sternals (fig. 25.31).

The humerus has a shallow sigmoid curvature, a long, low deltopectoral crest, and a prominent articular head (fig. 25.32). The ulna and radius are relatively straight, and the ulna has a large olecranon process in most species (slight in *C. dispar*). The distal ends of both bones are expanded.

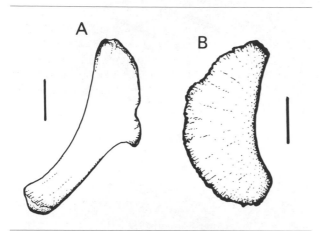

Fig. 25.30. Sternal bones. (A) *Iguanodon atherfieldensis* (after Norman 1986), (B) *Camptosaurus dispar*. Scale = 5 cm. (After Dodson and Madsen 1981.)

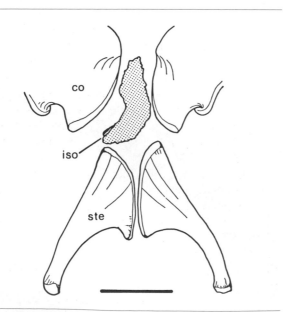

Fig. 25.31. *Iguanodon bernissartensis*. Intersternal ossification (stippled) in articulation with other pectoral elements. Scale = 20 cm. (After Norman 1980.)

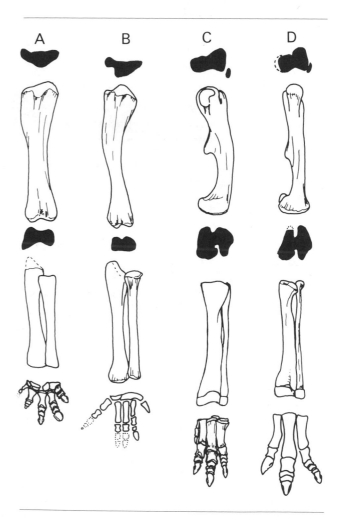

Fig. 25.32. Forelimb and hindlimb elements of *Camptosaurus* and *Ouranosaurus*. (A, C) forelimb and hindlimb elements of *C. dispar* (after Gilmore 1909), (B, D) forelimb and hindlimb of *O. nigeriensis* (after Taquet 1976).

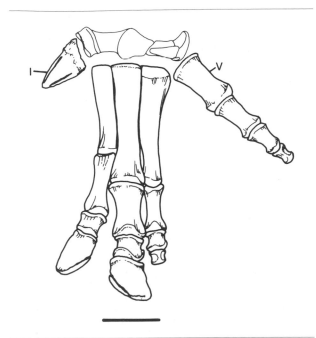

Fig. 25.33. *Iguanodon atherfieldensis.* Articulated manus and carpus in dorsal view. Scale = 5 cm. (After Norman 1980.)

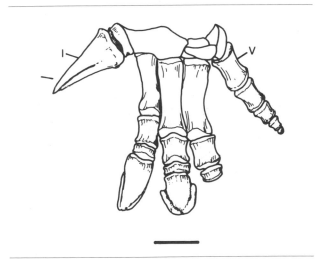

Fig. 25.34. *Iguanodon bernissartensis.* Articulated manus and carpus in dorsal view. Scale = 10 cm. (After Norman 1980.)

The carpals are unusually heavily ossified (figs. 25.33, 25.34). In addition, they are extensively fused, the first metacarpal with the radiale in all forms and the intermedium with the radiale in *C. dispar*, the intermedium with the ulnare in *O. nigeriensis*, and the intermedium with both the radiale and ulnare in species of *Iguanodon* (Dodson 1980). The distal carpals are vari-

ably ossified. Manual digits I and V diverge from the central axis of the hand. In *Iguanodon* and *Ouranosaurus,* digits II and IV are relatively long, slender, and may have been held together as a single unit. The digital formulae vary among iguanodonts (*Iguanodon atherfieldensis,* ?2–3–3–3–3; *I. bernissartensis,* 2–3–3–2–4; *Camptosaurus dispar,* 2–3–3–3–2; *Ouranosaurus nigeriensis,* ?1–3–?3–?3–?4). The manus is known in *Probactrosaurus* but has not been described to date. In *C. dispar,* the ungual of digit I is subconical (Sereno 1986). In species of *Iguanodon* and *Ouranosaurus,* the pollex is developed into a spike, but the situation is uncertain in *Muttaburrasaurus* and has not been reported in *Probactrosaurus.* Other features of the manus include the conversion of the remaining ungual claws into hooves and the ability of the digits to hyperextend to permit weight support on the forelimbs either while standing or walking.

Iguanodonts have a very conservative ilium with a long, deep preacetabular process with a slightly everted dorsal margin (fig. 25.35). The postacetabular portion of the ilium tends to be triangular, and its ventral edge forms a narrow brevis shelf. The public peduncle of the ilium is triangular in cross section, with a prominent lateral margin.

The pubis is rod shaped and runs parallel to the ischium. The prepubic process is deep and laterally flattened. There is no symphysis. The obturator notch is positioned proximally on the dorsal side of the pubis and is partly closed by contact with the ischium. The pubis itself is no more than about half the length of the ischium in nearly all iguanodonts. *Camptosaurus* differs in having a pubis that forms a symphysis and is equal in length to the ischium.

The ischium is long and decurved, with a moderately expanded distal end and a small leaf-shaped, proximally placed obturator process. The proximal end of the ischium is embayed where it forms the caudal margin of the acetabulum and has a very broad, robust iliac peduncle and a deep but narrow pubic peduncle.

In most taxa, the femur has a laterally flattened lesser trochanter lying next to the lateral and cranial margin of the greater trochanter (fig. 25.32). The proximal end of the femur, formed by the greater trochanter and the femoral head, is saddle-shaped. Distally, the femoral shaft is concave caudally beneath the crested fourth trochanter, and there is a noticeably sunken extensor groove.

The robust tibia bears a prominent cnemial crest. The narrower fibula contacts the calcaneum distally at a steplike junction, whereas the contact between the astragalus and tibia is more complexly stepped, re-

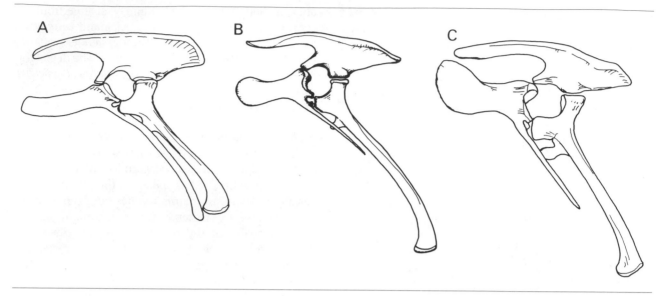

Fig. 25.35. Pelves of (A) *Camptosaurus dispar*, (B) *Iguanodon atherfieldensis*, and (C) *Ouranosaurus nigeriensis* in lateral view (after Gilmore 1909, Norman 1986, Taquet 1976, respectively).

inforced by a short ascending process. The distal tarsals are small, flat pads lodged against the proximal ends of the metatarsals. The pes has three functional toes (digits II-IV). The digital formula of the pes varies considerably among species (*Iguanodon atherfieldensis*, 0–3–4–5–0; *I. bernissartensis*, 0–3–4–5–0; *Camptosaurus dispar*, 2–3–4–5–0; *Ouranosaurus nigeriensis*, 0–3–4–5–0). In *Iguanodon*, the first digit is represented by a splintlike metatarsal alone. In *Camptosaurus*, the first metatarsal is considerably shorter than the others but sufficiently well developed to support the phalanges. The terminal phalanges in *Camptosaurus* form moderately sharp and slightly decurved claws. In all other iguanodonts, the unguals are dorsoventrally flattened and hooflike.

SYSTEMATICS

Very few detailed phylogenetic studies of iguanodonts have been published. Steel (1969) reviewed all iguanodonts then known and provided a dendrogram that indicated the relationships of what he considered to be the Iguanodontidae with all other ornithischians but did not provide critical evaluations within the group. In contrast, Galton (1972, 1974a) provided the first detailed phylogeny of ornithischians, including iguanodonts. These animals were defined as a family (the Iguanodontidae) based on a suite of primitive char-

acters, most of which came from limb proportions. In essence, iguanodontids consisted of those large ornithopods with so-called graviportal hindlimbs lacking the cranial specializations of hadrosaurids. As a consequence, iguanodontids became a polyphyletic aggregation of taxa (Dodson 1980; Norman 1984a, 1984b; Milner and Norman 1984), apparently iteratively derived from hypsilophodontids.

Taquet (1975) reviewed iguanodont relationships in light of the discovery of *Ouranosaurus nigeriensis* and concluded that the Iguanodontidae represent three discrete lineages, one that includes *Tenontosaurus tilletti* and relatives, a persistently primitive Cretaceous lineage derived from forms resembling camptosaurs of the Jurassic; a second lineage of robust iguanodonts (species of *Camptosaurus*, *Iguanodon bernissartensis*, and a so far unnamed species from Niger); and a third lineage of gracile iguanodonts (species of *Dryosaurus*, *I. atherfieldensis*, species of *Probactrosaurus*, and *Ouranosaurus nigeriensis*).

Last, Sereno (1984, 1986), Norman (1984a, 1984b), and Milner and Norman (1984) have provided the most recent assessments of the relationships of iguanodont ornithopods, among other ornithischians, using cladistic analyses. Norman accepted Dodson's (1980) referral of *Tenontosaurus* to the Hypsilophodontidae, separated the Dryosauridae from the Hypsilophodontidae *sensu stricto*, and placed the former as the sister taxon to the Iguanodontidae plus Hadrosauridae. The Iguanodontidae included *Camptosaurus, Iguanodon, Ouranosaurus, Probactrosaurus, Vectisaurus, Muttaburrasaurus,* and *Craspedodon.* By contrast, Sereno (1986) proposed that *Tenontosaurus* is the sister taxon of higher iguanodonts and hadrosaurids, the latter including, in serially closer order, *Dryosaurus, Camptosaurus, Probactrosaurus, Iguanodon,* and finally *Ouranosaurus* as successive outgroups to the Hadrosauridae. In the present work, the Iguanodontidae is greatly restricted, and several of Sereno's assignments are revised, notably, the position of *Probactrosaurus* and *Ouranosaurus* within the hierarchy. The revised character lists are derived from a detailed reassessment of Sereno (1986), dealt with in greater detail in Norman (in press).

As Sereno (1986) has documented, the iguanodont taxa considered in this chapter are a paraphyletic array of closely related forms. Within the group, species of *Camptosaurus* are the most primitive (forming with remaining iguanodonts and hadrosaurids a clade called the Ankylopollexia by Sereno [1986]). Second, species of *Iguanodon* and *Ouranosaurus* themselves form a clade termed the Iguanodontidae (this view differs from Sereno [1986], who views the relationship of these two genera as serially closer to hadrosaurids). Last, species of *Probactrosaurus* represent the most derived member of the iguanodont plexus and hence the sister group to the Hadrosauridae.

Looked at in detail (fig. 25.36; table 25.1), *Camptosaurus,* "higher" iguanodonts (i.e., Iguanodontidae and *Probactrosaurus*), and hadrosaurids differ from more primitive ornithopods (i.e., dryosaurids and hypsilophodontids, among others) in possessing dentary tooth crowns with a lingual surface having a distally offset and reduced primary ridge separated by a shallow, vertical trough from a low and broad secondary ridge (a small number of tertiary ridges are developed from the base of the marginal denticles); partial fusion of the carpals into two blocks, one associated with the distal radius, the other with the distal ulna; modification of the manual ungual into a spur; transverse flattening and broadening of the prepubic process of the pubis; and rhomboidal latticework of the ossified tendons. This large clade is termed the Ankylopollexia by Sereno (1986).

The Iguanodontidae is most closely related to *Probactrosaurus* and hadrosaurids. It shares with both groups a supraoccipital that is excluded from the dorsal margin of the foramen magnum; rod-shaped, caudolaterally directed processes on the sternals; distal expansion of the prepubic process of the pubis; reduction of the pubis; metacarpals II-IV bunched together, with

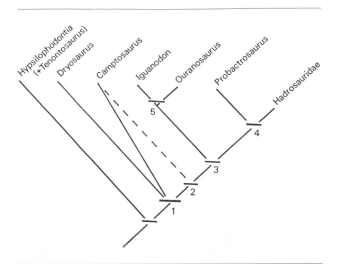

Fig. 25.36. Cladogram indicating relationships among iguanodonts based on the derived characters advocated in the text. Range limited to known and well-described taxa. Ambiguity over the position and status of *Camptosaurus* is indicated deliberately because the character states at node 2 are not regarded as particularly well founded.

metacarpals II and IV subequal in length and longer than metacarpal II; and unguals that are modified into hooves.

Finally, *Probactrosaurus* is most closely related to hadrosaurids and shares with the latter more than a single replacement tooth in each of the maxillary and dentary tooth families, ridges symmetrical on either side of the median primary ridge in maxillary teeth, a straight ischium, and loss of metatarsal I.

The Iguanodontidae (species of *Iguanodon* and *Ouranosaurus*) form a small clade uniquely characterized by an enlarged conical pollex ungual; a well-developed, obliquely projecting jugal process on the maxilla; a prominent lacrimal process of the maxilla that forms the rostral margin of the antorbital fenestra; a blocklike lacrimal that bears a distinctive pattern of recesses on its internal surface and forms the lateral wall of the tubular passage leading to the antorbital fenestra; and a narrow quadratojugal that is overlain by the jugal rostrally and embayed caudally where it forms the rostral margin of the paraquadrate foramen.

The affinities of *Muttaburrasaurus, Craspedodon,* and *Rhabdodon* are not well understood and these animals have not been included in the discussion of iguanodon relationships. They are considered Iguanodontia *incertae sedis* in table 25.1.

TABLE 25.1 Iguanodontidae and Related Ornithopoda

	Occurrence	Age	Material
Ornithopoda Marsh, 1881b			
Iguanodontia Dollo, 1882			
Camptosauridae Marsh, 1885			
Camptosaurus Marsh, 1885 (= *Brachyrophus* Cope, 1878g, *Cumnoria* Seeley, 1888, *Symphyrophus* Cope, 1878g)			
C. dispar (Marsh, 1879) (= *Camptonotus dispar* Marsh, 1879; including *Brachyrophus altarkansanus* Cope, 1878g, *Symphyrophus musculosus* Cope, 1878g, *Camptosaurus medius* Marsh, 1894, *C. nanus* Marsh, 1894, *C. browni* Gilmore, 1909)	Morrison Formation, Wyoming, Utah, Colorado, Oklahoma, United States	Kimmeridgian-Tithonian	25-30 disarticulated skull elements, some with associated postcrania, approximately 10 partial, articulated skeletons, juvenile to adult
C. amplus (Marsh, 1879c) (= *Camptonotus amplus* Marsh, 1879)	Morrison Formation, Wyoming, United States	Kimmeridgian-Tithonian	Pes
C. prestwichii (Hulke, 1880) (= *Iguanodon prestwichii* Hulke, 1880)	Kimmeridge Clay, Oxfordshire, England	Kimmeridgian	Fragmentary skull and skeleton
?*C. depressus* Gilmore, 1909	?Morrison Formation, Wyoming, Lakota Formation, South Dakota, United States	?Kimmeridgian-Tithonian; Barremian	Ilia and vertebrae
Iguanodontidae Cope, 1869b			
Iguanodon Mantell, 1825 (= *Iguanosaurus* Ritgen, 1828, *Heterosaurus* Cornuel, 1850)			
I. anglicus Holl, 1829 (including *Cetiosaurus brevis* Owen, 1842b, *Streptospondylus major* Owen, 1842b)	Wealden, West Sussex, England	Valanginian-Barremian	Teeth and postcranial fragments
I. atherfieldensis Hooley, 1924 (including *I. mantelli* Meyer, 1832, *Heterosaurus neocomiensis* Cornuel, 1850 partim, *Vectisaurus valdensis* Hulke, 1879a, *Sphenospondylus gracilis* Lydekker, 1888a)	Wealden, Isle of Wight, West Sussex, East Sussex, Surrey, Kent, Lower Greensand, Kent, England; Wealden, Hainaut, Belgium; Las Zabacheras Beds, Teruel, Spain; Wealden, Nordrhein-Westphalen, Federal Republic of Germany	Valanginian-Aptian	3 complete skulls and skeletons, partial skulls, isolated teeth, and postcranial remains, juvenile to adult
I. bernissartensis Boulenger, 1881 (including *I. seelyi* Hulke, 1882b)	Wealden, Isle of Wight, West Sussex, England; Wealden, Hainaut, Belgium; Las Zabacheras Beds, Capas Rojas, Teruel, Castellon, unnamed unit, Cuenca, Spain; Wealden, Nordrhein-Westphalen, Federal Republic of Germany	Valanginian-Albian	At least 26 associated skeletons and skulls, additional partial skeletal remains, teeth
I. dawsoni Lydekker, 1888a	Wealden, East Sussex, England	Valanginian-Barremian	2 partial skeletons
I. fittoni Lydekker, 1889e (including *I. hollingtoniensis* Lydekker, 1889e)	Wealden, East Sussex England	Valanginian-Barremian	3 partial skulls and jaws
I. hoggi Owen, 1874	Upper Purbeck Beds, Dorset, England	Berriasian	Mandible
I. lakotaensis Weishampel et Bjork, 1989	Lakota Formation, South Dakota, United States	Barremian	Skull, mandible, vertebrae
Unnamed iguanodontid (= **"Iguanodon"** *orientalis* Rozhdestvensky 1952a)	Khukhtekskaya Svita, Dundgov, Ovorhkangai, Shinekhudukskaya Svita, Dundgov, Mongolian People's Republic	Aptian-Albian	Complete skull, skull fragments, associated postcrania
Ouranosaurus Taquet, 1976			
O. nigeriensis Taquet, 1976	Elrhaz Formation, Agadez, Niger	late Aptian	Skull and postcrania, second skeleton
Unnamed iguanodontid (Chabli MS)	Elrhaz Formation Agadez, Niger	late Aptian	Partial skull, fragmentary postcranial skeleton
Unnamed taxon (= probactrosaurs)			
Probactrosaurus Rozhdestvensky, 1966			
P. gobiensis Rozhdestvensky, 1966	Dashuigou Formation, Nei Mongol Zizhiqu, People's Republic of China	Aptian-Albian	Skull and skeleton
P. alashanicus Rozhdestvensky, 1966	Dashuigou Formation, Nei Mongol Zizhiqu, People's Republic of China	Aptian-Albian	Fragmentary skull

Table 25.1, continued

Iguanodontia incertae sedis			
Craspedodon Dollo, 1883c			
C. lonzeensis Dollo, 1883c	Glauconie argileuse, Namur, Belgium	Santonian	3 teeth
Muttaburrasaurus Bartholomai et Molnar, 1981			
M. langdoni Bartholomai et Molnar, 1981	Griman Creek Formation, New South Wales, Mackunda Formation, Queensland, Australia	Albian	Skull and postcrania, fragmentary skeleton
Rhabdodon Matheron, 1869 (= *Mochlodon* Seeley, 1881)			
R. priscus Matheron, 1869 (including *Iguanodon suessi* Bunzel, 1871, *Oligosaurus adelus* Seeley, 1881, *Ornithomerus gracilis* Seeley, 1881, *Mochlodon robustus* Nopcsa, 1900, *Onychosaurus hungaricus* Nopcsa, 1902a, *Camptosaurus inkeyi* Nopcsa, 1900)	Sinpetru Beds, Hunedoara, Romania; Marnes Rouges inférieures, Aude, Grès de Saint-Chinian, Herault, Couches de Rognac, Bouches-du-Rhône, Grès à Reptiles, Var, Grès de Labarre, Ariège, France; unnamed unit, Soria, Lower "Garumnian" Beds, Lerida, unnamed unit, Lleida, Spain; Gosau Formation, Niederostereich, Austria	Campanian-Maastrichtian	Disarticulated skull and postcranial material, teeth, juvenile to adult
R. septimanicus Buffetaut et Le Loeuff, 1991 •	Grès à Reptiles, Herault, France	Campanian-Maastrichtian	Dentary

Nomina dubia	Material
Acantholpholis macrocercus Seeley, 1869 *partim*	Vertebrae, fragmentary skeletal elements
Acantholpholis stereocercus Seeley, 1869 *partim*	Vertebrae
Albisaurus scutifer Fritsch, 1905	Indeterminant scrap
Anoplosaurus curtonotus Seeley, 1878	Vertebrae, fragmentary skeletal elements
Anoplosaurus major Seeley, 1878	Vertebrae
Camptosaurus leedsi Lydekker, 1889b	Femur
(type of *Callovosaurus* Galton, 1980d)	
Eucercosaurus tanyspondylus Seeley, 1878	Vertebrae
Iguanodon exogirarus Fritsch, 1878	Indeterminant scrap
Iguanodon ottingeri Galton et Jensen, 1979b	Teeth
Loncosaurus argentinus Ameghino, 1898	Femur
Tichosteus lucasanus Cope, 1877f	Vertebra
Tichosteus aequifacies Cope, 1877f	Vertebrae

PALEOBIOLOGY AND PALEOECOLOGY

The earliest record of iguanodonts consists of indeterminate material (= *Callovosaurus leedsi*) from the Callovian of England, while the earliest diagnostic form is *Camptosaurus prestwichii* from the early Kimmeridgian, also of England. Other remains include isolated femora from the Kimmeridgian-Tithonian of Portugal (Galton 1980d) and the well-preserved *Camptosaurus dispar* from the Kimmeridgian of the United States. The latest-known taxon is *Rhabdodon priscus* from the Campanian-Maastrichtian of Europe.

From this record, it appears that these animals had a Euramerican distribution in the Jurassic. In the Early Cretaceous, however, iguanodonts had not only a wide European and North American distribution but also are known from Asia, Africa, and Australia. In the Late Cretaceous, iguanodonts are greatly restricted in distribution, being found only in Europe (*Craspedodon lonzeensis*).

Iguanodonts were primarily bipedal. The tail was long, muscular, and counterbalanced the front of the animal (Alexander 1985) so that the center of gravity was very close to the acetabulum. Position of the mass of the gut between the hindlimbs made possible through the reorganization of the pelvic bones also contributed significantly to the natural posture of the animals (Norman and Weishampel in press). The hindlimbs are also very large and powerfully built. The forelimbs tend to be about half the length of the hindlimb. Even though facultative bipedality seems to have been the norm among iguanodonts, several anatomical features indicate quadrupedal posture, in at least *Iguanodon bernissartensis*. In this animal, the forelimb girdle is robust, and an irregular intersternal ossification is developed between the sternals and coracoids. The forelimb is also relatively robust and elongate (70% of hindlimb length; Norman 1980). The carpus of *Iguanodon* (and *Ouranosaurus*), with its coossification and ligamentous connections, suggests weight bearing in conjunction with quadrupedality. The manual phalanges of digits II-IV terminate in flattened ungual hooves and were capable of hyperextension but not of opposition to form a grasping structure (Norman 1980).

In addition to the specializations of manual digits II to IV for weight support, the remaining digits are specialized for grasping and defense. Digit I appears to have been used as a stiletto-like, close-range weapon or for breaking into seeds and fruits, while digit V was the only one capable of some degree of opposition and hence prehension.

The forelimb of *Camptosaurus* differs from that of *Iguanodon* and *Ouranosaurus*, principally in that the digits are divergent. The metacarpals are short, and the digits terminate in narrow, pointed unguals. Once again, the manus is clearly capable of weight support.

Juveniles of *I. bernissartensis* have forelimb proportions that differ from those of adults (cf. 60 percent of hindlimb length in juveniles vs. 70 percent in adults). This relation contrasts with that in *Camptosaurus dispar*, in which forelimb proportions of juveniles are subequal to that in adults (Dodson 1980). Young individuals of *I. bernissartensis* may have been more thoroughgoing bipeds than adults; during ontogeny, disproportionate growth of the forelimb may have been associated with the subsequent tendency to adopt a quadrupedal gait while standing or during slow locomotion.

Footprint data relevant to interpretations of iguanodont locomotion are relatively rare. A trackway from southern England may show marks of the forefeet, but the hindfoot prints are very poor and do not allow a confident assignment of this trackway to *Iguanodon* (Norman 1980). Several trackways that indicate large ornithopods moving quadrupedally have been reported from the Lower Cretaceous of North America (Currie and Sarjeant 1979; Lockley 1986a). These appear to substantiate the anatomical argument for occasional quadrupedal locomotion.

Iguanodonts were probably low- to intermediate-grade subcursorial animals (Coombs 1978c), with an estimated maximal running speed of 24 km/hr. They were probably incapable of quadrupedal galloping, lacking as they do any ability to flex the vertebral column in the vertical plane, or of permitting scapular rotation to extend the stride of the forelimb.

Ossified tendons across the back formed tensile support for the vertebral column at the hips and presumably allowed the reduction of epaxial musculature within the trunk and proximal portion of the tail. The flexible attachment of these tendons to the neural spines by means of elastic ligaments provided a dynamic component to the system which accommodated vertical flexure of the vertebral column during normal terrestrial locomotion (Norman 1986).

Camptosaurus dispar is generally rare, although it is somewhat common at the Cleveland-Lloyd quarry in Utah (Madsen pers. comm.) and Como Bluff, Wyoming, United States (Gilmore 1909; Dodson, Behrensmeyer, Bakker, and McIntosh 1980). The same rarity is true of *Ouranosaurus nigeriensis* and the majority of other iguanodonts. The discovery of abundant remains of species of *Iguanodon* in southern England, Bernissart (Belgium), Nehden (Federal Republic of Germany), eastern Spain, and Portugal (Lapparent and Zbyszewski 1957; Santafe-Llopis et al. 1979, 1982; Norman 1980, 1986, 1987a, 1987b; Holder and Norman 1986; Norman et al. 1987), however, provides a rare opportunity to analyze the mode of preservations of the species of this genus and to speculate on its population density during the Early Cretaceous.

At least thirty-eight individuals were collected at Bernissart, but these appear not to have been the victims of a single catastrophic event as has been frequently suggested. Instead, the caracasses were deposited in groups on at least three separate occasions (Norman 1986, 1987a). That these accumulated in the marshy environment in considerable numbers over a geologically short time does not necessarily warrant the conclusion that these animals lived in herds. In contrast, the large number of *Iguanodon* skeletons from the very restricted Nehden locality may indicate gregarious behavior, since these carcasses are apparently associated with flash-flood conditions (Norman 1987b).

The role of species of *Iguanodon* in Early Cretaceous terrestrial communities was undoubtedly that of the dominant medium- to large-sized herbivore. *Iguanodon atherfieldensis* and *I. bernissartensis* are found consistently with smaller and apparently less abundant ornithopod herbivores (*Hypsilophodon foxii*, *Valdosaurus canaliculatus*), rare ankylosaurs (*Polacanthus foxii*), diplodocids, and exceedingly rare pachycephalosaurs (*Yaverlandia bitholus*).

 The success of iguanodonts as herbivores may well relate to the configuration of their skulls, more specifically, the pleurokinetic hinge system that allowed these animals to rotate their upper jaws laterally when the lower jaws were fully adducted (Norman 1984b). In existence since the evolution of hypsilophodontid ornithopods (Weishampel, 1984a; Norman and Weishampel 1985; see Sues and Norman, this vol.), pleurokinesis represents a solution to the problem of combining bilateral occlusion with a transverse power stroke. In addition, such a system reduces torsional stresses during isognathous occlusion in skulls (such as those of iguanodonts) that lack a rigid secondary palate.

The ability to produce a transverse power stroke and hence incorporate significant oral processing into the feeding repertoire of iguanodonts surely was important in view of the evolution of angiosperms in the Early Cretaceous. With their herbaceous, ground-cover life habit, these early angiosperms would have been easy targets for relatively low browsing iguanodonts (Wing and Tiffney 1987). In addition, foliage from the still-dominant, slow-growing gymnosperms were included as iguanodont foodstuffs. Although iguanodonts appear to track to a certain degree angiosperm radiation, profound evolutionary responses to the reorganization of the terrestrial plant realm did not take place until the Late Cretaceous, which involved a number of herbivorous groups, among them the iguanodont descendants, the hadrosaurids.

26

Hadrosauridae

DAVID B. WEISHAMPEL
JOHN R. HORNER

INTRODUCTION

More is known about hadrosaurids than virtually any other group of dinosaurs. Remains are often abundant and range from fully articulated skeletons (sometimes complete with sclerotic rings, stapes, hyoids, and ossified tendons) to disarticulated and isolated material. In addition, remains of eggs. hatchlings, and juveniles (Dodson 1975; once considered members of the Cheneosaurinae, Lull and Wright 1942), as well as footprints and trackways and skin impressions, have provided investigators glimpses of dinosaur biology which are not generally afforded for other groups of dinosaurs. Hadrosaurids, the so-called duck-billed dinosaurs, were large (7-10 m long, average adult body weight 3,000 kg) and had broad edentulous beaks, long and low skulls, and complex dentitions organized into dental batteries. Of this group, lambeosaurines are the most derived, principally through the radical modification of the nasal cavity, which is moved from the front of the face to a supraorbital position.

Hadrosaurids were the most diverse and abundant large vertebrates of Laurasia during the closing stages of the Late Cretaceous. They were also the last group of ornithopods to evolve in the Mesozoic (possibly as early as the Cenomanian [Lydekker 1888a; Rozhdestvensky 1977] but certainly by the Santonian [Kaye and Russell 1973; Carpenter 1982c]) and show a number of remarkable anatomical characteristics in their dentition and facial skeleton which have attracted considerable research attention relating particularly to feeding mechanisms and strategies and to social behavior.

Studies of hadrosaurids began with Leidy on rather poor material of *Trachodon mirabilis* and *Thespesius occidentalis* from Montana and South Dakota (Leidy 1856) and shortly thereafter on considerably better material pertaining to *Hadrosaurus foulkii* from New Jersey (Leidy 1858). Continuing through the nineteenth century, work by Cope and Marsh in both New Jersey and the western interior of the United States brought the study of hadrosaurids to what has been called the great North American dinosaur rush of the 1910s and 1920s. During this time, scientists such as Lambe, Brown, Parks, the Sternberg family, and Gilmore, increased our knowledge of North American hadrosaurids by virtually an order of magnitude, including the recognition of flat-headed, solid-crested, and hollow-crested forms. More recent studies of North American hadrosaurids include Langston (1960a), Ostrom (1961a, 1964), Hopson (1975b), Dodson (1975), Horner (1983, 1988, in press), Horner and Makela (1979), and Weishampel (1981a, 1981b, 1983, 1984a). Hadrosaurids were less abundant and diverse in eastern Asia but have been studied by Riabinin (1925, 1930b, 1939), Wiman (1929), Gilmore (1933a), Young

(1958a), Rozhdestvensky (1968a), and Maryańska and Osmólska (1981a, 1981b, 1984b). In Europe, two important hadrosaurid finds include indeterminate hadrosaurids (*Trachodon cantabrigiensis,* Lydekker 1888a; *Iguanodon hilli,* Newton 1892) and *Telmatosaurus transsylvanicus* (Nopcsa 1900, 1903). Last, recent work by Casamiquela (1964a), Brett-Surman (1979), and Bonaparte et al. (1984) details the presence of primitive hadrosaurids from South America.

In general, the most distinctive morphologic characters of hadrosaurids are concentrated in the skull. Taxonomic information has been harder to extract from the postcranium. There are, however, many details that are useful as criteria for taxonomic assignment as well as functional and paleoecologic differentiation among hadrosaurid taxa.

DIAGNOSIS

The Hadrosauridae may be diagnosed on the following features: dental batteries formed of closely packed tooth families, three to five replacement teeth per tooth position and from one to three functional teeth per position, supraorbitals fused with the dorsal rim of the orbit, elevation of the dorsal process of the maxilla and consequent displacement of the antorbital fenestra to the rostrodorsal surface of the maxilla, loss of the paraquadrate and surangular foramina, mesiodistal narrowing of maxillary teeth, reduction of the carpus, loss of manual digit I, large antitrochanter on the ilium, reduced pubic shaft, eight to ten sacral vertebrae.

ANATOMY

Skull and Mandible

Complete skulls are known for nearly all of the known genera and species of hadrosaurids (figs. 26.1-26.7), and virtually every established hadrosaurid taxon is based on at least some cranial material (see Lambe 1914a, 1920; Brown 1912, 1916a; Parks 1922; Gilmore 1924c; Sternberg 1926b; Lull and Wright 1942; Young 1958a; Ostrom 1961a; Heaton 1972; Maryańska and Osmólska 1981a; and Horner 1983 for detailed descriptions of various individual hadrosaurid taxa). In addition, information on cranial ontogeny from embryo to adult life stages is beginning to emerge from a number of recent studies (e.g., Dodson 1975; Horner and Makela 1979).

The traits best characterizing the hadrosaurid skull are the dental batteries, the alterations to the nasal cavity or arch, and the expanded rostrum or snout. The development of a complex dental battery results from the close packing of relatively small teeth with three to five replacement teeth per tooth position. Second, among many hadrosaurids, the facial skeleton is drastically reorganized to accommodate either a solid or hollow supracranial crest. The former occurs among brachylophosaurs and saurolophs and is formed by the nasals (and possibly the frontals), while the latter occurs in all lambeosaurines where it houses a hypertrophied nasal cavity and is made chiefly of the premaxillae and nasals with adjacent elements serving as the base and lateral supports. *Tsintaosaurus spinorhinus* presents a peculiar, dorsally projecting crest (fig. 26.1d) that appears to be made up entirely of the nasals (Young 1958a). The narrow space within the crest, if it is real, represents a cul de sac from the more conventionally placed nasal cavity within the facial skeleton. Last, the rostrum is laterally expanded and somewhat ducklike.

The skull of adult hadrosaurids is relatively long and reminiscent of that in horses or cows; most of the length of the skull is concentrated in the muzzle and cheek regions. In hatchlings and subadults, however, the skull is more box-like due to its relatively short muzzle. In dorsal view, the hadrosaurid skull is relatively narrow, widest along the margin of the beak, the jugal arch, the upper temporal arch, and the jaw joint (fig. 26.5c). All hadrosaurids have a relatively short, narrow braincase.

The premaxillae are edentulous and gently flare to form a blunt spatulate beak, broadly expanded in hadrosaurines but not in lambeosaurines and basal hadrosaurids as far as can be ascertained. In most hadrosaurids, the oral margin of the premaxilla is rugose and slightly serrated where it was covered by a horny rhamphotheca in life (Cope 1883; Versluys 1923; Sternberg 1935; Morris 1970). It is coarsely denticulate only in *Telmatosaurus.* Single or paired openings through the body of the premaxilla from the circumnarial fossa to the oral cavity appear to be found in all hadrosaurids except lambeosaurines (figs. 26.2a, 26.5c). The body of the premaxilla sends off two processes (the dorsal process [Pmx$_2$] and lateral processes [Pmx$_1$]) that together form the rear margin of both the external naris and surrounding circumnarial fossa, the latter often continuous onto the nasal bone. This fossa is present only in hadrosaurines (see Systematics section); it is shallow in *Prosaurolophus, Gryposaurus,* and *Maiasaura,* among others, but is deeply excavated in *Edmontosaurus regalis* and *Anatotitan copei.* In lambeosaurines, the circumnarial fossa (and hence the vestibule) is enclosed within the dorsal and lateral premaxillary processes

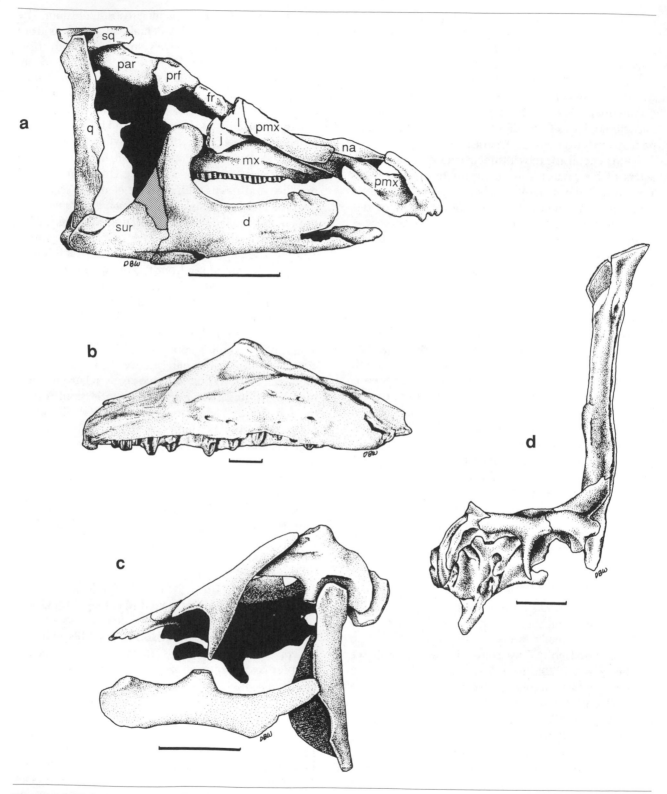

Fig. 26.1. Right lateral view of the skull of *Telmatosaurus transsylvanicus* (after Nopcsa 1900). b. Lateral view of the right maxilla of *Gilmoreosaurus mongoliensis*. c. Left lateral view of the skull of *Tanius sinensis* (after Wiman 1929). d. Right lateral view of the braincase and crest of *Tsintaosaurus spinorhinus* (after Young 1958a). Scale = 100 mm.

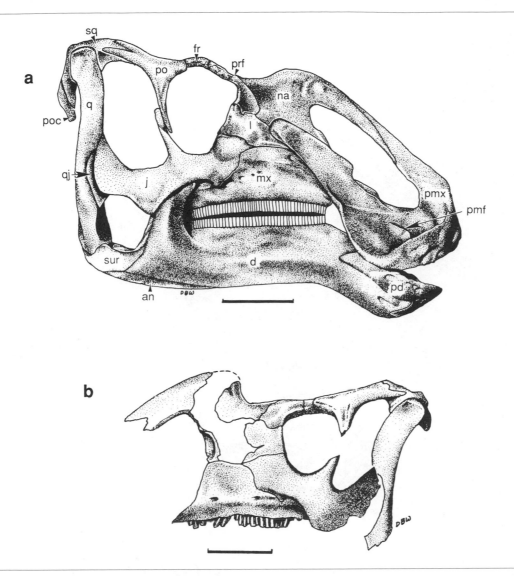

Fig. 26.2. a. Right lateral view of the skull of *Gryposaurus notabilis*. b. Left lateral view of the skull of *Aralosaurus* *tuberiferus* (after Rozhdestvensky 1968a). Scale = 100 mm.

(Hopson 1975a), such that the true external naris is internalized and shifted to a supraorbital position. In all nonlambeosaurine hadrosaurids, the dorsal and lateral premaxillary processes are not involved in shrouding the vestibule but rather extend caudally to accommodate a direct narial passageway through to the choanae at the back of the secondary palate. In these animals, the dorsal premaxillary process contacts the rostral process of the nasal directly above the external nares. However, in lambeosaurines in which the dorsal and lateral premaxillary processes are extremely folded and enlarged, the dorsal process makes up most of the hollow supraorbital crest (figs. 26.6b, 26.7). Primitively, the lateral premaxillary process is directed toward the

orbit in all hadrosaurids and terminates on the upper part of the lacrimal (shrouding the deeper nasal-maxilla contact) but above the orbit along the side of the hollow crest in lambeosaurines (the exceptions are species of *Parasaurolophus* whose lateral process terminates about the level of the highest point of the maxilla). In all lambeosaurines, the lateral premaxillary process houses the first part of the vestibule of the nasal cavity, while the dorsal process houses the remainder as well as part of the lateral diverticula. Only in lambeosaurines is there contact between the premaxilla and prefrontal where they make up the lateral wall and base of the crest.

Because of their intimate connection with the

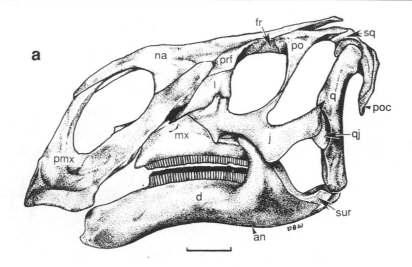

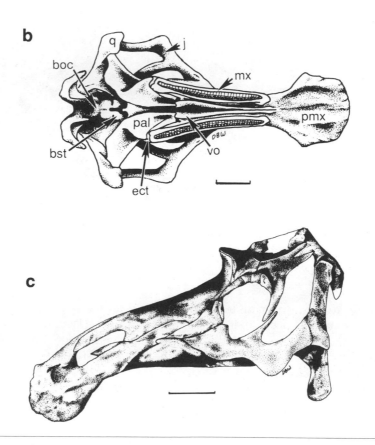

Fig. 26.3. a. Left lateral view of the skull of *Brachylophosaurus canadensis*. b. Palatal view of the skull of *Brachylophosaurus canadensis* (after Heaton 1972). c. Left lateral view of the skull of *Maiasaura peeblesorum* (after Horner 1983). Scale = 100 mm.

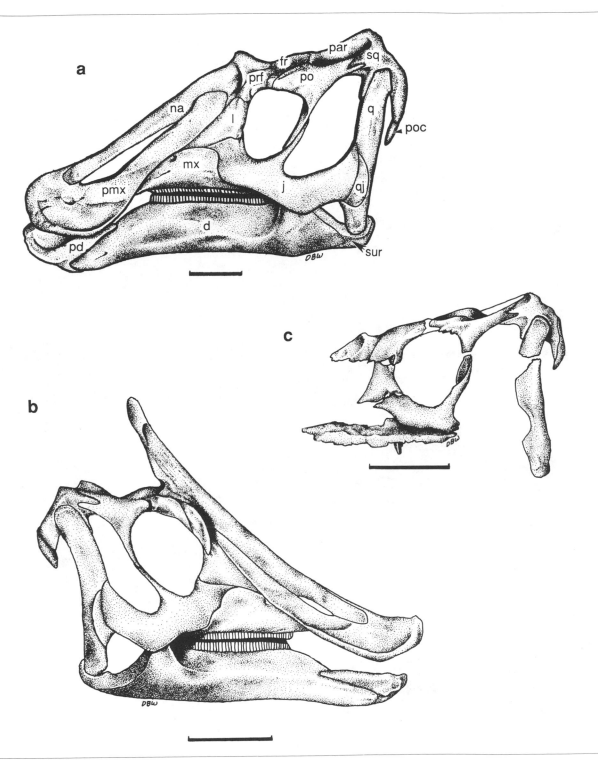

Fig. 26.4. a. Left lateral view of the skull of *Prosaurolophus maximus*. b. Right lateral view of the skull of *Saurolophus angustirostris* (after Maryańska and Osmólska 1981*a*). c. Left lateral view of the skull of *Lophorhothon atopus* (after Langston 1960*a*). Scale = 100 mm.

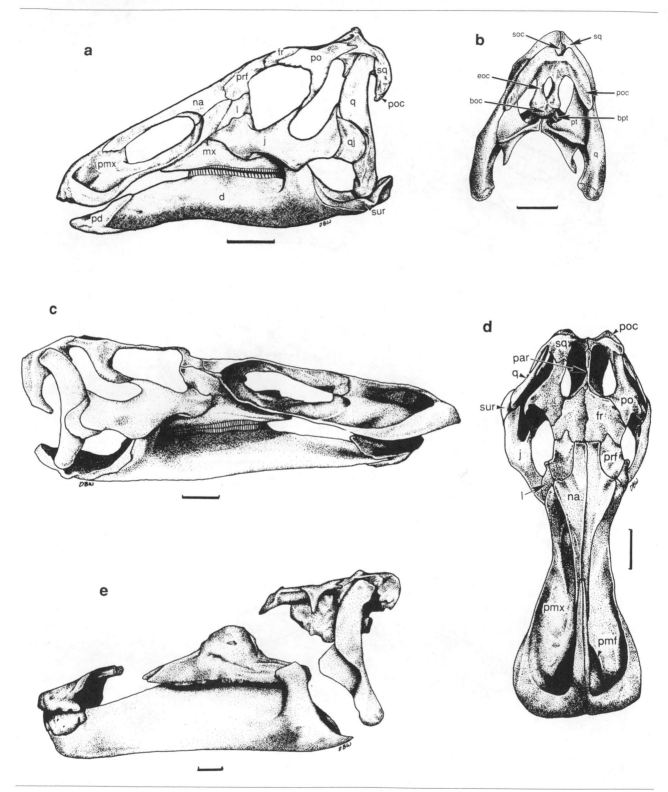

Fig. 26.5. a. Left lateral view of the skull of *Edmontosaurus saskatchewanensis*. b. Occipital view of the skull of *Edmontosaurus saskatchewanensis*. c. Right lateral view of the skull of *Anatotitan copei*. d. Dorsal view of the skull of *Anatotitan copei*. e. Left lateral view of the skull of *Shantungosaurus giganteus* (after Hu 1973). Scale = 100 mm.

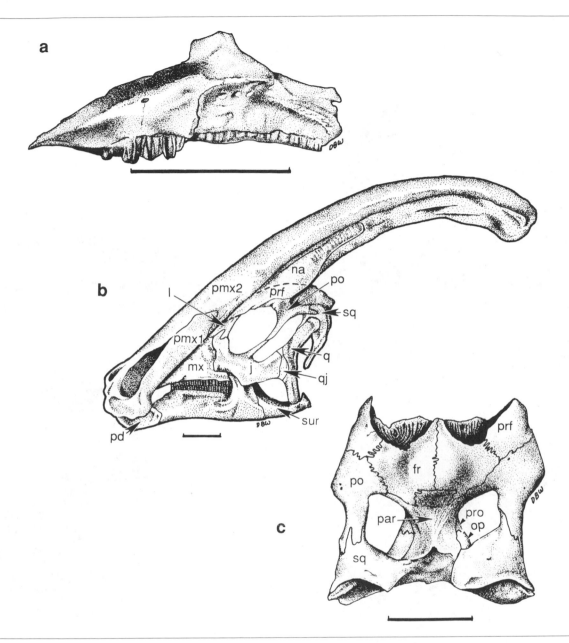

Fig. 26.6. a. Lateral view of the left maxilla of *Bactrosaurus johnsoni*. b. Left lateral view of the skull of *Parasaurolophus walkeri* (after Weishampel 1981*b*). c. Dorsal view of the skull roof of *Jaxartosaurus aralensis* (after Rozhdestvensky 1968*a*). Scale = 100 mm.

nasal cavity, the nasals of hadrosaurids are highly variable in their size, shape, and relationship. In nonlambeosaurine hadrosaurids, the nasal forms the dorsal margin of the external nares and continues caudally to contact the frontal dorsal to the orbit. Among some of these animals, the dorsal border of the nasal is arched above the external naris and in front of the orbit (fig. 26.2; "*Kritosaurus*" *incurvimanus, Gryposaurus notabilis, Aralosaurus tuberiferus*). In others (figs. 26.3, 26.4; *Saurolophus osborni, Maiasaura peeblesorum,* and *Brachylophosaurus canadensis, Prosaurolophus maximus, Lophorhothon atopus*), the caudal nasal region is expanded to form a solid crest dorsal to the orbit and often across the skull roof (the crest is supported from behind by upraised frontal buttresses in the first two taxa). In lambeosaurines, the nasal contributes in varying degrees to the lateral wall of the crest (figs. 26.6*b*, 26.7). In particular, the nasal contributes approximately the

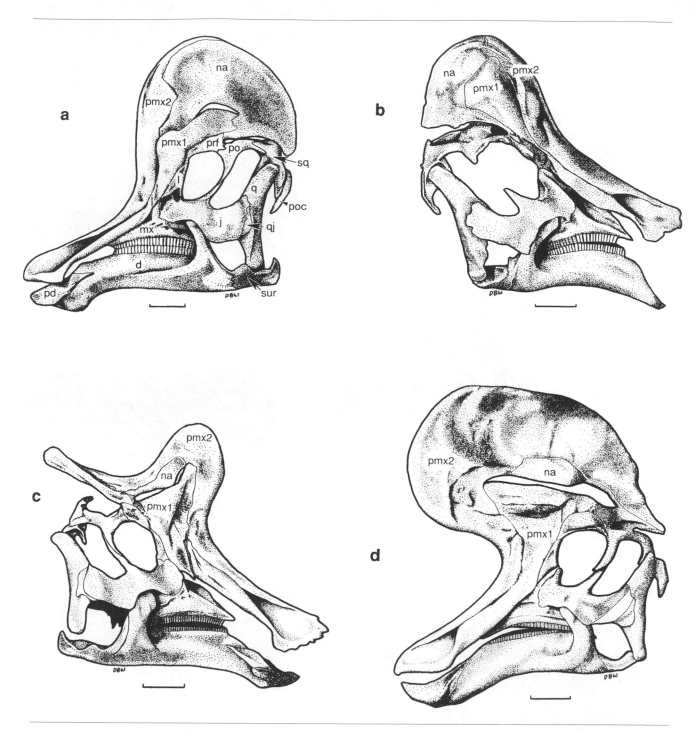

Fig. 26.7. a. Left lateral view of the skull of *Corythosaurus casuarius* (after Weishampel 1981*b*). b. Right lateral view of the skull of *Hypacrosaurus altispinus* (after Weishampel 1981*b*). c. Right lateral view of the skull of *Lambeosaurus lambei* (after Weishampel 1981*b*). d. Left lateral view of the skull of *Lambeosaurus magnicristatus* (after Weishampel 1981*b*). Scale = 100 mm.

same extent as the premaxilla in *Corythosaurus casuarius* and *Hypacrosaurus altispinus*, forming the caudal half of the crescentic crest. However, in species of *Parasaurolophus* and *Lambeosaurus*, they contribute only a very small, irregular portion of the crest wall (Weishampel 1981*b*). In all cases, the nasal makes up some part of the lateral wall of the lateral diverticula in adult lambeosaurines (juvenile and subadult lambeosaurines lack lateral diverticula; Weishampel 1981*b*). The nasals appear to form the entire crest in *Tsintaosaurus*, extending dorsally or slightly forward from the rostral margin of the skull roof (fig. 26.1d). A slight hollow space between the paired nasals is described by Young 1958*a*, but this passage is not an integral part of the nasal cavity as it is in lambeosaurines (Weishampel 1981*b*). Because of the hypertrophy of the nasal cavity, there is no nasal-lacrimal contact in lambeosaurines; in other hadrosaurids, the contact is found directly in front of the orbit as a butt or scarf joint.

The lacrimal is a short subrectangular bone on the rostral margin of the orbit. It appears wedge-shaped due to the overlapping lateral premaxillary process and the dorsal margin of the jugal. The lacrimal transmits the lacrimal canal from the rostroventral margin of the orbit to the lower portion of its medial surface. In addition, it forms a strongly concave buttress with the jugal at the orbital margin.

The prefrontal is found along the rostrodorsal portion of the orbital rim in all hadrosaurids. It is crescentic in lateral view and in lambeosaurines forms the lateral supports for the hollow crest. Coombs (1972) argued that the prefrontal originates either from the fusion of the supraorbitals with a greatly limited prefrontal or else as coalesced supraorbitals with no contribution by the prefrontal (which is itself lost). Work by Maryańska and Osmólska (1981*a*) and Horner (1983) indicates that the prefrontal is a coalescence of both supraorbitals and prefrontal. Additional supraorbitals fuse with the orbital margin of the frontal and sometimes with the postorbital.

The maxilla is one of the largest bones of the skull. Its triangular outline derives from a broad base formed by the tooth row and an apex formed by its dorsal process. The tooth row is inset from the lateral margin of the maxilla. The row is either straight or slightly concave in the buccal direction. The maxilla forms a transversely convex and broad articular surface for reception of the undersurface of the body and lateral process of the premaxilla. In species of *Edmontosaurus*, *Brachylophosaurus*, *Gryposaurus*, and *Anatotitan copei*, the rostral process of the maxilla makes a spheroidal contact with the premaxillary body (this condition may also occur in species of *Saurolophus*). Medial to this articulation is the median rostral process (Weishampel 1984*a;* anterior maxillary process of Heaton 1972, Horner 1983) that passes between the paired premaxillae in all nonlambeosaurine hadrosaurids; it is sufficiently long that it can be seen through the external naris in *Anatotitan, Edmontosaurus,* and *Brachylophosaurus*. In lambeosaurines, the premaxillary contact surface rises more abruptly than in the aforementioned species, most likely due to the reconfiguration of the vestibule of the nasal cavity. There is a medially oriented, convex shelf developed midway along the contact surface to support the premaxilla-maxilla articulation in lambeosaurines. Finally, in species of *Parasaurolophus*, the lateral margin of the premaxillary contact surface is slightly excavated, much like that seen in *Dryosaurus* and *Hypsilophodon*. The dorsal process of the maxilla is elevated, most probably in conjunction with the development of the maxillary dental battery. This elevation in turn accounts for the shift of the antorbital fenestra to the articular surface of the maxilla with the premaxilla (Weishampel 1984*a*). The lateral surface of the maxilla is dominated by the articular facet for the jugal. Here the crescentic contact surface is highly rugose and striated to support a strong union with the overlying jugal. Beneath the jugal articulation, there are as many as four foramina presumed to conduct nerves and vasculature to the cheek region (viz. Lull and Wright 1942 and Galton 1973*c* on hadrosaurid cheeks). Primitively, the lateral surface of the dorsal process of the maxilla forms the articulation for the lacrimal, while in lambeosaurines, the lacrimal fits into a small longitudinal groove below the dorsal edge of the process, giving the impression that the lacrimal is lapped laterally by the maxilla in articulated specimens. The medial maxillary surface is more or less flat, unlike the lateral surface, and features a series of special foramina (Edmund 1957) interconnected by a groove along the medial alveolar wall. The caudal extremity of the maxilla consists of a prominent longitudinal ectopterygoid shelf and a more dorsal ridge receiving the base of the palatine. Among lambeosaurines, the caudal half of the maxilla is truncated, giving the bone a disproportionately long forward section. The maxilla ends abruptly where it articulates in butt fashion with the pterygoid and ectopterygoid. The dorsal caudal process of the maxilla is found directly above the caudal terminus of the element; this process articulates with the ectopterygoid ramus of the pterygoid.

In virtually all hadrosaurids, the large jugal is thin, flat, and elongate. It forms the ventral margin of the orbit and infratemporal fenestra as well as the ventral half of the postorbital bar. It is thicker and in general more robust in *Edmontosaurus* and *Anatotitan* than in other hadrosaurid species. Rostrally, the jugal is expanded dorsoventrally, forming an extensive articulation with the maxilla in virtually all hadrosaurids. Exceptions include *Gilmoreosaurus, Bactrosaurus, and Tanius,* in which the maxillary contact surface of the jugal shows little or no dorsoventral expansion but is rather longitudinally elongate. In lambeosaurines, the rostral margin is broadly convex, while in remaining hadrosaurids, it is distinctly angular, extending along the maxilla-lacrimal contact (figs. 26.6, 26.7). This angular process is relatively large, long, and symmetrically triangular in species of *Maiasaura, Brachylophosaurus, Telmatosaurus* but slight and asymmetrical in *Prosaurolophus, Saurolophus, Edmontosaurus,* and *Anatotitan copei* (figs. 26.2, 26.3, 26.4, 26.5). The S-shaped ventral margin of the jugal is particularly accentuated in species of *Brachylophosaurus, Maiasaura peeblesorum, Gryposaurus notabilis, "Kritosaurus" incurvimanus,"* and species of *Parasaurolophus* (possibly also *Aralosaurus tuberiferus*), where it forms a flange immediately ventral to the infratemporal fenestra. The caudal half of the jugal overlies the adductor chamber and coronoid process and supported the cheek musculature in life (Galton 1973c). In *Brachylophosaurus* and *Maiasaura,* the caudal half of the element is incised along its ventral margin. The jugal meets the postorbital along a flat, slightly laterally facing surface (the joint surface is slightly excavated in both *Lophorhothon* and *Gilmoreosaurus*). This postorbital process is especially long in lambeosaurines. Medially, the caudal aspect of the jugal bears an angular facet for the reception of the rostral process of the quadratojugal. Immediately dorsal to the jugal-quadratojugal contact, the jugal laterally overlaps the quadrate to a very small extent.

The quadratojugal is a relatively small spacer between the jugal and quadrate. Like the jugal, it is a thin element, lateral to the adductor chamber. Roughly triangular in shape, the quadratojugal has a thickened, broadly crescentic caudal margin. The rostral process in all hadrosaurids is pointed, decidedly so in *Maiasaura.*

The postorbital defines the upper half of the caudal margin of the orbit and consists of a central body from which extend prefrontal, jugal, and squamosal processes (the second and last contribute to the postorbital bar and upper temporal arch, respectively). In addition, the postorbital contacts the frontal rostrally, the parietal medially, and laterosphenoid ventrally. In species of *Edmontosaurus, Anatotitan copei* and possibly *Shantungosaurus giganteus,* the body of the postorbital is enlarged and deeply excavated on its orbital surface (Ostrom 1961a). This modified postorbital flares laterally from the side of the orbit and greatly restricts the upper part of the infratemporal fenestra.

The squamosal makes up the caudal half of the upper temporal arch, forming an extensive scarf contact with the postorbital. The squamosal also articulates in front with the postorbital, medially with the parietal and supraoccipital, caudally with the exoccipital and opisthotic to form the paroccipital process, and ventrally with the dorsal head of the quadrate. The last joint is formed by a strong ellipsoidal cotylus that is synovial in nature. The squamosal-quadrate joint is bounded in front and back by stout pre- and postquadratic processes. The squamosals contact one another along the midline in an elevated occipital crest in all lambeosaurines. The internal surface of the squamosal acts as the attachment site for Mm. adductor mandibulae externus superficialis et medialis.

The quadrate is long and straight in lateral view. Extending rostromedially from the shaft, the pterygoid ramus of the quadrate broadly contacts the reciprocal quadrate ramus of the pterygoid. The pterygoid ramus is buttressed along the caudal margin of the quadrate shaft and is relatively broader among hadrosaurines than in lambeosaurines. On the rostral surface of the quadrate between the pterygoid flange and lateral wall, M. adductor mandibulae caudalis takes its origin. The dorsal quadrate head is hemispheric in lateral view and subtriangular in dorsal view; the apex of the triangle fits into the squamosal cotylus, while the base faces laterally. Extensive ridges and grooves on the rostral surface of the quadrate head opposite similar rugosities on the prequadrate process of the squamosal suggest the presence of very strong capsular ligaments around the synovial joint. Similar conditions are found between the caudal margin of the quadrate head and the postquadratic process. This caudal fibrous connection is marked by a protuberance on the rear surface of the quadrate head in virtually all hadrosaurids, but in lambeosaurines, it is either small or missing entirely. The shaft of the quadrate is nearly straight along its length, curving slightly caudally at its dorsal extreme, particularly in lambeosaurines. In its lower section, the quadrate contacts the quadratojugal along an emarginated, beveled articulation (in *Gilmoreosaurus,* there is an additional ventral buttress for the lower aspect of the quadratojugal). There is no paraquadrate foramen between the quadrate and quadratojugal. Immediately dorsal to the quadratojugal contact, the quadrate is overlapped to a small extent by the caudal extremity of the jugal. The mandibular condyle of the quadrate is formed chiefly by a large hemispheric lateral condyle and a much smaller, more dorsally placed medial condyle. In several forms (e.g., *Tsintaosaurus, Maiasaura*), the medial condyle is lost entirely, giving the articulation a totally rounded appearance. Finally, *Gilmoreosaurus, Bactrosaurus,* and *Telmatosaurus* have a more primitive transversely expanded ventral quadrate head.

The frontal is the largest element of the skull roof, extending broadly between the orbits and the nasal and parietal. It is rostrocaudally shorter in hadrosaurids with supracranial crests (i.e., lambeosaurines as well as *Prosaurolophus, Saurolophus, Brachylophosaurus,* and *Maiasaura*) than in flat-headed forms. In the first group (see fig. 26.6c), the rostrodorsal surface forms a broad, excavated base for the crest (contacting the nasals or lateral premaxillary process, the latter only in adult *Lambeosaurus lambei* and *L. magnicristatus*). Laterally, the frontal is exposed to a variable degree in the dorsal orbital rim in all hadrosaurids, except for lambeosaurines where it is excluded from the orbit entirely. Rostrally, the dorsal surface of the frontal is broadly excavated for reception of the more medial nasal and more lateral prefrontal, at least in hadrosaurines. It forms a complex interdigitate joint with the postorbital from the dorsal rim of the orbit to the supratemporal fenestra (often obscured in adult lambeosaurines due to the development of the crest). In addition, the frontal articulates with the parietal along a nearly transverse interdigitate joint, often accompanied by a median projection of the parietal into the interfrontal joint (the interparietal process of Lull and Wright 1942; see below). Lateral to the parietal contact, the frontal articulates along a coarsely interdigitate and rostrally trending joint with the postorbital. The paired frontals contact each other along a simple butt joint in juveniles which later becomes complexly interdigitate in adults, especially near the parietal. Ventrally, the frontal contacts the laterosphenoid and orbitosphenoid portions of the braincase.

The parietal is interposed between the squamosals and frontals and forms the medial wall of the supratemporal fenestrae. Although formed from right and left paired parietal elements, parietal fusion takes place well before the latest stages of embryogenesis. The midline of the parietal forms a sagittal crest in adults; this crest is only slightly developed in hatchlings and juveniles. The sagittal crest and the lateral wall of the parietal act as the attachment sites of Mm. adductor mandibulae externus profundus and pseudotemporalis. In dorsal view, the parietals have an hourglass shape. Rostrally, they contact the paired frontals along a coarsely interdigitate joint mediated along the midline by a short parietal process (the interparietal of Lull

and Wright 1942; found in *Jaxartosaurus, Corythosaurus, Lambeosaurus, Hypacrosaurus, Tanius, Shantungosaurus, Tsintaosaurus, Anatotitan copei,* and possibly *Edmontosaurus* and *Gryposaurus*). The short medial process of the postorbital forms a short, oblique interdigitate joint with the rostrolateral margin of the parietal. The ventral parietal margin articulates with the laterosphenoid, prootic, opisthotic, supraoccipital, and exoccipital.

Because it is not often prepared, the braincase in hadrosaurids is rather incompletely known. However, in all known cases, it is roughly quadrangular in lateral view, consisting of fused or well-sutured supraoccipital, exoccipitals, basioccipital in caudal view, and opisthotic, prootic, basisphenoid, parasphenoid, laterosphenoid, orbitosphenoid, and presphenoid bones along the lateral wall. The rostrodorsal end of the laterosphenoid is transversely expanded and forms a synovial joint with a reciprocal excavation on the undersurface of the frontal and postorbital bones. Caudal to this contact, the laterosphenoid forms the forward section of the lateral braincase wall. The basipterygoid processes project both ventrally and laterally from the base of the basisphenoid, between 35° and 55° in most hadrosaurids but as little as 15° to 30° in *Gilmoreosaurus, Edmontosaurus,* and *Tanius*. The basipterygoid process in *Tsintaosaurus* has a more depressed position relative to the base of the braincase and is shifted caudally about its long axis. The alar process on the lateral wall of the basisphenoid is found directly above the basipterygoid process and between the ophthalmic and combined maxillary and mandibular branches of the trigeminal nerve. Here, the basisphenoid forms a slight caudal projection that is moderately concave and rostrolaterally facing. It is interpreted as the origin of M. protractor pterygoideus (Galton 1974a; Norman 1984a; Weishampel 1984a). The lateral wall of the prootic and opisthotic bears a nearly horizontal and striated ridge that runs from the trigeminal foramen to the midsection of the paroccipital process. This ridge, the crista prootica, has been interpreted as the origin of M. levator pterygoideus (Norman 1984b; Weishampel 1984a). The occipital condyle is formed chiefly of the basioccipital, with two large exoccipital contributions to the occipital condyle flanking it along its lateral and dorsal aspect. The long, pendent paroccipital process is typically formed by the fused exoccipital and opisthotic. The tip of the paroccipital process contributes the origin for M. depressor mandibulae. Finally, the supraoccipital is excluded from the foramen magnum by the paired exoccipitals and is wedged between the overlying squamosals.

The bony palate extends from the premaxilla to directly beneath the orbits (fig. 26.3b). It consists of the vomers, palatines, pterygoids, and ectoptyergoid bones and is moderately vaulted in hadrosaurines and highly so in lambeosaurines due to modification of the nasal cavity.

The paired vomers are poorly known but lie adjacent to the midline of the rostral part of the palate. Each flat, triangular element sends a thin process forward to fit between the paired premaxillae. The elevated caudal margin articulates with the palatine.

The palatine is a trapezoidal element rigidly attached to the dorsal surface of the caudal process of the maxilla. From this position immediately caudal to the dorsal maxillary process, the palatine extends dorsally and slightly rostrally to articulate with the jugal. The caudodorsal palatine margin articulates with the pterygoid along a slightly movable joint (Weishampel 1984a).

The pterygoid bears three rami that diverge both rostrally and caudally from a central plate. The palatine ramus projects forward and dorsally and forms a rostrodorsal articulation with the palatine along a linear tongue and groove joint. The ectopterygoid ramus extends ventrally where it bluntly contacts the ectopterygoid. Last, the quadrate ramus with its caudal alar and caudoventral projections creates an extensive scarf joint with the pterygoid flange of the quadrate. The central plate is buttressed along a line from the palatine ramus to the caudoventral projection of the quadrate ramus that includes the medial wall of the pterygoid-basipterygoid joint. The central plate also contacts the maxilla along its rostrolateral surface immediately ventral to the base of the palatine-pterygoid joint. M. pterygoideus ventralis is thought to have its origin along the caudal aspect of the ectopterygoid ramus of the pterygoid.

The ectopterygoid is a flat, strap-shaped bone closely applied to the ectopterygoid shelf of the caudal process of the maxilla. It also fits into a fossa on the ectopterygoid ramus of the pterygoid. The dorsal surface of the ectopterygoid is the site of attachment for M. pterygoideus dorsalis.

In all hadrosaurids, the mandible is both long and massive, especially in species of *Edmontosaurus* and *Anatotitan copei*. In lambeosaurines, it is, however, relatively shorter and is ventrally deflected around its oral margin. The dentary amounts to more than 75 percent of the total mandibular length in *Anatotitan copei*. The dentary bears a conspicuously high, laterally offset coronoid process. This process extends into the adductor chamber medial to the jugal; it is also found lateral to the distal mandibular dentition. The dorsal and caudal surfaces of the coronoid process serve for the attachment of most of the adductor musculature (e.g., Mm. adductor mandibulae externus, pseudopterygoideus). The lateral dentary surface is relatively smooth and prominently convex along its length. There are several foramina for vessels and nerves to the face and buccal cavity in association with the cheeks and oral margin. The medial aspect of the dentary is slightly convex and marked by a series of special foramina at the base of the mandibular alveoli. The mandibular symphysis is linear and slightly inclined to the long axis of the mandible; well-striated grooves attest to the presence of well-developed symphyseal ligaments. The caudal margin of the Meckelian groove along the base of the medial dentary served as the insertion site for M. adductor mandibulae caudalis. The rostral region of the dentary is edentulous, shortest in lambeosaurines and basal hadrosaurids and longest in *Edmontosaurus, Shantungosaurus, Anatotitan copei, Saurolophus,* and *Prosaurolophus*. The mandibular dentition is emarginated from the lateral aspect of the dentary, forming a buccal cavity medial to the cheeks.

The unpaired predentary is fitted across the mandibular symphysis; it is a flat, scoop-shaped element bearing two lateral processes that contact the lateral margins of dentaries, a dorsal median process articulating with the dorsal margin of the mandibular symphysis, and a bilobate ventral median process that underlies the symphysis. The oral margin of the predentary is deeply serrated; directly behind the serrations, there are many vascular foramina and pits extending along its entire length. In life, this surface was surrounded by a cornified rhamphotheca (Cope 1883; Versluys 1923; Sternberg 1935; Morris 1970).

The surangular is the second largest mandibular element. It completes the lateral wall of the Meckelian fossa and, together with the more medial articular, forms the mandibular glenoid and the upturned retroarticular process. The mandibular glenoid (jaw joint) is ovoid and shallowly concave in shape and somewhat broader than the ventral quadrate head it accommodates. The surangular articulates with the more ventral and larger quadrate condyle. The lateral aspect of the retroarticular part of the surangular serves as the insertion for M. pterygoideus ventralis. There is no surangular foramen.

The articular is rectangular in medial view. A small portion of its medial wall articulates with the small dorsal condyle of the quadrate. Caudal to this position, the medial wall forms the insertion surface for Mm. depressor mandibulae and pterygoideus dorsalis.

The splenial is a small wedge-shaped bone on the medial surface of the mandible. Here, it roofs the caudal aspect of the Meckelian fossa. Likewise, the angular is a relatively small element, found on the ventromedial margin of the mandible. This long tapering element serves as the attachment site for M. mylohyoideus.

Hadrosaurid functional and replacement teeth are organized into a dental battery made up of as many as 60 closely packed tooth families (fig. 26.8; Weishampel 1984a). Each tooth family consists of from three to five successive teeth. Teeth in adjacent families interlock as they travel through the alveoli to their functional positions. At any given time, there can be one to three functioning teeth per family. All hadrosaurids have mesiodistally narrow maxillary teeth, while dentary teeth are similarly narrow in all reported hadrosaurids but *Telmatosaurus*. In comparisons between maxillary and dentary dentitions among species, maxillary teeth are slightly narrower mesiodistally than those in the dentary. As a consequence, there are a few more maxillary tooth positions than dentary positions in an individual. The number of tooth families increases during ontogeny as does tooth size but not in so pronounced a fashion as in other ornithopods. Lambeosaurines possess fewer tooth families per jaw than remaining hadrosaurids, but overall, there is no difference in absolute tooth density between the the groups. Tooth replacement is virtually continuous during life (but see Horner 1979). Z-spacing is approximately 2.2, while W ranges from 6.0 to 8.0 (Weishampel 1984a). In all hadrosaurids, the tooth rows converge mesially and are usually buccally concave along their long axis.

Crown morphology in hadrosaurids is strikingly different from that of other ornithopods. Taller than wide, the lanceolate crowns bear enamel only on a single side (buccal for maxillary teeth, lingual for dentary teeth). In the middle of this enameled face, there is a strong median carina. In *Telmatosaurus,* the carina on mesial dentary teeth is low and bisected by a shallow trough, producing in effect a double carina much like that of more primitive ornithopods. Marginal denticles are found in nearly all hadrosaurids, but these are strongly developed in *Claosaurus*. The near-cylindrical remainder of the tooth is formed of dentine. Dentary replacement teeth are relatively narrower and slightly less linearly arranged in lambeosaurines than in other hadrosaurids. In lambeosaurines, the median carina is sinuous and the crown-root angle is greater than 145°, in contrast to remaining hadrosaurids, in which the median carina is linear or only slightly curved and the crown-root angle ranges from 120° to 140° (Sternberg 1936; Langston 1960a; Horner in press).

Wear is virtually continuous along the entire tooth row. The occlusal plane is slightly undulatory along its length, steepest mesially and distally and shallowest in the middle of the tooth row. It varies from 40° to 55° from the horizontal in virtually all hadrosaurids (*Prosaurolophus* appears to have the steepest occlusal plane, commonly ranging to as much as 60°). Wear striae occur nearly transversely across the teeth (Weishampel 1983, 1984a). A slightly curved longitudinal groove can be observed in many mandibular dentitions. Given that tooth eruption is continuous through life, this groove appears to have formed because the maxillary dentition did not completely plane off the mandibular dentition during the transverse power stroke (Weishampel 1984a). In transverse section, the occlusal surface of the maxillary tooth row is virtually planar, while the dentary tooth row is always concave; in both cases, the occlusal surface cuts across two or more functional teeth to produce such curvatures.

Sclerotic rings are known in virtually all hadrosaurid taxa (Brown 1912; Edinger 1929; Russell 1940a; Ostrom 1961a). The hexagonal plates overlap in such a way as to create bilaterally symmetrical quadrants. There are 13 to 14 plates forming a single hadrosaurid sclerotic ring.

The stapes is known in both *Edmontosaurus regalis* and *Corythosaurus casuarius* (Versluys 1923; Colbert and Ostrom 1958). In both, it consists of a thin, nearly cylindrical rod that extends from the large notch between the paroccipital process and the head of the quadrate to the fenestra ovalis at the rostral border of the opisthotic. The footplate of the stapes appears to be embedded in the fenestra (Colbert and Ostrom 1958).

Of the hyoid elements, the first ceratobranchial is commonly preserved (Ostrom 1961a). The rostral end

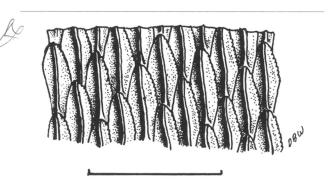

Fig. 26.8. Portion of the mandibular dentition (i.e., dental battery) of *Lambeosaurus lambei,* labial view. Scale = 50 mm.

is compressed, particularly where the paired elements oppose but do not appear to articulate with one another. Caudally, the ceratobranchial is cylindrical, highly tapered, and slightly sinuous along its length. The ossified hyoid lies under the mandible, approximately beneath the level of the coronoid process and supported the hyoid musculature and, presumably, the vocal organ.

Postcranial Skeleton

Axial Skeleton

In hadrosaurids, the number of presacral vertebrae varies from 29 to 32 for the more primitive hadrosaurids and from 30 to 34 for lambeosaurines. Of these, there are 12 to 13 cervicals in the former group and 13 to 15 in the latter. The natural curvature of the axial skeleton is well documented (Maryańska and Osmólska 1983, 1984b). All but the most cranial dorsals and the caudals are held nearly horizontal. The cranial dorsals show pronounced ventriflexion, while the cervicals are dorsiflexed.

The hadrosaurid atlas is composed of four parts that include the odontoid, hypocentrum, and sagitally split neural arch. An excellent description of the atlas and axis can be found in the detailed description of "*Kritosaurus*" *incurvimanus* by Parks (1920b) or Lull and Wright (1942). The odontoid appears to be a free element in all hadrosaurids. The axis in juvenile through adult stages is composed of a centrum and a neural arch that fuses to the centrum when the animal is about half-grown. The neural spine is large with a dorsally arching, craniodorsal flange that extends rostrally, where it separates the two halves of the neural arch of the atlas. The caudal end of the axis centrum is deeply concave for the reception of the strongly convex cranial end of the centrum of the third cervical vertebra. The third and remaining cervicals are all strongly opisthocoelous. The cranial cervical centra are dorsoventrally compressed, being wider than high. Near the caudal end of the series, the centra become deeper than wide and look more like the heart-shaped dorsal centra. The neural spines of all cervical vertebrae except for the atlas and axis are very small and restricted to a position between the pre- and postzygopophyses. Cervical ribs exist on all cervicals except for the atlas and axis and are progressively larger caudally.

Cervical vertebrae grade nearly imperceptibly into dorsals (Parks 1920b), but the latter bear distinct articular facets for the thoracic ribs whereas the former do not. Dorsal vertebrae number from 16 to 20 and possess similar centra and neural arches. The centra increase in size caudally and retain a heart-shaped cranial and caudal outline regardless of ontogenetic stage. Except for the first dorsal, which is slightly opisthocoelous, each dorsal centrum is amphiplatyan. The neural spines increase in height caudally and are generally tallest in lambeosaurines. Whereas the diapophyses are oriented nearly horizontally in lambeosaurines, these processes are angled strongly upward in more primitive hadrosaurids. The breadth of the spine compared to its height primitively ranges from approximately 1:1 to 1:3.5; in lambeosaurines, they range from 1:2 to 1:4.5.

The sacrum in adult hadrosaurids consists of 10 to 12 vertebrae, including single dorsosacral and caudosacral contributions. Fusion of the sacrum varies both ontogentically and between species. In *Gryposaurus notabilis*, there are nine fused sacrals, whereas *Maiasaura peeblesorum* has a sacrum formed of ten vertebrae, of which the first seven are fused and separated from a subsequent fusion of three. The seven cranial sacrals in both species possess parapophyses that extend laterally to support the iliac plate formed of the sacral ribs. This iliac plate, in turn, contacts the inner surface of the ilium; the first and second sacral ribs that contribute to the iliac plate also extend ventrally to provide additional contact with the iliac peduncle of the pubis (Maryańska and Osmólska 1983, 1984b). The eighth sacral, whether fused or unfused, possesses massive transverse processes that extend laterally to meet the caudal blade of the ilium. The remaining sacrals possess transverse processes that abruptly decrease in lateral extent caudally. These processes seldom reach the ilium. In *Saurolophus angustirostris,* the ends of the transverse processes are placed on the dorsal margin of the ilium (Maryańska and Osmólska 1984b). Neural spine height in the dorsals as well as sacrals is species dependent. In *Barsboldia sicinskii*, the neural spines are especially long and club-shaped at their termini (Maryańska and Osmólska 1981b). Sacral centra vary in shape, being more or less heart-shaped cranially, dorsoventrally compressed centrally, and rounded caudally. In juveniles, the dorsoventrally compressed sacrals possess an extremely large neural canal that becomes relatively increasingly smaller with age. Fusion of the sacrum begins at the junction of the second and third sacrals and progresses caudally. Fusion of the first and second sacrals apparently takes place at the same time as fusion of the sixth and seventh sacrals. The sacrum of all hadrosaurines has a median ventral groove that extends axially along at least the caudalmost four or five centra. The lambeosaurine sacrum has a ventral ridge (see Young 1958a).

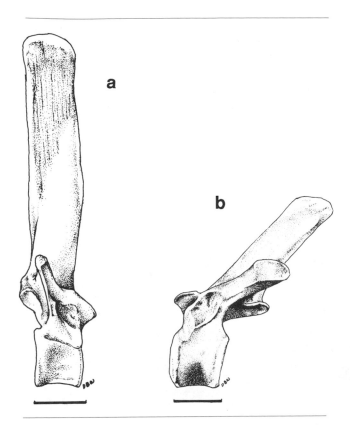

Fig. 26.9. Dorsal vertebrae from the caudal region. a. The lambeosaurine *Hypacrosaurus altispinus* (after Brown 1913). b. The gryposaur *Gryposaurus notabilis.* Scale = 100 mm. (After Lull and Wright 1942.)

The number of caudal vertebrae varies from 50 to 70 and is most likely species dependent. The first true caudal vertebra, located caudal to the end of the iliac blade, varies in shape but is usually axially compressed and possesses the tallest neural spine in the entire vertebral column (fig. 26.9). The cranial or caudal outline of the centra of nearly all caudals is hexagonal, with horizontal dorsal and ventral surfaces. Transverse processes on the first 15 or so caudal centra extend laterally and gradually decrease in size distally. Fusion of the transverse processes and neural arches of the caudals occurs when the animals are approximately a third grown. Neural spines are tallest and most angled in *Barsboldia* and *Hypacrosaurus,* lowest and least angled in *Edmontosaurus.* Chevrons begin distal to the second or third caudal in lambeosaurines and distal to the fourth or fifth in remaining hadrosaurids. With the exception of the first five chevrons, which are directed caudally and often touch one another, the chevrons maintain a caudoventral angle equivalent to the caudodorsal angle of the neural spines. Chevrons are attached

to nearly all of the caudal vertebrae except the proximal three or four and the distalmost centra, which bear little or no neural spine. Length and robustness of the chevrons varies considerably, some being longer and more slender than the neural spines, while others are equal in proportion to the neural spines. Some chevrons are transversely flattened, while others are cylindrical.

The dorsal ribs of hadrosaurids are very similar in form to those of other bipedal ornithischians. The shaft between the capitulum and tuberculum varies in length corresponding to the length of the diapophysis. Rib curvature is greater in the cranial region of the thorax than in the caudal region. Overall, however, the rib cage is generally laterally compressed, particularly compared to tetrapods such as ceratopsians. There are 16 to 20 ribs, the longest underlying the midshafts of the scapulae, then rapidly decreasing in length caudally. The distal ends of all ribs possess slightly enlarged processes for reception of the costal cartilages.

Ossified tendons in hadrosaurids are organized into a rhomboidal lattice along the lateral sides of the neural spines, extending from the middorsal region caudally to about the midsection of the tail. A detailed description of this lattice work is given by Parks (1920b). The tendons appear to have developed within the transversospinalis muscle group. The function of the tendons was most likely for additional support of the horizontal trunk and tail (Ostrom 1964; Galton 1970a; Norman 1980). The tendons apparently reduced ventral displacement of the tail and restricted its dorsoventral oscillation while probably allowing some degree of lateral motion.

Appendicular Skeleton

The hadrosaurid pectoral girdle and forelimb are distinctive among ornithopods yet vary to a large degree among species (fig. 26.10). The hadrosaurid scapula has an elongate and greatly expanded dorsal blade. A prominent acromion process extends caudodorsally from near the coracoid articular surface and gradually decreases in size. There is no evidence from either the ontogenetic growth series of *Maiasaura peeblesorum* or of an undescribed hypacrosaur from the Two Medicine Formation of Montana that the acromion process increases in relative length during growth, as has been claimed (Brett-Surman 1976). Little ontogenetic change is known for the scapula. The shape of the scapular blade varies considerably among hadrosaurids. Lambeosaurines possess a very narrow-waisted scapular neck in contrast to more primitive hadrosaurs, which have a thick scapular neck only slightly

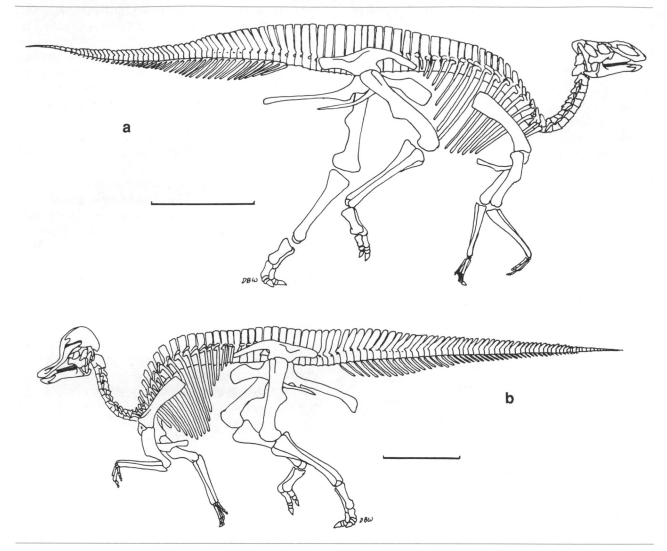

Fig. 26.10. a. The skull and postcranial skeleton of *"Kritosaurus" incurvimanus* (after Norman 1985). b. The skull and postcranial skeleton of *Corythosaurus casuarius* (after Paul 1987b). Scale = 1 m.

narrower than the dorsal border. In addition, the blade expands in width caudodorsally in lambeosaurines, while remaining hadrosaurines have scapulae with near-parallel sides.

The coracoid, although forming half of the glenoid fossa, is small in comparison to those of other ornithischians. The bone is expanded transversely for reception of the scapula and humerus and cranially becomes thin and bladelike. A caudoventral hook extends below the glenoid. A large foramen perforates the element in front of the glenoid. Sometimes the foramen opens into the glenoid-scapular contact, but other times it does not, regardless of ontogenetic stage.

The sternals are paired elements that appear to have been joined to one another by cartilage at their proximomedial ends. The bones have platelike proximal ends and slender shafts that extend caudoventrally along the distal ends of the cranial ribs. There are, however, no indications of attachment to the ribs, either by ligaments or cartilage (Horner pers. obs. from histologic sections). Surficial striations, directed medially, occur along the proximomedial bladelike surface, indicating the presence of either ligamentous or cartilaginous connections.

The humerus is distinctive in that it possesses a long robust deltopectoral crest that extends from the proximal end of the bone to about midshaft. Two morphs can be delimited on the basis of relative length and prominence of the deltopectoral crest. Primitively, the hadrosaurid humerus maintains a relatively short deltopectoral crest that projects very little from the shaft as compared to the lambeosaurine humerus. The humerus of *Maiasaura* represents the extreme in this regard; the deltopectoral crest is very small, projecting only half again the diameter of the humeral shaft. The deltopectoral crest in lambeosaurines projects more than the diameter of the humeral shaft. Muscle scars are prominent along the lateral border of the deltopectoral crest and caudomedial surface of the upper shaft in all hadrosaurids.

The ulna is straight, expanded only at the proximal end. A small olecranon process extends a short distance above the proximal articular surface. The proximal end is triangular in caudal view, possessing a grooved depression below the olecranon on the cranial face for the proximal end of the radius. Distally, the ulna is subtriangular in shape, with a flattened cranial face for the distal end of the radius. It appears that the relative length of the ulna is too variable to be a useful taxonomic character, as has been claimed (Lull and Wright 1942, Brett-Surman 1975).

The radius is a straight, slender bone with only slightly expanded proximal and distal ends. The radius is the shortest of the forelimb long bones.

The manus is composed of one or two small, subrounded carpals, the metacarpus, and four digits. Metacarpals II, III, and IV are relatively straight, elongate bones with only slight proximal and distal expansion. Metacarpals III and IV are nearly the same length, although metacarpal III is more massive. Metacarpal II is two-thirds the length of metacarpal III. Metacarpal V is usually about a third as long as metacarpal III. Each digit has three phalanges. With the exception of the unguals, the phalanges are elongate with expanded proximal ends. The unguals are small and hoof-shaped on digits II, III, and IV and ovoid on digit V. The unguals of digit II and IV are elongate, while that of digit III is broader and tends to resemble the ungual of pedal digit III of juvenile individuals.

Lull and Wright (1942) noted that the forelimb is about half the length of the hindlimb. However, there are a few notable exceptions. *"Kritosaurus" incurvimanus* has forelimbs that are two-thirds the length of their respective hindlimbs. Limb proportions vary considerably and most likely reflect differences in locomotor kinematics and weight support.

The pelvis and hindlimb are generally most robust in lambeosaurines among hadrosaurids (fig. 26.9). The pelvic elements are not fused, nor are the ilia coalesced with the sacral ribs. Rugose, pitted articular surfaces on the medial surface of the ilium and between the ilium, pubis, and ischium suggest that the unit was held together with masses of ligamentous connective tissue.

The ilium possesses a long, ventrally deflected, laterally compressed cranial process, a large antitrochanter, a deep, bladelike caudal process, a massive ischial peduncle, and a lightly constructed pubic peduncle. Hadrosaurines usually have an ilium that is dorsoventrally shallow, while lambeosaurines have a deep ilium (fig. 26.11). In addition, lambeosaurines generally have a much more massive antitrochanter and a more strongly arched cranial process. The ilium appears to be diagnostic to the generic level (Brett-Surman 1975).

The pubis is composed primarily of a cranial prepubic blade that extends forward beyond the ilium (fig. 26.11). The shape of the blade varies and may be useful in generic identification (Horner 1979). Some, like those of species of *Edmontosaurus*, *"E." minor*, *Anatotitan copei*, and *Shantungosaurus giganteus*, have a long, slender proximal shaft, with a short expanded blade. Others, including most of the lambeosaurines, have slightly waisted proximal blades that expand abruptly in the cranial direction. The pubic shaft is rounded and lightly constructed in all hadrosaurids and extends down the upper cranial face of the ischium a short distance past the obturator process.

The ischium is long and shaftlike. The ischium of hadrosaurines is very lightly constructed and has a long, narrow shaft, only very slightly expanded distally (fig. 26.11a). That of lambeosaurines is more massive and possesses a dilated, footlike structure distally (fig. 26.11b). In all species, this footlike structure expands at a 90° angle ventrally from the main shaft. The iliac process is always larger than the pubic process. Located a short distance distal to the pubic process is the obturator process that often coalesces with the base of the pubic process to form a foramen in older individuals.

The femur is a massive element (fig. 26.10), straight in lateral view and slightly bowed outwardly in cranial view. The femoral head is set off from the shaft by a distinct neck; the greater trochanter extends dorsally slightly above the head, usually forming all of the lateral border of the proximal femur. The lateral surface of the greater trochanter is flattened and striated for muscle attachment. The lesser trochanter is found on the craniolateral surface of the femur, immediately beneath the greater trochanter. The lesser trochanter, which is separated from the greater trochanter by an open cleft, varies among hadrosaurid genera in size, relative position, and distance from the greater trochanter. Specimens from the upper Santonian or lower Campanian, for example, possess lesser trochanters

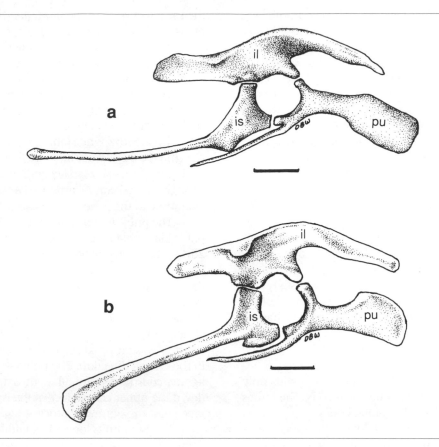

Fig. 26.11. a. Right lateral view of the pelvis of "*Kritosaurus*" *incurvimanus* (after Parks 1920*b*). b. Right lateral view of the pelvis of *Corythosaurus casuarius* (after Brown 1916*b*). Scale = 200 mm.

that are relatively large and separated from the greater trochanter by a wide, deep cleft (Kaye and Russell 1973). The fourth trochanter is found on the caudal surface of the femur at about midshaft level. It varies in overall shape (but tends to be triangular in outline) as well as in position. Some fourth trochanters are shaped like obtuse triangles, while others are more pendent and shaped like scalene triangles. Distally, the femoral condyles form broad contact surfaces for the proximal tibia. The condyles are slightly expanded cranially and greatly extended caudally. The larger medial condyle is flush with the inner surface of the femoral shaft, while the lateral condyle is smaller and separated from the outer surface of the shaft by a deep groove. Lull and Wright (1942) suggested that this groove was occupied by the distal tendon of M. iliofibularis. The craniolateral surface of the lateral condyle possesses moderately deep longitudinal grooves that radiate distally. Cranially, where the lateral and medial condyles terminate, there is often a contact of the condyles and the formation of an enclosed channel for passage of the

extensor tendons. Whether the two condyles meet and enclose a channel may be species dependent, but it is not age dependent. There are numerous muscle scars on the femur, and their precise locations vary considerably ontogenetically and interspecifically. There is always a deep muscle scar located on the medial surface of the fourth trochanter and adjacent shaft and an equally roughened area on the lateral surface of the shaft either just above, adjacent to, or immediately below the fourth trochanter. Particularly in lambeosaurines, a roughened area can be found on the craniodistal portion of the shaft, just above the lateral condyle. Both proximal and distal articular surfaces of the femur appear to have accommodated very large cartilaginous epiphyses (Lull and Wright 1942, Haines 1969).

The tibia is a very straight element, greatly expanded both proximally and distally. The proximal end possesses two condyles, the medial being the larger. The cnemial crest extends craniolaterally from the medial tibial condyle and wraps slightly around the proximal head of the fibula. Distally, the shaft twists outwardly at about 45° so that the distal end of the tibia as well as the pes are not oriented forward but rather are splayed to the side (viz. Mesaverde hadrosaurid trackways; Brown 1938). The distal craniolateral face of the tibia is flattened, while the caudomedial face is convex in outline. The distal craniolateral face is divided into two segments by a moderately deep depression that receives a small dorsal projection from the astragalus. The medial segment of the face receives the lower end of the fibula. Both the proximal and distal ends of the tibia, like the femur, appear to have accommodated large cartilaginous epiphyses. There are no deep muscle scars at any location on the tibia, although in some species small bumplike structures are located on the cranial side of the shaft, near the point of distal expansion.

The fibula is straight and shaftlike, slightly expanded both proximally and distally. The proximal expansion is crescentic, with its convex surface fitting into a concavity of the arcuate cnemial crest of the tibia. Distally, the shaft twists so that the distal expansion is oriented at about 90° from the proximal expansion. The caudomedial surface of the fibula is flattened where it meets the distal face of the tibia. The craniolateral surface is convex and thickest laterally where it expands to meet the calcaneum.

The pes is composed of three tarsals and three digits. The astragalus is crescentic and fits snugly over the distal part of the craniolateral aspect of the tibia. The astragalus never coalesces with the tibia; regardless, there is no indication of any movement between these two elements. The lateral side of the astragalus has a slight concave indentation for the medial side of the calcaneum. A distal tarsal is here reported for the first time in hadrosaurids (originally described as an undesignated element of the manus; Horner 1979). This element, articulating with the top of metatarsal IV, is circular in proximal view, flat on one side, and broadly elevated on the other.

The calcaneum, in lateral view, has a convex distal surface and a wide W-shaped proximal surface. The proximal depression on its caudal side fits into the inner craniolateral border of the tibia, while the cranial depression receives the distal end of the fibula. As noted by Lull and Wright (1942), the combined astragalus and calcaneum do not cover the entire width of the distal tibia. The lateral edge of the distal tibia forms a small portion of the metatarsal articulation.

With the notable exception of *Claosaurus* (Marsh 1872), all hadrosaurids possess only metatarsals II, III, and IV. *Claosaurus* apparently also possessed metatarsal I (Lull and Wright 1942). Metatarsal II is expanded craniocaudally at both proximal and distal ends. The distal third is deflected slightly medially. Metatarsal III is the largest and straightest. Its proximal end is subrounded with a notch on its medial side for the proximal end of metatarsal II. The distal end of metatarsal III is rectangular and slightly flattened craniocaudally. The distal third of metatarsal IV is deflected laterally. A substantial process projects from the proximal caudolateral surface of metatarsal IV. This process meets metatarsal III and apparently assisted in aligning metatarsal IV with metatarsal III. Divergence of both metatarsals II and IV away from metatarsal III prevented the toes from touching one another.

The phalangeal formula is 0−3−4−5−0. The first phalanges on digits II, III, and IV and the second phalanx on digit IV are dorsoventrally compressed and longer than wide. The second, third, and fourth phalanges on digit II and the second and third phalanges on digit III are dorsoventrally compressed and wider than long. All three ungual phalanges are spade-shaped. The plantar surfaces of the unguals are flattened except in *Maiasaura*, which has unguals with longitudinal plantar keels.

Impressions of integument are found with a number of articulated specimens (e.g., Osborn 1912*a*; Brown 1916*b*; Parks 1920*b*; Verluys 1923; Sternberg 1935; Lull and Wright 1942; Gilmore 1946*a*; Horner 1979; 1984*c*). Since there has been no comprehensive comparative study on the subject, it is unclear whether there are significant differences in the integument of the two major groups of hadrosaurids, or in fact among genera or species within groups. All hadrosaurids appear to have been covered by a pavement of small and large polygonal to ovate tubercles in a variety of arrangements extending over the entire body; large tubercles were distributed on a background of smaller tubercles (Osborn 1912*a*, figs. 7 and 8). Tubercles are coarser on the dorsal surface of the animal. Many species appear to have had a median integumentary fold or frill extending along the neck, back, and tail. Variation in this frill is evident from specimens of *Prosaurolophus maximus*, *"Kritosaurus" incurvimanus*, and an indeterminate hadrosaurid (Brown 1916*a*; Parks 1920*b*; Lull and Wright 1942; Horner 1984*c*). In several "mummified" specimens, webbing appears around the central digits of the manus. Although often thought to reflect at least some aspect of aquatic locomotion, this webbing recently has been interpreted as postmortem distortion of a digital pad (Bakker 1986), much like the pedal pad indicated by footprint morphology (Langston 1960*b*).

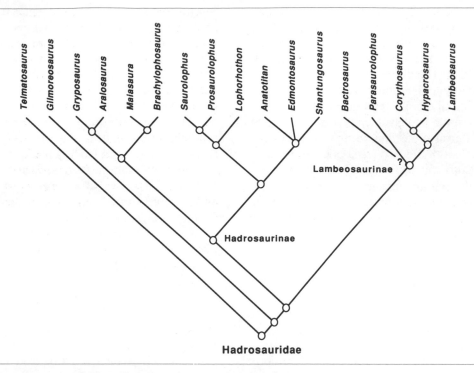

Fig. 26.12. Relationships among hadrosaurid taxa. Features characterizing each node are given in the text.

SYSTEMATICS

Brown (1914*a*), Lambe (1918*a*), Parks (1923), Lull and Wright (1942), Sternberg (1953, 1954), Ostrom (1961*a*), Rozhdestvensky (1966), Morris (1973*b*), Taquet (1975), Hopson (1975*b*), Brett-Surman (1979), and Horner (1983) have all considered to some extent the relationship of hadrosaurids to other ornithopod groups and/or the relationships of taxa within the Hadrosauridae. Most of these workers have associated hadrosaurid ancestry with one or more iguanodont ornithopods, ranging from *Camptosaurus dispar* to species of *Probactrosaurus.* More recently, Sereno (1984, 1986) recognized a close relationship between the iguanodont *Ouranosaurus nigeriensis* and the Hadrosauridae on the basis of a variety of characters, among them a prominent mandibular diastema, transverse expansion of the rostrum, dorsoventral expansion of the maxillary contact of the jugal, and loss of pedal digit I. Milner and Norman (1984) and Norman (1984*a*, 1986) argued that hadrosaurids are the sister group of a monophyletic Iguanodontidae based on an elongate, edentulous snout, exclusion of the supraoccipital from the foramen magnum, low and crested fourth trochanter, and loss of metatarsal I, among other features. Finally, in preliminary work, Horner (1985, 1988, in press)

suggested that hadrosaurs may have had a diphyletic origin from iguanodonts, such that hadrosaurines (Horner's Hadrosauridae) are the sister group of *Iguanodon,* based on a large, long supraoccipital, fusion of palpebrals with the dorsal margin of the orbit, and exclusion of the basioccipital from the foramen magnum, among other characters, and that lambeosaurines (Horner's Lambeosauridae) are more closely related to *Ouranosaurus* based on, among other things, a maxilla with a premaxillary shelf, loss of median rostral maxillary process, high crown-root angle, and expanded, short prepubic shaft and blade. The renewed vigor in dinosaur systematics exemplified by these three sets of studies has not only stimulated closer examination of taxonomic relationships of hadrosaurid taxa but also enabled the reevaluation of characters and phylogenetic importance for establishing hadrosaurid relationships.

This chapter considers the Hadrosauridae to be monophyletic (fig. 26.12), pending further research on the subject (see Horner 1990, in prep.). *Ouranosaurus + Iguanodon, Camptosaurus,* and *Dryosaurus* were used as successively more remote outgroups. Monophyly in hadrosaurids is based on the possession of three or more teeth per tooth position, elevation of the dorsal process of the maxilla and consequent reposition of the antorbital fenestra, loss of the surangular and paraquadrate foramina, fusion of the palpebral to the dorsal orbital margin, mesiodistal narrowing of maxillary teeth, reduction of the carpus, loss of manual digit I, large antitrochanter, reduction of the pubic shaft, and 8 to 10 sacral vertebrae. All of these character states exist primitively in hadrosaurids but are lacking in all outgroups.

The traditional version of the Hadrosaurinae is a paraphyletic taxon consisting of two major groups and at least two basal taxa (Table 26.1). The most primitive hadrosaurid is *Telmatosaurus transsylvanicus*, followed by *Gilmoreosaurus mongoliensis*. As far as can be ascertained, *Telmatosaurus* shares all of the hadrosaurid features described above but lacks those acquired by remaining hadrosaurid taxa. *Gilmoreosaurus* is very poorly known but appears to share two features with all higher hadrosaurids: relatively narrow dentary teeth and a strong single median carina on the enamel face of these same teeth. *Claosaurus agilis*, *Tanius sinensis*, and *Secernosaurus koerneri* may also belong among these basal hadrosaurids, but all are known from poor material that has yet to yield phylogenetically significant information.

Remaining hadrosaurids share a narrowing of the mandibular condyle of the quadrate and reduction or loss of the coarsely denticulate margin of the oral margin of the premaxilla. The first large clade related to this node is a revised Hadrosaurinae, reconstituted to included gryposaurs (*Gryposaurus notabilis*, *Aralosaurus tuberiferus*, *Hadrosaurus foulkii*), brachylophosaurs (*Maiasaura peeblesorum*, species of *Brachylophosaurus*), saurolophs (species of *Saurolophus*, *Prosaurolophus maximus*, ?*Lophorhothon atopus*), and edmontosaurs (species of *Edmontosaurus*, *Anatotitan copei*, *Shantungosaurus giganteus*). The Hadrosaurinae is recognized by the transverse expansion of the premaxilla and the development of a circumnarial depression, neither of which appears to be present in *Telmatosaurus*, *Gilmoreosaurus*, or lambeosaurines. Hadrosaurines may also be characterized by the presence of a ventrally grooved sacrum, but appropriate material demonstrating the ubiquity of this character within the subfamily and/or more basal hadrosaurids is not available.

Within the Hadrosaurinae, gryposaurs share a distinctively narrow preorbital nasal arch. Gryposaurs consist of *Gryposaurus* itself (note: *Gryposaurus* is here preferred to *Kritosaurus*, since the type of *Kritosaurus* [*K. navajovius* Brown, 1912] is based on very poor and questionably useful gryposaur material; see table 26.1), *Aralosaurus*, "*Kritosaurus*" *incurvimanus*, and *Hadrosaurus* (the last two taxa not indicated in fig. 26.12). The brachylophosaur group (*Brachylophosaurus*, *Maiasaura*)

share a symmetrical triangular maxillary process of the jugal (also seen in *Telmatosaurus*) and a transversely broad, solid nasofrontal crest. Gryposaurs and brachylophosaurs may be united on the basis of a prominent ventral flange of the jugal, but we do not consider this feature particularly strong. The other cluster of hadrosaurine taxa includes saurolophs and edmontosaurs. Saurolophs (*Saurolophus*, *Prosaurolophus*, *Lophorhothon*) are united on the basis of the caudal expression of a narrow, solid supraorbital crest formed by the nasal. Edmontosaurs (species of *Edmontosaurus*, *Anatotitan copei*, *Shantungosaurus giganteus*), however, themselves make up a small grouping characterized by complex excavation of the narial fossa, a massive jugal, and excavation of the orbital surface of the postorbital. Saurolophs and edmontosaurs share an elongate area between the dentary tooth row and predentary (a mandibular diastema), with a corresponding extension of the premaxilla.

The Lambeosaurinae consists of *Bactrosaurus johnsoni*, species of *Parasaurolophus*, *Corythosaurus casuarius*, *Hypacrosaurus altispinus*, and species of *Lambeosaurus*. These are diagnosed by the possession of a modified nasal cavity having a supraorbital position, a jugal with a truncated, rounded articulation with the maxilla, dentary teeth with a crown-root angle greater than 145°, elongate neural spines, robust appendicular elements, and a prominently footed ischium. *Bactrosaurus* and *Parasaurolophus* represent the most primitive lambeosaurines as a trichotomy with remaining members of the clade. *Lambeosaurus*, *Corythosaurus*, and *Hypacrosaurus* are united as derived lambeosaurines on the basis of the enlargement of the cavum nasi proprium portion of the nasal cavity. *Jaxartosaurus*, *Barsboldia*, and *Nipponosaurus* are also members of the Lambeosaurinae but are not included in figure 26.12 as the first two taxa are very poorly known and the last is recognized only from juvenile material.

The status of *Tsintaosaurus spinorhinus* deserves comment. Originally described as a hadrosaurine (Young 1958a), *Tsintaosaurus* has been allied with saurolophs (Hopson 1975b), referred to the Lambeosaurinae (Brett-Surman 1979), and excluded from the Lambeosaurinae (Weishampel 1981b). On further consideration, we considered this species a chimera based on a combination of lambeosaurine and hadrosaurine material, which has confounded attempts at properly classifying the animal. We are unconvinced that the nasal crest is hollow, and Taquet (pers. comm.) has suggested that it may be a displaced nasal. Be that as it may, we believe that the skull roof-neurocranium material, the premaxilla, the sacra with a ventral groove,

TABLE 26.1 Hadrosauridae

	Occurrence	Age	Material
Ornithopoda Marsh, 1881b			
Hadrosauridae Cope, 1869b			

Telmatosaurus Nopcsa, 1903 (=
    Limnosaurus Nopcsa, 1900,
    Hecatasaurus Brown, 1910)			

T. transsylvanicus (Nopcsa, 1900)
        (= *Limnosaurus transsylvanicus*
        Nopcsa, 1900)
	Sinpetru Beds, Hunedoara, Romania; Grès de Saint-Chinian, Herault, Grès à Reptiles, Var, France; unnamed unit, Lleida, Spain	Campanian-Maastrichtian	5-10 fragmentary skulls, some with associated postcrania, mixed age classes

Gilmoreosaurus Brett-Surman, 1979			

G. mongoliensis (Gilmore, 1933a)
        (= *Mandschurosaurus mongoliensis*
        Gilmore, 1933a)
	Iren Dabasu Formation, Nei Mongol Zizhiqu, People's Republic of China	Late Cretaceous ?Cenomanian, ?Maastrichtian)	Isolated skull material, associated, disarticulated postcrania, at least 8 individuals

Tanius Wiman, 1929			

T. sinensis Wiman, 1929
        (including *T. chingkankouensis*
        Young, 1958a)
	Wangshi Series, Shandong, People's Republic of China; Tsagayanskaya Svita, Heilongjiang, People's Republic of China	?Coniacian-?Maastrichtian	?3 disarticulated specimens, skull and postcranial material

Claosaurus Marsh, 1890a			
C. agilis (Marsh, 1872) (= *Hadrosaurus agilis* Marsh, 1872)	Niobrara Chalk, Kansas, United States	early Campanian	Single articulated postcranial skeleton with associated skull fragments
Secernosaurus Brett-Surman, 1979			
S. koerneri Brett-Surman, 1979	?Bajo Barreal Formation, Rio Negro, Argentina	?Maastrichtian	Associated disarticulated postcrania, braincase
Hadrosaurinae Lambe, 1918a			
Unnamed taxon (= gryposaurs)			
Gryposaurus Lambe, 1914a			
G. notabilis Lambe, 1914a	Judith River Formation, Alberta, Canada	late Campanian	Approximately 10 complete skull, 12 fragmentary skulls, associated postcrania
Hadrosaurus Leidy, 1858			
H. foulkii Leidy, 1858 (including *Ornithotarsus immanis* Cope, 1869c)	Woodbury Formation, New Jersey, United States	Campanian	Single articulated postcranial skeleton, teeth, isolated postcranial material
Unnamed gryposaur (= "*Kritosaurus*" *incurvimanus* Parks, 1920a)	Judith River Formation, Alberta, Canada	late Campanian	Single fully articulated skull and skeleton

Unnamed gryposaur (described as
        Hadrosaurus notabilis by Horner
        1979)
	Bearpaw Shale, Montana, United States	late Campanian	Nearly complete skull and postcrania

Undescribed gryposaur (unpublished)	Lower Two Medicine Formation, Montana, United States	Santonian	Several partial skulls and postcranial skeletons
Aralosaurus Rozhdestvensky, 1968a			
A. tuberiferus Rozhdestvensky, 1968a	Beleutinskaya Svita, Kazachskaya S.S.R., Union of Soviet Socialist Republics	?Turonian-early Santonian	Nearly complete, articulated skull

Unnamed gryposaur (= "*Kritosaurus*"
        australis Bonaparte, Franchi,
        Powell, et Sepulveda, 1984)
	Los Alamitos Formation, Rio Negro, Argentina	late Campanian-early Maastrichtian	Partial skulls with associated postcrania, approximately 5 individuals

?Kritosaurus navajovius Brown, 1910
        (type material is probably not
        diagnostic)
	Kirtland Shale, ?Fruitland Formation, New Mexico, ?Aguja Formation, Texas, United States	?Campanian-Maastrichtian	Fragmentary skull, associated (but unprepared) postcrania, referred material of 2-3 individuals

Unnamed taxon (= brachylophosaurs)			
Brachylophosaurus Sternberg, 1953			
B. canadensis Sternberg, 1953	Judith River Formation, Alberta, Canada	late Campanian	Complete skull and associated postcrania, partial skull
B. goodwini Horner, 1988	Judith River Formation, Montana, United States	late Campanian	Nearly complete skull and associated postcrania
Maiasaura Horner et Makela, 1979			
M. peeblesorum Horner et Makela, 1979	Upper Two Medicine Formation, Montana, United States	middle-late Campanian	More than 200 specimens, including articulated skull and postcrania, embryo to adult
Unnamed taxon (= saurolophs)			
Prosaurolophus Brown, 1916a			
P. maximus Brown, 1916a	Judith River Formation, Alberta, Canada; ?Judith River Formation, Montana United States	late Campanian	20-25 individuals, including at least 7 articulated skulls and associated postcrania,
Saurolophus Brown, 1912			
S. osborni Brown, 1912	Horseshoe Canyon Formation, Alberta, Canada	early Maastrichtian	Complete skull and skeleton, 2 complete skulls, disarticulated skull material
S. angustirostris Rozhdestvensky, 1952b	Nemegt Formation, Omnogov, White Beds of Khermeen Tsav, Nemegtskaya Svita, Bayankhongor, Mongolian People's Republic	?late Campanian or early Maastrichtian	At least 15 specimens, including articulated skull and postcranial skeleton

Table 26.1, continued

Lophorhothon Langston, 1960			
L. atopus Langston, 1960	Mooreville Chalk, Alabama, Black Creek Formation, North Carolina, United States	Campanian	Disarticulated skull and postcranial skeleton, isolated skeletal elements
Undescribed sauroloph (unpublished)	Upper Two Medicine Formation, Montana, United States	middle-late Campanian	Disarticulated, associated skull and postcrania pertaining to at least 4 individuals
Unnamed taxon (= edmontosaurs)			
Edmontosaurus Lambe, 1920			
E. regalis Lambe, 1917b (including *Trachodon atavus* Cope, 1871, *Agathaumas milo* Cope, 1874)	Horseshoe Canyon Formation, St. Mary River Formation, Scollard Formation, Alberta, Canada; Hell Creek Formation, Lance Formation, South Dakota, Hell Creek Formation, North Dakota, Lance Formation, Wyoming, Laramie Formation, Colorado, United States	early-late Maastrichtian	Approximately 7 fully articulated skull and associated postcrania, 5-7 articulated skulls, associated skull elements
E. annectens (Marsh, 1892b) (= *Claosaurus annectens* Marsh, 1892b, including *Thespesius edmontonensis* Gilmore, 1924c)	Scollard Formation, Alberta, Frenchman Formation, Saskatchewan, Canada; Hell Creek Formation, Montana, South Dakota, Lance Formation, Wyoming, Laramie Formation, Colorado, United States	late Maastrichtian	At least 5 articulated skull and associated postcranial skeletons, isolated skull material
E. saskatchewanensis (Sternberg, 1926b) (= *Thespesius saskatchewanensis* Sternberg, 1926b)	Frenchman Formation, Saskatchewan, Canada	late Maastrichtian	Complete skull, 3-4 partial skulls
Anatotitan Chapman et Brett-Surman, 1990 •			
A. copei (Lull et Wright, 1942) (= Anatosaurus copei Lull et Wright, 1942; possibly including *Trachodon longiceps* Marsh, 1897)	Hell Creek Formation, Montana Lance Formation, Hell Creek Formation, South Dakota, United States	late Maastrichtian	2-3 articulated skulls, associated postcrania, isolated skull elements
Shantungosaurus Hu, 1973			
S. giganteus Hu, 1973	Wangshi Series, Shandong, People's Republic of China	?Campanian	At least 5 individuals, consisting of disarticulated skull and postcranial elements
Unnamed edmontosaur (described as "*Hadrosaurus minor*", described by Colbert 1948b)	New Egypt Formation, New Jersey, United States	Maastrichtian	Partial hindlimb, vertebrae, ribs
Lambeosaurinae Parks, 1923			
Bactrosaurus Gilmore, 1933a			
B. johnsoni Gilmore, 1933a	Iren Dabasu Formation, Nei Mongol Zizhiqu, People's Republic of China	Late Cretaceous (?Cenomanian, ?Maastrichtian)	Disarticulated, associated skull and postcranial material pertaining to at least 6 individuals, juvenile to adult
Corythosaurus Brown, 1914a (including *Procheneosaurus* Matthew, 1920 *partim*)			
C. casuarius Brown, 1914a (including *C. excavatus* Gilmore, 1923, *C. intermedius* Parks, 1923, *Tetragonosaurus erectofrons* Parks, 1931, *C. bicristatus* Parks, 1935, *C. brevicristatus* Parks, 1935, *T. cranibrevis* Sternberg, 1935)	Judith River Formation, Alberta, Canada; ?Judith River Formation, Montana, United States	middle-late Campanian	Approximately 10 articulated skulls and associated postcrania, 10-15 articulated skulls, isolated skull elements, juvenile to adult
Hypacrosaurus Brown, 1913			
H. altispinus Brown, 1913 including *Cheneosaurus tolmanensis* Lambe, 1917a)	Horseshoe Canyon Formation, Alberta, Canada	early Maastrichtian	5-10 articulated skulls, some associated with postcrania, isolated skull elements, juvenile to adult
Lambeosaurus Parks, 1923 (including *Procheneosaurus* Matthew, 1920 *partim*, *Didanodon* Osborn, 1902)			
L. lambei Parks, 1923 (including *Tetragonosaurus praeceps* Parks, 1931, *Corythosaurus frontalis* Parks, 1935, *L. clavinitialis* Sternberg, 1935)	Judith River Formation, Alberta, Canada	late Campanian	Approximately 7 articulated skulls with associated postcrania, ?10 articulated skulls, isolated skull elements, juvenile to adult
L. magnicristatus Sterberg, 1935	Judith River Formation, Alberta, Canada	late Campanian	2 complete skulls, 1 with associated, articulated postcrania
?*L. laticaudus* Morris, 1981	"El Gallo" formation, Baja California, Mexico	Campanian	Isolated skull and postcranial elements
Parasaurolophus Parks, 1922			
P. walkeri Parks, 1922	Judith River Formation, Alberta, Canada; ?Hell Creek Formation, Montana, United States	late Campanian	Complete skull and postcranial skeleton
P. tubicen Wiman, 1931	Kirtland Shale, New Mexico, United States	Maastrichtian	Disarticulated, associated skull and postcrania, pertaining to at least 3 individuals
P. cyrtocristatus Ostrom, 1961b	Fruitland Formation, New Mexico, United States	late Campanian	Fragmentary skull, postcranial skeleton
Undescribed lambeosaurine (Currie pers. comm.)	Judith River Formation, Alberta, Canada	late Campanian	Skull

Table 26.1, continued

Undescribed lambeosaurine (unpublished)	Upper Two Medicine Formation, Montana, United States	late Campanian	Partial skull, isolated postcrania, many individuals
Lambeosaurinae incertae sedis			
Jaxartosaurus Riabinin, 1939			
J. aralensis Riabinin, 1939 (including *Procheneosaurus convincens* Rozhdestvensky, 1968a)	Dabrazinskaya Svita, Syderinskaya Oblast, Kazakhstan	?Turonian-Santonian	Isolated skull roof and braincase
Barsboldia Maryańska et Osmólska, 1981b			
B. sicinskii Maryanska et Osmólska, 1981b	Nemegt Formation, Omnogov, Mongolian People's Republic	Maastrichtian	Sacrum, pelvis
Nipponosaurus Nagao, 1936			
N. sachalinensis Nagao, 1936	Mh7 or Mh6 of Miho Group, Sachalinskaya Oblast, Russia	early Santonian or late Coniacian	Partial skull and associated postcrania, juvenile
Hadrosauridae incertae sedis			
Tsintaosaurus Young, 1958a			
T. spinorhinus Young, 1958a (including *Tanius laiyangensis* Zhen, 1976)	Wangshi Series, Shandong, People's Republic of China	?Campanian	Isolated skull and postcranial elements from at least 5 individuals

Nomina dubia:	Material
Arstanosaurus akkurganensis Suslov et Shilin, 1982	Caudal half of maxilla
Bactrosaurus prynadai Riabinin, 1939	Maxilla, dentary, both with teeth
Cionodon arctatus Cope, 1874	Fragmentary maxilla, vertebrae, fragmentary postcranial elements
Cionodon kysylkumensis Riabinin, 1931b	Fragmentary dentary, vertebrae, tibia
Cionodon stenopsis Cope, 1875	Fragmentary maxilla
Claorhynchus trihedrus Cope, 1892b	Fragmentary premaxillae, predentary
Claosaurus affinis Wieland, 1903	Pedal phalanx (lost)
Diclonius pentagonus Cope, 1876a	Fragmentary dentary with teeth
Diclonius calamarius Cope, 1876a	Teeth
Diclonius perangulatus Cope, 1876a	Teeth
Dysganus encaustus Cope, 1876a *partim*	Single tooth and 5 tooth fragments
Hadrosaurus breviceps Marsh, 1889a	Fragmentary dentary with teeth
Hadrosaurus cavatus Cope, 1871	Caudal vertebrae
Hadrosaurus minor Marsh, 1870	Dorsal vertebrae
Hadrosaurus paucidens Marsh, 1889a	Squamosal, maxilla
Hypsibema crassicauda Cope, 1869a	Caudal vertebrae, fragmentary humerus, fragmentary tibia, metatarsal II
Iguanodon hilli Newton, 1892	Tooth
Jaxartosaurus fuyunensis Wu, 1984	Edentulous dentary
Mandschurosaurus laosensis Hoffet, 1943	Ilium
Microhadrosaurus nanshiungensis Dong, 1979	Dentary with teeth
Orthomerus dolloi Seeley, 1883	Caudal vertebra, femur, tibia
Orthomerus weberi Riabinin, 1945	Fragmentary hindlimb elements
Neosaurus missouriensis Gilmore et Stewart, 1945 (Type of *Parrosaurus* Gilmore, 1945)	Caudal vertebrae, fragmentary dentary and predentary
Pteropelyx grallipes Cope, 1889a	Skeleton lacking skull
Sanpasaurus yaoi Young, 1946 *partim*	Fragmentary postcranium
Saurolophus kryschtofovici Riabinin, 1930b	Fragmentary ischium
Thespesius occidentalis Leidy, 1856	Caudal vertebrae, pedal phalanx
Trachodon altidens Lambe, 1902	Maxilla with teeth
Trachodon amurensis Riabinin, 1925 (Type of *Mandschurosaurus* Riabinin, 1930b)	Fragmentary skull and postcrania
Trachodon cantabrigiensis Lydekker, 1888a	Dentary tooth
Trachodon marginatus Lambe, 1902	Postcranial material, tooth fragments
Trachodon mirabilis Leidy, 1856	Dentary tooth
Trachodon selwyni Lambe, 1902	Dentary with teeth

and the ischia lacking prominent distal expansion are hadrosaurine, while the maxillae, the quadrates, most of the dentaries, the ventrally ridged sacra, and the prominently footed ischia are lambeosaurine. We further suggest that the lambeosaurine crest fragment (viz. Young 1958a:98, fig. 40) should be combined with this lambeosaurine material. Because of these difficulties, we treat *Tsintaosaurus spinorhinus* as a hadrosaurid *incertae sedis,* pending new and better associated material.

PALEOECOLOGY

From its outset, the study of hadrosaurid paleoecology has been largely one of functional typology: hadrosaurids did (or did not) live in a particularly uniform way, eat the same food items, and move around in the same manner. That this is not the case is clear from the variability of hadrosaurid skeletal elements and soft anatomy, biogeographic distribution, evolutionary patterns, and local environmental context, among other aspects of their biology. We here outline some of these distinctions in hadrosaurid paleoecology.

The fossil record of North American hadrosaurids is extremely good. The vast majority of all taxa are known from essentially complete skulls, almost all taxa are known from a number of individuals (including both juveniles and adults), and 60 to 70 percent have associated postcranial material, many of which are virtually complete postcranial skeletons (Dodson 1971; Sternberg 1970). Major biases among hadrosaurids include a strong tendency toward preservation of skulls without postcrania. Although preservation of solitary articulated specimens has been thought to be typical, recent work on bone beds, many of them monospecific, is reversing the historical trend (Horner 1984b; Hooker 1987). Juvenile hadrosaurid material, including both skulls and skeletons, is also becoming well known.

Hadrosaurids are known to have had a nearly worldwide distribution (North America, central and eastern Asia, Europe, and South America; Weishampel and Weishampel 1983; Weishampel, this vol.). To date, no hadrosaurid material has been found in Africa, India, Australia, or Antarctica. The more extreme hadrosaurid locations include the North Slope of Alaska, United States, and the Yukon Territory, Canada, where the animals lived as much as 70° to 85° N paleolatitude during the Maastrichtian (Rouse and Srivastava 1972; Russell 1984c; Davies 1987; Brouwers et al. 1987). The southern extreme is in Chubut Province, Argentina, at approximately 45° S paleolatitude (Bonaparte et al. 1984).

Much like those of other Late Cretaceous dinosaurs, hadrosaurid remains are found in diverse biotopes, including intermontane, terrestrial foredeep, and marine basins. More specifically, hadrosaurids are known from upper coastal plain deposits, lower coastal plain channels and overbank deposits, and delta plain sediments. Hadrosaurids are even known from a few, albeit rare, marine occurrences (*Hadrosaurus foulkii, Claosaurus agilis;* see Horner 1979 for other taxa known from marine localities).

In Montana, United States, and Alberta, Canada, where a considerable amount of work has been accomplished in the collection of Campanian-age hadrosaurid specimens, it appears that there is some habitat differentiation among hadrosaurid species (Russell 1967; Dodson 1971; Horner 1983; see also Lucas 1981 and Lehman 1987 on similar habitat partitioning in the Southwest of the U.S.). For instance, hadrosaurines such as *Gryposaurus notabilis, "Kritosaurus" incurvimanus,* and *Brachylophosaurus canadensis,* characterized by

their transversely expanded snouts and arched or inflated facial regions, are found almost exclusively in near-marine, deltaic sediments. The hadrosaurines *Prosaurolophus maximus* and *Brachylophosaurus goodwini* and all of the previously described lambeosaurines with the possible exception of *Parasaurolophus* (see Horner 1979) are found in lower coastal plain sediments deposited inland from the near-marine environments. A similar situation apparently occurred in Maastrichtian faunas, where the hadrosaurine *Edmontosaurus regalis* is found in near-marine environments, and *Saurolophus osborni* and the lambeosaurine *Hypacrosaurus altispinus* are found in marginal, more continental lowlands (Russell and Chamney 1967). Conservative species, such as the hadrosaurine *Maiasaura peeblesorum,* were apparently endemic to upper coastal plain environments during the Campanian (Horner 1983). To date, upper coastal plain hadrosaurines have not yet been found or recognized from Maastrichtian-age sediments of North America.

The coastal plain distribution of many hadrosaurine and lambeosaurine taxa is odd in that eggs and young individuals are absent or extremely rare for these same marginal and lower coastal plain habitats. Eggs and young of species living in upper coastal plain environments are very common, making up as much as 80 percent or more of total recovered specimens (Horner 1984b, 1987). Matthew (1915a) was the first to suggest that the absence of young and eggs from lower coastal plain environments reflected an isolated habitat for the young and egg laying. Matthew also believed that hadrosaurids (and ceratopsians) originated in upland habitats and that they had migrated upland to their ancestral habitat to nest. The young then remained for a time after hatching before returning to the lowlands. Although there have been numerous discoveries of hadrosaurid eggs, nests, and young since Matthew published these opinions, very little evidence from these findings has been brought to bear on Matthew's contention. Only recently have counterinterpretations been proposed. Carpenter (1982b) suggested that the environmental distribution of eggs reflects not preferred laying regions but rather local soil conditions in lowland environments. Here, the substrate would have been too acidic for preservation of calcareous eggs and young skeletons had they been there in the first place. Research to separate preservational constraints from biological preference and to discover whether particular lowland species entered and nested in upland environments is currently under way (Horner and Weishampel in prep.)

Hadrosaurids ranged in size from less than 1 kg at hatching to 2,000 to 4,000 kg in adulthood (estimates from hatchling *Maisaura peeblesorum* and an unnamed ?hadrosaurine [Barsbold and Perle 1983]; adult weight estimates for *Hypacrosaurus altispinus*, *Edmontosaurus regalis*, and *Anatotitan copei* [Anderson et al. 1985]). The largest hadrosaurids (e.g., *Shantungosaurus giganteus*) may have weighed as much as 16,000 kg, but these animals appear to have been relatively rare. Given their size and abundance during the Late Cretaceous, it is not surprising that these animals were the dominant terrestrial herbivores of their time. Hadrosaurids were clearly active foragers on ground cover and low arboreal foliage from conifers and deciduous trees and shrubs. Browsing appears to have been concentrated within the first 2 m above the ground, but the animals were capable of reaching possibly as high as 4 m to obtain relatively high-placed foliage or fructifications (Weishampel 1984c).

The landscape in which hadrosaurids foraged consisted of rich and diverse forests. Mega- and palynofloras indicate that gymnosperms were still dominant in the Late Cretaceous, with platanoid and magnoliid angiosperms making up the majority of the remainder of the plant realm (Wolfe 1976; Wing and Tiffney 1987). Many Late Cretaceous climax forests were composed of a scattering of emergent araucarian conifers and a low canopy of small angiosperm trees (Wing and Tiffney 1987). Swamp vegetation of the lower coastal plain was commonly dominated by taxodiacean conifers (Parker 1976). Remaining floral elements include various ferns, palm, and cycadophytes. It is on the foliage and fructifications of these suites of plant life that hadrosaurids must have fed.

The broad, complex occlusal surfaces and dental batteries of hadrosaurids, as well as large gut capacity, were well suited for a subsistence on low-quality, high-fiber vegetation (Farlow 1987b). Chewing in these animals involved bilateral occlusion, much like that in the more primitive hypsilophodontids and iguanodonts (see Sues and Norman, and Norman and Weishampel, this vol.). Hadrosaurids accomplished transverse tooth-tooth movement through a combination of orthal mandibular movement and pleurokinesis (i.e., slight lateral rotation of the maxillary region of the facial skeleton; Weishampel 1983, 1984a; Norman 1984a; Norman and Weishampel 1985). As a consequence, the entire hadrosaurid jaw system is formed into a seven-link mechanism that includes left and right maxillary units, counterpart mandibular units, a single muzzle unit, and two slightly streptostylic quadrate units. With such a jaw system, the broad, uninterrupted occlusal surface is available for compression and transverse tearing and cutting of plant items during the excursion of maxillary over dentary teeth. A possible record of hadrosaurid diet that comes from a published account of supposed gut contents is that of Kräusel (1922) from a "mummified" *Edmontosaurus annectens*. The contents consist of a good deal of fibrous material (conifer needles and branches) as well as deciduous foliage and numerous small seeds or fruits. Whether this material reflects the constituents of at least part of the hadrosaurid diet, as advocated by Ostrom (1964), or simply represents a fluvial concentration within the shadow of the hadrosaurid carcass, as suggested by Abel (1922), cannot be ascertained at present because the plant material has since been prepared away from the "mummified" specimen and broken up.

Once thought to be amphibious (Leidy 1858; Cope 1883), hadrosaurids are now viewed as facultative terrestrial bipeds, using quadrupedal stance during slow locomotion or when standing still, with a bipedal stance adopted for fast locomotion. The majority of the dorsal and all of the caudal series is ramrod straight and supported by ossified tendons; together with the slender pubic peduncle of the ilium, this condition indicates that the vertebral column was naturally held at or near a horizontal attitude, more like a ratite bird than a kangaroo (Galton 1970a). Hadrosaurids may have been able to reach 14 to 20 km/hr sustained running speeds (Thulborn 1982), although higher running velocities (approximately 50 km/hr; Coombs 1978c) may have been possible if used only rarely.

Although we agree that many aspects of hadrosaurid anatomy suggest that they were highly active on land, we are of the opinion that these animals may not have all been exclusively terrestrial foragers (see also Dodson 1971). Hadrosaurines found in near-marine environments, for example, may have been semi-aquatic, whereas those in more inland environments may have been more terrestrial. Other hadrosaurines such as *Maiasaura peeblesorum* were most likely fully terrestrial, as the anastomosing or braided fluvial sediments that have yielded *M. peeblesorum* material indicate very shallow depth. The postcranial morphology of *M. peeblesorum* also suggests a terrestrial existence as the limb elements are relatively short and very robust. Other features, the humerus and its deltopectoral crest, for example, vary considerably in size and robustness among species and genera, but their significance for differentiation of locomotor grade and habitat usage is yet to be ascertained.

The question as to whether any hadrosaurids spent any length of time in brackish or marine environments has yet to be adequately addressed. Although it has been presumed that those hadrosaurines found in marine sediments (see Horner 1979) were simply washed out to sea from inland habitats, there is no particularly good evidence for dismissing them from coastal marine habitats. Nearly all of the specimens from marine sediments, for example, are found articulated, and in at least one instance, the animal was entombed in skin (Horner 1979).

Hadrosaurid intraspecific behavior has been discussed in detail by Hopson (1975a), Dodson (1975), Horner and Makela (1979), Weishampel (1981a), and Horner (1982, 1984b). Colonial nesting is known in *Maiasaura peeblesorum* and may be widespread among hadrosaurids (see Barsbold and Perle 1983 for communal nesting in ?hadrosaurines from the Toogreeg beds of the Mongolian People's Republic). Parental care of the young seems to have existed for some period of time following hatching. Bonebed occurrences lend further information on possible gregariousness and migration. Hadrosaurids are commonly found in large bone beds, many of them exclusively monospecific, in both marginal lowland and upland sediments. Perhaps the best example comes from a bone bed surrounded by an ash-fall bentonite within the Two Medicine Formation, Montana, United States, that conservatively contains remains of at least 10,000 *M. peeblesorum* individuals ranging in size between 3 and 7.5 m in length (Hooker 1987). It is assumed that an aggregation of this size required migratory movement, most likely seasonal, to meet the energy demands of its members. Because *M. peeblesorum* specimens have not yet been found in lowland sediments, these animals likely migrated along a north-south route, as discussed by Hotton (1980; also including most other groups of dinosaurs). For these same reasons, it is likely that hadrosaurine and lambeosaurine species groups remained within their partitioned habitats (upper coastal plain, lower coastal plain, near-marine, etc.) during these seasonal migrations. Although movement out of these habitats may have occurred during nesting seasons, there is no evidence as yet that it actually occurred.

Several unusual cranial features in hadrosaurids likely evolved in the context of parent-offspring interactions, herding, and intraspecific social behavior. These structures center around the modification of the nasal cavity and associated lateral diverticula (Hopson 1975a; Weishampel 1981a, 1981b). Since there is good morphometric evidence for sexual dimorphism in the crests of two lambeosaurine species (Dodson 1975; also suspected in other hadrosaurids), these cranial structures are thought to have been used for species recognition, intraspecific combat, ritualized display, courtship, parent-offspring communication, and social ranking (Hopson 1975a; Molnar 1977). In doing so, they would have promoted successful matings within species by acting as premating genetic isolating mechanisms (Hopson 1975a) and enhanced offspring survival.

The accentuated nasal arch is thought to have been used as a weapon for broadside (in *Gryposaurus notabilis, Aralosaurus tuberiferus*) or head-pushing (in *Brachylophosaurus canadensis*) intraspecific combat. Fleshy lateral diverticula probably lay adjacent to the external nares; these were used for both visual and vocal communication. The enlarged solid crest of *Prosaurolophus maximus* and species of *Saurolophus* is thought to have supported elongate and inflatable paired diverticula. The structural modifications to produce a solid crest converted the weapon function of the nasal arch in *G. notabilis* and *B. canadensis* to that of a dominance rank symbol in *P. maximus, S. osborni,* and *S. angustirostris*. Although the nasal arch is not accentuated in *Anatotitan copei* and species of *Edmontosaurus,* their prominent circumnarial excavations are thought also to have housed inflatable diverticula. These diverticula are thought to have assumed a vocal function in these animals. Last, the hollow supracranial crests in lambeosaurines acted as vocal resonators as well as species-specific visual display organs (Wiman 1931; Hopson 1975a; Weishampel 1981a, 1981b).

MARGINOCEPHALIA

P. DODSON

The Marginocephalia (Sereno 1986) consists of the Ceratopsia plus the Pachycephalosauria, two groups of primarily Late Cretaceous herbivorous ornithischians that seemingly have little in common. Each represents a clade whose monophyletic status is highly corroborated, and each diverges somewhat from the ornithopod *Bauplan* but without the strikingly divergent features of the Ankylosauria and the Stegosauria. Both pachycephalosaurs and ceratopsians lack an obturator process on the ischium. The origins of each has long been sought among ornithopods (always without success), and it now appears that pachycephalosaurs and ceratopsians together represent the sister group to the Ornithopoda. A case has been made for uniting the two (Sereno 1984, 1986; Maryańska and Osmólska 1985; Sues and Galton 1987) as the Marginocephalia. Derived characters supporting the Marginocephalia involve especially the formation of a parietosquamosal shelf or incipient frill that overhangs the occiput. Sereno (1986) also cited a reduced pubis lacking a symphysis and the maxilla excluding the premaxilla from the internal naris, but the latter is plesiomorphic for the group. To this list, Maryańska and Osmólska (1985) added the greater separation of the acetabula than the dorsal borders of the ilia, a character that has the effect of spreading the hips.

As Sereno (1986) recognized, the monophyletic status of the Marginocephalia is a hypothesis that is by no means secure, although it seems preferable to any competing hypothesis. Most ceratopsians and pachycephalosaurs are of Late Cretaceous age and are highly derived. The earliest pachycephalosaurian, *Yaverlandia bitholus* (Galton 1971c) is of Early Cretaceous (Barremian) age, but its fragmentary nature permits no insight into possible outgroup relationships. Phyletic divergence between the Pachycephalosauria and primitive Ceratopsia is thus a matter of inference as basal marginocephalians are unknown.

One possible basal marginocephalian is *Stenopelix valdensis* from the Barremian of the Federal Republic of Germany. It consists of a single partial skeleton, unfortunately lacking a skull. Maryańska and Osmólska (1974) tentatively referred *Stenopelix* to the Pachycephalosauria on the basis of inferred exclusion of the pubis from the acetabulum and presence of strong caudal ribs. Sues and Galton (1982) assigned

S. valdensis to the Ceratopsia on the basis of the form of the ilium, reduced pubis, and decurved ischium. The ceratopsian status of *S. valdensis* is rejected by Sereno (this vol.); Maryańska (this vol.) similarly rejects it from the Pachycephalosauria. The uncertain status of *Stenopelix* seems best expressed by recognizing it as a basal marginocephalian, the sister group to the Pachycephalosauria and the Ceratopsia.

The biostratigraphic implication of the split of the Marginocephalia from the Ornithopoda is that basal marginocephalians must have existed since the earliest Jurassic, but their distinguishing characters must have been little evident. Their small size and generally nondescript character would have made them difficult to recognize prior to their definitive split into the Pachycephalosauria and Ceratopsia early in the Cretaceous.

Stenopelix

	Occurrence	Age	Material
Marginocephalia Sereno, 1986			
Stenopelix Meyer, 1857			
S. valdensis Meyer, 1857	Obernkirchen Sandstein, Niedersachsen, Federal Republic of Germany	Berriasian	Partial skeleton, no skull

27

Pachycephalosauria

TERESA MARYAŃSKA

INTRODUCTION

The Pachycephalosauria, often called dome-headed or thick-headed dinosaurs, is a group of bipedal ornithischians with thickened bones of the skull roof. Pachycephalosaurians retain a few primitive ornithischian features, including a short premaxilla, premaxillary dentition, more or less pronounced heterodonty, small leaflike cheek teeth, mandible with a low coronoid process and moderately developed retroarticular process, and ischium without an obturator process. The group is restricted to the Late Cretaceous, with the exception of an Early Cretaceous species, *Yaverlandia bitholus.* The Pachycephalosauria includes two families and fourteen genera. All but one of these genera are monospecific. The number of specimens per species varies from one in most species to several dozen in *Stegoceras validum* (table 27.1).

Although remains of the Pachycephalosauria are scarce, their geographic distribution is quite wide. The majority of Late Cretaceous species occur in the Northern Hemisphere (western North America and central Asa), but a single species, *Majungatholus atopus,* is reported from Gondwanaland (i.e. Madagascar).

The ancestry of the Pachycephalosauria is still an open question. As will be shown, several derived cranial and pelvic characters link the Pachycephalosauria with the Ceratopsia.

DIAGNOSIS

Members of the Pachycephalosauria are characterized by the following derived features: skull roof thickened, table-like or domed; jugal and quadratojugal strongly extended ventrally toward the articular surface of the quadrate; the jugal meets the quadrate ventral to the quadratojugal; rostral and medial walls of the orbit ossified; orbital roof with two supraorbitals, and contact between the supraorbital 1 and nasal excludes prefrontal-lacrimal contact; basicranial region strongly shortened sagittally, basal tubera thin and platelike, contact between prootic-basisphenoid plate and the quadrate ramus of the pterygoid; broad expansion of the squamosal on the occiput; external surfaces of skull bones strongly ornamented, prominent osteoderms on the rim of the squamosal; dorsal and caudal vertebrae with ridge-and-groove articulation between zygapophyses; long ribs on sacral and proximal caudal vertebrae; forelimb approximately 25 percent of the length of hindlimb; slightly twisted and bowed humerus with rudimentary deltopectoral crest; ilium with cranially broad and horizontal preacetabular process and a medial flange on the postacetabular process; pubis reduced and excluded from the acetabulum; ischium with a long dorsoventrally flattened cranial peduncle contacting the pubic peduncle of the ilium and sacral ribs, instead of the pubis; multiple rows of fusiform ossified tendons around the middle portion of the tail.

ANATOMY

The majority of pachycephalosaurians are known from isolated fragmentary skull caps. Exceptions include the Late Cretaceous *Stegoceras validum* and *Prenocephale prenes*, in both of which complete, single skulls have been found, and *Pachycephalosaurus wyomingensis, Homalocephale calathocercos, Tylocephale gilmorei* and *Goyocephale lattimorei,* all of which are known from single, largely complete skulls.

A complete postcranial skeleton has yet to be found. The most complete pertain to *H. calathocercos, G. lattimorei* and *S. validum.* Some postcranial fragments have also been found in several other Asian species. Incompleteness of postcranials is responsible for lack of any information concerning the number of cervicals (and thus neck length) and the structure of the manus.

The description of pachycephalosaurian cranial structure that follows is based largely on the skulls of *P. prenes, S. validum,* and *H. calathocercos,* while that of the postcranial skeleton is based mainly on *H. calathocercos* and *G. lattimorei.*

Skull and Mandible

The basic structure of the pachycephalosaurian skull is very peculiar, although generally based on a very uniform *Bauplan* (figs. 27.1, 27.2, 27.3). The skull is akinetic, generally comparatively short and high, with a well-ossified, thickened roof. The basicranium is short, while the occiput is very high with a broad expansion of the squamosal. Virtually all articulations are very firm; some of them are fused. Extreme thickening of the skull roof results in a domelike structure and correlates with the closure of the supratemporal fenestrae.

The premaxilla is short with a moderately developed lateral process wedged between the nasal and the maxilla. This condition, clearly visible in *Prenocephale* and *Goyocephale,* is also visible in *Stegoceras validum* (Sereno, pers. comm.). This process never reaches the lacrimal.

The maxilla is high and nearly vertical. The thick palatal process of the maxilla contacts its fellow medially. An extensive intramaxillary sinus penetrates the maxilla along nearly its entire length. A relatively small external antorbital fenestra is present at the suture of the maxilla and the lacrimal only in *P. prenes.*

The lacrimal varies in size; it is largest in *P. prenes* and smallest in *T. gilmorei.* In the vicinity of the antorbital fenestra, the lacrimal has a separate medial lamina that bounds the dorsomedial aspect of the intramaxillary sinus. Generally, the lacrimal does not reach rostrally as far as the nasal. There is no contact between the lacrimal and the prefrontal.

The jugal is extensive dorsoventrally, reaching close to the quadrate-mandibular joint ventrally, and it can form most of the caudal orbital rim. In all pachycephalosaurians, the jugal is strongly ornamented, especially on its caudoventral aspect. The quadratojugal is a teardrop-shaped bone. Its ventral margin is surrounded by small portions of the jugal and quadrate. The dorsal part of the quadrate shaft is always more or less deflected backward. The pterygoid ramus of the quadrate is broad.

The expanded squamosal overhangs the occiput. Here, it contacts the parietal medially along an extensive serrated suture. The squamosal is incorporated into the dome structure in *Prenocephale* and *Pachycephalosaurus,* while in *Stegoceras,* the form of squamosals varies in connection wtih the development of the dome. Together with the parietal, it forms a more or less prominent parietosquamosal shelf along the caudal skull margin. In *Homalocephale* and *Goyocephale,* which have comparatively large supratemporal fenestrae, the parietosquamosal contact is short. The squamosal bears a row of prominent osteoderms on its external margins.

The postorbital is very large and incorporated into the dome in *Prenocephale* and *Tylocephale.* Its share in the caudal orbital margin varies widely, being extensive in *P. prenes* and *H. calathocercos* and small in *S. validum.* In the last-named species and in *Pachycephalosaurus wyomingensis,* there is a continuation of the squamosal shelf along the postorbital. The bone is strongly ornamented.

Two distinctly separated ossifications, supraorbitals 1 and 2, roof the orbital cavity and form its heavy upper rim. The supraorbitals are present in both domed and flat-headed representatives of the group.

The shape of the nasal varies in the Pachycephalosauria. The transverse and sagittal profiles across the rostral part of the nasals are concave dorsally. The nasal is weakly convex at the caudal limit near its rather short contact with the frontal in *P. prenes.* A similar condition is found in *Pachycephalosaurus wyomingensis.* In *S. validum,* contact between the nasal and frontal is extensive, and the sagittal and transverse profiles through both nasals are convex.

The pterygoid has two rami. The palatal ramus has a nearly horizontal and broad caudal portion and a narrow, vertical rostral region. In most species, the palatal ramus meets its fellow medially. Dorsally, its ver-

TABLE 27.1 Pachycephalosauria

	Occurrence	Age	Material
Pachycephalosauria Maryańska et Osmólska, 1974			
Homalocephalidae Dong, 1978			
Goyocephale Perle, Maryańska, et Osmólska, 1982			
G. lattimorei Perle, Maryańska, et Osmólska, 1982	unnamed unit, Ovorkhangai, Mongolian People's Republic	?late Santonian or early Campanian	Fragmentary skull, nearly complete postcranial skeleton
Homalocephale Maryańska et Osmólska, 1974			
H. calathocercos Maryańska et Osmólska, 1974	Nemegt Formation, Omnogov, Mongolian People's Republic	?late Campanian or early Maastrichtian	Complete skull with associated, nearly complete postcranial skeleton
Wannanosaurus Hou, 1977			
W. yansiensis Hou, 1977	Xiaoyan Formation, Anhui, People's Republic of China	Campanian	Partial skull roof, mandible, fragments of postcranium
Pachycephalosauridae Sternberg, 1945			
Gravitholus Wall et Galton, 1979			
G. albertae Wall et Galton, 1979	Judith River Formation, Alberta, Canada	late Campanian	Frontoparietal dome
Majungatholus Sues et Taquet, 1979			
M. atopus Sues et Taquet, 1979	Grès de Maevarano, Majunga, Madagascar	Campanian	Partial skull
Ornatotholus Galton et Sues, 1983			
O. browni (Wall et Galton, 1979) (= *Stegoceras browni* Wall et Galton, 1979)	Judith River Formation, Alberta, Canada	late Campanian	Frontoparietal dome
Pachycephalosaurus Brown et Schlaikjer, 1943 (= *Tylosteus* Leidy, 1872)			
P. wyomingensis (Gilmore, 1931) (= *Troodon wyomingensis* Gilmore, 1931, including *Tylosteus ornatus* Leidy, 1872, *P. grangeri* Brown et Schlaikjer, 1943, *P. reinheimeri* Brown et Schlaikjer, 1943)	Lance Formation, Wyoming, Hell Creek Formation, South Dakota, Judith River Formation, Hell Creek Formation, Montana, United States	late Maastrichtian	Nearly complete skull, at least 3 skull fragments
Prenocephale Maryańska et Osmólska, 1974			
P. prenes Maryańska et Osmólska, 1974	Nemegt Formation, Omnogov, Mongolian People's Republic	?late Campanian or early Maastrichtian	Complete skull with associated partial postcranium
Stegoceras Lambe, 1902 (= *Troodon* Leidy, 1856, *sensu* Gilmore, 1924d *partim*)			
S. validum Lambe, 1918b (including *S. breve* Lambe, 1918b, *S. lambei* Sternberg, 1945, *Troodon sternbergi* Brown et Schlaikjer, 1943)	Judith River Formation, Alberta, Canada; Judith River Formation, Montana, United States	late Campanian	Complete skull with associated partial postcranium, several dozen frontoparietal domes
S. edmontonense (Brown et Schlaikjer, 1943) (= *Troodon edmontonensis* Brown et Schlaikjer, 1943)	Horseshoe Canyon Formation, Alberta, Canada; Hell Creek Formation, Montana, United States	Maastrichtian	2 skull fragments
Stygimoloch Galton et Sues, 1983 (= *Stenotholus* Giffin, Gabriel, et Johnson, 1987)			
S. spinifer Galton et Sues, 1983 (including *Stenotholus kohleri* Giffin, Gabriel, et Johnson, 1987)	Hell Creek Formation, Montana, Lance Formation, Wyoming, United States	late Maastrichtian	5 skull fragments
Tylocephale Maryańska et Osmólska, 1974			
T. gilmorei Maryańska et Osmólska, 1974	Barun Goyot Formation, Omnogov, Mongolian People's Republic	middle Campanian	Incomplete skull
Yaverlandia Galton, 1971c			
Y. bitholus Galton, 1971c	Wealden Marls, Isle of Wight, England	Barremian	Skull fragment
Unnamed pachycephalosaurid (= *"Troodon" bexelli* Bohlin, 1953)	Tsondolein Khuduk beds, Nei Mongol Zizhiqu, People's Republic of China	Campanian-Maastrichtian	Partial parietal
Pachycephalosauria incertae sedis			
Micropachycephalosaurus Dong, 1978			
M. hongtuyanensis Dong, 1978	Wang Formation, Shandong, People's Republic of China	Campanian	Partial mandible, associated postcranial fragments

Nomina dubia	Material
Heishansaurus pachycephalus Bohlin, 1953	Badly preserved cranial and postcranial fragments
Stegosaurus madagascariensis Pivetaut, 1926	Teeth

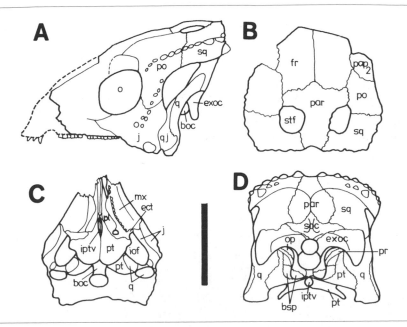

Fig. 27.1. A–D, *Homalocephale calathocercos:* skull in lateral (A), dorsal (B), palatal (C), and occipital (D) views (after Maryańska and Osmólska 1974). Scale = 9 cm.

tical portion underlies the caudal part of the palatine. The palatal ramus rises most steeply toward the midline in *P. prenes,* less steeply in *S. validum,* and only slightly in *H. calathocercos.* The quadrate ramus of the pterygoid broadly overlaps the basisphenoid rostrally to contact the prootic. The extensive caudal portion of this ramus is vertical and overlaps the pterygoid ramus of quadrate.

In *H. calathocercos* and *P. prenes,* there is a small bone directed rostrocaudally from the prootic at the contact of the latter bone with the laterosphenoid toward the quadrate wing of pterygoid. This bone was considered the epipterygoid by Maryańska and Osmólska (1984*a*). An ossified epipterygoid is absent in *S. validum* (Wall and Galton 1979).

The small ectopterygoid has a stout jugal process. The rostral part of the palatine tapers abruptly and is steeply inclined dorsomedially where it meets its fellow. Rostrally, both palatines continue above the vomer but do not meet each other. The long rostral portion of the vomer is wedged between the maxillae and does not reach the premaxillae. The secondary palate is very short.

The braincase is solidly constructed, and many of its elements are thick. The union of the neurocranium with the splanchnocranium is very strong. The small supraoccipital forms a small portion of the dorsal mar-

gin of the foramen magnum. The paroccipital process is formed of the exoccipital. In all the Pachycephalosauria, the exoccipital closely adheres along its dorsal margin to the squamosal and quadrate, except in *P. prenes,* where the exoccipital is separated from the quadrate and the squamosal at its lateral extreme.

The basioccipital forms most of the occipital condyle. The articular surface of the condyle faces ventrally. The condylar neck is very short. The basal tuber is flat and platelike and bounds a deep slit-like cavity, the rostral wall of which is formed by the prootic-basisphenoid plate. The prootic and basisphenoid are very strongly modified from the typical ornithischian pattern. Together, they produce a flat, transverse plate, formed dorsally by the prootic and ventrally by the basisphenoid. Laterally, the plate underlies the caudal surface of the quadrate ramus of the pterygoid. In this way, the basicranial region is separated from the palatal and suborbital regions. The part of the basisphenoid that forms the ventral portion of the basal tuber does not extend ventrally, yet nearly reaches the pterygoid in *H. calathocercos.* It is dorsoventrally shortest in *S. validum,* in which a comparatively wide space separates the basisphenoid from the pterygoid.

The parasphenoid is well ossified and distinct from the basisphenoid, overlapping the latter ventrally. The parasphenoidal rostrum is well developed. The lat-

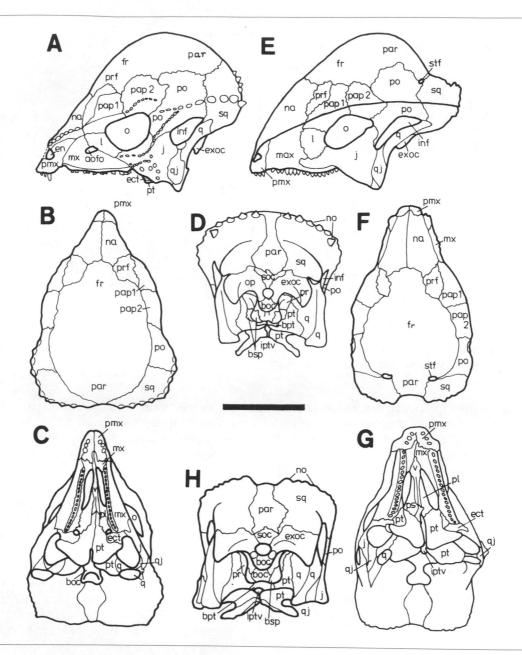

Fig. 27.2. A–D, *Prenocephale prenes;* E–H, *Stegoceras validum.* Skulls in lateral (A, E), dorsal (B, F), palatal (C, G), and occipital (D, H) views (after Maryańska and Osmólska 1974). Scale = 9 cm.

erosphenoid forms the dorsolateral wall of the braincase and is also seen as a small portion of the medial wall of the orbit. Contact between the presphenoid and parasphenoid on the lateral wall of the braincase seems to be absent in the Pachycephalosauria. Additional ossifications separate the two bones, and the peculiar compression of the basicranial region described above also prevents contact of the presphenoid and parasphenoid.

Endocranial morphology has been described only in *S. validum, Pachycephalosaurus wyomingensis,* and *Y. bitholus* (Hopson 1979; Wall and Galton 1979). The cerebral region of *Y. bitholus* differs from that of *S. validum* and *Pachycephalosaurus wyomingensis* in being shorter and broader with the hemispheres separated by a median sulcus. The olfactory peduncles are large and the olfactory lobes broad and divergent in *S. validum* and *Pachycephalosaurus wyomingensis.* In *Y. bitholus,* the

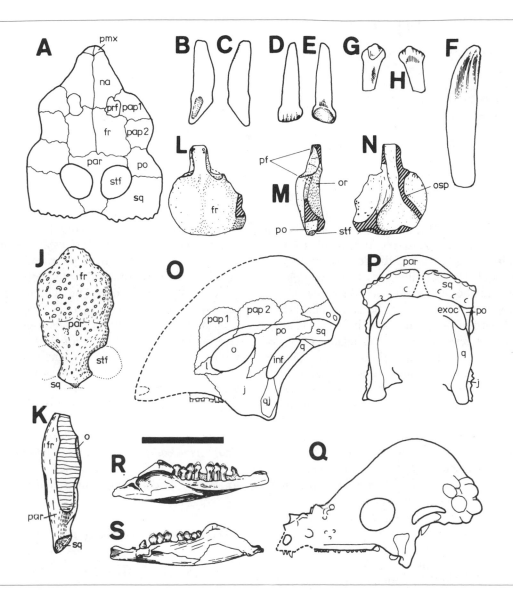

Fig. 27.3. A—H, *Goyocephale lattimorei:* skull in dorsal view (A); second left premaxillary tooth (B, C), a maxillary tooth (D, F), first mandibular tooth (F), distal mandibular tooth (G, H), in lingual (B, E, F, G) and labial (C, D, H) views, wear facets outlined. J—K, *Ornatotholus browni:* fragment of skull in dorsal (J) and lateral (K) views. L—N, *Yaverlandia bitholus:* fragment of skull in dorsal (L), lateral (M), ventral (N) views. O—P, *Tylocephale gilmorei:* fragment of skull in lateral (O) and occipital (P) views. Q, *Pachycephalosaurus wyomingensis:* skull in lateral view. R—S, *Wannanosaurus yansiensis:* left mandible in medial (R) and lateral (S) views. (J—N after Galton 1971c; O, P from Maryańska and Osmólska 1974; R, S after Hou 1977.) Scale = 9 cm (A, O, P), 1.5 cm (B—H), 4 cm (J, K), 5 cm (L—N), 24 cm (Q), 3 cm (R, S).

olfactory peduncles are slender and the olfactory lobes small and parallel.

The medial and rostral walls of the orbital cavity are completely ossified. Numerous additional small, well-sutured ossifications result in the complete bony separation of the nasal and orbital cavities.

The frontals and parietals are thick and occur in two different structural patterns. One, in which the uniformly thickened and flat frontals and parietals are suturally separated (including a clearly visible interfrontal suture), is exemplified by *H. calathocercos* and *G. lattimorei.* In these forms, the fused parietals make up the medial borders of the supratemporal fenestrae to produce an interfenestral bridge. The other pattern, with a more or less elevated and fused frontoparietal mass that forms a dome, is found in *P. prenes, S. validum,* and *Pachycephalosaurus wyomingensis.* In these forms, the supratemporal fenestrae are strongly reduced or absent. Each frontal is slightly domed in *Y. bitholus.*

Complete or fragmentary mandibles are known only in *S. validum, G. lattimorei, Wannanosaurus yansiensis, T. gilmorei,* and *Micropachycephalosaurus hongtuyanensis.* The mandible is generally shallow and slightly sigmoid in ventral view, with a moderately long retroarticular process and shallow coronoid eminence. The mandibular rami are not fused at the symphysis. The symphyseal surface is small. The predentary has never been found, but a very small grooved articular surface for its reception is present on the rostrolateral surface of the dentary. The dentary forms about two-thirds of the length of the mandible. The subtriangular splenial lies almost entirely on the lingual side of the mandible. The coronoid eminence is formed of the dentary, surangular, and coronoid. No mandible preserves a complete coronoid, but this bone seems to be rather small and does not project dorsally above the surangular. The postcoronoid portion of the mandible is short, the angular being the largest element in this region. The sutures between the prearticular and articular are obliterated. The prearticular rostrally extends at least to the middle of the mandibular fossa. Ornamentation covering the buccal and ventral surface of the mandible is most pronounced on the angular.

The pachycephalosaurian dentition is heterodont. All teeth are enameled on both sides. Three caniniform premaxillary teeth are present in *P. prenes* and *G. lattimorei,* the third being the largest. They are subconical, slightly recurved with denticulate distal edges, and oval in cross section. The crowns are distinct from the roots. They are worn along their lingual or linguodistal faces. In these species, there is a deep diastema between the premaxillary and maxillary series of teeth for the reception of the first dentary tooth. The premaxillary teeth in *S. validum* are pointed and slightly recurved distally with vertical wear facets along the lingual aspect of the crowns (Sues and Galton 1987). A small diastema is present between premaxillary and maxillary teeth in this species.

The maxillary teeth are small, being relatively largest in *T. gilmorei* and smallest in *P. wyomingensis.* The crowns are more or less triangular and transversely compressed. Generally, the lingual side of the crown is weakly convex vertically, while the buccal side is concave. The crowns have weak ridges that culminate in marginal denticles. The central denticle is the strongest. The teeth are arranged in a single row, and the distal margin of each tooth slightly overlaps buccally the mesial margin of the successive tooth.

Wear on maxillary teeth is variable. In *H. calatho-cercos,* extensive wear is found along the lingual surface of all maxillary teeth. These surfaces tend to form a confluent occlusal plane over the whole maxillary series. A similar pattern is observed in *G. lattimorei* and *T. gilmorei.* In *P. prenes,* there is a narrow wear surface along the ventral margin of each tooth. In *S. validum,* the oblique lingual wear facets are developed on the mesiolingual aspects of the crown (Sues and Galton 1987).

Mandibular teeth are only known in some species. Those of *S. validum* are generally similar to the teeth in the upper jaw. The mesial mandibular teeth are larger than distal. Mandibular heterodonty is more pronounced in *G. lattimorei,* in which the first tooth is much larger than successive ones and is caniniform. It is also much larger than all premaxillary teeth, polygonal in cross section, and bears a sharp, denticulate ridge on the distal edge. The remaining, closely arranged mandibular teeth of *G. lattimorei* have ridged, subtriangular crowns with pointed tips. The wear surface, almost parallel to the longitudinal axis of each tooth, is on the labial side of each crown. The mandibular teeth of *W. yansiensis* are similar to those in *G. lattimorei.*

The number of the teeth in pachycephalosaurian species is poorly known. The maxillae bear 16 teeth in *S. validum* and *G. lattimorei,* 17 in *P. prenes,* and 20 in *Pachycephalosaurus wyomingensis.* The number of dentary teeth varies from approximately 10 in *W. yansiensis* to 17 in *S. validum* and 18 in *G. lattimorei.*

Postcranial Skeleton

Axial Skeleton

The preserved part of the vertebral column of *H. calathocercos* (including 10 caudal dorsals, 6 sacrals, and 29 caudals) is the most complete known in the Pachycephalosauria, which allows the reconstruction of the natural curvature of the vertebral series in this species. The column rises slightly backward along the first three preserved dorsals, beyond which it is horizontal to the end of the sacrum. Behind the sacrum, the tail slopes rather steeply downward, beginning with the first caudal up to the thirteenth. More distally, the natural orientation of the tail is nearly horizontal.

The dorsal vertebrae are amphiplatyan, increasing in size caudally. The neural arches are about twice as high as the centra in the cranial dorsal vertebrae and about one and a half times as high in the caudal dorsals. In lateral view, the neural spines are subrectangular and robust, although relatively low and slightly inclined backward. The transverse processes are steeper caudally than cranially, except for the last dorsal, in which they are nearly horizontal. The zygapophyses of the dorsals are short and broad, and show a tongue-and-groove articulation with one another. The dorsal ribs are rod-like, slightly curved, and double-headed. Distance between the capitulum and tuberculum decreases caudally. A few last ribs are more slender and curved.

The sacrum in *H. calathocercos* was found in natural articulation with the pelvic girdle (fig. 27.4). It consists of six strongly-coossified, long, low vertebrae. There is a low keel along the ventral side of sacrals 2 through 4. The structure of the sacral ribs and their contacts with the pelvis are somewhat peculiar. Sacral 1 bears a normal transverse process fused with a broad and short sacral rib. The dorsoventrally-flattened end of the rib attaches loosely to the medial edge of the ilium. Sacral ribs 2 through 4 are stout and flattened craniocaudally. The rib of sacral 2 is directed caudally, while that of sacral 3 is angled cranially. Both ribs contact the cranial process of the ischium near the contact of the latter with the pubic peduncle of the ilium. The rib of sacral 4 is the most robust and forms a vertical plate. In its upper part, this plate contacts the ilium above and behind the acetabulum. The lower part of the rib attaches to the ischium just below the contact of the latter with the ischial peduncle of the ilium. The rib of sacral 5 is flattened craniocaudally, directed cranially, and meets the ilium at the articular facet for the fourth sacral rib. Sacral rib 6 is dorsoventrally flattened like that of the succeeding caudal rib, and it contacts the medial flange of the ilium. Sutures are visible between the sacral ribs and the transverse processes of sacrals 1, 4, 5, and 6. The sacrum and sacral ribs are similar in *P. prenes*. The sacrum of *G. lattimorei* is composed only of four weakly fused vertebrae. Distal contacts of sacral ribs 1 and 2 are not well known. That of sacral rib 3 is known to contact at least the upper border of the ilium as there is a facet for such a rib. In contrast to *H. calathocercos*, sacral rib 3 is caudally directed in *G. lattimorei*. Sacral rib 4 extends much less ventrally than in *H. calathocercos*, and it contacts only the ilium.

The caudal vertebrae in *H. calathocercos* and *G. lattimorei* are amphiplatyan and spool-like in lateral view. Caudals from the middle portion of the tail (starting with caudal 5) are shorter than either proximal and distal caudals and have deeply concave ventral profiles. The prezygapophyses of proximal caudals have concave articular surfaces that are directed dorsomedially. These become flatter and face more medially in more

distal caudals. Eight proximal caudals bear caudal ribs. A distinct suture is visible where each rib contacts the vertebra. The first four ribs are placed on the boundary between the centrum and neural arch. More distally, their position becomes more ventral. The ribs on the most proximal vertebrae are long, more than the length of three caudal centra. The neural spines of all preserved caudals are long and slightly inclined backward. The most proximal articular site of the chevrons, which are poorly known in pachycephalosaurians, is on caudal 5. Beginning with caudal 12, a basketwork of tendons surrounds the tail in *H. calathocercos*. These tendons were formed in both epaxial and hypaxial musculature. It is not known whether they extended to the very end of the tail. The tendons are fusiform and of different length and sizes. Similar ossified tendons were recovered together with skeletons of *P. prenes*, *G. lattimorei*, and *S. validum*. In the latter, the pieces of the tendons were described by Gilmore (1924d) as abdominal ribs.

Appendicular Skeleton

The shoulder girdle is known in *S. validum*. The scapula is relatively long and slender with a greatly expanded ventral end. The coracoid is imperfectly known but seems to be relatively large. The coracoid foramen is situated close to the coracoid-scapula contact.

Paired sternals are known in *H. calathocercos* and *G. lattimorei*. According to Sereno (pers. comm.), they are also present in *S. validum*. The sternals are elongated, transversely narrow with a slightly thickened distal articular end, and a medially expanded thin, proximal end.

The combined length of the humerus and radius of *S. validum*, *G. lattimorei*, and *W. yansiensis* is slightly greater than one-fourth the hindlimb (femur + tibia) length. The humerus is bowed with a twisted shaft (approximately 20°; fig. 27.4). The humeral head is weakly developed. The deltopectoral crest is low, and the distal condyles are only slightly separated. The ulna is about half the length of the humerus and has a very low olecranon process. The radius is stouter than the ulna. The structure of the manus is not known in the Pachycephalosauria.

The complete articulated pelvis is known only in *H. calathocercos* (fig. 27.4); some pelvic fragments are also known in *S. validum*, *P. prenes*, *G. lattimorei*, *W. yansiensis* and *M. hongtuyanensis*. The most characteristic features of the pelvis are its great transverse width and the almost complete exclusion of the pubis from the acetabulum. The last character is due to the addi-

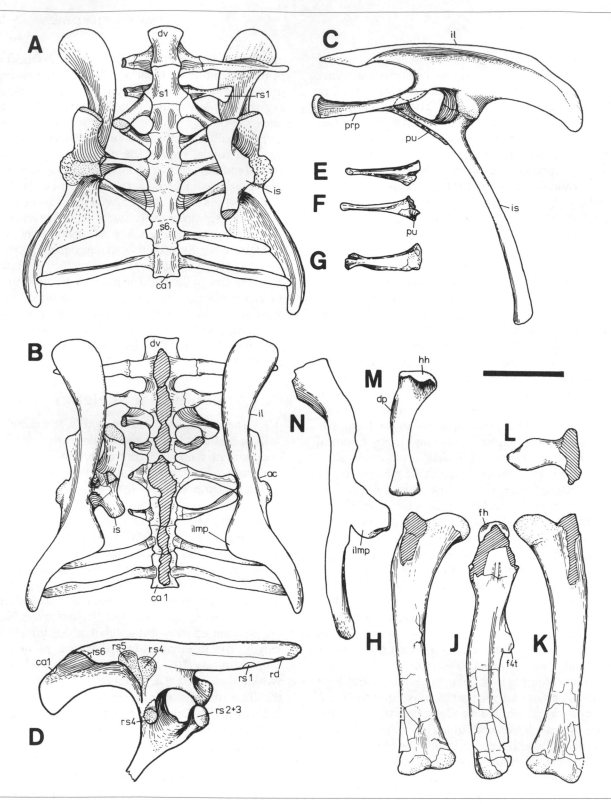

Fig. 27.4. A—L, *Homalocephale calathocercos:* sacropelvic region with the last dorsal and first caudal attached in ventral (A) and dorsal (B) views; reconstruction of pelvis (C) in lateral view; left ilium and ischium naturally articulated with facets for last dorsal, sacral, and first caudal ribs (D) in medial view; left prepubic process in dorsal (E), ventral (F), and medial (G) views; left femur in caudal (H), lateral (J), and cranial (K) views, articular head (L) in dorsal view. M—N, *Goyocephale lattimorei:* left humerus (M) in caudal view; left ilium (N) in dorsal view. (A—L after Maryańska and Osmólska 1974.) Scale = 6 cm.

tional, cranial contact of the ischium and ilium by means of the extended cranial (pubic) process of the ischium. This process closes the acetabulum ventrally. Additionally, the ventral (ischial) rim of the acetabulum is shifted medially with respect to the dorsal border of the ilium, which leaves the ventral half of the acetabulum unsupported.

The dorsoventrally-flattened preacetabular process of the ilium is broad and horizontal. The postacetabular process is rather high, vertical, and bears a horizontally expanded medial flange immediately behind the acetabulum. The shape of the postacetabular process varies in pachycephalosaurians in both height and curvature. In no case is there an antitrochanter. The pubic peduncle of the ilium is comparatively long and widened transversely. The small, medialmost corner of its articular surface contacts the pubic process of the ischium. The outer portion of this surface articulates with the pubis. The ischial peduncle of the ilium is also transversely broad, robust, and has a rough lateral protuberance. Consequently, the dorsal half of the acetabulum is wide, while its ventral surface faces somewhat laterally. On the medial border of the ilium, cranially to the medial flange, there are facets for the fourth and fifth sacral ribs.

An incomplete pubis is preserved in *H. calathocercos*. The prepubic process is flattened dorsoventrally near the acetabulum and laterally flattened distally. The dorsal margin of the prepubic process is raised cranially. Judging from the preserved portion, the pubis was very thin and short.

The ischium has a long and slender shaft that is weakly bowed and slightly curved inwardly toward its fellow. Immediately below the contact between the ischium and the pubic peduncle of the ilium, the long cranial process of the ischium bears a common articular facet for the distal ends of sacral ribs 2 and 3. Caudally, the ischium displays a flat articular facet for the lower portion of sacral rib 4.

The femur is preserved in *H. calathocercos, P. prenes, M. hongtuyanensis, W. yansiensis,* and *S. validum,* and in all cases the bone is recurved medially (fig. 27.4). The femoral head is comparatively long and compressed craniocaudally. The articular surface of the head faces dorsally and slightly cranially. A deep groove separates the head from the greater trochanter, but the entire trochanteric area is not preserved in any specimen. The fourth trochanter is moderately developed and slightly pendent. It lies entirely within the proximal half of the bone. In *M. hongtuyanensis,* the fourth trochanter is more proximal than in other species. Shallow extensor and flexor grooves separate the distal condyles. The lateral condyle is larger than the medial.

The tibia slightly exceeds the femur in length and has a slender shaft. Its lateral distal condyle is more prominent than the medial condyle. The fibula, well preserved in *S. validum,* is very slender. Of the proximal tarsals, only the astragalus is known (in *H. calathocercos*). Its upper margin rises slightly on the central aspect of the cranial face of the tibia. Two distal tarsals are preserved in *G. lattimorei.* Distal tarsal 1 is small, semilunar, and covers metatarsal II and a small fraction of metatarsal III. Distal tarsal 2 is subquadrate and caps metatarsal III, half of metatarsal IV, and medially a part of distal tarsal 1.

The metatarsals and pedal phalanges are poorly preserved in the Pachycephalosauria, but, as may be judged from the preserved parts in *S. validum, H. calathocercos,* and *G. lattimorei,* there are four metatarsals in these dinosaurs. The foot most probably had three functional digits. The first digit appears to be strongly reduced and the fifth apparently wanting. The pedal unguals are tapering but not distinctly recurved, with flat plantar sides. The ungual of the third toe is the most robust.

EVOLUTION AND SYSTEMATICS

Since the first skull of dome-headed dinosaur *Stegoceras validum* was described (Lambe 1902), the systematic position of pachycephalosaurians within the Ornithischia has often been discussed. They have been assigned to the Stegosauria (Lambe 1918*b*), Ceratopsia (Nopcsa 1904), Ankylosauria (Romer 1927; Nopcsa 1928*a,* 1929*a*), and mostly to the Ornithopoda (e.g., Gilmore 1924*d;* Brown and Schlaikjer 1943; Galton 1971*c*). In 1945, Sternberg formally erected the Pachycephalosauridae for all dome-headed dinosaurs as a family within the Ornithopoda. This systematic position was generally accepted (based primarily on their putative bipedality) although not unanimously. For instance, Rozhdestvensky (1964*a*) considered the Pachycephalosauridae as Ornithischia *incertae subordinis* and in 1973 assigned them to the Ankylosauria. The Pachycephalosauria was first assigned subordinal rank within the Ornithischia by Maryańska and Osmólska (1974) and considered descendants of the Ornithopoda. Lately, Sereno (1984, 1986) and Maryańska and Osmólska (1984*a,* 1985) suggested that the Pachycephalosauria is a sister group of the Ceratopsia, as both groups share in common a suite of derived characters (e.g., expansion of the parietals and squamosals overhanging the occiput, the vomer contacting the maxilla within the secondary palate as a result of exclusion of the premaxilla from the margin of the internal nares by the maxilla, strongly reduced pubis, pubic symphysis absent, iliac peduncles directed ventrolaterally, causing the acetabular portions of the ilia to be more widely separated than the dorsal portions of the ilia). Some resemblance of the postcranial skeleton of *S. validum* to that of the Ceratopsia was mentioned also by Brown and Schlaikjer (1943). This hypothesis was tentatively accepted by Sues and Galton (1987).

It is possible that divergence of the Pachycephalosauria and Ceratopsia took place in the Late Jurassic, as the most ancient known representative of the Pachycephalosauria occurs in the Barremian of Europe (Galton 1971*c*).

The first undoubted flat-headed pachycephalosaurian, *Homalocephale calathocercos*, was described in 1974 by Maryańska and Osmólska. In 1978, Dong erected the family Homalocephalidae for flat-headed pachycephalosaurians.

The presently known pachycephalosaurian species are divided by some authors (Dong 1978; Perle et al. 1982; Sues and Galton 1987) into two monophyletic families: the Homalocephalidae Dong, 1978, emend. Perle et al. 1982, so far exclusively Asian, and the Pachycephalosauridae Sternberg, 1945, with a much broader geographic distribution. The Homalocephalidae is distinguished by the completely flat, tablelike skull roof, whereas the Pachycephalosauridae is characterized by more or less extensive, domelike thickening of the skull roof bones. This idea is accepted here. The opposite hypothesis, that of a paraphyletic Homalocephalidae, was expressed by Sereno (1986).

Homalocephalidae

As here understood, the Homalocephalidae includes pachycephalosaurians characterized by a derived flat, tablelike skull roof (derived character) accompanied by relatively large, primitive supratemporal fenestrae. The Homalocephalidae comprises small to medium-sized dinosaurs, their total body length ranging from approximately 60 cm (*Wannanosaurus yansiensis*) to 150 cm (*Homalocephale calathocercos* and *Goyocephale lattimorei*). The family presently contains three monospecific genera, all from the Late Cretaceous of central Asia. Each species is represented by a type specimen only.

Homalocephale calathocercos, from the Upper Cretaceous Nemegt Formation, Mongolian People's Republic, is represented by the most complete cranial and postcranial material. The species is characterized by a rounded supratemporal fenestra, a broad interfenestral bridge approximately equal in size to the transverse width of each supratemporal fenestra, a moderately concave occiput, a roughly ornamented (i.e., pitted) cranial roof, a gently convex dorsal surface of the preacetabular process of the ilium, and a craniocaudally long medial iliac flange and downwardly curved postacetabular iliac process.

In *Goyocephale lattimorei*, from the Late Cretaceous, Mongolian People's Republic, the supratemporal fenestra is longitudinally oval, the bridge between the supratemporal fenestra is narrow (about a third the transverse width of the supratemporal fenestra), the occiput is weakly concave, the medial portion of the cranial roof is indistinctly ornamented, the dorsal surface of the preacetabular process of the ilium is flat and angularly bent along medial and lateral margins, and the medial iliac flange is craniocaudally short and the postacetabular process straight and subrectangular in lateral view.

The skull of *Wannanosaurus yansiensis*, from the Upper Cretaceous Xiaoyan Formation, People's Republic of China, is fragmentary but very small. Nevertheless it probably represents an adult individual judging from the obliterated sutures on the skull roof. The supratemporal fenestra in this species seems to be relatively larger than in *H. calathocercos* and *G. lattimorei*. The occiput is only slightly concave (similar to *G. lattimorei*), but the ornamentation of the caudal part of the frontal is coarser than on the caudal part of the squamosal. The pattern of skull roof ornamentation in *W. yansiensis* is granulated rather than pitted, hence differing from both *H. calathocercos* and *G. lattimorei*. Last, the humerus is strongly bowed.

Pachycephalosauridae

The genera including at least eleven species from the Cretaceous of Europe, North America, central Asia, and Madagascar are assigned to the Pachycephalosauridae. Almost all are known only from skull elements.

The outstanding feature of all of the members of this family is the domelike thickening of the bones of the skull roof, evidently a derived character. For this reason, the shape of the dome has been the main basis for the erection of almost all pachycephalosaurid genera and species.

The stratigraphically oldest known pachycephalosaurid, *Yaverlandia bitholus*, from the Barremian of England, was a small animal (length of the frontals approximately 45 mm). It is the only pachycephalosaurid that has two small domes, one on each frontal. The dorsal surface of the domes is pitted. Hopson (1979) doubted the relationship of *Y. bitholus* with other pachycephalosaurians because of the structure of its fragmentary endocranial activity. Although these differences in the endocranial cast are significant, it is also true that *Y. bitholus* shares with other pachycephalosauridae two characters not found in any other ornithischians: the thickening and doming of the skull roof and the textured nature of its dorsal surface. The species is thus considered here as representative of the Pachycephalosauridae.

Majungatholus atopus, from the ?Campanian of Madagascar, possesses a highly elevated single frontal dome, ornamented irregularly by nodes and furrows. A median depression is present on the parietals, which themselves are not domed. The supratemporal fenestrae are large.

Ornatotholus browni, from the Campanian Judith River Formation of Canada, shows thickening of both frontals and parietals but the dome is low and divided by a shallow depression into frontal and parietal parts. The dorsal surface of the dome is rough. The supratemporal fenestrae are well-developed like those in the species mentioned above.

In *Stegoceras validum*, from the Campanian Judith River Formation of Alberta, Canada, and Montana, United States, the frontoparietal dome is moderately to highly elevated, and its rostrocaudal length exceeds its width. The supratemporal fenestrae are either small or completely closed, particularly in those individuals with highly elevated domes. The parietosquamosal shelf is well developed, and its extent is inversely proportional to the development of the dome. The surface of the frontoparietal shelf is often rough. *S. edmontonense* differs from *S. validum* in lacking the pronounced nasofrontal elevation (Sues and Galton 1987).

The highest point of the strongly elevated dome in *Tylocephale gilmorei*, from the ?Campanian Barun Goyot Formation, Mongolian People's Republic, is situated far caudally. The postorbitals and supraorbitals are also incorporated into the dome. The parietosquamosal shelf itself is very narrow, and the supratemporal fenestrae are probably absent.

Gravitholus albertae, from the Campanian Judith River Formation, Canada, is a very poorly documented species, characterized by a high and very wide dome with a large depression on the parietals. The shape of its dome may be pathological, but the size of the braincase in this species is much smaller relative to dome size than that of large individuals of *S. validum* (Chapman et al. 1981). Hence, it is reasonable to maintain the validity of this genus and species until further material becomes available.

In *Prenocephale prenes*, from the Upper Cretaceous Nemegt Formation, Mongolian People's Republic, the prefrontals, supraorbitals, postorbitals, and a portion of the squamosals are incorporated into a very high dome. The parietosquamosal shelf is not developed, and the supratemporal fenestrae are completely closed. The surface of the dome is slightly roughened, while a row of conspicuous nodes is present along the caudolateral and caudal margins of the skull.

Pachycephalosaurus wyomingensis, from the Maastrichtian Lance and Hell Creek formations of the western interior of the United States, is the largest pachycephalosaurid (skull length over 600 mm) and shows extreme thickening of the skull roof and a closure of the supratemporal fenestrae. The parietosquamosal shelf is absent or very narrow. There is sutural contact between the squamosal and quadrate (Galton and Sues 1983). The snout is rather long, and ornamentation composed of large nodes is developed especially on the squamosals.

Stygimoloch spinifer, from the Maastrichtian Lance and Hell Creek formations of the western interior of the United States, is based only on parts of squamosals but can be characterized by a prominent squamosal shelf, ontogenic closure of the supratemporal fenestrae, and massive horn-cores on the squamosals (Galton and Sues 1983).

It should be noted here that there is a yet-to-be-established pachycephalosaurid genus from the Upper Cretaceous of Tsondolein Khuduk, Gansu Province, People's Republic of China. It was originally described as *Troodon bexelli* by Bohlin in 1953 and is represented by a badly preserved fragment of the parietal. The species differs significantly from all other described forms in its comparatively extensive parietosquamosal shelf, closed supratemporal fenestrae, and extensive participation of the parietal in the caudal margin of the skull. At the moment, this taxon should be considered "*Troodon*" *bexelli*, pending a redescription and renaming of the genus.

Micropachycephalosaurus hongtuyanensis, from the Upper Cretaceous Wang Formation, People's Republic of China, is often assigned to the Homalocephalidae (Dong 1978; Perle et al. 1982; Sues and Galton 1987). This species was established by Dong (1978) on poor material, including some skeletal fragments (part of a dentary with teeth, fragments of dorsal, sacral, and caudal vertebrae, small portions of the ilium, a fragmentary tibia, a well-preserved femur, and fragments of ossified tendons). Judging from the structure of all preserved skeletal elements, *M. hongtuyanensis* is obviously a member of the Pachycephalosauria, but as-

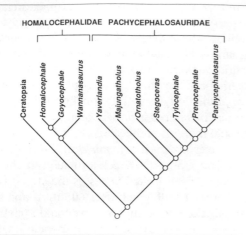

HOMALOCEPHALIDAE PACHYCEPHALOSAURIDAE

Fig. 27.5. Relationships of the Pachycephalosauria.

signation of this species to the Homalocephalidae or Pachycephalosauridae is not justified. However, *M. hongtuyanensis* differs from all other Asian pachycephalosaurian species by a more proximally positioned fourth trochanter and caudally tapering postacetabular process of the ilium, and is here treated as a valid species at the level of Pachycephalosauria *incertae sedis*.

The interrelationships within the Pachycephalosauria recently discussed by Sues and Galton (1987), based mainly on the structure of the skull roof, is here accepted in principal (fig. 27.5). The presumably oldest homalocephalid, the age of which is certainly Late Cretaceous and probably pre-Campanian (Perle et al. 1982), is *G. lattimorei*. Morphologically somewhat more primitive, *W. yansiensis* has comparatively larger supratemporal fenestrae and a more uncertain Late Cretaceous age (Hou 1978). The most derived homalocephalid with the respect to the relative size of supratemporal fenestrae and number of sacral vertebrae is the Late Campanian or Maastrichtian *H. calathocercos*.

The oldest-known pachycephalosaurid, and at the same time the oldest pachycephalosaurian, is the Barremian *Y. bitholus*. If pachycephalosaurid domes are arranged in a structural sequence, using the flat-headed homalocephalidae as an outgroup, the dome of *Y. bitholus* (having two low domes exclusively on the frontals and comparatively large supratemporal fenestrae) is the most primitive. The next would be that of *M. atopus* with a single, although prominent frontal dome, and the next would be that of *O. browni*, characterized by a very low dome divided between the frontals and parietals. The most derived are those domes consisting of a high, undivided frontoparietal mass. This type is found in *S. validum, T. gilmorei, P. prenes,*

and *Pachycephalosaurus wyomingensis*. The last two species display the most derived structure—the domes are very massive with obliterated supratemporal fenestrae. The domes of *M. atopus* and *O. browni* are derivable from that of *Y. bitholus*. It is likely that an early pachycephalosaurid species with an *O. browni*-like dome founded the Asiamerican evolutionary line, which climaxed in *Pachycephalosaurus* on the North American continent and *Prenocephale* on the Asian continent at the end of the Cretaceous. A dispersed line probably evolved on Gondwanaland which in the Campanian reached the advanced dome form exemplified by *Majungatholus*.

PALEOECOLOGY AND BEHAVIOR

The ecological habits of the Pachycephalosauria have been reconstructed by several authors, among them, Gilmore (1924*d*), Sternberg (1933), Galton (1970*c*, 1971*c*), Maryańska and Osmólska (1974), Galton and Sues (1983), and Sues and Galton (1987).

The majority of the pachycephalosaurian remains from western North America is restricted to skull caps. As mentioned by Sternberg (1933*b*), these domes are often water worn. The most abundant are isolated skull caps of *S. validum;* only one specimen of this species is represented by a well-preserved skull accompanied by a fragmentary postcranial skeleton. The total number of isolated pachycephalosaurian skull fragments in the Judith River Formation of western Canada suggests that pachycephalosaurians were abundant and constitute at least 10 percent of the total dinosaur fauna of this formation (Béland and Russell 1978). According to Dodson (1983), an intense taphonomic bias against preservation of articulated skeletons of this small dinosaur is responsible for preservation of only thickened skull caps in the sediments of the Judith River Formation. The abundantly represented, isolated pachycephalosaurian skull remains in the Judith River Formation were deposited in fluvial sediments under a seasonal, warm temperate or possibly subtropical climate (Dodson 1983). Similarly, European (*Y. bitholus*) and Malagasy (*M. atopus*) forms are known from skull fragments. In contrast, Asian specimens are commonly preserved as more or less complete skulls accompanied by at least some postcranial fragments, *T. gilmorei* and "*Troodon*" *bexelli* being exceptions.

The Barun Goyot Formation, which has yielded *T. gilmorei,* consists of alternating eolian dune and interdune deposits, sediments of small lakes and streams, and deposits of playas. These were probably deposited under conditions of a hot and semiarid climate (Gradzinski and Jerzykiewicz 1974; Gradzinski et al. 1977).

H. calathocercos and *P. prenes* were found in the Nemegt Formation. The sediments of this formation are typically fluvial. Dinosaur remains are commonly associated with channels. The rapid sedimentation of these sediments during seasonal floods created favorable conditions for preservation of bony material (Gradzinski 1970). The nearly complete skeletons of *H. calathocercos* and *P. prenes* and many other dinosaurs from the Nemegt Formation lack traces of long distance transport, indicating that these animals were entombed not very distant from their habitat. As was suggested by Gradzinski (1970), the petrographic composition of the deposits and the sedimentary features of the Nemegt Formation suggest a warm, seasonal, and rather humid climate.

Pachycephalosaurians were most probably herbivorous, as is indicated by their dentition. Despite their uniformity, relative size of the teeth and pattern of wear vary from species to species. This difference implies that they may have fed on different kinds of vegetation.

The very large orbits facing rostrolaterally and usually protected on their upper side by an extended orbital shelf suggests that vision played an important role in the life of these ornithischians and that it was partially stereoscopic.

The endocranial casts of the pachycephalosaurid skulls are characterized by very large olfactory lobes, which speak strongly for a good sense of smell in these animals.

The high morphologic diversity of the thickened frontoparietal mass has produced considerable discussion concerning the function of the dome. Colbert (1955) was the first to suggest that it might have served as a combat device. Subsequently, this idea was analyzed in detail by Galton (1970c, 1971c), Sues (1978b), Galton and Sues (1983), and Sues and Galton (1987). These authors concluded that the thick skull cap was used as a battering ram for intraspecific competition. This conclusion was based not only on the structure of bone of the dome but was also supplemented by some postcranial features, such as the ridge-and-groove articulation between zygapophyses of successive dorsal vertebrae, providing great rigidity of the vertebral column (Sues and Galton 1987). The flat-headed homalocephalids may have employed a different mode of intraspecific combat such as head-to-head shoving rather than battering (Galton and Sues 1983).

Several pachycephalosaurian features, for example, the shortening of basicranium, rostral inclination and vertical expansion of the occipital region, tongue-and-groove articulation between zygapophyses, reflect combat habits. The extremely strong medial extension of the caudodorsal wall of the acetabulum probably stabilized the hindlimbs during combat. The femoral head would have pressed caudodorsally against the acetabular wall, while the rigid vertebral column sloped down and forward. In this way, forces produced by impact would have been transmitted directly to the strengthened pelvic region.

Pachycephalosaurians were obligatory bipedal animals. The neck was probably thick and short. The weakly curved caudal dorsal ribs, very broad pelvis, widely spaced femora, and unusually long ribs on the proximal caudal vertebrae indicate that these animals had a bulky trunk and heavy tail. The tail had additional reinforcement through the basketwork of ossified tendons. The trunk was held rigid by the ridge-and-groove articulation of the zygapophyses of the dorsal and caudal vertebrae. Lateral flexibility of the tail was imparted only by the most proximal portion; the more distal part of the tail was rigid. This structure of the vertebral column indicates that it was held nearly horizontal while the animal walked. It is possible that the tail acted as a counterbalance to the front half of the body during walking.

The large available sample of skull caps referable to *S. validum* shows a high degree of intraspecific variability, which drew Brown and Schlaikjer's (1943) attention to possible sexual dimorphism in this species. They believed that the degree of development of the parietosquamosal shelf and the surface of the domed part of the skull roof are the characters that differentiate the sexes. Forms nearly lacking the shelf and having a smooth (but not necessarily smaller) frontoparietal dome were regarded as females, while forms with a roughened and extensive shelf were considered males.

Galton (1971c) noticed that the absence of the parietosquamosal shelf was a result of the more caudally extensive dome. He therefore reversed Brown and Schlaikjer's (1943) sex designations and suggested instead that forms with relatively larger domes were males. Morphometric analyses by Chapman et al. (1981) further support Galton's hypothesis of secondary sexual characteristics in this species. In particular, these analyses show that *S. validum* consisted of two morphs differentiated by the relative size of their domes and braincases. Those forms with larger, thicker, and more convex domes, which probably were used for head-ramming during intraspecific combat, are considered males, while females had less convex domes.

CERATOPSIA

P. DODSON

The Ceratopsia consists of the Psittacosauridae (Sereno, this vol.) and the Neoceratopsia (Dodson and Currie, this vol.). The term "Neoceratopsia" is a useful one coined by Sereno (1986) to designate the monophyletic assemblage that includes the Protoceratopsidae and the Ceratopsidae. This assemblage constitutes the Ceratopsia of earlier authors (e.g., Romer 1956, 1966). The inclusion of the Psittacosauridae within the Ceratopsia is a recent idea (Maryańska and Osmólska 1975) that has been corroborated and has achieved wide acceptance (e.g., Coombs 1982; Sereno 1986). The key to the interpretation of Psittacosauridae was the recognition of the rostral bone by Maryańska and Osmólska (1975). Autapomorphies of the monophyletic Ceratopsia (Sereno 1986) include a rostral bone, a skull with narrow beak and flaring jugals, a jugal that is deeper under the orbit than under the infratemporal fenestra and that has a dorosventral ridge, an incipient frill composed dominantly of parietals, and premaxillae with a strongly vaulted palatal surface. The Ceratopsia is a late group of dinosaurs, none being older than late Early Cretaceous. As presently known, these animals are restricted in distribution to Asia and western North America.

28

Psittacosauridae

PAUL C. SERENO

INTRODUCTION

Psittacosaurids, or the parrot-beaked dinosaurs, comprise a small group of herbivorous ornithischians, known only from Lower Cretaceous sediments of Asia. Not exceeding two meters in length, the psittacosaurid postcranium is remarkably primitive as compared to other ornithischians, with limb proportions of a facultative biped. The skull, in contrast, is highly modified, with a tall, parrotlike snout, which is proportionately shorter than the snout of any other ornithischian.

Despite a rich fossil record, psittacosaurs have received scant attention in the morphologic literature. Two brief notes on Mongolian psittacosaurs (Osborn 1923a, 1924b) and, similarly, two brief accounts on Chinese forms (Young 1958a; Chao 1962) constitute the primary sources for comparative information. Consequently, much of psittacosaurid systematics has been speculative; some authors recognize two genera and as many as four or five species (Young 1958a; Chao 1962), whereas others accept only a single species (Rozhdestvensky 1955).

Psittacosaurid remains were first uncovered in the Mongolian People's Republic (Outer Mongolia) in 1922 by the Third Asiatic Expedition of the American Museum of Natural History. Osborn (1923a) erected two genera, *Psittacosaurus mongoliensis* and *Protiguanodon mongoliensis*, based on two articulated skeletons (figs. 28.1, 28.2). In passing, Andrews (1932:223) mentioned the existence of fine juvenile material of *Psittacosaurus mongoliensis*, which remained unstudied until recently (Coombs 1982; Sereno 1987). In 1946, the Soviet-Mongolian Expeditions collected disarticulated bones of *P. mongoliensis* in Outer Mongolia (Rozhdestvensky 1955b). During the 1970s, the new Soviet-Mongolian Expeditions discovered further material of the same species, including several articulated skeletons.

The first well-preserved psittacosaurid remains from China were discovered during the 1950s and described as *P. sinensis* (Young 1958a) and *P. youngi* (Chao 1962; figs. 28.3, 28.4). Recently, two new species of Chinese psittacosaurids have been described, the first from northeastern China (Sereno and Chao 1988; fig. 28.5) and the second from northwestern China (Sereno et al. 1988; fig. 28.6).

Initial descriptions of the psittacosaurid cranium misinterpreted the rostromedian element of the snout as the premaxilla rather than the rostral bone, and until recently (Maryańska and Osmólska 1975; Coombs 1982), psittacosaurs were referred to the Suborder Ornithopoda. The rostral bone is the hallmark of the Suborder Ceratopsia, which traditionally included only small-bodied forms, such as *Protoceratops*, and larger-bodied ceratopsids. Now accepted as primitive ceratopsians, psittacosaurids play an important role in the diagnoses of the Ceratopsia and Neoceratopsia and the phylogenetic position of ceratopsians among other ornithischians (Sereno 1986).

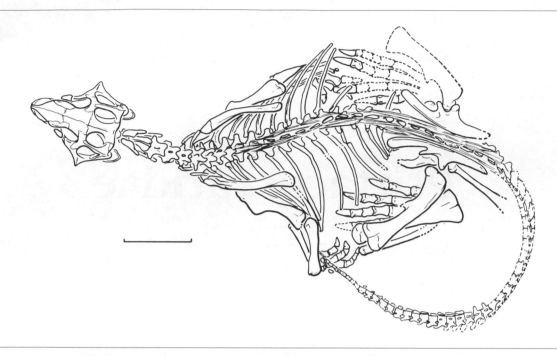

Fig. 28.1. Type skeleton of *Psittacosaurus mongoliensis* in dorsal view. Scale = 10 cm.

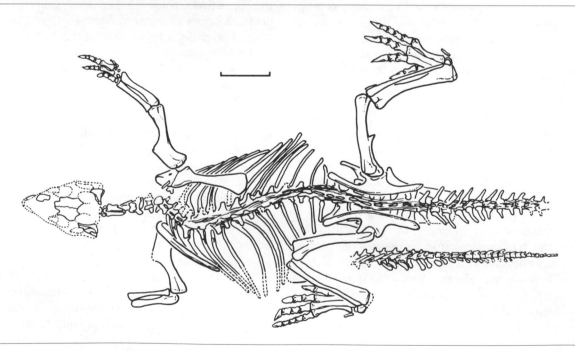

Fig. 28.2. Type skeleton of *Protiguanodon mongoliensis* in dorsal view. Distal tail section figured separately. Scale = 10 cm.

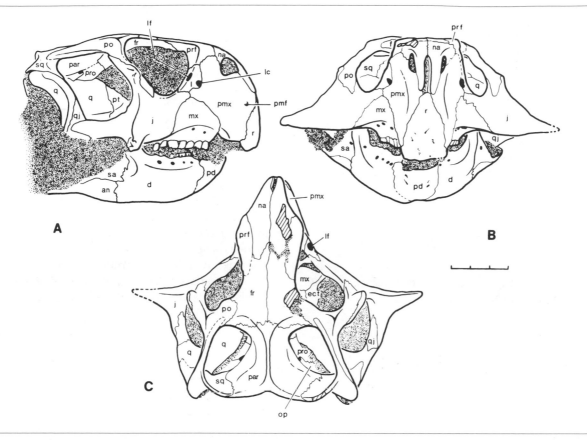

Fig. 28.3. Skull of *Psittacosaurus sinensis* in (a) lateral, (b) rostral, and (c) dorsal views. Scale = 3 cm.

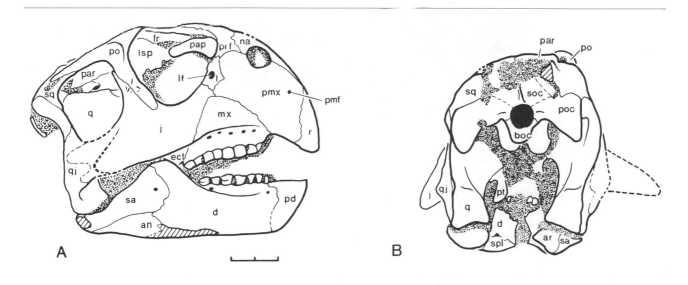

Fig. 28.4. Skull of *Psittacosaurus youngi* in (a) lateral and (b) caudal views. Scale = 2 cm.

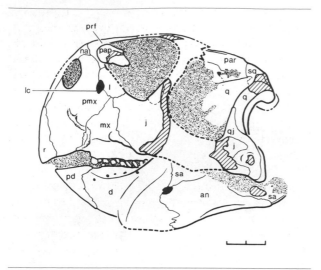

Fig. 28.5. Lateral view of the skull of *Psittacosaurus meile-yingensis*. Scale = 2 cm.

DIAGNOSIS

The Psittacosauridae is diagnosed on the following features: ceratopsians with a short preorbital skull segment (less than 40% of skull length); external naris positioned very high on the snout; nasal extending rostroventrally below the external naris, establishing contact with the rostral; caudolateral premaxillary process extremely broad, separating the maxilla from the external naris by a wide margin and extending dorsally to form a tall parrotlike rostrum; premaxilla, maxilla, lacrimal, and jugal sutures converge to a point on the snout; antorbital fenestra absent; antorbital fossa absent; unossified gap in the wall of the lacrimal canal; eminence on the rim of the buccal emargination of the maxilla near the junction with the jugal; elongate pterygoid mandibular ramus; dentary crown with bulbous primary ridge; manual digit IV with only one simplified phalanx; manual digit V absent.

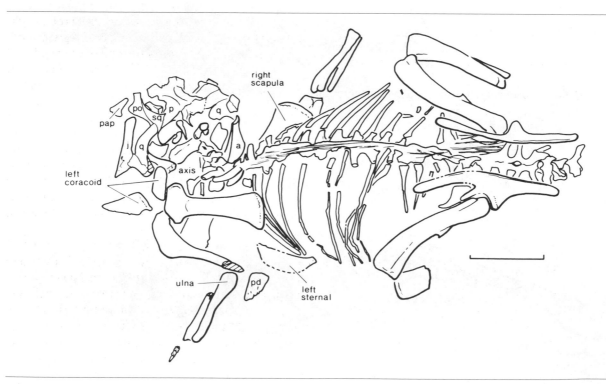

Fig. 28.6. Skeleton of *Psittacosaurus xinjiangensis*. Scale = 5 cm.

ANATOMY

Skull and Mandible

The psittacosaurid skull roof (figs. 28.3, 28.4, 28.5, 28.7) is constructed from relatively slender, thin elements, which bound broad supratemporal and infratemporal fenestrae. A noteworthy exception is the jugal, which is thickened in midsection to form a horn of variable size and shape. In dorsal view, the psittacosaurid cranium has a subtriangular contour due to the lateral projection of each jugal horn and the transverse compression of the beaklike snout.

In neoceratopsians (i.e., ceratopsians more derived than psittacosaurids), the rostral bone is strongly transversely compressed, extending rostrally to provide support for a ventrally curved beak. In psittacosaurids, in contrast, the rostral is thin in sagittal section and forms a transversely convex shield that caps a triangular surface on the conjoined premaxillae. The rostral also contacts the slender rostroventral processes of the nasals, a sutural contact absent among neoceratopsians. The tall, parrot-like rostrum is constructed almost entirely from the expansive caudolateral process of the premaxilla. The sheetlike premaxillary process is broadened to the extent that it no longer closely resembles the homologous pointed process in other ornithischians. In palatal view, the palatal process of the premaxilla arches ventrolaterally from the midline to the lateral bill margin. As in other ceratopsians and pachycephalosaurs, the caudal extension of the palatal process of the premaxilla does not reach the margin of the internal nares and does not establish contact with the footplate of the vomer.

A most unusual feature of the snout is the slender rostral process of the nasal, which extends ventral to the external naris past its usual termination on the internarial bar. The external naris is relatively small and positioned high on the snout, such that the ventral margin of the naris is situated dorsal to the ventral margin of the orbit, an unusual condition among ornithischians. The maxilla is strongly emarginated dorsal to the tooth row, and the antorbital fossa, the depression that primitively surrounds part or all of the antorbital fenestra, is absent. A small secondary depression may occur on the maxilla. In *P. mongoliensis*, for example, the secondary depression is small and subtriangular. The antorbital fenestra is absent in all species of psittacosaurids. The presence in psittacosaurids of a secondary maxillary depression should not be confused with the antorbital fossa in other dinosaurs.

A small subsidiary opening occurs on the sidewall of the psittacosaurid snout, usually between the premaxilla and lacrimal, and should not be confused with the antorbital fenestra. In psittacosaurids, the lateral wall of the lacrimal canal remains only partially ossified. An opening is present in the lateral wall of the canal about halfway along its passage from the margin of the orbit to the nasal cavity. The opening, which occurs in the lacrimal bone, is frequently bounded rostrally by the premaxilla, which overlaps the lacrimal.

The prefrontal forms the rostrodorsal margin of the orbit, which is turned laterally for articulation with the subtriangular palpebral. The form of the caudal margin of the skull roof in psittacosaurs bears some resemblance to the frilled margin in neoceratopsians. The parietal extends caudally over the occiput as a transversely broad shelf. The rostrocaudal proportion of the parietal shelf varies among psittacosaurid species, but the shelf constitutes at least one quarter of the length of the bone.

In ventral view of the skull, a transversely arched secondary palate is present rostrally, formed principally by the premaxillae. Caudal to the internal nares, the remainder of the palate is composed of the palatine, pterygoid, and ectopterygoid. The vomers, which fuse rostrally, arch in the midline from the secondary palate rostrally to the palatine and pterygoid caudally. The suborbital opening persists as a foramen between the palatal bones and the maxilla. An elongate flange of the pterygoid, the mandibular ramus, is directed caudoventrally toward the adductor fossa of the lower jaw.

The exoccipital and opisthotic are completely fused. The combined element borders the foramen magnum and sends a narrow paroccipital process laterally and slightly caudally. The foramen magnum is also

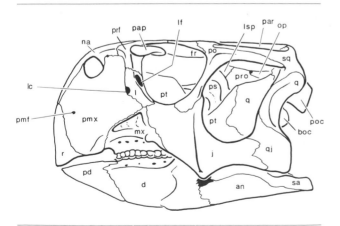

Fig. 28.7. Reconstruction of the skull of *Psittacosaurus mongoliensis* in lateral view.

bounded by the basioccipital below and the supraoccipital above. The plane of the occiput is nearly vertical.

The lower jaw is deep in lateral view. In *P. mongoliensis* and *P. meileyingensis,* a low flange hangs from the ventral margin of the dentary. The external mandibular fenestra is open in some species but closed in *P. sinensis* and *P. youngi.* The articular surface on the lower jaw for the mandibular condyle of the quadrate is peculiar; it is perfectly flat, rather than cupped as found in most archosaurs. The flat ventral surface of the jaw articulation may have facilitated rostrocaudal sliding of the quadrate condyles during mastication.

The slender stapes is preserved in two individuals. The length of the bony stapes equals the distance from the fenestra ovalis to the otic notch, which is formed rostrally by the quadrate shaft and caudally by the paroccipital process and caudal process of the squamosal. Gently curving, rod-shaped first ceratobranchials are preserved in *P. mongoliensis* and *P. sinensis.*

Premaxillary teeth do not occur during any growth stage in *Psittacosaurus.* Denticulate maxillary and dentary crowns are positioned along the tooth rows with a small amount of overlap along the crown edges. Both upper and lower tooth rows are inset from the lateral surface of the skull by a buccal emargination. Maxillary and dentary teeth are approximately equal in size and number in any single individual. Tooth count increases during growth from as few as five to as many as twelve in *P. mongoliensis.* Adult tooth count varies from eight to twelve among psittacosaurid species.

In all species of *Psittacosaurus,* the buccal surface of the maxillary crown is flatter than the lingual surface. The lateral surface is ornamented by a weak primary ridge, which terminates at the apex of the crown and is flanked by a few weak secondary ridges. The medial surface of the dentary crowns, in contrast, is dominated by a prominent median primary ridge. The primary ridge is quite bulbous near the crown base but tapers to the apex of the crown and is flanked to each side by flat crescentic surfaces with weak secondary ridges.

All crown surfaces are covered with enamel. The enamel is several times thicker on the lateral surface of the maxillary crowns and the medial surface of the dentary crowns than on opposing sides. This asymmetrical distribution of enamel in cheek crowns is now known to occur in a variety of ornithischians including ceratopsians, pachycephalosaurs, and ornithopods. The psittacosaurid dentition is characterized by broad planar wear surfaces with self-sharpening cutting edges. The wear surfaces of a single tooth row lie in approximately the same oblique plane (dorsomedial-ventrolateral) but do not form a continuous wear surface along the tooth row. Some crowns are truncated by single wear surfaces; others are truncated by two contiguous wear facets. A precise pattern of occlusion between the upper and lower tooth rows, such as alternate or tooth-to-tooth alignment, does not occur in *Psittacosaurus.*

Axial Skeleton

The principal divisions of the axial skeleton (figs. 28.1, 28.2, 28.8) are uniform across all species of *Psittacosaurus.* There are 21 presacrals, 6 sacrals, and approximately 45 caudals. In the presacral division, there may be as many as 8 or 9 cervical vertebrae in *P. mongoliensis* and *P. sinensis.* Distinction between cervical and dorsal vertebrae, in this case, is based on the position of the parapophysis. In dorsal vertebrae, the parapophysis is located above the neurocentral suture.

A proatlas has not been observed in *Psittacosaurus.* The atlas is composed of four discrete elements: a lozenge-shaped centrum, a U-shaped intercentrum, and two neural arches. The axial centrum, longest in the cervical series, is fused rostrally to a small, wedge-shaped axial intercentrum in the adult. The axial neural spine is well developed, extending caudodorsally at approximately 45°. The centra of the postaxial cervicals are transversely pinched and bear low ventral keels. The change in shape of the postaxial centra, from that of a trapezoid caudally to that of a parallelogram cranially, indicates that the neck assumed a gentle S-shaped curve in natural articulation.

As in the cervical centra, the articular surfaces in the dorsal centra are flat cranially and gently concave caudally. The dorsolateral angle of the transverse processes increases from approximately 35° in the cranial dorsals to approximately 40° in the mid-dorsals, before decreasing to the horizontal in the caudal dorsals.

The spool-shaped sacral centra are significantly longer than the adjacent dorsal and caudal centra. In adults, the sacral centra coossify but remain separate from the last dorsal and first caudal. The first sacral centrum is particularly robust. Sacral neural arches are characterized by nearly vertically oriented zygapophyses and neural spines that remain separate in the adult.

The caudal centra exhibit a regular decrease in length and height along the tail. The decrease in centrum height is stronger than the decrease in centrum length, resulting in elongate cylindrical centra at the

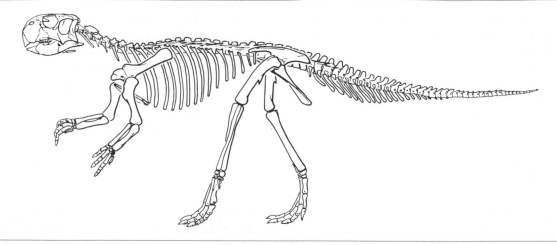

Fig. 28.8. Reconstruction of the skeleton of *Psittacosaurus mongoliensis* in lateral view (after Osborn 1924b).

distal end of the tail. The fused caudal ribs decrease rapidly in length from the first caudal distally, disappearing at midlength along the tail in *P. mongoliensis* and *P. sinensis*. The neural spines and zygapophyses decrease in height more gradually, the latter recognizable until the last few vertebrae.

The first chevron articulates between the second and third caudal centra. The chevrons, which are always longer than respective neural spines, decrease in length distally, disappearing near the end of the tail in *P. mongoliensis*.

Freely articulating ribs occur in the presacral series. The long atlantal rib, like the two or three caudalmost dorsal ribs, is holocephalous. The dichocephalous ribs of the axis and several of the cervicals that follow are are relatively short, extending only to the middle of the next centrum. The ribs of the caudal cervicals and cranial dorsals increase rapidly in length to a maximum between the fourth and seventh dorsals. Passing caudally from the first dorsal rib, the capitulum decreases in length as the parapophysis climbs the cranial edge of the transverse process. The last few dorsal ribs are short and rod shaped. The slender first sacral rib, fused proximally to the transverse process of the first sacral, attaches distally to the midsection of the preacetabular process of the ilium. The second to the sixth sacral ribs fuse to the sides of the sacral centra. Each rib is offset slightly cranially from its respective vertebra, such that the corner of each rib fuses to the caudal corner of the next sacral centrum.

Ossified epaxial tendons are present along nearly the entire dorsal and sacral column in *P. mongoliensis* and *P. xinjiangensis*. In the latter species, the tendons extend along at least the proximal half of the tail. In *P. sinensis*, in contrast, ossified tendons are absent. In *P. mongoliensis*, individual tendons span approximately four or five vertebrae, intertwining without apparent order.

Appendicular Skeleton

The scapula is longer than the humerus and has a relatively narrow blade with a prominent acromial process (figs. 28.1, 28.2, 28.6, 28.8). The blade is proportionately narrower in *P. sinensis*, as measured by the ratio between the length of the blade and the width behind the acromion. The scapula shares equally in the formation of the glenoid with the coracoid, a subquadrate bone with a well-developed cranioventral process.

P. mongoliensis has an ossified clavicle, as originally identified by Osborn (1924b, fig. 2). The supposed clavicle in *P. sinensis* (Young 1958a, fig. 52), however, appears to represent a displaced first ceratobranchial (Sereno 1987). In *P. mongoliensis*, the short, strap-shaped clavicle lies along the cranial margin of the coracoid and would not have reached the midline.

A pair of plate-like crescentic sternals are centered on the midline with a section of their thin medial edges in contact. The cranial end of each sternal is rounded and apparently nonarticular. The rectangular caudal end is scalloped for attachment to cartilaginous extensions of the cranial dorsal ribs.

The broadly expanded deltopectoral crest of the humerus forms a rectangular sheet of bone, which projects at an angle to the shaft. The poorly defined head is directed rostrodorsally toward the glenoid cavity. The radius and ulna are stoutly constructed with moderately expanded proximal and distal ends. The ulna, somewhat heavier than the radius, terminates proximally in a blunt olecranon process.

The carpus and manus are known primarily in *P. mongoliensis*. The carpus is composed of four carpals, which can be identified by their associations with forearm and metacarpal elements. The carpals include the radiale, intermedium, ulnare, and distal carpal 3, which are not tightly fitted to one another. The manus is asymmetrical, showing strong reduction of the fourth digit and loss of the fifth. The inner three digits are robust and terminate in broad unguals. The phalangeal formula is 2-3-4-1-0. The first digit and first metacarpal have stout proportions. The medial distal condyle of the first metacarpal is more proximal in position, which results in medial offset of the subsequent phalanges. During flexion, the phalanges of the first digit converge toward the center of the palm. The second metacarpal is longer than the first, and the phalanges of the second digit curve gently medially. Digit III is the longest digit and metacarpal three the longest metacarpal. Digit IV is reduced with only a small terminal phalanx. Metacarpal IV, in contrast to the inner three metacarpals, has a cylindrical shaft, which terminates distally without the development of paired condyles.

The ilium is low with a straplike preacetabular process. The postacetabular process, subequal in length to the preacetabular process, is proportionately deeper. The acetabulum is completely open without any development of a descending iliac flange. The narrow pubic peduncle projects cranioventrally. The ischial peduncle, in contrast, is robust and projects ventrolaterally. A distinct, ovate surface on the acetabular margin of the ischial peduncle articulates against the broad proximal end of the femur.

The two transversely compressed peduncles of the ischium join caudal to the acetabulum and pass distally as a flattened blade, angling ventromedially toward the midline at approximately 30° above the horizontal. Contact between right and left blades, which constitutes the ischial symphysis, is restricted to a small facet on the medial corner of the distal margin.

The pubis is small relative to the ilium and ischium but remains an integral part of the margin and articular surface of the acetabulum. The acetabular surface is located on the body of the pubis and faces caudolaterally and slightly dorsally. A short prepubic process projects craniolaterally, terminating before the tip of the preacetabular process of the ilium. The slender postpubic process, preserved only in *P. mongoliensis* and *P. sinensis*, projects caudomedially from the body of the pubis and passes along the ventral margin of the ischium. It does not join its opposite to form a pubic symphysis. In *P. mongoliensis*, the length of the process relative to the remainder of the pubis is nearly twice that in *P. sinensis*.

The femur is gently bowed in lateral view. The head, which arches medially from the shaft, shares with the greater trochanter the formation of a convex proximal articular surface. In lateral view, the greater trochanter is several times the craniocaudal width of the fingerlike lesser trochanter, which projects dorsally from the craniolateral side of the proximal shaft. Just above midshaft, a pendent fourth trochanter projects from the caudomedial side and is flanked medially by an oval depression. The distal condyles are separated by a shallow depression cranially and by a much deeper flexor groove caudally. The laterally curving cnemial crest dissipates distally along the tibial shaft. The compressed distal end of the tibia is rotated approximately 70° from the transversely compressed proximal end. The narrow distal articular surface is divided into a small lateral condyle, which extends behind the calcaneum, and a larger medial condyle, which is capped ventrally by the astragalus. The fibula is modestly expanded at both ends, which exhibit relative rotation to the degree observed in the tibia. The distal end lies against the flat cranial surface of the tibia and butts distally against the calcaneum.

The tarsus, known from articulated specimens in several species, consists of the astragalus, calcaneum, and lateral and medial distal tarsals. The cup-shaped astragalus tightly caps the medial distal condyle of the tibia. The short ascending process of the astragalus usually establishes a minor contact with the distal end of the fibula. The small wedge-shaped calcaneum is positioned on the cranial side of the tibia, articulating proximally with the fibula and distally with the lateral distal tarsal. The medial and lateral distal tarsals are tabular and positioned over the third and fourth metatarsals, respectively. The compact digitigrade pes, known largely in *P. mongoliensis* and *P. sinensis*, does not depart significantly from the primitive ornithischian condition. The first digit, for example, is significantly shorter than the second, third, and fourth digits, and the fifth digit is reduced to a metatarsal splint. The short fifth metatarsal appears to articulate proximally against the lateral side of the lateral distal tarsal. The phalangeal formula is 2-3-4-5-0. As in the manus, the unguals are broad, and all preceding phalanges have well-developed distal condyles.

EVOLUTION AND SYSTEMATICS

Ceratopsian Origins

Despite a long sojourn in the Ornithopoda, psittacosaurids are now generally regarded as primitive ceratopsians. Historically, the change of opinion favoring ceratopsian affinity appears to have hinged on the proper identification of the element at the front of the psittacosaurid skull. This element, the ceratopsian rostral bone, was tentatively identified for the first time in psittacosaurids by Romer (1956, 1968). Psittacosaurids exhibit a number of additional synapomorphies with other ceratopsians (i.e., neoceratopsians), which are enumerated below. These ceratopsian synapomorphies are important because they demonstrate the monophyly of the Ceratopsia, including psittacosaurids, despite the marked morphologic distance between psittacosaurids and neoceratopsians. The union of psittacosaurids and neoceratopsians in the Ceratopsia is based on cladistic, not phenetic (overall similarity), grounds, because psittacosaurids and other primitive ornithischians share many symplesiomorphies that are absent in neoceratopsians.

Recent agreement on the presence of the rostral bone in psittacosaurids has encouraged the proposal of many new ceratopsian synapomorphies that specifically would include psittacosaurids (Maryańska and Osmólska 1975; Norman 1984a; Cooper 1985). One of these additional characters is regarded here as a valid ceratopsian synapomorphy including psittacosaurids— jugals that project laterally well beyond the dorsal margin of the orbit (fig. 28.3; Maryańska and Osmólska 1975). In dorsal view of the ceratopsian skull, the distance between the frontal margin of the orbit and the lateralmost part of the jugal always exceeds the interorbital width across the frontals.

Additional ceratopsian synapomorphies have been identified in reference to a phylogeny with pachycephalosaurs and ornithopods as successive outgroups (Euornithopoda of Sereno 1986). The following synapomorphies, which unite Psittacosauridae and Neoceratopsia, reside entirely in the skull, because the psittacosaurid postcranium is remarkably primitive in comparison to the modified neoceratopsian postcranium. These synapomorphies include a flat subnarial margin between the narial fossa and ventral margin of the premaxilla (figs. 28.3, 28.4, 28.5, 28.7). In primitive ceratopsians, such as *Psittacosaurus, Leptoceratops, Bagaceratops,* and *Protoceratops,* the narial fossa surrounding the external nares is separated from the ventral margin of the premaxilla by a planar, near-vertical, surface of bone. In other ornithischians, the narial fossa surrounding the external nares extends close to the ventral margin of the premaxilla. This primitive condition also obtains in the Ceratopsidae, which is interpreted here as a reversal related to the hypertrophy of the ceratopsid narial fossa.

In lateral view of the ceratopsian skull, a dorsoventral crest traverses the lateral aspect of the jugal (figs. 28.3, 28.7). The crest divides the lateral aspect of the jugal into rostral and caudal surfaces ("biplanar lateral surface"; Sereno 1986). The clarity of division of the lateral surface is dependent on the prominence of the crest. Among primitive ceratopsians, such as *Psittacosaurus,* the dividing crest is prominent. In primitive neoceratopsians, such as *Leptoceratops, Bagaceratops, Protoceratops,* and *Montanoceratops,* it is accentuated by the epijugal, an accessory dermal cap unique to neoceratopsians. In the Ceratopsidae, in contrast, the jugal is relatively flat; the low, caudally displaced dividing crest, nonetheless, is discernible (e.g., Hatcher et al. 1907). There is no development of a similar crest on the lateral surface of the jugal in other ornithischians.

In ceratopsians, the infraorbital ramus of the jugal is deeper than the infratemporal ramus (figs. 28.3, 28.7). In psittacosaurids and primitive neoceratopsians, the infraorbital ramus ventral to the orbit is substantially deeper than the infratemporal ramus under the laterotemporal fenestra. In ceratopsids, the infraorbital ramus is only marginally deeper than the infratemporal ramus, which may be related to the substantial reduction in relative size of the infratemporal fenestra in this group. In contrast to ceratopsians, the infraorbital ramus of the jugal is significantly narrower than the infratemporal ramus in other ornithischians.

In all marginocephalians, the parietal and squamosal overhang the occiput. In the Ceratopsia, however, the composition of the overhanging shelf is unique. In members of this group, the parietal accounts for at least half of the caudal margin of the shelf or frill. The parietal contribution to the shelf in psittacosaurids is the least among ceratopsians, accounting for approximately half of the caudal margin. In neoceratopsians, the parietal dominates the caudal margin of the frill. Among outgroups to the Ceratopsia, a parietal-squamosal shelf occurs only in the Pachycephalosauria, in which the parietal contribution is limited to a narrow median band.

In ceratopsians, the premaxillary palate is vaulted. The premaxillary palate arches laterally from a dorsal position in the midline to sidewalls with a nearly vertical disposition. A sharp bill margin is formed from the ventral margin of the steeply inclined sidewalls. The vaulted premaxillary palate occurs in all ceratopsians, including psittacosaurids, primitive neoceratopsians such as *Protoceratops*, and in ceratopsids. The sidewall of the vaulted ceratopsian premaxillary palate is visible in medial view of a disarticulated premaxilla (e.g., *Triceratops*, Hatcher et al. 1907, fig. 28A, B). The premaxillary palate among pachycephalosaurs and ornithopods, in contrast, is either planar and horizontal or arched very gently away from the midline.

In ceratopsians, the ventral process of the predentary is broad and expands in transverse width toward its proximal end to equal, or exceed, one-half the maximum transverse width of the predentary. The ventral process of the predentary is very broad in primitive ceratopsians such as *Psittacosaurus, Leptoceratops,* and *Protoceratops*, and the greatest width of the process occurs proximally where the process joins the body of the predentary. The proximal end of the ventral process in these ceratopsians nearly equals the maximum transverse width of the predentary. In ceratopsids, the narrower ventral process, which also expands toward its base, is approximately half the maximum transverse width of the predentary. The lateral processes of the ceratopsian predentary insert into a groove of variable length on the dorsal margin of the dentary. In primitive ceratopsians, such as *Psittacosaurus* and *Leptoceratops*, the groove is well developed, extending the entire length of the contact with the short lateral process. In ceratopsids, the dentary groove receives only the distal end of the lateral process. The broad ventral process and slotted contact between the lateral predentary process and the dentary suggest that mobility at the dentary-predentary joint was minimal in ceratopsians ("immobile mandibular symphysis"; Sereno 1986).

In conclusion, psittacosaurids can be allied with other ceratopsians by these cranial synapomorphies. Numerous synapomorphies that unite neoceratopsians, however, are absent in psittacosaurids, which is an equally important conclusion. These symplesiomorphies specify the primitive ancestral condition of the ceratopsian clade. This is particularly significant for apomorphic characters that occur within the Neocera-

topsia as well as elsewhere within the Ornithischia. Within Neoceratopsia, for example, derived features of the masticatory apparatus include the prominent primary ridge on the maxillary crowns, close packing and increased replacement in the cheek dentition, distal extension of upper and lower tooth rows, a laterally displaced coronoid process, and the relative reduction of the postdentary elements. These same characters also occur within the Ornithopoda. Were it not for primitive sister taxa, such as psittacosaurids among ceratopsians, these parallelisms would be eligible for interpretation as synapomorphies, and the scheme of ornithischian phylogeny would be altered accordingly.

Psittacosaurid Systematics and Evolution

As originally conceived by Osborn in 1923*a*, the Psittacosauridae contained only the single species *Psittacosaurus mongoliensis*. Osborn's diagnosis mentioned the tall parrotlike rostrum, after which he had named the genus and family. More recent diagnoses of the family include several diagnostic (apomorphic) characters, but these have been mixed with many nondiagnostic (plesiomorphic) characters (e.g., Steel 1969). The revised diagnosis above lists only psittacosaurid synapomorphies. These characters, determined in reference to a hypothesis of relationships which specifies the Neoceratopsia and Pachycephalosauria as successive outgroups, constitute the basis for a monophyletic Psittacosauridae.

Psittacosaurid taxa in the descriptive literature are listed in table 28.1. Their taxonomic standing is briefly discussed below, and available evidence for their phylogenetic relationships is presented. On occasion, *Stenopelix*, a problematic taxon from the Lower Cretaceous of West Germany, has been referred to the Psittacosauridae or to the Ceratopsia (Romer 1956; Sues and Galton 1982). *Stenopelix*, however, does not exhibit any ceratopsian or psittacosaur synapomorphies and appears more closely related to the Pachycephalosauria (Maryańska and Osmólska 1974; Sereno 1987; Dodson this vol.).

Protiguanodon mongoliensis (fig. 28.2). The taxonomic status of *Protiguanodon mongoliensis* has remained controversial since its description by Osborn in 1923. Young (1985*a*) believed that the minor differences between *Protiguanodon mongoliensis* and *Psittacosaurus mongoliensis* only justify specific distinction, whereas others have suggested their specific synonymy (Rozhdestvensky 1955*b*, 1977; Coombs 1982). Osborn's

TABLE 28.1 Psittacosauridae

	Occurrence	Age	Material
Ceratopsia Marsh, 1890a			
Psittacosauridae Osborn, 1923a			
Psittacosaurus Osborn, 1923a (= *Protiguanodon* Osborn, 1923a)			
P. guyangensis Cheng, 1983	Lisangou Formation, Nei Mongol Zizhiqu, People's Republic of China	?Aptian-Albian	4 fragmentary individuals, one with partial skull
P. mongoliensis Osborn, 1923a (= *P. protiguanodonensis* Young, 1958a, including *Protiguanodon mongoliensis* Osborn, 1923a)	Khukhtekskaya Svita, Ovorkhangai, unnamed unit, Bayankhongor, Khulsyngolskaya Svita, Shinekhudukskaya Svita, Dundgov, Mongolian People's Republic, Jiufotang Formation, Liaoning, ?unnamed unit, Nei Mongol Zizhiqu, People's Republic of China; Shestakovskaya Svita, Gorno-Altayaskaya Avtonomnaya Oblast, Russia	Aptian-Albian	More than 75 individuals, including more than 15 skeletons
P. osborni Young, 1931 (including *P. tingi* Young, 1931)	Lisangou Formation, Xinpongnaobao Formation, Nei Mongol Zizhiqu, People's Republic of China	?Aptian-Albian	More than 3 individuals, jaw fragments, limb elements
P. sinensis Young, 1958a	Qingshan Formation, Shandong, People's Republic of China	?Aptian-Albian	More than 20 individuals, 5 complete skulls, 3 articulated skeletons
P. youngi Chao, 1962	Qingshan Formation, Shandong, People's Republic of China	?Aptian-Albian	Partial skeleton with skull
P. meileyingensis Sereno, Chao, Cheng, et Rao, 1988	Jiufotang Formation, Liaoning, People's Republic of China	?Aptian-Albian	4 individuals, 2 complete skulls
P. xinjiangensis Sereno et Chao, 1988	Lianmugin Formation, Xinjiang Uygur Zizhiqu, People's Republic of China	?Aptian-Albian	More than 10 individuals, including articulated skeleton with skull

supposed distinguishing characteristics, nevertheless, have never been challenged and are listed briefly below.

1. Maxillary teeth flattened with asymmetrical, trilobate form in *Psittacosaurus mongoliensis* but convex and symmetrical in *Protiguanodon mongoliensis*.

2. Teeth less compacted in *Protiguanodon mongoliensis*.

3. "Large occipital condyles" (*sic*) in *Psittacosaurus mongoliensis* but small in *Protiguanodon mongoliensis* (Osborn 1924b:8).

4. Epidermal ossifications on the throat and side of the face in *Psittacosaurus mongoliensis* but absent in *Protiguanodon mongoliensis*.

5. Five sacrals with pelvic articulations and 16 dorsals in *Psittacosaurus mongoliensis* but 6 sacrals and 15 dorsals in *Protiguanodon mongoliensis*.

6. Ribs, pelvic girdle, and hindlimbs more massive in *Psittacosaurus mongoliensis*.

7. Prepubic process much more robust in *Psittacosaurus mongoliensis*.

8. Wide geologic separation.

The reported difference between the maxillary crowns in the type specimens of *Psittacosaurus mongoli-* *ensis* and *Protiguanodon mongoliensis* is the result of misidentification of an isolated tooth associated with the latter. The tooth in question (Osborn 1924b, fig. 3B) exhibits the crown shape and bulbous primary ridge present in the dentary teeth of all species of *Psittacosaurus*. Osborn, however, identified it as a maxillary tooth and then observed differences in comparisons to maxillary teeth in *Psittacosaurus mongoliensis*. Because the dentary teeth in the type of *Psittacosaurus mongoliensis* were not exposed at the time of Osborn's writing and because maxillary crowns are not preserved in *Protiguanodon mongoliensis*, an appropriate comparison between the dentitions of these specimens was not possible. Comparison between the dentary teeth of these two specimens does not support their distinction. The supposed difference in spacing along the tooth row cannot be verified in a comparison of the type dentitions; slight overlap of adjacent crown edges is the consistent spacing.

Poor preservation prevents an accurate measure of the size of the occipital condyle in *Protiguanodon mongoliensis*, but there is no positive evidence of a significant difference in this part of the skull. The supposed epidermal ossifications on the side of the type skull of *Psittacosaurus mongoliensis* are sedimentary artifacts, as originally suggested by Granger (in Osborn 1924b:8). There is no bone structure in cross section, and additional nodules were found dispersed throughout the matrix, below the palate, and within the orbit.

In any case, the supposed absence of epidermal ossifications in *Protiguanodon mongoliensis* cannot be established given the poor preservation of the type skull.

Osborn's comparison of the number of dorsals and sacrals in the type skeletons of *Psittacosaurus mongoliensis* and *Protiguanodon mongoliensis* is not valid because he did not present any justification for the high dorsal counts of 15 or 16 and because the sacral region is not sufficiently exposed in *Psittacosaurus mongoliensis*. It is possible to compare presacral counts, however, and in this regard a valid difference emerges. In the type skeleton of *Protiguanodon mongoliensis*, there are 21 presacrals as compared to 22 in the type skeleton of *Psittacosaurus mongoliensis*. In both skeletons, the first sacral rib articulates with the base of the preacetabular process of the ilium. Two additional skeletons referred to *Psittacosaurus mongoliensis* have 21 presacrals, as in *Protiguanodon mongoliensis*. The difference in presacral count is significant but may be within the range of individual variation.

As Osborn noticed, the skeletal elements of *Psittacosaurus mongoliensis* are generally somewhat more robust than in *Protiguanodon mongoliensis*, although the difference is subtle. Some variation in size and in the exact proportions of individual elements is expected in any population sample, and such variation is evident in the enormous collection of psittacosaurid material now known from the Mongolian People's Republic. Osborn mentioned the more robust proportions of the prepubic process in the type skeleton of *Psittacosaurus mongoliensis*. The prepubic process in the type skeleton, however, appears to have been artifically deepened by post-mortem fracturing and crushing, which has been emphasized in Osborn's comparative pelvic reconstruction (Osborn 1924*b*, fig. 8).

Finally, stratigraphic and geographic separation are not valid criteria for taxonomic distinction. The detailed similarity between the type skeletons of *Psittacosaurus mongoliensis* and *Protiguanodon mongoliensis* argues in favor of their generic and specific synonymy.

Psittacosaurus mongoliensis (figs. 28.1, 28.8). All the characters in Osborn's original diagnosis of *P. mongoliensis* occur in other psittacosaur species, or, more generally, in other ornithischians. This species, however, can be diagnosed on cranial characteristics including the triangular maxillary depression and the up-turned lateral margin of the prefrontal.

Psittacosaurus osborni (fig. 28.9). Young (1931) described two species, *P. osborni* and *P. tingi*, that were distinguished primarily by their small size relative to *P.*

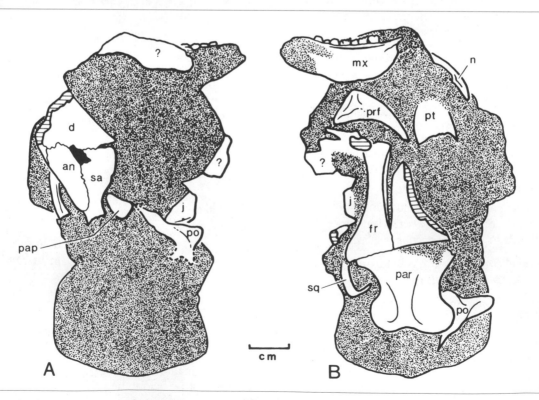

Fig. 28.9. Skull of *Psittacosaurus osborni* in (1) ventral and (2) dorsal views.

mongoliensis. Later, Young (1958*a*) reduced *P. tingi* to a junior synonym of *P. osborni*, which is based on a partial disarticulated skull. Obviously, small size alone is an insufficient basis for distinction, particularly in light of the size range in skeletal elements that occur at most psittacosaur localities (e.g., Coombs 1982). The strong median primary ridge on the dentary crowns, Young's second distinguishing feature, is now known to occur in all psittacosaur species. Cheng (1983) has recently referred additional small dental and postcranial remains to *P. osborni*.

Psittacosaurus sinensis (fig. 28.3). In 1958, Young described very complete psittacosaur material from eastern China as a new species, *P. sinensis*. Young's diagnosis was incomplete, listing only relatively small body size, a broadly proportioned skull, and fewer cheek teeth as distinguishing characters. Additional features are available in this species, such as the small jugal-postorbital horn core on the postorbital bar and the absence of the external mandibular fenestra. Contrary to earlier suggestions (Rozhdestvensky 1955*b*; Coombs 1982), the synonymy of *P. sinensis* as a junior synonym of *P. mongoliensis* cannot be upheld.

Psittacosaurus youngi (fig. 28.4). *P. youngi* was described from a partial skeleton discovered very near the localities that yielded *P. sinensis* (Chao 1962). *P. youngi* is extremely similar to *Psittacosaurus sinensis* and may be conspecific. A loose fragment of the nasal in the type skull has been mistaken as a nasal horn core (Maryańska and Osmólska 1975:172).

Psittacosaurus guyangensis. Recently, Cheng (1983) recognized another species, *P. guyangensis*, on the basis of a rostral skull fragment and referred postcrania. The distinguishing characteristics of this species, such as its intermediate size and the presence of nine maxillary teeth without a marked primary ridge, are difficult to assess in the absence of a detailed description of *P. mongoliensis*.

Psittacosaurus meileyingensis and *Psittacosaurus xinjiangensis*. The remains of these two new species were discovered in China recently, the first based on an excellent skull from the Jiufotang Formation near Chaoyoung, Liaoning Province, in northeastern China (Sereno et al. 1988; fig. 28.5) and the second based on an articulated skeleton from the Tugulu Group of the Junggar Basin in the Xinjiang Uygur Autonomous Region of northwestern China (Sereno and Chao 1988; fig. 28.6). The skull in *P. meileyingensis* has a nearly circular profile. The snout is relatively very short rostrocaudally; snout length, as measured from the front of the skull to the orbit margin, is approximately 27 percent of skull length, a smaller percentage of skull length than in any other ornithischian. Other distinguishing cranial features include a rugose protuberance on the quadratojugal and a prominent ventral flange on the dentary. Diagnostic aspects of *P. xinjiangensis* include a jugal horn with a flat rostral surface, a proportionately narrow iliac postacetabular process, and ossified tendons that extend distally at least midway along the tail.

Character information relevant to psittacosaurid interrelationships is scant; there are few derived characters that are distributed in more than one, but not in all, psittacosaurid species. The four best-known species include *Psittacosaurus mongoliensis*, *P. sinensis*, *P. meileyingensis*, and *P. xinjiangensis* (table 28.1). One potential synapomorphy favoring a subgroup of psittacosaurid species is the prominent, laterally projecting jugal horn. This character is shared by *P. sinensis* and *P. xinjiangensis* but is absent in *P. mongoliensis* and the new species from northeastern China. It is also absent in the nearest outgroup, the Neoceratopsia. The laterally projecting jugal horn, unfortunately, is the only available synapomorphy among psittacosaurid species free of missing data or uncertainty in character polarity.

PALEOECOLOGY

Psittacosaurids are known only from central Asia, broadly defined to include eastern and western China, Mongolia, and southern Siberia. The temporal distribution of psittacosaurs may be limited to the late Early Cretaceous (Aptian-Albian), but this has been dated by associated pollen and spores at only a single locality in Mongolia for *Psittacosaurus mongoliensis* (Bratzeva and Novodvorskaja 1975). An Early Cretaceous age is also supported by the associated invertebrate faunas (including ostracods and molluscs) and the vertebrate faunas, including freshwater fish (*Lycoptera*), turtles (*Hangiemys, Mongolemys*), dinosaurs (*Iguanodon, Shamosaurus*), and mammals (*Prokennalestes, Prozalambdalestes*) (Cockerell 1924; Kalandadze and Kurzanov 1974; Shuvalov 1975; Tumanova 1981). In China, the age of the beds that have furnished the remains of Chinese psittacosaurid species is poorly established but generally assumed to be Early Cretaceous. The associated faunas are poorly known, and only in Shandong Province are psittacosaurid-bearing deposits (*P. sinensis, P. youngi*) indisputably overlain by Upper Cretaceous sediments and faunas (e.g., the hadrosaurids *Tsintaosaurus* and *Tanius*; Young 1958*a*).

The slicing psittacosaurid dentition, with its self-sharpening cutting edges, is suited for mastication of plant material. Even the crowns of small hatchlings are truncated by wear. Polished gastroliths, exceeding fifty in number, are associated with two psittacosaurid skeletons and must have played a significant role in the breakdown of plant materials.

If limb length is estimated by addition of respective propodials, epipodials, and third metapodials, the length of the forelimb in *P. mongoliensis* is approximately 58 percent that of the hindlimb. This percentage of hindlimb length is somewhat higher than in *Hypsilophodon foxii*, which has a forelimb/hindlimb ratio of 51 percent. Proportions within the hindlimb in *Psittacosaurus* are very comparable to the presumed cursorial ornithopods, *Hypsilophodon*, *Dryosaurus*, and *Parkso-*

saurus (Janensch 1955; Galton 1973*d*) and the neoceratopsian *Microceratops* (Maryańska and Osmólska 1975). The tibia is slightly longer than the femur, as in *Dryosaurus* and *Microceratops*, but proportionately less elongate than in *Hypsilophodon* and *Parksosaurus*.

Given the proportions of the fore- and hindlimbs, it seems reasonable to suppose that *P. mongoliensis* was facultatively bipedal, that is, a capable and perhaps habitual biped at most speeds (fig. 28.8). The length of the forelimb and structure of the manus, however, do not preclude effective use of the forelimb in locomotion. The divergence of digit I suggests a limited grasping capability in the manus, which may have been used in procurement of vegetation.

29

Neoceratopsia

PETER DODSON

PHILIP J. CURRIE

INTRODUCTION

The Neoceratopsia (Sereno 1986) is an undoubtedly monophyletic (Norman 1984*a*; Maryańska and Osmólska 1985; Sereno 1986) assemblage of herbivorous ornithischians consisting of two families, the Protoceratopsidae and the Ceratopsidae. Monophyly of the Protoceratopsidae is in some doubt (Sereno 1986). Even if the protoceratopsids prove to be only a grade of primitive neoceratopsians, the name remains a useful one for purposes of communication. Subdivision of the Ceratopsidae into two monophyletic subfamilies, the Centrosaurinae and the Chasmosaurinae, seems advisable (Lehman in press).

Protoceratopsids were small (1 to 2.5 m long), relatively primitive herbivorous dinosaurs found in eastern and central Asia and western North America. Some of them appear to have been capable of rapid, bipedal progression (*Microceratops gobiensis*, Maryańska and Osmólska 1975; *Leptoceratops gracilis*, Coombs 1978*c*). Ceratopsids were medium to large (4 to 8 m long), somewhat rhinoceros-like quadrupeds with variably developed nasal and orbital horns and parietosquamosal frills. All neoceratopsians had large skulls in relation to body length. The skulls of protoceratopsids were relatively larger (Russell 1970*b*), but ceratopsids had the largest skulls of any terrestrial vertebrate (i.e., *Torosaurus latus,* 2.4 m, Colbert and Bump 1947). Neoceratopsians were greatly specialized in cranial anatomy, and taxonomy is based almost exclusively on the skull. They were rather conservative in the postcranial skeleton. It is generally assumed that quadrupedality was secondary (e.g., D. A. Russell 1970*b*; Coombs 1978*c*), given the bipedality of the successively more remote outgroups (Psittacosauridae, Pachycephalosauria, and Ornithopoda; Maryańska and Osmólska 1985; Sereno 1986). However, an obturator process is never seen in quadrupedal ornithischians, and a weak case may be made (Sereno 1984; Maryańska and Osmólska 1985) for uniting the Ceratopsia with the Thyreophora as persistently quadrupedal ornithischians. By this interpretation, quadrupedality is held to be a primitive ornithischian character that persisted in ceratopsians but was lost in the sister group Pachycephalosauria.

Ceratopsids developed complex dental batteries that paralleled those of hadrosaurids to some extent (Ostrom 1966). The development of frills and horns in neoceratopsians and the demonstration of sexual dimorphism (Kurzanov 1972; Dodson 1976; Lehman in press) suggests that neoceratopsians were gregarious, social animals. Their remains are often associated in monotypic bone beds (Currie 1981, 1987*b*; Currie and Dodson 1984; Lehman 1989), which also suggests herding. Ceratopsids had the largest brains relative to body size of all quadrupedal dinosaurs (Hopson 1977).

Neoceratopsians have one of the finest fossil records of any group of dinosaurs, both in terms of numbers and quality of specimens and in their contribution to contemporary faunas. However, most species are of late Campanian or Maastrichtian age, and so it has not been possible to trace lineages through time very convincingly.

The first ceratopsian material to be described includes species of *Agathaumas* Cope, 1872; *Polyonax* Cope, 1874; *Dysganus* Cope, 1876*a*; and *Monoclonius* Cope, 1876*a*. All of these are based on very incomplete material from which no concept of the ceratopsians could be derived. The first complete ceratopsian cranial material was that of *Triceratops* Marsh, 1889*b*. The first comprehensive study of ceratopsids was that by Hatcher et al. (1907). At that time, complete skulls were known for only *Triceratops* and *Torosaurus*; nevertheless, the 1907 monograph remains the most valuable single reference on ceratopsid anatomy. A sequel (Lull 1933) provides a useful review of many taxa described since 1907. *Leptoceratops* Brown, 1914*c*, was the first protoceratopsid to be described, but these animals were not well known until *Protoceratops* was discussed in a monograph by Brown and Schlaikjer (1940*a*). Further protoceratopsid diversity was revealed by renewed work in Mongolia (Maryańska and Osmólska 1975). No new ceratopsian taxa were named in North America between *Pachyrhinosaurus* Sternberg, 1950, and *Avaceratops* Dodson, 1986.

Although well-preserved neoceratopsian skulls are common, and most taxa are known from at least several specimens, the overwhelming majority consist of mature skulls with partial or complete closure of sutures. Complete skeletons are less common, and several ceratopsids have no reliably associated postcranials (e.g., *Monoclonius crassus*, *Torosaurus latus*, *Arrhinoceratops brachyops*). *Triceratops horridus* is known from more skulls than any other ceratopsid, but not a single complete skeleton is known; all reconstructions are composite (Gilmore 1919*b*; Osborn 1933; Erickson 1966). Discoveries and descriptions of certain skulls have thus had disproportionate influence on our understanding of certain details and homologies of neoceratopsian cranial anatomy. Noteworthy are *Brachyceratops montanensis* (Gilmore 1917; see also Lambe 1915 and Sternberg 1927*a*) and *Protoceratops andrewsi* (Brown and Schlaikjer 1940*a*). The anatomical diagnosis of the Ceratopsia by Romer (1956), which is taxonomically equivalent to the Neoceratopsia of Sereno (1986), is noteworthy for its depth and insight.

DIAGNOSIS

The Neoceratopsia can be diagnosed as follows: head very large relative to body, rostral and predentary sharply keeled and ending in a point, quadrate sloping rostrally, prominent broad parietal frill smoothly confluent with supratemporal fossa, exoccipitals exclude basioccipital from foramen magnum, predentary with bifurcated caudoventral process and laterally sloping triturating surface, teeth with ovate crowns in buccal view, maxillary teeth with prominent buccal primary ridge, cervicals 1 to 3 (or 4) fused to form syncervical, ischium gently decurved (Sereno 1986).

ANATOMY

Skull and Mandible

The skulls of neoceratopsians (figs. 29.1, 29.2, 29.3, 29.4) are broadly triangular in dorsal view, flaring widely across the jugals and tapering rostrally to a narrow, keeled, edentulous beak. A deep face is characteristic, and the tall premaxilla usually accounts for much of this height. The erect rostral that abuts the premaxilla and rounds out the profile of the face is a defining synapomorphy of the Ceratopsia (Maryańska and Osmólska 1975; Sereno 1986). The rostral has a rugose texture and was probably covered by horn in life. The ventral cutting edge is relatively thin and beveled. The rostral fuses to the premaxilla only in very mature individuals but frequently fails to be preserved in juveniles. The opposing predentary is longer and lower than the rostral. It terminates in a sharp point that rests inside the rostral. The bone is robust and the cutting edge is thick. Among ceratopsids, the cutting edge is relatively flat in chasmosaurines but strongly sloping in centrosaurines (Lehman 1990).

The premaxilla is prominent and in ceratopsids, may nearly equal the maxilla in length. A strap-like caudodorsally directed process often contacts the lacrimal in centrosaurines (e.g., *Centrosaurus apertus*, *Monoclonius crassus*), separating the nasal from the maxilla. A small, dorsal, elongate external naris is found in protoceratopsids, but in ceratopsids, the naris is enlarged, usually subcircular in shape, and surrounded by a depression in the premaxilla. In centrosaurines, a small process of the nasal and premaxilla projects into the caudal border of the external naris, but in chasmosaurines, a process projects dorsally from the rostral floor of the external naris. In chasmosaurines, there is either a depression or an interpremaxillary foramen in front

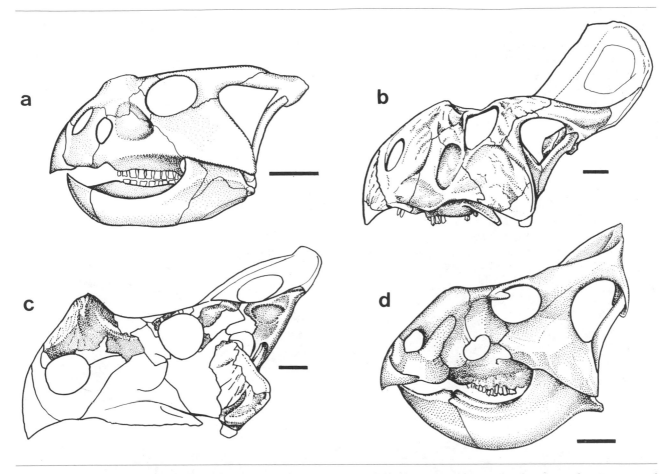

Fig. 29.1. Skulls of representative Protoceratopsidae. a, *Bagaceratops rozhdestvenskyi* (after Maryańska and Osmólska 1975); b, *Protoceratops andrewsi* (after Brown and Schlaikjer 1940*a*); c, *Montanoceratops cerorhynchus* (after Brown and Schlaikjer 1942); d, *Leptoceratops gracilis* (after D. A. Russell 1970*b*). Scale = 5 cm.

of the external naris (poorly developed in *Arrhinoceratops brachyops*, Tyson 1981). The ventral edge is sectorial (more so in centrosaurines than in chasmosaurines), occluding with the predentary. Medial flanges of the premaxillae are overlapped by premaxillary processes of the maxillae and together form a short hard palate. *Protoceratops andrewsi* is unique in the possession of two premaxillary teeth, either a primitive retention or a secondary specialization.

In protoceratopsids, the maxilla is tall and forms about two-thirds of the height of the face, while in ceratopsids, it is usually about half the height of the face. In protoceratopsids, there is fairly broad contact between the maxilla and jugal, but in ceratopsids, the contact is quite restricted. Here, a prominent longitudinal ridge on the dorsal maxilla marks the contact with the jugal. Among neoceratopsians generally, the dentigerous portion of the maxilla lies medial to the jugal ridge, and in ceratopsids, it continues caudad some distance me-

dial to the jugal. The caudal portion is rather delicate and is often lost in disarticulated specimens. In ceratopsids, the antorbital foramen is situated in a cleft on the dorsal maxilla between its jugal and lacrimal processes. There is great individual variability in the prominence of the foramen, and in some specimens it can barely be detected. In protoceratopsids, there is a prominent circular antorbital fossa with a small antorbital fenestra; in *Leptoceratops gracilis*, the fenestra is large. In *P. andrewsi* and *Bagaceratops rozhdestvenskyi*, there is a prominent sinus that communicates with the antorbital fossa but not the nasal cavity (Osmólska 1986).

The small quadrilateral lacrimal forms the rostral margin of the orbit between the jugal ventrally and the prefrontal dorsally. The nasal usually borders the lacrimal rostrodorsally. The lacrimal roofs the antorbital fenestra. Early fusion of the lacrimal frequently obscures its borders with adjacent elements.

In protoceratopsids, the nasal is relatively long

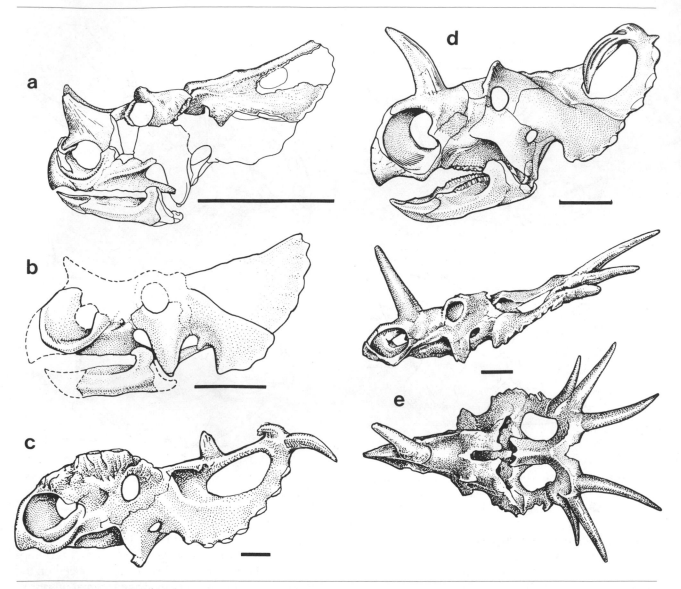

Fig. 29.2. Skulls of representative Centrosaurinae. a, *Brachyceratops montanensis* (after Gilmore 1917); b, *Avaceratops lammersi* (after Dodson 1986); c, *Pachyrhinosaurus canadensis*; d, *Centrosaurus apertus* (after Lull 1933); e, *Styracosaurus albertensis* (lateral and dorsal views, after Lull 1933). Scale = 20 cm.

and narrow, but in ceratopsids, it expands ventrally to become a major element of the face. Arching of the nasal suggesting an incipient nasal horn core is a general feature of the Protoceratopsidae (hence the name of the family), but only *B. rozhdestvenskyi* (otherwise very primitive) and *Montanoceratops cerorhynchus* show a differentiated nasal horn base. Among ceratopsids, the nasal horn core in centrosaurines is clearly formed by fusion of paired nasals; many specimens show this condition, first described in *Brachyceratops montanensis* (Gilmore 1917). In centrosaurines, nasal horn cores range from very low to 500 mm in height for *Styraco-*

saurus albertensis. In chasmosaurines, a separate center of ossification for the nasal horn core was claimed by Hatcher et al. (1907), disputed by Sternberg (1949), then reiterated by Ostrom and Wellnhofer (1986) and Lehman (in press). The "epinasal" bone described by Lambe (1915) in "*Eoceratops*" *canadensis* is almost certainly not a consistent separate center of ossification but only an anomaly characterizing one side of the nasal of a single specimen. The situation remains ambiguous due to the lack of convincing juvenile material (the smallest described skull of *Triceratops* [Schlaikjer 1935] is 1.38 m long). It is noteworthy, however, that

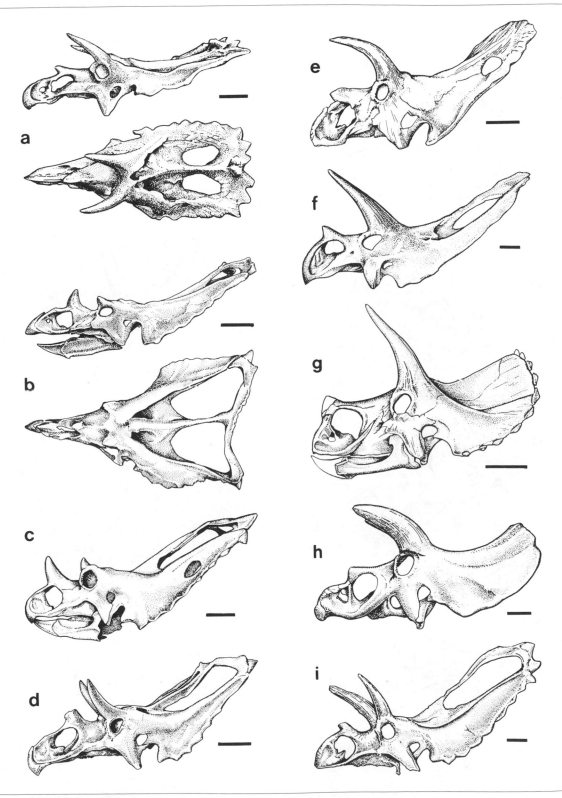

Fig. 29.3. Skulls of representative Chasmosaurinae. a, *Anchiceratops longirostris* (lateral and dorsal views, after Lull 1933); b, *Chasmosaurus belli* (lateral and dorsal views, after Lull 1933); c, *Chasmosaurus canadensis* (after Lull 1933); d, *Chasmosaurus belli* (after Lull 1933); e, *Arrhinoceratops brachyops* (after Lull 1933); f, *Torosaurus latus* (after Lull 1933); g, *Triceratops horridus* (after Hatcher et al. 1907); h, *Triceratops horridus* (after Lull 1933); i, *Pentaceratops sternbergii* (after Lull 1933). Scale = 20 cm.

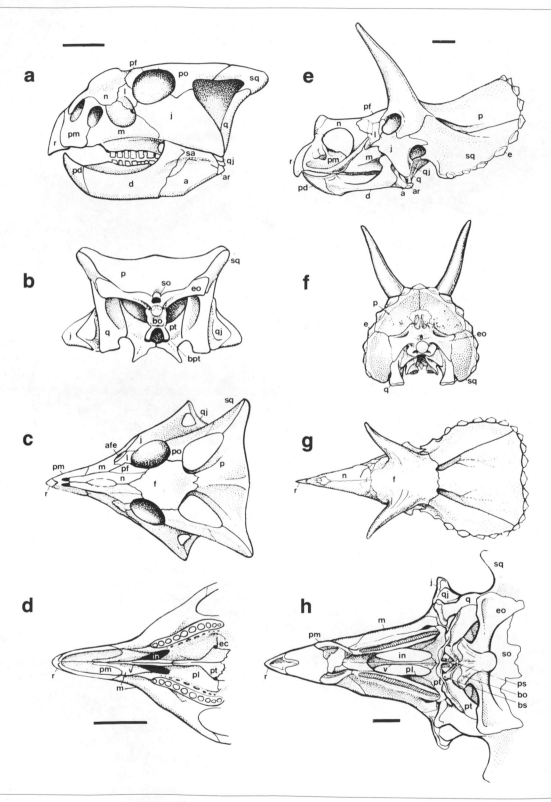

Fig. 29.4. Comparison of protoceratopsid and ceratopsid skulls in lateral (a, e), occipital (b, f), dorsal (c, g) and palatal (d, h) views. a-d. *Bagaceratops rozhdestvenskyi* (after Mary-ańska and Osmólska 1975). e, h. *Triceratops horridus* (after Hatcher et al. 1907). Scale for a, b, c, d = 5 cm, scale for e, f, g = 20 cm, scale for h = 10 cm.

the variable morphology of the nasal horn in species of *Triceratops* (Ostrom and Wellnhofer 1986; Lehman in press) is consistent with frequent loss of a loose osteoderm. In almost all ceratopsid skulls, left and right nasals are fused caudal to the nasal horn.

Bones of the postnasal midline series and of the dorsal orbital series remain little changed in protoceratopsids but are highly modified in ceratopsids. The relevant sutures were subject to relatively early fusion, and few skulls show them. Interpretations are largely based on *Styracosaurus albertensis* (Lambe 1913; Sternberg 1927*a*); "*Eoceratops*" *canadensis* (Lambe 1915); *Brachyceratops montanensis* (Gilmore 1917); *Monoclonius crassus* (Sternberg 1940*b*); and *Chasmosaurus mariscalensis* (Lehman 1989). Two salient changes involve the prefrontal and supraorbital (= palpebral of ornithopods and protoceratopsids). Medial expansion of the prefrontals results in midline contact of these elements, with consequent separation of the nasals from the frontals. In *C. mariscalensis*, the prefrontals are small and fail to meet on the midline. More problematic is the supraorbital (Coombs 1972). In protoceratopsids, the palpebral articulates loosely with the prefrontal and projects into the orbit. In ceratopsids, a supraorbital is never identifiable on the skulls of mature individuals but is described as being fused with the prefrontal and excluding the prefrontal from the orbital rim. Evidence for the supraorbital is seen in certain prefrontals (e.g., Sternberg 1927*a*; Horner pers. comm.) with an elaborate sutural pattern on the orbital rim, indicating a missing element, but such specimens are very rare.

In *Bagaceratops rozhdestvenskyi*, the postorbital is a relatively restricted element between the orbit and the infratemporal fenestra. With reduction in the size of the infratemporal fenestra, the postorbital is excluded from the fenestra (e.g., *Protoceratops andrewsi*). In ceratopsids, the postorbital is a major bone of the dorsal orbit and cheek. The orbital horn cores are outgrowths of this bone, and juvenile specimens and ontogenetic series (Gilmore 1917; Brown and Schlaikjer 1940*a*, 1940*b*; Currie 1981) show that only a single center of ossification is involved. However, certain centrosaurines including *Styracosaurus albertensis* and *Pachyrhinosaurus canadensis*, show irregular depressions in the orbital region surrounded by suture-like surfaces that suggest the loss of a separate center of ossification. Centrosaurine orbital horn cores are usually less than 150 mm in height, while orbital horn cores of large chasmosaurines may approach 1 m in length and have massive, hollow bases.

The enlarged jugal is highly derived in the Ceratopsia and ranks along with the rostral and parietal bones as the most characteristic bones of the infraorder. It has the form of an inverted triangle, the base of which forms the ventral rim of the orbit. A ventrolaterally flaring jugal boss is already evident in psittacosaurids, the sister group of the Neoceratopsia. The distal tip of the jugal is often thickened (particularly in *Pentaceratops sternbergii* and *Arrhinoceratops brachyops*), and this thickening may be further accentuated by an epidermal ossification, the epijugal. Jugal-squamosal contact is incipient in *Leptoceratops gracilis* and *Protoceratops andrewsi*. Broad jugal-squamosal contact dorsal to the infratemporal fenestra characterizes ceratopsids, as does development of a squamosal process ventral to the temporal opening.

In the ornithopod *Hypsilophodon foxii* (Galton 1974*a*) and in the neoceratopsian sister group, the psittacosaurids (Sereno, this vol.), the quadrate is tall, erect, and completely exposed laterally. In neoceratopsians, there is a progressive reorientation of the cheek. The ventral end of the quadrate is rotated forward, the infratemporal fenestra is pushed ventrally and compressed (it remains relatively unreduced in protoceratopsids), and the jugal, quadratojugal, and ventral quadrate are telescoped such as to lie side by side rather than in series front to back. The quadrate is relatively shorter in ceratopsids than in protoceratopsids and tends to be covered in lateral view by processes from the squamosal and jugal caudal to the infratemporal opening. The ventral end of the quadrate is robust and transversely expanded, but the dorsal end is surprisingly thin. The major stresses of mastication must have been transmitted to the skull roof through the quadratojugal and jugal rather than through the shaft of the quadrate. This arrangement seemingly precluded any streptostyly (Weishampel 1984*a*). A distinct pocket for the pterygoid on the pterygoid ramus of the quadrate is characteristic of ceratopsids. The quadratojugal is a spacer element interposed between the ventral jugal and the quadrate. It is thick ventrally and thins dorsally, wrapping to the caudal surface of the midquadrate shaft. Although the quadratojugal is largely obscured in lateral view by the jugal (consistently more exposed in *Triceratops horridus* than in most other ceratopsids by virtue of a relatively narrow jugal), its degree of exposure adjacent to the infratemporal opening is variable both individually and taxonomically.

In protoceratopsids, the paired frontals form a major portion of the cranium, border the orbit dorsally, and form the rostral limit of the supratemporal fenestrae. In *Protoceratops andrewsi* and *Bagaceratops rozhdestvenskyi*, a pair of modest frontoparietal depressions in adult specimens is associated with the rostral borders of the supratemporal fenestrae, reflecting expansion of the attachments of the jaw adductor musculature. In ceratopsids, the fused frontals are restricted in extent and bound neither the orbit nor the supratemporal fenestrae (Lambe 1915; Lehman 1990). A unique ceratopsid feature is the development of a secondary roof over the braincase (Brown and Schlaikjer 1940a). This secondary roof is the result of actual folding of the frontal bones to create a median sinus over the braincase (Gilmore 1919b). When the sinus is open dorsally, it is called a frontal fontanelle (Sternberg 1927a) or postfrontal fontanelle (Gilmore 1919b), even though in chasmosaurines the fontanelle is principally bounded by the postorbitals (Lehman 1990). When the fontanelle is closed over (as in *A. brachyops* and many specimens of *Triceratops horridus*), it forms a complete secondary roof with an enclosed frontal sinus. The fontanelle is irregularly shaped and deep. The frontal fontanelle bears no relationship to the supratemporal fenestrae and its associated musculature. Only in *Torosaurus latus* (Marsh 1892c; Hatcher et al. 1907) is there a communication between the supratemporal fenestrae and unique, paired frontal fontanelles. In chasmosaurines, the frontal is small and restricted to the rostral border of the fontanelle and the postorbitals form the sides of the fontanelle, but in centrosaurines, the frontals are extensive and bound the sides (Lehman 1990).

Incipiently in protoceratopsids and fully developed in ceratopsids, the parietosquamosal frill that extends behind the skull (and may exceed the length of the skull by as much as 50% in chasmosaurines; e.g., *Pentaceratops sternbergii*) is most characteristic of neoceratopsians. Primitively (e.g., *Leptoceratops gracilis, Bagaceratops rozhdestvenskyi*), the squamosal forms a simple bar with the postorbital that separates the infra- and supratemporal fenestrae and provides a cotylus for the head of the quadrate. In *Protoceratops andrewsi*, there is a distinct if modest postquadrate expansion of the squamosal. In all ceratopsids, the robust squamosal is expanded ventrally and the jugal is expanded caudally. As a result, the infratemporal fenestra is compressed and reduced in size, and jugal-squamosal contact excludes the postorbital from the infratemporal fenestra. Jugal-squamosal contact may also occur caudal to the infratemporal opening in ceratopsids (e.g., usual in *Centrosaurus*; variable in *Chasmosaurus*; no contact in *Triceratops*). Most characteristic of ceratopsids is a tremendous postquadrate expansion of the squamosal. In centrosaurines, the squamosal is short relative to the length of the frill (less than 500 mm in length), and the length of the postquadrate portion of the squamosal exceeds that of the rostral portion by as much as three to two. In chasmosaurines, the squamosal is both relatively and absolutely enormous, reaching the back of the frill. In the relatively small Judithian *Chasmosaurus belli*, the squamosal approaches 1 m in length, and in the Lancian *Torosaurus latus* (Colbert and Bump 1947), the squamosal measures 1.43 m. In chasmosaurines, the caudal portion may exceed the rostral portion by four to one. The ceratopsid squamosal is constricted opposite the quadrate articulation to form a jugal notch, beyond which it flares widely. On the medial surface of the squamosal, the attachment of the exoccipital is marked by a prominent ridge, ventral to which is a cotylus for the head of the quadrate (covered in lateral view). In chasmosaurines, squamosal fenestrae, usually unilateral, appear to be a common intraspecific variant in all genera. Possibly they are injury related (Lull 1933; Molnar 1977). The caudal free border of the squamosal is often ornamented either by undulations or by epoccipital bones.

Identification of the major median element of the frill in ceratopsids was problematic (Hay 1909; Huene 1911; Brown 1914c; Gilmore 1919b until the discovery and description of the protoceratopsid, *Protoceratops andrewsi* (Granger and Gregory 1923; Gregory and Mook 1925). It then became clear (Sternberg 1927a; Brown and Schlaikjer 1940a) that the element in question was the parietal, as asserted by Marsh (1891a) and Hatcher et al. (1907). In psittacosaurids, the near outgroup of neoceratopsians, the parietal roofs the brain and forms the medial border of the supratemporal fenestra (Sereno, this vol.). In neoceratopsians, the fused parietals are variably elongated and laterally expanded to form the central portion of the fan-shaped frill. In protoceratopsids, the parietal continues to roof the brain, but in ceratopsids, it barely participates in the braincase at all; here, it forms the caudal border of the frontal fontanelle. In *Leptoceratops gracilis*, the frill is unfenestrated, but in all other protoceratopsids (including *Bagaceratops rozhdestvenskyi*, Osmólska pers. comm.) and most ceratopsids, the frill is fenestrated. Parietal fenestrae are hypertrophied in species of *Chasmosaurus* and in *Pentaceratops sternbergii* but are secondarily closed in *Triceratops horridus* and in *Avaceratops lammersi*. The parietal is generally thickest sagittally and along its borders and is thinnest in areas lateral to the midline, regardless of whether or not there are parietal fenestrae. The free border of the parietal in ceratopsids often bears scallops, spikes, or dermal ossicles. Although areas adjacent to the supratemporal fenestrae have a smooth surface suggestive of muscle attachment, much of the surface of the parietal may have a pattern of vascular impressions that seems incompatible with such attachment and that suggests instead a closely adhering thick epidermis on both the dorsal and ventral surfaces.

The protoceratopsid palate (fig. 29.4) has recently been described (Osmólska 1986), but the palate of ceratopsids has received little attention (Hatcher et al. 1907; Lull 1933; Sternberg 1940b; Brown and Schlaikjer 1940a). Protoceratopsids have extensive palatines, well-ossified vomers, and ectopterygoids exposed on the surface of the palate. The internal nares (choanae) are positioned far forward and are oblique to the palatal plane, due to the narrowness and vaulting of the snout (Osmólska 1986). The secondary palate is short. In protoceratopsids (*Bagaceratops rozhdestvenskyi, Leptoceratops gracilis*), there is a palatine foramen situated between the pterygoid, ectopterygoid, and palatine. The palate is strongly vaulted both transversely and, in *B. rozhdestvenskyi* at least, longitudinally as well. In ceratopsids, the choanae are relatively enormous because of restriction of the palatines caudally. The palatine foramen is lost, and the ectopterygoids are restricted in size and eliminated from exposure on the ventral palate by contact between the maxilla and the palatine. The vomer is a straight median bar running between the pterygoids caudally and the palatal processes of the maxillae rostrally. The intervomerine suture remains prominent in *Protoceratops andrewsi* and *B. rozhdestvenskyi*. In protoceratopsids, the vomer rises steeply caudodorsally to meet the rostrodorsally inclined longitudinal process of the palatine. This orientation divides the nasal cavity into narrow, paired channels (Osmólska 1986). The vomer is rarely observed in ceratopsids despite the abundance of skulls. It has been reported only rarely in *Triceratops horridus* (Hatcher et al. 1907) and *Chasmosaurus belli* (Lull 1933). In these, the vomer runs horizontally. The distinctive palatine of neoceratopsians is restricted to the caudal half of the maxilla. A caudal transverse wing of the palatine achieves a vertical orientation to contact the lacrimal, prefrontal, and jugal medially in the vicinity of the antorbital foramen. In protoceratopsids, there is also a prominent parasagittal process that extends rostrodorsally to meet the vomers. Expansion of the palatine is accompanied by restricton of the ectopterygoid to a small, flattened bone on the dorsum of the caudal process of the maxilla. Contact of the ectopterygoid with the palatine has been lost in *Leptoceratops gracilis* (Sternberg 1949) and in *Centrosaurus apertus* (Sternberg 1940b). The pterygoid is a complex bone that retains its primitive role as a link between the quadrate and palate, with a brace on the braincase. In *P. andrewsi*, a concavity passes across the ventral surface of the pterygoid adjacent to the basipterygoid articulation and onto the quadrate ramus (Brown and Schlaikjer 1940a). This concavity seems to have been a precursor to the auditory tube (eustachian canal) that is characteristically channeled into the medial surface of the pterygoid in ceratopsids. The pterygoid is relatively smaller and the quadrate process of the pterygoid is more strongly developed in ceratopsids than in protoceratopsids.

The neoceratopsian braincase (fig. 29.5) shows few derived features. A rostral extension of the supraoccipital contacts the prootic and the laterosphenoid in all neoceratopsians. In protoceratopsids, the supraoccipital forms the dorsal border of the foramen magnum, but in ceratopsids, the supraoccipital is excluded from the foramen magnum by the exoccipitals. Distinct vertical elliptical depressions mark the attachment of M. rectus capitis on the supraoccipital of ceratopsids. In ceratopsids, a pair of strong rostrodorsal processes of the supraoccipital support the parietals at their junction with the frontals. In neoceratopsians, the occiput is broad, and the exoccipitals form long, slender, straplike processes that extend laterally to reach the squamosal and usually contribute to the support of the dorsal quadrate. In ceratopsids, the ends of the exoccipitals may become fan-shaped. In *P. andrewsi* (Brown and Schlaikjer 1940a) and *B. rozhdestvenskyi* (Maryańska and Osmólska 1975), there are three foramina, presumably for the exits of cranial nerves IX and X, XI, and XII, respectively. In ceratopsids, there are but two foramina, those for c. n. XI and XII being confluent. Contact between the exoccipital and the prootic is reduced in ceratopsids. In ceratopsids, the exoccipitals may show prominent depressions that indicate the attachment of M. longissimus capitis. The distal ends of the exoccipitals are moderately expanded in ceratopsids. In protoceratopsids, the exoccipitals do not roof the foramen magnum and make only a slight to moderate contribution to the occipital condyle. In ceratopsids, the exoccipitals meet above the foramen magnum and form two-thirds of the condyle. In neoceratopsians, the basioccipital is excluded from the foramen magnum by the exoccipitals, but only in ceratopsids is the basioccipital reduced to one-third of the ventral condyle. The large size and spherical form of the condyle with its distinct neck are diagnostic of neoceratopsians. In large ceratopsids, the occipital condyle is huge, measuring 96 mm in diameter in *Pachyrhinosaurus canadensis* (Langston 1975) and up to 102 mm in *Triceratops horridus* (Hatcher et al. 1907). The caudal face of the prominent basioccipital tubera are formed by the basioccipital, and a simple butt joint with the basisphenoid persists in all but the largest ceratopsids.

The opisthotic, prootic, and laterosphenoid retain

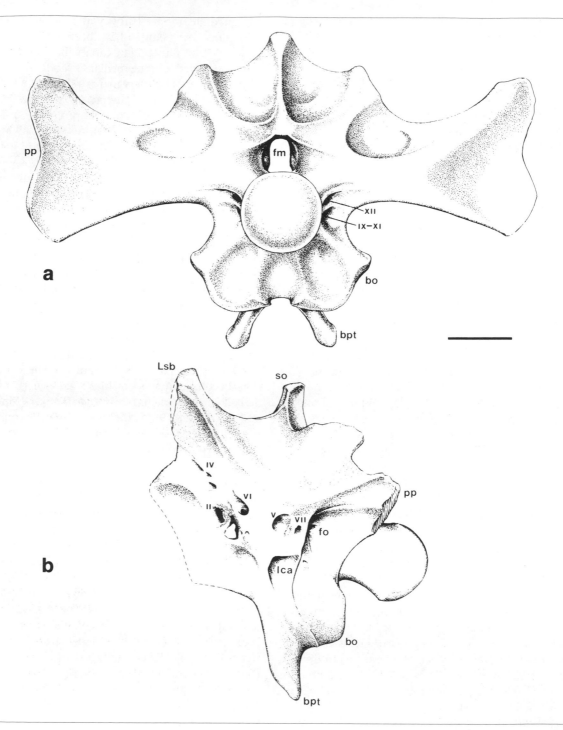

Fig. 29.5. Braincase of *Centrosaurus apertus* in occipital (a) and left lateral (b) views. Scale = 5 cm.

their usual morphology and contacts with adjacent bones and show the customary foramina. C. nn. I through IV exit the laterosphenoid, c. n. V passes out between the laterosphenoid and the prootic, and c. n. VII pierces the prootic. The orbitosphenoid is appar-ently ossified in *B. rozhdestvenskyi* and in ceratopsids, although it is rarely seen as separate from the latero-sphenoid due to early fusion of these elements. A pre-sphenoid is probably present as well. The basisphenoid is indistinguishable from the parasphenoid. The basicranial complex is pierced on the midline by a eustachian opening in neoceratopsians, except in *B. rozhdestenskyi*, in which the opening passes between the basioccipital and basisphenoid. The basipterygoid processes are shorter and more vertical in ceratopsids than in protoceratopsids. The rostral process of the parasphenoid is embraced by the pterygoids in *P. andrewsi* but contacts the vomer in ceratopsids.

The dentary is long, low, and straight (in *P. andrewsi* and *L. gracilis*, the dentary is deep and the ventral border is curved), and the coronoid process is pronounced. The coronoid process slopes up gradually from the ramus in protoceratopsids but is erect and strongly set off from the ramus, although relatively low, in ceratopsids. The apex of the coronoid process in ceratopsids is expanded rostrally and overhangs the ramus. The rostral, edentulous portion of the jaw is relatively short and terminates bluntly. A strong lateral ridge runs the length of the dentary and is confluent with the coronoid process; thus, the tooth row is strongly inset and passes medial to the coronoid process. The dentary is thickest at the base of the coronoid process, where it surrounds the Meckelian fossa in which M. intramandibularis is situated. The fossa is covered ventrally by an extensive splenial. The articular is of remarkable breadth in large ceratopsids such as *Triceratops horridus* and *Pentaceratops sternbergii*, corresponding to the breadth of the quadrate. The surangular is tall, forming half the coronoid process in protoceratopsids, and still reaching the apex of the coronoid process in ceratopsids.

The crowns of both maxillary and dentary teeth of neoceratopsians are leaf shaped, enameled on one side (buccally and lingually, respectively), and bear a strong median ridge, usually but not always asymmetrically placed (distad and mesiad on maxillary and dentary teeth, respectively). Secondary ridges are more prominent in protoceratopsids than in ceratopsids. The crown is set at a high angle to the root, resulting in an occlusal plane that is steep (greater than 60° in protoceratopsids) or vertical (ceratopsids; Ostrom 1966). In *Leptoceratops gracilis*, the dentition combines vertical shear with a horizontal shelf on the dentary teeth, such that the functions of shear and crushing are combined (Sternberg 1951*a*; Ostrom 1966). In protoceratopsids, tooth replacement is primitive; there is a functional tooth with a single replacement tooth at each position. In ceratopsids, there is a dental battery paralleling that of hadrosaurids, with three to five teeth in each vertical row. The tooth crowns interlock, while the roots that brace each tooth on its replacement are uniquely split, imparting great stability. A labile, vascular bone of attachment, similar to mammalian cement, further sta-

bilized each tooth in the battery. The number of maxillary tooth positions increases ontogenetically in protoceratopsids (e.g., from 8 to 15 in *P. andrewsi*). An increase from 20 to 28 positions is reported for *Chasmosaurus mariscalensis* by Lehman (1989), but ontogenetic data on tooth number are unknown for other ceratopsids. Total number of tooth positions generally follows absolute skull size, ranging from 10 in *B. rozhdestvenskyi* to 15 in *P. andrewsi*, 17 in *L. gracilis*, 20 in small and large *Brachyceratops montanensis*, 28 to 31 in *Centrosaurus apertus*, and 36 to 40 in *Triceratops horridus*. *P. andrewsi* and *Breviceratops kozlowskii* are unique among the Neoceratopsia in possessing premaxillary teeth.

Hyoid bones have been described for *Centrosaurus apertus* (Parks 1921), *Triceratops horridus* (Lull 1933), *P. andrewsi* (Colbert 1945), and *L. gracilis* (Sternberg 1951*a*). In *C. apertus* and *L. gracilis*, they take the form of simple slender rods, in *P. andrewsi* they have a flattened, bean shape, and in *T. horridus* a curved, irregular form.

Sclerotic rings are rarely preserved in neoceratopsians. A partial ring in *P. andrewsi* was described by Brown and Schlaikjer (1940*a*). It contains 12 complete and three partial plates, overlapping in typical fashion. A partial ring in "*Monoclonius nasicornus*" (= ?*Styracosaurus albertensis* female; Dodson in press) includes only five plates (Brown 1917); the thin plates have a fine rugose texture.

There is an interesting tendency in ceratopsids to fuse epidermal ossicles to the skull. Such osteoderms have been termed epinasal (Lambe 1915; Gilmore 1917; but of doubtful validity—Sternberg 1949); epijugal (Hatcher et al. 1907), and epoccipital (Marsh 1891*a*). Hatcher et al. (1907) pointed out the inappropriateness of the term *epoccipital*, as these ossicles adorn the squamosal and parietal but never the occipital bones, but the name has stuck. Possibly, the orbital horn of *Styracosaurus albertensis* (Lambe 1913, 1915) is another example of a unitary tendency among neoceratopsians to add epidermal ossifications to the skull. Modest epijugals are reported in *L. gracilis* (Sternberg 1951*a*), *P. andrewsi* (Brown and Schlaikjer 1940*a*), and *Montanoceratops cerorhynchus* (Brown and Schlaikjer 1942). Among ceratopsids, epijugals are more pronounced among chasmosaurines. In *Arrhinoceratops brachyops* (Parks 1925) and *Pentaceratops sternbergii* (Osborn 1923*b*), the epijugal forms a pronounced jugal "horn" whose thickness exceeds the width of the ventral end of the quadrate. However, epijugals do not appear to be very prominent in the chasmosaurine *Triceratops*, although they are seen as distinct in some material of *T. horridus* (Hatcher et al. 1907). Epoccipitals are found only in ceratopsids and appear to be more prominent among some chasmosaurines (e.g., species of *Chasmosaurus; P. sternbergii; Triceratops horridus*) than among any centrosaurine.

Postcranial Skeleton

Axial Skeleton

Although there are many very fine neoceratopsian specimens, very few have complete vertebral columns (fig. 29.6). Indeed, the type of "*Monoclonius nasicornus*" (Brown, 1917; = ?*Styracosaurus albertensis*), two specimens of *Leptoceratops gracilis* (Sternberg 1951*a*), and a skeleton of *Anchiceratops* "*longirostris*" (Lull 1933; = *A. ornatus*) may have the only complete vertebral columns in existence. The best skeleton of *Protoceratops andrewsi* described by Brown and Schlaikjer (1940*a*) was complete only to the level of caudal 32; they estimated the complete caudal count as greater than 40. There is more certainty about the presacral and sacral portions of the column. A typical ceratopsid vertebral formula consists of 10 cervicals, 12 dorsals, and 10 sacrals. Complete caudal counts, where known, range from 38 to 50.

Coalescence of the centra and neural arches of the first cervicals to form a syncervical (Ostrom and Wellnhofer 1986) or cervical bar (Langston 1975) characterizes all neoceratopsians but only in adults of the protoceratopsid *L. gracilis* (Sternberg 1951*a*). It has been controversial whether the syncervical incorporates the first three (Brown 1917) or the first four (Hatcher et al. 1907) cervicals; Romer (1956) accepted three or four cervicals. The consensus for many years has favored the interpretation of three cervicals (e.g., Parks 1921; Lull 1933; Brown and Schlaikjer 1940*a*; Sternberg 1951*a*; Langston 1975; Sereno 1986). In *L. gracilis*, the atlas is small, bears no neural spine, and three cervicals coossify only when large size is achieved (Sternberg 1951*a*). Ostrom and Wellnhofer (1986) reasonably point out the unlikelihood that the atlas had a separate neural arch in *Triceratops horridus*. Acceptance of this view implies that the atlas has been severely reduced in most neoceratopsians and that the small cranial element (complete ring in *T. horridus*, Hatcher et al. 1907; ventral intercentrum in *P. andrewsi*, Brown and Schlaikjer 1940*a*; intermediate in size in *Montanoceratops cerorhynchus*, Brown and Schlaikjer 1942) represents a vestigial atlas. The neoceratopsian syncervical, except in *L. gracilis*, thus apparently incorporates a fused atlas-axis complex as well as cervicals 3

and 4. This view is herein accepted, and vertebral counts are correspondingly adjusted.

Cervicals 5 to 10 in protoceratopsids are subequal in length, and the neural spines are reduced (Brown and Schlaikjer 1940*a*). In ceratopsids, these cervicals are relatively short and broad. The neural spines are uniform in height in *Triceratops horridus* (Hatcher et al. 1907; Ostrom and Wellnhofer 1986), but in other ceratopsids, the spines form a graded series increasing in height caudally. The cervical neural spines are relatively low in long-frilled *Chasmosaurus belli* (Sternberg 1927*b*) and *Pentaceratops sternbergii* (Wiman 1930) but apparently relatively high in *Anchiceratops ornatus* (Lull 1933). Adventitious fusion of cervicals 5 and 6 is reported in *Protoceratops andrewsi* (Brown and Schlaikjer 1940*a*) and *Styracosaurus albertensis* (Brown and Schlaikjer 1937) and of cervicals 6 and 7 in *Centrosaurus apertus* (Lull 1933). Cervical 10 is the last presacral vertebra with the capitular facet on the centrum (Brown 1917). Double-headed ribs are borne on all centra beginning with cervical 3. The tuberculum is prominent, and the shaft is short and straight. Elongation of the ribs usually begins with the sixth, except that in *Triceratops*, the eighth rib still is short (Ostrom and Wellnhofer 1986). The first long rib articulates with cervical 9 (considered dorsal 1 by Hatcher et al. 1907).

The usual number of free dorsals in ceratopsians is 12, but up to 3 additional dorsal centra may be incorporated into the sacrum as sacrodorsals. In protoceratopsids, dorsal centra are relatively long and low. In ceratopsids, they are short and high. Mid-dorsals in ceratopsids are significantly narrower than mid-cervicals, and this narrowing is more striking in chasmosaurines than in centrosaurines, corresponding to the relatively larger skulls in chasmosaurines. Faces of the centra are either round or pear-shaped. The neural canal decreases in diameter caudally down the dorsal series. Transverse processes are strongly elevated by dorsal 3 but decrease in prominence in the caudal half of the series, the region in which the zygapophyses increase in prominence. Neural spines are robust and generally show little variation in height, although there is somewhat more variation in *Chasmosaurus belli* (Sternberg 1927*b*) and *Pentaceratops sternbergii* (Wiman 1930) than in others. The tallest dorsal spine varies in position from dorsal 3 (*Leptoceratops gracilis*, Sternberg 1951*a*), to dorsal 5 to 7 (*C. belli, P. sternbergii*, Sternberg 1927*b*; Wiman 1930), to dorsal 7 to 8 (*Centrosaurus apertus*, Brown 1917, Lull 1933), to dorsal 9 (*Protoceratops andrewsi*, Brown and Schlaikjer 1940*a*), and back to dorsal 12 (*Montanoceratops cerorhynchus*, Brown and Schlaikjer 1942). Bundles of ossified epaxial tendons

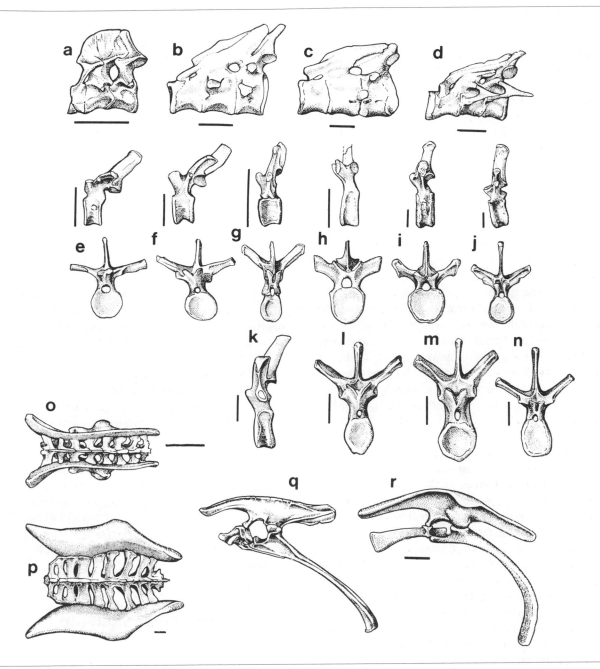

Fig. 29.6. Axial skeleton and pelvis of neoceratopsians. a, syncervical of *Protoceratops andrewsi* (left lateral view, after Brown and Schlaikjer 1940*a*); b, syncervical of *Styracosaurus albertensis* (left lateral view, after Langston 1975); c, syncervical of *Pachyrhinosaurus canadensis* (left lateral view, after Langston 1975); d, syncervical of *Triceratops horridus* (left lateral view, after Hatcher et al. 1907); e, cranial dorsal of *Centrosaurus apertus* (left lateral and cranial views, after Lull 1907); f, mid-dorsal of *Centrosaurus apertus* (left lateral and cranial views, after Lull 1933); g, dorsal vertebra of *Brachyceratops montanensis* (left lateral and cranial views, after Gilmore 1917); h, *Chasmosaurus canadensis* (left lateral and cranial views, after Tyson 1977); i, cranial dorsal of *Triceratops horridus* (left lateral and cranial views, after Hatcher et al. 1907); j, caudal dorsal of *Triceratops horridus* (left lateral and cranial views, after Hatcher et al. 1907); k, dorsal vertebra, *Centrosaurus apertus* (left lateral view, after Lull 1933); l, dorsal vertebra of *Styracosaurus albertensis* (cranial view, after Langston 1975); m, dorsal vertebra of *Pachyrhinosaurus canadensis* (cranial view, after Langston 1975); n, dorsal vertebra of *Triceratops horridus* (cranial view, after Hatcher et al. 1907); o, sacrum and ilia of *Montanoceratops cerorhynchus* (dorsal view, after Brown and Schlaikjer 1942); p, sacrum and ilia of *Triceratops horridus* (dorsal view, after Hatcher et al. 1907); q, pelvis of *Protoceratops andrewsi* (left lateral view, after Brown and Schlaikjer 1940*a*); r, *Chasmosaurus mariscalensis* (left lateral view, after Lehman 1989). Scales = 10 cm.

have been noted in the cranial dorsal region of *Styracosaurus albertensis* (Brown and Schlaikjer 1937), the mid-dorsal region of *Triceratops horridus* (Hatcher et al. 1907), and the dorsal and sacral regions of *L. gracilis* (Sternberg 1951*a*) and "*M. nasicornus*" (Brown 1917; = ?*S. albertensis*). In *L. gracilis*, the tendons lie at the bases of the neural arches, while in *T. horridus* they lie parallel at the summits of the arches. In "*M. nasicornus*," Brown described a "mass" of tendons, lacking any regular order. It is reasonable to assume that all ceratopsids had ossified tendons. Dorsal ribs are figured by Gilmore (1917), Brown and Schlaikjer (1942), and Ostrom and Wellnhofer (1986) and are described by Brown (1917). The second to sixth dorsal ribs are subequal in length, after which there is a variable shortening caudally. The twelfth rib may be quite weak. The cranial dorsal ribs are comparatively straight with a prominent tuberculum widely separated from the capitulum. In the middorsal region, the ribs are more curved and the tuberculum less prominent. Caudally, the tuberculum is weak and closer to the rib head.

In *L. gracilis*, there are six sacrals (Sternberg 1951*a*), in *Microceratops gobiensis*, there are seven (Maryańska and Osmólska 1975), and in *P. andrewsi* (Brown and Schlaikjer 1940*a*) and *Montanoceratops cerorhynchus* (Brown and Schlaikjer 1942) eight. Sacral fusion is less derived in protoceratopsids than in ceratopsids, but an acetabular bar, formed by the fusion of the distal ends of the sacral ribs, is present. Neural spines are tall and always separate from each other. The sacrum is not arched, and the ventral surface is not excavated. The sacrum of ceratopsids typically consists of 10 variably ankylosed centra (11 in *Pentaceratops sternbergii*; Wiman 1930), including three sacrodorsals, four sacrals with both transverse processes and sacral ribs, and three sacrocaudals. The degree of fusion among sacral vertebrae has a strong ontogenetic basis. In juvenile *Brachyceratops montanensis* (Gilmore 1917), there is no fusion, but eight centra are in sutural contact. A heavy, ventral acetabular bar typically runs half the length of the sacrum. Transverse processes are robust and abut the ilium; expanded distal ends may fuse (in *Triceratops horridus*, the cranial four and caudal two—Hatcher et al. 1907). In *T. horridus* and *P. sternbergii*, transverse processes 7 and 8 are the longest, and the following ones are shorter, thus imparting an "oval" shape to the sacrum in dorsal view. However, in *Centrosaurus apertus* (Lull 1933), the oval shape is scarcely discernible. The neural arches are low and broad and fused to a variable degree. In *P. sternbergii*, all spines are distinct; in *T. horridus*, spines 2 to 5 are fused; and in *Styracosaurus albertensis* (Brown 1917), spines 2 to 7 are fused. Sacral centra are wide and low, and the sacrum is arched dorsally,

with the result that the base of the tail is directed downward. The ventral surface of the sacrum is strongly excavated in chasmosaurines and less strongly so in centrosaurines (Lehman in press). In *B. montanensis*, sacral 1 is widest, while in *C. apertus*, sacrals 1 to 3 are wide, sacrals 4 to 7 are narrow, and the remaining sacrals widen again.

Complete caudal series are rarely found. The number of caudals in *S. albertensis* is 46 (Brown 1917), in *B. montanensis* 50 (Gilmore 1917), and in *Anchiceratops ornatus* 38 or 39 (Lull 1933). One individual of *Leptoceratops gracilis* has 38, and another 20 percent larger has 48 (Sternberg 1951*a*), suggesting ontogenetic increase. In *Pentaceratops sternbergii*, 28 caudals are preserved (Wiman 1930), and the last is so small that it is unlikely there were more than five others. The caudals are relatively simple, the size of all components decreasing steadily toward the end of the tail. Transverse processes terminate halfway down the tail (caudal 13 in *Protoceratops andrewsi*, 18 in *L. gracilis*, 19 in *Pentaceratops fenestratus*, 23 in *S. albertensis*, and 25 in *B. montanensis*) but reach relatively farther in ceratopsids than in protoceratopsids. Chevrons are about as long as corresponding neural spines but incline backward more strongly; they terminate about five segments from the end of the tail. A unique feature of *Protoceratops andrewsi* and *Montanoceratops cerorhynchus* is elongation of the proximal neural spines. In *P. andrewsi*, the neural spines increase in height until caudal 14, accounting for two-thirds the length of the tail, and then decrease to the end of the tail. The tail of *M. cerorhynchus* is apparently similar, although it is less completely known.

Appendicular Skeleton

The appendicular skeleton of neoceratopsians is generally conservative, with very few unique specializations (figs. 29.7, 29.8). A scapular spine is little developed in protoceratopsids but is more or less prominent in ceratopsids, usually crossing from the craniodorsal blade to the caudal supraglenoid ridge. The coracoid is moderately prominent, with a protracted caudal process that effectively limited the range of humeral extension. The coracoid typically fuses to the scapula in adults. In ceratopsids, the scapula forms two-thirds of the glenoid fossa and the coracoid only one-third. Ossified clavicles are known in *Protoceratops andrewsi* and *Leptoceratops gracilis* but have never been reported in ceratopsids. Paired, flattened, bean-shaped sternals are known throughout the neoceratopsians. There is al-

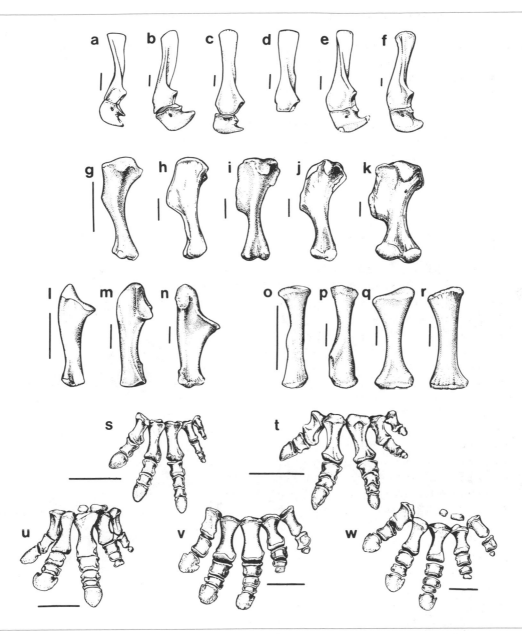

Fig. 29.7. Forelimb skeleton of neoceratopsians. a, left scapulocoracoid of *Leptoceratops gracilis* (lateral view, after Brown 1914c); b, left scapulocoracoid of *Centrosaurus apertus* (lateral view, after Lull 1933); c, left scapulocoracoid of *Centrosaurus apertus* (lateral view, after Parks 1921); d, left scapula of *Styracosaurus albertensis* (lateral view, after Langston 1975); e, left scapulocoracoid of *Monoclonius crassus* (lateral view, after Hatcher et al. 1907); f, left scapulocoracoid of *Triceratops horridus* (lateral view, after Hatcher et al. 1907); g, left humerus of *Leptoceratops gracilis* (lateral view, after Brown 1914c); h, left humerus of *Centrosaurus apertus* (lateral view, after Lull 1933); i, left humerus of *Centrosaurus apertus* (lateral view, after Parks 1921); j, left humerus of *Monoclonius crassus* (lateral view, after Hatcher et al. 1907); k, left humerus of *Triceratops horridus* (lateral view, after Hatcher et al. 1907); l, left ulna of *Leptoceratops gracilis* (caudal view, after Brown 1914c); m, left ulna of *Centrosaurus apertus* (caudal view, after Lull 1933); n, left ulna of *Triceratops horridus* (caudal view, after Hatcher et al. 1907); o, left radius of *Leptoceratops gracilis* (lateral view, after Brown 1914c); p, left radius of *Centrosaurus apertus* (lateral view, after Lull 1933); q, left radius of *Chasmosaurus belli* (lateral view); r, left radius of *Triceratops horridus* (lateral view, after Hatcher et al. 1907); s, left manus of *Protoceratops andrewsi* (dorsal view, after Brown and Schlaikjer, 1940a); t, left manus of *Leptoceratops gracilis* (dorsal view, after Brown and Schlaikjer 1940a); u, *Centrosaurus apertus*, left manus (dorsal view, after Lull 1933); v, *Centrosaurus apertus*, left manus (dorsal view, after Brown 1917); w, left manus of *Centrosaurus* sp. (dorsal view, after Brown, 1917). Scale = 10 cm.

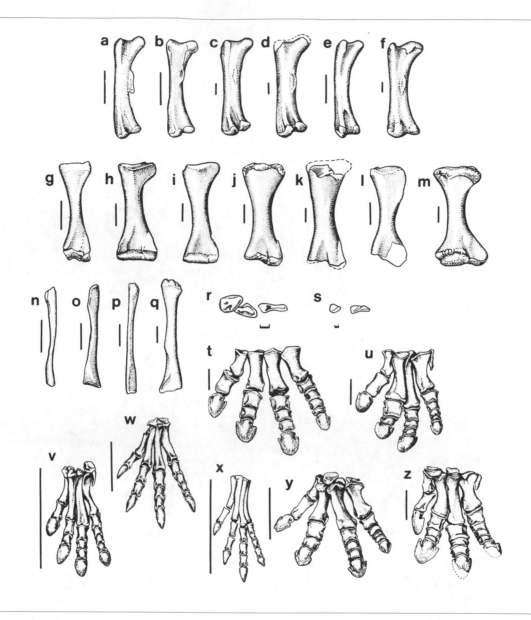

Fig. 29.8. Hindlimb elements of neoceratopsians. a, *Montanoceratops cerorhynchus* (caudal view of left femur, after Brown and Schlaikjer 1942); b, *Brachyceratops montanensis* (caudal view of left femur after Gilmore 1917); c, *Styracosaurus albertensis* (caudal view of left femur, after Langston 1975); d, *Pachyrhinosaurus canadensis* (caudal view of left femur, after Langston 1975); e, *Chasmosaurus belli* (caudal view of left femur); f, *Triceratops horridus* (caudal view of left femur, after Hatcher et al. 1907); g, left tibia, astragalus and calcaneum of *Montanoceratops cerorhynchus* (cranial view, after Brown and Schlaikjer 1942); h, left tibia, astragalus and calcaneum of *Centrosaurus apertus* (cranial view, after Lull 1933); i, left tibia, astragalus and calcaneum of *Monoclonius crassus* (cranial view, after Hatcher et al. 1907); j, left tibia of *Styracosaurus albertensis* (cranial view, after Langston 1975); k, left tibia of *Pachyrhinosaurus canadensis* (cranial view, after Langston 1975); l, left tibia of *Chasmosaurus belli* (cranial view) m, left tibia, astragalus and calcaneum of *Triceratops horridus* (cranial view, after Hatcher et al. 1907); n, left fibula of *Montanoceratops cerorhynchus* (lateral view, after Brown and Schlaikjer 1942); o, *Centrosaurus apertus*, left fibula (lateral view, after Lull 1933); p, *Monoclonius crassus*, left fibula (lateral view, after Hatcher et al. 1907); q, *Triceratops horridus* left fibula (lateral view, after Hatcher et al. 1907); r, left distal tarsals of *Brachyceratops montanensis* (proximal view, after Gilmore 1917); s, left distal tarsals of *Centrosaurus apertus* (proximal view, after Brown and Schlaikjer 1940a); t, left pes of *Centrosaurus* sp. (dorsal view, after Brown 1917); u, left pes of *Centrosaurus apertus* (dorsal view, after Brown 1917); v, left pes of *Protoceratops andrewsi* (dorsal view, after Brown and Schlaikjer 1940a); w, left pes of *Leptoceratops gracilis* (dorsal view, after Brown and Schlaikjer 1940a); x, left pes of *Microceratops gobiensis* (dorsal view, after Maryańska and Osmólska 1975); y, left pes of *Brachyceratops montanensis* (dorsal view, after Gilmore 1917); z, left pes of *Centrosaurus apertus* (dorsal view, after Lull 1933). r, s, scale = 1 cm; all other scales = 10 cm.

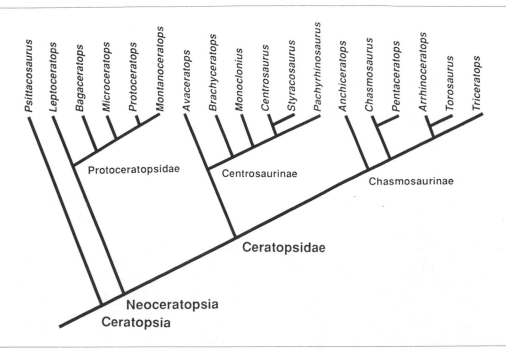

Fig. 29.9. Cladogram of the phylogenetic relationships of the Neoceratopsia.

ways a strong disparity in relative lengths of the fore-limbs and hindlimbs, forelimbs being typically only about 70 percent of the length of the hindlimbs. The limbs of ceratopsids are more massive than those of protoceratopsids. The head of the humerus is hemispheric and eccentric, extending onto the caudal proximal surface of humerus. In tiny *Microceratops gobiensis*, the humerus is light and elongate, the deltopectoral crest is very proximal, and the humeral head is strongly deflected medially (Maryańska and Osmólska 1975). The prominent deltopectoral crest extends distally half or more the length of the humerus in ceratopsids, in keeping with the increase in size and weight of large forms. Ceratopsids have a pronounced olecranon process on the ulna, longer in chasmosaurines than in centrosaurines (Lehman 1990). The manus is always smaller than the pes, but they are more equal in size in ceratopsids as compared with protoceratopsids. Five carpals are ossified in *P. andrewsi*, but only four are known in ceratopsids. The manus is comparatively broad. The phalangeal formula of the manus is 2-3-4-3-1 or 2. The unguals on the first three digits are somewhat blunt, and digits IV and V terminate in small nubbins of bone.

The ilium of *L. gracilis* is low and shows no eversion of the dorsal margin. In *P. andrewsi*, eversion of the preacetabular process is incipient and is slightly more derived in *Breviceratops kozlowskii* and *Montanoceratops cerorhynchus*. In ceratopsids, the entire dorsal margin is everted into a shelf-like structure. In protoceratopsids, the pubis is a small element. In ceratopsids, the prepubic process is pronounced, extending as far cranially as the tip of the ilium and flaring distally. In all neoceratopsians, the pubis itself is very short. In protoceratopsids, the ischium is long, slender, and relatively straight or gently decurved and lacks an obturator process. Ceratopsid ischia are more massive and strongly decurved, although the curve is stronger in chasmosaurines than in centrosaurines (Lehman 1990). In protoceratopsids, the tibia is longer than the femur, but in ceratopsids, the femur is always longer than the tibia. This is true even for juvenile ceratopsids (e.g., *Brachyceratops montanensis* Gilmore, 1917; *Avaceratops lammersi* Dodson, 1986). In ceratopsids, the greater and lesser trochanters tend to join. The fourth trochanter is pendent in protoceratopsids, including *M. cerorhynchus* (Brown and Schlaikjer 1942) but is reduced to a low prominence in ceratopsids. The distal condyles are more nearly equal in size than in protoceratopsids; however, the lateral condyle extends more distally than the medial one (Lehman 1982). From this, it may be inferred that the action of the femur was not precisely parasagittal. Calcanea are flattened and may be fused to the astragalus. Two distal tarsals are usually found, though three are reported in *Chasmosaurus belli* (Lull 1933) and *B. montanensis* (Gilmore 1917). The foot is compact in protoceratopsids, less so in ceratopsids. In *P. andrewsi* and *Microceratops gobiensis*, the foot is decidedly elongate. In all ceratopsids, the foot is relatively short. There are four functional metatarsals and frequently a small splint representing metatarsal V. The phalangeal formula in all neoceratopsians is 2-3-4-5-0. Unguals range from acute claws in *L. gracilis, M. gobiensis,* and *M. cerorhynchus* to blunt tapered in *P. andrewsi* and *Avaceratops lammersi* to the blunt-rounded, hoof-like unguals of typical ceratopsids.

Integument

Impressions of patches of skin have been described for several ceratopsids (Lambe 1914*b* on *Chasmosaurus belli*; Brown 1917 on "*Monoclonius nasicornus*" and "*M. cutleri*"; Sternberg 1925 on *Chasmosaurus belli*; Lull 1933 on "*M. cutleri*") but have not yet been reported in protoceratopsids. A large impression from the right pelvic area of *Chasmosaurus belli* (Sternberg 1925) shows a texture based on large circular plates up to 55 mm in diameter set in irregular rows and decreasing in size ventrally. The spacing between plates varies from 50 to 100 mm. Between the circular plates are fields of smaller, polygonal, nonimbricating scales. In "*M. cutleri*" (Brown 1917; Lull 1933, Pl. III), skin over the distal femur has the same general pattern as described for *Chasmosaurus belli,* except that the rounded plates are more widely separated.

PHYLOGENY AND EVOLUTION

Protoceratopsidae

Protoceratopsids (figs. 29.1, 29.4) represent a radiation of small dinosaurs (1 to 2.5 m long) that show at least incipient stages of many of the features that characterize the more derived family, the Ceratopsidae (table 29.1). All known protoceratopsids come from Upper Cretaceous sediments. They were successful dinosaurs independent of their presumed ancestry for ceratopsids, for they persisted through the Late Cretaceous. It is difficult to specify exact ages of the Mongolian protoceratopsids, because reliable correlations with European and North American marine and terrestrial sections remain elusive (Fox 1978; Lillegraven and McKenna 1986). Anomalously, protoceratopsids are unknown from the Nemegt Formation, the youngest of the Upper Cretaceous beds of Mongolia. The earliest protoceratopsid, *Microceratops gobiensis* Bohlin, 1953 (Santonian?), is somewhat poorly known, but its validity has been confirmed by Maryańska and Osmólska

(1975). It is a tiny animal, roughly 250 mm high at the hips, with highly cursorial hindlimbs suggestive of cursorial ornithopods, rather long front limbs, and a fenestrated frill. Late Maastrichtian *Leptoceratops gracilis,* one of the most primitive protoceratopsids, was one of the last dinosaurs in North America (Russell 1970*b*, 1984*c*). The skull of *Montanoceratops cerorhynchus* (Brown and Schlaikjer 1942; Sternberg 1951*a*) is poorly known, but *L. gracilis, Protoceratops andrewsi,* and *Bagaceratops rozhdestvenskyi* are each known from multiple specimens, with skulls spanning a significant size range (a factor of ten in *P. andrewsi*).

Monophyly of the Protoceratopsidae is questioned by Sereno (1986). There is a sharp discontinuity in size and in correlative allometric features between protoceratopsids and ceratopsids, and there is never confusion between members of one family and members of the other. However, Sereno (1986) believed that protoceratopsids consist of a group of successively more derived forms and that *Montanoceratops cerorhynchus,* the most derived member of the paraphyletic Protoceratopsidae, is the sister taxon of a monophyletic Ceratopsidae. The issue is unresolved on two accounts, as the skull of *M. cerorhynchus* is presently inadequately known, and so the hypothesized relationship between *M. cerorhynchus* and ceratopsids is purely hypothetical; the recent description of the palate of Asiatic protoceratopsids (Osmólska 1986) may demonstrate genuine autapomorphies of protoceratopsids. Characters that may ultimately establish a monophyletic Protoceratopsidae include a shallow, circular antorbital fossa, inclined parasagittal process of the palatine, and the maxillary sinus. The case for monophyly of protoceratopsids would be strengthened by the documentation of pre-Campanian ceratopsids, especially in Asia. Such evidence is presently lacking.

Protoceratopsid skulls show a mosaic of character states: there is no simple progression from primitive to derived, although the overall pattern is clear (fig. 29.9). All protoceratopsids have sharp, keeled, pendent rostrals that occlude with sharp, keeled predentaries, widely flaring jugals, quadrates that slope downward and forward, some degree of development of a parietal frill, large circular antorbital fossa with a small antorbital fenestra, vaulted palate with small rostral external nares, a syncervical, and a tibia equal to or longer than the femur. The ischium is nearly straight or gently decurved and lacks an obturator process. Unguals have the form of either acute or bluntly tapering claws. Trends that occur within the protoceratopsids include addition of vertebrae to the sacrum (*L. gracilis, B. rozhdestvenskyi,* six; *Microceratops gobiensis,* seven; *P. an-*

TABLE 29.1 Neoceratopsia

	Occurrence	Age	Material
Ceratopsia Marsh, 1890a			
Neoceratopsia Sereno, 1986			
Protoceratopsidae Granger et Gregory, 1923			
Bagaceratops Maryanska et Osmólska, 1975			
B. *rozhdestvenskyi* Maryańska et Osmólska, 1975	Red Beds of Khermeen Tsav, Omnogov, Mongolian People's Republic	middle Campanian	5 complete skulls, 17 fragmentary skulls, postcranial skeletons, juvenile to adult
Leptoceratops Brown, 1914c			
L. *gracilis* Brown, 1914c	Scollard Formation, Alberta, Canada Lance Formation, Wyoming, United States	late Maastrichtian	3 complete skulls, 2 partial skulls, skeletons
Microceratops Bohlin, 1953			
M. *gobiensis* Bohlin, 1953 (including M. *sulcidens* Bohlin, 1953)	Minhe Formation, Gansu, Nei Mongol Zizhiqu, People's Republic of China; Sheeregeen Gashoon Formation, Omnogov, Mongolian People's Republic	?Santonian Campanian-Maastrichtian	Partial skull, skeleton, several fragmentary specimens
Montanoceratops Sternberg, 1951a			
M. *cerorhynchus* (Brown et Schlaikjer, 1942) (= *Leptoceratops cerorhynchus* Brown et Schlaikjer, 1942)	St. Mary River Formation, Montana, United States	early Maastrichtian	Partial skull with associated skeleton, second articulated specimen
Protoceratops Granger et Gregory, 1923			
P. *andrewsi* Granger et Gregory, 1923	Djadochta Formation, Beds of Toogreeg, ?Beds of Alag Teg, Omnogov, Mongolian People's Republic; Minhe Formation, Gansu, Nei Mongol Zizhiqu, People's Republic of China	late Santonian or early Campanian	80 skulls, some skeletons, juvenile to adult
Breviceratops Kurzanov, 1990 •			
B. *kozlowskii* (Maryańska et Osmólska, 1975) (= ?*Protoceratops kozlowskii* Maryańska et Osmólska, 1975)	Barun Goyot Formation, Omnogov, Mongolian People's Republic	middle Campanian	2 partial juvenile skulls
Ceratopsidae Marsh, 1890a			
Centrosaurinae Lambe, 1915			
Avaceratops Dodson, 1986			
A. *lammersi* Dodson, 1986	Judith River Formation, Montana, United States	late Campanian	Partial skull, skeleton, juvenile
Brachyceratops Gilmore, 1914b			
B. *montanensis* Gilmore, 1914b	Two Medicine Formation, Montana, United States	Campanian-Maastrichtian	6 partial skulls, skeletons, subadult
Centrosaurus Lambe, 1904a (= *Eucentrosaurus* Chure et McIntosh, 1989 •)			
C. *apertus* (including *Monoclonius dawsoni* Lambe, 1902, M. *flexus* Brown, 1914d, M. *cutleri* Brown, 1917, C. *longirostris* Sternberg, 1940b)	Judith River Formation, Alberta, Canada	late Campanian	15 skulls, several skeletons, all adult; abundant bonebed material with rare juveniles and subadults
Monoclonius Cope, 1876a			
M. *crassus* Cope, 1876a (including M. *lowei* Sternberg, 1940b)	Judith River Formation, Montana, United States; Judith River Formation, Alberta, Canada	late Campanian	5 skulls, 1 complete
Pachyrhinosaurus Sternberg, 1950			
P. *canadensis* Sternberg, 1950	Horseshoe Canyon Formation, St. Mary River Formation, Waptiti Formation, Alberta, Canada	Maastrichtian	12 partial skulls, disarticulated cranial and postcranial material from bonebeds
Styracosaurus Lambe, 1913			
S. *albertensis* Lambe, 1913 (including S. *parksi* Brown et Schlaikjer, 1937, *Monoclonius nasicornus* Brown, 1917)	Judith River Formation, Alberta, Canada	late Campanian	2 skulls, 3 skeletons, additional material in bonebeds
S. *ovatus* Gilmore, 1930	Two Medicine Formation, Montana, United States	Campanian-Maastrichtian	3 skulls, cranial and postcranial material in bonebeds

Table 29.1, continued

Chasmosaurinae Lambe, 1915			
Anchiceratops Brown, 1914b			
A. ornatus Brown, 1914b (including *A. longirostris* Sternberg, 1929a)	Judith River Formation, Horseshoe Canyon Formation, Alberta, Canada	late Campanian-Maastrichtian	6 skulls, 1 complete skeleton
Arrhinoceratops Parks, 1925			
A. brachyops Parks, 1925	Horseshoe Canyon Formation, Alberta, Canada	Maastrichtian	Complete skull
Chasmosaurus Lambe, 1914a (= *Protorosaurus* Lambe, 1914b, *Eoceratops* Lambe, 1915)			
C. belli (Lambe, 1914a) (= *Monoclonius belli* Lambe, 1902, including *C. brevirostris* Lull, 1933)	Judith River Formation, Alberta, Canada	late Campanian	8 skulls, several skeletons
C. russelli Sternberg, 1940b	Judith River Formation, Alberta, Canada	late Campanian	3 complete or partial skulls
C. canadensis (Lambe, 1902) (= *Monoclonius canadensis* Lambe, 1902, including *Eoceratops canadensis* Lambe, 1915, *C. kaiseni* Brown, 1933)	Judith River Formation Alberta, Canada	late Campanian	4 complete or partial skulls
C. mariscalensis Lehman, 1989	Aguja Formation, Texas, United States	Campanian	12 disarticulated skull, postcrania, juvenile
Pentaceratops Osborn, 1923b			
P. sternbergii Osborn, 1923b (including *P. fenestratus* Wiman, 1930)	Fruitland Formation, Kirtland Shale, New Mexico, United States	?Campanian-Maastrichtian	8 complete or partial skulls, 1 complete and several partial skeletons
Torosaurus Marsh, 1891b			
T. latus Marsh, 1891b (including *T. gladius* Marsh, 1891, *Arrhinoceratops utahensis* Gilmore, 1946b)	Lance Formation, Wyoming, Hell Creek Formation, Montana, South Dakota, Laramie Formation, Colorado, North Horn Formation, Utah, Kirtland Shale, New Mexico, Javelina Formation, Texas, United States; Frenchman Formation, Saskatchewan, Canada	late Maastrichtian	5 partial skulls, isolated skull elements
Triceratops Marsh, 1889b (= *Sterrholophus* Marsh, 1891d, *Diceratops* Hatcher, 1905)			
T. horridus Marsh, 1889a (= *Ceratops horridus* Marsh, 1889b, including *T. flabellatus* Marsh, 1889b, *T. prorsus* Marsh, 1890b, *T. serratus* Marsh, 1890b, *T. elatus* Marsh, 1891b, *T. calicornis* Marsh, 1898b, *T. obtusus* Marsh, 1898b, *T. hatcheri* Lull, 1905, *T. brevicornus* Hatcher, 1905, *T. eurycephalus* Schlaikjer, 1935, *T. albertensis* Sternberg, 1949)	Lance Formation, Evanston Formation, Wyoming, Hell Creek Formation, Montana, South Dakota, Laramie Formation, Colorado, United States; Scollard Formation, Alberta, Frenchman Formation, Saskatchewan, Canada	late Maastrichtian	50 complete or partial skulls, some partial skeletons

Nomina dubia	Material
Agathaumas sylvestris Cope, 1872	Partial sacrum and pelvis
Asiaceratops salsopaludalis Nessov, Kaznyshkina, et Cherepanov, 1989 •	Teeth, cranial fragments, phalanx
Bison alticornis Marsh, 1887	Orbital horn cores
Ceratops montanus Marsh, 1888b	Occipital condyle, paired horn cores
Dysganus bicarinatus Cope, 1876a	Isolated teeth
Dysganus encaustus Cope, 1876a *partim*	Isolated teeth
Dysganus haydenianus Cope, 1876a	Isolated teeth
Dysganus peiganus Cope, 1876a	Tooth
Monoclonius fissus Cope, 1876a	Isolated pterygoid
Monoclonius recurvicornis Cope, 1889b	Braincase, 3 horns, isolated fragments
Monoclonius sphenocerus Cope, 1889b	Nasal horn, premaxilla
Notoceratops bonarelli Tapia, 1918	Edentulous dentary
Polyonax mortuarius Cope, 1874	Horn fragments, vertebrae
Triceratops galeus Marsh, 1889b	Nasal horn core
Triceratops ingens Lull, 1915b	Partial skull and skeleton
Triceratops maximus Brown, 1933	8 vertebrae, 2 ribs
Triceratops sulcatus Marsh, 1890a	Fragmentary skull
Turanoceratops tardabilis Nessov, Kaznyshkina, et Cherepanov, 1989 •	Fragmentary maxilla, ?horn core, ?rostral
Ugrosaurus olsoni Cobabe et Fastovsky, 1987	Partial premaxilla, nasal, isolated fragments

drewsi and *Montanoceratops cerorhynchus,* eight); eversion of the preacetabular process of the ilium (*?P. kozlowskii, M. cerorhynchus*); development of a nasal horn core (*B. rozhdestvenskyi, M. cerorhynchus*); presence of an epijugal (*L. gracilis, P. andrewsi, M. cerorhynchus?*); contact of the jugal and squamosal above the infratemporal opening (absent in *B. rozhdestvenskyi* and *?P. kozlowskii,* barely in *L. gracilis,* modest in *P. andrewsi,* unknown in *M. cerorhynchus*); fenestration of the parietal (*Microceratops gobiensis, B. rozhdestvenskyi, P. andrewsi*); and expansion of the postquadrate portion of the squamosal (*P. andrewsi* greatly derived over all others, as it is in relative size of the frill). A ventrally bowed dentary is unique to *L. gracilis* and *P. andrewsi.*

Protoceratops andrewsi once seemed extremely primitive (Granger and Gregory 1923). As diversity of the protoceratopsids has become better known (Maryańska and Osmólska 1975), it has become clear that *P. andrewsi* is derived within the protoceratopsids, the more so if its premaxillary teeth are viewed as a secondary specialization rather than as a primitive character. *Breviceratops kozlowskii* (Maryańska et Osmólska, 1975), from the Barun Goyot Formation is younger than *P. andrewsi* from the Djadochta Formation. Although it is based on juvenile specimens, it shows an interesting mosaic of character states. Primitive character states include the major contribution of the prefrontal to the orbital rim (as this character changes ontogenetically in *Bagaceratops rozhdestvenskyi,* it may also change in *Breviceratops kozlowskii*) and the wide separation of the jugal and squamosal. The low face and the abbreviated frill are also likely juvenile characters. Retained premaxillary teeth constitute a weak basis for referral to *Protoceratops.* For a very small animal, the skeleton is robust and the eversion of the preacetabular process of the ilium greater than that seen in *P. andrewsi. Microceratops gobiensis* is the oldest protoceratopsid and remains poorly known, although a good skeleton shows it to have had highly cursorial hindlimbs. Its frill is short but fenestrated. *Bagaceratops rozhdestvenskyi* comes from younger strata than *P. andrewsi* but is in most respects more primitive, except for the presence of a nasal horn. It is more primitive than *L. gracilis* in the separation of the squamosal from the jugal by the postorbital. A unique character is an auxiliary antorbital fenestra between the premaxilla and the maxilla. The parietal, originally described as solid, is now known to have fenestrae (Osmólska pers. comm.). Ten maxillary teeth in *B. rozhdestvenskyi* is the lowest tooth count known in the Neoceratopsia. The skull of *Montanoceratops cerorhynchus* is poorly known, but new discoveries in the St. Mary River Formation of Montana

(Weishampel pers. comm.) may change this situation. The squamosal and frill of *M. cerorhynchus* appear primitive, but an expanded external naris and reduced infratemporal opening would be considered derived characters if they can be corroborated. At present, it cannot yet be determined whether two independent evolutionary lineages existed in Asia and North America and whether ceratopsids originated *in situ* in North America or an Asian protoceratopsid gave rise to ceratopsids. A protoceratopsid discovered in the Santonian of Montana (Horner pers. comm.) may shed some light on questions such as these.

Ceratopsidae

The Ceratopsidae (figs. 29.2, 29.3, 29.4) forms a monophyletic assemblage of large (4 to 8 m) obligate quadrupeds, with large skulls (1 to 2.4 m in length), prominent parietosquamosal frills, and various combinations of nasal and postorbital horns. Ceratopsids may be diagnosed as follows: large heads, prominent parietosquamosal frills, variably but often strongly developed nasal or postorbital horns, enlarged external nares set in narial fossae, prefrontals meeting on midline to separate nasals from frontals, folding of frontals to form secondary skull roof expressed as frontal fontanelle, reduced infratemporal fenestrae, squamosal with strong postquadrate expansion, free border of squamosal often ornamented with or without epidermal ossifications, broad jugal-squamosal contact above the infratemporal fenestra, pocket on quadrate to receive wing of pterygoid, choanae enlarged, palatine reduced and longitudinal process lost, ectopterygoid reduced and eliminated from exposure on palate, palatine foramen lost, eustachian canal impressed on pterygoids, exoccipitals excluding supraoccipital from foramen magnum and forming two-thirds of occipital condyle, basioccipital forming one-third of the condyle, apex of coronoid process of dentary expanded, dental batteries, teeth with split roots, tendency to add epidermal ossifications to squamosal and parietal, cervical centra expanded in width for support of head, usually ten fused centra in sacrum, deltopectoral crest of humerus robust, dorsal border of ilium everted, ischium decurved, prepubis long and simple, postpubis short, femur considerably longer than tibia, unguals blunt, rounded, hooflike.

Ceratopsid remains have been claimed from South America and Asia. Neither claim is inherently unlikely, but to date, materials reported are fragmentary and ambiguous. *Notoceratops bonarellii* (Tapia 1918) is based on an incomplete edentulous dentary from the

Chubut region of Argentina. Huene (1929*a*) was convinced of its ceratopsian status, but weighed against this inconclusive fossil (which cannot be relocated, Bonaparte pers. comm.) is the complete lack of any other evidence for ceratopsians, including a dental record in microfaunal assemblages, in all of South America. The record is regarded as doubtful pending further discoveries. The probability of finding ceratopsids in Asia seems much greater, particularly in mesic lowland deposits. An isolated dermal bone from Lower Cretaceous strata at Baiying Bologai, Mongolian People's Republic, was identified by Gilmore (1933*b*) as an epoccipital of *Pentaceratops,* but Coombs (1987) referred the specimen to the Ankylosauria. If further material confirms the existence of ceratopsids in Asia, the question will be whether they show affinities with the ceratopsids of North America or represent an endemic radiation.

Centrosaurinae

Two distinct subfamilies of ceratopsids may be recognized (Lehman 1990), the Centrosaurinae and the Chasmosaurinae. Members of the less-derived Centrosaurinae (fig. 29.2) retain a relatively short high face and a frill that is usually shorter than the basal length of the skull (in *Styracosaurus albertensis,* hypertrophied spines from the caudal border of the parietal give the impression of a very long frill). The premaxillary nasal septum is simple, and a nasopremaxillary process may project into the caudal border of the external naris. The nasal horn is usually dominant over the paired orbital horns. A number of individuals show the horn to be formed of paired nasals that fuse relatively later in development in *Monoclonius crassus* than in *Centrosaurus apertus* (Sternberg 1940*b*). The premaxillary process may contact the lacrimal (intraspecific variation may be sexual in *Centrosaurus*—Dodson 1990). The premaxilla shears against the predentary with wide, inclined cutting surfaces (Lehman in press). Postorbital horn cores range from nonexistent (*Monoclonius crassus*) and apparently deciduous (or at least representing a separate center of ossification in *Styracosaurus albertensis* and *Pachyrhinosaurus canadensis*) to fully developed. There is great intraspecific variation in orbital and nasal horns, as there is in all aspects of cranial ornamentation. The long axis of the squamosal is declined caudoventral to the horizontal axis of the skull, as a consequence of the reorientation of the cheek and quadrate. The postquadrate expansion of the squamosal is modest and usually exceeds the length of the rostral portion in the ratio of 3 to 2. The squamosal rarely exceeds 500 mm in length and forms a relatively small part of the frill. It never reaches the back of the frill; hence the informal designation of centrosaurines as

short-squamosaled ceratopsids (Sternberg 1949). The parietal is relatively short, but is more derived than chasmosaurines in the degree of ornamentation of the caudal border with spikes, hooks, and processes (species of *Centrosaurus, Styracosaurus albertensis, Pachyrhinosaurus canadensis*). In *Avaceratops lammersi* and *Monoclonius crassus,* ornamentation of the frill is modest; in large *Brachyceratops montanensis* (Gilmore 1939), it is somewhat more pronounced. Parietal fenestrae are the rule, but *A. lammersi* is unique in apparently lacking them (Dodson 1986); *B. montanensis* too may have lacked parietal fenestrae. Gilmore (1917, 1939) assumed they were present, but no specimen confirms this (Dodson 1990). The tendency to form epidermal ossifications (epoccipitals) is not as strong as in chasmosaurines, although they are present. However, there is usually a strong primary pattern of scallops on the free border of the squamosal and parietal independent of any osteoderms. Centrosaurines are relatively small, generally not exceeding about 5 m in total length or 1.5 m in length of the skull. Perhaps correlating in part with this size, the olecranon was modest and the ischium only moderately curved (Lehman in press).

Centrosaurines are primarily Judithian (i.e., late Campanian, Eaton 1987) in age, and their remains are so far found principally in Montana and Alberta. *Pachyrhinosaurus canadensis* is the only Edmontonian (early Maastrichtian) member known, and it is also the most derived ceratopsid. It is by far the most massive centrosaurine; at 96 mm in diameter (Langston 1975), the size of its occipital condyle exceeds that of *Centrosaurus apertus* by 25 percent. Remains of *P. canadensis* are known from a number of bone beds in Alberta, and the recent discovery of a new site in central Alberta (Tanke 1988) makes it the most northerly adequately characterized dinosaur in Canada. *Pachyrhinosaurus* has also been identified on the North Slope of Alaska (Clemens pers. comm.). *P. canadensis* lacked typical ceratopsid horns but instead had a huge, pachyostotic nasofrontal boss (Sternberg 1950; Langston 1967, 1975) that has been interpreted as a battering ram. Currie (in prep.) believes it may alternatively have served as the base of an enormous, possibly keratinized, epidermal horn. The frill is incompletely known, but fragments are reminiscent of *Centrosaurus* or *Styracosaurus.* However, there was a unique unicornlike median horn on the parietal bar (fig. 29.2; Currie in prep.). *Pachyrhinosaurus canadensis* seems distinctly separate from known centrosaurines, but there is no justification for placing it in a separate family as advocated by Sternberg (1950). In every significant feature of its anatomy, its centrosaurine characters are clear.

Monoclonius has proved troublesome due to the construction of composite types (Cope 1876*a*, 1889*b*;

criticized by Hatcher et al. 1907 and Lambe 1915). However, the parietal of *Monoclonius crassus* is definitive, well preserved, and accessible for study (Dodson 1990). Brown (1914*d*, 1917) compounded the problem of *Monoclonius* by emending the diagnosis on the basis of well-preserved skulls of *Centrosaurus* from Alberta, and incidentally defined *Monoclonius crassus* out of the genus *Monoclonius* by emphasizing the thickening of the caudal border and the hooklike processes of the parietal of *Centrosaurus*, the very features that separate *Monoclonius* from *Centrosaurus* (Sternberg 1940*b*; Dodson 1990). As presently understood, there is only a single complete skull of *Monoclonius*, described as *M. lowei* (Sternberg 1940*b*); although this skull is imperfect, it is the sole specimen that includes nasal horn, orbital horn, squamosal, and parietal, the characters most valuable for ceratopsid taxonomy. Several incomplete skulls from Alberta are known.

Brachyceratops montanensis, from the Two Medicine Formation (Campanian) of Montana, is a second centrosaurine whose status has been uncertain. It was originally described on the basis of juvenile material (Gilmore 1914*b*, 1917), but a much larger specimen was later discovered (Gilmore 1939). It has been widely assumed that *B. montanensis* was a juvenile of *Monoclonius crassus* (e.g., Sternberg 1949), but Dodson (1990) has shown that the larger specimen cannot be referred to *M. crassus*, and thus it appears that *Brachyceratops montanensis* is a valid taxon.

Centrosaurus apertus is known from many skulls that show a wide range of intraspecific variation, previously encompassed in a series of species (Dodson 1990). In *Centrosaurus*, the caudal border of the parietal is thickened, a pair of hooklike structures project caudally, and a pair of horns project rostrally over the parietal fenestrae. *Styracosaurus albertensis* is very closely related to *Centrosaurus apertus* and is possibly congeneric. *Styracosaurus* is characterized by large hornlike processes projecting caudally from the border of the parietal. The specimen described by Brown (1917) as *Monoclonius nasicornus*, although having a typical *Centrosaurus* parietal, otherwise closely resembles *S. albertensis*, leading Dodson (1990) to conclude tentatively that *M. nasicornus* may be a female of *S. albertensis*. Gilmore (1930) erected a separate species from the Two Medicine Formation of Montana, *S. ovatus*, on the basis of a fragmentary parietal. Discovery of new *Styracosaurus* material from the Two Medicine Formation by J. R. Horner in 1986 appears to substantiate the validity of *S. ovatus*. *Monoclonius crassus* is also close to *Centrosaurus apertus* but probably not as close as is *Styracosaurus albertensis*, and it is more primitive than either of these genera in lacking elaborate parietal ornamentation. *Avaceratops lammersi* is apparently a small, solid-frilled, relatively primitive centrosaurine. The only specimen, a juvenile (Dodson 1986), measures 2.3 m in length, and adult size is estimated at 4 m. *Brachyceratops montanensis* is more derived than *A. lammersi* but less derived than *Monoclonius crassus* in having a low maxillary tooth count and either small parietal fenestrae or none at all (Dodson 1990). Phyletic relations within the Centrosaurinae are clear (fig. 29.9).

Chasmosaurinae

Members of the Chasmosaurinae are more highly derived than those of the Centrosaurinae in having a long, low face and a long frill with elongate squamosals that extend the full length of the long frill, which is typically longer than the basal length of the skull (figs. 29.3, 29.4). Chasmosaurines reach greater absolute sizes than do centrosaurines, with skulls ranging to 2.4 m long (*Torosaurus latus*; *Pentaceratops sternbergii* is nearly as large) and body lengths of 8 m. A small skull of *Triceratops horridus* measures 1.4 m and a large skull 2.1 m in length (Schlaikjer 1935). Paired postorbital horns predominate over the nasal horn. The face is relatively low compared to that of centrosaurines, and elongation of the premaxilla produces a distinctive rostrum. There is either a pit or an interpremaxillary foramen in front of the external naris, and a bony process projects into the naris from the rostral narial floor. The horizontal ventral edge of the premaxilla apparently forms a blunt cutting edge (Lehman in press), and the caudodorsal premaxillary process does not reach the lacrimal. Loss or breakage of the nasal horn core, as in specimens of *Triceratops* (Ostrom and Wellnhofer 1986) and *Arrhinoceratops brachyops* (Tyson 1981), is cited as evidence that the nasal horn core is a separate center of ossification, but the evidence is ambiguous (Sternberg 1949). The scarcity of juvenile specimens compounds the problem of interpretation. Postorbital horn cores are prominent, and in large skulls, the horn cores have massive bases. Sinuses extend into the bases of the orbital horns in *Triceratops horridus* and *Torosaurus latus*. Squamosals are very long (1 to 1.5 m) and are always concave dorsally, as if bent in the middle. This represents derivation from the more primitive centrosaurine condition, in which the axis of the squamosal is downward and backward. In chasmosaurines, the squamosal begins caudoventrally and then bends caudodorsally, to complement the erect parietal frill. The caudal postquadratic portion of the squamosal is longer than the rostral portion by as much as four to one. Squamosal fenestrae, unilateral or bilateral, are known from individuals of all species. Apart from its great length, the parietal is fairly conservative in chasmosaurines, and the caudal border always lacks the elaborate excrescences that some centrosaurines sport. Parietal fenestrae are moderate sized in *Anchiceratops ornatus, Torosaurus latus,* and *Arrhinoceratops brachyops.* In *Triceratops horridus,* the parietal is secondarily shortened and the parietal fenestrae are lost. In contrast, parietal fenestrae are hypertrophied in *Chasmosaurus belli* and *Pentaceratops sternbergii.* In these species, the frill consists essentially of a framework formed of a stout median bar and outer struts surrounding an open space. This arrangement would have offered little if any protection for the neck. Epoccipitals appear to be more prominent in chasmosaurines than in centrosaurines (although there are none in *Torosaurus latus* and *Arrhinoceratops brachyops*) and are even found on the parietal of *P. sternbergii, Anchiceratops ornatus,* and *Triceratops horridus.* The epijugal is prominent in *Arrhinoceratops brachyops* and is hypertrophied in *Pentaceratops sternbergii* (thickness of jugal plus epijugal nearly 150 mm in one individual) to form an extra set of "horns," as the name implies. In the postcranial skeleton of chasmosaurines, the olecranon is strong, the ischium is strongly decurved, and the longitudinal channel in the ventral surface of the sacrum is more pronounced than in centrosaurines (Lehman in press).

Chasmosaurines were very successful and are both Campanian and Maastrichtian in age (Judithian, Edmontonian, Lancian). They ranged from Texas to Alberta, and remains of ceratopsians from northern Mexico (e.g., Murray et al. 1960) and Alaska (Parrish et al. 1987) will likely prove to include chasmosaurines when diagnostic materials are found. Chasmosaurines coexisted with centrosaurines on the Judithian floodplain of Alberta, but apart from the Edmontonian centrosaurine *Pachyrhinosaurus canadensis,* they alone survived in the Maastrichtian and achieved very large body size. Most chasmosaurines are well characterized, based on excellent cranial material. Those that are based on inadequate material (e.g., *Ceratops montanus* Marsh, 1888*b; Ugrosaurus olseni* Cobabe et Fastovsky, 1987) are best regarded as *nomina dubia. Eoceratops canadensis* is probably *Chasmosaurus,* possibly a distinct species, *C. canadensis* (Lehman 1990). Many species have been described, and when modern biometric studies have been carried out, it is probable that growth, sexual dimorphism, and individual variation will be found to account for most of what was originally described as interspecific variation. For instance, in the judgment of Ostrom and Wellnhofer (1986), some 15 described species of *Triceratops* can be regarded as the single valid species, *T. horridus.* Lehman (in press) accepts the interpretation of a single species, *T. horridus,* but believes that sexual dimorphism, as expressed by size and orientation of the orbital horn cores, is evident in *Tri-*

ceratops. He posits three sympatric, sexually dimorphic, species of *Chasmosaurus,* including *C. belli, C. canadensis,* and *C. russelli,* from Alberta, and a fourth species, *C. mariscalensis,* from the Aguja Formation of Texas. In contrast, there is but a single species of the closely related (or possibly congeneric) *Pentaceratops, P. sternbergii,* from New Mexico. Additional material of *P. sternbergii* has been described recently (Rowe et al. 1981). *Arrhinoceratops brachyops* is known only from a single skull from Alberta (Tyson 1981). Five skulls of *Torosaurus* are reported, including two from Wyoming (Marsh 1892*c*) and one each from South Dakota (Colbert and Bump 1947), Montana (West pers. comm.), and Saskatchewan (Tokaryk 1986*a*). However, additional materials from Colorado (Carpenter pers. comm.), Utah (Gilmore 1946*b*), and Texas (Lawson 1976) have been referred to the genus as *T. utahensis. Chasmosaurus* seems the most geographically widespread genus, from Texas to Alberta, and possibly the most species-rich as well if Lehman's hypothesis is sustained.

Phyletic relations within the Chasmosaurinae are difficult to discern. It is very difficult to construct a consistent scheme that does not contain parallelisms and character-state reversals. *Anchiceratops ornatus* seems the most generalized chasmosaurine. *Triceratops horridus* is highly derived in developing a shortened solid frill, and *Pentaceratops sternbergii* is highly derived in the opposite sense of having a very long, very open frill. One possible phylogenetic scheme is presented in figure 29.9, but several alternative schemes can also be defended.

TAPHONOMY, PALEOECOLOGY, AND BIOGEOGRAPHY

The fossil record of neoceratopsians is very good; in fact, it is the best of any major group of dinosaurs. All genera are known from essentially complete skulls, almost all have multiple specimens, and about half have associated skeletal material, although complete postcrania are relatively uncommon. Juveniles form an important part of the record of protoceratopsids, but major biases in the fossil record of ceratopsids include a striking scarcity of juveniles (Dodson and Currie 1988) and subadults (Dodson 1986) and a strong bias toward preservation of skulls without postcrania. Although preservation of solitary articulated specimens is typical, there is also a tendency for ceratopsids to occur in monotypic bone beds, at least in Alberta (Currie and Dodson

1984), western Montana (Horner pers. comm.), and Texas (Lehman 1989). Fossils are found in a wide range of sedimentary environments, from semiarid upland floodplains in Mongolia (Osmólska 1980) to swampy deltaic lowland deposits of Alberta (Dodson 1971, 1983). Better-drained conditions in Mongolia, Montana, and Alberta seem to favor the preservation of juveniles, including nests (Gilmore 1917; Brown and Schlaikjer 1940a; Horner 1987).

Remains of *Protoceratops andrewsi* are found in areas where sedimentation took place in eolian dunes and small, intermittent lakes and streams (Gradzinski et al. 1977). The climate was hot, semiarid, and seasonal. It is significant that protoceratopsids are absent in the mesic Nemegt Formation of Mongolia which overlies the Djadochta and Barun Goyot formations. In North America, protoceratopsid remains are found in sediments laid down in upper coastal plain regions (Horner 1984b) that appear to have been seasonally dry. There is no evidence of protoceratopsids in the rich fauna of the contemporary lower coastal plain deposits of Alberta, Montana, or New Mexico. The coexistence of the very primitive *Leptoceratops gracilis* and the highly derived *Triceratops horridus* in terminal Cretaceous strata of Alberta is an interesting paradox that occurred in cooler, drier environments distant from the shores of a receding inland sea. *P. andrewsi* and other protoceratopsids appear to have been gregarious animals that nested communally. The discovery of a skeleton of *P. andrewsi* entangled with that of the dromaeosaurid *Velociraptor mongoliensis* (Kielan-Jaworwoska and Barsbold 1972; Barsbold and Perle 1983) suggests that the latter preyed on the former. Another association led to speculation (Osborn 1924a) that *Oviraptor philoceratops* included eggs of *P. andrewsi* in its diet. However, the apparent strength of the jaw muscles in oviraptorids (Barsbold et al., this vol.) makes it evident that they were capable of consuming much more challenging food as well.

Ceratopsids were very successful animals, typically forming 25 percent of the specimens in Campanian dinosaur communities (Dodson 1983, 1987) and 50 to 70 percent or more of the terminal Cretaceous communities, although at much reduced diversity. Ceratopsid remains are relatively common in Campanian and Maastrichtian beds of North America, where they inhabited coastal lowlands along the western shores of the inland sea. Ceratopsids ranged from the North Slope of Alaska (Parrish et al. 1987) at paleolatitudes of up to 85° N (Brouwers et al. 1987) and the Northwest Territories (Russell 1984c) south to northern Mexico (Murray et al. 1960). The existence of ceratopsids outside of western North America has not been confirmed, although isolated bones from South America and Asia have been claimed to be ceratopsid.

Monospecific bone beds are known for *Anchiceratops ornatus, Brachyceratops montanensis, Centrosaurus apertus, Chasmosaurus belli, C. mariscalensis, Monoclonius crassus, Pachyrhinosaurus canadensis, Styracosaurus albertensis,* and *S. ovatus* (Sternberg 1970; Currie and Dodson 1984; Tanke 1988). Minimum numbers of individuals may exceed 100 in some of these (Currie 1981, 1987b), and juveniles are included. Such remarkable concentrations of bones belonging to single species suggest that ceratopsids congregated in large, migratory herds and that these herds were susceptible to mass mortality. Bone beds in channel deposits indicate that flooding was responsible for mass drowning in some cases, but other bone beds are associated with quiet water environments. Drought and fire are also agents of both concentration and death. It is not yet possible to associate any species with a particular environment. Ceratopsids were large animals that moved through a wide range of habitats. It is inconceivable that half a dozen sympatric species of ceratopsid could coexist in herds during Judithian time (as richly documented at Dinosaur Provincial Park, Alberta; Dodson 1971, 1983; Béland and Russell 1978) unless each species had developed a distinct ecological niche. Such niche separation would be expected to include specific food preferences and microhabitat selection (e.g., Jarman and Sinclair 1979; Wing and Tiffney 1987; Coombs, this vol.), but such information is not readily extracted from the fossil record.

Several attempts have been made to analyze the feeding behavior of neoceratopsians. These studies suggest that the beak was used for grasping and plucking rather than for biting (Ostrom 1966). The head was held low, so herbaceous plants were probably favored as food (Tait and Brown 1928), although horns and beaks may have been used to push down taller plants. The close-packed batteries of teeth were adapted for shearing and cutting rather than grinding and were well suited for chopping up tough fibrous vegetation such as palms and cycads which grew in mesic lowland habitats (Ostrom 1966; Weishampel 1984c) but not in the semiarid habitats of Mongolia (Osmólska pers. comm.). No plant remains are preserved at Bayn Dzak, Mongolia, where the remains of *Protoceratops andrewsi* are found in abundance, but a xerophytic herbaceous vegetation may be postulated (Osmólska 1980). Various authors (e.g., Haas 1955; Ostrom 1966) have postulated that the enlarged frill of ceratopsians served as a framework for origin of hypertrophied muscles of mastication. In *Leptoceratops gracilis,* with its enlarged supratemporal fenestrae and incipient frill, this model seems plausible, but there are several reasons for doubting its generality. Typically, ceratopsids do not have very spacious supratemporal fenestrae. There is usually a well-defined area adjacent to the supratemporal fenestra which represents a modestly expanded attachment area for jaw adductor muscle, but major areas of the parietal may show a pronounced vascular texture that is incompatible with muscle attachment. What the possible advantage of a meter-long jaw muscle could be is unclear. It is well known that increase in length of a muscle does not increase its strength (e.g., Hill 1950; McMahon 1984). It is clear that the frill was sexually dimorphic in some species (Kurzanov 1972; Dodson 1976; Lehman in press) and that scallops, spikes, and spines on the frill served display or defensive purposes that were totally unrelated to mastication (although neither would they detract from masticatory ability). Wheeler (1978) drew attention to the potential of large surfaces with rich vascularization for thermoregulatory heat loss. Ceratopsid frills, some of which had surface areas of more than a square meter, fit this paradigm well.

Horns and frills provided a measure of protection from predators but were probably more useful for intraspecific display and competition (Farlow and Dodson 1975). Adventitious fenestrae in the cheeks and frills in *Arrhinoceratops, Centrosaurus, Chasmosaurus, Pentaceratops,* and *Triceratops* may be punctures that indicate that these animals fought with members of their own species. The intraspecific male display and combat of living ungulates (Geist 1966) provides a model for interpreting ceratopsid behavior (Farlow and Dodson 1975). One can easily imagine two chasmosaurine bulls approaching each other with their heads lowered and their crests elevated in a magnificent and colorful display. If bluffing failed to establish dominance, then the orbital horns were ideally suited for locking together for a relatively safe pushing and wrestling match. Centrosaurines had smaller crests to display, and the emphasis on the single nasal horn suggested a much higher potential for inflicting injury on a competing bull. Perhaps avoidance of injury was the reason for the development of the nasal boss in *Pachyrhinosaurus canadensis,* the most derived centrosaurine. Although centrosaurine crests were smaller, the addition of hooks, spikes, and other excrescences would have been just as effective. Possibly animals such as *Styracosaurus* engaged in side-to-side combat, with the nasal horns and spikes on the frill locking as the animals pushed against one another (Spassov 1979).

Given the conspicuous nature of the horns and crests of neoceratopsians, it is almost an a priori expectation that sexual dimorphism should be a significant cause of variability in neoceratopsians. There has been little rigorous biometric study of intraspecific variation in neoceratopsians, apart from *Protoceratops andrewsi* (Dodson 1976). To be convincing, studies of intraspecific variation need to take into account the broadest context possible, including such factors as population size and size range of the individuals as well geographic and stratigraphic distribution. Lehman (1989, in press) documented such variation in a Texas population of *Chasmosaurus mariscalensis* and posited that males are characterized by orbital horns that are larger, straighter, more erect, and vertical in orientation, while females are characterized by horns that are curved, procumbent, and laterally divergent. He generalized his results in a survey of chasmosaurines and postulates sexual dimorphism in all species. He posits that in *Triceratops horridus,* Lull's (1933) "lineage" of *Triceratops "calicornis"—T. "elatus"* represents males, and the "lineage" of *T. "brevicornus"—T. "horridus"* comprises females. How far this approach can be taken is not clear. Lehman tentatively accepts three sympatric, sexually dimorphic species of *Chasmosaurus* from the Judith River Formation of Alberta. While Lehman's interpretation may be correct, such a degree of ecological sympatry among congeners of similar size that probably competed for similar foods (Hutchinson 1959) seems questionable (Dodson 1975; Ostrom and Wellnhofer 1986). The problem can only be resolved by a biometrical mapping of intraspecific variability.

Future developments in the study of neoceratopsians will take a number of directions. New discoveries that will elucidate significant gaps in our knowledge may be anticipated in terrains where only fragments are presently known. These include the Santonian of Montana (Horner pers. comm.), the Campano-Maastrichtian of Alaska (e.g., Parrish et al. 1987), and the Upper Cretaceous of central Asia (Nessov 1982). Multivariate morphometry and phylogenetic analyses will map out intraspecific variability and rationalize sexual variability and species-level taxonomy. Biomechanical analyses of the dentition and masticatory system (e.g., Weishampel 1984*a,* 1984*c*) will allow a more precise understanding of the relationship between neoceratopsians and the plants that supported them (Coe et al. 1987; Wing and Tiffney 1987). Neoceratopsians will continue to provide fertile ground for the study of dinosaurian paleobiology.

Bibliography

Abel, O. 1919. Die Stämme der Wirbeltiere. De Gruyter, Berlin and Leipzig. 914 pp.

———. 1922. [Discussion of Kräusel 1922]. Palaeontol. Z. 4:87.

———. 1924. Die neuen Dinosaurierfunde in der Oberkreide Canadas. Naturwissenschaften 12:709–716.

———. 1930. Plastische Rekonstruktion des Lebensbildes von *Tyrannosaurus rex* Osborn. Palaeobiologica 3:103–130.

Abler, W. L. 1984. A three-dimensional map of a paleontological quarry. Contrib. Geol. Univ. Wyoming 23:9–14.

Adams, A. L. 1875. On a fossil saurian vertebra. *Arctosaurus osborni*, from the Arctic regions. Proc. Roy. Irish Acad. 2:177–179.

Adams, L. A. 1919. A memoir on the phylogeny of the jaw muscles in Recent and fossil vertebrates. Ann. New York Acad. Sci. 28:51–156.

Aguirrezabala, L. M., Angel Tores, J., and Viera, L. I. 1985. El Weald de Igea (Cameros—La Rioja). Sedimentologia, bioestratigrafia y paleoicnologia de grandes reptiles (Dinosaurios). Munibe 37:111–138.

Aguirrezabala, L. M., and Viera, L. I. 1983. Icnitas de dinosaurios en Santa Cruz de Yanguas (Soria). Munibe 35:1–13.

Aitken, W. G. 1961. Geology and palaeontology of the Jurassic and Cretaceous of southern Tanganyika. Bull. Geol. Surv. Tanganyika 31:1–144.

Akersten, W. A. 1985. Canine function in *Smilodon* (Mammalia; Felidae; Machairodontinae). Contrib. Sci. Nat. Hist. Mus. Los Angeles County 356:1–22.

Alcock, J. 1984. Animal Behavior: An Evolutionary Approach. 3d ed. Sinauer, Sunderland, Mass. 596 pp.

Alexander, R. M. 1976. Estimates of speeds of dinosaurs. Nature 261:129–130.

———. 1985. Mechanics of posture and gait of some large dinosaurs. Zool. J. Linn. Soc. London 83:1–25.

Allainé, D. D., Pontier, J. M., Lebreton, J. D., Trouvilliez, J., and Colbert, J. 1987. The relationship between fecundity and adult body weight in homeotherms. Oecologia 73:478–480.

Allen, P. 1975. Wealden of the Weald: A new model. Proc. Geol. Assoc. London 86:389–436.

Alonso, R. N. 1980. Icnitas de dinosaurios (Ornithopoda, Hadrosauridae) en el Cretacico superior de norte de Argentina. Acta Geol. Lilloana 15:55–63.

Alvarez, L. W., Alvarez, W., Asaro, F., and Michel, H. V. 1980. Extraterrestrial cause for the Cretaceous-Tertiary extinction. Science 208:1095–1108.

Alvarez, W., Alvarez, L. W., Asaro, F., and Michel, H. V. 1984. The end of the Cretaceous: Sharp boundary or gradual transition? Science 223:1183–1186.

Alvarez, W., Kauffman, E. G., Surlyk, F., Alvarez, L. W., Asaro, F., and Michel, H. V. 1984. Impact theory of mass extinctions and the invertebrate fossil record. Science 223:1135–1141.

Ambroggi, R., and Lapparent, A. F. de. 1954. Découverte d'empreintes de pas de Reptiles dans le Maestrichtien d'Agadir (Maroc). C. R. Soc. Géol. France 1954:50–54.

Ameghino, F. 1898. Sinopsis geológico-paleontológica. Segundo Censo Repúbl. Argent. Folia 1:112–255.

———. 1899. Sinopsis geológica paleontológica. Supplemento. Adiciones y corecciones. La Plata. 13 pp.

———. 1899. Nota preliminar sobre el *Loncosaurus argentinus* un represente de la familia de las Megalosauridae en la Republica Argentina. Ann. Soc. Cien. Argent. 47: 61–62.

Amprino, R. 1967. Bone histophysiology. Guy's Hospital Rep. 116: 51–69.

Anderson, F. M. 1958. Upper Cretaceous of the Pacific Coast. Mem. Geol. Soc. Am. 71: 1–378.

Anderson, H. M., and Anderson, J. M. 1970. A preliminary review of the biostratigraphy of the uppermost Permian, Triassic and Jurassic of Gondwanaland. Suppl. Palaeontol. afr. 13: 1–22.

Anderson, J. F., Hall-Martin, A., and Russell, D. A. 1985. Long-bone circumference and weight in mammals, birds and dinosaurs. J. Zool. A 207: 53–61.

Anderson, R. A., and Karasov, W. H. 1988. Energetics of the lizard *Cnemidophorus tigris* and life history consequences of food-acquisition mode. Ecol. Monogr. 58: 79–110.

Anderson, S. M. 1939. Dinosaur tracks in the Lakota Sandstone of the eastern Black Hills, South Dakota. J. Paleontol. 13: 361–364.

Anderton, R., Bridges, P. H., Leeder, M. R., and Sellwood, B. W. 1979. A Dynamic Stratigraphy of the British Isles. Allen and Unwin, London. 301 pp.

Andrews, C. W. 1921. On some remains of a theropodous dinosaur from the lower Lias of Barrow-on-Soar. Ann. Mag. Nat. Hist. (ser. 9) 8: 570–576.

Andrews, J. E., and Hudson, J. D. 1984. First Jurassic dinosaur footprint from Scotland. Scottish J. Geol. 20: 129–134.

Andrews, R. C. 1932. The New Conquest of Central Asia. The Natural History of Central Asia 1. Am. Mus. Nat. Hist., New York. 678 pp.

Andrews, R. M. 1982. Patterns of growth in reptiles. In: Gans, C., and Pough, F. H. (eds.). Biology of the Reptilia. Vol. 13: 273–320.

Andrews, R. M., and Pough, F. H. 1985. Metabolism of squamate reptiles: Allometric and ecological relationships. Physiol. Zool. 58: 214–231.

Anonymous 1978. New dinosaurian remains. Geol. Surv. India News 9(5): 4.

Antunes, M. T. 1976. Dinosaurios eocretacicos de Lagosteiros. Cienc. Terra 1: 1–35.

Antunes, M. T., and Pais, J. 1978. Notas sobre depositos de Taveiro. Estratigrafia, paleontologia, idade, paleoecologia. Cienc. Terra 4: 109–128.

Arambourg, C., and Wolff, R. G. 1969. Nouvelles données paléontologiques sur l'âge des "grès du Lubur" (Turkana grits) à l'Ouest du lac Rodolphe. C. R. Soc. Géol. France 1969: 190–191.

Archibald, J. D. 1987*a*. Late Cretaceous (Judithian and Edmontonian) vertebrates and geology of the Williams Fork Formation, N.W. Colorado. In: Currie, P. J., and Koster, E. H. (eds.). 4th Symp. Mesozoic Terr. Ecosyst. Short Pap. Tyrrell Mus. Palaeont., Drumheller, Alberta. Pp. 7–11.

———. 1987*b*. Stepwise and non-catastrophic extinctions in the Western Interior of North America: Testing observations in the context of an historical science. Mém. Soc. Géol. France 150: 45–52.

Arcucci, A. 1986. Nuevos materiales y reinterpretacion de *Lagerpeton chanarensis* Romer (Thecodontia, Lagerpetonidae nov.) del Triásico Medio de La Rioja, Argentina. Ameghiniana 23: 233–242.

Argast, S., Farlow, J. O., Gabet, R. M. and Brinkman, D. L. 1987. Transport-induced abrasion of fossil reptilian teeth: Implications for the existence of Tertiary dinosaurs in the Hell Creek Formation, Montana. Geology 15: 927–930.

Arid, F. M., and Vizotto, L. D. 1971. *Antarctosaurus brasiliensis*, um novo saurópode do Crétaceo superior do Sul do Brasil. An. Cong. Brasil. Geol. 1971: 297–305.

Armstrong, D. C., Castelin, C. A., and Lockett, N. H. 1967. Notes on the geology of the Kadsi River area, mid-Zambezi Valley. Detritus 2: 11–17.

Armstrong, E. 1983. Relative brain size and metabolism in mammals. Science 220: 1302–1304.

Armstrong-Ziegler, J. G. 1978. An aniliid snake and associated vertebrates from the Campanian of New Mexico. J. Paleontol. 52: 480–483.

———. 1980. Amphibia and Reptilia from the Campanian of New Mexico. Fieldiana Geol. 4: 1–39.

Astibia, H., Garcia-Garmilla, F., Orue-Extebarria, X., Rodriguez-Lazaro, J., Buscalioni, A. D., Sanz, J. L., and Jimenez-Fuentes, E. 1987. The Cretaceous-Tertiary boundary in a sector of the south limb of the Miranda-Trevino Synclinal: The first appearance of Chelonia and Archosauria in the Basque Country. Cretaceous Res. 8: 15–27.

Attridge, J., Crompton, A. W., and Jenkins, F. A. 1985. The Southern African Liassic prosauropod *Massospondylus* discovered in North America. J. Vert. Paleontol. 5: 128–132.

Auffenberg, W. 1980. The herpetofauna of Komodo, with notes on adjacent areas. Bull. Florida State Mus. Biol. Sci. 25: 39–156.

———. 1981. The Behavioral Ecology of the Komodo Monitor. Univ. Presses of Florida, Gainesville. 406 pp.

Avnimelech, M. A. 1962*a*. Dinosaur tracks in the Judean Hills. Proceedings, Israel Acad. Sci. Humanities Sect. Sci. 1: 1–19.

———. 1962*b*. Découverte d'empreintes de pas de Dinosaures dans le Cénomanien inférieur des environs de Jérusalem. C. R. Soc. Géol. France 1962: 233–235.

Awad, G. H., and Ghobrial, M. G. 1966. Zonal stratigraphy of the Kharga Oasis. General Organization, Govt. Printing offices, Cairo.

Axelrod, D. I., and Bailey, H. P. 1968. Cretaceous dinosaur extinction. Evolution 22: 595–611.

Baadsgaard, H., and Lerbekmo, J. F. 1980. A Rb/Sr age for the Cretaceous-Tertiary boundary (Z coal), Hell Creek, Montana. Can. J. Earth Sci. 17: 671–673.

Bachofen-Echt, A. 1925. *Iguanodon*-Fährten auf Brioni. Palaeontol. Z. 7: 172–173.

Baird, D. 1957. Triassic reptile footprint faunules from

Milford, New Jersey. Bull. Mus. Comp. Zool. 117: 449–520.

———. 1964. Dockum (Late Triassic) reptile footprints from New Mexico. J. Paleontol. 38: 118–125.

———. 1979. The dome-headed dinosaur *Tylosteus ornatus* Leidy, 1872 (Reptilia: Ornithischia: Pachycephalosauridae). Notulae Naturae 456: 1–11.

———. 1980. A prosauropod dinosaur trackway from the Navajo Sandstone (Lower Jurassic) of Arizona. In: Jacobs, L. L. (ed.), Aspects of Vertebrate History. Mus. No. Arizona Press, Flagstaff. Pp. 219–230.

———. 1984. No ichthyosaurs in the Upper Cretaceous of New Jersey. . .or Saskatchewan. Mosasaur 2: 129–134.

———. 1986*a*. *Halisaurus* and *Prognathodon*, two uncommon mosasaurs from the Upper Cretaceous of New Jersey. Mosasaur 3: 37–46.

———. 1986*b*. Upper Cretaceous reptiles from the Severn Formation of Maryland. Mosasaur 3: 63–85.

Baird, D., and Horner, J. R. 1977. A fresh look at the dinosaurs of New Jersey and Delaware. Bull. New Jersey Acad. Sci. 22: 50.

———. 1979. Cretaceous dinosaurs of North Carolina. Brimleyana 2: 1–28.

Bakker, R. T. 1968. The superiority of dinosaurs. Discovery 3: 11–22.

———. 1971*a*. The ecology of the brontosaurs. Nature 229: 172–174.

———. 1971*b*. Brontosaurs. McGraw-Hill Yearbook of Science and Technology. Pp. 179–181.

———. 1971*c*. Dinosaur physiology and the origin of mammals. Evolution 25: 636–658.

———. 1972. Anatomical and ecological evidence of endothermy in dinosaurs. Nature 238: 81–85.

———. 1974. Dinosaur bio-energetics—a reply to Bennett and Dalzell, and Feduccia. Evolution 28: 497–503.

———. 1975*a*. Dinosaur renaissance. Sci. Am. 232: 58–78.

———. 1975*b*. Experimental and fossil evidence for the evolution of tetrapod bioenergetics. In: Gates, D. M., and Schmerl, R. B. (eds.). Perspectives of Biophysical Ecology. Springer Verlag, New York. Pp. 365–399.

———. 1977. Tetrapod mass extinctions: A model of the regulation of speciation rates and immigration by cycles of topographic diversity. In: Hallam, A. (ed.), Patterns of Evolution. Elsevier, Amsterdam. Pp. 439–468.

———. 1978. Dinosaur feeding behavior and the origin of flowering plants. Nature 274: 661–663.

———. 1980. Dinosaur heresy—dinosaur renaissance. In: Thomas, R. D. K., and Olson, E. C. (eds.). A Cold Look at the Warm-Blooded Dinosaurs. Westview Press, Boulder. Pp. 351–462.

———. 1986. Dinosaur Heresies. William Morrow, New York: 481 pp.

———. 1987. The return of the dancing dinosaurs. In: Czerkas, S. J., and Olson, E. C. (eds.). Dinosaurs Past and Present. Nat. Hist. Mus. Los Angeles County. Vol. I. Pp. 38–69.

———. 1988. Review of the Late Cretaceous nodosauroid Dinosauria. *Denversaurus schlessmani*, a new armor-plated dinosaur from the latest Cretaceous of South Dakota, the last survivor of the nodosaurians, with comments on stegosaur-nodosaur relationships. Hunteria 1: 1–23.

———. MS. Dinosaurs and bioenergetic evolution: Quantitative test of the predator/prey model.

Bakker, R. T. and Galton, P. M. 1974. Dinosaur monophyly and a new class of vertebrates. Nature 248: 168–172.

Bakker, R. T., Williams, M., and Currie, P. 1988. *Nanotyrannus*, a new genus of pygmy tyrannosaur, from the latest Cretaceous of Montana. Hunteria 1: 1–30.

Ballerstedt, M. 1922. Zwei grosse, zweizehige Fährten hochbeiniger Bipeden aus dem Wealdensandstein bei Buckeburg. Z. Deutsch. Geol. Ges. 73: 76–91.

Banse, K., and Mosher, S. 1980. Adult body mass and annual production/biomass relationships of field populations. Ecol. Monogr. 50: 355–379.

Barbarena, M. C., Araujo, D. C., and Lavina, E. L. 1985. Late Permian and Triassic tetrapods of southern Brazil. Natl. Geogr. Res. 1:5–20.

Barbour, E. H. 1931. Evidence of dinosaurs in Nebraska. Bull. Nebraska State Mus. 21: 187–190.

Barsbold, R. 1972. [The taphonomy of the fauna of the Late Cretaceous of the MPR]. Akad. Nauk MNR. Geol. Inst. Pp. 53–60. (In Russian)

———. 1974. Saurornithoididae, a new family of small theropod dinosaurs from central Asia and North America. Palaeontol. Polonica 30: 5–22.

———. 1976*a*. [On a new Late Cretaceous family of small theropods (Oviraptoridae fam. n.) of Mongolia]. Doklady Akad. Nauk S.S.S.R. 226: 685–688. (In Russian)

———. 1976*b*. [On the evolution and systematics of the late Mesozoic dinosaurs]. Sovm. Sov.-Mong. Paleontol. Eksped. Trudy 3: 68–75. (Russian with English summary)

———. 1976*c*. [New data on *Therizinosaurus* (Therizinosauridae, Theropoda)]. Sovm. Sov.-Mong. Paleontol. Eksped. Trudy 3: 76–92. (In Russian with English summary)

———. 1977*a*. [Kinetism and peculiarity of the jaw apparatus of oviraptors (Theropoda, Saurischia)]. Sovm. Sov.-Mong. Paleontol. Eksped. Trudy 4: 37–47. (In Russian with English summary)

———. 1977*b*. [On the evolution of carnivorous dinosaurs]. Sovm. Sov.-Mong. Paleontol. Eksped. Trudy 4: 48–56. (In Russian with English summary)

———. 1981. [Toothless carnivorous dinosaurs of Mongolia]. Sovm. Sov.-Mong. Paleontol. Eksped. Trudy 15: 28–39. (In Russian with English summary)

———. 1983*a*. [Carnivorous dinosaurs from the Cretaceous of Mongolia]. Sov.-Mong. Paleontol. Eksped. Trudy 19: 1–117. (In Russian with English summary)

———. 1983*b*. [On some avian features in the morphology of carnosaurs]. Sovm. Sov.-Mong. Paleontol. Eksped.

Trudy 24: 96–103. (In Russian with English summary)

———. 1986. [Raubdinosaurier Oviraptoren]. In: Vorobyeva, E. I. (ed.). Herpetologische Untersuchungen in der Mongolischen Volsrepublik. Akad. Nauk S.S.S.R. Inst. Evolyucionnoy Morfologii i Ekologii Zhivotnykh im. A. M. Severtsova, Moskva. Pp. 210–223. (In Russian with German summary)

———. 1988. [New Late Cretaceous ornithomimid from the MPR]. Paleontol. Zh. 1988: 122–125. (In Russian).

Barsbold, R., Osmólska, H., and Kurzanov, S. M. 1987. On a new troodontid (Dinosauria, Theropoda) from the Early Cretaceous of Mongolia. Acta Palaeontol. Polonica 33: 49–52.

Barsbold, R., and Perle, A. 1979. [Modification of the saurischian pelvis and parallel evolution of the carnivorous dinosaurs]. Sovm. Sov.-Mong. Paleontol. Eksped. Trudy 8: 39–44. (In Russian with English summary)

———. 1980. Segnosauria, a new infraorder of carnivorous dinosaurs. Acta Palaeontol. Polonica 25: 185–195.

———. 1983. [On taphonomy of joint burying of juvenile dinosaurs and some aspects of their ecology]. Sovm. Sov.-Mong. Paleontol. Eksped. Trudy 24: 121–125. (In Russian with English summary)

———. 1984. [On first new find of a primitive ornithomimosaur from the Cretaceous of the M.P.R.] Paleontol. Zh. 1984: 121–123. (In Russian)

Bartholomai, A., and Molnar, R. E. 1981. *Muttaburrasaurus*, a new iguanodontid (Ornithischia: Ornithopoda) dinosaur from the Lower Cretaceous of Queensland. Mem. Queensland Mus. 20: 319–349.

Bassoulet, J. P. 1971. Découverte d'empreintes de pas de Reptiles dans l'Infralias de la région d'Ain-Sefra (Atlas saharien, Algérie). C. R. Soc. Géol. France 1971: 358–359.

Bassoullet, J.-P., and Iliou, J. 1967. Découverte de dinosauriens associés à des crocodiliens et des poissons dans le Crétacé inférieur de l'Atlas saharien (Algérie). C. R. Soc. Géol. France 1967: 294–295.

Bataller, J. R. 1960. Los vertebrados del Cretacico español. Notas Comun. Inst. Geol. Min. 60: 141–164.

Bazhanov, V. S. 1961. [The first discovery of dinosaur egg shells in the U.S.S.R.]. Trudy Inst. Zool. Akad. Nauk Kazakhskoi SSR 15: 177–181. (In Russian)

Bednarz, J. C. 1988. Cooperative hunting in Harris' hawks (*Parabuteo unicinctus*). Science 239: 1525–1527.

Beetschen, J.-C. 1985. Sur les niveaux à coquilles d'oeufs de dinosauriens de la région de Rennes-le-Château (Aude). In: Les Dinosaures de la Chine à la France. Mus. Hist. Nat. Toulouse, France. Pp. 113–126.

Behrensmeyer, A. K. 1975. Taphonomy and paleoecology of Plio-Pleistocene bone assemblages of Lake Rudolph, Kenya. Bull. Mus. Comp. Zool. 146: 473–578.

Behrensmeyer, A. K., Western, D., and Dechant-Boaz, D. E. 1979. New perspectives in vertebrate paleocology from a Recent bone assemblage. Paleobiology 5: 12–21.

Béland, P., and Russell, D. A. 1978. Paleoecology of Dinosaur Provincial Park (Cretaceous), Alberta, interpreted from the distribution of articulated remains. Can. J. Earth Sci. 15: 1012–1024.

———. 1979. Ectothermy in dinosaurs: Paleoecological evidence from Dinosaur Provincial Park, Alberta. Can. J. Earth Sci. 16: 250–255.

———. 1980. Dinosaur metabolism and predator/prey ratios in the fossil record. In: Thomas, R. D. K., and Olson, E. C. (eds.). A Cold Look at the Warm-Blooded Dinosaurs. Westview Press, Boulder. Pp. 85–102.

Bellair, P., and Lapparent, A. F. de. 1949. Le Crétacé et les empreintes de pas de Dinosauriens d'Amoura (Algérie). Bull. Soc. Hist. Nat. Afrique Nord 39: 168–175.

Benedetto, J. L. 1973. Herrerasauridae, nueva familia de saurisquios Triásicos. Ameghiniana 10: 89–102.

Bennett, A. F. 1973. Ventilation in two species of lizards during rest and activity. Comp. Biochem. Physiol. 46A: 653–671.

———. 1982. The energetics of reptilian activity. In: Gans, C., and Pough, F. H. (eds.). The energetics of reptilian activity. Biology of the Reptilia. Vol. 13: 155–199.

———. 1983. Ecological consequences of energy metabolism. In: Huey, R. B., Pianka, E. R., and Schoener, T. W. (eds.). Lizard Ecology: Studies of a Model Organism. Harvard Univ. Press, Cambridge. Pp. 11–23.

Bennett, A. F., and Dalzell, B. 1973. Dinosaur physiology: A critique. Evolution 27: 170–174.

Bennett, A. F. and John-Adler, H. B. 1984. The effect of body temperature on the locomotory energetics of lizards. J. Comp. Physiol. B 155: 21–27.

Bennett, A. F. and Ruben, J. A. 1979. Endothermy and activity in vertebrates. Science 206: 649–654.

———. 1986. The metabolic and thermoregulatory status of therapsids. In: Hotton, N., III, MacLean, P. D., Roth, J. J., and Roth, E. C. (eds.). The Ecology and Biology of Mammal-Like Reptiles. Smithsonian Inst., Washington, D.C. Pp. 207–218.

Bennett, A. F., Seymour, R. S., Bradford, D. F. and Webb, G. J. W. 1985. Mass-dependence of anaerobic metabolism and acid-base disturbance during activity in the salt-water crocodile, *Crocodylus porosus*. J. Exp. Biol. 118: 161–171.

Benton, M. J. 1982. The Diapsida: Revolution in reptile relationships. Nature 296: 306–307.

———. 1983*a*. Dinosaur success in the Triassic: A noncompetitive ecological model. Q. Rev. Biol. 58: 29–55.

———. 1983*b*. The Triassic reptile *Hyperodapedon* from Elgin: Functional morphology and relationships. Phil. Trans. Roy. Soc. London B 302: 605–717.

———. 1984*a*. The relationships and early evolution of the Diapsida. Zool. Soc. London Symp. 52: 575–596.

———. 1984*b*. Fossil reptiles of the German Late Triassic and the origin of the dinosaurs. In: Reif, W.-E., and Westphal, F. (eds.). 3d Symp. Mesozoic Terr. Ecosyst. Short Pap. Attempto Verlag, Tübingen. Pp. 13–18.

———. 1984c. Consensus on archosaurs. Nature 310: 101.

———. 1984d. Small comparisons for early dinosaurs. Nature 307: 111–112.

———. 1985a. Patterns in the diversification of Mesozoic non-marine tetrapods and problems in historical diversity analysis. Spec. Pap. Palaeontol. 33: 185–202.

———. 1985b. Classification and phylogeny of the diapsid reptiles. Zool. J. Linn. Soc. London 84: 97–164.

———. 1986. The Late Triassic reptile *Teratosaurus*—a rauisuchian, not a dinosaur. Palaeontology 29: 293–301.

Benton, M. J., and Clark, J. 1988. Archosaur phylogeny and the relationships of the Crocodylia. In: Benton, M. J. (ed.). The Phylogeny and Classification of the Tetrapods, Vol. I. Syst. Assoc. Spec. Vol. 35A: 289–332.

Benton, M. J., and Walker, A. D. 1985. Palaeoecology, taphonomy, and dating of Permo-Triassic reptiles from Elgin, northeast Scotland. Palaeontology 28: 207–234.

Berckhemer, F. 1938. Mitteilung über neue Fossilfunde aus dem Schwarzwalder Buntsandstein. Jahresh. Verein. Vaterländ. Naturkunde Württemberg 94: xxxv–xxxvi.

Bere, R. 1966. The African Elephant. Golden Press, New York. 96 pp.

———. 1970. Antelopes. Arco Publ., New York. 96 pp.

Berman, D. S., and Jain, S. L. 1982. The braincase of a small sauropod dinosaur (Reptilia: Saurischia) from the Upper Cretaceous Lameta Group, central India, with review of Lameta Group localities. Ann. Carnegie Mus. 51: 405–422.

Berman, D. S., and McIntosh, J. S. 1978. Skull and relationships of the Upper Jurassic sauropod *Apatosaurus* (Reptilia, Saurischia). Bull. Carnegie Mus. Nat. Hist. 8: 1–35.

———. 1986. Description of the lower jaw of *Stegosaurus* (Reptilia, Ornithischia). Ann. Carnegie Mus. 55: 29–40.

Bernier, P. 1984. Des dinosaures qui sautaient. La Recherche 15: 1438–1442.

Bernier, P., Barale, G., Bourseau, J.-P., Buffetaut, E., Demathieu, G., Gaillard, C., Gall, J.-C., and Wenz, S. 1984. Découverte de pistes de dinosaures sauteurs dans les Calcaires Lithographiques de Cérin (Kimméridgien supérieur, Ain, France)—implications paléoécologiques. Géobios Mém. Spéc. 8: 177–185.

Bertram, B. C. R. 1979. Serengeti predators and their ecosystems. In: Sinclair, A. R. E., and Norton-Griffiths, M. (eds.), Serengeti: Dynamics of an Ecosystem. Univ. Chicago Press, Chicago. Pp. 221–248.

Bessonnat, G., Lapparent, A. F. de, Montenat, C., and Ters, M. 1965. Découverte de nombreuses empreintes de pas de Reptiles dans le Lias inférieur de la Côte de Vendée. C. R. Acad. Sci. Paris 260: 5324–5326.

Beurlen, K. 1950. Neue Fährtenfunde aus der Fränkischen Trias. N. Jb. Geol. Palaeontol. Mh. 1950: 308–320.

Bickler, P. E., and Anderson, R. A. 1986. Ventilation, gas exchange, and aerobic scope in a small monitor lizard, *Varanus gilleni*. Physiol. Zool. 59: 76–83.

Bidar, A., Demay, L., and Thomel, G. 1972. *Compsognathus corallestris*, nouvelle espèce de dinosaurien théropode du Portlandien de Canjuers (Sud-Est de la France). Mus. Hist. Nat. Nice, Ann. 1: 1–34.

Biese, W. 1961. El Jurasico de Cerritos Bayos. Univ. Chile 19: 7–35.

Bird, R. T. 1944. A dinosaur walks into the museum. Nat. Hist. 47: 74–81.

———. 1985. Bones for Barnum Brown: Adventures of a Dinosaur Hunter. Texas Christian Univ. Press, Fort Worth. 225 pp.

Bishop, M. J., and Friday, A. E. 1988. Estimating the interrelationships of tetrapod groups on the basis of molecular sequence data. In: Benton, M. J. (ed.). The Phylogeny and Classification of the Tetrapods. Vol. 1. Syst. Assoc. Spec. Vol. 35A: 33–58.

Bjork, P. R. 1985a. A new iguanodontid dinosaur from the Lakota Formation of the northern Black Hills of South Dakota. Proc. Pacific Div. Am. Assoc. Adv. Sci. 4: 22.

———. 1985b. Preliminary report on the Ruby Site bone bed, Upper Cretaceous, South Dakota. Geol. Soc. Am. Rocky Mountain Sect. Abst. Prog. 17: 207.

Blake, D. K., and Loveridge, J. P. 1975. The role of commercial crocodile farming in crocodile conservation. Biol. Conservat. 8: 261–272.

Block, B. A., and Carey, F. G. 1985. Warm brain and eye temperature in sharks. J. Comp. Physiol. 156: 229–236.

Blows, W. T. 1987. The armoured dinosaur *Polacanthus foxi* from the Lower Cretaceous of the Isle of Wight. Palaeontology 30: 557–580.

Blueweiss, L., Fox, H., Kudzma, V., Nakashima, D., Peters, R., and Sams, S. 1978. Relationships between body size and some life history parameters. Oecologia 37: 257–272.

Bock, W. 1952. Triassic reptilian tracks and trends in locomotive evolution. J. Paleontol. 26: 395–433.

Bodily, N. M. 1969. An armored dinosaur from the Lower Cretaceous of Utah. Brigham Young Univ. Geol. Stud. 16: 35–60.

Bohlin, B. 1953. Fossil reptiles from Mongolia and Kansu. Sino-Swedish Exped. Publ. 37: 1–113.

Bohor, B. F., Foord, E. E., Modreski, P. J., and Triplehorn, D. M. 1984. Mineralogic evidence for an impact event at the Cretaceous-Tertiary boundary. Science 224: 867–869.

Bölau, E. 1952. Neue Fossilfunde aus dem Rhät Schonens und ihre paläogeographisch-ökologische Auswertung. Geol. Fören. Förh. 74: 44–50.

———. 1954. The first finds of dinosaurian skeletal remains in the Rhaetic-Liassic of N. W. Scania. Geol. Fören. Förh. 76: 501–502.

Bonaparte, J. F. 1969. Dos nuevas "faunas" de reptiles Triásicos de Argentina. I Gondwana Symp., Mar del Plata Ciencias Tierra 2: 283–306.

———. 1970. Annotated list of the South American Triassic tetrapods. Proc. Pap. Second Gondwana Symp. Pp. 665–682.

———. 1972. Los tetrápodos del sector superior de la Formación Los Colorados, La Rioja, Argentina (Triásico Superior). I Parte. Opera Lilloana 22: 1–183.

———. 1973. Edades/reptil para el Triásico de Argentina y Brasil. Actas Congr. Geol. Argent. 5: 93–129.

———. 1975a. The family Ornithosuchidae. Colloq. Internatl. C. N. R. S. 218: 485–502.

———. 1975b. Nuevos materiales de *Lagosuchus talampayensis* Romer (Thecodontia-Pseudosuchia) y su significado en el origen de los Saurischia. Chañarense inferior, Triásico medio de Argentina. Acta Geol. Lilloana, 13: 5–90.

———. 1976. *Pisanosaurus mertii* Casamiquela and the origin of the Ornithischia. J. Paleontol. 50: 808–820.

———. 1978a. El Mesozoico de América del Sur y sus Tetrápodos. Opera Lilloana 26: 1–596.

———. 1978b. *Coloradia brevis* n. g. et n. sp. (Saurischia Prosauropoda), dinosaurio Plateosauridae de la Formación Los Colorados, Triásico superior de La Rioja, Argentina. Ameghiniana 15: 327–332.

———. 1979a. Faunas y paleobiogeografia de los tetrápodos Mesozoicos de América del Sur. Ameghiniana 16: 217–238.

———. 1979b. Dinosaurs: A Jurassic assemblage from Patagonia. Science 205: 1377–1379.

———. 1980. Los vertebrados tetrápodos del limite Jurasico-Cretacico. Actas II Cong. Argent. Paleontol. Bioestrat. y I Congr. Latinoam. Paleontol. 5: 77–88.

———. 1982a. Faunal replacement in the Triassic of South America. J. Vert. Paleontol. 2: 362–371.

———. 1982b. Classification of the Thecodontia. Géobios Mém. Spéc. 6: 99–112.

———. 1984. Locomotion in rauisuchid thecodonts. J. Vert. Paleontol. 3: 210—218.

———. 1985. A horned Cretaceous carnosaur from Patagonia. Natl. Geogr. Res. 1: 149–151.

———. 1986a. History of the terrestrial Cretaceous vertebrates of Gondwana. Actas IV Congr. Argent. Paleontol. Bioestrat. 2: 63–95.

———. 1986b. The early radiation and phylogenetic relationships of the Jurassic sauropod dinosaurs, based on vertebral anatomy. In: Padian, K. (ed.). The Beginning of the Age of Dinosaurs. Cambridge Univ. Press, Cambridge. Pp. 247–258.

———. 1986c. Les Dinosaures (Carnosaures, Allosauridés, Sauropodes, Cétiosauridés) du Jurassique moyen de Cerro Condor (Chubut, Argentine). Ann. Paléontol. 72: 247–289, 325–386.

Bonaparte, J. F., and Bossi, G. 1967. Sobre la presencia de dinosaurios en la Formación Pirgua del Grupo Salta y su significado cronologico. Acta Geol. Lilloana 9: 25–44.

Bonaparte, J. F., Franchi, M. R., Powell, J. E., and Sepulveda, E. G. 1984. La Formación Los Alamitos (Campaniano-Maastrichtiano) del sudeste de Río Negro, con descripción de *Kritosaurus australis* n. sp. (Hadrosauridae). Significado paleogeográfico de los vertebrados. Asoc. Geol. Argent. Rev. 39: 284–299.

Bonaparte, J. F., and Gasparini, Z. B. de. 1979. Los sauropodos de los Grupos Neuquen y Chubut, y sus relaciones cronologicas. Actas VII Congr. Geol. Argent. Neuquen 2: 393–406.

Bonaparte, J. F., and Novas, F. E. 1985. *Abelisaurus comahuensis,* n.g., n. sp., Carnosauria del Cretacico tardio de Patagonia. Ameghiniana 21: 259–265.

Bonaparte, J. F., and Powell, J. E. 1980. A continental assemblage of tetrapods from the Upper Cretaceous beds of El Brete, northwestern Argentina (Sauropoda-Coelurosauria-Carnosauria-Aves). Mém. Soc. Géol. France 139: 19–28.

Bonaparte, J. F., Salfity, J. A., Bossi, G., and Powell, J. E. 1977. Hallazgo de dinosaurios y aves Cretácicas en la Formación Lecho de El Brete (Salta), próximo al límite con Tucumán. Acta Geol. Lilloana 14: 5–17.

Bonaparte, J. F., and Vince, M. 1979. El hallazgo del primer nido de Dinosaurios Triásicos (Saurischia, Prosauropoda), Triásico superior de Patagonia, Argentina. Ameghiniana 16: 173–182.

Bond, G. 1955. A note on dinosaur remains from the Forest Sandstone (Upper Karroo). Arnoldia 2: 795–800.

Bond, G., and Bromley, K. 1970. Sediments with the remains of dinosaurs near Gokwe, Rhodesia. Palaeogeogr. Palaeoclimatol. Palaeoecol. 8: 313–327.

Borner, M., FitzGibbon, C. D., Borner, M. M., Caro, T. M., Lindsay, W. K., Collins, D. A., and Holt, M. E. 1987. The decline of the Serengeti Thomson's gazelle population. Oecologia 73: 32–40.

Borsuk-Bialynika, M. 1977. A new camarasaurid sauropod *Opisthocoelicaudia skarzynskii,* gen. n. sp. n. from the Upper Cretaceous of Mongolia. Palaeontol. Polonica 37: 5–64.

Bossi, G. E., and Bonaparte, J. F. 1978. Sobre la presencia de un dinosaurio prosauropodo en la Formacion Quebrada del Barro, en el borde austral de la Cuenca de Marayes-El Carrizal (Triásico superior de San Juan). Acta Geol. Lilloana 15: 41–47.

Bouaziz, S., Buffetaut, E., Ghanmi, M., Jaeger, J. J., Martin, M., Mazin, J.-M., and Tong, H. 1988. Nouvelles découvertes de vertébrés fossiles dans l'Albien du Sud tunisien. Bull. Soc. Géol. France 1988: 335–339.

Bourcart, J., Lapparent, A. F. de, and Termier, H. 1942. Un nouveau gisement de dinosauriens jurassiques au Maroc. C. R. Acad. Sci. Paris 214: 120–122.

Bourgeois, J., and Dott, R. H. 1985. Stratigraphy and sedimentology of Upper Cretaceous rocks in coastal southwest Oregon: Evidence for wrench-fault tectonics in a postulated accretionary terrane. Bull. Geol. Soc. Am. 96: 1007–1019.

Bourlière, F. 1964. The Natural History of Mammals. 3d ed. Knopf, New York. 387 pp.

Bouvier, M. 1977. Dinosaur haversian bone and endothermy. Evolution 31: 449–450.

Bowen, C. F. 1915. The stratigraphy of the Montana Group. U.S. Geol. Surv. Prof. Pap. 90: 93–153.

———. 1918. The stratigraphy of the Hanna Basin, Wyoming. U.S. Geol. Suv. Prof. Pap. 108: 227–235.

Bradley, O. C. 1903. The muscles of mastication and the movement of the skull in Lacertilia. Zool. Jb. Anat. 18: 475–486.

Brady, L. F. 1960. Dinosaur tracks from the Navajo and Wingate sandstones. Plateau 32: 81–82.

Branisa, L. 1968. Hallazgo del amonite *Neolobites* en la caliza miraflores y de huellas de dinosaurios en la formacion El Molino y su significado para la determinacion de la edad del "Grupo Puca." Inst. Boliviano Petroleo Bol. 8: 16–29.

Bratzeva, T. M., and Novodvorskaja, I. M. 1975. [New conchostraca from the Lower Cretaceous of Mongolia]. [Fossil Flora and Fauna of Mongolia]. Sovm. Sov.-Mong. Paleontol. Eksped. Trudy 2: 205–209. (In Russian with English abstr.)

Breithaupt, B. 1982. Paleontology and paleoecology of the Lance Formation (Maastrichtian), east flank of Rock Springs Uplift, Sweetwater County, Wyoming. Contrib. Geol. Univ. Wyoming 21: 123–151.

———. 1985. Nonmammalian vertebrate faunas from the Late Cretaceous of Wyoming. Wyoming Geol. Assoc. Gdbk. 36: 159–175.

Brenner, K. 1973. Stratigraphie und Palaeogeographie des oberen Mittelkeupers in Südwest-Deutschland. Arbeiten Inst. Geol. Palaeontol. Univ. Stuttgart 68: 101–222.

Breton, G., Fournier, R., and Watte, J.-P. 1986. Le lieu de ponte de dinosaures de Rennes-le-Château (Aude). Ann. Mus. Havre 32: 1–12.

Brett, C. E., and Wheeler, W. H. 1961. A biostratigraphic evaluation of the Snow Hill Member, Upper Cretaceous of North Carolina. Southeastern Geol. 3: 49–132.

Brett-Surman, M. K. 1975. The appendicular anatomy of hadrosaurian dinosaurs. M.A. thesis, Univ. of California, Berkeley. 70 pp.

———. 1979. Phylogeny and palaeobiogeography of hadrosaurian dinosaurs. Nature 277: 560–562.

Brett-Surman, M. K., and Paul, G. S. 1985. A new family of bird-like dinosaurs linking Laurasia and Gondwanaland. J. Vert. Paleontol. 5: 133–138.

Brinkman, D. 1981. The origin of the crocodiloid tarsus and the interrelationships of thecodontian archosaurs. Breviora 464: 1–23.

Brinkman, D. B., and Sues, H.-D. 1987. A staurikosaurid dinosaur from the Upper Triassic Ischigualasto Formation of Argentina and the relationships of the Staurikosauridae. Palaeontology 30: 493–503.

Brinkman, D. L., and Conway, F. M. 1985. Textural and mineralogical analysis of a *Stegosaurus* (Reptilia: Ornithischia) plate. Compass (Sigma Gamma Epsilon) 63: 1–5.

Brinkmann, W. 1984. Erster Nachweis eines Hadrosauriers (Ornithischia) aus dem unteren Garumnium (Maastrichtium) des Beckens von Tremp (Provinz Lerida, Spanien). Palaeontol. Z. 58: 295–305.

Brion, P. E., and Dutuit, J. M. 1981. Figurations sedimentaires et traces d'activité au sol dans le Trias de la formation d'Argana et de l'Ourika (Maroc). Bull. Mus. Natl. Hist. Nat. (ser. 4) 3: 399–427.

Brodrick, H. 1909. Note on footprint casts from the Inferior Oolite near Whitby, Yorshire. Proc. Liverpool Geol. Soc. 10: 327–335.

Broin, F. de, Buffetaut, E., Koeniguer, J.-C., Rage, J.-C., Russell, D., Taquet, P., Vergnaud-Grazzini, C., and Wenz, S. 1974. Le faune de vertébrés continentaux du gisement d'In Beceten (Senonien du Niger). C. R. Acad. Sci. Paris 279: 469–472.

Broin, F. de, Grenot, C., and Vernet, R. 1971. Sur la découverte d'un nouveau gisement de vertébrés dans le Continental Intercalaire saharien: la Gara Samani (Algérie). C. R. Acad. Sci. Paris 272: 1219–1221.

Broom, R. 1904. On the occurrence of an opisthocoelian dinosaur (*Algoasaurus bauri*) in the Cretaceous beds of South Africa. Geol. Mag., ser. 5, 1: 445–447.

———. 1911. On the dinosaurs of the Stormberg, South Africa. Ann. S. Afr. Mus. 7: 291–308.

———. 1912a. On the remains of a theropodous dinosaur from the Northern Transvaal. Trans. Geol. Soc. S. Afr. 14: 82–83.

———. 1912b. Observations on some specimens of South African fossil reptiles preserved in the British Museum. Trans. Roy. Soc. S. Afr. 2: 19–25.

———. 1915. Catalogue of the type and figured specimens of fossil vertebrates in the American Museum of Natural History III. Permian, Triassic and Jurassic reptiles of South Africa. Bull. Am. Mus. Nat. Hist. 25: 105–164.

Broughton, P. L. 1981. Casts of vertebrate internal organs from the Upper Cretaceous of western Canada. J. Geol. 89: 741–749.

Broughton, P. L., Simpson, F., and Whitaker, S. H. 1978. Late Cretaceous coprolites from western Canada. Palaeontology 21: 443–453.

Brouwers, E. M., Clemens, W. A., Spicer, R. A., Ager, T. A., Carter, L. D., and Sliter, W. V. 1987. Dinosaurs on the North Slope, Alaska: high latitude, latest Cretaceous environments. Science 237: 1608–1610.

Brown, B. 1907. Gastroliths. Science 25: 392.

——. 1908. The Ankylosauridae, a new family of armored dinosaurs from the Upper Cretaceous. Bull. Am. Mus. Nat. Hist. 24: 187–201.

——. 1910. The Cretaceous Ojo Alamo beds of New Mexico with description of the new dinosaur genus *Kritosaurus*. Bull. Am. Mus. Nat. Hist. 28: 267–274.

——. 1912. A crested dinosaur from the Edmonton Cretaceous. Bull. Am. Mus. Nat. Hist. 31: 131–136.

——. 1913. A new trachodont dinosaur, *Hypacrosaurus*, from the Edmonton Cretaceous of Alberta. Bull. Am. Mus. Nat. Hist. 32: 395–406.

——. 1914a. *Corythosaurus casuarius*, a new crested dinosaur from the Belly River Cretaceous, with provisional classification of the family Trachodontidae. Bull. Am. Mus. Nat. Hist. 33: 559–565.

——. 1914b. *Anchiceratops*, a new genus of horned dinosaurs from the Edmonton Cretaceous of Alberta. With discussion of the origin of the ceratopsian crest and the brain casts of *Anchiceratops* and *Trachodon*. Bull. Am. Mus. Nat. Hist. 33: 539–548.

——. 1914c. *Leptoceratops*, a new genus of Ceratopsia from the Edmonton Cretaceous of Alberta. Bull. Am. Mus. Nat. Hist. 33: 567–580.

——. 1914d. A complete skull of *Monoclonius*, from the Belly River Cretaceous of Alberta. Bull. Am. Mus. Nat. Hist. 33: 549–558.

——. 1916a. A new crested trachodont dinosaur, *Prosaurolophus maximus*. Bull. Am. Mus. Nat. Hist. 35: 701–708.

——. 1916b. *Corythosaurus casuarius*: skeleton, musculature and epidermis. Bull. Am. Mus. Nat. Hist. 35: 709–716.

——. 1917. A complete skeleton of the horned dinosaur *Monoclonius*, and description of a second skeleton showing skin impressions. Bull. Am. Mus. Nat. Hist. 37: 281–306.

——. 1933. A new longhorned Belly River ceratopsian. Am. Mus. Novitates 669: 1–3.

——. 1935. Sinclair Dinosaur Expedition, 1934. Nat. Hist. 36: 3–15.

——. 1938. The mystery dinosaur. Nat. Hist. 41: 190–202, 235.

Brown, B., and Schlaikjer, E. M. 1937. The skeleton of *Styracosaurus* with the description of a new species. Am. Mus. Novitates 955: 1–12.

——. 1940a. The structure and relationships of *Protoceratops*. Ann. New York Acad. Sci. 40: 133–266.

——. 1940b. The origin of ceratopsian horn cores. Am. Mus. Novitates 1065: 1–8.

——. 1940c. A new element in the ceratopsian jaw with additional notes on the mandible. Am. Mus. Novitates 1092: 1–13.

——. 1942. The skeleton of *Leptoceratops* with the description of a new species. Am. Mus. Novitates 1169: 1–15.

——. 1943. A study of the troödont dinosaurs with the description of a new genus and four new species. Bull. Am. Mus. Nat. Hist. 82: 121–149.

Brunet, M., Coppens, Y., Pilbeam, D., Djallo, S., Behrensmeyer, K., Brillanceau, A., Downs, W., Dupéron, M., Ekodeck, G., Flynn, L., Heintz, E., Hell, J., Jehenne, Y., Martin, L., Mosser, C., Salard-Cheboldaeff, M., Wenz, S., and Wing, S. 1986. Les formations sedimentaires continentales du Crétacé et du Cenozoique camerounais: Premiers résultats d'une prospection paléontologique. C. R. Acad. Sci. Paris 303: 425–428.

Bryant, H. N., and Churcher, C. S. 1987. All sabretoothed carnivores aren't sharks. Nature 325: 488.

Bryant, L. J., Clemens, W. A., and Hutchison, J. H. 1986. Cretaceous-Tertiary dinosaur extinction. Science 234: 1172.

Buckland, W. 1824. Notice on *Megalosaurus* or great fossil lizard of Stonesfield. Trans. Geol. Soc. London 21: 390–397.

Budyko, M. I. 1984. [Evolution of the Biosphere]. Gidrometeoizdat, U.S.S.R. 488 pp. (In Russian)

Buffetaut, E. 1982. Mesozoic vertebrates from Thailand and their palaeobiological significance. Terra Cognita 2: 27–34.

——. 1984. Une vertèbre de dinosaurien sauropode dans le Crétacé du Cap de la Hève (Normandie). Actes Mus. Rouen Bull. Municipal Sci. 1984: 213–221.

——. 1985. The age of the Saint-Nicolas-de-Port vertebrate locality (Triassic of eastern France). Terra Cognita 5: 133.

Buffetaut, E., Bulow, M., Gheerbrant, E., Jaeger, J.-J., Martin, M., Mazin, J.-M., Milsent, C., and Rioult, M. 1985. Zonation biostratigraphique et nouveaux restes de vertébrés dans les "Sables de Glos" (Oxfordien supérieur, Normandie). C. R. Acad. Sci. Paris 330: 929–932.

Buffetaut, E., Cappetta, H., Gayet, M., Martin, M., Moody, R. T. J., Rage, J. C., Taquet, P., and Wellnhofer, P. 1981. Les vertébrés de la partie moyenne du Crétacé en Europe. Cretaceous Res. 2: 275–281.

Buffetaut, E., and Ingavat, R. 1983. Vertebrates from the continental Jurassic of Thailand. United Nations ESCAP CCOP Tech. Bull. 16: 68–75.

——. 1984. Un dinosaurien théropode de très petite taille dans le Jurassique supérieur du nord-est de la Thailande. C. R. Acad. Sci. Paris 298: 915–918.

——. 1985. The Mesozoic vertebrates of Thailand. Sci. Am. 253 (2): 80–87.

——. 1986a. The succession of vertebrate faunas in the continental Mesozoic of Thailand. Proc. Geol. Soc. Malaysia 1 (Bull. 19): 167–172.

——. 1986b. Unusual theropod dinosaur teeth from the Upper Jurassic of Phu Wiang, northeastern Thailand. Rev. Paléobiol. 5: 217–220.

Buffetaut, E., Ingavat, R., Sattayarak, N., and Suteetorn, V. 1985. Early Cretaceous dinosaur footprints from Phu Luang (Loei Province, northeastern Thailand) and their significance. Conference on Geology and Mineral Resources Development of Northeast Thailand. Khon Kaen Univ. Pp. 71–76.

Buffetaut, E., Marandat, B., and Sigé, B. 1986. Découverte de dents de deinonychosaures (Saurischia, Theropoda) dans le Crétacé supérieur du sud de la France. C. R. Acad. Sci. Paris 303: 1393–1396.

Buffetaut, E., Meijer, A. W. F., Taquet, P., and Wouters, G. 1985. New remains of hadrosaurid dinosaurs (Reptilia, Ornithischia) from the Maastrichtian of Dutch and Belgian Limburg. Rev. Paleobiol. 4: 65–70.

Buffetaut, E., Pouit, D., and Taquet, P. 1980. Une dent de dinosaurien ornithopode remaniée dans les faluns miocènes de Doué-Douces (Maine-et-Loire). C. R. Soc. Géol. France 1980: 200–202.

Buffetaut, E., and Wouters, G. 1986. Amphibian and reptile remains from the Upper Triassic of Saint-Nicolas-de-Port (eastern France) and their biostratigraphic significance. Modern Geol. 10: 133–145.

Buffrénil, V. de. 1980. Mise en evidence de l'incidence des conditions de milieu sur la croissance de *Crocodylus siamensis* (Schneider, 1801) et valeur des marques de croissance squelettiques pour l'évaluation de l'âge individuel. Arch. Zool. Exp. Gen. 121: 63–76.

Buffrénil, V. de, Farlow, J. O., and Ricqlès, A. de 1984. Histological data on structure, growth and possible functions of *Stegosaurus* plates. In: Reif, W.-E., and Westphal, F. (eds.). 3d Symp. Mesozoic Terr. Ecosyst. Attempto Verlag, Tübingen. Pp. 31–36.

———. 1986. Growth and function of *Stegosaurus* plates: Evidence from bone histology, Paleobiology 12: 459–473.

Bull, J. J. 1983. The Evolution of Sex-Determining Mechanisms. Benjamin Cummings, Menlo Park, Calif. 316 pp.

Bunzel, E. 1871. Die Reptilfauna der Gosauformation in der Neuen Welt bei Wiener-Neustadt. Abh. Geol. Reichsanst. Wien 5: 1–18.

Burggren, W. W. 1987. Form and function in reptilian circulations. Am. Zool. 27: 5–19.

Burghardt, G. M. 1977. Of iguanas and dinosaurs: Social behavior and communication in neonate reptiles. Am. Zool. 17: 177–190.

Buscalioni, A. D., and Sanz, J. L. 1987. Lista faunistica de los Vertebrados del Cretácico inferior del area de Galve (Teruel, España). In: Estud. Geol. Vol. Extraord. Galve-Tremp. Pp. 65–68.

Cabrera, A. 1947. Un saurópodo nuevo del Jurásico de Patagonia. Notas Mus. La Plata 12 Paleontol. 95: 1–17.

Cailliet, G. M., Natanson, L. J., Welden, B. A., and Ebert, D. A. 1985. Preliminary studies on the age and growth of the white shark, *Carcharodon carcharias*, using vertebral bands. Mem. So. Calif. Acad. Sci. 9: 49–60.

Calder, W. A., III. 1984. Size, Function, and Life History. Harvard Univ. Press, Cambridge. 431 pp.

Callison, G., and Quimby, H. M. 1984. Tiny dinosaurs: Are they fully grown? J. Vert. Paleontol. 3: 20–209.

Camp, C. L. 1935. Dinosaur remains from the province of Szechuan, China. Univ. Calif. Publ. Geol. Sci. 23: 467–472.

———. 1936. A new type of small bipedal dinosaur from the Navajo Sandstone of Arizona. Univ. Calif. Publ. Geol. Sci. 24: 39–56.

Camp, C. L., and Welles, S. P. 1956. Triassic dicynodont reptiles. I. The North American genus *Placerias*. Mem. Univ. Calif. 13: 255–348.

Cannon, G. L. 1907. Sauropodan gastroliths. Science 24: 116.

Caraco, T., and Wolf, L. L. 1975. Ecological determinants of group sizes of foraging lions. Am. Nat. 109: 343–352.

Carey, F. G., Casey, J. G., Pratt, H. L., Urquhart, D. and McCosker, J. E. 1985. Temperature, heat production and heat exchange in lamnid sharks. Mem. So. Calif. Acad. Sci. 9: 92–108.

Carey, F. G., Kanwisher, J. W., Brazier, O., Gabrielson, G., Casey, J. G., and Pratt, H. L., Jr. 1982. Temperature and activities of a white shark, *Carcharodon carcharias*. Copeia 1982: 254–260.

Carey, M. A., and Madsen, J. H., Jr. 1972. Some observations on the growth, function and differentiation of sauropod teeth from the Cleveland-Lloyd Quarry. Proc. Utah Acad. Sci. 49: 40–43.

Carpenter, K. 1979. Vertebrate fauna of the Laramie Formation (Maestrichtian), Weld County, Colorado. Contrib. Geol. Univ. Wyoming 17: 37–48.

———. 1982a. Skeletal and dermal armor reconstruction of *Euoplocephalus tutus* (Ornithischia: Ankylosauridae) from the Late Cretaceous Oldman Formation of Alberta. Can. J. Earth Sci. 19: 689–697.

———. 1982b. Baby dinosaurs from the Late Cretaceous Lance and Hell Creek formations and a description of a new species of theropod. Contrib. Geol. Univ. Wyoming 20: 123–134.

———. 1982c. The oldest Late Cretaceous dinosaurs in North America? Mississippi Geol. 3: 11–17.

———. 1984. Skeletal reconstruction and life restoration of *Sauropelta* (Ankylosauria: Nodosauridae) from the Cretaceous of North America. Can. J. Earth Sci. 21: 1491–1498.

———. 1987. Paleoecological significance of droughts during the Late Cretaceous of the Western Interior. In: Currie, P. J., and Koster, E. H. (eds.). 4th Symp. Mesozoic Terr. Ecosyst. Tyrrell Mus. Palaeontol., Drumheller, Alberta. Pp. 42–47.

Carpenter, K., and Breithaupt, B. 1986. Latest Cretaceous occurrence of nodosaurid ankylosaurs (Dinosauria, Ornithischia) in western North America and the gradual extinction of the dinosaurs. J. Vert. Paleontol. 6: 251–257.

Carpenter, K., and Parrish, M. 1985. Late Triassic vertebrates from Revuelto Creek, Quay County, New Mexico.

New Mexico Geol. Soc. Gdbk., 36th Field Conf. Pp. 197–198.

Carr, S. G., Oliver, J. G., Conor, C. H. H., and Scott, D. C. 1979. Andamooka opal fields. The geology of the precious stones field and the results of the subsidised mining program. Rept. Invest. Dept. Mines South Australia 51: 1–68.

Carrier, D. R. 1987. The evolution of locomotor stamina in tetrapods: circumventing a mechanical constraint. Paleobiology 13: 326–341.

Carroll, R. L. 1976. Eosuchians and the origin of archosaurs. In: Churcher, C. S. (ed.). Athlon. Roy. Ontario Mus., Toronto. Pp. 58–79.

Carroll, R. L., Belt, E. S., Dineley, D. L., Baird, D., and McGregor, D. C. 1972. Vertebrate paleontology of eastern Canada. XXIV Internatl. Geol. Congr., Gdbk. Field Excursion A59. Pp. 1–113.

Carter, D. R., and Hayes, W. C. 1977. Compact bone fatigue damage. I. Residual strength and stiffness. J. Biomech. 10: 323–337.

Casamiquela, R. M. 1963. Considéraciones acerca de *Amygdalodon* Cabrera (Sauropoda, Cetiosauridae) del Jurásico medio de la Patagonia. Ameghiniana 3: 79–95.

———. 1964*a*. Sobre un dinosaurio hadrosaurido de la Argentina. Ameghiniana 3: 285–308.

———. 1964*b*. Estudios icnológicos. Problemas y métodos de la icnología con aplication al estudio de pisadas mesozoicos (Reptilia, Mammalia) de la Patagonia. Ministerio Asuntos Sociales Province del Rio Negro. 229 pp.

———. 1964*c*. Sobre el hallazgo de dinosaurios triásicos en la Provincia de Santa Cruz. Argent. Austral. 35: 10–11.

———. 1967*a*. Los dinosaurios Chilenos. Mus. Nacl. Hist. Nat. (Chile) 134: 3–7.

———. 1967*b*. Un nuevo dinosaurio ornitisquio Triásico (*Pisanosaurus mertii;* Ornithopoda) de la Formación Ischigualasto, Argentina. Ameghiniana 4: 47–64.

———. 1980. La presencia del genero *Plateosaurus* (Prosauropoda) en el Triásico superior de la Formación El Tranquilo, Patagonia. Actas II Congr. Argent. Paleontol. Bioestrat. I Congr. Latinoam. Paleontol. 1: 143–158.

Casamiquela, R. M., Corvalan, J., and Franquesa, F. 1969. Hallazgo de dinosaurios en el Cretácico superior de Chile. Inst. Invest. Geol. Bol. 25: 1–31.

Casamiquela, R. M., and Fasola, A. 1968. Sobre pisadas de dinosaurios del Cretácico inferior de Colchagua (Chile). Univ. Chile Dept. Geol. Publ. 30: 1–24.

Casanovas, M. L., Santafé, J. L., Sanz, J. L., and Buscalioni, A. D. 1987. Arcosaurios (Crocodilia, Dinosauria) del Cretácico superior de la Conca de Tremp (Lleida, España). Estud. Geol. Vol. Extraord. Galve-Tremp. Pp. 95–110.

Casanovas-Cladellas, M. L., and Santafé-Llopis, J.-V. 1971. Icnitas de reptiles mesozoicos en la provincia de Logrono. Acta Geol. Hisp. 6: 139–142.

———. 1974. Dos nuevos yacimientos de icnitas de Dinosaurios. Acta Geol. Hisp. 9: 88–91.

Casanovas-Cladellas, M. L., Santafé-Llopis, J. V., and Sanz, J. L. 1984. Las Icnitas de "Los Corrales del Pelejon" en el Cretácico inferior de Galve (Teruel, España). Paleontol. Evol. 18: 173–176.

Casanovas-Cladellas, M. L., Santafé-Llopis, J. V., Sanz, J. L., and Buscalioni, A. 1985*a*. *Orthomerus* (Hadrosaurinae, Ornithopoda) del Cretácico superior del yacimiento de "Els Nerets" (Tremp, España). Paleontol. Evol. 19: 155–162.

———. 1985*b*. *Orthomerus* (Hadrosaurinae, Ornithopoda) du Crétáce supérieur du gisement de "Els Nerets" (Tremp, Lleida). In: Les Dinosaures de la Chine à la France. Mus. Hist. Nat., Toulouse, France. Pp. 99–111.

Casanovas-Cladellas, M. L., Perez-Lorente, F., Santafé-Llopis, J. V., and Fernandez-Ortega, A. 1985. Nuevos datos icnologicos del Cretácico inferior de la Sierra de Cameros (La Rioja, España). Paleontol. Evol. 19: 3–18.

Case, T. J. 1978*a*. On the evolution and adaptive significance of postnatal growth rates in the terrestrial vertebrates. Q. Rev. Biol. 53: 243–282.

———. 1978*b*. Speculations on the growth rate and reproduction of some dinosaurs. Paleobiology 4: 320–328.

Casey, R. 1973. The ammonite succession at the Jurassic-Cretaceous boundary in eastern England. In: Casey, R., and Rawson, P. R. (eds.). The Boreal Lower Cretaceous. Seel House Press, Liverpool. Pp. 193–266.

Casier, E. 1978. Les Iguanodons de Bernissart. Inst. Roy. Sci. Nat. Belgique, Brussels. 168 pp.

Chabli, S. 1985. Données nouvelles sur un "Dinosaurien" Jurassique moyen du Maroc: *Megalosaurus mersensis* Lapparent 1955, et sur les Mégalosauridés en général. In: Les Dinosaures de la Chine à la France. Mus. Hist. Nat. Toulouse, France. Pp. 65–72.

Chabreck, R. H. 1972. The foods and feeding habits of alligators from fresh and saline habitats in Louisiana. Proc. Ann. Conf. Southeast. Assoc. Game Fish Commissioners 25: 117–124.

Chakravarti, D. K. 1934. On a stegosaurian humerus from the Lameta beds of Jubbulpore. Q. J. Mineral. Metallurg. Soc. India, 30: 75–79.

———. 1935. Is *Lametasaurus indicus* an armored dinosaur? Am. J. Sci. (ser. 5) 30: 138–142.

Chao S. 1962. [Concerning a new species of *Psittacosaurus* from Layiyang, Shantung.] Vertebrata PalAsiatica 6: 349–360. (Chinese with Russian summary)

Chapman, R. E., Galton, P. M., Sepkoski, J. J., and Wall, W. P. 1981. A morphometric study of the cranium of the pachycephalosaurid dinosaur *Stegoceras*. J. Paleontol. 55: 608–616.

Chappell, M. A., and Ellis, T. M. 1987. Resting metabolic rates in boid snakes: Allometric relationships and temperature effects. J. Comp. Physiol. B 157: 227–235.

Charig, A. J. 1972. The evolution of the archosaur pelvis and hindlimb: an explanation in functional terms. In:

Joysey, K. A., and Kemp, T. S. (eds.). Studies in Vertebrate Evolution. Winchester, New York. Pp. 121–155.

——. 1976. "Dinosaur monophyly and a new class of vertebrates": A critical review. In: Bellairs, A. d'A., and Cox, C. B. (eds.). Morphology and Biology of Reptiles. Academic Press, New York. Pp. 65–104.

——. 1979. A New Look at the Dinosaurs. Heinemann, London. 160 pp.

——. 1980. A diplodocid sauropod from the Lower Cretaceous of England. In: Jacobs, L. L. (ed.). Aspects of Vertebrate History. Mus. No. Arizona Press, Flagstaff. Pp. 231–244.

Charig, A. J., Attridge, J., and Crompton, A. W. 1965. On the origin of the sauropods and the classification of the Saurischia. Proc. Linn. Soc. London 176: 197–221.

Charig, A. J., and Crompton, A. W. 1974. The alleged synonymy of *Lycorhinus* and *Heterodontosaurus*. Ann. S. Afr. Mus. 64: 167–189.

Charig, A. J., Krebs, B., Sues, H.-D. and Westphal, F. 1976. Thecodontia. Hdb. Paläoherpetol. 13: 1–137.

Charig, A. J., and Milner, A. C. 1986. *Baryonyx*, a remarkable new theropod dinosaur. Nature 324: 359–361.

Chatterjee, S. 1978*a*. A primitive parasuchid (phytosaur) reptile from the Upper Triassic Maleri Formation of India. Palaeontology 21: 83–127.

——. 1978*b*. *Indosuchus* and *Indosaurus*, Cretaceous carnosaurs from India. J. Paleontol. 52: 570–580.

——. 1982. Phylogeny and classification of thecodontian reptiles. Nature 295: 317–320.

——. 1984. A new ornithischian dinosaur from the Triassic of North America. Naturwissenschaften 71: 630–631.

——. 1985. *Postosuchus*, a new thecodontian reptile from the Triassic of Texas and the origin of tyrannosaurs. Phil. Trans. Roy. Soc. London B 309: 395–460.

——. 1986. The Late Triassic Dockum vertebrates: their stratigraphic and paleobiogeographic significance. In: Padian, K. (ed.). The Beginning of the Age of Dinosaurs. Cambridge Univ. Press, New York. Pp. 139–150.

——. 1987. A new theropod dinosaur from India with remarks on the Gondwana-Laurasia connection in the Late Triassic. In: McKenzie, G. D. (ed.). Gondwana 6: Stratigraphy, Sedimentology, and Paleontology. Geophys. Monogr. 41: 183–189.

Chatterjee, S., and Hotton, N., III. 1986. The paleoposition of India. J. Southeast Asian Earth Sci. 1: 145–189.

Chatterjee, S., and Majumdar, P. K. 1987. *Tikisuchus romeri*, a new rauisuchid reptile from the Late Triassic of India. J. Paleontol. 61: 787–793.

Chen P. 1983. A survey of the non-marine Cretaceous of China. Cretaceous Res. 4: 123–143.

Chen P., Li W., Chen J., Yea C., Wang Z., Shen Y., and Sun A. 1982. Stratigraphical classification of Jurassic and Cretaceous of China. Scientia Sinica B 25: 1227–1248.

Cheng Z. 1983. [Reptilia]. In: [The Mesozoic Stratigraphy and Paleontology of the Guyang Coal-Bearing Basin,

Nei Mongoll Autonomous Region, China]. Geology Press, Beijing. Pp. 123–136. (In Chinese)

Choi, H. I. 1985. Sedimentology and its implications for stratigraphic classifications of the Cretaceous Gyeongsang Basin. J. Geol. Soc. Korea 21: 26–37.

——. 1986. Sedimentation and evolution of the Cretaceous Gyeongsang Basin, S.E. Korea. Q. J. Geol. Soc. London 143: 29–40.

Chow M. 1951. Notes on the Late Cretaceous dinosaurian remains and the fossil eggs from Laiyang, Shantung. Bull. Geol. Soc. China 31: 89–96.

Chow M., and Rozhdestvensky, A. K. 1960. Exploration in Inner Mongolia—a preliminary account of the 1959 field work of the Sino-Soviet Paleontological Expedition (SSPE). Vertebrata PalAsiatica 4: 1–10.

Clark, G. A., Jr. 1964. Ontogeny and evolution in the Megapodes (Aves: Galliformes). Postilla 78: 1–37.

Clark, J. M., and Fastovsky, D. E. 1986. Vertebrate biostratigraphy of the Glen Canyon Group in northern Arizona. In: Padian, K. (ed.). The Beginning of the Age of Dinosaurs: Faunal Change across the Triassic-Jurassic Boundary. Cambridge Univ. Press, New York. Pp. 285–301.

Clark, J. M., and Hopson, J. A. 1985. Distinctive mammal-like reptile from Mexico and its bearing on the phylogeny of the Tritylodontidae. Nature 315: 398–400.

Clemens, W. A. 1980. Rhaeto-Liassic mammals from Switzerland and West Germany. Zitteliana 5: 51–92.

——. 1986. Evolution of the terrestrial vertebrate fauna during the Cretaceous-Tertiary transition. In: Elliott, D. K. (ed.). The Dynamics of Extinction. Wiley Interscience, New York. Pp. 63–86.

Cobabe, E. A., and Fastovsky, D. E. 1987. *Ugrosaurus olsoni*, a new ceratopsian (Reptilia: Ornithischia) from the Hell Creek Formation of eastern Montana. J. Paleontol. 61: 148–154.

Cobban, W. A. 1955. Cretaceous rocks of northwestern Montana. Billings Geol. Soc. Gdbk. 6th Ann. Field Conf. Pp. 107–119.

Cobban, W. A., and Reeside, J. B. 1952. Correlation of the Cretaceous formations of the western interior of the United States. Bull. Geol. Soc. Am. 63: 1011–1044.

Cockerell, T. D. A. 1924. Fossils in the Ondai Sair Formation, Mongolia. Bull. Am. Mus. Nat. Hist. 51: 129–144.

Coe, M. J., Bourn, D., and Swingland, I. R. 1979. The biomass, production and carrying capacity of giant tortoises on Aldabra. Phil. Trans. Roy. Soc. London B 286: 162–176.

Coe, M. J., Dilcher, D. L., Farlow, J. O., Jarzen, D. M., and Russell, D. A. 1987. Dinosaurs and land plants. In: Friis, E. M., Chaloner, W. G., and Crane, P. R. (eds.). The Origins of Angiosperms and their Biological Consequences. Cambridge Univ. Press, New York. Pp. 225–258.

Colbert, E. H. 1945. The hyoid bones in *Protoceratops* and *Psittacosaurus*. Am. Mus. Novitates 1301: 1–10.

———. 1947. The little dinosaurs of Ghost Ranch. Nat. Hist. 56: 392–399, 427–428.

———. 1948a. Triassic life in the southwestern United States. Trans. New York Acad. Sci. (ser. 2) 10: 229–235.

———. 1948b. A hadrosaurian dinosaur from New Jersey. Proc. Acad. Nat. Sci. Philadelphia 100: 23–37.

———. 1951. The Dinosaur Book: The Ruling Reptiles and Their Relatives. McGraw-Hill, New York: 156 pp.

———. 1955. Evolution of the Vertebrates. Wiley, New York. 479 pp.

———. 1961. Dinosaurs: Their Discovery and Their World. Dutton, New York. 300 pp.

———. 1964a. Relationships of saurischian dinosaurs. Am. Mus. Novitates 2181: 1–24.

———. 1964b. The Triassic dinosaur genera *Podokesaurus* and *Coelophysis*. Am. Mus. Novitates 2168: 1–12.

———. 1968. Men and Dinosaurs. Dutton, New York. 283 pp.

———. 1970. A saurischian dinosaur from the Triassic of Brazil. Am. Mus. Novitates 2405: 1–39.

———. 1974. The Triassic paleontology of Ghost Ranch. New Mexico Geol. Soc. Gdbk., 25th Field Conf. Pp. 175–178.

———. 1981. A primitive ornithischian dinosaur from the Kayenta Formation of Arizona. Bull. Mus. No. Arizona 53: 1–61.

———. 1983. Dinosaurs: An Illustrated History. Hammond, Maplewood, N.J. 224 pp.

———. 1989. The Triassic dinosaur *Coelophysis*. Bull. Mus. No. Arizona 57: 1–174.

Colbert, E. H., and Baird, D. 1958. Coelurosaur bone casts from the Connecticut Valley Triassic. Am. Mus. Novitates 1901: 1–11.

Colbert, E. H., and Bump, J. D. 1947. A skull of *Torosaurus* from South Dakota and a revision of the genus. Proc. Acad. Nat. Sci. Philadelphia 99: 93–106.

Colbert, E. H., Cowles, R. B., and Bogert, C. M. 1946. Temperature tolerances in the American alligator and their bearing on the habits, evolution and extinction of the dinosaurs. Bull. Am. Mus. Nat. Hist. 86: 327–373.

———. 1947. Rates of temperature increase in dinosaurs. Copeia 1947: 141–142.

Colbert, E. H., and Merrilees, D. 1967. Cretaceous dinosaur footprints from Western Australia. J. Roy. Soc. W. Australia 50: 21–25.

Colbert, E. H., and Ostrom, J. H. 1958. Dinosaur stapes. Am. Mus. Novitates 1900: 1–20.

Colbert, E. H., and Russell, D. A. 1969. The small Cretaceous dinosaur *Dromaeosaurus*. Am. Mus. Novitates 2380: 1–49.

Colinvaux, P. 1978. Why Big Fierce Animals Are Rare: an Ecologist's Perspective. Princeton Univ. Press, Princeton. 256 pp.

Congdon, J. D., Dunham, A. E. and Tinkle, D. W. 1982. Energy budgets and life histories of reptiles. In: Gans, C., and Pough, F. H. (eds.). Biology of the Reptilia. Vol. 13: 233–271.

Coombs, M. C. 1983. Large mammalian clawed herbivores: A comparative study. Trans. Am. Phil. Soc. 73: 1–96.

Coombs, W. P., Jr. 1971. The Ankylosauria. Ph.D. dissertation, Columbia Univ., New York.

———. 1972. The bony eyelid of *Euoplocephalus* (Reptilia, Ornithischia). J. Paleontol. 46: 637–650.

———. 1975. Sauropod habits and habitats. Palaeogeogr. Palaeoclimatol. Palaeoecol. 17: 1–33.

———. 1978a. The families of the ornithischian dinosaur order Ankylosauria. Palaeontology 21: 143–170.

———. 1978b. Forelimb muscles of the Ankylosauria (Reptilia, Ornithischia). J. Paleontol. 52: 642–658.

———. 1978c. Theoretical aspects of cursorial adaptations in dinosaurs. Q. Rev. Biol. 53: 393–418.

———. 1978d. An endocranial cast of *Euoplocephalus* (Reptilia, Ornithischia). Palaeontographica A 161: 176–182.

———. 1979. Osteology and myology of the hindlimb in the Ankylosauria (Reptilia, Ornithischia). J. Paleontol. 53: 666–684.

———. 1980a. Swimming ability of carnivorous dinosaurs. Science 207: 1198–1200.

———. 1980b. Juvenile ceratopsians from Mongolia—the smallest known dinosaur specimens. Nature 283: 380–381.

———. 1982. Juvenile specimens of the ornithischian dinosaur *Psittacosaurus*. Palaeontology 25: 89–107.

———. 1986. A juvenile ankylosaur referrable to the genus *Euoplocephalus* (Reptilia, Ornithischia). J. Vert. Paleontol. 6: 162–173.

———. 1987. Asiatic Ceratopsidae might be Ankylosauridae. In: Currie, P. J., and Koster, E. H. (eds.). 4th Symp. Mesozoic Terr. Ecosyst. Tyrrell Mus. Palaeontol., Drumheller, Alberta. Pp. 48–51.

Coombs, W. P., Jr., and Molnar, R. E. 1981. Sauropoda (Reptilia, Saurischia) from the Cretaceous of Queensland. Mem. Queensland Mus. 20: 351–353.

Cooper, M. R. 1980. The first record of the prosauropod dinosaur *Euskelosaurus* from Zimbabwe. Arnoldia 9: 1–17.

———. 1981a. The prosauropod dinosaur *Massospondylus carinatus* Owen from Zimbabwe: Its biology, mode of life and phylogenetic significance. Occasional Papers, Natl. Mus. Monuments Rhodesia (series B) 6: 689–840.

———. 1981b. Archosaur ankles: Interpretation of the evidence. S. Afr. J. Sci. 77: 307–309.

———. 1984. A reassessment of *Vulcanodon karibaensis* Raath (Dinosauria: Saurischia) and the origin of the Sauropoda. Palaeontol. afr. 25: 203–231.

———. 1985. A revision of the ornithischian dinosaur *Kangnasaurus coetzeei* Haughton, with a classification of the Ornithischia. Ann. S. Afr. Mus. 95: 281–317.

Cope, E. D. 1866. Discovery of a gigantic dinosaur in the Cretaceous of New Jersey. Proc. Acad. Nat. Sci. Philadelphia 18: 275–279.

———. 1867. An account of extinct reptiles that approach birds. Proc. Acad. Nat. Sci. Philadelphia 19: 234–235.

———. 1868. On some Cretaceous Reptilia. Proc. Acad. Nat. Sci. Philadelphia. 20: 237–242.

———. 1869a. [Remarks on *Eschrichtius polyporus, Hypsibema crassicauda, Hadrosaurus tripos,* and *Polydectes biturgidus*]. Proc. Acad. Nat. Sci. Philadelphia 21: 192.

———. 1869b. Synopsis of the extinct Batrachia, Reptilia, and Aves of North America. Trans. Am. Phil. Soc. 14: 1–252.

———. 1869c. [Remarks on *Holops brevispinus, Ornithotarsus immanis,* and *Macrosaurus proriger*]. Proc. Acad. Nat. Sci. Philadelphia 1869: 123.

———. 1871. Supplement to the synopsis of the extinct Batrachia and Reptilia of North America. Proc. Am. Phil. Soc. 12: 41–52.

———. 1872. On the existence of Dinosauria in the transition beds of Wyoming. Proc. Am. Phil. Soc. 12: 481–483.

———. 1874. Report on the stratigraphy and Pliocene vertebrate palaeontology of northern Colorado. Bull. U.S. Geol. Geogr. Surv. Territories 1: 9–28.

———. 1875. On the transition beds of the Saskatchewan district. Proc. Acad. Nat. Sci. Philadelphia 1875: 9–10.

———. 1876a. Description of some vertebrate remains from the Fort Union beds of Montana. Proc. Acad. Nat. Sci. Philadelphia 28: 248–261.

———. 1876b. On some extinct reptiles and Batrachia from the Judith River and Fox Hills beds of Montana. Proc. Acad. Nat. Sci. Philadelphia 28: 340–359.

———. 1877a. On a carnivorous dinosaurian from the Dakota beds of Colorado. Bull. U.S. Geol. Surv. Territories 3: 805–806.

———. 1877b. On a dinosaurian from the Trias of Utah. Proc. Am. Phil. Soc. 16: 579–584.

———. 1877c. On reptilian remains from the Dakota beds of Colorado. Proc. Am. Phil. Soc. 17: 193–196.

———. 1877d. On *Amphicoelias,* a genus of saurian from the Dakota epoch of Colorado. Proc. Am. Phil. Soc. 17: 242–246.

———. 1877e. On a gigantic saurian from the Dakota epoch of Colorado. Paleontol. Bull. 25: 5–10.

———. 1877f. On reptilian remains from the Dakota beds of Colorado. Proc. Am. Phil. Soc. 17: 193–196.

———. 1878a. A new genus of Dinosauria from Colorado. Am. Nat. 12: 188–189.

———. 1878b. A new opisthocoelous dinosaur. Am. Nat. 12: 406.

———. 1878c. A new species of *Amphicoelias.* Am. Nat. 12: 563–565.

———. 1878d. On the Vertebrata of the Dakota epoch of Colorado. Proc. Am. Phil. Soc. 17: 233–247.

———. 1878e. On some saurians found in the Triassic of Pennsylvania, by C. M. Wheatley. Proc. Am. Phil. Soc. 17: 231–232.

———. 1878f. Descriptions of new extinct *Vertebrata* from the upper Tertiary and Dakota formations. Bull. U.S. Geol. Geogr. Surv. Territories 1878: 379–396.

———. 1878g. On the saurians of the Dakota Cretaceous of Colorado. Nature 18: 476.

———. 1879. New Jurassic Dinosauria. Am. Nat. 13: 402–404.

———. 1883. On the characters of the skull in the Hadrosauridae. Proc. Acad. Nat. Sci. Philadelphia 35: 97–107.

———. 1887. The dinosaur genus *Coelurus.* Am. Nat. 21: 367–369.

———. 1889a. Notes on the Dinosauria of the Laramie. Am. Nat. 23: 904–906.

———. 1889b. The horned Dinosauria of the Laramie. Am. Nat. 23: 715–717.

———. 1889c. On a new genus of Triassic Dinosauria. Am. Natur. 23: 626.

———. 1892a. On the skull of the dinosaurian *Laelaps incrassatus* Cope. Proc. Am. Phil. Soc. 30: 241.

———. 1892b. Fourth note on the Dinosauria of the Laramie. Am. Nat. 26: 756–758.

Cope, J. C. W., Duff, K. L., Parsons, C. F., Torrens, H. S., Wimbledon, W. A., and Wright, J. K. 1980. Pt. 2: Jurassic. Geol. Soc. London Sp. Rept. 15: 1–109.

Cope, J. C. W., Getty, T. A., Howarth, M. K., Morton, N., and Torrens, H. S. 1980. Jurassic. Pt. 1: Introduction and Lower Jurassic. Spec. Rept. Geol. Soc. London 14: 1–73.

Corro, G. del. 1966. Un nuevo dinosaurio carnívoro del Chubut (Argentina). Communicaciones Mus. Argent. Cien. Nat. "Bernardino Rivadavia". Paleontol. 1: 1–4.

———. 1974. Un nuevo megalosaurio (Carnosaurio) del Cretácico de Chubut (Argentina). Communicaciones Mus. Argent. Ciencias Nat. "Bernardino Rivadavia" Paleontol. 1: 37–44.

———. 1975. Un nuevo saurópodo del Cretácico superior. *Chubutisaurus insignis* gen. et sp. nov. (Saurischia-Chubutisauridae nov.) del Cretácico superior (Chubutiano), Chubut, Argentina. Actas Congr. Argent. Paleontol. Bioestrat. 2: 229–240.

Cott, H. B. 1961. Scientific results of an inquiry into the ecology and economic status of the Nile crocodile (*Crocodylus niloticus*) in Uganda and Northern Rhodesia. Trans. Zool. Soc. London 29: 211–357.

———. 1971. Parental care in the Crocodilia, with special reference to *Crocodylus niloticus.* Crocodiles. Proc. of the First Meeting of Crocodile Specialists. Internatl. Union Conserv. Nature Nat. Res. Switzerland. Pp. 166–180.

Coulson, R. A. 1984. How metabolic rate and anaerobic glycolysis determine the habits of reptiles. Symp. Zool. Soc. London 52: 425–441.

Coulson, T. D., Coulson, R. A., and Hernandez, T. 1973. Some observations on the growth of captive alligators. Zoologica 58: 47–52.

Courtillot, V., Féraud, G., Maluski, H., Vandamme, D., Moreau, M. G., and Besse, J. 1988. Deccan flood basalts and the Cretaceous/Tertiary boundary. Nature 333: 843–846.

Cousminer, H. L., and Manspeizer, W. 1976. Triassic pollen date Moroccan High Atlas and the incipient rifting of Pangea as Middle Carnian. Science 191: 943–945.

Cowles, R. B. 1946. Fur and feathers: A response to falling temperature. Science 103: 74–75.

Cox, K. G. 1988. Gradual volcanic catastrophes? Nature 333: 802.

Cracraft, J. 1971. Caenagnathiformes: Cretaceous birds convergent on dicynodont reptiles. J. Paleontol. 45: 805–809.

———. 1986. The origin and early diversification of birds. Paleobiology 12: 383–399.

Crompton, A. W., and Attridge, J. 1986. Masticatory apparatus of the larger herbivores during Late Triassic and Early Jurassic times. In: Padian, K. (ed.). The Beginning of the Age of Dinosaurs. Cambridge Univ. Press, New York. Pp. 223–236.

Crompton, A. W., and Charig, A. J. 1962. A new ornithischian from the Upper Triassic of South Africa. Nature 196: 1074–1077.

Crompton, A. W., and Hiiemae, K. 1969. How mammalian molar teeth work. Discovery 5: 23–34.

Crompton, A. W., and Hotton, N., III. 1967. Functional morphology of the masticatory apparatus of two dicynodonts (Reptilia, Therapsida). Postilla 109: 1–51.

Crompton, A. W., Taylor, C. R., and Jagger, J. A. 1978. Evolution of homeothermy in mammals. Nature 272: 333–336.

Cruickshank, A. R. I. 1972. The proterosuchian thecodonts. In: Joysey, K. A., and Kemp, T. S. (eds.). Studies in Vertebrate Evolution. Oliver and Boyd, Edinburgh. Pp. 89–119.

———. 1975. The origin of sauropod dinosaurs. S. Afr. J. Sci. 71: 89–90.

———. 1979. The ankle joint in some early archosaurs. S. Afr. J. Sci. 75: 168–178.

Cruickshank, A. R. I., and Benton, M. J. 1985. Archosaur ankles and relationships of the thecodontian and dinosaurian reptiles. Nature 317: 715–717.

Currey, J. D. 1959. Differences in tensile strength of bone of different histological types. J. Anat. 93: 87–95.

———. 1962. The histology of the bone of a prosauropod dinosaur. Palaeontology 5: 238–246.

Currie, P. J. 1981. Hunting dinosaurs in Alberta's huge bonebed. Can. Geogr. J. 101: 32–39.

———. 1982. The osteology and relationships of *Tangasaurus mennelli* Haughton (Reptilia, Eosuchia). Ann. S. Afr. Mus. 86: 247–265.

———. 1985. Cranial anatomy of *Stenonychosaurus inequalis* (Saurischia, Theropoda) and its bearing on the origin of birds. Can. J. Earth Sci. 22: 1643–1658.

———. 1986. Dinosaur fauna. In: Naylor, B. G. (ed.), Dinosaur Systematics Symp. Field Trip Gdbk., Dinosaur Provincial Park. Pp. 17–23.

———. 1987a. Bird-like characteristics of the jaws and teeth of troodontid theropods (Dinosauria, Saurischia). J. Vert. Paleontol. 7: 72–81.

———. 1987b. New approaches to studying dinosaurs in Dinosaur Provincial Park. In: Czerkas, S. J. and Olson, E. C. (eds.). Dinosaurs Past and Present. Nat. Hist. Mus. Los Angeles County. Vol. II, pp. 101–117.

———. 1987c. Theropods of the Judith River Formation of Dinosaur Provincial Park, Alberta. In: Currie, P. J., and Koster, E. H. (eds.). 4th Symp. Mesozoic Terr. Ecosyst. Short Pap. Tyrrell Mus. Palaeontol., Drumheller, Alberta. 52–60.

Currie, P. J., and Dodson, P. 1984. Mass death of a herd of ceratopsian dinosaurs. In: Reif, W.-E., and Westphal, F. (eds.). 3d Symp. Mesozoic Terr. Ecosyst. Attempto Verlag, Tübingen. Pp. 61–66.

Currie, P. J., and Russell, D. A. 1988. Osteology and relationships of *Chirostenotes pergracilis* (Saurischia, Theropoda) from the Judith River (Oldman) Formation of Alberta, Canada. Can. J. Earth Sci. 25: 972–986.

Currie, P. J., and Sarjeant, W. A. S. 1979. Lower Cretaceous dinosaur footprints from the Peace River Canyon, British Columbia, Canada. Palaeogeogr. Palaeoclimatol. Palaeoecol. 28: 103–115.

Cuvier, G. 1812. Recherches sur les Ossemens Fossiles. Paris.

Cys, J. M. 1967. The inability of dinosaurs to hibernate as a possible key factor in their extinction. J. Paleontol. 41: 266.

Czerkas, S. A. 1987. A reevaluation of the plate arrangement on *Stegosaurus stenops*. In: Czerkas, S. J., and Olson, E. C. (eds.), Dinosaurs Past and Present. Nat. Hist. Mus. Los Angeles County. Vol. II. Pp. 82–99.

Czerkas, S. J., and Olson, E. C. (eds.) 1987. Dinosaurs Past and Present. Nat. Hist. Mus. Los Angeles County, Univ. Washington Press. Seattle. 2 vols. 161 pp., 149 pp.

Dagg, A. I., and Foster, J. B. 1976. The Giraffe: Its Biology, Behavior, and Ecology. Van Nostrand Reinhold, New York. 210 pp.

Dames, W. 1884. *Megalosaurus dunkeri*. Sitzungber. Ges. Naturforsch. Freunde Berlin 1884: 186–188.

Damon, R. 1884. Geology of Weymouth, Portland, and coast of Dorsetshire, from Swanage to Briport-on-the-Sea. Weymouth. 250 pp.

Darby, D. G., and Ojakangas, R. W. 1980. Gastroliths from an Upper Cretaceous plesiosaur. J. Paleontol. 54: 548–556.

DasGupta, H. C. 1931. On a new theropod dinosaur (*Orthogoniosaurus matleyi*, n. gen. et. n. sp.) from the Lameta beds of Jubbulpore. J. Asiatic Soc. Bengal. 26: 367–369.

Davies, K. L. 1987. Duck-bill dinosaurs (Hadrosauridae, Ornithischia) from the North Slope of Alaska. J. Paleontol. 61: 198–200.

Davitashvili, L. 1961. [The Theory of Sexual Selection]. Izdatel'stvo Akademia Nauk SSSR, Moscow. 538 pp. (In Russian)

———. 1969. [Causes of extinction of the organisms]. Nauka Press, Moscow. 440 pp. (In Russian)

———. 1978. [Evolutionary studies]. Metsniereba, vol. 2, 523 pp. Tbilisi. (In Russian)

Debénath, A., Raynal, J.-P., and Sbihi-Alaoui, F.-Z. 1979. Première fouille d'un site à dinosauriens dans le bassin de Taguelft (Atlas de Beni-Mellal, province d'Azilal, Maroc): Résultats et perspectives. C. R. Acad. Sci. Paris D 289: 899–902.

DeCourten, F. L. 1978. Non-marine flora and fauna from the Kaiparowits Formation (Upper Cretaceous) of the Paria River Amphitheater, southwestern Utah. Geol. Soc. Am. Abstr. Prog. 10: 102.

DeCourten, F. L., and Russell, D. A. 1985. A specimen of *Ornithomimus velox* (Theropoda, Ornithomimidae) from the terminal Cretaceous Kaiparowits Formation of southern Utah. J. Paleontol. 59: 1091–1099.

Dehm, R. 1935. Beobachtungen im oberen Bunten Keuper Mittelfrankens. Zbl. Mineral. Geol. Palaeontol. B 1935: 97–109.

Deitz, D. C. 1979. Behavioral ecology of young American alligators. Ph.D. dissertation, Univ. Florida, Gainesville.

Deitz, D. C., and Hines, T. C. 1980. Alligator nesting in north-central Florida. Copeia 1980: 249–258.

Delair, J. B. 1966. New records of dinosaurs and other fossil reptiles from Dorset. Proc. Dorset Nat. Hist. Archaeol. Soc. 87: 28–37.

———. 1973. The dinosaurs of Wiltshire. Wiltshire Archaeol. Nat. Hist. Mag. 68: 1–7.

———. 1980. Multiple dinosaur trackways from the Isle of Purbeck. Proc. Dorset Nat. Hist. Archaeol. Soc. 102: 65–67.

———. 1982. Notes on an armoured dinosaur from Barnes High, Isle of Wight. Proc. Isle of Wight Nat. Hist. Archaeol. Soc. 7: 297–302.

Delair, J. B., and Sarjeant, W. A. S. 1975. The earliest discoveries of dinosaurs. Isis 66: 5–25.

———. 1985. History and bibliography of the study of fossil vertebrate footprints in the British Isles: Supplement 1973–1983. Palaeogeogr. Palaeoclimatol. Palaeoecol. 49: 123–160.

Delany, M. F., and Abercrombie, C. L. 1986. American alligator food habits in north-central Florida. J. Wildl. Manage. 50: 348–353.

DeMar, R. 1972. Evolutionary implications of *Zahnreihen*. Evolution 26: 435–450.

Demment, M. W., and Van Soest, P. J. 1985. A nutritional explanation for body-size patterns of ruminant and nonruminant herbivores. Am. Nat. 125: 641–672.

Depéret, C. 1896. Note sur les dinosauriens sauropodes et théropodes du Crétacé supérieur de Madagascar. Bull. Soc. Géol. France, ser. 3, 24: 176–196.

Depéret, C., and Savornin, J. 1925. Sur la découverte d'une faune de vertébrés albiens à Timimoun (Sahara occidental). C. R. Acad. Sci. Paris 181: 1108–1111.

Desmond, A. J. 1975. The Hot-Blooded Dinosaurs. Blond and Briggs, London. 238 pp.

———. 1979. Designing the dinosaur: Richard Owen's response to Robert Edmond Grant. Isis 70: 224–234.

———. 1982. Archetypes and Ancestors. Univ. Chicago Press, Chicago. 287 pp.

Diaz, G. C., and Gasparini, Z. B. del. 1976. Los Vertebrados Mesozoicos de Chile y su aporte Geo-Paleontologico. Actas Sexto Congr. Geol. Argent. 23 pp.

Dickie, L. M., Kerr, S. R., and Boudreau, P. R. 1987. Size-dependent processes underlying regularities in ecosystem structure. Ecol. Monogr. 57: 233–250.

Digregorio, J. H. 1978. Estratigrafia de las acumulaciones Mesozoicas. VII Congr. Geol. Argent. Relatorio Geol. Recursos Nat. Neuquen. Pp. 37–66.

Diller, L. V., and Johnson, D. R. 1988. Food habits, consumption rates, and predation rates of western rattlesnakes and gopher snakes in southwestern Idaho. Herpetologica 44: 228–233.

Dobson, G. P., Wood, S. C., Daxboeck, C., and Perry, S. F. 1986. Intracellular buffering and oxygen transport in the Pacific blue marlin (*Makaira nigricans*): Adaptations to high-speed swimming. Physiol. Zool. 59: 150–156.

Dodson, P. 1971. Sedimentology and taphonomy of the Oldman Formation (Campanian), Dinosaur Provincial Park, Alberta (Canada). Palaeogeogr. Palaeoclimatol. Palaeoecol. 10: 21–74.

———. 1975. Taxonomic implications of relative growth in lambeosaurine hadrosaurs. Syst. Zool. 24: 37–54.

———. 1976. Quantitative aspects of relative growth and sexual dimorphism in *Protoceratops*. J. Paleontol. 50: 929–940.

———. 1980. Comparative osteology of the American ornithopods *Camptosaurus* and *Tenontosaurus*. Mém. Soc. Géol. France 139: 81–85.

———. 1983. A faunal review of the Judith River (Oldman) Formation, Dinosaur Provincial Park, Alberta. Mosasaur 1: 89–118.

———. 1984. Small Judithian ceratopsids, Montana and Alberta. In: Reif, W.-E., and Westphal, F. (eds.). 3d Symp. Mesozoic Terr. Ecosyst. Short Pap. Attempto Verlag, Tübingen. Pp. 73–78.

———. 1986. *Avaceratops lammersi*: A new ceratopsid from the Judith River Formation of Montana. Proc. Acad. Nat. Sci. Philadelphia 138: 305–317.

———. 1987. Microfaunal studies of dinosaur paleoecology, Judith River Formation of Alberta. In: Currie, P. J., and Koster, E. H. (eds.), 4th Symp. Mesozoic Terr. Ecosyst. Short Pap. Tyrrell Mus. Palaeontol., Drumheller, Alberta. Pp. 70–75.

———. 1990. On the status of the ceratopsids *Monoclonius* and *Centrosaurus*. In: Currie, P. J., and Carpenter, K. (eds.), Dinosaur Systematics: Perspectives and Approaches. Cambridge Univ. Press, New York. Pp. 231–243.

Dodson, P., Behrensmeyer, A. K., Bakker, R. T. and McIntosh, J. S. 1980. Taphonomy and paleoecology of the Upper Jurassic Morrison Formation. Paleobiology 6: 208–232.

Dodson, P., Behrensmeyer, A. K., and Bakker, R. T. 1980. Taphonomy of the Morrison Formation (Kimmeridgian-Portlandian) and Cloverly Formation (Aptian-

Albian) of the western United States. Mém. Soc. Géol. France 139: 87–93.

———. 1983. Paleoecology of the dinosaur-bearing Morrison Formation. Natl. Geogr. Soc. Res. Rept. 15: 145–156.

Dodson, P., and Currie, P. J. 1988. The smallest ceratopsid skull-Judith River Formation of Alberta. Can. J. Earth Sci. 25: 926–930.

Dodson, P., and Madsen, J. H., Jr. 1981. On the sternum of *Camptosaurus*. J. Paleontol. 55: 109–112.

Dollo, L. 1882. Première note sur les dinosauriens de Bernissart. Bull. Mus. Roy. Hist. Nat. Belgique 1: 161–180.

———. 1883*a*. Troisième note sur les dinosauriens de Bernissart. Bull. Mus. Roy. Hist. Nat. Belgique 2: 85–126.

———. 1883*b*. Quatrième note sur les dinosauriens de Bernissart. Bull. Mus. Roy. Hist. Nat. Belgique 2: 223–252.

———. 1883*c*. Note sur les restes de dinosauriens recontrés dans le Crétacé supérieur de la Belgique. Bull. Mus. Roy. Hist. Nat. Belgique 2: 205–221.

———. 1884. Cinquième note sur les dinosauriens de Bernissart. Bull. Mus. Roy. Hist. Nat. Belgique 3: 120–140.

———. 1903. Les dinosauriens de la Belgique. C. R. Acad. Sci. Paris 136: 565–567.

———. 1909. The fossil vertebrates of Belgium. Ann. New York Acad. Sci. 19: 99–119.

———. 1923. Le centenaire des iguanodons (1822–1922). Phil. Trans. Roy. Soc. London B 212: 67–78.

Dong Z. 1973*a*. [Dinosaurs from Wuerho]. Mem. Inst. Vert. Paleontol. Paleoanthropol. Acad. Sinica 11: 45–52. (In Chinese)

———. 1973*b*. [A fossil carnosaur tooth from Kuqa County, Xingjiang Autonomous Region.] Vertebrata PalAsiatica 11: 73. (In Chinese)

———. 1977. [On the dinosaurian remains from Turpan, Xinjiang]. Vertebrata PalAsiatica 15: 59–66. (In Chinese with English summary)

———. 1978. [A new genus of Pachycephalosauria from Laiyang, Shantung]. Vertebrata PalAsiatica 16: 225–228. (In Chinese with English summary)

———. 1979. [The Cretaceous dinosaur fossils in southern China]. In: Mesozoic and Cenozoic Red Beds in Southern China. Inst. Vert. Paleontol. Paleoanthropol. Nanjing Geol. Paleontol. Inst. Sci. Press, Beijing. Pp. 342–350. (In Chinese)

———. 1980. [The dinosaurian faunas of China and their stratigraphic distribution]. J. Strat. 4: 256–263.

———. 1984*a*. [A new prosauropod from Ziliujing Formation of Sichuan Basin]. Vertebrata PalAsiatica 22: 310–313. (In Chinese with English summary)

———. 1984*b*. [A new theropod dinosaur from the Middle Jurassic of Sichuan Basin]. Vertebrata PalAsiatica 22: 213–218. (In Chinese with English summary)

———. 1989. [On a small ornithopod (*Gongbusaurus wucaiwanensis* sp. nov.) from Kelamaili, Junggar Basin, Xinjiang, China.] Vert. PalAs. 27: 140–147. (In Chinese with English summary)

———. 1990. Stegosaurs in Asia. In: Currie, P. J., and Carpenter, K. (eds.). Dinosaur Systematics: Perspectives and Approaches. Cambridge Univ. Press, New York. Pp. 255–268.

Dong Z., Chang Y., Li X., and Zhou S. 1978. [Note on a new carnosaur (*Yangchuanosaurus shangyouensis* gen. et sp. nov.) from the Jurassic of Yangchuan District, Szechuan Province]. Kexue Tongbao 23: 298–302. (In Chinese)

Dong Z., Li X., Zhou S., and Chang Y. 1977. [On the stegosaurian remain from Zigong (Tzekung), Szechuan Province]. Vertebrata PalAsiatica 15: 307–312. (In Chinese with English summary)

Dong Z., and Tang Z. 1984. [Note on a new mid-Jurassic sauropod (*Datousaurus bashanensis*) from Sichuan Basin, China]. Vertebrata PalAsiatica 22: 69–74. (In Chinese with English summary)

———. 1985. [A new mid-Jurassic theropod (*Gasosaurus constructus* gen. et sp. nov.) from Dashanpu, Zigong, Sichuan Province, China]. Vertebrata PalAsiatica 23: 77–83. (In Chinese with English summary)

Dong Z., Tang Z., and Zhou S. 1982. [Note on the new Mid-Jurassic stegosaur from Sichuan Basin, China]. Vertebrata PalAsiatica 20: 83–87 (In Chinese with English summary)

Dong Z., Zhou S., and Zhang Y. 1983. [The dinosaurian remains from Sichuan Basin, China]. Palaeontologica Sinica 162: 1–145. (In Chinese with English summary)

Dorlodot, J. de. 1934. L'exploration du gîte à dinosauriens jurassiques de Damparis. La Terre et la Vie 4: 563–586.

Dorr, J. A. 1985. Newly found Early Cretaceous dinosaurs and other fossils in southeastern Idaho and westernmost Wyoming. Contrib. Mus. Paleontol. Univ. Michigan 27: 73–85.

Douglas, R. J. W., Norris, D. K., Thorsteinsson, R., and Tozer, E. T. 1963. Geology and petroleum potentialities of northern Canada. Geol. Surv. Can. Pap. 63–31: 1–28.

Douville, J. H. F. 1885. [Remarques sur *Halitherium* et sur un mégalosaurien]. Bull. Soc. Géol. France (sér 3) 13: 441.

Drickhamer, L. C., and Vessey, S. H. 1986. Animal Behavior: Concepts, Processes and Methods. Prindle, Weber & Schmidt. Boston. 619 pp.

Duffin, C. J., Coupatez, P., LePage, J. C., and Wouters, G. 1983. Rhaetian (Upper Triassic) marine faunas from "le Golfe du Luxembourg" in Belgium (preliminary note). Bull. Soc. Géol. Belgique 92: 311–315.

Dughi, R., and Sirugue, F. 1957. Les oeufs de dinosauriens du bassin d'Aix-en-Provence. C. R. Acad. Sci. Paris 245: 707–710.

———. 1958*a*. Sur les oeufs de dinosaures du bassin fluviolacustre d'Aix-en-Provence. C. R. 83d Congr. Nat. Soc. Sav. Aix. Pp. 183–205.

———. 1958*b*. Observations sur les oeufs de dinosaures du bassin d'Aix-en-Provence: Les oeufs à coquilles biostratifiées. C. R. Acad. Sci. Paris 246: 2271–2274.

———. 1966. Sur la fossilisation des oeufs de dinosaures. C. R. Acad. Sci. Paris 262: 2330–2332.

———. 1976. L'extinction des dinosaures à la lumière des gisements d'oeufs du Crétacé terminal du Midi de la France, principalement dans le bassin d'Aix-en-Provence. Paléobiol. Continentale 7: 1–39.

Duncan, R. A., and Pyle, D. G. 1988. Rapid eruption of the Deccan flood basalts at the Cretaceous/Tertiary boundary. Nature 333: 841–843.

Dunham, A. E., Overall, K. L., Porter, W. P., and Forster, C. A. 1989. Implications of ecological energetics and biophysical and developmental constraints for life history variation in dinosaurs. In: Farlow, J. O. (ed.), Paleobiology of the Dinosaurs. Geol. Soc. Am. Spec. Pap. 238: 1–19.

Dupont, E. 1878. Sur la découverte d'ossements d'*Iguanodon*, de poissons et de végétaux dans la fosse Sainte-Barbe du Charbonnage de Bernissart. Bull. Acad. Roy. Belgique (sér 2) 46: 387–408.

Dutuit, J. M. 1972. Découverte d'un dinosaure ornithischien dans le Trias supérieur de l'Atlas occidental marocain. C. R. Acad. Sci. Paris D275: 2841–2844.

———. 1974. Caractères généraux des gisements Triasiques Marocains. Ann. Geol. Surv. Egypt 4: 1–8.

Dutuit, J. M., and Ouazzou, A. 1980. Découverte d'une piste de dinosaure sauropode sur le site d'empreintes de Demnate (Haut Atlas marocain). Mém. Soc. Géol. France 59: 95–102.

Dwivedi, G. N., Mohabey, D. M., and Bandyopadhya, S. 1982. On discovery of vertebrate fossils in Infratrappean Lameta beds of Kheda District, Gujarat. IV Indian Geol. Congr. Curr. Trends Geol. 7: 79–87.

Earle, M. 1987. A flexible body mass in social carnivores. Am. Nat. 129: 755–760.

East, R. 1984. Rainfall, soil nutrient status and biomass of large African savanna mammals. Afr. J. Ecol. 22: 245–270.

Eaton, J. G. 1987. The Campanian-Maastrichtian boundary in the western interior of North America. Newsletter Stratigr. 18: 31–39.

Eaton, J. G., Kirkland, J. E., and Doi, K. 1989. Evidence of reworked Cretaceous fossils and their bearing on the existence of Tertiary dinosaurs. Palaios 4: 281–286.

Eaton, T. H., Jr. 1960. A new armored dinosaur from the Cretaceous of Kansas. Univ. Kansas Paleontol. Contrib. 25: 1–24.

Economos, A. C. 1981. The largest land mammal. J. Theor. Biol. 89: 211–215.

Edinger, T. 1929. Über knöcherne Scleralringe. Zool. Jb. 51: 163–226.

Edmund, A. G. 1957. On the special foramina in the jaws of many ornithischian dinosaurs. Roy. Ont. Mus. Div. Zool. Palaeontol. Contrib. 48: 1–14.

———. 1969. Dentition. In: Gans, C., Bellairs, A. d'A., and Parsons, T. S. (eds.). Biology of the Reptilia. Vol. 1: 117–200.

Edwards, M. B., Edwards, R., and Colbert, E. H. 1978. Car-nosaurian footprints in the Lower Cretaceous of eastern Spitsbergen. J. Paleontol. 52: 940–941.

Efremov, I. A. 1931. [Dinosauria in the red beds of middle Asia]. Trudy Paleozool. Inst. 1: 217–221. (In Russian)

Ehrlich, P. R., et al. 1983. Long-term biological consequences of nuclear war. Science 222: 1293–1300.

Eicher, D. L. 1960. Stratigraphy and micropaleontology of the Thermopolis Shale. Peabody Mus. Nat. Hist. 15: 1–126.

Eisenberg, J. F., O'Connell, M. A., and August, P. V. 1979. Density, productivity and distribution of mammals in two Venezuelan habitats. In: Eisenberg, J. F. (ed.). Vertebrate Ecology in the Northern Neotropics. Smithsonian Inst., Washington, D.C. Pp. 187–207.

El-Khashab, B. 1977. A review of the fossil reptile faunas of Egypt. Geol. Suv. Egypt Mining Authority Pap. 62: 1–9.

Ellenberger, P. 1965. Découverte de pistes de vértebrés dans le Permien, le Trias et le Lias inférieur, aux abords de Toulon (Var) et d'Anduze (Gard). C. R. Acad. Sci. Paris 260: 5856–5859.

———. 1970. Les niveaux paléontologiques de première apparition des mammifères primordiaux en Afrique du Sud et leur ichnologie. Etablissement de zones stratigraphiques detaillées dans le Stormberg du Lesotho (Afrique du Sud) (Trias supérieur à Jurassique). Proc. Pap. Second Gondwana Symp. Pretoria, S. Afr. Pp. 343–370.

Ellenberger, F., and Fuchs, Y. 1965. Sur la présence de pistes de vertébrés dans le Lotharingien marin de la région de Severac-le-Château (Aveyron). C. R. Soc. Geol. France 1965: 39–40.

Ellinger, T. U. H. 1950. *Camarasaurus annae*—A new American sauropod dinosaur. Am. Nat. 84: 225–228.

Elliott, D. K., ed. 1986. The Dynamics of Extinction. Wiley, New York. 294 pp.

Emery, S. H. 1985. Hematology and cardiac morphology in the great white shark, *Carcharodon carcharias*. Mem. So. Calif. Acad. Sci. 9: 73–80.

———. 1986. Hematological comparisons of endothermic vs. ectothermic elasmobranch fishes. Copeia 1986: 700–705.

Emmons, L. H. 1987. Comparative feeding ecology of felids in a Neotropical rainforest. Behavioral Ecol. Sociobiol. 20: 271–283.

Enay, R. 1980. Indices d'émersion et d'influences continentales dans l'Oxfordien supérieur-Kimméridgien inférieur en France. Interprétation paleogéographique et conséquences paléobiogéographiques. Bull. Soc. Geol. France (sér. 7) 22: 581–590.

Enlow, D. H. 1969. The bone of reptiles. In: Gans, C., Bellairs, A. d'A., and Parsons, T. S. (eds.). Biology of the Reptilia. Vol. 1: 45–80.

Enlow, D. H., and Brown, S. O. 1957. A comparative histological study of fossil and Recent bone tissue. Pt. II. Texas J. Sci. 9: 186–214.

———. 1958. A comparative histological study of fossil and Recent bone tissues. Pt. III. Texas J. Sci. 10: 187–230.

Erben, H. K. 1972. Ultrastrukturen und Dicken der Wand pathologischer Eischalen. Abhandl. Akad. Wiss. Lit. Mainz, Math.-Nat. Kl. 1972: 193–216.

———. 1975. Die Entwicklung der Lebenwesen; Spielregeln der Evolution. Verlag Piper, Munich.

Erben, H. K., Hoefs, J., and Wedepohl, K. H. 1979. Paleobiological and isotopic studies of eggshells from a declining dinosaur species. Paleobiology 5: 380–414.

Erickson, B. R. 1966. Mounted skeleton of *Triceratops prorsus* in the Science Museum. Sci. Publ. Sci. Mus. St. Paul, Minn. 1: 1–16.

Eriksen, L. 1979a. Telltale tracks. Pt. I. Museum Notes, Mus. Western Colorado 5: 2.

———. 1979b. Telltale tracks. Pt. II. Museum Notes, Mus. Western Colorado 5: 6–7.

Estes, R. 1964. Fossil vertebrates from the Late Cretaceous Lance Formation, eastern Wyoming. Univ. Calif. Publ. Geol. Sci. 49: 1–180.

Estes, R., Berberian, P., and Meszoely, C. A. M. 1969. Lower vertebrates from the Late Cretaceous Hell Creek Formation, McCone County, Montana. Breviora 337: 1–33.

Estes, R., and Sanchiz, F. 1982. Early Cretaceous lower vertebrates from Galve (Teruel), Spain. J. Vert. Paleontol. 2: 21–39.

Eudes-Deslongchamps, E. 1870. Note sur les reptiles fossiles appartenant à la famille des téléosauriens, dont les débris ont été recueillis dans les assises jurassiques de Normandie. Bull. Soc. Geol. France (ser. 2) 27: 299–351.

Eudes-Deslongchamps, J.-A. 1838. Mémoire sur le *Poekilopleuron bucklandi*, grand saurien fossile, intermédiaire entre les crocodiles et les lézards. Mém. Soc. Linn. Normandie 6: 37–146.

Evans, S. E. 1984. The classification of the Lepidosauria. Zool. J. Linn. Soc. London 82: 87–100.

———. 1988. The early history and relationships of the Diapsida. In: Benton, M. J. (ed.). The Phylogeny and Classification of the Tetrapods. Vol. I. Syst. Assoc. Spec. Vol. 35A: 221–260.

Ewer, R. F. 1965. The anatomy of the thecodont reptile *Euparkeria capensis* Broom. Phil. Trans. Roy. Soc. London B 248: 379–435.

———. 1973. The Carnivores. Cornell Univ. Press, Ithaca. 494 pp.

Exon, N. F., and Vine, R. R. 1970. Revised nomenclature of the 'Blythsdale' sequence. Queensland Govt. Mineral. J. 71: 48–52.

Fabre, J. 1977. Environment paléosédimentaire du gisement de vertébrés "berriasien" des Bessons, Petit Plan de Canjuers (Var). Sédimentologie fine d'une coupe effectuée dans une carrière des Bessons. C. R. Acad. Sci. Paris 284: 417–420.

Falconer, H. 1868. Memorandum on two remarkable vertebrae sent by Dr. Oldham from Jubbulpore-Spilsbury bed. Paleontological Memoirs and Notes of the late Hugh Falconer 1: 418–419.

Farlow, J. O. 1976a. Speculations about the diet and foraging behavior of large carnivorous dinosaurs. Am. Midl. Nat. 95: 186–191.

———. 1976b. A consideration of the trophic dynamics of a Late Cretaceous large-dinosaur community (Oldman Formation). Ecology 57: 841–857.

———. 1980. Predator/prey biomass ratios, community food webs and dinosaur physiology. In: Thomas, R. D. K., and Olson, E. C. (eds.). A Cold Look at the Warm-Blooded Dinosaurs. Westview Press, Boulder. Pp. 55–83.

———. 1981. Estimates of dinosaur speeds from a new trackway site in Texas. Nature 294: 747–748.

———. 1983. Dragons and dinosaurs. Paleobiology 9: 207–210.

———. 1987a. Lower Cretaceous dinosaur tracks, Paluxy River Valley, Texas. South Central Sect. Geol. Soc. Am., Baylor Univ. Pp. 1–50.

———. 1987b. Speculations about the diet and digestive physiology of herbivorous dinosaurs. Paleobiology 13: 60–72.

———. 1989. Ostrich footprints and trackways: Implications for dinosaur ichnology. In: Gillette, D. D., and Lockley, M. G. (eds.). Dinosaur Tracks and Traces. Univ. Cambridge Press, New York. Pp. 243–248.

Farlow, J. O., and Dodson, P. 1975. The behavioral significance of frill and horn morphology in ceratopsian dinosaurs. Evolution 29: 353–361.

Farlow, J. O., Thompson, C. V., and Rosner, D. E. 1976. Plates of *Stegosaurus:* Forced convection heat loss fins? Science 192: 1123–1125.

Fassett, J. E., and Hinds, J. S. 1971. Geology and fuel resources of the Fruitland Formation and Kirtland Shale of the San Juan Basin, New Mexico and Colorado. U.S. Geol. Surv. Prof. Pap. 676: 1–75.

Fastovsky, D. E. 1987. Paleoenvironments of vertebrate-bearing strata during the Cretaceous-Paleogene transition, eastern Montana and western North Dakota. Palaios 2: 282–295.

Fastovsky, D. E., and Dott, R. H., Jr. 1986. Sedimentology, stratigraphy, and extinctions during the Cretaceous-Paleogene transition at Bug Creek, Montana. Geology 14: 279–282.

Fastovsky, D. E., and McSweeney, K. 1987. Paleosols spanning the Cretaceous/Paleogene transition, eastern Montana and western North Dakota. Bull. Geol. Soc. Am. 99: 66–77.

Faul, H., and Roberts, W. A. 1951. New fossil footprints from the Navajo(?) Sandstone of Colorado. J. Paleontol. 25: 266–274.

Feduccia, A. 1980. The Age of Birds. Harvard Univ. Press, Cambridge. 196 pp.

Felder, W. M. 1975a. Lithostratigraphische Gliederung der Oberen Kreide in Süd-Limburg (Niederlande) und den Nachbargebieten. 1. Teil. Der Raum westlich der

Maas, Typusgebiet des "Maastricht". Publ. Naturhist. Genootsch. Limburg 24: 1–43.

———. 1975*b*. Lithostratigrafie van het Boven-Krijt en het Dano-Montien in Zuid-Limburg en het aangrenzende gebied. In: Toelichting bij geol. overzichtskaarten Nederland, Rijks Geol. Dienst Haarlem. Pp. 63–72.

Ferguson, M. W. J. 1985. Reproductive biology and embryology of the crocodilians. In: Gans, C., Billett, F., and Maderson, P. F. A. (eds.). Biology of the Reptilia. Vol. 14: 329–491.

Ferguson, M. W. J., and Joanen, T. 1982. Temperature of egg incubation determines sex in *Alligator mississippiensis*. Nature 296: 850–853.

Ferrusquia-Villafranca, I., Applegate, S. P., and Espinosa-Arrubarrena, L. 1980. Las huellas mas australes de dinosaurios en Norteamérica y su significación geobiológica. Actas II Congr. Argent. Paleontol. Bioestrat. I. Congr. Latinoam. Paleontol. 1: 249–263.

Finch, R. C. 1981. Mesozoic stratigraphy of central Honduras. Am. Assoc. Petrol. Geol. 65: 1320–1333.

Fiorillo, A. R. 1987. Significance of juvenile dinosaurs from Careless Creek Quarry (Judith River Formation), Wheatland County, Montana. In: Currie, P. J., and Koster, E. H. (eds.). Fourth Symposium Mesozoic Terr. Ecosyst. Tyrrell Mus. Palaeontol., Drumheller, Alberta. Pp. 88–95.

Fitzinger, L. J. 1843. Systema Reptilium. Vienna.

Fleay, D. H. 1937. Nesting habits of the Brush Turkey. Emu 36: 153–163.

Forster, C. A. 1985. A description of the postcranial skeleton of the Early Cretaceous ornithopod *Tenontosaurus tilletti*, Cloverly Formation, Montana and Wyoming. M.Sc. thesis, Univ. Pennsylvania, Philadelphia. 128 pp.

Fouch, T. D., Lawton, T. F., Nichols, D. J., Cashion, W. B., and Cobban, W. A. 1983. Patterns and timing of synorogenic sedimentation in Upper Cretaceous rocks of central and northeast Utah. In: Reynolds, M. W., and Dolly, E. D. (eds.). Mesozoic Paleogeography of west-central United States. Rocky Mountain Section, SEPM, Denver. Pp. 305–336.

Fox, L. R. 1975. Cannibalism in natural populations. Ann. Rev. Ecol. Syst. 6: 87–106.

Fox, R. C. 1978. Upper Cretaceous terrestrial vertebrate stratigraphy of the Gobi Desert (Mongolian People's Republic) and western North America. Geol. Assoc. Can. Spec. Pap. 18: 577–594.

Fox, W. 1866. Another Wealden reptile. Athenaeum 2014: 740.

Fraas, E. 1896. Die schwäbischen Triassaurier nach dem Material der Kgl. Naturalien-Sammlung in Stuttgart zusammengestellt. Mitth. königlichen Naturalien-Kabinett Stuttgart 5: 1–18.

———. 1908. Ostafrikanische Dinosaurier. Palaeontographica 55: 105–144.

———. 1913. Die neuesten Dinosaurierfunde in der schwäbischen Trias. Naturwissenschaften 45: 1097–1100.

Fraser, N. C., and Walkden, G. M. 1983. The ecology of a Late Triassic reptile assemblage from Gloucestershire, England. Palaeogeogr. Palaeoclimatol. Palaeoecol. 42: 341–365.

Fraser, N. C., Walkden, G. M., and Stewart, V. 1985. The first pre-Rhaetic therian mammal. Nature 314: 161–163.

Frazzetta, T. H. 1962. A functional consideration of cranial kinesis in lizards. J. Morphol. 111: 287–320.

Frederiksen, N. O. 1986. Reconnaissance biostratigraphy of *Expressipollis* and *Oculata* pollen groups in Campanian and Maastrichtian rocks of the North Slope, Alaska. Am. Assoc. Strat. Palynol., 19th Ann. Mtg. Prog. Abstr. P. 11.

Freytet, P. 1965. Découverte d'oeufs de dinosaures à Saint-André de Roquelongue (Aude). Bull. Soc. Études Sci. Aude 65: 121–123.

Friese, H. 1972. Die Dinosaurierfährten von Barkhausen im Wiehengebirge. Wittlager Heimathafte 5: 1–21.

———. 1979. Die Dinosaurierfährten von Barkhausen im Wiehengebirge. Veröffentl. Landkreises Osnabrück 1: 1–36.

Friis, E. M., Chaloner, W. G., and Crane, P. R. (eds.). 1987. The Origins of Angiosperms and Their Biological Consequences. Cambridge Univ. Press, Cambridge. 358 pp.

Frith, H. J. 1956. Temperature regulation in the nesting mounds of the Mallee-Fowl, *Leipoa ocellata* Gould. Commonw. Sci. Industr. Res. Org. Wildl. Res. 1: 79–95.

———. 1957. Experiments on the control of temperature in the sound of the Mallee-Fowl, *Leipoa ocellata* Gould (Megapodidae). Commonw. Sci. Industr. Res. Org. Wildl. Res. 2: 101–110.

———. 1962. The Mallee Fowl. Angus and Robertson, Sidney.

Fritsch, A. 1905. Synopsis der Saurier der Böhmischen Kreideformation. Sitz. könig. böhm. Ges. Wiss., II. Classe. 7 pp.

Gabunia, L. K. 1951. [Concerning the tracks of dinosaurs from the Lower Cretaceous deposits of western Georgia]. C. R. Acad. Sci. U.R.S.S. 81: 917–919. (In Russian)

———. 1969. [Extinction of ancient reptiles and mammals]. Metsniereba Press, Tbilisi. 234 pp. (In Russian with French summary)

———. 1984. [Extinction processes in the evolution of the vertebrates]. Trans. 27th Internatl. Geol. Congr., Moscow 1: 260. (In Russian)

Gallagher, W. B. 1984. Paleoecology of the Delaware Valley region. Pt. II: Cretaceous to Quaternary. Mosasaur 2: 9–44.

Galli, O., and Dingman, R. 1962. Cuadrangulos Pica, Alca, Matilla y Chacarilla. Carta geologica de Chile. Inst. Invest. Geol. Carta Geol. Chile 3.

Galton, P. M. 1969. The pelvic musculature of the dinosaur *Hypsilophodon*. Postilla 131: 1–64.

———. 1970*a*. The posture of hadrosaurian dinosaurs. J. Paleontol. 44: 464–473.

———. 1970*b*. Ornithischian dinosaurs and the origin of birds. Evolution 24: 448–462.

———. 1970*c*. Pachycephalosaurids-dinosaurian battering rams. Discovery 6: 23–32.

———. 1971*a*. The prosauropod dinosaur *Ammosaurus,* the crocodile *Protosuchus,* and their bearing on the age of the Navajo Sandstone of northeastern Arizona. J. Paleontol. 45: 781–795.

———. 1971*b*. Manus movements of the coelurosaurian dinosaur *Syntarsus* and opposability of the theropod hallux. Arnoldia 5: 1–8.

———. 1971*c*. A primitive dome-headed dinosaur (Ornithischia: Pachycephalosauridae) from the Lower Cretaceous of England, and the function of the dome in pachycephalosaurids. J. Paleontol. 45: 40–47.

———. 1971*d*. The mode of life of *Hypsilophodon,* the supposedly arboreal ornithopod dinosaur. Lethaia 4: 453–465.

———. 1972. Classification and evolution of ornithopod dinosaurs. Nature 239: 464–466.

———. 1973*a*. A femur of a small theropod dinosaur from the Lower Cretaceous of England. J. Paleontol. 47: 996–1001.

———. 1973*b*. On the anatomy and relationships of *Efraasia diagnostica* (v. Huene) n. gen., a prosauropod dinosaur (Reptilia: Saurischia) from the Upper Triassic of Germany. Palaeontol. Z. 47: 229–255.

———. 1973*c*. The cheeks of ornithischian dinosaurs. Lethaia 6: 67–89.

———. 1973*d*. Redescription of the skull and mandible of *Parksosaurus* from the Late Cretaceous with comments on the family Hypsilophodontidae (Ornithischia). Roy. Ontario Mus. Life Sci. Contrib. 89: 1–21.

———. 1974*a*. The ornithischian dinosaur *Hypsilophodon* from the Wealden of the Isle of Wight. Bull. Br. Mus. (Nat. Hist.) Geol. 25: 1–152.

———. 1974*b*. Notes on *Thescelosaurus,* a conservative ornithopod dinosaur from the Upper Cretaceous of North America, with comments on ornithopod classification. J. Paleontol. 48: 1048–1067.

———. 1975. English hypsilophodontid dinosaurs (Reptilia: Ornithischia). Palaeontology 18: 741–752.

———. 1976*a*. Prosauropod dinosaurs (Reptilia: Saurischia) of North America. Postilla 169: 1–98.

———. 1976*b*. *Iliosuchus,* a Jurassic dinosaur from Oxfordshire and Utah. Palaeontology 19: 587–589.

———. 1977*a*. On *Staurikosaurus pricei,* an early saurischian dinosaur from Brazil, with notes on the Herrerasauridae and Poposauridae. Palaeontol. Z. 51: 234–245.

———. 1977*b*. The ornithopod dinosaur *Dryosaurus* and a Laurasia-Gondwanaland connection in the Upper Jurassic. Nature 268: 230–232.

———. 1978. Fabrosauridae, the basal family of ornithischian dinosaurs (Reptilia: Ornithopoda). Palaeontol. Z. 52: 138–159.

———. 1980*a*. Partial skeleton of *Dracopelta zbyszewskii* n. gen. and n. sp., an ankylosaurian dinosaur from the Upper Jurassic of Portugal. Géobios 13: 451–457.

———. 1980*b*. Avian-like tibiotarsi of pterodactyloids (Reptilia: Pterosauria) from the Upper Jurassic of East Africa. Palaeontol. Z. 54: 331–342.

———. 1980*c*. Armored dinosaurs (Ornithischia: Ankylosauria) from the Middle and Upper Jurassic of England. Géobios 13: 825–837.

———. 1980*d*. European Jurassic dinosaurs of the families Hypsilophodontidae and Camptosauridae. N. Jb. Geol. Paläontol. Abhandl. 160: 73–95.

———. 1980*e*. *Priodontognathus phillipsii* (Seeley), an ankylosaurian dinosaur from the Upper Jurassic (or possibly Lower Cretaceous) of England. N. Jb. Geol. Paläontol. Mh. 1980: 477–489.

———. 1981*a*. *Craterosaurus pottonensis* Seeley, a stegosaurian dinosaur from the Lower Cretaceous of England, and a review of Cretaceous stegosaurs. N. Jb. Geol. Paläontol. Abhandl. 161: 28–46.

———. 1981*b*. A juvenile stegosaurian dinosaur, "*Astrodon pusillus*" from the Upper Jurassic of Portugal, with comments on Upper Jurassic and Lower Cretaceous biogeography. J. Vert. Paleontol. 1: 245–256.

———. 1981*c*. *Dryosaurus,* a hypsilophodontid dinosaur from the Upper Jurassic of North America and Africa. Postcranial skeleton. Palaeontol. Z., 55: 271–312.

———. 1982*a*. Juveniles of the stegosaurian dinosaur *Stegosaurus* from the Upper Jurassic of North America. J. Vert. Paleontol. 2: 47–62.

———. 1982*b*. The postcranial anatomy of the stegosaurian dinosaur *Kentrosaurus* from the Upper Jurassic of Tanzania, East Africa. Geol. Palaeontol. 15: 139–160.

———. 1982*c*. *Elaphrosaurus,* an ornithomimid dinosaur from the Upper Jurassic of North America and Africa. Palaeontol. Z. 265–275.

———. 1983*a*. The oldest ornithischian dinosaurs in North America from the Late Triassic of Nova Scotia, North Carolina and Pennsylvania. Geol. Soc. Am. Abstr. Prog. 15: 122.

———. 1983*b*. Armored dinosaurs (Ornithischia: Ankylosauria) from the Middle and Upper Jurassic of Europe. Palaeontographica A 182: 1–25.

———. 1983*c*. *Sarcolestes leedsi* Lydekker, an ankylosaurian dinosaur from the Middle Jurassic of England. N. Jb. Geol. Paläontol. Mh. 1983: 141–155.

———. 1983*d*. The cranial anatomy of *Dryosaurus,* a hypsilophodontid dinosaur from the Upper Jurassic of North America and East Africa, with a review of hypsilophodontids from the Upper Jurassic of North America. Geol. Palaeontol. 17: 207–243.

———. 1983*e*. A juvenile stegosaurian dinosaur, *Omosaurus phillipsi* Seeley from the Oxfordian (Upper Jurassic) of England. Géobios 16: 95–101.

———. 1984*a*. Cranial anatomy of the prosauropod dinosaur *Plateosaurus* from the Knollenmergel (Middle

Keuper, Upper Triassic) of Germany. I. Two complete skulls from Trossingen Württ. with comments on the diet. Geol. Palaeontol. 18: 139−171.

———. 1984b. An early prosauropod dinosaur from the Upper Triassic of Nordwürttemberg, West Germany. Stuttgarter Beitr. Naturk. B 106: 1−25.

———. 1985a. The poposaurid thecodontian *Teratosaurus suevicus* v. Meyer, plus referred specimens mostly based on prosauropod dinosaurs, from the Middle Stubensandstein (Upper Triassic) of Nordwürttemberg. Stuttgarter Beitr. Naturk. (ser. B) 116: 1−29.

———. 1985b. Cranial anatomy of the prosauropod dinosaur *Sellosaurus gracilis* from the Middle Stubensandstein (Upper Triassic) of Nordwürttemberg, West Germany. Stuttgarter Beitr. Naturk. B 118: 1−29.

———. 1985c. British plated dinosaurs (Ornithischia, Stegosauridae). J. Vert. Paleontol. 5: 211−254.

———. 1985d. Diet of prosauropod dinosaurs from the Late Triassic and Early Jurassic. Lethaia 18: 105−123.

———. 1985e. Cranial anatomy of the prosauropod dinosaur *Plateosaurus* from the Knollenmergel (Middle Keuper, Upper Triassic) of Germany. II. All the cranial material and details of soft-part anatomy. Geol. Palaeontol. 19: 119−159.

———. 1985f. Notes on the Melanorosauridae, a family of large prosauropod dinosaurs (Saurischia: Sauropodomorpha). Géobios 19: 671−676.

———. 1986a. Prosauropod dinosaur *Plateosaurus (= Gressylosaurus)* (Saurischia: Sauropodomorpha) from the Upper Triassic of Switzerland. Geol. Palaeontol. 20: 167−183.

———. 1986b. Herbivorous adaptations of Late Triassic and Early Jurassic dinosaurs. In: Padian, K. (ed.). The Beginning of the Age of Dinosaurs. Cambridge Univ. Press, Cambridge. Pp. 203−221.

———. 1988. Skull bones and endocranial casts of stegosaurian dinosaur *Kentrosaurus* Hennig, 1915 from Upper Jurassic of Tanzania, East Africa. Geol. Palaeontol. 22: 123−143.

Galton, P. M., and Bakker, R. T. 1985. The cranial anatomy of the prosauropod dinosaur "*Efraasia diagnostica*," a juvenile individual of *Sellosaurus gracilis* from the Upper Triassic of Nordwürttemberg, West Germany. Stuttgarter Beitr. Naturk. B 117: 1−15.

Galton, P. M., and Boine, G. 1980. A stegosaurian dinosaur femur from the Kimmeridgian Beds (Upper Jurassic) of the Cap de la Hève, Normandy. Bull. Trim. Soc. Géol. Normandie Amis Mus. Havre 87: 31−35.

Galton, P. M., Brun, R., and Rioult, M. 1980. Skeleton of the stegosaurian dinosaur *Lexovisaurus* from the lower part of the middle Callovian (Middle Jurassic) of Argences (Calvados), Normandy. Bull. Trim. Soc. Géol. Normandie Amis Mus. Havre 67: 40−60.

Galton, P. M., and Cluver, M. A. 1976. *Anchisaurus capensis* (Broom) and a revision of the Anchisauridae (Reptilia, Saurischia). Ann. S. Afr. Mus. 69: 121−159.

Galton, P. M., and Coombs, W. P. 1981. *Paranthodon africanus* (Broom), a stegosaurian dinosaur from the Lower Cretaceous of South Africa. Géobios 14: 299−309.

Galton, P. M., and Heerden, J. V. 1985. Partial hindlimb of *Blikanasaurus cromptoni* n. gen. and n. sp., representing a new family of prosauropod dinosaurs from the Upper Triassic of South Africa. Géobios 18: 509−516.

Galton, P. M., and Jensen, J. A. 1973a. Small bones of the hypsilophodontid dinosaur *Dryosaurus altus* from the Upper Jurassic of Colorado. Great Basin Nat. 33: 129−132.

———. 1973b. Skeleton of a hypsilophodontid dinosaur (*Nanosaurus(?) rex*) from the Upper Jurassic of Utah. Brigham Young Univ. Geol. Stud. 20: 137−157.

———. 1979a. A new large theropod dinosaur from the Upper Jurassic of Colorado. Brigham Young Univ. Geol. Stud. 26: 1−12.

———. 1979b. Remains of ornithopod dinosaurs from the Lower Cretaceous of North America. Brigham Young Univ. Geol. Stud. 25: 1−10.

Galton, P. M., and Powell, H. P. 1980. The ornithischian dinosaur *Camptosaurus prestwichii* from the Upper Jurassic of England. Palaeontology 23: 411−443.

———. 1983. Stegosaurian dinosaurs from the Bathonian (Middle Jurassic) of England, the earliest record of the Family Stegosauridae. Géobios 16: 219−229.

Galton, P. M., and Sues, H.-D. 1983. New data on pachycephalosaurid dinosaurs (Reptilia: Ornithischia) from North America. Can. J. Earth Sci. 20: 462−473.

Galton, P. M., and Taquet, P. 1982. *Valdosaurus*, a hypsilophodontid dinosaur from the Lower Cretaceous of Europe and Africa. Géobios 15: 147−159.

Gao R., Zhou S., and Huang D. 1986. [Discovery of a shoulder spine in stegosaurian material from Zigong]. Vertebrata PalAsiatica 24: 78−79. (In Chinese)

Gao S., Li B., and Dong G. 1981. [Footprint fossils found in Chabu District, Inner Mongolia, China]. Vertebrata PalAsiatica 19: 193. (In Chinese)

Garcia-Ramos, J. C., and Valenzuela, M. 1977a. Hallazgo de huellas de pisada de vertebrados en el Jurasico de la Costa Asturiana entre Gijon y Ribadesella. Breviora Geol. Asturica 21: 17−21.

———. 1977b. Huellas de pisada de vertebrados (Dinosaurios y otros) en el Jurasico Superior de Asturias. Estud. Geol. 33: 207−214.

Gardiner, B. G. 1982. Tetrapod classification. Zool. J. Linn. Soc. London 74: 207−232.

Garland, T., Jr. 1983a. The relation between maximal running speed and body mass in terrestrial mammals. J. Zool. 199: 157−170.

———. 1983b. Scaling the ecological cost of transport to body mass in terrestrial mammals. Am. Nat. 121: 571−587.

———. 1984. Physiological correlates of locomotory performance in a lizard. Am. J. Physiol. 247: R806−R815.

Garrick, L. D., and Lang, J. W. 1977. Social signals and behaviors of adult alligators and crocodiles. Am. Zool. 17: 225–239.

Garrick, L. D., and Garrick, R. A. 1978. Temperature influences on hatchling *Caiman crocodilus* distress calls. Physiol. Zool. 51: 105–113.

Garrick, L. D., Lang, J. W., and Herzog, H. A., Jr. 1978. Social signals of adult American alligators. Bull. Am. Mus. Nat. Hist. 160: 153–192.

Gasparini, Z. B. de. 1979. Comarios criticos sobre los vertebrados Mesozoicos de Chile. Segundo Congr. Geol. Chileno. Pp. H15–H32.

Gasparini, Z. de, Olivero, E., Scasso, R., and Rinaldi, C. 1987. Un ankylosaurio (Reptilia, Ornithischia) Campaniano en el continente Antartico. An. X Congr. Brasil. Paleontol. Pp. 131–141.

Gaudry, A. 1890. Les enchaînements du monde animal dans les temps géologiques. Fossiles secondaires. 323 pp.

Gauthier, J. 1984. A cladistic analysis of the higher systematic categories of the Diapsida. Ph.D. dissertation, Univ. California, Berkeley. 564 pp.

———. 1986. Saurischian monophyly and the origin of birds. In: Padian, K. (ed.). The Origin of Birds and the Evolution of Flight. Mem. Calif. Acad. Sci. 8: 1–55.

Gauthier, J., Estes, R., and de Queiroz, K. K. 1988. A phylogenetic analysis of Lepidosauromorpha. In: Estes, R., and Pregill, G. K. (eds.). The Phylogenetic Relationships of the Lizard Families. Stanford Univ. Press, Palo Alto. Pp. 15–98.

Gauthier, J., Kluge, A. G., and Rowe, T. 1988a. Amniote phylogeny and the importance of fossils. Cladistics 4: 105–208.

———. 1988b. The early evolution of the Amniota. In: Benton, M. J. (ed.). The Phylogeny and Classification of the Tetrapods. Vol. I. Syst. Assoc. Spec. Vol. 35A: 103–155.

Gauthier, J., and Padian, K. 1985. Phylogenetic, functional and aerodynamic analysis of the origin of birds. In: Hecht, M. K., Ostrom, J. H., Viohl, G., and Wellnhofer, P. (eds.). The Beginnings of Birds. Freunde des Jura-Museums, Eichstätt. Pp. 185–198.

Geiger, M. E., and Hopping, C. A. 1968. Triassic stratigraphy of the southern North Sea Basin. Phil. Trans. Roy. Soc. London B 254: 1–36.

Geist, V. 1966. The evolution of horn-like organs. Behaviour 27: 175–214.

Gervais, P. 1852. Zoologie et Paléontologie Françaises (Animaux Vertébrés). 1st ed. Paris. 271 pp.

———. 1859. Zoologie et Paléontologie Françaises. 2d ed. Paris 544 pp.

Geyer, O. F., and Gwinner, M. 1962. Der Schwäbische Jura. Sammlung Geol. Führer 40: 1–452.

Giffin, E. B., Gabriel, D. L., and Johnson, R. E. 1987. A new pachycephalosaurid skull from the Cretaceous Hell Creek Formation of Montana. J. Vert. Paleontol. 7: 398–407.

Gilinsky, N. L. 1986. Was there 26 Myr periodicity of extinctions? Nature 321: 533–534.

Gillette, D. D. 1987. A giant sauropod from the Jackpile SS member of the Morrison Fm (Upper Jurassic) of New Mexico. J. Vert. Paleontol. 7 (suppl.): 17A–18A.

Gillette, D. D., Pittman, J. G., and Thomas, D. A. 1985. Herding behavior and species diversity in Cretaceous sauropods of the American southwest. Geol. Soc. Am. Abstr. Prog. 17: 220.

Gilmore, C. W. 1907. The type of the Jurassic reptile *Morosaurus agilis* redescribed, with a note on *Camptosaurus*. Proc. U.S. Natl. Mus. 32: 151–165.

———. 1909. Osteology of the Jurassic reptile *Camptosaurus*, with a revision of the species of the genus, and a description of two new species. Proc. U.S. Natl. Mus. 36: 197–332.

———. 1913. A new dinosaur from the Lance Formation of Wyoming. Smithsonian Misc. Coll. 61: 1–5.

———. 1914a. Osteology of the armored Dinosauria in the United States National Museum, with special reference to the genus *Stegosaurus*. Bull. U.S. Natl. Mus. 89: 1–136.

———. 1914b. A new ceratopsian dinosaur from the Upper Cretaceous of Montana, with a note on *Hypacrosaurus*. Smithsonian Misc. Coll. 63: 1–10.

———. 1915a. Osteology of *Thescelosaurus*, an orthopodous dinosaur from the Lance Formation of Wyoming. Proc. U.S. Natl. Mus. 49: 591–616.

———. 1915b. A new restoration of *Stegosaurus*. Proc. U.S. Natl. Mus. 49: 355–357.

———. 1916. Contributions to the geology and paleontology of San Juan County, New Mexico. 2. Vertebrate faunas of the Ojo Alamo, Kirtland and Fruitland Formations. U.S. Geol. Surv. Prof. Pap. 98: 279–308.

———. 1917. *Brachyceratops*, a ceratopsian dinosaur from the Two Medicine Formation of Montana. U.S. Geol. Surv. Prof. Pap. 103: 1–45.

———. 1918. A newly mounted skeleton of the armored dinosaur *Stegosaurus stenops* in the United States National Museum. Proc. U.S. Natl. Mus. 54: 383–390.

———. 1919a. Reptilian faunas of the Torrejon, Puerco, and underlying Upper Cretaceous Formations of San Juan County, New Mexico. U.S. Geol. Surv. Prof. Pap. 119: 1–68.

———. 1919b. New restoration of *Triceratops*, with notes on the osteology of the genus. Proc. U.S. Natl. Mus. 55: 97–112.

———. 1920. Osteology of the carnivorous Dinosauria in the United States National Museum, with special reference to the genera *Antrodemus (Allosaurus)* and *Ceratosaurus*. Bull. U.S. Natl. Mus. 110: 1–154.

———. 1921. The fauna of the Arundel Formation of Maryland. Proc. U.S. Natl. Mus. 59: 581–594.

———. 1922. A new sauropod dinosaur from the Ojo Alamo Formation of New Mexico. Smithsonian Misc. Coll. 72: 1–9.

———. 1923. A new species of *Corythosaurus,* with notes on other Belly River Dinosauria. Can. Field-Nat. 37: 1–9.

———. 1924*a.* A new coelurid dinosaur from the Belly River Cretaceous of Alberta. Bull. Can. Dept. Mines Geol. Surv. 38: 1–12.

———. 1924*b.* A new species of *Laosaurus,* an ornithischian dinosaur from the Cretaceous of Alberta. Trans. Roy. Soc. Can. Sect. IV (ser. 3) 18: 1–6.

———. 1924*c.* A new species of hadrosaurian dinosaur from the Edmonton Formation (Cretaceous) of Alberta. Bull. Can. Dept. Mines Geol. Surv. 38: 13–26.

———. 1924*d.* On *Troodon validus,* an orthopodous dinosaur from the Belly River Cretaceous of Alberta, Canada. Bull. Dept. Geol. Univ. Alberta 1: 1–43.

———. 1925. A nearly complete articulated skeleton of *Camarasaurus,* a saurischian dinosaur from the Dinosaur National Monument. Mem. Carnegie Mus. 10: 347–384.

———. 1930. On dinosaurian reptiles from the Two Medicine Formation of Montana. Proc. U.S. Natl. Mus. 77: 1–39.

———. 1932*a.* On a newly mounted skeleton of *Diplodocus* in the United States National Museum. Proc. U.S. Natl. Mus. 81: 1–21.

———. 1932*b.* A new fossil lizard from the Belly River Formation of Alberta. Trans. Roy. Soc. Can. (ser. 3) 26: 117–120.

———. 1933*a.* On the dinosaurian fauna of the Iren Dabasu Formation. Bull. Am. Mus. Nat. Hist. 67: 23–78.

———. 1933*b.* Two new dinosaurian reptiles from Mongolia with notes on some fragmentary specimens. Am. Mus. Novitates 679: 1–20.

———. 1935. On the Reptilia of the Kirtland Formation of New Mexico, with descriptions of new species of fossil turtles. Proc. U.S. Natl. Mus. 83: 159–188.

———. 1936. Osteology of *Apatosaurus* with special reference to specimens in the Carnegie Museum. Mem. Carnegie Mus. 11: 175–300.

———. 1939. Ceratopsian dinosaurs from the Two Medicine Formation, Upper Cretaceous of Montana. Proc. U.S. Natl. Mus. 87: 1–18.

———. 1945. *Parrosaurus,* n. name, replacing *Neosaurus* Gilmore 1945. J. Paleontol. 19: 540.

———. 1946*a.* Notes on some recently mounted reptile fossil skeletons in the United States National Museum. Proc. U.S. Natl. Mus. 96: 195–203.

———. 1946*b.* Reptilian fauna of the North Horn Formation of central Utah. U.S. Geol. Surv. Prof. Pap. 210C: 1–52.

———. 1946*c.* A new carnivorous dinosaur from the Lance Formation of Montana. Smithsonian Misc. Coll. 106: 1–19.

Gilmore, C. W., and Stewart, D. R. 1945. A new sauropod dinosaur from the Upper Cretaceous of Missouri. J. Paleontol. 19: 23–29.

Ginsburg, L. 1964. Découverte d'un scélidosaurien (Dinosaure ornithischien) dans le Trias supérieur du Basutoland. C. R. Acad. Sci. Paris 258: 2366–2368.

Ginsburg, L., Lapparent, A. F. de, Loiret, B., and Taquet, P. 1966. Empreintes de pas de vertébrés tetrapodes dans les séries continentales à l'Ouest d'Agadès (République du Niger). C. R. Acad. Sci. Paris 263: 28–31.

Gittleman, J. L. 1985. Carnivore body size: Ecological and taxonomic correlates. Oecologia 67: 540–554.

Goodman, M., Weiss, M. L., and Czelusniak, J. 1982. Molecular evolution above the species level: Branching patterns, rates and mechanisms. Syst. Zool. 31: 35–42.

Gow, C. E. 1975*a.* A new heterodontosaurid from the Red Beds of South Africa showing clear evidence of tooth replacement. Zool. J. Linn. Soc. 57: 335–339.

———. 1975*b.* The morphology and relationships of *Youngina capensis* Broom and *Prolacerta broomi* Parrington. Palaeontol. afr. 18: 89–131.

———. 1981. Taxonomy of the Fabrosauridae (Reptilia, Ornithischia) and the *Lesothosaurus* myth. S. Afr. J. Sci. 77: 43.

Gow, C. E., and Raath, M. A. 1977. Fossil vertebrate studies in Rhodesia: Sphenodontid remains from the Upper Trias of Rhodesia. Palaeontol. afr. 20: 121–122.

Gradzinski, R. 1970. Sedimentation of dinosaur-bearing Upper Cretaceous deposits of the Nemegt Basin, Gobi Desert. Palaeontol. Polonica 21: 147–229.

Gradzinski, R., and Jerzykiewicz, T. 1974. Dinosaur- and mammal-bearing aeolian and associated deposits of the Upper Cretaceous in the Gobi Desert (Mongolia). Sed. Geol. 12: 249–278.

Gradzinski, R., Kazmierczak, J., and Lefeld, J. 1968. Geographical and geological data from the Polish-Mongolian palaeontological expeditions. Palaeontol. Polonica 19: 33–82.

Gradzinski, R., Kielan-Jaworowska, Z., and Maryanska, T. 1977. Upper Cretaceous Djadokhta, Barun Goyot and Nemegt formations of Mongolia, including remarks on previous subdivisions. Acta Geol. Polonica 27: 281–318.

Grambast, L., Marginez, M., Mattauer, M., and Thaler, L. 1967. *Perutherium altiplanense,* nov. gen., nov. sp., premier Mammifère mésozoïque d'Amerique du Sud. C. R. Acad. Sci. Paris 264: 707–710.

Granger, W., and Gregory, W. K. 1923. *Protoceratops andrewsi,* a pre-ceratopsian dinosaur from Mongolia. Am. Mus. Novitates 72: 1–9.

———. 1935. A revised restoration of the skeleton of *Baluchitherium,* gigantic fossil rhinoceros of central Asia. Am. Mus. Novitates 787: 1–3.

———. 1936. Further notes on the gigantic extinct rhinoceros, *Baluchitherium,* from the Oligocene of Mongolia. Bull. Am. Mus. Nat. Hist. 72: 1–73.

Gregory, H. E. 1950. Geology and Geography of the Zion

Park Region Utah and Arizona. U.S. Geol. Surv. Prof. Pap. 220: 1–200.

——. 1951. The Geology and Geography of the Paunsaugunt Region Utah. U.S. Geol. Surv. Prof. Pap. 226: 1–116.

Gregory, J. T. 1948. The type of *Claosaurus? affinis* Wieland. Am. J. Sci. 246: 29—30.

Gregory, W. K. 1951. Evolution Emerging. Macmillan, New York. Vols. 1, 2. 1749 pp.

Greigert, J., Joulia, F., and Lapparent, A. F. de. 1954. Répartition stratigraphique des gisements de vertébrés dans le Crétacé du Niger. C. R. Acad. Sci. Paris 239: 433–435.

Greppin, J. B. 1870. Description géologique du Jura bernois et de quelques districts adjacents. Beitr. Geol. Karte Schweiz 8: 1–357.

Gries, J. P. 1962. Lower Cretaceous stratigraphy of South Dakota and the eastern edge of the Powder River Basin. Wyoming Geol. Assoc. Gdbk. 17th Ann. Field Conf. Pp. 163–172.

Grigorescu, D. 1983*a*. A stratigraphic, taphonomic, and paleoecologic approach to a "forgotten land": The dinosaur-bearing deposits from the Haţeg Basin (Transylvania-Romania). Acta Palaeontol. Polonica 28: 103–121.

——. 1983*b*. Cadrul stratigrafic si paleoecologic al depozitelor continentale cu dinosauri din Basinul Haţeg. Sargetia 13: 37–47.

——. 1984. New paleontological data on the dinosaur beds from the Haţeg Basin. In: 75 Years of the Laboratory of Paleontology. Univ. Bucharest, Spec. Vol. Pp. 111–118.

——. 1987. Considerations of the age of the "Red Beds" continental formations in SW Transylvanian Depression. In: The Eocene from the Transylvanian Basin. Pp. 189–196.

Grigorescu, D., Hartenberger, J.-L., Radulescu, C., Samson, P., and Sudre, J. 1985. Découverte de mammifères et dinosaures dans le Crétacé supérieur de Pui (Roumanie). C. R. Acad. Sci. Paris 301: 1365–1368.

Grzimek, B. 1968. Grzimek's Animal Life Encyclopedia. Van Nostrand Reinhold Co., New York. 579 pp.

Guard, C. L. 1980. The reptilian digestive system: General characteristics. In: Schmidt-Nielsen, K., Bolis, L., Taylor, C. R., Bentley, P. J., and Stevens, C. E. (eds.). Comparative Physiology of Primitive Mammals. Cambridge Univ. Press, Cambridge. Pp. 43–51.

Guggisberg, C. A. W. 1972. Crocodiles, Their Natural History, Folklore and Conservation. Stackpole, Harrisburg, PA. 214 pp.

Gürich, G. J. E. 1926. Über Saurier-Fährten aus dem Etjo-Sandstein von Südwestafrika. Palaeontol. Z. 8: 112–120.

Guthrie, R. D. 1967. Differential preservation and recovery of Pleistocene large mammal remains in Alaska. J. Paleontol. 41: 243–246.

——. 1968. Paleoecology of the large-mammal community in interior Alaska during the Late Pleistocene. Am. Midl. Nat. 79: 346–363.

Haack, S. C. 1986. A thermal model of the sailback pelycosaur. Paleobiology 12: 450–458.

Haas, G. 1955. The jaw musculature in *Protoceratops* and in other ceratopsians. Am. Mus. Novitates 1729: 1–24.

——. 1969. On the jaw muscles of ankylosaurs. Am. Mus. Novitates 2399: 1–11.

Haines, R. W. 1969. Epiphyses and sesamoids. In: Gans, C., Bellairs, A. d'A., and Parsons, T. S. (eds.). Biology of the Reptilia. 1: 81–115.

Hallam, A. 1987. End-Cretaceous mass extinction event: Argument for terrestrial causation. Science 238: 1232–1244.

Hallett, M. 1987. Bringing dinosaurs to life. In: Czerkas, S. J., and Olson, E. C. (eds.). Dinosaurs Past and Present. Nat. Hist. Mus. Los Angeles County. Vol. I. Pp. 96–113.

Halstead, L. B., and Halstead, J. 1981. Dinosaurs. Blandford Press, Poole, Engl. 170 pp.

Hamley, T. In press. Functions of the tail in bipedal locomotion of lizards and dinosaurs. Mem. Queensland Mus.

Hao Y, and Guan S. 1984. The Lower-Upper Cretaceous and Cretaceous-Tertiary boundaries in China. Bull. Geol. Soc. Denmark 33: 129–138.

Hao Y., Su D., Yu J., Li P., Li Y., Wang N., Qi H., Guan S., Hu H., Liu X., Yang W., Ye L., Shou Z., Zang Q., et al. 1986. The Cretaceous System of China. Geol. Publishing House, Beijing. Pp. 1–341.

Haq, B. U., Hardenbol, J., and Vail, P. R. 1987. Chronology of fluctuating sea levels since the Triassic. Science 235: 1156–1167.

Hargens, A. R., Millard, R. W., Pettersson, K., and Johansen, K. 1987. Gravitational haemodynamics and oedema prevention in the giraffe. Nature 329: 59–60.

Harland, W. B., Cox, A. V., Llewellyn, P. G., Smith, A. G., and Walters, R. 1982. A Geologic Time Scale. Cambridge Univ. Press, Cambridge. 131 pp.

Harris, J. M., and Russell, D. A. MS. Preliminary notes on the occurrence of dinosaurs in the Turkana Grits of northern Kenya.

Harris, J. P., and Hudson, J. D. 1980. Lithostratigraphy of the Great Estuarine Group (Middle Jurassic), Inner Hebrides. Scottish J. Geol. 16: 231–250.

Harrison, C. J. O., and Walker, C. A. 1973. *Wyleyia:* A new bird humerus from the Lower Cretaceous of England. Palaeontology 16: 721–728.

——. 1975. The Bradycnemidae, a new family of owls from the Upper Cretaceous of Romania. Palaeontology 18: 563–570.

Harvey, P. H., and Zammuto, R. M. 1985. Patterns of mortality and age at first reproduction in natural populations of mammals. Nature 315: 319–320.

Hatcher, J. B. 1893. The Ceratops Beds of Converse County, Wyo. Am. J. Sci. (ser. 3) 14: 135–144.

————. 1901. *Diplodocus* (Marsh): Its osteology, taxonomy, and probable habits, with a restoration of the skeleton. Mem. Carnegie Mus. 1: 1–63.

————. 1902. Structure of the forelimb and manus of *Brontosaurus*. Ann. Carnegie Mus. 1: 356–376.

————. 1903*a*. A new sauropod dinosaur from the Jurassic of Colorado. Proc. Biol. Soc. Washington 16: 1–2.

————. 1903*b*. A new name for the dinosaur *Haplocanthus* Hatcher. Proc. Biol. Soc. Washington 16: 100.

————. 1903*c*. Osteology of *Haplocanthosaurus*, with description of a new species, and remarks on the probable habits of the Sauropoda and the age and origin of the *Atlantosaurus* beds. Mem. Carnegie Mus. 2: 1–72.

————. 1903*d*. Additional remarks on *Diplodocus*. Mem. Carnegie Mus. 2: 72–75.

————. 1905. Two new Ceratopsia from the Laramie of Converse County, Wyoming. Am. J. Sci. (ser. 4) 20: 413–419.

Hatcher, J. B., Marsh, O. C., and Lull, R. S. 1907. The Ceratopsia. U.S. Geol. Surv. Mongr. 49: 1–300.

Haughton, S. H. 1915. On some dinosaur remains from Bushmanland. Trans. Roy. Soc. S. Afr. 5: 259–264.

————. 1918. A new dinosaur from the Stormberg Beds of South Africa. Ann. Mag. Nat. (ser. 9) 2: 468–469.

————. 1924. The fauna and stratigraphy of the Stormberg series. Ann. S. Afr. Mus. 12: 323–497.

————. 1928. On some reptilian remains from the dinosaur beds of Nyassaland. Trans. Roy. Soc. S. Afr. 16: 67–75.

Hawthorne, J. M. 1987. The stratigraphy and depositional environments of Lower Cretaceous dinosaur track-bearing strata in Texas. Geol. Soc. Am. South-Central Sect. Abstr. Prog. 19: 169.

Hay, O. P. 1899. On the nomenclature of certain American fossil vertebrates. Am. Geol. 24: 345–349.

————. 1909. On the skull and brain of *Triceratops*, with notes on the brain-cases of *Iguanodon* and *Megalosaurus*. Proc. U.S. Natl. Mus. 36: 95–108.

Hayssen, V. 1984. Basal metabolic rate and the intrinsic rate of increase: An empirical and theoretical reexamination. Oecologia: 64: 419–421.

Hayssen, V., and Lacy, R. C. 1985. Basal metabolic rates in mammals: Taxonomic differences in the allometry of BMR and body mass. Comp. Biochem. Physiol. 81A: 741–754.

He X. 1979. [A newly discovered ornithopod dinosaur—*Yandusaurus* from Zigong, Sichuan]. In: [Contribution to International Exchange of Geology. Pt. 2. Stratigraphy and paleontology]. Geol. Publishing House, Beijing. Pp. 116–123. (In Chinese)

————. 1984. [The Vertebrate Fossils of Sichuan]. Sichuan Scientific and Technical Publishing House, Chengdu, Sichuan. 168 pp. (In Chinese)

He X., and Cai K. 1983. [A new species of *Yandusaurus* (hypsilophodont dinosaur) from the Middle Jurassic of Dashanpu, Zigong, Sichuan]. In: [Special paper on dinosaurian remains of Dashanpu, Zigong, Sichuan]. J. Chengdu College Geol. Suppl. 1. Pp. 5–14. (In Chinese with English summary)

————. 1984. [The Middle Jurassic dinosaurian fauna from Dashanpu, Zigong, Sichuan. Vol. 1. The ornithopod dinosaurs]. Sichuan Scientific and Technological Publishing House, Chengdu, Sichuan. Pp. 1–71. (In Chinese with English summary)

He X.-L., Li K., Cai K.-J., and Gao Y.-H. 1984. [*Omeisaurus tianfuensis*—a new species of *Omeisaurus* from Dashanpu, Zigong, Sichuan.] J. Chengdu College Geol. Suppl. 2: 13–32. (In Chinese with English summary)

Heaton, M. J. 1972. The palatal structure of some Canadian Hadrosauridae (Reptilia: Ornithischia). Can. J. Earth Sci. 9: 185–205.

Hecht, M. K., Ostrom, J. H., Viohl, G., and Wellnhofer, P. (eds.). 1985. The Beginnings of Birds. Freunde des Jura-Museums, Eichstätt. 382 pp.

Hecht, M. K., and Tarsitano, S. 1982. The paleobiology and phylogenetic position of *Archaeopteryx*. Géobios Mém. Spéc. 6: 141–149.

————. 1983. *Archaeopteryx* and its paleoecology. Acta Palaeontol. Polonica 28: 133–136.

Heerden, J. V. 1977. The comparative anatomy of the postcranial skeleton and the relationships of the South African Melanorosauridae (Saurischia: Prosauropoda). Ph.D. dissertation, Univ. Orange Free State, Bloemfontein. 175 pp.

————. 1979. The morphology and taxonomy of *Euskelosaurus* (Reptilia: Saurischia; Late Triassic) from South Africa. Navor. Nasl. Mus. Bloemfomtein 4: 21–84.

Heilmann, G. 1927. The Origin of Birds. Appleton, New York. 210 pp.

Heller, F. 1952. Reptilfährten-Funde aus dem Ansbacher Sandstein des Mittleren Keupers von Franken. Geol. Blätt. NO-Bayern. 2: 129–141.

Henneman, W. W., III. 1983. Relationship among body mass, metabolic rate, and the intrinsic rate of natural increase in mammals. Oecologia 56: 104–108.

————. 1984. Commentary. Oecologia 64: 421–423.

Hennig, E. 1912. Am Tendaguru. Leben und Wirken einer deutschen Forschungsexpedition zur Ausgrabung vorweltlicher Riesensaurier in Deutsch-Ostafrika. Stuttgart. 151 pp.

————. 1915*a*. Stegosauria. Fossilium Catalogus. I. Animalia. Pt. 9: 1–15.

————. 1915*b*. *Kentrosaurus aethiopicus*, der Stegosauride des Tendaguru. Sitzungsber. Ges. Naturforsch. Freunde Berlin 1915: 219–247.

————. 1916*a*. Zweite Mitteilung über den Stegosauriden vom Tendaguru. Sitzungsber. Ges. Naturforsch. Freunde Berlin 1916: 175–182.

————. 1916*b*. *Kentrurosaurus*, non *Doryphorosaurus*. Cbl. Mineral. Geol. Palaeontol. 1916: 578.

————. 1924. *Kentrurosaurus aethiopicus*. Die Stegosaurier-

Funde vom Tendaguru, Deutsch-Ostafrika. Palaeontographica (Suppl. 7) 1: 103–253.

———. 1936. Ein Dentale von *Kentrurosaurus aethiopicus* Hennig. Palaeontographica (Suppl. 7) 2: 309–312.

Henry, J. 1876. Étude stratigraphique et paléontologique de l'Infralias dans la Franche-Comté. Mém., Soc. Emulation Doubs, Besançon (ser. 4) 10: 285–476.

Hickey, L. J. 1981. Land plant evidence compatible with gradual, not catastrophic, change at the end of the Cretaceous. Nature 292: 529–531.

——— 1984. Changes in the angiosperm flora across the Cretaceous-Tertiary Boundary. In: Berggren, W. A., and Van Couvering, J. A. Catastrophes in Earth History. Princeton Univ. Press, Princeton. Pp. 279–314.

Hildebrand, S. F., Gilmore, C. W., and Cochran, D. M. 1930. Cold-Blooded Vertebrates. Smithsonian Sci. Series, New York. 375 pp.

Hill, A. V. 1950. The dimensions of animals and their muscular dynamics. Sci. Progress 38: 209–230.

Hinokuma, S. 1963. On the Cretaceous layers at Takashima, Nagasaki Pref. Bull. Kyushu Coal Mining 16: 14–20.

Hintze, L. F. 1973. Geologic history of Utah. Brigham Young Univ. Geol. Stud. 20: 1–181.

Hirsch, K. F., Stadtman, K. L., Miller, W. E., and Madsen, J. H., Jr. 1989. Upper Jurassic dinosaur egg from Utah. Science 243: 1711–1713.

Hitchcock, E. 1848. An attempt to discriminate and describe the animals that made the fossil footmarks of the United States, and especially of New England. Mem. Am. Acad. Arts Sci. 3: 129–256.

———. 1858. Ichnology of New England. White, Boston. 220 pp. 60 pl.

———. 1865. A supplement to the ichnology of New England. Wright and Potter, Boston. 96 pp.

Hoepen, E. C. N. van. 1916. De ouderdom der Transvaalsche Karroolagen. Verh. Geol.-mijnbouwkundig genootschap Nederland en kolonien (Geol. Ser.) 3: 107–117.

———. 1920. Contributions to the knowledge of the reptiles of the Karro Formation. 5. A new dinosaur from the Stormberg Beds. 6. Further dinosaurian material in the Transvaal Museum. Ann. Transvaal. Mus. 7: 77–141.

Hoffet, J. H. 1942. Description de quelques ossements du Sénoniene du Bas-Laos. C. R. Cons. Rech. Sci. Indochine 1942: 49–57.

———. 1943a. Description de quelques ossements de Titanosauriens du Sénonien du Bas-Laos. C. R. Cons. Rech. Sci. Indochine 1943: 1–8.

———. 1943b. Description des ossements les plus caractéristiques appartenant à des avipelviens du Sénonien du Bas-Laos. Bull. Cons. Rech. Sci. Indochine 1944: 179–186.

Hofman, M. A. 1983. Energy metabolism, brain size, and longevity in mammals. Q. Rev. Biol. 58: 495–512.

Hoffman, A. 1985. Patterns of family extinction depend on definition and geological time scale. Nature 315: 659–662.

Hoffstetter, R. 1957. Quelques observations sur les Stégosarinés. Bull. Mus. Natl. Hist. Nat. 29: 537–547.

Hohnke, L. A. 1973. Haemodynamics in the Sauropoda. Nature 244: 309–310.

Hölder, H., and Norman, D. B. 1986. Kreide-Dinosaurier im Sauerland. Naturwissenschaften 73: 109–116.

Holl, F. 1829. Handbuch der Petrefactenkunde. Pt. 1. Quedlinburg. 232 pp.

Holland, W. J. 1906. The osteology of *Diplodocus* Marsh. Mem. Carnegie Mus. 2: 225–278.

———. 1915a. Heads and tails; a few notes relating to the structure of the sauropod dinosaurs. Ann. Carnegie Mus. 9: 273–278.

———. 1915b. A new species of *Apatosaurus*. Ann. Carnegie Mus. 10: 143–145.

———. 1919. Report on Section of Paleontology. Ann. Rept. Carnegie Mus. 1919: 38.

———. 1924a. Description of the type of *Uintasaurus douglassi* Holland. Ann. Carnegie Mus. 15: 119–138.

———. 1924b. The skull of *Diplodocus*. Mem. Carnegie Mus. 9: 379–403.

Honkala, F. S. 1955. The Cretaceous-Tertiary boundary near Glacier National Park, Montana. Billings Geol. Soc. Guidebook, 6th Ann. Field Conf.: 124–128.

Hooijer, D. A. 1968. A Cretaceous dinosaur from the Syrian Arab Republic. Proc. K. Nederl. Akad. Wet. B 71 150–152.

Hooker, J. S. 1987. Late Cretaceous ashfall and the demise of a hadrosaurian "herd." Geol. Soc. Am. Rocky Mountain Sect. Abstr. Prog. 19: 284.

Hopson, J. A. 1975a. On the generic separation of the ornithischian dinosaurs *Lychorhinus* and *Heterodontosaurus* from the Stormberg Series (Upper Triassic) of South Africa. S. Afr. J. Sci. 71: 302–305.

———. 1975b. The evolution of cranial display structures in hadrosaurian dinosaurs. Paleobiology 1: 21–43.

———. 1977. Relative brain size and behavior in archosaurian reptiles. Ann. Rev. Ecol. Syst. 8: 429–448.

———. 1979. Paleoneurology. In: Gans, C. (ed.). Biology of the Reptilia. Vol. 9A: 39–146.

———. 1980a. Relative brain size in dinosaurs—implications for dinosaurian endothermy. In: Thomas, R. D. K., and Olson, E. C. (eds.). A Cold Look at the Warm-Blooded Dinosaurs. Westview Press, Boulder. Pp. 287–310.

———. 1980b. Tooth function and replacement in early Mesozoic ornithischian dinosaurs: Implications for aestivation. Lethaia 13: 93–105.

Horne, G. S., Attwood, M. G., and King, A. P. 1974. Stratigraphy, sedimentology, and paleoenvironment of Esquias Formation of Honduras. Am. Assoc. Petrol. Geol. Bull. 58: 176–188.

Horner, J. R. 1979. Upper Cretaceous dinosaurs from the Bearpaw Shale (marine) of south-central Montana with a checklist of Upper Cretaceous dinosaur remains from marine sediments in North America. J. Paleontol. 53: 566–578.

———. 1982. Evidence for colonial nesting and "site fidelity" among ornithischian dinosaurs. Nature 297: 675–676.

———. 1983. Cranial osteology and morphology of the type specimen of *Maiasaura peeblesorum* (Ornithischia: Hadrosauridae), with a discussion of its phylogenetic position. J. Vert. Paleontol. 3: 29–38.

———. 1984a. The nesting behavior of dinosaurs. Sci. Am. 250(4): 130–137.

———. 1984b. Three ecologically distinct vertebrate faunal communities from the Late Cretaceous Two Medicine Formation of Montana, with discussion of evolutionary pressures induced by interior seaway fluctuations. Montana Geol. Soc. 1984 Field Conf., Northwestern Montana. Pp. 299–303.

———. 1984c. A "segmented" epidermal tail frill in a species of hadrosaurian dinosaur. J. Paleontol. 58: 270–271.

———. 1985. Evidence for polyphyletic origination of the Hadrosauridae (Reptilia; Ornithischia). Proc. Pacific Div. Am. Assoc. Adv. Sci. 4: 31–32.

———. 1987. Ecological and behavioral implications derived from a dinosaur nesting site. In: Czerkas, S. J., and Olson, E. C. (eds.). Dinosaurs Past and Present. Vol. II. Nat. Hist. Mus. Los Angeles County. Pp. 51–63.

———. 1988. A new hadrosaur (Reptilia, Ornithischia) from the Upper Cretaceous Judith River Formation of Montana. J. Vert. Paleontol. 8: 314–321.

———. 1990. Evidence of diphyletic origination of the hadrosaurian (Reptilia; Ornithischia) dinosaurs. In: Currie, P. J., and Carpenter, K.(eds.). Dinosaur Systematics: Perspectives and Approaches. Cambridge Univ. Press, New York. Pp. 179–187.

Horner, J. R., and Makela, R. 1979. Nest of juveniles provides evidence of family structure among dinosaurs. Nature 282: 296–298.

Horner, J. R., and Weishampel, D. B. 1988. A comparative embryological study of two ornithischian dinosaurs. Nature 332: 256–257.

Hotton, N., III. 1955. A survey of adaptive relationships of dentition to diet in the North American Iguanidae. Am. Midl. Nat. 53: 88–114.

———. 1963. Dinosaurs. Pyramid Publications, New York. 192 pp.

———. 1980. An alternative to dinosaur endothermy: The happy wanderers. In: Thomas, R. D. K., and Olson, E. C. (eds.). A Cold Look at the Warm-Blooded Dinosaurs. Westview Press, Boulder. Pp. 311–350.

Hou L. 1977. [A new primitive Pachycephalosauria from Anhui, China]. Vertebrata PalAsiatica 15: 198–202. (In Chinese with English summary)

Hou L., Chao X., and Chu. 1976. [New discovery of sauropod dinosaurs from Sichuan]. Vertebrata PalAsiatica 14: 160–165. (In Chinese)

Hou L., Yeh H., and Zhao X. 1975. [Fossil reptiles from Fusui, Kwangshi]. Vertebrata PalAsiatica 13: 24–33. (Chinese with English summary)

Houston, D. C. 1979. The adaptations of scavengers. In: Sinclair, A. R. E., and Norton-Griffiths, M. (eds.). Serengeti: Dynamics of an Ecosystem. Univ. Chicago Press, Chicago. Pp. 263–286.

Hsü, K. J., and McKenzie, J. A. 1985. A Strangelove ocean in the earliest Tertiary. In: Sundquist, E. T., and Broecker, W. (eds.). The Carbon Cycle and Atmospheric CO_2: Natural Variation Archean to Present. Geophys. Monogr. Ser. 32: 487–492.

Hu C. 1973. [A new hadrosaur from the Cretaceous of Chucheng, Shantung]. Acta Geol. Sinica 2: 179–202. (In Chinese with English summary)

Hu S. 1964. [Carnosaurian remains from Alashan, Inner Mongolia]. Vertebrata PalAsiatica 8: 42–63. (In Chinese with English summary)

Huckriede, R. 1982. Die unterkretazische Karsthöhlen-Füllung von Nehden im Sauerland. 1. Geologische, paläozoologische und paläobotanische Befunde und Datierung. Geol. Palaeontol. 16: 183–242.

Huddleston, W. H. 1876. Excursion to Swindon and Faringdon. Proc. Geol. Assoc. London 4: 543–554.

Huene, F. von. 1904. *Dystropheus viaemalae* Cope in neuer Beleuchtung. N. Jb. Mineral. Geol. Palaeontol. Abhandl. 19: 319–333.

———. 1905. Über die Trias-Dinosaurier Europas. Z. Dtsch. Geol. Ges. Mb. 57: 345–349.

———. 1906. Ueber die Dinosaurier der aussereuropäischen Trias. Geol. Palaeontol. Abhandl. 8: 97–156.

———. 1907-08. Die Dinosaurier der europäischen Triasformation mit Berücksichtigung der aussereuropäischen Vorkommnisse. Geol. Palaeontol. Abhandl. Suppl. 1: 1–419.

———. 1909. Skizze zu einer Systematik und Stammesgeschichte der Dinosaurier. Zbl. Mineral. Geol. Palaeontol. 1900: 12–22.

———. 1910a. Ein primitiver Dinosaurier aus der mittleren Trias von Elgin. Geol. Palaeontol. Abhandl. 8: 317–322.

———. 1910b. Ueber den ältesten Rest von *Omosaurus* (*Dacentrurus*) im englischen Dogger. N. Jb. Mineral. Geol. Palaeontol. 1910: 75–78.

———. 1911. Beiträge zur Kenntnis des Ceratopsidenschädels. N. Jb. Mineral. Geol. Palaeontol. 1911: 146–162.

———. 1914a. Nachträge zu meinen früheren Beschreibungen triassischer Saurischia. Geol. Palaeontol. Abhandl. 13: 69–82.

———. 1914b. Saurischia et Ornithischia triadica ("Dinosauria" triadica). Fossilium Catalogus. I. Animalia 4: 1–21.

———. 1914c. Das natürliche System der Saurischia. Zbl. Mineral. Geol. Palaeontol. B 1914: 154–158.

———. 1914d. Ueber die Zweistämmigkeit der Dinosaurier, mit Beiträgen zur Kenntnis einiger Schädel. N. Jb. Mineral. Geol. Palaeontol. Beil.-Bd. 37: 577–589.

———. 1915. Beiträge zur Kenntnis einiger Saurischier der schwäbischen Trias. N. Jb. Mineral. Geol. Palaeontol. 1915: 1–27.

———. 1919. Kurze Übersicht über die Saurischia und ihre naturlichen Zusammenhange. Palaeontol. Z. 11: 269–273.

———. 1920. Bemerkungen zur Systematik und Stammesgeschichte einiger Reptilien. Z. Indukt. Abstamm. Vererb. Lehre 24: 162–166.

———. 1921. Reptilian and stegocephalian remains from the Triassic of Pennsylvania in the Cope Collection. Bull. Am. Mus. Nat. Hist. 44: 561–574.

———. 1922. Ueber einen Sauropoden im oberen Malm des Berner Jura. Eclogae Geol. Helvetiae 17: 80–94.

———. 1923. Carnivorous Saurischia in Europe since the Triassic. Bull. Geol. Soc. Am. 34: 449–458.

———. 1925. Ausgedehnte Karroo-Komplexe mit Fossilführung im nordöstlichen Südwestafrika. Cbl. Mineral. Geol. Palaeontol. 1925: 151–156.

———. 1926a. On serveral known and unknown reptiles of the order Saurischia from England and France. Ann. Mag. Nat. Hist. (ser. 9) 17: 473–489.

———. 1926b. The carnivorous Saurischia in the Jura and Cretaceous formations, principally in Europe. Revista Mus. La Plata 29: 35–167.

———. 1926c. Vollständige Osteologie eines Plateosauriden aus dem schwäbischen Trias. Geol. Palaeontol. Abhandl. 15: 129–179.

———. 1927. Sichtung der Grundlagen der jetzigen Kenntnis der Sauropoden. Eclogae Geol. Helvetiae 20: 444–470.

———. 1928. Lebensbild des Saurischier-Vorkommens im obersten Keuper von Trossingen in Württemberg. Palaeobiologica 1: 103–116.

———. 1929a. Los Saurisquios y Ornithisquios de Cretacéo Argentino. An. Mus. La Plata (ser. 2) 3: 1–196.

———. 1929b. Die Besonderheit der Titanosaurier. Cbl. Mineral. Geol. Palaeontol. 1929B: 493–499.

———. 1929c. Kurze Übersicht über die Saurischia und ihre natürlichen Zusammenhänge. Palaeontol. Z. 11: 269–273.

———. 1932. Die fossile Reptil-Ordnung Saurischia, ihre Entwicklung und Geschichte. Monogr. Geol. Palaeontol. (ser. 1) 4: 1–361.

———. 1934a. Ein neuer Coelurosaurier in der thüringischen Trias. Palaeontol. Z. 16: 145–170.

———. 1934b.Nuevos dientes de Saurios del cretáceo del Uruguay. Bol. Inst. Geol. Uruguay 21: 13–20.

———. 1941. Die Tetrapoden-Fährten im toskanischen Verrucano und ihre Bedeutung. N. Jb. Mineral. Geol. Palaeontol. 1941B: 1–34.

———. 1942. Die fossilen Reptilien des südamerikanischen Gondwanalandes. Ergebnisse der Sauriergrabungen in Südbrasilien 1928/29. Beck'sche Verlagsbuchhandlung, Munich. 332 pp.

———. 1948. Short review of the lower tetrapods. Roy. Soc. S. Afr. Spec. Publ. Robert Broom Commemorative Vol.: 65–106.

———. 1956. Paläontologie und Phylogenie der Niederen Tetrapoden. Fischer, Jena. 716 pp.

———. 1959. Paläontologie und Phylogenie der Niederen Tetrapoden. Nachträge und Ergänzungen. Fischer, Jena. 58 pp.

———. 1966. Ein Megalosauriden-Wirbel des Lias aus norddeutschem Geschiebe. N. Jb. Geol. Palaeontol. 1966: 318–319.

Huene, F. von, and Matley, C. A. 1933. The Cretaceous Saurischia and Ornithischia of the central provinces of India. Mem. Geol. Surv. India 21: 1–74.

Huene, F. von, and Maubeuge, P. L. 1954. Sur quelques restes de Sauriens du Rhétien et du Jurassique lorrains. Bull. Soc. Géol. France (sér. 6) 4: 105–109.

Hughes, B. 1963. The earliest archosaurian reptiles. S. Afr. J. Sci. 59: 221–241.

Hulke, J. W. 1869. Note on a large saurian humerus from the Kimmeridge Clay of the Dorset coast. Q. J. Geol. Soc. London 25: 386–389.

———. 1870. Note on a new and undescribed Wealden vertebra. Q. J. Geol. Soc. London 26: 318–324.

———. 1872. Appendix to a "Note on a new and undescribed Wealden vertebra." Q. J. Geol. Soc. London 28: 36–37.

———. 1874. Note on a very large saurian limb-bone adapted for progression upon land, from the Kimmeridge Clay of Weymouth, Dorset. Q. J. Geol. Soc. London 30: 16–17.

———. 1879a. Note (3rd) on (Eucamerotus Hulke) Ornithopsis H.G.Seeley = Bothriospondylus Owen = Chondrosteus magnus Owen. Q. J. Geol. Soc. London 36: 31–34.

———. 1879b. Note on Poikilopleuron bucklandi of Eudes-Deslongchamps (pere) identifying it with Megalosaurus bucklandi. Q. J. Geol. Soc. London 35: 233–238.

———. 1880a. Supplementary note on the vertebrae of Ornithopsis Seeley = Eucamerotus Hulke. Q. J. Geol. Soc. London 36: 31–35.

———. 1880b. Iguanodon prestwichii, a new species from the Kimmeridge Clay, distinguished from I. mantelli of the Wealden formation in the S.E. of England and the Isle of Wight by differences in the shape of the vertebral centra, by fewer than five sacral vertebrae, by the simpler character of its tooth-serrature, etc., founded on numerous fossil remains lately discovered at Cumnor, near Oxford. Q. J. Geol. Soc. London 36: 433–456.

———. 1881. Polacanthus foxii, a large undescribed dinosaur from the Wealden Formation in the Isle of Wight. Phil. Trans. Roy. Soc. London 178: 169–172.

———. 1882a. Note on the os pubis and ischium of Ornithopsis eucamerotus. Q. J. Geol. Soc. London 38: 372–376.

———. 1882b. Description of some Iguanodon-remains indicating a new species, I. seelyi. Q. J. Geol. Soc. London 38: 135–144.

———. 1887. Note on some dinosaurian remains in the collection of A. Leeds, Esq., of Eyebury, Northamptonshire. Q. J. Geol. Soc. London 43: 695–702.

Humphreys, W. F. 1979. Production and respiration in animal populations. J. Anim. Ecol. 48: 427–453.

Hut, P., Alvarez, W., Elder, W. P., Hansen, T., Kauffman, E. G., Keller, G., Shoemaker, E. M., and Weissman, P. R. 1987. Comet showers as a cause of mass extinctions. Nature 329: 118–126.

Hutchinson, G. E. 1959. Homage to Santa Rosalia, or why are there so many kinds of animals? Am. Nat. 93: 145–159.

Hutchison, J. H., and Archibald, J. D. 1986. Diversity of turtles across the Cretaceous/Tertiary boundary in northeastern Montana. Palaeogeogr. Palaeoclimatol. Palaeoecol. 55: 1–22.

Huxley, T. H. 1866. On some remains of large dinosaurian reptiles from the Stormberg Mountains, South Africa. Geol. Mag. 3: 563.

———. 1867. On *Acanthopholis horridus,* a new reptile from the Chalk-Marl. Geol. Mag. 4: 65–67.

———. 1869*a.* On *Hypsilophodon,* a new genus of Dinosauria. Abstr. Proc. Geol. Soc. London 204: 3–4.

———. 1869*b.* On the upper jaw of *Megalosaurus.* Q. J. Geol. Soc. London 25: 311–314.

———. 1870. On the classification of the Dinosauria with observations of the Dinosauria of the Trias. Q. J. Geol. Soc. London 26: 31–50.

Imlay, R. W. 1952. Correlation of the Jurassic formations of North America exclusive of Canada. Bull. Geol. Soc. Am. 63: 953–992.

Ingavat, R., Janvier, P., and Taquet, P. 1978. Découverte en Thaïlande d'une portion de fémur de dinosaure sauropode (Saurischia, Reptilia). C. R. Soc. Géol. France 1978: 140–141.

Ingavat, R., and Taquet, P. 1978. First discovery of dinosaur remains in Thailand. J. Geol. Soc. Thailand Bangkok 3: 1–6.

Iordansky, N. N. 1968. Cranial kinesis in lizards: A contribution to the problem of adaptive significance of skull kinesis. Zool. Zh. 45: 1398–1410.

Irish, E. J. W., and Havard, C. J. 1968. The Whitemud and Battle Formations ("Kneehills Tuff Zone")—a stratigraphic marker. Geol. Surv. Can. Pap. 67-63: 1–51.

Ishigaki, S. 1985. [Lower Jurassic dinosaur footprints from the central High Atlas Mountains, Morocco]. Nat. Stud. 32: 113–116. (In Japanese)

Iverson, J. B. 1980. Colic modifications in iguanine lizards. J. Morphol. 163: 79–93.

———. 1982. Adaptations to herbivory in iguanine lizards. In: Burghardt, G. M., and Rand, A. S. (eds.). Iguanas of the World: Their Behavior, Ecology and Conservation. Noyes Publ. Park Ridge, N.J. Pp. 60–76.

Jablonski, D. 1986. Causes and consequences of mass extinctions: A comparative approach. In: Elliott, D. K. (ed.). Dynamics of Extinction. Wiley Interscience, New York. Pp. 183–230.

Jacobs, L. L. 1980. Additions to the Triassic vertebrate fauna of Petrified Forest National Park, Arizona. J. Arizona-Nevada Acad. Sci. 15: 12.

Jacobs, L. L., and Murry, P. A. 1980. The vertebrate community of the Triassic Chinle Formation near St. Johns, Arizona. In: Jacobs, L. L. (ed.). Aspects of Vertebrate History. Mus. No. Arizona Press, Flagstaff. Pp. 55–72.

Jaekel, O. 1911. Die Wirbeltiere. Eine Übersicht über die fossilien und lebenden Formen. Borntraeger, Berlin. 252 pp.

———. 1914. Über die Wirbeltierfunde in der oberen Trias von Halberstadt. Palaeontol. Z. 1: 155–215.

Jain, S. L. 1980. The continental Lower Jurassic fauna from the Kota Formation, India. In: Jacobs, L. L. (ed.). Aspects of Vertebrate History. Mus. No. Arizona Press, Flagstaff. Pp. 99–123.

———. 1983. Spirally coiled 'coprolites' from the Upper Triassic Maleri Formation, India. Palaeontology 26: 813–829.

Jain, S. L., Kutty, T. S., Roy-Chowdhury, T., and Chatterjee, S. 1975. The sauropod dinosaur from the Lower Jurassic Kota Formation of India. Proc. Roy. Soc. London A 188: 221–228.

———. 1979. Some characteristics of *Barapasaurus tagorei,* a sauropod dinosaur from the Lower Jurassic of Deccan, India. Proc. IV Internatl. Gonwana Symp. Calcutta, Vol. 1: 204–216.

Jain, S. L., and Sahni, A. 1983. Some Upper Cretaceous vertebrates from central India and their palaeogeographic implications. In: Maheswari, H. K. (ed.). Cretaceous of India. Indian Assoc. Palynostratigraphers, Lucknow, India. Pp. 66–83.

———. 1985. Dinosaurian egg shell fragments from the Lameta Formation at Pisdura, Chandrapur District, Maharashtra. Geosci. J. 6: 211–220.

Janensch, W. 1914. Übersicht über die Wirbeltierfauna der Tendaguru-Schichten, nebst einer kurzen Charakterisierung der neu aufgeführten Arten von Sauropoden. Arch. Biontol. 3: 81–110.

———. 1920. Über *Elaphrosaurus bambergi* und die Megalosaurier aus den Tendaguru-Schichten Deutsch-Ostafrikas. Sitzungsber. Ges. Naturforsch. Freunde Berlin 1920: 225–235.

———. 1922. Das Handskelett von *Gigantosaurus robustus* und *Brachiosaurus brancai* aus den Tendaguru-Schichten Deutsch-Ostafrikas. Cbl. Mineral. Geol. Palaeontol. 1922: 464–480.

———. 1925*a.* Ein aufgestelltes Skelett des Stegosauriers *Kentrurosaurus aethiopicus* E. Hennig aus den Tendaguru-Schichten Deutsch-Ostafrikas. Palaeontographica (Suppl. 7) 1: 257–276.

———. 1925*b.* Die Coelurosaurier und Theropoden der Tendaguru-Schichten Deutsch-Ostafrikas. Palaeontographica (Suppl. 7) 1: 1–99.

———. 1926. Dinosaurier-Reste aus Mexico. Cbl. Mineral. Geol. Palaeontol. 1926: 192–197.

———. 1929*a.* Die Wirbelsäule der Gattung *Dicraeosaurus.* Palaeontographica (Suppl. 7) 2: 39–133.

———. 1929*b.* Material und Formengehalt der Sauropoden in der Ausbeute der Tendaguru-Expedition. Palaeontographica (Suppl. 7) 2: 1–34.

————. 1929c. Magensteine bei Sauropoden der Tendaguru-Schichten. Palaeontographica (Suppl. 7) 2: 135–144.

————. 1929d. Ein aufgestelltes und rekonstruiertes Skelett von *Elaphrosaurus bambergi* mit einem Nachtrag zur Osteologie dieses Coelurosauriers. Palaeontographica (Suppl. 7) 1: 279–286.

————. 1935-36. Die Schädel der Sauropoden *Brachiosaurus*, *Barosaurus* und *Dicraeosaurus* aus den Tendaguruschichten Deutsch-Ostafrikas. Palaeontographica. (Suppl. 7) 2: 147–298.

————. 1936. Über Bahnen von Hirnvenen bei Saurischiern und Ornithischiern, sowei einigen anderen Fossilen und rezenten Reptilien. Palaeontol. Z. 18: 181–198.

————. 1939. Der sakrale Neuralkanal einiger Sauropoden und anderer Dinosaurier. Palaeontol. Z. 21: 171–194.

————. 1947. Pneumatizität bei Wirbeln von Sauropoden und anderen Saurischiern. Palaeontographica (Suppl. 7) 3: 1–25.

————. 1950a. Die Wirbelsäule von *Brachiosaurus brancai*. Palaeontographica (Suppl. 7) 3: 27–93.

————. 1950b. Die Skelettrekonstruktion von *Brachiosaurus brancai*. Palaeontographica (Suppl. 7) 3: 97–103.

————. 1955. Der Ornithopode *Dysalotosaurus* der Tendaguruschichten. Palaeontographica (Suppl. 7) 3: 105–176.

————. 1961. Die Gliedmassen und Gliedmaszengürtel der Sauropoden der Tendaguru-Schichten. Palaeontographica (Suppl. 7) 3: 177–235.

Janis, C. 1976. The evolutionary strategy of the Equidae and the origins of rumen and cecal digestion. Evolution 30: 757–774.

————. 1986. Evolution of horns and related structures in hoofed mammals. Discovery 19: 9–17.

Jarman, P. J. 1974. The social organisation of antelope in relation to their ecology. Behaviour 58: 215–267.

Jarman, P. J., and Sinclair, A. R. E. 1979. Feeding strategy and the pattern of resource-partitioning in ungulates. In: Sinclair, A. R. E., and Norton-Griffiths, M. (eds.). Serengeti: Dynamics of an Ecosystem. Univ. Chicago Press, Chicago. Pp. 130–163.

Jarzen, D. M. 1982. Palynology of Dinosaur Provincial Park (Campanian), Alberta. Syllogeus 38: 1–69.

Jeletzky, J. A. 1960. Youngest marine rocks in western interior of North America and the age of the *Triceratops* beds; with remarks on comparable dinosaur-bearing beds outside North America. Proc. 21st Internatl. Geol. Congress 1960, Pt. 5: 25–40.

————. 1963. The allegedly Danian dinosaur-bearing rocks of the globe and the problem of the Mesozoic-Cenozoic boundary. J. Paleontol. 36: 1005–1018.

Jenkins, F. A., Jr. 1971. Limb posture and locomotion in the Virginia opossum (*Didelphis marsupialis*) and in other noncursorial mammals. J. Zool. 165: 303–315.

Jenkins, F. A., Jr., Crompton, A. W., and Downs, W. R. 1983. Mesozoic mammals from Arizona: New evidence on mammalian evolution. Science 222: 1233–1235.

Jenny, J., Jenny-Deshusses, C., Le Marrec, A., and Taquet, P. 1980. Découverte d'ossements de Dinosauriens dans le Jurassique inférieur (Toarcien) du Haut Atlas central (Maroc). C. R. Acad. Sci. Paris 290: 839–842.

Jenny, J., and Jossen, J.-A. 1982. Découverte d'empreintes de pas de dinosauriens dans le Jurassique inférieur (Pliensbachian) du Haut-Atlas central (Maroc). C. R. Acad. Sci. Paris 294: 223–226.

Jensen, J. A. 1966. Dinosaur eggs from the Upper Cretaceous North Horn Formation of central Utah. Brigham Young Univ. Geol. Stud. 13: 55–67.

————. 1970. Fossil eggs in the Lower Cretaceous of Utah. Brigham Young Univ. Geol. Stud. 17: 51–65.

————. 1985a. Uncompahgre dinosaur fauna: A preliminary report. Great Basin Nat. 45: 710–720.

————. 1985b. Three new sauropod dinosaurs from the Upper Jurassic of Colorado. Great Basin Nat. 45: 697–709.

Jerison, H. J. 1973. Evolution of the Brain and Intelligence. Academic Press, New York. 482 pp.

Johnson, J. H. 1931. The paleontology of the Denver Quadrangle, Colorado. Proc. Colorado Sci. Soc. 12: 355–378.

Johnston, C. 1859. Note on Odontography. Am. J. Dent. Sci. 9: 337–343.

Jordan, R. R. 1983. Stratigraphic nomenclature of non-marine Cretaceous rocks of inner margin of coastal plain in Delaware and adjacent states. Delaware Geol. Surv. Rep. Invest. 37: 1–43.

Julesz, B. 1971. Foundations of Cyclopean Perception. Univ. Chicago Press, Chicago. 406 pp.

Jurcśak, T. 1982. Occurrences nouvelles des Sauriens mesozoiques de Roumanie. Vert. Hungarica 21: 175–184.

Jurcśak, T., and Kessler, E. 1985. La paléofaune de Cornet—implications phylogénétiques et écologiques. Evolution et Adaptation II: 137–147.

Jurcśak, T., and Popa, E. 1979. Dinozaurieni ornitopozi din Bauxitele de la Cornet (Munţii Padurea Craiului). Nymphaea 7: 37–75.

————. 1983. La faune de dinosauriens du Bihor (Roumanie). In: Buffetaut, E., Mazin, J. M., and Salmon, E. (eds.). Actes Symp. Paléontol. Georges Cuvier. Montbéliard, France. Pp. 325–335.

Kaever, M., and Lapparent, A. F. de. 1974. Les traces de pas de dinosaures du Jurassique de Barkhausen (Basse Saxe, Allemagne). Bull. Soc. Géol. France 16: 516–525.

Kalandadze, N. N., and Kurzanov, S. M. 1974. [The Lower Cretaceous localities of terrestrial vertebrates in Mongolia]. Sovm. Sov.-Mong. Paleontol. Eksped. Trudy 1: 288–295. (In Russian with English summary)

Katz, J. L. 1971. Hard tissue as a composite material. 1. Bonds on the elastic behavior. J. Biomech. 4: 455–473.

————. 1981. Composite material models for cortical bone. In: Cowin, S. C. (ed.). Mechanical Properties of Bone. Am. Soc. Mech. Eng., New York. Pp. 171–184.

Kauffman, E. G. 1984. The fabric of Cretaceous marine extinctions. In: Berggren, W. A., and Van Couvering,

J. A. (eds.). Catastrophes and Earth History. Princeton Univ. Press, Princeton. Pp. 151–246.

Kaye, J. M., and Russell, D. A. 1973. The oldest record of hadrosaurian dinosaurs in North America. J. Paleontol. 47: 91–93.

Keefer, W. R. 1965. Stratigraphy and geologic history of the uppermost Cretaceous, Paleocene, and Lower Eocene rocks in the Wind River Basin, Wyoming. U.S. Geol. Surv. Prof. Pap. 495A: 1–77.

Keefer, W. R., and Troyer, M. L. 1956. Stratigraphy of the Upper Cretaceous and lower Tertiary rocks of the Shotgun Butte area, Fremont County, Wyoming. U.S. Geol. Surv. Oil Gas Invest. Chart OC-56.

Kellogg, R. 1929. The habits and economic importance of alligators. U.S. Dept. Agr. Tech. Bull. 147: 1–36.

Kemp, T. S. 1988. Haemothermia or Archosauria?: The interrelationships of mammals, birds, and crocodiles. Zool. J. Linn. Soc. London 92: 67–104.

Kennedy, W. J., Klinger, H. C., and Mateer, N. J. 1987. First record of an Upper Cretaceous sauropod dinosaur from Zululand, South Africa. S. Afr. J. Sci. 83: 173–174.

Kermack, D. 1984. New prosauropod material from South Wales. Zool. J. Linn. Soc. London 82: 101–117.

Kermack, K. A. 1951. A note on the habits of sauropods. Ann. Mag. Nat. Hist. (ser. 12) 4: 830–832.

Kérourio, P. 1981a. Nouvelles observations sur le mode de nidification et de ponte chez les dinosauriens du Crétacé terminal du Midi de la France. C. R. Soc. Géol. France 1981: 25–28.

———. 1981b. La distribution des "coquilles d'oeufs de dinosauriens multistratifiées" dans le Maestrichtien continental du sud de la France. Géobios 14: 533–536.

———. 1982. Un nouveau type de coquille d'oeuf présumé dinosaurien dans le Campanien et le Maestrichtien continental de Provence. Palaeovertebrata 12: 141–147.

Kielan-Jaworowska, Z., and Barsbold, R. 1972. Narrative of the Polish-Mongolian palaeontological expeditions 1967–1971. Palaeontol. Polonica 27: 5–13.

Kim, H. M. 1983. Cretaceous dinosaurs from Korea. J. Geol. Soc. Korea 19: 115–126.

———. 1986. New Early Cretaceous dinosaur tracks from Republic of Korea. In: Gillette, D. D. (ed.). 1st Internatl. Symp. Dinosaur Tracks Traces. Albuquerque, N. Mex. P. 17.

Kingham, R. F. 1962. Studies of the sauropod dinosaur Astrodon Leidy. Proc. Washington Jr. Acad. Sci. 1: 38–44.

Kiprijanow, W. 1883. Studien über die fossilen Reptilien Russlands. Th. 4. Ordnung Crocodilia Oppel. Indeterminirte fossile Reptilien. Acad. Sci. St. Petersb. Mem. (ser. 7) 31: 1–29.

Kitchell, J. A., and Estabrook, G. 1986. Was there 26 Myr periodicity in extinctions? Nature 321: 534–535.

Kitching, J. W. 1979. Preliminary report on a clutch of six dinosaurian eggs from the Upper Triassic Elliot Formation, northern Orange Free State. Palaeontol. afr. 22: 41–45.

———. 1981. Preliminary report on a clutch of six dinosaurian eggs from the Upper Triassic Elliot Formation, northern Orange Free State. Palaeont. afr. 24: 21.

Kitching, J. W., and Raath, M. A. 1984. Fossils from the Elliot and Clarens Formations (Karoo Sequence) of the Northeastern Cape, Orange Free State and Lesotho, and a suggested biozonation based on tetrapods. Palaeontol. afr. 25: 111–125.

Kiteley, L. W. 1983. Paleogeography and eustatic-tectonic model of late Campanian Cretaceous sedimentation, southwestern Wyoming and northwestern Colorado. In: Reynolds, M. W., and Dolly, E. D. (eds.). Mesozoic Paleogeography of west-central United States. Rocky Mountain Section, SEPM, Denver. Pp. 273–303.

Klein, G. deV. 1962. Triassic sedimentation, Maritime provinces of Canada. Bull. Geol. Soc. Am. 73: 1127–1146.

Koken, E. 1887. Die Dinosaurier, Crocodiliden und Sauropterygier des norddeutschen Wealden. Geol. Palaeont. Abhandl. 3: 311–420.

———. 1900. [Review of E. Fraas: Triassaurier]. N. Jb. Mineral. Geol. Palaeontol. 1900: 303.

Konishi, M., and Pettigrew, J. D. 1981. Some observations on the visual system of the oilbird (Steatornis caripensis). Natl. Geogr. Res. Rept. 1975: 439–449.

Kool, R. 1981. The walking speed of dinosaurs from the Peace River Canyon, British Columbia, Canada. Can. J. Earth Sci. 18: 823–825.

Kordos L. 1983. Fontosabb szorvanyleletek a mafi gerincesgyujtemenyeben (8. kozlemeny). M. All. Foldtani Intezet Evi Jelentese Az 1981: 503–511.

Koster, E. H. 1984. Sedimentology of a foreland coastal plain: Upper Cretaceous Judith River Formation at Dinosaur Provincial Park. Can. Soc. Petrol. Geol. 1984 Field Trip Gdbk. 115 pp.

Koster, E. H., and Currie, P. J. 1986. Sedimentological background. In: Naylor, B. G. (ed.). Dinosaur Systematics Symp. Field Trip Gdbk. Dinosaur Provincial Park. Tyrrell Mus. Palaeont. Drumheller, Alberta. Pp. 6–16.

Kotetishvilii, E. V. 1986. [Zonal stratigraphy of the Lower Cretaceous in Georgia and paleozoography of the Lower Cretaceous basins in the Mediterranean Province]. Trans. Geol. Inst. U.S.S.R. Acad. Sci. 91: 1–160. (In Russian)

Kramarenko, N. N. 1974. [On the works of the Joint Soviet-Mongolian Paleontological Expedition in 1969–1972]. Sovm. Sov.-Mong. Paleontol. Eksped. Trudy 1: 9–18. (In Russian with English summary)

Krassilov, V. A. 1981. Changes of Mesozoic vegetation and the extinction of the dinosaurs. Palaeogeogr. Palaeoclimatol. Palaeoecol. 34: 207–224.

———. 1985. [Cretaceous Period]. In: [Evolution of the Earth's Crust and Biosphere]. Nauka Press, Moscow. 240 pp. (In Russian)

———. 1987. Palaeobotany of the Mesophyticum: State of the art. Rev. Palaeobot. Palynol. 50: 221–234.

Kräusel, R. 1922. Die Nahrung von *Trachodon*. Palaeontol. Z. 4: 80.

Krebs, B. 1963. Bau und Funktion des Tarsus eines Pseudo-suchiers aus der Trias des Monte San Giorgio (Kanton Tessin, Schweiz). Palaeontol. Z. 37: 88–95.

———. 1976. Pseudosuchia. Hdb. Paläoherpetol. 13: 40–98.

Kues, B. S., Lehman, T. H., and Rigby, J. K., Jr. 1980. The teeth of *Alamosaurus sanjuanensis,* a Late Cretaceous sauropod. J. Paleontol. 54: 864–869.

Kuhn, O. 1938. Lebensbild des Wirbeltiervorkommens im Keuper von Ebrach. Palaeontol. Z. 19: 315–321.

———. 1939. Beiträge zur Keuperfauna von Halberstadt. Palaeontol. Z. 2: 258–286.

———. 1958*a*. Zwei neue Arten von *Coelurosaurischnus* aus dem Keuper Frankens. N. Jb. Geol. Palaeontol. Mh. 1958: 437–440.

———. 1958*b*. Die Fährten der vorzeitlichen Amphibien und Reptilien. Verlagshaus Meisenbach, Bamberg. 64 pp.

———. 1959. Ein neuer Microsaurier aus dem deutschen Rotliegenden. N. Jb. Geol. Palaeontol. Mh. 1959: 424–426.

———. 1961. Die Familien der rezenten und fossilen Amphibien und Reptilien. Meisenbach, Bamberg. 79 pp.

Kurtén, B. 1968. Pleistocene Mammals of Europe. Aldine, Chicago. 317 pp.

Kurzanov, S. M. 1972. [Sexual dimorphism in protoceratopsids]. Paleontol. Zh. 1972: 91–97. (In Russian)

———. 1976*a*. [Braincase structure of a carnosaur *Itemirus* gen. nov. and some problems of the cranial anatomy of dinosaurs]. Paleontol. Zh. 1976: 127–137. (In Russian)

———. 1976*b*. [A new Late Cretaceous carnosaur from Nogon-Tsav, Mongolia]. Sovm. Sov.-Mong. Paleontol. Eksped. Trudy 3: 93–104. (In Russian with English summary)

———. 1981. [On the unusual theropods from the Upper Cretaceous of Mongolia]. Sovm. Sov.-Mong. Paleontol. Eksped. Trudy 15: 39–50. (In Russian with English summary)

———. 1982. [Structural characteristics of the fore limbs of *Avimimus* (Avimimidae)]. Paleontol. Zh. 1982: 108–112. (In Russian)

———. 1983*a*. [*Avimimus* and the problem of the origin of birds]. Fossil Reptiles of Mongolia. Sovm. Sov.-Mong. Paleontol. Eksped. Trudy 24: 104–109. (In Russian with English summary)

———. 1983*b*. [New data on the pelvic structure in *Avimimus*]. Paleontol. Zh. 1983: 115–116. (In Russian)

———. 1987. [Avimimids and the problem of the origin of birds]. Sovm. Sov.-Mong. Paleontol. Eksped. Trudy 31: 5–95. (In Russian)

Kurzanov, S. M., and Bannikov, A. F. 1983. [A new sauropod from the Upper Cretaceous of Mongolia]. Paleontol. Zh. 1983: 91–97. (In Russian)

Kurzanov, S. M., and Tumanova, T. A. 1978. [On the structure of the endocranium in some ankylosaurs from Mongolia]. Paleontol. Zh. 1978: 90–96. (In Russian)

Kutty, T. S. 1969. Some contributions to the stratigraphy of the Upper Gondwana formations of the Pranhita-Godavari Valley, central India. J. Geol. Soc. India 10: 33–48.

Kutty, T. S., and Roy-Chowdhury, T. 1970. The Gondwana sequence of the Pranhita-Godavari Valley, India, and its vertebrate faunas. 2d Gondwana Symp. Proc. Pap. Pp. 303–308.

Laemmlen, M. 1956. Keuper. Lex. Stratigr., Internatl. I, Eur. 5: 1–335.

Lambe, L. M. 1902. New genera and species from the Belly River Series (mid-Cretaceous). Contrib. Canadian Palaeontol. Geol. Surv. Can. 3: 25–81.

———. 1904*a*. On the squamoso-parietal crest of two species of horned dinosaurs from the Cretaceous of Alberta. Ottawa Nat. 18: 81–84.

———. 1904*b*. On *Dryptosaurus incrassatus* (Cope) from the Edmonton Series of the Northwest Territory. Contrib. Can. Palaeontol. 3: 1–27.

———. 1910. Note on the parietal crest of *Centrosaurus apertus,* and a proposed new generic name for *Stereocephalus tutus*. Ottawa Nat. 14: 149–151.

———. 1913. A new genus and species from the Belly River Formation of Alberta. Ottawa Nat. 27: 109–116.

———. 1914*a*. On *Gryposaurus notabilis,* a new genus and species of trachodont dinosaur from the Belly River Formation of Alberta, with a description of the skull of *Chasmosaurus belli*. Ottawa Nat. 27: 145–155.

———. 1914*b*. On the fore-limb of a carnivorous dinosaur from the Belly River Formation of Alberta, and a new genus of Ceratopsia from the same horizon, with remarks on the integument of some Cretaceous herbivorous dinosaurs. Ottawa Nat. 27: 129–135.

———. 1914*c*. On a new genus and species of carnivorous dinosaur from the Belly River Formation of Alberta with a description of the skull of *Stephanosaurus marginatus* from the same horizon. Ottawa Nat. 28: 13–20.

———. 1915. On *Eoceratops canadensis,* gen. nov., with remarks on other genera of Cretaceous horned dinosaurs. Geol. Surv. Can. Mus. Bull. 12: 1–49.

———. 1917*a*. On *Cheneosaurus tolmanensis,* a new genus and species of trachodont dinosaur from the Edmonton Cretaceous of Alberta. Ottawa Nat. 30: 117–123.

———. 1917*b*. A new genus and species of crestless hadrosaur from the Edmonton Formation of Alberta. Ottawa Nat. 31: 65–73.

———. 1917*c*. The Cretaceous theropodous dinosaur *Gorgosaurus*. Mem. Geol. Surv. Can. 100: 1–84.

———. 1918*a*. On the genus *Trachodon* of Leidy. Ottawa Nat. 31: 135–139.

———. 1918*b*. The Cretaceous genus *Stegoceras* typifying a new family referred provisionally to the Stegosauria. Trans. Roy. Soc. Can. (ser. 3) 12: 23–36.

———. 1919. Description of a new genus and species (*Panoplosaurus mirus*) of armored dinosaur from the Belly River Beds of Alberta. Trans. Roy. Soc. Can. (ser. 3) 13: 39–50.

———. 1920. The hadrosaur *Edmontosaurus* from the Upper Cretaceous of Alberta. Mem. Can. Geol. Surv. 120: 1–79.

Lambert, D. 1983. A Field Guide to Dinosaurs. Avon Books, New York. 256 pp.

Lang, J. W., Whitaker, R., and Andrews, H. 1986. Male parental care in Mugger crocodiles. Natl. Geogr. Res. 2: 519–525.

Langston, W., Jr. 1959. *Anchiceratops* from the Oldman Formation of Alberta. Natl. Mus. Can. Nat. Hist. Pap. 3: 1–11.

———. 1960a. The vertebrate fauna of the Selma Formation of Alabama. Pt. VI. The Dinosaurs. Fieldiana Geol. Mem. 3: 313–361.

———. 1960b. A hadrosaurian ichnite. Natl. Mus. Can. Nat. Hist. Pap. 4: 1–9.

———. 1965. Fossil crocodilians from Columbia and the Cenozoic history of the Crocodilia in South America. Univ. Calif. Publ. Geol. Sci. 52: 1–157.

———. 1967. The thick-headed ceratopsian dinosaur *Pachyrhinosaurus* (Reptilia: Ornithischia) from the Edmonton Formation near Drumheller, Canada. Can. J. Earth Sci. 4: 171–186.

———. 1974. Nonmammalian Comanchean tetrapods. Geosci. Man 8: 77–102.

———. 1975. The ceratopsian dinosaurs and associated lower vertebrates from the St. Mary River Formation (Maestrichtian) at Scabby Butte, southern Alberta. Can. J. Earth Sci. 12: 1576–1608.

———. 1976. A Late Cretaceous vertebrate fauna from the St. Mary River Formation in western Canada. In: Churcher, C. S. (ed.). Athlon. Roy. Ont. Mus., Toronto. Pp. 114–133.

———. 1983. Lower Cretaceous dinosaur tracks near Glen Rose, Texas. In: Perkins, B. F., and Langston, W. (eds.). Field Trip Guide. Lower Cretaceous Shallow Marine Environments in the Glen Rose Formation: Dinosaur Tracks and Plants. Am. Assoc. Strat. Palynol. Pp. 39–61.

Langston, W., Jr., and Durham, J. W. 1955. A sauropod dinosaur from Colombia. J. Paleontol. 29: 1047–1051.

Lapparent, A. F. de. 1943. Les dinosauriens jurassiques de Damparis (Jura). Mém. Soc. Géol. France 47: 5–20.

———. 1946. Présence d'un dinosaurien sauropode dans l'Albien du Pays de Bray. Ann. Soc. Géol. Nord 66: 236–242.

———. 1947. Les dinosauriens du Crétacé supérieur du Midi de la France. Mém. Soc. Géol. France 56: 1–54.

———. 1955. Étude paléontologique des vertébrés du Jurassique d'El Mers (Moyen Atlas). Not. Mém. Serv. Géol. Maroc 124: 1–36.

———. 1958. Sur les Dinosauriens du "Continental intercalaire" du Sahara central. C. R. Acad. Sci. Paris 246: 1237–1240.

———. 1959. Describumiento de un yacimiento de huevos de dinosaurios en el Cretaceo superior de la depresion de Tremp (Provincia de Lerida). Not. Com. Inst. Geol. Min. Espan. 54: 51–53.

———. 1960a. Les dinosauriens du "Continental intercalaire" du Sahara central. Mém. Soc. Géol. France 88A: 1–57.

———. 1960b. Los dos dinosaurios de Galve (provincia de Teruel, España). Teruel 24: 177–197.

———. 1962. Footprints of dinosaur in the Lower Cretaceous of Vestspitsbergen-Svalbard. Arbok. Norsk. Polarinst. 1960: 14–21.

———. 1967. Les dinosaures de France. Sciences 51: 4–19.

Lapparent, A. F. de, and Aguirre, E. 1956. Algunos yacimientos de dinosaurios en el Cretácico superior de la Cuenca de Tremp. Estud. Geol. 12: 31–32.

Lapparent, A. F. de, and Davoudzadeh, M. 1972. Jurassic dinosaur footprints of the Kerman area, central Iran. Geol. Surv. Iran Rept. 26: 5–22.

Lapparent, A. F. de, Joncour, M. le, Mathieu, A., and Plus, B. 1965. Découverte en Espagne d'empreintes de pas de Reptiles mésozoïques. Bol. Soc. Espan. Hist. Nat. Sec. Geol. 63: 225–230.

Lapparent, A. F. de, and Lavocat, R. 1955. Dinosauriens. In: Piveteau, J. (ed.). Traîté de Paléontologie, Vol. 5: 785–962.

Lapparent, A. F. de, and Lucas, G. 1957. Vertèbres de dinosaurien sauropode dans le Callovien moyen de Rhar Rouban (frontière algéro-marocaine du Nord). Bull. Soc. Hist. Nat. Afr. Nord 48: 234–236.

Lapparent, A. F. de, and Montenat, C. 1967. Les empreintes de pas de reptiles de l'Infralias du Veillon (Vendée). Mém. Soc. Géol. France 107: 1–41.

Lapparent, A. F. de, and Oulmi, M. 1964. Une empreinte de pas de dinosaurien dans le Portlandien de Chassiron (île d'Oleron). C. R. Soc. Géol. France 1964: 232–233.

Lapparent, A. F. de, and Sadat, M. A. A. N. 1975. Une trace de pas de dinosaure dans le Lias de l'Elbourz, en Iran. Conséquences de cette découverte. C. R. Acad. Sci. Paris 280: 161–163.

Lapparent, A. F. de, and Stchepinsky, V. 1968. Les Iguanodons de la région de Saint-Dizier (Haute-Marne). C. R. Acad. Sci. Paris 266: 1370–1372.

Lapparent, A. F. de, and Stocklin, J. 1972. Sur le Jurassique et le Crétacé du Band-e-Turkestan (Afghanistan du Nord-Ouest). C. R. Soc. Géol. France 1971: 387–388.

Lapparent, A. F. de, and Zbyszewski, G. 1957. Les Dinosauriens du Portugal. Mem. Serv. Geol. Portugal 2: 1–63.

Lapparent, A. F. de, Zbyszewski, G., Moitinho, de Almeida, F., and Veiga Ferreira, O. da. 1951. Empreintes de pas de dinosauriens dans le Jurassique du Cap Mondego (Portugal). C. R. Soc. Géol. France 1951: 251–252.

Larkin, P. 1910. The occurrence of a sauropod dinosaur in the Trinity Cretaceous of Oklahoma. J. Geol. 18: 93–98.

Larsonneur, C., and Lapparent, A. F. de. 1966. Un dinosaurien carnivore, *Halticosaurus,* dans le Rhétien d'Airel (Manche). Bull. Soc. Linn. Normandie 7: 108–117.

Laugier, R. 1971. Le Lias inférieur et moyen du Nord-Est de la France. Sci. Terre Mém. 21: 1–300.

Lavigne, D. M. 1982. Similarity in energy budgets of animal populations. J. Anim. Ecol. 51: 195–206.

Lavocat, R. 1954. Sur les Dinosauriens du continental intercalaire des Kem-Kem de la Daoura. C. R. 19th Internatl. Geol. Congr. 1952: 65–68.

———. 1955a. Sur une portion de mandibule de théropode provenant du Crétáce supérieur de Madagascar. Bull. Mus. Natl. Hist. Nat. (sér. 2) 27: 256–259.

———. 1955b. Sur un membre antérieur du dinosaurien sauropode "*Bothriospondylus*" recueilli à Madagascar. C. R. Acad. Sci. Paris 240: 1795–1796.

———. 1955c. Les recherches de reptiles fossiles à Madagascar. Nat. Malgache 7: 203–207.

Laws. R. M. 1981. Experiences in the study of large mammals. In: Fowler, C. W., and Smith, T. D. (eds.). Dynamics of Large Mammal Populations. Wiley, New York. Pp. 19–45.

Lawson, D. A. 1976. *Tyrannosaurus* and *Torosaurus,* Maestrichtian dinosaurs from Trans-Pecos Texas. J. Paleontol. 50: 158–164.

Lawton, J. H. 1981. Moose, wolves, *Daphnia,* and *Hydra:* On the ecological efficiency of endotherms and ectotherms. Am. Nat. 117: 782–783.

Lee, W. T. 1912. Stratigraphy of the coal fields of northern central New Mexico. Bull. Geol. Soc. Am. 23: 471–686.

Lefeld, J. 1971. Geology of the Djadokhta Formation at Bayn Dzak (Mongolia). Palaeontol. Polonica 25: 101–127.

Lehman, T. M. 1982. A ceratopsian bonebed from the Aguja Formation (Upper Cretaceous), Big Bend National Park, Texas. M.A. thesis, Univ. Texas, Austin. 209 pp.

———. 1985a. Transgressive-regressive cycles and environments of coal deposition in Upper Cretaceous strata of Trans-Pecos Texas. Trans. Gulf Coast Assoc. Geol. Soc. 35: 431–438.

———. 1985b. Stratigraphy, sedimentology, and paleontology of Upper Cretaceous (Campanian-Maastrichtian) sedimentary rocks in Trans-Pecos Texas. Ph.D. dissertation, Univ. Texas, Austin. 299 pp.

———. 1987. Late Maastrichtian paleoenvironments and dinosaur biogeography in the Western Interior of North America. Palaeogeogr. Palaeoclimatol. Palaeoecol. 60: 189–217.

———. 1989. *Chasmosaurus mariscalensis,* n. sp., a new ceratopsian dinosaur from Texas. J. Vert. Paleontol. 9: 137–162.

———. 1990. The ceratopsian subfamily Chasmosaurinae: sexual dimorphism and systematics. In: Currie, P. J., and Carpenter, K. (eds.). Dinosaur Systematics: Perspectives and Approaches. Cambridge Univ. Press, New York. Pp. 211–229.

Leidy, J. 1856. Notice of remains of extinct reptiles and fishes, discovered by Dr. F. V. Hayden in the badlands of the Judith River, Nebraska Territory. Proc. Acad. Nat. Sci. Philadelphia 8: 72–73.

———. 1858. *Hadrosaurus foulkii,* a new saurian from the Cretaceous of New Jersey. Proc. Acad. Nat. Sci. Philadelphia 10: 215–218.

———. 1865. Memoir on the extinct reptiles of the Cretaceous formations of the United States. Smithsonian Contrib. Knowledge 14: 1–135.

———. 1868a. [Remarks on *Conosaurus* of Gibbs]. Proc. Acad. Nat. Sci. Philadelphia. 20: 200–202.

———. 1868b. [Remarks on a jaw fragment of *Megalosaurus*]. Proc. Acad. Nat. Sci. Philadelphia 1870: 197–200.

———. 1870. [Remarks on *Poicilopleuron valens, Clidastes intermedius, Leiodon proriger, Baptemys wyomingensis,* and *Emys stevensonianus*]. Proc. Acad. Nat. Sci. Philadelphia 1870: 3–5.

Lennier, G. 1870. Etude géologique et paléontologique sur l'embouchure de la Seine et les falaise de la Haute-Normandie. Le Havre. 254 pp.

Leonardi, G. 1977. On a new occurrence of tetrapod trackways in the Botucatu Formation in the State of São Paulo, Brazil. Dusenia, Curitiba 10: 181–183.

———. 1979a. Nota preliminar sobre seis pistas de dinosauros Ornithischia da Bacia do Rio do Peixe, em Souse, Paraiba, Brasil. An. Acad. Brasil. Cienc. 51: 501–516.

———. 1979b. New archosaurian trackways from the Tio do Peixe Basin, Paraiba, Brazil. Ann. Univ. Ferrara (Sezione IX) 5: 239–246.

———. 1980a. On the discovery of an abundant ichnofauna (vertebrates and invertebrates) in the Botucatu Formation *s. s.* in Araraguara, São Paulo, Brasil. An. Acad. Brasil. Cienc. 52: 559–567.

———. 1980b. Ornithischian trackways of the Corda Formation (Jurassic), Goias, Brazil. Actas II Congr. Argent. Paleontol. Bioestrat. I Congr. Latinoam. Paleontol. Pp. 215–222.

———. 1980c. Dez novas pistas de dinosauros (Theropoda Marsh, 1881) na bacia do Rio do Peixe, Paraiba, Brasil. Actas II Congr. Argent. Paleontol. Bioestrat. I Congr. Latinoam. Paleontol. Pp. 243–248.

———. 1981a. Ichnological data on the rarity of young in northeast Brazil dinosaurian populations. An. Acad. Brasil. Cienc. 53: 345–346.

———. 1981b. As localidades com rastos fosseis de tetraopodes na America Latina. An. II Congr. Latinoam. Paleontol. Pp. 929–940.

———. 1984. Le impronte fossili di dinosauri. In: Bonaparte, J. F., Colbert, E. H., Currie, P. J., Ricqlès, A. de., Kielan-Jaworowska, Z., Leonardi, G., Morello, N., and Taquet, P. Sulle Orme dei Dinosauri. Erizzo, Venice. C.N.P.Q. Brasile. Pp. 165–186.

———. 1989. Inventory and statistics of the South American dinosaurian ichnofauna and its paleobiological interpretation. In: Gillette, D. D., and Lockley, M. G.

(eds.). Dinosaur Tracks and Traces. Cambridge Univ. Press, New York. Pp. 165–178.

Leonardi, G., and Godoy, L. C. 1980. Novas pistas de Tetrapodes de Formacao Botucatu no Estado de Sao Paulo. An. XXXI Congr. Brasil. Geol. 5: 3080–3089.

Lillegraven, J. A., and McKenna, M. C. 1986. Fossil mammals from the "Mesaverde" Formation (Late Cretaceous, Judithian) of the Bighorn and Wind River basins, Wyoming, with definitions of Late Cretaceous North American land-mammal "ages." Am. Mus. Novitates 2840: 1–68.

Lipps, J. H. 1986. Extinction dynamics in pelagic ecosystems. In: Elliott, D. K. (ed.). Dynamics of Extinction. Wiley Interscience, New York. Pp. 87–106.

Llompart, C., Casanovas, M. L., and Santafe, J. V. 1984. Un nuevo yacimiento de icnitas de dinosaurios en las facies garumniensis de la Conca de Tremp (Lleida, España). Acta Geol. Hispanica 19: 143–147.

Llompart, C., and Krauss, S. 1982. Restos de moluscos y dinosaurios en formaciones estromatoliticas garumnienses al S del Montsee (Prov. de Lerida). Bol. Geol. Min. 93: 371–378.

Lockley, M. G. 1985. Vanishing tracks along Alameda Parkway. In: Chamberlain, C. K. (ed.). Environments of Deposition (and Trace Fossils) of Cretaceous Sandstones of the Western Interior. Soc. Econ. Paleontol. Min. Field Guide 2d Ann. Meeting, Golden, Colorado. Pt. 3: 131–142.

———. 1986a. A guide to dinosaur tracksites of the Colorado Plateau and American southwest. Univ. Colorado Denver Geol. Dept. Mag. Sp. Issue 1: 1–56.

———. 1986b. The paleobiological and paleoenvironmental importance of dinosaur footprints. Palaios 1: 37–47.

———. 1987. Dinosaur trackways and their importance in paleontological reconstruction. In: Czerkas, S. J., and Olson, E. C. (eds.). Dinosaurs Past and Present. Nat. Hist. Mus. Los Angeles County. Vol. I: 80–95.

Lockley, M. G., Houck, K. J., and Prince, N. K. 1986. North America's largest dinosaur trackway site: Implications for Morrison Formation paleoecology. Bull. Geol. Soc. Am. 97: 1163–1176.

Lockley, M. G., Young, B. H., and Carpenter, K. 1983. Hadrosaur locomotion and herding behavior: Evidence from footprints in the Mesaverde Formation, Grand Mesa Coal Field, Colorado. Mountain Geol. 2: 5–14.

Lohrengel, C. F. 1969. Palynology of the Kaiparowits Formation, Garfield County, Utah. Brigham Young Univ. Geol. Stud. 16: 61–180.

Longman, H. A. 1926. A giant dinosaur from Durham Downs, Queensland. Mem. Queensland Mus. 8: 183–194.

———. 1927. The giant dinosaur Rhoetosaurus brownei. Mem. Queensland Mus. 9: 1–18.

———. 1933. A new dinosaur from the Queensland Cretaceous. Mem. Queensland Mus. 10: 131–144.

Løvtrop, S. 1985. On the classification of the taxon Tetrapoda. Syst. Zool. 34: 463–470.

Lozinsky, R. P., Hunt, A. P., Wolberg, D. L., and Lucas, S. G. 1984. Late Cretaceous (Lancian) dinosaurs from the McRae Formation, Sierra County, New Mexico. New Mexico Geol. 6: 72–77.

Lucas, F. A. 1901. A new dinosaur, Stegosaurus marshi, from the Lower Cretaceous of South Dakota. Proc. U.S. Natl. Mus. 23: 591–592.

———. 1902a. A new generic name for Stegosaurus marshi. Science 16: 435.

———. 1902b. Paleontological notes: The generic name Omosaurus. Science 19: 435.

Lucas, S. G. 1981. Dinosaur communities of the San Juan Basin: A case for lateral variations in the composition of Late Cretaceous dinosaur communities. In: Lucas, S. G., Rigby, J. K., Jr., and Kues, B. S. (eds.). Advances in San Juan Basin Paleontology. Univ. New Mexico Press, Albuquerque. Pp. 337–393.

Lucas, S. G., Mateer, N. J., Hunt, A. P., and O'Neill, F. M. 1987. Dinosaurs, the age of the Fruitland and Kirtland Formations, and the Cretaceous-Tertiary boundary in the San Juan Basin, New Mexico. Geol. Soc. Am. Spec. Pap. 209: 35–50.

Lull, R. S. 1904. Fossil footprints of the Jura-Trias of North America. Mem. Boston Soc. Nat. Hist. 5: 461–557.

———. 1905. Restoration of the horned dinosaur Diceratops. Am. J. Sci. (ser. 4)20: 420–422.

———. 1910a. The armor of Stegosaurus. Am. J. Sci. (ser. 4) 29: 201–210.

———. 1910b. Stegosaurus ungulatus Marsh, recently mounted at the Peabody Museum of Yale University. Am. J. Sci. (ser. 4)30: 361–376.

———. 1911a. The reptilian fauna of the Arundel Formation. Lower Cretaceous Vol. Maryland Geol. Surv. Pp. 173–178.

———. 1911b. Systematic paleontology of the Lower Cretaceous deposits of Maryland: Vertebrata. Lower Cretaceous Vol. Maryland Geol. Surv. Pp. 183–211.

———. 1915a. Triassic life of the Connecticut Valley. Conn. State Geol. Nat. Hist. Surv. Bull. 24: 1–285.

———. 1915b. The mammals and horned dinosaurs of the Lance formation of Niobrara County, Wyoming. Am. J. Sci. (ser. 4)40: 319–348.

———. 1917. On the functions of the "sacral brain" in dinosaurs. Am. J. Sci. (ser. 4)44: 471–477.

———. 1919. The sauropod dinosaur Barosaurus Marsh. Mem. Conn. Acad. Arts Sci. 6: 1–42.

———. 1921. The Cretaceous armored dinosaur, Nodosaurus textilis Marsh. Am. J. Sci. (ser. 5)1: 97–127.

———. 1925. Organic Evolution. Macmillan, New York. 729 pp.

———. 1933. A revision of the Ceratopsia or horned dinosaurs. Peabody Mus. Nat. Hist. Bull. 3: 1–175.

———. 1942. Triassic footprints from Argentina. Am. J. Sci. 240: 421–425.

———. 1953. Triassic life of the Connecticut Valley. Rev. ed. Bull. Conn. State Geol. Nat. Hist. Surv. 81: 1–331.

Lull, R. S., and Wright, N. E. 1942. Hadrosaurian dinosaurs of North America. Geol. Soc. Am. Spec. Pap. 40: 1–242.

Lupton, C., Gabriel, D., and West, R. M. 1980. Paleobiology and depositional setting of a Late Cretaceous vertebrate locality, Hell Creek Formation, McCone County, Montana. Contrib. Geol. Univ. Wyoming 18: 117–126.

Lutcavage, M., and Lutz, P. L. 1986. Metabolic rate and food energy requirements of the leatherback sea turtle, *Dermochelys coriacea*. Copeia 1986: 796–798.

Lutz, T. M. 1985. The magnetic reversal record is not periodic. Nature 317: 404–407.

Lydekker, R. 1877. Notices of new and other Vertebrata from Indian Tertiary and Secondary Rocks. Rec. Geol. Surv. India 10: 30–43.

———. 1879. Indian pre-Tertiary Vertebrata. Fossil Reptilia and Batrachia. Palaeontol. Indica (ser. 4) 1: 1–36.

———. 1885. Indian pre-Tertiary Vertebrata. The Reptilia and Amphibia of the Maleri and Denwa groups. Palaeontol. Indica (ser. 4) 1: 1–38.

———. 1888a. Note on a new Wealden iguanodont and other dinosaurs. Q. J. Geol. Soc. London 44: 46–61.

———. 1888b. Catalogue of the fossil Reptilia and Amphibia in the British Museum. Pt. I. Containing the orders Ornithosauria, Crocodilia, Dinosauria, Squamata, Rhynchocephalia, and Proterosauria. Br. Mus. Nat. Hist. London 309 pp.

———. 1889a. *Orinosaurus capensis*. Geol. Mag. (ser. 3)6: 353.

———. 1889b. On the remains and affinities of five genera of Mesozoic reptiles. Q. J. Geol. Soc. London 45: 41–59.

———. 1889c. Catalogue of Fossil Reptilia and Amphibia. Pt. 4. Br. Mus. Nat. Hist. London. 295 pp.

———. 1889d. On a coelurid dinosaur from the Wealden. Geol. Mag. (ser. 3)6: 119–121.

———. 1889e. Notes on some points in nomenclature of fossil reptiles and amphibians, with preliminary notices of two new species. Geol. Mag. (ser. 3)6: 325–326.

———. 1890a. On the remains of small sauropodous dinosaurs from the Wealden. Q. J. Geol. Soc. London 46: 182–184.

———. 1890b. Note on certain vertebrate remains from the Nagpur District. Rec. Geol. Surv. India 23: 20–24.

———. 1890c. Contributions to our knowledge of the dinosaurs of the Wealden, and the Sauropterygia of the Purbeck and Oxford clay. Q. J. Geol. Soc. London 46: 36–53.

———. 1891. On certain ornithisaurian and dinosaurian remains. Q. J. Geol. Soc. London 47: 41–44.

———. 1893a. Contributions to the study of the fossil vertebrates of Argentina. I. The dinosaurs of Patagonia. An. Mus. La Plata Sec. Paleontol. 2: 1–14.

———. 1893b. On two dinosaurian teeth from Aylesbury. Q. J. Geol. Soc. London 49: 566–568.

———. 1893c. On the jaw of a new carnivorous dinosaur from the Oxford Clay of Peterborough. Q. J. Geol. Soc. London 49: 284–287.

———. 1895. On bones of a sauropodous dinosaur from Madagascar. Q. J. Geol. Soc. London 51: 329–336.

MacDonald, D. (ed.). 1984. The Encyclopedia of Mammals. Facts on File Publ., New York. 895 pp.

MacDonald, D. W. 1983. The ecology of carnivore social behavior. Nature 301: 379–384.

Madeira, J., and Dias, R. 1983. Novas Pistas de Dinossaurios no Cretacico Inferior. Comun. Serv. Geol. Portugal 69: 147–158.

Madsen, J. H., Jr. 1974. A new theropod dinosaur from the Upper Jurassic of Utah. J. Paleontol. 48: 27–31.

———. 1976a. *Allosaurus fragilis:* A revised osteology. Utah Geol. Min. Surv. Bull. 1091: 1–163.

———. 1976b. A second new theropod dinosaur from the Late Jurassic of East Central Utah. Utah Geol. 3: 51–60.

Madsen, J. H., Jr., and Stokes, W. L. 1963. New information on the Jurassic dinosaur *Ceratosaurus*. Geol. Soc. Am. Spec. Pap. 73: 90.

Maeda, N., and Fitch, W. M. 1981. Amino acid sequence of a myoglobin from the lace monitor lizard, *Varanus varius*, and its evolutionary implications. J. Biol. Chem. 256: 4301–4309.

Magnusen, W. E. 1980. Hatching and creche formation by *Crocodilus porosus*. Copeia 1980: 359–362.

Magnusen, W. E., Silva, E. V. da, and Lima, A. P. 1987. Diets of Amazonian crocodilians. J. Herpetol. 21: 85–95.

Maleev, E. A. 1952a. [A new family of armored dinosaurs from the Upper Cretaceous of Mongolia]. Doklady Akad. Nauk S.S.S.R. 87: 131–134. (In Russian)

———. 1952b. [A new ankylosaur from the Upper Cretaceous of Asia]. Doklady Akad. Nauk S.S.S.R. 87: 273–276. (In Russian)

———. 1954. [New turtle-like reptile in Mongolia]. Priroda 1954: 106–108. (In Russian)

———. 1955a. [New carnivorous dinosaurs from the Upper Cretaceous of Mongolia]. Doklady Akad. Nauk S.S.S.R. 104: 779–783. (In Russian)

———. 1955b. [Gigantic carnivorous dinosaurs of Mongolia]. Doklady Akad. Nauk S.S.S.R. 104: 634–637. (In Russian)

———. 1955c. [Carnivorous dinosaurs of Mongolia]. Priroda 1955: 112–115.

———. 1956. [Armoured dinosaurs from the Upper Cretaceous of Mongolia]. Trudy Paleontol. Inst. Akad. Nauk S.S.S.R. 62: 51–91. (In Russian)

———. 1965. [On the brain of carnivorous dinosaurs]. Paleontol. Zh. 1965: 141–143. (In Russian)

———. 1974. [Gigantic carnosaurs of the family Tyrannosauridae]. Sovm. Sov.-Mong. Paleontol. Eksped. Trudy 1: 132–191. (In Russian with English summary)

Malumian, N., Nullo, F. E., and Ramos, V. A. 1983. The Cretaceous of Argentina, Chile, Paraguay, and Uruguay. In: Moullade, M., and Nairn, A. E. M. (eds.). The Phanerozoic Geology of the World II. The Mesozoic, B. Elsevier, New York. Pp. 265–304.

Mantell, G. A. 1825. Notice on the *Iguanodon*, a newly discovered fossil reptile, from the sandstone of Tilgate Forest, in Sussex. Phil. Trans. Roy. Soc. London 115: 179–186.

———. 1833. Geology of the South East of England. London.

———. 1844. Medals of Creation; or first lessons in geology and in the study of organic remains. London.

———. 1850. On the *Pelorosaurus;* an undescribed gigantic terrestrial reptile. Phil. Trans. Roy. Soc. London 140: 379–390.

———. 1852. On the structure of the *Iguanodon* and on the fauna and flora of the Wealden Formation. Not. Proc. Roy. Inst. Gr. Br. 1: 141–146.

Marsh, O. C. 1870. [Remarks on *Hadrosaurus minor, Mosasaurus crassidens, Leiodon laticaudus, Baptosaurus,* and *Rhinoceros matutinus*]. Proc. Acad. Nat. Sci. Philadelphia 22: 2–3.

———. 1872. Notice on a new species of *Hadrosaurus*. Am. J. Sci. (ser. 3)3: 301.

———. 1877a. Notice of a new gigantic dinosaur. Am. J. Sci. (ser. 3)14: 87–88.

———. 1877b. Notice of some new dinosaurian reptiles from the Jurassic Formation. Am. J. Sci. (ser. 3)14: 514–516.

———. 1877c. New order of extinct Reptilia (Stegosauria) from the Jurassic of the Rocky Mountains. Am. J. Sci. (ser. 3)14: 513–514.

———. 1877d. Notice of some new vertebrate fossils. Am. J. Sci. (ser. 3)14: 249–256.

———. 1878a. Notice of new dinosaurian reptiles. Am. J. Sci. (ser. 3)15: 241–244.

———. 1878b. Principal characters of American Jurassic dinosaurs. Pt. I. Am. J. Sci. (ser. 3)16: 411–416.

———. 1879a. Principal characters of American Jurassic dinosaurs. Pt. II. Am. J. Sci. (ser. 3)17: 86–92.

———. 1879b. Additional characters of the Sauropoda. Am. J. Sci. (ser. 3)17: 181–182.

———. 1879c. Notice of new Jurassic reptiles. Am. J. Sci. (ser. 3)18: 501–505.

———. 1880. Principal characteristics of American Jurassic dinosaurs. Pt. III. Am. J. Sci. (ser. 3)19: 253–259.

———. 1881a. Principal characters of American Jurassic dinosaurs. Pt. IV. Spinal cord, pelvis, and limbs of *Stegosaurus*. Am. J. Sci. (ser. 3)21: 167–170.

———. 1881b. Principal characters of American Jurassic dinosaurs. Pt. V. Am. J. Sci. (ser. 3)21: 417–423.

———. 1881c. Classification of the Dinosauria. Am. J. Sci. (ser. 3)23: 81–86.

———. 1881d. A new order of extinct Jurassic reptiles (Coeluria). Am. J. Sci. (ser. 3)21: 339–340.

———. 1881e. The sternum in dinosaurian reptiles. Am. J. Sci. (ser. 3)19: 395–396.

———. 1883. Principal characters of American Jurassic dinosaurs. Pt. VI. Restoration of *Brontosaurus*. Am. J. Sci. (ser. 3)26: 81–85.

———. 1884a. Principal characters of American Jurassic dinosaurs. Pt. VIII. The Order Theropoda. Am. J. Sci. (ser. 3)27: 329–340.

———. 1884b. The classification and affinities of dinosaurian reptiles. Nature 31: 68–69.

———. 1884c. On the united metatarsal bones of ceratosaurs. Am. J. Sci. (ser. 3)28: 161–162.

———. 1884d. Principal characters of American Jurassic dinosaurs. Pt. VII. On the Diplodocidae, a new family of the Sauropoda. Am. J. Sci. (ser. 3)27: 161–167.

———. 1885. Names of extinct reptiles. Am. J. Sci. (ser. 3) 29: 169.

———. 1887a. Principal characters of American Jurassic dinosaurs. Pt. IX. The skull and dermal armor of *Stegosaurus*. Am. J. Sci. (ser. 3)34: 413–417.

———. 1887b. Notice of new fossil mammals. Am. J. Sci. (ser. 3)34: 323–331.

———. 1888a. Notice of a new genus of Sauropoda and other new dinosaurs from the Potomac Formation. Am. J. Sci. (ser. 3)35: 89–94.

———. 1888b. A new family of horned Dinosauria from the Cretaceous. Am. J. Sci. (ser. 3)36: 477–478.

———. 1889a. Notice of new American dinosaurs. Am. J. Sci. (ser. 3)37: 331–336.

———. 1889b. Notice of gigantic horned Dinosauria from the Cretaceous. Am. J. Sci. (ser. 3)38: 173–175.

———. 1889c. The skull of the gigantic Ceratopsidae. Am. J. Sci. (ser. 3)38: 501–506.

———. 1889d. Discovery of Cretaceous Mammalia. Am. J. Sci. (ser. 3) 38: 81–92.

———. 1890a. Additional characters of the Ceratopsidae with notice of new Cretaceous dinosaurs. Am. J. Sci. (ser. 3)39: 418–426.

———. 1890b. Description of new dinosaurian reptiles. Am. J. Sci. (ser. 3)39: 81–86.

———. 1891a. The gigantic Ceratopsidae, or horned dinosaurs, of North America. Am. J. Sci. (ser. 3)41: 167–177.

———. 1891b. Notice of new vertebrate fossils. Am. J. Sci. (ser. 3)42: 265–269.

———. 1891c. Restoration of *Stegosaurus*. Am. J. Sci. (ser. 3) 42: 179–181.

———. 1891d. Restoration of *Triceratops* [and *Brontosaurus*]. Am. J. Sci. (ser. 3) 41: 339–342.

———. 1892a. Restoration of *Claosaurus* and *Ceratosaurus*. Am. J. Sci. (ser. 3)44: 343–350.

———. 1892b. Notice of new reptiles from the Laramie Formation. Am. J. Sci. (ser. 3)43: 449–453.

———. 1892c. The skull of *Torosaurus*. Am. J. Sci. (ser. 3) 43: 81–84.

———. 1892d. Notes on Mesozoic vertebrate fossils. Am. J. Sci. (ser. 3)44: 171–176.

———. 1892e. Notes on Triassic Dinosauria. Am. J. Sci. (ser. 3) 43: 543–546.

———. 1894. The typical Ornithopoda of the American Jurassic. Am. J. Sci. (ser. 3)48: 85–90.

———. 1895. On the affinities and classification of the dinosaurian reptiles. Am. J. Sci. (ser. 3)50: 483–498.

———. 1896. The Dinosaurs of North America. U.S. Geol. Surv. 16th Ann. Rept. 1894–95: 133–244.

———. 1898a. On the families of sauropodous dinosaurs. Am. J. Sci. (ser. 4)6: 487–488.

———. 1898b. New species of Ceratopsia. Am. J. Sci. (ser. 4) 6: 92.

———. 1899. Footprints of Jurassic dinosaurs. Am. J. Sci. (ser. 4)7: 227–232.

Marshall, L. G., Muizon, C. de, Gayet, M., Lavenu, A., and Sigé, B. 1985. The "Rosetta Stone" for mammalian evolution in South America. Natl. Geogr. Res. 1: 274–288.

Martin, J. 1987. Mobility and feeding of Cetiosaurus (Saurischia: Sauropoda)—why the long neck? In: Currie, P. J., and Koster, E. H. (eds.). 4th Symp. Mesozoic Terr. Ecosyst. Short Pap. Tyrrell Mus. Palaeontol., Drumheller, Alberta. Pp. 154–159.

Martin, L. D. 1983a. The origin of birds and avian flight. Curr. Ornithol. 1: 105–129.

———. 1983b. The origin and early radiation of birds. In: Brush, A. H., and Clark, S. A. (eds.). Perspectives of Ornithology. Cambridge Univ. Press, New York. Pp. 291–338.

Martin, L. D., Stewart, J. D., and Whetstone, K. N. 1980. The origin of birds: Structure of the tarsus and teeth. Auk 97: 86–93.

Martin, R. B., and Burr, D. B. 1982. A hypothetical mechanism for the stimulation of osteonal remodeling by fatigue damage. J. Biomech. 15: 137–139.

Martin, R. D. 1981. Relative brain size and basal metabolic rate in terrestrial vertebrates. Nature 293: 57–60.

Martinez, R., Gimenez, O., Rodriguez, J., and Bochatey, G. 1986. Xenotarsosaurus bonapartei nov. gen et sp. (Carnosauria, Abelisauridae), un nuevo Theropoda de la Formacion Bajo Barreal, Chubut, Argentina. Actas IV Congr. Argent. Paleontol. Bioestrat. 2: 23–31.

Martinsson, G. G. 1975. [Towards the question about principles of stratigraphy and correlations of Mesozoic continental deposits]. Sovm. Sov.-Mong. Nauchnoissled. Geol. Eksped. Trudy 13: 7–24. (In Russian)

Martinsson, G. G., and Shuvalov, V. F. 1973. [Stratigraphic subdivision of the Upper Jurassic and Lower Cretaceous in southeastern Mongolia]. Acad. Nauk S.S.S.R. Izv. Ser. Geol. 10: 139–143. (In Russian)

Martinsson, G. G., Sochava, A. V., and Kolesnikov, C. M. 1971. [Fossil dinosaur eggs from the Gobi Desert]. Akademia Nauk S.S.S.R. Vestnik 7: 95–98. (In Russian)

Maryańska, T. 1969. Remains of armored dinosaurs from the uppermost Cretaceous in Nemegt Basin, Gobi Desert. Palaeontol. Polonica 21: 22–34.

———. 1971. New data on the skull of Pinacosaurus grangeri (Ankylosauria). Palaeontol. Polonica 25: 45–53.

———. 1977. Ankylosauridae (Dinosauria) from Mongolia. Palaeontol. Pol. 37: 85–151.

Maryańska, T., and Osmólska, H. 1974. Pachycephalosauria, a new suborder of ornithischian dinosaurs. Palaeontol. Polonica 30: 45–102.

———. 1975. Protoceratopsidae (Dinosauria) of Asia. Palaeontol. Polonica 33: 133–182.

———. 1981a. Cranial anatomy of Saurolophus angustirostris with comments on the Asian Hadrosauridae (Dinosauria). Palaeontol. Polonica 42: 5–24.

———. 1981b. First lambeosaurine dinosaur from the Nemegt Formation, Upper Cretaceous, Mongolia. Acta Palaeontol. Polonica 26: 243–255.

———. 1983. Some implications of hadrosaurian postcranial anatomy. Acta Palaeontol. Polonica 28: 205–208.

———. 1984a. Phylogenetic classification of ornithischian dinosaurs. 27th Internatl. Geol. Congr. 1: 286–287.

———. 1984b. Postcranial anatomy of Saurolophus angustirostris with comments on other hadrosaurs. Palaeontol. Polonica 46: 119–141.

———. 1985. On ornithischian phylogeny. Acta Palaeontol. Polonica 30: 137–150.

Mateer, N. J. 1976. New topotypes of Alamosaurus sanjuanensis Gilmore (Reptilia: Saurischia). Bull. Geol. Inst. Univ. Uppsala 6: 93–95.

———. 1981. The reptilian megafauna from the Kirtland Shale (Late Cretaceous) of the San Juan Basin, New Mexico. In: Lucas, S. G., Rigby, J. K., Jr., and Kues, B. S. (eds.). Advances in San Juan Basin Palaeontology. Univ. New Mexico Press, Albuquerque. Pp. 49–75.

———. 1987. A new report of a theropod dinosaur from South Africa. Palaeontology 30: 141–145.

———. 1989. Upper Cretaceous reptilian eggs from Zhejiang Province, China. In: Gillette, D. D., and Lockley, M. G. (eds.). Dinosaur Tracks and Traces. Cambridge Univ. Press, New York. Pp. 115–118.

Mateer, N. J., Hartman, J. H., Zhen S., Li J., Rao C., Chen J., and Reser, P. MS. Stratigraphy, palaeontology, and sedimentology of Cretaceous dinosaur-bearing strata in eastern China.

Mateer, N. J., Lucas, S. G., and Hunt, A. In press. Nonmarine Jurassic-Cretaceous boundary in western North America. In: Mateer, N. J., and Chen P. (eds.). Proc. 1st Internatl. Symp. Nonmarine Cretaceous Correlations. China Ocean Press, Beijing.

Mateer, N. J., and McIntosh, J. S. 1985. A new reconstruction of the skull of Euhelopus zdanskyi (Saurischia: Sauropoda). Bull. Geol. Inst. Univ. Uppsala 11: 125–132.

Matheron, P. 1869. Note sur les reptiles des dépôts fluviolacustres crétacés du bassin à lignite de Fuveau. Bull. Soc. Géol. France (sér. 2)26: 781–795.

Mathur, U. B., Pant, S. C., Mehra, S., and Mathur, A. K. 1985. Discovery of dinosaurian remains in Middle

Jurassic of Jailsalmer, Rajasthan, western India. Bull. Indian Geol. Assoc. 18: 59–65.

Mathur, U. B., and Srivastava, S. 1987. Dinosaur teeth from Lameta Group (Upper Cretaceous) of Kheda District, Gujarat. J. Geol. Soc. India 29: 554–566.

Matley, C. A. 1923. Note on an armored dinosaur from the Lameta Beds of Jubbulpore. Rec. Geol. Surv. India 55: 105–109.

Matsukawa, M., and Obata, I. 1985. Dinosaur footprints and other indentation in the Cretaceous Sebayashi Formation, Sebayashi, Japan. Bull. Natl. Sci. Mus. (ser. C)11: 9–36.

Matsumoto, T., and Obata, I. 1979. [Evaluation of ammonites and other fossils from the Cretaceous of Japan]. Kaseki 29: 43–58. (In Japanese)

Matsuo, H. 1967. A Cretaceous *Salvinia* from the Hashima Is. (Gunkan-Jima), outside of the Nagasaki Harbour, west Kyushu, Japan. Trans. Proc. Palaeontol. Soc. Japan 66: 49–55.

Matthew, W. D. 1905. The mounted skeleton of *Brontosaurus*. Am. Mus. J. 5: 62–70.

———. 1915*a*. Climate and evolution. Ann. New York Acad. Sci. 24: 171–318.

———. 1915*b*. Dinosaurs. Am. Mus. Nat. Hist. Hdbk. 5: 1–162.

Matthew, W. D., and Brown, B. 1922. The family Deinodontidae, with notice of a new genus from the Cretaceous of Alberta. Bull. Am. Mus. Nat. Hist. 46: 367–385.

———. 1923. Preliminary notices of skeletons and skulls of Deinodontidae from the Cretaceous of Alberta. Am. Mus. Novitates 89: 1–9.

Mawson, J., and Woodward, A. S. 1907. On the Cretaceous formation of Bahia (Brazil), and on vertebrate fossils collected therein. Q. J. Geol. Soc. London 63: 128–139.

McAllister, J. A. 1989. Dakota Formation tracks from Kansas: Implications for the recognition of tetrapod subaqueous traces. In: Gillette, D. D., and Lockley, M. G. (eds.). Dinosaur Tracks and Traces. Cambridge Univ. Press, New York. Pp. 343–348.

McAlpine, D. F. 1985. Size and growth of heart, liver, and kidneys in North Atlantic fin whales (*Balaenoptera physalus*), sei (*B. borealis*), and sperm (*Physeter macrocephalus*) whales. Can. J. Zool. 63: 1402–1409.

McCosker, J. E. 1985. White shark attack behavior: Observations and speculations about predator and prey strategies. Mem. So. Calif. Acad. Sci. 9: 123–135.

———. 1987. The white shark, *Carcharodon carcharias,* has a warm stomach. Copeia 1987: 195–197.

McGowan, C. 1984. Evolutionary relationships of ratites and carinates from the ontogeny of the tarsus. Nature 307: 733–735.

McIntosh, J. S. 1981. Annotated catalogue of the dinosaurs (Reptilia: Archosauria) in the collections of Carnegie Museum of Natural History. Bull. Carnegie Mus. Nat. Hist. 18: 1–67.

McIntosh, J. S., and Berman, D. S. 1975. Description of the palate and lower jaw of *Diplodocus* (Reptilia: Saurischia) with remarks on the nature of the skull of *Apatosaurus.* J. Paleontol. 49: 187–199.

McIntosh, J. S., and Williams, M. E. 1988. A new species of sauropod dinosaur, *Haplocanthosaurus delfsi* sp. nov., from the Upper Jurassic Morrison Fm. of Colorado. Kirtlandia 43: 3–26.

McKenna, M. C. 1980. Late Cretaceous and Early Tertiary vertebrate paleontological reconnaissance, Togwotee Pass area, northwestern Wyoming. In: Jacobs, L. L. (ed.). Aspects of Vertebrate History. Mus. No. Arizona Press, Flagstaff. Pp. 321–343.

McKenna, M. C., and Love, J. D. 1970. Local stratigraphic and tectonic significance of *Leptoceratops*, a Cretaceous dinosaur in the Pinyon Conglomerate, Northwestern Wyoming. U.S. Geol. Surv. Prof. pap. 700 D: D55–D61.

McKinney, M. L. 1987. Taxonomic selectivity and continuous variation in mass and background extinctions of marine taxa. Nature 325: 143–145.

McLachlan, I. R., and McMillan, I. K. 1976. Review and stratigraphic significance of southern Cape Mesozoic palaeontology. Trans. Geol. Soc. S. Afr. 79: 197–212.

McLean, D. M. 1978. A terminal Mesozoic "greenhouse": Lessons from the past. Science 201: 401–406.

———. 1985. Mantle degassing unification of the trans-K-T geobiological record. Evol. Biol. 19: 287–313.

McLoughlin, J. C. 1979. Archosauria: A New Look at the Old Dinosaur. Viking, New York. 117 pp.

McMahon, T. A. 1975. Using body size to understand the structural design of animals: Quadrupedal locomotion. J. Appl. Physiol. 39: 619–627.

———. 1984. Muscles, Reflexes, and Locomotion. Princeton Univ. Press, Princeton. 331 pp.

McMahon, T. A., and Bonner, J. T. 1983. On Size and Life. Sci. Am. Lib., New York. 255 pp.

McMannis, W. J. 1965. Resume of depositional and structural history of western Montana. Bull. Am. Assoc. Petrol. Geol. 49: 1801–1823.

McNab, B. K. 1970. Body weight and the energetics of temperature regulation. J. Exp. Biol. 53: 329–348.

———. 1978. The evolution of endothermy in the phylogeny of mammals. Am. Nat. 112: 1–21.

———. 1980. Food habits, energetics, and the population biology of mammals. Am. Nat. 116: 106–124.

———. 1983. Energetics, body size, and the limits to endothermy. J. Zool. 199: 1–29.

———. 1984. Commentary. Oecologia 64: 423–424.

———. 1986*a*. The influence of food habits on the energetics of eutherian mammals. Ecol. Monogr. 56: 1–19.

———. 1986*b*. Food habits, energetics, and the reproduction of marsupials. J. Zool. 208: 595–614.

———. 1987. The evolution of mammalian energetics. In: Calow, P. (ed.). Evolutionary Physiological Ecology. Cambridge Univ. Press, New York. Pp. 219–236.

McNab, B. K., and Auffenberg, W. 1976. The effect of large body size on the temperature regulation of the

Komodo dragon, *Varanus komodoensis*. Comp. Biochem. Physiol. 55A: 345–350.

McNaughton, S. J., and Georgiadis, N. J. 1986. Ecology of African grazing and browsing mammals. Ann. Rev. Ecol. Syst. 17: 39–65.

McNease, L., and Joanen, T. 1977. Alligator diets in relation to marsh salinity. Proc. Ann. Conf. SE Assoc. Wildl. Agencies 31: 36–40.

Medus, J. 1983. Des palynoflores de l'Infralias de Normandie (France). Géobios 16: 647–685.

Mehl, M. G. 1931. Additions to the vertebrate record of the Dakota Sandstone. Am. J. Sci. (ser. 5) 21: 441–452.

———. 1936. *Hierosaurus coleii* new aquatic dinosaur from the Niobrara Cretaceous of Kansas. J. Sci. Lab. Denison Univ. 31: 1–20.

Mellett, J. S. 1982. Body size, diet, and scaling factors in large carnivores and herbivores. Proc. 3d No. Am. Paleontol. Conf. 2: 371–376.

Melville, A. G. 1849. Notes on the vertebral column of *Iguanodon*. Phil. Trans. Roy. Soc. London 139: 285–300.

Mensink, H., and Mertmann, D. 1984. Dinosaurier-Fährten (*Gigantosauropus asturiensis* n. g. n. sp.; *Hispanosauropus hauboldi* n. g. n. sp.) im Jura Asturiens bei La Griega und Ribadesella (Spanien). N. Jb. Geol. Palaeontol. Mh. 1984: 405–415.

Messel, H., Vorlicek, G. C., Wells, A. G., and Green, W. J. 1981. Surveys of Tidal River Systems in the Northern Territory of Australia and their Crocodile Populations. Monograph 1. The Blyth-Cadell Rivers System Study and the Status of *Crocodylus porosus* in Tidal Waterways of Northern Australia: Methods for Analysis, and Dynamics of a Population of *C. porosus*. Pergamon, New York. 463 pp.

Meyer, H. von. 1832. Paleologica zur Geschichte der Erde. Frankfurt am Main. 560 pp.

———. 1837. Mitteilung an Prof. Bronn (*Plateosaurus engelhardti*). N. Jb. Mineral. Geol. Palaeontol. 1837: 317.

———. 1857. Beiträge zur näheren Kenntnis fossiler Reptilien. N. Jb. Mineral. Geol. Palaeontol. 1857: 532–543.

Miller, H. W. 1964. Cretaceous dinosaurian remains from southern Arizona. J. Paleontol. 38: 378–384.

Miller, W. E., Britt, B. B., and Stadtman, K. L. 1989. Tridactyl trackways from the Moenave Formation of southwestern Utah. In: Gillette, D. D., and Lockley, M. G. (eds.). Dinosaur Tracks and Traces. Cambridge Univ. Press, New York. Pp. 209–215.

Milner, A. R., and Norman, D. B. 1984. The biogeography of advanced ornithopod dinosaurs (Archosauria: Ornithischia)—a cladistic-vicariance model. In: Reif, W.-E., and Westphal, F. (eds.). 3d Symp. Mesozoic Terr. Ecosyst. Short Pap. Attempto Verlag, Tübingen. Pp. 145–150.

Minton, S. A., Jr., and Minton, M. R. 1973. Giant Reptiles. Scribner's, New York. 345 pp.

Mohabey, D. M. 1983. Note on the occurrence of dino-

saurian fossil eggs from Infratrappean Limestone in Kheda District, Gujarat. Curr. Sci. 52: 1194.

———. 1986. Note on dinosaur footprint from Kheda District, Gujarat. J. Geol. Soc. India 27: 456–459.

Molenaar, C. M. 1983. Major depositional cycles and regional correlations of Upper Cretaceous rocks, southern Colorado Plateau and adjacent areas. In: Reynolds, M. W., and Dolly, E. D. (eds.). Mesozoic Paleogeography of west-central United States. Rocky Mountain Section, SEPM, Denver. Pp. 201–223.

Molnar, R. E. 1974. A distinctive theropod dinosaur from the Upper Cretaceous of Baja California (Mexico). J. Paleontol. 48: 1009–1017.

———. 1977. Analogies in the evolution of display structures in ornithopods and ungulates. Evol. Theory 3: 165–190.

———. 1978. A new theropod dinosaur from the Upper Cretaceous of central Montana. J. Paleontol. 52: 73–82.

———. 1980a. An ankylosaur (Ornithischia: Reptilia) from the Lower Cretaceous of southern Queensland. Mem. Queensland Mus. 20: 77–87.

———. 1980b. Australian late Mesozoic terrestrial tetrapods: Some implications. Mém. Soc. Géol. France 139: 131–143.

———. 1980c. A dinosaur from New Zealand. 5th Internatl. Gondwana Symp. Pp. 11–16.

———. 1980d. An albertosaur from the Hell Creek Formation of Montana. J. Paleontol. 54: 102–108.

———. 1982. A catalogue of fossil amphibians and reptiles in Queensland. Mem. Queensland Mus. 20: 613–633.

———. 1984a. Ornithischian dinosaurs in Australia. In: Reif, W.-E., and Westphal, F. (eds.). 3d Symp. Mesozoic Terr. Ecosyst. Short Pap. Attempto Verlag, Tübingen. Pp. 151–156.

———. 1984b. Alternatives to *Archaeopteryx*: A survey of proposed early or ancestral birds. In: Hecht, M. K., Ostrom, J. H., Viohl, G., and Wellnhofer, P. (eds.). The Beginnings of Birds. Freunde des Jura-Museums, Eichstätt, Pp. 209–217.

Molnar, R. E., and Carpenter, K. In press. The Jordan theropod (Maastrichtian, Montana, U.S.A.) referred to the genus *Aublysodon*. Géobios.

Molnar, R. E., Flannery, T. F., and Rich, T. H. V. 1981. An allosaurid theropod dinosaur from the Early Cretaceous of Victoria, Australia. Alcheringa 5: 141–146.

———. 1985. Aussie *Allosaurus* after all. J. Paleontol. 59: 1511–1513.

Molnar, R. E., and Galton, P. M. 1986. Hypsilophodontid dinosaurs from Lightning Ridge, New South Wales, Australia. Géobios 19: 231–239.

Molnar, R. E., and Pledge, N. S. 1980. A new theropod dinosaur from South Australia. Alcheringa 4: 281–287.

Monbaron, M. 1978. Nouveaux ossements de dinosauriens de grande taille dans le bassin jurassico-crétacé de Taguelft (Atlas de Beni Mellal, Maroc). C. R. Acad. Sci. Paris 287: 1277–1279.

———. 1983. Dinosauriens du haut Atlas Central (Maroc). Actes Soc. Jurassienne Emulation 1983: 203–234.

Monbaron, M., Dejax, J., and Demathieu, G. 1985. Longues pistes de dinosaures bipèdes à Adrar-n-Ouglagal (Maroc) et répartition des faunes de grands Reptiles dans le domaine atlasique au cours du Mésozoïque. Bull. Mus. Natl. Hist. Nat. Paris (sér. 4)3: 229–242.

Monbaron, M., and Taquet, P. 1981. Découverte du squelette complet d'un cétiosaure (dinosaure sauropode) dans le bassin jurassique moyen de Tilougguit (Haut-Atlas central, Maroc). C. R. Acad. Sci. Paris 292: 243–246.

Mones, A. 1980. Nuevos elementos de la paleoherpetofauna del Uruguay (Crocodilia y dinosauria). Actas II Congr. Argent. Paleontol. Bioestrat. I. Congr. Latinoam. Paleontol. 1: 265–277.

Montanucci, R. R. 1968. Comparative dentition in four iguanid lizards. Herpetologica 24: 305–315.

Moody, R. 1977. A Natural History of Dinosaurs. Chartwell Books, Secaucus, N.J. 124 pp.

Mook, C. C. 1917a. Criteria for the determination of species in the Sauropoda with description of a new species of *Apatosaurus*. Bull. Am. Mus. Nat. Hist. 37: 815–819.

———. 1917b. The fore and hind limbs of *Diplodocus*. Bull. Am. Mus. Nat. Hist. 37: 355–358.

Morales, M. 1986. Dinosaur tracks in the Lower Jurassic Kayenta Formation near Tuba City, Arizona. In: Lockley, M. (ed.). Dinosaur Tracksites. Geol. Dept., Univ. Colorado, Denver. Pp. 14–16.

Morris, F. K. 1936. Central Asia in Cretaceous time. Bull. Geol. Soc. Am. 47: 1477–1534.

Morris, J. 1893. A Catalogue of British Fossils. Br. Mus. Nat. Hist. London.

Morris, W. J. 1970. Hadrosaurian dinosaur bills—morphology and function. Contrib. Sci. Nat. Hist. Mus. Los Angeles County 193: 1–14.

———. 1971. Mesozoic and Tertiary vertebrates in Baja California. Natl. Geogr. Soc. Res. Rept. 1965: 195–198.

———. 1973a. Mesozoic and Tertiary vertebrates in Baja California. Natl. Geogr. Soc. Res. Rept. 1966: 197–209.

———. 1973b. A review of Pacific coast hadrosaurs. J. Paleontol. 47: 551–561.

———. 1976. Hypsilophodont dinosaurs: A new species and comments on their systematics. In: Churcher, C. S. (ed.). Athlon. Roy. Ont. Mus., Toronto. Pp. 93–113.

———. 1981. A new species of hadrosaurian dinosaur from the Upper Cretaceous of Baja California—?*Lambeosaurus laticaudus*. J. Paleontol. 55: 453–462.

———. 1982. California dinosaurs. In: Bottjer, D. J., Colburn, I. P., and Cooper, J. D. (eds.). Late Cretaceous Depositional Environments and Paleogeography, Santa Ana Mountains, Southern California. Pacific Sect. SEPM, Field Trip Vol. Gdbk. Pp. 89–90.

Mourier, T., Jaillard, E., Laubacher, G., Noblet, C., Pardo, A., Sigé, B., and Taquet, P. 1986. Découverte de restes dinosauriens et mammaliens d'âge crétacé supérieur à la base des couches rouges du synclinal de Bagua (Andes nord-péruviennes): Aspects stratigraphiques sédimentologiques et paléogéographiques concernant la régression fin-crétacée. Bull. Soc. Géol. France 1986: 171–175.

Moussaye, G. de la. 1884. Sur une dent de *Neosodon* trouvée dans les sables ferrugineux de Wimille. Bull. Soc. Géol. France 1884: 51–54.

Muizon, C. de, Gayet, M., Lavenu, A., Marshall, L. G., Sigé, B., and Villaroel, C. 1983. Late Cretaceous vertebrates, including mammals, from Tiupampa, south-central Bolivia. Géobios 16: 747–753.

Murray, G. E., Boyd, D. R., Wolleben, J. A., and Wilson, J. A. 1960. Late Cretaceous fossil locality, eastern Parras Basin, Coahuila, Mexico. J. Paleontol. 34: 368–373.

Murry, P. A. 1986. Vertebrate paleontology of the Dockum Group, western Texas and eastern New Mexico. In: Padian, K. (ed.). The Beginning of the Age of Dinosaurs. Cambridge Univ. Press, New York. Pp. 109–137.

Nagao, T. 1936. *Nipponosaurus sachalinensis*, a new genus and species of trachodont dinosaur from Japanese Saghalien. J. Fac. Sci. Hokkaido Imperial Univ. (ser. 4) 3: 185–220.

Nagy, K. A. 1983. Ecological energetics. In: Huey, R. B., Pianka, E. R., and Schoener, T. W. (eds.). Lizard Ecology: Studies of a Model Organism. Harvard Univ. Press, Cambridge. Pp. 24–54.

———. 1987. Field metabolic rate and food requirement scaling in mammals and birds. Ecol. Monogr. 57: 111–128.

Nelson, M. E., and Crooks, D. M. 1987. Stratigraphy and paleontology of the Cedar Mountain Formation (Lower Cretaceous), eastern Emery County, Utah. In: Averett, W. R. (ed.). Paleontology and Geology of the Dinosaur Triangle. Mus. Western Colorado, Grand Junction, Colorado. Pp. 55–63.

Nessov, L. A. 1982. [Ancient mammals of the U.S.S.R.]. Ezhegodn. Vsesoyuzn. Paleontol. Obshchest. 25: 228–243. (In Russian)

Newell, N. D. 1967. Revolutions in the history of life. Geol. Soc. Am. Spec. Pap. 89: 62–92.

Newman, B. H. 1968. The Jurassic dinosaur, *Scelidosaurus harrisonii* Owen. Palaeontology 11: 40–43.

———. 1970. Stance and gait in the flesh-eating dinosaur *Tyrannosaurus*. Biol. J. Linn. Soc. London 2: 119–123.

Newton, E. T. 1892. Note on an iguanodont tooth from the Lower Chalk ("Totternhoe stone"), near Hitchin. Geol. Mag. (ser. 3)9: 49–50.

———. 1899. On a megalosauroid jaw from Rhaetic Beds near Bridgend (Glamorganshire). Q. J. Geol. Soc. London 55: 89–96.

Nicholls, E. L., and Russell, A. P. 1981. A new specimen of *Struthiomimus altus* from Alberta, with comments on the classificatory characters of the Upper Cretaceous ornithomimids. Can. J. Earth Sci. 18: 518–526.

———. 1985. Structure and function of the pectoral girdle and forelimb of *Struthiomimus altus* (Theropoda: Ornithomimidae). Palaeontology 28: 638–667.

Nichols, D. J., Jarzen, D. M., Orth, C. J., and Oliver, P. Q. 1986. Palynological and iridium anomalies at Cretaceous-Tertiary boundary, south-central Saskatchewan. Science 231: 714–717.

Nilsen, T. H., and Abbott, P. L. 1981. Paleogeography and sedimentology of Upper Cretaceous turbidites, San Diego, California. Am. Assoc. Petrol. Geol. Bull. 65: 1256–1284.

Nomina Anatomica. 1983. Williams and Wilkins, Baltimore. 86 pp.

Nomina Anatomica Veterinaria. 1983. World Assoc. Veterinary Anatomists, Cornell Univ., Ithaca. 218 pp.

Nopcsa, F. 1900. Dinosaurierreste aus Siebenbürgen. I. Schädel von *Limnosaurus transsylvanicus* nov. gen. et spec. Denkschr. Akad. Wiss. Wien 68: 555–591.

———. 1901. Synopsis und Abstammung der Dinosaurier. Földt. Közl. 31: 247–288.

———. 1902*a*. Dinosaurierreste aus Siebenbürgen. II. Schädelreste von *Mochlodon*. Denkschr. Akad. Wiss. Wien 72: 149–175.

———. 1902*b*. Notizen über cretacische Dinosaurier. 3. Wirbel eines südamerikanischen Sauropoden. Sitzungsber. K. Akad. Wiss. Wien 111: 108–114.

———. 1903. *Telmatosaurus*, new name for the dinosaur *Limnosaurus*. Geol. Mag. (ser. 4)10: 94–95.

———. 1905*a*. Notes on British dinosaurs. Pt. II: *Polacanthus*. Geol. Mag. (ser. 5)2: 241–250.

———. 1905*b*. Remarks on the supposed clavicle of the sauropodous dinosaur *Diplodocus*. Proc. Zool. Soc. London 1905: 289–294.

———. 1906. Zur Kenntnis des Genus *Streptospondylus*. Beitr. Palaeont. Geol. Österr.-Ung. 19: 59–83.

———. 1911*a*. Notes on British dinosaurs. Pt. IV. *Stegosaurus priscus*, sp. nov. Geol. Mag. (ser. 5)8: 109–115, 145–153.

———. 1911*b*. *Omosaurus lennieri*, un nouveau dinosaurien du Cap de la Hève. Bull. Soc. Géol. Normandie 30: 23–42.

———. 1912. Notes on British dinosaurs. Pt. V. *Craterosaurus* (Seeley). Geol. Mag. (ser. 5)9: 481–484.

———. 1915. Die Dinosaurier der siebenbürgischen Landesteile Ungarns. Mitteil. Jb. K. Ungar. Geol. Reichsanst. 23: 1–26.

———. 1916. *Doryphorosaurus* nov. nom. für *Kentrosaurus* Hennig. Cbl. Mineral. Geol. Palaeont. 1916: 511–512.

———. 1917. Über Dinosaurier. I: Notizen über die Systematik der Dinosaurier. Cbl. Mineral. Geol. Palaeont. 1917: 203–213.

———. 1918. *Leipsanosaurus* n. gen. ein neuer Thyreophore aur der Gosau. Földt. Közl. 48: 324–328.

———. 1923*a*. On the geological importance of the primitive reptilian fauna in the uppermost Cretaceous of Hungary. Q. J. Geol. Soc. London 74: 100–116.

———. 1923*b*. Die Familien der Reptilien. Fortschr. Geol. Palaeontol. 2: 1–210.

———. 1923*c*. Notes on British dinosaurs. Pt. VI. *Acanthopholis*. Geol. Mag. 60: 193–199.

———. 1926. A 4000 lb. insectivorous dinosaur. Illustr. London News 169: 466–467, 478.

———. 1928*a*. Paleontological notes on reptiles. Geol. Hungarica, Ser. Palaeontol. 1: 1–84.

———. 1928*b*. Dinosaurierreste aus Siebenbürgen. IV. Die Wirbelsäule von *Rhabdodon* und *Orthomerus*. Geol. Hungarica, Ser. Palaeontol. 1: 273–304.

———. 1928*c*. The genera of reptiles. Palaeobiologica, 1: 163–188.

———. 1929*a*. Dinosaurierreste aus Siebenbürgen V. Geol. Hungarica Ser. Paleontol. 4: 1–76.

———. 1929*b*. Sexual differences in ornithopodous dinosaurs. Palaeobiologica. 2: 187–201.

———. 1930. Zur Systematik und Biologie der Sauropoden. Palaeobiologica 3: 40–52.

Norman, D. B. 1980. On the ornithischian dinosaur *Iguanodon bernissartensis* from the Lower Cretaceous of Bernissart (Belgium). Inst. Roy. Sci. Nat. Belg. Mem. 178: 1–103.

———. 1984*a*. A systematic appraisal of the reptile order Ornithischia. In: Reif, W.-E., and Westphal, F. (eds.). 3d Symp. Mesozoic Terr. Ecosyst. Short Pap. Attempto Verlag, Tübingen. Pp. 157–162.

———. 1984*b*. On the cranial morphology and evolution of ornithopod dinosaurs. In: Ferguson, M. W. J. (ed.). The Structure, Development and Evolution of Reptiles. Symp. Zool. Soc. London 52: 521–547.

———. 1985. The Illustrated Encyclopedia of Dinosaurs. Crescent, New York. 208 pp.

———. 1986. On the anatomy of *Iguanodon atherfieldensis* (Ornithischia: Ornithopoda). Bull. Inst. Roy. Sci. Nat. Belg. Sci. Terre 56: 281–372.

———. 1987*a*. On the history of the discovery of fossils at Bernissart in Belgium. Arch. Nat. Hist. 14: 59–75.

———. 1987*b*. A mass-accumulation of vertebrates from the Lower Cretaceous of Nehden (Sauerland), West Germany. Proc. Roy. Soc. London 230: 215–255.

———. 1987*c*. Wealden dinosaur biostratigraphy. In: Currie, P. J., and Koster, E. H. (eds.). 4th Symp. Mesozoic Terr. Ecosyst. Short Pap. Tyrrell Mus. Palaeontol., Drumheller, Alberta. Pp. 165–170.

———. In press. A review of the Wealden iguanodontid dinosaur *Vectisaurus valdensis* Hulke 1874, with comments upon the status of the family Iguanodontidae. In: Currie, P. J., and Carpenter, K. (eds.). Dinosaur Systematics: Perspectives and Approaches. Cambridge Univ. Press, New York.

Norman, D. B., Hilpert, K.-H., and Hölder, H. 1987. Die Wirbeltierfauna von Nehden (Sauerland), Westdeutschland. Geol. Paläontol. Westphalens 8: 1–77.

Norman, D. B., and Weishampel, D. B. 1985. Ornithopod feeding mechanisms: Their bearing on the evolution of herbivory. Am. Nat. 126: 151–164.

———. In press. Feeding mechanisms in some small herbivorous dinosaurs: Processes and patterns. In: Rayner, J. M. V. (ed.). Biomechanics in Evolution. Cambridge Univ. Press, New York.

Northcutt, R. G. 1977. Elasmobranch central nervous system organization and its possible evolutionary significance. Am. Zool. 17: 411–429.

Novas, F. 1987. Un probable teropodo (Saurisquia) de la Formacion Ischigualasto (Triásico superior), San Juan, Argentina. IV Congr. Argent. Paleontol. Bioestrat. 2: 1–6.

Novodvorskaya, I. M. 1974. [Taphonomy of the Early Cretaceous vertebrate locality Khuren-Dukh, M.P.R.]. Sovm. Sov.-Mong. Paleontol. Eksped. Trudy 1: 305–313. (In Russian with English summary)

Nowak, R. M., and Paradiso, J. L. 1983. Walker's Mammals of the World. 4th ed. Johns Hopkins Univ. Press, Baltimore. 1362 pp.

Nowinski, A. 1971. *Nemegtosaurus mongoliensis* n. gen., n. sp. (Sauropoda) from the uppermost Cretaceous of Mongolia. Palaeontol. Polonica 25: 57–81.

Nurumov, T. N. 1964. [Finds of dinosaur remains in Kazakhstan]. In: [Problems of Herpetology]. Izd-vo LGU, Leningrad. Pp. 49–50. (In Russian)

Obata, I., Hasegawa, Y., and Otsuka, H. 1972. [Reptile fossils from the Cretaceous of Hokkaido]. Mem. Natl. Sci. Mus. 5: 213–222.

Obata, I., and Kanie, Y. 1977. Upper Cretaceous dinosaurbearing sediments in Majunga region, northwestern Madagascar. Bull. Natl. Sci. Mus. (ser. C)3: 161–172.

Officer, C. B., and Drake, C. L. 1983. The Cretaceous-Tertiary transition. Science 219: 1383–1390.

———. 1985. Terminal Cretaceous events. Science 227: 1161–1167.

Officer, C. B., Hallam, A., Drake, C. L., and Devine, J. D. 1987. Late Cretaceous and paroxysmal Cretaceous/Tertiary extinctions. Nature 326: 143–149.

Ogier, A. 1975. Étude de nouveaux ossements de *Bothriospondylus* (Sauropode) d'un gisement du Bathonien de Madagascar. Thèse de 3ᵉ cycle, Univ. Paris.

Olivero, E., Scasso, R., and Rinaldi, C. 1986. Revision del Grupo Marambio en la isla James Ross, Antartica. Contr. Inst. Antart. Argent. 331: 1–30.

Olsen, P. E. 1980*a*. A comparison of the vertebrate assemblages from the Newark and Hartford Basins (early Mesozoic, Newark Supergroup) of eastern North America. In: Jacobs, L. L. (ed.). Aspects of Vertebrate History. Mus. No. Arizona Press, Flagstaff. Pp. 35–54.

———. 1980*b*. Triassic and Jurassic formations of the Newark Basin. In: Manspeizer, W. (ed.). Field Studies of New Jersey Geology and Guide to Field Trips, 52nd Ann. Mt. N.Y. State Geol. Assoc. Pp. 2–39.

———. 1980*c*. Fossil great lakes of the Newark Supergroup in New Jersey. In: Manspeizer, W. (ed.). Field Studies in New Jersey Geology and Guide to Field Trips. 52nd Ann. Mt. N.Y. State Geol. Assoc. Pp. 352–398.

Olsen, P. E., and Baird, D. 1986. The ichnogenus *Atreipus* and its significance for Triassic biostratigraphy. In: Padian, K. (ed.). The Beginning of the Age of Dinosaurs. Cambridge Univ. Press, New York. Pp. 61–88.

Olsen, P. E., and Galton, P. M. 1977. Triassic-Jurassic tetrapod extinctions: Are they real? Science 197: 983–986.

———. 1984. A review of the reptile and amphibian assemblages from the Stormberg Group of southern Africa with special emphasis on the footprints and the age of the Stormberg. Palaeontol. afr. 25: 87–110.

Olsen, P. E., McCune, A. R., and Thomson, K. S. 1982. Correlation of the early Mesozoic Newark Supergroup by vertebrates, principally fishes. Am. J. Sci. 282: 1–44.

Olsen, P. E., Remington, C. L., Cornet, B., and Thomson, K. S. 1978. Cyclic change in Late Triassic lacustrine communities. Science 201: 729–733.

Olsen, P. E., Shubin, N. H., and Anders, M. H. 1987. New Early Jurassic tetrapod assemblages constrain Triassic-Jurassic tetrapod extinction event. Science 237: 1025–1029.

Olsen, P. E., and Sues, H.-D. 1986. Correlation of continental Late Triassic and Early Jurassic sediments, and patterns of Triassic-Jurassic tetrapod transition. In: Padian, K. (ed.). The Beginning of the Age of Dinosaurs. Cambridge Univ. Press, New York. Pp. 321–351.

Olshevsky, G. 1978. The Archosaurian Taxa (excluding Crocodylia). Mesozoic Meanderings 1: 1–50.

Olson, E. C. 1954. Fauna of the Vale and Choza: 8. Pelycosauria: *Dimetrodon*. Fieldiana Geol. 10: 205–210.

———. 1966. Community evolution and the origin of mammals. Ecology 47: 291–302.

———. 1983. Coevolution or coadaptation? Permo-Carboniferous vertebrate chronofauna. In: Nitecki, M. H. (ed.). Coevolution. Univ. Chicago Press, Chicago. Pp. 307–338.

Olsson, R. K. 1980. The New Jersey coastal plain and its relationship with the Baltimore Canyon Trough. In: Manspeizer, W. (ed.). Field Studies of New Jersey Geology and Guide to Field Trips. 52nd Ann. Mt. N.Y. State Geol. Assoc. Pp. 116–129.

Orth, C. J., Gilmore, J. S., Knight, J. D., Pillmore, C. L., Tschudy, R. H., and Fassett, J. E. 1981. An iridium anomaly at the palynological Cretaceous-Tertiary boundary in northern New Mexico. Science 214: 1341–1343.

Osborn, H. F. 1899*a*. A skeleton of *Diplodocus*. Mem. Am. Mus. Nat. Hist. 1: 191–214.

———. 1899*b*. Fore and hind limbs of carnivorous and herbivorous dinosaurs from the Jurassic of Wyoming. Bull. Am. Mus. Nat. Hist. 12: 161–172.

———. 1903. *Ornitholestes hermanni*, a new compsognathoid dinosaur from the Upper Jurassic. Bull. Am. Mus. Nat. Hist. 19: 459–464.

———. 1904. Manus, sacrum, and caudals of Sauropoda. Bull. Am. Mus. Nat. Hist. 20: 181–190.

———. 1905. *Tyrannosaurus* and other Cretaceous carnivorous dinosaurs. Bull. Am. Mus. Nat. Hist. 21: 259–265.

———. 1906. *Tyrannosaurus*, Upper Cretaceous carnivorous dinosaur (second communication). Bull. Am. Mus. Nat. Hist. 22: 281–296.

———. 1912a. Integument of the iguanodont dinosaur *Trachodon*. Mem. Am. Mus. Nat. Hist. 1: 33–54.

———. 1912b. Crania of *Tyrannosaurus* and *Allosaurus*. Mem. Am. Mus. Nat. Hist. 1: 1–30.

———. 1916. Skeletal adaptations of *Ornitholestes, Struthiomimus, Tyrannosaurus*. Bull. Am. Mus. Nat. Hist. 35: 733–771.

———. 1923a. Two Lower Cretaceous dinosaurs from Mongolia. Am. Mus. Novitates 95: 1–10.

———. 1923b. A new genus and species of Ceratopsia from New Mexico, *Pentaceratops sternbergii*. Am. Mus. Novitates 93: 1–3.

———. 1924a. Three new Theropoda, *Protoceratops* zone, central Mongolia. Am. Mus. Novitates 144: 1–12.

———. 1924b. *Psittacosaurus* and *Protiguanodon:* Two Lower Cretaceous iguanodonts from Mongolia. Am. Mus. Novitates 127: 1–16.

———. 1924c. Sauropoda and Theropoda of the Lower Cretaceous of Mongolia. Am. Mus. Novitates 128: 1–7.

———. 1933. Mounted skeleton of *Triceratops elatus*. Am. Mus. Novitates 654: 1–14.

Osborn, H. F., and Granger, W. 1901. Fore and hindlimbs of Sauropoda from the Bone Cabin Quarry. Bull. Am. Mus. Nat. Hist. 14: 199–208.

Osborn, H. F., and Mook, C. C. 1921. *Camarasaurus, Amphicoelias,* and other sauropods of Cope. Mem. Am. Mus. Nat. Hist. n.s. 3: 247–287.

Osmólska, H. 1976. New light on skull anatomy and systematic position of *Oviraptor*. Nature 262: 683–684.

———. 1979. Nasal salt glands in dinosaurs. Acta Palaeontol. Polonica 24: 205–215.

———. 1980. The Late Cretaceous vertebrate assemblages of the Gobi Desert, Mongolia. Mém. Soc. Géol. France 139: 145–150.

———. 1981. Coossified tarsometatarsi in theropod dinosaurs and their bearing on the problem of bird origins. Palaeontol. Polonica 42: 79–95.

———. 1982. *Hulsanpes perlei* n.g. n. sp. (Deinonychosauria, Saurischia, Dinosauria) from the Upper Cretaceous Barun Goyot Formation of Mongolia. N. Jb. Geol. Paläontol. Mh. 1982: 440–448.

———. 1986. Structure of the nasal and oral cavities in the protoceratopsid dinosaurs. Acta Palaeontol. Polonica 31: 145–157.

———. 1987. *Borogovia gracilicrus* gen. et sp. n., a new troodontid dinosaur from the Late Cretaceous of Mongolia. Acta Palaeontol. Polonica 32: 133–150.

Osmólska, H., and Roniewicz, E. 1970. Deinocheiridae, a new family of theropod dinosaurs. Palaeontol. Polonica 21: 5–19.

Osmólska, H., Roniewicz, E., and Barsbold, R. 1972. A new dinosaur, *Gallimimus bullatus* n. gen., n. sp. (Ornitho-mimidae) from the Upper Cretaceous of Mongolia. Palaeontol. Polonica 27: 103–143.

Ostrom, J. H. 1961a. Cranial morphology of the hadrosaurian dinosaurs of North America. Bull. Am. Mus. Nat. Hist. 122: 33–186.

———. 1961b. A new species of hadrosaurian dinosaur from the Cretaceous of New Mexico. J. Paleontol. 35: 575–577.

———. 1964. A reconsideration of the paleoecology of hadrosaurian dinosaurs. Am. J. Sci. 262: 975–997.

———. 1966. Functional morphology and evolution of the ceratopsian dinosaurs. Evolution 20: 290–308.

———. 1969a. A new theropod dinosaur from the Lower Cretaceous of Montana. Postilla 128: 1–17.

———. 1969b. Osteology of *Deinonychus antirrhopus,* an unusual theropod from the Lower Cretaceous of Montana. Peabody Mus. Nat. Hist. Bull. 30: 1–165.

———. 1970a. Stratigraphy and paleontology of the Cloverly Formation (Lower Cretaceous) of the Bighorn Basin area, Wyoming and Montana. Peabody Mus. Nat. Hist. Bull. 35: 1–234.

———. 1970b. Terrestrial vertebrates as indicators of Mesozoic climates. N. Am. Paleontol. Conv. Proc. D: 347–376.

———. 1972a. Were some dinosaurs gregarious? Palaeogeogr. Palaeoclimatol. Palaeoecol. 11: 287–301.

———. 1972b. Dinosaur. In: McGraw-Hill Yrbk. Sci. Tech. pp. 176–179.

———. 1973. The ancestry of birds. Nature 242: 136.

———. 1974a. The pectoral girdle and forelimb function of *Deinonychus* (Reptilia: Saurischia): A correction. Postilla 165: 1–11.

———. 1974b. *Archaeopteryx* and the origin of flight. Q. Rev. Biol. 49: 27–47.

———. 1975a. The origin of birds. Ann. Rev. Earth Planet. Sci. 3: 55–77.

———. 1975b. On the origin of *Archaeopteryx* and the ancestry of birds. Cent. Natl. Rech. Sci. 218: 519–532.

———. 1976a. On a new specimen of the Lower Cretaceous theropod dinosaur *Deinonychus antirrhopus*. Breviora 439: 1–21.

———. 1976b. *Archaeopteryx* and the origin of birds. Biol. J. Linn. Soc. London 8: 81–182.

———. 1978a. The osteology of *Compsognathus longipes* Wagner. Zitteliana 4: 73–118.

———. 1978b. *Leptoceratops gracilis* from the "Lance" Formation of Wyoming. J. Paleontol. 52: 697–704.

———. 1980a. *Coelurus* and *Ornitholestes:* Are they the same? In: Jacobs, L. L. (ed.). Aspects of Vertebrate History: Essays in Honor of Edwin Harris Colbert. Mus. No. Arizona Press, Flagstaff. Pp. 245–256.

———. 1980b. The evidence for endothermy in dinosaurs. In: Thomas, R. D. K., and Olson, E. C. (eds.). A Cold Look at the Warm-Blooded Dinosaurs. Westview Press, Boulder. Pp. 15–54.

———. 1981. *Procompsognathus*—theropod or thecodont? Palaeontographica A 175: 179–195.

———. 1985. The meaning of *Archaeopteryx*. In: Hecht, M. K., Ostrom, J. H., Viohl, G., and Wellnhofer, P. (eds.). The Beginnings of Birds. Freunde des Jura-Museums Eichstätt. Pp. 161–176.

———. 1986. Social and unsocial behavior in dinosaurs. In: Nitecki, M. H., and Kitchell, J. A. (eds.). Evolution of Animal Behavior. Oxford Univ. Press, New York. Pp. 41–61.

Ostrom, J. H., and McIntosh, J. S. 1966. Marsh's Dinosaurs. Yale Univ. Press, New Haven. 388 pp.

Ostrom, J. H., and Wellnhofer, P. 1986. The Munich specimen of *Triceratops* with a revision of the genus. Zitteliana 14: 111–158.

Owen, M. R., and Anders, M. H. 1988. Evidence from cathodo-luminescence for non-volcanic origin of shocked quartz at the Cretaceous/Tertiary boundary. Nature 334: 145–146.

Owen, R. 1840-1845. Odontography. London. 655 pp.

———. 1841. A description of a portion of the skeleton of *Cetiosaurus*, a gigantic extinct saurian occurring in the Oolitic Formation of different parts of England. Proc. Geol. Soc. London 3: 457–462.

———. 1842a. Deuxième rapport sur les reptiles fossiles de la Grande-Bretagne. Institut 10: 11–14.

———. 1842b. Report on British fossil reptiles. Pt. II. Rept. Br. Assoc. Adv. Sci. 11: 60–204.

———. 1854. Descriptive catalogue of the fossil organic remains of Reptilia contained in the Museum of the Royal College of Surgeons of England. Br. Mus. Nat. Hist. London. 184 pp.

———. 1859. Monograph on the fossil Reptilia of the Wealden and Purbeck Formations. Suppl. II. Crocodilia (*Streptospondylus* etc.). Palaeontogr. Soc. Monogr. 11: 20–44.

———. 1860. Palaeontology. Edinburgh. 420 pp.

———. 1861a. A monograph of the fossil Reptilia of the Lias Formations. I. *Scelidosaurus harrisonii*. Palaeontogr. Soc. Monogr. 13: 1–14.

———. 1861b. Monograph on the fossil Reptilia of the Wealden and Purbeck Formations. Pt. V. Lacertilia. Palaeontogr. Soc. Monogr. 12: 31–39.

———. 1863. A monograph of the fossil Reptilia of the Lias Formations. II. *Scelidosaurus harrisonii* Owen of the lower Lias. Palaeontogr. Soc. Monogr. 14: 1–26.

———. 1872. Monograph on the fossil Reptilia of the Wealden and Purbeck formations. Suppl. IV. Dinosauria (*Iguanodon*) (Wealden). Palaeontogr. Soc. Monogr. 25: 1–15.

———. 1874. Monograph on the fossil Reptilia of the Wealden and Purbeck formations. Suppl. V. *Iguanodon*. Palaeontogr. 27: 1–18.

———. 1875. Monographs of the fossil Reptilia of the Mesozoic formations (Pt. II) (genera *Bothriospondylus, Cetiosaurus, Omosaurus*). Palaeontogr. Soc. Monogr. 29: 15–93.

———. 1876a. Descriptive and illustrated catalogue of the fossil Reptilia of South Africa in the collection of the British Museum. British Mus. Nat. Hist., London 88 pp.

———. 1876b. Monograph of the fossil Reptilia of the Wealden and Purbeck Formations. Suppl. 7. Crocodilia (*Poikilopleuron*), Dinosauria (*Chondrosteosaurus*). Palaeontogr. Soc. Monogr. 30: 1–7.

———. 1877a. Monograph of the fossil Reptilia of the Mesozoic formations. Pts. II and III. Genera *Bothriospondylus, Cetiosaurus, Omosaurus*). Palaeontogr. Soc. Monogr. 29: 15–94.

———. 1877b. Monograph of the fossil Reptilia of the Mesozoic formations (*Omosaurus*, continued). Palaeontogr. Soc. Monogr. 31: 95–97.

———. 1884. A history of British fossil reptiles. Cassell, London.

Packard, G. C., Tracy, C. R., and Roth, J. J. 1977. The physiological ecology of reptilian eggs and embryos, and the evolution of viviparity within the class Reptilia. Biol. Rev. 52: 71–105.

Padian, K. 1982. Macroevolution and the origin of major adaptations: Vertebrate flight as a paradigm for the analysis of patterns. Proc. 3d No. Am. Paleontol. Conv. Vol. 2: 387–392.

———. 1983. Osteology and functional morphology of *Dimorphodon macronyx* (Buckland) (Pterosauria: Rhamphorhynchoidea) based on new material in the Yale Peabody Museum. Postilla 189: 1–44.

———. 1986. On the type material of *Coelophysis* Cope (Saurischia: Theropoda) and a new specimen from the Petrified Forest of Arizona (Late Triassic: Chinle Formation). In: Padian, K. (ed.). The Beginning of the Age of Dinosaurs. Cambridge Univ. Press, New York. Pp. 45–60.

Padian, K., Alvarez, W., Birkelund, T., Futterer, F. T., Hsu, K. J., Lipps, J. H., McLaren, D. J., Raup, D. M., Shoemaker, E. M., Smit, J., Toon, O. B., and Wetzel, A. 1984. The possible influences of sudden events on biological radiations and extinctions, group report. In: Holland, H. D., and Trendall, A. F. (eds.). Patterns of Change in Earth Evolution. Springer Verlag, New York. Pp. 77–102.

Padian, K., and Clemens, W. A. 1985. Terrestrial vertebrate diversity: Episodes and insights. In: Valentine, J. W., (ed.). Phanerozoic Diversity Patterns. Princeton Univ. Press, Princeton. Pp. 41–96.

Padley, D. 1985. Do the life history parameters of passerines scale to metabolic rate independently of body mass? Oikos 45: 285–287.

Paladino, F. V., Spotila, J., Dodson, P. and Hammond, J. K. 1989. Temperature dependent sex determination for reptiles, and the implications for dinosaur population dynamics and possible extinction. Geol. Soc. Am. Spec. Pap. 238: 63–70.

Paris, J.-P., and Taquet, P. 1973. Découverte d'un fragment de dentaire d'hadrosaurien (reptile dinosaurien) dans le Crétáce supérieur des Petites Pyrénées (Haute-

Garonne). Bull. Mus. Natl. Hist. Nat., Sci. Terre 22: 17–27.

Parker, L. R. 1976. The paleoecology of the fluvial coal-forming swamps and associated floodplain environments in the Blackhawk Formation (Upper Cretaceous) of central Utah. Brigham Young Univ. Stud. Geol. 22: 99–116.

Parkinson, J. 1930. The Dinosaur in East Africa. Witherby, London. 192 pp.

Parks, W. A. 1920a. Preliminary description of a new species of trachodont dinosaur of the genus *Kritosaurus, Kritosaurus incurvimanus*. Trans. Roy. Soc. Can. (ser. 3) 13: 51–59.

———. 1920b. The osteology of the trachodont dinosaur, *Kritosaurus incurvimanus*. Univ. Toronto Stud. (Geol. Ser.) 11: 1–74.

———. 1921. The head and fore limb of a specimen of *Centrosaurus apertus*. Trans. Roy. Soc. Can. (ser. 4) 15: 53–63.

———. 1922. *Parasaurolophus walkeri*, a new genus and species of crested trachodont dinosaur. Univ. Toronto Stud. (Geol. Ser.) 13: 1–32.

———. 1923. *Corythosaurus intermedius*, a new species of trachodont dinosaur. Univ. Toronto Stud. (Geol. Ser.) 13: 1–32.

———. 1924. *Dyoplosaurus acutosquameus*, a new genus and species of armored dinosaur; with notes on a skeleton of *Prosaurolophus maximus*. Univ. Toronto Stud. (Geol. Ser.) 18: 1–35.

———. 1925. *Arrhinoceratops brachyops*, a new genus and species of Ceratopsia from the Edmonton Formation of Alberta. Univ. Toronto Stud. (Geol. Ser.) 19: 5–15.

———. 1926a. *Thescelosaurus warreni*, a new species of ornithopodous dinosaur from the Edmonton Formation of Alberta. Univ. Toronto Studies (Geol. Ser.) 21: 1–42.

———. 1926b. *Struthiomimus brevitertius*, a new species of dinosaur from the Edmonton Formation of Alberta. Trans. Roy. Soc. Can. (ser. 3)20: 65–70.

———. 1928a. *Albertosaurus arctunguis*, a new species of theropodous dinosaur from the Edmonton Formation of Alberta. Univ. Toronto Stud. (Geol. Ser.) 25: 3–42.

———. 1928b. *Struthiomimus samueli*, a new species of Ornithomimidae from the Belly River Formation of Alberta. Univ. Toronto Stud. (Geol. Ser.) 26: 1–24.

———. 1931. A new genus and two new species of trachodont dinosaurs from the Belly River Formation of Alberta. Univ. Toronto Stud. (Geol. Ser.) 31: 1–11.

———. 1933. New species of dinosaurs and turtles from the Belly River Formation of Alberta. Univ. Toronto Stud. (Geol. Ser.) 34: 1–33.

———. 1935. New species of trachodont dinosaurs from the Cretaceous formations of Alberta, with notes on other species. Univ. Toronto Stud. (Geol. Ser.) 37: 1–45.

Parrish, J. M. 1986. Locomotor adaptations in the hindlimb and pelvis of the Thecodontia. Hunteria 1: 1–35.

Parrish, J. M., and Carpenter, K. 1986. A new vertebrate fauna from the Dockum Formation (Late Triassic) of eastern New Mexico. In: Padian, K. (ed.). The Beginning of the Age of Dinosaurs. Cambridge Univ. Press, New York. Pp. 151–160.

Parrish, J. M., Parrish, J. T., Hutchinson, J. H., and Spicer, R. A. 1987. Late Cretaceous vertebrate fossils from the North Slope of Alaska and implications for dinosaur ecology. Palaios 2: 377–389.

Patrulius, D., Marinescu, F., and Baltres, A. 1983. Dinosauriens ornithopodes dans les Bauxites néocomiennes de l'unité de Bihor (Monts Apuseni). Anuarul Inst. Ului Geol. Geofiz. Bucarest: 59: 109–118.

Patterson, C., and Smith, A. B. 1987. Is the periodicity of extinctions a taxonomic artefact? Nature 330: 248–251.

Paul, G. S. 1984a. The archosaurs: A phylogenetic study. In: Reif, W.-E., and Westphal, F. (eds.). 3d Symp. Mesozoic Terr. Ecosyst. Short Pap. Attempto Verlag, Tübingen. Pp. 175–180.

———. 1984b. The segnosaurian dinosaurs: Relics of the prosauropod-ornithischian transition? J. Vert. Paleontol. 4: 507–515.

———. 1987a. Predation in the meat-eating dinosaurs. In: Currie, P. J., and Koster, E. H. (eds.). 4th Symp. Mesozoic Terr. Ecosyst. Short Pap. Tyrrell Mus. Palaeontol., Drumheller, Alberta. Pp. 171–176.

———. 1987b. The science and art of restoring the life appearance of dinosaurs and their relatives. In: Czerkas, S. J., and Olson, E. C. (eds.). Dinosaurs Past and Present. Vol. II. Nat. Hist. Mus. Los Angeles County. Pp. 5–49.

———. 1988. The small predatory dinosaurs of the mid-Mesozoic: The horned theropods of the Morrison and Great Oolite—*Ornitholestes* and *Proceratosaurus*—and the sickle-claw theropods of the Cloverly, Djadokhta and Judith River—*Deinonychus, Velociraptor* and *Saurornitholestes*. Hunteria 2: 1–9.

Paul, L. W., and Juhl, J. H. 1966. The Essentials of Roentgen Interpretation. Harper, New York. 902 pp.

Pauwels, F. 1948. Die Bedeutung der Bauprinzipien des Stütz- und Bewegungsapparates für die Beanspruchung der Röhrenknochen. Z. Anat. Entwicklungsgesch. 114: 129–180.

Peabody, F. E. 1961. Annual growth zones in living and fossil vertebrates. J. Morph. 108: 11–62.

Pedley, T. J. 1987. How giraffes prevent oedema. Nature 329: 13–14.

Peng G.-Z. 1990. [A new species of small ornithopod type from Zigong, Sichuan.] J. Zigong

Perle, A. 1977. [On the first finding of *Alectrosaurus* (Tyrannosauridae, Theropoda) in the Late Cretaceous of Mongolia]. Probl. Geol. Mong. 3: 104–113. (In Russian)

———. 1979. [Segnosauridae—a new family of theropods from the Late Cretaceous of Mongolia]. Sovm. Sov.-Mong. Paleontol. Eksped. Trudy 8: 45–55. (In Russian)

———. 1981. [A new segnosaurid from the Upper Cretaceous of Mongolia]. Sovm. Sov.-Mong. Paleontol. Eksped. Trudy 15: 50–59. (In Russian)

———. 1982. [On a new finding of the hindlimb of *Therizinosaurus* sp. from the Late Cretaceous of Mongolia]. Probl. Geol. Mong. 5: 94–98. (In Russian)

———. 1985. [Comparative myology of the hip region in the bipedal dinosaurs]. Paleontol. Zh. 1985: 108–112. (In Russian)

Perle, A., Maryańska, T., and Osmólska, H. 1982. *Goyocephale lattimorei* gen. et sp. n., a new flat-headed pachycephalosaur (Ornithischia, Dinosauria) from the Upper Cretaceous of Mongolia. Acta Palaeontol. Polonica 27: 115–127.

Peters, R. H. 1983. The Ecological Implications of Body Size. Cambridge Univ. Press, New York. 329 pp.

Petersen, K., Isakson, J. I., and Madsen, J. H., Jr. 1972. Preliminary study of the paleopathologies in the Cleveland-Lloyd dinosaur collection. Proc. Utah Acad. Sci. 49: 44–47.

Peterson, F., and Turner-Peterson, C. E. 1987. The Morrison Formation of the Colorado Plateau: Recent advances in sedimentology, stratigraphy, and paleotectonics. Hunteria 2: 1–18.

Peterson, O. A., and Gilmore, C. W. 1902. *Elosaurus parvus,* a new genus and species of Sauropoda. Ann. Carnegie Mus. 1: 490–499.

Peterson, W. 1924. Dinosaur tracks on the roof of coal mines. Nat. Hist. 24: 388–391.

Petri, S., and Campanha, V. A. 1981. Brazilian continental Cretaceous. Earth-Sci. Rev. 17: 69–85.

Petri, S., and Mendes, J. C. 1983. Brazil. In: Moullade, M., and Nairn, A. E. M. (eds.). The Phanerozoic Geology of the World II. The Mesozoic, B. Elsevier, New York. Pp. 151–179.

Petters, S. W. 1981. Stratigraphy of Chad and Iullemmeden basins (west Africa). Eclogae Geol. Helv. 74: 139–159.

Pettigrew, J. D. 1986. The evolution of binocular vision. In: Pettigrew, J. D., Sanderson, K. J., and Lewick, W. R. (eds.). Visual Neuroscience. Cambridge Univ. Press, New York. Pp. 208–222.

Pettingill, O. S., Jr. 1970. Ornithology in Laboratory and Field. 4th ed. Burgess Publ. Co., Minneapolis.

Peyer, B. 1944. Über Wirbeltierfunde aus dem Rhät von Hallau (Kt. Schaffhausen). Eclogae Geol. Helv. 36: 260–263.

Phillips, J. 1870. [Notice of some specimens of megalosaurian bones in the Oxford Museum]. In: Huxley, T. H. 1870. Further evidence on the affinity between dinosaurian reptiles and birds. Q. J. Geol. Soc. London 26: 13–16.

———. 1871. Geology of Oxford and the Valley of the Thames. 529 pp.

Pianka, E. R. 1986. Ecology and Natural History of Desert Lizards. Princeton Univ. Press, Princeton. 208 pp.

Pidancet, J., and Chopard, S. 1862. Note sur un saurien gi-

gantesque appartenant aux Marnes irisées. C. R. Acad. Sci. Paris 54: 1259–1262.

Pillmore, G. L., Tschudy, R. H., Orth, C. J., Gilmore, J. S., and Knight, J. D. 1984. Geologic framework of nonmarine Cretaceous-Tertiary boundary sites, Raton Basin, New Mexico and Colorado. Science 223: 1180–1183.

Pipiringos, G. N., and O'Sullivan, R. B. 1978. Principal unconformities in Triassic and Jurassic rocks, western interior United States—a preliminary survey. U.S. Geol. Surv. Prof. Pap. 1035A: 1–29.

Plaziat, J.-C. 1961. Présence d'oeufs de dinosauriens dans le Crétacé supérieur des Corbières et existence d'un niveau marin dans le Thanétien aux environs d'Albas (Aude). C. R. Soc. Géol. France 1961: 196.

Pleijel, C. 1975. Nya dinosauriefotspar fran Skanes Rat-Lias. Fauna och flora 3: 116–120.

Plieninger, T. 1846. Über ein neues Sauriergenus und die Einreihung der Saurier mit flachen, schneidenden Zähnen in eine Familie. Jahresh. Ver. Naturk. Württemberg 2: 148–154.

Polis, G. A., and Myers, C. A. 1985. A survey of intraspecific predation among reptiles and amphibians. J. Herpetol. 19: 99–107.

Ponomarenko, A. G., and Popov, Y. A. 1980. [Paleobiocoenoses of Early Cretaceous Mongolian lakes]. Paleont. Zhur 1980: 3–13. (In Russian)

Popenoe, W. P., Imlay, R. W., and Murphy, M. A. 1960. Correlation chart of the Cretaceous formations of the Pacific Coast of the United States and northwestern Mexico. Bull. Geol. Soc. Am. 71: 1491–1540.

Pough, F. H. 1980. The advantages of ectothermy for tetrapods. Am. Nat. 115: 92–112.

———. 1983. Amphibians and reptiles as low-energy systems. In: Aspey, W. P., and Lustick, S. I. (eds.). Behavioral Energetics: The Cost of Survival in Vertebrates. Ohio State Univ. Press, Columbus. Pp. 141–188.

Pough, F. H., and Andrews, R. M. 1984. Individual and sibling-group variation in metabolism of lizards: The aerobic model for the origin of endothermy. Comp. Biochem. Physiol. 79A: 415–419.

Powell, J. E. 1978. Contribución al conocimiento geológico del extremo sur de la Sierra de La Candelaria, Provincia de Salta, República Argentina. Acta Geol. Lilloana 15: 49–58.

———. 1979. Sobre una asociación de dinosaurios y otras evidencias de vertebrados del Cretácico superior de la región de La Candelaria, Prov. de Salta, Argentina. Ameghiniana 16: 191–204.

———. 1980. Sobre la presencia de una armadura dermica en algunos dinosaurios titanosáuridos. Acta Geol. Lilloana 15: 41–47.

———. 1986. Revisión de los titanosáuridos de América del Sur. Ph. D. dissertation. Univ. Nac. Tucumán, Argentina. 493 pp.

Prange, H. D., Anderson, J. F., and Rahn, H. 1979. Scaling of skeletal mass to body mass in birds and mammals. Am. Nat. 113: 103–122.

Prasad, K. N. 1968. Some observations on the Cretaceous dinosaurs of India. Mem. Geol. Soc. India 2: 248–255.

Prasad, K. N., and Verma, K. K. 1967. Occurrence of dinosaurian remains from the Lameta Beds of Umrer, Nagpur District, Maharashtra. Curr. Sci. 36: 547–548.

Price, L. E. 1947. Sedimentos Mesozoicos na eaia de Sao Marcos, Estado do Maranhao. Not. Prelim. Estud. Dept. Nacl. Prod. Min. 40: 1–8.

———. 1951. Um ovo de dinossaurio na Formacao Bauru, do Cretacico do Estado de Minas Gerais. Not. Prelim. Estud. Dept. Nacl. Prod. Min. 53: 1–6.

———. 1960. Dentes de Theropoda num Testemunho de Sonda no Estado do Amazonas. An. Acad. Brasil. Cienc. 32: 79–84.

Preim, F. 1914. Sur les vertébrés de Cretácé et de l'Eocène d'Egypte. Bull. Soc. Géol. France (sér. 4)14: 366–382.

Quenstedt, F. A. 1858. Der Jura. Tübingen. 842 pp.

Quinn, J. F. 1987. On the statistical detection of cycles in extinctions in the marine fossil record. Paleobiology 13: 465–478.

Quinn, J. H. 1973. Arkansas dinosaur. Geol. Soc. Am. S.-Cent. Sect. Abst. Progr. 5: 276–277.

Raath, M. A. 1967. Notes on an occurrence of fossil bone in the Sipolilo-Musengezi area, Rhodesia. Detritus 2: 18–20.

———. 1969. A new coelurosaurian dinosaur from the Forest Sandstone of Rhodesia. Arnoldia 4: 1–25.

———. 1972. Fossil vertebrate studies in Rhodesia: a new dinosaur (Reptilia: Saurischia) from near the Triassic-Jurassic boundary. Arnoldia 5: 1–37.

———. 1974. Fossil vertebrate studies in Rhodesia: further evidence of gastroliths in prosauropod dinosaurs. Arnoldia 7 (5): 1–7.

———. 1977. The anatomy of the Triassic theropod *Syntarsus rhodesiensis* (Saurischia: Podokesauridae) and a consideration of its biology. Ph.D. dissertation, Rhodes Univ., Grahamstown, S. Afr.

———. 1980. The theropod dinosaur *Syntarsus* (Saurischia: Podokesauridae) discovered in South Africa. S. Afr. J. Sci. 76: 375–376.

———. 1985. The theropod *Syntarsus* and its bearing on the origin of birds. In: Hecht, M. K., Ostrom, J. H., Viohl, G., and Wellnhofer, P. (eds.). The Beginnings of Birds. Freunde des Jura-Museums Eichstätt. Pp. 219–227.

Rampino, M. R., and Strothers, R. B. 1988. Flood basalt volcanism during the past 250 million years. Science 241: 663–668.

Rampino, M. R., and Volk, T. 1988. Mass extinctions, atmospheric sulphur and climactic warming at the K/T boundary. Nature 332: 63–65.

Rao, L. R. 1932. On a reptilian vertebra from the south Indian Cretaceous. Am. J. Sci. (ser. 5)24: 221–224.

———. 1956. Recent contributions to our knowledge of the Cretaceous rocks of south India. Proc. Indian Acad. Sci. 44: 185–245.

Raup, D. M. 1986. The Nemesis Affair. Norton, New York. 220 pp.

Raup, D. M., and Sepkoski, J. J., Jr. 1984. Periodicity of extinctions in the geologic past. Proc. Natl. Acad. Sci. U.S.A. 81: 801–805.

———. 1986. Periodic extinction of families and genera. Science 231: 833–836.

———. 1988. Testing for periodicity of extinction. Science 241: 94–96.

Rawson, P. F., Curry, D., Dilley, F. C., Hancock, J. M., Kennedy, W. J., Neale, J. W., Wood, C. J., and Worssam, B. C. 1978. Cretaceous. Geol. Soc. London Spec. Rept. 9: 1–70.

Ray, C. E. 1965. Variation in the number of marginal tooth positions in three species of iguanid lizards. Breviora 236: 1–15.

Ray, G. E. 1941. Big for his day. Nat. Hist. 48: 36–39.

Reeds, C. A. 1932. A review of the work of the Central Asiatic Expeditions. In: Reeds, C. A. (ed.). The New Conquest of Central Asia. Natural History of Central Asia 1. Am. Mus. Nat. Hist., New York. Pp. 553–574.

Reeside, J. B. 1924. Upper Cretaceous and Tertiary formations of the western part of the San Juan Basin, Colorado and New Mexico. U.S. Geol. Surv. Prof. Pap. 134: 1–70.

Reeside, J. B., Applin, P. L., Colbert, E. H., Gregory, J. T., Hadley, H. D., Kummel, B., Lewis, P. J., Love, J. D., Maldonado-Koerdell, M., McKee, E. D., McLaughlin, D. B., Muller, S. W., Reinemund, J. A., Rodger, J., Sanders, J., Silberling, N. J., and Waage, K. 1957. Correlation of the Triassic formations of North America exclusive of Canada. Bull. Geol. Soc. Am. 68: 1451–1513.

Regal, P. J. 1975. The evolutionary origin of feathers. Q. Rev. Biol. 50: 35–66.

———. 1978. Behavioral differences between reptiles and mammals: An analysis of activity and mental capabilities. In: Greenburg, N., and MacLean, P. D. (eds.). Behavior and Neurology of Lizards. U.S. Dept. Health, Ed., Welfare. Washington, D.C. Pp. 183–202.

Regal, P. J., and Gans, C. 1980. The revolution in thermal physiology: implications for dinosaurs. In: Thomas, R. D. K., and Olson, E. C. (eds.). A Cold Look at the Warm-Blooded Dinosaurs. Westview Press, Boulder. Pp. 167–188.

Rehnelt, K. 1950. Ein Beitrag über Fährtenspuren im unteren Gipskeuper von Bayreuth. Berichte Naturforsch. Ges. Bayreuth 6: 27–36.

———. 1952. Ein weiterer dinosauroider Fährtenrest aus dem Benkersandstein von Bayreuth. Geol. Blätter NO-Bayern 2: 39–40.

Reid, R. E. H. 1981. Lamellar-zonal bone with zones and annuli in the pelvis of a sauropod dinosaur. Nature 292: 49–51.

———. 1983. High vascularity in bones of dinosaurs, mammals and birds. Geol. Mag. 120: 191–194.

———. 1984a. The histology of dinosaurian bone, and its possible bearing on dinosaurian physiology. In: Ferguson, M. J. W. (ed.). The Structure, Evolution, and

Development of Reptiles. Symp. Zool. Soc. London 52: 629–663.

———. 1984*b*. Primary bone and dinosaur physiology. Geol. Mag. 121: 589–598.

———. 1984*c*. Lamellar-zonal bone with zones and annuli in the pelvis of a sauropod dinosaur. Nature 292: 49–51.

———. 1987*a*. Review of R. T. Bakker, The Dinosaur Heresies. Modern Geol. 11: 271–280.

———. 1987*b*. Bone and dinosaur "endothermy." Modern Geol. 11: 133–154.

Reig, O. A. 1963. La presencia de dinosaurios saurisquios en los "Estratos de Ischigualasto" (Mesotriásico superior) de las Provincias de San Juan y La Rioja (Republica Argentina). Ameghiniana 3: 3–20.

———. 1970. The Proterosuchia and the early evolution of the archosaurs: An essay about the origin of a major taxon. Bull. Mus. Comp. Zool. 139: 229–292.

Reinhardt, J., Gibson, T. G., Bybell, L. M., Edwards, L. E., Fredericksen, N. O., Smith, C. C., Sohl, N. F., and Schwimmer, D. R. 1981. Upper Cretaceous and Lower Tertiary Geology of the Chattahoochee River Valley, western Georgia and eastern Alabama. 16th Ann. Field Trip, Georgia Geol. Soc.

Retallack, G. J., and Leahy, G. D. 1986. Cretaceous/Tertiary dinosaur extinction. Science 234: 1170–1171.

Retallack, G. J., Leahy, G. D., and Spoon, M. D. 1987. Evidence from paleosols for ecosystem changes across the Cretaceous/Tertiary boundary in eastern Montana. Geology 15: 1090–1093.

Reynolds, S. H. 1939. A collection of reptile bones from the Oolite near Stow-on-the-Wold, Gloucestershire. Geol. Mag. 76: 193–214.

Rhodin, A. G. J. 1985. Comparative chondro-osseus development and growth of marine turtles. Copeia 1985: 752–771.

Riabinin, A. N. 1914. [Report on a dinosaur from Transbaikalia]. Trudy Muz. Petra Velikogo 8: 133–140. (In Russian)

———. 1925. [A mounted skeleton of the gigantic reptile *Trachodon amurense* nov. sp.]. Izvest. Geol. Komissaya 44: 1–12. (In Russian with English summary)

———. 1930*a*. [Towards a problem of the fauna and age of the dinosaur beds on the Amur River]. Zapiski Russ. Min. Obshchestva (ser. 2)59: 41–51. (In Russian with English summary)

———. 1930*b*. [*Mandshurosaurus amurensis* nov. gen., nov. sp., a hadrosaurian dinosaur from the Upper Cretaceous of Amur River]. Monogr. Russkogo Paleontol. Obshchestva 2: 1–36. (In Russian with English summary)

———. 1931*a*. [Two dinosaurian vertebrae from the Lower Cretaceous of Transcaspian Steppes]. Zapiski Russkogo Min. Obshchestva (ser. 2) 60: 110–113. (In Russian with English summary)

———. 1931*b*. [Dinosaurian remains from the Upper Cretaceous of the lower Amudaria River]. Zapiski Rus-

skogo Min. Obshchestva (ser. 2)60: 114–118. (In Russian with English summary)

———. 1937. [The discovery of crested dinosaurs in the Upper Cretaceous beds of south Kazakhstan]. Priroda 1937: 91. (In Russian)

———. 1938. [Some results of the studies of the Upper Cretaceous dinosaurian fauna from the vicinity of the Station Sary-Agach, south Kazakhstan]. Probl. Paleontol. 4: 125–135. (In Russian with English summary)

———. 1939. [The Upper Cretaceous vertebrate fauna of south Kazakhstan. I. Reptilia. Pt. 1. Ornithischia]. Tsentral. Nauchno-issled. Geol. Inst. Trudy 118: 1–40. (In Russian with English summary)

———. 1945. [Dinosaurian remains from the Upper Cretaceous of the Crimea]. Vsesoyuznyy Nauchno-issled. Geol. Inst. Mat. Paleontol. Stratigr. 4: 4–10. (In Russian with English summary)

Rice, D. D., and Shurr, G. W. 1983. Patterns of sedimentation and paleogeography across the Western Interior Seaway during time of deposition of Upper Cretaceous Eagle Sandstone and equivalent rocks, northern Great Plains. In: Reynolds, M. W., and Dolly, E. D. (eds.). Mesozoic Paleogeography of west-central United States. Rocky Mountain Section, SEPM, Denver. Pp. 337–358.

Rich, T. H. V., Molnar, R. E., and Rich, P. V. 1983. Fossil vertebrates from the Late Jurassic of Early Cretaceous Kirkwood Formation, Algoa Basin, southern Africa. Trans. Geol. Soc. S. Afr. 86: 281–291.

Rich, T. H. V., and Rich, P. V. 1989. Polar dinosaurs and biotas of the Early Cretaceous of southeastern Australia. Natl. Geogr. Res. 5: 15–53.

Richmond, N. D. 1965. Perhaps juvenile dinosaurs were always scarce. J. Paleontol. 39: 503–505.

Ricklefs, R. E. 1979. Ecology. Chiron Press, New York. 966 pp.

Ricqlès, A. J. de. 1967. La paléontologie de terrain: un bilan international. Atomes 243: 337–341.

———. 1968*a*. Quelques observations paléohistologiques sur le dinosaurien sauropode *Bothriospondylus*. Ann. Univ. Madagascar 6: 157–209.

———. 1968*b*. Recherches paléohistologiques sur les os longs des tétrapodes. I. Origine du tissu osseux plexiforme des dinosaurien sauropodes. Ann. Paléontol. (Vert.) 54: 133–145.

———. 1969. L'histologie osseuse envisagée comme indicateur de la physiologie thermique chez les tétrapodes fossiles. C. R. Acad. Sci. Paris 268: 782–785.

———. 1974. Evolution of endothermy: Histological evidence. Evol. Theory 1: 51–80.

———. 1976. On bone histology of fossil and living reptiles, with comments on its functional and evolutionary significance. In Bellairs, A. d'A. and Cox, C. B. (eds.). Morphology and Biology of Reptiles. Linn. Soc. Symp. 3: 123–150.

———. 1980. Tissue structures of dinosaur bone: Functional significance and possible relation to dinosaur physiology. In: Thomas, R. D. K., and Olson, E. C. (eds.). A Cold Look at the Warm-Blooded Dinosaurs. Westview Press, Boulder. Pp. 103–139.

———. 1983. Cyclical growth in the long limb bones of a sauropod dinosaur. Acta Palaeontol. Pol. 28: 225–232.

Rieber, H. 1985a. Saurier und andere Fossilien von Frick. In: Frick Gestern und Heute. A. Frick AG, Frick, Switzerland. 16 pp.

———. 1985b. Der Plateosaurier von Frick. Unizürich 6: 3–5.

Rigby, J. K., Jr. 1982. *Camarasaurus* cf. *supremus* from the Morrison Formation near San Ysidro, New Mexico—the San Ysidro Dinosaur. New Mex. Geol. Soc. Gdbk. 33d Field Conf. Pp. 271–272.

Rigby, J. K., Jr., Newman, K. R., Smit, J., Van der Kars, S., Sloan, R. E., and Rigby, J. K. 1987. Dinosaurs from the Paleocene part of the Hell Creek Formation, McCone County, Montana. Palaios 2: 296–302.

Riggs, E. S. 1901a. The largest known dinosaur. Science 13: 549–550.

———. 1901b. The dinosaur beds of the Grand River Valley of Colorado. Field Columbian Mus. Geol. 1: 267–274.

———. 1901c. The fore leg and pectoral girdle of *Morosaurus*. Publ. Field Columbian Mus. Geol. 1: 275–281.

———. 1903a. *Brachiosaurus altithorax*, the largest known dinosaur. Am. J. Sci. (ser. 4)15: 299–306.

———. 1903b. Structure and relationships of opisthocoelian dinosaurs. Pt. I. *Apatosaurus* Marsh. Publ. Field Columbian Mus. Geol. 2: 165–196.

———. 1904. Structure and relationships of opisthocoelian dinosaurs. Pt. II. The Brachiosauridae. Publ. Field Columbian Mus. Geol. 2: 229–248.

Riley, H., and Stutchbury, S. 1836. A description of various fossil remains of three distinct saurian animals discovered in the Magnesian Conglomerate near Bristol. Proc. Geol. Soc. London 2: 397–399.

———. 1840. A description of various fossil remains of three distinct saurian animals, recently discovered in the Magnesian Conglomerate near Bristol. Trans. Geol. Soc. London (ser. 2)5: 349–357.

Rixon, A. E. 1968. The development of the remains of a small *Scelidosaurus* from a Lias nodule. Mus. J. 67: 315–321.

Romer, A. S. 1923. The pelvic musculature of saurischian dinosaurs. Bull. Am. Mus. Nat. Hist. 48: 605–617.

———. 1927. The pelvic musculature of ornithischian dinosaurs. Acta Zool. 8: 225–275.

———. 1933. Vertebrate Paleontology. Univ. Chicago Press, Chicago. 491 pp.

———. 1956. Osteology of the Reptiles. Univ. Chicago Press, Chicago. 772 pp.

———. 1966. Vertebrate Paleontology. 3d ed. Univ. Chicago Press, Chicago. 468 pp.

———. 1968. Notes and Comments on Vertebrate Paleontology. Univ. Chicago Press, Chicago. 304 pp.

———. 1971. The Chañares (Argentina) Triassic fauna. X. Two new but incompletely known long-limbed pseudosuchians. Breviora 378: 1–10.

———. 1972a. The Chañares (Argentina) Triassic reptile fauna. XV. Further remains of the thecodonts *Lagerpeton* and *Lagosuchus*. Breviora 394: 1–7.

———. 1972b. The Chañares (Argentina) Triassic reptile fauna. XVI. Thecodont classification. Breviora 395: 1–24.

Roth, J. J., and Roth, E. C. 1980. The parietal-pineal complex among paleovertebrates. In: Thomas, R. D. K., and Olson, E. C. (eds.). A Cold Look at the Warm-Blooded Dinosaurs. Westview Press, Boulder. Pp. 189–231.

Rothschild, B. M. 1988. Stress fracture in a ceratopsian phalanx. J. Paleontol. 62: 302–303.

Rouse, G. E., and Srivastava, S. K. 1972. Palynological zonation of Cretaceous and early Tertiary rocks of the Bonnet Plume Formation, northeastern Yukon, Canada. Can. J. Earth Sci. 9: 1163–1179.

Rowe, T. 1986. Homology and evolution of the deep dorsal thigh muscles in birds and other Reptilia. J. Morphol. 189: 327–346.

———. 1988. Definition, diagnosis and origin of Mammalia. J. Vert. Paleontol. 8: 241–264.

———. 1989. A new species of the theropod dinosaur *Syntarsus* from the Early Jurassic Kayenta Formation of Arizona. J. Vert. Paleontol. 9: 125–136.

Rowe, T., Colbert, E. H., and Nations, J. D. 1981. The occurrence of *Pentaceratops* (Ornithischia: Ceratopsia) with a description of its frill. In: Lucas, S., Rigby, J. K., Jr., and Kues, B. S. (eds.). Advances in San Juan Basin Paleontology. Univ. New Mexico Press. Albuquerque. Pp. 29–48.

Rozhdestvensky, A. K. 1952a. [Discovery of an iguanodont in Mongolia]. Doklady Akad. Nauk S.S.S.R. 84: 1243–1246. (In Russian)

———. 1952b. [A new representative of the duck-billed dinosaurs from the Upper Cretaceous deposits of Mongolia]. Doklady Akad. Nauk S.S.S.R. 86: 405–408. (In Russian)

———. 1955a. [The first discovery of a dinosaur in the U.S.S.R. in its original place]. Bull. Soc. Nat. Moscou Sect. Geol. 60: 118. (In Russian)

———. 1955b. [New data concerning psittacosaurs, Cretaceous ornithopods]. Voprosy Geol. Azii, Izdatelstvo Akad. Nauk S.S.S.R., Moskva. Vol. 2: 783–788. (In Russian)

———. 1957. [On the Upper Cretaceous dinosaur localities on the Amur River]. Vertebrata PalAsiatica 1: 285–291. (In Russian with English summary)

———. 1960. [Locality of Lower Cretaceous dinosaurs in the Kuznetz Basin]. Paleontol. Zh. 1960: 165. (In Russian)

———. 1964a. [Family Pachycephalosauridae]. In: Orlov, Y. A. (ed.). Osnovy Paleontologii. Izdatelstvo Nauka. Moskva. Pp. 558–589. (In Russian)

———. 1964*b*. [New data on the dinosaur localities of Kazakhstan Territory and in central Asia]. Tashkentskiy Gosudarstvennyy Univ. V. I. Lenina 234: 227–242. (In Russian)

———. 1965. [Growth changes and some problems of systematics of Asian dinosaurs]. Paleontol. Zh. 1965: 95–109. (In Russian)

———. 1966. [New iguanodonts from Central Asia. Phylogenetic and taxonomic relationships between late Iguanodontidae and early Hadrosauridae]. Paleontol. Zh. 1966: 103–116. (In Russian)

———. 1968*a*. [Hadrosaurs of Kazakhstan]. In: Tatarinov, L. P., et al. (eds.). [Upper Paleozoic and Mesozoic Amphibians and Reptiles]. Akademia Nauk S.S.S.R. Moscow. Pp. 97–141. (In Russian)

———. 1968*b*. [Find of a gigantic dinosaur]. Priroda 1968: 115–116. (In Russian)

———. 1970*a*. [Mesozoic and Cenozoic terrestrial vertebrate complexes of central Asia and adjoining regions of Kazakhstan and their stratigraphic position]. In: Biostratigraficheskie i Paleobiofatial'nye Issledovaniya i ikh Prakticheskoe Znachenie. Pp. 50–58. (In Russian)

———. 1970*b*. [On the gigantic unguals of some enigmatic Mesozoic reptiles]. Paleontol. Zh. 1970: 131–141. (In Russian)

———. 1971. [Investigations of the Mongolian dinosaurs and their role for the subdivision of the terrestrial Mesozoic]. Sovm. Sov.-Mong. Paleontol. Eksped. Trudy 3: 21–31. (In Russian)

———. 1972. [Development of dinosaur faunas in Asia and in other continents and paleogeography of the Mesozoic]. Izvest. Akad. Nauk S.S.S.R. 12: 115–133. (In Russian)

———. 1973*a*. [Animal Kingdom in Ancient Eurasia]. 159 pp. (In Japanese)

———. 1973*b*. [The study of Cretaceous reptiles in Russia]. Paleontol. Zh. 1973: 90–99. (In Russian)

———. 1974*a*. [A history of the dinosaur faunas in Asia and other continents and some problems of paleogeography]. Sovm. Sov.-Mong. Paleontol. Eksped. Trudy 1: 107–191. (In Russian)

———. 1974*b*. [Investigations of the Mesozoic reptiles in Russia and future prospects]. Paleontol. Zh. 1974: 26–32. (In Russian)

———. 1977. The study of dinosaurs in Asia. J. Palaeontol. Soc. India 20: 102–119.

Rozhdestvensky, A. K., and Khozatsky, L. I. 1967. [Late Mesozoic land vertebrates from the Asiatic part of the U.S.S.R.]. In: Martinson, G. G. (ed.). [Stratigraphy and Paleontology of Mesozoic and Paleogene-Neogene Continental Deposits of Asia and Adjoining Regions of Kazakhstan and their Stratigraphic Position]. Nauka Press, Leningrad. Pp. 82–92. (In Russian)

Rubey, W. W., Oriel, S. S., and Tracey, J. I. 1961. Age of the Evanston Formation, western Wyoming. U.S. Geol. Surv. Prof. Pap. 424B: 153–154.

Rudwick, M. J. S. 1985. The Meaning of Fossils. Univ. Chicago Press, Chicago. 288 pp.

Rühle von Lilienstern, H., Lang, M., and Huene, F. von. 1952. Die Saurier Thüringens. Gustav Fischer, Jena. 40 pp.

Russell, D. A. 1967. A census of dinosaur specimens collected in western Canada. Natl. Mus. Can. Nat. Hist. Pap. 36: 1–13.

———. 1969. A new specimen of *Stenonychosaurus* from the Oldman Formation (Cretaceous) of Alberta. Can. J. Earth Sci. 6: 595–612.

———. 1970*a*. Tyrannosaurs from the Late Cretaceous of western Canada. Natl. Mus. Nat. Sci. Publ. Palaeontol. 1: 1–34.

———. 1970*b*. A skeletal reconstruction of *Leptoceratops gracilis* from the upper Edmonton Formation (Cretaceous) of Alberta. Can. J. Earth Sci. 7: 181–184.

———. 1972. Ostrich dinosaurs from the Late Cretaceous of western Canada. Can. J. Earth Sci. 9: 375–402.

———. 1973. The environments of Canadian dinosaurs. Can. Geogr. J. 87: 4–11.

———. 1977. A Vanished World: The Dinosaurs of Western Canada. Natl. Mus. Nat. Sci. Natl. Mus. Can., Ottawa. 142 pp.

———. 1979. The enigma of the extinction of the dinosaurs. Ann. Rev. Earth Planet. Sci. 7: 163–182.

———. 1980. Reflections of the dinosaurian world. In: Jacobs, L. L. (ed.). Aspects of Vertebrate History. Mus. No. Arizona Press, Flagstaff. Pp. 275–268.

———. 1982. The mass extinctions of the late Mesozoic. Sci. Am. 246: 58–65.

———. 1984*a*. Terminal Cretaceous extinctions of large reptiles. In: Berggren, W. A., and Van Couvering, J. A. (eds.). Catastrophes and Earth History. Princeton Univ. Press, Princeton. Pp. 383–384.

———. 1984*b*. The gradual decline of dinosaurs—fact or fallacy? Nature 307: 360–361.

———. 1984*c*. A check list of the families and genera of North American dinosaurs. Syllogeus 53: 1–35.

———. 1987. A note on the terminal Cretaceous nodosaurids of North America. J. Vert. Paleontol. 7: 102.

Russell, D. A., and Béland, P. 1976. Running dinosaurs. Nature 264: 486.

Russell, D., Béland, P., and McIntosh, J. S. 1980. Paleoecology of the dinosaurs of Tendaguru (Tanzania). Mém. Soc. Géol. France 139: 169–175.

Russell, D. A., and Chamney, T. P. 1967. Notes on the biostratigraphy of dinosaurian and microfossil faunas in the Edmonton Formation (Cretaceous), Alberta. Natl. Mus. Can. Nat. Hist. Pap. 35: 1–22.

Russell, D. A., and Séguin, R. 1982. Reconstruction of the small Cretaceous theropod *Stenonychosaurus inequalis* and a hypothetical dinosauroid. Syllogeus 37: 1–43.

Russell, D., Russell, D., Taquet, P., and Thomas, H. 1976. Nouvelles récoltes de vertébrés dans les terrains continentaux du Crétacé supérieur de la région de Majunga (Madagascar). C. R. Soc. Géol. France 1976: 205–208.

Russell, L. S. 1935. Fauna of the upper Milk River Beds, southern Alberta. Trans. Roy. Soc. Can. (ser. 3)29: 115–127.

———. 1940a. The sclerotic ring in the Hadrosauridae. Paleontol. Contrib. Roy. Ont. Mus. 3: 1–7.

———. 1940b. Edmontonia rugosidens (Gilmore), an armored dinosaur from the Belly River series of Alberta. Univ. Toronto Stud. (Geol. Ser.) 43: 3–28.

———. 1948. The dentary of Troodon, a genus of theropod dinosaur. J. Paleontol. 22: 625–629.

———. 1949. The relationships of the Alberta Cretaceous dinosaur "Laosaurus" minimus Gilmore. J. Paleontol. 23: 518–520.

———. 1964. Cretaceous non-marine faunas of northwestern North America. Roy. Ont. Mus. Life Sci. Contrib. 61: 1–24.

———. 1965. Body temperature of dinosaurs and its relation to their extinction. J. Paleontol. 39: 497–501.

———. 1966. Dinosaur hunting in western Canada. Roy. Ont. Mus. Life Sci. Contrib. 70: 1–37.

———. 1968. A dinosaur bone from Willow Creek beds in Montana. Can. J. Earth Sci. 5: 327–329.

———. 1970. Correlation of the Upper Cretaceous Montana Group between southern Alberta and Montana. Can. J. Earth Sci. 7: 1099–1108.

———. 1975. Mammalian faunal succession in the Cretaceous system of western North America. Geol. Assoc. Can. Spec. Pap. 13: 137–161.

———. 1987. Biostratigraphy and palaeontology of the Scollard Formation, Late Cretaceous and Paleocene of Alberta. Roy. Ont. Mus. Life Sci. Contrib. 147: 1–23.

Rütimeyer, L. 1856a. (Dinosaurus gresslyi). Biblio. Universelle Genève Arch. Sept. 1856: 53.

———. 1856b. Reptilienknochen aus dem Keuper. Allg. Schweiz. Ges. Gesamt. Naturwiss. Verh. 41: 62–64.

Saha, S. 1982. The dynamic strength of bone and its relevance. In: Ghista, D. N. (ed.). Osteoarthromechanics. Hemisphere Publ. Corp., Washington, D.C. Pp. 1–43.

Sahni, A. 1972. The vertebrate fauna of the Judith River Formation, Montana. Bull. Am. Mus. Nat. Hist. 147: 321–412.

Sahni, A., Rana, R. S., Kumar, K., and Loyal, R. S. 1984. New stratigraphic nomenclature for the Intertrappean beds of the Nagpur region, India. Geosci. J. 5: 55–58.

Sahni, A., Rana, R. S., and Prasad, G. V. R. 1984. SEM studies of thin egg shell fragments from the Intertrappeans (Cretaceous-Tertiary transition) of Nagpur and Asifabad, peninsular India. J. Palaeontol. Soc. India. 29: 26–33.

Said, R. 1962. The Geology of Egypt. Elsevier Publishing Co., Amsterdam. 377 pp.

Saito, T., Yamanoi, T., and Kaiho, K. 1986. End-Cretaceous devastation of terrestrial flora in the boreal Far East. Nature 323: 253–255.

Salomon, J. 1982. Les formations continentales du Jurass-

ique supérieur-Crétacé inférieur en Espagne du Nord (Chaîne Cantabrique et NW Ibérique). Mém. Géol. Univ. Dijon 6: 1–228.

Sander, P. M. 1987. Taphonomy of the Lower Permian Geraldine Bonebed in Archer County, Texas. Palaeogeogr. Palaeoclimatol. Palaeoecol. 61: 221–236.

Santa Luca, A. P. 1980. The postcranial skeleton of Heterodontosaurus tucki from the Stormberg of South Africa. Ann. S. Afr. Mus. 79: 159–211.

———. 1984. Postcranial remains of Fabrosauridae (Reptilia: Ornithischia) from the Stormberg of Southern Africa. Palaeont. Afr. 25: 151–180.

Santa Luca, A. P., Crompton, A. W., and Charig, A. J. 1976. A complete skeleton of the Late Triassic ornithischian Heterodontosaurus tucki. Nature 264: 324–328.

Santafé-Llopis, J. V., Casanovas-Cladellas, M. L., Sanz-Garcia, J. L., and Calzada-Badia, S. 1979. Los dinosaurios de Morella (Nota preliminar). Acta Geol. Hisp. 13: 149–154.

———. 1982. Geología y Paleontología (Dinosaurios) de las Capas Rojas de Morella (Castellón, España). Diputación Prov. Castellón Diputación Barcelona. Pp. 1–169.

Sanz, J. L. 1982. A sauropod dinosaur tooth from the Lower Cretaceous of Galve (Province of Teruel, Spain). Géobios 15: 943–949.

———. 1983. A nodosaurid ankylosaur from the Lower Cretaceous of Salas de los Infantes (Province of Burgos, Spain). Géobios 16: 615–621.

———. 1985. Nouveaux gisements de dinosaures dans le Crétacé Espagnol. In: Les Dinosaures de la Chine à la France. Mus. Hist. Nat. Toulouse, France. Pp. 81–88.

Sanz, J. L., and Buscalioni, A. D. 1987. New evidence of armoured dinosaurs in the Upper Cretaceous of Spain. In: Currie, P. M., and Koster, E. H. (eds.). 4th Symp. Mesozoic Terr. Ecosyst. Short Pap. Tyrrell Mus. Palaeont., Drumheller, Alberta. Pp. 199–204.

Sanz, J. L., Buscalioni, A. D., Casanovas, M. L., and Santafé, J. V. 1987. Dinosaurios del Cretácico inferior de Galve (Teruel, España). Estud. Geol. Vol. Extraord. Galve-Tremp. Pp. 45–64.

Sanz, J. L., Casanovas, M. L., and Santafé, J. V. 1984. Iguanodontidos (Reptilia, Ornithopoda) del yacimiento del Cretácico inferior de San Cristobal (Galve, Teruel). Acta Geol. Hisp. 19: 171–176.

Sanz, J. L., Moratalla, J. J., and Casanovas, M. L. 1985. Traza icnologica de un dinosaurio iguanodontido en el Cretácico inferior de Cornago (La Rioja, España). Estud. Geol. 41: 85–91.

Sanz, J. L., Santafé, J.-V., and Casanovas, L. 1983. Wealden ornithopod dinosaur Hypsilophodon from the Capas Rojas Formation (lower Aptian, Lower Cretaceous) of Morella, Castellon, Spain. J. Vert. Paleontol. 3: 39–42.

Sarjeant, W. A. S. 1970. Fossil footprints from the Middle Triassic of Nottinghamshire and the Middle Jurassic of Yorkshire. Mercian Geol. 3: 269–282.

―――. 1974. A history and bibliography of the study of fossil vertebrate footprints in the British Isles. Palaeogeogr. Palaeoclimat. Palaeoecol. 16: 265–378.

Sastry, M. V. A., Rao, B. R. J., and Mamgain, V. D. 1969. Biostratigraphy zonation of the Upper Cretaceous formation of Trichinopoly district, south India. Mem. Geol. Soc. India 2: 10–17.

Sattayarak, N. 1983. Review of the continental Mesozoic stratigraphy of Thailand. In: Workshop on Stratigraphy Correlation of Thailand and Malaysia. Geol. Soc. Thailand, Geol. Soc. Malaysia, Bangkok. Pp. 127–148.

Sauvage, H. E. 1874. Mémoire sur les dinosauriens et les crocodiliens des terrains jurassiques de Boulogne-sur-Mer. Mém. Soc. Géol. France (sér. 2)10: 1–58.

―――. 1876a. Notes sur les reptiles fossiles. Bull. Soc. Géol. France (sér. 3)4: 435–442.

―――. 1876b. De la présence du type dinosaurien dans le Gault du nord de la France. Bull. Soc. Géol. France (sér. 3)4: 439–442.

―――. 1880. Synopsis des poissons et reptiles des terraines jurassiques de Boulogne-sur-Mer. Bull. Soc. Géol. France (sér. 3)8: 524–547.

―――. 1882. Recherches sur les reptiles trouvés dans le Gault de l'est du bassin de Paris. Mém. Soc. Géol. France (sér. 3)2: 1–42.

―――. 1888. Sur les reptiles trouvés dans le Portlandien supérieur de Boulogne-sur-Mer. Bull. Soc. Géol. France (sér. 3)16: 623—632.

―――. 1897. Notes sur les reptiles fossiles. Bull. Soc. Géol. France (sér. 3)25: 864–875.

―――. 1897–98. Vertébrés fossiles du Portugal. Contributions à l'étude des poissons et des reptiles du Jurassique et du Crétacique. Direction Trauvaux Geolog. Portugal. Mem. Comm. Serv. Geol. Portugal 1897–98: 1–46.

―――. 1907. Vertébrés. In: Thiéry, P., Sauvage, H.-E., and Crossman, M. (eds.). Note sur l'Infralias de Provenchere-sur-Meuse 365: 6–17.

Scarlett, R. J., and Molnar, R. E. 1984. Terrestrial bird or dinosaur phalanx from the New Zealand Cretaceous. New Zealand J. Zool. 11: 271–275.

Schaffer, W. M., and Reed, C. A. 1972. The co-evolution of social behavior and cranial morphology in sheep and goats (Bovidae, Caprini). Fieldiana Zool. 61: 1–88.

Schaller, G. B. 1972. The Serengeti Lion. Univ. Chicago Press, Chicago. 480 pp.

Schlaikjer, E. M. 1935. Contributions to the stratigraphy and paleontology of the Goshen Hole area, Wyoming. II. The Torrington member of the Lance Formation and a study of a new *Triceratops*. Bull. Mus. Comp. Zool. 76: 31–68.

Schlüter, T., and Schwarzhans, W. 1978. Eine Bonebed-Lagerstätte aus dem Wealden Süd-Tunesiens (Umgebung Ksar Krerachfa). Berliner Geowissen. Abh. A 8: 53–75.

Schmidt, H. 1969. *Stenopelix valdensis* H.v.Meyer, der kleine Dinosaurier des norddeutschen Wealden. Palaeontol. Z. 43: 194–198.

Schmidt, R. G. 1978. Rocks and mineral resources of the Wolf Creek area, Lewis and Clark and Cascade Counties, Montana. Bull. U.S. Geol. Surv. 1441: 1–91.

Schmidt-Nielsen, K. 1984. Scaling. Why Is Size So Important? Cambridge Univ. Press, New York. 241 pp.

Schmitz, O. J., and Lavigne, D. M. 1984. Intrinsic rate of increase, body size, and specific metabolic rate in marine mammals. Oecologia 62: 305–309.

Schopf, T. J. M. 1982. Extinction of the dinosaurs: A 1982 understanding. Geol. Soc. Am. Spec. Pap. 190: 415–422.

Seeley, H. G. 1869. Index to the fossil remains of Aves, Ornithosauria and Reptilia, from the Secondary system of strata arranged in the Woodwardian Museum of the University of Cambridge. Cambridge. 143 pp.

―――. 1870. On *Ornithopsis*, a gigantic animal of the pterodactyle kind from the Wealden. Ann. Mag. Nat. Hist. (ser. 4) 5: 279–283.

―――. 1871. On *Acanthopolis platyus* (Seeley), a pachypod from the Cambridge Greensand. Ann. Mag. Nat. Hist. (ser. 4) 8: 305–318.

―――. 1874. On the base of a large lacertian cranium from the Potton Sands, presumably dinosaurian. Q. J. Geol. Soc. London 30: 690–692.

―――. 1875. On the maxillary bone of a new dinosaur (*Priodontognathus phillipsii*), contained in the Woodwardian Museum of the University of Cambridge. Q. J. Geol. Soc. London 31: 439–443.

―――. 1876. On *Macrurosaurus semnus* (Seeley), a long-tailed animal with procoelous vertebrae from the Cambridge Upper Greensand, preserved in the Woodwardian Museum of the University of Cambridge. Q. J. Geol. Soc. London 32: 440–444.

―――. 1879. On the Dinosauria of the Cambridge Greensand. Q. J. Geol. Soc. London 35: 591–635.

―――. 1881. On the reptile fauna of the Gosau Formation preserved in the Geological Museum of the University of Vienna. Q. J. Geol. Soc. London 37: 620–707.

―――. 1882. On *Thecospondylus horneri*, a new dinosaur from the Hastings Sand, indicated by the sacrum and the neural canal of the sacral region. Q. J. Geol. Soc. London 38: 457–460.

―――. 1883. On the dinosaurs from the Maastricht beds. Q. J. Geol. Soc. London 39: 246–253.

―――. 1887a. On *Aristosuchus pusillus* Owen, being further notes on the fossils described by Sir R. Owen as *Poikilopleuron pusillus*, Owen. Q. J. Geol. Soc. London 43: 221–228.

―――. 1887b. On the classification of the fossil animals commonly called Dinosauria. Proc. Roy. Soc. London 43: 165–171.

―――. 1888a. The classification of the Dinosauria. Rept. Br. Assoc. Adv. Sci. 1887: 698–699.

―――. 1888b. *Thecospondylus* (*Thecocoelurus*) *horneri*. Q. J. Geol. Soc. London 44: 79.

———. 1889. Note on the pelvis of *Ornithopsis*. Q. J. Geol. Soc. London 45: 391–397.

———. 1891. On *Agrosaurus mcgillivrayi,* a saurischian reptile from the northeast coast of Australia. Q. J. Geol. Soc. London 47: 164–165.

———. 1893. *Omosaurus phillipsi.* Ann. Rept. Yorkshire Phil. Soc. 1892: 52–57.

———. 1894a. On *Euskelosaurus browni* (Huxley). Ann. Mag. Nat. Hist. (ser. 6)14: 317–340.

———. 1894b. On *Hortalotarsus skirtopodus,* a new saurischian fossil from Barkly East, Cape Colony. Ann. Mag. Nat. Hist. (ser. 6)13: 411–419.

———. 1895. On the type of the genus *Massospondylus,* and on some vertebrae and limb-bones of *M.* (?) *browni.* Ann. Mag. Nat. Hist. (ser. 6)15: 102–125.

———. 1898. On large terrestrial saurians from the Rhaetic beds of Wedmore Hill, described as *Avalonia sanfordi* and *Picrodon herveyi.* Geol. Mag. (ser. 4)5: 1–6.

———. 1901. In: Huene, F. von. 1901. Notizen aus dem Woodwardien-Museum in Cambridge. Cbl. Min. Geol. Palaeontol. 1901: 718.

Seemann, R. 1933. Das Saurischierlager in den Keupermergeln bei Trossingen. Jahresh. Ver. Vaterl. Naturk. Württemberg 89: 129–160.

Seidensticker, J. 1976. On the ecological separation of tigers and leopards. Biotropica 8: 225–234.

Seitel, M. R. 1979. The osteoderms of the American alligator and their functional significance. Herpetologica 35: 375–380.

Sepkoski, J. J., Jr. 1982. Mass extinctions in the Phanerozoic oceans: A review. Geol. Soc. Am. Spec. Pap. 190: 283–289.

Sepkoski, J. J., Jr., and Raup, D. M. 1986. Periodicity in marine extinction events. In: Elliott, D. K. (ed.). Dynamics of Extinction. Wiley Interscience, New York. Pp. 3–36.

Sereno, P. C. 1984. The phylogeny of the Ornithischia: A reappraisal. In: Reif, W.-E., and Westphal, F. (eds.). 3d Symp. Mesozoic Terr. Ecosyst. Short Pap. Attempto Verlag, Tübingen. Pp. 219–226.

———. 1986. Phylogeny of the bird-hipped dinosaurs (Order Ornithischia). Natl. Geogr. Soc. Res 2: 234–256.

———. 1987. The ornithischian dinosaur *Psittacosaurus* from the Lower Cretaceous of Asia and the relationships of the Ceratopsia. Ph.D. dissertation, Columbia Univ., New York. 554 pp.

Sereno, P. C., and Chao S. 1988. *Psittacosaurus xinjiangensis* (Ornithischia: Ceratopsia), a new psittacosaur from the Lower Cretaceous of northwestern China. J. Vert. Paleontol. 8: 363–365.

Sereno, P. C., Chao S., Cheng Z., and Rao C. 1988. *Psittacosaurus meileyingensis* (Ornithischia: Ceratopsia), a new psittacosaur from the Lower Cretaceous of northeastern China. J. Vert. Paleontol. 8: 366–377.

Seymour, R. S. 1976. Dinosaurs, endothermy and blood pressure. Nature 262: 207–208.

———. 1979. Dinosaur eggs: Gas conductance through the shell, water loss during incubation and clutch size. Paleobiology 5: 1–11.

Sheehan, P. M., and Hansen, T. A. 1986. Detritus feeding as a buffer to extinction at the end of the Cretaceous. Geology 14: 868–870.

Shen Y. 1981. [Fossil conchostracans from the Chijinpu Formation (Upper Jurassic) and the Xinminpu Group (Lower Cretaceous) in Hexi Corridor, Gansu]. Acta Palaeontol. Sinica 20: 266–272. (In Chinese with English summary)

Shen Y., and Mateer, N. J. In press. An outline of the Cretaceous System in northern Xinjiang. In: Mateer, N. J., and Chen P. (eds.). Proc. 1st Internatl. Symp. Nonmarine Cretaceous Correlations. China Ocean Press, Beijing.

Shepherd, J. D., Galton, P. M., and Jensen, J. A. 1977. Additional specimens of the hypsilophodontid dinosaur *Dryosaurus altus* from the Upper Jurassic of western North America. Brigham Young Univ. Geol. Stud. 24: 11–15.

Shikama, T. 1942. Footprints from Chincou, Manchoukuo, of *Jeholosauripus,* the Eo-Mesozoic dinosaur. Bull. Cent. Nat. Mus. Manchoukuo 3: 21–31.

Shilin, P. V., and Suslov, Y. V. 1982. [A hadrosaur from the northeastern Aral region]. Paleontol. Zh. 1982: 131–135. (In Russian)

Shine, R. 1986. Food habits, habitats and reproductive biology of four sympatric species of varanid lizards in tropical Australia. Herpetologica 42: 346–360.

Shipman, P. 1981. Life History of a Fossil: An Introduction to Taphonomy and Paleoecology. Harvard Univ. Press, Cambridge. 222 pp.

Shuvalov, V. F. 1974. [On the geology and age of Khobur and Khuren-Dukh localities in Mongolia]. Sovm. Sov.-Mong. Paleontol. Eksped. Trudy 1: 296–304. (In Russian with English summary)

———. 1975. [Buylyasutuin-Khuduk, a new Early Cretaceous dinosaur locality]. Sovm. Sov.-Mong. Paleontol. Eksped. Trudy 2: 210–213. (In Russian with English summary)

———. 1976. [The upper Senonian of southeastern Mongolia]. Akad. Nauk S.S.S.R. Izv. Ser. Geol. 1976: 58–62. (In Russian)

Shuvalov, V. F., and Chkhikvadze, V. M. 1975. [New data about Late Cretaceous turtles of south Mongolia]. Sovm. Sov.-Mong. Paleontol. Eksped. Trudy 2: 214–229. (In Russian with English summary)

Sigé, B. 1968. Dents de micromammifères et fragments de coquilles d'oeufs de dinosauriens dans la faune de vertébrés du Crétacé supérieur de Laguna Umayo (Andes péruviennes). C. R. Acad. Sci. Paris 267: 1495–1498.

Sigogneau-Russell, D. 1983. A new therian mammal from the Rhaetic locality of Saint-Nicolas-de-Port (France). Zool. J. Linn. Soc. London 78: 175–186.

Sill, W. D. 1974. The anatomy of *Saurosuchus galilei* and the relationships of the rauisuchid thecodonts. Bull. Mus. Comp. Zool. 146: 317–362.

Simionescu, J. 1913. *Megalosaurus* aus der Unterkreide der Dobrogea. Cbl. Min. Geol. Palaeontol. 1913: 686–687.

Simmons, D. J. 1965. The non-therapsid reptiles of the Lufeng Basin, Yunnan, China. Fieldiana Geol. 15: 1–93.

Sinclair, A. R. E., and Norton-Griffiths, M. 1979. Serengeti: Dynamics of an Ecosystem. Univ. Chicago Press, Chicago. 389 pp.

Skipp, B., and McGrew, L. W. 1972. The Upper Cretaceous Livingston Group of the western Crazy Mountains Basin, Montana. Montana Geol. Soc. Gdbk. 21st Field Conf. Pp. 99–106.

Slijper, E. J. 1962. Whales. Basic Books, New York. 475 pp.

Sloan, R. E., Rigby, J. K., Jr., Van Valen, L. M., and Gabriel, D. 1986. Gradual dinosaur extinction and simultaneous ungulate radiation in the Hell Creek Formation. Science 232: 629–633.

Smith, E. N. 1979. Behavioral and physiological thermoregulation of crocodilians. Am. Zool. 19: 239–247.

Smith, K. K. 1982. An electromyographic study of the function of the jaw adducting muscles in *Varanus exanthematicus* (Varanidae). J. Morphol. 173: 137–158.

Sochava, A. V. 1969. [Dinosaur eggs from the Upper Cretaceous of the Gobi Desert]. Paleontol. Zh. 1969: 517–527. (In Russian)

Sohn, I. G. 1979. Nonmarine ostracodes in the Lakota Formation (Lower Cretaceous) from South Dakota and Wyoming. U.S. Geol. Surv. Prof. Pap. 1069: 1–22.

Sollas, W. J. 1879. On some three-toed footprints from the Triassic conglomerate of South Wales. Q. J. Geol. Soc. London 35: 511–516.

Spassov, N. B. 1979. Sexual selection and the evolution of horn-like structures of ceratopsian dinosaurs. Paleontol. Stratigr. Lithol. 11: 37–48.

———. 1982. The "bizarre" dorsal plates of *Stegosaurus*: Ethological approach. C. R. Acad. Bulg. Sci. 35: 367–370.

Spicer, R. A., Burnham, R. J., Grant, P., and Glicken, H. 1985. *Pityrogramma calomelanos*, the primary post-eruption colonizer of Volcán Chichonal, Chiapas, Mexico. Am. Fern J. 75: 1–5.

Spotila, J. R. 1980. Constraints of body size and environment on the temperature regulation of dinosaurs. In: Thomas, R. D. K., and Olson, E. C. (eds.). A Cold Look at the Warm-Blooded Dinosaurs. Westview Press, Boulder. Pp. 233–252.

Spotila, J. R., and Gates, D. M. 1975. Body size, insulation, and optimum body temperature of homeotherms. In: Gates, D. M., and Schmerl, R. B. (eds.). Perspectives of Biophysical Ecology. Springer-Verlag, Berlin. Pp. 291–301.

Spotila, J. R., Lommen, P. W., Bakken, G. S., and Gates, D. M. 1973. A mathematical model for body temperatures of large reptiles: Implication for dinosaur ecology. Am. Nat. 107: 391–404.

Spotila, J. R., and Standora, E. A. 1985. Environmental constraints on the thermal energetics of sea turtles. Copeia 1985: 694–702.

Srivastava, S. 1983. Excavation of dinosaur bones in the Lameta beds near Rahioli, Kheda District, Gujarat. S. S. I. News W. Region 2: 1–13.

Srivastava, S., Mohabey, D. M., Sahni, A., and Pant, S. C. 1986. Upper Cretaceous dinosaur egg clutches from Kheda District (Gujarat, India). Palaeontographica A 193: 219–233.

Standora, E. A., and Spotila, J. R. 1982. Regional endothermy in the sea turtle, *Chelonia mydas*. J. Thermal Biol. 7: 159–165.

Stanley, S. M. 1984a. Mass extinctions in the oceans. Sci. Am. 250: 64–72.

———. 1984b. Marine mass extinction: A dominant role for temperature. In: Nitecki, M. H. (ed.). Extinctions. Univ. Chicago Press, Chicago. Pp. 69–117.

———. 1987. Extinction. Sci. Am. Library, New York. 242 pp.

Stapel, S. O., Leunissen, J. A. M., Versteeg, M., Wattel, J., and DeJong, W. W. 1984. Ratites as oldest offshoot of avian stem—evidence from alpha-crystallin A sequences. Nature 311: 257–259.

Stearns, S. C. 1983. The influence of size and phylogeny on patterns of covariation among life-history traits in the mammals. Oikos 41: 173–187.

Steel, R. 1969. Ornithischia. Hdb. Paläoherpetol. 15: 1–82.

———. 1970. Saurischia. Hdb. Paläoherpetol. 14: 1–87.

Stephenson, L. W., King, P. B., Monroe, W. H., and Imlay, R. W. 1942. Correlation of the outcropping Cretaceous formations of the Atlantic and Gulf coastal plain and Trans-Pecos, Texas. Bull. Geol. Soc. Am. 53: 435–448.

Sternberg, C. H. 1909. The Life of a Fossil Hunter. Holt, New York. 286 pp.

———. 1917. Hunting Dinosaurs in the Bad Lands of the Red Deer River, Alberta, Canada. World Co. Press, Lawrence, Kan. 232 pp.

Sternberg, C. M. 1921. A supplementary study of *Panoplosaurus mirus*. Trans. Roy. Soc. Can. (ser. 3)15: 93–102.

———. 1925. Integument of *Chasmosaurus belli*. Can. Field-Nat. 39: 108–110.

———. 1926a. Dinosaur tracks from the Edmonton Formation of Alberta. Can. Dept. Mines Bull. 44: 85–87.

———. 1926b. A new species of *Thespesius* from the Lance Formation of Saskatchewan. Can. Field-Nat. 38: 66–70.

———. 1927a. Homologies of certain bones of the ceratopsian skull. Trans. Roy. Soc. Can. (ser. 3)21: 135–143.

———. 1927b. Horned dinosaur group in the National Museum of Canada. Can. Field-Nat. 41: 67–73.

———. 1928. A new armored dinosaur from the Edmonton Formation of Alberta. Can. Field-Nat. 22: 93–106.

———. 1929a. A new species of horned dinosaur from the Upper Cretaceous of Alberta. Bull. Natl. Mus. Can. 54: 34–37.

———. 1929*b*. A toothless armored dinosaur from the Upper Cretaceous of Alberta. Bull. Natl. Mus. Can. 54: 28–33.

———. 1932*a*. Two new theropod dinosaurs from the Belly River Formation of Alberta. Can. Field-Nat. 46: 99–105.

———. 1932*b*. Dinosaur tracks from Peace River, British Columbia. Ann. Rept. Natl. Mus. Can. 1930: 59–85.

———. 1933*a*. A new *Ornithomimus* with complete abdominal cuirass. Can. Field-Nat. 47: 79–83.

———. 1933*b*. Relationships and habitat of *Troodon* and the nodosaurs. Ann. Mag. Nat. Hist. (ser. 10) 11: 231–235.

———. 1935. Hooded hadrosaurs from the Belly River Series of the Upper Cretaceous. Bull. Natl. Mus. Can. 77: 1–37.

———. 1936. The systematic position of *Trachodon*. J. Paleont. 10: 652–655.

———. 1937. Classification of *Thescelosaurus*: A description of a new species. Proc. Geol. Soc. Am. 1936: 375.

———. 1940*a*. *Thescelosaurus edmontonensis*, n. sp., and classification of the Hypsilophodontidae. J. Paleontol. 14: 481–494.

———. 1940*b*. Ceratopsidae from Alberta. J. Paleontol. 14: 468–480.

———. 1945. Pachycephalosauridae proposed for dome-headed dinosaurs, *Stegoceras lambei* n. sp., described. J. Paleontol. 19: 534–538.

———. 1949. The Edmonton fauna and description of a new *Triceratops* from the Upper Edmonton member; phylogeny of the Ceratopsidae. Bull. Natl. Mus. Can. 113: 33–46.

———. 1950. *Pachyrhinosaurus canadensis*, representing a new family of Ceratopsia. Bull. Natl. Mus. Can. 118: 109–120.

———. 1951*a*. Complete skeleton of *Leptoceratops gracilis* Brown from the Upper Edmonton member on the Red Deer River, Alberta. Bull. Natl. Mus. Can. 123: 225–255.

———. 1951*b*. The lizard *Chamops* from the Wapiti Formation of northern Alberta: *Polyodontosaurus grandis* not a lizard. Bull. Natl. Mus. Can. 123: 256–258.

———. 1953. A new hadrosaur from the Oldman Formation of Alberta: Discussion of nomenclature. Bull. Dept. Res. Dev. Can. 128: 1–12.

———. 1954. Classification of American duck-billed dinosaurs. J. Paleontol. 28: 382–383.

———. 1955. A juvenile hadrosaur from the Oldman Formation of Alberta. Ann. Rept. Natl. Mus. Can. 1953–54 Bull. 138: 120–122.

———. 1970. Comments on dinosaurian preservation in the Cretaceous of Alberta and Wyoming. Natl. Mus. Can. Publ. Nat. Hist. 4: 1–9.

Sternberg, R. M. 1940. A toothless bird from the Cretaceous of Alberta. J. Paleontol. 14: 81–85.

Sternfeld, R. 1911. Zur Nomenklatur der Gattung *Gigan-tosaurus* Fraas. Sitzungsb. Gesellsch. Naturforsch. Freunde Berlin 1911: 398.

Stigler, S. M., and Wagner, M. J. 1987. A substantial bias in nonparametric tests for periodicity in geophysical data. Science 238: 940–945.

———. 1988. Testing for periodicity of extinction. Science 241: 96–99.

Stipanicic, P. N. 1983. The Triassic of Argentina and Chile. In: Moullade, M., and Nairn, A. E. M. (eds.). The Phanerozoic Geology of the World II. The Mesozoic, B. Elsevier, New York. Pp. 181–199.

Stipanicic, P. N., and Bonetti, M. I. R. 1970*a*. Posiciones estratigraficas de las principales floras Jurasicas Argentinas. I. Floras Liasicas. Ameghiniana 7: 57–78.

———. 1970*b*. Posiciones estratigraficas y edades de las principales floras Jurasicas Argentina II. Floras Doggerianas y Malmicas. Ameghiniana 7: 101–118.

Stokes, W. L. 1944. Jurassic dinosaurs from Emery County, Utah. Proc. Utah Acad. Sci. 21: 11.

———. 1978. Animal tracks in the Navajo-Nugget Sandstone. Contrib. Geol. Univ. Wyoming 16: 103–107.

———. 1985. The Cleveland-Lloyd Dinosaur Quarry—Window to the Past. U.S. Govt. Printing Office, Washington, D.C. 27 pp.

———. 1987. Dinosaur gastroliths revisited. J. Paleontol. 61: 1242–1244.

Stokes, W. L., and Bruhn, A. F. 1960. Dinosaur tracks from Zion National Park and vicinity, Proc. Utah Acad. Sci. 37: 75–76.

Storer, J. E. 1975. Dinosaur tracks, *Columbosauripus ungulatus* (Saurischia: Coelurosauria), from the Dunvegan Formation (Cenomanian) of northeastern British Columbia. Can. J. Earth Sci. 12: 1805–1807.

Stott, D. F. 1975. The Cretaceous System in northeastern British Columbia. Geol. Assoc. Can. Spec. Pap. 13: 441–467.

Stout, W., and Service, W. 1981. The Dinosaurs. Bantam, New York.

Stovall, J. W. 1938. The Morrison of Oklahoma and its dinosaurs. J. Geol. 46: 583–600.

Stovall, J. W., and Langston, W., Jr. 1950. *Acrocanthosaurus atokensis*, a new genus and species of Lower Cretaceous Theropoda from Oklahoma. Am. Midl. Nat. 43: 696–728.

Stromer, E. 1915. Wirbeltier-Reste der Baharije-Stufe (unterstes Cenoman). 3. Das Original des Theropoden *Spinosaurus aegyptiacus* nov. gen. nov. spec. Abh. K. Bayer. Akad. Wissensch., Math.-Phys. Kl. 28: 1–32.

———. 1931. Wirbeltier-Reste der Baharije-Stufe (unterstes Cenoman). 10. Ein Skelett-Rest von *Carcharodontosaurus* nov. gen. Abh. Bayer Akad. Wissensch. Math.-naturwiss. Abt. 9: 1–23.

———. 1932. Wirbeltierreste der Baharije-Stufe (unterstes Cenoman). 11. Sauropoda. Abh. Bayer. Akad. Wissensch. Math.-naturwiss. Abt. 10: 1–21.

———. 1934*a*. Wirbeltierreste der Baharije-Stufe (unterstes

Cenoman). 13. Dinosauria. Abh. Bayer. Akad. Wissensch. Math.-naturwiss. Abt. 22: 1–79.

———. 1934b. Die Zähne des *Compsognathus* und Bemerkungen über das Gebiss der Theropoda. Cbl. Min. Geol. Palaeontol. B 1934: 74–85.

Stromer, E., and Weiler, W. 1930. Beschreibung von Wirbeltier-Resten aus dem nubischen Sandsteine Oberägyptens und aus ägyptischen Phosphaten nebst Bemerkungen über die Geologie der Umgegend von Mahamid in Oberagypten. Abh. Bayer. Akad. Wissensch. Math.-naturwiss. Abt. 7: 1–42.

Suarez-Vega, L. C. 1974. Bibliografia parcialmente comentada del Jurasico de Asturias. Cuad. Geol. Iber. 2: 581–588.

Sues, H.-D. 1978a. A new small theropod dinosaur from the Judith River Formation (Campanian) of Alberta, Canada. Zool. J. Linn. Soc. London 62: 381–400.

———. 1978b. Functional morphology of the dome in pachycephalosaurid dinosaurs. N. Jb. Geol. Paläontol. Mh. 1978: 459–472.

———. 1980a. Anatomy and relationships of a new hypsilophodontid dinosaur from the Lower Cretaceous of North America. Palaeontographica A 169: 51–72.

———. 1980b. A pachycephalosaurid dinosaur from the Upper Cretaceous of Madagascar and its paleobiogeographical implications. J. Paleontol. 54: 954–962.

Sues, H.-D., and Galton, P. M. 1982. The systematic position of *Stenopelix valdensis* (Reptilia: Ornithischia) from the Wealden of north-western Germany. Palaeontographica A 178: 183–190.

———. 1987. Anatomy and classification of the North American Pachycephalosauria (Dinosauria: Ornithischia). Palaeontographica A 198: 1–40.

Sues, H.-D., Olsen, P. E., and Shubin, N. H. 1987. A diapsid faunule from the Lower Jurassic of Nova Scotia, Canada. In: Currie, P. M., and Koster, E. H. (eds.). 4th Symp. Mesozoic Terr. Ecosyst. Short Pap. Tyrrell Mus. Palaeontol., Drumheller, Alberta. Pp. 203–205.

Sues, H.-D., and Taquet, P. 1979. A pachycephalosaurid dinosaur from Madagascar and a Laurasia-Gondwanaland connection in the Cretaceous. Nature 279: 633–635.

Sukhanov, V. B. 1964. [Subclass Testudinata]. In: Orlov, Y. A. (ed.). Osnovy Paleontologii. Izdatel'stvo Navka, Moskva. Pp. 354–438. (In Russian)

Sullivan, R. M. 1987. A reassessment of reptilian diversity across the Cretaceous-Tertiary boundary. Nat. Hist. Mus. Los Angeles County Contrib. Sci. 391: 1–26.

Sun A., Cui G., Li Y., and Wu X. 1985. [A verified list of Lufeng saurischian fauna]. Vertebrata PalAsiatica 23: 1–12. (In Chinese with English summary)

Sunquist, M. E. 1981. The social organization of tigers (*Panthera tigris*) in Royal Chitawan National Park, Nepal. Smithsonian Contrib. Zool. 336: 1–98.

Suslov, Y. V. 1982. [Dromaeosaurid unguals from the Upper Cretaceous deposits of the Kzyl-Orda region]. Akad.

Nauk Kazokhskoy S.S.R. Inst. Zool. Mat. istorii fauny flory Kazakhstana 8: 5–16. (In Russian)

———. 1983. [The locality of *Psittacosaurus* in Chamrin-Us (East Gobi, MPR)]. Sovm. Sov.-Mong. Paleontol. Eksped. Trudy 24: 118–120. (In Russian)

Swinton, W. E. 1946. The Isle of Wight dinosaur. Illustr. London News 209: 278.

———. 1947. New discoveries of *Titanosaurus indicus* Lyd. Ann. Mag. Nat. Hist. (ser. 11) 14: 112–123.

———. 1950. Fossil eggs from Tanganyika. Illustr. London News 217: 1082.

Taigen, T. L. 1983. Activity metabolism of anuran amphibians: Implications for the origin of endothermy. Am. Nat. 121: 94–109.

Tait, J., and Brown, B. 1928. How the Ceratopsia carried and used their head. Trans. Roy. Soc. Can. (ser. 3)22: 13–23.

Talbot, M. 1911. *Podokesaurus holyokensis,* a new dinosaur from the Triassic of the Connecticut Valley. Am. J. Sci. (ser. 4)31: 469–479.

Tan L. 1983. [Stratigraphy]. In: [The Mesozoic Stratigraphy and Paleontology of the Guyang Coal-Bearing Basin, Nei Monggol Autonomous Region, China]. Geol. Press, Beijing. Pp. 4–30.

Tanke, D. 1988. Ontogeny and dimorphism in *Pachyrhinosaurus* (Reptilia, Ceratopsidae), Pipestone Creek, N.W. Alberta, Canada. J. Vert. Paleontol. 8 (suppl.): 27A.

Tapia, A. 1918. Una mandibula de dinosaurio procedente de Patagonia. Physis 4: 369–370.

Taquet, P. 1975. Remarques sur l'évolution des iguanodontidés et l'origine des hadrosauridés. Colloq. Internatl. C.N.R.S. 218: 503–511.

———. 1976. Géologie et paléontologie du gisement de Gadoufaoua (Aptien du Niger). Cahiers Paléontol. C.N.R.S. Paris. Pp. 1–191.

———. 1977. Les découvertes récentes de dinosaures du Jurassique et du Crétacé en Afrique, au Proche et Moyen-Orient et en Inde. Mém. Soc. Géol. France 8: 325–330.

———. 1984. Une curieuse spécialisation du crâne de certains dinosaures carnivores du Crétacé: Le museau long et étroit des Spinosauridae. C. R. Acad. Sci. Paris 299: 217–222.

———. 1985. Les découvertes récentes de dinosaures au Maroc. In: Les Dinosaures de la Chine à la France. Mus. Hist. Nat. Toulouse, France. Pp. 39–43.

Taquet, P., and Welles, S. P. 1977. Redescription du crâne de dinosaure théropode de Dives (Normandie). Ann. Paléontol. (Vertébrés) 63: 191–295.

Tarsitano, S. 1983. Stance and gait in theropod dinosaurs. Acta Palaeontol. Pol. 28: 251–264.

Tarsitano, S., and Hecht, M. K. 1980. A reconsideration of the reptilian relationships of *Archaeopteryx.* Zool. J. Linn. Soc. London 69: 149–182.

Tatarinov, L. P. 1966. [Basipterygoid articulation and vidian canal in therapsids]. Paleontol. Zh. 1966: 101–115. (In Russian)

———. 1985. [New data on a collection of tetrapod teeth from the Rhaetian of Hallau (Switzerland)]. Paleontol. Zh. 1985: 137–138. (In Russian)

Taylor, C. R., Caldwell, S. L., and Rowntree, V. L. 1972. Running up and down hills: Some consequences of size. Science 178: 1096–1097.

Taylor, J. A. 1979. The foods and feeding habits of subadult *Crocodylus porosus* in northern Australia. Austral. Wildl. Res. 6: 347–359.

Tener, J. S. 1965. Muskoxen in Canada, a Biological and Taxonomic Review. Queen's Printer, Ottawa. 166 pp.

Thaler, L. 1962. Empreintes de pas de dinosaures dans les dolomies du Lias inférieur des Causses. C. R. Soc. Géol. France 1962: 190–192.

Thenius, E. 1974. Niederösterreich. Verh. Geol. Bundesanst., Vienna. 280 pp.

Théobald, N., Blanc, M., and David, E. 1967. Découverte d'ossements de dinosaure *Plateosaurus* cf. *poligniensis* (Pidancet et Chopard) dans les Marnes irisées supérieures des environs de Salins (Jura). Ann. Sci. Univ. Besançon (ser. 3) 3: 21–25.

Thévenin, A. 1907. Paléontologie de Madagascar. IV. Dinosauriens. Ann. Paléontol. 2: 121–136.

Thomas, D. A., and Gillette, D. D. 1985. Ornithopod and theropod ichnofauna in the Dakota Formation, Clayton Lake State park, northeastern New Mexico. Geol. Soc. Am. Rocky Mt. Sect. Abst. Prog. 17: 267.

Thomas, R. D. K., and Olson, E. C. (eds.). 1980. A Cold Look at Warm-Blooded Dinosaurs. Westview Press. Boulder. 514 pp.

Thompson, S. D. 1987. Body size, duration of parental care, and the intrinsic rate of natural increase in eutherian and metatherian mammals. Oecologia 71: 201–209.

Throckmorton, G. S. 1976. Oral food processing in two herbivorous lizards, *Iguana iguana* (Iguanidae) and *Uromastix aegyptius* (Agamidae). J. Morphol. 148: 363–390.

Thulborn, R. A. 1970a. The systematic position of the Triassic ornithischian dinosaur *Lycorhinus angustidens*. Zool. J. Linn. Soc. 49: 235–245.

———. 1970b. The skull of *Fabrosaurus australis*, a Triassic ornithischian dinosaur. Palaeontology 13: 414–432.

———. 1971a. Tooth wear and jaw action in the Triassic ornithischian dinosaur *Fabrosaurus*. J. Zool. 164: 165–179.

———. 1971b. Origins and evolution of ornithischian dinosaurs. Nature 234: 76–78.

———. 1972. The post-cranial skeleton of the Triassic ornithischian dinosaur *Fabrosaurus australis*. Palaeontology 15: 29–60.

———. 1973a. Teeth of ornithischian dinosaurs from the Upper Jurassic of Portugal. Serv. Geol. Portugal Mem. 22: 89–134.

———. 1973b. Thermoregulation in dinosaurs. Nature 245: 51–52.

———. 1974. A new heterodontosaurid dinosaur (Reptilia: Ornithischia) from the Upper Triassic red beds of Lesotho. Zool. J. Linn. Soc. London 55: 151–175.

———. 1977. Relationships of the Lower Jurassic dinosaur *Scelidosaurus harrisonii*. J. Paleontol. 51: 725–239.

———. 1978. Aestivation among ornithopod dinosaurs of the African Trias. Lethaia 11: 185–198.

———. 1980. The ankle joints of archosaurs. Alcheringa 4: 241–261.

———. 1982. Speeds and gaits of dinosaurs. Palaeogeogr. Palaeoclimatol. Palaeoecol. 38: 227–256.

———. 1984a. The avian relationships of *Archaeopteryx*, and the origin of birds. Zool. J. Linn. Soc. London 82: 119–158.

———. 1984b. Preferred gaits of bipedal dinosaurs. Alcheringa 8: 243–252.

Thulborn, R. A., and Wade, M. 1979. Dinosaur stampede in the Cretaceous of Queensland. Lethaia 12: 275–299.

———. 1984. Dinosaur trackways in the Winton Formation (mid-Cretaceous) of Queensland. Mem. Queensland Mus. 21: 413–517.

Tokaryk, T. T. 1985. Fossil vertebrates in the Eastend Museum. Blue Jay 43: 126–127.

———. 1986a. Ceratopsian dinosaurs from the Frenchman Formation (Upper Cretaceous) of Saskatchewan. Can. Field-Nat. 100: 192–196.

———. 1986b. Collection of fossil vertebrates for the Saskatchewan Museum of Natural History, 1985. Blue Jay 44: 150–152.

Tongiorgi, M. 1980. Orme di tetrapodi dei Monti Pisani. In: I vertebrati fossili italiani. Catalogo della Mostra, Verona. Pp. 77–84.

Tracy, C. R. 1976. Tyrannosaurs: Evidence for endothermy? Am. Nat. 110: 1105–1106.

Tracy, C. R., Turner, J. S., and Huey, R. B. 1986. A biophysical analysis of possible thermoregulatory adaptations in sailed pelycosaurs. In: Hotton, N., MacLean, P. D., Roth, J. J., and Roth, E. C. (eds.). The Ecology and Biology of Mammal-like Reptiles. Smithsonian Inst. Press, Washington, D.C. Pp. 195–206.

Tracy, J. I., and Oriel, S. S. 1959. Uppermost Cretaceous and lower Tertiary rocks of the Fossil Basin. Intermontane Assoc. Petrol. Geol. 10th Ann. Field Conf. Gdbk. Pp. 126–130.

Tricas, T. C., and McCosker, J. E. 1984. Predatory behavior of the white shark (*Carcharodon carcharias*), with notes on its biology. Proc. Cal. Acad. Sci. 43: 221–238.

Trivers, R. L. 1972. Parental investment and sexual selection. In: Campbell, B. (ed.). Sexual Selection and the Descent of Man. Aldine, Chicago. Pp. 136–179.

———. 1974. Parent-offspring conflict. Am. Zool. 14: 249–264.

Tschudy, R. H., Pillmore, C. L., Orth, C. J., Gilmore, J. S., and Knight, J. D. 1984. Disruption of the terrestrial plant ecosystem at the Cretaceous-Tertiary boundary, Western Interior. Science 225: 1030–1032.

Tschudy, R. H., and Tschudy, B. D. 1986. Extinction and survival of plant life following the Cretaceous/Tertiary boundary event, Western Interior, North America. Geology 14: 667–670.

Tschudy, R. H., Tschudy, B. D., and Craig, L. C. 1984. Palynological evaluation of Cedar Mountain and Burro Canyon formations, Colorado Plateau. U.S. Geol. Surv. Prof. Pap. 1281: 1–21.

Tsybin, Y. I., and Kurzanov, S. M. 1979. [New data on Upper Cretaceous localities of vertebrates of Baishin-Tsav region]. Sovm. Sov.-Mong. Paleontol. Eksped. Trudy 8: 108–112. (In Russian with English summary)

Tucker, M. E., and Burchette, T. P. 1977. Triassic dinosaur footprints from South Wales: Their context and preservation. Palaeogeogr. Palaeoclimatol. Palaeoecol. 22: 195–208.

Tumanova, T. A. 1977. [New data on the ankylosaur *Tarchia gigantea*]. Paleontol. Zh. 1977: 92–100. (In Russian)

———. 1981. [The morphological peculiarities of ankylosaurs]. Paleontol. Zh. 1981: 124–128. (In Russian)

———. 1983. [The first ankylosaur from the Lower Cretaceous of Mongolia]. Sovm. Sov.-Mong. Paleontol. Eksped. Trudy 24: 110–120. (In Russian with English summary)

———. 1985. Skull morphology of the ankylosaur *Shamosaurus scutatus* from the Lower Cretaceous of Mongolia. In: Les Dinosaures de la Chine à la France. Mus. Hist. Nat. Toulouse, France. Pp. 73–79.

———. 1987. [The armored dinosaurs of Mongolia]. Sovm. Sov.-Mong. Paleontol. Eksped. Trudy 32: 1–80. (In Russian)

Turco, R. P., Toon, O. B., Ackerman, T. P., Pollack, J. B., and Sagan, C. 1983. Nuclear winter: Global consequences of multiple nuclear explosions. Science 222: 1283–1292.

Turner, J. S., and Tracy, C. R. 1986. Body size, homeothermy, and control of heat exchange in mammal-like reptiles. In: Hotton, N., MacLean, P. D., Roth, J. J., and Roth, E. C. (eds.). The Ecology and Biology of Mammal-like Reptiles. Smithsonian Inst. Press, Washington, D.C. Pp. 185–194.

Tverdochlebov, V. P., and Tsybin, Y. I. [Genesis of the Upper Cretaceous dinosaur localities Tugrikin-Us and Alag-Teg]. Sovm. Sov.-Mong. Paleontol. Eksped. Trudy 1: 314–319. (In Russian with English summary)

Tyson, H. 1981. The structure and relationships of the horned dinosaur *Arrhinoceratops* Parks (Ornithischia: Ceratopsidae). Can. J. Earth Sci. 18: 1241–1247.

Urlichs, M. 1966. Zur Fossilführung und Genese der Feuerletten, der Rät-Lias-Grenzschichten und des unteren Lias bei Nürnberg. Erlanger Geol. Abh. 64: 1–42.

Van Gelder, R. G. 1969. Biology of Mammals. Scribner's, New York. 197 pp.

Van Tyne, J., and Berger, A. J. 1959. Fundamentals of Ornithology. Wiley, New York. 624 pp.

Van Valen, L. M. 1988. Paleocene dinosaurs or Cretaceous ungulates in South America. Evol. Monogr. 10: 1–79.

Van Valkenburgh, B. 1985. Locomotor diversity within past and present guilds of large predatory mammals. Paleobiology 11: 406–428.

———. 1988. Incidence of tooth breakage among large, predatory mammals. Am. Nat. 131: 291–302.

Vaughan, T. A. 1978. Mammalogy. 2d ed. Saunders, Philadelphia. 522 pp.

———. 1986. Mammalogy. 3d ed. CBS (Saunders) College Publishing, Philadelphia. 522 pp.

Versluys, J. 1910. Streptostylie bei Dinosauriern, nebst Bemerkungen über die Verwandtschaft der Vögel und Dinosaurier. Zool. Jb., Anat. 30: 175–260.

———. 1923. Der Schädel des Skelettes von *Trachodon annectens* im Senckenberg-Museum. Abh. Senckenb. Naturf. Ges. 38: 1–19.

Vézina, A. F. 1985. Empirical relationships between predator and prey size among terrestrial vertebrate predators. Oecologia 67: 555–565.

Vianey-Liaud, M., Jain, S. L., and Sahni, A. 1987. Dinosaur eggshells (Saurischia) from the Late Cretaceous Intertrappean and Lameta formations (Deccan, India). J. Vert. Paleontol. 7: 408–424.

Viele, G. W., and Harris, F. G. 1965. Montana Group stratigraphy, Lewis and Clark County, Montana. Am. Assoc. Petrol. Geol. Bull. 49: 379–417.

Viera, L. I., and Aguirrezabala, L. M. 1982. El Weald de Munilla (La Rioja) y sus icnitas de dinosaurios. Munibe 34: 245–270.

Viera, L. I., and Angel Torres, J. 1979. El Wealdico de la zona de Enciso (Sierra de los Cameros) y su fauna de grandes reptiles. Munibe 31: 141–157.

Villatte, J., Taquet, P., and Bilotte, M. 1985. Nouveaux restes de dinosauriens dans le Crétacé terminal de l'anticlinal de Dreuilhe. État des connaissances dans le domaine Sous-Pyrénéen. In: Les Dinosaures de la Chine à la France. Mus. Hist. Nat. Toulouse, France. Pp. 89–98.

Vine, R. R., and Day, R. W. 1965. Nomenclature of the Rolling Downs Group, northern Eromanga Basin, Queensland. Queensland Govt. Min. J. 66: 416–421.

Virchow, H. 1919. Atlas and Epistropheus bei den Schildkröten. Sitzungsber. Ges. Naturforsch. Freunde Berlin 1919: 303–332.

Visser, J. N. J. 1984. A review of the Stormberg Group and Drakensberg volcanics in Southern Africa. Palaeontol. afr. 25: 5–27.

Voss-Foucart, M. F. 1968. Paléoprotéines des coquilles fossiles d'oeufs de dinosauriens du Crétacé supérieur de Provence. Comp. Biochem. Physiol. 24: 31–36.

Wade, M. 1989. The stance of dinosaurs and the Cossack Dancer syndrome. In: Gillette, D. D., and Lockley, M. G. (eds.). Dinosaur Tracks and Traces. Cambridge Univ. Press, New York. Pp. 73–82.

Wagner, A. 1861. Neue Beiträge zur Kenntnis der urweltlichen Fauna des lithographischen Schiefers. V. *Compsognathus longipes* Wagn. Abh. Bayer. Akad. Wiss. 9: 30–38.

Waldman, M. 1974. Megalosaurids from the Bajocian

(Middle Jurassic) of Dorset. Palaeontology 17: 325–339.

Walker, A. 1981. Diet and teeth—dietary hypotheses and human evolution. Phil. Trans. Roy. Soc. London B 292: 57–64.

Walker, A. D. 1964. Triassic reptiles from the Elgin area: *Ornithosuchus* and the origin of carnosaurs. Phil. Trans. Roy. Soc. London B 248: 53–134.

———. 1972. New light on the origin of birds and crocodiles. Nature 237: 257–263.

Walker, C. A. 1981. A new subclass of birds from the Cretaceous of South America. Nature 292: 51–53.

Wall, W. P., and Galton, P. M. 1979. Notes on pachycephalosaurid dinosaurs (Reptilia: Ornithischia) from North America, with comments on their status as ornithopods. Can. J. Earth Sci. 16: 1176–1186.

Walls, G. L. 1942. The Vertebrate Eye and its Adaptive Radiation. Cranbrook Inst. Sci., Bloomfield Hills, Mich. 785 pp.

Wang S. et al. 1985. The Jurassic System of China. Geol. Publ. House, Beijing.

Wang Y., and Sun D. 1983. A survey of the Jurassic System of China. Can. J. Earth Sci. 20: 1646–1656.

Ward, P. 1983. The extinction of the ammonites. Sci. Am. 249: 136–147.

Ward, P., Wiedmann, J., and Mount, J. F. 1986. Maastrichtian molluscan biostratigraphy and extinction patterns in a Cretaceous/Tertiary boundary section exposed at Zumaya, Spain. Geology 14: 899–903.

Ward, P. D., and Signor, P. W. 1983. Evolutionary tempo in Jurassic and Cretaceous ammonites. Paleobiology 9: 183–198.

Warrington, G., Audley-Charles, M. G., Elliot, R. E., Evans, W. B., Ivimey-Cook, H. C., Kent, P. E., Robinson, P. L., Shotton, F. W., and Taylor, F. M. 1980. Triassic. Spec. Rept. Geol. Soc. London 13: 1–78.

Weaver, J. C. 1983. The improbable endotherm: The energetics of the sauropod dinosaur *Brachiosaurus*. Paleobiology 9: 173–182.

Webb, G. J. W., Buckworth, R., and Manolis, S. C. 1983. *Crocodylus johnstoni* in the McKinlay River, N.T. VI. Nesting biology. Austral. Wild. Res. 10: 607–637.

Webb, G. J. W., and Gans, C. 1982. Galloping in *Crocodylus johnstoni*—a reflection of terrestrial activity? Rec. Austral. Mus. 34: 607–618.

Weishampel, D. B. 1981a. Acoustic analysis of potential vocalization in lambeosaurine dinosaurs (Reptilia: Ornithischia). Paleobiology 7: 252–261.

———. 1981b. The nasal cavity of lambeosaurine hadrosaurids (Reptilia: Ornithischia): Comparative anatomy and homologies. J. Paleontol. 55: 1046–1057.

———. 1983. Hadrosaurid jaw mechanics. Acta Palaeontol. Pol. 28: 271–280.

———. 1984a. Evolution of jaw mechanisms in ornithopod dinosaurs. Adv. Anat. Embryol. Cell Biol. 87: 1–110.

———. 1984b. Trossingen: E. Fraas, F. von Huene, R. Seemann and the "Schwäbische Lindwurm" *Plateosau-*

rus. In: Reif, W.-E., and Westphal, F. (eds.). 3d Symp. Mesozoic Terr. Ecosyst. Short Pap. Attempto Verlag, Tübingen. Pp. 249–253.

———. 1984c. Interactions between Mesozoic plants and vertebrates: Fructifications and seed predation. N. Jb. Geol. Paläontol. Abhandl. 167: 224–250.

Weishampel, D. B., and Bjork, P. R. 1989. The first indisputable remains of *Iguanodon* (Ornithischia: Ornithopoda) from North America: *Iguanodon lakotaensis* n. sp. J. Vert. Paleontol. 9: 56–66.

Weishampel, D. B., and Horner, J. R. 1986. The hadrosaurid dinosaurs from the Iren Dabasu fauna (People's Republic of China, Late Cretaceous). J. Vert. Paleontol. 6: 38–45.

———. 1987. Dinosaurs, habitat bottlenecks, and the St. Mary River Formation. In: Currie, P. J., and Koster, E. H. (eds.). 4th Symp. Mesozoic Terr. Ecosyst. Short Pap. Tyrrell Mus. Palaeont., Drumheller, Alberta. Pp. 222–227.

Weishampel, D. B., and Jensen, J. A. 1979. *Parasaurolophus* (Reptilia: Hadrosauridae) from Utah. J. Paleontol. 53: 1422–1427.

Weishampel, D. B. and Norman, D. B. 1987. Dinosaur-plant interactions in the Mesozoic. In: Currie, P. J., and Koster, E. H. (eds.). 4th Symp. Mesozoic Terr. Ecosyst. Short Pap. Tyrrell Mus. Palaeont., Drumheller, Alberta. Pp. 228–233.

———. 1989. Vertebrate herbivory in the Mesozoic: jaws, plants, and evolutionary metrics. Geol. Soc. Am. Spec. Pap. 238: 87–100.

Weishampel, D. B., and Weishampel, J. B. 1983. Annotated localities of ornithopod dinosaurs: Implications to Mesozoic paleobiogeography. Mosasaur 1: 43–88.

Weishampel, D. B., and Westphal, F. 1986. Die Plateosaurier von Trossingen. Ausstellungskat. Univ. Tübingen 19: 1–27.

Weiss, W. 1934. Eine Fährtenschicht im mittelfränkischen Blasensandstein. Jb. Mitteil. Oberrhein. Geol. Ver. 23: 5–11.

Welles, S. P. 1952. A review of the North American Cretaceous elasmosaurs. Univ. Calif. Publ. Geol. Sci. 29: 47–143.

———. 1954. New Jurassic dinosaur from the Kayenta Formation of Arizona. Bull. Geol. Soc. Am. 65: 591–598.

———. 1970. *Dilophosaurus* (Reptilia, Saurischia), a new name for a dinosaur. J. Paleontol. 44: 989.

———. 1971. Dinosaur footprints from the Kayenta Formation of northern Arizona. Plateau 44: 27–38.

———. 1983. Two centers of ossification in a theropod astragalus. J. Paleontol. 57: 401.

———. 1984. *Dilophosaurus wetherilli* (Dinosauria, Theropoda). Osteology and comparisons. Palaeontographica A 185: 85–180.

Welles, S. P., and Long, R. A. 1974. The tarsus of theropod dinosaurs. Ann. S. Afr. Mus. 64: 191–218.

Wellman, H. W. 1959. Divisions of the New Zealand Cretaceous. Trans. Roy. Soc. N. Zealand 87: 99–163.

Welty, J. C. 1975. The Life of Birds. 2d ed. Saunders, London. 623 pp.

————. 1982. The Life of Birds. 3d ed. Saunders College Publ., Philadelphia. 745 pp.

Werner, D. I., Baker, E. M., Gonzalez, E. del, and Sosa, I. R. 1987. Kinship recognition and grouping in hatchling green iguanas. Behavioral Biol. Sociobiol. 21: 83–89.

Western, D. 1979. Size, life history and ecology in mammals. Afr. J. Ecol. 17: 185–204.

————. 1983. Production, reproduction and size in mammals. Oecologia 59: 269–271.

Wetmore, A. 1960. A classification for the birds of the world. Smithsonian Misc. Coll. 139: 1–37.

Wheeler, P. E. 1978. Elaborate CNS cooling structures in large dinosaurs. Nature 275: 441–443.

Whitaker, R., and Whitaker, Z. 1984. Reproductive biology of the mugger (*Crocodylus palustris*). J. Bombay Nat. Hist. Soc. 81: 297–316.

White, T. E. 1958. The braincase of *Camarasaurus lentus* (Marsh). J. Paleontol. 32: 477–494.

————. 1973. Catalogue of the genera of dinosaurs. Ann. Carnegie Mus. 44: 117–155.

Whybrow, P. J. 1981. Evidence for the presence of nasal salt glands in the Hadrosauridae (Ornithischia). J. Arid Env. 4: 43–57.

Wieland, G. 1903. Notes on the marine turtle *Archelon*. II. Associated fossils. Am. J. Sci. (ser. 4)15: 211–216.

————. 1909. A new armored saurian from the Niobrara. Am. J. Sci. (ser. 4)27: 250–252.

————. 1911. Notes of the armored Dinosauria. Am. J. Sci. (ser. 4)31: 112–124.

Wieser, W. 1985. A new look at energy conversion in ectothermic and endothermic animals. Oecologia 66: 506–510.

Wiffen, J., and Molnar, R. E. MS. An Upper Cretaceous ornithopod from New Zealand.

Wild, R. 1973. Die Triasfauna der Tessiner Kalkalpen. XXIV. *Tanystropheus longobardicus* (Bassani) (Neue Ergebnisse). Schweiz. Paläontol. Abh. 94: 1–162.

————. 1978a. Ein Sauropoden-Rest (Reptilia, Saurischia) aus dem Posidonienschiefer (Lias, Toarcium) von Holzmaden. Stuttgart. Beitr. Naturk. (Ser. B)41: 1–15.

————. 1978b. Die Flugsaurier (Reptilia, Pterosauria) aus der oberen Trias von Cene bei Bergamo, Italien. Boll. Soc. Paleontol. Italiana 17: 176–256.

Wilford, J. N. 1986. The Riddle of the Dinosaur. Alfred A. Knopf, New York. 304 pp.

Williams, D. L. G., Seymour, R. S., and Kerourio, P. 1984. Structure of fossil dinosaur eggshell from the Aix Basin, France. Palaeogeogr. Palaeoclimatol. Palaeoecol. 45: 23–37.

Williston, S. W. 1905. A new armored dinosaur from the Upper Cretaceous of Wyoming. Science 22: 503–504.

Wilson, E. O. 1975. Sociobiology: The New Synthesis. Belknap Press, Cambridge. 697 pp.

Wilson, M. C., and Currie, P. J. 1985. *Stenonychosaurus inequalis* (Saurischia: Theropoda) from the Judith River Formation of Alberta: new findings on metatarsal structure. Can. J. Earth Sci. 22: 1813–1817.

Wiman, C. 1929. Die Kreide-Dinosaurier aus Shantung. Palaeontol. Sinica (ser. C)6: 1–67.

————. 1930. Über Ceratopsia aus der oberen Kreide in New Mexico. Nova Acta Reg. Soc. Sci. Upsaliensis (ser. 4)7: 1–19.

————. 1931. *Parasaurolophus tubicen*, n. sp. aus der Kreide in New Mexico. Nova Acta Reg. Soc. Sci. Upsaliensis (ser. 4)7: 1–11.

Wing, S. L., and Tiffney, B. H. 1987. The reciprocal interaction of angiosperm evolution and tetrapod herbivory. Rev. Palaeobot. Palynol. 50: 179–210.

Winkler, D. A., Murry, P. A., Jacobs, L. L., Downs, W. R., Branch, J. R., and Trudel, P. 1988. The Proctor Lake dinosaur locality, Lower Cretaceous of Texas. Hunteria 2: 1–8.

Wolbach, W. S., Lewis, R. S., and Anders, E. 1985. Cretaceous extinctions: Evidence for wildfires and search for meteorite material. Science 230: 167–170.

Wolbach, W. S., Gilmour, I., Anders, E., Orth, C. J., and Brooks, R. R. 1988. Global fire at the Cretaceous-Tertiary boundary. Nature 334: 665–669.

Wolfe, J. A. 1976. Stratigraphic distribution of some pollen types from the Campanian and lower Maestrichtian rocks (Upper Cretaceous) of the Middle Atlantic States, U.S. Geol. Surv. Prof. Pap. 977: 1–18.

————. 1987. Late Cretaceous-Cenozoic history of deciduousness and the terminal Cretaceous event. Paleobiology 13: 215–226.

Wolfe, J. A., and Upchurch, G. R. 1986. Vegetation, climatic and floral changes at the Cretaceous-Tertiary boundary. Nature 324: 148–152.

————. 1987. North American non-marine climates and vegetation during the Late Cretaceous. Palaeogeogr. Palaeoclimatol. Palaeoecol. 61: 33–77.

Woodward, A. R., Hines, T. C., Abercrombie, C. L., and Nichols, J. D. 1987. Survival of young American alligators on a Florida lake. J. Wildl. Manage. 51: 931–937.

Woodward, A. S. 1895. Note on megalosaurian teeth discovered by Mr. J. Alstone in the Portlandian of Aylesbury. Proc. Geol. Assoc. London 14: 31–32.

————. 1901. On some extinct reptiles from Patagonia, of the genera *Meiolania*, *Dinilysia*, and *Genyodectes*. Proc. Zool. Soc. London 1901: 169–184.

————. 1905. On parts of the skeleton of *Cetiosaurus leedsi*, a sauropodous dinosaur from the Oxford Clay of Peterborough. Proc. Zool. Soc. London 1905: 232–243.

————. 1906. On a tooth of *Ceratodus* and a dinosaurian claw from the Lower Jurassic of Victoria, Australia. Ann. Mag. Nat. Hist. (ser. 7)103: 1–3.

————. 1908a. Note on *Dinodocus mackesoni*, a cetiosaurian from the Lower Greensand of Kent. Geol. Mag. (ser. 5) 5: 204–206.

————. 1908b. Note on a megalosaurian tibia from the lower Lias of Wilmcote, Warwickshire. Ann. Mag. Nat. Hist. (ser. 8)1: 257–265.

Wright, J. C. 1986. Effects of body temperature, mass, and activity on aerobic and anaerobic metabolism in juvenile *Crocodylus porosus*. Physiol. Zool. 59: 505–513.

Wu S. 1973. [A fossil of *Jaxartosaurus* is discovered in the Xinjiang]. Vertebrata PalAsiatica 11: 217–218. (In Chinese)

———. 1984. In: [The locations of ancient organisms in the northwest regions, Xinjiang Uygur Autonomous Region Edition III. Mesozoic, Cenozoic]. Geol. Res. Div. Xinjiang Oil Admin. Bureau, Cartogr. Grp. Xinjian Geol. Bureau. Geol. Publ., Beijing.

Xia W., Li X., and Yi Z. 1983. [The burial environment of dinosaur fauna in Lower Shaximiao Formation of Middle Jurassic at Dashanpu, Zigong, Sichuan]. J. Chengdu Coll. Geol. 2: 46–59. (In Chinese with English summary)

Xie J. 1980. [Discovery of dinosaur fossils in Tongwei County, Gansu Province]. Vertebrata PalAsiatica 18: 178.

Yabe, H., Inai, Y., and Shikoma, T. 1940. Discovery of dinosaurian footprints from the Cretaceous (?) Yangshan, Chinchou. Preliminary note. Proc. Imp. Acad. Tokyo 16: 560–563.

Yadagiri, P., and Ayyasami, K. 1979. A new stegosaurian dinosaur from Upper Cretaceous sediments of south India. J. Geol. Soc. India 20: 521–530.

Yagadiri, P., Prasad, K. N., and Satsangi, P. P. 1979. The sauropod dinosaur from the Kota Formation of Pranhitagodavari Valley, India. 4th Internatl. Gondwana Symp. Pp. 199–202.

Yeh H. 1975. Mesozoic Redbeds of Yunnan. Academia Sinica, Beijing.

Yorath, C. J., and Cook, D. G. 1981. Cretaceous and Tertiary stratigraphy and paleogeography, northern interior plains, District of Mackenzie. Geol. Surv. Can. Mem. 398: 1–76.

Young C.-C. 1930. On a new Sauropoda, with notes on other fragmentary reptiles from Szechuan. Bull. Geol. Soc. China 19: 279–315.

———. 1931. On some new dinosaurs from western Suiyan. Inner Mongolia. Bull. Geol. Surv. China 2: 159–166.

———. 1935a. Dinosaurian remains from Mengyin, Shantung. Bull. Geol. Soc. China 14: 519–533.

———. 1935b. On a new nodosaurid from Ninghsia. Palaeontol. Sinica (ser. C)11: 1–28.

———. 1937. A new dinosaurian from Sinkiang. Palaeontol. Sinica (ser. C)2: 1–25.

———. 1939. On a new Sauropoda, with notes on other fragmentary reptiles from Szechuan. Bull. Geol. Soc. China 19: 279–315.

———. 1941a. A complete osteology of *Lufengosaurus huenei* Young (gen. et sp. nov.). Palaeontol. Sinica (ser. C)7: 1–53.

———. 1941b. *Gyposaurus sinensis* (sp. nov.), a new Prosauropoda from the Upper Triassic Beds at Lufeng, Yunnan. Bull. Geol. Soc. China 21: 205–253.

———. 1942a. *Yunnanosaurus huangi* (gen. et sp. nov.), a new Prosauropoda from the Red Beds at Lufeng, Yunnan. Bull. Geol. Soc. China 22: 63–104.

———. 1942b. Fossil vertebrates from Kuangyuan, N. Szechuan, China. Bull. Geol. Soc. China 22: 293–309.

———. 1944. On the reptilian remains from Weiyuan, Szechuan, China. Bull. Geol. Soc. China 24: 187–209.

———. 1947. On *Lufengosaurus magnus* (sp. nov.) and additional finds of *Lufengosaurus huenei* Young. Palaeontol. Sinica 12: 1–53.

———. 1948a. Further notes on *Gyposaurus sinensis* Young. Bull. Geol. Soc. China 28: 91–103.

———. 1948b. On two new saurischians from Lufeng, Yunnan. Bull. Geol. Soc. China 28: 75–90.

———. 1948c. Notes on the occurrence of sauropod remains from N. Kweichow, China. Sci. Rec. 2: 200–206.

———. 1951. The Lufeng saurischian fauna in China. Palaeontol. Sinica (ser. C) 13: 1–96.

———. 1954. On a new sauropod from Yiping, Szechuan, China. Acta Palaeontol. Sinica 2: 355–369.

———. 1958a. [The dinosaurian remains of Laiyang, Shantung]. Palaeontol. Sinica (ser. C)16: 1–138. (Chinese and English)

———. 1958b. New sauropods from China. Vertebrata PalAsiatica 2: 1–28. (In English with Chinese summary)

———. 1959. On a new Stegosauria from Szechuan, China. Vertebrata PalAsiatica 3: 1–8. (In English with Chinese summary)

———. 1960. Fossil footprints in China. Vertebrata PalAsiatica 4: 53–66. (In English with Chinese summary)

———. 1965. [Fossil eggs from Nanshiung, Kwangtung and Kanchou, Kiangsi]. Vertebrata PalAsiatica 9: 141–170. (In Chinese with English summary)

———. 1982a. Selected works of Yang Zhungjian (Young Chung-Chien). Sci. Press, Beijing. 219 pp. (In Chinese)

———. 1982b. [On a new genus of dinosaur from Lufeng, Yunnan]. In: Yang Zhong Jian Wen Ji [Collected Papers of Yang Zhong Jian (Young, C.-C.)]. Pp. 38–42. (In Chinese)

Young, C.-C., and Chao, H. C. 1972. *Mamenchisaurus hochuanensis* sp. nov. Inst. Vert. Paleontol. Paleoanthropol. Monogr. (Ser. A)8: 1–30.

Zakharov, S. A. 1964. [The Cenomanian dinosaur whose tracks were found in the Shirkent River Valley]. In: Reiman, V. M. (ed.). [Paleontology of Tadzhikistan]. Akad. Nauk Tadzhik S.S.R. Press, Dushanbe. Pp. 31–35. (In Russian with English summary)

Zborzewski, A. 1834. Aperçu des recherches physiques rationnelles, sur les nouvelles curiosités Podolie-Volhyniennes, et sur leurs rapports géologiques avec les autres localités. Bull. Soc. Nat. Moscou 7: 224–254.

Zbyszewski, G. 1946. Les ossements d'*Omosaurus* découverts près de Baleal (Peniche). Serv. Geol. Portugal Com. 28: 135–144.

Zeng D., and Zhang J. 1979. [On the dinosaurian eggs from the western Dongting Basin, Hunan]. Vertebrata PalAsiatica 17: 131–136. (Chinese with English summary)

Zhai R., Zheng J., and Tong Y. 1978. [Stratigraphy of the mammal-bearing Tertiary of the Turfan Basin, Sinkiang]. Mem. Inst. Vert. Paleontol. Paleoanthropol. 13: 68–81. (In Chinese)

Zhang Y., Yang D., and Peng G. 1984. [New materials of *Shunosaurus* from the Middle Jurassic of Dashanpu, Zigong, Sichuan]. J. Chengdu College Geol. Suppl. 2: 1–12. (In Chinese with English summary)

Zhao Z. 1975. [The microstructure of fossil dinosaur eggs from Nanxiong County, Guandong Province, and issues in their classification]. Vertebrata PalAsiatica 13: 105–117. (In Chinese)

———. 1979. [Discovery of the dinosaurian eggs and footprint from Neixiang County, Henan Province]. Vertebrata PalAsiatica 17: 304–309. (Chinese with English summary)

Zhao Z., and Ding S. 1976. [Discovery of dinosaurian eggshells From Alxa, Ningxia Hui Autonomous Region]. Vertebrata PalAsiatica 14: 42–44. (In Chinese)

Zhen S. 1976. [A new species of hadrosaur from Shandong]. Vertebrata PalAsiatica 14: 166–168. (In Chinese with English summary)

Zhen S., Li J., and Rao C. 1986. [Dinosaur footprints of Jinning, Yunnan]. Mem. Beijing Nat. Hist. Mus. 33: 1–18. (In Chinese with English summary)

Zhen S., Li J., and Zhen B. 1983. [Dinosaur footprints of Yuechi, Sichuan]. Mem. Beijing Nat. Hist. Mus. 25: 1–19. (In Chinese with English summary)

Zhen S., Li J., Zhang B., Chen W., and Zhu S. 1987. Bird and dinosaur footprints from the Lower Cretaceous of Emei County, Sichuan. 1st Internatl. Symp. Nonmarine Cretaceous Correlations Abst. Pp. 37–38.

Zhen S., Rao C., Li J., Mateer, N. J., and Lockley, M. G. 1989. A review of dinosaur footprints in China. In: Gillette, D. D., and Lockley, M. G. (eds.). Dinosaur Tracks and Traces. Cambridge Univ. Press, New York. Pp. 187–198.

Zhen S., Zhen B., Mateer, N. J., and Lucas, S. G. 1985. The Mesozoic reptiles of China. Bull. Geol. Inst. Univ. Uppsala 11: 133–150.

Zheng X. 1982. [Fossil footprints]. In: [Paleontological Handbook of Hunan Province]. Geol. Press, Beijing. Pp. 485–489. (In Chinese)

Zherychin, V. V. 1978. [Development and change in Cretaceous and Cenozoic faunal communities of Tracheates and Chelicerates]. Trudy Paleont. Inst. Akad. Nauk. S.S.S.R. 165: 198. (In Russian)

———. 1987. [Community regulation of evolutionary processes]. Paleont. Zhur. 1987: 3–12. (In Russian)

Zhou S. 1984. [The Middle Jurassic dinosaurian fauna from Dashanpu, Zigong, Sichuan, Vol. 2. Stegosaurs]. Sichuan Sci. Tech. Publ. House, Chengdu. 51 pp. (In Chinese with English summary)

Zug, G. R., Wynn, A. H., and Ruckdeschel, C. 1986. Age determination of loggerhead sca turtles, *Caretta caretta*, by incremental growth marks in the skeleton. Smithsonian Contrib. Zool. 427: 1–34.

Contributors

Rinchen Barsbold
Department of Palaeontology and Stratigraphy
Geological Institute
Mongolian Academy of Sciences
Ulan Bator
Mongolian People's Republic

Michael J. Benton
Department of Geology
University of Bristol
Bristol B58 1RJ
England

Walter P. Coombs, Jr.
Department of Biology
Western New England College
Springfield, Massachusetts 01119
U.S.A.

Philip J. Currie
Tyrrell Museum of Palaeontology
P.O. Box 7500
Drumheller, Alberta T0J 0Y0
Canada

Peter Dodson
Laboratories of Anatomy
School of Veterinary Medicine
University of Pennsylvania
Philadelphia, PA 19104
U.S.A.

Dong Zhiming
Institute of Vertebrate Paleontology and
    Paleoanthropology
Academia Sinica
Beijing
People's Republic of China

James O. Farlow
Department of Earth and Space Sciences
Indiana University-Purdue University
Fort Wayne, IN 46805
U.S.A.

Peter M. Galton
Department of Biology
University of Bridgeport
Bridgeport, CT 06610
U.S.A.

Jacques A. Gauthier
Herpetology
California Academy of Sciences
Golden Gate Park
San Francisco, CA 94118
U.S.A.

John R. Horner
Museum of the Rockies
Montana State University
Bozeman, MT 59717
U.S.A.

Sergei M. Kurzanov
Palaeontological Institute
U.S.S.R. Academy of Sciences
Profsoyuznaya, 113
117321 Moscow
U.S.S.R.

Teresa Maryańska
Muzeum Ziemi
Polska Akademia Nauk
Al. Na Skarpie 20/26
00−488 Warszawa
Poland

683

John S. McIntosh
Science Tower
Wesleyan University
Middletown, CT 06457
U.S.A

Ralph E. Molnar
Queensland Museum
P.O. Box 836
Fortitude Valley
Queensland 4006
Australia

David B. Norman
Sedgwick Museum
University of Cambridge
Downing Street
Cambridge CB2 3EQ
England

Halszka Osmólska
Zakład Paleobiologii
Polska Akademia Nauk
Al. Zwirki i Wigury 93
02–089 Warszawa
Poland

John H. Ostrom
Department of Vertebrate Paleontology
Peabody Museum of Natural History
Yale University
New Haven, CT 06511
U.S.A.

Timothy Rowe
Department of Geological Sciences
University of Texas
Austin, TX 78713
U.S.A.

Paul C. Sereno
Department of Organismal Biology and Anatomy
University of Chicago
Chicago, IL 60637
U.S.A.

Hans-Dieter Sues
Vertebrate Palaeontology
Royal Ontario Museum
Toronto, Ontario M5S 2C6
Canada

Leonid P. Tatarinov
Palaeontological Institute
U.S.S.R. Academy of Sciences
Profsoyuznaya 123
Moscow 117321
U.S.S.R.

David B. Weishampel
Department of Cell Biology and Anatomy
The Johns Hopkins University
School of Medicine
Baltimore, MD 21205
U.S.A.

Lawrence M. Witmer
Department of Anatomy
New York College of Osteopathic Medicine
Old Westbury, NY 11568
U.S.A.

The Dinosauria: Index to Genera and Species

NOTES: References to tables are in *italics*; references to illustrations are in **boldface**. The subheading "provenance" denotes references to the dinosaurian distribution chapter, pages 63–139. The subheading "anatomy" denotes references to anatomical discussions that are primarily comparative or descriptive. The subheading "phylogeny" denotes references to discussions that are often anatomical but deal primarily with evolution and phylogeny. The subheading "taxonomy" denotes references to discussions that are often anatomical but deal primarily with taxonomy. Synonymies employed in the text are expressed as *see also* references in this index.

Agrosaurus macgillivrayi, 336
  provenance, 75
Alamosaurus, 351, 396
  cranial elements, lack of, 354
  paleoecology, 404
  phylogeny, 401
  taxonomy, 384, 395, 397,
    397–98, 399
Alamosaurus sanjuanensis,
    351, 396
  caudal vertebra, **363**
  humerus, **366**
  ischium, **371**
  provenance, 120, 121, 123
  radius, **368**
  scapulocoracoid, **364**
  sternal plate, **365**
Alamosaurus sp., provenance,
    118, 121
Albertosaurus, 189. See also
    Deinodon; Gorgosaurus
  abundance, 60
  behavior, 34
  biogeography, 211
  braincase, **175**
  phylogeny, **187**
  taxonomy, 199
  teeth, 34
Albertosaurus arctunguis, 189.
    See also *Albertosaurus
    sarcophagus*
Albertosaurus libratus, 189. See
    also *Deinodon horridus;
    Dryptosaurus kenabekides;
    Gorgosaurus libratus;
    Gorgosaurus sternbergi;
    Laelaps falculus; Laelaps
    hazenianus; Laelaps
    incrassatus*
  anatomy, 169–87
  behavior, 216, 220, 223
  hatchling, **220**
  lacrimal, **193**
  manus, **182**
  metatarsus, **188, 198**
  ontogeny, 219–20
  provenance, 112, 116, 121
  skull, **170**
  taphonomy, 211–21
  taxonomy, 193
Albertosaurus megagracilis, 189
  provenance, 117
Albertosaurus periculosus, 190. See
    also *Tarbosaurus bataar*
  provenance, 133
Albertosaurus sarcophagus, 189.
    See also *Albertosaurus
    arctunguis*
  provenance, 113, 116, 119
Albertosaurus sp., provenance,
    113, 114, 115, 118, 119,
    121, 123, 124
Albisaurus scutifer, 531
  provenance, 128
Alectrosaurus, 190
  biogeography, 211
Alectrosaurus olseni, 190
  anatomy, 171–78
  metatarsus, **188**

paleoecology, 415
phylogeny, **187**
provenance, 130, 133
taxonomy, 193, 198, 199
tibia, **186**
Algoasaurus bauri, 352
  provenance, 94
Alioramus, 190
  biogeography, 211
Alioramus remotus, 190
  anatomy, 171–78
  braincase, **175**
  phylogeny, **187**
  provenance, 130
  skull, **170**
  taxonomy, 193, 199
Aliwalia, 147
  taxonomy, 18
Aliwalia rex, 147
  known material, 145–46
  provenance, 73
Allosaurus, 189. See also *Antro-
    demus; Creosaurus; Epan-
    terias; Labrosaurus;
    Saurophagus*
  anatomy, 34, 156, 160–
    61, 290
  behavior, 38, 217
  taphonomy, 34, 454
  taxonomy, 196–198
Allosaurus ferox, 189. See also *Al-
    losaurus fragilis*
Allosaurus fragilis, 189. See also
    *Allosaurus ferox; Al-
    losaurus lucaris; Creo-
    saurus atrox; Epanterias
    amplexus; Hypsirophus dis-
    curus; Labrosaurus ferox;
    Laelaps trihedrodon; Sau-
    rophagus maximus*
  anatomy, 169–87
  axial neural spine, **194**
  behavior, 212–17, 222–23
  biogeography, 210
  braincase, **174, 175**
  chevron, **195**
  cranial kinesis, **212**
  extensor groove of femur, **197**
  femoral head, **196**
  femur, **185**
  forelimb elements, **181**
  hindlimb elements, **186**
  ilium and sacrum, **196**
  injuries, 224
  jaw joint, **194**
  lacrimal, **193**
  manus, **182**
  metatarsus, **198**
  ontogeny, 219–20
  pelvis, **183**
  pes, **188**
  phylogeny, 187, **187**
  provenance, 86, 87, 88
  skulls, **170, 172, 222**
  taphonomy, 211–12
  taxonomy, 193, 194, 197–
    98, 199, 201, 202–203,
    205, 313, 314–15

vertebrae, **179, 215**
Allosaurus lucaris, 189. See also
    Allosaurus fragilis
Allosaurus medius, 308. See also
    Dryptosaurus medius
  provenance, 98
  taxonomy, 311
Allosaurus sibiricus, 189. See also
    Chilantaisaurus sibiricus
  taxonomy, 200
Allosaurus tendagurensis, 189
  biogeography, 210
  provenance, 94
Allosaurus sp.
  biogeography, 210–11
  provenance, 111
Alocodon, taxonomy, 423
Alocodon kuehnei, 424
  provenance, 84
Altispinax, 307, 307
Altispinax dunkeri, 307, 307. See
    also *Megalosaurus dunkeri*
  provenance, 98, 100, 103
Alvarezsaurus, 282
Alvarezsaurus calvoi, 282
  provenance, 137
Ammosaurus, 335
  anatomy, 327–33
  pelvis, **332**
  taphonomy, 341
  taxonomy, 29
Ammosaurus major, 335, 340. Se
    also *Anchisaurus major;
    Anchisaurus solus*
  provenance, 77
  taxonomy, 339
Ammosaurus cf. *major,* prove-
    nance, 76
Ammosaurus sp., prove-
    nance, 75
Amphicoelias, 349, 391
Amphicoelias altus, 349, 391.
    See also *Amphicoelias
    fragillimus*
  femur, **372**
  provenance, 88
Amphicoelias fragillimus, 349. See
    also *Amphicoelias altus*
Amphicoelias latus, 348. See also
    Camarasaurus supremus
Amphisaurus, 335. See also
    Anchisaurus
Amtosaurus, 475
Amtosaurus magnus, 475, 480
  provenance, 130
Amygdalodon, 347, 378
  taxonomy, 378–79
Amygdalodon patagonicus,
    347, 378
  provenance, 85
Anatotian copei, 557. See also *Tra-
    chodon longiceps*
  anatomy, 535–46, 551
  behavior, 561
  paleoecology, 560
  phylogeny, **554**
  provenance, 118, 122
  skull, **540**
  taxonomy, 555

Anchiceratops, 612
  behavior, 37–38
  phylogeny, **609**
Anchiceratops longirostris, 612. See
    also *Anchiceratops ornatus*
  provenance, 114
  skull, **597**
Anchiceratops "*longirostris,*" anat-
    omy, 604
Anchiceratops ornatus, 612, 616.
    See also *Anchiceratops
    longirostris*
  anatomy, 604–606
  provenance, 113, 114
  taphonomy, 617
Anchisaurus, 335. See also *Amphi-
    saurus; Megadactylus;
    Yaleosaurus*
  anatomy, 323–33
  braincase, **325**
  manus, **329**
  paleoecology, 342–44
  pelvis, **332**
  pes, **332**
  phylogeny, **19, 333**
  skull, **324**
  taphonomy, 341
  taxonomy, 23, 29, 318
Anchisaurus colurus, 335. See also
    Anchisaurus polyzelus
Anchisaurus major, 335. See also
    Ammosaurus major
Anchisaurus polyzelus, 335, 339.
    See also *Anchisaurus colu-
    rus; Megadactylus polyzelus*
  provenance, 77
Anchisaurus solus, 335. See also
    Ammosaurus major
Anchisaurus sp., provenance, 75
Ankylosaurus, 475
  phylogeny, **478**
Ankylosaurus magniventris,
    475, 480
  anatomy, 457–65
  hindlimb elements, **472**
  humerus, **470**
  ischium, **471**
  paleoecology, 483
  phylogeny, 481
  provenance, 114, 118
  skull, **458, 461**
  tail club, **467**
  teeth, **464**
  vertebrae, **466**
Ankylosaurus cf. *magniventris,*
  provenance, 119
Ankylosaurus sp., provenance,
    115
Anodontosaurus, 475
Anodontosaurus lambei, 475. See
    also *Euoplocephalus tutus*
Anoplosaurus curtonotus, 531
  provenance, 100
Anoplosaurus major, 531
  provenance, 100
Anserimimus, 226
Anserimimus planinychus, 226
  provenance, 130
Antarctosaurus, 351–52, 397

The Dinosauria: Stratigraphic Index

NOTES: References to tables are in *italics*. Main references are to the dinosaurian distribution chapter, pages 63–139.

The Dinosauria: Subject Index

NOTES: References to tables are in *italics*; references to illustrations are in **boldface**. The dinosaurian distribution chapter, pages 63–139, is already arranged systematically by period and locality and, except for the maps, is not indexed here. "Anatomy" subheadings denote references to anatomical discussions that are primarily comparative or descriptive. The subheading "phylogeny" denotes references to discussions that are often anatomical but deal primarily with evolution and phylogeny. The subheading "taxonomy" denotes references to discussions that are often anatomical but deal primarily with taxonomy. As a general rule, because the text is already organized systematically, the figures are indexed here in greater detail than is the text.

Designer: Marvin Warshaw
Compositor: G&S Typesetters, Inc.
Text: 10/12 Meridien
Display: Helvetica
Printer: Malloy Lithographing, Inc.
Binder: Malloy Lithographing, Inc.